STEEL AND COMPOSITE STRUCTURES

BALKEMA – Proceedings and Monographs
in Engineering, Water and Earth Sciences

PROCEEDINGS OF THE 3rd INTERNATIONAL CONFERENCE ON STEEL AND COMPOSITE STRUCTURES (ICSCS07), MANCHESTER, UK, 30 JULY–1 AUGUST 2007

Steel and Composite Structures

Editors

Y.C. Wang
The University of Manchester, UK

C.K. Choi
Korea Advanced Institute of Science & Technology, Korea

Taylor & Francis
Taylor & Francis Group

LONDON / LEIDEN / NEW YORK / PHILADELPHIA / SINGAPORE

Taylor & Francis is an imprint of the Taylor & Francis Group, an informa business

© 2007 Taylor & Francis Group, London, UK

Typeset by Charon Tec Ltd (A Macmillan Company), Chennai, India
Printed and bound in Great Britain by TJ International Ltd, Padstow Cornwall

Published by: Taylor & Francis/Balkema
                P.O. Box 447, 2300 AK Leiden, The Netherlands
                e-mail: Pub.NL@tandf.co.uk
                www.taylorandfrancis.co.uk/engineering, www.crcpress.com

ISBN: 978-0-415-45141-3

Steel and Composite Structures – Wang & Choi (eds)
© 2007 Taylor & Francis Group, London, ISBN 978-0-415-45141-3

Table of contents

Material, beams, columns and frames

Connections

Bridges

Fire resistance

Dynamics, earthquake and impact

Fracture and Fatigue

Thin-walled structures

Structural analysis

Space, shell and hybrid structures

Steel and Composite Structures – Wang & Choi (eds)
© 2007 Taylor & Francis Group, London, ISBN 978-0-415-45141-3

Preface

The volume contains a total of 151 papers, including 7 keynote papers, 7 invited papers in a special session on codification, and 137 contributed papers presented at the 3rd International Conference on Steel and Composite Structures (ICSCS'07) held on 30 July–1 August 2007, in Manchester, UK. The conference was organised by the University of Manchester, in collaboration with "Steel & Composite Structures, *An International Journal*. We are grateful to Professor David Nethercot for giving us permission to use this conference to honour his 60th birthday. This series of conferences was founded by Professor C.K. Choi and the first two conferences were successfully held in Pusan, Korea (ICSCS'01) and Seoul, Korea (ICSCS'04) respectively.

These conferences bring together leading researchers and engineers from around the world to discuss and disseminate the latest advances in the analysis, design and construction of steel and composite structures under different loading and environmental conditions.

We would like to express our gratitude to the 7 keynote presenters whose support for the conference was essential to its success. Their papers cover areas of great interest to steel and composite structures. They include state of the art reports on advanced steel and composite structural analysis and design methods and steel and composite structural behaviour and design under fire and extreme loading conditions. We would like to thank Professor S.L. Chan for organising the special session on codifications and the 7 invited speakers to share with us their considerable expertise. However, ultimately, it is contributions from all participants of the conference that made this conference a memorable one. Thanks are also due to members of the International Scientific Committee for promotion of this conference and for reviewing the papers. Organising a major international conference would not have been possible without a dedicated team of local organisers and we would like to acknowledge the meticulous work of staff members of the conference centre at the University of Manchester. Dr. Y.C. Wang would also like to acknowledge his appreciation of the University of Manchester, particularly, School of MACE for allowing him to spend time on organising this conference.

The editors hope that the papers contained in this volume provide some inspiration to researchers and designers in their further endeavours to enable more innovative, elegant, safe and economic steel and composite structures to be built.

Editors

Dr. Y.C. Wang, *University of Manchester*
Professor C.K. Choi, *Korea Advanced Institute of Science and Technology*

Steel and Composite Structures – Wang & Choi (eds)
© 2007 Taylor & Francis Group, London, ISBN 978-0-415-45141-3

Organisation

Co-chairs

Y.C. Wang	UK
C.K. Choi	Korea
S.L. Chan	Hong Kong, China
D.A. Nethercot	UK

International scientific committee

C.G. Bailey	UK
F.S.K. Bijlaard	The Netherlands
R. Bjorhovde	USA
C. Both	The Netherlands
M. Bradford	Australia
I.W. Burgess	UK
W.F. Chen	USA
L. Da Silva	Portugal
J.B. Davison	UK
L. Dezi	Italy
D. Dubina	Romania
C.J. Earls	USA
J.C. Ermopoulos	Greece
E.S. Easterling	USA
J.M. Franssen	Belgium
L. Gardner	UK
Y. Goto	Japan
L.H. Han	China
J.E. Harding	UK
G.Q. Li	China
R. Liew	Singapore
E.M. Lui	USA
V. Kodur	USA
T. Krauthammer	USA
V. Kulyabko	Ukraine
D. Lam	UK
J. Loughlan	UK
M. Mahendran	Australia
P. Makelainen	Finland
D.B. Moore	UK
A. Orton	UK
R.J. Plank	UK
K. Rusmassen	Australia
P. Schaumann	Germany
N.E. Shanmugam	Singapore
Z.H. Shi	Japan
J.G. Teng	Hong Kong, China
B. Uy	Australia
F. Wald	Czech
B. Zhao	France
X.L. Zhao	Australia

Keynote papers

Steel and Composite Structures – Wang & Choi (eds)
© 2007 Taylor & Francis Group, London, ISBN 978-0-415-45141-3

Semi-rigid and partial strength joint action: Then, now, when?

David A. Nethercot

Department of Civil and Environmental Engineering, Imperial College London, UK

ABSTRACT: Research conducted during the past 100 years into the influence of semi-rigid and partial strength joint action on the behaviour of steel and composite frames is selectively reviewed. The emphasis is on the static behaviour of systems comprising conventional open section members. Comments regarding the maturity of the subject – in particular the extent to which the principles may be applied in practice – are made. Some suggestions for the relative lack of take up by Industry and a possible way to overcome this are offered.

1 INTRODUCTION

The earliest studies of the strength and restraining characteristics of structural steelwork beam to column connections took place almost a century ago (1) Laboratory tests on a variety of riveted connections clearly demonstrated the possession of finite degrees of moment capacity and rotational stiffness i.e. partial strength and semi rigidity. Despite this, steelwork design has remained firmly based on the two extremes of "pin" and "rigid" until the present time – why? By reviewing important developments and setting these against the background of practical need, an attempt is made to answer the question. The topic is then taken further by suggesting what additional developments might be necessary in order that the more logical, more rigorous and potentially more competitive alternative of semi-continuous design might become firmly established.

Undoubtedly the simplest assumption for the behaviour of the beam to column connections when designing regular steel frames is to assume pin joints. These are assumed to be incapable of developing any worthwhile moment and to permit free rotation between adjacent members. Some variants recognise that there might be a small degree of moment transfer – essentially due to the eccentricity of the line of action of the beam reactions compared with the column centreline. The method is very attractive in that it effectively requires no structural analysis, with a simple statical apportionment of loads, meaning that individual members may essentially be designed in isolation from one another. Because of the assumption of pin connections it is, of course, necessary for the structure to possess a separate lateral restraint system to withstand horizontal loading. In many cases lift shafts, service cores or specially designed braced bays fulfil this role.

The alternative assumption of rigid beam to column connections means that joints must possess a moment capacity at least equal to that of the adjacent members and should be sufficiently stiff for the initial angle between members to remain essentially unchanged under load. This then permits the frame to be analysed using conventional rigid frame techniques; nowadays this implies the use of standard software packages. Because of the interaction between components, the frame must be considered as a whole i.e. changing the properties of one member will influence the distribution of forces around the structure and thus potentially have an influence on the design of every other member. By combining member design requirements with the rigid frame analysis, recent American and Australian Codes have developed the so called "Advanced Analysis" approach to design which hands over all the structural calculations to a single computerised operation.

The third possibility – semi-continuous construction – has actually been permitted and recognised in certain specialist applications for several decades. For example, some clauses covering its implementation may be found in the earliest editions of BS449; these were subsequently revised and have been included in each revision of BS5950. The so called Wind Moment Method appears in technical papers and design documents from many parts of the world, certain practical design methods e.g. the systems employed by the manufacturers of light gauge purlin systems, implicitly rely on semi-rigid joint action but without making this apparent to the user. More recently, both EC3 and EC4 have included specific provisions for semi-continuous construction, linked to considerations of moment capacity, rotational stiffness and rotation capacity of steel and composite connections. In the United States Partially Restrained (PR) construction is a feature of the current AISC Code, following inclusion of more specialised Type 2 construction methods in previous documents.

The paper will review selected important contributions to the development of a growing understanding of semi-rigid and partial strength (SRAPS) joint design, will identify both barriers and encouragements to the use of semi-continuous construction and will close by suggesting what more remains to be done.

2 THEN – EARLY DAYS, PRE-1945

The introductory paper to the first ever edition of the Journal of Constructional Steel Research by Lord Baker (2) refers to his early experience with the Steel Structures Research Committee (SSRC – not to be confused with the American Structural Stability Research Council) and of locating "Only six papers were considered worthy of record" during a search in 1929. Of these "four addressed Structural Analysis"; he did not state whether one of the remaining two was the 1917 Illinois Engineering Experiment Station Bulletin by Wilson and Moore (1) that would appear to be the first report of laboratory tests on the restraint characteristics of Steelwork connections. This early study measured the moment transmitted by the joint and the rotation of the beam relative to the column, plotting the results in the form of a moment – rotation or $M - \varphi$ curve. It thus established the fundamental connection property, equivalent to the stress-strain characteristic for material or the load – end shortening relationship for a stub column. During the subsequent 90 years researchers worldwide have added hundreds of measured $M - \varphi$ curves to those reported by Wilson and Moore.

Following Baker's initial review the SSRC itself conducted tests on connections – both in the laboratory and on site using London buildings under construction (3–5). The former added to the pool of $M - \varphi$ data, further establishing the point that practical joints conform to neither of the classical assumptions of pins or fixed, whilst the site measurements demonstrated that the column joints transmitted non-negligible moments from the beams into the columns. Despite devising a design method that properly recognised this, the SSRC approach failed to dislodge the deceptively simple and computationally easy "Simple Design" approach as the normal method for designing non-sway steel frames in the UK.

On the far side of the Atlantic in the United States designers were more interested in sway frames i.e. a system in which lateral forces are resisted by frame action rather than by a separate lateral bracing arrangement, so concentrated on joints having rotational stiffness and moment capacity. Despite focussing on "Moment Frames", the recognition that real joints functioned in an intermediate fashion led

to the derivation of the Wind Connection Method (now referred to in the UK as the Wind Moment Method), in which joints are assumed to function as pins under the gravity loading but as rigid when resisting lateral loads. Specific types of connections were subsequently developed and the method promoted by the American Institute of Steel Construction under the title Type 2 Construction (more recently renamed Partially Restrained Construction).

Further studies followed the lead of the SSRC – notably those by Young and Jackson (6) and by Rathburn (7), plus a comprehensive study by Hechtman and Johnston (8) published in 1947. However the scene was changing, bolts and welds would rapidly replace rivets as the principal form of fastening, scientific research in Universities was set to expand, computing was soon to radically change our ability to conduct significant volumes of calculation, the benefits of composite construction were starting to be recognised and National Design Codes were providing a focus for delivering the findings from research.

3 GRADUAL REALISATION, 1945–1985

The initial thrust of the effects of semi-rigid joint action – starting with the original work of reference 1 and subsequently developed independently by Baker (4, 5) and by Rathburn (7) was on the elastic response of beams using the slope – deflection method. This was subsequently extended and adapted to embrace moment distribution and matrix stiffness techniques as summarised in the review by Jones et al (9). This then permitted the behaviour of complete frames to be addressed, including the overall response for sway frames.

However, a rather more informative development, again arising from the work of the SSRC – this time by Batho (4, 5) – was the beam line approach. Figure 1 illustrates the concept. Essentially, using moment – area principles the end restraint moment M_E may be determined as a function of the connection rotation from

$$M_E = M^F - \frac{2EI}{L} \phi_b \qquad (1)$$

in which M^F is the fixed end moment ($wL^2/12$ for a uniform load w).

By plotting the connection's curve on the same diagram the intersection point P gives the value of moment and rotation. Beam line methods were later developed by Kennedy (10) and Nethercot (11) and the concept used as the basis for the codified method given in a supplement to BS449 (12) and later adapted by Roberts (13) to work with the ingenious variable stiffness approach to column design proposed by

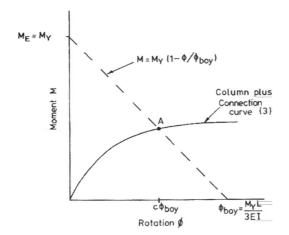

Figure 1. Beam line representation of equilibrium of beam and connection.

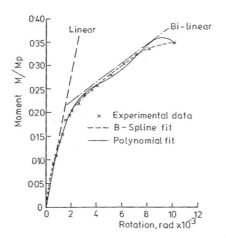

Figure 2. Representation of $M - \varphi$ Curves.

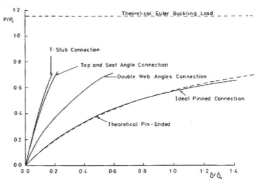

Figure 3. Load vs deflection curves columns.

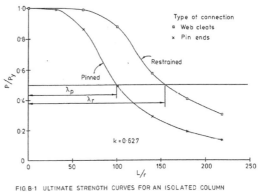

FIG.B·1 ULTIMATE STRENGTH CURVES FOR AN ISOLATED COLUMN

Figure 4. Extraction of effective length column curves.

Wood [14]. This technique, which requires only simple calculations, provides very useful insights into the interplay of strength, stiffness and rotation – the three key features for appreciating SRAPS design.

Work also took place on the effects of the end restraint provided by the combination of beams and connections to columns [15]–[17]. By suitably mathematically representing selected connection curves of the type shown in figure 2, analyses were set up to trace the column's load-deflection response up to failure as indicated in figure 3. The results could then be used to devise relationships between the column effective length factors and the degree of end restraint [18] of the type shown in figure 4 that permitted semi-rigid joint effects to be included in conventional column design using column curves. In principle, any arrangement

could now be treated in a far more rigorous fashion than previously.

The realisation that the proper representation of connection $M - \varphi$ data was an essential component of research into SRAPS action meant that considerable attention was devoted to devising appropriate representations [19]. This, in turn, meant that the test data needed to be collated. Fortunately, computing had advanced sufficiently that collections that had been laboriously assembled could now be handled in this way, thus greatly facilitating their usage in validation and comparative studies.

During the 1970s work in Delft [20] had identified the removal of expensive column stiffeners in moment transmitting beam to column joints as potentially economically very beneficial. The "difficulty" was that the resulting connections were now definitely semi-rigid and partial strength so member and frame design had to recognise this [21]. From these studies emerged the so-called component method of joint design, in which the response of a particular connection is modelled by combining the responses of each of its key components e.g. end plates in bending, bolts in

5

tension etc. This highlighted the three key measures of joint performance as:

- Strength (moment capacity) – principally governing beam strength.
- Stiffness – principally governing beam deflections and column strength.
- Rotation capacity – principally governing beam/frame strength when redistribution of moments is required.

Both these three properties and the component method feature strongly in the approach to joints and the link between joint behaviour and plane frame response taken by EC3.

Although the earliest recorded use of composite construction dates back to the late 19th Century, with the first Patent being filed in 1926, it was not until the 1960s that both research and practice became well established. Having addressed the basics of shear connection, beam and column design, it was natural for the interest in bare steel joints to migrate across to composite construction. Johnson and Hope-Gill (22) appear to have been the first to conduct tests on semi-rigid composite connections. The motivation arose because of the difficulties identified with design for full continuity; since the largest moments tended to occur at the joints where the cross-section was less efficient than for the mid-span in sagging bending, redistribution of moments was required for economic design. This meant that the lower part of the cross-section in the support region i.e. the steel beam's web and bottom flange, was placed in compression with the result that either local buckling reduces ductility or tight geometrical limits were required so as to ensure adequate rotation capacity, thereby eliminating several standard sections from consideration. Accommodating the support rotations necessary to redistribute moments satisfactorily within the joints themselves was therefore an attractive possibility. However, take-up was surprisingly slow e.g. Zandonini (23) in his 1989 review identified only 6 studies covering a total of 25 composite connections tests, by 1995 (24) this number had become 118. The real development of this topic therefore belongs in the next section.

Another important feature to appear in the 1970s was the use of numerical (FE) analyses of joints as a supplement to laboratory testing. Early studies were handicapped by lack of computing power, as well as by the absence of satisfactory techniques to handle bolts, friction, contact issues, bolt preload, lack of fit etc. All of these were, of course, simply challenges for the analysts so that by the 1990s studies resolving these problems were appearing quite frequently.

During this period interest started in areas in addition to the basic in-plane action of bare steel connections between I-section members. Studies of the Flexibility of welded tubular joints (previously

Table 1. Issues largely solved.

In-plane $M - \varphi$ test data for H sections
End restraint and column curves
2-D Frame Analysis
3-D response of columns in braced frames
Prediction of M_c, K and ϕ_u
Composite connection properties and $M - \varphi$ characteristics
Moment redistribution and rotation capacity requirement
Influence of connection restraint on beam deflections
Explicit inclusion of SRAPS provisions in Codes
Behaviour of connections under repeated and reversed loading

assumed rigid), joints in scaffolding systems (previously neglected), out of plane and 3-dimensional response (previously considered too difficult) all appeared. Most continue as research topics to the present day.

4 RESEARCH BOOM LATE 20TH/EARLY 21ST CENTURY

Table 1 list 10 topics which had, by this time, received sufficient attention for them to be regarded as mature and reasonably well understood. Thus the opportunity existed to exploit SRAPS knowledge in more realistic and more competitive design approaches. In particular, most of the issues listed in Table 1 that are of a fundamental nature had matured to the extent that:

- Conducting the exploratory tests.
- Identifying key features of behaviour.
- Developing appropriate theory and prediction methods, including numerical modelling.
- Simplifying the findings.

In terms of SRAPS Applications to the Static Behaviour of Bare Steel and Composite Frame Structures could be said to be largely completed. Of course, it is always possible to identify specific items not yet fully resolved but a combination of imaginative interpretation of existing knowledge plus carefully focussed additional studies should be sufficient to cover variants of these topics.

The exception in the table is response under seismic loading, for which significant progress (25, 26) – not discussed explicitly here as the topic is not within the author's field of expertise – has been made that there is a sense that work needs to be done before a similar degree of maturity is reached. This arises because a consensus of the benefits of recognising and including SRAPS in available approaches to seismic design is not yet clear. It is, however, of interest to observe particular developments e.g. the concept of joints or joint regions as devices to deliver ductility (25, 26), the use of potential connections (27) recognition that

Table 2. Growth of SRAPS papers in the JCSR.

Period	No. of SRAPS papers
1980–84	4
1985–89	17
1990–94	36
1995–99	52
2000–04	42
2005 date	37

frames with semi-rigid (rather than rigid) joints pick up load rather differently – and possibly less seriously – in their interaction with the ground (28. 29), the idea that braced frames with semi-rigid joints provide attractive alternatives to moment frames with complex joints that involve significant on site fabrication etc.

A simple yet interesting way of monitoring the growth in SRAPS related work is to note the increase in SRAPS papers appearing in the Journal of Constructional Steel Research. Table 2 shows how for each 5 year period since 1980–84 this has risen to the point where it is regularly 10 times the original figure. The most recent and as yet incomplete quinquennium looks set to significantly beat this.

5 WORK IN PROGRESS

It is still possible to find interesting and potentially valuable papers appearing on the basic issues i.e. prediction of joint moment capacity, techniques to assess the response of unbraced frames, but the suspicion is that the real challenges for developing and adopting SRAPS lie in new situations. Some interesting and as yet not fully resolved clues may be obtained from the past e.g. tubular joints, scaffolding, cold formed frames etc, but it seems likely that the principal interest will be in including SRAPS in contemporary issues such as:

• Behaviour in fire
• Dynamic response – i.e. floor vibrations
• More work on seismic resistance
• Progressive Collapse
• Response to impact or blast loading

Work on each of these issues according to the list of four features outlined above is in many cases at an early stage e.g. no studies are known of the important practical problems of the dynamic response of either composite floors or grandstands that include SRAPS considerations, Imperial work on progressive collapse is currently limited by lack of detailed data on the response of joints to the sort of load regimes that occur during a progressive collapse etc.

This suggests that SRAPS will continue to be a research issue for the foreseeable future but that, as always, the research community will need to be ambitious in the problems it tackles and imaginative in the approaches it follows.

6 ADOPTION

It of interest to reflect why – given the very substantial research effort over the past decades, culminating in numerous attempts to present the findings in a straightforward and usable fashion – there has not been greater adoption of SRAPS in practice. Interest in the design of structural connections per se has, as reviewed by the author (30, 31), radically changed over the past couple of decades, with the introduction of the BCSA/SCI Green Book Design Guides bringing about a substantial change in practice. Recently, the author (32) summarised simple design approaches for dealing with non-sway and sway bare steel and composite frames based on his own contributions; each are comparable in terms of calculation volume with traditional methods. Numerous other parallel references may be found dealing with alternative approaches. Reference 32 listed 7 "improvements" and 3 "conclusions" related to SRAPS, from which possible reasons for the limited uptake may be stated as:

• It is hard to improve on results obtained from the "simple construction" approach given that it combines effective lengths of less than the storey height and no significant column moments.
• Many forms of connection that actually behave as SRAPS are still regarded as rigid and full strength and design based on this presumption.
• There is a perception that using SRAPS requires more calculations than do traditional approaches.
• Its use requires a better understanding of the actual structural behaviour.

Set against this the benefits might be stated as:

(i) More design options and therefore more opportunities to identify the most appropriate solution.
(ii) Genuine economies compared with traditional approaches.
(iii) The move to limit states design and the current interest in performance based approaches in codes mean that greater consistency between the design approach and the resulting structural behaviour is expected.
(iv) Conducting calculations should no longer be an issue.
(v) A new generation need not be wedded to the old ideas.
(vi) Growing number of designs governed by serviceability conditions, for which SRAPS offers potentially greater benefits than for the ultimate condition.

All of these points to straightforward reluctance/inertia to adopt something new and to change established practices as being the essential reason. The solution:- why not introduce these concepts into structural teaching? After all, 50 years ago plastic theory did not feature in undergraduate university courses yet now who would teach a coherent degree course in Civil Engineering without the concept of plastic hinges featuring in the structural lectures? Pausing to think of the incidental benefits, SRAPS illustrates fundamental concepts such as:

- Resistance
- Stiffness
- Rotation Capacity
- Restraint
- Moment Redistribution
- Compatibility
- Load – Deformation Characteristics

No doubt others could add more but the list is a sufficiently convincing argument for using the concept of SRAPS as a way of illustrating in a realistic fashion many important structural principles. When might we see aspects of SRAPS become as commonplace as plastic hinge theory, column effective length, Merchant – Rankine Formula, P-Δ effects and other regular features of the interface between the behaviour of steel structures and approaches adopted in design?

7 CONCLUSION

Nearly 100 years of Research Activity into the effects of semi-rigid and partial strength joint action on the performance of components and structures has been selectively reviewed. The findings suggest: That a good understanding of the fundamental principles has been available for something over 20 years, that many of the findings have been distilled into workable design practices that compare favourably with existing approaches but that take up remains disappointing – largely due to inertia and general resistance to change. It is suggested that one way to resolve this would be to introduce the principles and simple applications into undergraduate steel design courses; it is argued that this would also have the benefit of reinforcing many key concepts in Structural Engineering.

ACKNOWLEDGEMENTS

Whilst responsibility for the statements made in this paper rest solely with the author, many of the ideas and thoughts are the result of collaborations with a number of talented and industrious research students and colleagues over many years. To all these grateful thanks is extended.

REFERENCES

1. Wilson, W. M. and Moore, H.F., "Tests to Determine the Rigidity of Riveted Joints in Steel Structures. University of Riveted joints in Steel Structures", University of Illinois, Engineering Experiment Station, Bulletin No. 104, Urbana, USA, 1917.
2. Baker, J. F., "Early *Steelwork Research*", Journal of Constructional Steel Research, Vol.1. No.1, 1980, pp. 3–9.
3. Steel Structures Research Committee, First Report, Department of Sc and Industrial Research, HMSO, London. 1931.
4. Steel Structures Research Committee, Second Report, Department of Scientific and Industrial Research, HMSO, London, 1934.
5. Steel Structures Research Committee, Final Report, Department of Scientific and Industrial Research, HMSO, London, 1936.
6. Young, C. R. and Jackson K.B., "The Relative Rigidity of Welded and Riveted Connections." Canadian Journal of Research, 1934, **II**, No. 1. 62–100 and **II**, No. 2, 101–34.
7. Rathbun, J.C., "Elastic Properties of Riveted Connections." Transactions of American Society of Civil Engineers, 1936, 101, 524–53.
8. Hechtman, R. A. and Johnston, B. J., "Riveted Semi-Rigid Beam to Column Building Connections. Progress Report No. 1, Committee of Steel Structures Research, AISC, November 1947.
9. Jones, S.W., Kirby, P.A. and Nethercot, D.A., "The Analysis of Frame with Semi-Rigid Connection – A State-of-the-Art Report", Journal of Constructional Steel Research, Vol.3., No. 2, 1983, pp. 2–33.
10. Kennedy, D. J. L. "Moment-rotation characteristics of shear connections", *Engineering Journal*, American Institute of Steel Construction 6, No 4, October 1969, pp. 105–115.
11. Nethercot, D.A., "Joint Action and the Design of Steel Frames" The Structural Engineer, Vol. 63A, No. 12, Dec, 1984, pp. 371–379.
12. PD 3343 Supplement No. 1 to BS 449: Part I: *The use of structural steel building*, London, British Standards Institution, 1971.
13. Roberts, E. H., "Semi-rigid design using the variable stiffness method of column design" *Joints in structural steelwork*, ed., Howlett, J.H., Jenkins W.M. and Stainsy, R., London, Pentech Press, 1981, pp. 5.36–5.49.
14. Wood R. H., *A new approach to column design*, London HMSO, 1974.
15. Lui, E. M. and Chen, W.F., "Strength of H-columns with small end restraints", *Structural Engineer,* 61B, No. 2, March 1983, pp. 17–26.
16. Jones S. W, Kirby, P. K. and Nethercot, D. A., "Columns with semi-rigid joints", *Journal of the Structural Division,* ASCE, 108, , No. ST2, April 1982, pp361–372.
17. Jones, S.W., Kirby, P.A. and Nethercot, D.A., "Effect of Semi-Rigid Connections on Steel Column Strength", *Journal of Constructional Steel Research*, Vol.1, No. 1, Sept, 1980, pp.38–40.
18. Lui, E. M, and Chen, W. F., "Strength of H-columns with small end restraints", *The Structural Engineer*, 61B, No. I, March 1983, pp. 17–26.

19. Nethercot, D.A. and Zandoninin, R., "Methods of Predictions of Joint Behaviour: Beam-to-Column Connections", in "Structural Connections" Stability and Strength", ed. Narayanan, R., *Elsevier Applied Science*, 1989, pp.23–62.

20. Zoetemeijer, P., "Influence of Joint Characteristics on Structural Response of Frames", in "Structural Connections" Stability and Strength", ed. Narayanan, R., *Elsevier Applied Science*, 1989, pp. 121–152.

21. Zoetemeijer, P., "A design method for the tension side of statically loaded bolted beam-to-column connections, *Heron* 20, No. , Delft, 1974.

22. Johnson, R.P. and Hope-Gill, M.C., "Semi-Rigid Joints in Composite Frame", *IABSE Ninth Congress*, Preliminary Report, Amsterdam, May 1972, pp. 133–144.

23. Zandonini, R., "Semi-Rigid Composite Joints" In Structural Connections, Stability and Strength, ed R Narayanan, *Elsevier Applied Science*, 11989, pp. 63–120.

24. Nethercot, D. A., "Semi-rigid Joint Action and the Design of Non-sway Composite Frames", *Engineering Structures*, Vol.17, No. 8, 1995, pp. 554–567.

25. FEMA 350 (2000). Recommended Seismic Design Criteria for New Steel Moment-Frame Buildings. *Federal Emergency Management Agency*, Washington, DC.

26. FEMA 355D (2000). State of the Art on Connection Performance. *Federal Emergency Management Agency*, Washington, DC.

27. ANSI/AISC 358-05 (2005). Pre-qualified Connections for Special and Intermediate Steel Moment Frames for Seismic Applications. *American Institute of Steel Construction*, Chicago, Illinois.

28. Elnashai, A. S. and Elghazouli, A. Y. (1994) Seismic Behaviour of Semi-Rigid Steel Frames: Experimental and Analytical Investigations, *Journal of Constructional Steel Research*, 29. pp. 149–174.

29. Leon, R. T and Kim, D. H. (2004). Seismic Performance of PR Frames in Zones of Infrequent Seismicity. *13th World Conference on Earthquake Engineering*, Vancouver, BC, Canada, Paper No. 2696.

30. Nethercot, D. A. "Towards a Standardisation of the Design and Detailing of Connection", *Journal of Constructional Steel Research,* Vol. 46, 1998, pp. 3–4 (Abstract, full paper available on CD Rom).

31. Nethercot, D. A., "Connection Research and its Impact on Practice During the Dowling Era", *Journal of Constructional Steel Research,* Vol. 62, 2006, pp. 1165–1170.

32. Nethercot, D. A., "How to Benefit by Using Semi-Rigid Construction". *South African Institute of Steel Construction 50th Anniversary Conference,* Nov 2006.

Steel and Composite Structures – Wang & Choi (eds)
© 2007 Taylor & Francis Group, London, ISBN 978-0-415-45141-3

From 30% to 70% market share – how did it happen in the UK?

D.B. Moore & D. Tordoff
The British Constructional Steelwork Association Ltd., UK

ABSTRACT: The reasons behind the growth in steel's market share in the UK are complex and comprise a diverse range of factors. These include productivity and capability improvements by steelwork contractors and steelmakers; 'design & build'; the demise of the 'main contractor'; CAD/CAM; the acceptance by clients of the benefit early return of capital invested; metal decking; intumescent paint; lighter foundations, longer spans, more lettable space; safety; a vibrant industry with strong leading companies and market segmentation of the capabilities of the companies in the industry; the introduction by Corus Group plc (then called the British Steel Corporation) of its Regional Structural Advisory service; commercial, technical, contractual and market development activities and the weakness of concrete marketing. The market share growth has taken decades to achieve and maintain.

This paper highlights, in chronological order, some of the key influences, initiatives and activities which have taken place, all contributing in one way or another to the success of the UK steel construction industry. It would not be possible to cover all topics in a single paper and hence this paper focuses on the industry collaborative actions by BCSA.

1 INTRODUCTION

To fully appreciate the reasons behind the success of the use of steel in construction in the UK it is necessary to look back at where the industry started and to understand how it developed, what the influences acting upon it were, eg competition from concrete; the impact of social, economic and political change; together with the infrastructure supporting the industry in terms of industry collaboration, technical and marketing activities, etc.

The historical development of steelwork in construction is a subject which has never been precisely recorded because it came about as part of the general development in building techniques dating back to before the industrial revolution. Cast iron beams and columns are recorded as having been incorporated in the building of a five-storey mill at Shrewsbury as far back as 1797 and, by the middle of the nineteenth century, developments in the use of wrought iron had made it acceptable to Brunel and Stevenson in the construction of many famous railway bridges still in use today. The demand for ironwork in buildings then grew very quickly and, under its impetus, technical developments soon made supplies of the new Bessemer and Open Hearth steels available. The first rolled sections in this new completely material to be used in a steel-framed building in the British Isles were in a furniture emporium in County Durham, built in 1900.

The British Constructional Steelwork Association Ltd was formed in 1906 and is the national organization for the steel construction industry; its Member companies undertake the design, fabrication and erection of steelwork for all forms of construction in building and civil engineering; its Associate Members are those principal companies involved in the purchase, design or supply of components, materials, services, etc, related to the Industry. The principal objectives of the Association are to promote the use of structural steelwork, to assist specifiers and clients, to ensure that the capabilities and activities of the industry are widely understood and to provide members with professional services in technical, commercial, contractual and certification matters. The services provided by BCSA work both for the overall benefit of the industry and the direct benefit of individual companies.

The turnover of the UK steel construction industry is currently approximately £5,000 million pa with 15,000 direct employees and a further 50,000 indirect employees. The UK steel construction industry is a world-leader and steel is the leading construction material in the United Kingdom. Steel's market share of non-domestic multi-storey building construction (buildings of two or more storeys) has increased from 33% in 1980 to an all time record 70% today. Steel maintains a 98% share of single storey non-domestic construction.

2 RETROSPECT

The history of Structural Steelwork, in common with the history of practically any other subject, does not have a particular starting point, and an arbitrary threshold must be chosen. After thousands of years of building in timber and masonry, it is only in the last quarter of the second millennium that metal has been used; indeed, it is only just over a hundred years since any significant steel structures appeared.

Cast iron, wrought iron and steel have developed one from the other and each has been incorporated into buildings and bridges using continuously improving techniques. It may well be worthwhile, therefore, to look briefly at the beginnings of the structural use of metal and follow its progress over what is, after all, a comparatively short period of time.

2.1 *Material*

The so called 'Industrial Revolution', it could be said, had its origins in the accidental coincidence of a number of happenings. One of great importance was Abraham Darby's discovery in the early years of the eighteenth century, that coke could be used in place of charcoal for smelting iron in a blast furnace. Cast iron became plentiful and cheap, finding endless uses domestically and industrially. It was easy to mould, had reasonable resistance to corrosion and had a formidable compressive strength, but it was unfortunately brittle and had poor tensile qualities. It must have seemed strange that, by using the same raw materials in another process, wrought iron could be made, exhibiting quite different characteristics. It was tough, malleable, had good tensile properties and could be welded simply by hammering pieces together at white heat. Unfortunately the production process was slow, output was limited and the end product was, consequently, very expensive.

So the alchemists went to work to find the philosopher's stone that would turn the now abundant supply of cast iron into a material with these very desirable qualities, in much larger quantities and at a more reasonable price. There were a number of false starts and claims that could not be substantiated but credit for the invention that stimulated a huge increase in the production of wrought iron is generally accorded to Cort who, in 1783, developed the puddling furnace. He also made another significant contribution in the invention of grooved rolls which enabled all manner of shapes to be produced with economy. Some were decorative but the greatest importance to the fabricator was the rolling of structural sections, initially angles and tees. The puddling process was still, however, highly labour intensive and was limited by what a man could manipulate from the furnace to the hammer, usually about 100 lbs. These small blooms could be combined by forging or rolling but even in the middle of the nineteenth century, it was exceptional to build up ingots weighing as much as a ton.

There was an urgent need for a better method and Bessemer, who was neither iron maker nor metallurgist, actually found something that he was not really looking for – mild steel. He was an able inventor and was trying to devise improved ways of producing wrought iron and carbon steel not, it must be said, for structural purposes, but to replace the brittle cast iron used in gun barrels. It was unfortunate that after the tremendous excitement created by the publication of Bessemer's work in 1856, the process proved to be unreliable and it took two further years of experiment to establish that good quality steel could only be made from iron that had been smelted from low phosphorus ore. In the meantime the iron masters, not easily persuaded that all was now well, had lost interest which encouraged Bessemer to set up his own plant in Sheffield. Output expanded quickly, allowing him to fulfill his original purpose, since his steel was used to make guns for both sides in the Franco-Prussian war.

The industry then flourished and wrought iron, which had held sway for most of the century, fell into decline. It is appropriate that in the final act, before the curtain fell, wrought iron was chosen in 1887 to build one of its finest structures – world famous – the first building to reach a height of 300 m – the Eiffel Tower.

The huge demand for wrought iron had led to the formation of many companies engaged in its production. Apart from the ironworks, where each blast furnace might serve as many as twenty puddling furnaces, there were also many wrought iron makers who bought their pig iron and built their own furnaces. After all, the technology was fairly simple and the capital cost of setting up was not enormous. Not all of them, of course, converted to steel making because the demand for wrought iron continued, although in continuous decline, right up until the 1950s. Nevertheless, after Bessemer and also later Siemens had done their work, it was not long before there were in excess of two hundred steel makers in England and Wales. The success of wrought iron production, indeed the world leadership, led to an unfortunate complacency and steel making in Great Britain started with problems which took nearly a century to resolve. There were too many companies operating on too small a scale, many with too wide a range of products. They were all fiercely independent. They suffered from nepotism where management was a matter of relationship rather than ability and they seemed blind to the fact that both production and efficiency were, in the USA and Europe, rapidly overtaking them. Bankruptcies, liquidations and amalgamations reduced the numbers over a period of time, but they were never sufficiently profitable to undertake the research and development that was necessary. Even so, their problems were by no means all of their own making. As an industry which relied on continuous volume production, it was the most vulnerable to any economic downturn, leading to serious ills from fierce price cutting. Also, transport had so much improved by this time that there was an international trade in steel and our manufacturers were further embarrassed by low priced imports from countries which protected their own industries by imposing tariffs.

As the years went by, unbridled competition, poor management, violent trade cycles, wars and political interference all impinged on the British steel industry which makes it all the more astonishing that late in the twentieth century, it emerged as a single, efficient public company.

2.2 *Structure*

The early structural use of cast iron could well have been unrecorded props and lintels around the ironworks themselves but Smeaton claimed to have used cast iron beams in the floor of a factory in 1755, and overcame the disparate values of tensile and compressive performance by designing asymmetrical sections, where the area of the bottom flange was several times that of the top. Another early use of the material was in the columns which supported the galleries in St Anne's Church, Liverpool in 1772 while in 1784 John Rennie built Albion Mill in London with a frame entirely of cast iron.

From the middle of the eighteenth century, the ironmasters, in their enthusiasm, turned their hand to making everything that they could from a material which was now becoming plentiful. 'Iron-mad Wilkinson', one of the most famous, made a cast iron boat and confounded the scoffers when it actually floated. He went on to build a church for his work people where the door and window frames were cast iron, as too was the pulpit. His only failure was in the casting of his own coffin, which sadly could not be used since prosperity had substantially increased his girth. But his greatest contribution, at least in the eye of the structural engineer, was as an enthusiastic promoter of the iron bridge, built by Telford in 1779, and which gave the name to the small town which surrounds it.

Perhaps the greatest demonstration of the advantages of mass production and prefabrication was presented by the construction of the Crystal Palace in which was housed the Great Exhibition of 1851. Here, on a grand scale, the processes of design, fabrication and erection were co-ordinated, allowing the whole structure to be completed in an extremely short time to the astonishment of the general public. Perhaps the mass production aspects of cast iron influenced the builders and would doubtless keep the cost down,

Figure 1. The first large-span bridge in cast iron – Coalbrookdale, 1779.

Figure 2. The Transept of the Great Exhibition of 1851.

but it is surprising that more wrought iron was not used. Even so, Prince Albert was so impressed that he ordered a prefabricated 'iron' ballroom which was duly erected at Balmoral.

Although the use of self-supporting steel frames became the norm for industrial buildings, there seems to have been reluctance, in the UK at least, to take advantage of the benefits that could be gained in commercial use. Offices, hotels and shops continued to be built traditionally although steel beams made possible greater unobstructed floor space and their use as lintels opened up the ground floors and created the street scenes that we know today. In some ways it is surprising that the urban American experience was so different. They had far more space than European cities and it might have been supposed that there would have been less urgency to build vertically. However, they were the architects who developed the potential of the metal frame in high rise buildings.

Wire rope made possible the design, by Elisha Otis, of the passenger elevator in the 1850s which, in turn,

13

Figure 3.   The Ritz Hotel, London.

Figure 4.   Section books.

made multi-storey buildings acceptable to the general public. However, the complete metal frame did not appear until the nine-storey Home Insurance Building in Chicago was erected in 1883. Wrought iron was still in vogue and it is only the four topmost storeys that were framed in steel, which soon became, and still remains, the preferred material for tall structures. Even so, the metal frame was not immediately or universally accepted and multi-storey buildings in Chicago continued, for a while, to use load bearing walls, but as these approached seven feet in thickness at ground level, the supporters of the system conceded defeat.

The Ritz hotel in London is often quoted as London's first steel frame but in 1906 it would not have been permitted to act as a complete steel skeleton. It must therefore be said that there were a good many steel framed buildings in various British cities which pre-dated it. A good example is the Midland Hotel in Manchester built in 1903.

If it is true that fashions are set in the capital city, then it is hardly surprising that steel framed buildings developed slowly. It was not until 1909 that London County Council acknowledged that the thickness of external walls might safely be reduced should a steel skeleton be introduced.

2.3   Section books

Dorman Long produced their first section book in 1887, setting out the properties of all the profiles that they rolled at that time including beams up to 18 ins deep. Others followed, although there was no standard applicable and each company rolled sections which it considered the most saleable. Some time later, a number of fabricators also produced books of section tables incorporating all manner of useful data. Before any sort of standard or regulation appeared, these were the sources of design information and contained recommendations on stresses, factors of safety and loadings

for various categories of buildings as well as formulae for the design of beams and columns. Roof trusses, compound beams and columns with safe loads over a range of spans and heights, wind pressures, details of sheeting and glazing and the design of gutters and downpipes all were included as people vied with each other to put together the most sought after handbook which would keep their name before the architects and engineers in whose hands lay the appointment of contractors.

It cannot be said that the last fifteen years of the 19th century was a period of continuous economic prosperity. However, neither were trade cycles as extreme as they had been, nor unemployment as severe or as long lasting. Demands for steel increased both at home and abroad, shipbuilding enjoyed some good years and the steel makers earned a little respite from the uncertainties of the previous decade but, although the UK's national production increased, its proportion of world output continued to decline as it was overtaken by both Germany and the USA.

Society was not totally devoid of social conscience. The first hesitant steps were taken in legislation to provide education for those who could not afford to pay, to prevent the worst abuses of child labour and to provide compensation for employees killed or injured at work, but seemingly nothing could be done to regulate the continuing trade cycles which were the root cause of so many industrial disputes. After a hundred and fifty years of industrialization we were no nearer to finding an answer to the problems in the relationship between employer and employee.

So the 19th century ended with many problems but with some hope and with one certainty – the structural steelwork industry was firmly established with a proven capability for the design and fabrication of contracts large and small, anywhere in the world.

2.4   The fabricating industry

More than most industries, the fabricators suffered from the peaks and troughs of trade cycles and in the interest of efficiency, most of them reduced their 'all things to all men' approach and concentrated on one or another aspect of the trade. The simple truth was that there were far too many companies competing in the structural fabrication industry. The decline in the use of cast iron and wrought iron naturally led foundries and forges into the use of steel, with varying degrees of efficiency and not, in the overall, with much success. The effect, however, of only a small number of competitors who were short of work or, as has frequently been the case, were not aware of their own costs, was to drive the pricing structure of the whole market downwards.

Of course the actions of the steel mills were closely watched by the fabricators because they knew that in spite of vociferous protests to the contrary, the mills were selling simple fabrication at cut prices directly to the general contractors. Equally, the mills knew that in spite of vigorous denials, the fabricators were, at times, buying steel from abroad. In fact this stand-off sufficed to keep both parties more or less in line with only occasional transgressions. There was, nevertheless, a deep rooted suspicion, which persisted until the 1980s when the steel industry finally sold off its fabricating capacity, that competition was unfair because, it was claimed, those companies which were owned by the steel mills received their materials at a considerable discount.

2.5 *Trade associations*

Against this background the fabricators found it very hard to make a living. Not only did they suffer the self-inflicted wounds of under-pricing their work to meet the fierce competition, they also found themselves, more often than not, in the position of sub contractor with the attendant hazards of 'Dutch auctioning', a device used by many main contractors to achieve rock bottom prices. To add to their woes, some of the main contractors were notoriously slow payers while a few proved to be unstable and in the event of their failure, the fabricator suffered a financial loss with no redress.

Observing that the steel makers were benefiting from co-operation, involving some degree of price fixing, and that labour rates were fairly consistent across the industry, the fabricators made tentative moves towards creating an organization for their protection. In 1906, five of the larger fabricators in the Manchester area put their heads together and the Steelwork Society was formed. The impact was not immediate since there were many competing firms outside the Society, but gradually these people found that membership was to their advantage and thirty years later the organization could muster a total of forty companies in the Northern Counties. Similar groups formed in other regions of the country, ultimately to be amalgamated into The British Constructional Steelwork Association in 1936.

The minutes of the early meetings of the Society are anything but revealing. The only thing regularly and accurately reported is the passing for payment of stationery and postage charges, which is hardly an accurate measure of their activity. The substance of the meetings is only hinted at and it is clear that the members were uncertain of their legal position to the extent of giving themselves code names and numbers. Reading between the lines, it would seem that their activities were fairly harmless, concerned mainly with exchange of information about wage rates and conditions and discussion on the effect of various pieces of legislation. Clearly, meeting and getting to know each other helped them to exchange small favours, but such arrangements were obviously outside the official business.

Tentative moves towards promoting steel construction were made by the propaganda committees of the regional groups without significant progress until, in 1928, The British Steelwork Association was established, supported by both fabricators and steel makers, with the declared intention of promoting the use of steel for constructional purposes, by means of publicity, marketing, information and technical research.

The activities and authority of the BCSA grew rapidly and were merged into the new BCSA organization on its formation in 1936. Although its life was short, the excellent work done by the BSA should never be overlooked. Its initiative in research brought real economies to the design of structures while, at the same time, its promotional efforts, which included some excellent literature, were widely applauded. In these respects, the high standards set an excellent example for the BCSA to follow.

On the outbreak of war in 1939, the total capability of BCSA was offered to the relevant government departments who on many occasions, during the following six years, called for assistance in fulfilling urgent requirements. At one time, the whole BCSA drawing office was engaged in converting ship builders' drawings into those more suitable to the fabricators' workshops. BCSA then arranged for its members to fabricate thirty thousand tons of components for dispatch to the shipyards. The fabrication of bridges and barges, hangars and hulls for armoured cars and the various works for Mulberry harbour, were handled by BCSA which matched requirements to capacity and capability in the allocation of work.

After the war, a number of important technical books were published by BCSA eg on plastic theory, and in 1955 alone 22,000 technical publications were distributed.

2.6 *Design standards*

In the first years of the 20th century, design was left very much to the discretion and skill of the engineer or architect. There were no universally agreed permissible stresses or factors of safety, nor were design methods in any way mandatory. Reliance was

Figure 5. Floating bridge.

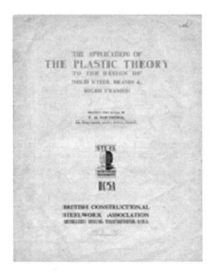

Figure 6. Technical books – Plastic Theory.

Figure 7. Building with Steel magazine, Feb 1960.

placed entirely upon the integrity of the designer and it must be said that this confidence was very seldom misplaced. The incidence of collapse was rare and, as throughout the history of metal construction, the period of instability during construction was by far the most hazardous time. Nevertheless, when submitted to modern methods of analysis, some old designs, particularly the connections between members, have been found wanting, but the understandable ignorance of those responsible remains hidden by the factors of safety and by the forgiving nature of a ductile material.

The first steps towards regulation came with the foundation of the British Standards Institution. On 26 April 1901 the first meeting of the Engineering Standards Committee took place. As a result, BS 1 came about through which the variety of sizes of structural steel sections was reduced from 175 to 113 and the number of gauges of tramway rails was reduced from 75 to 5. This brought estimated savings in steel production costs at the time of £1 million a year. Steel merchant's costs were reduced due to fewer varieties. This made steel cheaper for the users so everyone benefited.

The first regulations controlling design came in the London County Council (General Powers) Act of 1909 which gave detailed rules on permissible stress and loading and also, very significantly, made it lawful to erect '... buildings wherein the loads and stresses are transmitted through each storey to the foundations by a skeleton framework of metal...'.

3 A NEW PROMOTION CAMPAIGN

At the start of the 1960s it was realised that a step change was needed and a first serious attempt was made at growing the steel construction market.

BCSA, together with the steelmakers, launched a new 'Joint Propaganda Project', involving new staff, new brochures, a magazine, a programme of works visits, etc. 'Building With Steel' commenced publication in February 1960 with a circulation of 15,000 copies.

The tacit agreement between 'steel' and 'concrete' not to cite the other materials in their 'propaganda' started to break down in 1961 when a leaflet 'Why Choose Concrete' drew comparisons between concrete and steel.

A group of architects was taken by BCSA to Paris in 1962 to see steel framed buildings being erected and the extent to which no fire protection was being applied. By 1962 BCSA's annual promotion spend

Figure 8.   Structural Steel Design Awards Trophy.

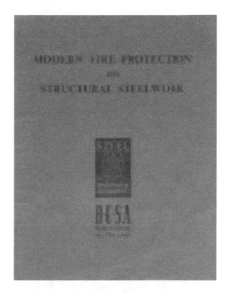

Figure 9.   Modern Fire Protection for Structural Steelwork, 1962.

had reached £96,000 equivalent to £1.4million at 2006 values. In 1963 BCSA published 62,000 technical brochures, 58,000 pamphlets for engineers and architects and 86,000 copies of 'Building With Steel'.

However by 1965 unfavorable comparisons were being drawn by members and others to the increasing activities of the Cement & Concrete Association (C & CA); designs were being converted from steel to concrete on the grounds of cost and speed of erection. 'BCSA is unable to come to terms with basic problems and its obvious lack of imagination' said one member.

An Export Group was established in 1966 with an Export Director who made extensive overseas visits to Conferences and Exhibitions to promote the British structural steelwork industry. What was to become the industry's ongoing flagship promotion event, the Structural Steel Design Awards Scheme, was established in 1968 and 40 entries were received in the first year.

3.1   Technical developments

In the late 1950s and early 1960s, three major developments helped the industry:

- the introduction of high strength bolts – one result of the combined effect of high strength bolts and arc welding was to virtually eliminate rivetting, both in the fabrication shop and on site.
- the rolling of Universal Beam and Column sizes – when introduced in 1962 immediately increased the ability to compete in overseas markets.
- the development in the use of digital computing in engineering – the computer provided a powerful tool to both designer and draughtsman. In the beginning

this was used for the analysis of rigid frames, thus enabling the true potential of electric arc welding to be fully appreciated.

The new section range in 1962 together with increased yield strengths necessitated the development and publication of a new set of safe load tables. Further new technical publications continued to be produced, eg 'Deflection of Portal Frames'.

The results of an investigation into the economics of alternative methods of fire protection were published in 1962, viz 'Modern Fire Protection for Structural Steelwork' and BCSA encouraged the formation of the Structural Steelwork Fire Protection Association in order to promote lightweight protection systems. A series of full scale fire tests was carried out at the Fire Research Station in 1964.

BCSA recognised the benefits of University links and in 1966 sponsored a programme of research at the Universities of Cambridge, Swansea and Newcastle and at Imperial College. During the 1960s BCSA developed a bridge department which established a world wide reputation for its design advisory service. In 1967 a heavy involvement took place in the changeover to the metric system and this necessitated the production and printing in metric of many publications, eg the Safe Load Tables and 'Metric Practice', plus courses for members. A special working party was established in 1968 to determine the types of weathering steel which could be made available. By 1969 BCSA had established a Computer Design Service through which members could 'obtain scantlings for single bay portal frames under various loading conditions within 24 hours'.

Figure 10.   Black books.

4   THE 1970'S

With the economic downturn BCSA's fortunes suffered at the end of the 1960s along with those of the industry and it became necessary to reduce the levy paid by members by 40%. Costs needed to be cut and, following discussions with the British Steel Corporation, CONSTRADO (Constructional Steel Research and Development Organization) was formed in the early part of 1971 as a new 'independent' technical organization (to be funded by the British Steel Corporation). Certain of the BCSA's technical staff, including the Technical Director, were transferred to CONSTRADO. Whilst some of BCSA's technical activities, eg education, bridge design department, sponsorship of University research, etc were transferred to CONSTRADO it was recognized that BCSA needed to retain a core technical expertise to represent the industry on technical issues such as Building Regulations, welding, computers, etc and to advise members on technical matters.

BCSA's cost study of multi-storey buildings project resulted in the devising of a method of construction for application to a range of office buildings of up to nine storeys in height, incorporating a low cost fire protection system.

The Appraisal Rules issued by the Merrison Committee in the early 1970s after the collapse during erection of certain box girder bridges gave rise to considerable concern amongst the bridgework fabricators; the arduous requirements of the Rules led to the establishment of a working party on tolerances. Subsequently BCSA became extensively involved in the drafting of the new limit state rules in BS 5400 for design and fabrication to ensure that steel bridges continued to be safe and economical to construct.

Similarly the Association was involved in the middle of the decade in the early stages of the revision of BS 449 (later to become BS 5950) 'so that the thinking in that regard could be considered and in order also to ensure that the proposed adoption of limit state design was kept within practical bounds'.

In 1977 the ECCS Recommendations on the Design of Constructional Steelwork were published and it was proposed by the European Commission that they 'be adopted as the model for the new Eurocodes'.

BCSA's Industrial Training Advisor regularly visited member companies to advise members on training courses and to advise members on grants from the Engineering Industry Training Board.

4.1   *Contractual*

Changes to the way in which steel construction contracts were placed were vital for improving the fortunes and standing of the industry. At the start of the decade BCSA published its Conditions of Sale for Structural Steelwork for use by members incorporating improved payment terms, in particular, progress payments for black steel and for fabricated material stored at works. These, and also BCSA's new model fluctuation clauses, were registered with the Register of Restrictive Trading Agreements. The first edition of the Members' Contractual Handbook was published in 1973. Payment terms were an ongoing concern and in 1976 BCSA reached agreement with the government to increase the proportion of the value of payment for materials delivered to site to 97%. Continued pressure was also maintained for the introduction of provisions allowing for the early release of retentions to steelwork contractors who frequently had to wait months, even years, before their entitlement is paid over. During 1979 BCSA introduced its Professional Indemnity insurance scheme for members.

4.2   *Specify steels*

In 1975 a new promotion campaign was launched called 'Specify Steel' to help remedy the fall-off in members' order books. The campaign opened with a co-ordinated advertisement in the Financial Times on 25 March together with an editorial supplement. The core of the new campaign was a series of promotional publications in several languages, incorporating the campaign logo and demonstrating members' achievements in various market sectors. 20,000 copies were produced in Farsi, Arabic, Spanish and Portuguese and distributed to embassies, government agencies and specifiers overseas.

Figure 11.   Specify Steel Campaign.

Figure 12.   BCSA News.

A BCSA display stand was also developed for use at exhibitions, together with car stickers, a 'Specify Steel' leaflet and regional specifier seminars.

The British Steel Corporation (now Corus) launched its highly successful, and still ongoing, Structural Advisory Service and programme of seminars and sponsored research. To this day regional advisory engineers keep in regular contact and visit designers and specifiers to show them, by alternative designs if necessary, how their construction projects can best be built in steel. The research and seminar programme was particularly successful in changing attitudes to steel's performance in fire. This programme was one of the main factors behind steel's success and it continues to move with the times, today featuring sustainability, safety in design and construction, etc.

5   THE 1980'S

A result of BCSA lobbying, the Secretary of State for Transport agreed in the early 1980s that twelve bridges would go out to tender based on two designs – a concrete design and a steel design. Prior to this most designs were done in concrete; steel won the vast majority of the tenders and this led to the growth in the use of steel for bridgework.

With considerable support from the British Steel's technical marketing campaigns, steel's popularity continued to grow. In the mid-1980s the overall efficiency, competitiveness and quality of steel-framed construction became increasingly recognized. 1985 saw a large resurgence of interest in steel construction which was achieved not only because of fundamental economic factors, but also because of various technical

innovations, such as profiled steel sheet decking, lightweight fire protection, and the introduction of the BCSA/DTI sponsored FASTRAK 5950 suite of computer programs.

1985 also saw the industry's efforts to put its house in order, by way of improvements in quality, responsible wage settlements, quick and reliable on-site erection. These were rewarded when, for the first time, steel superseded in-situ concrete as the most popular form of construction for multi-storey buildings. By the end of the 1980s British Steel Corporation had been de-nationalised and had become an efficient steel producer by world standards of productivity.

5.1   *Market development*

In 1984 BCSA restarted a promotion magazine with its new quarterly 'BCSA News'. The previous 'Building with Steel' had been taken over by the British Steel Corporation and then soon ceased its publication. In 1986 the title of the new BCSA magazine was changed to 'Steel Construction' and by 1988 was being issued six times per annum, with a distribution list of 13,000.

Press publicity began to be increased with special 'steel construction' supplements in the trade and regional press, eg in the 'Yorkshire Post' and in the 'Birmingham Post'.

BCSA took an active role in the National Economic Development Organisation's (NEDO) Constructional Steelwork Economic Development Committee which was active in the fields of research, education, promotion and relations with government. A NEDO report

Figure 13. Fastrak 5950.

on the 'Efficiency of Multi-storey Buildings' was published in 1985.

5.2 *Contractual*

In 1984 BCSA launched its Liability Insurance Scheme for members and in 1985 published its Model Tendering Terms and Conditions for Steelwork Fabrication. In addition to the regular local and regional members' meetings held throughout the country in 1986 the programme of regular National Meetings was introduced with one of the first being on 'The Invisible Costs of Contracting' at which guest speakers discussed the fabrication industry's contractual problems. Unreasonable warranties were a major problem and hence BCSA published its own standard form of connection warranty which resulted in many clients agreeing to remove some of the more objectionable clauses.

5.3 *Computer applications*

At the start of the 1980s BCSA acquired several of the new microcomputers and instigated a loan scheme whereby members could borrow them 'to demonstrate their capability as aids in the design, detailing and fabrication shop information'. During 1982 over 40 member companies took part on the scheme. The

Figure 14. Steel Construction Quality Assurance Scheme.

FASTRAK 5950 suite of programs commenced in 1984 under BCSA's direction for the design and estimating of multi-storey and portal framed buildings designed to BS 5950. In 1987 BCSA conceived and instigated the pan-European Eureka project which eventually resulted in the publication of the Cimsteel integration standards.

5.4 *Technical*

Demand for technical publications remained buoyant and many new books were published, eg 'Manual on Connections', 'Fabrication of Steel Bridges', 'Erector's Manual', 'Erection of Structural Steelwork', 'International Structural Steelwork Handbook' and 'Historical Structural Steelwork Handbook'. The new limit state BS 5950 to supersede the permissible stress based BS 449 was finally published in August 1985, however, it was to be another fifteen years before it effectively replaced BS 449.

In 1986 BCSA developed and launched the Quality Assurance Certification Scheme under the direction of an independent Governing Board. CONSTRADO was 'privatised' and became the Steel Construction Institute, an independent centre of technical excellence. SCI went on to play a major role in helping to provide the technical tools to facilitate steel's market share growth. The BCSA launched the National Structural Steelwork Specification for Building Construction (the 'Black' Book) with the aim of achieving greater uniformity in contractual project specifications and eliminating the plethora of conflicting requirements which were faced by the industry. This necessitated extensive consultation and research, eg into the standards and level of inspection of welds.

6 THE 1990'S

The early 1990s saw UK economy and the construction industry in recession. Not only was there a dramatic

Figure 15. The Register of Qualified Steelwork Contractors.

drop in the production of structural steelwork from c 1,400,000 tonnes in 1989 to 780,000 tonnes in 1992, prices also collapsed. All companies suffered greatly and many ceased trading. The industry, the backbone of construction, faced a difficult situation – attacked from all sides, unable to control its present and unable to plan its future. Most notably companies in the steel construction sector suffered severely when several large London property developers got into trouble. But even in the difficult times the industry continued to move forward; in the home market steel's market share of buildings of two or more storeys continued to increase. Exports saw a dramatic leap in the early 1990s with companies proving that the UK has a world leading steel construction industry by winning orders all around the globe. In anticipation for the Single European Market, member companies looked towards mainland Europe to develop their market potential and to increase their share of exported steelwork; to assist BCSA produced a detailed Guide to Exporting Steelwork to Europe

A major initiative in 1995 was the development and launch of the Register of Qualified Steelwork Contractors Scheme. This was more than just a list of companies as each applicant company must qualify by being audited by specialist auditors who check the company's financial and technical resources and track record. The Highways Agency quickly gave its endorsement by including in its tender documentation a requirement that only firms listed on the Register

for the type and value of work to be undertaken will be employed for the fabrication and erection of bridgeworks

6.1   *Contractual*

In 1992 an amendment to SMM7 came into force; this consisted of changes negotiated by BCSA to improve the steelwork section of the standard method. By 1994 it was generally felt that an improvement in the industry's fortunes was on its way. Sir Michael Latham's independent review of procurement and contractual arrangements commenced and promised to be a watershed for the entire construction industry – starting a return back to good practices that could only benefit the industry. Throughout 1995 members responded to the campaign to lobby their MPs to legislate against contractual and payment abuses; the success of this campaign led to the inclusion of construction legislation in the Queen's Speech in November 1995. Legislation was introduced into the House of Lords in February 1996 as part of the Housing, Grants, Construction and Regeneration Bill and received Royal Assent in July 1996.

The late 1990's saw improvements to the commercial environment with the elimination of cash retentions on steel construction contracts when BCSA registered an agreement with the Office of Fair Trading whereby members agreed that they would not accept the deduction of cash retentions.

6.2   *Promotion*

In 1991 BCSA launched its Steel Construction Challenge for Schools and new brochures were prepared to promote the industry. The Association's magazine was combined with SCI's magazine into a new bi-monthly publication 'New Steel Construction' with a circulation list of 10,000. The first National Steel Construction Week took place in October 1992 when over 3,000 delegates took part in 50 separate events around the country. In 1997 the first 'Directory for Specifiers and Buyers' was published to explain, not only the capabilities of member companies, but also the competitive advantages of steel in construction.

In 1999 over 1,000 delegates from 26 countries came to London for BCSA's International Steel Construction Conference and Exhibition, culminating with a Millennium Banquet held at Guildhall with HRH The Princess Royal as the Principal Guest.

6.3   *Technical*

In 1991 the Construction Products Regulations were published to facilitate the introduction of CE marking – although it was to be another 15 years before this started to make an impact on the industry. New books continued to be produced, for example 'Moment Connections' in 1995. A Commentary on the National

Figure 16.   Directory for Specifiers and Byers.

Figure 17.   Mind the Gap brochure.

Structural Steelwork Specification was first published in 1996 and the first in a series of Health & Safety booklets. A series of government sponsored research projects were carried out looking at the future structure and direction of the industry.

The Association was active in a wide range of activities, including: wind loadings, stability during erection, simplification of EC3, fire, health & safety, CDM regulations, CAD/CAM, connections group, welding group, CNC users group, EPA group, etc. BCSA worked closely with its colleagues in Corus to develop the steel construction market and BCSA, Corus and SCI together started planning the design guides which will be necessary for a smooth transition to implement the Eurocodes.

6.4 *Commercial*

Overseas missions took place to Japan, China, South Africa, Italy and Brazil to learn from sister industries worldwide and to seek out new export opportunities. As part of the Association's strategic review it was concluded that that the two major aims to be pursued in the future would be to help improve members' business competence and their profitability.

A particular focus has always been placed on developing good relationships with government Ministers and officials and in 1999, the Construction Minister said 'Without hesitation I can say that your sector – and more specifically BCSA – is held in high regard both within government and outside. The constructional steelwork sector is a vital element of the UK construction industry. Your sector has much to be proud of and I know you continue to look for

opportunities for further innovation and development – developments in your products – developments in your methods of working. I cannot stress too highly how government benefits from having a constructive dialogue with an industry that is innovative and forward looking – an industry that both responds to and informs customers' interests here in the UK and overseas. Your staff and members make an invaluable contribution to the development and implementation of new codes and standards'.

7   2000 ONWARDS

The new millennium got off to a good start with steel showing its dominance as a framing material for multi-storey buildings continuing with a market share at 69%. Steel was strong in all sectors, but showed its greatest dominance in the multi-storey industrial (92%), leisure (79%), retail (75%) and office (70%) fields; in no area did its share of the market fall significantly below 50%. In the non-domestic single-storey building market steel enjoyed a market share of 90%. Speed of construction remained the number one reason for choosing steel, with 'lowest overall cost' coming second.

2004 was a turbulent year for the steel construction industry with frequent sharp increases in the price of our basic raw material - steel. The price increases were driven by global pressures on iron ore, coke and transport costs. The increases not only applied to steel

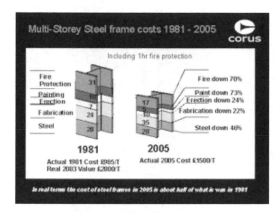

Figure 18. Multi-storey steel frame costs.

Figure 19. Safer Steel Construction.

sections, plate and strip, but also to reinforcing steel. During 2004 steel sections increased in price by around 50%. However the price of rebar increased by 50% in the first three months of the year alone. Consequently steel's competitive position over concrete for construction remained; for example the cost of a steel beam and composite slab floor building frame was then around £90/m^2 compared to £150/m^2 for a reinforced concrete frame and flat slab and £175/m^2 for a concrete frame with a post tensioned flat slab.

In real terms, the cost of steel frames is currently about half of their cost in 1981.

Despite the dramatic rise in steel prices during 2004, industry output and market share both increased and forward orders continued to be very healthy.

In 2005 steel prices were more stable with price rises for steelwork in the order of 5 to 7%. Demand in UK was nearly as high as the all-time peak of 1.4 million tonnes in 1988/89: in 2004 output was 8% up on the previous year at 1.3 million tonnes. The industry continued to gain market share in a number of key sectors such as residential, hospitals and educational.

By 2006 BCSA members' forward order books were in a healthy situation and the latest independent report showed that steel's market share had reached an all time record high. Steel's share of the multi storey non-residential buildings market had reached 70% for the first time; over the past 25 years steel's market share had steadily increased from 30% to 70%.

The latest independent annual cost comparison study between steel frames and concrete once again proved that steel provides the most advantageous and economical framing solution. Although all construction products are facing raw material and energy price increases steel, with the mills facing universal power/energy and raw material cost increases, is not the worst affected material. These pressures, combined with higher steel demand levels, lower stocks and higher prices in other regions of the world resulted

in the steel mills indicating that conditions were right for them to introduce higher market prices for structural sections. The section price increases are expected to result in the price of fabricated steelwork increasing by c 15% over 2006.

7.1 Contractual

Through its contacts in Government, BCSA was instrumental in obtaining a statement in the March 2004 Budget that a review would take place of the operation of the adjudication and payment provisions in the Construction Act in order to identify what improvements can be made to change the situation regarding the unreasonable delays in payment which members suffer in construction contracts. It is hoped that this Review will deliver results: BCSA is pleased that the government accepts that there is a problem as far as certainty of payment is concerned and that action is needed to ensure that the vital lifeblood of the industry – cash – flows more smoothly in future. Money is not everything but it is extremely difficult to live without it.

Timely and detailed project information is another issue which affects all and together with other organisations – such as the Association for Consultancy and Engineering – BCSA is aiming to produce a new document for steelwork alongside the National Structural Steelwork Specification which will set out guidance on information requirements.

7.2 Health and safety

A key theme of BCSA's is health and safety. The reality of life is that regrettably accidents will and do happen, but all must recognise this and work together to reduce, if not eliminate, the risk of accidents, no matter where they are likely to occur. Safety is our collective responsibility. The health and safety of all the people involved in the industry was therefore placed at the top of BCSA's agenda. To this end BCSA was active in a

Figure 20. Steel Construction News.

Figure 21. University CD pack.

variety of areas, both on and off site – publishing new guides, liaising with HSE and researching the causes of accidents.

In 2002 the Minister with responsibility for Health and Safety in Construction launched BCSA's "Safer Steel Construction" programme. This was a wide package of measures, including the Safe Site Handover Certificate, which quickly gained acceptance amongst the leading main contractors as the standard setting requirement for the safe sites. The following year saw the launch the new BCSA "Erector Cards", jointly with the Construction Skills Certification Scheme.

In 2004 new best practice guides were prepared for the safe erection of low rise buildings; metal decking and stud welding; bridges; and erecting steelwork in windy conditions. The Steel Construction Certification Scheme extended its scope to certification services for health and safety, in addition to quality management and environmental management.

Two more guides are being published in 2006 – one on the erection of Multi Storey Buildings and one on the Loading and Unloading of Trailers. A new 'Safety in Steel Construction' service is being launched in October 2006.

7.3   Technical

With regard to safety in use, it is a fact that structural frames, whether they be steel, concrete or timber, need to be designed to resist fire and by 2003 an extensive programme of full scale fire tests had been successfully carried out on the steel frame at Cardington. BCSA put forward the view that all new forms of construction for frames and floors in all materials should be extensively fire tested.

The Construction Products Regulations and CE marking for steel sections and plate were implemented in September 2006 and this has significant implications for steelwork contractors with regard to traceability. CE marking for fabricated steelwork will probably be introduced in 2007. BCSA is issuing guidance to members on this topic.

With the forthcoming Eurocodes BCSA is doing all it can, together with its partners Corus and SCI, to help with implementation in order to ensure steel structures continued to be easy and economic to design. For example the National Annex for Eurocode 3 is being drafted by BCSA and design aids for Eurocode 3 and 4 are under development. However it is vital for the construction industry as a whole and its clients that British Standards should be maintained until such time as the new Eurocodes and their supporting documents have been demonstrated to be user friendly, unambiguous and resulted in safe and economic structures. It is unlikely that the full package of Eurocode documents will be available until 2007/8 – hence the planned withdrawal of British Standards in 2010 is far too early and needed to be extended.

7.4   Marketing

A second magazine, "Steel Construction News", has been launched, jointly published by BCSA and Corus. The circulation of which is over 100,000 copies per

Sustainability
in Steel Construction

Figure 22. Sustainability.

Figure 23. Special Report on Steel Construction.

issue – the biggest of any publication in the UK construction industry – distributed by way of inserts into the leading trade magazines.

In 2002 a highly successful bridge conference attended by 240 delegates was held at the Institution of Civil Engineers to launch the new 'Steel Bridges' book. BCSA continued to disseminate best practice guidance and in 2003 over 300 delegates attended the Steel Buildings Conference and Exhibition. At the Conference a new comprehensive book on 'Steel Buildings' was published. Increased promotion took place during 2003/4 with full page advertisements in the press showing examples of steel framed hospitals, multi-story residential buildings, schools and car parks. A further Conference and Exhibition took place in 2005 with the publication of a new 'Steel Details' book.

In 2004 the first of a new annual "Steel Construction – Be Part Of It" pack was sent to almost 10,000 university undergraduates, comprising an introductory booklet about steel construction together with explanatory CDs.

7.5   Co-operation

Through the Specialist Engineering Contractors Group (SECG) BCSA co-operates with the other specialist engineering contracting sectors to develop better payment and contractual terms; the review of the Construction Act is one result of this co-operation. BCSA is leading the discussions between the UK government's Olympic Delivery Authority and SECG for the construction of the stadia and other facilities for the 2012 Olympics. Through the Metals Forum (an umbrella body covering 500,000 employees) BCSA works with other trade associations to raise the profile of the metals manufacturing and processing industries and to identify new European legislation which will impact on member companies.

7.6   Sustainability

In 2005 the Steel Construction Sustainability Charter was launched by BCSA, the objective of which

was to develop steel as a sustainable form of construction in terms of economic viability, social progress and environmental responsibility. BCSA requires that Sustainability Charter members make a formal declaration to operate their businesses in efficient and financially sustainable ways in order to undertake contracts that satisfy clients and add value for stakeholders. BCSA is auditing Charter member companies and plans to develop and publish key performance indicators that benchmark the development of sustainable steel construction generally and that permit individual Sustainability Charter members to measure their own progress.

7.7   BCSA Centenary

The Association's Centenary Dinner was held at the Savoy Hotel, London on 7 March 2006 attended by 450 members, their guests and guests of the Association. The Principal Guest, The Minister for Industry, Construction and the Regions, congratulated the BCSA on its Centenary, saying that the industry's achievements are all around us and that "steel is now more than ever the material of choice".

The Prime Minister, Tony Blair MP, wrote:

'Over the past 100 years steel construction has become an indispensable part of the built environment. The proof is there for us all to see in our daily lives – through new steel-framed hospitals, schools, transport terminals, power stations, bridges, water tanks, factories, offices, residential buildings and sports stadia. You have a history to be proud of, from Tower Bridge in London, Osaka airport terminal in Japan, the giant telescope in Hawaii, water tanks in India, industrial structures in Israel, to power stations in China, to give just a few examples. BCSA members are those whose technical knowledge and capabilities are regarded as amongst the best in the world. They have embraced

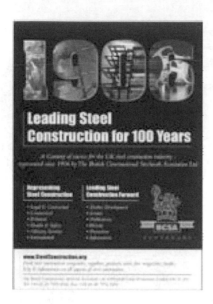

Figure 24. BCSA Centenary.

innovation in all aspect of their operations. Their contribution will continue to be vital to achieving what we all seek – a world-class built environment, built by a world-class construction industry.'

In association with BCSA's Centenary, the 'Financial Times' published a Special Report on Steel Construction: http://news.ft.com/reports/steelconstruction 2006. Over 130,000 copies of the report were printed and included in the newspaper on 8th March 2006.

The Association's Centenary Banquet was held at Blenheim Palace on 16 June 2006 attended by 460 member company representatives, Past Presidents, Fellows and staff.

Looking forward, the hosting of the 2012 Olympics should prove to be of great benefit to the industry and increase demand by about 5% over each of the next five years.

8   FURTHER READING

For further background reading on the development of the use of steel in construction in the UK see: www.steelconstruction.org/CenturyOfSuccess.

Steel and Composite Structures – Wang & Choi (eds)
© 2007 Taylor & Francis Group, London, ISBN 978-0-415-45141-3

Performance of steel in full-scale dynamic tests

Reidar Bjorhovde

The Bjorhovde Group, Tucson, Arizona, USA

ABSTRACT: The major findings of an investigation into the performance of steel under very high-demand dynamic tests are presented. The initial aim of the study was to develop criteria for certain practical beam-to-column connections with high ductility demands. The full-scale test specimens utilized common beam and column sizes as well as welded flanges and cover plates and bolted webs. The test loading was either quasi-static or dynamic. The first part of the study involved testing of a series of connections with columns that had been cold-straightened using various forms of straightening protocols. Quasi-static testing was used for some of the tests; it is adequate for seismic purposes. However, the high-cycle (1 Hz) dynamic testing that was used for the other connections in the project is much more severe and certainly closer to those imposed by an earthquake. The results show that the straightening has no effect, positive or negative, on the performance of the connections. All specimens performed very well; some of them exceeded the requirements by wide margins. The special tests of very large shapes under high-demand through-thickness conditions demonstrated that even under these, most extreme conditions, steel material fractures did not occur. The results show that the anticipated through-thickness failures do not occur; it is therefore advisable to re-examine all such code criteria.

1 INTRODUCTION

Steel has been the primary construction material for a very large number of buildings, bridges and other structures for almost 150 years. Its elastic and inelastic response characteristics make it a material with predictable and reliable behavior under a wide range of service conditions.

The most complex elements of structures in any kind of construction are the connections. For steel buildings this is particularly true for beam-to-column connections, where structural details and fabrication processes combine to produce three-dimensional demand conditions. To facilitate construction and allow for economical usage, certain connection types became common over the years. They proved their adequacy through service in many buildings, designers were confident about design methods and details, and fabricators produced high-quality structures. In particular, beam-to-column connections utilizing welded joints between the beam and column flanges and a bolted beam web were used extensively. Tests and analyses showed that these connections were capable of developing appropriate levels of strength, and they also exhibited desirable deformation characteristics.

Use of steel structures in areas of high seismicity was considered especially advantageous, due to the inherent inelastic deformation capacity of the material. Design and fabrication practices proved their worth

through a number of minor and major earthquakes. However, the understanding of seismic effects and structural behavior has advanced significantly over the past ten to twenty years, and efficient design tools have facilitated increasingly fine-tuned designs. Prompted by owners and architects, structural systems also changed, to allow for differing working and living space arrangements. As a result, structures in some ways have become simpler, with fewer primary load-carrying elements, but at the cost of reduced redundancy. The effects of events such as earthquakes would therefore have to be accommodated by fewer structural members and especially connections.

The 1994 Northridge earthquake had a significant effect on state-of-the-art structural thinking in the United States about ductile structural response. A number of steel-framed structures were found to have cracks in their beam-to-column connections, and it was thought that the earthquake had caused these failures. It is now clear that a number of the cracks had occurred before the earthquake, and that such cracking also has taken place in structures where no seismicity had been present.

Furthermore, several cases of what appeared to be fracture of the steel in the through-thickness (TT) direction introduced a novel limit state that might have to be incorporated into the design codes.

As a specific example of the non-earthquake-related cracking incidences, the fabricator for a large project

27

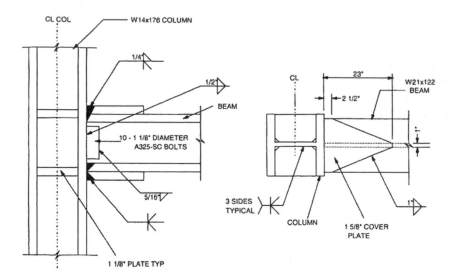

Figure 1. Original connection for the test program.

in California experienced cracking in the column of beam-to-column assemblies during the shop fabrication. The connections were of the welded flange, bolted web variety, but also utilized welded cover plates as well as continuity plates (stiffeners) for the column. Details of the connection are shown in Fig. 1. Although the beam and column sizes shown in the figure are not identical to those of the project in question, they are representative of what was used.

The cracks in the as-built (original) connection were found in the web of the column, in the region of the cross section now commonly referred to as the "k-area". This is a small area of the web of the wide-flange shape surrounding the location where the transition fillet from the flange enters the web. The k-dimension measures the distance from the outside of the flange to the end of the transition fillet in the web. It was subsequently determined that the cracking had occurred as a result of inadequate drying and improper storage of the welding electrodes; this prompted hydrogen cracking in the weld metal.

Originally the earthquake- and fabrication-related cracks found in the Northridge structures were thought to have taken place in part as a result of inadequate material properties. Much discussion took place about yield stresses significantly higher than the specified minimum values, and some investigators tended to state that current steel mill practices therefore were faulty. Additional problems evolved as subsequent examinations found that the k-area of H-shapes was prone to exhibit high strength and hardness as well as low ductility and fracture toughness. These properties were related to the fact that many sizes of wide-flange shapes are commonly rotary straightened in the steel mill in order to meet the straightness requirements of

ASTM (ASTM, 2007). This is typical practice for mills all over the world, but is nevertheless a phenomenon that merited study. These issues and events form the background for the studies that are presented here.

2 SCOPE OF TESTING PROGRAM

2.1 Testing of beam-to-column connections

All specimens used W14 × 176 (approximately H350 × 264) columns and W21 × 122 (approximately H530 × 183) beams in ASTM A572 Grade 50 or the comparable A992 steel. This material has a specified minimum yield stress of 50 ksi (355 N/mm²). A total of 17 full-scale beam-to-column connections were fabricated and tested. Details of the test program are given by Bjorhovde et al. (2000).

Eight of the test connections were the same as those of the California structure where the original cracking had been detected. This is the connection that is shown in Fig. 1. The connections used cover-plated, complete joint penetration (CJP) welds to attach the beam to the column, and CJP welds were used for the continuity plates that were placed in the web area of the column. The beam web connection was provided by ten 1-1/8 inch (28 mm) diameter ASTM A325 high strength bolts; this grade is the same as the European grade 8.8. Further, it was decided to examine the potential influence of the column straightening protocol that had been used by the steel producer to achieve members that would meet the steel standard's straightness criteria (ASTM, 2007). Thus, five of the eight specimens had rotary straightened columns, two had gag straightened columns, and one used an unstraightened member.

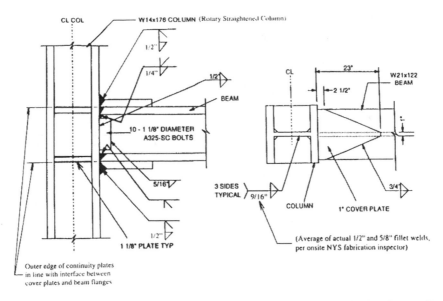

Figure 2. Type 3 connection (note especially the repositioned continuity plates (stiffeners) for the column web).

Six specimens utilized a revised connection, with thinner (1 inch (25 mm)) cover plates, and fillet welds instead of the CJP welds for the continuity plates. The use of fillet welds, in particular, was a major departure from current practice. Five of the specimens used rotary straightened columns; one had an unstraightened member.

Of the remaining three specimens, one was identical to the original specimens, with the only change being a 1/2 inch (12 mm) transition fillet weld between the cover plates and the column flange. It was felt that this would provide a better force and deformation transfer path for the connection in an area where cracks had been proven to initiate the eventual failures. The column was rotary straightened.

The final two specimens were further revised (referred to as Type 3) from the other designs. It was identical to the Type 1 revised connections, but also used the 1/2 inch (12 mm) transition fillet weld that was utilized for Type 2. In addition, the continuity plates were repositioned, to allow for an improved load and fracture path for the connection. The columns were rotary straightened. Figure 2 shows the details of the Type 3 connection.

Figure 3 shows the test frame assembly and the installation of a subassemblage with the connection.

2.2 Through-thickness testing of heavy shapes

Based on the performance of the connections in this testing program, as well as the presumed failures of certain columns in the through-thickness (TT) direction during the Northridge earthquake, an additional program of tests was devised (Dexter and Melendrez,

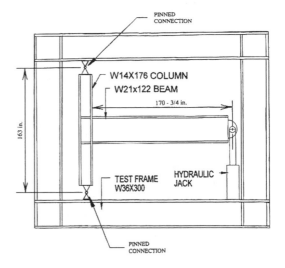

Figure 3. Test frame with an installed connection assembly.

2000). The aim was to determine whether the fractures that had been found were the result of unacceptable steel material properties in the TT direction. Further, these tests would also be conducted under very high-demand, albeit static, loads, emulating the most severe seismic conditions.

Figures 4 and 5 show some of the details of the specimen configuration that was designed to determine the governing limit state(s) of the assembly. The shapes were W14 × 426 (approximately H475 × 639, with a flange thickness of 3.04 inches = 77 mm), and W14 × 605 (approximately H530 × 908, with a flange thickness of 4.16 inches = 106 mm) in ASTM A992

29

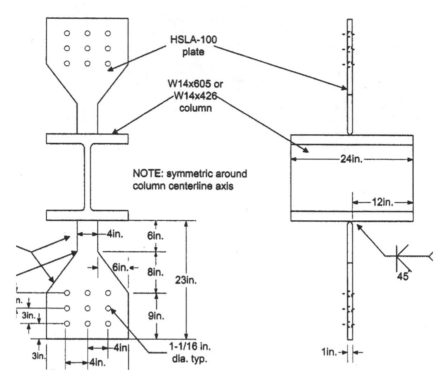

Figure 4. Details of through-thickness test specimen.

steel. This material has a specified minimum yield stress of 50 ksi (355 N/mm²), a specified maximum yield stress of 65 ksi (460 N/mm²), and a maximum yield-to-tensile strength ratio of 0.85.

The "pull-plates" that were welded to the flanges in the TT direction (see Fig. 4) were of the ASTM A514 grade, with a specified minimum yield stress of 100 ksi (700 N/mm²). For the first group of these tests the cross section of the plates was 4 in² (approximately 2580 mm²), with a tested yield strength of 400 kips (1,780 kN). The second series, which was recently started, uses a pull plate cross section of 6 in² (3,870 mm²), to accommodate a cyclic tensile load going from 0 to 600 kips (2,670 kN) and back to 0 in 1 Hz. The plates were purposely oversized, to ensure that any failure would not take place in these components.

3 CONNECTION DESIGN

The connections were designed in accordance with the criteria of the AISC steel design code and the AISC seismic code in force at the time the programs were developed (the 2005 versions are given in the list of references). To achieve optimal seismic performance

of structures and their elements, current US principles utilize the "strong column, weak beam" concept, where plastic hinges will form in the beams at the ultimate limit state. This provides for improved structural redundancy and ductile failure modes for a structure as a whole.

However, for these tests it was decided to impose the most demanding conditions possible on the column material, primarily since the cracking that had been observed occurred in the columns. It was felt that this would represent a worst case scenario. The test specimens therefore reflect assemblies with strong columns and weak beams as well as weak panel zones. The strength of the panel zone relative to the beams is approximately $0.66 M_p$, where M_p is the fully plastic moment of the beam.

4 LOADING PROTOCOLS

A number of beam-to-column connection tests have been conducted in past research projects. Many of these used slowly increasing or effectively static loads. Recognizing the importance of dynamic and especially seismic response characteristics, the US Applied Technology Council (ATC) has developed testing criteria

Figure 5. Through-thickness test specimen assembled in the testing machine.

that are based on cyclic loads (ATC, 1992). These are often referred to as quasi-static testing conditions, since it is not attempted to model earthquake loading input. Rather, using a displacement control approach, the cyclic load is applied in alternate directions, with increasing amplitudes of the load application point. This is the quasi-static loading protocol that was utilized for eight of the seventeen connection tests.

More recent studies have emphasized the need for the test loading to simulate seismic conditions as closely as possible. This led to the development of criteria that focused on loads applied at certain loading or strain rates, to mimic the earthquake response of the structure. Although opinions still differ as to whether dynamic loads impose more realistic conditions than quasi-static loads, the former in all likelihood reflects a worst case scenario. It was therefore decided to test the second group of nine connections specimens dynamically, using a frequency of 1 Hz.

5 COLUMN STRAIGHTENING PROTOCOLS

One of the issues that led to the decision to perform this program of connection tests was the performance of the steel itself in the web of the column. Specifically, cracks had developed in the k-area of some columns during welding. It was also noted that the k-area is deformed significantly because of the rotary straightening procedure that is used in steel mills for small to medium size shapes to meet the straightness requirements of the material delivery standard (ASTM, 2007). As a result, the k-area steel of rotary straightened shapes tends to have higher strength and hardness than other areas of the cross section, but also lower ductility and toughness. Since this form of straightening is applied continuously, the localized areas of changed material characteristics appear along the full length of the column.

Heavier shapes are straightened by the application of concentrated loads at discrete points; this is termed gag straightening. Gag straightening therefore does not introduce continuous areas of high strength and hardness in the manner of rotary straightening; it is only a localized phenomenon for such shapes.

To investigate the possible effect of straightening *per se*, two of the specimens were also fabricated with unstraightened columns. However, unstraightened shapes are not commercially available products.

6 COMPUTATION OF DISPLACEMENTS AND ROTATIONS

6.1 *Measure of performance*

The key measure for the performance of a connection is its plastic rotation capacity or angle, θ_p. This measures the ability of a connection to sustain plastic deformations prior to failure, and is therefore regarded as a criterion by which the connection can be evaluated for seismic performance and suitability.

6.2 *Major assumptions*

If the test frame (see Fig. 3) is infinitely stiff, the true beam tip displacement would be defined as the vertical deflection of the beam end, measured relative to its undeformed original position. However, the frame is not rigid, and deformations of some magnitude will take place. If the specimen displacements are measured in relation to the test frame, the frame deformations would be added to the true specimen displacements in some fashion, resulting in incorrect beam tip and other deformations. To account for these effects during the tests, vertical and horizontal displacements were measured at the centers of the top and bottom column supports (pins), the center of the pin at the beam end, and at the center of the column

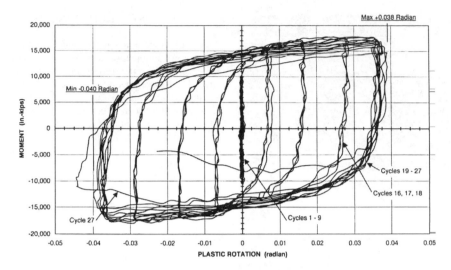

Figure 6. Moment vs. plastic rotation hysteresis loops for a Type 3 connection.

panel zone. These displacements were used to calculate the actual beam tip deflections and to eliminate the effects of the test frame flexibility.

In addition to the vertical translations of the column end supports, horizontal movements occur at both column ends as well as at the panel zone center. The sense of these displacements will indicate whether the entire test specimen rotates as a rigid body around a point on the column close to the bottom pin. Using an additional specimen for verification of the testing system performance, it was found that the top column pin moved horizontally, in direct proportion to the beam end deflection, with magnitudes up to ±8 mm. The lower column pin also moved horizontally, but in the opposite direction; the magnitudes of this translation were never larger than ±1 mm. It was decided to treat this deformation as insignificant. The entire specimen therefore rotated as a rigid body about the bottom column pin support.

6.3 Computation of rotations and translations

During an earthquake, the deformation demand is partly accommodated by the elastic displacements of the frame. Additional deformations have to be provided by the structure in the form of plastic hinge rotations in the beams and by plastic deformations in the column panel zones. The guidelines (FEMA, 1995) that were developed recommended that new steel-framed construction should be able to accommodate plastic rotations of at least 0.030 radians in the connection regions. Minimum rotation capacities of 0.025 radians were recommended for connections of retrofitted structures.

For connection testing, either as proof of performance of existing construction or for the acceptance of new designs, the minimum plastic rotation capacities indicated above must be sustained for at least one full cycle of loading. It is noted that the FEMA guideline recommendations for connection testing are based on a quasi-static testing protocol; no criteria address the issues associated with dynamic testing.

6.4 Cumulative and normalized cumulative plastic rotations

Most research on connections has limited the presentation of the results to the requisite plastic rotations and the accompanying number of cycles. Included are also hysteresis loops and observations of failure modes. However, beyond the appearance of the hysteresis loops there is nothing provided to permit an analysis of the important measure of *energy absorption capacity* of a connection. For seismic performance this is a key measure of suitability.

For the past number of years Japanese research reports have included data on cumulative plastic rotations and normalized cumulative plastic rotations (Nakashima et al., 1998). The former is the sum of the plastic rotations associated with each cycle of loading until failure occurs; the latter is a relative measure of the same. However, it is evident that the cumulative plastic rotation for a connection reflects its energy absorption capacity and therefore provides key information on its performance ability. The survival of a connection for one cycle says little about its potential response under sustained seismic activity.

The cumulative plastic rotation, $\Theta_p = \Sigma \theta_p$, is defined as the sum of the individual plastic rotations occurring during each complete half cycle of the test. The quantity also includes the excursion amount

32

occurring at failure. The normalized cumulative plastic rotation, η, is:

$$\eta = \Theta_p/\theta_y = (\Sigma\theta_p)/\theta_y \qquad (1)$$

The term θ_y is defined as the elastic rotation of the beam cross section for a moment equal to the fully plastic moment.

7 MATERIALS TESTING

Extensive testing was performed for the column materials in the connection specimens, since the cracking and eventual connection failure would occur within these members. The steel grade was ASTM A572 Grade 50 and subsequently A992, and the tensile property and chemical analysis tests showed that the steel in all of the specimens was satisfactory. The tension specimens were taken from flanges, web and k-area material. As expected, the tensile properties of the web and flange steel met and reasonably exceeded the minimum requirements. The k-area of the rotary straightened columns had higher yield and tensile strengths and lower ductility. The gag straightened and unstraightened members showed nearly uniform strength and ductility properties at all locations. These results were all as expected.

Charpy V-Notch (CVN) specimens for impact testing were taken from the flanges as well as the web, core and k-region of the columns. As expected, the flange and web materials in the rotary straightened columns exhibited excellent toughness; the core and especially the k-area steel was much less tough. Gag straightened and unstraightened columns did not display the k-region decreases in toughness.

8 PERFORMANCE OF CONNECTIONS

The preceding chapters have outlined the scope and many of the details that played a role in the execution of the connection tests. It would require too much space to discuss the results for all of the connections; instead, only the performance of certain connections will be examined here.

8.1 Connection design and materials

Figures 1 shows the original ("as-built") connection, and Fig. 2 details the final (Type 3) revised connection. The heavy cover plates and complete joint penetration welds of the as-built connection were the same as those of the structure that prompted the study. The continuity plates were located with their mid-thickness at the level of the interface between the beam flange and the cover plate.

For the Type 3 connection in Fig. 2, of particular interest are (1) the smaller thickness (1 inch (25 mm)) cover plate; (2) the 1/2 inch (12 mm) transition fillet weld between the top and bottom cover plates and column flanges; (3) the fillet welded continuity plates; and (4) the repositioned continuity plates. Following the first tests, it was decided that a repositioning of the continuity plates would allow for an improved fracture path following crack initiation. The continuity plates were placed with their outside edges in line with the interface between the beam cover plate and the beam flange.

8.2 Selected test results

Only a selection of the most important results will be presented in this paper. Complete test data are given in Bjorhovde et al. (2000).

Figure 6 shows the moment vs. plastic rotation hysteresis loops for one of the Type 3 connections. It had a rotary straightened column, and the testing protocol was dynamic. The connection maintained full stiffness and integrity for all of the ±0.030 radians cycles as well as the three first ±0.038 radians cycles (nos. 19–21). The stiffness and beam tip load decreased slightly and uniformly for each cycle after no. 21, although the connection maintained its integrity, with no observed cracking. The test continued with plastic rotations of ±0.038 radians and with complete integrity during 26-1/2 cycles. The specimen eventually failed due to cracking in the top flange cover plate to column flange region. Local buckling did not occur.

By the end of the 26 cycles the maximum and minimum plastic rotations were +0.038 and −0.040 radians; during the first half of the 27th cycle the plastic rotation was +0.039 radians. The connection underwent cumulative plastic rotations of +0.986 and −0.985 radians. These results compare very closely to the cumulative plastic rotation of 1.004 radians for the other Type 3 specimen.

8.3 Connection failure characteristics

The performance of the Type 3 connections warrants additional comments. For the test described here, a crack initiated at the weld root between the top cover plate and the beam flange. This crack eventually extended through the column flange and intersected with a secondary crack that had initiated at the weld access hole adjacent to the toe of the continuity plate weld. The crack at the access hole had the appearance of stable growth, before it intersected with the flange crack. Following intersection of the cracks, the fracture extended from the top continuity plate towards the bottom continuity plate.

It was found that the location of the secondary crack initiation was at the continuity plate weld that coincided with the k-area. The fracture markings indicated that the crack started at the intersection of the continuity plate fillet weld and the k-area. Additional fractographic evaluation showed that there were numerous (7 in total) crack arrest marks. This indicates that the crack had been propagating for at least 7 cycles, meaning that the initiation occurred before the 20th cycle. This is consistent with the results for other specimens. These results also indicate that the k-area cracking at the weld for the continuity plate is a secondary fracture, and therefore not a primary failure location. Finally, although the crack was propagating in the k-area, the arrest marks demonstrate that this was a slow and ductile cracking phenomenon.

All of the connections performed well to very well. In particular, the findings for the two Type 3 connections point to a number of significant results. Primary among these is the fact that crack initiation did not occur in the k-area, but at the toe of the cover plate weld. A secondary crack did initiate in the k-area, but this propagated in a stable manner until it intersected with the crack from the toe of the cover plate weld. The stable crack growth in the k-area negates the perceived problem that the k-area and its high hardness and low toughness will always result in unstable brittle fracture.

The pronounced effect of the relocation of the continuity plates further emphasizes the critical nature of the connection detailing. By providing for improved load paths, the rotation capacity and endurance of the assembly were increased very significantly. Finally, two identical and very complex assemblies were tested, for which a large number of parameters could influence the final results. The close correlation between the two tests attests to the quality of the materials and the connection design and fabrication as a whole.

9 THROUGH-THICKNESS TESTS

9.1 *Introduction*

In addition to the tests reported by Dexter and Melendrez (2000), further studies on the performance of steel in the through-thickness direction have been conducted at the University of Minnesota (Dexter et al., 2001; Dexter and Bergson, 2004) and at the Canadian Frontier Engineering Research (C-FER) group of the Alberta Research Council (Bjorhovde, 2006).

A total of 58 tests aimed at establishing the failure modes and their parameters of influence for the flanges of heavy to very heavy hot-rolled shapes that were loaded in the through-thickness direction. The loads were applied statically for 57 specimens; for the most recent (single) test the load was applied dynamically with a 1 Hz frequency. The test specimen is shown in Figs. 4 and 5.

The testing program will re-commence this year, to ensure that all conceivable parameters of influence have been addressed insofar as the dynamic effects are concerned.

9.2 *Failure characteristics*

All of the 58 test specimens failed by fracture in the 100 ksi (700 N/mm^2) pull plates that were welded to the W-shape flanges as shown in Fig. 4. The stress level at failure in the pull plates was approximately 125 ksi (880 N/mm^2) for all of the specimens. There were no through-thickness failures of any kind in the flanges of the heavy shapes, including for the one specimen that was tested dynamically.

There was no yielding of the flange material in these tests, indicating that there is no significant demand for ductility in the TT direction. This is contrary to most technical opinions on the subject, although these opinions have not been supported by the results of full-scale tests such as these.

9.3 *Some other observations*

Steel chemistry and especially the sulfur contents were also examined for any influence on the TT performance of the first 57 test specimens (Dexter et al., 2001). The results show clearly that there is but a very minor influence of the percentage of sulfur; no other chemical elements appear to have any effect.

10 FINAL OBSERVATIONS

10.1 *Connection tests*

Overall, the test results demonstrate that there are no significant performance differences between assemblies using rotary straightened, gag straightened and unstraightened columns.

The specimens included flange and cover-plate complete joint penetration welded beam to column joints with bolted web connections, and CJP-welded continuity plates. One such connection also had a fillet welded transition between the cover plate and the column flange. Other specimens used thinner cover plates and fillet welded continuity plates. Finally, two specimens used the thinner cover plates and fillet welded continuity plates, in addition to the transition fillet weld, and the continuity plates were repositioned.

Eight of the connection specimens were tested under quasi-static, displacement-controlled loading conditions. The other nine specimens were tested under 1 Hz dynamic loading, also using displacement-controlled amplitudes. The dynamic loading is a much more severe condition, but the differences in performance between otherwise identical specimens are not significant. Since dynamic testing is more difficult to perform and also demands testing equipment of much higher hydraulic capacity, quasi-static testing appears to be the most practical approach for full-scale connection testing.

The cumulative and normalized cumulative plastic rotation capacities are significantly better and more realistic measures of performance, especially since they measure energy absorption.

The Type 3 connections in particular demonstrated excellent plastic rotation and energy absorption capacities. It was found that although cracks developed and propagated through portions of the column material, the propagation was slow and stable, with numerous crack arrest events during the tests. This also occurred for the cracks that propagated into the k-area of the columns, demonstrating that a crack in this region will propagate in stable fashion, given appropriate connection details and fracture paths. Finally, fabrication is much easier with the thinner cover plates, and the fillet welded and repositioned continuity plates.

10.2 *Through-thickness tests*

The TT tests represent novel findings that are likely to have a significant impact on steel construction economy in all areas, but especially where seismic conditions prevail. The TT failure that was found in one connection in the Northridge Earthquake could not be replicated by any specimens in this extensive series of full-scale tests. This includes the one very high-demand test that used a 1 Hz frequency.

It is therefore recommended that design codes do not need to have criteria for checking the through-thickness strength of welded beam-to-column connections.

REFERENCES

American Society for Testing and Materials (ASTM) (2007), "Specification for General Requirements for Rolled Structural Steel Bars, Plates, Shapes and Sheet Piling", ASTM Standard A6/A6M, ASTM International, Conshohocken, Pennsylvania.

American Institute of Steel Construction (AISC) (2005), "Specification for Structural Steel Buildings", ANSI/AISC Standard 360-05, AISC, Chicago, Illinois.

American Institute of Steel Construction (AISC) (2005), "Seismic Provisions for Structural Steel Buildings", ANSI/AISC Standard 341–05 and 341s1–05, AISC. Chicago, Illinois.

Applied Technology Council (ATC) (1992), "Guidelines for Cyclic Seismic Testing of Components of Steel Structures", ATC Guideline No. ATC-24. ATC, Redwood City, California.

Bjorhovde, R., Goland, L. J. and Benac, D. J. (2000), "Performance of Steel in High-Demand Full-Scale Connection Tests", Proceedings, North American Steel Construction Conference, AISC, Chicago, Illinois.

Bjorhovde, R. (2006), "Dynamic Testing for Through-Thickness Strength of a Heavy W-Shape Flange", Internal Report, The Bjorhovde Group, March.

Dexter, R. J. and Melendrez, M. I. (2000), "Through-Thickness Properties of Column Flanges in Welded Moment Connections", *Journal of Structural Engineering*, ASCE, Vol. 126, No. 1 (pp. 24–31).

Dexter, R. J., Prochnow, S. D. and Perez, M. I. (2001), "Constrained Through-Thickness Strength of Column Flanges of Various Grades and Chemistries", *Engineering Journal*, AISC, Vol. 38, No. 4 (pp. 181–198).

Dexter, R. J. and Bergson, P. M. (2004), "T-Tests to Determine Through-Thickness Strength of Column Flanges", Internal Research Report, Department of Civil Engineering, University of Minnesota, March.

Federal Emergency Management Agency (FEMA) (1995), "Interim Guidelines: Evaluation, Repair, Modification and Design of Steel Moment Frames", FEMA Report No. FEMA-267. FEMA, Washington, D.C.

Nakashima, M., Suita, K., Morisako, K. and Maruoka, Y. (1998), "Tests of Welded Beam-Column Assemblies. I: Global Behavior", *Journal of Structural Engineering*, ASCE, Vol. 124, No. 11 (pp. 1236–1244).

Steel and Composite Structures – Wang & Choi (eds)
© 2007 Taylor & Francis Group, London, ISBN 978-0-415-45141-3

Restrained distortional buckling in continuous composite beams

M.A. Bradford

Centre for Infrastructure Engineering & Safety, School of Civil & Environmental Engineering,
The University of New South Wales, Sydney, Australia

ABSTRACT: Composite steel-concrete beams are usually in configurations for which the steel joist is subjected to tension and the concrete slab is subjected to compression, but bridge girders continuous over an internal support or composite beams in the region of a beam-to-column connection have these internal stress resultants reversed, so that the steel component is subjected to compression. The compression in the steel may therefore lead to potential instability failure by local or overall buckling. The overall buckling, that is considered in this paper, is not covered by Vlasov theory because the restraint provided by the slab necessitates that the cross-section distorts in its plane during buckling; this buckling mode is known as a restrained distortional buckle. The prebuckling distribution of stress in the steel component is complicated because both the bending moment and the axial compression vary along the length of the component so that it resembles a beam-column with varying bending moment as well as axial force. This paper reviews studies of distortional buckling in composite beams and highlights that the complexity of the phenomenon precludes simple yet accurate design procedures. A spline finite strip buckling model is presented to analyse the problem, and some numerical solutions are presented.

1 INTRODUCTION

Steel-concrete composite beams are an efficient and widely adopted design solution for flexural members in positive bending because the steel joist is mostly subjected to tension and the concrete slab is subjected to compression. In continuous bridge beams at an internal support and in the region of a beam-to-column connection in a frame structure, a reverse situation is encountered because the negative bending at these locations results in compressive action in the steel joist and tensile action in the concrete slab, and both steel and concrete are not best suited to these actions (Oehlers & Bradford 1995, 1999). Tension in the slab causes tensile cracking, and so longitudinal reinforcement is used in the slab to provide the tensile load path, and compression in the steel joist can lead to possible failure by buckling. Another situation in which the steel joist is subjected to compressive action is in a compartment fire in a building frame, for which the restraint of the columns induces significant axial compression in the joist (Bradford 2006, Luu & Bradford 2007).

Buckling of bare steel members in frames is generally treated by considering local buckling and lateral-torsional buckling as being separate instability modes (Trahair *et al.* 2001) and design is based on this (Woolcock *et al.* 1999), coupled nonlinear buckling modes are usually considered only in thin-walled cold formed

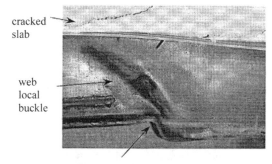

cracked slab

web local buckle

flange local buckle

Figure 1. Inelastic local buckling of composite beam in negative bending (Loh *et al.* 2004).

design (Hancock 1998). Local buckling in the negative moment region of a composite beam (Figure 1) has been fairly widely researched (Climenhaga & Johnson 1972, Johnson & Hope-Gill 1976, Johnson & Bradford 1983, Bradford & Johnson 1987) from the standpoint of moment redistribution and the development of a ductile failure mechanism, and it can usually be treated in design by using the same limiting plate slenderness ratios that are used for the design of bare steel structures (Bradford 1986, Azhari & Bradford 1993). A typical local buckle in a composite beam is shown in Figure 1 (Loh *et al.* 2004).

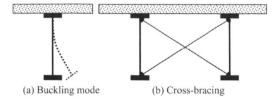

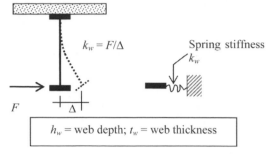

(a) Buckling mode   (b) Cross-bracing

Figure 2. Restrained distortional buckling (RDB) in a composite beam.

h_w = web depth; t_w = web thickness

Figure 3. U-frame approach.

Lateral-torsional buckling (Trahair 1993) is based on Vlasov's assumption that the cross-section remains rigid during buckling; this assumption cannot be made for the overall buckling of a composite beam that is shown in Figure 2 because of the near-to-rigid connection of the steel joist to the comparatively very stiff slab. Since this restraint (if the slab is completely rigid) means that either the steel joist does not buckle at all in an overall mode, or that the overall buckling mode takes the distortional shape shown in Figure 2a, Vlasov's assumption cannot be used and the buckling mode of the joist is a so-called lateral-distortional buckle (RDB). Predicting the RDB response of steel beams is considerably more difficult than the lateral buckling response of bare steel beams, and it can be eliminated in practice by the use of cross-bracing in the negative bending region, as shown in Figure 2b.

Distortional buckling, which can be thought of as a linear interaction between local and lateral-torsional buckling, has been studied quite extensively. Bradford (1992) presented a comprehensive review of studies of the topic prior to the 1990's. It can be reasoned that the buckling load for a beam restrained as shown in Figure 2a is higher than that of its counterpart which is unrestrained at along its top flange, and because of this, designing for RDB on the basis of lateral buckling is very conservative and uneconomic. The British Bridge code BS5400 (1978) made use of the concept of an "inverted U-frame" (Figure 3) for determining the buckling strength of a composite beam in negative bending, for which the flexible web is assumed to restrain the flange, that is treated as a uniformly compressed strut restrained continuously along its edge with a stiffness per unit length of $k_w = Et_w^3/[4(1 - v^2)h_w]$. This idealisation is overly simplistic, mainly because the flange is subjected to a variation of axial stress along its length and because the deformation of the web also contributes to the distortional buckle. Design models were proposed by Svensson (1985), Goltermann & Svensson (1988), Williams and Jemah (1987), Williams et al. (1993) and others based on this concept. Implicitly, these models recognised fundamentally that the steel joist is a beam-column with varying axial force as well as bending moment along its length, with continuous elastic restraint of one of its flanges. The Eurocode

EC4 (BSI 2004) provides design guidance for the joist of a composite beam restrained in this way, in which it considers the flexibility of the bottom flange during buckling ($1/k$) to be influenced by both the flexibility of the web ($1/k_w$) and the twist flexibility of the concrete slab during buckling ($1/k_c$) by (Bradford 1998) $1/k = 1/k_c + 1/k_w$. The EC4 does not, however, provide formulae for the elastic buckling moment M_{cr} of the steel joist, but design formulations such as that of Ng & Ronagh (2004) and Vrcelj & Bradford (2006a) can be used to determine the elastic moment to cause RDB based on a rational distortional buckling analysis.

The lateral buckling strength of a beam is usually determined from an interaction between the elastic buckling moment and the fully plastic moment; this concept has also been used for the distortional buckling strength of a beam (Oehlers & Bradford 1995). In the EC 4, the same relationship between the elastic buckling moment and plastic moment that exists for lateral buckling (as given in EC3 2006) is used for the RDB of the steel joist. This assumption has been the subject of some research (Weston & Nethercot 1987, Bradford 1989, Gao & Bradford 1992, Vrcelj & Bradford 2006b) and it appears that more research is needed to substantiate the assumption, or to provide an alternative relationship.

Owing to the complexities in modelling RDB in the inelastic range of material response, recourse is needed to numerical techniques, which can also be quite difficult to formulate accurately (Ronagh & Bradford 1994, Poon & Ronagh 2004). General-purpose plate and shell finite elements provide a useful framework of analysis, but they lose the generic modelling of the structural mechanics that describes the buckling and some have been shown to produce erroneous results for RDB (Vrcelj 2004). The spline finite strip method, which can accommodate local, distortional and lateral buckling modes, is an attractive technique for studying the RDB of composite beams, with a review of the technique being given by Vrcelj (2004). This method is used herein to study the inelastic RDB of composite beams. The spline finite strip method uses basic B_3 spline functions to model the buckling shape

38

lengthwise, and so is able to include a multiplicity of end boundary conditions. Each strip, into which the steel member is discretised in the conventional spline finite strip method, is assumed to buckle into a cubic profile across its width defined by buckling the deformations and rotations at its two edges. It has been shown (Azhari & Bradford 1994, Azhari *et al.* 2000) that augmenting the transverse cubic displacement with a "bubble term" across the domain Ω that comprises of a linear combination of symmetric shapes of the form $(1 + \eta)^n(1 - \eta)^n/2^{2n}$ (where $\eta \in \Omega = [-1, 1]$ and $n = 2,3,...$) improves the accuracy and convergence of the spline finite strip significantly. This representation is used in the present study.

This paper presents a spline finite strip treatment for RDB in composite beams using bubble functions to augment the flexural buckling deformations in order to capture accurately the plate buckling component of the distortional buckling mode. Because calculating the stresses in the finite strips into which the steel joist is discretised is complex, an in-plane analysis is undertaken first in which cracking of the slab is included in negative bending. This analysis is followed by an out-of-plane analysis, in which the RDB is determined using the spline finite strip method and the residual stresses induced in welded steel beams are included. The technique is validated with methods reported elsewhere, and is then used to present some illustrative numerical studies.

2 SPLINE FINITE STRIP METHOD

2.1 General

This section describes the bubble augmented spline finite strip method for the inelastic RDB of continuous composite beams. The analysis of continuous beams consists of two parts: the first part is an in-plane analysis using the well-known force or flexibility method to determine the moment and shear force distribution along the member length, while the second part is an out-of-plane buckling analysis using the spline finite strip method. The buckling is considered as being a bifurcation from a straight pre-buckled configuration, so that the in-plane and buckling analyses are uncoupled.

The stiffness matrix is comprised of constants in elastic buckling analysis and the geometric stiffness matrix contains terms proportional to the buckling load factor, and the addition of the elastic and geometric stiffness matrices leads to a formulation of the equilibrium equation that may be solved by standard eigenvalue procedures. However, for inelastic buckling analysis, the stiffness matrix must be modified to include the effects of the altered stiffness properties of the steel associated with the plastic deformation prior

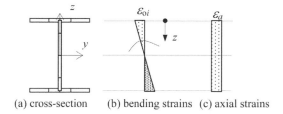

(a) cross-section   (b) bending strains   (c) axial strains

Figure 4. Applied strains in cross-section.

to buckling. This modification is effected by altering the elastic out-of-plane stiffness matrices so that they contain coefficients which depend on the state of plasticity in the plates and therefore on the state of stress. The modification adopted depends on the inelastic plate buckling theory being used.

It is assumed here that the steel is linear-elastic until the yield stress f_y is reached. Beyond the elastic range, two methods of representation are generally employed. In one, a finite relationship is considered between stress and strain, with the elastic modulus E being replaced by the secant modulus E_s, which depends on the state of stress. The other method is an incremental one, which employs the tangent modulus E_t, which also varies with the stress. Commonly, the theory involving finite laws is called deformation theory, whilst that involving infinitesimal laws is called flow theory. Both types of theory assume that the plastic law proposed applies during loading, while unloading takes place elastically. In order to define the plasticity theories in mathematical forms, the assumption that the principal axes of stress coincide with the principal axes of plastic part of the strain or its increment should be made. The analysis here assumes that there is full interaction between the slab and joist.

2.2 Prebuckling (in-plane) analysis

The first stage in the buckling analysis requires a calculation of the distribution of the strains applied to the member prior to invoking the RDB bifurcation analysis. Firstly, an appropriate residual stress model needs to be selected, and then an initial axial strain and curvature (Figure 4) are applied as would occur when the member is subjected to axial and bending actions. The residual stresses which occur in welded sections, such as plate girders, are included in this analysis; residual stresses in hot-rolled members are quantified in Bradford & Trahair (1985) not considered further here. The welding residual stresses illustrated in Figure 5 are based on Cambridge University research, summarised by Kitipornchai & Wong-Chung (1987) and Bradford (1988). The constitutive curve for the structural steel is shown in Figure 6. It represents a trilinear idealisation, with a plastic plateau and a constant strain hardening modulus E_{st}. The stress-strain curve assumes that

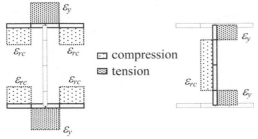

(a) residual strains in flanges

(b) residual strains in web

Figure 5. Cambridge residual strain model for welded section.

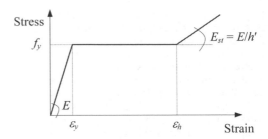

Figure 6. Stress-strain idealisation for steel.

the shear force does not influence the yielding of the member, although in theory this could easily be included using Von Mises' yield surface with an effective stress. For simplicity, the concrete is assumed to carry compressive stress only and the reinforcement to be elastic-fully plastic.

The bending moment, shear and axial force distributions in the member are determined at each Gauss point prior to the buckling analysis. For the purpose of this analysis, the length of the composite beam L is divided into m sections of equal length h, as described by Vrcelj & Bradford (2004); this discretisation in the steel joist forms the basis of the finite strip analysis of Section 2.3.

An iterative method that is described in Vrcelj (2004) is used to determine the distribution of the bending moment and shear force along the statically indeterminate member, since the flexural rigidity EI_y is initially unknown because it depends on the extent of the sagging and hogging regions of the beam. The determination of major axis flexural rigidity EI_y is more complicated than that used in elastic analysis due to the variation of the degree of yielding (and cracking) along the beam. For the first iteration, the values of the flexural rigidity $EI_y^{(1)}$ at each Gauss point are assumed as being their elastic (non-cracked) values based on transformed area principles. The redundant reactions are then calculated using the force method and simple statics is employed to determine the moment and shear force distribution along the member. By defining $\varepsilon_{oi}^{(i)}$ as the strain at the top of the section at a given Gauss point i, and $\kappa^{(i)}$ as the curvature, as illustrated in Figure 4, the strain at any point z below the top fibre of the section can be expressed as

$$\varepsilon = \varepsilon_{oi}^{(i)} + \varepsilon_r - z\kappa^{(i)}, \qquad (1)$$

where ε_r is the residual strain (specified in Figure 5). The value of z in Equation 1 can be adjusted

iteratively by employing the Newton-Raphson procedure, because the pure bending condition for the composite beam that

$$N^{(i)} \equiv 0 \qquad (2)$$

must be enforced at every cross-section. The axial force $N^{(i)}$ and moment $M^{(i)}$ at the given value of strain $\varepsilon_{oi}^{(i)}$ and curvature $\kappa^{(i)}$ for each Gauss point are then obtained by numerical integration over the composite cross-section as

$$N^{(i)} = \int_A \sigma(y,z)\mathrm{d}A \quad \text{and} \quad M^{(i)} = \int_A \sigma(y,z)z\,\mathrm{d}A, \qquad (3)$$

where $\sigma(y,z)$ is the stress calculated at strain ε, obtained from the constitutive relationship

$$\sigma(y,z) = \begin{cases} E\varepsilon & \varepsilon < \varepsilon_y \\ |\varepsilon| f_y / \varepsilon & \varepsilon_y \le |\varepsilon| \le \varepsilon_h \\ E_{st}\varepsilon & \varepsilon > \varepsilon_h \end{cases} \qquad (4)$$

for the steel joist or reinforcement, or from

$$\sigma(y,z) = \begin{cases} E_c\varepsilon & \varepsilon > 0 \\ 0 & \varepsilon \le 0 \end{cases} \qquad (5)$$

for the concrete, where E_c is its elastic modulus. The integrations in Equations 3 to 5 are carried out numerically by subdividing the flanges and web into a number of rectangles that distinguish elastic, yielded and strain-hardened regions around the section and using a trapezoidal integration technique, and by formal integration through the slab thickness.

The major axis flexural rigidities at each Gauss point $EI_y^{(2)}$ are recalculated using the secant modulus as

$$\kappa^{(i)} = \frac{M^{(i)}}{EI_y^{(2)}} \qquad (5a)$$

with known values of curvature $\kappa^{(i)}$ and moment distribution $M^{(i)}$ along the member. The calculated

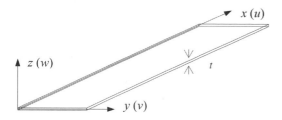

Figure 7.   Finite strip used in buckling study.

flexural rigidity values $EI_y^{(2)}$ are then compared with the assumed values of $EI_y^{(1)}$ and if the normalised Euclidean norm

$$\left\| EI_y^{(n)} / EI_y^{(n-1)} \right\| \le c \qquad (6)$$

for the two sets of values $EI_y^{(2)}$ and $EI_y^{(1)}$ is less than some predetermined accuracy c, then these values are accepted. Otherwise, the procedure is repeated n times until the condition in Equation 6 is satisfied.

2.3   Finite strip (out of plane) analysis

For the finite strip method used here that employs both spline interpolation functions and a bubble term, the flanges and web are treated as an assemblage of longitudinal strips subjected to membrane stresses σ_x, σ_y and τ determined from the in-plane analysis, and these are increased by a load factor λ under proportional loading until buckling occurs. This involves assembling the flexural stiffness and stability (geometric stiffness) matrices $\mathbf{k}_F$ and $\mathbf{g}_F$, and the membrane stiffness and stability matrices $\mathbf{k}_M$ and $\mathbf{g}_M$ for each strip. A typical finite strip is depicted in Figure 7, where y and z are the local strip axes.

The membrane buckling displacements u and v for a finite strip are assumed to be

$$\begin{aligned} u = &\left(N_1\psi_{-1}\alpha_{-1} + N_1\psi_o\alpha_o + ... + N_1\psi_{m+1}\alpha_{m+1} \right)_i \\ &+ \left(N_2\psi_{-1}\alpha_{-1} + N_2\psi_o\alpha_o + ... + N_2\psi_{m+1}\alpha_{m+1} \right)_j \quad \text{and} \end{aligned} \qquad (7)$$

$$\begin{aligned} v = &\left(N_1\psi_{-1}\beta_{-1} + N_1\psi_o\beta_o + ... + N_1\psi_{m+1}\beta_{m+1} \right)_i \\ &+ \left(N_2\psi_{-1}\beta_{-1} + N_2\psi_o\beta_o + ... + N_2\psi_{m+1}\beta_{m+1} \right)_j , \end{aligned} \qquad (8)$$

or more concisely in matrix form as

$$u = \left\langle \mathbf{M}_1 \quad \mathbf{M}_2 \right\rangle \left\langle \boldsymbol{\alpha}_i \quad \boldsymbol{\alpha}_j \right\rangle^{\mathrm{T}} ; \quad v = \left\langle \mathbf{M}_1 \quad \mathbf{M}_2 \right\rangle \left\langle \boldsymbol{\beta}_i \quad \boldsymbol{\beta}_j \right\rangle^{\mathrm{T}} . \qquad (9)$$

In these equations,

$$\begin{aligned} \boldsymbol{\alpha}_i &= \left\langle \alpha_{i-1}, \alpha_{i0}, \alpha_{i1}...,\alpha_{im-2}, \alpha_{im-1}, \alpha_{im}, \alpha_{im+1} \right\rangle^{\mathrm{T}}, \\ \boldsymbol{\alpha}_j &= \left\langle \alpha_{j-1}, \alpha_{j0}, \alpha_{j1}...,\alpha_{jm-2}, \alpha_{jm-1}, \alpha_{jm}, \alpha_{jm+1} \right\rangle^{\mathrm{T}} \quad \text{and} \end{aligned} \qquad (10)$$

$$\begin{aligned} \boldsymbol{\beta}_i &= \left\langle \beta_{i-1}, \beta_{i0}, \beta_{i1}...,\beta_{im-2}, \beta_{im-1}, \beta_{im}, \beta_{im+1} \right\rangle^{\mathrm{T}}, \\ \boldsymbol{\beta}_j &= \left\langle \beta_{j-1}, \beta_{j0}, \beta_{j1}...,\beta_{jm-2}, \beta_{jm-1}, \beta_{jm}, \beta_{jm+1} \right\rangle^{\mathrm{T}} \end{aligned} \qquad (11)$$

represent vectors of kernel displacements at nodal lines i and j,

$$N_1 = 1-\eta \quad \text{and} \quad N_2 = \eta \qquad \text{(where } \eta = y/b) \qquad (12)$$

are linear interpolation functions across the strip and

$$\begin{aligned} \mathbf{M}_1 &= \left\langle N_1\psi_{-1}, N_1\psi_o, N_1\psi_1,..., N_1\psi_{m+1} \right\rangle; \\ \mathbf{M}_2 &= \left\langle N_2\psi_{-1}, N_2\psi_o, N_2\psi_1,..., N_2\psi_{m+1} \right\rangle \end{aligned} \qquad (13)$$

contain the spine functions defined by

$$\psi_i(x) = \frac{1}{6h^3} \begin{cases} 0 & x < x_{i-2} \\ (x-x_{i-2})^3 & x_{i-2} \le x \le x_{i-1} \\ h^3 + 3h^2(x-x_{i-1}) + 3h(x-x_{i-1})^2 - 3(x-x_{i-1})^3 & x_{i-1} \le x \le x_i \\ h^3 + 3h^2(x_{i+1}-x) + 3h(x_{i+1}-x)^2 - 3(x_{i+1}-x)^3 & x_i \le x \le x_{i+1} \\ (x_{i+2}-x)^3 & x_{i+1} \le x \le x_{i+2} \\ 0 & x_{i+2} \le x. \end{cases} \qquad (14)$$

that result from the spline function representation of the function $f(x)$ as

$$f(x) = \sum_{i=-1}^{m+1} a_i \psi_i(x). \qquad (15)$$

Similarly, the flexural displacements of the strip are assumed to be

$$\begin{aligned} w = &\left(N_3\psi_{-1}\gamma_{-1} + N_3\psi_o\gamma_o + ... + N_3\psi_{m+1}\gamma_{m+1} \right)_i \\ &+ \left(N_4\psi_{-1}\delta_{-1} + N_4\psi_o\delta_o + ... + N_4\psi_{m+1}\delta_{m+1} \right)_i \\ &+ \left(N_3\psi_{-1}\gamma_{-1} + N_3\psi_o\gamma_o + ... + N_3\psi_{m+1}\gamma_{m+1} \right)_j \\ &+ \left(N_4\psi_{-1}\delta_{-1} + N_4\psi_o\delta_o + ... + N_4\psi_{m+1}\delta_{m+1} \right)_j \\ &+ \left(N_5\psi_{-1}\omega_{-1} + N_5\psi_o\omega_o + ... + N_5\psi_{m+1}\omega_{m+1} \right)_B \end{aligned} \qquad (16)$$

or in matrix format as

$$w = \left\langle \mathbf{M}_3 \; \mathbf{M}_4 \; \mathbf{M}_3 \; \mathbf{M}_4 \; \mathbf{M}_5 \right\rangle \left\langle \boldsymbol{\gamma}_i \; \boldsymbol{\delta}_i \; \boldsymbol{\gamma}_j \; \boldsymbol{\delta}_j \; \boldsymbol{\omega}_B \right\rangle^{\mathrm{T}} , \qquad (17)$$

where the vectors γ_i, δ_i and γ_j, δ_j are kernel freedoms associated with nodal lines i and j, and the flexural cubic interpolation functions are

$$\begin{aligned} N_{3i} &= 1-3\eta^2 + 2\eta^3, & N_{4i} &= \eta\left(1-2\eta+\eta^2\right), \\ N_{3j} &= 3\eta^2 - 2\eta^3, & N_{4j} &= \eta\left(\eta^2 - 1\right). \end{aligned} \qquad (18)$$

Equation 16 contains the bubble function

$$N_5 = \frac{1}{2^{2n}}\eta^n(1-\eta)^n \qquad (n = 2, 3, ...) \qquad (19)$$

while Equation 17 contains the associated bubble freedoms

$$\boldsymbol{\omega}_B = \left\langle \omega_{B-1}, \omega_{B0}, \omega_{B1}..., \omega_{Bm-2}, \omega_{Bm-1}, \omega_{Bm}, \omega_{Bm+1} \right\rangle^{\mathrm{T}} . (20)$$

The displacement fields for the strip in Equations 9 and 17 are expressed in terms of kernel or coefficient degrees of freedom which need to be transformed into nodal degrees of freedom defined at the section knots, prior to the assembly of the strips, in order to satisfy the compatibility and equilibrium conditions. This procedure can be effected by the transformation represented by (Vrcelj & Bradford 2007)

$$\mathbf{u} = \left\langle u \quad v \quad w \right\rangle^{\mathrm{T}} = \boldsymbol{\psi}\boldsymbol{\Delta} \tag{21}$$

in which ψ is a matrix of interpolations containing the spline terms, linear and cubic terms as well as the bubble interpolation, and Δ is a vector of assemblable freedoms that may be prescribed in a similar fashion to conventional finite element treatments as, for example, $0 \equiv$ 'fixed' or $1 \equiv$ 'free'. The formulation of Vrcelj & Bradford (2007) that allows the buckling displacements to be written in the form of Equation 21 circumvents the loss of versatility that arises when conventional amended splines are needed to prescribe boundary conditions.

The finite strip buckling formulation results in the well-known eigenproblem

$$\left[\mathbf{K}(\lambda) - \mathbf{G}(\lambda)\right]\mathbf{Q} = \mathbf{0} \tag{22}$$

where $\mathbf{Q}$ is the vector of assemblable buckling freedoms for the joist, $\mathbf{K}$ is the stiffness matrix for the joist and $\mathbf{G}$ is the stability matrix for the joist, assembled from their component matrices for each strip given by

$$\mathbf{k}_F(\lambda) = \int_V \mathbf{B}_F{}^{\mathrm{T}} \mathbf{D}_F(\lambda) \mathbf{B}_F \mathrm{d}V \quad \text{and} \tag{23}$$

$$\mathbf{k}_M(\lambda) = \int_V \mathbf{B}_M{}^{\mathrm{T}} \mathbf{D}_M(\lambda) \mathbf{B}_M \mathrm{d}V \tag{24}$$

for the stiffness matrix and

$$\mathbf{g}_F(\lambda) = \int_V \boldsymbol{\varepsilon}_N{}^{\mathrm{T}} \begin{bmatrix} \sigma_x(\lambda) & \sigma_y(\lambda) \\ \sigma_y(\lambda) & \tau_{xy}(\lambda) \end{bmatrix} \boldsymbol{\varepsilon}_N \mathrm{d}V \quad \text{and} \tag{25}$$

$$\mathbf{g}_M(\lambda) = \int_0^L \int_0^b \begin{Bmatrix} \partial u/\partial x \\ \partial v/\partial x \end{Bmatrix} \sigma_x(\lambda) \begin{Bmatrix} \partial u/\partial x \\ \partial v/\partial x \end{Bmatrix}^{\mathrm{T}} t \, \mathrm{d}y \, \mathrm{d}x \tag{26}$$

for the stability matrix in which the property matrices $\mathbf{D}_F$ and $\mathbf{D}_M$ contain terms associated with the plastic assumptions for the buckling analysis and are functions of the applied load factor λ, $\mathbf{B}_F$ and $\mathbf{B}_M$

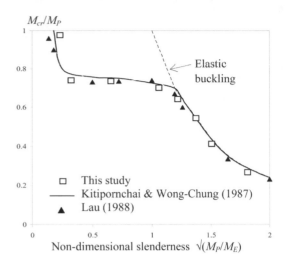

Figure 8. Comparisons of inelastic buckling moments for section shown in Figure 9.

are strain matrices derived by appropriate differentiation of the interpolating functions, ε_N is the vector of nonlinear strains and σ_x, σ_y and τ_{xy} are the stresses that depend on the level of applied loading. The property matrices were developed from the formulation of Dawe & Kulak (1984) that is similar to that of Bradford (1986).

2.4 Solution of buckling problem

Initially, a regime of transverse loading is applied to the beam, and for the in-plane analysis that is nonlinear, it is assumed that the state of stress is not path-dependent. Because of this, this regime or base loading is defined at a sufficiently low loading level to be a small proportion of that to cause the plastic moment in the joist being reached. This base load is then increased monotonically by a load factor λ under proportional loading.

At each value of λ, the in-plane analysis described in Section 2.2 is invoked to enable the yielded regions of the joist to be defined, as well as the distributions of axial force, moment and shear in the steel joist that allow the stresses $\sigma_x(\lambda)$, $\sigma_y(\lambda)$ and $\tau_{xy}(\lambda)$ to be determined. The matrices in Equations 23 to 26 can then be assembled and the matrix $\mathbf{A}(\lambda) = \mathbf{K}(\lambda) - \mathbf{G}(\lambda)$ in Equation 22 is then inspected. The base loading is chosen so that $\mathbf{A}$ is positive definite for the starting load factor $\lambda = 1$, and the process is repeated at small increments of λ until the matrix $\mathbf{A}$ becomes negative definite. Once bracketed, the buckling load factor can be determined using the method of bisections and the buckling load, current state of stress and buckling eigenshapes ($\mathbf{Q}$ in Equation 22) may then be found.

3 NUMERICAL VALIDATION OF THEORY

The accuracy of the spline finite strip method of inelastic buckling analysis described in the previous section was investigated by comparing the buckling loads and modes computed with some existing theoretical values. It should be noted that Vrcelj (2004) reported that standard research software such as ABAQUS (2003) can produce erroneous results when modelling RDB. Kitipornchai & Wong-Chung (1987) and Lau (1988) investigated the inelastic buckling of welded monosymmetric I-beams subjected to uniform bending moment, and the current method of inelastic buckling analysis of the joist was used to model their studies. The yield stress was $f_y = 250$ N/mm^2 and the stress-strain curve of the material was assumed to be the same as that in Figure 6 with $E_{st} = 0$. The critical moments reported by Kitipornchai & Wong-Chung (1987) and Lau (1988) are compared with the present method in Figure 8 for the T-section shown in Figure 9. In Figure 8, M_{cr}, M_E and M_P are the inelastic critical moment, elastic critical moment and fully plastic moment respectively. In Lau's analysis, the flange outstand and web plate were subdivided into four and ten strips respectively, and because of symmetry, only four spline sections were required longitudinally for half of the length of a beam. However, in the analysis developed in this study only two strips were required to model the flange outstand and four strips were used to model the web plate. The computed critical moments are within 5% of those obtained by Kitipornchai & Wong-Chung (1987) and Lau (1988).

4 NUMERICAL STUDIES

4.1 General

The spline finite strip technique has been applied to study the inelastic RDB of composite beams subjected to transverse loading. The steel joist shown in Figure 10 was used in the study and the number of sections used in the spline function representation is shown in Figure 11 as a function of the beam slenderness; this discretisation was found to be optimal in terms of computer time and accuracy. The studies presented are symmetrical and use was made of this symmetry in the modelling of only half of the beam with the appropriate buckling boundary support conditions being implemented.

4.2 Through-girder beam

A simply supported composite beam for which the unrestrained flange is subjected to compressive stresses throughout was studied; this case being representative of a half-through girder bridge. It was assumed that the shear stress in the web was uniform

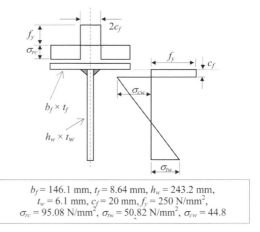

b_f = 146.1 mm, t_f = 8.64 mm, h_w = 243.2 mm, t_w = 6.1 mm, c_f = 20 mm, f_y = 250 N/mm^2, σ_{rc} = 95.08 N/mm^2, σ_{tw} = 50.82 N/mm^2, σ_{cw} = 44.8

Figure 9. Cross-section and properties used for validation.

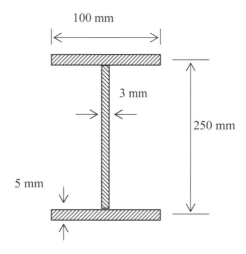

$2c_f$ = 20 mm, σ_{rc} = 62.5 N/mm^2
E = 200,000 N/mm^2 E_c = 25,000 N/mm^2
E_{st} = $E/33$, ε_h = $11\varepsilon_y$, ν = 0.3

Figure 10. Material properties and dimensions of joist used in numerical buckling studies.

($\tau = V/(h_w t_w)$ where V is the total shear resisted by the composite cross-section) and that the shear distribution in the flanges was linear (Trahair *et al.* 2001).

Figure 12 shows the inelastic buckling moments for simply supported plain steel and steel-concrete composite beams subjected to a uniformly distributed load. In the figure, the inelastic buckling moment M_{cr} is normalised with respect to plastic moment M_P and is plotted against the modified slenderness $\sqrt{(M_P/M_E)}$. It is evident from the figure that the buckling behaviour

43

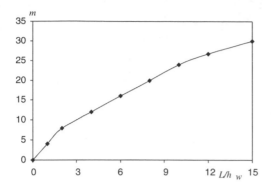

Figure 11. Number of sections versus slenderness of joist used in numerical buckling studies.

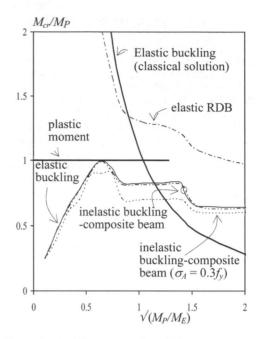

Figure 12. Buckling moments for half-through girder.

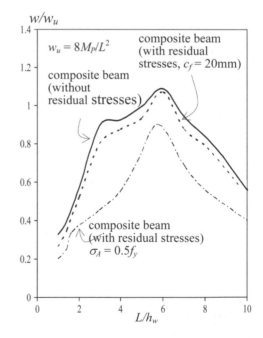

Figure 13. Buckling loads for half-through girder.

for the cross-section shown in Figure 10, for a range of different slenderness ratios, is elastic. The presence of the residual stresses ($c_f = 10\,mm$) shown in Figure 12 as a dashed line coinciding with the full line that represents inelastic and/or elastic buckling for the beam, is insignificant for this configuration of geometry and loading. Nevertheless, the presence of the uniform axial force, which was adopted as 30% of the total yield moment for the steel I-section shown in Figure 10, is quite notable. The figure also shows the elastic critical buckling moments for plain steel beams when the assumption that the cross-section remains rigid is valid, as well as the RDB moments for composite beams when this assumption is no longer applicable. The RDB values, derived by the analysis method developed by Vrcelj & Bradford (2004), do not account for any local buckling instability of the cross-sectional plates. The figure clearly illustrates significant reductions in the buckling capacity caused by the cross-sectional distortions, such as local buckling, RDB and the combination of the two, in spite of the beneficial effects of the rigid restraint provided by the concrete at the tension flange of the steel I-section joist.

The results in Figure 13 show the reductions of the beam ultimate load capacity for simply supported steel-concrete composite beams. In this figure, the normalised load capacity is plotted as a function of beam slenderness L/h_w. The axial stress was calculated as 50% of the bending stress on the onset of yielding for the given cross-section. Noteworthy reductions in the buckling capacity due to the presence of the uniform axial force in the beam, which inevitably reallocates the neutral axis closer to the restrained flange, are obvious from the graphs. Significant reductions in the ultimate load capacity are also observed due to the residual stresses ($c_f = 20$) and it is observable that the coupling of local and distortional buckling is a governing failure mode for the cross-sections of this geometry.

4.3 Continuous I-section beam

Figure 14 shows a two-span continuous beam with equal span lengths that is simply supported and subjected to a uniformly distributed load, and Figure 15

44

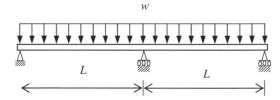

Figure 14.   Two-span continuous beam.

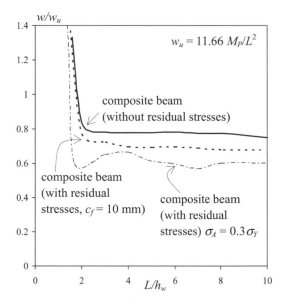

Figure 15.   Buckling loads for continuous composite beam.

shows the buckling load capacities w, normalised with respect to the beam ultimate load capacity w_u plotted as a function of the beam slenderness L/h_w.

An extensive study was undertaken by Vrcelj & Bradford (1999) to investigate the influences of the axial stress on the buckling behaviour of a continuous steel-concrete composite beam. It was concluded that the presence of the axial stresses is of particular significance in propped construction, where the ratio between axial and bending stresses in the compression flange is in the range $0.2 \sim 0.3$. This ratio is somewhat less for unpropped construction, where the self-weight of the beam contributes to an increased bending stresses. It was further shown that in two-span continuous composite beams, the axial force varies longitudinally with prevailing compressive stresses in the vicinity of the internal support that can be even higher than $20 \sim 30\%$ of the bending stress. Therefore, in this analysis the axial stress was assumed to vary longitudinally as a sixth order polynomial function, based on the conclusions of Vrcelj & Bradford (1999).

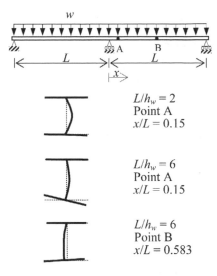

Figure 16.   Buckling modes for continuous beam.

The presence of the axial residual stresses in the beam results in a considerable reduction of the buckling load carrying capacity and contributes to further instability of the cross-section in the negative moment region where the coupling of localised and RDB modes is significant. On the other hand, the presence of the rigid restraint provided by the concrete medium contributes to an ample increase in the capacity when compared to that of a bare steel section beam, as shown in Figure 15.

Generally, when the span ratio is low the governing buckling mode was found to be local buckling, or local buckling coupled with distortional buckling, whilst for high span ratios, owing to the rigid restraint provided at the top flange, the governing buckling mode was found to be distortional, as illustrated in Figure 16.

5   CONCLUDING REMARKS

This paper has provided a review of the buckling of continuous composite beams for which the top flange of the steel I-section joist is restrained by the concrete flange, and for which some portion of the bottom unrestrained flange is subjected to compression. The buckling problem is similar to that of a beam-column for which both the bending moment and axial force vary along the length of the beam, and for which the beam has continuous partial restraint. It was reasoned that the buckling mode must be accompanied by cross-sectional distortion because of the restraint provided by the slab. Modelling this buckling mode in design is difficult, and some prescriptive design techniques of questionable accuracy and applicability appear in design codes of practice. Surprisingly, little or no experimental work seems to have been reported that addresses RDB in composite beams, and it seems that this is a consequence of contemporary design practice "over-designing" beams against RDB by using Class 1, 2 or 3 cross-sections to prevent local buckling.

A spline finite strip technique was described which is able to predict the interaction of the restrained distortional and local buckling modes for the inelastic buckling of composite girders. The method is efficient and it can handle a number of boundary conditions in order to provide comprehensive solutions to this complicated but important buckling problem. This technique can be used to develop further accurate design proposals, for which further research work and especially experimental verification is needed.

ACKNOWLEDGEMENT

The work in this paper was supported by the Australian Research Council through a Federation Fellowship awarded to the author.

REFERENCES

ABAQUS Standard User's Manual Version 5.8. 1988. Pawtucket, RI: Hibbit, Karlsson and Sorensen Inc.

Azhari, M. & Bradford, M.A. 1993. Local buckling of composite tee-beams with longitudinal stiffeners. *Canadian Journal of Civil Engineering* 20(6): 923–930.

Azhari, M. & Bradford, M.A. 1994. Local buckling by the complex finite strip method using bubble functions. *Journal of Engineering Mechanics, ASCE* 120(1): 43–57.

Azhari, M., Hoshdar, S. & Bradford, M.A. 2000. On the use of bubble functions in local buckling analysis of plate structures by the spline finite strip method. *International Journal for Numerical Methods in Engineering* 48(4): 583–593.

Bradford, M.A. 1986. Local buckling analysis of composite beams. *Civil Engineering Transactions, I.E. Australia* CE28(4): 312–317.

Bradford, M.A. 1988. Buckling strength of deformable monosymmetric I-beams. *Engineering Structures* 10: 167–173.

Bradford, M.A. 1989. Buckling strength of partially restrained I-beams. *Journal of Structural Engineering, ASCE* 115(5): 1272–1276.

Bradford, M.A. 1992. Lateral-distortional buckling of steel I-section members. *Journal of Constructional Steel Research* 23: 97–116.

Bradford, M.A. 1998. Inelastic buckling of I-beams with continuous elastic tension flange restraint. *Journal of Constructional Steel Research* 48: 63–77.

Bradford, M.A. 2006. Generic model for a composite T-beam at elevated temperature. *Proceedings, 4th International Workshop on Structures in Fire*, Aveiro, Portugal, 805–812.

Bradford, M.A. & Gao, Z. 1992. Distortional buckling solutions for continuous composite beams. *Journal of Structural Engineering, ASCE* 118(1): 73–89.

Bradford, M.A. & Johnson, R.P. 1987. Inelastic buckling of composite bridge girders near internal supports. *Proceedings of The Institution of Civil Engineers London*, Part 2 83(2): 143–159.

Bradford, M.A. & Trahair, N.S. 1985. Inelastic buckling of beam-columns with unequal end moments. *Journal of Constructional Steel Research* 5: 195–212.

British Standards Institution. 1978. *BS5400-2:1978 Steel, concrete and composite bridges*. London: BSI.

British Standards Institution 2004. *Eurocode 4: Design of composite steel and concrete structures. Part 1-1: General rules and rules for buildings BS EN 1994-1-1*. London: BSI.

Climenhaga, J.J. & Johnson, R. P. 1972. Local buckling in continuous composite beams. *The Structural Engineer* 50(9): 367–375.

Dawe, J.L. & Kulak, G.L. 1984. Plate instability of W-shapes. *Journal of the Structural Division, ASCE* 110(ST6): 1278–1291.

Goltermann, P. & Svensson, S.E. 1988. Lateral distortional buckling – predicting the elastic critical stress. *Journal of Structural Engineering, ASCE* 114(7): 1606–1625.

Hancock, G.J. 1998. *Design of Cold-Formed Steel Structures*. 3rd edn, Sydney: Australian Institute of Steel Construction.

Johnson, R.P. & Bradford, M.A. 1983. Distortional lateral buckling of stiffened composite bridge girders. *International Conference on Instability and Plastic Collapse of Steel Structures*, Manchester, 569–580.

Johnson, R.P. & Hope-Gill, M.C. 1976. Tests on three three-span continuous composite beams. *Proceedings of The Institution of Civil Engineers*, Part 2 61(2): 367–381.

Kitipornchai, S. & Wong-Chung, A.D. 1987. Inelastic buckling of welded monosymmetric I-beams. *Journal of Structural Engineering, ASCE* 113(4): 740–756.

Lau, S.C.W. 1988. Distortional buckling of thin-walled columns. *PhD Thesis*, University of Sydney, Australia.

Loh, H.Y., Uy, B. & Bradford, M.A. 2004. The effects of partial shear connection in the hogging moment regions of composite beams. Part I – Experimental study. *Journal of Constructional Steel Research* 60: 897–919.

Luu T.K. & Bradford, M.A. Thermal-induced restrained distortional buckling of composite steel-concrete beams. *Proceedings International Conference on Steel and Aluminium Structures*, Oxford, UK.

Ng, M.L.H. & Ronagh, H.R. 2004. An analytical solution for the elastic lateral-distortional buckling of I-section beams. *Advances in Structural Engineering* 7(2): 189–200.

Oehlers, D.J. & Bradford, M.A. 1995. *Composite Steel and Concrete Structural Members: Fundamental Behaviour*. Oxford: Pergamon Press.

Oehlers, D.J. & Bradford, M.A. 1999. *Elementary Behaviour of Composite Steel and Concrete Structural Members*. Oxford: Butterworth-Heinemann.

Poon, C.P & Ronagh, H.R. 2004. Distortional buckling of I-beams by finite element method. *Advances in Structural Engineering* 7(1): 71–80.

Ronagh, H.R. & Bradford, M.A. 1994. Some notes on finite element buckling formulations for beams. *Computers and Structures* 52(6): 1119–1126.

Svensson, S.E. 1985. Lateral buckling of beams analysed as elastically supported columns subjected to varying axial force. *Journal of Constructional Steel Research* 5: 179–193.

Trahair, N.S. 1993. *Flexural-Torsional Buckling of Structures*. London: E&FN Spon.

Trahair, N.S., Bradford, M.A. & Nethercot, D.A. 2001. *The Behaviour and Design of Steel Structures to BS5950*. 3rd edn (British), London: Spon Press.

Vrcelj, Z. 2004. Buckling modes in continuous composite beams. *PhD Thesis*, The University of New South Wales, Sydney, Australia.

Vrcelj, Z. & Bradford, M.A. 2004. Bubble functions in the analysis of plates and plate assemblies by the spline finite strip method. *UNICIV Report No. R*-431, The University of New South Wales, Sydney, Australia.

Vrcelj, Z. & Bradford, M.A. 2006a. Elastic distortional buckling of continuously restrained I-section beam-columns. *Journal of Constructional Steel Research* 62(3): 223–230.

Vrcelj, Z. & Bradford, M.A. 2006b. Inelastic restrained distortional instabilities in continuous composite T-*beams. Proceedings, XI International Conference on Metal Structures*, Rzeszow, Poland, 375–383.

Vrcelj, Z. & Bradford, M.A. 2007. A simple method for the inclusion of external and internal supports in the spline finite strip method (SFSM) of analysis. *Computers and Structures*, in press.

Weston, G. & Nethercot, D.A. 1987. Continuous composite bridge beams – stability of the steel compression flange in hogging bending. *Proceeding of ECCS Colloquium on Stability of Plate and Shell Structures*, Ghent, Belgium, 47–52.

Williams, F.W. & Jemah, A.K. 1987. Buckling curves for elastically supported columns with varying axial force, to predict lateral buckling of beams. *Journal of Constructional Steel Research* 7(2): 133–147.

Williams, F.W., Jemah, A.K. & Lam, D.H. 1993. Distortional buckling curves for composite beams. *Journal of Structural Engineering, ASCE* 119(7): 2134–2149.

Woolcock, S.T., Kitipornchai, S. & Bradford, M.A. 1999. *Design of Portal Frame Buildings*. 3rd edn, Sydney: Australian Institute of Steel Construction.

Steel and Composite Structures – Wang & Choi (eds)
© 2007 Taylor & Francis Group, London, ISBN 978-0-415-45141-3

Structural steel connections and their effect on progressive collapse

Krauthammer, T.
Center for Infrastructure Protection and Physical Security, University of Florida, USA

ABSTRACT: This paper describes a multi-year study on the behavior of structural steel connections under blast loads, and the implementation of the findings for the simulation of progressive collapse in multi-story steel framed structures. It was observed that different types of connections can lead to different behavior sequences during progressive collapse simulations.

1 INTRODUCTION AND BACKGROUND

Progressive collapse became an issue following the Ronan Point incident in 1968, when a gas explosion in a kitchen on the 18th floor of a 22-story precast building caused extensive damage to the entire corner of that building. The failure investigation of that incident resulted in important changes in the UK building code. It requires to provide a minimum level of strength to resist accidental abnormal loading by either comprehensive 'tying' of structural elements, or (if tying is not possible) to enable the 'bridging' of loads over the damaged area (the smaller of 15% of the story area, or $70\,m^2$), or if bridging is not possible to insure that key elements can resist $34\,kN/m^2$. Although many in the UK attribute the very good performance of numerous buildings subjected to blast loads to these guidelines, it might not be always possible to quantify how close those buildings were to progressive collapse.

Progressive collapse is a failure sequence that relates local damage to large scale collapse in a structure. The local failure can be defined as a loss of the load-carrying capacity of one or more structural components that are part of the whole structural system. Preferably, once any structural component fails, the structure should enable an alternative load-carrying path. After the load is redistributed through a structure, each structural component will support different loads. If any load exceeds the load-carrying capacity of any member, it will cause another local failure. Such sequential failures can propagate through the structure. If a structure loses too many members, it may lead to partial or total collapse. This type of collapse behavior may occur in various types of framed structures.

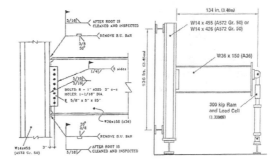

Figure 1. Connection Test Models (Engelhardt and Sabol 1994).

2 ANALYSIS OF STEEL CONNECTIONS UNDER QUASI-STATIC LOAD

The first step of the research approach required one to validate the selected computational connection model. This was achieved by using a previously tested WUF-B connection (Engelhardt and Sabol 1994), as shown in Figure 1.

The aforementioned structure was modelled by using the finite element code ABAQUS/Explicit (ABAQUS, 2005), and detailed three dimensional models, as shown in Figure 2(a). All geometric material, loading, and support information from the test were incorporated in the simulations.

Tip displacement results from the simulations and tests were compared, as shown in Figure 3. The comparisons showed that the simulation using the proposed model provided were within 10 percents of experimental data. Therefore, this validated simulation approach was adopted for studying the behavior of steel connection under blast loads.

However, since it is almost impossible to include hundreds of detailed structural steel connections in practical numerical simulations of three-dimensional multi-story frame analyses, the detailed connection model had to be simplified, as shown in Figure 2(b). Connection properties are generally described using moment-rotation curves that can be extracted from push-over simulations of the detailed connection models, as shown in Figure 4.

These curves define the mechanical properties of the simplified connector elements (e.g., initial stiffness, yield and/or ultimate strength, deformations, etc.). This approach enables one to replace the detailed finite element connection model with representative connector elements, as shown in Figure 2(b). Push-over analyses were then carried out for the simplified models, and the results were compared with the same data from the detailed push-over simulations. According to comparisons of the tip displacement results in

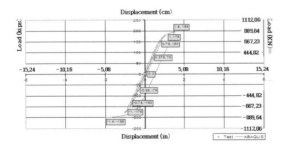

(a) Detailed model   (b) Simplified model

Figure 2.   Connection Simulation Models.

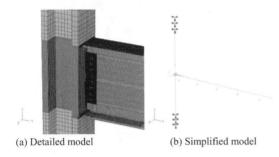

Figure 3.   Code and Model Validation Results.

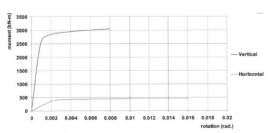

Figure 4.   Moment-Rotation Curves.

Figure 5, the simpler models were able to reproduce the behaviors quite accurately.

These results showed that this finite element analysis approach could clearly provide accurate connection characteristics that could be effectively applied to simplified frame models.

3   ANALYSIS OF STEEL CONNECTIONS UNDER BLAST LOADS

Currently, U.S. design guidelines for steel connections in structures subjected to blast loads are based on recommendations in TM 5-1300 (Department of the US Army 1990). The approach involves various simplifications, and address single storied steel frames under dead loads from the self weight of the structure. The actual effect of blast and dead loads on real steel connections may exceed the margins predicted by TM 5-1300. Consequently, finite element simulations using ABAQUS/Explicit 6.5 (ABAQUS Inc. 2005) were employed to assess the behaviours of 16 selected steel moment connections under such loads in more detail (Yim et al. 2006a). The maximum rotational capacities of the connections were then compared against values derived with the TM 5-1300 approach. Target connections were placed between beam and column at the ground floor of a multi-story building (Figure 6). The general dimensions of beams and columns were taken from (Engelhardt et al. 1994). For each connection type, four different load cases were studied, corresponding to explosions at the centre of the floor: the two side walls of the room either could be frangible, venting the internal pressure, or survive the shock load and reflect it back into the internal space. Reference maximum blast pressures were calculated based upon the TM 5-1300 criteria. For the finite element calculations, the computer codes SHOCK and FRANG (Naval Civil Engineering Lab. 1988, and Wager and Connett, 1989) were used to compute equivalent shock and gas pressures (Table 1), assuming the detonation of 8.5 kg TNT located at the centre of the floor. This effective charge was 20%

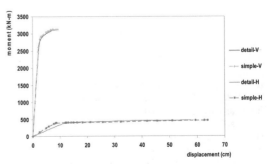

Figure 5.   Push-Over Response Comparisons.

smaller than required to fail the structure, as per TM 5-1300.

In the simulations, the computed blast pressures were applied as spatially uniform surface loads on the sidewalls, and were transferred to the beams and the column of the connection. Dead loads, applied on the top flanges of the beams and axially on the top cross section of the column, correspond to those for a 10-story office building. The numerical model was analyzed with and without dead loads, and with and without dynamic increase factors (DIF) to evaluate their influence on connection response (Figure 7).

The analysis was performed using reinforced moment connection that had been modified by adding flange cover plates and web stiffeners, as described below (Engelhardt et al. 1994). The flanges of beam and flange cover plates were welded to the face of the column using groove welds. The cover plates helped reduce stresses on the beam flange groove welds and also theoretically relocated the plastic hinge away from the face of the column. Stiffeners were provided for the column web to prevent local web buckling in the panel zone. As a material property,

an isotropic elasto-plastic model was used for each connection component (Yim et al. 2006a). Yield and ultimate strengths were increased to account for strain rate effects using dynamic increase factors (DIF) as recommended in TM 5-1300. Related to the DIF, the recommended dynamic design stress is given by: $f_{dy} = c \cdot a \cdot f_y$, where f_{dy} is a dynamic yield stress, c is a dynamic increase factor on the yield stress, a is an average strength increase factor, and f_y is a static yield stress based on uniaxial tensile stress. Since brittle fracture on the weld connections was anticipated under the blast loads, the shear failure model was adopted; ABAQUS removes elements from the mesh as they failed. The shear failure model was based on the value of the equivalent plastic strain at an element integration point and used to represent brittle fracture of the welds. The finite element models were created using predominantly 8-noded continuum brick elements with reduced integration, with wedge elements

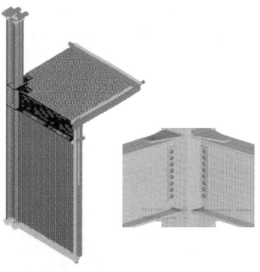

Figure 7. Representative 3D Finite Element Model of Steel Connection.

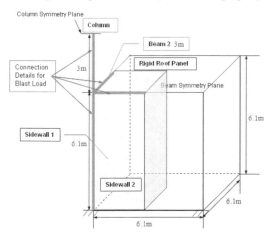

Figure 6. 3D Frame Model.

Table 1. Loading Data.

Case	Member	Shock Pressure		Gas Pressure	
		Peak P (MPa)	Time (msec)	Peak P (MPa)	Time (msec)
Two walls failed	Walls	1.07	1.81	0.19	29.95
	Floor	0.2	7.19	0.19	31.08
Wall 1 failed, Wall 2 reflects	Walls	1.07	2.23	0.19	43.5
Wall 2 failed, Wall 1 reflects	Floor	0.2	9.06	0.19	44.52
Two walls reflect	Walls	1.07	2.65	0.19	5864
	Floor	0.2	11.17	0.19	5866

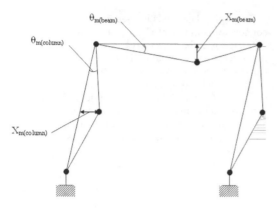

Figure 8.   Structural Deformation Model (TM5-1300).

Table 2.   Theoretical Response to Blast Loads, based on TM 5-1300.

Member	Maximum Deflection, (cm)	Ductility Ratio	Rotation deg.(rad.)	Rotational Limit, deg. (rad.)
Beam	1.73	2.2	0.32 (0.0056)	2 (0.0349)
Column	6.98	1.8	1.31 (0.0228)	

used as required by the geometry. The responses and failure criteria based on TM 5-1300 criteria are shown in Table 2 and Figure 8, and indicate that the representative room could withstand the loads from the explosive charge.

Finite element results when blast pressures (from Table 1) were applied to floor and sidewalls are summarized in Tables 3 through 6. Maximum global and local rotations, associated with the plastic hinge that is formed in the beam, are shown in Figure 9. The predicted global rotations of the beams are close to the TM 5-1300 results for the frangible wall cases. However, the beams where the reflecting walls were located rotated much more than TM 5-1300 computation predicted: greater impulse and energy were transferred to the beam and column in the room when no venting occurs. All local rotations for the different cases exceeded the limit of 2 degrees, as specified in TM 5-1300.

Also it was noted that the beams twisted more severely in the horizontal direction, since the realistic internal blast pressures in three; the horizontal rotations clearly exceeded the TM 5-1300 limit criteria. These findings indicate that one should expect severe damage in the connections from not only the vertically applied pressure but also the blast propagation in

three dimensions. Deformation data of beams and column in the various conditions indicate that dead loads and DIFs enhanced the structural strength, but the beam cross sections twisted additionally due to dead loads. The column rotations indicate that the columns did not significantly affect the connection damage. According to the stress and strain results, components in all connections yielded for all simulated cases.

The findings from these blast analyses show the value of investigating structural connections using the more realistic conditions available in high-resolution finite element programs. For example, a steel moment connection judged safe based on TM 5-1300 criteria failed in the finite element simulations. Moreover, TM 5-1300 criteria may need to be revised on the basis or more complex behaviors. Especially, when attached walls do not fail, beams showed much higher rotations than in the cases where the wall failed.

4   PROGRESSIVE COLLAPSE OF STEEL FRAME STRUCTURES

Progressive collapse is a failure sequence in which local damage leads to large scale collapse in a structure. This type of intrinsically transient nonlinear phenomenon is very difficult to understand, model, or design against without finite element analysis.

Ten-story 3D moment frames with rigid and semi-rigid connections were studied for their sensitivity to failure of specific columns (Yim et al. 2006b). Three failure modes were considered: material, column buckling, and connection failures; only material and buckling failures with semi-rigid connections have been studied extensively elsewhere (Aristizabal-Ochoa 1997, Ermopoulos 1991 & Liew et al. 2000). Experiments have shown that a real steel connection is neither rigid nor pinned; furthermore, the relationship between the moment and the beam column rotation is nonlinear (Kameshki & Saka 2003). In this study, the moment-rotation relationship of the 10 story frame was obtained by extensive preliminary 3D finite element simulation of steel connections.

Six initial failures with rigid and semi-rigid connections were used to analyze the frames for progressive collapse of five story frames. Frame columns were based on a simple LRFD design procedure manual (AISC 1994); other details of the finite element model are described in (Yim et al. 2006b). Both ideal (rigid plus hinge) and semi-rigid connections were adopted for the progressive collapse analyses.

Analyses were performed up to seven seconds after the initial failure, modeled by instantaneous removal of a designated column. The initial column failure cases

Table 3. Maximum Displacements and Rotations of Beams (Two walls failed) [unit: degree (radian)].

	Beam 1				Beam 2			
	DL DIF	DL NO DIF	NO DL DIF	NO DL NO DIF	DL DIF	DL NO DIF	NO DL DIF	NO DL NO DIF
Vertical								
Global	0.455	0.609	0.439	0.648	0.572	0.602	0.502	0.565
	(0.008)	(0.011)	(0.008)	(0.011)	(0.010)	(0.011)	(0.009)	(0.010)
Local	1.164	1.868	1.016	1.705	1.558	1.707	1.707	1.634
	(0.020)	(0.033)	(0.018)	(0.030)	(0.027)	(0.030)	(0.030)	(0.029)
Horizontal								
Global	3.267	3.336	3.197	3.241	0.196	0.232	0.181	0.226
	(0.057)	(0.058)	(0.056)	(0.057)	(0.003)	(0.004)	(0.003)	(0.004)
Local	12.540	14.60	12.17	14.534	1.340	1.718	1.718	1.701
	(0.219)	(0.255)	(0.212)	(0.254)	(0.023)	(0.030)	(0.030)	(0.030)

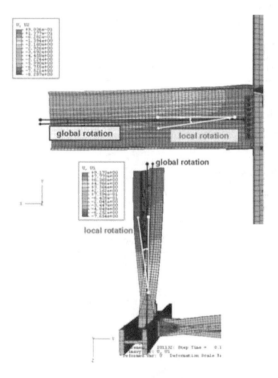

Figure 9. Global and Local Rotations (Vertical and Horizontal).

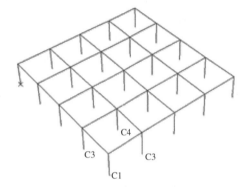

Figure 10. Initial Column Failure Cases.

are shown in Figure 10. Two types of steel connections were used in the frame analyses: ideal connections and semi-rigid connections. The ideal connections were composed of hinge and rigid connections. The hinge connection is incapable of transferring moments from one member to another, while a rigid connection can transfer moments applied one member to the other. The semi-rigid connection is defined in which

moment in the connection is proportional to rotation up to certain limits. These connections were designed as part of the building and the moment-rotation relationships were extracted from connection analyses. These relationships were adopted for the progressive collapse analyses of the 10-story steel frame building. Only Case 6, where three columns were removed, caused total collapse of the building. Figure 11 shows the result for Case 6 of the building with ideal connections. Case 6 with semi-rigid connections also collapsed, but differently, as shown in Figure 12. The first failure was initiated at a connection, as shown in Figure 12-(b). As connections failed, the floors above the removed columns started to fall to the ground, and it caused columns buckling in the 6th floor, as shown in Figure 12-(c). These columns buckling cases initiated horizontal failure propagation in the 6th floor, so that the whole floor failed. After that, the columns in the first floor started to buckle because the floors collapsed and it leaded to the total collapse of the building.

The 10-story frame, designed for gravity and lateral loads, showed fairly good performance in the

Table 4. Maximum Displacements and Rotations of Beams (Wall 1 failed, Wall 2 reflects) [unit: degree (radian)].

	Beam 1				Beam 2			
	DL DIF	DL NO DIF	NO DL DIF	NO DL NO DIF	DL DIF	DL NO DIF	NO DL DIF	NO DL NO DIF
Vertical								
Global	0.320	0.427	0.338	0.288	3.825	4.813	3.485	4.154
	(0.006)	(0.007)	(0.006)	(0.005)	(0.067)	(0.084)	(0.061)	(0.073)
Local	0.757	1.295	0.880	1.088	6.294	11.47	5.920	7.981
	(0.013)	(0.023)	(0.015)	(0.019)	(0.110)	(0.200)	(0.103)	(0.139)
Horizontal								
Global	3.853	3.508	3.700	3.513	0.713	1.002	0.674	0.821
	(0.067)	(0.061)	(0.065)	(0.061)	(0.012)	(0.017)	(0.012)	(0.014)
Local	9.155	8.701	9.094	10.062	5.563	6.373	5.762	5.531
	(0.160)	(0.152)	(0.159)	(0.176)	(0.097)	(0.111)	(0.101)	(0.097)

Table 5. Maximum Displacements and Rotations of Beams (Wall 1 reflects, Wall 2 failed) [unit: degree (radian)].

	Beam 1				Beam 2			
	DL DIF	DL NO DIF	NO DL DIF	NO DL NO DIF	DL DIF	DL NO DIF	NO DL DIF	NO DL NO DIF
Vertical								
Global	2.893	3.590 (F)	2.626	3.266	0.279	0.477 (F)	0.330	0.387
	(0.050)	(0.063)	(0.046)	(0.057)	(0.005)	(0.008)	(0.006)	(0.007)
Local	2.073	9.395 (F)	5.084	9.290	0.573	0.905 (F)	0.641	0.793
	(0.036)	(0.164)	(0.089)	(0.162)	(0.010)	(0.016)	(0.011)	(0.014)
Horizontal								
Global	2.064	2.504 (F)	1.929	2.427	0.253	0.217 (F)	0.227	0.213
	(0.036)	(0.044)	(0.034)	(0.042)	(0.004)	(0.004)	(0.004)	(0.004)
Local	21.137	18.09 (F)	21.53	20.84	1.020	0.997 (F)	0.942	1.050
	(0.369)	(0.316)	(0.376)	(0.364)	(0.018)	(0.017)	(0.016)	(0.018)

Table 6. Maximum Displacements and Rotations of Beams (Two walls reflect) [unit: degree (radian)].

	Beam 1				Beam 2			
	DL DIF	DL NO DIF	NO DL DIF	NO DL NO DIF	DL DIF	DL NO DIF	NO DL DIF	NO DL NO DIF
Vertical								
Global	2.254	3.104	1.932	2.936	3.066	4.155	2.883	4.020
	(0.039)	(0.054)	(0.034)	(0.051)	(0.054)	(0.073)	(0.050)	(0.070)
Local	2.018	4.010	2.131	3.804	4.574	7.981	4.392	7.486
	(0.035)	(0.070)	(0.037)	(0.066)	(0.080)	(0.139)	(0.077)	(0.131)
Horizontal								
Global	1.523	1.746	1.647	1.675	0.529	0.821	0.478	0.751
	(0.027)	(0.030)	(0.029)	(0.029)	(0.009)	(0.014)	(0.008)	(0.013)
Local	18.461	16.62	16.317	15.33	4.458	5.531	4.522	5.293
	(0.322)	(0.290)	(0.285)	(0.268)	(0.078)	(0.097)	(0.079)	(0.092)

DL: dead load; DIF: dynamic increase factor; (F): connection failure.

finite element simulations. Even though the ideal and semi-rigid connection cases both caused total collapse for Case 6, they showed very different qualitative behavior. The collapse of the semi-rigid connection case was caused by a cascade of local failures, such as connection failures and columns buckling. However, the collapse of the ideal connection case was caused by column buckling in the first floor. These different

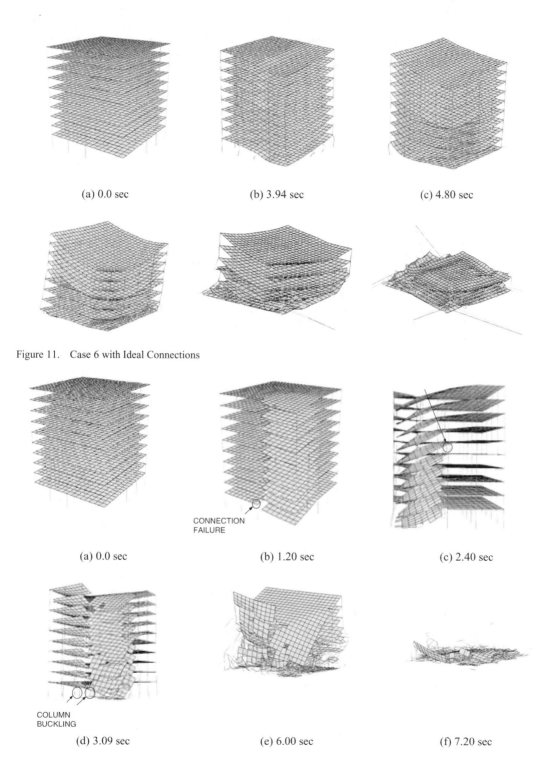

(a) 0.0 sec   (b) 3.94 sec   (c) 4.80 sec

Figure 11.   Case 6 with Ideal Connections

CONNECTION
FAILURE

(a) 0.0 sec   (b) 1.20 sec   (c) 2.40 sec

COLUMN
BUCKLING

(d) 3.09 sec   (e) 6.00 sec   (f) 7.20 sec

Figure 12.   Case 6 with Semi-rigid Connections

failure mechanisms are quite apparent in the nonlinear finite element results. Progressive collapse analyses of 3D frames also showed that once failure propagation was initiated (i.e. horizontal column buckling), it would not stop until it caused total collapse (or almost total collapse). If a structure has to be protected against progressive collapse, horizontal column buckling propagation appears to be the most critical factor to be controlled.

5 CONCLUSIONS

Finite element simulations can be used to model many important phenomena associated with structural behavior under blast, with or without addressing computational fluid dynamics simulations of the explosion. A fairly wide range of blast wave loading models can be incorporated in explicit finite element structural calculations, ranging from transient uniform loads to complex spatially varying propagating loads. Critical aspects of structural behavior, such as the details of steel connection designs, can not only be studied using finite element analysis, but the results may be used to define macroscopic structural connection behavior used in larger models. The work cited in this paper showed that connection behavior under blast loading can vary significantly from standard design criteria. Importantly, the design criteria may not be conservative for the studied cases. Further, the design criteria themselves may require refinement and revision in light of nonlinear transient effects, such as progressive collapse, which can be revealed only through computational analysis, or prohibitively costly experiments. The progressive collapse analyses discussed here combined fairly complex nonlinear mechanics with the nonlinear macroscopic models of connection behavior derived from the models of the connection geometry. Finite element analysis of progressive collapse due to blast effects also revealed qualitative information about structural failure such as, in these cases, the sensitivity to connections types.

ACKNOWLEDGEMENTS

This study was conducted under contract with the U.S. Army Engineer and Development Center (ERDC), Vicksburg, MS. The authors wish to acknowledge both the generous support provided by the sponsor, and the active involvement of ERDC technical personnel in this program. Also, the author wishes to thank Dr. H.C. Yim, of the Center for Infrastructure Protection and Physical Security at the University of Florida, Dr. J.H. Lim of WSP Cantor Seinuk, New York, and Dr. J. Cipolla of ABAQUS Inc. for their cooperation and assistance.

REFERENCES

ABAQUS Inc. 2005. *ABAQUS Analysis User's Manual*, Version 6.5.

American Institute of Steel Construction. 1994. *Load and Resistance Factor Design.*

Aristizabal-Ochoa, J.D. 1997. Elastic Stability of Beam-Columns with Flexural Connections under Various Conservative End Axial Forces, *Journal of Structural Engineering, Vol. 123, No. 9, pp. 1194–1200, September 1997.*

Cipolla, J. 2006. Generalized incident wave loading on acoustic and solid elements, *Proc. 77th Shock and Vibration Symposium, 2006.*

Department of the US Army. 1990. *Structures to Resist the Effects of Accidental Explosions* TM 5-1300.

Department of the US Army. 1992. *Fundamentals of protective design for conventional weapons* TM 5-855-1.

Engelhardt, M.D., Sabol, T.A., Aboutaha, R.S. & Frank, K.H. 1994. *Northridge Moment Connection Test Program Report for AISC.*

Ermopoulos, J. CH. 1991. Buckling Length of Framed Compression Members with Semi-rigid Connections, *Journal of Constructional Steel Research, Vol. 18, pp. 139–154, 1991.*

Griffiths, H., Pugsley, A. & Saunders, O. 1968. *Collapse of Flats at Ronan Point, Canning Town*, Her Majesty's Stationery Office, London, UK.

Kameshki, E.S. & Saka, M.P. 2003. Genetic Algorithm Based Optimum Design of Nonlinear Planar Steel Frames with Various Semi-Rigid Connections, *Journal of Constructional Steel Research 59, pp. 109–134, 2003.*

Naval Civil Engineering Lab. 1988. *SHOCK User's Manual*, Port Hueneme, CA, USA.

Liew, J.Y.R., Chen, W.F. & Chen, H. 2000. Advanced Inelastic Analysis of Frame Structures, *Journal of Constructional Steel Research 55, pp. 245–265, 2000.*

Ripley, R.C., Donahue, L. & Zhang, F. 2005. Modeling Complex Blast Loading in Streets, *6th Asia-Pacific Conference on Shock & Impact Loads on Structures, 7–9 December, 2005.* Perth, Australia.

Wager, P. & Connett, J. 1989. *FRANG User's Manual*, Naval Civil Engineering Lab., Port Hueneme, CA. USA.

Yim, H.C., Starr, C., Krauthammer, T & Lim, J.H. 2006b. Assessment of Steel Moment Connections for Blast Loads, *Proc. 2nd International Conference on Design and Analysis of Protective Structures 2006.* 13–15 November 2006, Singapore.

Yim, H.C., Krauthammer, T., Lim, J.H. & Kyung, K.H. 2006a. Progressive Collapse Studies of Steel Structures, *Proc. 32nd DoD Explosive Safety Seminar, 22–24 August 2006.* Philadelphia, PA USA.

Steel and Composite Structures – Wang & Choi (eds)
© *2007 Taylor & Francis Group, London, ISBN 978-0-415-45141-3*

Structural fire engineering of steel framed buildings

C.G. Bailey

*School of Mechanical, Aerospace and Civil Engineering, The University of Manchester,
Manchester, United Kingdom, UK*

ABSTRACT: There are a number of approaches to ensure the safe design of steel structures under fire conditions. These range from a simple elemental prescriptive approach to a more advanced structural fire engineering approach. In the simple approach, realistic structural and fire behaviour are ignored and optimum design solutions in terms of safety and economy are impossible to obtain. It is also assumed that any ignored beneficial effects outweigh any ignored detrimental effects experienced in fires within real buildings. This paper presents various methods of achieving fire safety for steel-framed buildings and introduces the benefits of using performance-based approaches. By considering the actual fire and structural behaviour, through more advanced methods, any 'weak-links' within the design can be identified, and rectified, allowing safer, more robust, and possibly more economical buildings to be constructed.

1 INTRODUCTION

Steel has a high thermal conductivity and, compared to concrete and masonry, will rise in temperature very quickly in a fire. Similar to all materials steel will lose strength and stiffness at elevated temperature (Fig. 1). Therefore steel-framed buildings will need to be designed to ensure they achieve the required safety during a fire.

The minimum legislative level of safety for structural fire design aims to provide an acceptable risk associated with the safety of the building occupants, fire fighters and people in the proximity of the building. Life safety requirements are covered by Regulations which may be functional or prescriptive.

Figure 1. Structural collapse of steel structure following a fire.

For example, the Building Regulations in England and Wales (2000) provide the following functional objectives relating to structural aspects of fire safety.

- The building shall be designed and constructed so that in the event of fire its stability will be maintained for a reasonable period.
- To inhibit the spread of fire within the building it shall be divided with fire resisting construction to an extent appropriate to the size and intended use of the building.

To meet these life safety requirements either a performance-based approach or the simple prescriptive rules, as outlined in the approved documents (ADB, 2000) or guidance (Bailey, 2007), could be adopted.

Due to the perceived disadvantage of steel structures in fire there has been extensive research activity, in understanding the behaviour of steel structures in fire, resulting in a number of practical design approaches. These are discussed in this paper.

2 APPLIED FIRE PROTECTION

The traditional and still the most popular approach to achieve the required levels of safety is to apply fire protection to all exposed areas of steel. The use of fire protection can be in the form of proprietary materials comprising sprays, boards or intumescent coatings, or generic materials comprising concrete,

Figure 2. Typical methods of applied fire protection.

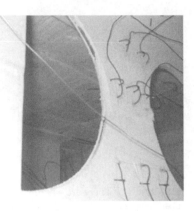

Figure 3. Poor application of intumescent coatings.

brick, block, gypsum plaster and certain types of plasterboard (Fig. 2).

Typically the specification of fire protection thicknesses to steel elements has been based on ensuring that the steel does not exceed a maximum temperature of 550°C for columns, and 620°C for beams supporting concrete floors, for a given fire resistance period tested in a standard furnace. These temperatures are based on the assumption that a fully-stressed member at ambient conditions will lose its design safety margin when it reaches 550°C. The maximum temperature for beams supporting concrete floors is increased to 620°C, since the top flange is at a lower temperature compared to the web and bottom flange, due to the concrete floor acting as a heat sink. Generally the 550/620°C maximum temperatures are considered conservative since the members are not fully stressed at ambient temperature, the stress-strain-temperature relationship of steel at elevated temperatures (used to derive the 550/620°C values) is too simplistic, and in practice the structural elements do not behave in isolation.

It is possible to reduce the protection thickness based on the design of steel members during the fire limit state using current codes of practice. However, this has been found to be very difficult in practice due to the reluctance of protection manufacturers to release the required thermal properties of their materials. It is possible to work with individual manufacturers to specify protection thicknesses based on the fire design of the steel members on a project-by-project basis, but this increases the design time and is rarely a method adopted by designers. There are some exceptions; noticeably manufacturers of cellular beams in the UK are specifying failure temperatures (and in some cases protection thicknesses) for their beams.

The use of intumescent coatings has increased recently, especially when applied off-site which reduces construction time and arguably increases quality. However, the coatings are easily damaged during transportation, are not always applied adequately (Fig. 3) and due to pressures of construction time

are sometimes delivered to site when they are not sufficiently cured.

3 PARTIAL FIRE PROTECTION

It is possible to adopt forms of construction (Bailey et al, 1999a) which eliminate the need for additional passive fire protection. The common forms of beams and columns that utilise partial protection are described below. The construction systems have generally been developed based on standard fire resistance tests (BSEN1363, 2005) and the basic principles given in fire design codes (BSEN1993-1-2, 2005 & BSEN1994-1-2, 2005).

By placing a significant portion of the beam within the depth of the supported concrete slab it is possible to specify steel beams without the need to apply additional fire protection. Some systems, known as slim-floor beams, are constructed such that the beam is encased in the supporting concrete slab with only the bottom flange or plate exposed to any fire (Fig. 4). SCI and Corus have promoted systems known as 'Slimflor®' and 'Slimdek®' where beams can readily achieve 60 minutes fire resistance without the need to protect the exposed bottom flange or plate. The Slimdek® system, incorporating an asymmetrical beam, has been tested at full-scale (Bailey, 2003) and shown to perform extremely well when subjected to a severe fire.

Another form of slim-floor beam is the 'Deltabeam', which can readily achieve 2 hours fire resistance. The systems available have their own associated advantages and disadvantages and it is difficult to provide generalised information on the choice of the best system. The use of the systems should be assessed on a project-by-project basis considering ultimate, serviceability and fire design together with buildability issues.

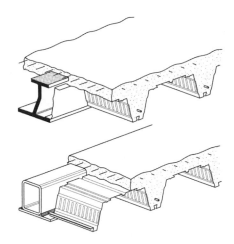

Figure 4.   Slim-floor beams.

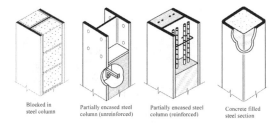

Blocked in steel column   Partially encased steel column (unreinforced)   Partially encased steel column (reinforced)   Concrete filled steel section

Figure 5.   Partially protected columns.

For typical downstand 'I' beams the floor slab can be supported by shelf-angles, with the legs of the angles pointing upwards into the slab; a form of construction commonly referred to as 'shelf angle beams'. The part of the steel section and supporting angles embedded in the supporting slab can allow 30 minutes fire resistance to be readily obtained for this type of member. It is possible to increase the fire resistance to 60 minutes although the required thickness of the concrete slab may make this form of construction uneconomical. Another form of construction, where the beam is partially protected by concrete, consists of filling the area between the flanges and web with reinforced concrete. This type of construction is popular in continental Europe where the cost of proprietary fire protection materials is relatively high. The system can readily achieve 2 hours fire resistance and has the advantage of being resistant to impact damage, although the increase in self-weight of the structure and buildability issues can be seen as a disadvantage.

As with steel beams it is possible to enhance the fire resistance of steel columns by adopting systems where concrete and masonry materials provide sufficient, although partial, protection. A simple method is to place aerated concrete blocks between the inner faces of the flanges (Fig. 5). For columns, of size $203 \times 203 \times 46$ UC and greater, 30 minutes fire resistance can be achieved. Another method, developed by SCI (Bailey, 1999a), involves filling between the flanges with unreinforced concrete. Welded plates and shear fixings allow the load to be transferred from the steel section to the concrete during a fire. This system can achieve 60 minutes fire resistance and can resist impact damage. A variation of the SCI infill column is to provide reinforcement to the infill concrete (Fig. 5). Similar to the infill beams, this system is popular in continental Europe and can achieve 2 hours fire resistance.

The most common form of steel column that generally does not use applied proprietary fire protection materials is concrete-filled hollow steel sections (Fig. 5). The infill concrete may be unreinforced or reinforced depending on the required load-capacity and fire resistance. For reinforced concrete infill columns, 2 hours fire resistance can readily be obtained. To achieve the most efficient concrete filled column, in terms of fire resistance, the section should be designed in the cold state such that the load-carrying capacity of the steel shell is low compared to the load-carrying capacity of the concrete core.

4   DESIGN OF FLOOR PLATES

Tensile membrane action in concrete (or steel and concrete composite) slabs has been observed in full-scale fire tests (Bailey *et al* 1999b) (Fig. 6) and in real buildings following actual fires. Simple design procedures given in current fire design codes for floor slabs are based on flexural action and ignore any beneficial effect of membrane action. However, provided they follow general engineering principles, the codes do allow advanced design methods to be used but they present no guidance on these methods.

In 2000 a simple pragmatic fire design method (Bailey & Moore 2000, Newman *et al* 2000), which utilised the behaviour of tensile membrane action, was published in a tabulated form for composite floor slabs supported by a grillage of steel beams. The method was also published in more detail in 2001(Bailey 2001), allowing designers to use the method to its full potential. The design method is based on the assumption that the floorplate is divided into a collection of horizontally unrestrained slabs spanning over unprotected beams and supported around their perimeter by protected steel beams. By utilising the membrane action of the floor slab a large proportion of the steel beams within a given floorplate can be left unprotected as shown in Fig. 7.

The original design approach was recently updated in 2006 (Newman *et al* 2006) to include more efficient reinforcement patterns and the practical use of natural fires. However, even with these recent updates,

Figure 6. Large displacements of the floor following fire test.

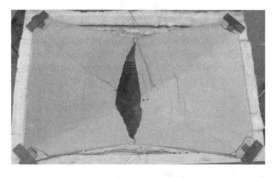

Figure 7. Use of membrane action allows unprotected beams within a floorplate.

[Figure 8 image]

Figure 8. Tests of concrete slabs to understand tensile membrane action (loading removed).

the original method still had a number of assumptions which could only be refined by extensive testing. This led to a test programme looking at the tensile membrane action of horizontally unrestrained slabs at both ambient and elevated temperatures, which has just been completed at the University of Manchester. The slabs incorporated welded mesh reinforcement, which is typically used in composite floors in the UK. The tests (Fig. 8) comprised 22 tests at ambient temperature which were repeated at elevated temperatures. The results from the tests have led to a refinement of the original design method which has included a more accurate estimate of the in-plane stress distribution and a limit on the load-carrying capacity due to crushing of the concrete in the corners of the slab.

5 FIRE ENGINEERING

A comprehensive 'full' performance-based approach to fire safety engineering in buildings is an extremely complex multi-disciplinary design procedure. The 'full' approach will involve consideration of active and passive measures, movement of smoke and fire, detection systems, fire safety management, structural response and risk analysis.

It is possible to carry out a *structural* fire performance-based approach, which will allow the designer to understand and explain how buildings perform, should they be subjected to severe fires. The main advantage of a performance-based approach is that actual acceptable performance criteria can be defined and the level of safety for each part of the design can be assessed. This is a significant improvement on the acceptance criteria underlying the prescriptive approach which relates to stability, insulation and integrity defined in an unrealistic small-scale standard fire test.

The acceptable criteria within a performance-based fire design should be based on the global fire strategy for the building. The following points should be considered when considering the acceptable structural response:

- The structure should remain stable for a reasonable worst case fire scenario. If natural fire curves are used the effect of the cooling stage of the fire, on the behaviour of the structure, should be considered. For example, for steel-framed structures a significant proportion of the connections should be able to reasonably accommodate large tensile forces without loss of vertical shear capacity.
- Both vertical and horizontal compartmentation should be maintained for the duration of the reasonable worst case fire scenario. Vertical displacement of the floor slabs and beams in the proximity of the compartment walls should be considered particularly when more advanced methods are being adopted. These displacements can be an order higher than those experienced at ambient temperature.

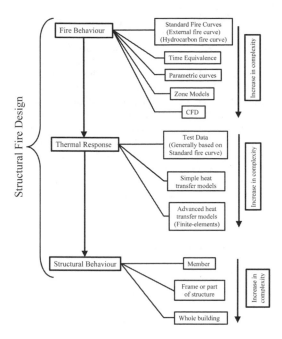

Figure 9. Available methods to define the fire, heat transfer and structural response.

- All escape routes, especially for phased evacuation, should remain tenable for a reasonable period of time.
- Fire-fighting shafts should not be compromised for the duration of the reasonable worst case fire scenario.
- By consultation with specialist suppliers, the effect of large structural movements on any applied fire protection, fire stopping, penetration seals, and the integrity of ducts and dampers should be considered for the reasonable worst case fire scenario.
- If identified as a critical fire scenario, the risk and consequence of fire spread up the building, through windows, should be considered within the structural fire design strategy.

A performance-based structural fire design consists of defining the fire behaviour, the transfer of heat to the structure and high temperature structural analysis. Fig. 9 shows the available methods covering these three aspects of the design.

The choice of the design approach will depend on:

- The defined requirements and objectives.
- The experience of the designer.
- The potential economical return.
- The need to consider higher levels of safety above the regulatory requirement.

It is possible to use any permutation of the methods shown in Fig. 9 to define the fire behaviour, heat transfer and structural response. The following, general, guidance is provided when considering different permutations.

- The accuracy of the design as a whole should be considered. For example the designer would need to consider the effect, and validity, of using the simple standard temperature-time relationships with advanced heat transfer and structural response models, when carrying out a deterministic approach. The Eurocodes do allow such a design approach but it must be noted that there is little to be gained in predicting the heat transfer and structural response to a high level of accuracy when the prediction of the fire is crude and bears little resemblance to reality. However, this combination may be appropriate when carrying out a comparative approach. An example would be the case where a standard fire is used and advanced analysis is used to compare the relative performance of a simple compliant structure with that of a more complex structure when test or prescriptive design data is not available.
- If there is reliable thermal test data, relevant to the assumed fire behaviour, then this may be sufficient to replace the need for a thermal analysis for input into structural finite element analysis.
- The knowledge and experience of the designer should be considered. The use of zone models, CFD and finite-element heat-transfer and structural models requires specialist knowledge and should only be used by suitably experienced personnel.
- The accuracy and availability of the data representing the fuel load, ventilation, and thermal properties of the compartment boundaries, heat release rates, material properties and applied static loads.
- Availability of software for zone, CFD and finite-element models.
- Available time to carry out the design.
- Capital cost of the project. For a low cost project the use of advanced fire models may not be justified.
- The importance of considering the structural behaviour during the cooling phase of the fire. If the structural behaviour during the cooling stage is considered to be important then standard fires cannot be used.

5.1 Fire Behaviour

The factors influencing the severity of a fire in a compartment are:

- Fire load type, density and distribution.
- Combustion behaviour of the fire load.
- Compartment size and geometry.
- Ventilation conditions of the compartment.
- Thermal properties of the compartment boundaries.

The occurrence of flashover, in a compartment fire, defines a transition in the fire development.

Figure 10. CFD modelling of a compartment fire.

Fire model	Nominal fires	Time equivalences	Compartment fires		Zone models		CFD / field models
			Parametric	Localised	One-zone	Two-zone	
Complexity	Simple	Intermediate			Advanced		
Fire behaviour	Post-flashover fires			Pre-flashover fires	Post-flashover fires	Pre-flashover/localised fires	Pre-flashover temperature-time relationships
Temperature distribution	Uniform in whole compartment			Non-uniform along plume	Uniform	Uniform in each layer	Time and space dependent
Input parameters	- Fire type - No physical parameters	- Fire load - Ventilation conditions - Thermal properties of boundary - Compartment size		- Fire load & size - Height of ceiling	- Fire load - Ventilation conditions - Thermal properties of boundary - Compartment size - Detailed input for heat & mass balance of the system		Detailed input for solving the fundamental equations of the fluid flow
Design tools	←———————— EN1991-1-2 ————————→				COMPF2 OZone SFIRE-4	CCFM CFAST OZone	FDS SMARTFIRE SOFIE
	PD7974-1	PD7974-1					
	Simple equations for hand calculations	Spreadsheet		Simple equations	Computer models		

Figure 11. Options for modelling compartment fires.

Therefore, many fire models are classified as pre- or post-flashover models, except for computational fluid dynamic (CFD) models, which can model all stages of the fire (Fig. 10). There are a number of options available to calculate the fire severity, as shown in Fig. 9. A summary of the fire models, their complexity, predicted fire behaviour, input parameters and design tools are summarised in Fig. 11.

The simplest approach is to use the standard fire curves but, as explained previously, these time-temperature relationships are not based on any physical parameters and do not consider the cooling stages of a fire, which can be extremely important when considering the structural behaviour. The time-equivalence is a simple approach that tries to relate the actual temperature of a structural member from an anticipated fire severity, to the time taken for the same member to attain the same temperature when subjected to the standard fire curve. There are a number of time-equivalence methods which take into account the amount of fuel load, compartment size, thermal characteristics of the compartment boundaries and ventilation conditions. Although simple to use, the time-equivalence is a crude approximate method of modelling real fire behaviour and the limitations of the method should be fully understood.

Parametric fire curves allow the time-temperature relationship to be estimated over the duration of the anticipated fire. Compartment size, boundary characteristics, fuel load and ventilation are considered. The approach is simple to use and, with the aid of simple spreadsheets, fire predictions can be easily derived.

Zone models are simple computer models that divide the considered fire compartment into separate zones, where the condition in each zone is assumed to be uniform. The simplest model is a one-zone model where the conditions within the compartment are assumed to be uniform and represented by a single temperature. A more sophisticated modelling technique is the use of Computational Fluid Dynamics (CFD) to predict fire growth and compartment temperatures. CFD has been shown to be successful in the modelling of smoke movement and has recently been applied to the modelling of fires. Similar to the use of any computer model, both the zone and CFD models require expertise in defining the correct input data and in assessing the feasibility of the calculated results.

5.2 Heat transfer to the structure

The temperature distribution through structural members is dependent on the radiation and convective heat transfer coefficients at the member's surface and conduction of heat within the member. The available methods are summarised in Fig. 12.

For materials with a high thermal conductivity, such as steel, it may be sufficiently accurate to ignore thermal gradients within members and assume a uniform temperature. This assumption is valid provided the member is not in contact with a material of low thermal conductivity which will act as a heat-sink and thus create a thermal gradient through the member. Simple design equations exist to predict the temperatures of steel members which are fully exposed to fire or steel members that support a concrete floor slab and are exposed on three sides.

Estimating the heat transfer in materials with a low thermal conductivity and moisture retention, such as concrete and masonry, becomes extremely complex due to the high thermal gradients. To carry out a performance-based approach, which investigates the structural response of the building, it is extremely important to obtain an accurate estimate of the temperature gradient through the structural members. Simple design charts are given in codes and design guides defining the temperature distribution through members, which have been derived from standard fire tests. These charts can only be used if the standard fire curve is assumed to define the fire behaviour.

If parametric curves, zone models or CFD are used to model the fire behaviour then either simple or advanced heat transfer models should be used. Careful attention should be given to the modelling of moisture if simple heat transfer models are adopted.

5.3 Structural Response

The available structural design methods are summarised in Fig. 13. The simplest method of predicting

Model	Design charts / Test data	Simple formulae	Advanced models
Complexity	Simple	Intermediate	Advanced
Heat transfer modes	Conduction		Convection Radiation Conduction
Analysis ability	- Test results - Standard fire conditions	- Empirical solutions - Standard fire conditions	- Accurate solutions - Any fire conditions
Member types	- Dependent on available test data	- Mainly steel members	- Any material & construction methods
Input parameters	- Construction type - Member geometry	- Heat flux or fire curves - Boundary conditions - Member geometry - Material thermal properties	
Solutions	- Cross-sectional temperature charts - Tabulated thermal data	- Simple cross-sectional temperature profile	- One- to three-dimensional time & space dependent temperature profile
Design tools	- Fire part of Eurocodes - Test/Research reports	- Fire part of Eurocodes - Design guides	- Finite element package
	Design charts/tables	*Spreadsheet*	*Computer models*

Figure 12. Options for estimating the heat transfer.

Model	Simple element	Sub-models	Advanced computer finite-element models
Complexity	Simple	Intermediate	Advanced
Input parameters	Temperature through the cross-section. Material strength and stiffness reduction. Applied static load. Simplified boundary conditions.	Temperature through the cross-section and along the member. Material strength and stiffness reduction. Applied static load. Boundary conditions.	Temperature through and along the cross-section. Full material stress-strain-temperature relationship. Applied static load. Boundary conditions. Element type and density.
Accuracy	Ignores real behaviour but assumed to be conservative. Ultimate strength calculation.	Begins to consider actual load paths and restraint. Ultimate strength calculation	Predicts internal stresses, displacements, and rotations for all members throughout the duration of the fire. Localised behaviour is not modelled accurately in whole building modelling.
Design tools	Simple equations for hand calculations	Simple equations for hand calculations. Plastic design, redistribution of moments. Simple computer models.	Commercially available or purpose written computer software.

Figure 13. Options for structural analysis.

the structural response of buildings in fire is to analyse individual members. The design adopts relevant partial safety factors which provides realistic estimates of the likely applied load at the time of the fire and the likely material resistance of the member. The approach of designing individual members has evolved from results and observations from standard fire tests.

The use of member fire design at the FLS, as covered by current codes of practice, utilises principles which closely follow the approach used to check members at the Ultimate Limit State (ULS). The main differences between ULS and FLS is that for fire design different partial safety factors for load and material resistance are used (to represent an accidental limit state) and the strength and stiffness of the member is reduced based on the temperature distribution through the cross-section.

Although member and frame design at FLS is a significant improvement on the prescriptive approaches, allowing designers to obtain some indication (although limited) of the actual behaviour of buildings in fire, recent fire tests on full-scale buildings have shown that member design is generally not realistic. To most designers this will come as no surprise since member design methods at ULS and SLS are only an approximation of the real behaviour of buildings. However, provided this approximation is conservative, resulting in safe, usable and economic buildings then the design approach is acceptable.

At present, research is on-going looking at the modelling of whole building behaviour in fire with specialist companies using commercially available or purpose-written software. The main use of such software is in the modelling of steel-framed structures where significant savings in fire protection can be obtained. The main disadvantage of using sophisticated models is that they are seen as a 'black-box', which makes checking of designs difficult. In addition, the models are not able to simulate localised failure to a sufficient level of accuracy, particularly reinforcement fracture in the slab and connection failure when considering whole building behaviour. At present designers make conservative assumptions by restricting the maximum allowable strains in the reinforcement and specifying ductile connections that have been shown to retain their vertical shear capacity following a fire.

6 ADVANTAGES OF USING PERFORMANCE BASED APPROACHES

The concept of fire resistance, and the standard fire test, has the considerable advantage of being easily understood by designers and checking bodies. To-date it has generally been shown to be an adequate approach for ensuring a minimum level of fire safety in buildings. However, the fire resistance test has generally been considered to stifle the understanding of how buildings behave in fire. In addition, the use of the standard fire test as a means of meeting the regulatory requirement has resulted in designers and manufacturers concentrating on ensuring that their system/product performs well in the test, without any thought on how it will perform in a real structure under a real fire. As we increase our knowledge, striving towards improving the economy and use of buildings, different forms of construction and longer spans are being utilised. For example, 15 m span steel beams are now the norm in buildings whereas our fire tests are still on 4.5 m span beams. How can we be sure that these long span beams will be adequate in fire?

The question of how 'robust' the designed structure is, should a fire occur, cannot be answered unless a fire engineering approach is adopted. This is because the building does not represent a collection of individual elements working independently of each other as tested in a standard furnace. The interaction between structural elements in a fire has both possible beneficial and detrimental effects on the survival of the building as a whole. Beneficial effects are generally due to the formation of alternative load-path mechanisms such as compressive and tensile membrane action, catenary action and possible rotational restraint from connections. Evidence of tensile membrane action has been provided by large-scale fire tests

Figure 14. Failure of non-loadbearing compartment wall due to deformation of the structure.

Figure 15. Fracture of end-plates and shear failure of bolts in various connection types following a fire.

and was used to develop the design method utilising tensile membrane action discussed previously.

The detrimental effect of a collection of structural elements acting as a complete building can be due to the restraint of thermal expansion resulting in large compressive forces being induced into elements (particularly vertical elements) leading to instability.

Another detrimental effect can be the behaviour of walls, which in a standard fire test may be shown to perform adequately but in a real building the movement of the heated structure around the wall may result in premature collapse. This effect was shown for a non-loadbearing compartment wall in one of the fire tests on the steel-framed building at Cardington (Fig. 14). In this test the wall was placed off-grid and the deflection of the unprotected beams caused significant deformation of the wall, leading to compartmentation failure.

A common mode of detrimental structural behaviour observed during full-scale fire tests and following real fires, consists of fracture of connections in steel framed buildings which occurs during the cooling phase of the fire. The high tensile forces induced into the connections during cooling led to fracture of the connection's welds and bolts (Fig. 15). For an end-plate type of connection it was shown that they performed adequately, maintaining vertical shear under high tensile forces, even though fracture of the end-plate occurred. However, in the fin-plate type of connection the vertical shear was lost following failure of the bolts.

A further significant disadvantage of the standard fire test, and thus the prescriptive approach, is that the time-temperature relationship does not represent a real fire. There are generally three distinct phases to a real fire comprising a growth, steady burning and cooling phase. The severity of the fire is governed by the geometry of the compartment, the amount of combustible material, the ventilation conditions and the thermal characteristics of the compartment boundary. Different types of fire can result in different structural behaviour. For example a short duration high temperature fire will affect the strength of any unprotected steelwork whereas a long duration low temperature fire will result in a higher average temperature in protected members and composite floors resulting in greater thermal expansion and a greater overall reduction in strength.

7 CONCLUSIONS

There are a number of design approaches available to ensure structural fire safety of buildings. The simplistic elementary prescriptive approach of specifying forms of construction, which will achieve the required fire resistance periods, are commonly used. However, by using these simple approaches the designer cannot assess the actual levels of fire safety, robustness of the building and whether the optimum economical design solution has been achieved. The elemental prescriptive approach also ignores any detrimental effects observed from full-scale fire tests due to the building acting in its entirety.

By carrying out a fire engineering approach the actual structural behaviour and realistic fire scenarios are considered and any 'weak-points', identified within the design. Any identified 'weak-points' can be easily, and typically cheaply, rectified to obtain a more overall robust building. The performance-based approach can also form a part of a risk analysis to consider multiple extreme loading events, such as earthquakes and fire, or explosions and fire, with the aim of reducing the overall probability of loss of life and financial loss.

REFERENCES

Approved Document B – Fire Safety 2000. The Stationary Office Limited, London UK. ISBN 185 112351 2.

Bailey C.G., Newman G.M and Simms W.I., *Design of Steel Framed Buildings without Applied Fire Protection.* SCI Publication 186. The Steel Construction Institute, Ascot. 1999a. ISBN 1 85942 062 1.

Bailey, C.G., Lennon, T. and Moore, D.B.(1999b), The behaviour of full-scale steel framed buildings subjected to compartment fires, *The Structural Engineer*, Vol. 77, No. 8, April 1999. pp. 15–21.

Bailey C.G. and Moore D.B (2000) structural behaviour of steel frames with composite floorslabs subject to fire: Part 2: Design. The Structural Engineer Vol. 78 No. 11 June 2000 pp. 28–33.

Bailey C.G. (2001) Structures supporting composite floor slabs: design for fire. BRE Digest 462. December 2001. ISBN 1 86081 527 8.

Bailey C.G. (2003) Large scale fire test on a composite slim-floor system. *Steel and Composite Structures.* Volume 3, Number 3, pp.153–168.

Bailey C.G (2007). *One Stop Shop in Structural fire Engineering.* www.structuralfiresafety.org. The University of Manchester.

BSEN1363-1:1999. Fire resistance tests – General requirements. British Standards Institution, London, 1999.

BSEN1993-1-2: Eurocode 3. Design of steel structures Part 1.2. General rules. Structural fire design. British Standards Institution, London, 2005.

BSEN1994-1-2. Eurocode 4. Design of composite steel and concrete structures. Part 1.2. General rules. Structural fire design. British Standards Institution, London, 2005.

Newman G.M., Robinson J,T. and Bailey C.G., (2000) *Fire Safe design: A New Approach to Multi-Storey Steel-Framed Buildings.* SCI Publication P288. The Steel Construction Institute, Ascot.

Newman G.M., Robinson J,T. and Bailey C.G.(2006), *Fire Safe design: A New Approach to Multi-Storey Steel-Framed Buildings (Second Edition).* SCI Publication P288. The Steel Construction Institute, Ascot. 2006.

The Building Regulations 2000, The Stationary Office Limited, London, UK. ISBN 011 099897 9.

Steel and Composite Structures – Wang & Choi (eds)
© *2007 Taylor & Francis Group, London, ISBN 978-0-415-45141-3*

Concept and application of member-based linear and system-based nonlinear analysis in steel structure design

S.L. Chan

Department of Civil and Structural Engineering, The Hong Kong Polytechnic University, Hong Kong, China

ABSTRACT: The term "Advanced Analysis" and "Second-order Analysis" appear in code and other documents for decades but they still remain basically a check for the effect of sway moment or a tool for academic research. In practical design, engineers only use the method for checking of the magnitude of sway moments in slender frames. In fact, the direct design application of the method provides a valuable tool for practical design. Their underlying principle also carries a very different philosophy to the conventional design approach. This paper outlines the similarities and differences between the two design concepts, namely the linear effective length method and the second-order elastic or plastic analysis and design method. The era for re-thinking our design philosophy seems to be approaching. In this paper, the experience and theory on using the new design concept and method are also described with worked example demonstrated.

1 INTRODUCTION

In conventional design, prescriptive rules, featured by extensive uses of empirical parameters and assessment are followed and the design is principally member-based. The failure of a member is normally considered as the failure of the complete structure and the design is mainly limited to the formation of the first-plastic hinge and the analysis method is confined to the elastic and small deflection range.

Research on nonlinear frame analysis has been carried out for many decades. Recent work more related to practical applications include the advanced analysis by Liew et al. (1994, 2000), Izzuddin and Smith (1996) and White and Hajjar (1997). The works have also been summarised and reported in detail by Yang and Kuo (1994) and Chan and Chui (2000). Applications to special structures like transmission towers are also demonstrated by Albermani and Kitipornchai (2003) and Chan and Cho (2005) and performance of various nonlinear numerical methods is also studied by Clarke and Hancock (1990). The theoretical works have been extensively researched with good progress made, but the existing method is still difficult to apply to real engineering problems. For example, most codes require consideration of member imperfection in buckling strength determination but the straight cubic element without initial curvature for simulation of member imperfection is widely used whilst modelling of a member by several elements is inconvenient to adopt. Practicing engineers most commonly use the

elastic second-order analysis for finding of P-Δ sway moment which in fact has limited contribution to the an improved design in terms of safety, economy and design efficiency.

In modern design codes, the requirements for progressive collapse lead engineers to go beyond the elastic limit and the first plastic hinge. Further, the second-order analysis or the direct analysis has been more widely coded. The new Eurocode-3 (2005) and Code of practice for the structural uses of steel Hong Kong (2005) even place the chapter for design method based on second-order analysis in front of the chapter for conventional design, indicating the preference of the new method by the code drafters.

During the past decade or so, the use of higher grade steel, commonly grade S355 or higher series, becomes more popular in industry. In pace with the advocacy of the advanced design method making use of the true ultimate strength of a structural system rather than the first plastic hinge of one of its members, the behaviour of a structure at ultimate or collapse limit state is required to be more precisely understood.

This paper outlines the problem of the older member-based method against the advantages of the new method. A drift of research direction more on the system-based performance is recommended to researchers in steel, composite and in fact to any type of structures made of different materials. The detailed element formulation can be found in a two-part paper by Chan and Zhou (2004).

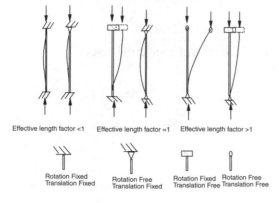

Effective length factor <1   Effective length factor =1   Effective length factor >1

Rotation Fixed
Translation Fixed

Rotation Free
Translation Fixed

Rotation Fixed
Translation Free

Rotation Free
Translation Free

Figure 1.   The effective length factors.

2   THE LINEAR ANALYSIS WITH THE EFFECTIVE LENGTH METHOD

Second-order effects exist whenever a structure is in compression or the structure deforms. Generally speaking, the effects are more important for slender structures and vice versa. The effective length method to reduce the buckling resistance is to account for this effect. Fairly speaking, this approach is indispensable in the pre-computer age, but becomes more difficult to handle contemporary structures of complex geometry and made of high strength steel with buckling being more critical.

In the linear design using the effective length method, the critical problem will be on the assessment of the buckling strength and the assumption of effective length. Below are the typical values for effective length factor (L_e/L).

The above member-based design is only for members under isolation. To account for system or frame stiffness, some design codes like the Eurocode-3 (2005) use the classification method.

A building frame is classified as non-sway, sway sensitive and sway-ultra-sensitive frames, according to the value of elastic critical load factor, λ_{cr}, as,

$$\lambda_{cr} = \frac{F_N}{F_V} \frac{h}{\delta_N} \qquad (1)$$

in which in which F_V is the factored dead plus live loads on and above the floor considered, F_N is the notional horizontal force taken typically as 0.5% of F_V for building frames, h is the storey height and δ_N is the notional horizontal deflection of the upper storey relative to the lower storey due to the notional horizontal force F_N.

For different ranges of elastic buckling load factor, λ_{cr}, the codes recommend varied charts to determine

the effective length factor (see, for example, Appendix E in BS5950 [2000]).

In the Hong Kong Code (2005), an amplification factor as follows is further required to ensure the moment amplified along a member is properly accounted for.

$$\lambda_{cr} = \frac{\pi^2 EI}{F_c L_e^2} \qquad (2)$$

in which L_e is the effective length, EI is the flexural constant and F_c is the member compression force due to design loads.

Selecting a smaller λ_{cr} from equations 1 and 2, the amplified moment, M_{amp}, is calculated as,

$$M_{amp} = \frac{\lambda_{cr}}{\lambda_{cr} - 1} M_L \qquad (3)$$

When we know the effective length, we can determine the slenderness ratio (L_e/r) and then we can use the charts, tables or formulae in a code to find the buckling strength p_c and buckling resistance P_c as,

$$P_c = p_c A \qquad (4)$$

in which p_c is the buckling strength, P_c is the buckling resistance and A is the cross sectional area.

The above effective length method is not applicable for design of slender frames because "effective length" is based on undeformed geometry and the end nodes of columns in general deflect with the loads and the structure is deforming as illustrated in Figure 2. If the structure does not deform, it has no displacement, no strain, no stress and no resistance against external loads and therefore the assumption of using the original geometry for checking is not logical. The stiffness or sway sensitivity of a frame is then used for assessment of the effective length factor (L_e/r) and amplified moment in Equation 3, but the actual amplification for bending moment is linked to the deflection of the frame in a nonlinear manner and therefore the method is limited to rather stocky and regular structures which may not be the case in practice.

The underlying principle for the effective length method is to reduce the resistance of a structure and this is conceptually unacceptable since buckling is due to external load and deflection. The mixed use of reducing buckling resistance of a member in Equation 3 and amplifying the moment in Equation 4 is another inconsistency.

An important message from the simple portal frame in Figure 2 is the fact that the second-order moment is related to the lateral deflection and the axial force as the P-delta moment, the simplified use of effective length takes no account of the moment induced by lateral force which varies continuously with the lateral

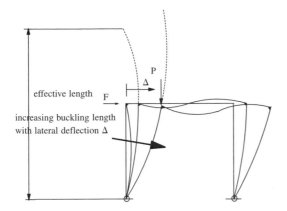

Figure 2. Effective length is taken as the buckling length before deformation.

Figure 3. The system behaviour of a braced portal.

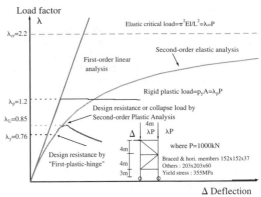

Figure 4. Types of structural analysis.

drift. Therefore the basic assumption with the effective length method of no pre-buckling deformation until it buckles cannot reflect correctly the second-order effects.

As can be seen in Figure 3, the stiffness of one bay increases significantly the buckling strength of the complete frame and the buckling resistance of all vertical columns is affected by the "system" stiffness that it should not be assessed in isolation. Therefore we should not design members in isolation.

3 TYPE OF ANALYSIS

An over view of design and analysis method is shown graphically in Figure 4. Various types of buckling loads and analysis represent different levels of accuracy in modelling. Naturally, a more simple method gives a less accurate solution and leaving the part of safety check to the design part.

4 IMPERFECTIONS

A valid second-order analysis must consider imperfections at element and frame levels. Ignoring either one of them will provide an under-designed solution to a design exercise. To this, the following two aspects of imperfections are required for consideration in an analysis and design software to code requirement.

4.1 Global P-Δ imperfection

Linear analysis uses the moment amplification to enlarge the linear moment for sway effect, which can be due to wind load or notional force normally taken as 0.5% for permanent structures and 1% for temporary structures.

In second-order analysis, wind load or notional force are still used, but an alternative and more reliable and convenient method is to use the elastic buckling mode as the imperfection mode with amplitude set equal to the out-of-plumbness normally taken as one-200th of the building height for permanent structures or other justified values. Other imperfection mode assumption like use of post-buckling shape can be used. When designing a structure of unconventional shape, the second method of using elastic buckling mode as imperfection mode is more rational since the locations and direction of application of notional force become difficult or controversial to determine.

4.2 Member P-δ imperfection

Linear analysis uses various buckling curves to represent different values of member imperfections and this leads to the production of different design tables in various notional codes, with the exception of the LRFD (1986) which uses only one buckling curve.

In a second-order or P-Δ–δ elastic analysis, the initial curvature can be pre-set to the code values which can be calibrated from the buckling curves. The member curvature can be varied with an increasing axial force such that the P-δ effect can be considered correspondingly (see example 1 of this paper).

69

4.3 Structural resistance check

In the codified linear analysis and design, a member is required for checking against member buckling and sectional strength. In the second-order analysis for design, only the section capacity check is required and checking can be exercised in the following expression.

$$\frac{P}{p_y A} + \frac{(M_y + P\Delta_y + P\delta_y)}{M_{Ry}} + \frac{(M_z + P\Delta_z + P\delta_z)}{M_{Rz}} = \varphi \le 1 \tag{5}$$

in which p_y is the design strength, A is the cross sectional area, M_y M_z M_{Ry} and M_{Rz} are respectively the design and resistant moments about the principal axes, P is the axial force, Δ and δ are the second-order delta effects and φ is the section capacity factor.

In dealing with members with possibility of lateral torsional buckling, the minor axis moment can include the component of the major axis moment (Trahair and Chan, 2003) and the twist angle. Its research is still being undertaken and refined.

5 PLASTIC ANALYSIS

In the first plastic hinge design approach where the design load is limited to the load causing the formation of the first plastic hinge, the checking of Equation 5 is carried out for all members and the design resistance for the complete structure is then checked and monitored. The design load should not cause any member to possess a plastic hinge. When using the plastic "advanced analysis", the analysis is continued even after the first plastic hinge and the design load is limited only by the load causing the complete structure to collapse as indicated in the geometry and material nonlinear load vs. deflection plot, which may occur after formation of a series of plastic hinges. Equation 5 is used again for checking and insertion of plastic hinge in a member. Nowhere along a member is allowed to have a sectional capacity factor greater than 1 in equation 5.

The use of one-element per member with a smooth extension of elastic to plastic analysis represents a consistent design procedure covering the linear and elastic analysis and design to an advanced design without use of different models, techniques and assumptions. The consistent design philosophy is much more easily accepted by the profession.

6 EXAMPLES

In Hong Kong, software agencies commonly claim their programs to have the capability to do a second-order analysis. Examples are therefore essential to

Figure 5. Buckling and advanced analysis for the elastic buckling and design buckling resistant loads of a pin-fix column.

confirm their claims to prevent the first structure designed by an improper second-order analysis against collapse.

6.1 Example 1 Verification example for testing of element and software

The first example is to demonstrate the performance of a curved element to simulate the buckling strength and behaviour of a column under compression. The success of the element in capturing the buckling behaviour of a column implies a consistent extension on the use of a structural model for linear analysis to second-order analysis. Modeling of a member by several elements to simulate initial imperfection results not only in unnecessary requirement of longer computer time, but also in a complication in modelling and an inconsistency in linear and nonlinear structural models.

The following example is a column of cross section CHS88.9 × 3.2, length 5 m and under an axial force. The analytical Euler's buckling load and design resistance from the code BS5950 (2000) are respectively 131 kN and 109 kN, compared well with the solutions for buckling and design resistance by the proposed method of 131 kN and 108 kN using only one element in the model.

6.2 Example 2 Collapse analysis of a portal

Check the structural adequacy of the following portal. The section is 686 × 254 × 140 UB of grade S355 steel. The frame is restrained out-of-plane, rigid-jointed and pin-supported with dimensions shown in Figure 6 below. The complete linear, second-order and advanced analyses are carried out for demonstration of their variations. The frame is designed by the linear, the second-order and the advanced analysis for direct comparison on their efficiency and accuracy.

The maximum bending moment at top of column is 500 kN-m.

The section capacity check in a linear analysis is carried out as follows.

$$\frac{F_c}{P_y} + \frac{M_x}{M_{cx}} + \frac{M_y}{M_{cy}} = \frac{1000 \times 10^3}{355 \times 17800} + \frac{500}{1573} = 0.476$$

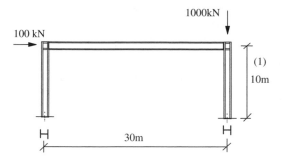

1000kN

100 kN

(1)

10m

30m

Figure 6. The portal designed by different methods.

The member check using the effective length method

$$\frac{F_c}{P_c}+\frac{m_x\overline{M}_x}{M_{cx}}+\frac{m_y\overline{M}_y}{M_{cy}}= \quad <1, \text{O.K.}$$

$$\frac{1000}{2800}+\frac{0.6\times500}{1573}=0.548$$

Member buckling check to non-sway mode effective length under amplified moment.

$$\frac{F_c}{\overline{P}_c}+\frac{m_xM_x}{M_{cx}}+\frac{m_yM_y}{M_{cy}}\leq1$$

For $\overline{P}_c$, effective length factor (Le/L) =1

$$\frac{L_e}{L}=1.0$$

L$_e$=10m

$$\frac{L_e}{r}=\frac{10000}{277}=36.1$$

From design table, the buckling strength $p_c = 327.9\,\text{N/mm}^2$, $P_c = 327.9 \times 17800/10^3 = 5837\,\text{kN}$.

The amplified moment at joint B and C due to $P-\Delta$ effect is obtained as follows.

$$\lambda_{cr} = smaller \ \ of \ \ \frac{F_N}{F_V}\frac{h}{\delta_N} \ \ and \ \ \frac{\pi^2EI}{F_cL_E^2} = smaller$$

of 6.49 and 26.9, use 6.49

$$M = \overline{M}\frac{\lambda_{cr}}{\lambda_{cr}-1}=500\times\frac{6.49}{6.49-1}=591kN-m$$

$$\frac{F_c}{\overline{P}_c}+\frac{m_xM_x}{M_{cx}}+\frac{m_yM_y}{M_{cy}}= \frac{1000}{5837}+\frac{0.6\times591}{1573}=0.397$$
<1, O.K.

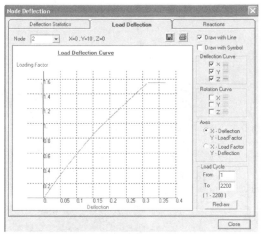

Figure 7. The load vs. deflection plot using advanced analysis.

By second-order P-Δ–δ analysis
The section capacity factor is 0.544. This is close to the most critical value above.

Advanced analysis – elastic-plastic analysis with initial imperfection.
The complete load vs. deflection plot of the portal is shown in Figure 7 below. It can be seen that the collapse load factor is about 1.54. The advanced analysis, when applied to a larger structure, normally requires longer computing time as the load steps are required to be smaller. From our experience, it is advisable that only the critical load case is worthy of detailed investigation by an advanced analysis while most less critical load cases may only require second-order elastic analysis with one load cycle for the resistance check under the design loads.

6.3 *Example 3 Design of a glass supporting structure*

The structure shown below is under the consideration of static, wind, live and seismic loads. As can be seen in the design process, the design checking was completed by a fraction of time required for a manual design with accuracy improved.

The computer model and a shot of the structure under construction are shown in Figures 8 and 9 above. Five combined load cases have been assumed for the structural safety check and the complete design process was completed in minutes. The elastic-plastic load vs. deflection plot is also shown in Figure 10. It can be seen that the first member fails at a load factor (i.e. *Load Resistance/Design Load*) equal to 1.65 in one of the tension brace while the collapse load in a particular load case is 2.1.

71

Figure 8. The structure under construction.

Figure 9. The modelled structure.

7 CONCLUSIONS

To date, the second-order and advanced analyses is essentially a tool for secondary check of buckling strength of a structure. Most engineers only use the

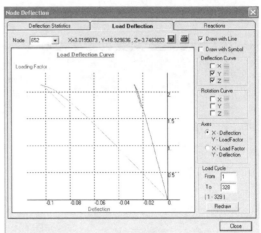

Figure 10. The load vs deflection plot allowing for plastic hinge and buckling effects.

method for finding of the sway P-Δ moment and little real-case application is given in literatures. This paper proposes an effective approach for this type of advanced design with its advantages illustrated. It is evidenced that the key issues for the practicality of the design method lie on (1) acceptable accuracy of using one element to model one member with initial curvature set equal to initial imperfection of a member specified in various codes and (2) automatic use of buckling mode as initial imperfection mode. Both these considerations have been allowed for in our software NIDA which has been used for design of a number of practical structures with enhanced safety and significant saving in design time and cost.

ACKNOWLEDGEMENT

The authors acknowledge the financial support by the Research Grant Council of the Hong Kong Special Administrative Region Government under the projects "Advanced analysis of steel frames and trusses of non-compact sections using the deteriorating plastic hinge method (PolyU 5117/06E)" and "Second-Order and Advanced Analysis of Wall-Framed Steel Structures (PolyU 5115/05E)".

REFERENCES

AISC. Load and resistance factor design specification for structural steel buildings. American Institute of Steel Construction Chicago; 1986.

Albermani, F.G.A. and Kitipornchai, S. (2003), "Numerical simulation of structural behaviour of transmission towers", Thin-walled Structures, vol. 41, no.2–3, pp. 167–177.

BSI. Structural use of steelwork in building – Part 1: Code of practice for design – Rolled and welded sections, BS5950, BSI, London; 2000.

Chan, S.L. and Chui, P.P.T. (2000),"Non-linear Static and Cyclic analysis of semi-rigid steel frames", Elsevier Science, pp.336.

Chan, S.L. and Zhou, Z.H. (2004), "Elastoplastic and large deflection analysis of steel frames by one element per member. Parts 1 and 2, Journal of Structural Engineering, ASCE, April, vol. 130, No.4, pp.538–553.

Chan, S.L. and Cho, S.H. (2005), "Second-order P-Δ–δ analysis and design of angle trusses allowing for imperfections and semi-rigid connections", International Journal of Advanced Steel Construction, 1(1), pp. 157–172.

Clarke, M.J. and Hancock, G.J. (1990), "A study of incremental-iterative strategies for nonlinear analysis", International Journal for Numerical Methods in Engineering, 29, pp.1365–1391.

Eurocode 3, "Design of steel structures", European Standard, 2005.

Hong Kong Code of Practice for the Structural Uses of Steel by the Limit State Approach, 2005.

Izzuddin, B.A. and Smith, D.L. (1996), Large displacement analysis of elso-plastic thin-walled frames Parts 1 and 2, Journal of Structural Engineering, ASCE, 122(8), pp. 905–925.

Liew, J.Y.R. and Chen, W.F. (1994), "Implications of using refined plastic hinge analysis for load and resistance factor design", Thin-walled Structures, 20 (1-4), pp.17–47.

Liew, J.Y.R., Chen, W.F. and Chen, H. (2000), Advanced inelastic analysis of frame structures, Journal of Constructional Steel Research, 55(1–3), pp. 245–265.

NIDA (2007). Non-linear Integrated Design and Analysis user's manuel, NAF-NIDA series, Version VII, Department of Civil and Structural Engineering, The Hong Kong Polytechnic University, Hong Kong. (http://www.nida-naf.com)

Trahair, N.S. and Chan, S.L (2003), "Out-of-plane advanced analysis of steel structures", 25, Engineering Structures, pp.1627–1637.

White, D.W. and Hajjar, J.F. (1997). Design of Steel Frames Without Consideration of Effective Length, Engineering Structures, 19(10), 797–810.

Yang, Y,B. and Kuo, S.R., (1994), "Theory and analysis of nonlinear framed structures", Prentice Hall, N.Y.

Special session on codification (Invited papers)

Steel and Composite Structures – Wang & Choi (eds)
© 2007 Taylor & Francis Group, London, ISBN 978-0-415-45141-3

Buckling strength of angle trusses in experiment, by code and by second-order analysis

S.L. Chan
Department of Civil and Structural Engineering, The Hong Kong Polytechnic University, Hong Kong, China

S.H. Cho & W.F. Chen
Department of Civil Engineering, Hawaii University, Hawaii, USA

ABSTRACT: Angle members are widely used in light-weight steel skeleton and they are commonly under high axial force with eccentricity. Interestingly, different design codes recommend varied load resistances and second-order analysis widely used in design of steel frames of doubly symmetrical sections is seldom reported. This paper proposes a practical second-order analysis and design method for trusses composed of angles sections. Realistic modeling of semi-rigid connections associated with one- and two- bolt end connections with flexible gusset plate and member imperfections such as initial curvatures and residual stresses is made. Load eccentricity is also simulated. The proposed method can be readily applied to reliable, robust, efficient and effective design of angle trusses and frames without the uncertain assumption of effective length.

1 GENERAL INSTRUCTIONS

Single angle members have a broad range of applications, such as web members in roof trusses, members of transmission towers and other bracing members. Most angle members are slender and therefore relatively weak in compression resistance compared with other steel sections, but angle sections are widely used because of its light weight and the L-shaped section making the angles easy for storage, transportation and fabrication.

The design of angle trusses is complicated by the structural behavior as follows. Firstly, since angles are asymmetric or mono-symmetric, their principal axes are always inclined to the plane of truss or frame. Secondly, it is not uncommon to bolt or weld an angle member to another member directly or to a gusset plate at its end through their legs. Therefore, in practice, an angle member is loaded eccentrically through one leg. As a result, an angle member is subject to an axial force as well as a pair of end moments at its ends.

Since single angle web members may be attached to the chord members on the same side or on alternate or opposite sides, this affects the directions of the moments. Twisting may also appear simultaneously as the shear centre of the cross-section is located at the point of intersection of the two legs away from the centroid. Finally, the connection at each end provides some degree of end fixity which is beneficial to

the compression capacity of the angle members. This further complicates the analysis of angle members.

These mentioned features are almost unique to angle sections making the design of single angle members controversial for some time. In a rational design procedure, the adverse effect of the end eccentricity and the beneficial effect due to the end restraint on the compression capacity should be considered. However, in most conventional design methods and codes widely used today, the design procedure is overly simplified with many assumptions not valid. For example, the load eccentricity and the end restraint may be neglected during the analysis.

Tests of angle struts have been carried out by Trahair et al. (1) Although this situation is an extreme case and rarely occurs in reality, it represents the worst scenario and provides us an aspect of the future research works. Adluri and Madugula [2] compared results of some experimental data on eccentrically loaded single angle members free to rotate in any directions at the ends from the available literature with AISC LRFD [3] and AISC ASD [4] specifications.

The experimental investigations, which covered a wide spectrum of single angle struts, were carried out by Wakabayashi and Nonaka [5], Mueller and Erzurumlu [6] and Kitipornchai and Lee (7) tested the angle struts. Kitipornchai and Woolcock (8) further investigated the buckling strength of angle trusses with web members placed on the same and on the opposite sides of the truss and reported that the first arrangement leads to a smaller eccentric moment and therefore the buckling resistance is stronger.

The tested results were summarized and concluded that the interaction formulae given in AISC LRFD [3] and AISC ASD [4] are highly conservative when applied to eccentrically loaded single angle members. It is because these interaction formulas were derived primarily for doubly symmetric sections and the moment ratios in these formulas are evaluated for the case of maximum stresses about each principal axis. This practice does not pose a problem on doubly symmetric sections such as I-sections because the four corners are critical for moments about both principal axes simultaneously.

However, for angle sections, as they are monosymmetric or asymmetric, the points having maximum bending stress about both principal axes sometimes do not coincide. As a consequence, the loading capacities of the sections calculated from these interaction equations are underestimated [2]. In order to eliminate the unnecessary discrepancy between the actual failure load and the design load, Adluri and Madugula [2] suggested that the moment interaction factors given in AISC LRFD [3] should be revised. Bathon et al. [8] carried out 75 full-scale tests which covered a slenderness ratio ranging from 60 to 210. The test specimens were unrestrained against rotation at the end supports. It was noted that the ASCE Manual 52 [9] under-predicted the capacities of single angle struts.

The above-mentioned research did not include the effect due to end connection details, which may also affect the buckling resistances of the angle struts. Elgaaly et al. [10] conducted an experimental program to investigate the structural behavior of non-slender single angle struts as part of three-dimensional trusses. The specimens cover a range of slenderness ratio from 60 to 120 including single-bolted and double-bolted conditions. Results show that both the ASCE Manual 52 [9] and AISC LRFD [3] are inadequate for single angle members with low slenderness ratio.

A review of literature shows there is a lack of research work on testing of single slender angle struts in the form of a truss. Comparisons with the experimental results and the predicted results by the traditional simplified method instead of the axial force-moment interaction method are also inadequate. In the investigation, a series of laboratory tests of angle trusses were conducted including single-bolted and double-bolted end conditions and web members on the same side and on alternate sides.

This paper describes the experimental program and results. Comparison among the test results will be made between the experimental results against those predicted by the design rules. A proposed second-order analysis and design method is also introduced and the method is validated by the test results. The concept and application of the method to general structures with members of symmetrical cross sections have been discussed by Liew et al. [11], among others, and this paper extends the work to members with asymmetrical cross section and under eccentric moments.

2 PROPOSED SECOND-ORDER ANALYSIS

The conventional design procedure based on first order linear analysis is traditionally used during the pre-computer age, when the computer time was expensive or when the computer speed was slow. To date, the prevalence of low-cost personal computers and the growing importance of environmental and economical concerns provide a natural choice to develop a practical second-order analysis method.

This method has been well-researched by Chen and Chan [15], Chan and Chui [16], Chan and Cho [17] with the second-order effects included during the analysis through update of geometry. In other words, the member deflection (δ) and the global displacement (Δ) are taken into account so that section capacity check is adequate for strength design as follows:

$$\frac{P}{A_g p_y} + \frac{M_y + P(\delta_z + \Delta_z)}{Z_y p_y} + \frac{M_z + P(\delta_y + \Delta_y)}{Z_z p_y} = \phi \le 1$$
(1)

in which p_y is the design strength, P is the external force applied to the section, A_g is the cross-sectional area, M_y and M_z are the external moments about the y and z axes respectively, Z_y and Z_z are the section modulus about the y and z axes respectively. $P(\delta_z + \Delta_z)$ and $P(\delta_y + \Delta_y)$ are the collective moments about the y and z axes respectively due to the change of member stiffness under load and large deflection effects of which the consideration allows for the effect of "effective length" automatically.

In other words, there is no need to reduce the compressive strength of the member to account for the $P - \delta$ and $P - \Delta$ effects. Moreover, the characteristics of realistic structure (e.g. initial imperfection and residual stresses) are also considered in the analysis so the design is completed readily with the analysis.

To determine the design buckling load of a structure, approximately 1% to 10% of the predicted failure load is applied incrementally until the sectional capacity factor, φ in Eq. (1) is equal to 1. The Newton-Raphson method combined with the minimum residual displacement iterative scheme [18] is utilized.

For designing single angle struts, a method was previously proposed by Cho and Chan [19] based on the aforementioned second-order analysis and design concept. Using the software NIDA (structural analysis software "Nonlinear Integrated Design and Analysis" version 7) [20], initial curvature is imposed along the member so that bending can be triggered at the instance that the load is applied instead of axial shortening. The values of the initial curvature are calculated based on the compressive strength curve given in BS5950 [13] which are $2.8 \times 10^{-3}L$ for equal angles and $2.0 \times 10^{-3}L$ for unequal angles.

The results computed by NIDA agree well with BS5950 [13]. The Code of Practice for the Structural Use of Steel 2005 [21] further gives explicit values of imperfections for angles. However, this method cannot truly reflect the end condition that a practical angle is exposed to. The method was modified by Chan and Cho [22]; the end condition is symbolized by a rotational spring element inserted at each end of the member. The value of the rotation spring stiffness is calculated from the dimensions of the gusset plates and its material properties. Therefore, the rotational stiffness due to the double-bolted connection can be considered at the early stage of the analysis rather than at the design stage as in the linear analysis and effective length design method. Only the rotational deformation of the connection spring element is considered for design because the effects of the axial and shear forces in the connection deformations are small when compared with that of bending moments.

However, this modified method is still inadequate to consider the directions of the end moments. Under some circumstances, these end moments would be advantageous to the overall structure. In this paper, the method is further refined here. The end moments due to load eccentricity are considered by connecting the angle web members at each end to the chord members by rigid arms. The rigid arm will be the element joining the centroid and the point of load application so that the magnitude and the direction of the end moments due to load eccentricity can be taken into account immediately during the analysis.

For single-bolted connection, the connection joints are allowed to rotate freely. For double-bolted connection, rotational springs are inserted to the joints connecting rigid arm elements to the angle web member element in the in-plane direction so that the couples due to the double-bolted connection can be considered. The merit of this approach over the purely equivalent imperfection approach is that it considers the direction of the end moments so that the aforementioned effect can be reflected during analysis. The spring stiffness of a rotational spring can be calculated as follows:

The couple, M formed by the pair of bolts is given by:

$$M = F \cdot d = k\theta \tag{2}$$

The shear stress, τ across the cross-section, A, of the bolt is:

$$\tau = \frac{F}{A} \tag{3}$$

in which A is the shear area and can be taken as 0.9 of the cross sectional area recommended in most design codes like the Hong Kong Steel Code 2005 [21].

The shear strain, γ of the bolt shank is:

$$\gamma = \frac{2\delta}{l} = \frac{d \cdot \theta}{l} = \frac{\tau}{G} \tag{4}$$

Rearranging terms, the rotational stiffness, k, due to the double-bolted connection will be given by:

$$k = \frac{GAd^2}{l} \tag{5}$$

where F is the shear force exerted on the bolt; d is the distance between the centroid of the two bolts; θ is the rotation of the bolt group; δ is the displacement of the bolt; l is the length of the bolt shank; G is the shear modulus of elasticity.

To account for the rotation stiffness of the spring element in the analysis, the following incremental tangent stiffness matrix is superimposed to the element stiffness matrix.

$$\begin{bmatrix} M_e \\ M_i \end{bmatrix} = \begin{bmatrix} S_c & -S_c \\ -S_c & S_c \end{bmatrix} \begin{bmatrix} \theta_e \\ \theta_i \end{bmatrix} \tag{6}$$

in which M_e and M_i are the incremental external and internal moments at two ends of a connection. The external node refers to the one connected to the global node and the internal node is joined to the angle element. The stiffness of the connection S_c can be related to relative rotations at the two ends of the connection spring as:

$$S_c = \frac{M_e}{\theta_e - \theta_i} = \frac{M_i}{\theta_i - \theta_e} \tag{7}$$

in which θ_e and θ_i are the conjugate rotations for the moments M_e and M_i.

3 THE INCREMENTAL AND ITERIATIVE PROCEDURE FOR ANALYSIS OF ANGLE TRUSSES AND FRAMES

The basic equations for incorporating the end connection stiffness are considered both in the tangent and the secant stiffness matrix equations as follows.

79

3.1 Tangent stiffness matrix

The incremental force is assumed in the software and the incremental displacement is solved. The basic element stiffness is modified by addition of the tangent stiffness of the connection spring modeled as a dimensionless spring element in a computer analysis as:

$$\begin{bmatrix} S_1 & -S_1 & 0 & 0 \\ -S_1 & k_{11}+S_1 & k_{12} & 0 \\ 0 & k_{21} & k_{22}+S_2 & -S_2 \\ 0 & 0 & -S_2 & S_2 \end{bmatrix} \begin{bmatrix} \theta_{1e} \\ \theta_{1i} \\ \theta_{2i} \\ \theta_{2e} \end{bmatrix} = \begin{bmatrix} M_{1e} \\ M_{1i} \\ M_{2i} \\ M_{2e} \end{bmatrix} \qquad (8)$$

in which S_1 and S_2 are the spring stiffness for simulation of semi-rigid connections at ends; k_{ij} is the stiffness coefficients of the element; θ_{1e}, θ_{1i}, θ_{2e} and θ_{2i} are respectively the rotations at two sides for the two ends of an element shown in Fig. 3 below.

Assembling the element matrices, the global stiffness matrix for the angle frame and truss is formed and stored in a one dimensional array in the computer analysis. The incremental displacement vector is solved and added to the last displacement and used for determination of resistance as follows.

3.2 Resistance determination

The resistance is determined as the sum of resisting forces for all elements as,

$$[R] = \sum [k_e][L_S][u] \qquad (9)$$

in which $[R]$ is the structural resistance of the angle frame, $[k_e]$ is the secant stiffness, $[L_s]$ is the transformation matrix for connection spring stiffness and $[u]$ is the accumulated element displacement transformed to the element local axis.

The resistance of the angle frame is determined as the maximum load not violating Eq. (1). This is based on the conventional use of load causing the formation of the first plastic hinge.

4 COMPARISON WITH CODES

4.1 Experimental program

Four single angle struts were tested as web members of a two-dimensional truss as shown in Fig. 1. In the first set, the web members of the truss are connected to the chord members on the same side. In the second set, the tests are repeated with the web members connected to the chord members on alternate sides. The specimens are of Grade S275 and two meters long making the slenderness ratio around 150. The leg length-to-thickness ratio meets the BS5950 [13] requirements

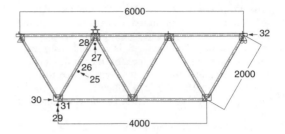

Figure 1. The layout of the tested truss.

so that local buckling can be ignored. Each end of the member is connected to a gusset plate.

The test included single and double bolted connections. The trusses were loaded in pair and sufficient lateral restraints were provided to ensure out-of-plane buckling at connecting nodes between chords and webs is avoided. Load was applied at the upper joint of the target failure member through a hydraulic jack. At the targeted member, two displacement transducers were placed in in-plane and out-of-plane directions and transducers were also used to monitor the movements of the top and the bottom joints of the targeted member so that its movement of the target member relative to the truss can be measured.

At the load where the targeted member buckled or failed, the deformations of the remaining parts of the truss were small and reversible. Thus, after each test, the failed member was replaced by a new specimen so that the next test could be conducted under almost the same conditions. The testing of the specimen is indicated in Fig.2.

4.2 Comparison between test, code and proposed method

The theoretical predicted loads, the test result and the coded design loads for the tested truss are tabulated in Fig.3 below. It can be seen that the proposed method provides a consistent more accurate prediction than the code predictions. The use of imperfection of length/200 is found to provide us a safe estimate of the design resistance of an angle truss.

The difference between the tested result and the proposed method ranges from 14 to 21% with the largest difference between the theory and the test occurs at the first sample of single bolt with webs connected on the same side of the truss. It is probably due to the web being more controlled by buckling behavior which depends heavily on the actual imperfections but unknown of the member. The actual imperfection cannot be measured directly in the sample as it includes not only the geometrical imperfection, but also the material imperfection in the form of residual stress which is included in the present equivalent imperfection.

Figure 2. The buckled web member.

Web location	End restraint	Test load (kN)	Test Load/ BS5950	Test Load/NIDA
Same Side	Single bolt	67.2	1.618	1.209
Same Side	Double bolts	78.4	1.145	1.172
Opposite Side	Single bolt	57.5	1.459	1.141
Opposite Side	Double bolts	72.1	1.104	1.161

Figure 3. Comparison between code and proposed method.

5 CONCLUSIONS

This paper proposes a second-order analysis and design method free from any assumption of effective length. Using this method, the second-order effects (e.g. initial curvatures, residual stresses, $P - \delta$ and $P - \Delta$) are explicitly included in the analysis so that effective length is not required to be assumed. The proposed design method is validated by laboratory tests of truss using single angle members of slenderness ratio about 150 as web members and the test results agreed well with the computed results.

ACKNOWLEGEMENT

The authors acknowledge the financial support by the Research Grant Council of the Hong Kong Special Administrative Region Government under the projects "Advanced analysis of steel frames and trusses of non-compact sections using the deteriorating plastic hinge method (PolyU 5117/06E)" and "Second-Order and Advanced Analysis of Wall-Framed Steel Structures (PolyU 5115/05E)".

REFERENCES

[1] Trahair NS, Usami T, Galambos TV. Eccentrically Loaded Single Angle Columns. Research Report No. 11. Structural Division, Civil and Environmental Engineering Department, School of Engineering and Applied Science, Washington University, St. Louis, Missouri, USA; 1969.

[2] Adluri SMR, Madugula MKS. Eccentrically loaded steel angle struts. Engineering Journal AISC 1992; 31(3):59–66.

[3] AISC. Load and resistance factor design specification for structural steel buildings. American Institute of Steel Construction Chicago; 1986.

[4] AISC. Specification of allowable stress design. AISC Inc, Chicago; 1989.

[5] Wakabayashi M, Nonaka T. On the buckling strength of angles in transmission towers. Bulleting of the Disaster Prevention Research Institute, Kyoto University, Japan, 1965; 15(2):1–18.

[6] Mueller WH, Erzurumlu H. Behavior and strength of angles in compression: an experimental investigation. Research Report of Civil-Structural Engineering. Division of Engineering and Applied Science. Portland State University, Oregon, USA; 1983.

[7] Kitipornchai S, Lee H.W. Inelastic Experiments on Angle and Tee Struts. Journal of Constructional Steel Research, 1986; 6 (3): 219–236.

[8] Woolcock S.T, Kitipornchai S. Design of Single Angle Web Struts in Trusses. Journal of Structural Engineering, ASCE, 1986; 112 (6): 1327–1345.

[9] ASCE. Manuals and reports on engineering practice no 52. Guide for Design of Steel Transmission Towers, ASCE, New York; 1988.

[10] Elgaaly M, Davids W, Dagher H. Non-slender single angle struts. Engineering Journal, AISC 1992; 31(3):49–59.

[11] Liew J.Y.R, Chen, W.F, Chen, H. Advanced inelastic analysis of frame structures, Journal of Constructional Steel Research, 55(1-3), 2000, pp. 245–265.

[12] AISC. Load and Resistance Factor Design Specification for Structural Steel Buildings, AISC, INC, Chicago; 1999.

[13] BSI. Structural use of steelwork in building – Part 1: Code of practice for design – Rolled and welded sections, BS5950, BSI, London; 2000.

[14] CEN. Eurocode 3 Design of steel structures – Part 1-1: General rules and rules for building, BS EN 1993-1-1. CEN, BSI, London; 2005.

[15] Chen WF, Chan SL. Second-Order Inelastic Analysis of Steel Frames Using Element with Midspan and End Springs. Journal of Structural Engineering 1995; 121(3):530–41.

[16] Chan SL, Chui PPT. Non-linear static and cyclic analysis of semi-rigid steel frames. Elsevier Science; 2000.

[17] Chan SL, Cho SH. Design of steel frames using calibrated design curves for buckling strength of hot-rolled members. In: Chan SL, Teng JG, Chung KF, editors. Proceedings of Advances in Steel Structures. Elsevier; 2002, 1193–1199.

[18] Chan SL. Geometric and material nonlinear analysis of beam-columns and frames using the minimum residual displacement method. International Journal for Numerical Methods in Engineering 1988; 26:2657–69.

[19] Cho SH, Chan SL. Practical second-order analysis and design of single angle trusses by an equivalent imperfection approach. Steel and Composite Structures 2005;5(6):443–58.

[20] NIDA. Non-linear Integrated Design and Analysis user's Manuel, NAF-NIDA series, Version 7. Department of Civil and Structural Engineering, The Hong Kong Polytechnic University, Hong Kong. (http://www.nida-naf.com)

[21] Code of Practice for Structural Use of Steel 2005, Buildings Department, Hong Kong SAR Government, 2005.

[22] Chan SL, Cho SH. Second-order $P-\Delta-\delta$ Analysis and Design of Angle Trusses Allowing for Imperfections and Semi-rigid Connections. Advanced Steel Construction 2005;1(1):169–83.

Steel and Composite Structures – Wang & Choi (eds)
© 2007 Taylor & Francis Group, London, ISBN 978-0-415-45141-3

The Direct Analysis Method: A new AISC approach to stability design

Donald W. White
Georgia Institute of Technology, Atlanta, Georgia, USA

ABSTRACT: The 2005 edition of the American Institute of Steel Construction (AISC) Specification for Structural Steel Buildings introduces a new approach for the stability design of steel structures, called the Direct Analysis Method (DM). This paper provides an overview of the DM and its relationship to more traditional stability design approaches.

1 INTRODUCTION

In the context of unbraced moment frames, the Direct Analysis Method (DM) is a fundamental new alternative to more traditional AISC Effective Length Method (ELM) design procedures. However, this new approach relates very closely to prior procedures in the 1999 AISC LRFD Specification for the stability design of braced frames and column bracing systems. Major advantages of the DM include:

1 No effective length (K) factor calculations are required,
2 The internal forces are represented more accurately at the ultimate strength limit state than in traditional design approaches, and
3 The method applies in a logical and consistent way for all types of structures including braced frames, moment frames and combined framing systems.

The provisions for the DM are contained in Appendix 7 of the 2005 AISC Specification. In addition, AISC (Griffis & White 2007) and AISC/MBMA (Kaehler et al. 2007) design guides that address the detailed application of the DM are in the process of final review and publication. Because of the advantages of the DM, AISC is considering a move of its DM provisions to the main body of the Specification in 2010 combined with a shift of its ELM provisions to an appendix.

2 OVERVIEW OF THE DM

The DM entails the use of a second-order elastic analysis that includes a nominally reduced stiffness and a nominal initial out-of-plumbness of the structure. For simply-connected braced structures, the DM requires two modifications to a conventional elastic second-order analysis:

1 A uniform out-of-plumbness of $\Delta_o = L/500$ is included in the analysis. This out-of-plumbness accounts for the influence of initial geometric imperfections, initial load eccentricities and other related effects on the internal forces under ultimate strength loadings. For rectangular structures, this out-of-plumbness effect may be modeled by applying an equivalent notional lateral load of

$$N_i = 0.002\, Y_i \qquad (1)$$

at each level of the structure, where Y_i is the factored gravity load acting at the ith level. Explicit modeling of out-of-plumbness is easier to automate in computer-based design. However, unless automated methods of specifying out-of-plumbness are available in analysis software, it is often easier to apply the above notional loads rather than modify the structure geometry. The above nominal out-of-plumbness is equal to the maximum tolerance specified in the AISC Code of Standard Practice (AISC 2005b).

2 The elastic stiffnesses of all the components in the structure are reduced by a uniform factor of 0.8. This factor accounts for the influence of partial yielding of the most critically loaded component(s) due to applied load and residual stress effects, as well as uncertainties with respect to the overall displacements and stiffness of the structure at the strength limit states.

For moment frames, the DM requires one additional modification to the elastic analysis: for members in which the axial force αP_r exceeds $0.5P_y$, an additional inelastic stiffness reduction of

$$\tau = 4\left(1 - \alpha P_r / P_y\right)\alpha P_r / P_y \qquad (2)$$

is applied to the member flexural rigidity EI, or in other words,

$$\bar{EI}_e = 0.8\tau EI \qquad (3)$$

where $\bar{EI}_e$ is the member effective flexural rigidity, αP_r is the required axial force at the ultimate strength load level ($\alpha = 1.6$ for Allowable Strength Design (ASD) and 1.0 for Load and Resistance Factor Design (LRFD)) and P_y is the member cross-section yield load $A_g F_y$. This modification is required to account for the more severe impact of distributed yielding on the flexural (versus the axial or shear) deformations in certain situations. Distributed yielding has a greater effect on the flexural rigidity particularly in cases such as weak-axis bending of I-section members.

The above adjustments to the elastic analysis model, combined with an accurate calculation of the second-order effects, provide an improved representation of the second-order inelastic forces at the ultimate strength limit. Due to this improvement, the AISC DM bases the member axial resistance P_n on the actual unsupported length for all types of framing systems. By adopting the fundamental approach of including both a nominal out-of-plumbness as well as a nominal stiffness reduction effect in the structural analysis, the analysis and design of all types of structures and their components is placed on a single logical and consistent footing. Engineers can approach the overall design of braced, unbraced or combined framing systems from the single perspective of providing a sufficient combination of sidesway stiffness and component resistances to satisfy the internal strength demands. The designer is able to avoid the complexities of determining appropriate buckling solutions for the member P_e values (or K factors) for general frame geometries and configurations.

The above modifications are for the assessment of strength. In contrast, serviceability limits are checked using the ideal geometry and the nominal (unreduced) elastic stiffness. In addition, the Specification component resistance equations are always expressed in terms of the unreduced nominal elastic stiffnesses, i.e., $E = 200,000$ MPa (29,000 ksi). Furthermore, it should be noted that a uniform factor of 0.8, applied to all the stiffness contributions, influences only the second-order effects in the structural model. That is, for structures in which the second-order amplification is small, the stiffness reduction has a negligible effect on the system internal forces.

The rationale for the above modifications is discussed in detail by Surovek-Maleck & White (2004a) and White et al. (2006). The reader is referred to Maleck (2001), Martinez-Garcia (2002), Deierlein (2003 & 2004), Surovek-Maleck & White (2003 & 2004b), Nair (2005), Martinez-Garcia & Ziemian (2006), White et al. (2007a, b), White & Griffis (2007),

Griffis & White (2007) and Kaehler et al. (2007) for other detailed discussions as well as validation and demonstration of the DM concepts.

In contrast, traditional ELM procedures are based on analysis of the ideal, geometrically-perfect, nominally-elastic structure. The traditional ELM procedures account for residual stress and geometric imperfection effects on the sidesway stability of unbraced moment frames solely in an implicit fashion via:

1 The calculation of an effective length factor K, usually greater than 1.0, or equivalently a column elastic flexural buckling load P_e, usually smaller than the Euler buckling load $P_{eL} = \pi^2 EI/L^2$, combined with
2 The use of the Specification column strength curve in determining the member nominal axial resistances P_n.

3 BASIC ILLUSTRATION OF THE DM

One of the simplest illustrations of the DM and its relationship to traditional ELM approaches is the solution for the design strength of a fixed-base cantilever. Figure 1 shows a W254 x 89 (W10 x 60) cantilever subjected to a vertical load P and a proportional horizontal load of $H = 0.01P$, adapted from Deierlein (2004). The bending is about the major-axis and the member is braced out-of-plane such that its in-plane resistance governs. The column slenderness in the plan of bending is $L/r_x = 40$ based on the member's actual length and $KL/r_x = 80$ based on an effective length factor of $K = 2$.

Figure 2 shows plots of the axial load versus the moment at the column base determined using three approaches:

1 The DM,
2 The traditional AISC ELM, and
3 A rigorous second-order Distributed Plasticity Analysis (DPA).

Load and Resistance Factor Design (LRFD) is used with $\phi_b = \phi_c = 0.9$ in each of these solutions. The rigorous DPA accounts explicitly for the spread of yielding through the cross-section and along the length of the member as the loads are increased.

The specific DPA used here is based on a factored stiffness and strength of $0.9E$ and $0.9F_y$, and out-of-plumbness and out-of-straightness on the geometry of $0.002L$ and $0.001L$ respectively (oriented in the same direction as the bending due to the applied loads), the Lehigh residual stress pattern (Galambos & Ketter 1959) with a maximum compressive residual stress at the flange tips of $0.3(0.9F_y) = 0.27F_y$, and an assumed elastic-perfectly plastic material stress-strain response. A small post-yield stiffness of $0.001E$ is

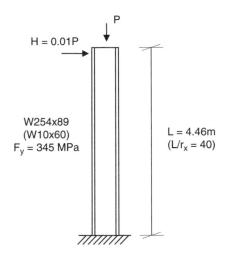

Figure 1. Example cantilever beam-column.

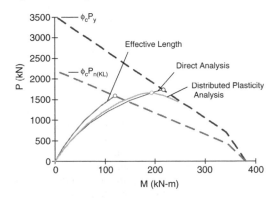

Figure 2. Results for example cantilever beam-column.

Table 1. Summary of calculated design strengths, cantilever beam-column example.

	P_{max} (kN)	M_{max} (kN-m)	M_{max}/HL	$P_{max}/P_{max(DPA)}$
ELM	1590	119	1.68	0.96
DM	1730	217	2.79	1.05
DPA	1650	198	2.68	

analysis. Overlaid on the force-point traces are the beam-column strength envelopes (Equations (H1-1) of the AISC Specification) for the ELM and the DM procedures. The strength envelopes are drawn using dashed lines in the figure. The $\phi_c P_n$ anchor points for the strength envelopes are $\phi_c P_{n(KL)} = 2210\,$kN for the ELM and $\phi_c P_y = 3540\,$kN for the DM.

The design strengths, determined as the combined P and M at the intersection of the force-point traces with the in-plane beam-column strength curves are summarized in Table 1. The ratios of the maximum base moments $M_{max} = HL + P\,(\Delta + \Delta_o)$ to the primary moment HL indicate the magnitude of the second-order effects. The axial load at the DM strength limit, which is representative of the strength in terms of the total applied load, is five percent higher than obtained from the DPA. Conversely, the axial load at the ELM beam-column strength limit is four percent smaller than that obtained from the DPA. Both of these estimates are within the target upper bound of five percent unconservative error relative to refined inelastic analysis established in the original development of the AISC LRFD beam-column strength equations (ASCE 1997; Surovek-Maleck & White 2004a).

The difference in the calculated internal moments is much larger. This difference is expected since the ELM compensates for its underestimation of the physical moments by reducing the value of the axial resistance term P_n (using $K = 2$). Conversely, the DM imposes additional requirements on the analysis to obtain an improved estimate of the physical internal moments. This more accurate calculation of the internal moments also influences generally the design of restraining members and their connections. For instance, in this example, the column base moments from the DM are more representative of the actual moments required to support the applied loads corresponding to the calculated member resistance. In this regard, the DM provides a direct estimate of the true required strengths for all of the structural components. Conversely, supplementary requirements are needed to account for this behavior in the context of the traditional ELM procedures. The 2005 AISC Specification implements these supplementary requirements as:

1 A minimum lateral load equal to the notional load in Equation (1) is applied with all gravity-only load combinations, and

used for numerical stability purposes. These are established parameters for calculation of benchmark design strengths in ASCE (1997) and in the other background references cited at the end of the previous section. The use of $0.9E$ and $0.9F_y$ in the DPA solution produces identical results for the calculated design strength limit to the use of the nominal E and F_y values followed by a posteriori factoring of the nominal strengths by 0.9.

In Figure 2, the reader should note that the internal moment from the DM is larger than that obtained from the traditional ELM. This is due to the use of a reduced stiffness of $0.8EI_x$, as well as the influence of the nominal initial out-of-plumbness of $0.002L$. The axial load anchor point for the DM beam-column strength curve is actually taken as $\phi_c P_n = \phi_c P_y = 3540\,$kN based on the cross-section axial load resistance ($L = 0$). The rationale for and limits on the use of this extension of the 2005 AISC provisions are discussed subsequently. The solid-line force-point traces in Figure 2 show that the DM internal moments are very similar to the internal moments calculated by the rigorous inelastic

85

2 The usage of the ELM is restricted to frames having a sidesway amplification $\Delta_{2nd}/\Delta_{1st} \leq 1.5$.

4 ADDITIONAL DM CONSIDERATIONS

In certain instances, the 2005 AISC Specification allows the designer to relax some of the base DM requirements outlined above in Section 2. Specifically, AISC (2005) allows the following simplifications:

1 If $\Delta_{2nd}/\Delta_{1st} \leq 1.5$ based on the nominal elastic stiffness, or equivalently if $\Delta_{2nd}/\Delta_{1st} \leq 1.71$ based on the reduced stiffness, the nominal out-of-plumbness only needs to be included for gravity-only load combinations. This simplification is based on the fact that the out-of-plumbness effects tend to be small relative to the lateral load effects for all ASCE 7-05 (ASCE 2005) lateral load combinations when the sidesway amplification is smaller than the above limits.

2 In structures where some of the members are subjected to axial loads $\alpha P_r > 0.5P_y$, an additional out-of-plumbness of 0.001 L or the corresponding additional notional load of $0.001Y_i$ may be applied in lieu of reducing the flexural rigidity based on Equations (2) and (3). This out-of-plumbness or the equivalent notional lateral load must be applied regardless of whether the above $\Delta_{2nd}/\Delta_{1st} \leq 1.5$ limit is satisfied or not. It accounts approximately for the τ effect and avoids the need to adjust the member flexural rigidities as a function of the applied loads.

3 Where an initial out-of-plumbness smaller than $0.002L$ is justified, the base out-of-plumbness of $0.002L$ may be reduced. One instance where this can occur is in building frames taller than about seven stories, where the AISC (2005b) envelope within which exterior column work points must fall will always limit the total out-of-plumbness of the structure to a smaller value. Griffis & White (2007) provide recommendations for modeling of the nominal out-of-plumbness for these types of frames. Also, if tighter controls are applied to the constructed geometry than specified in the AISC (2005b) Code of Standard Practice, a Δ_o smaller than $0.002L$ would be justified.

White et al. (2006), Griffis & White (2007) and Kaehler et al. (2007) recommend another simplification of the AISC DM requirements that is useful for certain types of structures. The background research to the DM indicates that $P_n = QP_y$ gives a close approximation to rigorous benchmark solutions involving flexural stability limit states when either:

1 $\alpha P_r \leq 0.10P_{eL}$, where $P_{eL} = \pi^2 EI/L^2$ in the plane of bending, or

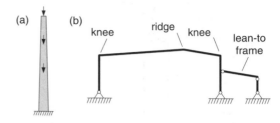

Figure 3. Cases where the definition of member length in the plane of bending becomes unclear.

2 An appropriate nominal out-of-straightness (e.g., $0.001L$) is included in the analysis.

(Q being the form factor associated with local buckling under uniform axial compression in the AISC Specification). The first of the above limits is satisfied by the columns in most low-rise steel frames. This simplification is particularly useful for special situations where the definition of an appropriate member length in the plane of bending is unclear. Figure 3 shows two of these cases.

Figure 3a is a single tapered cantilever column with axial loads applied at various locations along its length. This type of configuration is common for structures such as towers for microwave antennas. Questions arise in this case as to whether the member length is the distance between the different axial loads or if it is the overall physical length of the column. Figure 3b shows an unsymmetrical gabled portal frame with a shallow pitch of the rafters and a lean-to frame attached to the column on the right-hand side. Questions arise as to whether the length(s) of the rafter(s) in this frame are the distances from the knees to the ridge, or the overall distance along the rafters between their intersections with the columns at the knee joints. Also, since the lean-to frame in Figure 3b involves only gravity framing with no resistance to sidesway, is the right-hand column two members, one from the column base to the connection of the lean-to frame and one from the lean-to connection to the knee joint, or is it a single member from the base to the knee joint? In these cases, P_n may be taken equal to QP_y with Q determined using an axial stress $f = F_y$ when either of the conditions:

1 $P \leq 0.10P_{eL}$ based on a conservatively assumed member length, or
2 An appropriate nominal out-of-straightness is included in the analysis.

is satisfied. Using this approach, the consideration of member flexural stability is moved entirely from the resistance side to the analysis side of the design problem. However, when the axial loads are large, this approach generally requires the consideration of nominal imperfections having an affinity with potential buckling modes of the structure in the second-order analysis.

In the specific case of gabled rafter(s) where the ridge work point is offset from the rafter chord in a direction opposite to the direction of the transverse loads applied to the member, such as in Figure 3b, Kaehler et al. (2007) recommend that the P_n corresponding to the *in-plane* flexural buckling limit state may be taken equal to QP_y without the consideration of the above limits. This is because the offset of the ridge work point for these types of members nullifies the importance of any out-of-straightness relative to the chord between the ends of the on-slope length of the rafters.

Fortunately, for most cases, the length L pertaining to the member flexural stability limit states is clear, P_{nL} (the member axial compressive resistance calculated using $KL = L$) is only slightly smaller than QP_y, and $P_r = P_u \ll \phi_c P_{nL}$ (LRFD) or $P_r = P_a \ll P_{nL} / \Omega_c$ (ASD). Therefore, the AISC Specification approach of using $P_n = P_{nL}$ is a simple and practical option. The basic approach of using $P_n = P_{nL}$ is specifically to guard against a non-sway "braced" flexural buckling mode of failure in any members that are highly-slender and relatively weak compared to the rest of the structural system. Also, the use of $P_n = P_{nL}$ is consistent with the philosophy stated in the Commentary to Section C2.b of the 2005 AISC Specification that, in most cases, members should not be designed for $K < 1$. This practice is recommended because of the potential implications of the column stability behavior on beam moments and brace forces when the column axial loads are greater than the axial resistance based on $K = 1$. Furthermore, in all of the above cases, the member limit states associated with torsional, torsional-flexural and lateral-torsional buckling still must be checked. These limit states involve certain member lengths.

5 CONCLUDING REMARKS

This paper has provided on overview of the new Direct Analysis Method for stability design provided in (AISC 2005). Due to its advantages, it is believed that the DM will become the preferred approach in future editions of the AISC Specification.

ACKNOWLEGEMENTS

The concepts in this paper have benefited greatly from discussions with the members of the AISC Technical Committee 10 (TC10), the Structural Stability Research Council (SSRC) Task Group 4 on frames, and the former SSRC Task Group 29 on Inelastic Analysis for Frame Design. Professor J. Yura of the University of Texas at Austin, Dr. S. Nair of Teng and Associates, Inc. and Prof. G.G. Deierlein of Stanford University are thanked for their guidance in the TC10 efforts toward the implementation of the 2005 AISC provisions. In addition, special recognition goes to Mr. Larry Griffis of Walter P. Moore and Associates, Inc. and Mr. Richard Kaehler of Computerized Structural Design, S.C., for their lead authorship of new AISC and AISC/MBMA Design Guides addressing the detailed application of the DM to various types of framing systems. The opinions, findings and conclusions expressed in this paper are the author's and do not necessarily reflect the views of the above individuals, groups and organizations.

REFERENCES

AISC (1999). *Load and Resistance Factor Design Specification for Structural Steel Buildings*, American Institute of Steel Construction, Chicago, IL.

AISC (2005). *Specification for Structural Steel Buildings*, ANSI/AISC 360–05, American Institute of Steel Construction, Chicago, IL.

AISC (2005b). *Code of Standard Practice for Steel Buildings and Bridges*, American Institute of Steel Construction, Chicago, IL.

ASCE (2005). *Minimum Design Loads for Buildings and Other Structures*, SEI/ASCE 7–05, ASCE, Reston, VA.

ASCE (1997). Effective Length and Notional Load Approaches for Assessing Frame Stability: Implications for American Steel Design, Task Committee on Effective Length, Technical Committee on Load and Resistance Factor Design, Structural Engineering Institute, American Society of Civil Engineers, 442 pp.

Deierlein, G. (2003). "Background and Illustrative Examples on Proposed Direct Analysis Method for Stability Design of Moment Frames," Report on behalf of AISC TC10, July 13 2003, Dept. of Civil and Environmental Engineering, Stanford University, 17 pp.

Deierlein, G. (2004). "Stable Improvements: Direct Analysis Method for Stability Design of Steel-Framed Buildings," *Structural Engineer*, November, 24–28.

Galambos, T.V. & Ketter, R.L. (1959). "Columns Under Combined Bending and Thrust," Journal of the Engineering Mechanics Division, ASCE, 85(EM2), 135–152.

Griffis, L.G. & White, D.W. (2007). *Stability Design of Steel Buildings*, Steel Design Guide, American Institute of Steel Construction, to appear.

Kaehler, R.C., White, D.W. & Kim, Y.D. (2007), *Frame Design Using Web-Tapered Members*, Steel Design Guide, Metal Building Manufacturers Association and American Institute of Steel Construction, to appear.

Maleck, A. (2001). "Second-Order Inelastic and Modified Elastic Analysis and Design Evaluation of Planar Steel Frames," Doctoral dissertation, School of Civil and Environmental Engineering, Georgia Institute of Technology, Atlanta, GA, 579 pp.

Martinez-Garcia, J.M. (2002). "Benchmark Studies to Evaluate New Provisions for Frame Stability Using Second-Order Analysis," M.S. Thesis, School of Civil Engineering, Bucknell Univ., 241 pp.

Martinez-Garcia, J.M. & Ziemian, R.D. (2006). "Benchmark Studies to Compare Frame Stability Provisions," *Proceedings*, SSRC Annual Technical Sessions, 425–442.

Nair, R.S. (2005). "Stability and Analysis Provisions of the 2005 AISC Specification for Steel Buildings," *Proceedings*, Structures Congress 2005, ASCE, 3 pp.

Surovek-Maleck, A.E. & White, D.W. (2003). "Direct Analysis Approach for the Assessment of Frame Stability: Verification Studies," *Proceedings*, SSRC Annual Technical Sessions, 18 pp.

Surovek-Maleck A.E. & White, D.W. (2004a). "Alternative Approaches for Elastic Analysis and Design of Steel Frames. I: Overview," *Journal of Structural Engineering*, ASCE, 130(8), 1186–1196.

Surovek-Maleck A.E. & White, D.W. (2004b). "Alternative Approaches for Elastic Analysis and Design of Steel Frames. II: Verification Studies," *Journal of Structural Engineering*, ASCE, 130(8), 1197–1205.

White, D.W. & Griffis, L.G. (2007). "Stability Design of Steel Buildings: Highlights of a New AISC Design Guide," *Proceedings*, North American Steel Construction Conference, April, American Institute of Steel Construction, Chicago, IL, to appear.

White, D.W., Surovek, A.E., Alemdar, B.N., Chang, C.J., Kim, Y.D., Kuchenbecker, G.H., (2006), "Stability Analysis and Design of Steel Building Frames Using the AISC 2005 Specification," *International Journal of Steel Structures*, 6, 71–91.

White, D.W., Surovek, A.E. & Kim, S.-C. (2007a). "Direct Analysis and Design using Amplified First-Order Analysis: Part 1 – Combined Braced and Gravity Framing Systems," *Engineering Journal*, American Institute of Steel Construction, Chicago, IL, to appear.

White, D.W., Surovek, A.E. & Chang, C.-J. (2007b). "Direct Analysis and Design using Amplified First-Order Analysis: Part 2 – Moment Frames and General Framing Systems," *Engineering Journal*, American Institute of Steel Construction, Chicago, IL, to appear.

Steel and Composite Structures – Wang & Choi (eds)
© *2007 Taylor & Francis Group, London, ISBN 978-0-415-45141-3*

Evaluation of stress concentration and deflection of simply supported box girder

E. Yamaguchi
Kyushu Institute of Technology, Kitakyushu, Japan

T. Chaisomphob & J. Sa-nguanmanasak
Thammsat University, Pathumthani, Thailand

C. Lertsima
Asian Engineering Consultants Corp., Bangkok, Thailand

ABSTRACT: The shear lag has been studied for many years. Nevertheless, existing research gives a variety of stress concentration factors. Unlike the elementary beam theory, the application of load is not unique in reality. For example, concentrated load can be applied as point load or distributed load along the height of the web. This non-uniqueness may be a reason for the discrepancy of the stress concentration factors in the existing studies. Many researchers employed the finite element method for studying the effect of the shear lag. However, very few researches have taken into account the influence of the finite element mesh on the shear lag phenomenon, although stress concentration can be quite sensitive to the mesh employed in the finite element analysis. This may be another source for the discrepancy of the stress concentration factors. It also needs to be noted that much less studies seem to have been conducted for the shear lag effect on deflection while some design codes have formulas. The present study evaluates stress concentration and deflection of a simply supported box girder by the three-dimensional finite element method. The whole girder is modeled by shell elements, and extensive parametric study with respect to the geometry of a box girder is carried out. The results are compared with those due to design equations.

1 INTRODUCTION

In the elementary beam theory, the normal stress in the longitudinal direction produced by bending deformation is assumed to be proportional to the distance from the neutral axis and therefore uniform across the flange width. However, as a flange gets wider, this assumption becomes invalid: the normal stress distribution is not uniform in the wide flange, but the stress takes the maximum value at the flange-web intersection in general, decreasing toward the middle of the flange. This phenomenon is called the shear lag.

The shear lag has been studied for many years. Timoshenko and Goodier (1970) have documented one of the earliest researches due to von Karman. The best-known achievement in the past is probably the one due to Reissner (1941; 1946). While these early researches are analytical, a numerical means, the finite element method in particular, is often utilized in the recent studies. A concise but excellent literature review of research on the shear lag is available in Tenchev (1996).

Although much research has been done for the problem and several design codes have already provided formulas to account for the shear lag effect (British 1982; Japan 2002; CEN 2003), discrepancies in numerical results are observed in the literature. This seems to be attributable to the factors that have considerable influence on numerical results but have been overlooked. For example, concentrated load can be applied as point load or distributed load along the height of the web. Such non-uniqueness in loading can be a reason for the discrepancy of the stress concentration factors in the existing studies. It is also noteworthy that shear lag effect on deflection has not been studied much while some design codes have formulas for it (British 1982; Japan 2002).

The three-dimensional finite element analysis of a simply supported box girder by shell elements is carried out to investigate the shear lag effect in the present study. Two loading conditions of concentrated load at the mid-span and uniformly distributed load along the beam length are employed. Multiple ways to apply those loads are considered. Much attention is paid to

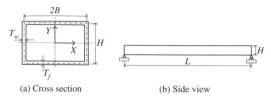

(a) Cross section   (b) Side view

Figure 1.   Structural geometry of box girder.

Table 1. K_c in literature.

| Literature | K_c | |
	Concentrated load	Distributed load
Tenchev (1996)	1.31	1.07
British (1982)	1.39	1.05
Japan (2002)	1.23	1.09
Eurocode 3 (2003)	–	1.05
Song et al. (1990)[)]	1.34	1.04
Sedlacek et al. (1993)	1.35	1.05
Tahan et al. (1997)	1.58	1.05
Lee et al. (2000)	1.56	1.05

finite element mesh as well, so as to minimize discretization error. Note that not many researchers seem to have carefully considered the way of applying load and the finite element mesh.

The normal stress in the longitudinal direction in the flange is of interest for investigating the shear lag effect on stress. The stress in the mid-span cross section is focused on in particular, since the largest stress is expected. The vertical displacement at the mid-span is also computed to see the shear lag effect on deflection. An extensive parametric study is conducted, and the results are compared with those due to design equations. In all the analyses, a finite element program, MARC (1994), is used.

2 ANALYSIS MODEL

Simply supported box girders under concentrated load at mid-span or uniformly distributed load are analyzed. The symbols employed in the present study for describing the structural geometry are illustrated in Figure 1. For those box girders, the stress concentration factors at the mid-span can be evaluated by the formula given in the design codes (British 1982; CEN 2003; Japan 2002). The factors can be obtained also by the results in the literature (Lee & Wu 2000; Sedlacek & Bild 1993; Song et al. 1990; Tahan et al. 1997; Tenchev 1996). For a box girder with $H/L = 1.0$, $B/H = 1.0$ and $T_f/T_w = 1.0$, those values are summarized in Table 1 where K_c stands for the stress concentration factor defined by the ratio of the maximum normal stress in the flange to that of the elementary beam theory.

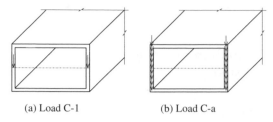

(a) Load C-1   (b) Load C-a

Figure 2. Concentrated load.

Significant discrepancy is recognized, especially under concentrated load. As Table 1 may prove, it is not an easy task to evaluate a rigorous stress distribution having the shear lag phenomenon. Generally, the largest normal stress in a flange occurs at the edge and sharp decrease is observed toward the middle of the flange. Because of the large stress gradient, K_c can be very sensitive to a mathematical model set up for the analysis.

The existing studies may be classified into two groups based on the type of analysis: analytical approach (Reissner 1941, 1946; Sedlacek & Bild 1993; Song et al. 1990; Tahan et al. 1997; Timoshenko & Goodier 1970) and finite element approach (Lee & Wu 2000; Moffatt and Dowling 1975; Tenchev 1996). The former often needs to introduce assumptions so as to simplify a problem and yield a solution. On the other hand, the finite element method requires few assumptions in principle. However, due to the limitation of computer capacity, the shear lag is investigated in the two-dimensional framework of a plane stress problem in some finite element analysis (Lee & Wu 2000; Tenchev 1996) while the behavior in accordance with the elementary theory of bending is assumed in some other finite element analysis (Moffatt and Dowling 1975).

In general, the less the assumptions imposed on the analysis are, the closer to the reality the mathematical model becomes. To this end, in the present study, an entire box girder is modeled as it is, using 4-node shell elements: the shear lag problem is not reduced to a plane-stress problem and no beam assumptions are implemented.

In the three-dimensional finite element model, unique load in the beam theory can be applied in various ways. Herein the load applications that may cause local effects on the stress distribution in the flange are avoided. As for concentrated load, therefore, two loading models shown in Figure 2 are adopted: Load C-1 is concentrated load at the middle of the web and Load C-2 is uniformly distributed load along the height of the web. Two loading models shown in Figure 3 are considered for distributed load: Load D-1 is uniformly distributed load along the centerline of the web and Load D-2 is uniformly distributed load not only along the beam axis but also along the web height of every

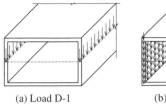
(a) Load D-1

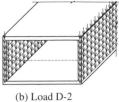
(b) Load D-2

Figure 3.   Distributed load.

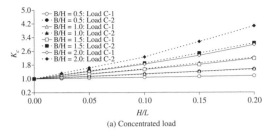
(a) Concentrated load

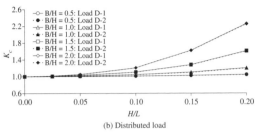
(b) Distributed load

Figure 4.   Variation of K_c with respect to H/L $(T_f/T_w = 1.0)$.

cross section. To be noteworthy, although the stress concentration may be influenced by the way the load is applied, researchers other than Tenchev (1996) have not described their loading condition explicitly, to the best knowledge of the authors.

3   NUMERICAL EVALUATION

The structural model described in the previous chapter is analyzed by the finite element method, using shell elements. Although the finite element method is very versatile and powerful, the results may depend largely on finite element mesh employed in the analysis, which is especially so when stress concentration is dealt with. In this conjunction, the extrapolation method called "multimesh extrapolation" (Cook et al. 1989) is utilized to enhance accuracy in the present study. This method is based on a theoretical observation that strain error is proportional to element size (Cook et al. 1989). To carry out the extrapolation, multiple finite element analyses are inevitable for each box girder. The same method is utilized for the evaluation of deflection as well. Note that deflection error is proportional to the square of element size (Cook et al. 1989).

4   PARAMETRIC STUDY

Based on the modeling and the numerical evaluations described above, three-dimensional finite element analysis is conducted so as to reveal the influence of the parameters that characterize the geometry of a box girder. In particular, the following values are considered: $H/L = 0.025, 0.05, 0.10, 0.15, 0.20$; $B/H = 0.5$, 1.0, 1.5, 2.0; $T_f/T_w = 0.5, 1.0, 1.5, 2.0$. The combination of all these values results in 80 box girders different from each other in geometry. In this paramaetric study, as explained earlier, multiple loadings are applied to each girder and multiple finite element meshes are used to eliminate discretization error by the multimesh extrapolation method for every girder under a specific loading condition.

4.1   Stress concentration factor K_c

As a typical example of the present numerical results, Figure 4 shows the variation of K_c with respect to

H/L for the cross sections with $T_f/T_w = 1.0$ under concentrated load (Loads C-1 and C-2) and uniformly distributed load (Loads D-1 and D-2). Load C-2 induces larger K_c than Load C-1 consistently and the difference is considerable. Load D-1 yields larger K_c than Load D-2, but the difference appears insignificant. Thus, it is decided that while both Loads C-1 and C-2 are applied in the case of concentrated load, only Load D-1 needs to be considered for distributed load in the present finite element analyses. To show the influences of B/H and T_f/T_w on K_c, Figures 5 and 6 are presented.

The general trends regarding the effect of the geometrical parameters on K_c observed in Figures 4 to 6 can be summarized as follows:

(a) K_c tends to grow with the increase of H/L.
(b) The influence of H/L on K_c is very small for B/H equal to 0.5 under Loads C-2 and D-1.
(c) K_c tends to grow with the increase of B/H.
(d) The influence of B/H on K_c is very small for H/L equal to 0.025 in the case of concentrated load and equal to or smaller than 0.05 in the case of distributed load.
(e) K_c tends to grow with the increase of T_f/T_w.
(f) The influence of T_f/T_w on K_c is very small for H/L equal to or smaller than 0.05 in the case of concentrated load. Under distributed load, the influence of T_f/T_w on K_c is very small for the entire range of T_f/T_w considered herein.

Figure 7 presents K_c based on those due to Tenchev (1996), British (1982), Japan (2002), CEN (2003) and Lee et al. (2000) together with the present numerical results. Discrepancy is recognized in concentrated load: while Tenchev (1996) and Japan (2002) yield

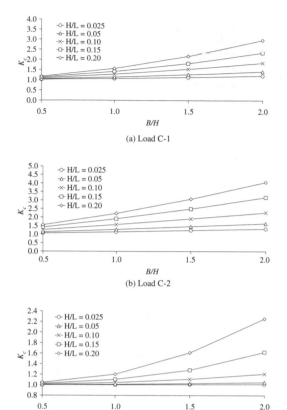

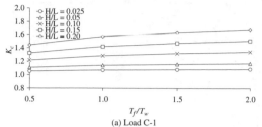

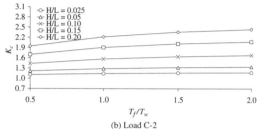

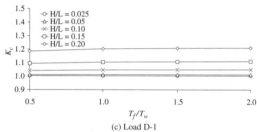

Figure 5. Variation of K_c with respect to B/H ($T_f/T_w = 1.0$).

Figure 6. Variation of K_c with respect to T_f/T_w ($B/H = 1.0$).

K_c close to that of the present results with Load C-1, K_c due to British (1982) lies between the two sets of the present results. In the distributed load, discrepancy is smaller, although K_c due to Japan (2002) becomes larger than that due to the others, as B/H increases.

4.2 Deflection magnification factor D_m

Figure 8 shows a typical variation of D_m with respect to H/L under concentrated load (Loads C-1 and C-2) and uniformly distributed load (Loads D-1 and D-2). Although Load C-2 yields larger D_m than Load C-1, the values due to the two loadings are close to each other. Load D-2 gives larger D_m than Load D-1, but the difference is very small. Thus, the way of applying load is insignificant for D_m for both concentrated and distributed loads. In the following numerical analyses, therefore, only Loads C-2 and D-2 are considered. For the influence of B/H and T_f/T_w on D_m, Figures 9 and 10 are presented.

The general trends observed in Figures 8 to 10 can be summarized as follows:

(a) D_m tends to grow with the increase of H/L.
(b) D_m tends to grow with the increase of B/H.
(c) The influence of B/H is very small for H/L equal to or smaller than 0.05.
(d) D_m tends to grow with the increase of T_f/T_w.
(e) The influence of T_f/T_w is very small for H/L equal to or smaller than 0.05.

Some design codes (British 1982; Japan 2002) provide formulas for the shear lag effect on deflection. Figure 11 presents D_m calculated through those formulas together with the present numerical results. While D_m due to the design formulas exhibits the similar tendencies to that obtained by the present finite element analysis, discrepancy is evident: D_m due to the design formulas considerably underestimates D_m by the finite element analysis. The difference between the D_m values due to the formulas in the two design codes is also obvious: D_m due to Japan (2002) is consistently larger than that due to British (1982).

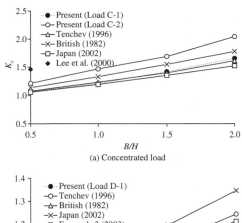

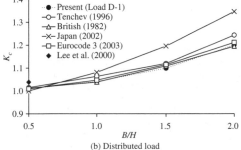

Figure 7. Comparison of K_c ($H/L = 0.1$, $T_f/T_w = 0.5$).

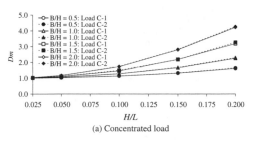

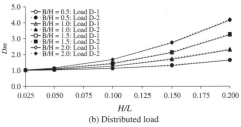

Figure 8. Variation of D_m with respect to H/L ($T_f/T_w = 1.0$).

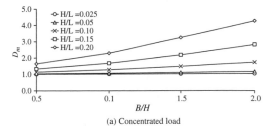

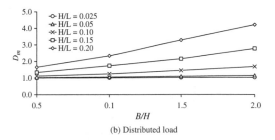

Figure 9. Variation of D_m with respect to B/H ($T_f/T_w = 1.0$).

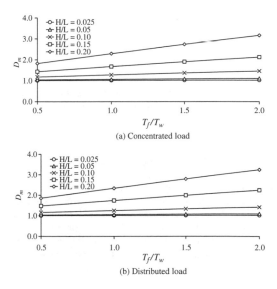

Figure 10. Variation of D_m with respect to T_f/T_w ($B/H = 1.0$).

5 CONCLUDING REMARKS

Extensive parametric study with respect to the geometry of a box girder has been carried out by the three-dimensional finite element analysis so as to evaluate the stress concentration and deflection of a simply supported box girder. Shell elements have been used to model the entire box girder. The effect of the way of applying load and the dependency of the stress concentration on the finite element mesh have been carefully treated, while many existing research works seem to have paid little attention to these aspects of analysis modeling.

The present study indicates that the real stress and deflection can be much larger than those due to the beam theory. It is also shown that the existing studies

93

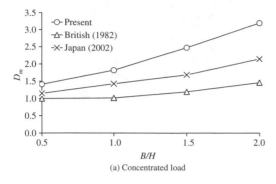

(a) Concentrated load

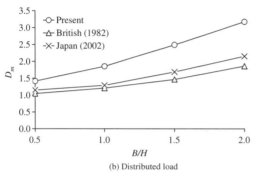

(b) Distributed load

Figure 11. Comparison of D_m ($H/L = 0.2$, $T_f/T_w = 0.5$).

including design codes may yield the stress and deflection quite different from each other and the present result.

ACKNOWLEDGEMENTS

The present research was partially supported by the Thailand Research Fund through the Royal Golden Jubilee Ph.D. Program (Grant No. PHD/0028/2544, PHD/0055/2543) and by the Ministry of Education, Science, Sports and Culture of Japan through Grant-in-Aid for Scientific Research (B) (No. 15360244). It is also the outgrowth of the academic agreement between Sirindhorn International Institute of Technology, Thammasat University and Faculty of Engineering, Kyushu Institute of Technology, and has been partially supported by the two academic bodies. These supports are gratefully acknowledged.

REFERENCES

British Standards Institution 1982. Steel, concrete and composite bridges, BS 5400, Part 3. *Code of Practice for Design of Composite Bridges.*

CEN 2003. *Eurocode 3: Design of Steel Structures*, prEN 1993 1–5.

Cook, R.D., Malkus, D.S. & Plesha, M.E. 1989. *Concepts and Applications of Finite Element Analysis*, 3rd Edition. John Wiley & Sons.

Japan Road Association 2002. *Design Specifications for Highway Bridges*, Part II Steel Bridges. Maruzen.

Lee, C.K. & Wu, G.J. 2000. Shear lag analysis by the adaptive finite element method 1: Analysis of simple plated structures. *Thin-Walled Struct.* 38: 285–309.

MARC Analysis Research Corporation 1994. *MARC Manuals,* Vol. A-D, Rev. K.6.

Moffatt, K.R. & Dowling, P.J. 1975. Shear lag in steel box girder bridges. *Struct. Engng.* 53(10): 439–448.

Reissner, E. 1941. Least work solutions of shear lag problems. *J. Aeronaut. Sci.* 8: 284–291.

Reissner, E. 1946. Analysis of shear lag in box beam by the principle of minimum potential energy. *Q. Appl. Math.* 6(3): 268–278.

Sedlacek, G. and Bild, S. 1993. A simplified method for the determination of the effective width due to shear lag effects. *J. Construct. Steel Research* 24: 155–182.

Song, Q.G. & Scordelis, A.C. 1990. Formulas for shear-lag effect of T-, I-, and box beams. *J. Struct. Engng., ASCE* 116(5): 1306–1318.

Tahan, N., Pavlovic, M.N. & Kotsovos, M.D. 1997. Shear-lag revisited: The use of single Fourier series for determining the effective breadth in plated structures. *Comput. & Struct.* 63(4): 759–767.

Tenchev, R.T. 1996. Shear lag in orthotropic beam flanges and plates with stiffeners. *Int. J. Solids Struct* 33(9): 1317–1334.

Timoshenko, S.P. & Goodier, J.N. 1970. *Theory of Elasticity*, 3rd Edition. McGraw-Hill.

Steel and Composite Structures – Wang & Choi (eds)
© *2007 Taylor & Francis Group, London, ISBN 978-0-415-45141-3*

Advanced analysis of building frames against blast and fire

J.Y. Richard Liew & Y. Hang

Department of Civil Engineering, National University of Singapore, Singapore

ABSTRACT: The lesson learned from the terrorist attacks on buildings is a need to assure that the structure is capable of sustaining localised damage without catastrophic failure. This paper presents a numerical approach for analyzing steel frame structures subject to localized damage caused by blast load and subsequently investigating their vulnerability under fire attack. The proposed numerical method adopts an adaptive mixed element approach for modeling large-scale framework and it is proven sufficiently accurate for capturing the detailed behavior of member and frame instability associated with the effects of high strain rate and fire temperature. A multi-storey steel building frame is analysed so that the complex interaction effects of blast and fire can be understood and quantified. The frame is found to be vulnerable, as it possesses little fire resistance due to the deformation on key structural elements caused by the high blast load.

1 INTRODUCTION

A significant amount of research work has been carried out in the past decades to assess the fire resistance of structures. Simplified method has been proposed by Kruppa (1979) to calculate the fire resistance of a beam based on the plastic mechanism method. Wong (2001) developed the plastic analysis method for fire resistant analysis of simple framed structures. Liew et al. (1998, 2004) and Chan & Chan (2001) proposed second order plastic hinge methods for large displacement analysis of semi-rigid steel frames under elevated temperature. All these methods described above vary in their applicability and degree of sophistication; however, they cannot be applied to situations in which the fire is a direct consequence of an explosion. Blast load is considered a severe hazard because it can deliver huge amount of energy to the structure causing potential damages to structural members. Research works have been carried out to predict the blast response of structures in order to assess the damages and to mitigate the risk. Krauthammer et al. (1999) studied the behaviour of structural connections under explosion load using the computational approaches. Ettouney et al. (1996) studied the vulnerability of different structural features of commercial building subjected to blast loading. Kaliszky & Logo (2006) presented three methods to determine the optimal layout of structures subject to extreme loading.

The combined effects of blast and fire could lead to progressive failure of structure as in the case of terrorist attack on the World Trade Center buildings. The first attempt to perform an integrated blast and fire analysis is by Izzuddin et al. (2000). Explicit and implicit dynamic analysis techniques are integrated to solve the blast and fire loading. Liew & Chen (2004) investigated the interaction of blast and fire load on the performance of columns and two-dimensional steel frames using ABAQUS implicit dynamic solver. Subsequently, they (Liew & Chen 2004) extended the work to three-dimensional steel frames. In all these works, simple blast loading models with variable peak pressure value are used instead of realistic blast and fire loadings. A monotonic increase of the temperature with assumed temperature gradient is used for the fire analysis. The study only restricted to simple steel frames. These examples are meant to demonstrate the solution technique; the numerical results do not usually represent the real structural behaviour.

Depending on the difference between the structure's natural period and the duration of the blast load, the structure may display three types of responses when subjected to blast and fire. The effect of peak blast pressure and duration vary with the types. (When the duration of the blast load is very small compared to the natural period of the structure, the peak blast pressure is less important. When the load duration is similar to the natural period of the structure, both peak force and the impulse will be important to determine the structural response. However, in case of accidental fire, the duration of the load is much longer than the natural period of the structure, and the loading is termed quasi-static.)

To capture the combined effect of fire and blast in an integrated analysis, high-strain rate caused by an explosion, softening of materials due to fire and dynamic and thermal responses of structure are to be assessed in a time domain. Considering the computational resources, an adaptive mixed element approach is adopted to achieve a realistic modelling of the overall framework subject to localized explosion and fire. The influence of blast loading on the fire resistance of a three-dimensional (3D) steel frame is studied. Although integrated analyses are not adopted at present, the current work forms the basis for further studies in which the explosion and fire analyses may be coupled.

2 STRAIN RATE EFFECT DUE TO BLAST LOAD

The mechanical properties of steel are affected noticeably by the rate at which strain takes place. If the mechanical properties under static loading are considered as a basis, the effects of increasing strain rate on steel strength can be illustrated in Figure 1. The yield strength (σ_y) increases substantially to a dynamic yield strength (σ_{dy}). Experimental evidences (Bassim & Panic 1999, Ogawa 1985) show that when the strain rate is very high, the dynamic yield strength may increase beyond that of the ultimate strength as illustrated in Figure 1. The elastic modulus generally remains insensitive to the loading rate. The ultimate tensile strength increases slightly; however, the percentage increase is less than that for the yield strength, and the elongation at rupture either remains unchanged or is slightly reduced due to the increased strain rate. High strain rate may affect cross section classification due to the increase in yield strength. Compact steel section may be down-graded to slender section leading to occurrence of local buckling at high strain rate (Chen & Liew 2005). Table 1 shows the plate slenderness limits in accordance with EC3: Part 1.1, but with yield strength replaced by the dynamic yield strength calculated from the Cowper and Symonds model.

The estimation of the strain rate and the calculation of the increase of the yield and ultimate strengths may be also obtained from design manuals such as the TM5-130037. The increase of strength due to strain rate effect is obtained by multiplying the static strength by a dynamic increase factor (DIF). The DIF values are acquired empirically and are based on the average strain rates of the steel members. The following observations must be taken into account when using the recommendations in TM5-130037:

1. The strain rates used are relatively low and thus the DIF values adopted are conservative.
2. The average strain rate is used as opposed to peak strain rate value, which is exceptionally high near the beam's ends and in the connections. High strain rate effect may cause local buckling of cross sections, which may not be identified through analytical approaches such as those recommended in various design manuals. Thus, numerical methods are essential to capture detailed behavioural effect at high strain rate locations.

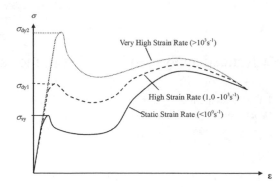

Figure 1. Stress-strain relationship of steel material under high-strain rate.

Table 1. Section re-classification under high strain rate based on Cowper and Symonds model.

Strain rate s^{-1}	Plate slender limit		
	Flange of I-shaped rolled beam	Flange of I-shaped welded beam	Webs in flexural compression
0*	15ε	14ε	124ε
10^{-5}	13.5ε	12.6ε	112.0ε
10^{-4}	13.4ε	12.5ε	110.5ε
10^{-3}	13.1ε	12.2ε	108.3ε
10^{-2}	12.7ε	11.9ε	105.1ε
10^{-1}	12.2ε	11.3ε	100.5ε
10^{0}	11.4ε	10.6ε	94.3ε
10^{1}	10.5ε	9.8ε	86.5ε
10^{2}	9.3ε	8.7ε	77.3ε
10^{3}	8.1ε	7.6ε	67.3ε

* Corresponding to Class 3 plate slender limit (Eurocode 3).

3. In cases when the DIF values from the TM5-1300 are used, the yield strength may exceed that of the ultimate strength. For design, it is conservative to use the lower of the two strength values. However, for detailed numerical modelling, the enhancement in yielding strength at peak strain rate should be modelled to capture and the possibility of local buckling of cross section and to assess the vulnerability of the blast-affected structure.

3 ANALYSIS OF STRUCTURES SUBJECTED TO BLAST LOAD

Blast analysis methods can be varied from single-degree of freedom (SDOF) or multi-degree of freedom (MDOF) dynamic models to more sophisticated finite element methods (FEM) based on either beam element model or shell element model.

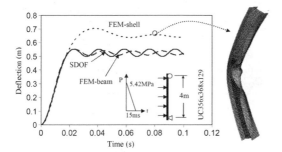

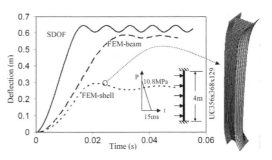

Figure 2. Comparison of the column mid-height defection with time predicted by SDOF, FEM-beam model and
FEM-shell model- column with pin-roller ends.

Figure 3. Comparison of the column mid-height deflection with time predicted by SDOF, FEM-beam model and
FEM-shell model- column with fix-fix ends.

The SDOF method is applied to single member,
which involves simplification of a continuous structural member into a dynamic system with concentrated equivalent mass and stiffness. The responses of
multi-storey frames can be represented by MDOF systems where the floors are represented by concentrated
masses. Solutions to the SDOF/MDOF systems can be
found by referring to the design charts or using classical dynamic solution methods. FEM (Finite Element
Method) based on beam element model considers the
interaction effects between individual members and
the overall framework.

One major shortcoming of the beam element model
is that individual member capacity is calculated based
on plastic hinge theory. However, under rapid blast
loading, member failure may be triggered by high
shear deformation rather than by flexural deformation.
There is also a possibility of local buckling of steel sections due to high strain rates, which will increase the
effective yield strength of materials while the Young's
modulus remains almost unchanged. When the strain-
rate is varied from 1.0 to $10^3 \sec^{-1}$, the yield strength is
increased by a factor of 1.47 to 2.2. That means that the
semi-compact plate slenderness limit is downgraded
by around 18 to 37% (see Table 1) as the plate slenderness criteria are functions of $\varepsilon = \sqrt{235/f_y}$ according
to EC3: Part 1.1.

The local buckling and high strain rate problems
can be solved by using FEM based on shell element
model together with appropriate constitutive models
accounting for the strain-rate effect. If more refined
solid element is used, it is possible to simulate spalling
and cracking of concrete or brittle failure of materials.
However, refined model also means a large amount of
labour in building the model and the need of intensive
computation resources.

A case study is performed to compare the blast
response obtained from the SDOF method and the
FEM using beam element and shell element. A steel
column with section size 356×368 UC129 and height
4.0m is subjected to uniform lateral pressure due to

blast effect. Material Grade S355 ($\sigma_y = 355 \, \text{N/mm}^2$)
is used and the column section is classified as non-
slender for both the flange and web plates. Two types
of boundary conditions are considered: pin-roller ends
and fixed ends. The SDOF analysis is performed for
steel members under accidental loading.

The finite element analysis is performed using
ABAQUS program with both beam and shell element
models. The blast pressure is applied on one of the
flange plate and forcing the member to bend about its
major axis. Geometrical initial imperfection is applied
by appropriate scaling of the member's buckling shape
with maximum initial deflection of span/1000 at the
mid-height.

Figure 2 and Figure 3 show the mid-height lateral
displacement of the columns with pin-roller ends and
fixed ends. Triangular blast loading with zero rise time
is used. Based on plastic-hinge theory, the stiffness of
the column with pin-roller ends can be represented
by a linear elastic-idealistically plastic curve. With
appropriate choice of the shape function, the SDOF
method gives almost exactly the same result as the
FEM results using beam elements. FEM using shell
elements predicts larger displacement because local
buckling reduces the effective stiffness of the section
as shown in the insert figure of Figure 2.

The column with fixed ends has undergone three
stages of deformation: (1) linear elastic regime, (2)
inelastic transition regime where yielding occurs at
the beam ends and (3) plastic regime where a plastic
mechanism is formed. When an elastic fully plastic
model is used, predicted. Figure 3 shows that the initial
stiffness predicted by the SDOF method is higher than
the other two. However, the SDOF method predicts
larger maximum deflection. This is because of the error
in the estimation of the stiffness in the hardening phase.
For the FEM shell element model, the flange plate
deforms toward the web upon the impingement of the
blast pressure as shown in the insert figure of Figure 3.
As the blast pressure is applied perpendicular to the
flange plate, the blast pressure is dissipated with the

deformation of the flange. Therefore, the deflection from the shell model is smaller than the SDOF method and the FEM with beam element model.

This example demonstrates the complexity of modeling the response of structure subjected to blast loading. The behaviour of structure depends on the difference between the blast duration and the structural natural period. The SDOF method predicts similar response behavior as the FEM-beam model provided that local buckling (or distortion) of cross section does not occur and appropriate representation for the inelastic stiffness can be made in the SDOF method.

4 MATERIAL PROPERTIES AT ELEVATED TEMPERATURE

Test evidence shows that typical stress-strain relationship of steel at elevated temperature does not exhibit a distinctive yield plateau. A more relaxed effective yield strength based on 2% strain is considered applicable due to the non-linearity of the stress-strain curve at elevated temperature. In the present analysis, an effective yield stress corresponding to 2% strain as recommended by EC3: Part1.2 is adopted for numerical analysis. The nonlinear stress-strain curve is modelled as multi-linear curves to be specified for the uniaxial representation of the constitutive law of steel at elevated temperature. Von-Mises yield criterion is then used to extrapolate a yield surface in 3D principle stress space. The metal plasticity model is characterized as associated flow with isotropic hardening.

When plastic hinge theory is used, the cross section plastic strength will have to be reduced. The plastic strength interaction curve at elevated temperature is represented by a constricting yield surface and the detailed description of second-order plastic hinge formulation considering the thermal effect is given in Ma & Liew (2004). One major short coming of the plastic hinge analysis method is that it cannot capture the behavoural effects associated with local/distorsional buckling of cross sections. The compactness of cross section is affected by the degradation of Young's modulus and yield strength. The classification of plate slenderness ratio which is a function of E_t/σ_{yt} is shown in Table 2. The rolled beam flange slenderness limit for class 3 plate changes from 15ε at 20°C to 13.6ε at 800°C in accordance with Eurocode 3. Therefore, a compact cross section at ambient temperature may become a slender section at elevated temperature.

5 ANALYSIS OF STRUCTURES SUBJECT TO FIRE

Fire analysis methods must be able to take into account the change of material properties and thermal

Table 2. Section re-classification under high temperature.

Temperature (°C)	Plate slender limit		
	Flange of I-shaped rolled beam	Flange of I-shaped welded beam	Webs in flexural compression
20*	15ε	14ε	124ε
100	15ε	14ε	124ε
200	14.2ε	13.3ε	117.6ε
300	13.4ε	12.5ε	110.9ε
400	12.6ε	11.7ε	103.8ε
500	13.2ε	12.3ε	108.8ε
600	12.2ε	11.4ε	100.7ε
700	11.3ε	10.5ε	93.2ε
800	13.6ε	12.7ε	112.2ε
900	15.9ε	14.9ε	131.5ε

* Corresponding to Class 3 plate slender limit (Eurocode 3).

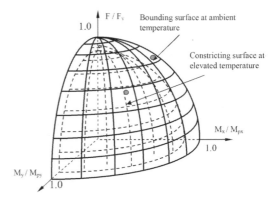

Figure 4. Constricting plastic strength surface at elevated temperature.

expansion of materials with increase of temperature. Liew and his associates (Liew et al. 1998, Ma & Liew 2004) proposed the use of the plastic hinge based method to model bare steel frames subject to fire. The proposed beam model can capture accurately the ultimate behaviour of beam-column using only one element per member. The beam element displacement fields are derived from the exact solution of the fourth order differential equation for a beam-column subjected to end forces (Liew & Chen 2004). Material non-linearity is modeled by yield hinges allowing them to form at any point along the length of the member and at the ends of the member. The plastic strength of cross section is a function of temperature. A constricting yield surface, as shown in Figure 4, is used to model the cross sectional strength at elevated temperature.

A more rigorous analysis is to model the member using fibre elements. The member is divided into small

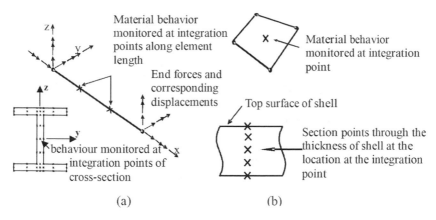

Figure 5. Modelling of (a) Beam Element (b) Shell Element.

fibers and the section forces and moments are calculated by integration of the stress at each fiber in the section. The spread of plasticity is captured by monitoring the stress-strain relationship at each fiber in accordance with the temperature in each fibre. For a more accurate representation of detailed localized effects such as shear and local deformation of plate components in the section, shell element modeling is required.

6 ANALYSIS OF STRUCTURES SUBJECT TO BLAST AND FIRE

An integrated analysis of structures subject to blast and fire imposes a greater challenge to numerical solution technique because

1. Blast load is rapid and intense; the corresponding analysis method is characterized by the inertia effect and strain-rate effect. The material model should include strain-rate effect and preferably material failure criterion by specifying maximum strain in which material fracture is expected to take place.
2. Fire is a relatively milder and prolonged process and nonlinear static analysis is preferred. Thermal creep effect is normally considered implicitly in the constitutive model, which considers degradation of the Young's modulus and effective yield strength at elevated temperatures. Restrained stresses induced by thermal expansion of heated members must be included in the analysis.

Methods for solving the dynamic and impulsive equilibrium equations are broadly characterized as implicit or explicit. Explicit scheme obtains values for dynamic quantities at time $t + \Delta t$ based entirely on available values at time t. The benefit of explicit solver is that the governing equilibrium equation can be solved directly without the need to transpose the stiffness matrix. The main drawback is that it requires very small time steps to maintain stability, normally in 10^{-6}s to 10^{-5}s for dynamic analysis of structural members.

Implicit scheme removes the constraint of using small time step by solving for the dynamic quantities at time $t + \Delta t$ based not only on values at t, but also on these same quantities at $t + \Delta t$. Because they are implicit, nonlinear equations must be solved and the global matrix needs to be transposed. Since a global set of equilibrium equations need to be solved in each increment, the cost of performing incremental analysis in the implicit scheme is expected to be greater than that of explicit analysis scheme. However, for highly nonlinear problems, smaller time steps should be used to meet convergence requirements. It is generally accepted that implicit solver is more appropriate for fire analysis while explicit solver is most commonly used for blast analysis.

In this paper, a mixed-element technique is used for adapting the mesh refinement of the structural model to the expected response. The structure model consists of different zones where the members may have different level of discretization (Fig. 5). It is proposed that frame members that are not directly subjected to blast and fire to be modeled using 2-node beam elements. Structural members that are under the direct action of blast and fire are modeled using 4-node shell elements to capture the local buckling, distortion of steel section plates and the inelasticity effects associated with high shear caused by blast loads.

The beam-column model is shown in Figure 5a. Spread of plasticity is modelled along the member length. The cross-sectional stiffness is calculated by integration of stresses and strains over the sub-areas of the cross-section. The cross-sectional stiffness is then integrated along the member length to obtain

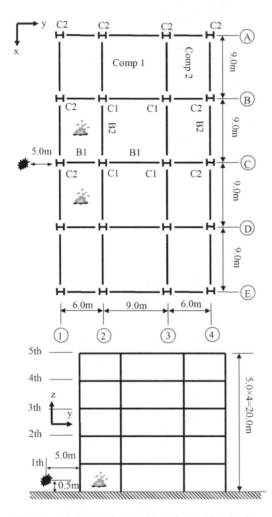

Figure 6. Five-storey frame subjected to accident loads.

Table 3. Structural Members Section.

	C1	C2	Beam
Level 1	UC356 × 202	UC356 × 153	B1:
Level 2	UC356 × 153	UC356 × 153	UB305 × 48
Level 3	UC356 × 153	UC305 × 118	B2:
Level 4	UC305 × 118	UC305 × 118	UB305 × 54
Level 5	UC203 × 60	UC203 × 60	

flange of steel sections and should be chosen based on convergence study.

7 MULTI-STOREY BUILDING FRAME SUBJECT TO ACCIDENTAL LOADS

This example illustrates the use of adaptive mixed element model to perform scenario-based fire and explosion analysis of a multi-storey frame. The first step is to identify the various possible accident scenarios that might occur during the service life of the structure. Numerical simulations are then carried out to evaluate the likelihood and consequences of such scenarios. This is essential that the structure could withstand and survive the blast load without the risk of total collapse.

Figure 6 shows a five storey steel building frame with solid concrete slab subject to the actions of blast and fire. All the steel members are Grade S275 steel and slab is Grade 30 normal weight concrete of 120 mm thick. The total floor load at accidental limit state is 6.25 kN/m^2. The steel section sizes and dimensions are summarised in Table 3.

In the numerical analysis, members that are affected by the explosion and fire loads are modelled using 4-node doubly curved shell elements with reduced integration and finite membrane strain (see Fig. 6). Gradual yielding is captured by evaluating the stresses and strains at each section point located at the integration point through the thickness of the shell. Along the length of the member, fine meshes with aspect ratios close to 1.0 are used in the regions with high stresses under the combined actions of high bending moments, shear and axial forces, and coarser meshes are used for in the region with lower stresses. The other non-critical members are modeled using the 2-noded beam-column elements.

7.1 Fire Accident Scenarios

The frame is first analysed for the fire accident scenario occurred at the lower compartment as shown in Figure 6. It is assumed that all the members in the compartment are engulfed in ISO834 standard fire. The temperature distribution in the heated members is assumed to be uniformly distributed over the member

the member stiffness. Comparison studies and bench-marking of beam-column approach versus shell element approach have been reported by Jiang et al. (2002) in ambient temperature and by Liew (2004) in fire conditions. However, with the inherit assumption of plane section remaining plane after deformation, the beam-column element approach cannot model the behavioural effects of cross-section associated with distortion and local buckling of component plates and significant shear yielding (Chen & Liew 2005).

The 4-node doubly curved shell element with reduced integration and finite membrane strains is shown in Figure 5b. Gradual yielding is captured by evaluating the stresses and strains at each section point located at the integration point through the thickness of the shell. A number of shell elements are used through the depth of the web and across the width of

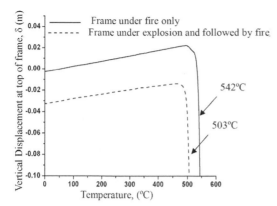

Figure 7.   Response of frame under the effect of explosion and fire.

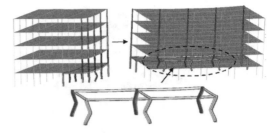

Figure 8.   Collapse mode of frame under fire attack.

lengths and across the sections. The Eurocode 3 Part 1–2 stress-strain relationship of steel at elevated temperature is used to model the steel structure under fire. Effect of fire is found to have negligible influence on the concrete slab because of its low heat conductivity up to the limiting temperature of the frame.

Figure 7 shows the temperature-vertical displacement curve at the top of the frame. During the initial heating phase, the compartment columns expand upwards due to the increase of temperature (upward displacement is shown as positive value). After a period of heating, the columns begin to shorten under the constant floor load because of the reduction of stiffness of steel members at elevated temperature. When fire temperature reaches 542°C, the axial displacement in z direction increases rapidly and progressive collapse occurs. The critical temperature of the frame is predicted to be $T_{max} = 542$°C. Classification of cross-section compactness is a function of yield strength and modulus of elasticity, which are affected by elevated temperature. The section is re-classified using the critical temperature using Table 2. The class 3 plate slender limits of the flange and beam web change to 12.8ε and 105.4ε, respectively and they are classified as compact. Hence, local buckling is not observed in the affected members. The deformed configuration of the frame at collapse is shown in Figure 8.

7.2   Blast and Fire Scenarios

The frame is then analysed for a possible accident scenario of blast load occurred at a stand off distance of 5 m from the front surface of the building. The edge columns at the building front are directly subject to the effect of blast load. The TNT equivalent weight of the explosive is assumed to be 1000 kg. The blast pressures applied on the columns at the front surface of the building are calculated from CONWEP program.

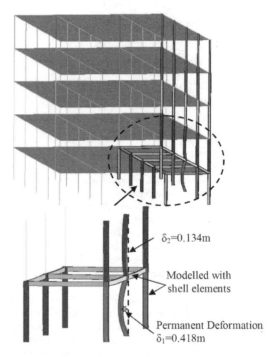

Figure 9.   Response of frame subjected to explosion load.

Plastic strains and deformations are observed for the columns located at grid line 1-C of level 1 and 2. The permanent deformation at the mid-height of the columns are $\delta_1 = 0.418$ m and $\delta_2 = 0.134$ m, which are higher than the nominal out-of-straightness value of $\delta_0 = L/1000 = 0.004$ m. The peak effective strain rates of the columns reach 104.4 s^{-1} and 46.3 s^{-1} at the bottom of columns. The section re-classification can be obtained from Table 1. When the effective strain rate of 1-C column in level 1 reaches 104.4 s^{-1}, the plate slender limits of the flange of I-shaped rolled beam and webs in flexural compression change to 9.3ε and 77.1ε, respectively. When the effective strain rate of 1-C column in level 2 reaches 46.3 s^{-1}, the plate slender limits of the flange of I-shaped rolled beam and webs in flexural compression change to 9.7ε and 80.5ε, where ε equals to $\sqrt{235/275} = 0.924$. Accordingly, the

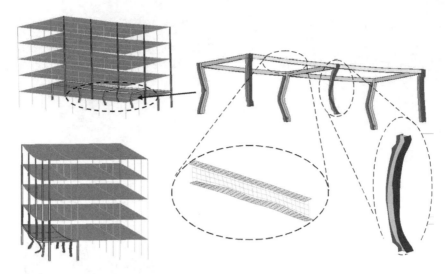

Figure 10. Collapse mode of frame subjected to explosion load and followed by fire attack.

width/thickness of the column flange is 8.65, which is between 9.7ε and 9.3ε, and the width/thickness of the column web is 26.07, which is smaller than 77.1ε. Therefore, local buckling occurred only at the bottom of the column in 1st level. The blast response of structure is shown in Figure 9.

The explosion triggers a fire occurrence in the ground floor compartment. After the explosion is over, the deformed frame is subjected to fire attack with increasing temperature. The same fire scenario as previous part is adopted.

When the outer face columns are subjected to blast load, they deform laterally causing some of the floor load to be transferred to the internal columns. The fire resistance of the internal columns reduces because of higher load ratio. After a period of heating, the internal columns began to buckle about the weak axes and the column loads are redistributed to the adjacent columns, and part of the loads are resisted by the external columns. With the combined effect of the increased axial force, the degradation of strength and stiffness caused by elevated temperature, permanent deformation due to blast, the external columns buckle. The inability of the structural system to redistribute the loads causes the progressive failure of the building frame. The frame failure mode is shown in Figure 10. The critical temperature of the frame is predicted to be 503°C, which is reduced by 7.2% compared to that under fire only.

8 CONCLUSIONS

Columns subject to blast loading may fail by shear yielding rather than by flexural deformation. There is also a possibility of local buckling of steel sections due to high strain rates, which increase the effective yield strength of materials while the modulus of elasticity remains almost unchanged. These local failure phenomena can only be capture by refined element model together with the appropriate constitutive models accounting for the strain-rate effect. For members subjected to blast and fire loads, detailed finite element modeling is necessary to improve the accuracy of predicting the limiting behaviour of structural members in post-blast fire situation. Initial deformation in the form of permanent deflection caused by blast loads decreases the load carrying capacity of the columns in fire.

To model large-scale steel frames subject to blast and fire, it is proposed that a mixed-element technique be adopted to allow for adapting the refinement of the structural model to the expected response. Frame members that are not directly subjected to explosion and fire can be economically modelled using the beam element approach. Structural members that are subject to direct extreme loads are best modelled using shell elements to capture localized effects associated with shear yielding, strength hardening due to high stain rate and degradation of strength and stiffness at elevated temperature.

An example of multistory frame subjected to explosion and fire was discussed, in which permanent deformation, local buckling and plastification caused by explosion reduced the fire resistance of the column and the frame. The interaction between explosion load and fire resistance may not be significant for isolated columns, but the fire resistance of multistory frame may be reduced more after the occurrence of an explosion depending on the magnitude of blast load. This is attributed to the stability interaction between the deformed members and the overall frame system. Blast load may cause local buckling and permanent deformation of the affected members, which will affect the overall stability of the frame when it is later exposed to fire. Hence, the fire resistance of steel structure is expected to be lower after it was subject to high impulsive loading causing plastic permanent deformation to critical members. The vulnerability and survivability of various framing systems exposed to blast and fire loads should be further studied to develop an effective strategy to protect building structures against willful attacks.

REFERENCES

Bassim, M.A. & Panic, N. 1999. High Strain Rate Effects on the Strain of Alloy Steels. Journal of Materials Processing Technology 92–93: 481–485.

Chan, S.L. & Chan, B.H.M. 2001. Refined plastic hinge analysis of steel frames under fire. Steel and Composite Struct 1(1): 111–130.

Chen, H. & Liew, J.Y.R. 2005. Explosion and fire analysis of steel frames using mixed element method. J. of Engineering Mechanics, ASCE 131(6): 606–616.

Ettouney, M. Smilowitz, R. & Rittenhouse, T. 1996. Blast Resistant Design of Commercial Buildings. Practice Periodical on Structural Design and Construction 1(1): 31–39.

Izzuddin, B.A., Song, L. Elnashi, A.S. & Dowling, P.J. 2000. An integrated adaptive environment for fire and explosion analysis of steel frames – Part II: verification and application. J. Constr. Steel Res. 53: 87-111.

Jiang, X.M., Chen, H. & Liew, J.Y.R. 2002. Spread-of-plasticity analysis of three dimensional steel frames. J. Constr. Steel Res. 58(2): 193–212.

Kaliszky, S. & Logo, J. 2006. Optimal design of elasto-plastic structures subjected to normal and extreme loads. Computers and Structures 84(28): 1770–1779.

Krauthammer, T. 1999. Blast-resistant structural concrete and steel connections. International Journal of Impact Engineering 22(9–10): 887–910.

Kruppa, J. 1979. Collapse temperature of steel structures. J. Struct. Div., ASCE; 105(ST9): 1769–1788.

Liew, J.Y.R. 2004. Performance based fire safety design of structures – A multi-dimensional integration. Advances in Structural Engineering, Multi-Science Publisher, U.K 7(4): 111–133.

Liew, J.Y.R. & Chen, H. 2004. Explosion and fire analysis of steel frames using fiber element approach. J. of Structural Engineering, ASCE 130(7): 991–1000.

Liew, J.Y.R. & Chen, H. 2004. Direct Analysis for Performance-based Design of Steel and Composite Structures. Progress in Structural Engineering and Materials. John Wiley, UK 6(4): 213–228.

Liew, J.Y.R., Tang, L.K., Holmaas, T. & Choo, Y.S. 1998. Advanced analysis for the assessment of steel frames in fire. J Constr. Steel Res 47(1–2): 19–45.

Liew, J.Y.R., Chen, H., Shanmugam, N.E. & Chen, W.F. 2000. Improved nonlinear plastic hinge analysis of space frame structures. Engrg. Struct. 22: 1324–1338.

Ma, K.Y. & Liew, J.Y.R. 2004. Nonlinear plastic hinge analysis of 3-D steel frames in fire. J. of Struc. Engrg., ASCE 130(7): 981–990.

Ogawa, K. 1985. Mechanical Behaviour of Metals Under Tension – Compression Loading at High Strain Rates. International Journal of Plasticity 1:347–358

Wong, M.B. 2001. Elastic and plastic methods for numerical modelling of steel structures subject to fire. J Constr. Steel Res 57(1): 1–14.

Steel and Composite Structures – Wang & Choi (eds)
© *2007 Taylor & Francis Group, London, ISBN 978-0-415-45141-3*

A practical design method for semi-rigid composite frames under vertical loads

G.-Q. Li
School of Civil Engineering, Tongji University, Shanghai, P.R. China

J.-F. Wang
School of Civil Engineering, Tsinghua University, Beijing, P.R. China

ABSTRACT: Although the benefits of semi-rigid connection and composite action of slab are extensively documented in the design of steel frames, they are not widely used much in practice. The primary cause is lack of appropriate practical design methods. In this paper, a practical method suitable for the design of semi-rigid composite frames under vertical loads is proposed. The proposed method provides the design of the connections, beams and columns for semi-rigid composite frames at the limit states of load-bearing and serviceability. The rotational stiffness of beam-to-column connections for calculating the deflection of the frame beams and the effective length factor of columns are also determined. The proposed method not only takes into account the actual behavior of the beam-to-column connections and its influence on the behavior of the overall structures, but also it is simple and convenient for designer to use in engineering practice.

1 INTRODUCTION

Semi-rigid composite frames are a novel structural system being developed to better utilize the composite floor slabs and flexible connections (Leon et al., 1987). This structural system extends the beneficial aspects of composite action to the negative (hogging) moment region of continuous beams by providing reinforcements across frame columns in slabs. The resultant system offers significant gains in stiffness and strength not only on the frame beams themselves but also at the connections.

Traditional approaches to the design of composite frames would regard the beam-to-column connections as being either notionally pinned or rigid. One of the approaches is to assume no continuity at beam-to-column connections and the composite beams are designed as simply supported between columns with the beam-to-column connections being required to transmit only the shears at beam ends. With rigid connections, full continuity is assumed and there is no relative rotation between the beam and the adjacent column. However, in a semi-rigid composite frame, moment resisting connections attract some hogging moment to the beam end, thereby reducing the sagging moment that the frame beam must support. The use of connections that provide reasonable degree of continuity can be beneficial to frameworks (Nethercot, 1995). Fig 1 gives three types of beam-to-column connections. The moment distributions of

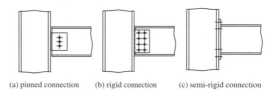

(a) pinned connection   (b) rigid connection   (c) semi-rigid connection

Figure 1. Types of beam-to-column connections.

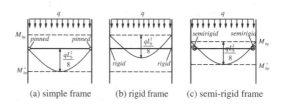

(a) simple frame   (b) rigid frame   (c) semi-rigid frame

Figure 2. Moment distribution of frames under uniformly distributed load.

the corresponding frames employing the above three types of connections under uniformly distributed load are shown in Fig.2.

The structural benefits of the semi-rigid composite frames are widely recognized, which may result in: (1) optimization of the moment distribution in the frames and improvement of the behavior of the frame beams, (2) reduction of cost and time for the construction of the frames, (3) good earthquake-resistance.

Nowadays there are a lot of studies on semi-rigid connections (Chen and Lui, 1991; Fang et al, 1999; Wang, 1999; Ahmed and Nethercot, 1997; Ahmed et al., 1997; Kemp and Nethercot, 2001; Cabrero and Bayo, 2005), but some key problems have not been solved effectively for the practical design of semi-rigid composite frames. Those problems mainly include:

(1) How to determine the effective rotational stiffness of beam-to-column connections at the ultimate load-bearing limit state or at the serviceability limit state?

(2) How to check the strength, stiffness and stability requirements for the semi-rigid composite frames at the ultimate load-bearing limit state or at the serviceability limit state?

The objective of this work is to propose a practical design method for semi-rigid composite frames under vertical loads, based on previous experimental and theoretical research (Wang, 2005). The proposed method provides the design requirements on the connections, beams and columns at the ultimate load-bearing limit state or at the serviceability limit state.

2 CONNECTION DESIGN

2.1 Moment capacity check of connections

The moment capacity of the semi-rigid beam-to-column composite connection should be checked with

$$M_u \leq M_{bp}^- \qquad (1)$$

where M_{bp}^- is the hogging moment capacity of the composite beam as shown in Fig.3.

The hogging moment capacity of the composite beam specified by GB50017 (2003) is expressed as

$$M_{bp}^- = (S_1 + S_2)f + A_{st}f_{st}(y_3 + y_4/2) \qquad (2)$$

where f is the design strength of steel; S_1 and S_2 are the area integral upper and lower the centriod axis of the steel beam, respectively; A_{st} is the cross-sectional area of the longitudinal reinforcing bars in the effective width of the slab for the composite beam; f_{st} is the design strength of the reinforcing bars; y_3 is the distance of the centriod axis of the longitudinal reinforcing bars to the neutral axis of the composite beam; y_4 is the distance of the centriod axis of the steel beam to the neutral axis of the composite beam. When the netural axis of the composite beam is within the web of the steel beam, $y_4 = A_{st}f_{st}/(2t_wf)$; when the netural axis of the composite beam is within the upper flange of the steel beam, y_4 is equal to the distance of the centriod axis of the steel beam to the web upper edge of the steel beam.

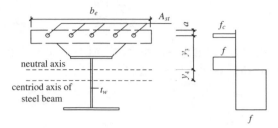

Figure 3. Negative moment capacity of composite beam.

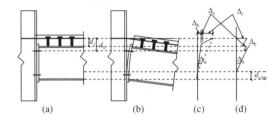

Figure 4. Available rotation capacity model of beam-to-column connection (a) connection before deformation; (b) connection after deformation; (c) deformation of components; (d) representation of deformations for rotation calculation.

2.2 Rotation capacity check of connections

The required rotation capacity of the connection should be checked with

$$\theta_a \leq \theta_u \qquad (3)$$

where θ_a is the required rotation capacity of the connection; and θ_u is the available rotation capacity of the connection.

2.2.1 Available rotation capacity of the connection

Ahmed and Nethercot (1997) proposed a simple model to calculate the available rotation capacity of the composite flush endplate connection. The available rotation capacity model accounts for the plastic deformations of the components, shown as in Fig.4.

The available rotation capacity of the semi-rigid composite connection, θ_u, can be obtained with

$$\theta_u = \frac{\Delta_r}{L_r - d_{c,bw}} + \frac{\Delta_s}{L_r - d_{c,bw} - d_c} + \frac{\Delta_b}{L_r - d_{c,bw} - d_b} \qquad (4)$$

where Δ_r is the elongation of the rebars; Δ_s is the slip of the studs; Δ_b is the elongation of the bolts; L_r is the distance of the reinforcements to the centerline of the bottom flange of the steel beam; $d_{c,bw}$ is the depth of the compression area in the steel beam; d_c is the distance of the steel beam to the centriod axis of the rebars; and d_b is the distance of the bolts to the rebars.

2.2.2 Required rotation capacity of the connection

For earthquake resistance, the required rotation capacity of the connection is proposed as (AISC,1997)

$$\theta_a = 0.03 \text{rad} \qquad (5)$$

3 BEAM DESIGN

3.1 Effective stiffness

In analysis of composite frames, the composite effect of steel beam and concrete slab on the frame behavior should be considered. However, since the ability of concrete to bear tension is ignorable, the composition of concrete slab with steel beam should not be considered at the location of the beam where the moment makes the concrete slab in tension. Because the moment is various at different locations in the beam of a frame, the effective stiffness of the composite beam may also varies with the location where the moment makes the concrete slab in compression or in tension. Despite this apparent complexity, the effective second moment of inertia to predict an acceptable behavior of the beam in a composite frame has been proposed by Ammerman and Leon (1990), which is given by

$$I_b = 0.6I_1 + 0.4I_2 \qquad (6)$$

where I_b is the effective moment of the inertia of the composite beam; I_1 and I_2 are the moments of inertia of the fully composite beam and the steel beam without composition, respectively.

3.2 Check for serviceability limit state

The check of the deflection of the frame beam for the serviceability limit state is specified by GB50017 (2003) as

$$\delta \le L_b / 400 \qquad (7)$$

where δ is the maximum deflection of the beam; and L_b is the span of the beam.

For calculating the deflections for a frame beam, it should consider the rotational restraints at the ends of the beam. The spring stiffness represents the stiffness of the connection itself, plus that of the adjoining structure.

For a frame beam with semi-rigid connections, the deflection of the beam, δ, can be determined with the following three portions (seen in Fig.5): (1) defection of the corresponding simple beam under uniformly distributed load, δ_s, (2) defection of the simple beam under the moment at the end 1, δ_{M1}, (3) defection of

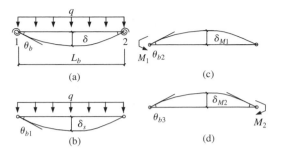

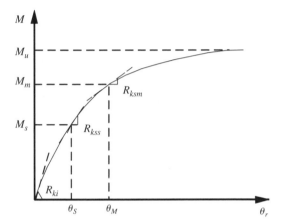

Figure 5. Deflections of the frame beam.

Figure 6. Rotational stiffness of semi-rigid connection.

the simple beam under the moment of at the end 2, δ_{M2}. So, the deflection of the frame beam under the uniformly distributed load can be expressed as

$$\delta = \delta_s + \delta_{M1} + \delta_{M2} \qquad (8)$$

According to the integral method, the deflection of the beam in a semi-rigid composite frame can also be approximately calculatedly by

$$\delta_{sr} = \delta_{rigid} + (1 - \omega)(\delta_{pin} - \delta_{rigid}) \qquad (9)$$

where δ is the deflection of the beam in the semi-rigid composite frame under vertical loads; δ_{rigid} is the deflection of the beam in the corresponding rigid composite frame under vertical loads; δ_{pin} is the deflection of the corresponding simply supported beam under vertical loads; and ω is a no-dimension factor, defined as

$$\omega = \frac{M}{M_{rigid}} = \frac{1}{1 + \frac{M_{rigid}}{\theta_{pin} R_k}} \qquad (10)$$

107

Table 1. Value of factor $\eta_1(n=1)$.

α_0 β	0	0.01	0.02	0.04	0.08	0.1	0.25	0.5	0.75	1	2	10
0	0.748	0.729	0.711	0.678	0.625	0.604	0.522	0.479	0.464	0.456	0.443	0.434
0.1	0.748	0.732	0.717	0.688	0.641	0.621	0.536	0.488	0.470	0.461	0.446	0.434
0.25	0.748	0.735	0.723	0.699	0.658	0.641	0.556	0.501	0.479	0.468	0.450	0.435
0.5	0.748	0.738	0.729	0.711	0.678	0.663	0.583	0.522	0.494	0.479	0.456	0.436
0.75	0.748	0.740	0.732	0.718	0.690	0.678	0.604	0.540	0.508	0.491	0.462	0.437
1	0.748	0.741	0.735	0.723	0.699	0.688	0.621	0.556	0.522	0.501	0.464	0.438
1.25	0.748	0.742	0.737	0.726	0.706	0.696	0.635	0.570	0.534	0.512	0.474	0.440
1.5	0.748	0.743	0.738	0.729	0.711	0.702	0.646	0.583	0.545	0.522	0.479	0.441
2	0.748	0.744	0.740	0.732	0.718	0.711	0.663	0.604	0.566	0.540	0.491	0.443
3	0.748	0.745	0.742	0.737	0.726	0.721	0.684	0.635	0.598	0.570	0.512	0.448
4	0.748	0.745	0.743	0.739	0.731	0.727	0.697	0.655	0.621	0.594	0.531	0.453
5	0.748	0.746	0.744	0.741	0.734	0.730	0.706	0.670	0.639	0.613	0.548	0.458
10	0.748	0.747	0.746	0.744	0.740	0.738	0.725	0.704	0.684	0.666	0.609	0.482

where M is the moment at the end of the beam in the semi-rigid composite frames under vertical loads; M_{rigid} is moment at the end of the beam in the corresponding rigid composite frames under vertical loads; θ_{pin} is the rotation of the corresponding simply supported beam under vertical loads; R_{ks} is the rotational stiffness of the connection(see Fig.6).When the beam-to-column connection is pined, $\omega=0$; when it is rigid, $\omega=1$; and when it is semi-rigid, $0<\omega<1$.

For composite beam with semi-rigid connections under uniformly distributed load, the rotational stiffness of the connection at the serviceability limit state of the beam, R_{ks}, can be expressed as

$$R_{ks} = \eta_1 R_{ki} \tag{11}$$

where η_1 is a factor, obtained with

$$\left[1-\eta_1^{\,n}\right]^{1/n} = \frac{0.5+\dfrac{\alpha_0}{\eta_2}+\beta}{1+2\dfrac{\alpha_0}{\eta_1}+2\beta} \tag{12}$$

where $\eta_2=(1 - -0.667^n)^{(n+1)/n}$, $\alpha_0=EI_b/R_{ki}L_b$, $\beta=EI_b/k_cL_b$.

Generally, the shape parameter, n, in Eq.(12) is equal to 1.5, the stiffness factor η_1 is then obtained in Table 1. or Fig.7.

3.3 Check of ultimate limit state

Traditional design procedures for a steel fame is first assume the member sizes according to previous experience, and analyze the frame to obtain the internal forces in the frame caused by applied loads. According to the calculated internal forces, the predetermined members may be checked to see whether they are

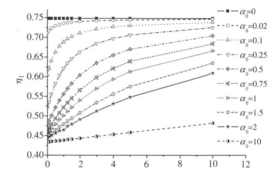

Figure 7. Value of factor $\eta_1(n=1.5)$.

appropriate for the applied load. If they are satisfactory, the design is complete, otherwise, the member sizes have to be adjusted to better match the internal forces and the above procedure must be repeated until the member sizes are satisfactory for the frame to resist the applied loads.

In the above traditional design procedure, the structural analysis of the whole structure of the frame is the most time-consuming part of the work. However, for semi-rigid composite frames under vertical loads, the structural analysis can be bypassed by using the plastic analysis Wang (2005).

At the ultimate limit state, the safety of the beam with semi-rigid connections is satisfied with

$$M_d \leq M_{bp}^+ \tag{13}$$

where M_d is the maximum sagging moment of the composite beam due to the vertical loads; M_{bp}^+ is the sagging plastic moment of the composite beam.

When the neutral axis of the composite beam is within the concrete slab, i.e. $A_s f \leq b_e h_{c1} f_c$ (seen

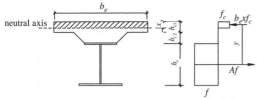

neutral axis

(a) neutral axis of the composite beam is within the concrete slab

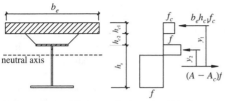

neutral axis

(b) neutral axis of the composite beam is within the steel beam

Figure 8.   Ultimate sagging moment capacity of composite beam.

in Fig.8a.), M_{bp}^+ can be determined by GB50017 (2003)

$$M_{bp}^+ = b_e x f_c y \qquad (14)$$

in which

$$x = A_s f / (b_e f_c) \qquad (15)$$

$$y = h_{c1} + h_{c2} + h_s/2 - x/2 \qquad (16)$$

where A_s is the cross-sectional area of the steel beam; x is the depth of the concrete slab in compression; b_e is the effective width of concrete slab; f_c is the compressive design strength of concrete;
and when the neutral axis the composite beam is within the steel beam, i.e. $A_s f > b_e h_{c1} f_c$ (seen in Fig.8b.),

$$M_{bp}^+ = b_e h_{c1} f_c y_1 + A_c' f y_2 \qquad (17)$$

in which

$$A_c' = 0.5(A - b_e h_{c1} f_c / f) \qquad (18)$$

$$y_1 = h_{c1}/2 + h_{c2} + h_s/2 \qquad (19)$$

$$y_2 = h_s/2 \qquad (20)$$

where A_c' is the cross-sectional area of the part of the steel beam in compression; y_1 is the distance of the centriod of the concrete in compression to the centriod of the steel beam.

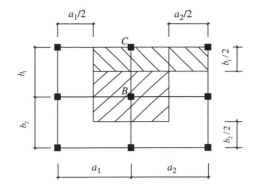

Figure 9.   Load area of column.

4   COLUMN DESIGN

4.1   Analysis of internal forces

4.1.1   Moments
The moments of the upper column and the lower column to a beam-to-column connection may be estimated by summing up the contributions of the moments from the adjacent beams as

$$M_{c1} = \left[\frac{EI_{c1}/L_{c1}}{EI_{c1}/L_{c1} + EI_{c2}/L_{c2}} \right] \times M_{b\,max} \qquad (21a)$$

$$M_{c2} = \left[\frac{EI_{c2}/L_{c2}}{EI_{c1}/L_{c1} + EI_{c2}/L_{c2}} \right] \times M_{b\,max} \qquad (21b)$$

where M_{c1}, M_{c2} are the moments of the upper column and the lower column at a beam-to-column connection, respectively; EI_{c1}, EI_{c2} are the stiffness of the upper column and the lower column, respectively; L_{c1}, L_{c2} are the length of the upper column and the lower column, respectively; M_{bmax} is the unbalanced moment of the composite beams joining the connection under ultimate state.

4.1.2   Axis force
The axis force of each column in a frame may be estimated by the load on the floor area assigned for the column to support approximately. In Fig.9, the load area of column A, B and C are $a_2 b_1/4, (a_1 + a_2)(b_1 + b_2)/4$, $(a_1 + a_2)(b_1 + b_2)/4$, respectively.

4.2   Check of columns

According to GB50017 (2003), linear interaction formulas have been adopted for the capacity checking of frame columns against failures in section yielding, in-plane instability and out-of-plane bucking.

The section strength check of the column is expressed as

$$\frac{N}{A} + \frac{M_x}{\gamma_x W_{nx}} \le f \tag{22}$$

where γ_x is the plasticity factor of the section; N, M_x are the design values of the axial compression force and the maximum moments of the column, respectively; A is cross-sectional area of the column; W_{nx} is net section module of the column.

The check of in-plane stability is of the column can be made with

$$\frac{N}{\varphi_x A} + \frac{\beta_{mx} M_x}{\gamma_x W_{1x}\left(1 - 0.8 N / N'_{Ex}\right)} \le f \tag{23}$$

where β_{mx} is the equivalent moment factor; W_{1x} is the section module of the column; φ_x is the in-plane strength reduction factor of the column due to axial bucking; and λ_x is the slenderness ratio of the column; $N'_{Ex} = \pi^2 EA/(1.1\lambda_x^2)$.

The checking against lateral-torsional bucking of the column is performed by

$$\frac{N}{\varphi_y A} + \eta \frac{\beta_{tx} M_x}{\varphi_b W_{1x}} \le f \tag{24}$$

where φ_x, φ_y is the out-of-plane strength reduction factor of the column due to axial buckling; β_{tx} is the equivalent moment factor; φ_b is the strength reduction factor of the column due to bend bucking η is the section effective factor.

5 EXAMPLE

An example is presented to demonstrate the application of the design procedure presented hereinabove for semi-rigid composite frames under vertical loads. The geometry of the frame for example is shown in Fig. 10. The dead load on the beam, g_k, is 35.16 kN/m, and the live load, q_k, is 14.84 kN/m. The design load of: for checking serviceability limit state is, $g_k + q_k = 50$ kN/m; and for checking ultimate limit state is, $1.4 g_k + 1.6 q_k = 72.97$ kN/m.

The section of HW250 × 250 × 9 × 14 and HN300 × 150 × 6.5 × 9 are selected for the columns and beams of the frame, respectively. The steel beams are connected to the column flanges by means of flush end plates of 14 mm in thickness and two rows of M22 Grade 10.9 bolts. The slab reinforcement ratio of each specimen is 0.9%. The material of steel for the frame is Q235, with a modulus of elasticity of 206,000 MPa and yield stress of 235 Mpa. The material of concrete is C30, with a modulus of elasticity of 30,000 MPa and $f_c = 14.3$ Mpa. The material of rebar is HRB335, with a modulus of elasticity of 200,000 MPa and $f = 300$ MPa.

For comparison, the design results for the steel frame with semi-rigid connections are compared to those with pinned and rigid connections, which are given in Table2. The cost for the frames designed is obtained from major Chinese steel fabricators. The result shows that semi-rigid composite design is the most cost-effective for this kind of low-rise steel frames.

6 CONCLUSION

The benefits of semi-rigid connections are extensively documented in the design of steel frames. Semi-rigid composite structures result in better efficiency and economy. However, owing to lack of appropriate practical design methods, they are not widely used much in practice. A Practical design method for semi-rigid composite frames under vertical loads has been proposed on the basis of previous research achievements. The applicability of the proposed method is demonstrated by a design example and cost estimations for

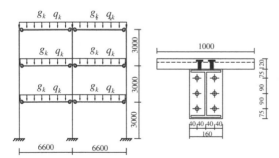

Figure 10. Example frame (unit = mm).

Table 2. Comparison of steel and composite frames with various connections.

Structural type	Beam	Column	Total Cost estimation (RMB)	Cost comparison
Simple steel frame	HN450 × 200 × 9 × 14	HW200 × 200 × 8 × 12	37,304,000	150%
Rigid steel frame	HN400 × 150 × 8 × 13	HW250 × 250 × 9 × 14	36,440,000	146%
Simi-rigid steel frame	HN400 × 150 × 8 × 13	HW200 × 200 × 8 × 12	30,768,000	124%
Simple composite frame	HN350 × 175 × 7 × 11	HW200 × 200 × 8 × 12	28,912,000	116%
Rigid composite frame	HN300 × 150 × 6.5 × 9	HW250 × 250 × 9 × 14	30,576,000	123%
Simi-rigid composite frame	HN300 × 150 × 6.5 × 9	HW200 × 200 × 8 × 12	24,896,000	100%

the design example have been also carried out. The semi-rigid composite frame designed has proved as the most cost effective solution in comparison to the design results of the frames using traditional types of connections.

The proposed method provides the design formulas for check of connections, beams and columns in semi-rigid composite frames at either serviceability limit state or ultimate state under vertical loads. With the development of the design method for resisting horizontal loads, the design and application of semi-rigid composite frames will be eventually introduced in everyday practice.

ACKNOWLEDGEMENTS

The authors would like to acknowledge the assistance of Dr Qing-ping Liu of Tongji University to help to fulfill the work presented in this paper. The support from the National Science Foundation of China through the Outstanding Yong Scholars Project (No.50225825) awarded to the first author is gratefully acknowledged.

REFERENCES

[1] Leon R T, Ammerman D J, Lin J, Mccauley R D. (1987). Semi-rigid composite steel frames. *Engineering J.* 24:4, 147–156.

[2] Nethercot D A. (1995). Semirigid joint action and the design of nonsway composite frame. *Engineering Structures* 17:8, 554–567.

[3] Chen W F, Lui E M. (1991). *Stability Design of Steel Frames*, CRC Press, Boca Raton.

[4] Fang L X, Chan S L, Wong Y L. (1999). Strength analysis of semi-rigid steel-concrete composite frames. *Journal of Constructional Steel Research* 52, 269–291.

[5] Wang J Y. (1999). *Nonlinear analysis of semi-rigid composite joints under lateral loading experimental and theoretical study*, Ph.D thesis, the Hong Kong Polytechnic University.

[6] Cabrero J M, Bayo E. (2005). Development of practical design methods for steel structures with semi-rigid connections. *Engineering Structures* 27, 1125–1137.

[7] Kish N, Chen W F. (1990). Moment-rotation relations of semi-rigid connections with angles. *J.Struct. Engrg.*, ASCE, 16:7, 1813–1834.

[8] Xiao Y, Choo B S, Nethercot D A. (1996). Composite connections in steel and concrete part 2- moment capacity of end plate beam-to-column connections. *J.Construct. Steel. Res.* 37:1, 63–90.

[9] Ahmed B, Nethercot D A. (1997). Design of flush endplate connections in composite beams. *Struct Engr* 75:14, 233–244.

[10] Ahmed B, Li T Q, Nethercot D A. (1997). Design of composite finplate and angle cleated connections, *J.Construct.Steel Res.* 41:1, 1–29.

[11] [GB50017.(2003). *Code for Design of Steel Structures*, China Plan Press, China.

[12] Ahmed B, Nethercot D A. (1997). Prediction of initial stiffness and available rotation capacity of major-axis flush end-plate connections. *J. Construct. Steel. Res.* 41:1, 31–60.

[13] Nethercot D A, Li T Q, Choo B S. (1995). Required rotations and moment redistribution for ccmposite frames and continuous beams. *J. Construc. Steel. Res.* 35:2, 121–64.

[14] Kemp A R, Nethercot D A. (2001). Required and available rotations in continuous composite beams with semi-rigid connections. *J. Construct. Steel. Res.* 57,375–400.

[15] Ammerman D J, Leon R. (1990). Unbranced frames with semirigid composite connections. *Eng.J.*, AISC, 27:1, 1–10.

[16] Wang J F. (2005). *The practical design approach of semi-rigidly connected composite frames under vertical loads*, Ph.D thesis, Tongji University.

[17] CEN, Eurocode 3.(2005). *Design of steel structures -Part 1-8: Design of joints*. ENV 1993–1–8.Brussels: CEN.

[18] CEN, Eurocode 4. (2004) *Design of Composite Steel and Concrete Structures - Part 1–1: General rules and rules for buildings*. ENV 1994–1–1.Brussels: CEN.

[19] AISC(2005) *Seismic Provisions for Structural Steel Buildings*.

Steel and Composite Structures – Wang & Choi (eds)
© *2007 Taylor & Francis Group, London, ISBN 978-0-415-45141-3*

Eurocode 3: Design of steel structures "Ready for practice"

F.S.K. Bijlaard

Department of Civil Engineering & Geosciences, Delft University of Technology, Delft, The Netherlands

ABSTRACT: The European Standards Organisation (CEN) has taken over the initiative from the European Commission to develop a set of harmonized European standards for Structural Design, the Eurocodes. Together with standards for fabrication and erection and harmonized product standards these standards form the bases for the design of structures and structural products. After a period of experimental use of the ENV (European Pre Standard)-versions of the Eurocodes, these are now official EN's (European Standards).

Eurocode 3 covers design of steel structures. This paper gives an overview of the context in which Eurocode 3 exists and of the accompanying codes. The various individual parts of Eurocode 3 are mentioned. Attention will be paid on the statistical evaluation of test results that form the basis for the determination of the recommended values for the safety elements like the partial safety factors for strength and stability. The National Annexes are discussed and finally an opinion is presented on the possibilities and difficulties the practitioners will experience using the Eurocode 3 and on the future developments that are needed.

1 HISTORY AND CONTEXT OF EUROCODE 3

In the early eighties of previous century the European Commission took the initiative to ask several experts to develop a complete series of structural design codes, the Eurocodes. The first draft of Eurocode 3 was prepared in 1983/1984, and published by the CEC in 1985. Extensive and detailed comments on the 1985 draft were received from the twelve member states of the EEC in 1987. At that time the responsibility for further development of the first generation Eurocodes was mandated to CEN were than handed over to the European Standards Organisation CEN to further develop these documents towards formal CEN Standards. To achieve this goal CEN has set up a Technical Committee TC 250: "Structural Eurocodes", which within CEN is solely responsible for all structural design codes. This TC has nine Subcommittees (SC), each responsible for one volume, see figure 1.

The Eurocode programme is aiming at two-dimensional harmonization:

(1) Harmonization across the borders of the European Countries;
(2) Harmonization between different construction materials, construction methods and types of building and civil engineering works to achieve full consistency and compatibility of the various

codes with each other and to obtain comparable safety levels. This is covered in Eurocode 0 Basis of Design.

Basis of Design (EN1990) covers general design philosophy, basis of design, structural reliability, common non material-related aspects and common terminology and symbols.

Eurocode 1 covers general actions (effects of loadings) applicable to all structures in Part 1 and additional actions for specific structures in further parts. Eurocode 1 has the following Parts:

Densities, self-weight, imposed loads (EN1991-1-1), Actions exposed to fire (EN1991-1-2), Snow loads (EN1991-1-3), Wind actions (EN1991-1-4), Thermal actions (EN1991-1-5), Actions during execution (EN1991-1-6), Accidental actions due to impact or explosion (EN1991-1-7), Traffic loads on bridges (EN1991-2), Actions on silos and tanks (EN1991-3) and Actions induced by cranes and machinery (EN1991-4)

The Eurocodes – being design standards – are to be used in combination with standards for fabrication and erection (in Euro-lingo called: execution) and product standards. The Eurocodes on Geotechnical Design and on Earthquake are also of importance the material

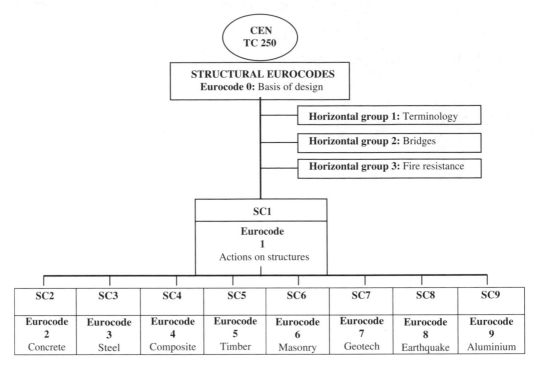

Figure 1. Organization of CEN/TC250.

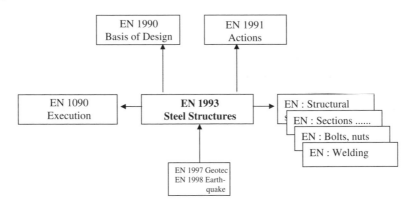

Figure 2. Relation of EN1993: Eurocode 3 with other European standards.

related structural Eurocodes. For Eurocode 3 this relation is illustrated in figure 2. These related standards will be discussed briefly.

Execution:
The design procedures in the Eurocodes are only valid if adequate workmanship criteria during fabrication and erection are satisfied. For example, the levels of initial geometric imperfections assumed in many of the strength rules in Eurocode 3 are directly related to these criteria and are therefore invalid if they are exceeded.

A separate CEN committee, TC135 "Execution of Steel Structures" has drafted the fabrication and erection rules in close contact with CEN TC250/SC3.

These rules for fabrication and erection are given in ENV1090, to be converted in EN1090.

The main reasons for developing a European Standard for execution of steel structures are:

– To transfer the requirements set during design from the designer to the constructor, i.e. to be a link between design and execution.

– To give instructions to the constructor on how to execute the physical work (fabrication, welding, bolting, erection, protective treatment) as well as to give requirements for accuracy of the work. The standard will thus serve as a document, which gives standardized technical requirements when ordering a steel structure.

– To inform and serve as a checklist for the designer with respect to information which needs to be specified in the project specification for the particular project. It is foreseen and required that each project shall have a project specification, which defines the technical requirements for that project. Such a project specification could be a single drawing for a minor project or a comprehensive package of documents for a complicated structure.

Products:
Steel structures are normally fabricated from standardized industrial products. In the development of the design rules assumptions had to be made on the values and statistical variations of the geometrical and physical properties of materials and products. In the Eurocodes these assumptions are based on European product standards (EN's) and therefore Eurocodes refer to EN product standards. These standards are mainly equal with or derived from existing Euronorms or ISO-standards. The referenced product standards for example concern the following categories of products:

Structural steel, Sections and plates, Bolts, nuts and washers, Welding consumables, Rivets and Corrosion protection. They include classification, product range, dimensions, and tolerances, physical and chemical properties.

Basis of Design and Actions on structures. These codes have already been discussed.

Geotechnical Design and Earthquake:
Eurocode 7 deals in Part 1 with geotechnical design for structures and in Parts 2 and 3 with geotechnical field tests and laboratory tests to assist design. This code is especially important for soil-structure interaction.

Eurocode 8 covers design of structures for earthquake resistance and is organised on a similar basis as the material-related Eurocodes.

2 STRUCTURE OF EUROCODE 3 STEEL STRUCTURES

The set-up of EN 1993 Steel Structures [1], in terms of Parts and Sections, resulted from policy decided by SC3 (with respect of the Parts) and TC250 (with respect of the Model Sections within each Part) and is as follows. The parts EN 1993-1-1 to EN 1993-1-12 are the so-called General Documents and contain rules that are applicable to more than one type of steel

structure. The parts EN 1993-2 to EN 1993-6 are the so-called application parts. In an application part reference is made to those clauses of the general parts that are to be taken into account for that application. Part EN 1993-1-1 is an exception to this system because that part contains besides the general rules also the specific rules for buildings.

EN 1993: Eurocode 3 - Design of Steel Structures consists of the following general parts:

EN 1993- Part 1.1: General rules and rules for buildings

Part 1.1 gives rules for the design of Steel Structures in general together with specific provisions for Buildings. Provisions in Part 1.1 specific to buildings have been placed at the end of clauses and are indicated with a capital B behind the clause number. The intention being to make clear what is specific to buildings.

This part pays attention to the global analysis of the structure and the imperfections to be taken into account. Several Ultimate Limit States are considered like Resistances of Cross-Sections; Buckling Resistance of Members; Uniform Built-up Compression Members.

EN 1993- Part 1.2: Structural fire design

This part is concerned with the determination of resistance of steel structures to fire. It provides information on how to determine the fire resistance based on the classical method with the standard fire curve but also on the much more advanced method "the Natural Fire Concept".

EN 1993- Part 1.3: Supplementary rules for cold formed members and sheeting

This part is concerned with the strength and stability design of cold formed members and sheeting and is especially important for building products. Special attention is paid on the testing procedures of structural elements as well as op complete assemblies.

EN 1993- Part 1.4: Supplementary rules for stainless steels

This part is concerned with stainless steel that has become popular for use in buildings, chimneys and tanks. A very important item is the selection of the material dependent on the circumstances of use.

EN 1993- Part 1.5: Plated structural elements (in-plane loaded)

This part is in particular important for bridge structures and other structures where plate stability plays an important role.

EN 1993- Part 1.6: Strength and stability of Shells
The design of structures like chimneys, tanks and pipelines is based on the theory of shells. In this

part attention is paid effects of the boundary conditions, the influence of the geometrical imperfections and the design concepts for the limit state design.

EN 1993- Part 1.7: Strength and Stability of Planar Plated Structures subject to out-of-plane loading

This part is in particular important for silo and container structures and other structures where there is interaction between plate stability plays and plates in bending.

EN 1993- Part 1.8: Design of Joints

This part contains information on the strength capacity of individual fasteners and groups of fasteners. It also provides information about welded connections. Modelling of beam-to-column joints and the influence of joints on the global analysis is treated. Detailed information is provided on the design and verification of structural joints connecting H or I sections based on the component method. Detailed information is also given about hollow section joints.

EN 1993- Part 1.9: Fatigue

This part is about the fatigue verification.

EN 1993- Part 1.10: Material toughness and through-thickness properties

This part provides information about maximum permitted thickness value. Furthermore it provides an evaluation using fracture mechanics. It gives a selection procedure for the selection of materials for through-thickness properties.

EN 1993- Part 1.11: Design of structures with tension elements

Aspects like design situations and partial factors, strength of steels and wires, length and fabrication tolerances, corrosion protection of each individual wire and corrosion protection of the rope/strand/cable interior are treated. Attention is paid to transient design situations during the construction phase, persistent design situation during service and non-linear effects from deformations. Also the various Ultimate and Serviceability Limit States are dealt with like strength vibrations and fatigue.

EN 1993- Part 1.12: Additional rules for the extension of EN1993 up to steel grades S700

This part contains additions to EN 1993-1-1 to EN 1993-1-11, additions to application parts EN 1993-2 to EN 1993-6 and additions to EN 1090 from which it can be concluded if the rules can be used for steel grades up to S700.

EN 1993: Eurocode 3 - Design of Steel Structures consists of the following application parts:
Steel Bridges (EN 1993- Part 2), Towers and Masts (EN 1993- Part 3.1), Chimneys (EN 1993- Part 3.2), Silos (EN 1993- Part 4.1), Tanks (EN 1993- Part 4.2), Pipelines (EN 1993- Part 4.3), Piling (EN 1993- Part 5) and Crane supporting structures (EN 1993- Part 6).

All these application parts contain Sections with clauses giving specific information to be taken into account for the design of that application. These clauses either refer to clauses in the various general parts and / or provides additional information specific for that application.

3   SAFETY LEVEL

In applying the rules in Eurocode 3 a structural safety is reached of not less than the reliability index β equal 3,8. Because the member states are entitled to choose their own safety level for structures, the rules are set up such that they contain safety elements of which the value can be chosen by the individual member state. These safety elements are for instance the partial (safety) factors for the resistance (limit states) of structural elements. For these safety elements in the Eurocodes so-called recommended values are given in notes accompanying the clauses containing these safety elements. To promote harmonization of design rules throughout Europe the Commission strongly advises to choose the recommended values for these safety elements.

The procedure to determine the partial safety factors for the resistance of structural elements can be explained in gross terms by the following steps:

1. The rules in the Euro-codes are based on limit state design format

   - Effects of Actions $E_d \leq R_d$ Design Resistance
   - Effect of Actions
   - $E_d = E_k \cdot \gamma_E$
   - where E_k is the Characteristic Value for the Effects of Actions
   - and γ_E is the Load Factor
   - Design Resistance
   - $R_d = R_k / \gamma_M$
   - where R_k is the Characteristic Value for the Resistance
   - and γ_M is the partial safety factor or Model Factor

2. Splitting the Action side from the Resistance side leads to a simplified procedure to determine the Partial Safety Factors γ_E and γ_M by using the sensitivity factor $\alpha_E = 0.7$ for the effects of actions side and $\alpha_R = 0.8$ for the resistance side

3. Start with "ideal" assumptions:

- (a) The strength function is a product function of independent variables (e.g. bolt in tension: $F_t = A_s \cdot f_u$ where: A_s is the stress area of the threaded part and f_u is the tensile strength of the bolt material);
- (b) A large number of test results is available;
- (c) All actual geometrical and material properties are measured;
- (d) All variables have a log-normal distribution;
- (e) The design function is expressed in the mean values of the variables;
- (f) There is no correlation between the variables of the strength function.

4. The *Standard Procedure* to determine γ_M:

- *Step 1*: Develop a "Strength Function" for the strength capacity of the structural member or detail considered,

$$r = g_R(X_i)$$

The strength function includes all relevant basic variables X_i which control the resistance in the limit state.

All variables should be measured for each test specimen i (assumption C).

- *Step 2*: Compare experimental and theoretical values.

The experimental values r_{ei} are known from the tests.

Using the relevant strength function and putting the actual properties into the formula, leads to the theoretical values r_{ti}.

- *Step 3*: Check whether the correlation between the experimental and theoretical values is sufficient.

If the strength function is exact and complete, all points (r_{ti}, r_{ei}) lie on the bisector of the angle between the axes of the diagram and the correlation coefficient $\rho = 1.0$

The correlation coefficient r needs to be determined. If the value $\rho \geq 0.9$, than the correlation is considered to be sufficient. In general the points (r_{ti}, r_{ei}) will scatter (See figure 3).

- *Step 4*: Determine the mean value correction via the correction terms $b_i = r_{ei}/r_{ti}$
The mean value correction

$$b_{mean} = \frac{1}{n}\sum_{i=1}^{n} b_i$$

The corrected strength function is:
$r_m(X_m) = b_{mean} \cdot r_t(X_m) = b_{mean} \cdot g_R(X_m)$ (See figure 4)

- *Step 5*: Determine the coefficient of variation V_δ of the error terms δ_i

The error terms δ_i of each experimental value r_{ei} with respect to each theoretical, mean value corrected, resulting in $b_{mean} \cdot r_{ti}$, is determined as follows:

$$\delta_i = \frac{r_{ei}}{b_{mean} \cdot r_{ti}}$$

In most cases the coefficient of variation V_δ is relatively small

$$\delta_{mean} = \frac{1}{n}\sum_{i=1}^{n}\delta_i$$

The standard deviation of the error terms δ_i is σ_δ and the coefficient of variation is $V_\delta = \sigma_\delta/\delta_{mean}$

- *Step 6*: Determine the coefficient of variation V_{X_i} of the basic variables X_i in the strength function

The coefficient of variation of all basic variables may only be determined from the test-data assumed that the test population is fully representative for the variation in the actual situation. This is normally not the case, so the coefficients of variation have to be determined from pre-knowledge.

- *Step 7*: Determine the characteristic value, the 5%-fractile of the strength.

For the log-normal distribution (assumption D) the characteristic strength follows from:

$$r_k = r_m(X_m) \cdot e^{-k_s \cdot \sigma_{\ln r} - 0.5\sigma_{\ln r}^2}$$

where $\sigma_{\ln r} = \sqrt{\ln(V_r^2 + 1)} \approx V_r$
k_s is the fractile coefficient for the 5% fractile
See figure 5.

- *Step 8*: Determine the design value of the strength r_d and the partial safety factor (Model Factor) γ_M

The design value for the strength is related to the chosen reliability index $\beta = 3.8$
k_s is replaced by $k_d = \alpha_R \cdot \beta$ and
$r_d = r_m(X_m) \cdot e^{-k_d \cdot \sigma_{\ln r} - 0.5\sigma_{\ln r}^2}$
with $k_d = \alpha_R \cdot \beta = 0.8 \cdot 3.8 = 3.04$

See figure 6.

The design value of the strength function is obtained by dividing the characteristic strength function by the partial safety factor or so-called model factor containing all uncertainties in the strength function.

$$r_d = \frac{r_k}{\gamma_M}$$

The characteristic strength and the design strength is known so that the partial safety or model factor can be calculated.

$$\gamma_M = \frac{r_k}{r_d} = \frac{r_m(X_m) \cdot e^{-k_s \cdot \sigma_{\ln r} - 0.5\sigma_{\ln r}^2}}{r_m(X_m) \cdot e^{-k_d \cdot \sigma_{\ln r} - 0.5\sigma_{\ln r}^2}}$$

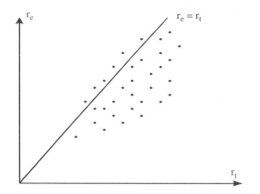

Figure 3. Scatter of the points (r_{ti}, r_{ei}) in the theory – experiment diagram.

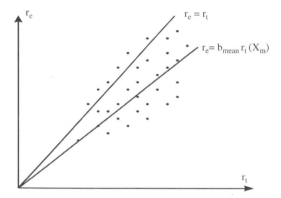

Figure 4. Mean value correction in the theory – experiment diagram.

This procedure, to determine the partial safety factors for the resistance of structural elements by a statistical evaluation of strength functions against experimental test results, is developed by Bijlaard in 1987 in close cooperation with Prof. Dr.-Ing. G. Sedlacek, RWTH-Aachen, Germany and Prof. Ir. J.W.B. Stark, Delft University of Technology, The Netherlands [2]. The procedure contains adjustments for deviations from the "ideal" assumptions such as "a variable is defined as having another statistical distribution that a log-normal one, a limited number of test results is available, not all actual properties are measured and that the strength function is non linear with respect to the variables and may contain additions of the variables. The procedure is further developed to take account of the so-called "tail"-effect of the statistical distributions by the research team of Prof. Sedlacek and forms now Annex D of EN 1990 "Basis of Design".

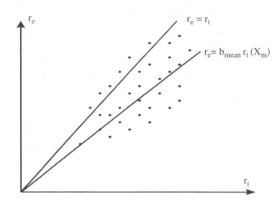

Figure 5. Characteristic values r_k in the theory – experiment diagram.

4 NATIONAL ANNEXES

Guidance Paper L [3] is the lead document for the content and format of the Structural Eurocodes and the National Annexes. In paragraph 2.3 of Guidance Paper L guidance is given about the purpose and format of National Annexes. From these statments in Guidance Paper L about the National Annexes the following can be concluded [4], [5]: The National Standards Bodies should normally publish a National Annex, on behalf of and with the agreement of the national competent authorities.

The preparation of National Annexes is likely to lead to different results per member state for the choice of Nationally Determined Parameters (NDP's) regarding:

– choice of values of symbols or classes where options are given in the Eurocode,
– determination of country specific data (geographical, climatic, etc),
– choice of procedures where alternatives are given, and also for:
– decisions on the use of informative Annexes,
– references to "non-contradictory complementary information" (NCCI).

The objective of CEN-TC250 is to provide a mechanism by which a convergence, where relevant, of the NDP's can be achieved, so that further harmonization can be realized.

The NDP's selected by the different Member States will need to be collected for each Eurocode in order to identify the variety of these NDP's and any differences of these NDP's from the recommended values in the Eurocodes, especially for those NDP's which are not related due to safety, geographical or climatic reasons. Additionally, gaps in information in a Eurocode part may be identified by, for example,

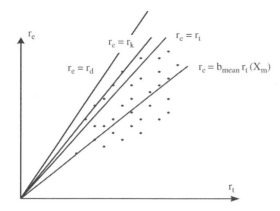

Figure 6. Design values r_d in the theory – experiment diagram.

gathering non-contradictory complementary information (NCCI). Note: Only references to NCCI are allowed to be given in National Annexes, not the actual guidance.

For this purpose the Joint Research Centre (JRC) at Ispra, Italy, has built a database. After this database is filled with the data from the National Annexes of the Member States an analysis will be made on which a further development and harmonization program for the Eurocodes can be based.

Based on the forgoing statments a National Annex may contain the following aspects:

(1) A National Annex provides values for the NDP's where in the text of the Eurocode via a Note explicitly an opening is given for the National Annex to do so. These NDP's can be values for the safety elements γ_M, a choice for a method or providing further information. It is the right of the Member States to make these choices, but the European Commission invites the Member States to make the choice for the recommended values when provided in the Eurocode. If a Member State deviates from the recommended values, that Member State will be invited to explain why that deviation is made.

(2) A National Annex has to determine the status of the Informative Annexes to the Eurocode parts. The possibilities are:

  – An Informative Annex may be completely rejected and set aside;
  – An Informative Annex may be left as an Informative Annex;
  – An Informative Annex may be left as it is and than be made Normative;
  – An Informative Annex may (partly) be modified and than be made Normative;

(3) A National Annex may refer to Non-Contradictory Complementairy Information (NCCI). That NCCI can be of the following nature:

  – NCCI filling a gap in the Eurocode. Such NCCI may be made Normative;
  – NCCI filling a gap in the Eurocode but made informative;
  – NCCI providing information about the use of the rules in the Eurocode. Such NCCI has to be informative.

5 INTRODUCTION OF EUROCODE 3 IN THE DESIGN PRACTICE

To introduce the Eurocodes in practice a lot is to be done to make the designers familiar with the Eurocode rules. Many initiatives are under way to organise courses to provide worked examples, to adjust education material for technical schools to the Eurocodes and to draft back ground reports to the rules in Eurocodes. The advantage of Eurocode 3 for the designer is that this code provides extensive information about how to calculate the structural behaviour of components like columns, beams and joints. However, many times it is said that the Eurocode is too complex for use in day-to-day practice. In the opinion of the author this is not the case but it is admitted that working with the Eurocode is a lot of work. And it is true that the designer has not much time to do his job in a commercial and competitive surrounding. Therefore it is necessary that user-friendly software is available to the designer to take the time consuming rules of the code to determine the joint behaviour out of his hands. In that situation the designer can spend his time to his profession being a designer looking for alternative structural solutions to reach a final design that reaches minimum integral costs (design + material + fabrication + erection + end-of-life + re-use) and leave the number crunching to the computer using adequate software. In that respect a warning should be made in using so-called expert-systems from the market. The designer should be very alert on the correctness of the software itself and on the correct use of that software. The term "expert-system" just means that only experts should use this software. In that case we can stop saying "Simple rules sell steel" and replace that by saying "Simple TOOLS sell Steel".

6 CONCLUSIONS

  – The process of harmonization of design standards of the member countries of CEN will take a period of about three decades. Compared to the "life time" of an existing code in a country of about 15 years, for the Eurocodes this period is not so bad.

119

- Eurocode 3 "Design of Steel Structures" comprises a fairly complete set of design codes for uniquely designed structures and for a wide range of structural steel products.
- The introduction of the Eurocodes in the design practice needs great care. Design examples, guide lines, design tools (special software) should be developed in the various countries. Explanations of differences and the justification for these changes should be supplied to support the acceptance of the Eurocodes.
- To support these local activities in the various member states, background documents need to be drafted on which local design tools and examples need be based.

REFERENCES

[1] EN 1993: 2004 "Eurocode 3 : Design of Steel Structures, Parts 1 to 6, CEN Central Secretariat, Rue de Stassart 36, B-1050 Brussels, BELGIUM

[2] Bijlaard, F.S.K., Sedlacek, G. and Stark, J.W.B. (1987), "Procedure for the determination of design resistance from tests – Background report to Eurocode 3", TNO Building and Construction Research, Delft, The NETHERLANDS

[3] GUIDANCE PAPER L (concerning the Construction Products Directive – 89/106/EEC) "APPLICATION AND USE OF EUROCODES" (issued following consultation of the Standing Committee on Construction at the 53rd meeting on 19 December 2001 and written procedure ended on 25 January 2002, as document CONSTRUCT 01/483 Rev.1), Rue de la Loi 200, B-1049 Bruxelles – Belgique Bureau/office SC15. Téléphone/switchboard: (32) 2 299 11 11. Télécopieur/fax (32) 2 296 10 65. Document publicly available at http://europa.eu.int/comm/enterprise/construction/index.htm

[4] CEN/TC250 N518 "Nationally Detemined Parameters and National Annexes for ENEurocodes"

[5] CEN/TC250 N544A National Annexes - Paper for discussion

Steel and Composite Structures – Wang & Choi (eds)
© *2007 Taylor & Francis Group, London, ISBN 978-0-415-45141-3*

Adopting the Eurocodes in the UK: Lessons for the steel structure community worldwide

D.A. Nethercot

Department of Civil Engineering, Imperial College London, London, UK

ABSTRACT: After 35 years in preparation the Structural Eurocodes are now becoming a reality. The suite of documents is set to replace National Codes for construction work in all member states. Views are expressed on exactly what this means for the Structural Engineering profession, covering: likely changes in working practices, replacement of the current portfolio of supporting Code-related design material, timescale and cost. Although based on the UK situation and experience, it is believed that much of the material is directly relevant to those contemplating major Code changes, wherever the location.

1 INTRODUCTION

As a result of its signing up to the Treaty of Rome in the 1970's, the UK Government committed the country to the eventual adoption of the package of Structural Eurocodes as a replacement for its National Standards for Construction works. This followed directly from the Commission's desire to remove "all artificial barriers to trade"; National Standards for Structural Design were seen as one such barrier. Some 30 years later – far longer than was anticipated – these documents are finally being released, with the enthusiasm with which they are being embraced as replacements for National Standards varying widely between the different European countries. Put simply, for those nations without a strong portfolio of up-to-date, comprehensive National Codes, they are seen as a clear improvement; for those nations with well developed National Standards plus all the associated supporting design infrastructure i.e. Design Guides, Manufacturers' Design Literature, Textbooks, Computer Software etc, the situation is somewhat more equivocal. The UK is firmly in the second camp.

However, no matter how unpopular the Structural Eurocodes might be in some quarters – and the pages of Verulum in the Structural Engineer provide ample evidence of the strength of feeling – the UK had already agreed that they will eventually replace the well known British Standards documents. Thus we must embrace them, manage the transition from a design environment based on BS 5950, BS 8110, BS 5400 etc to one based on EC3, EC2, EC4 etc, and ensure that we take advantage of the opportunities and both recognise and restrict the risks. As the IStructE Report (1)

prepared for the Office of the Deputy Prime Minister to advise on what should be done to support this transition stated:

- "The introduction of the Structural Eurocodes represents the biggest challenge ever faced by the Structural Engineering Community".

Remember that this transition covers all Codes in all materials, including loading, and as a consequence all the supporting design infrastructure – an investment of many hundreds or possibly many thousands of man-years. Perhaps we should take comfort from this quote:

- "The onset of new or revised Regulations invariably heralds a trying period of the unfortunate people who have to work such regulations. This applies both to those who have to comply with, and those who have to administer, such regulation."

A recent quote from Verulum? No, it relates to the introduction of the 1938 Edition of BS449, a document regarded with much affection and seen as a model for what we should be doing now by some rose tinted spectacle wearers.

It is, of course, impossible to do justice to an operation that has occupied several decades and involved the efforts of hundreds of experts from many different countries in a single conference presentation. This paper is thus, of necessity, selective in its coverage and personal in its treatment. Its aim is to convey something of the current situation in the UK (and by implication in many other European countries and, indeed, in any country facing a programme of major change to its Structural Codes – such as New Zealand, for which many of the lessons learnt are regarded as particularly appropriate due to the impending revision of the Steel Structures Standard, NZS 3404), especially the key issues that must be faced and the measures being adopted to cope. Naturally, it focuses on the position for Steel and Composite Construction.

2 THE CURRENT POSITION

Figure 1 shows the timetable required by CEN (the European body with overall responsibility for producing the Structural Eurocodes, although actual publication will be by the National Standards bodies) for managing the transition. Given that the 10 Eurocodes comprise some 56 separate parts (with EC3 consuming 17 of these), the process will be staggered over several years. But there is some desire in certain quarters to see all the withdrawals completed by 2010. Indeed, sections of the Concrete Community are pressing for the basic parts of BS8110 to be replaced by the equivalent parts of EC2 by 2008. For most of EC3 and EC4 the situation is that of being just above the line i.e. the technical content of every part is finalised, work is in progress on the National Annexes and publication is "imminent". Only when the Code is published with its associated National Annex (that contains some "Nationally determined parameters" and, possibly, some special provisions related to sections where alternative procedures are given) is it available for use as an alternative to the National Standards within the legal Building Regulations framework. Before this the technical provisions may only be considered and referred to in the manner of any other potentially helpful and relevant technical source. Readers interested in the process, especially the precise role of Eurocodes in the UK technical approvals process, are referred to either the ODPM Report (1) or to the summary publication in the August 2004 issue of The Structural Engineer (2). The position in other European countries will, of course, be rather different.

3 PARTICULAR FEATURES

Preparation of the Structural Eurocodes has been an enormous undertaking, involving large numbers of people, each from different backgrounds. Therefore certain rules had to be established, either upfront or in a more evolutionary way during the actual writing process:

3.1 *Principles and Application Rules*

In the original thinking "Principles" were defined as those rules that must be observed, whilst "Application Rules" provided one way (but not necessarily the only way) of satisfying the relevant principle. However, certainly for the Steel and Composite Eurocodes, this

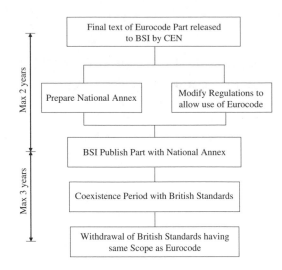

Figure 1. Implementation timetable.

concept has largely lapsed during the actual writing process, with the result that the documents do not really adhere to it and that very few actual "Principles" remain. One interesting aside which results from this concept is the idea of "equivalent application rules" i.e. precisely how is an alternative to the stated application rule defined? Academically satisfying but (probably) practically unworkable phrases such as "An equivalent application rule must be shown to lead to the same level of safety as the stated application rule" have been promulgated at various times but the whole concept is, at least for the present, largely dormant.

3.2 *No duplication of Material*

Although this appears efficient in theory – since it means that documents will not use up space repeating rules available elsewhere – it actually makes usage more difficult. Users of Codes normally complain at having to consult more than a single document or to refer to separate sections within the same document in order to locate all the material required for a particular task – although they are prepared to accept a limited amount of this e.g. the separation of information on loading and that on response. However, the concept has been enforced so rigidly that someone designing a composite bridge needs that part of the composite Code that deals with bridge structures, plus the more general parts of that Code, plus the equivalent material from the Concrete Codes, plus the equivalent material from the Steel Codes etc.

3.3 *Axis System*

That selected matches the arrangement commonly used by computer software; it is therefore different to the one normally used when specifying the major and minor principal axes for steel sections. For those used to a different arrangement, this is, of course, potentially confusing – they will simply have to learn to work with the new arrangement. Moreover, all published design guidance will, of course, need revision – for this reason alone!

3.4 Language

Several new terms and expressions have been introduced as a result of the need for the various clauses to have precisely the same meaning in each of the major European languages. Indeed, an exercise has been undertaken for each document to translate the original working draft (each of which was originally prepared in English) into French and then into German and then back into English as a way of checking consistency. Probably the best known example of this is the use of "Actions" for "Loads"; careful examination will reveal that this is actually a more appropriate term since certain "Loads" such as temperature change are not strictly speaking loads at all.

3.5 Arrangement

This is essentially by phenomena rather than by task i.e. collecting together all the rules on shear resistance rather than all those needed to design a beam. Users will simply have to learn where to look and thus need to know what to look for. There is clearly a role here for Design Guides and Computer Software to collect together all the Clauses required for particular tasks and to re-present the material in a more user friendly fashion. Interestingly, some 10 years ago at an earlier Pacific Structural Steel Conference, the author (3) suggested how this feature might be one aspect of a computer based implementation of Structural Codes that would facilitate a new way of working.

3.6 Code not "Textbook"

Different views prevail throughout Europe on what is appropriate content for a Code. In the UK, in common with most other English speaking countries e.g. USA, Australia etc, the tradition has been to provide a significant amount of guidance to the user. The majority of Mainland Europe, however, regard much of this material as "Textbook" and has traditionally only provided essential points within the Code itself. One of the clearest manifestations of this is the process for designing against lateral torsional buckling in EC3, for which knowledge of the elastic critical moment is required to initiate the process. No guidance is provided on the calculation of the theoretical elastic critical moment – a quality similar in principle to the Euler Load for a strut but significantly more complex to determine.

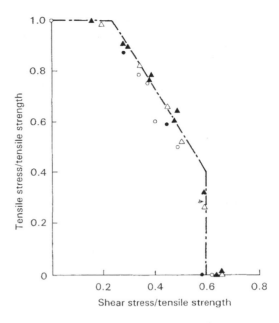

Figure 2. Trilinear interaction curve for bolts under combined tension and shear; comparison with test data.

3.7 Elegance vs. Convenience

This topic has already been the subject of a separate presentation by the author (4). For several topics, for which various design approaches are possible, an elegant presentation rather than an easy to use process has normally been preferred. One such example is illustrated in Figure 2 which shows test data for bolts in combined shear and tension and two alternative design rules. The more elegant corresponds to the arc of a circle and leads to a simple expression; the alternative comprising three straight lines is less easily presented but has the great practical virtue that for the majority of the design situations one of the two load components will be sufficiently small that the full resistance for the other mode may be used. EC3 uses the arc of a circle.

4 PRACTICAL ISSUES

4.1 Withdrawal

Referring to figure 1, the first key event in the publication/adoption/withdrawal sequence is approval and publication of the final text – known as the date of availability (DAV). BSI then has two years in which to publish the document with its associated National Annex (see below). Only then does it become available for use as an alternative to the National Standard. A period of up to three years of co-existence is then permitted whilst related documents are published. At the end of this time BSI is required to withdraw all conflicting National Standards. Of course, given that there is not an exact match between BSs and the Eurocodes, some judgement will be required in deciding precisely which documents are deemed to be in conflict. It is also worth noting that the term "withdrawn" has a specific meaning with BSI; it means no longer maintained and listed, it does not mean outlawed.

4.2 *National Annexes*

No change to the text prepared by CEN is permitted. However, it is recognised that certain safety and local issues are a National matter. Therefore a limited number of Nationally determined parameters and National choices may be covered through the provision of a National Annex for each Code. Such documents may also make reference to non-conflicting complimentary information (NCCI) of the type defined below. It is expected that the National Annexes will be brief.

4.3 *NCCI*

The technical provisions of the Eurocodes and the relevant BSs are not an exact match e.g. the Committee responsible for BS5950 identified 64 items in BS5950: Part 1 that were not replicated in EC3 Part 1.1. Of course, these varied in importance from relatively trivial advisory items to significant technical provisions such as the comprehensive guidance on aspects of portal frame design given in the BS. Clearly some of this material needs to be preserved and made available to Designers. This is the sort of content deemed suitable for an NCCI; other material not presently included within a National Code but dependent on it and widely used by the Industry might also be treated in a similar fashion. Thus the BCSA/SCI "Green Books" (5–7) on connection design could be viewed in this way.

4.4 *Implementation Issues*

It should be recognised that there are different groups within the Construction Sector who will each have rather different requirements in terms of assisting them with the transition. Table 1 reproduces material from reference 2 covering the identified needs of: Non-technical users, designers and regulators, academia and those providing training and software producers.

5 PRACTICAL ISSUES

5.1 *Publication Timetable*

Arguably the single most vital information for those working in Practice is accurate and up to date information on the publication schedule of the Eurocodes, augmented by similar information on related matters such as Design Guides, Educational

Course, Software releases etc. It is hoped that Eurocode Expert (8) can fulfil this role and the IStructE Committee is therefore working closely to provide advice and guidance. Much can, of course, be done through direct links to the relevant parts of the sites of other organisations active in supporting their sector e.g. BSI for its publication schedule, SCI, Corus and BCSA for steel related material.

5.2 *Comments*

Back in 1975 the official European Community view of the Structural Eurocodes was that they would:

- Provide common design criteria.
- Ensure a common understanding.
- Facilitate exchanges of people and products.
- Facilitate marketing across boarders.
- Provide a common basis for R&D.
- Increase competitiveness.

A few years later the European Economic Association listed:

- Provide a framework for harmonised technical specifications.
- Be a means for demonstrating compliance.
- Provide the basis for specifying contracts.

A more recent and personal view is presented in Table 2. This suggests that for every potential benefit there may well be a broadly equivalent potential downside. A successful transition clearly requires that the former be maximised and the latter be minimised. This argues for a coordinated programme of migration. A rather starker justification comes from simple economics. In preparing a presentation for the Annual BCSA Conference in 2003, the author (9) secured an estimate of the likely transition costs to the UK Structural Steelwork community that suggested:

- £2,500 million – Annual value of UK Steel Construction Industry activity.
- £2000 million – total design costs.
- £100 million – fraction of design that is directly Code related.
- £10 million – suggested efficiency loss.
- ? consequences of a misapplication failure.

Accepting that the actual figures are to some extent subjective, the message is, nonetheless, clear: Migration from a design environment based on National Standards to one based on the Eurocodes is a very substantial undertaking. For this reason the IStructE Report (1) stated:

- The Structural Eurocodes are the most wide-ranging change to codification of structural design ever experienced in the UK.

Table 1. Issues related to implementation.

Issue	Comment
Designers	Designers are unlikely to adopt the structural Eurocodes until they see a competitive advantage, are required to do so by clients or there is effectively no alternative because the British Standards have been withdrawn and reference in the Approved Documents to the Building Regulations removed. The investment required to purchase the Eurocodes themselves and supporting guidance documents as well as provide the necessary training for staff will be significant, As an illustration, it has been estimated that the cost (including loss of productivity) of adopting the structural Eurocodes within a consultancy with 16 fee earning staff will be approximately £0.25 M. With their fee levels already under pressure, some designers are likely to resist making this expenditure for as long as possible.
Residual British Standards	At the end of the coexistence period BSI is required to withdraw those British Standards having the same scope as the European Package. However, the scope of each of the Eurocodes is such that in the vast majority of cases their content will not match precisely that of the corresponding British Standards. For each British Standard that will eventually be withdrawn a decision needs to be reached as to whether the material within it that is not superseded by the appropriated into another publication such as a handbook. These documents will need to be available at the time of publication of the relevant Eurocode and its National Annex, or as soon as possible thereafter. The National Annex of each Eurocodes is permitted to contain reference to Non-conflicting complementary information (NCCI) with will assist designers in applying it in the UK. Much of the NCCI that it will be desirable to reference is likely to be from existing British Standards that will be withdrawn. If the finalisation of the National Annexes is not to be delayed, decisions on the fate of the material from these standards need to be reached quickly.
Non-conflicting complementary information sources other than British Standards	It is likely that the National Annexes will include reference to a wide variety of drawn from sources other than current British Standards. Much of the necessary material will be updated documents from the various bodies of industry who regularly publish authoritative guidance. However, finalisation of the National Annexes will require that these organisations decide which documents will be updated to reflect the Eurocodes and to what time they will be produced.
National Annexes	Considerable work still remains to be carried out to calibrate the Eurocodes. Thus far most of the emphasis in preparing the National Annexes appears to have been on buildings but it is essential that the necessary work in connection this bridges and other types of structures is completed. Providers of the guidance material and software need the National Annexes to be finalised before they will be able to do most of their work. Consequently it will not be for the majority of such material to be published until some time after the appropriate National Annex is finalised.
Background Information and	Production of guidance documents and software will be greatly assisted if information related to the background of the provisions, for example the basis of formulae, of the Eurocodes could be gathered together and made available. Authoritative interpretation of some sections of the codes will be necessary to enable them to be used as intended by their authors.
BSI	Within the UK the copyright of the Eurocodes will rest with the BSI. In order to be useful, guidance documents, in both paper and electronic format, will need to be able to quote extracts from them. It is also likely that material from the existing British Standards not republished as residual standards will need to be incorporated in other forms of NCCI. It is therefore essential that arrangements are made which allow reproduction of what is required without unnecessary bureaucracy or prohibitive cost.

Table 2. Potential benefits and perceived risks.

Benefits – Technically the most advanced	
Claim	Risk
Based on most up to date view on topics covered	Unfamiliar to users – too complex?
Consensus of expert views across Europe	Political "horse trading" opportunistic influences
Aligned with modern approaches e.g. computer based methods	Too sophisticated for "simple applications"
Benefits – Breadth of Coverage	
Claim	Risk
More comprehensive than all previous codes	Too broad coverage of minority items at expense of fuller treatment of frequently used topics
Suitable for wide applications	Arrangement of material makes harder to use for straightforward topics
No duplication	Need to consult several documents for even simplest task e.g. composite construction
Benefits – Wide Acceptance	
Claim writing process/legal position ensures acceptance by all members	Still some "local practices"; risk of little relevance if only working locally
Aids export of designs and products	Also aids others importing
Common practice e.g. design based on testing	May invalidate existing earlier testing not in accordance with new procedures

Table 3. Implementation in the UK – needs of industry.

Designers and regulators
- Need access to a clear jargon free programme with dates of
  - publication of the Eurocodes
  - publication of the National Annexes
  - availability of guidance documents
  - availability of design aids
  - withdrawal of British Standards

Academia and those providing training
- Require
  - Teaching Notes
  - Access to a concise version of the Eurocodes
  - Textbooks
  - Background information

Software Producers
- Final drafts of
  - Eurocodes
  - National Annexes
  - Residual Standards
  - NCCI
- Interpretation issues resolved

- They are in many ways the most technically advanced suite of Structural Design Codes anywhere in the world.
- They should provide sufficient opportunities for export of UK design expertise and products.

It went on to identify several key issues and then to define the assistance required by different sectors of the community. Table 3, gives some examples.

6 SOLUTIONS

Clearly the most important technical requirement for a smooth transition is recognition that direct use of the published Eurocodes as the basis for every day design is impractical and that Structural Engineers will expect to work with a far more user friendly arrangement. This could take one or more of the following forms:

- Design Guides that repackage Code material into an easier to use format.
- Design Manuals that extract material required for the most commonly conducted tasks and present them more simply.
- A "Concise" version of key documents, containing only that material required for the most commonly conducted tasks.

Each of these requires skilled resources for their production and does, of course, suffer from the common problem with any printed document: Any revision to some aspects of the underlying material necessitates revising and reprinting. No doubt some will be produced but longer term the future must lie with electronic provisions. It is for this reason that the main SCI vehicle for supporting the introduction of EC3 and EC4 is access – steel (10). This is a website, the result of a Pan-European collaborative project led by the SCI, designed to assist Eurocode implementation across all of the participating countries. Specifically, it assists in the transition to EC3 and EC4 based Structural Steel Design by providing:

- Unified design paths for routine buildings.
- Simple and complete Design Guidance.

Table 4. Contents of access-steel.

	Multi-Storey	Single Storey	Residential	Fire	TOTAL
Case Study	8	4	11	12	35
Scheme Development	17	11	12	18	58
Flow Chart	17	14	6	12	48
NCCI	30	17	–	3	50
Data		1		7	8
Static Examples	18	10	7	17	52
Interactive Examples	11				11

Table 5. Comparison of Eurocode 3 (EN 1993–1–1) with BS449 and BS 5950–1–200 for Simplified Design.

Design an unrestrained UB in S275 steel — the steps commonly used in practice		
Eurocode 3 kept simple	**BS 449**	**BS 5950 kept simple**
Choose UB size	Choose UB size	Choose UB size
Look up h/t_w, i_z and f_y in tables	Look up D/T and r_y in tables	Look up D/T, r_y and f_y in tables
Calculate slenderness $\dfrac{l}{i_y}$	Calculate slenderness $\dfrac{l}{r_y}$	Calculate slenderness $\dfrac{L_E}{r_y}$
Look up $\chi_{LT} f_y / \gamma_M$, the allowable buckling stress in a table such as Table 2 on p26 (values need to be tabulated). Calculate buckling resistance moment $M_{b,Rd} = \mathit{k}_{\sigma} f_y / \gamma_{M1} W_{pl}$ Compare applied moment with $M_{b,M}$	Look up allowable buckling stress p_{bc} in a Table 3a and check thickness in table 2. Calculate applied stress $f_{bc} = \dfrac{M}{Z}$ Compare f_{bc} with p_{bc}	Look up allowable buckling stress p_b from Table 20 assuming $\beta_W = 1$. Calculate buckling resistance moment $M_b = p_b S_x$ Compare applied moment with M_b
Check shear	Check shear	Check shear
Look up h and t_w in tables	Look up D and t in tables	Look up D and t in tables
Calculate the shear resistance $V_{c,Rd} = (f_y / \gamma_{M0} / \sqrt{3}) h t_w$ (Note that γ_{M0} is expected to be 1.0) Check applied shear ≤ $1.5 V_{c,Rd}$	Calculate the shear stress $f_q = Q / (Dt)$ Compare f_q with p_t from Table 11	Calculate the shear resistance $P_v = 0.6 p_y D$ Check applied shear ≤ $0.6 P_v$

- Material for inexperienced Designers.
- 250 plus harmonised resources.
- Multilingual (English, French, Spanish and German).
- Links to Eurocode Clauses.
- Quality Assured and printable
- User friendly IT systems: Fast structured search
- And most importantly, free registration.

At present it contains 262 separate contact items, distributed across the various forms of structure and types of material as detailed in Table 4. Interested readers are encouraged to access the site on www.access-steel.com.

There are, of course, ample opportunities to assist in a more simplified implementation of certain of the more widely used Eurocode procedures. For example, King (11) has recently produced a short article explaining how lateral-torsional buckling may be approached in terms familiar to UK Designers. This is a potentially difficult topic (at least at first sight) for the UK because our traditional approach for designing laterally unrestrained beams appears quite different. This is due to the form of presentation adopted in the Eurocode – both the specific rules for checking LTB and the separation of the complete set of design checks required by topic rather than by task. However, Table 5, which compares EC3, BS5950: Part 1 and the former BS449, reveals a remarkable similarity between them. Careful study shows that EC3 contains an LTB check that is actually very similar to that of BS5950: Part 1 but the form of presentation in terms of M_{cr} rather than λ_{LT} makes its direct implementation more difficult – since M_{cr} is first required and EC3 provides no assistance. New Zealand readers of the Eurocode should note this approach is actually very similar to that used in NZS 3404 – with the important difference that comprehensive guidance on calculating M_0 (M_{cr}) is given in the NZ Standard. To implement the procedure of the first column requires tables of the form of Table 6 that are not in the Code but which could well appear in Design Guides. Reference 11 explains that

Table 6. Design table required to support simpler EC3 implementation.

Lateral torsional buckling resistance: Part table of stress $x_{LT}f_y/\gamma_{M1}$

UBs, S275 steel, thickness not exceeding 16 mm

Slenderness		h/t_f									
l/i_y	$\bar{\lambda}$	5	10	15	20	25	30	35	40	45	50
30	0.35	275	275	275	275	275	275	275	275	275	275
35	0.40	275	275	275	275	275	275	275	275	275	275
40	0.48	275	275	275	275	275	274	274	274	274	274
45	0.52	275	275	272	271	270	269	269	269	268	268
50	0.58	275	273	268	266	265	264	263	263	263	263
210	2.42	238	193	159	135	118	107	98	92	87	83
220	2.53	236	189	154	130	113	102	93	87	82	78
230	2.65	233	185	150	125	109	97	89	83	78	74
240	2.78	223	175	140	117	101	90	82	76	71	68
250	2.88	229	178	141	117	101	89	81	75	70	67

adoption of this procedure does impose some restrictions in terms of not fully exploiting all the possibilities of EC3. But similar trade-offs – more rapid calculations in exchange for slightly lower design resistances – is an everyday feature of design. Surely, Designers are asked to exercise this sort of judgment all the time. Thus if the design is to be conducted in such a way that the most refined solution is necessary then recourse must be made to the full procedure — but even then the use of a simplified approximate procedure to obtain a close first estimate of the likely solution is helpful.

7 DESIGN

Information on aspects of the technical background to some of the material contained in EC3 and EC4, together with comment on how this and the actual rules differ from current UK practice, is provided in the series of textbooks being published by Thomas Telford. These are intended as both teaching aids and as reference material (12–15).

8 CONCLUSIONS

An attempt has been made to summarise key features of the process currently being experienced within the UK as Structural Design migrates from a climate based on National Standards to one based on the forthcoming Eurocodes. The principal conclusion is that the magnitude of the task is such that unless it is planned, resourced and orchestrated properly, not only will a major opportunity be lost but the existing position may well be compromised. A number of specific aspects of the necessary campaign have been identified and

described. It is suggested that similar issues face any country undertaking a programme of major change in its Structural Codes. A particular feature of the Eurocode operation is, of course, the number of countries involved and thus the limited extent to which the views (no matter how soundly based) of one nation can prevail. Whilst international scientific collaboration undoubtedly leads to a greater volume of knowledge and a better understanding, the translation of this into working procedures is itself a significant task yet to reach maturity. For the foreseeable future custom and practice has a powerful influence on the expectations and needs of the user in the local community.

REFERENCES

Institution of Structural Engineers, "National Strategy for Implementation of the Structural Eurocodes: Design Guidance, report for the Office of the Deputy Prime Minister, IStructE 2004. (www.istructe.org.uk/eurocodes).

Institution of Structural Engineers, "Implementation of the Structural Eurocodes", The Structural Engineer, 3 August 2004, pp. 21–24.

Byfield, M.P. and Nethercot, D.A., "Can Codes of Practice be both Comprehensive and User-Friendly?" Fourth Pacific Structural Steel Conference, Singapore, Vol. 1. Steel Structures, ed. Shanmugan, N.E. and Choo, Y.C., Pergamon, 1995, pp. 29–40.

Nethercot, D.A.,"Structural Steel Design Codes": Vehicles for Improving practice or for Implementing Research?" ICASS 05, ed., Shen, Z.Y, Li, G.L. and Chan, S.L., Shanghai, June 2005.

BCSA/SCI, "Joints in Simple Construction", Vol. 1, Design Methods, Second Edition, SCI, 1993. .

BCSA/SCI, "Joints in Steel Construction, Moment Connections", SCI, 1995,

BCSA/SCI, "Joints in Steel Construction, Composite Construction", SCI Pub. 213, 1998.

www.eurocodes.ac.uk, "Eurocodes Expert".

Nethercot, D.A., "Eurocode 3: A Personal View", New Steel Construction, March –April 2004, pp. 26–27.

Owens. G.W. and Roszykiewicz, C., "Maximising Opportunities for the Eurocodes", in "Steel; A New and Traditional Material for Building", ed Dubina, D. and Ungureanu, V, Taylor and Francis, London, 2006. pp 55 – 65.

King, C. "Steel Design can be Simple Using EC3", New Steel Construction, April 2005, pp. 25–27.

Gardner, L. and Nethercot, D.A., "Designer's Guide to EN 1993-1-1 Eurocode 3: Design of Steel Structures, General Rules and Rules for Bridges", Thomas Telford Ltd, 2005.

Moore, D.B et all, "Designers Guide to EN 1991-1-2, EN 1993-1-2 and EN1996-1-2: Fire Engineering Actions on Steel and Composite Structures", Thomas Telford Ltd, 2006.

Hendry, C.R. and Johnson, R.P., "Designer's Guide to Eurocode 4: Design of Composite Structures: Part 2. Rules for Bridges", Thomas Telford Ltd, 2006.

Johnson, R.P. and Anderson, D. "Designer's Guide to EN 1994–1–1: Eurocode 4: Design of Composite Steel and Concrete Structures", Thomas Telford Ltd, 2004.

Material, beams, columns and frames

Steel and Composite Structures – Wang & Choi (eds)
© *2007 Taylor & Francis Group, London, ISBN 978-0-415-45141-3*

Influence of the Bauschinger effect on the deflection behaviour of cambered steel and steel concrete composite beams

Hauke Grages & Jörg Lange
Institute for Steel Structures and Materials Mechanics, Darmstadt University of Technology, Germany

Norbert Sauerborn
Stahl+Verbundbau gmbh Dreieich, Germany

ABSTRACT: Aim of the project was to measure deflections of standard steel concrete composite beams, used in office and industrial buildings, assess and analyse the reasons for those deformations and develop simple design aids for a realistic calculation of the deflections. This research was necessary, since experience has shown that the calculated results do frequently not match the deflections measured on site. This can reduce the serviceability of a structure and lead to high rectification costs. The focus was on the differences between calculations and in-situ behaviour by analysing the measured deflections of beams in structures under construction.

During the measurements it appeared that cambered beams deform more than straight beams. This effect was analysed in tests and it was found that the reason for this behaviour lies in the Bauschinger Effect, rather than in residual stresses. It will be shown when and how this effect should be considered.

1 INTRODUCTION

Steel concrete composite structures are common practice in todays office and industrial buildings. The advantages of the materials steel and concrete lead to a very economic alternative for wide spanned slabs, especially in terms of high bearing capacity and small dimensions as well as time efficiency due to pre-fabrication.

Steel concrete composite structures are widely used for bridges and industrial buildings as well as for parking lots and multi storey buildings.

By prefabricating the steel substructure time for construction on site is reduced tremendously. Using composite slabs reduces the time for construction further since they replace parts of the reinforcement and are used as formwork. An alternative for composite slabs are partly or full pre-cast concrete elements that also reduce the construction time. The focus of this research project was on commonly used standard steel concrete composite beams as shown in Figure 1.

A building is designed for a special purpose. On one side the verification of the bearing capacity ensures the structural integrity; on the other hand the design for serviceability ensures a proper utilization in terms of the users' demands.

To design a buildings bearing capacity theoretical loads are taken into account. They are usually larger than the real and expected loads. They include a Safety factor. To design the buildings serviceability

Figure 1. Standard section of a composite beam.

it is of special necessity to work with realistic loads to calculate the real deflections and dynamic response.

Due to the development to longer spans and the ability of composite structures to bridge those spans the design for serviceability becomes more relevant. Often the calculated deflections do not match those seen on site. The knowledge about the real deflections of composite beams can reduce high rectification costs

In the beginning of the project we analyzed buildings that were under construction. The deflections were measured in all relevant situations starting from the fabrication of the steel beam, via the assembly of the structure until the building was finished and major long term deflections had occurred. The focus was on the history and structural details of the buildings erection, e.g. whether the beams are supported during casting of the concrete or the deflections are influenced by other structural details like stiff connections. Also material uncertainties resulting from the concrete and residual stresses were taken into consideration.

2 STATE OF THE ART

Commonly the deflection of a steel concrete composite beam is calculated for the following steps of loading with the corresponding second moments of area: The construction dead load of the construction, i. e. the steel beam and the concrete slab under consideration of temporarily propping. Second the superimposed dead load (cladding, finishing, mechanical services) third the live load and fourth long term influences of the concrete(creep and shrinkage).

Most of the commonly used calculation programs are following this pattern. When all influences are considered and calculated a camber will be assumed for the known dead load and parts of the live load, creep and shrinkage. For those assumptions the beam is assumed to be straight after its history of loading.

Reality shows, that the calculated deformations frequently do not match the real deflections. Reasons for this discrepancy can be that the assumptions made in the calculation are wrong, the theory for calculating the deflection has mistakes or further influences on the deflection behaviour of the beam are not taken into account. For this reason it became necessary to compare the calculations with the deflections of composite beams in real buildings to exclude that any influences are overseen by assembling a composite beam in a laboratory.

3 RESULTS OF SITE MEASUREMENT

Basis for the analysis of the deflection of the beams are site measurements. In three parking lots build in composite construction the deflection of the beams were measured. The shape of the beams was measured

– after cambering the beams,
– after welding of the shear studs,
– before and after galvanising, (if the beams were galvanised)
– after erecting the steel structure,
– after concreting,
– after removing of the construction supports, if the dead load shall act on the composite beam,
– frequently after finishing the structure for long term deformation.

During the measurement of the beams on site it appeared, that especially the deflections of highly cambered beams were larger than calculated.

4 THE BAUSCHINGER EFFECT

4.1 Production of the camber

The camber of all beams was fabricated by hydraulic presses as shown in Figure 2.

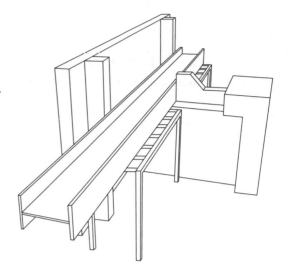

Figure 2. Hydraulic press.

The press deforms the beam plastically in the middle of the machine's supports. After that the beam is moved about 30–50 cm and the beam is locally deformed again. In that manner the entire beam is cambered until it has an ideal shape according to the previously calculated deflection.

At first it was assumed, that the residual stresses implemented by cambering are responsible for the increase of deflections. Therefore tests were accomplished to measure the strains of the beam while cambering. In addition a Finite Element Analysis was performed that simulated the process of cambering. The calculated strains fitted properly to the results of the tests. In the analysis it was simulated that the beam was loaded in the opposite direction of the camber. Just before the load bearing capacity was reached, the deformations increased slightly compared to a non cambered beam. The amount of camber had no influence on the deflection of the beams. All cambered beams showed a very similar load-deflection-behaviour (Fig. 3)

Still the measurements on site gave reason to the assumption that the increase of deformation was depending on the process of cambering.

4.2 Pre-Test to the BAUSCHINGER Effect

Since the previously described considerations and analysis did not lead to the intended results pre tests were performed to justify further investigation.

Following HOFF & FISCHER tests were executed to gain more information of the influence of the BAUSCHINGER Effect on the deflection of beams.

Five 1 m long pieces of IPE 100 in grade S355 were loaded above the yield point. The amount of the

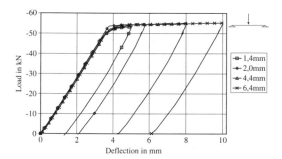

Figure 3. Calculated M-κ-diagram for bending beams after varied plastic pre-strain.

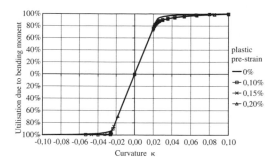

Figure 4. Plastic deformation – first load.

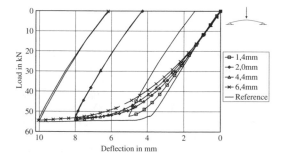

Figure 5. Plastic deformation – reversal load.

deflection and thus the plastic deformation was varied. Afterwards the specimen were loaded against the previous deformation again until the yield point was reached (Figs. 4–5).

While the load-deflection graphs of the first loading are nearly identical and bilinear, the course of the graphs changes tremendously when loaded in the opposite direction. Starting at a load of 30% of the bearing capacity the graphs branch out and show a three times larger deflection when reaching the bearing capacity, compared to the reference graph of a non cambered beam. The deflection increases depending on the pre-strain already at small loads. The bearing capacity seems not to be influenced by the camber.

The observed increase of deflection at the measurement on site seems to be initiated by the BAUSCHINGER Effect, which results form cambering the beams.

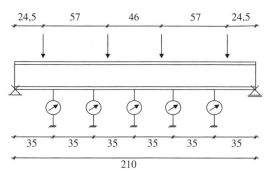

Figure 6. Load and measurement of deflection.

Table 1. List of specimens.

Group of specimen	Number of specimens	Average camber (1/100 mm)
1	3	0
2	3	828
3	2	893

on the pre-strain already at small loads. The bearing capacity seems not to be influenced by the camber.

The observed increase of deflection at the measurement on site seems to be initiated by the BAUSCHINGER Effect, which results form cambering the beams.

4.3 Bending Tests to the BAUSCHINGER Effect

The pre-tests have shown that the BAUSCHINGER Effect affects also standard steel used in structural steelwork and is not limited to high alloyed steel used in mechanical engineering. Still no description could be found that describes the influence of the BAUSCHINGER Effect on bending beams. Thus it is essential to analyse the material behaviour to simulate the BAUSCHINGER Effect at bending beams.

For the tests 18 m long IPE 220 in steel grade S355 were used. The beams were cambered with a hydraulic press by different amounts. Afterwards the beams were sawn into eight sections. The two partly non deformed sections at the ends could not be used. It was assumed that the inner six sections of 2300 mm had equal curvature and stresses. The beams were loaded in four points and the deflections were measured (Fig. 6).

According to table 1 three groups of beams with different camber were combined. In Figure 7 the load-deflection-graphs for the tests are shown. It is obvious that the cambered specimen deform stronger than the non cambered beams. As expected the deflection of the beams of group three are larger than those of group two due to a larger amount of cambering.

135

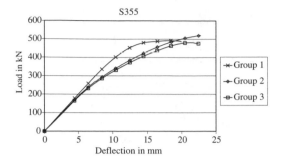

Figure 7.   Load-deflection-diagram of cambered beams.

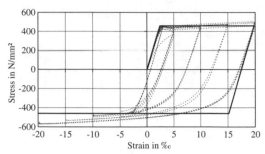

Figure 8. σ–ε–diagram for first and reversal specimes out of S355.

4.4   Material analysis

It was previously proven that the BAUSCHINGER Effect is responsible for the increase of the deflection of cambered beams. To simulate this effect a material model has to be developed that can be implemented in an analysis program to calculate the resulting deflections. Therefore a material analysis is performed with the steel of the bending specimen.

All tests were performed strain controlled with pre strain of 3, 5, 10, 15 and 20‰. The results of the tests with S355 are shown in figure 8.

The bold line gives the elastic-ideal-plastic material model as reference. The influence of the BAUSCHINGER Effect can be seen from a pre-strain of 5‰. At a strain of 20‰ the deformations become larger already while unloading the specimen. Thus the BAUSCHINGER Effect affects the deformation of a cambered steel beam from the first moment of its load history.

4.5   Material Model

Most of the existing attempts to find a material model that describes the BAUSCHINGER Effect are, according to YAMADA & TSUJI, oriented at the material model of RAMBERG and OSGOOD. The shape of stress-strain-relation is usually considered as a multilinear graph. Thus for each pre-strain a different material model must be specified. This is most inconvenient for the use in an analysis program or in an FEM simulation.

A material model has to be developed, that simulates the material behaviour from the reversal point of the load depending on the previous strain. According to the existing models initially the material model from RAMBERG-OSGOOD is used, that describes the stabilized cyclic σ–ε–curve:

$$\varepsilon_a = \frac{\sigma_a}{E} + \left(\frac{\sigma_a}{K'}\right)^{1/n'} \tag{1}$$

Some modifications are necessary since only the first load cycle is relevant and not the stabilized

cyclic curve. Due to the distinctive LÜDERS plateau of the tested steel the stress-strain-relation up to the reversal point is assumed to be elastic-plastic. Under consideration of the material model of MASING the material model of RAMBERG-OSGOOD can be changed to describe the material behaviour from the reversal point. Thus:

$$\Delta\varepsilon = \frac{\Delta\sigma}{E} + \left(\frac{\Delta\sigma}{2(1-n')\,K'}\right)^{1/n'} \tag{2}$$

With this equation a relation is given to describe the material behaviour for the reversal curve in dependence of the factors n' and K'. The tests show that the shape of the reversal curve is depending on the pre-strain of the specimen. For the described material model factors had to be searched for each curve so that it corresponds properly with the experimental results.

For illustration the point of origin is defined at the reversal point of the load. Since the reversal curve is nearly identical with the elastic line in the following diagrams only the area from $\Delta\sigma = 440$ N/mm² to $2\Delta\sigma = 880$ N/mm² is shown. In the first picture this area is shown hatched for a pre-strain $\varepsilon_{pre} = 20$‰. The test results are displayed as dotted lines, the calculation results as fat line and the thin line shows the elastic-ideal-plastic reference (Fig. 9)

To calculate the curves the following parameters for n' and K' were chosen:

Since the solidification exponent n' is constant for this material the curvature of the reversal graph is only depending on the solidification coefficient K'. The maximal possible stress is set to the yield stress. Especially for small pre-strain the stresses would become too big otherwise. If K' is displayed over the pre-strain ε_{pre}, it is possible to describe this curve with the equation:

$$K' = b_1 + \frac{1}{\left(\varepsilon_{pre}\right)^{b_2}} \tag{3}$$

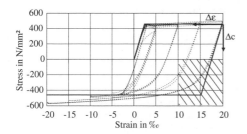

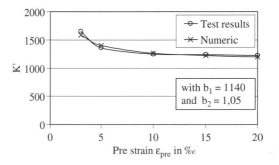

Table 2. Parameters for n' and K'.

Pre strain ε_{pre} (in ‰)	n'	K'
3	0,175	1650
5	0,175	1360
10	0,175	1250
20	0,175	1220

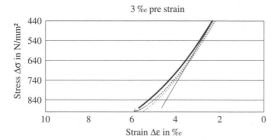

3 ‰ pre strain

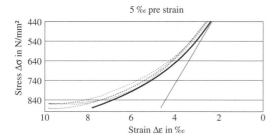

5 ‰ pre strain

Figure 10. K'-ε_{pre}-Diagram for reversal loaded specimen.

and the variables b_1 and b_2. The amount of pre-strain ε_{pre} must be inserted as decimal number. The parameters b_1 and b_2 are chosen similar to n' and K' until the graphs are similar. Since the BAUSCHINGER Effect appears only when the steel was plastically deformed the graph begins when ε_{pre} is larger than $f_{y,k}/E$ (Fig. 10).

It is now possible to describe the material behaviour from the reversal point just by knowing the previous strain. This material model can easily be implemented into a computer program.

A fibre model was developed to calculate the deflection of a cambered steel beam, considering the BAUSCHINGER Effect. At first the process of cambering the beam has to be simulated. Depending on the distance of the supports and the distance of pressure points this calculation has to be done until the entire beam is cambered. While deforming the beam with the hydraulic press the material model is elastic-ideal-plastic. While releasing the pressure the previously described material model will be activated if the deformation is plastic.

In the upper picture of figure 11 the strains and stresses in the different fibres are shown while the beam is deformed plastically in the press. After releasing the load there are still strains and stresses in the section. They were calculated with the developed material model. This procedure has to be repeated until the entire beam is cambered to the required amount. After this the pre-strain of every fibre is known as well as the residual stresses due to cambering. Thus it is clear which material model has to be chosen for

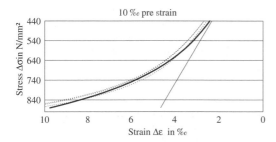

10 ‰ pre strain

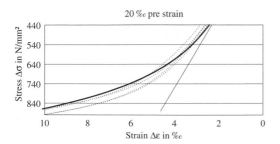

20 ‰ pre strain

Figure 9. σ ε diagram reversal load after plastic deformation.

137

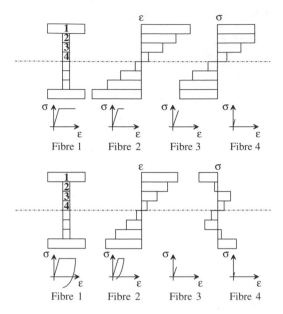

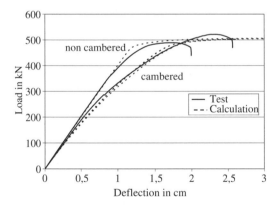

Figure 11. Assignement of a material model depending on the pre-strain.

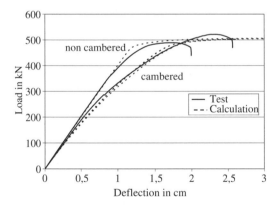

Figure 12. Comparison of tests and numerical simulation.

every fibre when the beam is loaded in the opposite direction.

4.6 Comparison of tests and simulation

To verify the material model and the simulation by a fibre model, the results are compared with the results of the performed tests. First the camber has to be simulated. The 18 m original beam of group 2 had a camber of 475 mm. For the tests a 2100 mm long part of the beam was loaded. After simulating the camber of the 18000 mm beam the loading of a 2100 mm long section according to the performed tests is calculated.

In figure 12 the results of the calculation are compared with the test results.

The calculation of the deflection was performed with different combinations of distance of supports and distance of pressure points, while the camber was kept constant. The dotted lines showing the calculation results are close to each other, thus independent of the chosen varieties. Obviously only the amount of camber has major influence. The calculations match the measured deflection good.

Thus it is possible to calculate the increase of deflection due to the BAUSCHINGER Effect with the presented material model.

5 CONCLUSION

It is common practice to camber wide spanned steel beams in steel and composite construction. The process of cambering the beams with a hydraulic press influences the deflection behavior of the beams depending on the amount of camber.

Tests were performed to analyze the material behavior to gain more information about the BAUSCHINGER Effect. A material model was developed that is able to consider the BAUSCHINGER Effect depending on the amount of pre-strain in every fibre of the steel beams section. With the material data of the tests and the developed material model it is possible to calculate the increase of deformation due to the BAUSCHINGER Effect.

With regard to the site measurements the BAUSCHINGER Effect should be considered in steel and composite construction when the camber exceeds a value of L/200. This is approximately equivalent to a pre-strain of the beams flanges of 4‰ and therefore is susceptible to the BAUSCHINGER Effect.

REFERENCES

Bauschinger, J. 1881. Über die Veränderung der Elasticitäts-grenze und des Elasticitätsmoduls. *Civiling.N.F.* (27)
Hoff, H. & Fischer, G. 1958. Beobachtungen über den Bauschinger-Effekt an weichen und mittelharten Stählen. *Stahl und Eisen 78*
Yamada & Tsuji. 1993. Stoffgesetze des Baustahls – Zur Trag-analyse des Stahltragwerkes. *Festschrift Udo Vogel*
Graves, H., in Press, 2007. Beitrag zur Verformungsanalyse von Verbundträgern. PhD.-Thesis, Darmstadt

138

Steel and Composite Structures – Wang & Choi (eds)
© 2007 Taylor & Francis Group, London, ISBN 978-0-415-45141-3

New full-range stress-strain model for stainless steels

W.M. Quach
Department of Civil and Environmental Engineering, University of Macau, Macau, China

J.G. Teng & K.F. Chung
Department of Civil and Structural Engineering, The Hong Kong Polytechnic University, Hong Kong, China

ABSTRACT: Advanced numerical modelling of cold-formed stainless steel members, from manufacturing to ultimate failure under loading, requires knowledge of the stress-strain relationship of the material over a wide range of strains. Although a number of stress-strain models exist, they are only capable of accurate predictions either over a limited strain range or for the tensile stress-strain behaviour only. This paper presents a three-stage stress-strain model for stainless steels which is capable of accurate predictions over the full range of both tensile and compressive strains. The new stress-strain model is defined by the basic Ramberg-Osgood parameters (E_0, $\sigma_{0.2}$ and n) and is based on a careful interpretation of existing experimental data. The accuracy of the proposed model is demonstrated by comparing its predictions with experimental stress-strain curves.

1   INTRODUCTION

In addition to material anisotropy, stainless steel alloys are characterized by nonlinear stress-strain behaviour and different mechanical properties in tension and compression. The nonlinear stress-strain behaviour of stainless steels is commonly described by the Ramberg-Osgood relationship. This relationship is defined using the initial elastic modulus E_0, the 0.2% proof stress $\sigma_{0.2}$ and the strain-hardening exponent n. In practice, the definition of the strain-hardening exponent n requires the Ramberg-Osgood curve to match the measured stress-strain curve exactly at the 0.01% proof stress $\sigma_{0.01}$ and the 0.2% proof stress $\sigma_{0.2}$, so that the Ramberg-Osgood expression can closely approximate the measured stress-strain curve up to $\sigma_{0.2}$. However, the use of the Ramberg-Osgood expression for higher strains can lead to overestimation of stresses with serious inaccuracy as indicated by numerous previous studies (Macdonald et al. 2000, Rasmussen 2003, Gardner & Nethercot 2004).

A number of studies (Macdonald et al. 2000, Mirambell & Real 2000, Olsson 2001, Rasmussen 2003, Gardner & Nethercot 2004) have been conducted on the modelling of the stress-strain behaviour of stainless steels for higher strains. Although a number of stress-strain models have resulted from these studies, each of them is capable of accurate predictions either for a limited strain range or for the tensile stress-strain behaviour only. Moreover, most of them (Maconald et al. 2000, Mirambell & Real 2000,

Olsson 2001, Gardner & Nethercot 2004) require the knowledge of not only the basic Ramberg-Osgood parameters (E_0, $\sigma_{0.2}$ and n), but also some additional parameters which are generally not specified in existing design codes (e.g. AS/NZS 2001, ASCE 2002) and need to be found from experimental stress-strain curves.

Rasmussen (2003) proposed a stress-strain model for stainless steels which describes the stress-strain curve over the entire strain range using only the basic Ramberg-Osgood parameters; the model was deemed to be applicable to all stainless steel alloys in both tension and compression. The agreement between experimental stress-strain curves (mostly from tension coupon tests) and the predictions of Rasmussen's (2003) model was shown to be very good overall. However, in Rasmussen's (2003) model, the expressions for the ultimate stress σ_u and the ultimate strain ε_u were developed by interpreting the test data of tension coupons and the expression for the strain-hardening exponent for the strain range of $\varepsilon > \varepsilon_{0.2}$ (where $\varepsilon_{0.2}$ is the total strain at the 0.2% proof stress $\sigma_{0.2}$) was obtained by trial and error using stress-strain curves which were mostly from tension coupon tests. Hence, the applicability of this full-range model to stainless steels in compression has not been appropriately demonstrated.

Indeed, when the full-range stress-strain curves predicted by Rasmussen's (2003) model are compared with experimental stress-strain curves from the published literature (Korvink et al. 1995, Macdonald et al. 2000, Rasmussen et al. 2002, Gardner & Nethercot 2004), Rasmussen's (2003) model is generally found to provide very close predictions for tension coupon tests but underestimate stresses at strains $\varepsilon > \varepsilon_{0.2}$ for most compression coupon tests (see Figs 3 and 4); the difference between the experimental results and the predictions increases with the strain.

The above review indicates that all existing stress-strain models for stainless steels suffer from some limitations. This paper thus presents a new stress-strain model for stainless steels, which is capable of accurate predictions over the full ranges of both tensile and compressive strains. The new stress-strain model is defined using only the basic Ramberg-Osgood parameters (E_0, $\sigma_{0.2}$ and n) and is based on a careful interpretation of existing experimental data.

It should be noted that although a stress-strain model that is accurate for strains of small and intermediate values is often sufficient for modelling structural responses under loading, a stress-strain model that is accurate over the full ranges of both tensile and compressive strains is needed in the numerical modelling of the manufacturing process of cold-formed stainless steel members. The present work was conducted to provide an accurate stress-strain model for use in the finite element simulation of the manufacturing process and in the subsequent buckling analysis of cold-formed stainless steel members (Quach 2005).

2 THREE-STAGE FULL-RANGE STRESS-STRAIN MODEL

2.1 Expression for the stress-strain model

The present stress-strain model describes the stress-strain behaviour in three stages. As in other models (Mirambell & Real 2000, Rasmussen 2003, Gardner & Nethercot 2004), for the first stage from zero stress to the 0.2% proof stress, the Ramberg-Osgood expression (see Equation 5) is adopted.

For the second stage, covering the strain range from $\varepsilon_{0.2}$ to $\varepsilon_{2.0}$ (where ε_a is the total strain at the a% proof stress σ_a), Gardner and Nethercot's (2004) model with a simple modification is adopted, as they have shown their model to be accurate up to a tensile strain of approximately 10% and a compressive strain of approximately 2%. In deriving their expression, the point of the 1% proof stress $\sigma_{1.0}$ was chosen as a calibration point, although it can be easily shown that the final expression defined by Gardner and Nethercot's (2004) does predict a curve that does not pass exactly through this point. Although the associated errors are negligible, the following modified version of Gardner and Nethercot's (2004) expression which ensures that the curve passes through the point of $\sigma_{1.0}$ exactly

is adopted here for the second stage to maintain mathematical preciseness:

$$\varepsilon = \frac{\sigma - \sigma_{0.2}}{E_{0.2}} + \left[0.008 + \left(\sigma_{1.0} - \sigma_{0.2} \right)\left(\frac{1}{E_0} - \frac{1}{E_{0.2}} \right) \right]$$
$$\times \left(\frac{\sigma - \sigma_{0.2}}{\sigma_{1.0} - \sigma_{0.2}} \right)^{n'_{0.2,1.0}} + \varepsilon_{0.2}, \quad \sigma_{0.2} < \sigma \leq \sigma_{2.0} \tag{1}$$

where $n'_{0.2,1.0}$ is a strain-hardening exponent representing a curve that passes through both points of $\sigma_{0.2}$ and $\sigma_{1.0}$.

According to Olsson (2001), the true stress-nominal strain curve of stainless steels can be approximated as a straight line for strains exceeding $\varepsilon_{2.0}$. In the present model, this straight-line approximation is adopted for the third stage covering the range of $\varepsilon > \varepsilon_{2.0}$; that is, the true stress-nominal strain curve for this range is assumed to be a straight line passing through the points of the 2% proof stress at $\varepsilon_{2.0}$ and the ultimate stress at ε_u. The true stress-nominal strain relationship for $\varepsilon > \varepsilon_{2.0}$ can then be expressed as

$$\sigma_t = a + b\varepsilon \tag{2}$$

in which σ_t is the true stress, ε is the nominal strain, a and b are constants which can be obtained from values at the boundary points of the third stage ($\sigma_{t2.0}$, $\varepsilon_{2.0}$) and (σ_{tu}, ε_u). $\sigma_{t2.0}$ and σ_{tu} are respectively the 2% true proof stress and the true ultimate stress.

The nominal stress σ and the nominal strain ε can be converted to the true stress σ_t and the true strain ε_t using the following relationships for both cases of tension and compression (Chakrabarty 2000):

$$\sigma_t = \sigma(1 \pm \varepsilon) \quad \text{and} \quad \varepsilon_t = \pm \ln(1 \pm \varepsilon) \tag{3}$$

where the upper sign corresponds to tension and the lower sign to compression, and σ, ε, σ_t and ε_t are absolute values for both tension and compression.

By substituting Equation 3 into Equation 2 and using the boundary values of stresses and strains, the nominal stress-strain relationship and the constants a and b can be obtained as

$$\varepsilon = \frac{\sigma - a}{b \mp \sigma}, \quad \sigma > \sigma_{2.0} \tag{4a}$$

with

$$a = \sigma_{2.0}(1 \pm \varepsilon_{2.0}) - b\varepsilon_{2.0} \tag{4b}$$

$$b = \frac{\sigma_u(1 \pm \varepsilon_u) - \sigma_{2.0}(1 \pm \varepsilon_{2.0})}{(\varepsilon_u - \varepsilon_{2.0})} \tag{4c}$$

Finally, a three-stage nominal stress-strain model for the full ranges of both tensile and compressive strains is given by the following equations:

$$
\varepsilon = \begin{cases}
\dfrac{\sigma}{E_0} + 0.002\left(\dfrac{\sigma}{\sigma_{0.2}}\right)^n, & \sigma \le \sigma_{0.2} \\[3mm]
\dfrac{\sigma - \sigma_{0.2}}{E_{0.2}} + \left[0.008 + \left(\sigma_{1.0} - \sigma_{0.2}\right)\left(\dfrac{1}{E_0} - \dfrac{1}{E_{0.2}}\right)\right] \\[3mm]
\quad \times \left(\dfrac{\sigma - \sigma_{0.2}}{\sigma_{1.0} - \sigma_{0.2}}\right)^{n'_{0.2,1.0}} + \varepsilon_{0.2}, & \sigma_{0.2} < \sigma \le \sigma_{2.0} \\[3mm]
\dfrac{\sigma - a}{b \mp \sigma}, & \sigma > \sigma_{2.0}
\end{cases} \tag{5}
$$

For ease of practical application, it is desirable to characterize the above stress-strain model using only the basic Ramberg-Osgood parameters (E_0, $\sigma_{0.2}$ and n) or ($e = \sigma_{0.2}/E_0$ and n), since the values of other parameters including $\sigma_{1.0}$, $\sigma_{2.0}$, $n'_{0.2,1.0}$, σ_u and ε_u, especially their values for compression, are generally not available in existing design specifications. Therefore, these additional parameters need to be expressed in terms of the basic Ramberg-Osgood parameters. Such expressions are presented below.

2.2 Expression for $\sigma_{2.0}$

As Gardner and Nethercot's (2004) model is adopted to describe the second stage of the stress-strain curve ($\sigma_{0.2} < \sigma \le \sigma_{2.0}$) in the present model, $\sigma_{2.0}$ can be determined by imposing the boundary values ($\sigma_{2.0}$, $\varepsilon_{2.0}$) on Equation 1, which leads to the following expression:

$$
\sigma_{2.0} = \sigma_{0.2}
$$

$$
+ \left(\sigma_{1.0} - \sigma_{0.2}\right) A^{1/n'_{0.2,1.0}} \left[1 - \left(\frac{1}{E_{0.2}} - \frac{1}{E_0}\right)\frac{\sigma_{2.0}}{B}\right]^{1/n'_{0.2,1.0}} \tag{6a}
$$

$$
A = B/\left[0.008 + e\left(\sigma_{1.0}/\sigma_{0.2} - 1\right)\left(1 - E_0/E_{0.2}\right)\right] \tag{6b}
$$

$$
B = 0.018 + e\left(E_0/E_{0.2} - 1\right) \tag{6c}
$$

Since $\sigma_{2.0}$ appears on both sides of Equation 6a, an explicit expression for $\sigma_{2.0}$ cannot be obtained from Equation 6a. An approximation of $\sigma_{2.0}$ by an explicit expression is thus desirable. As $|(1/E_{0.2} - 1/E_0)\sigma_{2.0}/B| < 1$ and $n'_{0.2,1.0} > 1$ for the typical ranges of the basic parameters e and n, the binomial expansion of the second term in Equation 6a results in an infinite convergent series:

$$
\left[1 - \left(\frac{1}{E_{0.2}} - \frac{1}{E_0}\right)\frac{\sigma_{2.0}}{B}\right]^{\frac{1}{n'_{0.2,1.0}}} = 1 - \left(\frac{1}{n'_{0.2,1.0}}\right)\left(\frac{1}{E_{0.2}} - \frac{1}{E_0}\right)
$$

$$
\times \frac{\sigma_{2.0}}{B} + \frac{1}{2}\left(\frac{1}{n'_{0.2,1.0}}\right)\left(\frac{1}{n'_{0.2,1.0}} - 1\right)\left(\frac{1}{E_{0.2}} - \frac{1}{E_0}\right)^2 \frac{\sigma_{2.0}^2}{B^2} - \cdots \tag{7}
$$

By keeping only the first two terms of the infinite series of Equation 7 and then substituting it into Equation 6a, an approximation for $\sigma_{2.0}$ can be obtained as

$$
\sigma_{2.0} \approx \frac{1 + \left(\sigma_{1.0}/\sigma_{0.2} - 1\right)A^{1/n'_{0.2,1.0}}}{1 + e\left(E_0/E_{0.2} - 1\right)\left(\sigma_{1.0}/\sigma_{0.2} - 1\right)\dfrac{A^{1/n'_{0.2,1.0}}}{n'_{0.2,1.0}B}} \sigma_{0.2} \tag{8}
$$

in which $E_{0.2}/E_0 = 1/(1 + 0.002n/e)$, $\sigma_{1.0}/\sigma_{0.2}$ and $n'_{0.2,1.0}$ are given by Equations 9 and 10 presented later.

According to existing design codes, typically, the values of e vary from 0.001 to 0.003 and the values of n vary from 3 to 16. Within these ranges, the maximum relative error due to the approximation given by Equation 8 is about 3% for austenitic and duplex alloys and about 5% for ferritic alloys. Such small errors can be considered as negligible. If higher accuracy is desired, a more accurate value of $\sigma_{2.0}$ can be obtained by substituting the approximation obtained from Equation 8, into the right-hand-side of Equation 6a.

2.3 Expression for $\sigma_{1.0}$ and $n'_{0.2,1.0}$

In order to establish the expressions for $\sigma_{1.0}$ and $n'_{0.2,1.0}$ in terms of the parameters e and n, experimental data from coupon tests were analysed for their relationships. Among the available test data, only the test data of austenitic and duplex alloys reported by the Steel Construction Institute (SCI 1991) and Gardner & Nethercot (2004) contain values of both the 1% proof stress $\sigma_{1.0}$ and the basic Ramberg-Osgood parameters (E_0, $\sigma_{0.2}$ and n), and only the test data of austenitic alloys reported by Gardner & Nethercot (2004) contain values of $n'_{0.2,1.0}$. Hence, the test data from both the Steel Construction Institute (SCI 1991) and Gardner & Nethercot (2004) (127 sets of test data for tension coupons and 133 sets for compression coupons) were used to develop the expression for $\sigma_{1.0}$, while the test data from Gardner & Nethercot (2004) (15 sets for tension coupons and 16 sets for compression coupons) were used to develop the expression for $n'_{0.2,1.0}$.

The expressions for $\sigma_{1.0}$ and $n'_{0.2,1.0}$ were obtained from the best-fit curves shown in Figures 1 and 2, and they are given by Equations 9 and 10 respectively:

$$
\sigma_{1.0}/\sigma_{0.2} = 0.542/n + 1.072 \quad \text{for tension coupons} \tag{9a}
$$

$$
\sigma_{1.0}/\sigma_{0.2} = 0.662/n + 1.085
$$
$$
\text{for compression coupons} \tag{9b}
$$

and

$$
n'_{0.2,1.0} = 12.255\left(E_{0.2}/E_0\right)\left(\sigma_{1.0}/\sigma_{0.2}\right) + 1.037
$$
$$
\text{for tension coupons} \tag{10a}
$$

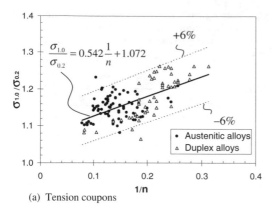

(a) Tension coupons

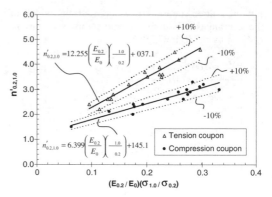

Figure 2. Relationship between $n'_{0.2,1.0}$ and $(E_{0.2}/E_0)(\sigma_{1.0}/\sigma_{0.2})$.

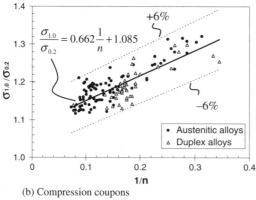

(b) Compression coupons

Figure 1. Relationship between $\sigma_{1.0}/\sigma_{0.2}$ and $1/n$.

$$n'_{0.2,1.0} = 6.399\left(E_{0.2}/E_0\right)\left(\sigma_{1.0}/\sigma_{0.2}\right) + 1.145$$
$$\text{for compression coupons (10b)}$$

Equations 9 and 10 provide predictions with maximum deviations from the test data being about $\pm 6\%$ and $\pm 10\%$ respectively. In Equation 10, $E_{0.2}/E_0 = 1/(1 + 0.002n/e)$ and the value of $\sigma_{1.0}/\sigma_{0.2}$ is given by Equation 9.

Ferritic alloys are characterized by n values ($= 6.5 \sim 16$) which are generally higher than those for austenitic and duplex alloys. Although the database used for developing Equations 9 and 10 covers only austenitic and duplex alloys, the n values of the test data range from 2.9 to 13.7, which also cover the major practical range of n values for ferritic alloys. Hence, Equations 9 and 10 are believed to provide reasonably close predictions for ferritic alloys as well.

2.4 Expression for σ_u and ε_u

The empirical expressions for σ_u and ε_u proposed by Rasmussen (2003), in terms of the basic parameters e

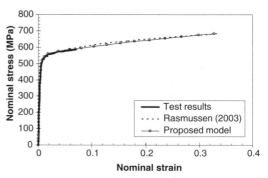

(a) Full-range stress-strain curves

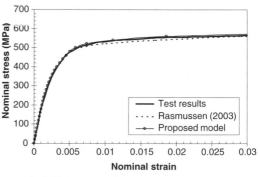

(b) Initial stress-strain curves

Figure 3. Nominal stress-strain curves for the tension coupon cut from the thick lipped channel section tested by Macdonald et al. (2000).

and n, are accepted in the present model for stainless steels in tension but the use of them for compressive behaviour is not justifiable, since these expressions were developed by interpreting the test data of tension coupons. Therefore, in the proposed model,

Rasmussen's expressions for σ_u and ε_u are adopted only for tensile strains:

$$\sigma_u^{ten}/\sigma_{0.2}^{ten} = \left(0.2 + 185\,e^{ten}\right)^{-1}$$

for austenitic and duplex alloys (11a)

$$\sigma_u^{ten}/\sigma_{0.2}^{ten} = \frac{1 - 0.0375\left(n^{ten} - 5\right)}{0.2 + 185\,e^{ten}}$$ for all alloys (11b)

$$\varepsilon_u^{ten} = 1 - \sigma_{0.2}^{ten}/\sigma_u^{ten} \qquad (11c)$$

in which the superscript *"ten"* refers to tension.

There has been sufficient experimental evidence to suggest that the macroscopic stress-strain curve of metal in simple compression coincides with that in simple tension when the true stress is plotted against the true strain (Cottrell 1964, Kalpakjian 1991, Chakrabarty 2000). Therefore, it is reasonable to assume that the true stress-strain curve of a stainless steel alloy in uniaxial compression coincides with that in uniaxial tension at sufficiently large strains such as the ultimate tensile strain. This assumption means that the true ultimate compressive stress σ_{tu}^{com} and the true ultimate compressive strain ε_{tu}^{com} can be approximated by the corresponding true ultimate tensile stress σ_{tu}^{ten} and true ultimate tensile strain ε_{tu}^{ten}.

Making use of this approximation (i.e. $\sigma_{tu}^{com} \approx \sigma_{tu}^{ten}$ and $\varepsilon_{tu}^{com} \approx \varepsilon_{tu}^{ten}$) and Equation 3, the nominal ultimate compressive stress σ_u^{com} and the nominal ultimate compressive strain ε_u^{com} can be obtained as

$$\sigma_u^{com} \approx \sigma_u^{ten}\left(1 + \varepsilon_u^{ten}\right)^2 \text{ and } \varepsilon_u^{com} \approx 1 - \left(1 + \varepsilon_u^{ten}\right)^{-1} \quad (12)$$

in which σ_u^{ten} and ε_u^{ten} are given by Equation 11. Finally, σ_u and ε_u in Equation 4c are given by σ_u^{ten} and ε_u^{ten} (Equation 11) for tensile strains and by σ_u^{com} and ε_u^{com} (Equation 12) for compressive strains.

3 COMPARISON WITH TEST DATA

A total of 39 experimental stress-strain curves available in the existing literature (Korvink et al. 1995, Macdonald et al. 2000, Rasmussen et al. 2002, Gardner & Nethercot 2004) were used to assess the accuracy of the proposed three-stage stress-strain model. The test data included results from both tension and compression coupon tests on different alloys: 31 tests on austenitic alloys, 4 tests on duplex alloys and 4 tests on ferritic alloys.

In obtaining stress-strain curves using the proposed three-stage stress-strain model for comparison, the values of the basic Ramberg-Osgood parameters reported in the existing literature were used. Representative comparisons between the proposed stress-strain model and the experimental stress-strain curves are shown in

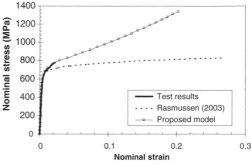

(a) Full-range stress-strain curves

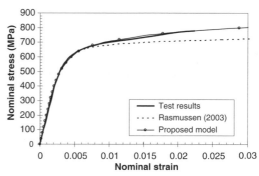

(b) Initial stress-strain curves

Figure 4. Nominal stress-strain curves for the transverse compression coupon cut from the duplex stainless steel plate tested by Rasmussen et al. (2002).

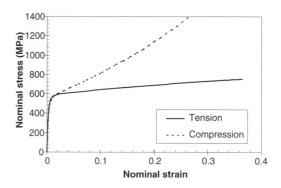

Figure 5. Typical nominal stress-strain curves for stainless steels defined by the proposed stress-strain model.

Figures 3 and 4. The whole set of comparisons can be found elsewhere (Quach 2005, Quach et al. 2007). Figures 3 and 4 show clearly that Rasmussen's (2003) model underestimates stresses at strains $\varepsilon > \varepsilon_{0.2}$ for compression coupon tests; the difference between the experimental results and the predictions increases with the strain.

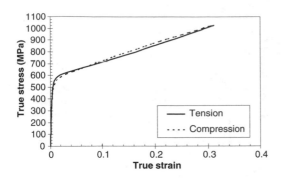

Figure 6. Typical true stress-strain curves for stainless steels defined by the proposed stress-strain model.

The proposed three-stage stress-strain model is generally in close agreement with the experimental stress-strain curves over wide ranges of both tensile and compressive strains. Typical nominal stress-strain curves for stainless steels predicted by the proposed stress-strain model are shown in Figure 5 for the full ranges of tensile and compressive strains, while the corresponding true stress-strain curves are shown in Figure 6.

4 CONCLUSIONS

This paper has presented a new full-range stress-strain model for stainless steel alloys, which is capable of accurate predictions over the full ranges of both tensile and compressive strains, as is required by advanced finite element simulations of cold-formed stainless steel sections covering both the manufacturing process and behaviour under loading. At the first instance, the stress-strain curves of stainless steels over full ranges of strains need to be characterized using not only the basic Ramberg-Osgood parameters but also certain additional parameters which may not be available in existing design specification or cannot be obtained from tests due to experimental limitations (e.g. ultimate compressive stress). In the proposed stress-strain model, these additional parameters are also defined using the basic Ramberg-Osgood parameters, based on careful interpretations of existing experimental data. As a result, the new full-range stress-strain model is defined using the basic Ramberg-Osgood parameters alone. The proposed stress-strain model has been compared with available experimental stress-strain curves, and in general, very close agreement has been demonstrated over wide ranges of both tensile and compressive strains. In particular, the new model is much more accurate than Rasmussen's (2003) model in predicting the full-range compressive stress-strain behaviour.

ACKNOWLEDGEMENTS

The authors would like to thank The Hong Kong Poly-technic University (Project No. G-V864), University of Macau (Ref. No. RG070/05-06S/QWM/FST) and the Research Grants Council of the Hong Kong S.A.R. (Project No. PolyU5056/02E) for their financial support.

REFERENCES

ASCE 2002. *Specification for the Design of Cold-Formed Stainless Steel Structural Members, SEI/ASCE 8–02.* New York: American Society of Civil Engineers.

AS/NZS 2001. *Cold-Formed Stainless Steel Structures, AS/NZS 4673:2001.* Sydney: Australian/New Zealand Standard, Standards Australia.

Chakrabarty, J. 2000. *Applied Plasticity.* New York: Springer-Verlag, Inc.

Cottrell, A.H. 1964. *The Mechanical Properties of Matter.* New York: Wiley.

Gardner, L. & Nethercot, D.A. 2004. Experiments on stainless steel hollow sections – Part 1: Material and cross-sectional behaviour. *Journal of Constructional Steel Research* 60: 1291–1318.

Kalpakjian, S. 1991. *Manufacturing Processes for Engineering Materials, 2nd edition.* United States: Addison-Wesley, Inc.

Korvink, S.A., Van den Berg, G.J. & Van den Merwe, P. 1995. Web crippling of stainless steel cold-formed beams. *Journal of Constructional Steel Research* 34: 225–248.

Macdonald, M., Rhodes, J. & Taylor, G.T. 2000. Mechanical properties of stainless steel lipped channels. *Proceedings of the Fifteenth International Specialty Conference on Cold-formed Steel Structures,* St. Louis, Missouri, U.S.A., 19–20 October 2000: 673–686.

Mirambell, E. & Real, E. 2000. On the calculation of deflections in structural stainless steel beams: an experimental and numerical investigation. *Journal of Constructional Steel Research* 54: 109–133.

Olsson, A. 2001. *Stainless Steel Plasticity – Material Modelling and Structural Applications,* Doctoral Thesis. Sweden: Division of Steel Structures, Luleå University of Technology.

Quach, W.M. 2005. *Residual Stresses in Cold-Formed Steel Sections and Their Effect on Column Behaviour,* PhD Thesis. Hong Kong: Department of Civil and Structural Engineering, The Hong Kong Polytechnic University.

Quach, W.M., Teng, J.G. & Chung, K.F. 2007. Three-stage full-range stress-strain model for stainless steels, in preparation.

Rasmussen, K.J.R. 2003. Full-range stress-strain curves for stainless steel alloys. *Journal of Constructional Steel Research* 59: 47–61.

Rasmussen, K.J.R., Burns, T., Bezkorovainy, P. & Bambach, M.R. 2002. *Numerical Modelling of Stainless Steel Plates in Compression, Research Report No. R813, March 2002.* Sydney: Department of Civil Engineering, University of Sydney.

SCI (1991). *Tests on Stainless Steel Materials, Report No. SCI-RT-251.* London: Steel Construction Institute.

Steel and Composite Structures – Wang & Choi (eds)
© *2007 Taylor & Francis Group, London, ISBN 978-0-415-45141-3*

Shear resistance of elliptical hollow sections

L. Gardner, T.M. Chan & C. Ramos
Department of Civil and Environmental Engineering, Imperial College London, UK

ABSTRACT: Following the introduction of hot-rolled elliptical hollow sections to the construction industry, recent research has been performed to develop supporting structural design guidance. This paper focuses on shear resistance. Twenty four shear tests were performed on hot-rolled steel elliptical hollow section members. The shear tests were arranged in a three-point bending configuration with span-to-depth ratios ranging from 1 to 8. This enabled the study of cross-section resistance in shear and the interaction between shear and bending. Measurements were taken of cross-section geometry, local and global initial geometric imperfections and material properties in tension. Key test results and sample load-deformation histories are presented. These results have been used to verify proposed design expressions for shear resistance and resistance under combined shear and bending.

1 INTRODUCTION

The introduction of hot-rolled elliptical hollow sections (EHS) to the construction industry opens a new chapter in the use of structural hollow sections. Their aesthetic appeal, complemented by sound structural efficiency, offers a promising alternative for engineers and architects to fulfil their design visions. This has been demonstrated by a number of recent projects including the Honda Central Sculpture in Goodwood, UK, the Jarrold Department Store in Norwich, UK (Corus 2006a) and the Society Bridge in Braemar, UK, shown in Figure 1.

Gardner & Chan (2007) have recently proposed cross-section classification criteria for EHS in compression, bending about both principal axes and combined compression plus bending. The study included the development of measures of slenderness for EHS and slenderness limits, providing efficient design rules for cross-section resistance in compression and in-plane bending. However, there remains a lack of verified design guidance for other structural phenomena. Development of such guidance is currently underway, and this paper focuses on the scenario for shear. Detailed experimental studies are described herein and design recommendations are made.

Initial studies of EHS connections have been made by Bortolotti et al. (2003), Choo et al. (2003), Pietrapertosa & Jaspart (2003) and Willibald et al. (2006).

2 EXPERIMENTAL STUDIES

A series of precise full-scale laboratory tests on EHS (grade S355), manufactured by Corus Tubes (Corus

Figure 1. Use of elliptical hollow sections for the Society Bridge in Braemar, UK.

2006b), was performed at Imperial College London. To date, the test programme has included tensile coupon tests, stub column tests, in-plane bending tests, shear tests and column buckling tests. Results from the stub column tests and bending tests have been reported by Chan & Gardner (2007) and Chan & Gardner (submitted-a) respectively. The EHS had an aspect ratio of two, with the overall cross-sectional dimensions being 150×75 mm, with thicknesses of 4 mm, 5 mm and 6.3 mm. A total of six tensile coupon tests and twenty four shear tests were conducted.

Table 1. Mean measured dimensions of tensile coupons.

Specimen	Width b (mm)	Thickness t (mm)
150 × 75 × 4-TC1	19.99	4.15
150 × 75 × 4-TC2	20.03	4.16
150 × 75 × 5-TC1	20.06	5.10
150 × 75 × 5-TC2	20.15	5.08
150 × 75 × 6.3-TC1	19.90	6.43
150 × 75 × 6.3-TC2	19.93	6.36

Table 2. Key results from tensile coupon tests.

Specimen	Young's Modulus E (N/mm^2)	Yield stress f_y (N/mm^2)	Ultimate tensile stress f_u (N/mm^2)
150 × 75 × 4-TC1	217400	380	512
150 × 75 × 4-TC2	217700	373	514
150 × 75 × 5-TC1	216900	374	506
150 × 75 × 5-TC2	217200	364	503
150 × 75 × 6.3-TC1	217700	381	509
150 × 75 × 6.3-TC2	215200	400	515

2.1 Tensile coupon tests

The basic stress-strain behaviour of the material for each of the tested section sizes was determined through tensile coupon tests carried out in accordance with EN 10002–1 (2001).

Two parallel coupons, each with the nominal dimensions of 320 × 20 mm, were machined longitudinally along the centreline of the flattest portions of each of the tested cross-sections. The tensile tests were performed using an Amsler hydraulic testing machine. To prevent slippage of the coupons in the jaws of the testing machine, holes were drilled and reamed 20 mm from each end of the coupons for pins to be inserted.

Linear electrical strain gauges were affixed at the midpoint of each side of the tensile coupons. Load, strain, displacement and input voltage were all recorded using the data acquisition equipment DATASCAN and logged using the DSLOG computer package. Mean measured dimensions and the key results from the six tensile coupon tests are reported in Tables 1 and 2.

2.2 Shear tests

The shear tests were arranged in a three-point bending configuration (see Figures 2 and 3) providing a

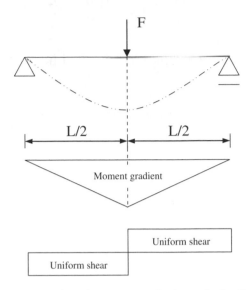

Figure 2. Schematic arrangement for three-point bending test.

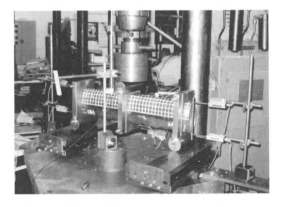

Figure 3. Three-point bending test setup.

bending moment gradient and uniform shear loading. A range of spans was tested to investigate the interaction between shear and bending. For bending about the minor axis (loading in the y-y direction), the span-to-depth ratio ranged from 2 to 8, whilst for major axis bending (loading in the z-z direction), the span-to-depth ratio ranged from 1 to 4. The tested beams were loaded at mid-span using a 100 T Amsler hydraulic actuator. The vertical displacement was measured with an LVDT, whilst two additional LVDTs were positioned at each end of the beam to measure end rotation. Four linear electrical resistance strain gauges were affixed to the extreme tensile and compressive fibres of the section at a distance of 50 mm either side of the mid-span of the beam. Load, strain, displacement, and input voltage were all recorded using the

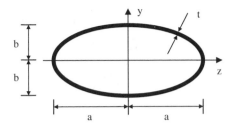

Figure 4. Geometry of an elliptical hollow section.

Table 3. Mean measured cross-sectional dimensions for shear specimens.

Specimen	Major axis outer diameter $2a$ (mm)	Minor axis outer diameter $2b$ (mm)	Thickness t (mm)
$150 \times 75 \times 4.0$-S1	150.42	75.64	4.24
$150 \times 75 \times 5.0$-S1	150.01	75.90	5.20
$150 \times 75 \times 6.3$-S1	148.91	75.84	6.36
$150 \times 75 \times 4.0$-S2	150.51	75.60	4.24
$150 \times 75 \times 5.0$-S2	150.53	75.63	5.17
$150 \times 75 \times 6.3$-S2	148.71	75.77	6.31
$150 \times 75 \times 4.0$-S3	150.39	75.44	4.22
$150 \times 75 \times 5.0$-S3	150.44	75.45	5.26
$150 \times 75 \times 6.3$-S3	148.74	75.83	6.31
$150 \times 75 \times 4.0$-S4	150.53	75.67	4.22
$150 \times 75 \times 5.0$-S4	150.11	75.85	5.23
$150 \times 75 \times 6.3$-S4	148.74	75.83	6.30
$150 \times 75 \times 4.0$-S5	150.65	75.47	4.19
$150 \times 75 \times 5.0$-S5	150.41	75.69	5.08
$150 \times 75 \times 6.3$-S5	148.57	76.01	6.31
$150 \times 75 \times 4.0$-S6	150.51	75.65	4.20
$150 \times 75 \times 5.0$-S6	149.91	75.91	5.16
$150 \times 75 \times 6.3$-S6	148.75	76.05	6.31
$150 \times 75 \times 4.0$-S7	150.53	75.27	4.19
$150 \times 75 \times 5.0$-S7	150.71	75.26	5.26
$150 \times 75 \times 6.3$-S7	148.51	75.89	6.29
$150 \times 75 \times 4.0$-S8	150.49	75.56	4.17
$150 \times 75 \times 5.0$-S8	150.21	75.65	5.06
$150 \times 75 \times 6.3$-S8	148.67	75.85	6.30

Table 4. Summary of results of shear tests.

Specimen	Length L (mm)	Direction of loading	Ultimate load F_u (kN)
$150 \times 75 \times 4.0$-S1	600	y-y	103
$150 \times 75 \times 5.0$-S1	600	y-y	119
$150 \times 75 \times 6.3$-S1	600	y-y	148
$150 \times 75 \times 4.0$-S2	600	z-z	189
$150 \times 75 \times 5.0$-S2	600	z-z	237
$150 \times 75 \times 6.3$-S2	600	z-z	303
$150 \times 75 \times 4.0$-S3	450	y-y	124
$150 \times 75 \times 5.0$-S3	450	y-y	157
$150 \times 75 \times 6.3$-S3	450	y-y	192
$150 \times 75 \times 4.0$-S4	450	z-z	236
$150 \times 75 \times 5.0$-S4	450	z-z	301
$150 \times 75 \times 6.3$-S4	450	z-z	393
$150 \times 75 \times 4.0$-S5	300	y-y	173
$150 \times 75 \times 5.0$-S5	300	y-y	212
$150 \times 75 \times 6.3$-S5	300	y-y	265
$150 \times 75 \times 4.0$-S6	300	z-z	308
$150 \times 75 \times 5.0$-S6	300	z-z	403
$150 \times 75 \times 6.3$-S6	300	z-z	527
$150 \times 75 \times 4.0$-S7	150	y-y	204
$150 \times 75 \times 5.0$-S7	150	y-y	270
$150 \times 75 \times 6.3$-S7	150	y-y	355
$150 \times 75 \times 4.0$-S8	150	z-z	426
$150 \times 75 \times 5.0$-S8	150	z-z	551
$150 \times 75 \times 6.3$-S8	150	z-z	686

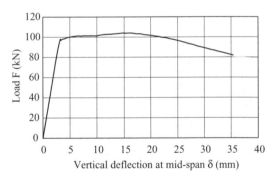

Figure 5. Load versus mid-span deflection curve for $150 \times 75 \times 4.0$-S1 (minor axis).

data acquisition equipment DATASCAN and logged using the DSLOG computer package.

The geometry of an elliptical hollow section is depicted in Figure 4 and the mean measured dimensions and the key results from the shear tests are summarised in Tables 3 and 4. Geometric properties for EHS are defined in EN 10210–2 (2006).

Figures 5 and 6 show typical load versus mid-span deflection curves for shear tests about the minor and major axes respectively. Full test results have been reported by Chan & Gardner (submitted-b).

3 SHEAR RESISTANCE

According to EN 1993–1–1 (2005), the design value of the shear force V_{Ed} at each cross section should satisfy Equation 1.

$$\frac{V_{Ed}}{V_{c,Rd}} \leq 1 \tag{1}$$

where $V_{c,Rd}$ is the design shear resistance, which may be evaluated on the basis of an elastic or a plastic shear

147

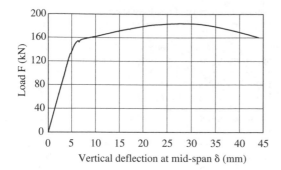

Figure 6. Load versus mid-span deflection curve for 150 × 75 × 4.0-S2 (major axis).

stress distribution. These two cases are considered in the following sub-sections.

3.1 *Elastic shear resistance*

To remain elastic, the design shear stress τ_{Ed} should satisfy the following criterion at all points in the cross-section:

$$\frac{\tau_{Ed}}{f_y/(\sqrt{3}\gamma_{M0})} \leq 1 \qquad (2)$$

where γ_{M0} is a partial factor for cross-section resistance, generally taken equal to unity. The elastic shear stress distribution can be approximated by the fundamental shear formula given by Equation (3).

$$\tau_{Ed} = \frac{V_{Ed}S}{It} \qquad (3)$$

where V_{Ed} is the design shear force, S is the first moment of area above the level at which the shear stress is being evaluated, I is the second moment of area of the whole cross-section and t is the thickness at the examined point. Thus, for a specific EHS under a given design shear force V_{Ed}, I and t are constant, and the elastic shear stress, which is directly proportional to S, varies with the distance from the neutral axis. For a typical EHS of dimensions $2a = 150\,\text{mm}$, $2b = 75\,\text{mm}$ and $t = 4\,\text{mm}$ under transverse shear loading in the y-y and z-z directions, the elastic shear stress distributions are shown in Figure 7.

Similarly to other doubly symmetric sections subject to shear in the y-y and z-z directions, the elastic shear stress distribution of Figure 7 varies parabolically with depth from zero shear stress at the extremities of the cross-section to a maximum shear stress (limited to the yield stress in shear τ_y) at the neutral axis. The minor difference in shear stress distribution between the two loading directions relates to the section geometry.

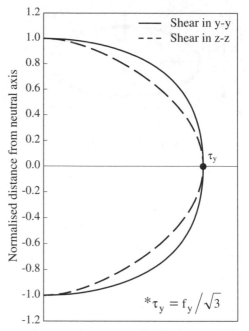

Average shear-stress distribution τ_{Ed} (N/mm2)

Figure 7. Elastic shear stress distribution for transverse load in the y-y and z-z directions.

3.2 *Plastic shear resistance*

In EN 1993–1–1 (2005), the design plastic shear resistance $V_{pl,Rd}$ of a cross-section is given by Equation 4.

$$V_{pl,Rd} = \frac{A_v\,(f_y/\sqrt{3})}{\gamma_{M0}} \qquad (4)$$

where A_v is the shear area, f_y is the material yield strength. For circular hollow sections (CHS) and tubes of uniform thickness, it is suggested that the shear area is defined as $2A/\pi$, where A is the cross-sectional area. This formula can be derived by considering an infinitesimal area of a CHS as shown in Figure 8.

Assuming that the shear stress τ_0 is distributed uniformly around the cross-section, the transverse load V is given by:

$$V = 2 \times \int_0^\pi (r \times d\theta \times t) \times (\tau_0 \times \sin\theta) \qquad (5)$$

$$V = 4 \times r \times t \times \tau_0 \qquad (6)$$

$$V = \frac{2A\tau_0}{\pi} \qquad (7)$$

$$\therefore A_v = \frac{V}{\tau_0} = \frac{2A}{\pi} \qquad (8)$$

148

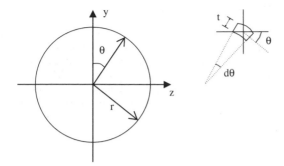

Figure 8. Geometry of a circular hollow section.

The same approach may also be applied to elliptical hollow sections. For an EHS of uniform thickness subject to transverse loading along the y-y axis, assuming a uniform shear stress distribution around the cross-section, the transverse load is balanced by the summation of the vertical component of the shear stress $\times$ dA. This is found to be equal to the product of the vertical projection of the area (given by $[(2b-t)+(2b-t)]t = 2(2b-t)t$) with shear stress. Likewise, when the transverse load is applied in the z-z direction, the corresponding projected area is equal to $2(2a-t)t$. Therefore, for an elliptical hollow section of constant thickness, the shear area A_v is defined by Equations 9 and 10.

$$A_v = (4b - 2t)t \quad \text{for load in the y-y direction} \quad (9)$$

$$A_v = (4a - 2t)t \quad \text{for load in the z-z direction} \quad (10)$$

3.3 Shear buckling

The shear resistance of hot-rolled elliptical hollow sections of current commercial proportions is dominated by yielding in shear, and hence the above formulations may be applied. However, for sections of more slender proportions, shear buckling may occur. Derivation of the elastic critical shear buckling stress for EHS and the slenderness limits beyond which shear buckling should be considered are currently underway at Imperial College London.

4 DESIGN GUIDANCE

The design shear resistance for elliptical hollow sections has been discussed in Section 3 and proposals for shear area A_v (Equations 9 and 10) to be used in the general formulation for plastic shear resistance (Equation 4) have been made. The influence of shear on bending moment resistance, together with associated design recommendations, is described in this section. Table 5 summarises the key results of the

Table 5. Summary of normalised results of shear tests.

Specimen	$V_u/V_{pl,Rd}$	$M_u/M_{pl,Rd}$ or $M_u/M_{el,Rd}$
150 × 75 × 4.0-S1*	0.39	1.40
150 × 75 × 5.0-S1*	0.38	1.39
150 × 75 × 6.3-S1*	0.37	1.41
150 × 75 × 4.0-S2	0.35	1.21
150 × 75 × 5.0-S2	0.37	1.29
150 × 75 × 6.3-S2	0.37	1.33
150 × 75 × 4.0-S3*	0.47	1.27
150 × 75 × 5.0-S3*	0.50	1.37
150 × 75 × 6.3-S3*	0.49	1.38
150 × 75 × 4.0-S4	0.44	1.14
150 × 75 × 5.0-S4	0.47	1.22
150 × 75 × 6.3-S4	0.49	1.30
150 × 75 × 4.0-S5*	0.67	1.19
150 × 75 × 5.0-S5*	0.69	1.26
150 × 75 × 6.3-S5*	0.67	1.27
150 × 75 × 4.0-S6	0.58	1.00
150 × 75 × 5.0-S6	0.63	1.11
150 × 75 × 6.3-S6	0.65	1.16
150 × 75 × 4.0-S7*	0.79	0.70
150 × 75 × 5.0-S7*	0.86	0.79
150 × 75 × 6.3-S7*	0.90	0.85
150 × 75 × 4.0-S8	0.80	0.70
150 × 75 × 5.0-S8	0.88	0.77
150 × 75 × 6.3-S8	0.85	0.75

*Results normalised by $M_{el,Rd}$.

shear tests where the ultimate test shear V_u $(=F_u/2)$ has been normalised by the plastic shear resistance given by Equation 4 and the ultimate test moment M_u $(=F_u/2 \times L/2)$ has been normalised by elastic moment resistance $M_{el,Rd} = f_y W_{el}$ (where W_{el} is the elastic section modulus) or plastic moment resistance $M_{pl,Rd} = f_y W_{pl}$ (where W_{pl} is the plastic section modulus), depending on the cross-section classification. According to the slenderness limits proposed by Gardner & Chan (2007), all tested sections are deemed Class 3 when bending about the minor axis (loading in the y-y direction) and Class 1 when bending about the major axis (loading in the z-z direction). Therefore, for loading in the z-z direction, the ultimate test moment has been normalised by $M_{pl,Rd}$ and for loading in the y-y direction, the ultimate test moment has been normalised by $M_{el,Rd}$. The normalised test results are also plotted in Figure 9.

The results demonstrate that where the shear force is less than half the plastic shear resistance, the effect of shear on the bending moment resistance is negligible. Conversely, for high shear force (greater than 50% of $V_{pl,Rd}$), there is a reduction in bending moment resistance.

To quantify this reduction in bending moment resistance due to the presence of shear, Eurocode 3 Part 1-1 (2005) proposes a reduced bending moment

149

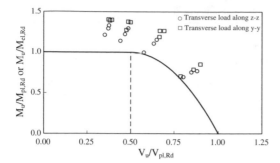

Figure 9. Interaction between shear and moment.

resistance $M_{V,Rd}$ based upon a reduced yield strength f_{yr} (Equation 11), that should be applied to the shear area A_v.

$$f_{yr} = (1-\rho)f_y \tag{11}$$

where $\rho = \left(\dfrac{2V_{Ed}}{V_{pl,Rd}} - 1 \right)^2 \tag{12}$

In Figure 9, the reduced yield strength f_{yr}, based on the reduction factor ρ, has been applied to the full cross-sectional area. This approach may be seen provide a design resistance under combined bending moment and shear that closely reflects the observed test behaviour. For cases where the applied shear force is greater than 50% of the shear resistance (defined by Equations 4, 9 and 10), it is therefore recommended that Equations 11 and 12 be adopted to calculate the reduced bending moment resistance for elliptical hollow sections.

5 CONCLUSIONS

As part of the development of comprehensive structural design rules for elliptical hollow sections, this paper describes a study of shear resistance. Results from six tensile coupon tests and twenty four shear tests in a three-point bending arrangement have been presented. Geometric properties and the key findings from the shear tests have been reported, including sample load-deformation histories. Elastic and plastic shear resistance functions have been proposed. The interaction between shear and bending moment has been discussed and results have demonstrated the appropriateness of adopting the reduced bending moment resistance formula in Eurocode 3.

ACKNOWLEDGEMENTS

The authors are grateful to the Dorothy Hodgkin Postgraduate Award Scheme for the project funding, and would like to thank Corus for the supply of test specimens and for funding contributions, Eddie Hole and Andrew Orton (Corus Tubes) for their technical input and Ron Millward and Alan Roberts (Imperial College London) for their assistance in the laboratory works.

REFERENCES

Bortolotti, E., Jaspart, J. P., Pietrapertosa, C., Nicaud, G., Petitjean, P. D. & Grimault, J. P. 2003. Testing and modelling of welded joints between elliptical hollow sections. *Proceedings of the 10th International Symposium on Tubular Structures, September 2003, Madrid.*
Chan, T. M. & Gardner, L. 2007. Compressive resistance of hot-rolled elliptical hollow sections. *Engineering Structures.* In press.
Chan, T. M. & Gardner, L. submitted-a. Bending strength of hot-rolled elliptical hollow sections. *Journal of Structural Engineering, ASCE.*
Chan, T. M. & Gardner, L. submitted-b. Shear resistance of elliptical hollow sections. *Journal of Constructional Steel Research.*
Choo, Y. S., Liang, J. X. & Lim, L. V. 2003. Static strength of elliptical hollow section X-joint under brace compression. *Proceedings of the 10th International Symposium on Tubular Structures, September 2003, Madrid.*
Corus 2006a. *Celsius® 355 Ovals*, Corus Tubes – Structural & Conveyance Business.
Corus 2006b. *Celsius® 355 Ovals – Sizes and Resistances Eurocode Version*, Corus Tubes – Structural & Conveyance Business.
EN 10002-1 2001. *Metallic materials – Tensile testing – Part 1: Method of test at ambient temperature*, CEN.
EN 10210–2 2006. *Hot finished structural hollow sections of non-alloy and fine grain steels – Part 2: Tolerances, dimensions and sectional properties*, CEN
EN 1993-1-1 2005. *Eurocode 3: Design of steel structures – Part 1-1: General rules and rules for buildings*, CEN.
Gardner, L. & Chan, T. M. 2007. Cross-section classification of elliptical hollow sections. *Steel and Composite Structures.* In press.
Pietrapertosa, C. & Jaspart, J. P. 2003. Study of the behaviour of welded joints composed of elliptical hollow sections. *Proceedings of the 10th International Symposium on Tubular Structures, September 2003, Madrid.*
Willibald, S., Packer, J. A. & Martinez-Saucedo, G. 2006. Behaviour of gusset plate connections to ends of round and elliptical hollow structural section members. *Canadian Journal of Civil Engineering*, **33(4)**, 373–383.

150

Steel and Composite Structures – Wang & Choi (eds)
© 2007 Taylor & Francis Group, London, ISBN 978-0-415-45141-3

Vertical shear resistance of a composite slim floor beam

P. Mäkeläinen & J. Zhang
Laboratory of Steel Structures, Department of Civil and Environmental Engineering,
Helsinki University of Technology, Espoo, Finland

S. Peltonen
Peikko Group, Lahti, Finland

ABSTRACT: A new application of headed studs will be presented. The use of headed studs is to increase the vertical shear resistance of the Finnish composite slim floor beam. This type of composite beam can be analyzed by strut-and-tie model (CEB-FIP Model Code 1990). The purpose of using the headed studs is to take advantage of its tensile resistance when embedded in concrete and make it to behave as "tie". The tests were arranged to get the real behavior of the single stud. The analysis procedure and results of the experiments will be compared and discussed. As a conclusion, headed studs can be applied into this composite beam to enhance the vertical shear resistance. More research will be carried and the more reliable and economic solutions are expected.

1 INTRODUCTION

Headed studs are widely used in composite building members behaving as shear connectors to provide composite actions between steel and concrete. There are also some applications in which the headed studs are used to increase the vertical shear resistance of concrete I-Beams (Grayed & Ghali, 2004). In this application, the vertical shear resistances are increased by utilizing the tensile strength of the steel of the heads studs.

In this article, a new method will be developed and applied to increase the vertical shear resistance of one typical Finnish slim floor beam named as Deltabeam. The significant feature of this new method is to take advantage of the tensile strength of the concrete when the studs inside are under tension load and pull-out.

Deltabeam is a hollow steel-concrete composite beam made from welded steel plates with holes in the sides. The box section of the steel beam is concreted at site during construction. After the concrete has hardened, Deltabeam acts as a composite beam working together with hollow-core slabs, composite slabs, or in-situ concrete slabs, as a uniform load-bearing structure (see Figure 1). The shear connection for Deltabeam is achieved by the dowel action between the steel web and concrete appearing in the openings through the steel section, after the hardening of the concrete (Peltonen & Leskelä, 2004)

Deltabeam has structural and economical advantages like other composite beams; it has considerable

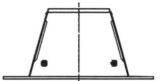
(a) The initial steel section of the beam

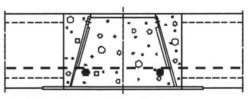
(b) Concrete cast and the slabs fixed

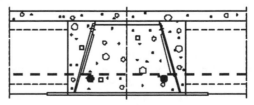
(c) The finished ones—slim floor composite beam

Figure 1. Deltabeam phases in different construction stages (Peltonen & Leskelä, 2004).

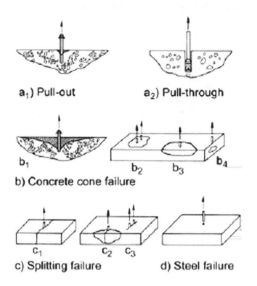

a$_1$) Pull-out   a$_2$) Pull-through

b) Concrete cone failure

c) Splitting failure   d) Steel failure

Figure 2. Tension failure modes (Eligehausen et al. 2006).

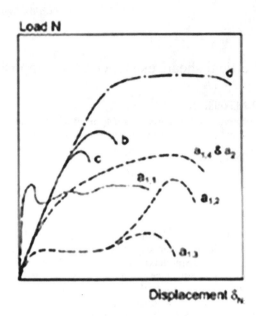

Figure 3. Idealized load-displacement curves for tensile loaded anchors exhibiting various failure modes (Fuchs et al. 1995).

good load-bearing capacity and fire resistance. The strut-and-tie Model (CEB, 1993) can be applied for analysis and design of Deltabeam.

Anyhow, when thin bottom plates are used, the vertical shear resistance of the Deltabeam becomes critical; the strut force in the concrete content enclosed in the steel sections exceeds the resistance of the bottom plate.

The idea of applying the headed studs to enhance the vertical shear resistance comes from that it is possible to safely exploit the tensile capacity of the concrete when design anchors, given that a sufficient conservative factor of safety is used (Eligehausen, et al. 2006). When the concrete struts tend to push out, the headed studs inside concrete are under the tension load, the internal tensile force can be transferred into the concrete by means of bearing interlock between the studs and the base material. In this way, tensile strength of the concrete around the stud can be used. It provides the possibility of using the headed studs in the Deltabeam to reinforce the concrete against the premature failures due to push out.

2   THE HEADED STUDS

2.1   Literature review

The headed studs used in this research are deformed steel bars, short relative to the length of concrete members, and provided with forged head for anchorage at one end. The diameter of the shaft is about 1/3 times of the diameter of the forged head. The diameter of the stud shaft used in this study is 20 mm, with the steel grade of A500HW.

Anchors typically exhibit four possible failure modes, shown in Figure 2, when loaded in tension. Each of modes is characterized by a unique load-displacement behavior. Figure 3 depicts the idealized load-displacement curves for various tension failure modes for fasteners.

Anchors, in general exhibit steel failure, concrete cone failure or concrete splitting. Pull-out failure will occur only if the mechanical interlock (bearing surface) is too small. The experiments have shown that an anchor head area equal to 9 or 10 times the cross-sectional area of the stem can provide secure mechanical anchorage with negligible slip (Ghali & Youakim, 2005). The pull-out failure does not happen to the studs discussed in this article.

Steel failure is ductile failure and happens by yielding of headed stud before any break-out of concrete occurs (Fuchs et al. 1995). The load-displacement curve is shown in Figure 3, d.

Concrete cone failure is a brittle failure mode characterized by the formation of a cone-shaped fracture surface in the concrete (Figure 2, b$_1$). The full tensile capacity of the concrete is utilized. Headed studs with an adequately large bearing surface will generate concrete cone breakout failure if the steel capacity is not exceeded. The load-displacement curve is shown in Figure 3, b.

Splitting failure occurs when dimensions of the concrete component are limited or the stud is installed closed to an edge (Figure 2, c$_1$ and c$_2$) or a line of

152

Figure 4. Typical concrete breakout cone obtained in the tests (Eligehausen et al. 1997).

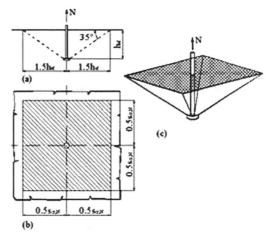

Figure 5. Idealized concrete cone for individual fastening under tensile loading after CCD method.

studs is installed in close proximity to each other (Figure 2, c_3).

In engineering practice, headed studs are indeed often used to transfer loads into reinforced concrete members. Provided the steel strength of the stud as well as the load bearing area of the head is large enough, a headed stud subjected to a tensile load normally fails by cone shaped concrete breakout. A typical concrete cone observed in experiments is shown in Figure 4 (Eligehausen et al. 1997). The failure is due to the failure of concrete in tension by forming a circumferential crack.

2.2 Concrete cone resistance

A concrete cone failure in a headed stud under tensile load is characterized by the formation of a conical fracture surface initiating and radiating from the top of the anchor head, provided there are sufficient distances from edges and corners. The initiation of crack starts already under very small loads. The crack propagation is small until about 90% of the bearing capacity is reached, and then it increases considerably (Sawade, 1994). The main influence factors of the bearing capacity are the mechanical properties of the concrete, the area of the failure surface and the embedment depth of the anchor.

Several models exist to describe the concrete cone resistance of anchors, such as:

- The Concrete Capacity Design (CCD) Model
- The Fracture Toughness Model
- The Sawade Model
- The Tensile Strength Model

The CCD (Fuchs et al. 1995) model predicts best the mean concrete cone failure load. It also has the advantage of simplicity and suitability for design purposes. The CCD model was developed to provide

an easily used model which lends itself to analyzing group and edge effects (Bazant, 1984). The fracture cone is idealized as a pyramid, assuming a quadratic base length of three times the embedment depth h_{ef}, (see Figure 5, a). The failure surface, therefore, corresponds to $(3h_{ef})^2$. In general, the following values can be taken: $s_{cr,N} = 3.0h_{ef}$ and $c_{cr,N} = 1.5h_{ef}$.

This fracture mechanics theory is adopted in CEB design guide for fastenings in concrete (CEB, 1995), the limitation of the minimum anchorage depth of the studs and the cracking condition of the base material are also taken into account in this design guide.

The concrete properties under tension are modeled by means of the tensile strength only, which is calculated from the concrete compressive strength using empirical formulas. The characteristic resistance $N^0_{Rk,c}$ of a single headed stud without edge and spacing effects, anchored in cracked concrete, is calculated as (CEB, 1995):

$$N^0_{Rk,c} = k_1 \cdot f_{ck}^{0.5} \cdot h_{ef}^{1.5} [N] \qquad (1)$$

$$k_1 = 7.5 N^{0.5} / mm^{0.5} \qquad (2)$$

The characteristic resistance $N_{Rk,c}$ of a headed stud or a group of headed studs in the case of concrete failure is given by the following equation (CEB, 1995):

$$N_{Rk,c} = N^0_{Rk,c} \cdot \Psi_{A,N} \cdot \Psi_{s,N} \cdot \Psi_{ec,N} \cdot \Psi_{re,N} \cdot \Psi_{ucr,N} \qquad (3)$$

$\Psi_{s,N}$ = the factor to take into account the influence of edges of the concrete member on the distribution of stresses in the concrete.

$\Psi_{ec,N}$ = the factor to take into account a group effect when different tension loads are acting on the individual anchors of a group.

153

The above two factors are not necessary to be taken into account in this study.

$\Psi_{re,N}$ = the shell spalling factor

$\Psi_{ucr,N}$ = the cracking factor (for base materials)

For this study, the anchorage depth is larger enough and the shell spalling factor can be taken as $\Psi_{re,N} = 1.0$; and the concrete might be assumed as cracked, then $\Psi_{ucr,N} = 1.0$.

So, the most important factor in equation (3) is $\Psi_{A,N}$. It is the factor to take into account the geometric effects of spacing and edge distance either for a single stud or a group anchors.

For a single anchor, the failure surface might not be as the base of an idealized quadratic pyramid with length equal to $3h_{ef}$ due to the edge effects, which means the break-out cone may be truncated if the stud located close to an edge. For a group of studs, each anchor in the group develops the failure surface of a single anchor, which may overlap with the failure surface from its neighbor, depending on the spacing between them. The resulting failure surface is equal to the enclosed area of the failure surfaces of all anchors. The factor $\Psi_{A,N}$ is calculated by the following formula (CEB, 1995):

$$\Psi_{A,N} = \frac{A_{c,N}}{A_{c;N}^0} \qquad (4)$$

$A_{c,N}^0$ = area of concrete cone of an individual anchor with a large spacing and edge distance at the concrete surface, idealizing the concrete cone as a pyramid with a height equal to $s_{cr,N}$.

$A_{c,N}$ = actual area of concrete cone of the anchorage at the concrete surface. It is limited by overlapping concrete cones of adjacent anchors ($s < s_{cr,N}$) as well as by edges of the concrete member ($c < c_{cr,N}$). It may be deduced from the idealized failure cone.

2.3 Assumption about the failure surface

In the above review, it is shown that the concrete cone failure load is mainly influenced by the actual area of concrete cone, the tensile stress of concrete and the anchor embedded depth h_{ef}. As a consequence, the failure surface of a single stud cast inside the Deltabeam-shaped concrete block is an important subject to be studied in this research.

The cross-section of the Deltabeam is a trapezoidal cross-section. Thus, the pull-out behavior of the single stud should be investigated in such shaped trapezoidal section. Assume the length of the specimen is enough for a single stud to develop the failure along the longitudinal direction of it. Obviously, the edge effects in transverse direction of the beam must be significant, while there are not edge effects along the longitudinal direction, see Figure 6(b), compared with the idealized failure cone in Figure 6(a). The assumed failure surface

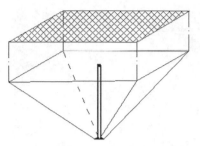

(a) Idealized concrete failure cone and correspoonding failure surface

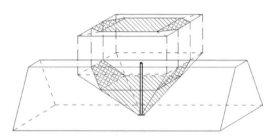

(b) Assumed concrete failure cone for the Deltabeam-shaped concrete block and corresponding failure surface

Figure 6.   Cooperation of the failure surface.

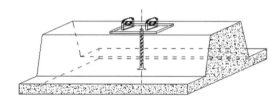

Figure 7.   The shape of specimens for single studs pulled out from Deltabeam-shaped concrete block.

in Figure 6(b) is hatched by shadows; the area of the shadows can be calculated according to geometrical dimensions of the beam.

3   TESTS

In order to inspect and verify the failure mode, failure surface assumed in previous part, to get the inclined angle of breaking out cone and to obtain the maximum failure load of a single headed stud, a series of commonly used size of Deltabeam-shaped concrete blocks were cast with the corresponding studs inside (see Figure 7). The length of the concrete specimens is 1 meter; the dimensions of the cross-section are shown in Figure 8.

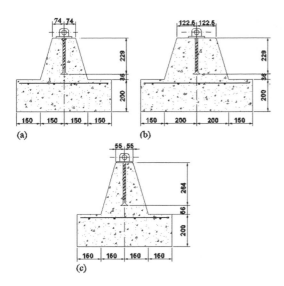

Figure 8. The shape of specimens for single studs pulled out from Deltabeam-shaped concrete block.

Table 1. The dimensions of the specimens.

	D26–300	D26–400	D32–300
Width of bottom flange (mm)	300	400	300
Height of specimen (mm)	260	260	320
Length of stud (mm)	229	230	265
Diameter of the stud (mm)	20		

The concrete grade is C30 (cubic)/7-days, the steel grade of the studs is A500 HW.

Three specimens were tested. The specimens are named with D26–300, D26–400 and D32–300. The dimensions of the specimens are shown in Table 1.

The setup of the tests are carefully arranged to make sure the accuracy of the results, see Figure 9. The bottom part of the specimens with thickness of 200 mm and the cantilever parts along the specimens are used to fix the specimens along the longitudinal direction and avoid the bending moments in the middle point at the top surface. Additionally, the specimens are fixed at middle points of the cantilever parts from both sides. The steel reinforcement bars are added at top level of the bottoms of the specimens, avoiding the shear cracking at the concrete corners.

The loading speed in the load history was 10 kN/s. One displacement measure was arranged at the top surface of the steel plates, where the studs were welded underside; the other two displacement measures were

Figure 9. The setup of the tests.

arranged at the top surface in the both ends of the specimens.

The failure patterns got from the tests were considerable fulfilled the CCD failure mode; see Figure 10(a), (b), and (c). The corresponding inclinations between the failure surface and the surface of the concrete members were 32.0°, 31.0° and 33.5°. These results are quite close to the angle 35° assumed in the CCD method and CEB design code.

The failure loads are listed in Table 2. The failure loads according to the assumptions shown in Figure 6(b) can be calculated according to equation (1), (2) and (3). The results are listed in Table 3.

There are obvious variations between the two groups of loads in Table 2 and Table 3, the variation is about 38% for D26-300, 22% and for D26-400 and 40% for D32-300.

The reason for these variations is that real failure surface is not exactly as the assumptions in Figure 6(b). Theoretically, the summit of the failure cone starts from the stems of the studs, and the cracks will continuously grow with an angle of 35° to the top surface of the concrete until the cracks reaching the edges. The lowest points of the cracks observed from the failure specimens are almost at the same level of the heads of the studs. This phenomenon indicates that the cracks in transverse direction of the specimens did not develop accurately with an angle of 35° to the top surface as in the CCD model. Another phenomenon should also be noticed: the cracks develop first horizontally along the longitudinal direction of the specimens in some certain distance, and after that they start to develop at the top surface of the concrete and form the break-out cones.

The above two phenomena cause the differences between the actual area of failure surface and the theoretical assumptions. The real areas of cracking surfaces measured from the failure specimens are about 18–35% more than the theoretical ones. It gives good explanations for the variations between the results in Table 2 and Table 3.

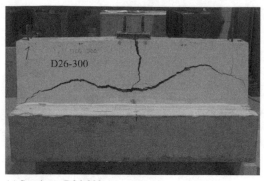

(a) Specimen D26-300

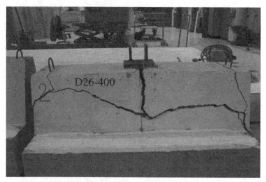

(b) Specimen D26-400

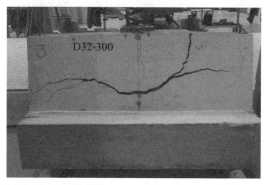

(c) Specimen D32-300

Figure 10. Failure shapes of the specimens.

As a conclusion, the failure mode in the above tests can be cataloged into concrete cone failure, the inclined angles of breaking out concrete cone are quite close to 35° and the areas of failure surfaces assumed based on the theory are smaller than the actual ones, which indicate that this assumption is conservative and safe.

Table 2. The failure loads in the tests.

Test No.	Name of the beam	F_{max_test}(kN)
1	D26-300	105.54
2	D26-400	125.84
3	D32-300	104.53

Table 3. The failure loads according to the calculations.

Calculation No.	Name of the beam	F_{max_cal}(kN)
1	D26-300	65.20
2	D26-400	97.57
3	D32-300	62.10

4 VERTICAL SHEAR RESISTANCE OF DELTABEAM

The Deltabeam is one type of hollow slim floor composite beam; the concrete body inside the steel sections provides the possibility of using truss model (CEB, 1993) when analyzing the flow of internal forces. The vertical shear resistance of the Deltabeam can be evaluated by considering a strut-and-tie system, which will develop in the concrete inside the boxed steel section (Leskelä, 1995). The resistance is defined by the component failure.

It is assumed that before the onset of shear failure the concrete is cracked diagonally in the region of the maximum shear force, and there are two main reasons, which will lead to the development of the shear failure: (1) compression failure of the concrete in the diagonal struts, the failure load expressed by $D_{R,S}$; (2) yielding of the webs between the web holes, the failure load expressed by $D_{R,W}$.

Then the minimum value of the $D_{R,S}$ and $D_{R,W}$ in vertical direction provides the limitation of the ultimate shear resistance of the beam.

The shear resistance of Deltabeam can be contributed by several parts: (1) the force anchored by web holes due to the dowel action; (2) the force anchored by the bottom flange; (3) the force anchored by other possible elements.

If the bottom steel flange is thick enough, the ultimate resistance can be developed unconditionally (Leskelä, 1999). However, when the bottom flange is not thick enough or the bottom flanges have lost their actions (such as fire situation), the contribution of steel bottom flange would be considerable small or neglected. The left anchorage forces could be not sufficient to prevent the concrete strut crushed out or the failure of the steel webs between the holes. The shear resistance of the beam will decrease.

As the tests shown in previous part, the steel headed studs can employ the local capacity of the concrete

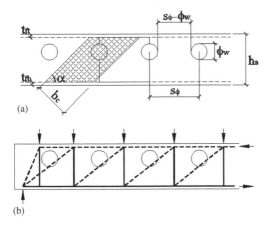

(a)

(b)

(a) Geometry for considering the diagonal cracking inside the steel section of the Deltabeam
(b) Truss model according to the design assumption of the Deltabeam

Figure 11. Truss model analysis for Deltabeam (Leskelä, 1999).

to carry tensile stresses. They can be arranged into Deltabeam. When the beam is under the vertical load, the concrete struts have tendency to crush down, as a consequence, the reaction forces drive the studs in tensile loads.

5 SUMMARY AND PROSPECTION

According to the above research work, it is sure that the headed studs can be applied into Deltabeam to provide the contribution to the vertical shear resistance.

More work will be carried out in the following research. The behaviors of the studs together with Deltabeam under vertical load should be analyzed and tested. The work will bring other two interesting topics: (1) the principle of arrangement of the studs into Deltabeam; (2) the composition action when studs applied into this composite beam.

The principle of arrangement of studs into Deltabeam is an optimizing problem, since the analysis of the beam is based on the strut-and-tie model, which provides the flexibility to adjust the struts angle to get more reliable and economic solutions.

The study of the composite action can get more efficient usage of all the materials included in this composite beam.

REFERENCES

Bazant, Z. P. 1984 Size effect in blunt fracture: Concrete, Rock, Metal. *Journal of Engineering Mechanics*, Vol. 101, No. 4.
Comité Euro-International du Béton (CEB), *CEB-FIP Model Code 1990*, Thomas Telford, Ltd., London, 1993, 437 p.
Comité Euro-International du Béton (CEB): COMITE EURO-INTERNATIONAL DU BETON 1995, *Design of Fastenings in Concrete*.
Eligehausen, R. Mallée, R. & Rehm, G. 1997 *Befestigungstechnik*. Berlin: Ernst and Sohn.
Eligehausen, R. Mallée, R. & Silva, J. F. 2006 *Anchorage in Concrete Construction*. Berlin: Ernst and Sohn.
Fuchs, W. Eligehausen, R. & Breen, J. B. 1995 Concrete Capacity Design (CCD) Approach for Fastening to Concrete. *ACI: Structural Journal* VOL. 92, NO.1.
Gayed, R.B. & Ghali, A. 2004 Double-Head Studs as Shear Reinforcement in Concrete I-Beams. *ACI: Structural Journal* VOL. 101, NO.4.
Ghali, A. & Youakim, S. A. 2005 Headed Studs in Concrete: State of the Art. *ACI: Structural Journal* VOL. 102, NO.5.
Leskelä, M.V. 1995 Vertical Shear Resistance of Deltabeam – Normal Temperature Design*. A report paper for Deltatek, Lahti, Oulu, Finland.
Leskelä, M.V. 1999 General Principle for Evaluating of the Vertical Shear Resistance of the Deltabeam—Normal Temperature Design and Fire Design*. Lahti, Oulu, Finland.
Peltonen, S. & Leskelä, M. V. 2004 Connection Behaviour of a Concrete Dowel in a Circular Web Hole of a Steel Beam: *Composite Construction in Steel and Concrete V*, South Africa.
Sawade, G. 1994 Ein energetisches Materialmodell zur Berechnung des Tragverhaltens von zugbeanspruchtem Beton, Dissertation, University of Stuttgart, Germany

* Not public papers

Steel and Composite Structures – Wang & Choi (eds)
© *2007 Taylor & Francis Group, London, ISBN 978-0-415-45141-3*

Nonlinear analysis of steel frames using element with midspan plastic hinge

Yu-shu Liu, Guo-qiang Li & Ya-mei He

School of Civil Engineering, Tongji University, Shanghai, China

ABSTRACT: A nonlinear analysis method of steel frames using element with midspan plastic hinge is proposed. This method can analyze the frame component imposed with laterally-distributed loads only using one element even that a plastic hinge appears within the component. By dividing the component into two parts at the location of the maximum moment, the incremental stiffness matrix of the two parts from time t to $t + dt$ are derived, then the beam element stiffness equation with midspan plastic hinge after the static condensation can be obtained. What's more, this method also considers the influences of some geometrical and material non-linear factors including second-order effect of axial forces, shear deformation, cross-sectional plastification, residual stress and initial imperfection. This method not only overcomes the time-consuming defect of plastic zone method of frame components because of the fine mesh discretization but also makes up for the problem of the traditional plastic hinge element that plastic hinges must form at the elemental ends. Analysis results show that the proposed method is satisfactory.

1 INTRODUCTION

The members of a steel frame may be subjected to laterally-distributed loads, so plastic hinges will be formed within the members. The common method[1~4] to treat this case for analysis of the frame is to arrange a node at the location of the plastic hinge within the member to divide the original one element into two or more elements representing the member. This will increase the number of nodes and degrees of freedom for analysis of the frame. Moreover, the traditional element with plastic hinge formed at the end(s) must fix the locations of nodes in advance, which can not suit the case that the locations of the plastic hinge within the member with laterally-distributed loads may vary during the loading process. In this paper, an approach for nonlinear analysis of steel frames using element with midspan plastic hinge is proposed. This approach can use one element to simulate one member in a frame even plastic hinge may form within the member.

2 NONLINEAR ANALYSIS METHOD OF STEEL FRAMES

2.1 *Refined Plastic Hinge Column Element*

Li and Shen[5] presented an improved plastic hinge model, which considered the cross-section plastifica-tion. Using this model the elasto-plastic incremental stiffness equation of the frame column element is given by

$$[k_P]\{\Delta\delta\} = \{\Delta f\} \tag{1}$$

Where, $\{\Delta\delta\}$ and $\{\Delta f\}$ refer to the incremental nodal displacements and forces, respectively, $[k_P]$ refers to the elasto-plastic stiffness matrix, and takes the following form

$$[k_p] = [k_e] - [k_e][G][E][L][E]^T[G]^T[k_e] \tag{2}$$

where

$$[L]^{-1} = [E]^T[G]^T([k_e] + [k_n])[G][E]$$

$$[k_n] = \text{diag}[\alpha_1 k_{e11}, \ \alpha_1 k_{e22}, \ \alpha_1 k_{e33},$$
$$\alpha_2 k_{e44}, \ \alpha_2 k_{e55}, \ \alpha_2 k_{e66}]$$

$$[E] = \begin{bmatrix} 1 & 1 & 1 & 0 & 0 & 0 \\ 0 & 0 & 0 & 1 & 1 & 1 \end{bmatrix}^T$$

$$[G] = \text{diag}\left[\frac{\partial x_1}{\partial N_1}, \ 0, \ \frac{\partial x_1}{\partial M_1}, \ \frac{\partial x_2}{\partial N_2}, \ 0, \ \frac{\partial x_2}{\partial M_2}\right]$$

$[k_e]$ represents the elastic stiffness matrix of the beam element accounting for the second order effect and shearing deformation[5]. In matrix $[G]$, $x_i (i = 1, 2)$

denotes the ultimate yield surface function of the section. In this paper, the ultimate yield surface function for I type section given by reference [2] and [6] is used here and can be written as

$$x_i = \left(\frac{N}{N_y}\right)^{1.3} + \frac{M}{M_p} = 1 \tag{3}$$

In matrix $[k_n]$, $\alpha_i(i = 1, 2)$ denotes the elasto-plastic hinge parameter of the two end sections, represents the plastification extent of the two end sections and can be expressed as

$$\alpha_i = \frac{r_i}{1 - r_i} \tag{4}$$

Where, $r_i(i = 1, 2)$ is the restoring force parameter of the two end sections and takes the form as

$$r_i = \begin{cases} 1 & M \leq M_{sN} \\ 1 - \dfrac{M - M_{sN}}{M_{pN} - M_{sN}}(1 - \beta) & M_{sN} \leq M \leq M_{pN} \\ \beta & M \geq M_{pN} \end{cases}$$

M, M_{sN} and M_{pN} represent the moment of the section, the initial yield moment the ultimate yield moment under the axial force N, respectively. β represents the material strain hardening coefficient, for normal low carbon steel and low alloyed steel, β can take 0.01~0.02, and M_{pN} can be given by equation (3).

The initial yield surface equation[2,6] without accounting for the influences of residual stress is expressed as

$$\frac{N}{N_y} + \frac{\gamma M}{M_p} = 1.0 \tag{5}$$

and $$M_{sN} = (1.0 - \frac{N}{N_y})M_p / \gamma \tag{6}$$

The initial yield surface equation[2,6] accounting for the influences of residual stress is expressed as

$$\frac{N}{0.8N_y} + \frac{\gamma M}{0.9M_p} = 1.0 \tag{7}$$

and $$M_{pN} = 0.9(1.0 - \frac{N}{0.8N_y})M_p / \gamma \tag{8}$$

where, γ is the plastification coefficient of the section.

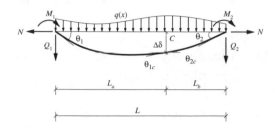

Figure 1. Beam element with midspan plastic hinge

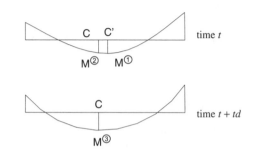

Figure 2. Position of maximum moment of different load step.

2.2 Refined Plastic Hinge Beam Element with Midspan Plastic Hinge

Figure 1 shows the beam element with midspan plastic hinge. Referring to Fig.1, an internal node C between elemental ends is inserted so that the element is divided into two parts, the lengths of which are L_a and L_b respectively. Assume the maximum bending moment, $\mathbf{M}^{①}$, at time t is at position C' and the maximum bending moment, $\mathbf{M}^{③}$, at time $t + \mathrm{d}t$ is at position C (see Fig. 2). For derivation of incremental stiffness matrix of the element during $t \rightarrow t + \mathrm{d}t$, a virtual state of moment, $\mathbf{M}^{②}$, is conceived, which is the bending moment at the same position of $\mathbf{M}^{③}$ at the time t. The incremental stiffness relationship of each part of the element can be expressed as the standard form as for the part of L_a

$$\begin{Bmatrix} dQ_1 \\ dM_1 \\ dQ_{1c} \\ dM_{1c} \end{Bmatrix} = [K_{pa}] \begin{Bmatrix} d\delta_1 \\ d\theta_1 \\ d\delta_{1c} \\ d\theta_{1c} \end{Bmatrix} = \begin{bmatrix} a_{11} & a_{12} & a_{13} & a_{14} \\ & a_{22} & a_{23} & a_{24} \\ & & a_{33} & a_{34} \\ & & & a_{44} \end{bmatrix} \begin{Bmatrix} d\delta_1 \\ d\theta_1 \\ d\delta_{1c} \\ d\theta_{1c} \end{Bmatrix}$$

$$\tag{9a}$$

$$\begin{Bmatrix} dQ_{2c} \\ dM_{2c} \\ dQ_2 \\ dM_2 \end{Bmatrix} = [K_{pb}] \begin{Bmatrix} d\delta_{2c} \\ d\theta_{2c} \\ d\delta_2 \\ d\theta_2 \end{Bmatrix} = \begin{bmatrix} b_{11} & b_{12} & b_{13} & b_{14} \\ & b_{22} & b_{23} & b_{24} \\ & & b_{33} & b_{34} \\ & & & b_{44} \end{bmatrix} \begin{Bmatrix} d\delta_{2c} \\ d\theta_{2c} \\ d\delta_2 \\ d\theta_2 \end{Bmatrix}$$

$$\tag{9b}$$

160

where $[K_{pa}]$ and $[K_{pb}]^{[5]}$ are the elasto-plastic stiffness matrices for the parts of L_a and L_b of the element respectively, a_{ij} and b_{ij} ($i,j=1,2,3,4$) are the corresponding elements in such matrices.

It can be seen from Fig.1 that the two parts of the elements share the same deformation components at their junction, namely $d\delta_{1c}=d\delta_{2c}=d\delta_c$ and $d\theta_{1c}=d\theta_{2c}=d\theta_c$. Combining Eq. (9a) and Eq. (9b), one has

$$
\begin{Bmatrix} dQ_1 \\ dM_1 \\ dQ_2 \\ dM_2 \\ dQ_{1c}+dQ_{2c} \\ dM_{1c}+dM_{2c} \end{Bmatrix} = \begin{bmatrix} a_{11} & a_{12} & 0 & 0 & a_{13} & a_{14} \\ & a_{22} & 0 & 0 & a_{23} & a_{24} \\ & & b_{33} & b_{34} & b_{13} & b_{23} \\ & & & b_{44} & b_{14} & b_{24} \\ & & & & a_{33}+b_{11} & a_{34}+b_{12} \\ & & & & a_{44}+b_{22} \end{bmatrix} \begin{Bmatrix} d\delta_1 \\ d\theta_1 \\ d\delta_2 \\ d\theta_2 \\ d\delta_c \\ d\theta_c \end{Bmatrix}
$$

$$(10)$$

For the purpose of static condensation to eliminate the freedom degree of the displacements of internal node, above stiffness matrix is partitioned into internal and external degrees of freedom as

$$
\begin{Bmatrix} df_e \\ df_i \end{Bmatrix} = \begin{bmatrix} k_{ee} & k_{ei} \\ k_{ei}^T & k_{ii} \end{bmatrix} \begin{Bmatrix} d\delta_e \\ d\delta_i \end{Bmatrix}
$$

$$(11)$$

where $\{df_e\}$ and $\{df_i\}$ are the elemental end and internal force vectors respectively, $\{d\delta_e\}$ and $\{d\delta_i\}$ are elemental end and internal deformation vectors respectively. Their expressions are as follows

$$\{df_e\} = [dQ_1,\ dM_1,\ dQ_2,\ dM_2]^T,$$
$$\{d\delta_e\} = [d\delta_1,\ d\theta_1,\ d\delta_2,\ d\delta_2]^T,$$
$$\{df_i\} = [dQ_{1c}+dQ_{2c},\ dM_{1c}+dM_{2c}]^T,$$
$$\{d\delta_i\} = [d\delta_c,\ d\theta_c]^T,$$

$$
k_{ee} = \begin{bmatrix} a_{11} & a_{12} & 0 & 0 \\ a_{12} & a_{22} & 0 & 0 \\ 0 & 0 & b_{33} & b_{34} \\ 0 & 0 & b_{34} & b_{44} \end{bmatrix},
$$

$$
k_{ei} = \begin{bmatrix} a_{13} & a_{14} \\ a_{23} & a_{24} \\ b_{13} & b_{23} \\ b_{14} & b_{24} \end{bmatrix}, k_{ii} = \begin{bmatrix} a_{33}+b_{11} & a_{34}+b_{12} \\ a_{34}+b_{12} & a_{44}+b_{22} \end{bmatrix}
$$

$$(12)$$

Since no external forces are applied at internal node C, namely $\{df_i\}=\{0\}$, $\{d\theta_i\}$ in Eq. (11) can be expressed with $\{d\theta_e\}$. The stiffness equation condensed off internal displacement vector is as

$$\left(k_{ee}-k_{ei}k_{ii}^{-1}k_{ei}^T\right)\{d\delta_e\}=\{df_e\}$$

$$(13)$$

In above derivation, it is assumed that the internal plastic hinge occurs at position of C at time t, and

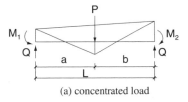

(a) concentrated load

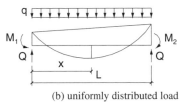

(b) uniformly distributed load

Figure 3. Load patterns within beam span.

the moment increases from $\mathbf{M}^{②}$ at t to $\mathbf{M}^{②}$ at $t=dt$. But actually in the duration $t \to t+dt$, the moment change should have been from $\mathbf{M}^{①}$ at position of C' to $\mathbf{M}^{③}$ at position of C. A stiffness matrix modification ($[k_{ee}-k_{ei}k_{ii}^{-1}k_{ei}^T]_{C,t}-[k_{ee}-k_{ei}k_{ii}^{-1}k_{ei}^T]_{C',t}$) may be superimposed to approximately take the effect from position change of internal plastic hinge into account. The subscripts in the stiffness matrix modification indicate the position and the time of maximum bending moment.

Assume the internal plastic hinge occurs at the position of maximum bending moment between two ends. The position of the maximum bending moment between the two ends of the element, position C, varies in loading process. Hence, the rational way to trace the internal plastic hinge is to calculate the position of the maximum bending moment at each loading step after elemental yielding. Two common internal loading patterns for beam elements are concentrated load and uniformly distributed load, as shown in Fig. 3.

If one concentrated load is applied within the beam span (see Fig.3a), the position of the maximum moment within span is certainly the loading position. But, if a uniformly distributed load is applied(see Fig.3b), the position of the maximum moment within span is changeable. The condition of the maximum moment within the beam span is that

$$\frac{dM(x)}{dx}=0 \quad \text{or} \quad Q(x)=0 \quad (14)$$

The shear force at end 1 can be expressed as

$$Q_1 = \frac{M_1-M_2}{L}+\frac{1}{2}qL \quad (15)$$

161

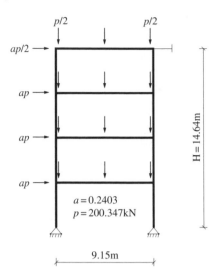

Figure 4. Four-story frame with concentrated loads at beams.

$$a = 0.2403$$
$$p = 200.347\text{kN}$$

9.15m

Table 1. Dimensions of all the components of four-storey steel frame.

Section	H (mm)	B (mm)	t_w (mm)	t_f (mm)	A (mm²)	I (×10⁶ mm⁴)
W16 × 40	406.7	177.5	7.9	12.7	7610	215
W10 × 60	259.6	256	10.7	17.3	11400	142
W12 × 79	314.5	306.8	11.9	18.8	15000	276

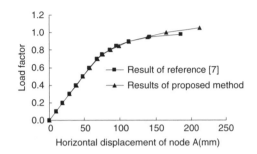

Figure 5. Load-displacement curve of four-storey steel frame.

And letting the shear force be equal to zero yields the position of the maximum moment desired

$$x = \frac{M_1 - M_2}{qL} + \frac{1}{2}L \qquad (16)$$

As for the beam element with both concentrated load and uniformly distributed load within span, one can divide this element into two segments at the position where the concentrated load applied. The maximum moment position of each segment can be determined according the method for the uniformly distributed load case as above-mentioned. With comparison of the maximum moments of two segments of the element induced by the uniformly distributed load and the bending moment where the concentrated load applied, the real maximum moment of this beam element can be obtained with the maximum of the above three moments.

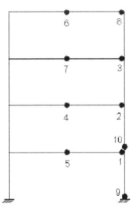

Figure 6. Appearing sequence of plastic hinges of four-storey steel frame.

3 NUMERICAL EXAMPLES

The structure examined is a four-story frame with mid-span concentrated loads as shown in Fig. 4. Table 1 gives the frame member size. The material elastic modulus E of steel is 206 kN/mm², the sections of all the beams are W16 × 40, the sections of the first storey columns are W12 × 79 and the sections of the other columns are W10 × 60.

The horizontal displacement versus load factor curves both obtained by analysis with the elements with internal hinge proposed in this paper and with the

normal elements through dividing the frame beam into two elements[7] are shown in Fig. 5. The ultimate load factor obtained with the proposed elements is $\lambda = 1.03$ while that with normal elements[7] is $\lambda = 0.99$. The sequence of plastic hinges formed in the frame is illustrated in Fig. 6.

Vogel six-story frame [8] usually appears in benchmark study of planar steel frames. The frame size and load information are illustrated in Fig. 7 and the frame member sizes are listed in Table 2. The material elastic modulus E of steel is 206 kN/mm² and the yield

Table 2. Dimensions of all the components of Vogel six-storey steel frame.

Section	d (mm)	b_f (mm)	t_w (mm)	t_f (mm)	A (mm²)	I ($\times 10^6$ mm⁴)	S ($\times 10^3$ mm³)
HEB300	300	300	11.0	19.0	14,900	251.7	1869
IPE240	240	120	6.2	9.8	3,910	38.92	367
IPE300	300	150	7.1	10.7	5,380	83.56	628
IPE330	330	160	7.5	11.5	6,260	117.7	804
IPE360	360	170	8.0	12.7	7,270	162.7	1019
IPE400	400	180	8.6	13.5	8,450	231.3	1307
HEA340	330	300	9.5	16.5	13,300	276.9	1850
HEB160	160	160	8.0	13.0	5,430	24.92	354
HEB200	200	200	9.0	15.0	7,810	56.96	643
HEB220	220	220	9.5	16.0	9,100	80.91	827
HEB240	240	240	10.0	17.0	10,600	112.6	1053
HEB260	260	260	10.0	17.5	11,800	149.2	1283

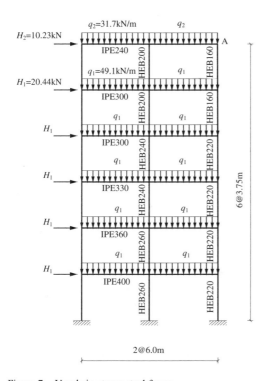

Figure 7. Vogel six-storey steel frame.

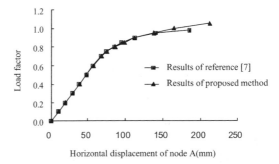

Figure 8. Load-displacement curve of Vogel six-storey steel frame.

and moment diagrams in the ultimate state are shown in Fig. 9, where the final plastic hinge distribution is dotted in the moment diagram.

4 CONCLUSION

An approach for nonlinear analysis of steel frames using element with midspan plastic hinge is proposed in this paper. This approach can use just one element to simulate one member in a frame even plastic hinge may form within the member when subjected to laterally-distributed loads. What's more, this approach also considers the influences of some geometrical and material nonlinear factors including second-order effect of axial forces, shear deformation, cross-section plastification, residual stress, initial geometrical imperfection. The numerical results show that the proposed approach is efficient and satisfactorily accurate, and is suitable for the nonlinear analysis of steel frames.

strength f_y is 235 N/mm². The horizontal displacement of right-upper corner (Node A) versus load factor curve by the elasto-plastic hinge model presented in this paper is compared with the results in reference [8] with plastic zone method in Fig. 8. The ultimate load factor obtained by the method proposed is $\lambda = 1.15$ while that by reference [8] is $\lambda = 1.18$. The axial force

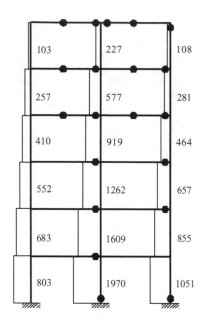

103 | 227 | 108

257 | 577 | 281

410 | 919 | 464

552 | 1262 | 657

683 | 1609 | 855

803 | 1970 | 1051

(a)axial forces(kN)

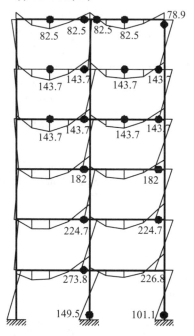

78.9

82.5 82.5 82.5 82.5

143.7 143.7 143.7 143

143.7 143.7 143.7 143

182 182

224.7 224.7

273.8 226.8

149.5 101.1

Figure 9.   Internal forces of Vogel six-storey steel frame.

REFERENCES

S. E. Kim and W. F. Chen, Practical advanced analysis for unbraced steel frames design, Journal of Structural Engineering, ASCE, Vol.122(11): 1259-1265, 1996

W. S. King, D. W. White and W. F. Chen, A modified plastic hinge method for second-order inelastic analysis of rigid frames. Structural Engineering Review, 4(1): 31-41, 1992

J. Y. R. Liew, Second-order refined plastic hinge analysis of frame design Part 1, Journal of Structural Engineering, Vol.119: 3196-3216, 1993

J. Y. R. Liew, Second-order refined plastic hinge analysis of frame design Part 2, Journal of Structural Engineering, Vol.119: 3217-3237, 1993

LI Guo-qiang and SHEN Zu-yan. Elastic and elasto-plastic analysis and computational theory for steel frame systems. Shanghai: Shanghai Science and Technology Press, 1998. (in Chinese)

L. Duan and W. F. Chen, Design interaction equation for steel beam-columns. Journal of Structural Engineering, Vol.115(5): 1225-1243, 1989

W. F. CHEN Analysis and design of beam-column, volume(I), planar problem characteristic and design. Beijing: China Communications Press, 1997. (in Chinese)

S.L. Chan and P.P.T. Chui. Non-linear static and cyclic analysis of steel frames with semi-Rigid connections£¬Oxford: Elesvier, 2000

Steel and Composite Structures – Wang & Choi (eds)
© *2007 Taylor & Francis Group, London, ISBN 978-0-415-45141-3*

Experimental and finite element studies on tapered steel plate girders

N.E. Shanmugam & W.H.M. Wan Mohtar

Department of Civil and Structural Engineering, Universiti Kebangsaan Malaysia, Bangi, Malaysia

ABSTRACT: The object of this paper is to investigate the behavior of tapered plate girders, primarily subjected to shear loading; experimental as well as finite element results obtained from the studies are presented in this paper. 11 medium-scale girders, one of uniform section and 10 tapered, were tested to failure and all girders were analysed by finite element method using the computer package ABAQUS. The results are compared and the accuracy of the finite element modeling established. Web thickness, loading direction – inclined flange in tension or compression and taper angle were varied and, the elastic and ultimate load behaviour studied.

1 INTRODUCTION

Tapered plate girders in which web depth is varied along the span can be effectively used to provide space for services thus maximizing the headroom available in high-rise buildings. Engineers are reluctant to use such girders in construction because of complexities associated with the taper angle. The tapered web alters the stress distribution within the member and will, in most cases, influence the collapse behavior. Even though extensive studies have been made on plate girders of uniform section, information available on tapered plate girders is limited.

Rockey and Skaloud (1971, 1972) showed that for plate girders having proportions similar to those employed in civil engineering construction, the ultimate load carrying capacity is influenced by the flexural rigidity of the flanges. They conducted ultimate load tests on three series of plate girder models, in each of which only the size of the flanges, and therefore their flexural rigidity, was varied. A more generally applicable mechanism was proposed by Porter et al.(1975, 1978), and after exhaustive comparisons with test data obtained from various sources the accuracy of these methods have been established. It is assumed in this method that failure will occur when a certain region of the web yields as a result of the combined effect of the inclined tensile membrane stress field and the web buckling stress when four plastic hinges form in the flanges. Further studies to investigate the effects of pure bending, combined action of shear and bending, edge or patch loading on the behaviour of plate girders have been carried out by other researchers (Owen et al., 1970; Rockey, 1968; Calladin, 1973; Roberts et al, 1979, 1981 and 1983).

Development of membrane tension in the web is denoted as tension field action and enables the web

to sustain loads well in excess of the elastic critical load. A consequence of the membrane tension in the web is the inward pulling of the flanges, under increasing loads. Eventually, plastic hinges are formed in the flanges leading to collapse of the girder. The collapse of a plate girder could, therefore, be assumed to consist of three contributing factors viz. (i) elastic critical load, (ii) resistance by tension field and (iii) contribution by flanges.

The elastic and inelastic behavior of plate girders having uniform cross-section along the span is well understood; a number of theoretical as well as experimental investigations on the ultimate load behavior have been reported and design methods proposed. However, there is no sufficient information available in the literature to understand the ultimate load behaviour of tapered plate girders. Only a limited number of published work could be traced on steel girders (Davis and Mandal, 1979) and on aluminum girders (Sharp. et al.,1971; Roberts and Newark, 1996; 1997). Therefore, an experimental investigation of tapered plate girders with different taper angles will be helpful to understand the behavior and load-carrying capacity of these girders.

In this paper, an experimental study on steel tapered plate girders carried out to study the ultimate load behaviour is reported. The experimental work deals with the elastic and ultimate load behavior of the girders under a point load applied at the mid span. Eleven full-scale girders, one of uniform cross-section and the rest with tapered webs, were tested to failure. The girders were anlaysed using finite element computer code ABAQUS and the results obtained are compared with the corresponding values from the experiments. The accuracy of the finite element modeling is thus established. The study also provides an insight into the elastic and inelastic behaviour of tapered girders.

2 EXPERIMENTAL PROGRAM

Two series of tests were carried out on plate girders fabricated by welding steel plates. One of the flanges was kept horizontal whilst the second was inclined following the depth of the web plate. In one series (A series), girders were tested with the horizontal flange in tension and, in the other (B series) the horizontal flange remained in compression. In each of the two series, five girders with different taper angles and different web thickness were considered. The test specimens were designed such that they represented plate girders that occur in practice. The dimensions measured for all specimens are summarized in Table 1. In the table, H refers to the depth at the support, t_w web thickness, t_{s1} thickness of stiffener at the mid-span, t_{s2} thickness of stiffeners near the mid-span and t_{s3} thickness of stiffeners at the support.

2.1 Fabrication of the test specimens

All steel plates used for the specimens are hot-rolled of Grade 43A. Web depth was varied along the span with larger depth at the mid-span and smaller one at the support. The depth at the mid-span of all the girders was kept as 800 mm and at the support it was varied as shown in Table 1 so as to get the required taper angle. The various components were first assembled by means of tack welds. The assembled plates were then clamped against out-of-plane deformations before providing intermittent welds. Sufficient time was allowed for cooling before final continuous fillet welds were applied. Welding was carried out at different or alternate locations and the specimens were allowed to cool to minimize the effect of initial distortion caused by excessive heating at particular locations. The test specimens are identified in the text as, for example, 200-30-A, the first field viz. 200 referring to the web thickness of 2 mm the second field 30 referring to the taper angle equal to 30 degrees and the last field A or B referring to the horizontal flange in tension or in compression in the test.

2.2 Material properties

The steel plates used had a nominal yield stress equal to 280 MPa. Three coupons from each of the stock plates were tested. The coupons were tested under displacement control in an Instron universal testing machine. The average material properties such as yield stress, ultimate stress, modulus of elasticity and Poisson's ratio for series A and B are respectively 290 N/mm², 323 N/mm², 215 kN/mm² and 0.29 for the web-plate and flange-plates; the corresponding values for the stiffeners were 280 N/mm², 304 N/mm², 205 kN/mm² and 0.27, respectively.

2.3 Test setup

A test rig built on a strong floor and capable of applying up to a maximum load of 1000 kN was used to test the specimens. The test rig was designed to support test specimens by means of a horizontal rectangular frame. Under maximum loading, the deflection of the support frame is negligible so as to provide a perfect, non-deflecting surface. The supp- port frame rests on the main lateral loading frame by means of a cradle. The vertical loading was applied by means of an Instron actuator capable of applying a maximum load of 1000 kN. The point load was transferred to the specimen via a ball and socket bearing so that differential movement at failure can be prevented. The rollers sandwiched between two hardened steel raceways were inserted between the ends of the specimen and the rig to ensure a simply supported condition. A typical testing arrangement with a test specimen mounted onto the rig is shown in Fig. 1. When the applied load reached a high value, it is possible for the beam to deflect in the lateral direction. The beam would, therefore, fail by lateral buckling before reaching the ultimate load. In order to prevent the beam from such a failure, it was restrained laterally by means of two tie rods: one end of the tie rod was connected to the stiffeners at the mid-span of the girder and the other end to a vertical steel member. One such typical bracing arrangement is shown in Fig. 2.

Table 1.   Measured dimensions of the test specimens.

Specimen	H (mm)	t_w (mm)	t_{s1} (mm)	t_{s2} (mm)	t_{s3} (mm)
200-0	800	2.12	16.24	10.12	10.25
200-10-A	649	2.11	15.86	10.24	10.47
200-10-B	658	1.98	16.14	9.48	10.15
200-20-A	496	2.11	16.26	9.38	9.86
200-20-B	480	1.94	15.84	9.89	10.21
200-30-A	305	1.96	16.47	9.83	9.86
200-30-B	296	2.02	15.74	9.89	9.56
300-20-A	486	3.08	26.32	20.58	12.35
300-20-B	491	3.17	26.84	20.32	12.85
400-20-A	483	4.19	25.78	19.58	11.48
400-20-B	480	3.95	26.04	19.72	11.54

Figure 1.   A typical testing arrangement.

2.4 Instrumentation

The vertical mid-span deflections of the girder were measured by linear variable displacement transducers (LVDT) with a measurement range of 50 to 200 mm. Transducers were placed at mid-span, support and quarter span along the centerline of the girder. Transducers at quarter spans on either side of the mid-span were used to monitor the symmetry in the behavior of the girder. A transducer (LVDT) was mounted onto specimens 200-20-A and 200-20-B to measure lateral deflection corresponding to the center point of the diagonal line of web-plate. Hori-zontal deflections were also measured at the corner of the web plate in the case of the girders 400-20-A and 400-20-B.

Strains along the flange plate were measured by electrical resistance strain gauges and post-yield rosettes were employed to monitor the strains on web plates. In the front surface, 5 strain rosettes were fixed and, in the back surface at the same location as those on the front side 5 rosettes were mounted. Thus, the stress distribution along the web would be monitored. Also, 3 strain gauges were placed longitudinally in the compression flange along the width, and another 3 strain gauges in the tension flange. From these strain values, the axial tension or compression force developed in flanges could be computed. Readings from the strain gauges and transducers were recorded on floppy diskette using a data-logger (TD-301) connected to a computer.

2.5 Test procedure

The specimen was mounted on the test rig with the mid-point of the girder located centrally under the jack, carefully checking by plumb. Care was taken to ensure that the specimen sits on the roller bearing and its centroidal line coincided with the centerline of the actuator applying the load. All electrical strain gauges and transducers were connected to a data acquisition system programmed to record the output on a floppy

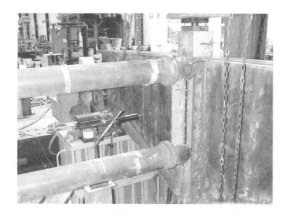

Figure 2. A typical lateral bracing arrangement.

diskette with simultaneous print out of the data at every load increment of 5 kN. Load-displacement and load-strain curves were plotted through a computer and it was thus possible to identify the onset of failure in the specimen. Before the actual test, a small preload of 50 kN was applied slowly on the specimen and then removed. This operation was repeated twice in order to remove any settlement in the support system and to ensure that the specimen was properly seated on the supports. At the same time, the readings of strain gauges and transducers were checked to ascertain that they functioned properly.

After ensuring that all the instruments were working satisfactorily, the strain gauges and transducers were initialized. The load was then applied on the specimen in a predetermined increment at a constant rate. After every 50 kN, the load was kept constant to allow the specimen to stabilize. At every 5 kN load increment, the strain gauges and transducer readings were recorded; the plate girder was examined for yielding. Close observations were made to locate the severe deformation of flange to form a hinge. Testing was terminated when the load dropped to one-third of the ultimate load. The ultimate load and mode of failure were recorded for each specimen. The test procedure adopted was same for all girders.

3 FINITE ELEMENT MODELLING

The finite element program ABAQUS version 5.4 (Hibbert, Karlsson, and Sorenson Inc. 1994), was employed to simulate the behaviour of the tapered plate girders. The analysis includes both material and geometric non-linearities. The plasticity model consists of von-Mises yield surface combined with Prandtl-Reuss flow rule. The standard Newton method is used for solving the non-linear equilibrium equations. Riks Arc length method of control is used in the analyses. This method is useful in solving unstable static equilibrium problems, where the unloading path needs to be monitored. For all the analyses described, an automatic time step size was used.

Since out-of-plane buckling deformations are not present in perfectly flat plates under in-plane loading conditions, a small variation from flatness of $d/1000$ is introduced in order to initiate out-of-plane deformations. The shapes of initial deformations were based on the buckling mode shapes. The geometry of the test specimens, loading and boundary conditions are symmetrical about the vertical line through the mid-span. The analyses were, therefore, carried out on one half of the girder and appropriate boundary conditions were used at the section of symmetry.

A surface model consisting of 8-noded (S8R5) reduced integration quadrilateral shell elements was used to avoid any ill conditioning in the analysis. The S8R5 element that allows for changes in the

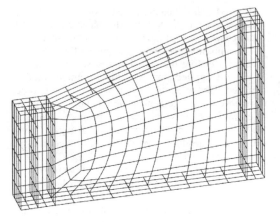

Figure 3. A typical finite element mesh for a girder.

Table 2. Summary of FEM ultimate loads.

Girder no.	FEM results P_{FEM}, kN	Test result P_{Test}, kN	P_{FEM}/P_{Test}
200-00	335	327	1.02
200-10-A	337	350	0.96
200-20-A	373	355	1.05
200-30-A	377	359	1.05
300-20-A	463	450	1.03
400-20-A	596	574	1.04
200-10-B	349	320	1.09
200-20-B	338	322	1.05
200-30-B	314	325	0.97
300-20-B	411	400	1.03
400-20-B	595	578	1.03

thickness as well as finite membrane strains was chosen. A typical finite element mesh, chosen based on a series of convergence studies carried out, is shown in Fig. 3. After the mesh generation, all mesh data are converted to standard ABAQUS input file for analysis and the elastic buckling analysis is carried out first. Then, the incremental-iterative load steps and required output commands need to be added in accordance with ABAQUS/Standard language. In the result file, information such as reaction, buckling or ultimate displacements can be extracted for evaluation of the buckling or ultimate load. The non-linear and large deflection analyses are used to compute the ultimate load of the plate girder. The results obtained from the analyses are extensive; however, only selected relevant ones are presented herein.

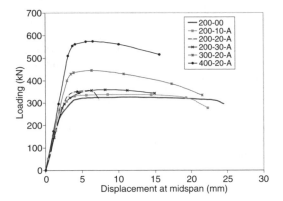

Figure 4. Load-deflection plots for A series.

4 RESULTS AND DISCUSSION

Results obtained from the experimental investigation and the finite element analyses are presented in the form of tables and figures. Experimental ultimate loads along with the corresponding values obtained from finite element analyses are summarized in Table 2 for all the tested specimens. Load-vertical displacement curves obtained from the tests are plotted against the applied load in Figs. 4 and 5 for series A and B, respectively. Figure 6 shows the load deflection plots in which the results obtained from the experiments and finite element analyses are presented for typical girders. It can be seen from Table 2 and Figure 6 that there is reasonable agreement between the two sets of results and agreement is close at the initial stages of loading. The finite element results overestimate the ultimate loads for all speci- mens except for specimens 200-10-A and 200-30-B.

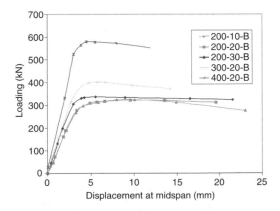

Figure 5. Load-deflection plots for B series.

The average ratio of FEM loads to experimental failure loads for the specimens tested is 1.03. This shows that the finite element modelling is capable of predicting the ultimate load-carrying capacity of tapered plate girders with reasonable accuracy in all cases.

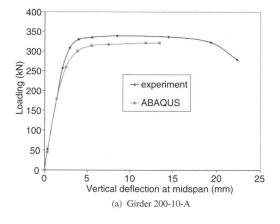

(a) Girder 200-10-A

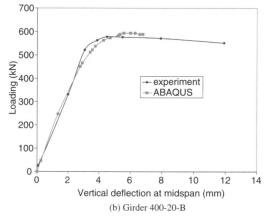

(b) Girder 400-20-B

Figure 6.   Comparison of FEM and Test Results.

(a) Experiment

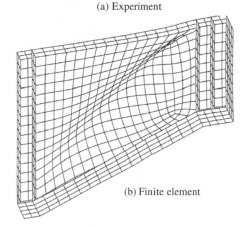

(b) Finite element

Figure 7.   View after failure – Girder 300-20-B.

The load-deflection behavior and failure modes have been observed to be similar for all the specimens. The web plate resisted shear up to the elastic critical load in all cases. It can be seen from the load – displacement plots that the curves remained linear at the initial stages of loading. When the load approached approximately one-third of the ultimate load, the web started to wrinkle slowly, and the load-deflection curve began to deviate from linear behavior. Any further increase in load did not cause a rapid collapse of the girder but resulted in the formation of buckles in a waveform parallel to the direction of the tension field. After the onset of buckling in the web, a further increase in load led the waveform spreading along the web, and caused severe bending of the flanges due to inward pulling at four positions, two of these were located on upper flanges, the rest on the lower flanges. When the load reached the ultimate value, the web was found to wrinkle extensively with the flange bending towards the web. Eventually, plastic hinges were formed in the flanges and the girder collapsed.

As the load increased beyond the bifurcation point of the web plate, the diagonal tension developed along with the bending stresses resulting from out-of-plane deformations. With further increments of loading, material yielding begins along the main diagonal tension line under the coupling action of the diagonal tension and bending, and four hinges on the upper and lower flanges formed. Depending upon the geometry, the web plate is capable of carrying additional loads in excess of that at which the web starts to buckle, due to the post-buckling reserve strength. While the load was increased, the yielding zone of the web-plate was observed to form along a diagonal band to serve as a chord, and the diagonal band kept spreading along the web with further load increase. The bending stress together with membrane stress acts on the web, causing the material on the front and back surfaces along the diagonal line to yield consecutively.

Views after failure of typical girders are shown along with the corresponding finite element predictions in Figures 7 and 8. Generally good agreement is

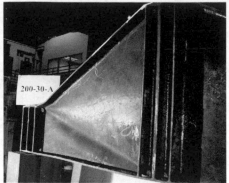

(a) Experiment

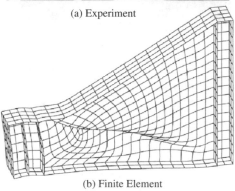

(b) Finite Element

Figure 8.   View after failure – Girder 200-30-A.

observed between the deflected web profiles observed in the experiment and those obtained from the finite element analyses. Similar observations could be made in respect of deformations in flanges also.

5   CONCLUSIONS

Experimental and Finite Element studies on the behaviour and load carrying capacity of tapered plate girders have been described in this paper. Based on the results presented herein it can be concluded that the proposed three dimensional finite element modelling using the software package ABAQUS is reliable in predicting the ultimate strengths of the tapered plate girders with sufficient accuracy. If inclined flanges are under tension, the taper angle has a moderate influence over shear capacity of the tapered girder. Compared to the shear capacity of a girder of uniform section, a 5 % loss in shear capacity would occur. Whilst inclined flanges are subjected to compression, the ultimate load of tapered girder becomes higher than that of the straight girder by more than 5%.

ACKNOWLEDGEMENTS

The investigation described in this paper has been carried out at the Department of Civil Engineering, National University of Singapore. Support by Yongnam Engineering (Singapore) Pte. Ltd. who supplied the test specimens is gratefully acknowledged.

REFERENCES

ABAQUS Theory Manual Version 5.4, Hibbit, Karlsson & Sorensen, Inc., USA., 1994.
ABAQUS User's Manual Version 5.4, Hibbit, Karlsson & Sorensen, Inc., USA., 1994.
ABAQUS/POST Manual Version 5.4, Hibbit, Karlsson & Sorensen, Inc., USA., 1994.
Calladine, C. R. Plastic theory for Collapse of Plate Girders under Combined Shearing force and Bending Moment, Structural Engineer, Volume 51, Issue 4, pp.147–154. 1973.
Davis, G and Mandal, S. N. The collapse behaviour of Tapered Plate Girders Loaded within the Tip. Proc. Institution of Engineers, Part 2, Vol. 67, pp.65–80. 1979.
Evans, H.R., Porter, D.M. and Rockey, K.C. The Collapse Behavior of Plate Girders Subjected to Shear and Bending. Int. Ass. Bridge Struct. Eng, Periodical, Proc. P-18/78. 1978.
Owen, D.R.J., Rockey, K.C. and Skaloud, M. Ultimate Load Behavior of Longitudinally Reinforced Web Plate Subjected to Pure Bending. IABSE Publications, Vol. pp.113–148, 1970.
Porter, D. M., Rockey, K, C, and Evans, H. R., The collpase behaviour of plate girders loaded in shear. Strut. Engnr. 53, 313–325, 1975.
Roberts, T.M. and Rockey, K.C. A Mechanism Solution for Predicting the Collapse Loads of Slender Plate Girders when Subjected to In-Plane Patch Loading. Proc. Instn. Civ. Engrs, Part 2, 67, pp.155–175. 1979.
Roberts, T.M. Slender Plate Girders Subjected to Edge Loading. Proc. Instn. Civ. Engrs., Part 2, 71, pp.805–819. 1981.
Roberts, T.M. and Markovic, N., Stocky Plate Girders Subjected to Edge Loading. Proc. ICE, Part 2, V. 75, pp.539–550. 1983.
Roberts, T. M., Newark, A.C. Shear strength of tapered aluminium plate girders, Bicentenary conference on thin-walled structures, London, 1996.
Roberts, T.M and Newark, A.C.B. Shear Strength of Tapered Aluminium Plate Girders. Thin-Walled Structures, Volume 29, Issues 1–4, pp.47–58. 1997.
Rockey, K. C. Factors influencing ultimate behaviour of plate of plate girders. Conference on Steel Bridges, June 24th–30th, British Constructional Steelwork Association, Institution of Civil Engineers, London, 1968.
Rockey, K.C., & Skaloud, M., The ultimate load behaviour of plate girders loaded in shear. Journal, The Structural Engineer, Vol. 50, No. 1, , pp.29–48, January, 1972.
Skaloud, M. and Rockey, K.C. Ultimate Load Behaviour of Plate Girders Loaded in Shear. IABSE Proc., London. Design of Plate and Box Girders for Ultimate Strength, pp.1–19. 1971.
Sharp, M. L., and Clark, J. W. Thin Aluminum shear Web. J. Struct. Div., ASCE, 1021–1038, 1971.

Steel and Composite Structures – Wang & Choi (eds)
© 2007 Taylor & Francis Group, London, ISBN 978-0-415-45141-3

Continuous composite beams with large web openings

T. Weil & J. Schnell
Kaiserslautern University of Technology, Kaiserslautern, Germany

ABSTRACT: Continuous composite beams can be designed according to the plastic hinge theory. This design method keeps many questions open in case beams with large web openings are regarded. At the Kaiserslautern University of Technology a research project was started to clarify these questions. In this paper some investigations of this project and its results are explained. In addition, the conclusion, which can be drawn from the experiments and the numerical parameter study, is outlined.

1 INTRODUCTION

Continuous composite beams can be designed according to the plastic hinge theory, which is explained by Bode (1998) in detail. With this method the plastic reserves of cross section and system can be used to full capacity. This design method keeps many questions open if beams with large web openings are regarded. At the Institute for Concrete Structures and Structural Design of the Kaiserslautern University of Technology a research project was started to solve these problems. Within this project six large-scale tests were arranged. Furthermore a comprehensive numerical parameter study was accomplished to follow up the unsolved questions.

In this paper problems to the mentioned topic are explained and the experimental and computed investigations, which were arranged, are presented.

2 PLASTIC HINGE THEORY

According to the plastic hinge theory in a statically indeterminate system plastic hinges are formed one after another until a kinematic mechanism is generated in the ultimate limit state. Then it is no longer possible to increase the load. The use of the plastic hinge method depends on the plastic structural behaviour of the cross-sections and its capacity of rotation.

A plastic moment hinge is generated, as soon as the plastic bearing capacity (Figure 1) of the cross section is accomplished at one location. In case rotation is possible at this location. The bending moment operates at the plastic moment level; an increase of this moment is no longer possible. The potential rotation at the location of the first plastic hinge enables the system that the overall load can be increased beyond this loading

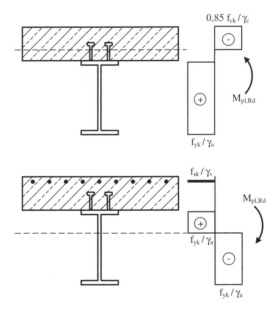

Figure 1. Plastic tension distribution of a composite cross section (bearing capacity for positive und negative bending moment).

state. The load is carried across other system areas. In this case other plastic hinges are generated. After occurring the last plastic hinge the system becomes kinematic; the load bearing capacity of the system is accomplished.

At continuous beams without openings plastic moment hinges are normally formed above the middle support and in midspan. At continuous beams with large web openings plastic ranges can be generated in the area near the opening. This means that under certain circumstances local plastic moment hinges are formed in the four corners of the opening. These local plastic hinges all together result in a global plastic hinge for shear forces.

The basic differences in the design according to the plastic hinge method between continuous composite beams with and without web openings are described by Schnell & Weil (2004). The effect on the stress resultants, which is caused by the opening, is shown there. In this context the following unsolved questions can be raised among other objectives:

– Which terms and conditions are needed, that the capacity of rotation of the partial cross sections is high enough to form a shear force plastic hinge?
– What is the structural and deformation behaviour of such a plastic hinge for shear forces, whilst it acts as a global mechanism with one degree of freedom?
– After the formation of the plastic hinge for shear forces the rearrangement of the stress resultants possibly enables an extra plastic moment hinge, during increase of loading. At which condition does this combination of plastic hinges occur and what are the structural and deformation properties for this case as a global mechanism with two degrees of freedom?
– In case the area of the opening is dimensioned in such way that there is no global plastic hinge possible, the deformation ability at this location is bigger. Then plastic hinges are generated above the middle support and in the field: What is the influence of the deformation ability for the evidence of the needed capacity of rotation in these plastic hinges?

3 EXPERIMENTAL INVESTIGATIONS

3.1 Test beams

Six large-scale tests were arranged in the Laboratory for Construction Engineering at the Kaiserslautern University of Technology to answer the mentioned questions. All investigated test beams are built with the same material strength. For every experiment a material test was accomplished. The resulting values are differing, which can be explained by the statistical spread of the material strength.

In the first test, which is named V1-T350, a two-span composite beam was examined. The cross section of this beam is shown in Figure 2, the corresponding elevation in Figure 3. The height of the used shear studs amounts to 10,5 cm. The span of the symmetrical system is 3,50 m. The location of the opening is in the zero point of moments, which is calculated elastically based on an equally sized beam without opening. A concentrated load operates in every midfield of the system.

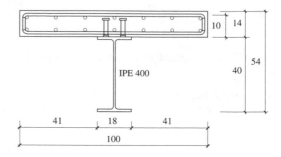

Figure 2. Cross section of the test beam V1-T350 (measurements in cm).

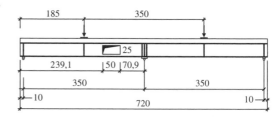

Figure 3. Elevation of the test beam V1-T350 (measurements in cm).

The second experiment is named V2-G400. It complies with the first test V1-T350 except for the length of span. The distance between the supports is raised from 3,50 m up to 4,00 m.

The third test beam V3-DL400 has an enforced concrete slab. To achieve this enforcement four shear rails are placed at the area of the opening inside the slab. Furthermore, the height of the shear studs is increased from 10 cm up to 12,5 cm along the total beam.

The fourth test beam V4-S400 was stressed with a uniformly distributed load. For all other properties V4-S400 fits with the examined beam V2-G400.

The experiments V5-DL400N and V6-DL400P vary the parameter "location of the opening". For the test V5-DL400N the opening is placed in the negative range of the bending moment. The opening of the test beam V6-DL400P is located in the positive range of the bending moment. Both test beams match the test beam V3-DL400 and are built with shear rails, too.

3.2 Test results

The test results are described and analysed in detail in Schnell, J. & Weil, T. (2007). In this section only some significant results are shown. These summarizing results refer to "deformation and behaviour in bending", "expansion behaviour at the area of the opening", "stress resultants" and "failure mode".

A load-deflection-diagram of both fields was created for every test. The diagram of the experiment

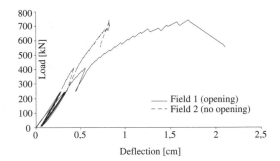

Figure 4. Load-deflection-diagram of both fields for test V1-T350.

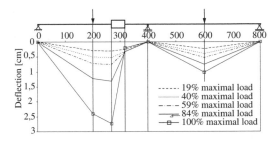

Figure 5. Bending line during the experiment V2-G400.

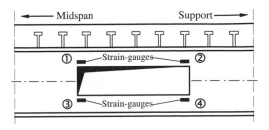

Figure 6. Definition for the corners of the opening.

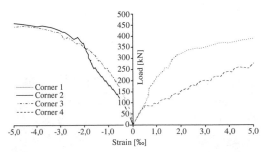

Figure 7. Strain-curves of the web at the area of the opening (V2-G400, locations of the strain-gauges in Figure 6).

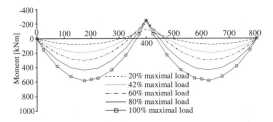

Figure 8. Bending moment lines during the test V4-S400.

V1-T350 is given in Figure 4. Comparison of both curves shows clear differences. The value of the deflection in midspan, the strength of which is weakened with the opening (field 1), is bigger than in the other field (field 2). This occurs already short-time after the start of the test. This difference becomes even more definite during the further test procedure.

In principle the characteristics of the load-deflection-diagram, which means the difference in both fields, are similar in all experiments. But looking on the amount of deflection the other tests exhibit bigger deflections because of the wider span.

Exemplary bending lines for the test V2-G400 are shown in Figure 5. It must be notified that the biggest ratio of deflection in field 1 is in the area of the opening. This becomes particularly clear at ultimate limit state. Moreover, it indicates a formation of a global plastic hinge for shear forces.

The analysis of the strain-gauges in the corners of the opening shows the expected influence the opening causes in the entire system. In Figure 7. the performance data of one exemplary strain-gauge is represented for every corner of the opening from V2-G400. The yield strain of the used steel is at 1,58‰ for this test. The results show a formation of plastic ranges in all four corners of the opening. The four corners are defined in Figure 6.

The bending moment curves during the test V4-S400 can be seen in Figure 8. In this picture

the redistribution of bending moments from the middle support to midfield becomes clear. Between 80% and 100% of the maximum load there is just a large increase of the moment in mid-span and a small decrease of the moment above the support.

The experiment V4-S400 is the only test, which was accomplished with a uniformly distributed load. All other tests were superimposed with two concentrated loads. These tests show also a redistribution of bending moments and an accompanied adjustment of the zero point of moments. The bending moment curves of the test V3-DL400 are pictured in Figure 9.

The mechanism of the collapse was somehow different at the experimental investigations. The test beams V1-T350 and V2-G400 failed basically identical. The kind of collapse was a distinct shear failure in the concrete slab above the opening (Figure 10). In spite of the plastification of the steel girder the shear force between the opening and the middle support could be increased. That was enabled by the bearing capacity of

173

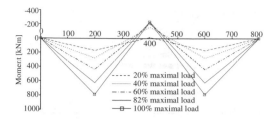

Figure 9. Bending moment lines during the test V3-DL400.

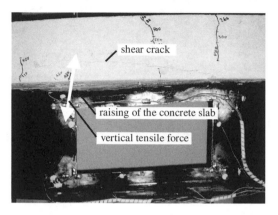

Figure 10. Area of the opening short-time before reaching ultimate limit state (V2-G400).

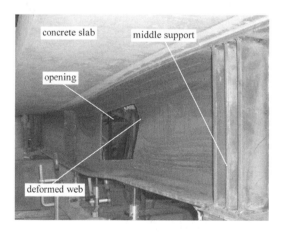

Figure 11. Field with opening at ultimate limit state (V3-DL400).

the concrete slab. For this reason the formation of the plastic moment hinge in the midspan began later than it was expected. But there was no fully developed formation of plastic hinges, because the efficiency of the second degree of freedom in the first-degree statically indeterminate system was limited by another mode. The shear force near the opening even caused a shear failure at the concrete slab. So the shear force and also

the moment, which was caused by the adjustment of the elastic calculated zero point of moments towards the middle support, could not be transported over the opening because of the originated shear crack. The occurring bending moment above the middle support does not come close to the plastic bending moment. The bending moment in midspan allowed only minor plastic deformations.

The other experimental investigations show a different failure. The ultimate limit state was not combined with an abrupt collapse. In the corners of the opening distinctive yield zones were generated. This can be called a global plastic hinge for shear forces. Furthermore, in midspan a plastic moment hinge was formed. This plastic hinge mechanism defined the ultimate load of the system (Figure 11). The load could not be increased any more. The deformation, which was raised after reaching the maximal load, induced again a gaping shear crack at the concrete slab.

4 CALCULATIVE PARAMETER STUDY

4.1 Basic model and modified parameter

A complex model of a continuous beam was analysed with the software package ANSYS working with the finite element method. This model was calibrated on the results of the accomplished large-scale tests. Based on this a comprehensive parameter study was performed. The whole results of the numerical investigations are mentioned in Schnell, J. & Weil, T. (2007). In this section a couple of interesting test results is presented.

Starting from a basic model the parameters "field length", "thickness and width of the concrete slab", "kind of load", "location of opening" and "kind of steel girder" were varied. Altogether 80 beams were calculated to investigate the varied parameter. Part of this were also calculations with equivalent beams but without an opening as well as systems, where two openings had been placed symmetrically to the middle support. The dimensions of the basic model are shown in Figure 12 and conform to the first test beam (Figure 2 and Figure 3).

4.2 Results of the calculations

The deformed shape of one of the calculated models is shown in Figure 13. The grey scales display the longitudinal stress. It becomes evident that the highest stress in the whole beam acts in the corners of the opening because the shear force is divided there and the secondary moments are at maximum.

The deflection in the left field is maximal. It is representatively affected by the shear force, which is transported over the opening. The influence of the

Elevation:

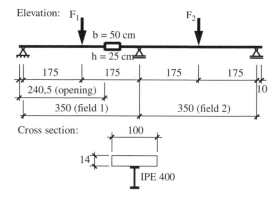

Cross section:

Figure 12. System and loads of the basic model (measurements in cm).

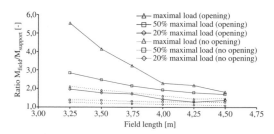

Figure 13. Deformed shape and longitudinal stress of the basic model at ultimate state.

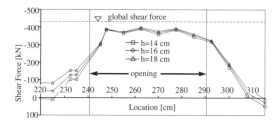

Figure 14. Moment ratio with the modified parameter "field length".

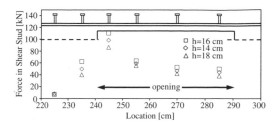

Figure 15. Shear force in the concrete slab above the opening with modified slab thickness ($V_{global} = 430$ kN).

deformation due to the bending moment has just a minor relevance.

The moment ratio with the modified parameter "field length" is pictured in Figure 14. It is obvious, that a short field length causes a higher bending moment in the field than over the middle support. The

Figure 16. Tensile Force in the shear studs with modified slab thickness ($V_{global} = 430$ kN).

global influence of the opening then decreases with a growing field length.

The partial shear force of the concrete slab in the area around the opening is shown in Figure 15. The three lines of the shear force display different slab thicknesses. Over 90% of the shear force is conducted through the concrete slab, although the remaining web is still 35% of the unweakened web. The transport of the shear force into the concrete slab occurs in concentrated form at the beginning of the opening (Figure 16). Independent of the concrete slab thickness above the opening the partial shear force is kept at an almost constant level. The partial shear force behind and in front of the opening grows up, in case the thickness of the slab increases.

In Figure 16 the tensile forces in the shear studs, which are active at the same location, are shown. The tensile force in the shear studs increases, in case the concrete slab becomes thinner and the length of the shear studs remains unchanged. Thereby the partial shear force in a thin concrete slab above the opening can correspond to the partial shear force in a thick concrete slab.

5 CONCLUSION

The capacity of rotation in the corners of the opening is always enough to form a local plastic moment hinge. But sometimes it is not possible to generate the plastic hinge in a fully developed way. In fact, the initiation of yielding is detected in all four corners, but the shear bearing capacity of the concrete slab limits the ultimate state of the total system. In this case a shear failure occurs. On account of this behaviour, no more re-distribution of moments is possible. The shear bearing capacity of the concrete slab must be increased to get a definitive global plastic hinge for shear forces. This condition was accomplished with such three investigated test beams, which were built with shear rails inside the concrete slab.

The opening has a great influence on the deformation. The increase of the deformation, which is shown in the field with the opening, is mainly achieved at the area of the opening. The comparison of the deformations to an unweakened beam is only equal to the test beams with the enforced concrete slab.

The structural behaviour of a plastic hinge for shear forces as a global mechanism with one degree of freedom is especially defined by the shear bearing capacity of the concrete slab above the opening. A collapse of the single statically indeterminate system is solely possible by the behaviour in the area of the opening.

A combination of a global plastic hinge for shear forces and moments was not detected during the investigations. The analysis of the test results shows that it is unlikely to generate such a combination. A global negative moment induces a tensile force in the upper part of the cross section. Thereby the shear bearing capacity of the concrete slab is reduced. To form a plastic moment hinge the moment has to be increased. For this reason the tensile force in the concrete slab also increases. For the generation of a combination of plastic hinges in the area of the opening the shear force has to be increased even more. This is unimaginable because the shear bearing capacity becomes smaller with a raising negative bending moment. In this case the system would fail by a shear collapse at the concrete slab before the formation of a plastic moment hinge.

The investigated beams had no stiffeners around the opening. The opening can be fitted with welded stiffeners in order to make the formation of a global plastic hinge impossible. No beams with stiffeners near the opening were investigated. If there is a collapse at the area of the opening and a clear formation of a plastic moment hinge mechanism possible, can not be answered now.

Furthermore the most interesting cognitions discovered by the investigations are as follows:

- The shear bearing capacity of the concrete slab at the opening has an important implication for the generation of the plastic hinge mechanism.

- Beams with an opening have always a larger redistribution of the bending moments than beams without an opening.
- The positioning of an opening at the elastic calculated zero point of moments causes an additional positive bending moment occurring already at a low loading level.
- The global influence of the opening decreases with increasing field length.
- The biggest stresses in the entire beam are activated in the corners of the opening. This is caused by the local moments, which are generated by the global shear force.
- Whilst the concrete slab becomes thinner and the shear studs remain equal the tensile force in the shear studs increases.

ACKNOWLEDGEMENT

This research was supported by the DFG (Deutsche Forschungsgemeinschaft).

REFERENCES

Bode, H. 1998. Euro-Verbundbau, Konstruktion und Berechnung. 2. völlig neu bearbeitete Auflage. Düsseldorf: Werner Verlag.

Schnell, J. & Weil, T. 2004. Zweifeldrige Verbundträger mit großer Stegöffnung. Erfahrungen und Zukunft des Bauens, Festschrift zum 70. Geburtstag von Gert König: pp. 527–538. Leipzig: Institut für Massivbau und Baustofftechnologie, Universität Leipzig.

Schnell, J. & Weil, T. 2007. Forschungsbericht zum DFG-Forschungsvorhaben SCHN771/1-1 u. 2, Stegöffnungen in Verbunddurchlaufträgern. Kaiserslautern: Fachgebiet Massivbau und Baukonstruktion, Technische Universität Kaiserslautern.

Steel and Composite Structures – Wang & Choi (eds)
© *2007 Taylor & Francis Group, London, ISBN 978-0-415-45141-3*

Full-scale composite beam tests on an Australian type of trapezoidal steel decking

S. Ernst & A. Wheeler

School of Engineering, University of Western Sydney, Australia

ABSTRACT: The effect of trapezoidal decking on the composite shear strength has recently been investigated by numerous authors and various experimental methods for determining the stud strength presented. These methods have included various configurations of push-out tests and the appropriateness of these tests to determine the strength of connectors in slabs has been discussed. In this paper the results from a number of full scale composite secondary beam tests carried are presented. The specimens tested included beams that used both traditional reinforcement and beams that utilised types of reinforcement to suppress particular failure modes. An accompanying set of push-out tests were also conducted and the results compared to the full scale beam tests. The results of the beam tests and push-out tests demonstrated a good agreement in both the modes of failure and the ultimate load capacity.

1 INTRODUCTION

Traditionally in Australia the type of decking that has been used in steel concrete composite decks was the so-called closed form deck. These closed form decks have relatively low and narrow ribs that are ignored when considering the stud strength in both primary and secondary beams. This practice is reflected in the Australian Composite Structures Standard (AS 2003), which currently does not allow for open rib (trapezoidal) decking.

In recent years a number of international and new trapezoidal decking profiles have been introduced into the Australian market. These decks result in longer un-propped spans with significant voids being formed in the floor system. However, when trapezoidal decking is orientated perpendicular to the beams (secondary beam) the mode of deformation and failure differed significantly to that of a solid slab (Fisher (1970) and Robinson (1969)).

A number of studies into the effect of the trapezoidal decking have been carried and various methods of design proposed (Grant et al. 1977, Johnson Yuan 1998b, Rambo-Roddenberry et al. (2000)). These deign methods have typically be derived from test results carried out on a limited number of decking profiles. The profile of a deck, i.e. the depth and width of ribs, has a significant influence on the shear connection behaviour and the resulting shear strength. It is therefore critical to have a thorough understanding of the effects of decking geometry on the stud behaviour.

To ascertain the shear capacity of connectors in a concrete slab with a particular decking profile to-date is only possible through testing, and the most accurate method to use is a full–scale composite beam test. Although consideration must be made as to how the boundary conditions are modelled, and the stud strength is determined. It is recognised that the costs of full-scale tests are extremely high, and the general practice is to utilise a push test to determine the shear behaviour, were only a small section of the composite member is tested. However, various types of push tests are used worldwide (Roik an Hanswille (1987), Hicks and McConnel (1996), Johnson Yuan 1998b, Rambo-Roddenberry et al. (2000), Bradford (2006)), and there has been considerable discussion the appropriateness of the boundary conditions and how these tests are carried out and how they relate to real structures.

An extensive research programme has been undertaken at the University of Western Sydney looking at various aspects of the effect of trapezoidal decking on the performance of welded studs and at the relationship between beam tests and push test. In this paper the results for a full scale composite beam test will be presented, along with the results from a companion push tests series.

The beam test were carried out using a typical trapezoidal deck from Australia and tested as both internal and edge beams. In both the beam test and the push tests traditional reinforcement, as well as reinforcement aimed at improving the shear strength stud performance were utilised. Measurement instrument

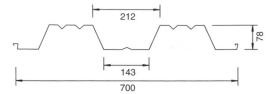

Figure 1. Geometry of trapezoidal steel decking.

were also placed along the beams to enable the accurate assessment of the behaviour at any given shear connection in the specimens.

2 EXPERIMENTAL PROGRAMME

2.1 Beam Tests

The tests carried out in this study were all secondary beams with the steel decking running transversely to the steel section. The tests series comprised two full-scale beams, one an internal beam and the other an edge beam. To maximize both the cost and the amount of information from each specimen, the beams were divided into two halves. One half of the beam was reinforced in accordance with Eurocode 4 (CEN 2004), while the other half had a single layer of reinforcement coupled with novel reinforcing components. The dividing of the composite beams in two halves has been used previously (Patrick et al. 1995) and provided the advantage of a direct comparison of the shear connection behaviour with different layouts while maintaining constant material properties. In this test program, special test procedures were developed to ensure sufficient levels of slip were experienced at both ends.

The effective span of both beams was 4200 mm between the end supports, this length ensured a sufficient number of shear connectors in each shear span The test specimens comprised of a hot-rolled steel beam I-section (250UB31.3) with a nominal yield stress of 300 MPa and a 150 mm thick slab with a target concrete strength of 32 MPa. For the internal beam (Beam 1), the concrete slab was 1800 mm wide, while for the edge beam (Beam 2), the concrete slab had an outstand of 300 mm and an overall width of 1200 mm. The same trapezoidal steel decking was used in both beams and had a base metal thickness of 0.75 mm and a nominal yield stress of 550 MPa. The geometry of the decking is shown in Figure 1.

All studs were 19 mm diameter headed studs placed as pairs diagonally either side of the centre embossment and welded to the beam with a specified height of 127 mm after welding. The spacing between transverse centres was 80 mm. Initial trials demonstrated that the confidence of full stud welding could not be guaranteed when welded through the decking so, all

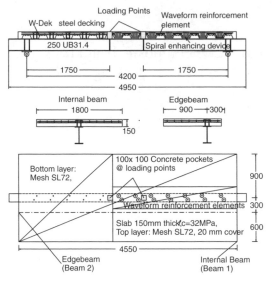

Figure 2. Beam Test Layouts.

studs were welded directly to the beam through holes placed in the sheeting.

The beams were designed using the partial shear connection assumptions with the minimum requirements given in Eurocode 4 (CEN 2004) provided. Generally loading of beam tests is through a number of points on top of the concrete slab (Rambo-Roddenberry et al., 2000), in this study loading was through a single point directly on to the steel beam. This method prevents any clamping effects of the concrete slab which might be present if the top surface of the concrete slab is loaded directly and models the type of loading you would expect from a hanging beam. A general layout of the specimens is shown in Figure 2. This single point loading provided the maximum possible shear span without the local influence of applied loads; thus allowing for a simpler investigation of the redistribution of the stud forces between the individual shear connections.

Slab reinforcement (shown in Figure 2) included a bottom layer of SL72 reinforcement mesh on the weak side of each beam specimen (left side), placed directly on top of the steel decking. The top reinforcement layer for the entire slab was SL72 mesh with 20 mm cover. The waveform reinforcement elements used on the strong side of the beam were as detailed in Figure 3, with four longitudinal bars bent to match the deck with 20 mm cover. The four longitudinal bars were spaced at 150 mm intervals and welded to the transverse bars as shown in Figure 3.

This waveform reinforcement was coupled with stud enhancement devices (see also Ernst 2006). These devices consist of a round steel wire which spirals around the stud and simply clipped into position over

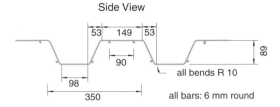

Side View

all bends R 10

all bars: 6 mm round

Figure 3. Detail waveform reinforcement element.

Figure 4. Spiral enhancing device.

Figure 5. Counterweight for Beam2 specimen.

the head of the shear connectors after the stud is welded into position (Figure. 4). The purpose of this device is to confine the concrete around the base of the stud connector, thus reducing bending of the stud shank, and minimising the effects of localized failure.

The concrete slabs were cast on the steel beams that were continuously propped, along their lengths. Sufficient concrete cylinders to monitor the concrete compressive strength at regular intervals were also poured. All specimens were stored under moist conditions until testing resumed.

To fully monitor the behaviour of each specimen, twenty three linear potentiometers were used to determine the displacements. The vertical deflections of the bottom flange of the steel section were monitored at the loading points and at mid-length of each shear span. The horizontal slips at the location of each pair of shear connection were measured relative to the slab, along with the total slip of the concrete slab at both ends of the beam. The additional five potentiometers measure the relative uplift of the concrete slab from the steel section along the length of the beam.

The strains in the steel section and top of the concrete slab were measured at six different locations along each beam. It was expected that the stresses in the steel beam in the cross-sections close to the supports would remain in the elastic range, so two strain gauges, one on the top and one on the bottom of the steel beam were deemed to be sufficient. At the cross-sections closest to the loading points where yielding

of the steel section was expected, a total of ten strain gauges were used, eight being attached to the steel section and two to the top surface of the concrete slab.

Once the concrete reached its target strength, the specimens were lifted into position with the ends placed on the roller supports. For Beam 2, the eccentric concrete slab lay-out, a counterweight G was used in order to shift the centre of gravity to the centreline of the steel section (Figure 5).

During the test the load was applied directly to the steel beam top flange. Initially, the weak (conventionally reinforced) side of the specimen was tested (Test A) and the load was cycled ten times between 10 kN and 40% of the expected composite beam strength. Further cycles were also undertaken at 60% and 80% of the expected strength. The specimen was then loaded until the shear connections at the weak side of the beam reached their maximum strength and initial failure was observed. The test was then stopped at this point.

Once the initial test was completed the load was removed and the composite beam was moved to the position that the load could be applied to the loading point at the strong side of the beam (Test B). Again the load was cycled in a similar fashion as in Test A between 10 kN and 40%, 60% and 80% of the expected composite beam strength.

For Beam 1 the load was increased until disproportionate increases in slip on the weak side shear connections was identified. At this point the end slip on the weak site was prevented and the test continued. The load was increased until large end slips on the strong side were observed. At this point the loads and the restraints on the weak side were removed. The weak side was the tested (Test C) to determine the post peak behaviour of the weak side of the specimen.

For Beam #2 after testing of the weak side end restraints were placed prior to testing of the strong side to prevent slip on the weak side. Additional tie down was also required the Beam #2 test due to the concrete

Table 1. Push test specimens.

Specimen no.	Width (mm)	Stud reinforcement*	Max stud strength (kN)
SDM1	1800	Nil	69.8
SDM2	1800	WR/SR	86.4
SDM3	1200	Nil	65.9
SDM4	1200	WR/SR	85.7

* WR – wave reinforcement, SR – stud reinforcement.

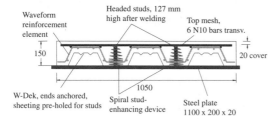

Figure 6. Push out specimen Details.

failure mode that occurred in the initial test. Again once the strong side tests were stopped, the specimens were moved so that the load could be applied to the weak side and Test C was carried out.

A full description of the beam test setup, beam specimens and results may be found in Ernst (2007).

2.2 Push tests

To quantify the relationship between the push tests and the full scale beam test, a series of companion push tests were fabricated and tested in parallel to the full scale tests. A total of 4 push tests were carried out in this investigation with configurations being identical to those in the beam tests.

The geometry and the reinforcing layout of the specimens are found in Table 1. Stud dimension and placement were identical to the beam tests and the specimens poured at the same time as the slab to ensure consistence of materials.

The push test specimens were all 1050 mm long and 150 mm deep, the setup also ensured that two ribs with shear connectors were engaged, see Figure 6. A layer of top reinforcement was placed to prevent the back breaking failure.

The instrumentation of the specimens included strain gauges in bolts to quantify the uplift loads in the bolts. Both horizontal displacements (slip) and vertical displacements at the location of the shear connecters and the horizontal slip at the end of the specimen we measured.

The specimens were fastened to the purpose built push out rig (Ernst 2006) and the vertical loads applied.

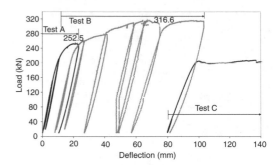

Figure 7. Load-deflection behaviour of Beam 1.

The specimens were first cycled to a proof load and then tested to failure. A full description of the push test specimens and results may be found in Ernst (2007).

3 TEST RESULTS AND FAILURES

In all the tests results the self-weight of the composite beams and any counterweight are included in the load calculations.

3.1 Beam 1

The load deflection behaviour of beam 1 is shown in Figure 7. The initial test on the waekside (Test A) was stopped at a load of 252.5kN when the load-deflection curve had softened significantly. A longitudinal concrete crack at the concrete slab surface propagating along the centreline of the slab from the concrete pocket towards the end of the slab was the only visible damage. At this point four of the five shear connections had already experienced significant plastic deformations with slips of approximately 2mm at the three shear connections closest to the support.

In the strong side tests (Test B), the load applied increased to a maximum load of 316.6 kN. Immediately after this maximum was recorded, the slip measurements indicated that the weak side may be limiting the capacity of the section. As previously indicated the weak end was restrained and the test resumed. At an end slip of close to 8 mm on the strong side of the specimen a consistent load of approximately 314 kN was observed. A longitudinal concrete crack propagated along the centreline of the concrete slab in a similar fashion to Test A was observed. A slight bulging of the steel decking in the direction of the longitudinal shear force was also observed in some of the ribs.

When loading on the weak side of the composite beam resumed in Test C, the specimen did not reach its earlier strength (Test A). This indicated that the

Figure 8. Stud pull-out failures in Beam 1.

Figure 9. Strong side Failure Modes – Beam 1.

movement of the weak side during Test B played a roll in reducing the shear connection strength.

The ultimate failure modes observed in the weak side was stud pull out failure out as shown in Figure 8, with the slab being easily lifted from the decking.

The failure modes of the strong side are shown in Figure 9. This figures shows the concrete slab after a cut was made and the slab removed adjacent to the beam. Horizontal stud pull-out cracks are clearly visible and had developed in most of the concrete ribs. The restraint of the additional shear reinforcement in preventing a pull out failure is demonstrated. Diagonal cracks between the base of the shear connector and the top corner of the concrete rib had also developed in most of the connections indicating the formation of a wedge breaking out of the concrete rib, i.e. the onset of rib punch-through failure.

3.2 Beam 2

The load deflection behaviour of beam test 2 is shown in Figure 10. Test A was stopped when the first horizontal rib shearing crack started to appear in the concrete rib closest to the loading point at 237.9 kN. At this point, the shear connections closest to the support had slips of approximately 1.5 mm. A longitudinal crack, propagating from the concrete pocket along the centreline of the concrete slab towards the weak end of the slab, similar to the one observed in the previous Beam 1 tests was also obseved.

In the strong side tests (Tests B), the total load was increased to 295.3 kN when a failure of the concrete

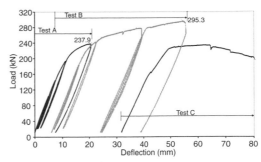

Figure 10. Load-deflection behaviour of Beam 2.

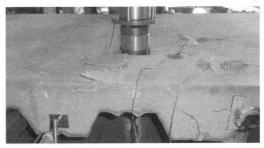

Figure 11. Concrete compressive zone failure.

Figure 12. Cracking of the concrete slab at weak side.

compressive zone on the narrow side of the concrete slab appeared, see Figure 11.

During the strong side tests the horizontal rib-shearing crack on the weak side continued to propagate and was restrained using a clamping system. Once the slab has failed the clamping system was removed the beam moved and the weak side tested for the post peak behaviour (Test C). The ultimate failure mode for the weak side was the rib shearing carks as shown in Figure 12.

3.3 Push tests

The results from the push test are presented in Figure 13 and 14 for the internal and edge beam respectively. The results presented show the average force per stud and the measured end slip of the specimens. The

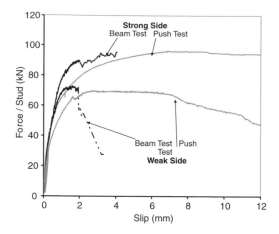

Figure 13.   Internal Stud Force – Slip Relationship.

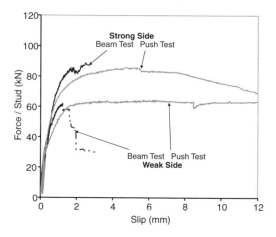

Figure 14.   Edge Stud Force – Slip Relationship.

failure modes observed in the push tests were stud pull out in the internal beam tests while rib shearing failures were observed in the edge beam tests. The peak failure loads per stud are presented in Table 1.

The effect of the additional reinforcement in the push test is evident with significant increases in the strength of these studs. In additional tests carried out by Ernst (2006) with alternative decking profiles and stud configuration resulted in significant increases in both the strength and ductility of the studs.

3.4   Comparison between beam and push tests

In the beam tests the forces in the shear connection in each shear span were determined by the variation in the strain measurements across the steel cross-section at individual locations (Ernst 2006). By taking the average of the shear connections investigated, the average force per stud verses the slip was determined

and plotted against the push test in Figure 13 and Figure 14.

Both the maximum shear strength and the initial stiffness of the shear connections are in agreement for both the push and the beam test. However, the connection in the beam test demonstrate stiffer behaviour after 60% of the maximum strength is reached.

4   CONCLUSION

Two secondary composite beams were tested, and divided in two different halves where the shear connections on one side were conventionally reinforced fulfilling the design provisions of Eurocode 4 (CEN 2004) and the connections on the other side included waveform reinforcing elements and spiral enhancing devices. The beams were tested and instrumented to allow for the determination of the loads being transferred to the studs throughout the test. A set of companion push tests were also prepared and tested to enable a direct comparison between the results from beam tests and push tests.

The test results showed that the concrete-related premature failure modes of stud pull-out, rib shearing and rib punch-through, experienced in earlier push-out tests, also occurred in a full-scale composite beam applications. The behaviour of the conventionally reinforced shear connections which experienced stud pullout failures in the internal beam specimen and rib shearing failures in the edge beam specimen appeared to be of an even more brittle nature than in the accompanying push-out tests. The use of the reinforcing elements on the other hand overcame the brittle effects of these failure modes by reinforcing the potential failure surfaces. The composite beam load and deflection capacities were significantly increased as the shear connections generally experienced increased stud strengths and deformation capacities which subsequently improved the force redistribution between individual stud connections.

The shear connection strengths and failure modes experienced in the composite beam tests were found to generally compare very well with the results obtained from similar push-out test specimens. In the full scale beam test both the rib shear failure and the stud pull out failure modes were evident. It therefore follows, that when conducting small scale tests, such as push tests, measures that suppress these failure modes should be avoided.

REFERENCES

Bradford, M.A, Filonov, A., Hogan, T.J, Uy, B. and Ranzi, G. (2006) Strength and ductility of shear connection in composite T-beams with trapezoidal steel decking. *Proceedings of 8th International Conference on Steel, Space and Composite Structures*, Kuala Lumpur, Malaysia, pp. 15–26

CEN (2004). "Design of composite steel and concrete structures, Part 1–1: General rules and rules for buildings." Eurocode 4, European Committee for Standardization (CEN).

Ernst S (2006) "Factors Affecting the Behaviour of the Shear Connection of Steel-Concrete Composite Beams" PhD Thesis, University of Western Sydney.

Ernst, S., Patrick, M., Bridge, R. Q., and Wheeler, A. (2004). "Reinforcement requirements for secondary composite beams incorporating trapezoidal decking." *Proceedings of Composite Construction in Steel and Concrete V Conference*, Kruger Park, South Africa, ASCE, 236–246.

Ernst, S., Bridge, R. Q., and Patrick, M. (2006). "Behaviour of the longitudinal shear connection in secondary composite beams incorporating trapezoidal steel decking." *Proceedings of the Australasian Conference on the Mechanics of Structures and Materials.* Christchurch, New Zealand.

Ernst, S., Bridge, R. Q., and Wheeler, A. (2007). "Correlation of beam tests with push-out test in steel concrete composite beams." ASCE *Journal of the Structural Division*, in publication.

Fisher, J. W. (1970). "Design of composite beams with formed metal deck." *AISC Engineering Journal*, 3(7), 88–96.

Grant, J. A., Fisher, J. W., and Slutter, R. G. (1977). "Composite Beams with Formed Steel Deck." *AISC Engineering Journal*, 14, 24–43.

Hicks, S. J., and McConnel, R. E. (1996). "The shear resistance of headed studs used with profiled steel sheeting." *Proceedings of Composite Construction in Steel and Concrete III Conference*, Irsee, Germany, ASCE, 325–338.

Johnson, R. P., and Yuan, H. (1998a). "Existing rules and new tests for stud shear connectors in throughs of profiled steel sheeting." *Proceedings of the Institution of Civil Engineers: Structures & Buildings*, 128, 244–251.

Johnson, R. P., and Yuan, H. (1998b). "Models and design rules for stud shear connectors in troughs of profiled sheeting." *Proceedings of the Institution of Civil Engineers: Structures & Buildings*, 128, 252–263.

Patrick, M., Dayawansa, P. H., Eadie, I., Watson, K. B., and van der Kreek, N. (1995). "Australian composite structures Standard AS 2327, Part 1: Simply-Supported Beams." *Journal of the Australian Institute of Steel Construction*, 29(4), 2–40.

Rambo-Roddenberry, M., Lyons, C., Easterling, W. S., and Murray, T. M. (2000). "Performance and strength of welded shear studs." *Proceedings of Composite Construction in Steel and Concrete IV Conference*, Banff, Canada, ASCE, 458–469.

Robinson, H. (1969). "Composite beam incorporating cellular steel decking." *Journal of the Structural Division*, 95(3), 355–380.

Roik, K., and Hanswille, G. (1987). "Zur Dauerfestigkeit von Kopfbolzendübeln bei Verbundträgern." *Der Bauingenieur*, 62, 273–285.

Standards Australia (2003). "Composite structures, Part 1: Simply supported beams." AS 2327.1–2003, Standards Australia.

Steel and Composite Structures – Wang & Choi (eds)
© *2007 Taylor & Francis Group, London, ISBN 978-0-415-45141-3*

A mixed co-rotational formulation for 3D beam element

Z.X. Li

Department of Civil Engineering, Zhejiang University, Hangzhou, China

ABSTRACT: A 3-node co-rotational element formulation for 3D beam undergoing large displacements and large rotations is presented, and a simple element-independent formulation is achieved. Besides translational nodal variables, vectorial rotational variables are defined to replace traditional angular rotational variables, resulting in all nodal variables additive in an incremental solution procedure. To alleviate or eliminate the membrane and shear locking phenomena, Hellinger-Reissner functional is introduced, where assumed membrane strains and shear strains are adopted to replace part of the corresponding conforming strains, and the internal force vector is calculated by enforcing stationarity upon the Hellinger-Reissner functional. Furthermore, by differentiating the internal force vector with respect to the nodal variables, a symmetric element tangent stiffness matrix is achieved. Finally, several examples of elastic beams with large displacements and large rotations are analyzed; the results demonstrate the computational efficiency and convergence of the present formulation.

1 INTRODUCTION

Developing an efficient beam element formulation for large displacement analysis of framed structures has been an issue of many researchers. There already exist various formulations to meet this requirement, Hsiao *et al.*(1987) had divided them into three categories: Total Lagrangian formulation, Updated Lagrangian formulation and co-rotational formulation. For convenience, these formulations can also be classified into two groups: formulations with asymmetric element tangent stiffness matrices and formulations with symmetric element tangent stiffness matrices. Due to the non-commutativity of spatial rotations, most co-rotational formulations belong to the first category. For an asymmetric tangent stiffness matrix, more storage is occupied to store all its components, and the computational efficiency is poor. Simo and Vu-quoc (1986) denoted that in a conservative system, although their developing tangent stiffness matrix is always asymmetric, it can become symmetric once the incremental loading process arrives at an equilibrium level. Crisfield (1990; 1996), Crsisfield and his co-workers(1996) had also found this phenomenon, so they symmetrized artificially the element tangent stiffness by excluding the non-symmetric term. This treatment can improve the computational efficiency greatly. Crisfield (1996) and Simo (1992) also predicted that a symmetric tangent stiffness matrix in a co-rotational framework could be achieved if a certain set of additive rotational variables are employed.

In this paper, the author defined a set of vectorial rotational variables and proposed an advanced co-rotational framework for 3D beam element, and achieved symmetric tangent stiffness matrices both in the local and global coordinate systems. In this formulation, several basic assumptions were adopted: (1) all elements are straight at the initial configuration; (2) the shape of the cross-section does not distort with element deforming; (3) the cross-section of all elements are bisymmetric; (4) restrained warping effects are ignored.

2 DESCRIPTION OF THE CO-ROTATIONAL FRAMEWORK

The local and global coordinate systems of the beam element are illustrated in Figure 1, where three local coordinate axes run along two principal axes of the cross-section at the internal node and their cross-product, and translate and rotate with the element rigid-body motion, but does not deform with the element. An auxiliary point in one of the symmetry plane of the beam element is employed in defining local coordinate axes (see Point A in Figure 1).

$\mathbf{e}_{x0}, \mathbf{e}_{y0}, \mathbf{e}_{z0}$ are the normalized orientation vectors of local x-, y- and z-axes, respectively. They are calculated from

$$\mathbf{e}_{x0} = \frac{\mathbf{v}_{120}}{|\mathbf{v}_{120}|} \qquad (1a)$$

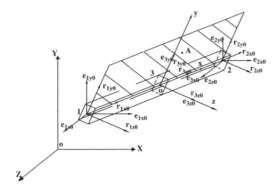

Figure 1. Definition of local and global coordinate systems.

$$\mathbf{e}_{z0} = \frac{\mathbf{V}_{120} \times \mathbf{V}_{3A0}}{|\mathbf{V}_{120} \times \mathbf{V}_{3A0}|} \tag{1b}$$

$$\mathbf{e}_{y0} = \mathbf{e}_{z0} \times \mathbf{e}_{x0} \tag{1c}$$

where,

$$\mathbf{V}_{120} = \mathbf{X}_{20} - \mathbf{X}_{10}, \quad \mathbf{V}_{3A0} = \mathbf{X}_{A0} - \mathbf{X}_{30} \tag{2}$$

$\mathbf{X}_{i0}$ $(i = 1,2,3,A)$ is the global coordinates at Node i.

The orientation vectors $\mathbf{e}_{ix}$, $\mathbf{e}_{iy}$, $\mathbf{e}_{iz}$ at Node i in the deformed configuration are calculated from the rotational variables directly in an incremental solution procedure. In particular, at Node 3 (the internal node of the beam element), $\mathbf{e}_{3x}$, $\mathbf{e}_{3y}$, $\mathbf{e}_{3z}$ are coincident with the orientation of local coordinate axes,

$$\mathbf{e}_{3x} = \mathbf{e}_x, \quad \mathbf{e}_{3y} = \mathbf{e}_y, \quad \mathbf{e}_{3z} = \mathbf{e}_z \tag{3}$$

and at the initial configuration,

$$\mathbf{e}_{3x0} = \mathbf{e}_{x0}, \quad \mathbf{e}_{3y0} = \mathbf{e}_{y0}, \quad \mathbf{e}_{3z0} = \mathbf{e}_{z0} \tag{4}$$

however, the initial orientation vectors of two end nodes are defined by

$$\mathbf{e}_{ix0}^T = \langle 1,0,0 \rangle, \quad \mathbf{e}_{iy0}^T = \langle 0,1,0 \rangle, \quad \mathbf{e}_{iz0}^T = \langle 0,0,1 \rangle \; i = 1,2 \tag{5}$$

In the global coordinate system, each element employs 18 degrees of freedom,

$$\mathbf{u}_G^T = \langle U_1 \, V_1 \, W_1 \, e_{1y,n_1} \, e_{1y,m_1} \, e_{1z,n_1} \cdots U_3 \, V_3 \, W_3 \, e_{3y,n_3} \, e_{3y,m_3} \, e_{3z,n_3} \rangle \tag{6}$$

where, (U_i, V_i, W_i) are global nodal translational displacements; $(e_{iy,ni}, e_{iy,mi}, e_{iz,ni})$ are vectorial nodal rotational variables, they are three independent components of $\mathbf{e}_{iy}$ and $\mathbf{e}_{iz}$, respectively.

In the local coordinate system, each element has 12 degrees of freedom, and each end node 6 freedoms,

$$\mathbf{u}_L^T = \langle u_1 \, v_1 \, w_1 \, r_{1y,n_1} \, r_{1y,m_1} \, r_{1z,n_1} \, u_2 \, v_2 \, w_2 \, r_{2y,n_2} \, r_{2y,m_2} \, r_{2z,n_2} \rangle \tag{7}$$

where, (u_i, v_i, w_i) are local nodal translational displacements, and $(r_{iy,ni}, r_{iy,mi}, r_{iz,ni})$ are the vectorial rotational variables at Node i, they are three components of nodal orientation vectors $\mathbf{r}_{iy}$ and $\mathbf{r}_{iz}$ in the local coordinate system.

The rotational variables $(e_{iy,ni}, e_{iy,mi}, e_{iz,ni})$ are defined according to the following procedure:

Firstly, assumed that

$$|e_{iy,l_i}| \ge |e_{iy,m_i}|, \quad |e_{iy,l_i}| \ge |e_{iy,n_i}| \quad (l_i, m_i, n_i \in \{1,2,3\}) \tag{8}$$

at the preceding incremental loading step: Case 1: if

$$|e_{iz,l_i}| \ge |e_{iz,m_i}|, \quad |e_{iz,l_i}| \ge |e_{iz,n_i}| \tag{9}$$

then, at the end of the current incremental load step, other components of $\mathbf{e}_{iy}$ and $\mathbf{e}_{iz}$ can be calculated from

$$e_{iy,l_i} = s_1 \sqrt{1 - e_{iy,n_i}^2 - e_{iy,m_i}^2} \tag{10a}$$

$$e_{iz,m_i} = \frac{-e_{iy,m_i} e_{iy,n_i} e_{iz,n_i} + s_2 e_{iy,l_i} \sqrt{1 - e_{iy,n_i}^2 - e_{iz,n_i}^2}}{1 - e_{iy,n_i}^2} \tag{10b}$$

$$e_{iz,l_i} = s_3 \sqrt{1 - e_{iz,m_i}^2 - e_{iz,n_i}^2} \tag{10c}$$

where, $\{n_i, m_i, l_i\}$ is a circular permutation; s_1 and s_3 take a numeric value of 1 or -1, they have the same signs as $e_{iy,li}$ or $e_{iz,li}$ at the last incremental load step; s_2 is also such a kind of constants, and it is conditioned on $\mathbf{e}_{iy}\mathbf{e}_{iz} = 0$.

Case 2: if

$$|e_{iz,m_i}| \ge |e_{iz,l_i}|, \quad |e_{iz,m_i}| \ge |e_{iz,n_i}| \tag{11}$$

then, at the end of the current incremental load step, other components of $\mathbf{e}_{iy}$ and $\mathbf{e}_{iz}$ can be calculated from

$$e_{iy,l_i} = s_1 \sqrt{1 - e_{iy,n_i}^2 - e_{iy,m_i}^2} \tag{12a}$$

$$e_{iz,l_i} = \frac{-s_1 \sqrt{1 - e_{iy,m_i}^2 - e_{iy,n_i}^2} e_{iy,n_i} e_{iz,n_i} + s_2 e_{iy,m_i} \sqrt{1 - e_{iz,n_i}^2 - e_{iz,n_i}^2}}{1 - e_{iy,n_i}^2} \tag{12b}$$

$$e_{iz,m} = s_3 \sqrt{1 - e_{iz,n}^2 - e_{iz,l}^2} \tag{12c}$$

where, s_1, s_2, s_3 are the same kind of constants as those in Case 1.

Vector $\mathbf{e}_{ix}$ is the cross-product of Vectors $\mathbf{e}_{iy}$ and $\mathbf{e}_{iz}$,

$$\mathbf{e}_{ix} = \mathbf{e}_{iy} \times \mathbf{e}_{iz} \tag{13}$$

The definition of local vectorial rotational variables $(r_{iy,ni}, r_{iy,mi}, r_{iz,ni})$ follows the same route as that of $(e_{iy,ni}, e_{iy,mi}, e_{iz,ni})$.

Rigid-body motion contributes nothing to strains, so it is excluded in advance to achieve an element-independent formulation. The relationships between local and global nodal variables are given as,

$$\mathbf{t}_i = \mathbf{R}(\mathbf{d}_i - \mathbf{d}_3) + (\mathbf{R} - \mathbf{R}_0)\mathbf{v}_{i0} \tag{14a}$$

$$\mathbf{r}_{iy} = \mathbf{R}\mathbf{R}_i^{\mathrm{T}}\mathbf{e}_{y0} \tag{14b}$$

$$\mathbf{r}_{iz} = \mathbf{R}\mathbf{R}_i^{\mathrm{T}}\mathbf{e}_{z0} \tag{14c}$$

where,

$$\mathbf{t}_i = \begin{Bmatrix} u_i \\ v_i \\ w_i \end{Bmatrix}, \qquad \mathbf{R}_0 = \begin{Bmatrix} \mathbf{e}_{x0}^{\mathrm{T}} \\ \mathbf{e}_{y0}^{\mathrm{T}} \\ \mathbf{e}_{z0}^{\mathrm{T}} \end{Bmatrix} \tag{15a}$$

$$\mathbf{R} = \begin{Bmatrix} \mathbf{e}_x^{\mathrm{T}} \\ \mathbf{e}_y^{\mathrm{T}} \\ \mathbf{e}_z^{\mathrm{T}} \end{Bmatrix}, \qquad \mathbf{R}_i = \begin{Bmatrix} \mathbf{e}_{ix}^{\mathrm{T}} \\ \mathbf{e}_{iy}^{\mathrm{T}} \\ \mathbf{e}_{iz}^{\mathrm{T}} \end{Bmatrix} \quad (i=1,2) \tag{15b}$$

Especially, at the internal node,

$$\mathbf{t}_3^{\mathrm{T}} = \langle 0\,0\,0 \rangle, \quad \mathbf{r}_{3y}^{\mathrm{T}} = \langle 0\,1\,0 \rangle, \quad \mathbf{r}_{3z}^{\mathrm{T}} = \langle 0\,0\,1 \rangle \tag{16}$$

$\mathbf{v}_{i0}$ is the relative vector oriented from Node 3 to Node i,

$$\mathbf{v}_{i0} = \mathbf{X}_{i0} - \mathbf{X}_{30} \quad (i=1,2) \tag{17}$$

3 KINEMATICS OF A 3-NODE ISO-PARAMETRIC BEAM ELEMENT

In the present 3-node iso-parametric beam element, the local coordinates, displacements and vectorial rotations at any point of the element central line are interpolated by using Lagrangian shape functions. The initial and current local coordinates at any point in the element can be depicted as

$$^0\mathbf{g} = \sum_{i=1}^{3} \mathbf{h}_i\,\mathbf{x}_{i0} + y_l \sum_{i=1}^{3} \mathbf{h}_i\,\mathbf{r}_{iy0} + z_l \sum_{i=1}^{3} \mathbf{h}_i\,\mathbf{r}_{iz0} \tag{18}$$

$$\mathbf{g} = \sum_{i=1}^{3} \mathbf{h}_i(\mathbf{t}_i + \mathbf{x}_{i0}) + y_l \sum_{i=1}^{3} \mathbf{h}_i(\mathbf{r}_{iy} - \mathbf{r}_{iy0})$$

$$+ z_l \sum_{i=1}^{3} \mathbf{h}_i(\mathbf{r}_{iz} - \mathbf{r}_{iz0}) \tag{19}$$

where, $\mathbf{h}_i$ is the Lagrangian shape function at Node i; $\mathbf{x}_{i0}$ is the initial local coordinates at Node i; y_l and z_l are the relative coordinates of any point in the element to its central line.

Considering the possibility of large displacements and large rotations, Green strain measure is introduced to describe the strain-displacement relationship. For beam element, the relationship of strain-displacement is given as,

$$\varepsilon_{xx} = \frac{1}{2}\left(\mathbf{g}_x^{\mathrm{T}}\mathbf{g}_x - {}^0\mathbf{g}_x^{\mathrm{T}}{}^0\mathbf{g}_x\right) \tag{20a}$$

$$\varepsilon_{xy} = \frac{1}{2}\left(\mathbf{g}_x^{\mathrm{T}}\mathbf{g}_y - {}^0\mathbf{g}_x^{\mathrm{T}}{}^0\mathbf{g}_y\right) \tag{20b}$$

$$\varepsilon_{xz} = \frac{1}{2}\left(\mathbf{g}_x^{\mathrm{T}}\mathbf{g}_z - {}^0\mathbf{g}_x^{\mathrm{T}}{}^0\mathbf{g}_z\right) \tag{20c}$$

where,

$$\mathbf{g}_x = \frac{\partial \mathbf{g}}{\partial x}, \qquad {}^0\mathbf{g}_x = \frac{\partial^0 \mathbf{g}}{\partial x} \tag{21}$$

For convenience, (20a–c) can be rewritten as,

$$\varepsilon_{xx} = \varepsilon_{xx}^{(0)} + y_l\varepsilon_{xx}^{(1)} + z_l\varepsilon_{xx}^{(2)} + y_l z_l\varepsilon_{xx}^{(3)} + y_l^2\varepsilon_{xx}^{(4)} + z_l^2\varepsilon_{xx}^{(5)} \tag{22a}$$

$$\varepsilon_{xy} = \varepsilon_{xy}^{(0)} + y_l\varepsilon_{xy}^{(1)} + z_l\varepsilon_{xy}^{(2)} \tag{22b}$$

$$\varepsilon_{xz} = \varepsilon_{xz}^{(0)} + y_l\varepsilon_{xz}^{(1)} + z_l\varepsilon_{xz}^{(2)} \tag{22c}$$

4 ELEMENT FORMULATION

To alleviate or eliminate membrane and shear locking problems, Hellinger–Reissner partial mixed functional (To and Wang 1999) are employed, where part of strains are replaced by assumed strains,

$$\pi_{\mathrm{HR}} = \mathrm{E}\int_V \varepsilon_{xx}^a \varepsilon_{xx}\,\mathrm{d}V - \frac{1}{2}\mathrm{E}\int_V (\varepsilon_{xx}^a)^2\,\mathrm{d}V$$

$$+ \mathrm{k}_0\mathrm{G}\int_V \gamma_{xy}^a \gamma_{xy}\,\mathrm{d}V - \frac{1}{2}\mathrm{k}_0\mathrm{G}\int_V (\gamma_{xy}^a)^2\,\mathrm{d}V$$

$$+ \mathrm{k}_0\mathrm{G}\int_V \gamma_{xy}^a \gamma_{xy}\,\mathrm{d}V - \frac{1}{2}\mathrm{k}_0\mathrm{G}\int_V (\gamma_{xy}^a)^2\,\mathrm{d}V - \mathrm{W}_e \tag{23}$$

where,

$$\varepsilon_{xx}^a = \mathbf{P}^{\mathrm{T}}\boldsymbol{\alpha} + y_l\varepsilon_{xx}^{(1)} + z_l\varepsilon_{xx}^{(2)} + y_l z_l\varepsilon_{xx}^{(3)} + y_l^2\varepsilon_{xx}^{(4)} + z_l^2\varepsilon_{xx}^{(5)} \tag{24a}$$

$$\gamma_{xy}^a = \mathbf{P}^{\mathrm{T}}\boldsymbol{\beta} + y_l\gamma_{xy}^{(1)} + z_l\gamma_{xy}^{(2)} \tag{24b}$$

$$\gamma_{xz}^a = \mathbf{P}^{\mathrm{T}}\boldsymbol{\chi} + y_l\gamma_{xz}^{(1)} + z_l\gamma_{xz}^{(2)} \tag{24c}$$

$$\mathbf{P}^{\mathrm{T}} = \langle 1,r \rangle, \qquad \boldsymbol{\alpha}^{\mathrm{T}} = \langle \alpha_1,\alpha_2 \rangle \tag{24e}$$

$$\boldsymbol{\beta}^{\mathrm{T}} = \langle \beta_1,\beta_2 \rangle, \qquad \boldsymbol{\chi}^{\mathrm{T}} = \langle \chi_1,\chi_2 \rangle \tag{24f}$$

E and G are the Young's modulus and the shear modulus, respectively; k_0 the shear factor of the cross-section; V the element volume; W_e the work done by the external force $\mathbf{f}_{ext}$.

By enforcing the variation upon Hellinger–Reissner functional π_{HR} with respect to $\mathbf{u}_L$, α, β and χ,

$$\delta\pi_{HR} = E\int_V \varepsilon_{xx}^a \delta\varepsilon_{xx}\, dV + E\int_V \varepsilon_{xx}\delta\varepsilon_{xx}^a\, dV$$

$$-E\int_V \varepsilon_{xx}^a \delta\varepsilon_{xx}^a\, dV + k_0 G\int_V \gamma_{xy}^a \delta\gamma_{xy}\, dV$$

$$+k_0 G\int_V \gamma_{xy}\delta\gamma_{xy}^a\, dV - k_0 G\int_V \gamma_{xy}^a \delta\gamma_{xy}^a\, dV$$

$$+k_0 G\int_V \gamma_{xz}^a \delta\gamma_{xz}\, dV + k_0 G\int_V \gamma_{xz}\delta\gamma_{xz}^a\, dV$$

$$-k_0 G\int_V \gamma_{xz}^a \delta\gamma_{xz}^a\, dV - \delta W_e \tag{25}$$

where,

$$\delta\varepsilon_{xx}^a = \mathbf{P}^T\delta\boldsymbol{\alpha} +$$
$$\left(y_i \mathbf{B}_{xx}^{(1)} + z_i \mathbf{B}_{xx}^{(2)} + y_i z_i \mathbf{B}_{xx}^{(3)} + y_i^2 \mathbf{B}_{xx}^{(4)} + z_i^2 \mathbf{B}_{xx}^{(5)}\right)\delta\mathbf{u}_L \tag{26a}$$

$$\delta\varepsilon_{xx} = \left(\mathbf{B}_{xx}^{(0)} + y_i \mathbf{B}_{xx}^{(1)} + z_i \mathbf{B}_{xx}^{(2)} +\right.$$
$$\left.y_i z_i \mathbf{B}_{xx}^{(3)} + y_i^2 \mathbf{B}_{xx}^{(4)} + z_i^2 \mathbf{B}_{xx}^{(5)}\right)\delta\mathbf{u}_L \tag{26b}$$

$$\delta\gamma_{xy}^a = \mathbf{P}^T\delta\boldsymbol{\beta} + \left(y_i \mathbf{B}_{xy}^{(1)} + z_i \mathbf{B}_{xy}^{(2)}\right)\delta\mathbf{u}_L \tag{26c}$$

$$\delta\gamma_{xy} = \left(\mathbf{B}_{xy}^{(0)} + y_i \mathbf{B}_{xy}^{(1)} + z_i \mathbf{B}_{xy}^{(2)}\right)\delta\mathbf{u}_L \tag{26d}$$

$$\delta\gamma_{xz}^a = \mathbf{P}^T\delta\boldsymbol{\chi} + \left(y_i \mathbf{B}_{xz}^{(1)} + z_i \mathbf{B}_{xz}^{(2)}\right)\delta\mathbf{u}_L \tag{26e}$$

$$\delta\gamma_{xz} = \left(\mathbf{B}_{xz}^{(0)} + y_i \mathbf{B}_{xz}^{(1)} + z_i \mathbf{B}_{xz}^{(2)}\right)\delta\mathbf{u}_L \tag{26f}$$

$$\delta W_e = \mathbf{f}_{ext}\delta\mathbf{u}_L \tag{26g}$$

Considering the independence of $\delta\alpha$, $\delta\beta$, $\delta\chi$ and $\delta\mathbf{u}_L$, meanwhile, assumed that the cross-section of the beam element is bisymmetric, then the local internal force vector $\mathbf{f}$ of the element can be determined by

$$\mathbf{f} = \mathbf{f}_{ext} = EA\left(\int_L \mathbf{B}_{xx}^{(0)T}\mathbf{P}^T\, dx\right)\boldsymbol{\alpha}$$

$$+Ak_0 G\left[\left(\int_L \mathbf{B}_{xy}^{(0)T}\mathbf{P}^T\, dx\right)\boldsymbol{\beta} + \left(\int_L \mathbf{B}_{xz}^{(0)T}\mathbf{P}^T\, dx\right)\boldsymbol{\chi}\right]$$

$$+I_y\left[E\int_L\left(\mathbf{B}_{xx}^{(0)T}\varepsilon_{xx}^{(4)} + \mathbf{B}_{xx}^{(4)T}\varepsilon_{xx}^{(0)} + \mathbf{B}_{xx}^{(1)T}\varepsilon_{xx}^{(1)}\right)dx\right.$$

$$\left.+k_0 G\int_L\left(\mathbf{B}_{xy}^{(1)T}\gamma_{xy}^{(1)} + \mathbf{B}_{xz}^{(1)T}\gamma_{xz}^{(1)}\right)dx\right]$$

$$+I_z\left[E\int_L\left(\mathbf{B}_{xx}^{(0)T}\varepsilon_{xx}^{(5)} + \mathbf{B}_{xx}^{(5)T}\varepsilon_{xx}^{(0)} + \mathbf{B}_{xx}^{(2)T}\varepsilon_{xx}^{(2)}\right)dx\right.$$

$$\left.+k_0 G\int_L\left(\mathbf{B}_{xy}^{(2)T}\gamma_{xy}^{(2)} + \mathbf{B}_{xz}^{(2)T}\gamma_{xz}^{(2)}\right)dx\right]$$

$$+I_{wx}\int_L \mathbf{B}_{xx}^{(4)T}\varepsilon_{xx}^{(4)}\, dx + I_{wy}\int_L \mathbf{B}_{xx}^{(5)T}\varepsilon_{xx}^{(5)}\, dx$$

$$+I_{wyz}E\int_L\left(\mathbf{B}_{xx}^{(3)T}\varepsilon_{xx}^{(3)} + \mathbf{B}_{xx}^{(5)T}\varepsilon_{xx}^{(4)} + \mathbf{B}_{xx}^{(4)T}\varepsilon_{xx}^{(5)}\right)dx \tag{27}$$

where,

$$\mathbf{H} = A\int_L \mathbf{P}\mathbf{P}^T\, dx, \qquad \mathbf{F}_1 = A\int_L \mathbf{P}\varepsilon_{xx}^{(0)}\, dx \tag{28a}$$

$$\mathbf{F}_2 = A\int_L \mathbf{P}\gamma_{xy}^{(0)}\, dx, \qquad \mathbf{F}_3 = A\int_L \mathbf{P}\gamma_{xz}^{(0)}\, dx \tag{28b}$$

$$\boldsymbol{\alpha} = \mathbf{H}^{-1}\mathbf{F}_1, \quad \boldsymbol{\beta} = \mathbf{H}^{-1}\mathbf{F}_2, \quad \boldsymbol{\chi} = \mathbf{H}^{-1}\mathbf{F}_3 \tag{28c}$$

The element tangent stiffness matrix in the local coordinate system can be calculated by differentiating the local internal force vector $\mathbf{f}$ of the element with respect to $\mathbf{u}_L$,

$$\mathbf{k}_T = EA\left(\int_L \frac{\partial\mathbf{B}_{xx}^{(0)T}}{\partial\mathbf{u}_L^T}\mathbf{P}^T\, dx\right)\boldsymbol{\alpha} + Ak_0 G\left[\left(\int_L \frac{\partial\mathbf{B}_{xy}^{(0)T}}{\partial\mathbf{u}_L^T}\mathbf{P}^T\, dx\right)\boldsymbol{\beta}\right.$$

$$\left.+\left(\int_L \frac{\partial\mathbf{B}_{xz}^{(0)T}}{\partial\mathbf{u}_L^T}\mathbf{P}^T\, dx\right)\boldsymbol{\chi}\right]$$

$$+EA^2\int_L \mathbf{B}_{xx}^{(0)T}\mathbf{P}^T\, dx\left(\mathbf{H}^{-1}\int_L \mathbf{P}\mathbf{B}_{xx}^{(0)}\, dx\right)$$

$$+A^2 k_0 G\left[\int_L \mathbf{B}_{xy}^{(0)T}\mathbf{P}^T\, dx\left(\mathbf{H}^{-1}\int_L \mathbf{P}\mathbf{B}_{xy}^{(0)}\, dx\right)\right.$$

$$\left.+\int_L \mathbf{B}_{xz}^{(0)T}\mathbf{P}^T\, dx\left(\mathbf{H}^{-1}\int_L \mathbf{P}\mathbf{B}_{xz}^{(0)}\, dx\right)\right]$$

$$+EI_y\left[\int_L\left(\mathbf{B}_{xx}^{(4)T}\mathbf{B}_{xx}^{(0)} + \frac{\partial\mathbf{B}_{xx}^{(0)T}}{\partial\mathbf{u}_L^T}\varepsilon_{xx}^{(4)} + \mathbf{B}_{xx}^{(0)T}\mathbf{B}_{xx}^{(4)} + \mathbf{B}_{xx}^{(1)T}\mathbf{B}_{xx}^{(1)}\right)dx\right.$$

$$\left.+\int_L\left(\frac{\partial\mathbf{B}_{xx}^{(4)T}}{\partial\mathbf{u}_L^T}\varepsilon_{xx}^{(0)} + \frac{\partial\mathbf{B}_{xx}^{(1)T}}{\partial\mathbf{u}_L^T}\varepsilon_{xx}^{(1)}\right)dx\right]$$

188

$$+ k_0 GI_y \left[\int_L \left(\mathbf{B}_{xy}^{(1)\mathrm{T}} \mathbf{B}_{xy}^{(1)} + \frac{\partial \mathbf{B}_{xy}^{(1)\mathrm{T}}}{\partial \mathbf{u}_L^\mathrm{T}} \gamma_{xy}^{(1)} \right) \mathrm{d}x \right.$$

$$\left. + \int_L \left(\mathbf{B}_{xz}^{(1)\mathrm{T}} \mathbf{B}_{xz}^{(1)} + \frac{\partial \mathbf{B}_{xz}^{(1)\mathrm{T}}}{\partial \mathbf{u}_L^\mathrm{T}} \gamma_{xz}^{(1)} \right) \mathrm{d}x \right]$$

$$+ EI_z \left[\int_L \left(\mathbf{B}_{xx}^{(5)\mathrm{T}} \mathbf{B}_{xx}^{(0)} + \frac{\partial \mathbf{B}_{xx}^{(0)\mathrm{T}}}{\partial \mathbf{u}_L^\mathrm{T}} \varepsilon_{xx}^{(5)} + \mathbf{B}_{xx}^{(0)\mathrm{T}} \mathbf{B}_{xx}^{(5)} + \mathbf{B}_{xx}^{(2)\mathrm{T}} \mathbf{B}_{xx}^{(2)} \right) \mathrm{d}x \right.$$

$$\left. + \int_L \left(\frac{\partial \mathbf{B}_{rr}^{(5)\mathrm{T}}}{\partial \mathbf{u}_L^\mathrm{T}} \varepsilon_{xx}^{(0)} + \frac{\partial \mathbf{B}_{xx}^{(2)\mathrm{T}}}{\partial \mathbf{u}_L^\mathrm{T}} \varepsilon_{xx}^{(2)} \right) \mathrm{d}x \right]$$

$$+ k_0 GI_z \left[\int_L \left(\mathbf{B}_{xy}^{(2)\mathrm{T}} \mathbf{B}_{xy}^{(2)} + \frac{\partial \mathbf{B}_{xy}^{(2)\mathrm{T}}}{\partial \mathbf{u}_L^\mathrm{T}} \gamma_{xy}^{(2)} \right) \mathrm{d}x \right.$$

$$\left. + \int_L \left(\mathbf{B}_{xz}^{(2)\mathrm{T}} \mathbf{B}_{xz}^{(2)} + \frac{\partial \mathbf{B}_{xz}^{(2)\mathrm{T}}}{\partial \mathbf{u}_L^\mathrm{T}} \gamma_{xz}^{(2)} \right) \mathrm{d}x \right]$$

$$+ EI_{wyz} \left[\int_L \left(\mathbf{B}_{xx}^{(3)\mathrm{T}} \mathbf{B}_{xx}^{(3)} + \mathbf{B}_{xx}^{(4)\mathrm{T}} \mathbf{B}_{xx}^{(5)} + \mathbf{B}_{xx}^{(5)\mathrm{T}} \mathbf{B}_{xx}^{(4)} \right) \mathrm{d}x \right.$$

$$\left. + \int_L \left(\frac{\partial \mathbf{B}_{xx}^{(3)\mathrm{T}}}{\partial \mathbf{u}_L^\mathrm{T}} \varepsilon_{xx}^{(3)} + \frac{\partial \mathbf{B}_{xx}^{(5)\mathrm{T}}}{\partial \mathbf{u}_L^\mathrm{T}} \varepsilon_{xx}^{(4)} + \frac{\partial \mathbf{B}_{xx}^{(4)\mathrm{T}}}{\partial \mathbf{u}_L^\mathrm{T}} \varepsilon_{xx}^{(5)} \right) \mathrm{d}x \right]$$

$$+ EI_{wy} \int_L \left(\mathbf{B}_{xx}^{(4)\mathrm{T}} \mathbf{B}_{xx}^{(4)} + \frac{\partial \mathbf{B}_{xx}^{(4)\mathrm{T}}}{\partial \mathbf{u}_L^\mathrm{T}} \varepsilon_{xx}^{(4)} \right) \mathrm{d}x$$

$$+ EI_{wz} \int_L \left(\mathbf{B}_{xx}^{(5)\mathrm{T}} \mathbf{B}_{xx}^{(5)} + \frac{\partial \mathbf{B}_{xx}^{(5)\mathrm{T}}}{\partial \mathbf{u}_L^\mathrm{T}} \varepsilon_{xx}^{(5)} \right) \mathrm{d}x \qquad (29)$$

Due to the commutativity of the local nodal variables in calculating the second derivatives of the Hellinger–Reissner partial mixed functional with respect to $\mathbf{u}_L$, it is obvious that $\mathbf{k}_T$ is symmetric.

The global internal force vector $\mathbf{f}_G$ can be calculated from the local internal force vector $\mathbf{f}$,

$$\mathbf{f}_G = \mathbf{T}^\mathrm{T} \mathbf{f} \qquad (30)$$

where, $\mathbf{T}$ is the transformation matrix from global coordinate system to local coordinate system, it is calculated from

$$\mathbf{T} = \frac{\partial \mathbf{u}_L}{\partial \mathbf{u}_G^\mathrm{T}} \qquad (31)$$

The global tangent stiffness matrix is calculated by differentiating $\mathbf{f}_G$ with respect to $\mathbf{u}_G$,

$$\mathbf{k}_{TG} = \frac{\partial \mathbf{f}_G}{\partial \mathbf{u}_G^\mathrm{T}} = \mathbf{T}^\mathrm{T} \frac{\partial \mathbf{f}}{\partial \mathbf{u}_G^\mathrm{T}} + \frac{\partial \mathbf{T}^\mathrm{T}}{\partial \mathbf{u}_G^\mathrm{T}} \mathbf{f}$$

$$= \mathbf{T}^\mathrm{T} \frac{\partial \mathbf{f}}{\partial \mathbf{u}_L^\mathrm{T}} \frac{\partial \mathbf{u}_L}{\partial \mathbf{u}_G^\mathrm{T}} + \frac{\partial \mathbf{T}^\mathrm{T}}{\partial \mathbf{u}_G^\mathrm{T}} \mathbf{f} = \mathbf{T}^\mathrm{T} \mathbf{k}_T \mathbf{T} + \frac{\partial \mathbf{T}^\mathrm{T}}{\partial \mathbf{u}_G^\mathrm{T}} \mathbf{f} \qquad (32)$$

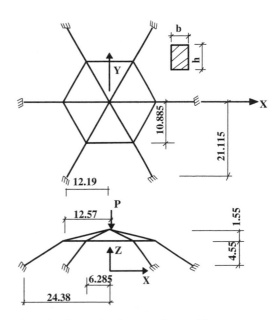

Figure 2. Geometry of a space dome subject to an apex concentrated load.

It is obvious that the first term in the right side of (32) is symmetric. The second term includes the second derivatives of local nodal variables with respect to global nodal variables, where the global nodal variables are commutative, thus the second term is also symmetric, resulting in a symmetric element tangent stiffness matrix $\mathbf{k}_{TG}$ in the global coordinate system.

5 EXAMPLES

5.1 Space dome subject to a concentrated load at apex

A space dome is fixed at its six boundary nodes, and subjected to a concentrated load at the apex (see Figure 2). All its members have the same rectangular cross sections $b \times h = 0.76\,\mathrm{m} \times 1.22\,\mathrm{m}$, and the same material properties $E = 2.069 \times 10^{10}\,\mathrm{N/m^2}$ and $G = 8.83 \times 10^9\,\mathrm{N/m^2}$.

In numerical analysis, each member is treated as one element. The deflection curve at the apex of the dome is depicted in Figure 3. It is in well agreement with the results from Teh and Clarke (1999) and Izzuddin (2001).

5.2 A space arc frame subject to vertical and horizontal concentrated loading

This arc frame consists of two groups of members (see Figure 4). The cross-section properties of the members in the arc frame planes are $A_1 = 0.5$,

189

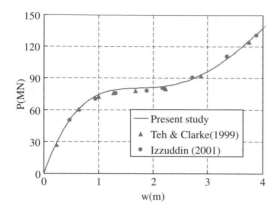

Figure 3. Load-deflection curve at the apex of the space dome.

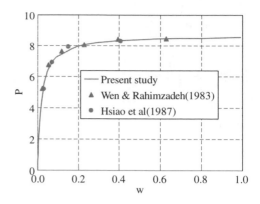

Figure 5. Response of space arc frame under ultimate concentrated loading.

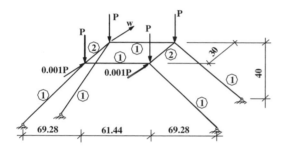

Figure 4. Space arc frame.

$I_{y1} = 0.4$ and $I_{z1} = 0.133$, respectively, and for the rib members, $A_2 = 0.1$, $I_{y2} = 0.05$ and $I_{z2} = 0.05$, respectively. The material properties are $E = 4.32 \times 10^5$ and $G = 1.66 \times 10^5$. This frame is pinned at four boundary nodes. In addition to four vertical concentrated loads P, the structure is also subjected to two lateral concentrated loads 0.001P (see Figure 4).

In numerical analysis, each member is treated as one element. The sideward deflection curve at a loading point (Figure 4) of the arc frame is presented in Figure 5. It is in close agreement with the solution from Hsiao et al.(1987) and Wen & Rahimzadeh (1983).

6 CONCLUSION

Compared with the existing 3-D beam element formulations for framed structures with large displacements and large rotations, there are several features in the proposed procedure: (1) the element tangent stiffness matrix is symmetric, so it ensures the computational efficiency and the storage source saving; (2) vectorial rotational variables are defined, so all nodal variables of freedoms are additive, and 'correction matrix' is

avoided; (3) these variables can be used to describe large member deformation. Through several examples test, the proposed procedure demonstrates satisfying accuracy and efficiency in large displacement analyses of framed structures.

ACKNOWLEDGEMENTS

This work is supported by National Natural Science Foundation of China (50408022) and the visiting scholarship from the Future Academic Star Project of Zhejiang University. In addition, this research also benefits from the financial supports of the Scientific Research Foundation for the Returned Overseas Chinese Scholars, provided respectively by State Education Ministry and Zhejiang Province.

REFERENCES

Crisfield, M.A. 1990. Consistent co-rotational formulation for non-linear, three-dimensional, beam-elements. *Computer Methods in Applied Mechanics and Engineering* 81: 131–150.
Crisfield, M.A. 1996. *Nonlinear Finite Element Analysis of Solid and Structures* Vol.2. Chichester: John Wiley & Sons.
Crisfield, M.A. & Moita, G.F. 1996. A Unified co-ratational framework for solids, shells and beams. *International Journal of Solids and Structures* 33: 2969–2992.
Hsiao, K.M., Horng, H.J. & Chen, Y.R. 1987. A corotational procedure that handles large rotations of spatial beam structures. *Computers and Structures* 27: 769–781.
Izzuddin, B.A. 2001. Conceptual issues in geometrically nonlinear analysis of 3D framed structures. *Computer Methods in Applied Mechanics and Engineering* 191: 1029–1053.
Simo, J.C. 1992. (Symmetric) Hessian for geometrically nonlinear models in solid mechanics. Intrinsic definition and

geometric interpretation. *Computer Methods in Applied Mechanics and Engineering* 96: 189–200.

Simo, J.C. & Vu-quoc, L. 1986. A three-dimensional finite-strain rod model. Part II: computational aspects. *Computer Methods in Applied Mechanics and Engineering* 58: 79–116.

Teh, L.H. & Clarke, M.J. 1999. Symmetry of tangent stiffness matrices of 3D elastic frame. *Journal of Engineering Mechanics –ASCE* 125: 248–251.

To, C.W.S. & Wang, B. 1999. Hybrid strain based geometrically nonlinear laminated composite triangular shell finite elements. *Finite Elements in Analysis and Design* 33(2): 83–124.

Wen, R.K. & Rahimzadeh, J. 1983. Nonlinear elastic frame analysis by finite element. *Journal of Structural Engineering –ASCE* 109: 1951–1971.

Steel and Composite Structures – Wang & Choi (eds)
© 2007 Taylor & Francis Group, London, ISBN 978-0-415-45141-3

In-plane inelastic buckling and strength of steel arches

Y.-L. Pi, Mark Andrew Bradford, Francis Tin-Loi
Centre for Infrastructure Engineering and Safety, School of Civil & Environmental Engineering, The University of New South Wales, Sydney, Australia

ABSTRACT: The in-plane inelastic buckling and strength of circular steel arches are investigated using a rational finite element model for the nonlinear inelastic analysis of arches. The finite element model determines the elastic-plastic behaviour of an arch by taking into account the effects of large deformations, material nonlinearities, initial geometric imperfection, and residual stresses. Radial loads uniformly distributed around the arch axis, concentrated loads, and distributed loads along the horizontal projection of an arch are studied, which induce either uniform compression or combined bending and compression in the arch. The effects of initial imperfection, rise-to-span ratio, residual stresses and the end support conditions on the in-plane inelastic stability and strength of steel arches are included in the study. Useful design equations against in-plane failure are proposed for steel arches under uniform compression and under combined bending and compressive actions.

1 INTRODUCTION

This paper is concerned with the in-plane nonlinear buckling behaviour and strength design of steel circular arches that are subjected to transverse loads, as shown in Fig. 1. Arches resist general loading by a combination of predominant axial compressive as well bending actions. Under these actions, an arch, which is adequately braced by lateral restraints so that its out-of-plane failure is fully prevented, may suddenly buckle and fail in the in-plane of loading.

Many commonly-used design codes (*AS4100* 1998, *BS5950* 1998, *LRF* 2000, Galambos 1988) do not give explicit methods for designing steel arches against their in-plane failure. The few that do use methods are essentially based on the strength of steel beam-columns. Despite their widespread use, research that reports on the strength and design of steel arches against their in-plane failure appears to be quite limited and has concentrated on pin-ended arches. Verstappen et al. (1998) proposed that the in-plane buckling design strength of non-shallow circular arches should be determined by using a linear interaction design equation that was developed for straight beam-columns. Pi and Trahair (1996, 1999) studied the in-plane elasto-plastic behaviour of pin-ended circular steel arches, while Pi and Bradford (2004) investigated the in-plane elasto-plastic behaviour of fixed circular steel arches.

The aims of this paper are to use an advanced nonlinear inelastic finite element (FE) model, developed by the authors (Pi et al. 2007), to investigate the in-plane elasto-plastic failure behaviour and strength

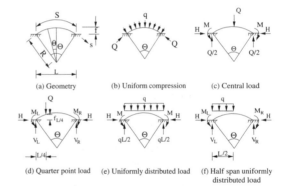

Figure 1. Arch and loading.

(a) Geometry   (b) Uniform compression   (c) Central load

(d) Quarter point load   (e) Uniformly distributed load   (f) Half span uniformly distributed load

of circular steel arches, and to propose design equations for steel arches against in-plane failure based on this numerical investigation.

2 ELASTIC IN-PLANE BUCKLING

The elastic buckling load of arches under uniform compression can be used as a reference load in the formulation for the in-plane strength design. Pi et al. (2002) studied the elastic buckling of arches under uniform compression and obtained the buckling load for arches as in the following.

The expressions for the elastic buckling loads of shallow and deep arches are different. When an arch has an included angle $\Theta \leq 90°$, it is considered to be a shallow arch.

For a shallow pin-ended arch with $3.88 \leq \lambda \leq 9.38$, where λ is the shallowness of the arch and defined by

$$\lambda = \frac{S\Theta}{2r_x} \tag{1}$$

with r_x being the radius of gyration of the cross-section about its major principal axis, its buckling load can be approximated as

$$N_{acr} = (0.15 + 0.006\lambda^2)N_{cr} \tag{2}$$

where N_{cr} is the second mode elastic flexural buckling load of a pin-ended column.

For a shallow pin-ended arch with $\lambda > 9.38$,

$$N_{acr} = \left(0.26 + 0.74\sqrt{1 - 0.63\,\pi^4/\lambda^2}\right)N_{cr}. \tag{3}$$

For shallow fixed arches with $9.87 \leq \lambda \leq 18.6$

$$N_{acr} = (0.36 + 0.00111\lambda^2)N_{cr} \tag{4}$$

where N_{cr} is the second mode elastic flexural buckling load of a fixed column. N_{cr} is given by

$$N_{cr} = \frac{\pi^2 EI}{(kS)^2} \tag{5}$$

where $k = 0.5$ for pin-ended arches and $k = 0.35$ for fixed arches.

For shallow fixed arches with $\lambda > 18.6$

$$N_{acr} = \left(0.6 + 0.4\sqrt{1 - 3.109\,\pi^4/\lambda^2}\right)N_{cr}. \tag{6}$$

When an arch has an included angle $\Theta > 90°$, it is a deep arch, its elastic buckling load is given by

$$N_{acr} = \left[1 - (\Theta/\pi)^2\right]N_{cr} \quad \text{for pin-ended arch,} \tag{7}$$

$$N_{acr} = \left[1 - (\Theta/1.4304\pi)^2\right]N_{cr} \quad \text{for fixed arch.} \tag{8}$$

3 NONLINEAR ELASTO-PLASTIC FINITE ELEMENT MODEL

A nonlinear inelastic FE model for arches, developed elsewhere by the authors (Pi et al. 2007), is used in this paper to investigate the in-plane strength of steel arches. The formulation of the FE model is based on the following assumptions and considerations:

1. Use of the Euler-Bernoulli theory of bending,

2. Application of nonlinear strain-displacement relationships that allow for large displacements and rotations,
3. Inclusion of the linear and nonlinear geometric effects of the load position on the strains and the tangent stiffness,
4. Inclusion of distributions of longitudinal normal residual stresses due to manufacturing and curving, as are described subsequently,
5. Inclusion of initial geometric imperfections.

Full details of the FE model are given in Pi et al. (2007), and the effectiveness and accuracy of this model has been verified extensively and is therefore not reproduced here.

4 RESIDUAL STRESSES, STRESS-STRAIN CURVE

Residual stresses may be induced in the manufacturing and curving processes of steel arches. It has been shown that the residual stresses and the initial geometric imperfections have significant effects on the strength of steel arches, and so need to be considered in the development of the strength design equations for steel arches.

The residual stresses σ_r consist of two components: the manufacturing and the cold rolling (or curving) residual stresses shown in Figs 2(c)–2(d). In Fig. 2(c), $\sigma_{rt} = 0.5\sigma_y$ is typically the tensile residual stress at the flange-web junctions and $\sigma_{rc} = 0.35\sigma_y$ is typically the compressive residual stress at the flange tips, where σ_y is the yield stress.

Rolling residual stresses are induced in cold curving the arch. It is assumed that, for example, a straight I-section member is bent elastic-plastically to obtain an overbent arch, and then released so that the overbent arch relaxes elastically into its final shape with a permanent set. Although the depth of the plastic zone after overbending depends on the radius of the arch, and so also do the final rolling residual stresses, the final rolling flange residual stresses σ_{rrf} are quite small after the relaxation. Hence, for convenience, the rolling residual stresses are assumed here to be independent of the arch radius.

The maximum rolling residual stresses occur in the web near the neutral axis, and so their effects on inelastic buckling are quite small as inelastic buckling is controlled primarily by flange yielding.

The maximum rolling flange residual stresses σ_{rrf} are assumed to be given by

$$\sigma_{rrf} = \sigma_y\left(\frac{Z_p}{Z_e} - 1\right) \tag{9}$$

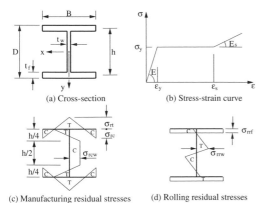

(a) Cross-section   (b) Stress-strain curve

(c) Manufacturing residual stresses   (d) Rolling residual stresses

Figure 2. Cross-section, stress-strain curve and residual stresses.

where Z_{px} and Z_{ex} are the plastic and elastic moduli of the cross-section about its major principal axis, respectively. The maximum rolling web residual stresses σ_{rrw} are assumed to be given by $\sigma_{rrw} = 0.9\sigma_y$.

The residual stresses shown in Fig. 2(c) and 2(d) satisfy the in-plane bending equilibrium condition

$$\int_A \sigma_r y \, dA = 0, \tag{10}$$

while the axial force equilibrium condition

$$\int_A \sigma_r dA = 0 \tag{11}$$

is used to determine the mid-web residual stress σ_{rcw}, where y is the coordinate in the principal axes of the cross-section, and A is the area of the cross-section.

The residual stresses defined in this section, and the steel I-section and the tri-linear stress-strain relationship shown in Fig. 2 are used throughout this study. The dimensions of the I-section are: the depth of the cross-section $D = 0.2613$ m, the flange width $B = 0.151$ m, the flange thickness $t_f = 0.0123$ m, and the web thickness $t_w = 0.0077$ m. The material properties for the tri-linear stress-strain curve are: the yield stress $\sigma_y = 250$ MPa, the corresponding yield strain $\varepsilon_y = 0.00125$, Young's modulus of elasticity $E = 200,000$ MPa, the strain at which the strain hardening starts is assumed to be 11 times the yield strain, the maximum strain is assumed to be 71 times the yield strain. The strain hardening modulus $E_s = 6,000$ MPa. The shear modulus of elasticity is $G = 80,000$ MPa.

5 STRENGTH IN UNIFORM COMPRESSION

The design proposal for the in-plane strength capacity N_{ac} of a steel arch that is subjected to uniform compression is based on the methodology in *AS4100* (1998) for

compression members, but modified accordingly for arches. Thus

$$N_{ac} = \phi \alpha_{ac} N_Y \leq N_Y \tag{12}$$

where ϕ is a capacity reduction factor ($\phi = 0.9$ is suggested and used in *AS4100* (1998)), N_Y is the squash load of the cross-section and given by

$$N_Y = A\sigma_y \tag{13}$$

and α_{ac} is the in-plane slenderness reduction factor of an arch and given within the methodology of *AS4100* (1998) by

$$\alpha_{ac} = \xi_a \left[1 - \sqrt{1 - \left(\frac{90}{\xi_a \lambda_{ag}} \right)^2} \right] \tag{14}$$

with

$$\xi_a = \frac{(\lambda_{ag}/90)^2 + 1 + \eta_a}{2(\lambda_{ag}/90)^2} \tag{15}$$

in which the modified slenderness λ_{ag} is defined as

$$\lambda_{ag} = \frac{L_e}{r_x} \sqrt{\frac{\sigma_y}{250}} \tag{16}$$

with the effective length $L_e = kS/\sqrt{k_{ac}}$; and the imperfection parameter η_a is given by

$$\eta_a = 0.00326(\lambda_{ag} - 13.5) \geq 0. \tag{17}$$

In Eqn (13), σ_y is expressed in units of MPa and the factor k_{ac} is given by $k_{ac} = N_{acr}/N_{cr}$.

To verify the proposed in-plane strength of an arch in uniform compression given by Eqn (10), the FE model was used to investigate the in-plane strength N_{ac} of arches in uniform compression. In the FE analysis, the in-plane geometric imperfections were assumed to be

$$v_0 = v_{0,S/4} \sin \frac{2\pi s}{S} \tag{18}$$

for pin-ended arches and

$$v_0 = \frac{v_{0,S/4}}{2} \left(1 - \cos \frac{2\pi s}{S/2} \right) \tag{19}$$

for fixed arches where $v_{0,S/4}$ is the initial in-plane crookedness at the quarter point of the arch length and s is the coordinate along the arch axis. The expression

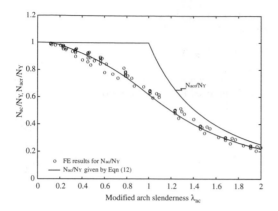

Figure 3. In-plane strength of pin-ended arches subjected to uniform compression.

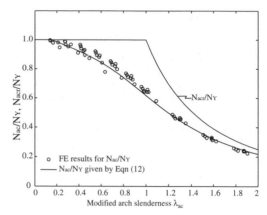

Figure 4. In-plane strength of fixed arches subjected to uniform compression.

for the imperfection satisfies the kinematical boundary conditions.

The central crookedness implied by the Australian steel structures design code *AS4100* (1998) for the major axis strength of steel columns was obtained by Bild and Trahair (1987), which is used in this study as the initial crookedness $v_{0,S/4}$ at the quarter point of the arch length, so that

$$\frac{1000v_{0,S/4}}{S/2} = 2.5\left(1 - \frac{1}{1+2\lambda_c}\right) \tag{20}$$

where λ_c is the modified slenderness of the corresponding column given by

$$\lambda_c = \sqrt{\frac{N_Y}{N_{cr}}} \ . \tag{21}$$

The variations of the dimensionless in-plane axial compressive strength N_{ac}/N_Y with the modified in-plane slenderness λ_{ac} obtained from the FE results are shown in Figure 3 for pin-ended arches and in Figure 4 for fixed arches, where the modified in-plane slenderness λ_{ac} is defined as

$$\lambda_{ac} = \sqrt{\frac{N_Y}{N_{acr}}} \tag{22}$$

Also shown in Figures 3 and 4 is the variation of the dimensionless proposed in-plane strength of fixed arches N_{ac}/N_Y with the modified in-plane slenderness λ_{ac}. It can be seen that proposed equation provides good predictions for the in-plane strengths of arches in uniform compression.

It worth pointing out that for a pin-ended arch with $\lambda < 3.88$ or a fixed arch with $\lambda < 9.87$ that is subjected to a radial load uniformly distributed around it, the arch

becomes very shallow and can be treated as a beam with initial in-plane geometric imperfection. Hence, the proposed Eqn (12) is not suitable in this case.

6 STRENGTH IN COMBINED BENDING AND COMPRESSION

6.1 First-order actions in elastic arches

Under general loading, arches are subjected to combined bending and axial compressive actions. Whether or not the compressive action or the bending action are relatively significant to each other very much depends on the loading condition.

To investigate the effects of the loading condition on the bending and compressive actions in an arch, first-order elastic analyses were performed by using the flexibility method. Four loading cases were investigated: a central concentrated load, a quarter point concentrated load, a vertical load uniformly distributed over the entire span of an arch, and a vertical load uniformly distributed over a half of the span of an arch (Fig. 1).

The results for the maximum compressive action N and the maximum bending action M are shown in Fig. 5 as the relationship between the ratio $(N/N_Y)/(N/N_Y + (M/M_p))$ and the included angle Θ for arches with length $S = 10$ m, where M_p is the full plastic moment of the cross-section. When an arch is subjected to a vertical load uniformly distributed over its entire span, the axial compressive action is relatively high and the bending action is relatively low. In addition, the compressive action of arches with a moderate included angle Θ is relatively higher than that of arches with a small or large included angle Θ. When an arch is subjected to a vertical load uniformly distributed over a half of its span, to a central concentrated

196

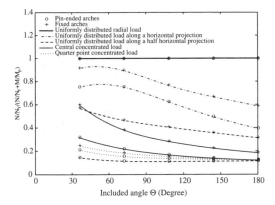

Figure 5. Maximum first-order bending moment and axial compressive force in arches.

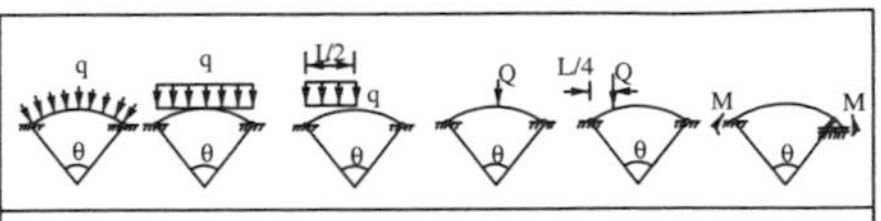

load, or to a quarter point concentrated load, the axial compressive action is relatively low and the bending action is relatively high. When the included angle Θ of an arch is very small, the bending action is relatively significant for all the four loading cases.

6.2 In-plane design strength

The in-plane strength of a steel arch subjected to combined bending and axial compressive actions is related to a number of factors, such as the buckling behaviour, yielding, the initial curvature, the included angle, the slenderness, the shallowness, residual stresses, initial in-plane geometric imperfections, and loading conditions. It is therefore difficult to develop simple and accurate equations for the design of arches that are subjected to the combined bending and axial compressive actions against in-plane failure. Instead, a lower bound interaction equation for the in-plane strength of arches has been sought based on the accurate numerical results.

The in-plane strength capacity design check for a steel arch that is subjected to combined bending and axial compressive actions is proposed as

$$\frac{N^*}{\phi\alpha_{an}N_{ac}} + \frac{M^*}{\phi\alpha_{am}M_p} \leq 1 \tag{23}$$

where ϕ is the same capacity reduction factor for uniform compression, N_{ac} is the in-plane axial compression capacity of an arch in uniform compression given by Eqn (12), N^* is the maximum axial compression obtained by a first-order in-plane elastic analysis, M^* is the maximum moment given by

$$M^* = \delta_b M_m \quad \text{with} \quad \delta_b = \frac{1}{1 - N^*/N_{acr}} \tag{24}$$

Arch						
α_{an}	1.00	1.00	1.00	1.10	1.10	–
α_{am} $\lambda \leq 10$	–	$1.13+0.137\lambda$	$1+0.02\lambda$	$1.35+0.065\lambda$	$1.1+0.05\lambda$	1.00
α_{am} $\lambda > 10$	–	2.50	1.20	2.00	1.60	1.00

Figure 6. Factors for in-plane strength of pin-ended arches.

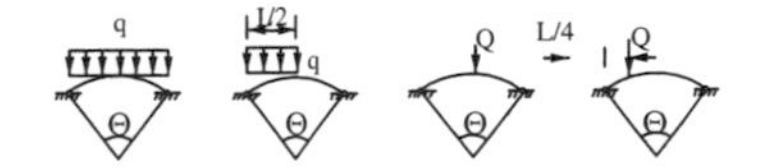

Arch				
α_{an}	1.00	1.00	1.10	1.10
α_{am} $\lambda < 9.38$	$1.13+0.146\lambda$	$1+0.021\lambda$	$1.35+0.069\lambda$	$1.1+0.053\lambda$
α_{am} $\lambda \geq 9.38$	2.50	1.20	2.00	1.60

Figure 7. Factors for in-plane strength of fixed arches.

where M_m is the maximum moment obtained by a first-order in-plane elastic analysis and N_{acr} is the in-plane elastic buckling load of an arch in uniform compression and given by Eqns (2)–(4) and (6)–(8), and α_{an} and α_{am} are the axial compression and moment modification factors for the in-plane strength.

The modification factor α_{am} accounts for the combined effects of the initial curvature, the included angle and the slenderness of the arch, and the non-uniform distribution of bending moment over the arch length, while the modification factor α_{an} accounts for the non-uniform distribution of axial compressive force over the arch length because the effects of the initial curvature. The effects of the included angle and the slenderness of the arch have already been accounted in the in-plane axial compression capacity N_{ac}. The values of α_{an} and α_{am} are given in Figure 6 for pin-ended arches and in Figure 7 for fixed arches.

It is recommended that if the amplification factor $\delta_b > 1.4$, then a second-order in-plane elastic analysis should be carried out to obtain M^* and N^*. This recommendation is consistent with that of AS4100 (1998) for frames consisting of straight members.

The predictions of the proposed interaction equation (23) are compared with the FE results (taking $\phi = 1$) for four load cases: a central concentrated load, a quarter point concentrated load, a uniformly distributed load over the entire arch, and a uniformly distributed load over a half of the arch (Fig. 1). In the FE analysis, the initial in-plane imperfections defined by Eqns (18)–(20) were used.

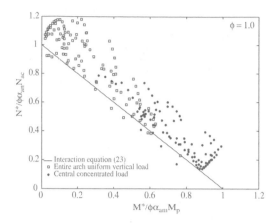

Figure 8.   In-plane strength of pin-ended arches subjected to symmetric loading.

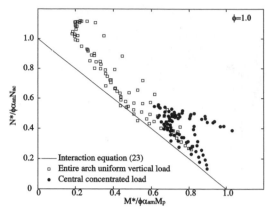

Figure 10.   In-plane strength of fixed arches subjected to symmetric loading.

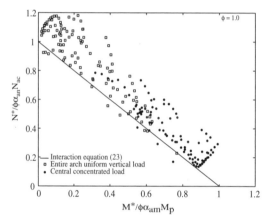

Figure 9.   In-plane strength of pin-ended arches subjected to asymmetric loading.

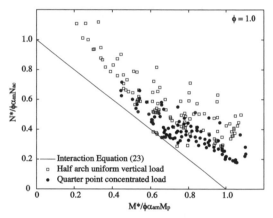

Figure 11.   In-plane strength of fixed arches subjected to asymmetric loading.

Comparisons of the predictions of the proposed equation (23) with the FE results are shown in Figures 8 and 9 for pin-ended arches that are subjected to symmetric and asymmetrical loads respectively, and in Figures 10 and 11 for fixed arches that are subjected to symmetric and asymmetric loads respectively. It can be seen that the interaction equation (23) provides good lower bound predictions that are not unduly conservative for the in-plane strength of steel arches that are subjected to the combined bending and axial compressive actions.

7   CONCLUSIONS

Design equation for the in-plane strength of steel arches in uniform compression has been proposed which provides good predictions for the in-plane strength. An interaction equation for the in-plane design strength of steel arches that are subjected to the combined bending and axial compressive actions has also been proposed. It considers the effects and interactions of various factors, such as in-plane buckling, yielding, the included angle, the slenderness, the shallowness, residual stresses, initial in-plane geometric imperfections, non-uniform distributions of bending moments and axial compressive forces, loading conditions. The proposed design interaction equation is simple and easy to use, is consistent with the Australian steel structures design code *AS4100* (1998), and may easily be formulated for other steel structure design codes. Generally, a first-order or second-order in-plane elastic analysis is sufficient for the design. The proposed design equation provides good lower bound predictions for the in-plane strength of steel arches that are subjected to bending and axial compressive actions.

ACKNOWLEDGEMENT

This work has been supported by the Australian Research Council through Discovery Projects awarded to the authors and a Federation Fellowship to the second author.

REFERENCES

AS4100, Steel structures (1998). Sydney: Standards Australia.

Bild S, Trahair NS. (1988). Steel column strength models. *Journal of Constructional steel Research*, 11, 13–26. *BS5950, Structural use of steel in building, part 1, code of practice for design in simple and continuous construction: hot-rolled sections.* 1998). London: British Standards Institution.

Galambos TV. (1988). *Guide to stability design criteria for metal structures*, 4th ed, New York: John Wiley & Sons. *Load and resistance factor design specification for structural steel buildings,* (2000). Chicago: American Institute of Steel Construction.

Pi YL, Bradford MA. (2004). In-plane strength and design of fixed steel I-section arches. *Engineering Structures*, 26(3): 291–301.

Pi YL, Bradford MA, Tin-Loi, F, Gilbert, RI. (2007). Geometric and material nonlinear analysis of elastically restrained arches. *Engineering Structures*, 29(3):283–295.

Pi YL, Bradford MA, Uy B. (2007). A rational elasto-plastic spatially curved thin-walled beam element. *International Journal for Numerical Methods in Engineering,* (published on internet).

Pi Y-L, Bradford MA, Uy B. (2002). In-plane stability of arches. *International Journal of Solids & Structures,* 39(2): 105–125.

Pi Y-L , Trahair NS. (1996). In-plane inelastic buckling and strength of steel arches. *Journal of Structural Engineering*, ASCE, 122(7): 734–747.

Pi Y-L, Trahair NS. (1999). In-plane buckling and design of steel arches. *Journal of Structural Engineering*, ASCE, 125(9), 1291–1298.

Verstappen I., Snijder HH, and Bijlaard FSK, Steembergen HMGM. (1998). Design rules for steel arches-in-plane stability. *Journal of Constructional Steel Research,* 46(1–3), 125–126.

Steel and Composite Structures – Wang & Choi (eds)
© *2007 Taylor & Francis Group, London, ISBN 978-0-415-45141-3*

Initial imperfections and buckling analysis for beam-columns: A comparative study of EC-3 proposals

D.J. Yong
University of Piura, Piura, Peru

D. Fernandez-Lacabe, A. López & M.A. Serna
Tecnun (University of Navarra), San Sebastian, Spain

ABSTRACT: As established by Eurocode 3, the verification of member stability should be carried out considering imperfections and second order effects. Two procedures may be used for this purpose: (1) second order analysis of members with local imperfections, and (2) buckling resistance analysis. If second order analysis is carried out, applied moments and moments due to initial bow imperfection are properly amplified in the presence of compression forces. Consequently, only cross-section resistance checks are needed. Eurocode 3 provides two ways for defining design values of initial imperfections: the traditional e_0/L coefficient and the theoretical expression used to obtain EC3 buckling curves. On the other hand, buckling resistance of members subjected to bending and axial compression may be obtained using interaction formulae. In this procedure initial imperfection is taking into account through a compression load reduction factor, and the amplification of bending moments is approximated by means of interaction factors. This paper presents a comparative study of both procedures for cases where analytical second order solution is possible. Results show how interaction factors may underestimate member resistance in some cases, and how member slenderness affects differences in the resistance given by the two procedures.

1 INTRODUCTION

A beam-column is a structural member subjected to combined axial compression and bending. In fact, most members in a building frame may be considered as beam-columns. Beam-columns may develop different types of buckling: flexural buckling, torsional buckling, flexural-torsional buckling and lateral-torsional buckling. The occurrence of any one of these types of buckling will depend on member slenderness, flexural rigidity, torsional rigidity and geometry of the cross section.

Eurocode 3 (2005) establishes that "the verification of the stability of frames and their parts should be carried out considering imperfections and second order effects" (EC3–5.2.2 (2)). To accomplish this analysis requirement, two procedures are possible. In the first modus operandi, second order effects in individual members and relevant member imperfections are totally accounted for in the global analysis of the structure. As a consequence, no individual stability check is necessary and only cross-section resistance must be checked. EC3 makes it possible to define the design values of initial local bow imperfections in two different ways: (1) by means of a simple table of e_0/L

values which are independent of member slenderness, and (2) by using a closed form equation which depends upon cross-section properties and member slenderness. In addition, cross-section resistance is checked by means of interaction formulae, and two approaches are offered: (1) to use a general conservative approximation based on a lineal summation of the utilization ratios, and (2) to apply specific tailored expressions which depends on cross-section shape and bending axis.

The second procedure does not introduce local imperfections nor perform second order analysis at member level. In this case, individual member stability must be checked following the appropriate equations which incorporate reduction factors for buckling resistances and interaction factors.

The first version of Eurocode 3 (1992) incorporated a first set of interaction formulae for members subjected to combined axial compression and bending. These first interaction formulae often provided conservative strength estimates, leading to moderately uneconomical designs, especially in members subjected to biaxial bending and compression (Lindner 2003; Gonçalves & Camotim 2004). In order to correct this undesirable situation, Technical Committee 8 of the European Convention for Constructional Steel-work (ECCS – TC 8), led by Prof. Lindner, presented a design methodology based on two levels of interaction formulae which were called Level 1 and Level 2 (Boissonnade et al. 2006). The Level 1 formulae were developed by an Austrian-German research team (Greiner et al. 1999; Greiner & Lindner 2000; Greiner & Ofner 1999; Greiner & Lechner 2002; Greiner 2001; Lindner 2001) whose approach was based on theoretical investigations combined with the test results originally available at the time. The Austrian-German research team carried out numerical simulations of the elastic-plastic behavior of single members, taking into account geometric nonlinearity, initial imperfections, residual stresses and plastic zones along the length of the member (Ofner 1999, 2003). The results were evaluated statistically (Lindner 2001). The physical effects of the instability problems were taken into account by means of a compact coefficient which makes the Level 1 formulae simple and user-friendly for the practical design (Greiner 2001). The Level 2 formulae, which were developed by a French-Belgian research team (Boissonnade et al. 2002, 2004) and are based on a second-order in-plane elastic theory, have been progressively extended to spatial and elastic-plastic behavior. Each constitutive coefficient is normally associated with a single physical effect of instability, which gives the Level 2 formulae a physical transparency (Boissonnade et al. 2004). The two levels of interaction formulae were included in the Eurocode 3 as Method 1 for Level 2 and Method 2 for Level 1. Results obtained by these two methods for some particular cases may be found in Yong et al. (2006).

This paper presents a comparative study of the two procedures outlined above. The comparison is made for in plane bending and linear moment distribution, since these simple cases allow the analytical solution of the differential equation produced by the second order analysis. A medium size rolled section, HEB 240, has been chosen for this study. Results are presented for two values of relative slenderness, 1 and 2, with bending about either the strong axis or the weak axis, and three moment diagrams: constant moment; zero moment at one end; and opposite moments at member ends. For the procedure based on second order analysis, the two values of imperfections and the two cross-section resistance interaction formulae are considered. As a consequence, each comparison will include six cases.

The paper first presents a short scheme of Eurocode 3 beam-column analysis. Next, the analytical solution for the beam-column with imperfection and lineal moment distribution is given. The comparative study is then presented by means of interaction diagrams. Finally, the main conclusions of the paper are summarized.

Table 1. Design values of initial local bow imperfections.

Buckling curve	Elastic analysis e_0/L	Plastic analysis e_0/L
a_0	1/350	1/300
a	1/300	1/250
b	1/250	1/200
c	1/200	1/150
d	1/150	1/100

2 EUROCODE 3 BEAM-COLUMN ANALYSIS

According to Eurocode 3, stability of beam-columns must be checked taking into account both second order effects and imperfections. If member imperfections are introduced in the model and second order analysis is performed, only cross-section checks are required. On the other hand, when structural analysis is carried out without introducing member imperfections, a second step is needed to check member stability. This step is referred to as buckling analysis.

2.1 Member imperfections

Table 1 presents the design values of initial local bow imperfection established by EC3. As an alternative to these values, the sine shape imperfection used to obtain EC-3 buckling curves may be considered. This imperfection is given by:

$$e = e_0 \sin \frac{\pi x}{L} = \alpha \left(\bar{\lambda} - 0.2 \right) \frac{W}{A} \sin \frac{\pi x}{L} \qquad (1)$$

where α is the imperfection factor, $\bar{\lambda}$ is the relative slenderness, and W and A are the resistance modulus and the cross-section area, respectively.

2.2 Resistance of cross-section

To check cross-section resistance for classes 1 to 3, EC-3 gives the following conservative approximation:

$$\frac{N_{Ed}}{N_{Rd}} + \frac{M_{y,Ed}}{M_{y,Rd}} + \frac{M_{z,Ed}}{M_{z,Rd}} \le 1 \qquad (2)$$

where N_{Ed}, $M_{y,Ed}$ and $M_{z,Ed}$ are the design normal force and bending moments, respectively, and N_{Rd}, $M_{y,Rd}$ and $M_{z,Rd}$ are the related design resistances.

EC-3 also provides more accurate cross-section resistance formulae for rectangular and doubly symmetrical I- and H-sections. Particularly, for standard rolled I- or H-sections the following approximation can be used:

$$M_{N,y,Rd} = M_{pl,y,Rd} (1-n)/(1-0.5a) \le M_{pl,y,Rd} \qquad (3)$$

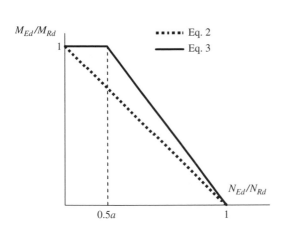

Figure 1. Strong axis cross-section resistance diagram.

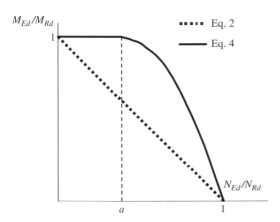

Figure 2. Weak axis cross-section resistance diagram.

$$for\ n \leq a: \quad M_{N,z,Rd} = M_{pl,z,Rd}$$

$$for\ n > a: \quad M_{N,z,Rd} = M_{pl,z,Rd}\left[1-\left(\frac{n-a}{1-a}\right)^2\right] \quad (4)$$

where

$$n = N_{Ed}/N_{pl,Rd}$$
$$a = \left(A-2bt_f\right)/A \leq 0.5 \quad (5)$$

Figures 1 and 2 show the interaction diagrams corresponding to the strong axis and the weak axis.

2.3 Buckling analysis

According to Eurocode 3, when second order analysis with member imperfections is not carried out,

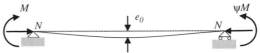

Figure 3. Beam-column subjected to linear moment diagram.

members which are subjected to combined bending and axial compression should satisfy:

$$\frac{N_{Ed}}{N_{b,y,Rd}} + k_{yy}\frac{M_{y,Ed}}{M_{b,y,Rd}} + k_{yz}\frac{M_{z,Ed}}{M_{z,Rk}\Big/\gamma_{M1}} \leq 1 \quad (6)$$

$$\frac{N_{Ed}}{N_{b,z,Rd}} + k_{zy}\frac{M_{y,Ed}}{M_{b,y,Rd}} + k_{zz}\frac{M_{z,Ed}}{M_{z,Rk}\Big/\gamma_{M1}} \leq 1 \quad (7)$$

where $N_{b,y,Rd}$ and $N_{b,z,Rd}$ are the design buckling resistance forces, $M_{b,y,Rd}$ is the design buckling resistance moment, γ_{M1} is the buckling resistance factor, and coefficients k are the interaction factors. These interaction factors approximate the bending amplification effect due to the axial compression. On the other hand, buckling resistances are obtained by using reduction factors.

As indicated above, EC3 (Eurocode 3, 2005) provides two methods to compute the interaction factors used in equations (6) and (7): Method 1, also called the French-Belgian approach, and Method 2, also referred to as the German-Austrian Method.

3 SECOND ORDER ANALYTICAL SOLUTION

Let us considered the beam column pictured in Figure 3, subjected to a compression force and a lineal bending moment diagram. The coefficient Ψ is always equal to or less than one.

Initial bow is given by:

$$e(x) = e_0 \sin\frac{\pi x}{L} \quad (8)$$

The elastic curve differential equation is:

$$EIv'' + Nv = -Ne_0 \sin\frac{\pi x}{L} - \frac{L-(1-\Psi)x}{L}M \quad (9)$$

where v is the elastic deflection.

The solution to Equation 9 is given by:

$$v = \frac{N}{N_{cr} - N} e_0 \sin \frac{\pi x}{L} - \frac{M}{N} \frac{L - (1 - \Psi)x}{L} +$$

$$+ \frac{M}{N} \frac{\sin\left[\pi \sqrt{\frac{N}{N_{cr}}}\left(1 - \frac{x}{L}\right)\right] + \Psi \sin\left(\sqrt{\frac{N}{N_{cr}}} \frac{\pi x}{L}\right)}{\sin\left(\pi \sqrt{\frac{N}{N_{cr}}}\right)} \quad (10)$$

where N_{cr} is the elastic critical force.

Using the elastic deflection given by Equation 10, the total second order bending moment can be obtained:

$$M_{SO} = \frac{1}{1 - \frac{N}{N_{cr}}} N e_0 \sin \frac{\pi x}{L} +$$

$$+ M \frac{\sin\left[\pi \sqrt{\frac{N}{N_{cr}}}\left(1 - \frac{x}{L}\right)\right] + \Psi \sin\left(\sqrt{\frac{N}{N_{cr}}} \frac{\pi x}{L}\right)}{\sin\left(\pi \sqrt{\frac{N}{N_{cr}}}\right)} \quad (11)$$

Since M_{SO} has been obtained using second order analysis and local bow imperfection, only cross-section resistance is needed at the point where M_{SO} is maximum. This point will change with the value of Ψ, and may be located at the end of the beam.

4 COMPARATIVE RESULTS FOR STRONG AXIS BENDING

The comparative study has been performed for an HEB 240 rolled section. Two relative slenderness, 1 and 2, and three different moment diagrams have been used. Figures 4 to 9 show results for bending with respect to the strong axis. In all Figures, curves M1 and M2 refer to Method 1 and Method 2. All other curves correspond to second order analysis. SR and GR indicates specific cross-section resistance (Equations 3 and 4), and General resistance (Equation 2), respectively; finally, A and B letters stand for imperfection given by Equation 1 and by Table 1, respectively.

As expected, second order analysis with imperfection A always gives the biggest member resistance. For uniform bending moment diagram all procedures give similar results, as can be seen in Figures 4 and 5.

For a triangular bending moment diagram (Figures 6 and 7), differences in resistance start to be appreciable. For relative slenderness equal to 2 differences may be as high as 15%.

Finally, discrepancies are more significant for bending diagrams with opposite end moments. This

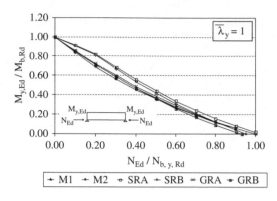

Figure 4. Strong axis interaction diagram ($\Psi = 1$).

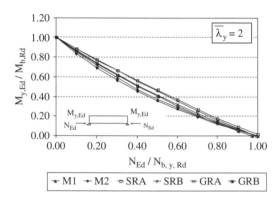

Figure 5. Strong axis interaction diagram ($\Psi = 1$).

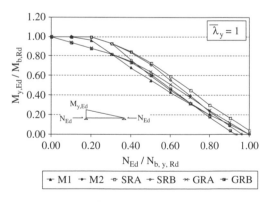

Figure 6. Strong axis interaction diagram ($\Psi = 0$).

is particularly noticeable for the case depicted in Figure 8, where Method 2 turns out to be excessively conservative (up to 40%) with respect to the resistance given by other procedures.

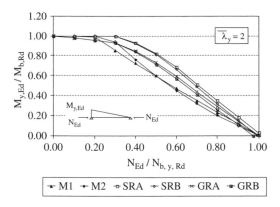

Figure 7. Strong axis interaction diagram ($\Psi = 0$).

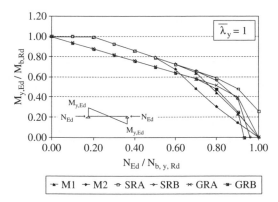

Figure 8. Strong axis interaction diagram ($\Psi = -1$).

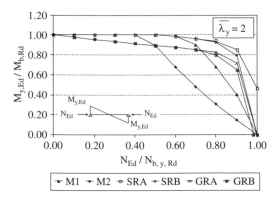

Figure 9. Strong axis interaction diagram ($\Psi = -1$).

5 COMPARATIVE RESULTS FOR WEAK AXIS BENDING

Figures 10 to 15 give the results obtained when bending is with respect to the weak axis. Even though there

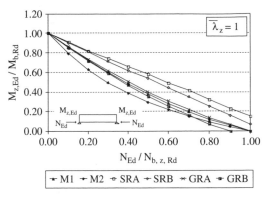

Figure 10. Weak axis interaction diagram ($\Psi = 1$).

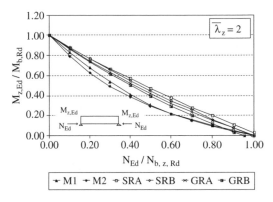

Figure 11. Weak axis interaction diagram ($\Psi = 1$).

is general agreement for uniform bending moment diagram (Figures 10 and 11), differences among the resistance given by the different procedures are significantly bigger than those obtained when bending with respect to the strong axis.

Figures 12 and 13 show results for a triangular bending diagram. It can be seen that the use of a specific cross-section resistance leads to a significant increase in member resistance.

Finally, Figures 14 and 15 present the interaction resistance curves for an opposite end moments diagram. It is clear that second order analysis with a specific cross-section resistance check gives higher member resistance than the methods based on buckling analysis formulae.

As with the results obtained for bending with respect to the strong axis, Method 2 appears to be too conservative. It is also remarkable how the use of specific cross-section resistance allows a significant bending moment at both member ends even when the axial force reaches the general buckling force, which is computed assuming general cross-section resistance.

As a final remark, it is also worth noticing the fact that the initial local bow imperfection given by Table 1

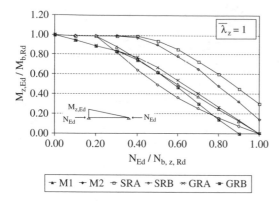

Figure 12. Weak axis interaction diagram ($\Psi = 0$).

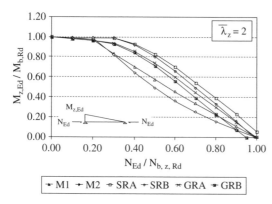

Figure 13. Weak axis interaction diagram ($\Psi = 0$).

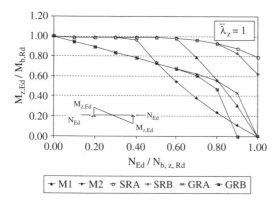

Figure 14. Weak axis interaction diagram ($\Psi = -1$).

leads to lower member resistance than Equation 1. However, its effect is significantly lower than that of formulae used to compute the cross-section resistance, as can clearly be seen in Figure 14.

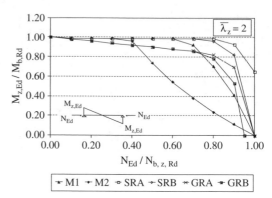

Figure 15. Weak axis interaction diagram ($\Psi = -1$).

6 CONCLUSIONS

This paper has presented a comparative study of all possibilities offered by Eurocode 3 to determine the stability resistance of members subjected to bending and axial compression. Results for different cases of linear moment distributions show that second order analysis and buckling analysis lead to quite similar results for constant moment distribution ($\Psi = 1$). However, significant differences may appear when the member is subjected to opposite end moments ($\Psi = -1$). Member resistances computed using analytical second order solutions and Equation 1 imperfections are the highest for all cases. As expected, the use of specific cross-section resistance formulae lead to an increase in member resistance. On the other hand, Method 1 has shown to be very efficient in approximating analytical results for most cases, with lower efficiency for non uniform moment distributions. Method 2 is also very efficient and simple to use. However, its simplicity may lead to an excess of conservatism for some cases. Finally, differences due to the method of computing the initial local bow imperfection have turned out to be less important than those related to cross-section resistance determination.

ACKNOWLEDGEMENTS

The authors would like to acknowledge the support of this research by the ArcelorMittal Chair of the University of Navarra.

REFERENCES

Boissonnade, N., Greiner, R., Jaspart, J.-P. & Lindner, J. 2006. *Rules for Member Stability in EN 1993-1-1. Background documentation and design guidelines*. European Convention for Constructional Steelwork.

Boissonnade, N., Jaspart, J.-P., Muzeau J.-P. & Villette M. 2002. Improvement of the interaction formulae for beam columns in Eurocode 3. *Computers and Structures*, 80, 2375–2385.

Boissonnade, N., Jaspart, J.-P., Muzeau J.-P. & Villette M. 2004. New interaction formulae for beam-columns in Eurocode 3: The French-Belgian approach. *Journal of Constructional Steel Research*, 60, 421–431.

Eurocode 3, 1992. European Committee for Standardization. ENV 1993-1-1. Eurocode 3: Design of steel structures. Part 1-1: General Rules and rules for buildings. Brussels.

Eurocode 3, 2005. European Committee for Standardization. prNV 1993-1-1. Eurocode 3: Design of steel structures. Part 1-1: General Rules and rules for buildings (draft). Brussels.

Gonçalves, R. & Camotim, D. 2004. On the application of beam-column interaction formulae to steel members with arbitrary loading and support conditions. *Journal of Constructional Steel Research*, 60, 433–450.

Greiner, R. 2001. Background information on the beam-column interaction formulae at Level 1. ECCS TC 8. Paper N° TC 8-2001. Technical University Graz.

Greiner, R. & Lechner, A. 2002. Elastic-plastic beam-column behaviour within structural systems. ECCS TC 8. Paper N° TC 8-2002-19. Graz University of Technology.

Greiner, R. & Lindner, J. 2000. Proposal for buckling resistance of members: Flexural and lateral torsional buckling. ECCS-Validation Group, Report 7.

Greiner, R. & Lindner, J. 2006. Interaction formulae for members subjected to bending and axial compression in Eurocode 3- the Method 2 approach. *Journal of Constructional Steel Research*, 62, 757–770.

Greiner, R. & Ofner, R. 1999. Validation of design rules for member stability of European Standards-Proposal for buckling rules. In: D. Dubina & M. Iványi (eds), *Stability and Ductility of Steel Structures*. Timisoara, Romania: Elsevier.

Greiner, R., Ofner, R. & Salzgeber, G. 1999. Lateral torsional buckling of beam-columns: Theoretical background. ECCS-Validation Group, Report 5.

Lindner, J. 2001. Evaluation of interaction formulae al Level 1 approach with regard to ultimate load calculations and test results: Flexural buckling and lateral torsional buckling. Report 2144E, TU Berlin.

Lindner, J. 2003. Design of beams and beam columns. *Progress in Structural Engineering and Materials*, 5, 38–47.

Ofner, R. 1999. Results of a parametric study of steel beams under axial compression and biaxial bending – Comparisons with code regulations. *Eurosteel Conference*, Prag.

Yong, D.J., López, A. & Serna, M.A. 2006. A Comparative study of AISC LRFD and EC3 approaches to beam-column buckling resistance. In D. Camotin et al. (eds), *Stability and Ductility of Steel Structures; Proceedings of International Colloquium on Stability and Ductility of Steel Structures*, Lisbon, Portugal, vol 2, pp 1109–1116, September 6–8.

Steel and Composite Structures – Wang & Choi (eds)
© 2007 Taylor & Francis Group, London, ISBN 978-0-415-45141-3

Numerical modelling of uniform stainless steel members in bending and axial compression

N. Lopes & P.M.M. Vila Real
University of Aveiro, Portugal

L. Simões da Silva
University of Coimbra, Portugal

ABSTRACT: In this paper the accuracy and safety of the interaction formulae for the evaluation of the resistance of stainless steel beam-columns from Eurocode 3 are evaluated. Two new methods for the design of carbon steel beam-columns at room temperature are proposed on Part 1.1 of Eurocode 3. The possibility of using these two procedures with stainless steel elements will be checked. New formulae for the safety evaluation of stainless steel columns will be also proposed. Its influence on the beam-column interaction formulae will be taken into account.

1 INTRODUCTION

Stainless steel has countless desirable characteristics for a structural material (Estrada, 2005), (Gardner, 2005) and (Euro Inox and SCI, 2006). Although its use in construction is rapidly increasing, it is still necessary to develop the knowledge of its structural behaviour.

Regarding combined bending and axial compression of stainless steel members, Part 1.4 of Eurocode 3 (CEN, 2005b) has two notes saying that the national annexes may give other interaction formulae and others interaction factors, which suggests that the beam-columns formulae and the interaction factors were not well established for stainless steel members, at the time of the conversion from ENV to EN.

Stainless steels are known by their non-linear stress-strain relationships with a low proportional stress and an extensive hardening phase. Well defined yield strength does not exist, the conventional limit of elasticity at 0.2% is usually considered.

The final version of Part 1-1 of Eurocode 3, EN 1993-1-1 (CEN, 2005a), introduced several changes in the design formulae for carbon steel beam-columns, when compared with the previous versions of Eurocode 3. These modifications took place during the conversion of Eurocode 3 from ENV (CEN, 1992) to EN status.

Two new methods for the design of carbon steel beam-columns at room temperature are proposed in Part 1.1 of Eurocode 3 (CEN, 2005a), which are the result of the work carried out by two working groups that followed different approaches (Boissonnade et al, 2006), a French-Belgian team and an Austrian-German one. In this paper it will be checked if these two procedures can also be used with stainless steel elements.

In order to study the possibility of having, in parts 1-1 and 1-4 of the EN versions of Eurocode 3, the same approach for beam-columns, a numerical investigation was carried out and is presented in this paper.

In a companion paper, the authors made a proposal for the safety evaluation of stainless steel columns (Lopes et al, 2007), which is more accurate than the approach in Eurocode 3, and is safe when compared with the numerical results. This new proposal necessarily affects the behaviour of the interaction formulae for beam-columns and its influence can be seen in figures 5 to 8 on the curves with "NP".

The aim of this work is to achieve a better understanding of stainless steel beam-columns behaviour, throughout numerical simulations on H-columns. These simulations have shown that the design resistance formulae from the Eurocode 3, with the proposed safety factor $\gamma_{M1} = 1.1$ are too much conservative. The numerical results were obtained with the finite element program SAFIR (Franssen, 2005), which has been adapted to be able to model stainless steel structures according the material properties defined in EN 1993-1-4. More details on the Program can be found in Lopes et al, 2007.

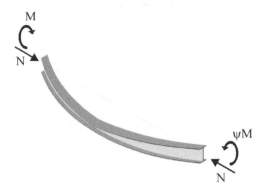

Figure 1. Stainless steel beam-column.

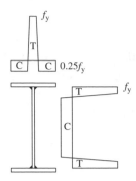

Figure 2. Residual stresses: C – compression; T – tension.

2 CASE STUDY

A beam-column with axial compression and bending moments applied in the extremities, as shown in figure 1 was chosen. It was considered bending moment in the strong axis and in the weak axis with a ψ ratio of 1, which corresponds to uniform bending diagram.

The equivalent welded HEA 200 section of the stainless steel grade 1.4301 (also known as 304) was used in the numerical simulations. Because of the lack of space, only the results for beam-columns with of 3 and 7 meters spans will be shown in the paper. However more lengths were analyzed, showing the same type of behavior.

The possibility of buckling around the strong axis (yy) and around the weak axis (zz) was considered.

In the numerical simulations, a lateral geometric imperfection given by the following expression was considered:

$$y(x) = \frac{l}{1000} \sin\left(\frac{\pi x}{l}\right) \tag{1}$$

where l is the length of the beam. Also, an initial rotation around the beam axis with a maximum value of $l/1000$ radians at mid span was considered.

The adopted residual stresses follow the typical patterns for carbon steel welded sections, considered constant across the thickness of the web and flanges. The distribution is shown in figure 3, and has the maximum value of f_y (yield strength) (Chen and Lui, 1991), (Gardner and Nethercot, 2004) and (Greiner et al, 2005).

3 BEAM-COLUMNS SAFETY EVALUATION

Part 1-1 and Part 1-4 are the parts from Eurocode 3 dedicated respectively to the design of carbon steel and stainless steel structural elements. In this section a brief description of the methods prescribed in those parts of the Eurocode 3, including a new proposal from the authors will be presented.

3.1 Eurocode 3 carbon steel interaction curves

3.1.1 General format

The project team involved in the conversion phase from ENV to EN of Part 1.1 of Eurocode 3 has revised the interaction formulae for carbon steel beam-columns safety check (CEN, 2005a). The reasons for this change were the complexity and lack of a physical rationale behind the interaction formulae of ENV 1993-1-1 (CEN, 1992). Two alternative proposals (Boissonnade et al, 2006) were adopted specifically implementing the concepts of amplification factor and equivalent uniform bending moment.

The procedure for the determination of the interaction factors for "method 1" is reported in Annex A of Part 1.1 of Eurocode 3 and was developed by a French-Belgian team by combining theoretical rules and numerical calibration to account for all the differences between the real model and the theoretical one. "Method 2" is described in Annex B of Part 1.1 of Eurocode 3 and results from an Austrian-German proposal that attempted to simplify the verification of the stability of beam-columns, all interaction factors being obtained by means of numerical calibration.

These two methods were changed here to have into account the reduction factor for flexural buckling of stainless steel columns (see the companion paper, Lopes et al, 2007).

According to EN 1993-1-1 (CEN, 2005a), the stability of beam-columns (classes 1 and 2), in the case of bending around the strong and weak axis, is checked in accordance with the following interaction formulae:

$$\frac{N_{Ed}}{\chi_y \dfrac{N_{Rk}}{\gamma_{M1}}} + k_{yy} \frac{M_{y,Ed} + \Delta M_{y,Ed}}{\chi_{LT} \dfrac{M_{y,Rk}}{\gamma_{M1}}} \leq 1 \tag{2}$$

and

$$\frac{N_{Ed}}{\chi_z \dfrac{N_{Rk}}{\gamma_{M1}}} + k_{zz} \frac{M_{z,Ed} + \Delta M_{z,Ed}}{\dfrac{M_{z,Rk}}{\gamma_{M1}}} \leq 1 \qquad (3)$$

where:

$$N_{Rk} = A f_y, \quad M_{z,Rk} = W_{pl,z} f_y$$
$$\text{and} \quad M_{y,Rk} = W_{pl,y} f_y \qquad (4)$$

For class 1 and class 2 sections $\Delta M_{y,Ed}$, $\Delta M_{z,Ed}$ are equal to zero. k_{yy} and k_{zy} are the interaction factors. The partial safety coefficient γ_{M1} is equal to 1.

3.1.2 Determination of the interaction factors using Method 1

The procedure for the determination of the interaction factors for "method 1" is reported in Annex A of Part 1.1 of Eurocode 3. Without attempting to explain the background of this proposal, the interaction factors are expressed by the following relations:

$$k_{yy} = c_{my} \frac{\mu_y}{1 - \dfrac{N_{Ed}}{N_{cr,y}}} \frac{1}{c_{yy}} \qquad (5)$$

$$k_{zz} = c_{mz} \frac{\mu_z}{1 - \dfrac{N_{Ed}}{N_{cr,z}}} \cdot \frac{1}{c_{zz}} \qquad (6)$$

where:

$$c_{my} = c_{my,0} \quad \text{e} \quad c_{mz} = c_{mz,0} \qquad (7)$$

and

$$c_{my,0} = 0.79 + 0.21\psi + 0.36(\psi - 0.33)\frac{N_{Ed}}{N_{cr,y}} \qquad (8)$$

$$c_{mz,0} = 0.79 + 0.21\psi_z + 0.36(\psi_z - 0.33)\frac{N_{Ed}}{N_{cr,z}} \qquad (9)$$

The other coefficients not described here can be obtained in Annex A of Part 1.1 of Eurocode3.

3.1.3 Determination of the interaction factors using Method 2

"Method 2" (Boissonnade et al. 2006) is described in Annex B of Part 1.1 of Eurocode 3 According to this approach, the interaction factors are expressed by the following relations

$$k_{yy} = c_{my}\left(1 + (\overline{\lambda}_y - 0.2)\frac{N_{Ed}}{\chi_y \dfrac{N_{Rk}}{\gamma_{M1}}}\right) \qquad (10)$$

$$\text{but} \quad k_{yy} \leq c_{my}\left(1 + 0.8\frac{N_{Ed}}{\chi_y \dfrac{N_{Rk}}{\gamma_{M1}}}\right)$$

and

$$k_{zz} = c_{mz}\left(1 + (2\overline{\lambda}_z - 0.6)\frac{N_{Ed}}{\chi_z \dfrac{N_{Rk}}{\gamma_{M1}}}\right) \qquad (11)$$

$$\text{but} \quad k_{zz} \leq c_{mz}\left(1 + 1.4\frac{N_{Ed}}{\chi_z \dfrac{N_{Rk}}{\gamma_{M1}}}\right)$$

with

$$c_{my} = c_{mLT} = 0.6 + 0.4\psi \geq 0.4 \qquad (12)$$

3.2 Eurocode 3 stainless steel interaction curves

Part 1-4 of Eurocode 3, which is the part from the Eurocode dedicated to stainless steel structures, gives the following expressions

$$\frac{N_{Ed}}{N_{b,i,Rd}} + k_i \frac{M_{i,Ed}}{W_{pl,i} \dfrac{f_y}{\gamma_{M1}}} \leq 1 \qquad (13)$$

where

$$k_i = 1.0 + 2(\overline{\lambda}_i - 0.5)\frac{N_{Ed}}{N_{b,Rd,i}} \qquad (14)$$

$$\text{but} \quad 1.2 \leq k_i \leq 1.2 + 2\frac{N_{Ed}}{N_{b,Rd,i}}$$

where i refers to the y or z axis.

Although Eurocode 3 suggests for carbon steel a safety factor $\gamma_{M1} = 1.0$, for stainless steel the value suggested for this factor is $\gamma_{M1} = 1.1$. Figures 3 and 4 show that with this safety factor the Eurocode 3 gives

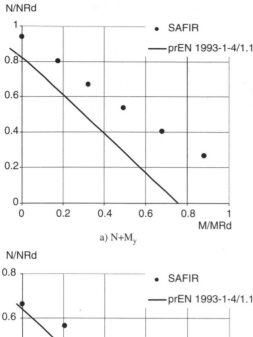

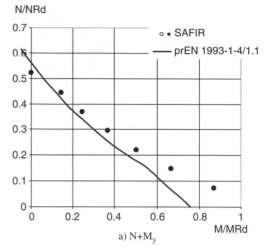

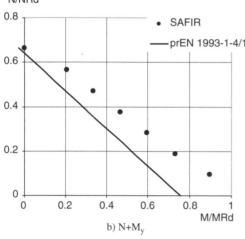

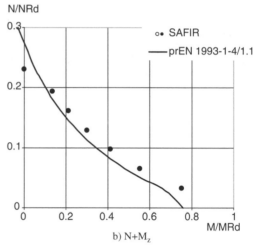

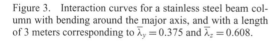

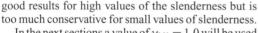

Figure 3. Interaction curves for a stainless steel beam column with bending around the major axis, and with a length of 3 meters corresponding to $\bar{\lambda}_y = 0.375$ and $\bar{\lambda}_z = 0.608$.

Figure 4. Interaction curves for a stainless steel beam column with bending around the major axis, and with a length of 7 meters corresponding to $\bar{\lambda}_y = 0.875$ and $\bar{\lambda}_z = 1.419$.

good results for high values of the slenderness but is too much conservative for small values of slenderness.

In the next sections a value of $\gamma_{M1} = 1.0$ will be used so that comparison between the formulae from Part 1.4 of Eurocode 3 and the other procedures analyzed in this paper could be made.

3.3 Proposal for the stainless steel interaction curves

In the design method prescribed in Eurocode 3, there is a minimum limit value for the interaction factor k_i. as shown in equation (14).This means that when the

axial force becomes zero the interaction factor takes the value 1.2, leading to a resistant moment smaller than the resistant moment for pure bending prescribed in Eurocode (see curves prEN 1993-1-4 and prEN 1993-1-4 NP in figures 5 to 8).

In the parametric study presented in this paper the possibility of not limiting the interaction factor k_i to a minimum value of 1.2 was tested so that the plastic moment could be reached when no axial force is acting, i. e. $k_i = 1.0$ (see curve prEN 1993-1-4 NP + NK in figures 5 to 8).

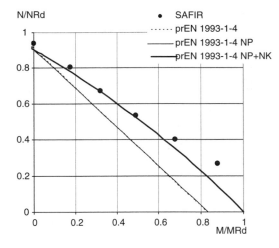

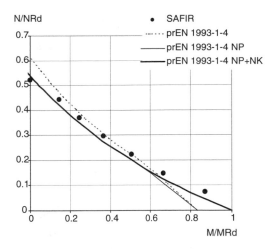

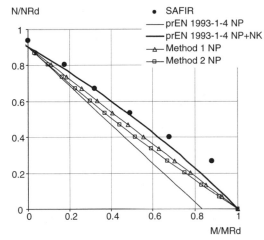

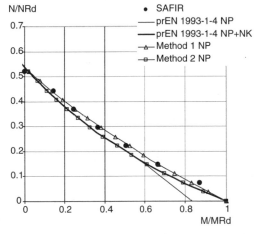

Figure 5. Interaction curves for a stainless steel beam column with bending around the major axis, and with a length of 3 meters corresponding to $\overline{\lambda}_y = 0.375$.

Figure 6. Interaction curves for a stainless steel beam column with bending around the major axis, and with a length of 7 meters corresponding to $\overline{\lambda}_y = 0.875$.

4 PARAMETRIC STUDY

Figures 5 to 8 show the comparison carried out in this study. The interaction curves chosen for this parametric study were obtained using: (i) Eurocode 3 "EN 1993-1-4"; (ii) Eurocode 3 with the new proposal made for stainless steel columns (Lopes et al, 2007) "EN 1993-1-4 NP"; (iii) the proposal made in this paper without the minimum limit value of 1.2 for the interaction factor k_i "EN 1993-1-4 NP+NK"; and (iv) the formulae from Part 1.1 of Eurocode 3 for carbon steel beam-columns with the new proposal for stainless steel columns "Method 1 NP" and "Method 2 NP".

The results for stainless steel beam columns with bending around the major axis are shown in figures 5

and 6. Figures 7 and 8 present the results for stainless steel beam-columns with bending around the minor axis.

5 CONCLUSION

The method that approximates better the numerical results from SAFIR is the method "prEN1993-1-4 NP+NK". This method corresponds to the Part 1.4 of Eurocode 3 with the new proposal for stainless steel columns (Lopes et al, 2007) and not considering the minimum limit value of 1.2 for the factor k_i as it is described in section 3.3 of this paper.

The "method 1 NP" and "method 2 NP" adapted from the formulae from Part 1.1 of Eurocode 3 for carbon steel and the new proposal for stainless steel

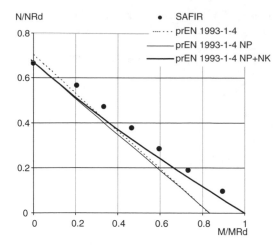

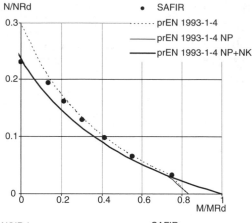

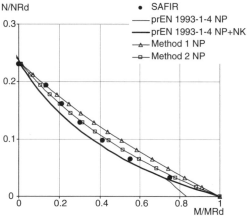

Figure 8. Interaction curves for a stainless steel beam column with bending around the minor axis, and with a length of 7 meters corresponding to $\overline{\lambda}_z = 1.419$.

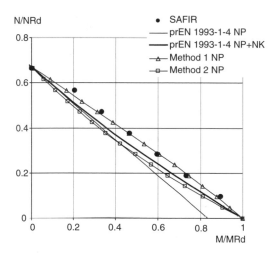

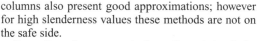

Figure 7. Interaction curves for a stainless steel beam column with bending around the minor axis, and with a length of 3 meters corresponding to $\overline{\lambda}_z = 0.608$.

columns also present good approximations; however for high slenderness values these methods are not on the safe side.

Regarding the proposal from Part 1.4 of the Eurocode 3 it was shown that a safety factor $\gamma_{M1} = 1.1$ is really needed but the minimum limit of 1.2 for the interaction factor k_i should be withdrawn. Although the value of 1.1 for the safety factor γ_{M1} should be considered for high slenderness values it was also shown that for small values of the slenderness the Eurocode 3 is too much conservative.

It should be pointed out that as the parametric study shown in this paper contemplates only few cases, more studies considering other types of stainless steel

(austenitic-ferritics and ferritics) and other cross sections should be made to check the accuracy of the presented proposal (prEN1993-1-4 NP+NK).

ACKNOWLEDGEMENTS

The authors wish to acknowledge to the Foundation Calouste Gulbenkian (Portugal) for its supports through the scholarship given to the first author.

REFERENCES

Boissonnade N.; Greiner, R.; Jaspart, J.P; Lindner J. 2006. Rules for member stability in EN 1993-1-1. *Background documentation and design guidelines*, ECCS.
CEN European Committee for Standardisation 2005a. EN 1993-1-1, Eurocode 3, Design of Steel Structures – Part

1-1. General rules and rules for buildings. Brussels, Belgium.

CEN European Committee for Standardisation 2005b. prEN 1993-1-4, Eurocode 3, Design of Steel Structures – Part 1-4. General rules – Supplementary Rules for Stainless Steels. Brussels, Belgium.

CEN European Committee for Standardisation 1992. ENV 1993-1-1 Eurocode 3: Design of Steel Structures - Part 1-1: General rules and rules for buildings. Brussels, Belgium.

Chen W. F. and Lui E. M. 1991. *Stability design of steel frames*, CRC Press.

Euro Inox and SCI, Steel Construction Institute. 2006. Designers Manual for Structural Stainless Steel.

Estrada, I. 2005. Shear Design of Stainless Plate Girders. *PhD Thesis*, Universitat Politècnica de catalunya, Barcelona.

Franssen, J.-M. 2005. SAFIR. A Thermal/Structural Program Modelling Structures under Fire. *Engineering Journal*, A.I.S.C., Vol. 42, No. 3, pp. 143–158.

Franssen, J.-M. 1989. Modelling of the residual stresses influence in the behaviour of hot-rolled profiles under fire conditions, (in French), *Construction Métallique*, Vol. 3, pp. 35–42.

Gardner, L. 2005. The use of stainless steel in structures. *Prog. Struct. Engng Mater.*

Gardner, L., Nethercot, D. A. 2004. Numerical Modeling of Stainless Steel Structural Components – A consistent Approach. *Journal of Constructional Engineering*, ASCE, pp. 1586–1601.

Greiner, R., Hörmaier, I., Ofner, R., Kettler, M., 2005. Buckling behaviour of stainless steel members under bending. *ECCS Technical Committee 8 – Stability*.

Lopes, N., Vila Real, P., Silva, L. 2007. Numerical modelling of the Flexural buckling of axially loaded stainless steel members, proceedings of the third International Conference on Steel and Composite Structures ICSCS07, Manchester, United Kingdom.

Steel and Composite Structures – Wang & Choi (eds)
© 2007 Taylor & Francis Group, London, ISBN 978-0-415-45141-3

The analysis of deflection of composite steel-concrete beams with partial shear connection

J. Valivonis & G. Marčiukaitis
Department of Reinforced Concrete and Masonry Structures, Vilnius Gediminas Technical University

ABSTRACT: Results of experimental and theoretical investigation of deflection of composite steel and concrete beams with different shear connectors are analyzed in this article. A composite beam made of I profile steel beam with 50 mm wide concrete slab with different quantity and shape of shear connectors were tested. A method for calculation of deflection of composite beam, which theoretically estimates the stiffness of the shear connection, based on "built-up bars theory", is also proposed. Theoretical results have good coincidence with experimental results.

1 INTRODUCTION

The composite steel and concrete structures are widely used in civil and industrial buildings and bridges. The composite floors in the most cases consist of steel beams of I section with a reinforced concrete slab installed above them. The reinforced concrete slab with the steel beam in such composite floor can either be connected by the means of the shear connectors or not. Depending on the number and the type of the shear connectors, the layers in the composite steel and concrete floor can be considered fully or partially connected.

In the case when additional shear connectors are not used, effect of reinforced concrete slabs in analysis of the steel beams is neglected. In the case of the full shear connection between the reinforced concrete slab and the steel beam, the floor beams in their analysis are considered as the composite members. Slip at the interface between reinforced concrete slabs and steel beams are neglected.

Experimental and theoretical investigations performed by many authors (Loh, H.Y et al. 2004, Hosain, M.U. et al.1992, Jurkiewiez, B. & Hottier, J.M. 2005, Lawson, R.M. 1992, Lawson, R.M. 1992, Jurkiewiez, B. & Braymand, S. 2007 Jurkiewiez, B. & Braymand, S. 2007, Lam, D. & Ellobody, E. 2005, Motak, J. & Machacek, J. 2004) indicate that it is complicated to amount the full shear connection between the reinforced concrete layer and the steel members. In serviceability stage, especially at high stresses, slip between the layers takes place. It results in the change of the stiffness of the shear connection. This phenomenon affects flexural stiffness of a composite steel and concrete member, which directly influences deflection of the floor. It means that deflection of the floor increases with the slip in the shear connection between the steel and concrete members.

In analysis (Ranzi, G. et al. 2003, Gurkšnys, K. et al. 2005, Salari, M.R. et al. 1998, Wang, Y.C. 1998, Marčiukaitis G. et al. 2006) of the composite steel and concrete floors with the steel beams, it is desirable to take into account the partial shear connection between the layers. Substantially more accurate results of analysis can be obtained and it is possible to assess actual action of the composite floor in various stages of its behavior when the shear partial stiffness is appreciable.

2 ANALYSIS OF DEFLECTIONS OF BEAMS IN A COMPOSITE FLOOR

In analysis of deflection of a composite steel and concrete flexural member, it is desirable to allow for the shear stiffness of the shear connection between the concrete and the steel layers. In this case better description of the flexural stiffness of the composite member is obtained. The theory of built-up bars (Rzanitsyn, A. 1986) can be adopted for analysis of deflection of the composite steel and concrete beams. This theory gives opportunity to take into account not only the flexural stiffness of individual layers of the composite member but and the stiffness of the shear connectors as well.

The vertical deformations y of flexural composite steel and concrete structures subjected to uniformly distributed load can be described by differential equation (Rzanitsyn, A. 1986):

$$\frac{dy^4}{dx^4} - \lambda^2 \frac{d^2 y}{dx^2} = \frac{\lambda^2 p \cdot x}{2}\left(\frac{l-x}{E_{eff}I_{eff}}\right) + \frac{p}{E_{c.eff}\cdot I_c + E_p I_p} \quad (1)$$

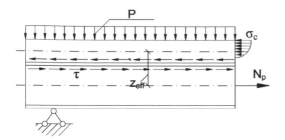

Figure 1. Diagram for analysis of a composite member.

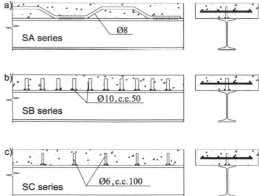

Figure 2. General view of the beams.

After solution of this equation and with allowance for boundary conditions $y(0) = 0$; $y(l) = 0$; $dy^2(0)/dx^2 = 0$ and $dy^2(l)/dx^2 = 0$, the vertical displacement (deflection) at the middle of the span of the composite member is expressed by:

$$y\left(\frac{l}{2}\right) = \delta = \frac{5 \cdot p \cdot l^4}{384 E_{eff} I_{eff}} + \frac{p}{\lambda^4 \cdot D}\left(\frac{1}{ch(0,5 \cdot \lambda \cdot l)} + \frac{\lambda^2 l^2}{8} - 1\right) \quad (2)$$

The quantity describing flexural stiffness of the composite member can be determined by

$$\frac{1}{D} = \frac{1}{E_{c,eff} \cdot I_c + E_p \cdot I_p} - \frac{1}{E_{eff} \cdot I_{eff}} \quad (3)$$

Characteristics of the stiffness for the shear connection between the concrete and the steel:

$$\lambda = \sqrt{\alpha \cdot \gamma} $$

Characteristic of stiffness for the composite beam is determined by:

$$E_{eff} \cdot I_{eff} = E_{c,eff} \cdot I_c + E_p I_p + \frac{E_{c,eff} \cdot A_c \cdot E_p \cdot A_p \cdot z_{eff}^2}{E_{c,eff} A_c + E_p A_p} \quad (4)$$

where: $E_{c,eff}, E_p$ = elasticity modules of the concrete and the steel; I_c, I_p, A_c, A_p = moments of inertia and areas of cross-sections of the concrete layer and of the structural steel section; z_{eff} = distance between centers of gravity of the concrete layer and of the structural steel section.

The stiffness factor γ for flexural composite steel and concrete structures is determined by:

$$\gamma = \frac{1}{E_{c,eff} \cdot A_c} + \frac{1}{E_p A_p} + \frac{z_{eff}^2}{E_{c,eff} \cdot I_c + E_p I_p} \quad (5)$$

The stiffness factor for the shear connection between the concrete and the steel:

$$\alpha = \frac{b_w \cdot G_w}{z_{eff}} \quad (6)$$

where: b_w = width of the connection between the concrete and the structural steel section; G_w = shear stiffness characteristic for the concrete and the structural steel section.

The vertical displacement of composite steel and concrete members due to the action of two concentrated loads can be determined by:

$$\frac{dy^4}{dx^4} - \lambda^2 \frac{dy^2}{dx^2} = \lambda^2 \frac{M}{E_{eff} \cdot I_{eff}} \quad (7)$$

After solution of the differential equation, deflection of a flexural member at the middle of the span can be expressed by:

$$\delta = M\left(\frac{l^2}{8 \cdot E_{eff} I_{eff}} + \frac{1}{D}\left(\frac{ch(0,5\lambda l) - 1}{\lambda^2 ch(0,5\lambda l)}\right)\right) \quad (8)$$

3 EXPERIMENTAL INVESTIGATION

The composite steel and concrete beams were investigated in this research. Cross-section of the experimental composite beams is shown in Fig. 2. The specimens were made of 100 m high structural steel section of I cross-section. The concrete layer was 50 mm in thickness and 200 mm in width. The cube strength of the concrete layer $f_{c,cube} = 23{,}1$ MPa. The concrete layer in the middle of its cross-section was provided with reinforcing fabrics made of S240 class reinforcing steel Ø 6 mm in diameter. Modulus of elasticity of the concrete $E_{cm} = 24{,}9 \cdot 10^3$ MPa. The concrete layer and the structural steel section were jointed by the shear connectors.

Six 2,0 m long composite beams were manufactured and tested. The beams were grouped in three groups: SA, SB, SC. The type of the shear connectors between the concrete and the structural steel section

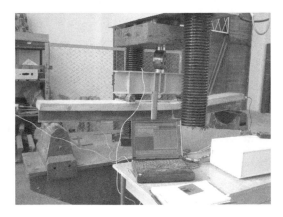

Figure 3.   General view of the set up for the test of the beams.

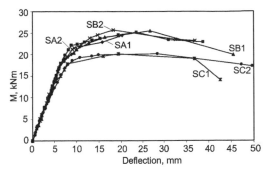

Figure 4.   Experimental deflections of the composite beams.

was varied in the individual groups. In the beams *SA* the shear connectors were made of a zigzag shape steel bar and situated at the middle of the top flange of the structural steel section along its full length (Fig. 2a). One side of the bar was welded to the structural steel section. The bar was Ø 8 mm in diameter. The beams *SB* were provided with the vertical stud shear connector in diameter of Ø 10 mm the end of which was welded to the structural steel section and spacing of connectors was 200 mm (Fig. 2b). The beams *SC* were provided with the vertical stud shear connector in diameter of Ø 6 mm and with spacing of 200 mm (Fig. 2c). The stud shear connectors were manufactured of reinforcing steel of S240 class.

Manufactured specimens differ in the type and intensity of the shear connectors seeking to examine the stiffness of the shear connection between the concrete and the structural steel section.

In parallel specimens for experimental investigations in the stiffness of the shear connection between the concrete and the structural steel section were manufactured.

The beams were tested by two concentrated forces (Fig. 3). The load was increased in steps. Forces and deflections of the beam were recorded with electronic gauges "ALMEMO".

4   RESULTS OF EXPERIMENTAL INVESTIGATION

During the tests deflection of the beam was recorded at each load step. The graphs of deflections of the composite beams are shown in Fig. 4. Experimental investigations showed that deflections of the composite steel and concrete beams depend on the shear stiffness of the connection between the layers of the concrete and the steel. Stiffness of the connection is determined by the stiffness of the shear connectors provided between the layers.

Experimental investigation indicated that in behavior of the tested composite beams three stages in development of deflections can be distinguished. The first stage is the stage of elastic behavior of the composite beam. Experimental investigations (Fig. 4) pointed out that the first stage reaches $M \sim 0{,}65 M_R$. The concrete layer in this stage acts jointly with the structural steel section. The slip in the contact between the concrete and the structural steel members was small and it can be ignored.

It has no influence on the flexural stiffness of the composite steel and concrete beams. In this stage of behavior the growth of deflection was proportional with the load. It shows that composite beams are with full shear connection. The second stage of composite steel and concrete beam is stage of its elastic plastic behavior. It occurs when the bending moment exceeds $0{,}65 M_R$ and continues up to $M \approx (0{,}85..0{,}9) M_R$. During this stage of behavior the slip in the shear connection between the concrete and the steel of the composite beams increase. The flexural stiffness of the composite beams decreases, deflections increase out of proportion (Fig. 4). Nevertheless, in this stage the shear connection of the composite beams is not destroyed, there are no cracks in the concrete layer.

The third stage begins when $M \approx (0{,}85..0{,}9) M_R$. It is the stage of failure of the composite steel and concrete beams. At this stage of behavior of the composite beams the concrete layer and the structural steel section separate from each other. The concrete and the steel layers act separately. Deflection of the beams increases continuously. Failure of the element commences. In the beams of SA group longitudinal crack in the reinforced concrete slab opened.

Analysis of results of experimental investigations showed that the beams of *SA* and *SB* groups were of the highest stiffness (their deflection was the smallest), while the shear stiffness of these beams was the highest. The shear connectors in the beams of *SA* group

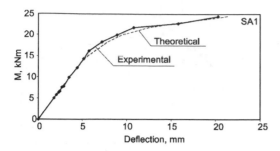

Figure 5. Comparison of experimental and theoretical results for SA1 beams.

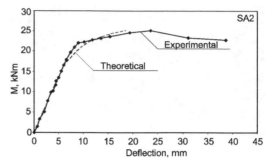

Figure 6. Comparison of experimental and theoretical results for SA2 beams.

were made of a steel bar bent in the shape of a zigzag; the beams of *SB* group were provided with sufficiently closely spaced the stud shear connectors in diameter of Ø10 mm. The lowest stiffness was of the beams which were provided with the shear connectors of the lowest stiffness (*SC*).

Comparison of investigation results pointed out that when the beams behaved elastically (in the first stage of behavior) ($M = 10\,\text{kN} \cdot \text{m}$) for all the beams deflections were similar in their value (Fig. 4). In the second stage of behavior the rate of growth of deflection of the composite steel and concrete beams depends on the stiffness of connection between the layers. The rate of growth of deflection for the beams of *SC* group is substantially greater than that for the beams of *SA* and *SB* groups. Under the action of the same load in its value ($M = 18\,\text{kN·m}$) deflection of the beams of SC group is by 25% greater than that of the beams of SA and SB groups. It is conditioned by the different stiffness of the shear connections between the layers.

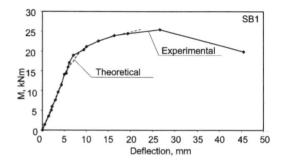

Figure 7. Comparison of experimental and theoretical results for SB1 beams.

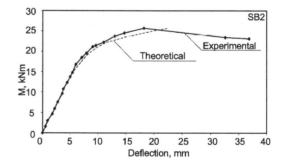

Figure 8. Comparison of experimental and theoretical results for SB2 beams.

5   COMPARISON OF EXPERIMENTAL AND THEORETICAL RESULTS

Using characteristics of materials of experimental composite steel and concrete beams and the method of analysis presented above, calculation of deflections for the beams was performed. The method proposed gave opportunity to take account of the actual stiffness of the shear connection between the steel and concrete members of the composite steel and concrete beams up to the failure stage.

Experimental and theoretical investigations pointed out that the shear stiffness of connection between the concrete and the steel members of the composite beams can be defined the stiffness characteristic G_w. This characteristic depends on the shear stiffness of shear connectors provided in the connection and on concrete deformations due to the bearing stresses.

In the elastic (the first) stage of behavior of composite steel and concrete beams while in the connection between the concrete and the structural steel members only elastic deformations occur, the shear connection between the concrete and the structural steel section is full. In this stage deflection of the beams can be calculated as for elastic structure using known methods for calculation of deflection for building structures.

The method of analysis proposed by us can be used as well, taking stiffness characteristic for connection between the layers, G_w, equal to the shear modulus of the concrete. In calculations $G_w = 9960\,\text{MPa}$

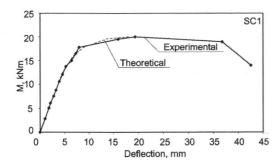

Figure 9. Comparison of experimental and theoretical results for SC1 beams.

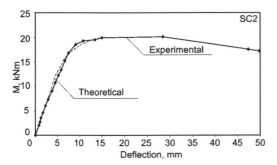

Figure 10. Comparison of experimental and theoretical results for SC2 beams.

was taken. For the elastic plastic stage of behavior of the composite steel and concrete beams, as $M = 0.65$ $M_R \ldots 0.90\,M_R$, the stiffness characteristic for connection between the layers, G_w (in MPa) was determined by empirical expression

$$G_w = \left(\alpha - \beta \cdot \frac{M}{M_R} \right)^2 10^3 \qquad (9)$$

where: for the beams of SA and SB group $\alpha = 4.1$, for the beams of SC group $\alpha = 4.25$ and for beams SA, SB and SC, $\beta = 4$.

Experimental calculations of deflections for composite steel and concrete beams demonstrated sufficiently good agreement between experimental and theoretical deflections (Figs 5–10).

6 CONCLUSIONS

Flexural stiffness of the composite steel and concrete structures depends greatly on the stiffness of the shear connection between the steel and concrete members. The stiffness of the shear connection depends on amount and type of the shear connectors and on deformability of the concrete subjected to bearing.

Experimental investigations of the composite steel and concrete beams made it possible to distinguish 3 stages in their behavior: elastic, elastic plastic and plastic. In the elastic stage the shear connection steel and concrete members is full. In the elastic plastic stage the deflections of the structure are greatly effected by the shear stiffness of the connection between the steel and concrete. The partial shear connection between steel and concrete members is in this stage.

In the elastic plastic stage it is recommended calculation of deflection of the composite steel and concrete structures to accomplish according to the proposed method based on the theory of built up bars. In calculations the shear stiffness of the connection between the layers is defined by the characteristic G_w.

This characteristic is determined in experimental way and depends on the type of the connection, its stiffness, and stresses.

REFERENCES

Loh , H.Y & Uy, B. & Bradford, M.A. 2004. The effects of partial shear connection in the hogging moment regions of composite beams. Part II-Analytical study. *Journal of Constructional Steel Research* 60: 921–962.

Hosain, M.U. & Chien, E.Y.L. & Kennedy, D.J.L. 1992. New Canadian provisions related to the design of composite beams. In Easterling W.S. & Roddis W.M.K (eds), Composite constructions in steel and concrete II; Proc. intern. symp. Trout Lodge Potosi, Missouri, 14–19 June 1992. New York.

Jurkiewiez, B. & Hottier, J.M. 2005. Static behaviour of a steel-concrete composite beam with an innovative horizontal connection. *Journal of Constructional Steel Research* 61: 1286–1300.

Lawson, R.M. 1992. Shear connection in composite beams. In Esterling W.S. & Roddis W.M.K. (eds), Composite construction in steel and concrete II; Proc. intern. symp.Tourt Lodge Potosi, Missouri 14–19 June 1992. New York.

Ellobody, E. & Young, B. 2006. Performance of shear connection in composite beams with profiled steel sheeting. *Journal of Constructional Steel Research* 62: 682–694.

Jurkiewiez, B. & Braymand, S. 2007. Experimental study of a pre-cracked steel-concrete composite beam. *Journal of Constructional Steel Research* 63: 135–144.

Gurkšnys, K. & Kvedaras, A. & Kavaliauskas, S. 2005. Behaviour evaluation of "sleeved" connectors in composite timber-concrete floors. *Journal of Civil Engineering and Management XI(4)*: 277–282.

Lam, D. & Ellobody, E. 2005. Behaviour of headed stud shear connectors in composite beam. *Journal of Structural Engineering, ASCE 131(1)*: 96–107.

Motak, J. & Machacek, J. 2004. Experimental behaviour of composite girders with steel undulating web and thin – walled shear connector's hilti stripcon. *Journal of Civil Engineering and Management X(1)*: 45–49.

Ranzi, G. & Bradford, M.A. & Uy, B. 2003. A general method of analysis of composite beams with partial interaction. *Steel & Composite Structures 3(3)*: 169–184.

Salari, M.R. & Spacone, E. & Shing, P.B.& Frangopol, D.M. 1998. Nonlinear analysis of composite beams with deformable shear connectors. *Journal of Structural Engineering 124(10)*: 1148–1158.

Wang, Y.C. 1998. Deflection of steel-concrete composite beams with partial shear interaction. *Journal of Structural Engineering 124(10)*: 1159–1164.

Marčiukaitis G. & Jonaitis, B. & Valivonis, J. 2006. Analysis of deflection composite slabs with profiled sheeting up to the ultimate moment. *Journal of Constructional Steel Research 62*:820–830.

Rzanitsyn, A. 1986. Built-up Bars and Plates. Moscow: Strojizdat (in Russian).

Steel and Composite Structures – Wang & Choi (eds)
© *2007 Taylor & Francis Group, London, ISBN 978-0-415-45141-3*

Simplified method for flexural failure analysis of steel-concrete composite beams post-tensioned with external tendons

A. Dall'Asta & A. Zona
Dipartimento di Progettazione e Costruzione dell'Ambiente, University of Camerino, Ascoli Piceno, Italy

L. Ragni
Dipartimento di Architettura, Costruzioni e Strutture, Marche Polytechnic University, Ancona, Italy

ABSTRACT: The ultimate capacity of beams with external prestressing cables slipping at saddle points cannot be evaluated by a local analysis of the critical sections and a nonlinear analysis of the whole structure is required. In the past some simplified approaches were proposed for reinforced concrete beams. On the other hand no simplified procedures are available for steel-concrete composite beams. In this work a new simplified method is introduced for evaluating the traction increment in the cable and consequently the flexural strength of composite beams. The proposed approach is described in detail and some applications of practical interest to simply supported beams are discussed.

1   INTRODUCTION

External prestressing in steel-concrete composite beams is an interesting technique for the design of new structures as well as for the reinforcement and repair of existing constructions. Many advantages are offered by external prestressing (Saadatmanesh et al. 1989), such as: enlarged elastic range of behaviour, increased ultimate capacity, improved fatigue behaviour, limitation of deflections, control of concrete slab cracking in the hogging regions, reduction of structural weight with benefits on construction economy and aesthetic value. The design and analysis of externally prestressed beams call for the attention to some specific issues. Differently from conventional bonded prestressing, the ultimate capacity of beams with external prestressing cables slipping at saddle points cannot be evaluated by a local analysis of the critical sections. A nonlinear analysis of the whole structure is required to correctly evaluate the cable traction at failure.

Being this type of analysis quite complex simplified approaches are an important support for engineers in the design process of external prestressed structures. In the past simplified formulas have been proposed in order to estimate the stress increment of the cable at collapse. However these formulas are limited to reinforced concrete beams prestressed with unbonded internal and external tendons (e.g., ACI 1995). In 1991 an extensive review of the existing equations for the determination of the ultimate stress in unbonded tendons in concrete beams was conducted by Naaman and

Alkhairi (Naaman & Alkhairi 1991a). It was concluded that at the time of the study, none of the published equations combined sufficient accuracy and simplicity to be recommended for code prediction. In the same 1991 Naaman and Alkhairi (Naaman & Alkhairi 1991b) proposed a new methodology for evaluating the stress increment at collapse in concrete beams prestressed by unbonded tendons. Their formulas were incorporated in the AASHTO code in 1994 (AASHTO 1994). In 1996 Aparicio and Ramos (Aparicio & Ramos 1996) carried out a wide parametric analysis on concrete bridges using their own finite element beam model. They concluded that the American formulas (including those of Naaman and Alkhairi) are not consistent with the actual behaviour of externally prestressed bridges and that in some cases these formulas recommend too high increments of stress (American formulas were obtained by interpolating a large number of experimental tests of prestressed beams without attempting to describe their collapse modalities). On the other hand the European code (CEN 1995) is too conservative allowing no increase of prestressing stress at ultimate unless a nonlinear analysis is performed. In the same work Aparicio and Ramos (Aparicio & Ramos 1996) proposed an alternative estimation of the stress increment in the tendon, based on their parametric analysis. Their prediction of cable stress increment depends on the beam static scheme, cross section shape and construction method. More recently the Authors (Dall'Asta et al. 2007) proposed a new simplified method for the evaluation of the flexural

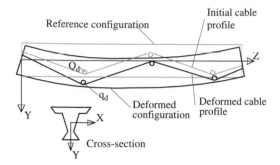

Reference configuration

Initial cable profile

Q_{d}

q_d

Deformed configuration

Deformed cable profile

Cross-section

Figure 1. Beam-tendon structural system.

strength of concrete beams prestressed with external cables (cable traction increment, position of the critical section where collapse occurs, collapse load multiplier) which is based on a simplified description of the deformation at collapse. Compared to previously presented methods, this approach shows a better approximation of the results obtained by a finite element nonlinear analysis and a clear understanding of collapse modalities.

In this paper the Authors extend the method previously presented for concrete beams (Dall'Asta et al. 2007) to steel-concrete composite beams. A comparison between the results deriving from the proposed method and the numerical solution provided by a nonlinear finite element model, is illustrated. The comparison is developed for some simple cases of practical interest.

2 PROPOSED METHOD

2.1 Review of the adopted analytical model

A beam with symmetrical cross section is considered. An orthogonal reference frame $\{O; X,Y,Z\}$ is introduced: the Z-axis is parallel to the beam axis and the vertical plane YZ is the plane of geometric, material and load symmetry of the structure (Fig. 1). Unit vectors $\mathbf{i}$, $\mathbf{j}$, $\mathbf{k}$ are parallel to X, Y and Z respectively. The cross section of the beam is rigid in its own plane and remains plane and orthogonal to the beam axis after deformation. Perfect bond acts between concrete and reinforcements. External prestressing cables are symmetrically disposed with respect to the symmetry plane YZ. Each couple of tendons is considered as one single resultant cable with path contained in the symmetry plane. For the sake of simplicity only one resultant cable is considered in the following.

The previous assumptions lead to the following description of the displacement $\mathbf{u}$ of a point of the beam (i.e., only in-plane bending occurs):

$$\mathbf{u}(y,z) = v(z)\mathbf{j} + [w(z) - yv'(z)]\mathbf{k} \tag{1}$$

where w is the axial displacement of a reference fiber of ordinate zero and v is the vertical displacement. The only non-zero strain component in the beam is the axial strain ε

$$\varepsilon(y,z) = \varepsilon_0(z) + y\theta(z) \tag{2}$$

where $\varepsilon_0(z) = w'(z)$ is the axial strain at the reference fiber of ordinate zero in the cross-section and $\theta(z) = -v''(z)$ is the curvature. Once that general nonlinear constitutive laws are introduced for concrete and reinforcement steel, resisting internal forces resultants of the stress in the beam are the axial force N_R and the bending moment M_R:

$$N_R(\varepsilon_0,\theta) = \int_A \sigma(\varepsilon)dA \tag{3}$$

$$M_R(\varepsilon_0,\theta) = \int_A y\sigma(\varepsilon)dA \tag{4}$$

where the bending moment is computed with respect to the reference fiber of ordinate zero.

The path of the cable is assigned by $D+1$ points where the points from 1 to $D-1$ locate the positions of the intermediate deviators and the points 0 and D locate the positions of the end anchorages:

$$\mathbf{Q}_d = y_d\mathbf{j} + z_d\mathbf{k} \tag{5}$$

where y_d and z_d $(d = 0, \dots D)$ are the coordinates of the d-th deviator of the cable. It is assumed that the cable traces a rectilinear line between two subsequent saddles. The total length of the cable in the undeformed state is given by

$$L_t = \sum_{d=1}^{D} |\mathbf{Q}_d - \mathbf{Q}_{d-1}| = \sum_{d=1}^{D} \sqrt{(\Delta_d y)^2 + (\Delta_d z)^2} \tag{6}$$

where

$$\Delta_d(\bullet) = \bullet(z_d) - \bullet(z_{d-1}) \tag{7}$$

After the deformation of the beam the deviators assume the new positions

$$\mathbf{q}_d = [y_d + v(z_d)]\mathbf{j} + \{z_d + w(z_d) - y_d v'(z_d)\}\mathbf{k} \tag{8}$$

where w_d and v_d are the components of the displacement of the d-th deviator, and the total length of the cable in the deformed state is given by:

$$l_t = \sum_{d=1}^{D} |\mathbf{q}_d - \mathbf{q}_{d-1}| = \\ = \sum_{d=1}^{D} \sqrt{(\Delta_d y + \Delta_d v)^2 + (\Delta_d z + \Delta_d w - \Delta_d (yv'))^2} \tag{9}$$

Assumed that the cable can slip with negligible friction at the saddle points, its strain can be calculated by the ratio between the global stretching of the cable path and its initial length, so that it depends on the displacement field of the whole beam. The hypothesis of negligible friction is satisfied in many real cases, as for example for individually-coated single-strand tendons (Conti et al. 1993). Since the displacements considered in the beam description are assumed to be very small, the following linear expression is adopted for the deformed cable length, in order to obtain a consistent formulation and to avoid making the problem non-linear by introducing negligible terms (Dall'Asta & Dezi 1993, 1998):

$$\varepsilon_t = \frac{l_t - L_t}{L_t} \cong$$
$$\cong \frac{1}{L_t} \sum_{d=1}^{D} \{\alpha_d [\Delta_d w - \Delta_d (yv')] + \beta_d \Delta_d v\} \tag{10}$$

where

$$\alpha_d = \frac{\Delta_d(z)}{\sqrt{[\Delta_d(z)]^2 + [\Delta_d(y)]^2}} \tag{11}$$

$$\beta_d = \frac{\Delta_d(y)}{\sqrt{[\Delta_d(z)]^2 + [\Delta_d(y)]^2}} \tag{12}$$

and the cable traction is

$$T(\varepsilon_t) = T_0 + A_t \sigma_t(\varepsilon_t) \tag{13}$$

with T_0 the traction in the reference configuration, A_t the cable cross-section area and σ_t the cable stress (σ_t is a nonlinear function of the cable strain).

The balance condition can be obtained from the Virtual Work Theorem

$$\int_0^L \int_A \sigma \, \delta \hat{\varepsilon}_z \, dA \, dz + L_t T \, \delta \hat{\varepsilon}_t = \int_0^L \mathbf{f} \cdot \delta \hat{\mathbf{u}} \, dz$$
$$\forall [\delta \hat{\mathbf{u}}, \delta \hat{\varepsilon}, \delta \hat{\varepsilon}_t] \in U \tag{14}$$

where $U =$ space containing all the displacements and strain fields compatible with internal and external constraints (Dall'Asta & Dezi 1993, 1998).

2.2 Basic assumptions of the proposed simplified method

The proposed method aims at evaluating in a simplified way the flexural strength of reinforced concrete beams with external prestressing. The method is based on the observation that at collapse the distribution of

the strain $\varepsilon_{0u}(z)$ at the assigned reference point and the distribution of the curvature $\theta_u(z)$ along the beam axis do not notably change under given conditions. In particular the shapes of these functions obviously depend on the cable path, on the load distribution and on the static scheme which control the bending moment diagram shape. However they do not remarkably depend on other parameters such as beam geometry (span length, span-to-depth ratio, and slab width), material characteristics, and cable initial traction. Once that approximated shape functions for $\varepsilon_{0u}(z)$ and $\theta_u(z)$ are found in a satisfactory range of approximation, the problem can be transformed from an analytical formulation involving the whole beam into an algebraic formulation related to the strain of the critical section, located at z_{cr}, where failure condition occurs. The failure condition consists in attaining a maximum value of strain (ultimate strain) in the concrete slab or in the steel beam. Denoting with y_F the position of the fiber where the limit strain ε_{Fu} is reached, it is possible to derive the strain distribution in the form

$$\varepsilon = \varepsilon_{Fu} + (y - y_F)\theta_u \tag{15}$$

which is alternative of equation (2) and depends on θ_u only. The strain

$$\varepsilon_{0u} = \varepsilon_F - y_F \theta_u \tag{16}$$

at the reference can consequently be derived.

Aiming at approximating strain by using shape functions, the problem can be reformulated describing the axial strain and the curvature by means of the functions

$$\eta(z) = \theta(z)/\theta_u \tag{17}$$

$$\gamma(z) = \varepsilon_0(z)/\varepsilon_{0u} \tag{18}$$

The rotation and displacements, excluding rigid-body motions which do not influence the strain, can be obtained from the following expressions:

$$v' = -\theta_u \int_0^z \eta(\xi) d\xi = \theta_u h L \tag{19}$$

$$v = -\theta_u \int_0^z \int_0^\zeta \eta(\xi) d\xi d\zeta = \theta_u k L^2 \tag{20}$$

$$w = \varepsilon_{0u} \int_0^z \gamma(\xi) d\xi = \varepsilon_{0u} g L \tag{21}$$

where

$$h = -\frac{1}{L} \int_0^z \eta(\xi) d\xi \tag{22}$$

225

$$k = -\frac{1}{L^2}\int_0^z\int_0^\zeta \eta(\xi)d\xi d\zeta \qquad (23)$$

$$g = \frac{1}{L}\int_0^z \gamma(\xi)d\xi \qquad (24)$$

are non-dimensional quantities and L is the total length of the beam span. Substituting Equations 19-21 in Equation 10, the strain of the cable becomes:

$$\varepsilon_t(\theta_u) \cong \frac{L}{L_t}\left\{(\varepsilon_F - y_F\theta_u)\sum_{d=1}^D \alpha_d\Delta_d(g) + \right.$$
$$\left. + \theta_u\sum_{d=1}^D[\beta_d L\Delta_d(k) - \alpha_d\Delta_d(yh)]\right\} \qquad (25)$$

Once that $h_d = h(z_d)$, $k_d = k(z_d)$ and $g_d = g(z_d)$ have been evaluated for each deviator, the position of all deviators at collapse and related increment of strain in the cable are functions of θ_u only. Consequently the cable traction is given by

$$T(\theta_u) = T_0 + A_t\sigma_t(\varepsilon_t(\theta_u)) = T_0 + \Delta T(\theta_u) \qquad (26)$$

where the traction increment due to the cable deformation with respect to the reference configuration is denoted with ΔT.

Thus the ultimate curvature can be evaluated using the nonlinear equation

$$N_R(\theta_u) + T_0 + \Delta T(\theta_u) = 0 \qquad (27)$$

that can easily be solved with an iterative procedure.

The load multiplier λ_u at collapse which is related to the position z_{cr} of the critical section, can be obtained by solving the following system

$$M_R(\theta_u) = \lambda_u M_{\bar{q}}(z_{cr}) - [T_0 + \Delta T(\theta_u)]e(z_{cr}) \qquad (28)$$

$$\lambda_u\left.\frac{dM_{\bar{q}}(z)}{dz}\right|_{z=z_{cr}} - [T_0 + \Delta T(\theta_u)]\left.\frac{de(z)}{dz}\right|_{z=z_{cr}} = 0 \qquad (29)$$

in which M_q is the bending moment due to the reference load and $e(z)$ is the cable eccentricity at the considered section.

It can be observed that Equation 28 is linear with respect to λ_u, which can be consequently expressed as:

$$\lambda_u = \frac{M_{Ru}(\theta_u)}{M_{\bar{q}}(z_{cr}) - e(z_{cr})[T_0 + \Delta T_u(\theta_u)]} \qquad (30)$$

and substituted in Equation 29, obtaining an equation in the only unknown z_{cr}:

$$M_{Ru}(\theta_u)\left.\frac{dM_{\bar{q}}(z)}{dz}\right|_{z=z_{cr}} +$$
$$- \{M_{\bar{q}}(z_{cr})[T_0 + \Delta T_u(\theta_u)]$$
$$- e(z_{cr})[T_0 + \Delta T_u(\theta_u)]^2\}\left.\frac{de(z)}{dz}\right|_{z=z_{cr}} = 0 \qquad (31)$$

whose complexity is related to the type of loading and to the tendon path.

2.3 Explicit formulas for simply supported beams with one intermediate deviator

A symmetric simply supported beam of length L with draped tendon with one intermediate deviator at mid-span, under uniform load is considered. The tendon eccentricity for the first half of the span is given by:

$$e(z) = 2e_0 z/L \quad \text{for } 0 \leq z \leq L/2 \qquad (32)$$

where e_0 is the eccentricity at mid-span. Consequently the tendon strain given by Equation 25 becomes:

$$\varepsilon_t(\theta_u) =$$
$$= \frac{L}{L_t}\{(\varepsilon_F - y_F\theta_u)[\alpha(g_2 - g_0)] + \theta_u[2L\beta k_1]\} \qquad (33)$$

with

$$\alpha = \alpha_1 = \alpha_2 = \frac{L/2}{\sqrt{\dfrac{L^2}{4} + e_0^2}} \qquad (34)$$

$$\beta = \beta_1 = -\beta_2 = \frac{e_0}{\sqrt{\dfrac{L^2}{4} + e_0^2}} \qquad (35)$$

being $y_0 = y_2 = 0$, $h_1 = 0$ and $h_0 = -h_2$ due to symmetry, $k_0 = k_2 = 0$ due to assigned end supports.

Equation 31 giving the position of the critical section is a second order algebraic equation that can easily be solved by hand calculation:

$$az_{cr}^2 + bz_{cr} + c = 0 \qquad 0 \leq z_{cr} \leq L/2 \qquad (36)$$

with:

$$a = \frac{e_0 q}{L}T_u(\theta_u) \qquad (37)$$

$$b = \frac{4e_0^2}{L^2}[T_u(\theta_u)]^2 - M_R(\theta_u)q - e_0qT_u(\theta_u) \qquad (38)$$

$$c = M_R(\theta_u)\frac{qL}{2} \qquad (39)$$

226

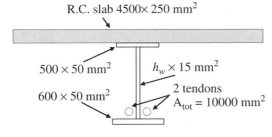

R.C. slab 4500× 250 mm²

500 × 50 mm² $h_w \times 15$ mm²

600 × 50 mm²   2 tendons
                $A_{tot} = 10000$ mm²

Figure 2.   Cross section of the composite beams tested.

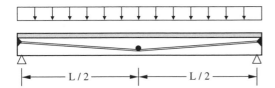

Figure 3.   Tendon path and loading condition of the composite beams tested.

2.4   Determination of the coefficients related to the shape functions

In evaluating the traction force in the cable the values g_d, h_d and k_d only are involved while the complete expressions of the shape functions are not required. The values of the previous quantities can be deduced from the displacements and rotation of deviators obtained by a nonlinear analysis (Dall'Asta & Zona 2005). These values necessarily vary for different static configurations, cables profiles, loading distributions. However, similar values were observed by varying, within the range of practical interest, the beam length L, the depth-length ratio h/L, the slab width and the initial prestressing force. This makes it possible to determine approximated values of g_d, h_d and k_d which can be applied for obtaining sufficiently accurate results for beams with the same cable profile, static constraints and load distribution.

3   APPLICATION EXAMPLE

3.1   Steel-concrete composite beams tested

In this work the attention is limited to simply supported composite beams with draped tendon path defined by one intermediate deviator at mid span and with uniform distributed load. The geometric characteristics of the beam are reported in Figures 2–3.

The steel reinforcement in the concrete slab are $(A_{s,sup} + A_{s,inf})/A_c = 0.5\%$ where A_c is the cross-section area of the slab, and $A_{s,sup} = A_{s,inf}$ are the area of reinforcement in the top and bottom of the slab with

Table 1.   Composite beams tested.

Beam	L (m)	h_w (mm)	h/L
B3020	30	1150	1/20
B3025	30	850	1/25
B3030	30	650	1/30
B4020	40	1650	1/20
B4025	40	1250	1/25
B4030	40	970	1/30
B5020	50	2150	1/20
B5025	50	1650	1/25
B5030	50	1300	1/30

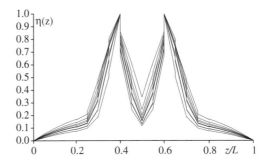

Figure 4.   Shape functions of curvature at collapse.

an assumed cover of 30 mm. A set of beams with length spanning from 30 m to 50 m and span-to-depth ratio varying from 20 to 30 are analyzed in order to consider beams usually used in bridge constructions (Table 1). The initial prestressing force under self-weight only is $T_0 = 11500$ kN for all beams.

3.2   Results from finite element nonlinear analysis

The set of beams in Table 1 was analyzed by using a finite element model (Dall'Asta & Zona 2005) previously introduced and validated by comparisons with experimental tests. The reader is referred to (Dall'Asta & Zona 2005) for every detail on nonlinear constitutive laws and other numerical aspects. The analysis sequence consists of three phases: I) self weight only in the unpropped beam; II) jacking of the external tendons up to the prescribed initial force T_0; III) application of an increasing uniform load up to collapse.

The trends of the shape functions $\eta(z)$ and $\gamma(z)$ at collapse are shown in Figure 4 and Figure 5 respectively for the nine beams considered. It is observed that the shape functions do not remarkably change as a result of the variation of the beam length and of the span-to-dept ratio. This is particularly true for the shape functions of the curvature.

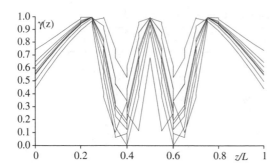

Figure 5. Shape functions of axial strain at collapse.

Table 2. Shape coefficients from nonlinear analysis.

Beam	k_1	h_0	h_2	g_1	g_2
B3020	0.0541	0.1548	−0.1548	−0.1723	−0.3445
B3025	0.0586	0.1684	−0.1684	−0.2299	−0.4598
B3030	0.0628	0.1808	−0.1808	−0.3208	−0.6416
B4020	0.0444	0.1261	−0.1261	−0.1189	−0.2379
B4025	0.0506	0.1446	−0.1446	−0.1526	−0.3052
B4030	0.0549	0.1576	−0.1576	−0.1943	−0.3887
B5020	0.0384	0.1082	−0.1082	−0.0948	−0.1897
B5025	0.0437	0.1241	−0.1241	−0.1169	−0.2338
B5030	0.0495	0.1413	−0.1413	−0.1464	−0.2929

In Table 2 the coefficient h_d, k_d, and g_d ($d = 1,2,3$) obtained from the integration of the shape functions are reported ($h_1 = 0$ and $h_0 = -h_2$ due to symmetry, $k_0 = k_2 = 0$, $g_0 = 0$ due to the assigned restraints). As expected there are not large variations of these coefficients and the fluctuations of their values are regular (the absolute values of h_d, k_d, and g_d decrease with the increasing of the beam length, while they increase with the decreasing of the span-to-depth ratio). These observations agree with the results previously obtained for the case of reinforced concrete beams (Dall'Asta et al. 2007).

3.3 Comparison between finite element nonlinear analysis and simplified method

The proposed simplified method is applied to the nine beams considered by using as shape coefficients for each beam the mean values ($k_1 = 0.0508$, $g_2 = -0.3438$). The results obtained from nonlinear analysis and simplified method are compared in Table 3.

The simplified method gives a good approximation of the results from the nonlinear analysis (the maximum difference in the tendon traction at collapse is 4.82%, the maximum difference in the increment of the tendon traction at collapse is 18.17%).

Table 3. Comparison between nonlinear analysis and simplified method.

Beam	Nonlinear analysis		Simplified method	
	T_u (kN)	ΔT_u (kN)	T_u (kN)	ΔT_u (kN)
B3020	16279.02	4779.02	15737.85	4237.85
B3025	15778.74	4278.74	15018.25	3518.25
B3030	15129.38	3629.38	14470.10	2970.10
B4020	16756.43	5256.43	16742.18	5242.18
B4025	16403.37	4903.37	15993.02	4493.02
B4030	16060.78	4560.78	15320.18	3820.18
B5020	17062.99	5562.99	17569.17	6069.17
B5025	16761.46	5261.46	16751.37	5251.37
B5030	16457.52	4957.52	16120.85	4620.85

4 CONCLUSIONS

A simplified method for the analysis of externally prestressed concrete beams was developed in order to calculate the traction increment in the cable at failure without performing a nonlinear analysis involving the whole beam. The method furnishes a direct relationship between the curvature of the beam section where failure occurs, and the increment of the cable force, by means of approximated shape functions describing the strain at collapse along the beam. The proposed approach was applied to some cases of practical interest where it furnished a good approximation of the results obtained by a nonlinear finite element analysis.

REFERENCES

AASHTO American Association of State Highway and Transportation Officials. 1994. *AASHTO LRFD Bridge Design Specification.* Washington D.C.
ACI American Concrete Institute. 1995. *ACI 318-95 Building code requirements for reinforced concrete.* Detroit, Michigan.
Aparicio, A.C. & Ramos, G. 1996. Flexural strength of externally prestressed concrete bridges. *ACI Structural Journal* 93(5): 512–523.
CEN Comité Européen de Normalization. 1995. *ENV 1992-1-5 Eurocode 2, Design of concrete structures, Part 1.5 General rules – Structures with unbonded and external prestressing tendons.* Brussels.
Conti, E., Tardy, R. & Virlogeux, M. 1993. Friction losses in some externally prestressed bridges in France. *Proceedings of the Workshop on Behaviour of External Prestressing in Structures*, Saint-Rémy-lès-Chevreuse, France.
Dall'Asta, A. & Dezi, L. 1993. Nonlinear analysis of beams prestressed by unbonded cables. *Journal of Engineering Mechanics ASCE* 119(4): 720–732.
Dall'Asta, A. & Dezi, L. 1998. Nonlinear behavior of externally prestressed composite beams: analytical model. *Journal of Structural Engineering ASCE* 124(5): 588–597.
Dall'Asta, A., Ragni, L. & Zona, A. 2007. Simplified method for failure analysis of concrete beams prestressed with

external tendons. *Journal of Structural Engineering ASCE* 133(1): 121–131.

Dall'Asta, A. & Zona, A. 2005. Finite element model for externally prestressed composite beams with deformable connection. *Journal of Structural Engineering ASCE* 131(5): 706–714.

Naaman, A. & Alkhairi, M. 1991a. Stress at ultimate in unbonded post-tensioning tendons: part 1 – evaluation of the state of the art. *ACI Structural Journal* 88(5): 641–651.

Naaman, A. & Alkhairi, M. 1991b. Stress at ultimate in unbonded post-tensioning tendons: part 2 – proposed methodology. *ACI Structural Journal* 88(6): 683–692.

Saadatmanesh, H., Albrecht, P. & Ayyub, B.M. 1989. Guidelines for flexural design of prestressed composite beams. *Journal of Structural Engineering ASCE* 115(11): 2944–2961.

Steel and Composite Structures – Wang & Choi (eds)
© 2007 Taylor & Francis Group, London, ISBN 978-0-415-45141-3

A steel-concrete composite beam model with partial interaction including the shear deformability of the steel component

G. Ranzi

School of Civil Engineering, The University of Sydney, Sydney, Australia

A. Zona

Dipartimento di Progettazione e Costruzione dell'Ambiente, University of Camerino, Ascoli Piceno, Italy

ABSTRACT: This paper presents an analytical model for the analysis of steel-concrete composite beams with partial shear interaction including the shear deformability of the steel component. This model is obtained by coupling an Euler-Bernoulli beam for the reinforced concrete slab to a Timoshenko beam for the steel beam. The composite action is provided by a continuous shear connection which enables relative longitudinal displacements to occur between the two components. The balance conditions are derived using the principle of virtual work and the numerical solution of the problem is obtained by means of the finite element method. Extensive numerical simulations are carried out on realistic three-span continuous composite beams to evaluate the effects of the shear deformability of the steel member on the overall structural response.

1 INTRODUCTION

The use of steel-concrete composite beams has gained popularity in the last century thanks to its ability to well combine the advantages of both steel and concrete. Composite members exhibit enhanced strength and stiffness when compared to the contribution of their components acting separately, and represent a competitive structural solution in many civil engineering applications, such bridges and buildings. In the 40s and 50s of the last century the first studies on composite beam behavior underlined that the relative displacement between the steel beam and the reinforced concrete slab requires to be included in the beam model for an adequate representation of the composite action. One of the earliest papers dealing with this problem (Newmark et al. 1951) proposed a model usually referred to as the Newmark model. This model couples two Euler-Bernoulli beams, i.e. one for the reinforced concrete slab and one for the steel beam, by means of a deformable shear connection distributed at their interface. The shear connection enables relative longitudinal displacements to occur between the two components (partial shear interaction) while preventing their vertical separation. Since then, many researchers have extended the Newmark model to include vertical separation (e.g., Adekola 1968), material nonlinearities (e.g., Arizumi et al. 1981), time dependent behavior of concrete (e.g., Tarantino & Dezi 1992), shear-lag effects (e.g., Dezi et al. 2001),

geometric nonlinearities (e.g., Cas et al. 2004), and out-of-plane bending as well as torsion (Dall'Asta 2001). It is beyond the scope of this paper to provide a review of the state of the art; useful reviews were presented in other works (e.g., Leon & Viest 1996, Spacone & El-Tawil 2004).

The objective of this paper is to propose a further modification of the original Newmark beam model to include the shear deformability of the steel beam. The Newmark model neglects the shear deformation of the concrete slab and of the steel beam, being both components modeled as Euler-Bernoulli beams (i.e., beams with infinite shear stiffness), and this hypothesis might not be appropriate in some cases, e.g., for composite beams with reduced span-to-depth ratio and for wide flange cross sections with thin webs. The simplest beam model that incorporates the effects of shear deformability is the Timoshenko model. The proposed model for composite beams is formulated by coupling an Euler-Bernoulli beam for the reinforced concrete slab with a Timoshenko beam for the steel member. The composite action is provided by a continuous shear connection which, as in the Newmark model, enables longitudinal relative displacements to occur between the two components (partial shear interaction). This model is referred throughout this paper as the Euler-Bernoulli-Timoshenko composite beam model (EB-T model). Such model was preferred to a model coupling two Timoshenko beams for two reasons: (i) the shear deformability of the slab is commonly very small due

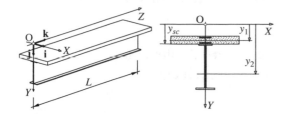

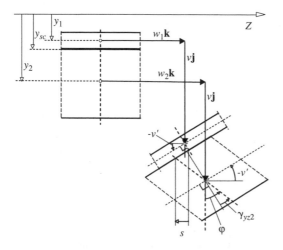

Figure 1. Typical composite beam and cross-section.

Figure 2. Displacement field of the EB-T beam model.

to its flexural slenderness while the shear deformability might not be negligible for the steel beam; (ii) it permits a more practical extension to include material nonlinearities in the analysis as a biaxial constitutive law for the concrete is not required.

2 ANALYTICAL MODEL

2.1 Model assumptions

A prismatic composite beam made of a reinforced concrete slab and a steel beam is considered (Fig. 1).

In its undeformed state, the composite beam occupies the cylindrical region $V = A \times [0,L]$ generated by translating its cross-section A, with regular boundary ∂A, along a rectilinear axis orthogonal to the cross-section and parallel to the Z axis of an ortho-normal reference system $\{O;X,Y,Z\}$; $\mathbf{i}, \mathbf{j}, \mathbf{k}$ are the unit vectors of axis X,Y,Z. The composite cross-section domain is made by the slab, referred to as A_1, and by the steel beam, referred to as A_2. Both A_1 and A_2 are assumed to be symmetric about the YZ plane. Also loads are symmetric with respect to the YZ plane which represents the plane of bending. No torsion and out-of-plane flexure are considered.

The slab component is modelled by the Euler-Bernoulli beam theory, i.e., small displacements and strains, plane sections perpendicular to the beam axis remain plane, rigid and perpendicular to the beam axis after deformation. Perfect bond occurs between reinforcement and concrete. The steel beam component is modelled by the Timoshenko beam theory, i.e., small displacements and strains, plane sections perpendicular to the beam axis remain plane and rigid but not necessarily perpendicular to the beam axis after deformation, being their rotation independent from the slope of the beam. The composite action between the two components is provided by a continuous deformable interface along a rectilinear line Λ at the interface between the two layers, whose domain consists of the points in the YZ plane with $y = y_{sc}$ and $z \in [0,L]$, y_{sc} being defined in Figure 1. The connection is assumed to permit only discontinuities parallel to the beam axis; thus no vertical separation can occur between the two layers.

2.2 Displacement and strain fields

The displacement field of a generic point $P(x,y,z)$ of the composite beam is defined by vector $\mathbf{d}$:

$$\mathbf{d}(y,z;t) = \begin{cases} \mathbf{d}_1(y,z;t) \quad \forall (x,y) \in A_1, \ z \in [0,L] \\ \mathbf{d}_2(y,z;t) \quad \forall (x,y) \in A_2, \ z \in [0,L] \end{cases} \quad (1)$$

with

$$\mathbf{d}_1(y,z;t) = v(z;t)\mathbf{j} + [w_1(z;t) - (y - y_1)v'(z;t)]\mathbf{k} \quad (2)$$

$$\mathbf{d}_2(y,z;t) = v(z;t)\mathbf{j} + [w_2(z;t) + (y - y_2)\varphi(z;t)]\mathbf{k} \quad (3)$$

where $w_1(z;t)$ and $w_2(z;t)$ are the axial displacements of the reference fibres of the two beam components located at y_1 and y_2 respectively (Fig. 2); $v(z;t)$ represents the deflection of both components as no vertical separation can take place; $\varphi(z;t)$ is the rotation of the bottom layer. For generality, the reference fibres of the two components are located at user-defined positions referred to as y_1 and y_2. Translations are positive when directed towards their relevant positive axes, rotations when counter clockwise. Variable t is time or another ordering parameter in the problem considered (e.g., the time from casting of the slab concrete). The prime represents the derivative with respect to z.

The slip between the two components, which represents the discontinuity of axial displacements at their interface, is given by the vector $\mathbf{s}$ as

$$s(z;t) = s(z;t)\mathbf{k} = \mathbf{d}_2(0,y_{sc},z;t) - \mathbf{d}_1(0,y_{sc},z;t) = \\ = [w_2(z;t) - w_1(z;t) - h_2\varphi(z;t) + h_1v'(z;t)]\mathbf{k} \quad (4)$$

232

where $h_1 = y_{sc} - y_1$ and $h_2 = y_2 - y_{sc}$.

Based on the assumed displacement field, the non zero components of the strain field are the axial strain:

$$\varepsilon_z(y,z;t) = \frac{\partial \mathbf{d}}{\partial z} \cdot \mathbf{k} = \begin{cases} \varepsilon_{z_1}(y,z;t) & \forall (x,y) \in A_1, z \in [0,L] \\ \varepsilon_{z_2}(y,z;t) & \forall (x,y) \in A_2, z \in [0,L] \end{cases} \quad (5)$$

with

$$\varepsilon_{z_1}(y,z;t) = w'_1 - (y - y_1)v'' \quad (6)$$

$$\varepsilon_{z_2}(y,z;t) = w'_2 + (y - y_2)\varphi' \quad (7)$$

and the shear deformation

$$\gamma_{yz2}(y,z;t) = \frac{\partial \mathbf{d}_2}{\partial z} \cdot \mathbf{j} + \frac{\partial \mathbf{d}_2}{\partial y} \cdot \mathbf{k} = \\ = v' + \varphi \quad \forall (x,y) \in A_2, z \in [0,L] \quad (8)$$

2.3 Balance conditions

The principle of virtual work for the beam model proposed is:

$$\sum_{\alpha=1}^{2} \int_L \int_{A_\alpha} \sigma_{z\alpha} \hat{\varepsilon}_{z\alpha} dAdz + \int_L \int_{A_2} \tau_{yz2} \hat{\gamma}_{yz2} dAdz + \int_L g_{sc} \hat{s} dz = \\ = \sum_{\alpha=1}^{2} \int_L \int_{A_\alpha} \mathbf{b} \cdot \hat{\mathbf{d}} dA\, dz + \sum_{\alpha=1}^{2} \int_L \int_{\partial A_\alpha} \mathbf{t} \cdot \hat{\mathbf{d}} ds\, dz + \sum_{\alpha=1}^{2} \int_{A_{\alpha 0,L}} \mathbf{t} \cdot \hat{\mathbf{d}} dA$$

$$\forall \hat{\mathbf{d}} \quad (9)$$

in which $\sigma_{z\alpha}$ (axial stress), τ_{yz2} (shear stress) and g_{sc} (interface shear force) represent the active stresses (i.e.. stresses that produce internal work) computed from the relevant strains $\varepsilon_{z\alpha}$, γ_{yz2} and s_{sc} once the constitutive laws are introduced, $\mathbf{b}$ and $\mathbf{t}$ are the body and surface forces respectively, and the third integral on the right hand-side of the equation represents the work done by the surface forces applied at the cross-sections $A_{\alpha 0,L}$ at the beam ends, i.e. $z = 0, L$. All applied loads can vary in the time domain with quasi-static rate (i.e., no dynamical effects are included). Virtual displacements and strains are identified by means of a hat "^" placed above the variables considered. Integrals over the slab cross section A_1 include the contributions of both concrete and reinforcement. The solution of the problem is then sought in the spaces of regular functions fulfilling the kinematic boundary conditions.

Based on the weak formulation specified in Equation 9, the stress resultant entities, which are duals of the kinematic entities derived from the assumed displacement field, are the axial forces

$$N_\alpha = \int_{A_\alpha} \sigma_{z\alpha} dA_\alpha \quad \alpha = 1,2 \quad (10)$$

duals of the axial strains $\varepsilon_1 = w_1'$ and $\varepsilon_2 = w_2'$, the bending moments

$$M_\alpha = \int_{A_\alpha} \sigma_{z\alpha}(y - y_\alpha) dA_\alpha \quad \alpha = 1,2 \quad (11)$$

duals of the curvatures $\theta_1 = -v''$ of the Bernoulli beam and $\theta_2 = \varphi'$ of the Timoshenko beam respectively, the shear force

$$V_2 = \int_{A_2} \tau_{yz2} dA_2 \quad (12)$$

dual of the shear strain, and the interface shear force g_{sc} dual of the interface slip s.

2.4 Constitutive laws

Linear elastic behavior is assumed for the steel beam, shear connection, and reinforcement. The time-dependent behavior of the concrete slab is represented by an integral-type linear viscous-elastic constitutive model (CEB 1984).

3 FINITE ELEMENT FORMULATION

3.1 Displacement-based finite element approach

Due to the complexity of the analytical formulation, numerical solutions for the proposed model are sought by means of a displacement-based finite element formulation. A step-by-step procedure for the time-dependent analysis is adopted (CEB 1984). Three new finite elements are derived based on different shape functions for the displacement field. In the following, it is discussed how undesired locking problems can be avoided with a careful selection of these shape functions.

3.2 Locking problems

Numerical problems may occur in structural modelling when two or more displacement fields are coupled and when the solution is sought in finite dimensional spaces, as those used in finite element approximations (Reddy 2004). The accuracy of the solution depends on some characteristic parameters involved in the coupled terms and, for limit values, further relations between unknowns might develop reducing the dimensions of the solution space. In some cases, the dimension goes to zero and the model completely "locks" or, in general, a stiffer response and spurious strains can be observed when the phenomenon occurs. Examples of these problems include shear locking (Reddy 1997, Yunhua 1998) that may develop in Timoshenko beam elements by varying the shear stiffness, the eccentricity issue (Blaauwendraad 1972, Gupta & Ma 1977, Crisfield 1991) that may affect the Euler–Bernoulli beam when varying the origin of the reference system, and slip locking (Dall'Asta & Zona 2004) that may occur in Newmark composite beams with deformable shear connection. In all of these cases, the generalized strains are functions of various components of the generalized displacements or of their derivatives. For example, in the Timoshenko beam model the shear deformation depends on the first derivative of the deflection and on the rotation, in the Euler-Bernoulli beam model the axial deformation is determined based on derivatives of both axial displacement and deflection, while in the Newmark model the interface slip is calculated from axial displacements and from the first derivative of the deflection. In these cases, locking problems can be avoided when consistent contributions of the generalized displacements or their derivatives, i.e. same order polynomials from each contribution, are provided to the generalized strains (Dall'Asta & Zona 2004, Reddy 2004).

The particularity of the proposed EB-T composite beam model is that the three types of locking problems previously outlined, i.e., eccentricity issue, shear locking and slip locking, may occur simultaneously in the same finite element. Thus the following functions need to be approximated with polynomials of the same order: (i) axial displacement w_1 and first derivative of deflection v' in the Bernoulli component; (ii) axial displacement w_2 and rotation φ in the Timoshenko component; (iii) rotation φ and first derivative of deflection v' in the Timoshenko component; and (iv) axial displacements w_1 and w_2 as well as the two rotations v' and φ.

3.3 Simplest displacement-based finite element

The simplest element which can be derived based on the EB-T model has 10 degrees-of-freedom (DOF) that are the least DOF required for describing the problem under consideration. Its shape functions consist of a cubic function for the deflection and a linear function for the axial displacements of the Euler-Bernoulli component. Since no vertical separation can occur between the two layers, the same cubic function is used to approximate the deflection of the Timoshenko component despite the fact that an isolated Timoshenko beam would require a linear function for the deflection. Conversely the shape functions for the axial displacement and rotation are the simplest possible for an isolated Timoshenko beam (i.e., linear functions). Based on the previous considerations, this simple 10DOF finite element does not fulfill the consistency

conditions between the different displacement fields coupled in the problem; in fact, conditions (i), (iii) and (iv) discussed in the previous sub-paragraph are not satisfied. The use of this element leads to poor and unsatisfactory results (Ranzi & Zona 2007), similar to those observed for 8DOF element of the Newmark model (Dall'Asta & Zona 2004).

3.4 Refined displacement-based finite elements

The simplest finite element fulfilling the consistency conditions of the displacement field is the 13DOF which enhances the order of the approximated polynomials for the axial displacements and rotations of the Timoshenko component to parabolic functions while using the same polynomials of the 10DOF element for the deflection.

In this study a second higher order consistent 21DOF element is preferred for numerical applications to obtain more accurate results. Its shape functions are fifth order polynomials for the deflection, fourth order polynomials for the axial displacements of both components, and fourth order polynomials for the rotation of the Timoshenko component.

4 PARAMETRIC ANALYSIS

4.1 Parametric analysis objectives

An extensive parametric study is proposed to investigate the influence of the shear deformability of the steel beam on the overall response using realistic composite members based on Australian design charts (Rapattoni et al. 1998). This investigation is carried out considering approximately 100 three-span continuous beams (external spans with length equal to 0.8 of the internal one), subjected to a uniformly distributed load. The extreme cases of low ($\alpha L = 1$) and high ($\alpha L = 50$) shear connection stiffness are considered, with the non dimensional parameter αL defined in (Dall'Asta & Zona 2004).

Particular attention is given in this study to identify under which conditions the proposed formulation yields a structural response significantly different from the one obtained using the Newmark model. It is worth observing that the inclusion in this study of high and low stiffness values for the shear connection gives qualitative information on the effect of the nonlinear behavior of the shear connection itself. In fact the shear connection strongly reduces its stiffness even with small values of the interface slip, as happens in the service state conditions when the other materials are still in the linear elastic range. Thus the two cases of high and low shear connection stiffness approximately represent the upper and lower bounds of the expected response in the case of a service state analysis including the nonlinear behavior of the shear connection.

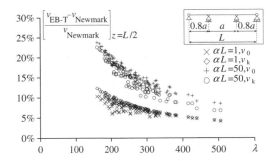

Figure 3. Differences for the mid-span deflection.

4.2 Parametric analysis main results

The results presented are obtained with the proposed 21DOF element for the EB-T beam model and with the 16DOF element (Dall'Asta & Zona 2002) for the Newmark model. The two subscript "EB-T" and "Newmark" are introduced to highlight whether their values are calculated using the EB-T or Newmark models respectively. The following response quantities were monitored: vertical deflection, rotations and curvatures for the two components, slip between slab and steel beam, and axial force and moment resisted by the two components. To clearly evaluate the differences between the two models, only the non-dimensional variations between their results are reported for both short- and long-term values separately; for this purpose, two additional subscripts "0" and "k" are used to specify whether a variable is calculated based on an instantaneous analysis (at time t_0) or a long-term one (at time $t_k = 25550$ days $= 70$ years) respectively. Results are plotted as a function of the dimensionless parameter λ defined as

$$\lambda = \frac{GA_{w2}}{EJ_2} L^2 \qquad (13)$$

where GA_{w2} and EJ_2 are respectively the elastic shear stiffness and bending stiffness of the steel beam and L represents the distance between two points of contraflexure, taken as 0.7 of the internal span for the continuous systems. As the proposed parametric study is concerned with H-shaped sections only, the area resisting shear actions is assumed to include the web and its prolongations into the flanges (i.e. area with width equal to the web thickness and height equal to the steel section depth).

Due to space limitations, only selected results are shown hereafter. More details as well as results relevant to other structural configurations (simply supported beams) can be found in (Ranzi & Zona 2007).

The differences in the values calculated for the mid-span deflection of the internal span leads to large

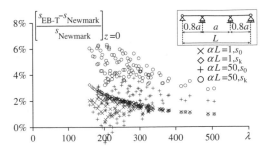

Figure 4. Differences for the beam end slip.

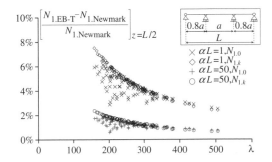

Figure 5. Differences for the slab axial force.

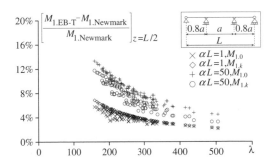

Figure 6. Differences for the slab bending moment.

differences with peaks of 25% and 13% for high and low shear connection stiffness respectively (Fig. 3); long-term effects tend to reduce these differences for high levels of shear connection stiffness while leading to an opposite result for low connection stiffness. Similar comments are applicable for rotations and curvatures not shown here (Ranzi & Zona 2007).

Figure 4 shows the variations in the calculation of the slip at the end supports. These differences are more significant for rigid connections in which case time effects increase the peak differences from 4% to 6%. After inspecting the slip distribution along the member length for the bridge arrangements considered, it is noted that these differences are mostly localised near the end supports and remain in general below 3%.

The internal actions, i.e. axial forces and bending moments resisted by the two components calculated at mid-span of the internal span, are significantly affected by the shear connection stiffness, with peak variations of 8% and 16% for rigid shear connections for the axial forces (Fig. 5) and for the bending moment in the top component (Fig. 6).

Based on the results shown and other results obtained (Ranzi & Zona 2007), the differences between the response numerical prediction of the EB-T and Newmark models are significant and highlight the need to carefully evaluate the influence of shearing deformations on the overall structural response and on the determination of deformations and stress resultants.

5 CONCLUSIONS

A new model for the analysis of steel-concrete composite beams with partial shear interaction including the shear deformability of the steel component is presented. The reinforced concrete slab is described by means of the Euler-Bernoulli beam theory while the steel member is described by means of the Timoshenko beam theory. The two components are coupled together by means of a continuously distributed interface allowing interlayer slip while preserving contact. The analytical formulation is derived by means of the principle of virtual work. Due to the complexity of the governing system of differential equations, the numerical solution is obtained by using the finite element method. Accurate locking-free finite elements are introduced. Although in this study attention is limited to the linear viscous-elastic case, the proposed model can easily be extended to material nonlinear analysis.

Using the proposed finite element formulation, an extensive parametric study, based on approximately 100 three-span continuous steel-concrete composite beams, is illustrated in order to evaluate under which conditions shearing deformations become important and need to be considered in the structural analysis. The parametric analysis focuses on the service state condition and includes the effects of the partial interaction and of the time dependent behavior of concrete. Response variables monitored (deflections, rotations, curvatures, interlayer slip and internal forces) indicate that significant differences might exist between the results calculated by means of the composite beam model ignoring shear deformation of the steel beam and the proposed model.

REFERENCES

Adekola, A.O. 1968. Partial interaction between elastically connected elements of a composite beam. *International Journal of Solids and Structures* 4: 1125–1135.

Arizumi, Y., Hamada, S. & Kajita, T. 1981. Elastic-plastic analysis of composite beams with incomplete interaction by finite element method. *Computers and Structures* 14(5–6): 453–462.

Blaauwendraad, I.J. 1972. Realistic analysis of reinforced concrete framed structures. *Heron* 18(4): 1–31.

Cas, B., Saje, M. & Planinc, I. 2004. Non-linear finite element analysis of composite planar frames with an interlayer slip. *Computers & Structures* 82: 1901–1912.

CEB Comité Euro-International du Béton. 1984. *CEB Design manual on structural effects of time-dependent behaviour of concrete*. Saint-Saphorin, Switzerland.

Crisfield, M.A. 1991. The eccentricity issue in the design of plate and shell elements. *Communications in Applied Numerical Methods* 7: 47–56.

Dall'Asta, A. 2001. Composite beams with weak shear connection. *International Journal of Solids and Structures* 38: 5605–5624.

Dall'Asta, A. & Zona, A. 2002. Non-linear analysis of composite beams by a displacement approach. *Computers and Structures* 80(27–30): 2217–2228.

Dall'Asta, A. & Zona, A. 2004. Slip locking in finite elements of composite beams with deformable shear connection. *Finite Elements in Analysis and Design* 40(13–14): 1907–1930.

Dezi, L., Gara, F., Leoni, G. & Tarantino, A.M. 2001. Time dependent analysis of shear-lag effect in composite beams. *Journal of Engineering Mechanics ASCE* 127(1): 71–79.

Gupta, A.K. & Ma, P.S. 1977. Error in eccentric beam formulation. *International Journal for Numerical Methods in Engineering* 11: 1473–1483.

Leon, R.T. & Viest, I.M. 1996. Theories of incomplete interaction in composite beams. *Proceedings of the Composite Construction in Steel and Concrete*, Irsee Germany.

Newmark, N.M., Siess. C.P. & Viest, I.M. 1951. Tests and analysis of composite beams with incomplete interaction. *Proceedings of the Society of Experimental Stress Analysis* 9(1): 75–92.

Ranzi, G. & Zona, A. (2007). A steel-concrete composite beam model with partial interaction including the shear deformability of the steel component. *Engineering Structures*, in print.

Rapattoni, F., Eastwood, D., Bennett, M. & Cheung, H. 1998. *Composite steel road bridges – concepts and design charts*. BHP Integrated Steel, Australia.

Reddy, J.N. 1997. On locking-free shear deformable beam finite elements. *Computer Methods in Applied Mechanics and Engineering* 149: 113–132.

Reddy, J.N. 2004. *An Introduction to Nonlinear Finite Element Analysis*. Oxford University Press, UK.

Spacone, E. & El-Tawil, S. 2004. Nonlinear analysis of steel-concrete composite structures: state-of-the-art, *Journal of Structural Engineering ASCE* 130(2): 159-168.

Tarantino, A.M. & Dezi, L. 1992. Creep effects in composite beams with flexible shear connectors. *Journal of Structural Engineering ASCE* 118(8): 2063–2081.

Yunhua, L. 1998. Explanation and elimination of shear locking and membrane locking with field consistency approach. *Computer Methods in Applied Mechanics and Engineering* 162: 249–269.

Steel and Composite Structures – Wang & Choi (eds)
© *2007 Taylor & Francis Group, London, ISBN 978-0-415-45141-3*

Numerical modelling of the flexural buckling of axially loaded stainless steel members

N. Lopes & P.M.M. Vila Real
University of Aveiro, Portugal

L. Simões da Silva
University of Coimbra, Portugal

ABSTRACT: The use of structural elements in stainless steel has been increased in the last years, due to its higher corrosion resistance and fire resistance, when compared with the carbon steel. In this work a geometrical and material non linear computer code has been used to determine the buckling resistance of stainless steel columns, considering that the material model behaves according to the hypotheses of Part 1.4 of Eurocode 3. The numerical results are compared with the results obtained using the formulae presented in this Eurocode and a new proposal is made for a simple model that ensures conservative results.

1   INTRODUCTION

For more than three decades, there has been an enormous research effort by the steel industry and the academics, to investigate the behaviour of steel structures, resulting in the development of a number of design rules (CEN, 2005a), which were incorporated within the structural Eurocodes. More recently, research works have been oriented towards stainless steel structures.

The use of stainless steel for structural purposes has been limited to projects with high architectural value, where the innovative character of the adopted solutions is a valorization factor for the structure. The high initial cost of stainless steel, associated with: (i) limited design rules, (ii) reduced number of available sections and (iii) lack of knowledge of the additional benefits of its use as a structural material, are some of the reasons that force the designers to avoid the use of the stainless steel in structures (Estrada, 2005) and (Gardner, 2005). However, a more accurate analysis shows a good performance of the stainless steel when compared with the conventional carbon steel.

The biggest advantage of stainless steel is its higher corrosion resistance. However, its aesthetic appearance, easy maintenance, high durability and reduced life cycle costs are also important characteristics. It is known that the fire resistance of stainless steel is bigger than the carbon steel usually used in construction. Due to the higher ductility provided by stainless steel when compared with carbon steel, it is to be expected that it will have also a better performance in case of earthquakes.

The high corrosion resistance of stainless steel in most of the aggressive environments has been the reason for its use in structures located near the sea, and also in oil-producing, chemical, nuclear, residual waters and food storage facilities. Its corrosion resistance results in a well adherent and transparent layer of oxide rich in chromium that forms itself spontaneously on the surface in the presence of air or any other oxidant environment. In case it is crossed, or has some cut damage, the superficial layer regenerates itself immediately in the presence of oxygen.

Although its use in construction is increasing, it is still necessary to develop the knowledge of its structural behaviour. Stainless steels are known by its non-linear stress-strain relationships with a low proportional stress and an extensive hardening phase. There is not a well defined yield strength, being usually considered for design at room temperature the 0.2% proof strength, $f_y = f_{0.2proof}$.

Table 1 shows some mechanical properties of stainless steel 1.4301 (also known as 304) and compare them to the ones from carbon steel S235, at room temperature (CEN, 2005a, b).

Figure 1 shows the stress-strain relationships of carbon steel S235 and stainless steel 1.4301.

Codes of practice are aimed at providing safe, competitive and, as far as possible, simple procedures for

Table 1. Mechanical properties of the stainless steel 1.4301 and carbon steel S235.

Mechanical properties	Carbon steel S235	Stainless steel 1.4301
Ultimate strength (MPa)	360	520
Yield strength (MPa)	235	210
Ultimate strain	>15%	40%

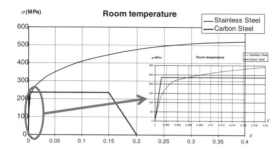

Figure 1. Stress-strain relationships of carbon steel S235 and stainless steel 1.4301.

the design of structures. Drafting and implementing a consistent set of Structural Eurocodes involving a large number of groups of experts is naturally a recursive task where each part must reflect the scientific advances and design options of all other related parts.

The prEN 1993-1-4 "Supplementary rules for stainless steels" gives design rules for stainless steel structural elements at room temperature. In this paper the accuracy and safety of the currently prescribed formulae for columns are evaluated. These evaluations were carried out by performing numerical simulations on equivalent welded stainless steel H-sections.

A comparison between the numerical results obtained with the program SAFIR (Franssen, 2005), which was adapted to model the stainless steel (CEN, 2005b, c), and the buckling curves given by the expressions from Part 1.4 Eurocode 3 for stainless steel structural elements subjected to axial compression has been made. Based on this comparison a proposal for the flexural buckling resistance of columns is made. It will shown that this new proposal is safer then the formulae from the Eurocode 3. Finally, a parametric study between the different proposals and the numerical results will be made.

2 NUMERICAL MODEL

The program SAFIR has been adapted according to the material properties defined in prEN 1993-1-4 (CEN, 2005c) and EN 1993-1-2 (CEN, 2005b), to model the behaviour of stainless steel structures (Lopes et al,

2005). This program, widely used by several investigators, has been validated against analytical solutions, experimental tests and numerical results from other programs (Franssen et al, 1994)., and has been used in several studies that led to proposals for safety evaluation of structural elements, already adopted in Eurocode 3.

In the numerical simulations, geometrical imperfections and residual stresses were considered.

The stainless steel stress-strain relationships, at room temperature and at high temperatures, were introduced in the program SAFIR, so that the material non-linearity could be considered, instead of the perfect elasto-plastic behaviour.

For the numerical simulations a two dimensional beam element has been used, based on the following formulations and hypotheses:

- Displacement type element in a total co-rotational description;
- Prismatic element;
- The displacement of the node line is described by the displacements of the three nodes of the element, two nodes at each end supporting three degrees of freedom, two translations and one rotation, plus one node at the mid-length supporting one degree of freedom, namely the non-linear part of the longitudinal displacement;
- The Bernoulli hypothesis is considered, i.e., in bending, plane sections remain plane and perpendicular to the longitudinal axis and no shear deformation is considered;
- No local buckling is taken into account, which is the reason why only Class 1 and Class 2 sections can be used (CEN, 2005a), allowing for a fully plastic stress distribution on the cross-section;
- The strains are small (von Kármán hypothesis), i.e.

$$\frac{1}{2}\frac{\partial u}{\partial x} \ll 1 \qquad (1)$$

- where u is the longitudinal displacement and x is the longitudinal co-ordinate;
- The angles between the deformed longitudinal axis and the undeformed but translated longitudinal axis are small, i. e., $\sin \varphi \cong \varphi$ and $\cos \varphi \cong 1$ where φ is the angle between the arc and the chord of the translated beam finite element;
- The longitudinal integrations are numerically calculated using Gauss' method;
- The cross-section is discretised by means of triangular or quadrilateral fibres. At every longitudinal point of integration, all variables, such as temperature, strain, stress, etc., are uniform in each fibre;
- The tangent stiffness matrix is evaluated at each iteration during the convergence process (pure Newton-Raphson method);

Figure 2.   Column subjected to axial compression.

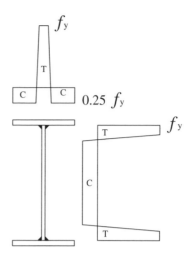

Figure 3.   Residual stresses: C – compression; T – tension.

- Residual stresses are considered by means of initial and constant strains (Franssen, 1989);
- The material behaviour in case of strain unloading is elastic, with the elastic modulus equal to the Young modulus at the origin of the stress-strain curve. In the same cross-section, some fibres that have yielded may therefore exhibit a decreased tangent modulus because they are still on the loading branch, whereas, at the same time, some other fibres behave elastically;
- The collapse criterion of the structure is defined as the instant when the stiffness matrix becomes not positive definite, becoming impossible to establish the equilibrium of the structure.

3   CASE STUDY

A column (see figure 2), subjected to axial compression was studied.

The following welded cross-sections were used: HEA 200 and HEB 280. The stainless steel grades 1.4301 and 1.4401 (also known as 316) for each cross-section were studied. These cross sections are class 2 and 1 for axial compression, therefore the local buckling problems were not considered. Also, it was considered the possibility of buckling around the strong axis (yy) and around the weak axis (zz).

In the numerical simulations, a lateral geometric imperfection given by the following expression was considered:

$$v(x) = \frac{l}{1000} \sin\left(\frac{\pi x}{l}\right) \qquad (2)$$

where l is the length of the beam.

An initial rotation around the beam axis with a maximum value of $l/1000$ radians at mid span was also considered.

The adopted residual stresses follow the typical patterns for carbon steel welded sections, considered constant across the thickness of the web and flanges. The distribution is shown in figure 3, and has the maximum value of f_y (yield strength) (Chen and Lui, 1991), (Gardner and Nethercot, 2004) and (Greiner et al, 2005), considered constant across the thickness of the web and flanges.

4   FLEXURAL BUCKLING SAFETY EVALUATION OF STAINLESS STEEL ELEMENTS

4.1   Formulae prescribed in Eurocode 3

According to prEN 1993-1-4 (CEN, 2005c), the resistance to axial compression of class 1 and class 2 cross section, is determined through:

$$N_{b,Rd} = \chi A f_y \frac{1}{\gamma_{M1}} \qquad (3)$$

Part 1.4 of Eurocode 3 recommends the use of the value 1.1 for the partial safety factor γ_{M1}. However in the study made here it was used a value of 1.0 for the partial safety coefficient.

Depending on the buckling axis, two reduction factors are defined χ_y and χ_z, given by:

$$\chi_y = \frac{1}{\phi_y + \sqrt{\phi_y^2 - \bar{\lambda}_y^2}} \text{ with } \chi_y \leq 1 \qquad (4)$$

Table 2. Values of the factors α and $\bar{\lambda}_0$.

Cross section	α	$\bar{\lambda}_0$
Open welded section (strong axis)	0.49	0.2
Open welded section (weak axis)	0.76	0.2

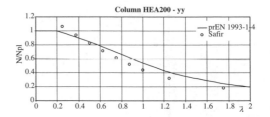

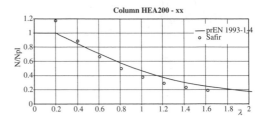

Figure 4. Comparison between the curve from Eurocode 3 and the numerical results: buckling around the yy axis and the zz axis.

$$\chi_z = \frac{1}{\phi_z + \sqrt{\phi_z^2 - \bar{\lambda}_z^2}} \quad \text{with } \chi_z \leq 1 \tag{5}$$

When is possible to occur buckling in the two axes, χ becomes equal to the minimum value between χ_y and χ_z. It should be noted that here the possibility of torsional buckling and torsional-flexural buckling is not taken in to account.

The coefficient ϕ is determined with the expression

$$\phi = \frac{1}{2}\left[1 + \alpha\left(\bar{\lambda} - \bar{\lambda}_0\right) + \bar{\lambda}^2\right] \tag{6}$$

where the factors α (imperfection factor) and $\bar{\lambda}_0$ are defined in table 2.

The adimensional slenderness of Class 1, 2 and 3 sections is determined with

$$\bar{\lambda} = \sqrt{\frac{Af_y}{N_{cr}}} \tag{7}$$

where N_{cr} is the Euler elastic critical force for flexural buckling.

Eurocode 3 states that the buckling effects may be ignored for

$$\bar{\lambda} \leq \bar{\lambda}_0 \quad \text{or} \quad \frac{N_{Ed}}{N_{cr}} \leq \bar{\lambda}_0^2 \tag{8}$$

Considering as reference, for example, a column with the cross section HEA 200, in the stainless steel grade 1.4301, the graphics from figure 4 were obtained. In this figure the numerical results are compared with the results obtained with the formulae from Part 1.4 of Eurocode 3 (expressions (2) (6), without considering the partial safety factor γ_{M1}). In this figure

$$N_{pl} = Af_y \tag{9}$$

or

$$N / N_{pl} = \chi \tag{10}$$

making $\gamma_{M1} = 1.0$ in equation (2).

It can be observed in figure 4 that for values of slenderness higher than 0.5, Eurocode 3 gives values for the reduction factor for flexural buckling higher

than the results obtained with the program SAFIR. Taking the numerical values as true values, it can be stated that the expressions form the Eurocode are not safe. This is reason why the EN version of part 1.4 of the Eurocode 3 suggests the value of 1.1 for the safety factor γ_{M1}.

4.2 Proposal for improving the Eurocode 3 formulae

With the purpose of achieving a better approximation to the numerical results, in compressed columns with high slenderness, a proposal to improve the Eurocode 3 formulae for the buckling of stainless steel elements is made in this section.

This proposal was developed, in order to minimize the changes in Eurocode 3. Thus, as for Part 1.1 of Eurocode 3 for the design of beams in carbon steel with lateral-torsional buckling (CEN, 2005a) and (Boissonnade et al, 2006), it is suggested here the introduction of a factor β with the value of 1.5, in the equations used to determining the reduction factor χ and the coefficient ϕ, according to the following formulae.

$$\chi = \frac{1}{\phi + \sqrt{\phi^2 - \beta\bar{\lambda}^2}} \quad \text{with } \chi \leq 1 \tag{11}$$

$$\phi = \frac{1}{2}\left[1 + \alpha\left(\bar{\lambda} - \bar{\lambda}_0\right) + \beta\bar{\lambda}^2\right] \tag{12}$$

It will be shown in the following point that this proposal results in a great improvement in the behaviour of the buckling curves.

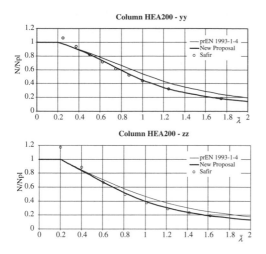

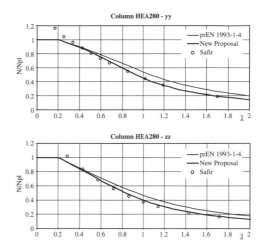

Figure 5. Column with cross section HEA200 in stainless steel grade 1.4301.

Figure 7. Column with cross section HEB280 in stainless steel grade 1.4301.

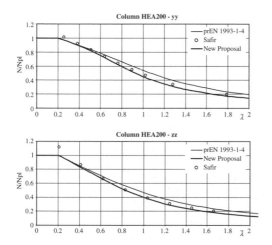

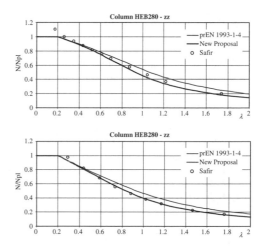

Figure 6. Column with cross section HEA200 in stainless steel grade 1.4401.

Figure 8. Column with cross section HEB280 in stainless steel grade 1.4401.

5 PARAMETRIC STUDY

In Figures 5 to 8, the results for stainless steel columns subjected to flexural buckling, around the strong axis (yy) and around the weak axis (zz) are shown.

A comparison is made between the curve obtained through the formulae from part 1.4 of Eurocode 3, described on the section 4.1 of this paper ("prEN 1993-1-4" in the graphics), the curve based on the proposal made in point 4.2 (noted "New Proposal"), and the numerical results determined with program SAFIR, for two cross sections and two different stainless steel grades.

Figures 5 to the 8 show that the introduction of the factor β, improves the behaviour of the buckling curves obtained with the formulae prescribed in the Part 1.4 of Eurocode 3.

As Part 1.4 of Eurocode 3 recommends the use of a partial safety factor γ_{M1} equal to 1.1, the comparisons between the design resistances are not simple. It is known that the partial safety coefficient γ_{M1} is used to guarantee that the design values present a probability lower than 10^{-3} not to be verified during the useful life of the structure (CEN, 2002), related to the dispersions resulted from the material and geometrical properties and of the calculation model, being given by

$$\gamma_{M1} = \gamma_m \gamma_{Rd} \tag{13}$$

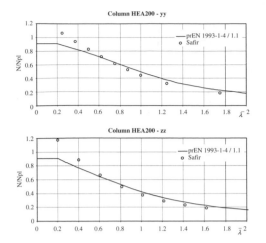

Figure 9. Comparison between the buckling curve defined in equation (13) and the numerical results: for the buckling around the yy axis zz axis.

where γ_m represents the material variability and γ_{Rd} the calculation model variability and geometry variability. The results presented in this work only allow to conclude on the quality and dispersion of the calculation model, since the numerical simulations were made with the nominal properties of the stainless steel. Thus, it is not possible to directly compare the results presented in this paper with the design values given by the equation (3), because the values of γ_m and γ_{Rd} are not given. Any way, Part 1.4 of the Eurocode 3 presents also unsafe values, as it can be verified in figure 9, where it is represented the buckling curve achieved with the partial safety coefficient γ_{M1} equal to 1.1, in the determination of the reduction factor χ.

$$\chi = \frac{1}{1.1} \cdot \frac{1}{\phi + \sqrt{\phi^2 - \bar{\lambda}^2}} \qquad (14)$$

The curve corresponded to this reduction factor is represented in figure 9 as "prEN 1993-1-4/1.1".

6 CONCLUSIONS

The unsafe nature observed in the buckling curves from Part 1.4 of Eurocode 3, justifies the presentation of the proposal made in this paper. This study has considered two types of columns with equivalent HEA200 and HEB280 cross-sections, two stainless steel grades, the 1.4301 and the 1.4401 and the possibility of buckling around the two axis yy and zz for each cross section.

It is worth noting that as the parametric study presented here only considered few cases and that studies with other types of stainless steel (austenitic-ferritics and ferritics) and with other cross sections (class 3 and class 4) have not been made yet, it is expected that future work, based on statistical treatment of the results, may suggest improvements in the proposal presented here in order to contemplate the influence of the steel grade as well as the influence of the cross section in the flexural buckling of compressed elements.

REFERENCES

Boissonnade N.; Greiner, R.; Jaspart, J.P; Lindner J. 2006. Rules for member stability in EN 1993-1-1. Background documentation and design guidelines, ECCS.
CEN European Committee for Standardisation 2005a. EN 1993-1-1, Eurocode 3, Design of Steel Structures – Part 1-1. General rules and rules for buildings. Brussels, Belgium.
CEN European Committee for Standardisation 2005b. EN 1993-1-2 Eurocode 3: Design of Steel Structures - Part 1–2: General rules – Structural fire design. Brussels, Belgium.
CEN European Committee for Standardisation 2005c. prEN 1993-1-4, Eurocode 3, Design of Steel Structures – Part 1-4. General rules – Supplementary Rules for Stainless Steels. Brussels, Belgium.
CEN European Committee for Standardisation 2002. EN 1990,– Eurocode, Basis of structural design, Brussels, Belgium.
Chen W. F. and Lui E. M. 1991. Stability design of steel frames, CRC Press.
Estrada, I. 2005. Shear Design of Stainless Plate Girders. PhD Thesis, Universitat Politècnica de catalunya, Barcelona.
Franssen, J.-M., Schleich, J.-B., Cajot, L.-G., Talamona, D., Zhao, B., Twilt, L. and Both, K., 1994. A comparison between five structural fire codes applied to steel elements, Fourth International Symposium on Fire Safety Science, 1125–1136, Ottawa, Canada.
Franssen, J.-M. 2005. SAFIR. A Thermal/Structural Program Modelling Structures under Fire. Engineering Journal, A.I.S.C., Vol. 42, No. 3, pp. 143–158.
Franssen, J.-M. 1989. Modelling of the residual stresses influence in the behaviour of hot-rolled profiles under fire conditions, (in French), Construction Métallique, Vol. 3, pp. 35–42.
Gardner, L. 2005. The use of stainless steel in structures. Prog. Struct. Engng Mater.
Gardner, L., Nethercot, D. A. 2004. Numerical Modeling of Stainless Steel Structural Components - A consistent Approach. Journal of Constructional Engineering, ASCE, pp. 1586–1601.
Greiner, R.; Hörmaier, I.; Ofner, R.; Kettler, M., 2005. Buckling behaviour of stainless steel members under bending. ECCS Technical Committee 8 – Stability.
Lopes, N., Vila Real, P.M.M., Piloto, P., Mesquita, L. e Simões da Silva, L., 2005. Numerical Modelling of the Lateral-torsional Buckling of Stainless Steel I-beams under High Temperatures (in portuguese). Congreso de Métodos Numéricos en Ingeniería, Granada, Spain.

Steel and Composite Structures – Wang & Choi (eds)
© *2007 Taylor & Francis Group, London, ISBN 978-0-415-45141-3*

Weak axis buckling of steel-concrete composite columns under strong axis bending

J.M. Rotter

Institute for Infrastructure and Environment, University of Edinburgh, Edinburgh, UK

ABSTRACT: Steel concrete composite columns in the form of encased rolled steel sections are normally assumed to susceptible only to in-plane buckling, to in-plane failure under combined bending and axial load, or to cross-section failure under biaxial bending and axial load. These columns have a high torsional stiffness, so failure by lateral-torsional buckling is eliminated. This paper discusses the failure mode of weak axis buckling failure when a composite column is subjected to axial load and major axis bending alone. This failure mode does not appear to have been studied before, so this paper sets out the mechanics behind this failure mode and presents numerical predictions of column buckling in this mode. The paper concludes that most cases of lateral buckling under major axis bending can be treated by identifying a specific slenderness of column at which lateral buckling overtakes in-plane strength failure as the critical failure mode. The study is equally applicable to reinforced concrete columns, which are also susceptible to this mode of failure.

1 INTRODUCTION

The failure modes of a steel-concrete composite column are commonly assumed to in-plane buckling, in-plane failure under combined bending and axial load, or cross-section failure under biaxial bending and axial load. These columns have a high torsional stiffness, so failure by lateral-torsional buckling is eliminated.

When a slender column made of material that displays a non-linear stress-strain curve in compression is subjected to axial compression, the stiffness of the cross-section progressively declines. Under pure axial compression, the buckling load of a perfect column reduces from the Euler load to the tangent modulus load, as described by Considère (1889), Engesser (1889) and Shanley (1947). A discussion of the well-known controversy concerning the tangent and double modulus theories is beyond the scope of this paper but will be familiar to most readers.

The tangent modulus load provides a strong indicator of the true strength of most columns, and has been widely used even for steel columns by recognising the role of residual stresses in reducing the cross-section stiffness even when the tangent modulus of the material is not progressively reducing (ideal elastic-plastic material). In steel columns, the residual stresses cannot play a very strong role, so there has naturally been much debate about the role of geometric imperfections, leading to Ayrton & Perry (1886) descriptions of buckling strength. These

imperfections play a much lesser role when the stress-strain curve itself is significantly nonlinear, as is the case for concrete.

Most applications of tangent modulus theory have discussed buckling in terms of the stress state in the cross-section, paying less attention to the critical factor of the integral of the tangent moduli leading to a declining value in the tangent flexural rigidity. It is the reduced flexural rigidity which leads to a reduced buckling strength, and this flexural rigidity is not only reduced by axial compression (where the stress state is uniform so the calculation is relatively trivial) but also under combined bending and axial compression, where the tangent modulus of the material varies across the cross-section according to the stress state at each point.

Thus, buckling of a pin-ended column of length L occurs about the weak Y axis when the axial load N is given by

$$N_{cr} = \pi^2 \frac{EI_{TY}}{L^2} \qquad (1)$$

where EI_{TY} is the tangent flexural rigidity, reduced by both axial force N and strong axis bending moment M_X, both of which are inducing inelastic stress states. If the strong axis bending moment M_X is relatively uniform throughout the column length (small P-Δ effects), a simple first estimate of the lateral buckling strength can be made.

This paper describes and analysis of this weak axis buckling failure mode when a pin-ended steel-concrete composite column is subjected to axial load and uniform major axis bending alone. This failure mode does not appear to have been studied before. The paper describes a study which is set out in more detail in Rotter (1977).

2 CROSS-SECTION ANALYSIS

A composite steel-concrete column cross-section is shown in Figure 1, subject to axial load N and major axis bending moment M_X. The steel stress-strain curve is characterised by an ideal elastic-plastic behaviour with strain-hardening and the concrete by the CEB (1970) stress-strain curve (Fig. 2). Residual stresses were assumed in the classic simple pattern (Galambos & Ketter, 1959) with a magnitude of $0.3\sigma_y$ at the flange tips, together with zero tensile strength and no shrinkage in the concrete.

Using the rapid exact inelastic cross-section analysis based on Green's theorem (Rotter, 1985), the precise values of the axial force N, biaxial bending moments M_X and M_Y, and tangent stiffnesses of the cross section may be directly evaluated for any given strain state $(\varepsilon_a, \phi_X, \phi_Y)$ to yield

$$N = \iint\sigma \mathrm{d}x\mathrm{d}y \quad M_X = \iint\sigma y\mathrm{d}x\mathrm{d}y \quad M_Y = \iint\sigma x\mathrm{d}x\mathrm{d}y$$

$$T_{11} = \iint E_T \mathrm{d}x\mathrm{d}y \quad T_{22} = \iint E_T y^2\mathrm{d}x\mathrm{d}y \quad T_{33} = \iint E_T x^2\mathrm{d}x\mathrm{d}y$$

$$T_{12} = T_{21} = \iint E_T y\mathrm{d}x\mathrm{d}y \quad T_{13} = T_{31} = \iint E_T x\mathrm{d}x\mathrm{d}y$$

$$T_{23} = T_{32} = \iint E_T xy\mathrm{d}x\mathrm{d}y \tag{2}$$

where incremental changes in the stress resultants $\mathrm{d}N$, $\mathrm{d}M_X$ and $\mathrm{d}M_Y$ are related to incremental changes in the axial strain ε_a and curvature about each axis ϕ_X and ϕ_Y by

$$\begin{pmatrix} \mathrm{d}N \\ \mathrm{d}M_X \\ \mathrm{d}M_Y \end{pmatrix} = \begin{bmatrix} T_{11} & T_{12} & T_{13} \\ T_{21} & T_{22} & T_{23} \\ T_{31} & T_{32} & T_{33} \end{bmatrix} \begin{pmatrix} \mathrm{d}\varepsilon_a \\ \mathrm{d}\phi_X \\ \mathrm{d}\phi_Y \end{pmatrix} \tag{3}$$

Inverting this matrix leads to

$$\begin{pmatrix} \mathrm{d}\varepsilon_a \\ \mathrm{d}\phi_X \\ \mathrm{d}\phi_Y \end{pmatrix} = \begin{bmatrix} U_{11} & U_{12} & U_{13} \\ U_{21} & U_{22} & U_{23} \\ U_{31} & U_{32} & U_{33} \end{bmatrix} \begin{pmatrix} \mathrm{d}N \\ \mathrm{d}M_X \\ \mathrm{d}M_Y \end{pmatrix} \tag{4}$$

and applying the condition that the axial load N is constant during buckling ($\mathrm{d}N = 0$), leads to

$$\begin{pmatrix} \mathrm{d}\phi_X \\ \mathrm{d}\phi_Y \end{pmatrix} = \begin{bmatrix} U_{22} & U_{23} \\ U_{32} & U_{33} \end{bmatrix} \begin{pmatrix} \mathrm{d}M_X \\ \mathrm{d}M_Y \end{pmatrix} \tag{5}$$

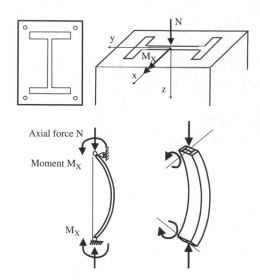

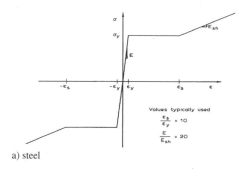

Figure 1. Composite column under axial load and uniform strong axis bending.

Figure 2. Adopted stress strain curves for steel and concrete.

a) steel

b) concrete (CEB, 1970)

which may then be inverted to give the tangent flexural rigidities at constant axial load N

$$\begin{pmatrix} \mathrm{d}M_X \\ \mathrm{d}M_Y \end{pmatrix} = \begin{bmatrix} EI_{TX} & EI_{TXY} \\ EI_{TYX} & EI_{TY} \end{bmatrix} \begin{pmatrix} \mathrm{d}\phi_X \\ \mathrm{d}\phi_Y \end{pmatrix} \tag{6}$$

The tangent flexural rigidities EI_{TX} and EI_{TY} control the stability of the column for buckling in each plane, so studies of these properties under different stress resultants are of interest.

The description in the remainder of this paper must be conducted in the context of an example problem, but the results are made dimensionless to illustrate general behaviour that may be expected in any similar cross-section. The chosen cross-section is a $254 \times 146 \times 37$ Universal Beam with yield stress $\sigma_Y = 250$ MPa encased in 360×248 mm of concrete with maximum stress (Fig. 2b) of $\sigma_c = 27.6$ MPa.

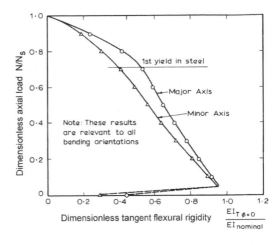

Figure 3. Loss of tangent flexural rigidity at small curvature caused by axial loads.

3 TANGENT FLEXURAL RIGIDITY REDUCTION WITH AXIAL LOAD AND BENDING MOMENT

As the axial load N on a composite column increases, the tangent flexural rigidity for small curvatures $EI_{T\phi} = 0$ about each principal axis declines. This effect leads to a classical tangent modulus theory buckling load about each axis.

The reducing flexural rigidity $EI_{T\phi} = 0$ is compared in Figure 3 with the traditional reference value $EI_{nominal}$ for each axis of bending, which is evaluated assuming no cracking and the initial tangent modulus for each material as

$$EI_{Xnominal} = \iint E_0 y^2 dx dy$$

$$EI_{Ynominal} = \iint E_0 x^2 dx dy \qquad (7)$$

The loss of tangent flexural rigidity under axial load alone is progressive and is shown in Figure 3 at different levels of axial load N relative to the squash load N_S

$$N_S = A_s \sigma_Y + A_c \sigma_c \qquad (8)$$

where A_s is the total steel area and A_c the total concrete area (not gross).

However, at any fixed axial load, if the strong axis bending moment is increased, there is also a progressive loss of tangent flexural rigidity about both the strong and weak axes, and if the weak axis tangent value falls to the point where Equation 1 is satisfied, then the column will buckle about the weak axis, despite the fact that it is being bent about the strong axis.

A progressive loss of tangent flexural rigidity about each axis, relative to the value for small curvatures shown in Figure 3, is caused by increasing strong axis bending moments M_X. The bending moment M_X is here made dimensionless by using the ultimate moment M_{UX} of the cross-section at the axial load N.

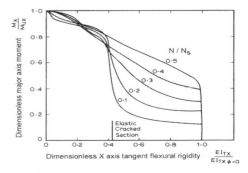

a) Strong axis tangent EI

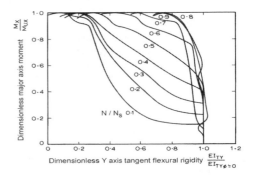

b) Weak axis tangent EI

Figure 4. Loss of tangent flexural rigidity caused by strong axis bending at different axial loads.

At low axial loads, concrete cracking intercedes rapidly to reduce the strong axis tangent flexural rigidity (Fig. 4a) towards the elastic cracked section value, with the moment at first cracking clearly closely related to the level of axial load on the section. From approximately half the ultimate moment M_{UX}, the tangent flexural rigidity falls rapidly towards zero at ultimate. This kind of information has long been known.

By contrast, the effect on the weak axis tangent flexural rigidity (Fig. 4b) is less well understood: the loss of concrete section with cracking has a dramatic effect, rapidly reducing what was already a smaller flexural rigidity than the strong axis value. However, the weak axis value is still finite as the ultimate moment is reached, indicating that the column will indeed fail in strong axis bending if it reaches this moment!

4 BUCKLING ABOUT THE WEAK AXIS UNDER STRONG AXIS BENDING

Consider an isolated column bent about its strong axis and pinned at its ends for weak axis bending. If the major axis moments are small and the column is almost axially loaded, minor axis flexural buckling will always be the failure mode. If the column is stocky and subject to large major axis moments, major axis failures can always be induced. This latter condition is always possible because the minor axis tangent flexural rigidity is positive when the major axis moment M_X reaches its ultimate value M_{UX} (Fig. 4b) when $EI_{TX} = 0$.

Considering the loading path of increasing major axis moments at a fixed axial load N, the tangent flexural rigidities both decline as the major axis moment increases. Although initially $EI_{TY} < EI_{TX}$, by the time the ultimate moment is reached, $EI_{TY} > EI_{TX}$, so between these two extremes the condition $EI_{TY} = EI_{TX}$ is satisfied. At this point, the column is equally ready to buckle about both axes. This point therefore provides a criterion by which out-of-plane minor axis buckling may be distinguished from in-plane major axis failure. The value $EI_{TY} = EI_{TX}$ is here termed the intersection value of tangent flexural rigidity EI_{TI}: its value depends only on the axial load present. Values of EI_{TI} at different axial loads are shown in Figure 5, where error bars are shown because the cross-over is rather ill-conditioned.

From zero axial load to balanced failure ($N/N_s = 0.28$), EI_{TI} falls (Fig. 5), but it rises steadily thereafter until it approaches the falling tangent flexural rigidity for small moments $EI_{TY\phi} = 0$ (Fig. 3), at which point minor axis buckling dominates anyway ($N/N_s = 0.74$). At axial loads below $N/N_s = 0.74$, minor axis buckling can be induced by a combination of axial load and major axis moments, and the intersection value of tangent flexural rigidity may be approximately and conservatively estimated as

$$\frac{EI_{TY}}{EI_{Y\text{nominal}}} = 0.41\left(\frac{N}{N_S}\right) \tag{9}$$

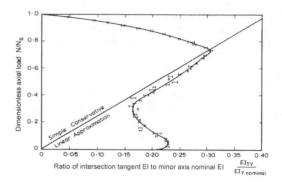

Figure 5. Intersection value of tangent flexural rigidity EI_{TI} relative to weak axis nominal $EI_{Y\text{nominal}}$ at different axial loads.

which may be combined with Equation 1 to give

$$L_{cr} = \pi\sqrt{0.41\frac{EI_{Y\text{nominal}}}{N_S}} \tag{10}$$

Columns longer than this all buckle about the weak axis before reaching the major axis ultimate moment, whilst columns shorter than this will attain the major axis ultimate moment.

The nominal minor axis radius of gyration of the composite column is

$$r_Y = \sqrt{\frac{E_S I_{YS} + E_C I_{YC}}{E_S A_S + E_C A_C}} = \sqrt{\frac{EI_{Y\text{nominal}}}{\Sigma(EA)}} \tag{11}$$

which transforms Equation 10 into

$$\left(\frac{L}{r_Y}\right)_{cr} = 2.0\sqrt{\frac{EA}{N_S}} \tag{12}$$

which is a particularly simple criterion to apply.

In a fuller exposition, this analysis has been verified using complete nonlinear analyses of columns under biaxial bending (Rotter, 1977).

5 CONCLUSIONS

This mode of failure in both steel-concrete composite and reinforced concrete columns does not appear to have been studied before, so this paper has briefly outlined the mechanics behind this failure mode and presented numerical predictions of column buckling in this mode. The conclusion was drawn that most cases of lateral buckling under major axis bending can be treated by identifying, for a given cross-section, a specific length of column at which lateral buckling overtakes in-plane strength failure as the critical failure mode. This analysis is equally applicable to reinforced concrete columns, which are also susceptible to this mode of failure.

ACKNOWLEDGEMENTS

This paper was written in Hong Kong whilst the author was Royal Society Kan Tong Po Visiting Professor. The author most grateful to the Royal Society, the Kan Tong Po fund and the Hong Kong Polytechnic University for their generous support.

REFERENCES

Ayrton, W.E. & Perry, J. 1886. On struts *The Engineer* 62 (10): 464.

CEB 1970. International Recommendations for Design and Construction of Concrete Structures, *Comité European de Beton*, Paris, FIP.

Considère, A. 1889. Resistance des pièces comprimées, *Congrès Internationale de Procédés de Construction, Exposition Universale Internationale de 1889*, Paris, 3: 371.

Engesser, F. 1889. Uber die Knickfestigkeit gerader Stäbe, *Zeitschrift fur Architektur und Ingenieurwesen*, 35: 455.

Galambos, T.V. & Ketter, R.L. 1959. Columns under combined bending and thrust, *Journal of the Engineering Mechanics Division, American Society of Civil Engineers*, 85 EM2 1–30.

Shanley, F.R. 1947. Inelastic column theory. *Journal of Aerospace Science*, 14 5: 261–267.

Rotter, J.M. 1977. *The Behaviour of Continuous Composite Columns*, PhD Thesis, Univ. Sydney, Australia.

Rotter, J.M. 1985. Rapid Inelastic Biaxial Bending Analysis, *Journal of the Structural Division, American Society of Civil Engineers*, 111 ST12: 2659–2674.

Steel and Composite Structures – Wang & Choi (eds)
© *2007 Taylor & Francis Group, London, ISBN 978-0-415-45141-3*

Behavior of slender steel reinforced concrete composite columns

G.T. Zhao
School of Architecture & Civil Engineering, Inner Mongolia University of Science & Technology, China

M.X. Zhang
Department of Civil Engineering, Shanghai University, China

ABSTRACT: An experimental study of the behavior of steel reinforced concrete columns and results of a nonlinear numerical analysis are presented. Eight slender steel reinforced concrete composite columns with rectangular section were tested under axial and eccentric loading conditions. Effects of various geometric and material parameters such as concrete strength, slenderness of columns and eccentricity of the applied axial load were studied. The load bearing capacity is reduced with increased slenderness ratio and eccentricity. Significant gains in load capacity are obtained with increased concrete strength for column subjected to axial load, but the capacity is not strongly influenced by the strength of concrete for column subjected to eccentric load.

1 INTRODUCTION

In steel reinforced concrete (SRC) structures , the introduction of shaped steel has made it possible to design columns more slender. However, although they can reduce column size, the columns become less ductile due to the brittleness of high-strength concrete, another the capacity of slender column is strongly influenced by the second order displacement. Therefore, the structural performance of slender steel reinforced concrete composite columns has recently become a major concern for design engineers.

Many aspects, such as effect of slenderness, amount of shaped steel, ductility, compressive axial force ratio, and eccentricity of the applied load, have to be investigated in order to understand the structural behavior of the columns completely. Some researchers, for example Morino S., Matsui C. and Watanabe H. (1984), Mirza S.A. and Skrabek B.W. (1989), Lakshmi B. and Shanmugam N.E. (2000), have studied short steel reinforced concrete composite columns of normal and high-strength concrete subject to axial load. Only a few have studied full-scale composite columns under eccentrically applied axial loading, for example Mirza S.A. and Tikka T.K. (1999), Mirza S.A. and Lacroix, E.A. (2003). However, the behavior of slender steel reinforced concrete composite columns is not yet fully understood. ACI318-95 and Chinese standard JGJ138-2001 (Technical specification for steel reinforced concrete composite structures) permit a moment magnifier approach for experimental investigation. These material properties were incorporated into a model in which the material model for concrete

was based on design of slender composite columns. This approach is strongly influenced by the effective flexural stiffness of the column that varies due to cracking, creep, nonlinearity of the concrete stress-strain curve, slenderness of column, and eccentricity of the applied axial load. This is expected because the ACI and JGJ138 equations were developed for reinforced concrete columns subjected to high axial loads but were modified, without any further investigation, for use in steel reinforced concrete column designs. So this paper describes experimental and theoretical investigations on the behavior of 8 slender steel reinforced concrete composite columns under axial and eccentric loading conditions. Details of the experimental investigation including description of the test columns, testing arrangements, failure modes and mechanism, strain characteristics, load-deformation responses and effects of various geometric and material parameters are presented. In this study, the three parameters varied were the concrete strength, slenderness of columns and eccentricity of the applied axial load. Then this paper presents a numerical method for the analysis of pin–ended slender columns, producing axial force or axial force combined with symmetrical single-curvature bending, this method is applicable for determining the material failure load or buckling failure load of a slender steel reinforced concrete composite column. In the method, both material and geometric nonlinearties are taken into account. In addition, the mechanical properties, such as the compressive concrete and steel strength, the modulus of elasticity, were measured In the nonlinear fracture mechanics. This model was, in turn, used in a nonlinear numerical

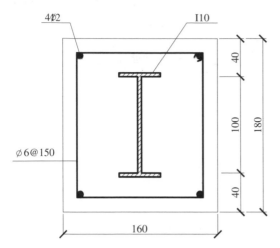

4Φ2   I10

40

100   180

Φ6@150

40

160

Figure 1. Geometry and details of configurations.

Table 1. Details of test columns.

No.	Section dimension h × b (mm)	Concrete strength f_{cu}(MPa)	Eccentricity (mm)	Length (mm)	Ratio of length to width
A1	180 × 160	65.6	0	2800	18
A2	180 × 160	59.8	0	2800	18
A3	180 × 160	55.7	0	3500	22
A4	180 × 160	50.7	0	3500	22
A5	180 × 160	53.8	0	4100	26
A6	180 × 160	67.0	0	4100	26
E1	180 × 160	43.3	30	3200	18
E2	180 × 160	46.6	40	3200	18

Table 2. Mechanical properties of structural steel and reinforcement bar.

Type	f_y(MPa)	f_u(MPa)	E(MPa)
Structural steel	379	507	2.058×10^5
Reinforcement bar	358	471	2.24×10^5

program in order to predict the responses of the slender concrete columns. Observations of the failure mechanisms during the tests and the results of the analysis, as well as some reasons for the failure of the columns under axial and eccentric compressive loading, are presented. The predicted failure loads are closer to the experimental values.

2 EXPERIMENTAL PROGRAM

2.1 Geometry and configuration

The objectives of the studies were to research the structural behavior of long, slender steel reinforced concrete composite columns subjected axial and eccentric axial loading. The test series consisted of 8 slender columns, with compressive cube(150 mm) strength of concrete from 40.3 Mpa to 67.0 Mpa,and shaped steel I10 encased in concrete. The lengths of the columns were 2.8 m, 3.2 m, 3.5 m, or 4.1 m, with rectangular cross sections as shown in Fig 1. The thickness of the concrete cover, measured to the outer edge of the stirrup, was 15 mm, and measured to the outer edge of the shaped steel flange, was 40 mm, for all of the columns. The stirrup spacing was 150 mm; its diameter was 6 mm. The longitudinal reinforcing bar s diameter was 12 mm. The length-to-width ratios, defined as the ratio of the column length, L, to the cross-section dimension, b (applied axial load), or h (subjected eccentric axial loading), were 18, 22 or 26. The eccentricity of applied loading was from 0 to 60 mm.The parameters varied in the tests reported here were the concrete strength, eccentricity, and the slenderness of the columns. Table 1 shows the parameters of the test columns.

2.2 Material properties

The concrete mixes, designed with target compressive cube (150 mm) strength of 40 Mpa, 50 Mpa or 60 Mpa, were produced at the structural engineering laboratory at Inner Mongolia University of Science and Technology. Superplasticizer was used in the concrete mixes to obtain high strength and work ability. The strengths of hardened concrete at 28 days are given in Table 1. The column specimens were cast horizontally in steel forms. The concrete was thoroughly vibrated by means of an internal vibrator. The columns were remolded after approximately seven days and cured under laboratory conditions until tested. The mechanical properties of the shaped steel and longitudinal reinforcement bar are presented in Table 2.

2.3 Test setup

The columns were hinged at the ends and were applied with an axial or an initial eccentric load at both ends. A curved plate of steel and a steel circular bar formed the hinge. The buckling length of the simply supported column is the distance between the bearings at each support. All of the tests were carried out in a vertical hydraulic long column-testing machine with a capacity of 5000 KN. The load, which was determined by measurements from an oil pressure gauge calibrated, was increased at a constant rate without interruption. When the load approached the calculated maximum load, the oil pressure gauge was used to indicate how the deformation should be increased in order to capture the post peak curve.

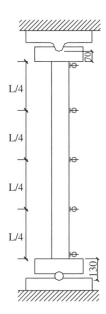

Figure 2. Test setting.

The deflection in the bending direction was measured by displacement gauge at five locations in order to determine the deflected shape of the column (Fig. 2). Another dial gauge was used to check for possible biaxial bending in the perpendicular direction. The vertical displacement of the lower movable plate of the column-testing machine was measured in relation to the laboratory floor by a displacement transducer. For each specimen, vertical 60-mm-long strain gauges were glued onto the concrete at each side of the column at mid-height, 6-mm-long strain gauges were glued onto reinforcement bars and each flange or web of the shaped steel at mid-height. To ensure that the failure would occur in the instrumented region of the columns, the ends of the test specimens were further confined with stirrups spaced apart 50 mm.

3 TEST RESULTS

All columns failed due to crushing of the concrete cover at the mid-height for test specimens subjected to the axial load, spalling of concrete cover and tensile cracks were not observed before reaching the maximum load. When reaching the maximum load, deflection at the mid-height was increased rapidly, spalling of concrete cover, few tensile cracks were observed. The columns exhibited a rather sudden and explosive type of failure, especially the columns with a length-to-width ratio of 26. For the columns subjected to the eccentric load, the tensile cracks were observed about reaching 90% the maximum load, the failure is due to the fact that the compressive concrete spalled at the mid-height. The column failure location varied

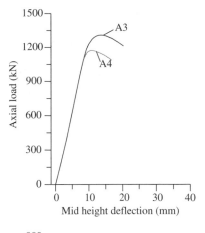

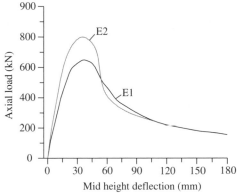

Figure 3. The measured load versus mid-height deflection relation.

from mid-height to an extreme of 500 mm below or up mid height. Further more, a characteristic feature of the failure surface was that the tension face exhibited a horizontal bending crack or cracks across the thickness of concrete cover. On the compressed side, a concrete section, the width of the side and approximately 2.5 time this in length, the vertical cracks were observed at maximum load. The loss of the protecting cover combined with a high load level finally led to buckling of the flange of shaped steel at the end of the test.

In Fig. 3, the measured load versus mid-height deflection relations are shown. The ascending load branches, the mid-height deflection of the columns was a little, but it was increased with the increasing of load and it was almost linear, before reaching 90% of maximum load. About reaching 90% of maximum load, the mid-height deflection value become to go up rapidly, the mid-height deflection- load curve slope was decreased. That is to say, the stiffness of column became reduced. Compared with the measured load

251

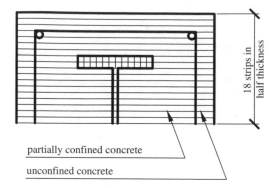

Figure 4. Discretization of composite one-half cross section for computing strength.

Figure 5. The modeling of a composite column.

versus mid-height deflection relation curves, the influence of the column's physical defect is more strongly; the buckling capacity is decreased increasing the slenderness. As can be observed from the Figure 3, the mid-height deflection is approximately the same at the measured maximum load for the same cross section length-to-width ratio, and initial defect.

4 NUMERICAL ANALYSIS OF THE COMPOSITE COLUMNS

For computations of strength, the following assumptions were used: (1) the strain in the cross section was proportional to the distance from the neutral axis, (2) there was no slip between the structural steel or reinforcing bars and the surrounding concrete, (3) the concrete and steel stresses were calculated as functions of the strains, (4) the effects of residual stresses in structural steel section were included, (5) the concrete confinement provided by lateral ties and structural steel section flanges was considered. The composite steel-concrete cross section was assumed to consist of four different materials, each represented by a different stress-strain curve. These materials include the unconfined concrete outside the transverse tie reinforcement, the partially confined concrete with the transverse ties, the longitudinal reinforcement, and the structural steel section as indicated in Fig. 4.

The modeling of a composite steel-concrete column subjected to symmetrical single curvature bending is illustrated in Fig. 5. For the purpose of analysis, the symmetry about the mid-length permitted the use of an equivalent column that was one-half the length of the original column. A column was loaded by introducing a small-applied axial load and a small bending moment at the top node, reflecting the end eccentricity used in the physical test of the column specimen. The applied axial load and bending moment were then increased

in increments of constant proportions, using a second-order analysis procedure, until failure occurred. For the analysis of columns subjected to pure axial load, an imperfection was added to the initially straight element model in order to ensure a smooth transition from column stability to column instability.

The failure strength of a column was defined as the peak strength reached on the load-deflection response curve when the column was subjected to pure axial force or axial force combined with bending moment. Thus, this analysis can be applied to short up to very long steel reinforced concrete composite columns, which may fail in material failure or buckling failure.

The elastic-perfectly plastic stress-strain curves defined by measured values of yield strength and modulus of elasticity were used for structural steel section and reinforcing steel bars. The descriptions of both the unconfined and the partially confined concretes in compression outside and inside the lateral ties, respectively, were taken from Park et al (1982). Discretization of composite one-half cross section is shown in Fig. 4 for computing strength.

5 ANALYSIS AND DISCUSSION OF RESULTS

The proposed analytical method above was applied to investigate the behavior of pin-ended rectangular slender steel reinforced concrete composite columns; the different ratios of length-to-width, concrete strengths, and eccentricities of applied loading were analyzed. The analytical values are compared with the corresponding experimental results in order to assess the accuracy of the proposed method.

Table 3. Comparison of tested and computed strength.

No.	A1	A2	A3	A4	A5	A6	E1	E2
N_u^o/kN	1900	1457	1270	1183	1040	1330	820	678
N_u^c/kN	1633	1341	1067	993	863	1233	739	647
N_u^o/N_u^c	1.16	1.09	1.19	1.19	1.20	1.08	1.11	1.05

5.1 Comparisons of tested and computed strength

The tested strengths and the calculate strengths by non-linear analysis method are presented in table 3. The calculated values in all cases are close to the corresponding experimental results. The mean value of the ratio of experimental load N_u^o to calculated load N_u^c is 1.13, it's standard deviation is 0.05, therefore it can be concluded that the proposed method provides an accurate solution.

5.2 Load-deflection response

To enable a comparative study of the influence of different eccentricity ratios, length-to-width ratios and concrete strengths on columns behavior, the three different eccentricity ratios, length-to-width ratios and concrete strengths were chosen. Fig. 6(a) presents the relation between axial and mid height deformation obtained by the proposed analytical method for the columns subjected to different initial load eccentricities, the eccentricity varies from e/h = 0.1 to 0.3 about the major axis. These columns of length 3.6 m pinned at the ends are subjected to single curvature bending. It is clear that column capacity is strongly affected by the amount of eccentricity. Fig. 6 (b) shows the results of the simulations for three length-to-width ratios columns subjected to an initial eccentricity of 18 mm. When the length-to-width ratio increases, the load bearing capacity is decreased. Fig. 6(c) shows the relations between the load and the mid-height deflection for three different concrete strength columns subjected to an initial eccentricity of 18 mm. The advantage of using a higher compressive concrete strength is not obvious. Comprehensive parametric studies show that the two main parameters that influenced the force-deflection response are column slenderness and eccentricity of the applied load. It can be seen from the Figure 3 (a) that the load-deflection response remains very stiff even up to the ultimate load in the case of columns subjected to axial load. As the eccentricity increases the load-carrying capacity drops significantly and the initially stiffness shown decreases.

Table 4 shows the comparisons of the experimental and predicted strength for 10 columns. The failure loads predicted by the proposed method show a good agreement with the experimental values. They

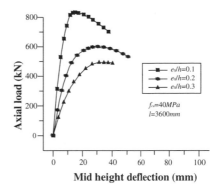

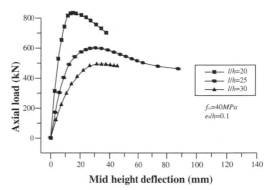

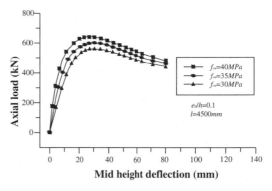

Figure 6. Computed relation of axial and mid height deflection.

are generally closer to the experimental data than the predictions by ACI318-05(2005) and Technical specification for steel reinforced concrete composite structures (China Specification 2002). The average ratios of Pexpt/Pcalc are 1.22 (ACI318-05), 0.97 (EC2), 1.21 (China Specification 2002) and1.127 (proposed method), respectively. On average, ACI318-05 gives more conservative predictions than this method, whereas EC2 predictions are slightly less conservative. Note that the coefficient of variation of Pexpt/Pcalc of this method is generally smaller than that of the three codes, as shown in Table 4.

Table 4. Comparison of Test and Computed Results.

No.	L/r	e (mm)	Pexpt (kN)	Pexpt/Pcalc			
				ACI318	EC2	China	Proposed method
A1	61	0	1900	1.19	0.98	1.14	1.16
A2	61	0	1457	0.98	1.02	1.12	1.09
A3	76	0	1270	1.41	0.99	1.23	1.19
A4	76	0	1183	1.07	0.96	1.17	1.19
A5	89	0	1040	1.21	0.97	1.31	1.20
A6	89	0	1330	1.47	0.94	1.37	1.08
E1	62	30	820	1.43	0.87	1.05	1.11
E2	62	40	678	1.05	1.05	1.08	1.05
Mean				1.22	0.97	1.21	1.127
Standard deviation				0.168	0.052	0.112	0.06

6 CONCLUSIONS

The influence of different eccentricity ratios, length-to-width ratios and concrete strengths on slender steel reinforced concrete column strength was studied. The results of tests and the numerical analysis on the columns presented here allow the following conclusions to be drowned. The buckling capacity is reduced with the slenderness ratio increase. For the eccentrically loaded columns, load carrying capacity is found to drop significantly with an increase of eccentricity, but that is not strongly influenced by the concrete strength. For column subjected to axial load, the buckling capacity is increased with the increase of concrete strength. An analytical method to compute the ultimate strength of slender steel reinforced concrete composite columns has been proposed. Comparison of ultimate strength, which was obtained by using the proposed method with the corresponding experimental results for column, has proven the accuracy of the proposed method.

REFERENCES

ACI committee 318, 1995, *Building code requirements for structural concrete (ACI 318-95) and commentary (ACI 318R-95)*, American Concrete Institute, Mich :Farmington Hills.

China standard, 2002, *Technical specification for steel reinforced concrete composite structures(JGJ138-2001)*, Beijing: China Architecture & Building Press.

Morino, S., Matsui, C. and Watanabe, H., 1984, Strength of biaxial loaded SRC columns. *Composite and Mixed Construction, ASCE*, New York, 241–253.

Mirza, S.A. and Skrabek, B.W. 1991, Reliability of short composite beam-column strength interaction. *Journal of Structural Engineering ASCE VID*, No.8, 2320–2334.

Lakshmi, B. and shanmugam, N.E., 2000, Behavior of steel-concrete composite columns, *Pro. 6th ASCCS Int. Conf. on Steel and Concrete Composite Structures, Association for International Cooperation and Research in Steel-Concrete Composite Structures (ASCCS)*, Los Angeles, 449–456.

Mirza, S.A. and Tikka, T.K. 1999, Flexural stiffness of composite columns subjected to major axial bending, *ACI Structural Journal*, V.96, No.1 January-February, 19–28.

Mirza, S.A. and Lacroix, E. A. 2003, Finite element analysis of composite steel-concrete columns, *Symposium on The Art and Science of Structural Concrete Design*, 185–205.

Park, R., Priestley, M. J. N. and Gill, W. D. 1982, Ductility of square confined concrete columns, *Journal of the Structural Division, ASCE*, V.108,No.ST4, 929–950.

Wang, Y.C. 1999, Tests on slender composite columns, *Journal of Construction Steel Research*, 49, 25–41.

Steel and Composite Structures – Wang & Choi (eds)
© *2007 Taylor & Francis Group, London, ISBN 978-0-415-45141-3*

Preload effect on the axial capacities of concrete-filled columns

J.Y. Richard Liew & De-Xin Xiong
Department of Civil Engineering, National University of Singapore, Singapore

ABSTRACT: *Concrete-filled steel tubes* (CFST) are usually used as the primary columns in high-rise construction. The steel tubes support the construction load and the weight of concrete slab and steel frame, and they are used as permanent formwork for infilling concrete. Due to the construction procedure, the steel tubes are usually preloaded by construction loads and the weight of the upper structures. The preload produces initial stress and deformation in the steel tube and will affect the stiffness and strength of the composite column. In this paper, a modified Eurocode approach is proposed to evaluate the preload effect on the axial resistance of CFST. The proposed approach is verified by experimental results and finite element analysis results. It is showed that the proposed method could give good estimation of the preload effect on the capacities of CFST.

1 INTRODUCTION

Concrete-filled steel tubes (CFST) are economic composite columns, and they have been widely used in many countries. The occurrence of the local buckling of steel tube is delayed by the restraint afforded by the concrete, and the strength and ductility of concrete is enhanced by the confining effect provided by the steel tube (Liang et al, 2006). The steel tubes serve as permanent formwork for multistorey construction in which concrete may be pumped upwards from the base or cast downwards from the top of the columns. Therefore, CFST are becoming popular for high rise construction (Liew, 2004).

There exists a problem of preloading in the practical construction of CFST structures. The hollow steel tubes are usually installed first with steel floor beams and metal floor decking, and they are connected to the bracing structures. The floor decks are then cast and concrete is pumped into the hollow steel tubes after several floors have been constructed. Therefore, before the infilled concrete has been hardened, the tubes are usually pre-loaded by the weight of unsolidified core-concrete, self-weight of steelwork and floor slabs, working load, etc. These loads bring some initial stresses and deformations in the steel tubes which could affect the capacity of the composite columns. Based on experimental studies and numerical analyses, there have been some reports which address the importance of considering the influence of the preload (Huang et al, 1996; Zha, 1996; Zhang et al, 1997; Han & Yao, 2003; Xiong & Zha, 2007).

Past investigations on the preload effect on the behavior of CFST columns are mainly experimental studies and numerical analyses which result in some empirical formulae proposed for estimating the preload effect. These empirical formulae are obtained based on experimental calibration, and they depend on the specimens used in the experiments and the models used in numerical simulations. Therefore, the formulae proposed by different researchers are different from each other, and they are not convenient for designers to use.

A comprehensive investigation, including theoretical analyses, experimental study and FE analysis, has been primarily done on the preload effect on the axial capacities of CFST in this paper. Theoretical analyses are carried out and approximate design method is proposed based on the modified Eurocode's approach. Experimental study is carried out on CFST with various preload ratios, material strengths and column lengths. The test results together with other published data are used to validate the proposed design method. FE analysis is adopted to provide systematic verification for the proposed approach.

2 SIMPLIFIED APPROACH TO CONSIDER THE PRELOAD EFFECT

When a compressive axial force N acts on a column with initial out-of-straightness δ_0 at the mid-height, the maximum total deflection δ_0 at the mid-height can be approximately given by (Chen and Atusta, 1976)

$$\delta = \frac{\delta_o}{1 - \frac{N}{N_{cr}}} \tag{1}$$

where $N_{cr} = \pi^2 EI/L^2$ is the Euler buckling load of composite column, and L is the effective column length. The maximum moment at the mid-height of the column is obtained as

$$M = N\delta = N\dfrac{\delta_o}{1 - \dfrac{N}{N_{cr}}} \tag{2}$$

and the maximum equivalent nominal stress is

$$\sigma_{max} = \dfrac{N}{A} + \dfrac{M}{S} = \dfrac{N}{A}\left(1 + \dfrac{\delta_o y}{r^2}\dfrac{1}{1 - \dfrac{N}{N_{cr}}}\right) \tag{3}$$

where A is the cross-sectional area, S is the elastic section modulus, y is the maximum distance from the neutral axis of cross-section to the outside edge, r is the radius of gyration for the cross-section.

Assuming that the column reaches its maximum resistance when $\sigma_{max} = N_{pl,Rk}/A$ where $N_{pl,Rk}$ is the characteristic value of plastic resistance of composite cross-section to compression, N reaches the ultimate axial load N_{Ek}, and then we can get

$$N_{pl,Rk} = A\sigma_{max} = N_{Ek}\left(1 + \dfrac{\delta_o y}{r^2}\dfrac{1}{1 - \dfrac{N_{Ek}}{N_{cr}}}\right) \tag{4}$$

From the above equation, the slenderness reduction factor χ is derived as

$$\chi = \dfrac{N_{Ek}}{N_{pl,Rk}}$$

$$= \dfrac{1}{\left[\left(\dfrac{1 + \rho + \lambda^2}{2}\right) + \sqrt{\left(\dfrac{1 + \rho + \lambda^2}{2}\right)^2 - \lambda^2}\,\right]} \tag{5}$$

$$= \dfrac{1}{\varphi + \sqrt{\varphi^2 - \lambda^2}}$$

where $\rho = \delta_0 y/r^2$, $\phi = 0.5(1 + \rho + \lambda^2)$, and λ is the non-dimensional slenderness given by

$$\lambda = \sqrt{\dfrac{N_{pl,Rk}}{N_{cr}}} \tag{6}$$

Taking $\rho = \alpha(\lambda - 0.2)$, the Equation (5) is the same as the formula recommended in Eurocode 4 to calculate the slenderness reduction factor, where α is an imperfection factor corresponding to the appropriate buckling curve (for concrete filled tubes, buckling curve "a" is adopted in Eurocode 4, $\alpha = 0.21$).

Equation (5) can be further modified to consider the preload effect. When a preload N_{pre} acts on the bare steel tube, the maximum deflection δ_1 at the mid-height of the column is given by

$$\delta_1 = \dfrac{\delta_o}{1 - \dfrac{N_{pre}}{N_{a,cr}}} \tag{7}$$

where $N_{a,cr} = \pi^2 E_a I_a/L^2$ is the Euler buckling load of the bare steel tube. Equation (7) gives the initial deflection of a preloaded composite column before composite action is achieved. After the infilled concrete has gained sufficient strength, δ_1 could be treated as the initial out-of-straightness deflection of the composite column just like δ_0 while an equivalent Euler buckling load $(N_{cr} - N_{pre})$ should be used for considering the action of the existing preload. When a superimposed load N_{add} is applied to the composite column with initial deformation δ_1, the maximum total deflection δ_2 at the mid-height of the composite column can be approximated as

$$\delta_2 = \dfrac{\delta_1}{1 - \dfrac{N_{add}}{N_{pre,cr} - N_{pre}}}$$

$$= \dfrac{\delta_o}{\left(1 - \dfrac{N_{pre}}{N_{a,cr}}\right)\left(1 - \dfrac{N_{add}}{N_{cr} - N_{pre}}\right)} \tag{8}$$

$$= \dfrac{1 - N_{pre}/N_{cr}}{1 - N_{pre}/N_{a,cr}}\dfrac{\delta_o}{1 - \dfrac{N_{pre} + N_{add}}{N_{cr}}}$$

where $N_{cr} = \pi^2 (EI)_{eff}/L^2$ is the equivalent Euler buckling load of composite column.

Then the maximum total moment is

$$M = (N_{add} + N_{pre})\delta_2 \tag{9}$$

Therefore, after the superimposed load N_{add} is applied on the composite column, the maximum equivalent nominal stress is

$$\sigma_{max} = \dfrac{N_{pre} + N_{add}}{A} + \dfrac{M}{S} \tag{10}$$

Assuming that the composite column reaches its maximum capacity when $\sigma_{max} = N_{pl,Rk}/A$, and the total compressive load $N = N_{pre} + N_{add}$ reaches its ultimate axial load $N_{pre,Ek}$ with preload effect, it obtains

$$N_{pl,Rk} = A\sigma_{max}$$

$$= N_{pre,Ek}\left[1 + \dfrac{1 - N_{pre}/N_{cr}}{1 - N_{pre}/N_{a,cr}}\rho\dfrac{1}{1 - \dfrac{N_{pre,Ek}}{N_{cr}}}\right] \tag{11}$$

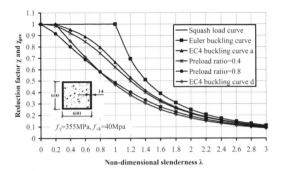

Figure 1. Buckling curves with preload effect.

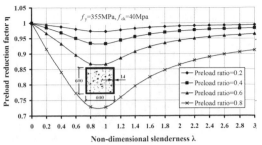

Figure 2. Preload reduction factor under different preload ratios.

Therefore, considering the preload effect, the slenderness reduction factor χ_{pre} is

$$\chi_{pre} = \frac{N_{pre,Ek}}{N_{pl,Rk}} = \frac{1}{\varphi_{pre} + \sqrt{\varphi_{pre}^2 - \lambda^2}} \qquad (12)$$

where

$$\varphi_{pre} = 0.5[1 + \alpha(\xi\lambda - 0.2) + \lambda^2] \qquad (13)$$

In Equation (11), $(1 - N_{pre}/N_{cr})/(1 - N_{pre}/N_{a,cr})$ represents the preload influence based on Euler buckling load. Actually, for non-slender composite columns, the capacities are much smaller than Euler buckling load. Therefore, in order to consider the preload influence effectively, it is more reasonable to adopt the actual ultimate axial load to substitute Euler buckling load, i.e.

$$\xi = \frac{1 - N_{pre}/N_{Ek}}{1 - N_{pre}/N_{a,Ek}} \qquad (14)$$

where ξ is the preload effect factor, N_{Ek} is the ultimate axial capacity of composite column without preload effect, $N_{a,Ek}$ is the ultimate axial capacity of bare steel tube.

Let β denotes the preload ratio, i.e.

$$\beta = N_{pre}/N_{a,Ek}, \text{ but } \beta < 1.0 \qquad (15)$$

Figure 1 shows an example for the design curves under different preload ratio values.

Normally, the preload ratio is not greater than 0.8. For convenience and according to Figure 1, it is conservative to adopt buckling curve d recommended in Eurocode 4 to design preloaded composite columns.

For preloaded CFST, the ultimate axial resistance is

$$N_{pre,Ek} = \chi_{pre} N_{pl,Rk} \qquad (16)$$

Table 1. Specimen details and results.

No.	L mm	β	f_y MPa	f_{ck} MPa	N_{exp} kN	N_{cal} kN	$\frac{N_{cal}}{N_{exp}}$
CS1	708.4	0.252	300	37.2	3677	3056	0.831
CS2	708.4	0	300	107.5	5410	5264	0.973
CS3	708.4	0.250	300	107.2	4667	5224	1.120
CI1	1728	0.299	405	44.4	3648	3043	0.834
CI2	1728	0	405	98.5	4977	4693	0.943
CI3	1728	0.305	405	112.6	5278	4941	0.936
CI4	1728	0.380	405	138.6	5437	5561	1.020
CL1	3078	0.306	393	48.7	3160	2725	0.862
CL2	3078	0	393	99.8	4204	3960	0.942
CL3	3078	0.310	393	110.7	4580	3919	0.856
CL4	3078	0.399	393	125.2	4827	4035	0.836

For non-preloaded CFST, the ultimate axial resistance is

$$N_{Ek} = \chi N_{pl,Rk} \qquad (17)$$

The preload reduction factor η is defined as

$$\eta = \frac{N_{pre,Ek}}{N_{Ek}} = \frac{\chi_{pre} N_{pl,Rk}}{\chi N_{pl,Rk}} = \frac{\chi_{pre}}{\chi} \leq 1.0 \qquad (18)$$

The smaller the preload reduction factor η, the greater is the preload effect. Figure 2 shows an example for preload reduction factor under different preload ratios.

3 EXPERIMENT INVESTIGATION

3.1 Test details

The column specimens were fabricated from hot-rolled circular steel tubes of 219 mm diameter and 6.3 mm thick. The details of the specimens are summarized in Table 1. The steel tubes were preloaded by means of the pre stressing strands anchored between the two end plates attached to the specimens. The pre-stressing

Figure 3. Specimen CL3 before and after failure.

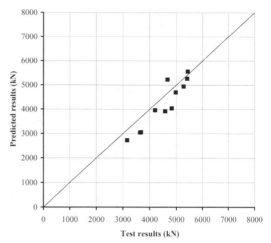

Figure 4. Test and corresponding predicted results.

Table 2. Comparison with test results dong by Zha (1996).

No.	L mm	β	f_{cube} MPa	f_y MPa	N_{exp} kN	kN	$\dfrac{N_{cal}}{N_{exp}}$
ZS1-1	1862	0	42.2	325	895	891	0.995
ZS1–2	1862	0	42.2	325	872	891	1.020
ZS2	1862	0.305	42.2	325	882	868	0.985
ZS3	1862	0.436	42.2	325	715	853	1.190
ZM1-1	2793	0	42.2	325	743	745	1.000
ZM1–2	2793	0	42.2	325	682	745	1.090
ZM2	2793	0.311	42.2	325	748	718	0.960
ZM3	2793	0.498	42.2	325	800	690	0.863

strands were stressed to achieve the desired amount of preloads. One of the specimens is shown in Figure 3.

3.2 Comparison with the results from the proposed design method

A total of eight preloaded specimens were tested. All the results are tabulated in Table 1 and shown in Figure 4. N_{cal} and N_{exp} respectively indicate the predicted and the experimental values. N_{cal} was calculated by the proposed equations. f_y values corresponded to the actual stresses in the steel tubes at the ultimate loads of the composite columns. For non-preloaded composite columns, the capacities are predicted by means of the equations adopted in Eurocode 4, just as the same as the results calculated from the proposed formulae provided that the preloads are zero. Compared with the test results, it shows that the predicted results calculated by proposed formulae are a little conservative as well as the results predicted by Eurocode 4.

3.3 Verification with other published test results

The predicted results from the proposed formulae are compared with available test results from the work

done by Zha (1996). The circular cross-section of tube is 133mm diameter and 4.5mm thick. The concrete strengths in the experiments are measured by the cube strengths f_{cube} and then the corresponding cylinder strengths f_{ck} are calculated and used in the proposed equations. Eight circular cross-sectional specimens, with the results from both the tests and proposed formulae, are shown in Table 2 and shown in Figure 5. The predicted results are in good agreement with the experimental results.

The predicted results from the proposed equations are also compared with another available test results from the work done by Han and Yao (2003). The square cross-section of tube is 120 mm breadth and 2.65 mm in thick. Six square cross-sectional specimens, with the results from both the tests and proposed formulae, are shown in Table 3 and shown in Figure 5. The specimens of H1-2 and H2-2 are short struts, so the preloads have no influence on the capacities. It shows that the predicted results also agree well with the experimental results.

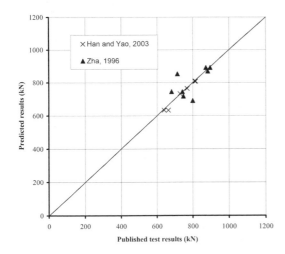

Figure 5. Published test and corresponding predicted results.

Table 3. Comparison with test results done by Han (2003).

No.	L mm	β	f_{cube} MPa	f_y MPa	N_{exp} kN	N_{cal} kN	$\dfrac{N_{cal}}{N_{exp}}$
HS1-1	360	0	20.1	340	640	633	0.990
HS1-2	360	0.499	20.1	340	664	633	0.954
HS2-1	360	0	36.0	340	816	808	0.991
HS2-2	360	0.499	36.0	340	812	808	0.996
HM1-1	1400	0	36.0	340	769	764	0.994
HM1-2	1400	0.499	36.0	340	730	733	1.000

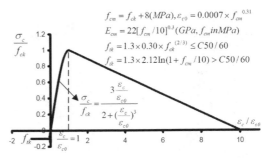

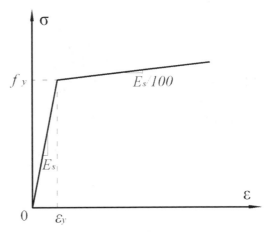

Figure 6. Stress-strain curve for concrete.

Figure 7. Stress-strain curve for steel.

4 FINITE ELEMENT ANALYSES

4.1 The finite element modelling

The general purpose finite element package ABAQUS is adopted for the numerical simulation. The continuum solid element C3D8R and the conventional shell element S4R are used to model core-concrete and square steel tube respectively. The stress-strain relationship curves adopted for concrete and steel are shown in Figure 6 and Figure 7.

In case of preloading, the whole loading process is divided into two steps. The preload is applied on the top of tube in the first step. Then the core-concrete is added in the model and it works compositely with the tube in the second step. In order to get the ultimate capacity in the second step, the vertical displacement of the column top is controlled and then the vertical reaction at the column bottom reveals the total axial load subjected by the column. Perfect bonding between tube and concrete is assumed. Half length composite column, fixed at bottom and free at the top, is modeled.

And maximum initial deflection of $L/1000$ is considered as the geometrical imperfection, which is shown in Figure 8.

4.2 FE analysis results

Three groups of specimens with different lengths are analyzed by considering various preload values. The square cross-section of steel tube is 600 mm diameter and 14 mm thick. The concrete strength is $f_{ck} = 40$ MPa, and the steel strength is $f_y = 355$ MPa. The curves of axial load–vertical displacement at the column tops are shown in Figure 9. And the numerical results with corresponding predicted results are tabulated in Table 4 and shown in Figure 10.

It can be seen from Table 4 and Figure 9 that the results predicted by the proposed formulae, compared with the results from FE analyses, are a little conservative. Therefore, it is safe enough to use the proposed formulae for the design of preloaded concrete-filled composite columns.

259

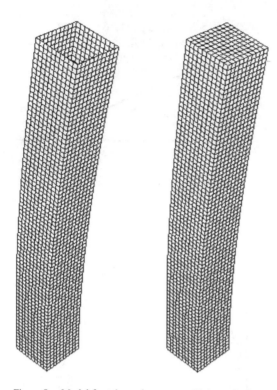

Figure 8. Model for tube and concrete with imperfection.

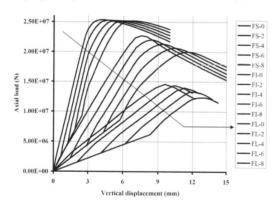

Figure 9. Axial load-vertical displacement curves.

5 CONCLUSION

A modified column buckling formula is derived to consider the preload effect on the axial capacities of CFST, and it is verified by test and FE analysis results. It shows that the proposed formula can provide sufficient estimation of the preload effect. The preload effect is influenced by the preload ratios, slenderness ratios, and the cross-sectional configuration

Table 4. FE analysis results compared with predicted results.

No.	L mm	β	N_{FE} kN	N_{cal} kN	$\dfrac{N_{cal}}{N_{FE}}$	η_{FE}	η_{cal}
FS-0	3600	0	25355	24541	0.968	1	1
FS-2	3600	0.2	25277	24374	0.964	0.997	0.993
FS-4	3600	0.4	25225	24102	0.955	0.995	0.982
FS-6	3600	0.6	25191	23576	0.936	0.994	0.961
FS-8	3600	0.8	25174	22135	0.879	0.993	0.902
FI-0	9000	0	22705	22108	0.974	1	1
FI-2	9000	0.2	22007	21686	0.985	0.969	0.981
FI-4	9000	0.4	21295	21028	0.987	0.938	0.951
FI-6	9000	0.6	20627	19858	0.962	0.908	0.898
FI-8	9000	0.8	19994	17151	0.858	0.881	0.776
FL-0	18000	0	14589	13013	0.892	1	1
FL-2	18000	0.2	14383	12725	0.885	0.986	0.978
FL-4	18000	0.4	13958	12287	0.880	0.957	0.944
FL-6	18000	0.6	13296	11535	0.868	0.911	0.886
FL-8	18000	0.8	12348	9887	0.801	0.846	0.760

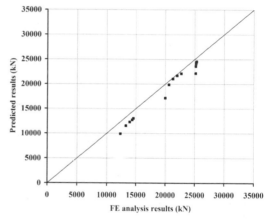

Figure 10. FE analysis and corresponding predicted results.

of the composite column such as relative wall thickness of steel tube, material strengths, etc. The preload has little influence on the short struts and very slender columns, but it has significant effect on intermediate length columns which shouldn't be ignored. A reduction may exceed 20% if the preload is quite large. For convenience, it is conservative to use the buckling curve d recommended in Eurocode 4 for the design of preloaded concrete-filled columns if the preload ratio is not greater than 0.8.

REFERENCES

Chen W.F. & Atsuta T. 1976. Theory of Beam-Columns, Vol. I: In-Plane Behavior and Design. McGraw-Hill, Inc. USA.
Chen W.F. & Liew J.Y.R. 2003. Composite Steel-Concrete Structures, in the Civil Engineering Handbook, Second

Edition, CRC Press, Boca Raton, Florida, 2003, chapter 51, 51–1 to 51–60

Eurocode 4. 2004. Design of Composite Steel and Concrete Structures, Part 1.1: General Rules and Rules for Buildings. Brussels: Commission of European Communities

Han L.H. & Yao G.H. 2003. Behaviour of concrete-filled hollow structural steel (HSS) columns with pre-load on the steel tubes. *Journal of Constructional Steel Research* 59(12): 1455–1475

Huang S.J., Zhong S.T. et al. 1996. Experimental Research of Prestress Effect on Bearing Capacity of Concrete Filled Steel Tubular Axial Compressive Members. *Journal of Harbin University of Architecture and Engineering* 29(6): 44–50

Liang Q.Q., Uy B. & Liew J.Y.R. 2006. Nonlinear analysis of concrete-filled thin-walled steel box columns with local buckling effects. *Journal of Constructional Steel Research* 62(6): 581–591

Liew J.Y.R. 2004. Buildable design of multi-storey and large span steel structures. *J Steel Structures, Korean Society of Steel Structures* 4(2): 53–70

Saw H.S. & Liew J.Y.R. 2000. Assessment of current methods for the design of composite columns in buildings. *Journal of Constructional Steel Research* 53(2): 121–147

Xiong D.X. & Zha X.X. 2007. A Numerical Investigation on the Behavior of Concrete-Filled Steel Tubular Columns under Initial Stresses. *Journal of Constructional Steel Research* 63(5): 599–611

Zha X.X. 1996. Investigation on the behavior of concrete filled steel tubular compression-bending-torsion members under the initial stress. *Harbin University of C.E. Architecture*

Zhang X.Q., Zhong S.T. et al. 1997. Experimental Study about the Effect of Initial Stress on Bearing Capacity of Concrete Filled Steel Tubular Members Under Eccentric Compression. *Journal of Harbin University of C.E. Architecture* 30(1): 50–56

Steel and Composite Structures – Wang & Choi (eds)
© *2007 Taylor & Francis Group, London, ISBN 978-0-415-45141-3*

An experimental and numerical study of high strength concrete filled tubular columns

J.M. Portoles
Universitat Jaume I, Castellon, Spain

M.L. Romero, J.L. Bonet & D. Hernandez-Figueirido
Universidad Politécnica de Valencia, Valencia, Spain

ABSTRACT: In recent years an increment in the utilization of concrete tubular columns was produced due to its high stiffness, ductility and fire resistance. On the other hand the use of high strength concrete (HSC) is more common due to the advances in the technology. The use of this material presents different advantages, mainly in elements subjected to high compressions as building supports or bridge columns.

However, there is a notably lack of knowledge in the behavior of high strength concrete filled tubular columns which produces that the existing simplified design models for normal strength concretes are not valid for them. This paper presents the results of a research project, where a numerical and an experimental study of high strength CFT's is performed.

1   INTRODUCTION

The utilization of hollow steel sections is well-known in multi-story buildings because a substantial reduction of the cross section is obtained. Moreover, the high strength concrete is spent more and more mainly for precast concrete structures, but the influence of the high strength on this type of columns (CFT's) is not well studied.

The actual simplified design methods are similar to the methods for reinforced concrete, assuming the steel section as an additional layer of reinforcement.Each country has its different code of design for composite sections (Japan, Australia, Canada, United States, Europe, etc.). The design code of the Euro-Code 4 (1992) (allows only the utilization of concrete with strength lower than 50 MPa (cylinder strength); therefore for high strength concretes the method and the interaction diagrams are not valid. Furthermore, as the section is reduced, for an equal length of the element, the slenderness is increased and the buckling is more relevant.

1.1   *Normal strength concrete*

The utilization in Europe of normal strength concrete filled tubular columns is well-known some decades ago when appeared the first monograph from the CIDECT (1970) simplifying its applicability for practical engineers. Later research works gave rise to the monograph n°5, CIDECT (1979). All this documents were the base to make the Euro-Code 4 (1992), with a special section for CFT's. In Spain, the ICT (Instituto de la Construcción Tubular) published a practical monograph, CIDECT (1998), with the idea to make easier the design of this type of sections.

1.2   *High strength concrete*

However, there are not a lot of investigations regarding high performance materials for CFT's, focused in the buckling. The research on high strength concrete (HSC) has demonstrated that the tensile capacity does not increase in the same proportion as the compression capacity. For hollow sections filled with concrete the tension problem is not as much important because the concrete cannot split of. Therefore in this type of section is where more advantage is taken.

1.2.1   *Experimental tests*

The more important contributions are concentrated in the last 5–10 years:

Grauers [1993] performed experimental tests over 23 short columns and 23 slender columns, stating that the methods of the different codes were valid but should be extended in order to analyze the effect of other parameters. Bergman (1994) studied the confinement mainly for normal strength concrete and partially for high strength concrete, but applying only axial load.

Aboutaba et al. (1999) compared classical columns of HSC with concrete filled tubular columns with HSC. Those ones presented more lateral stiffness and ductility.

Certainly, the research of Rangan & Joyce (1992) and Kilpatrick & Rangan (1999) has advanced a lot in the field. They presented experimental results from 9 columns for uniaxial bending and 24 columns for double curvature. However they used small sections and proposed a simplified design method regarding the Australian code, which does not follow the same hypothesis than the Eurocode-4, as i.e. the buckling diagrams.

Liu et al. (2003) compared experimentally the capacity of 22 rectangular sections with the different codes (AISC, ACI, EC4) and they concluded that the Euro-Code 4 was on the unsafe side while other codes over-designed the sections.

Varma et al. (2002, 2004) studied initially the behavior of the square and rectangular tubular columns and recently they have investigated cyclic loads and the existence of plastic hinges.

Gourley et al. (2001) have published a research report about the state of the art of concrete filled tubular columns.

1.2.2 Numerical models

Concerning the numerical models, it can be stated that there are not a lot of specific studies that applied the finite element method or sectional analysis to this type of structure. Most of them as Hu et al. (2003), Lu et al. (2000), and Shams & Saadeghvaziri (1999) study normal strength concretes. Only, recently Varma et al. (2005) have implemented a fibber model applied to square tubular sections but for short columns, without taking into account the buckling.

There is also an important research work from Johansson and Gylltoft (2001) where the the effects of three ways to apply a load to a columns is investigated numerically and experimentally.

If a good sectional characterization (moment-curvature) was performed, it can be inferred that the actual simplified methods are valid as a first approach to study the strength of these supports.

Few months ago, Zeghiche and Chaoui (2005) have published a small study for circular sections following this procedure. They affirmed that more numerical and experimental tests should be performed to check the validity of the buckling design methods of the EC4 for high strength concrete and double curvature.

Due to that the authors are performing a research project to study the effect of high strength concrete in the buckling. It has three parts: experimental study, one-dimensional numerical model and three-dimensional model. In this paper the initial results of the experimental and one-dimensional (1-D) part is presented.

For the 1-D part a nonlinear finite element numerical model for circular concrete filled tubular sections will be presented. The method has to be computationally efficient and must represent the behaviour of such columns, taking into account the effect of high strength concrete and second order effects.

Also an experimental study of circular concrete filled tubular sections is presented. The experimental tests selected corresponds to circular tubular columns filled with concrete (CFT) with pinned supports at both ends subjected to axial load and uniaxial bending. In these tests the eccentricity of the load at the ends is fixed and the maximum axial load of the column is evaluated.

2 NUMERICAL MODEL

In this section the numerical model based on the finite element method is illustrated.

2.1 Formulation

The finite element selected is a classical one-dimensional 13 degrees of freedom (d.o.f.'s.) element. It has 6 d.o.f.'s at each node (three displacement and three rotations), and a longitudinal degree of freedom in the mid-span to represent a non-constant strain distribution to represent the cracking. The Navier-Bernoulli hypothesis is accepted for the formulation. Also perfect bond between the concrete and structural steel is assumed. In the model the local buckling of the hollow steel section is neglected by stating at least the minimum thickness pointed out in the EC-4 (1992) (Art 4.8.2.4.):

$$D/t \le 90 \cdot \varepsilon^2 \qquad (1)$$

where: D external diameter of the circular section
t thickness of the circular hollow steel section

$\varepsilon = \sqrt{235/f_y}$

f_y yielding stress of the structural steel (MPa)

The model includes the second order effects by the formulation of large strains of the element (using the nonlinear deformation matrix $-B_L-$ and the geometric stiffness matrix $-K_g-$) and large displacements (by stating the force equilibrium in the deformed shape). The model automatically obtains the maximum load by using the total potential energy analysis "V" of the structure, Gutiérrez et al (1983); detecting if the structure reaches an stable equilibrium position ($\delta^2 V > 0$), unstable equilibrium ($\delta^2 V < 0$) or instability ($\delta^2 V = 0$). Moreover, the Newton-Raphson method was selected to solve the nonlinear system of equations for a known load level. An arc-length method,

Crisfield (1981), was used for displacement control. A more extensive definition of the finite element model can be found at Romero et al. (2005).

2.2 Constitutive equations of the materials

A bilinear (elastic-plastic) stress-strain diagram for the structural steel was assumed, EC-4 (1992) (Art 3.3.4)). The equation proposed by the CEB (1990) for the Model Code was used for the columns filled with normal strength concretes ($f_c \leq 50$ MPa); for the cases of high strength concrete ($f_c > 50$ MPa) was selected the equation from the CEB-FIP (1995). The tension-stiffening effect was considered by a gradual unload method, Bonet (2001). The permanent deformations due to cyclic loads were not included in the model. Therefore it is assumed that the maximum load is not dependent on the adopted path.

2.3 Cross section integration

The classical section integration of a circular (or annular) section is performed by a decomposition of the cross section into layers. However, this integration procedure has a high computational cost because it needs a large amount of information to characterize the section and a lot of calculations to obtain an admissible error. Also, due to the lack of an exact adjustment of the layers to the geometry, some convergence problems in the nonlinear iteration solution could appear due to the sharp variation of the neutral axis.

Because of that, this paper proposes a numerical algorithm to integrate the stress field of tubular circular sections filled with normal and high strength concrete (CFT) using the Gauss-Legendre quadrature.

The internal forces from the section integration are obtained as the addition of the concrete and structural steel internal forces:

$$N = N_c + N_a$$
$$M_y = M_{cy} + M_{ay} \qquad (2)$$
$$M_z = M_{cz} + M_{az}$$

where: (N_c,M_{cy},M_{cz}) concrete internal forces
(N_a,M_{ay},M_{az}) structural steel internal forces

On the other hand, the tangent stiffness matrix of the section (D_t) is obtained adding the constitutive matrix of the concrete (D_{ct}) and the steel constitutive matrix (D_{at}).

The steps followed in the numerical algorithm are: decompose the section in "wide" layers, Figure 1a, transform them in a path integral and later on to evaluate the integral by a Gauss-Legendre quadrature. To evaluate the internal forces of the cross section or the constitutive matrix, the terms of the concrete and steel are performed separately.

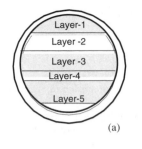

(a)

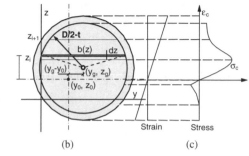

(b)   (c)

Figure 1. Numerical integration of concrete. Circular section (a) Decomposition in wide layers (b) Transformation into a path integral (c) Concrete stress field.

2.3.1 Concrete stress integration
The internal forces from the stresses of a circular concrete section can be obtained by the next equations (Figure 1b):

$$N_c = \iint_{A_c} \sigma_c(y,z)dA_c = \int_{z_g+D/2-t-x}^{z_g+D/2-t} b(z) \cdot \sigma_c(z)dz$$

$$M_{cy} = \iint_{A_c} \sigma_c(y,z) \cdot (z-z_0)dA_c = \qquad (3)$$

$$= \int_{z_g+D/2-t-x}^{z_g+D/2-t} b(z) \cdot (z-z_0) \cdot \sigma_c(z)dz$$

where: A_c is the area of the concrete integration region; (y_0, z_0) are the coordinates of the stress reference centre; (y_g, z_g) are the circular centre of gravity coordinates; $\sigma_c(z)$ is the concrete stress in a fibber with a coordinate "z"; b(z) is the width of the section in terms of "z":

$$b(z) = 2 \cdot \sqrt{(D/2-t-z+z_g) \cdot (D/2-t+z-z_g)} \quad (4)$$

The y-axis has been chosen parallel to neutral axis in the equations. As the stress field has a predominant direction (perpendicular to the neutral axis); the integral of the stresses in the concrete compression zone A_c can be transformed into a path integral in terms of z.

The components of the concrete tangent stiffness matrix (D_{ct}) can be computed in the same way, where

E_{ct} is the tangent elastic modulus of the concrete. For example:

$$D_{ct}(1,1) = \iint_{A_c} E_{ct}(y,z) \cdot dA_c = \int_{z_g+D/2-t-x}^{z_g+D/2-t} b(z) \cdot E_{ct}(z) \cdot dz \qquad (5)$$

2.3.2 Stress integration using the gauss-lengedre quadrature

Consequently, the integral of stresses over the integration are of concrete can be reduced to a path integral using the Green's theorem:

$$\iint_{A_c} f(y,z) \cdot dy \cdot dz = \int_{a_1}^{a_2} h(z)dz \qquad (6)$$

To perform each integral, the Gauss-Legendre method is used. The next coordinate transformation is required (Figure 1b):

$$z = \frac{x}{2} \cdot (\xi + 1) + \left[z_g + \frac{D}{2} - t - x \right] \quad \Rightarrow \quad dz = \frac{x}{2}d\xi \qquad (7)$$

Hence, the relating integrals (equation 3 and 5) are computed using the next expressions:

$$\iint_{A_c} f(y,z) \cdot dy \cdot dz = \int_{a_1}^{a_2} h(z)dz = \frac{x}{2} \cdot \sum_{k=1}^{npg} \omega_k \cdot h(\xi_k) \qquad (8)$$

where: ξ_k is the value of the curvilinear coordinate of the Gauss point "k", ω_k is the weight associated to this Gauss point and "npg" is in the number of Gauss points used in t he integration.

2.3.3 Stress Integration using "wide layers"

The accuracy of the integral depends on: the number of Gauss points used, the shape of the concrete constitutive equation and on the shape of the section (circular, in this case).

When the mathematical function to integrate $(h(z))$ is not adjusted to a small order polynomial or it is defined by branches, it is necessary to use a large amount of Gauss point to achieve an acceptable accuracy (Bonet et al. 2005). For these cases is better to subdivide the integration area into wide layers parallel to the neutral axis. Thus, for instance, the implementation of this method for the typical stress-strain concrete relationship could be performed using five wide layers (Figure 1a).

In this case, for each wide layer the coordinate transformation is performed using next equation (Figure 1b):

$$z = \frac{(z_{i+1} - z_i)}{2} \cdot (\xi + 1) + z_i \quad \Rightarrow \quad dz = \frac{(z_{i+1} - z_i)}{2}d\xi \qquad (9)$$

Therefore, the internal forces and the constitutive matrix of the concrete section are obtained as the addition of the integrals of each layer.

$$\iint_{A_c} f(y,z) \cdot dy \cdot dz = \int_{a1}^{a2} h(z)dz =$$
$$= \sum_{i=1}^{n°layers} \left[\frac{(z_{i+1} - z_i)}{2} \cdot \sum_{k=1}^{npg} \omega_k \cdot h(\xi_k) \right]_i \qquad (10)$$

2.3.4 Stress integration of the structural steel

The internal forces (N_a, M_{ay}, M_{az}) and the constitutive matrix of the tubular section (Dat) are computed by superposition of two circular sections with a radius (D/2) and (D/2-t) respectively.

The next procedure is followed for each of the subsections: to decompose the section into wide layers, to transform it in a path integral and to evaluate the integral using the Gauss-Legendre quadrature.

3 EXPERIMENTAL PROGRAM

Twenty-four test specimens of NSC and HSC columns will be studied in this experimental program; see Table 1. They are designed to investigate the effects of four main parameters on their behavior: slenderness (Length/Diameter), ratio D/t (Diameter/ thickness), strength of concrete (Fck) and eccentricity. The column lengths, L, are 2135 and 3135 mm and the cross-sections are circular with a 100, 125, 159 and 200 mm of outer diameter. The thickness of the steel tubes are 3, 4 and 6 mm. The strength of concrete varies from 30, 60 to 90 MPa and the axial load is applied with two eccentricities: 20 and 50 mm. All of the tests are going to be performed in the laboratory of the Department of Mechanical Engineering and Construction of the Universitat Jaume I in Castellon, Spain.

3.1 Materials

3.1.1 Concrete

All columns were cast using concrete batched in the laboratory. The concrete compressive strength fck was determined both from the cylinder and cubic compressive tests at 28 days. The specimens were cast in a inclined position.

3.1.2 Steel

The steel used was S275JR. The yield strength fy, the ultimate strength fu, the strain at hardening, the ultimate strain and modulus of elasticity E of the steel were obtained from the Eurocode 2.

266

Table 1. Tests for CFT columns, and numerical results.

Test	D (mm)	t (mm)	D/t	L (Mm)	L/D	fck (MPa)	e (mm)	e/D	Numerical Pmax (kN)	Numerical dy (m)
1	100	3	33.33	2135	21.35	17	20	0.20	**173.6**	**0.015**
2	100	3	33.33	2135	21.35	32	50	0.50	**127.7**	**0.022**
3	100	3	33.33	2135	21.35	60	20	0.20	**250.7**	**0.019**
4	100	3	33.33	2135	21.35	60	50	0.50	**146.7**	**0.023**
5	101.6	3	33.33	2135	21.35	90	20	0.20	**287.1**	**0.022**
6	101.6	3	33.33	2135	21.35	90	50	0.50	**158.3**	**0.023**
7	100	3	33.33	3135	31.35	30	20	0.20	**139.4**	**0.021**
8	100	3	33.33	3135	31.35	30	50	0.50	**94.3**	**0.032**
9	100	3	33.33	3135	31.35	60	20	0.20	**160.7**	**0.021**
10	100	3	33.33	3135	31.35	60	50	0.50	**110.8**	**0.039**
11	100	3	33.33	3135	31.35	90	20	0.20	**173.9**	**0.022**
12	100	3	33.33	3135	31.35	90	50	0.50	**115.4**	**0.034**
13	101.6	5	16.67	2135	21.35	30	20	0.20	**273.0**	**0.016**
14	101.6	5	16.67	2135	21.35	30	50	0.50	**175.4**	**0.023**
15	101.6	5	16.67	2135	21.35	60	20	0.20	**318.7**	**0.018**
16	101.6	5	16.67	2135	21.35	60	50	0.50	**198.3**	**0.023**
17	101.6	5	16.67	2135	21.35	90	20	0.20	**353.8**	**0.021**
18	101.6	5	16.67	2135	21.35	90	50	0.50	**212.6**	**0.025**
19	101.6	5	16.67	3135	31.35	30	20	0.20	**195.7**	**0.021**
20	101.6	5	16.67	3135	31.35	30	50	0.50	**133.9**	**0.032**
21	101.6	5	16.67	3135	31.35	60	20	0.20	**230.0**	**0.034**
22	101.6	5	16.67	3135	31.35	60	50	0.50	**146.1**	**0.033**
23	101.6	5	16.67	3135	31.35	90	20	0.20	**245.4**	0.038
24	101.6	5	16.67	3135	31.35	90	50	0.50	**153.5**	**0.033**

3.2 Test setup

The specimens are tested in a special 2000 kN capacity testing machine. The eccentricity of the applied compressive load was equal at both ends, so the columns are subjected to single curvature bending. It was necessary to built up special assemblages at the pinned ends to apply the load eccentrically. Figure 2 presents a general view of the test. Five LVDTs were used to measure symmetrically the deflection of the column at midlength(L/2), and also at four additional levels. The strains were measured at the mid-span section using strain gages.

4 RESULTS

All the 1-D numerical tests are accomplished, Table 1, but on the date of the congress only the half of the tests (one per week) will be carried out.

Also twelve experimental tests can be presented right now (March 2007), Table 2.

The accuracy degree is obtained from the next equation:

$$\xi = \frac{P_{u,test}}{P_{u,NS}} \tag{11}$$

where $P_{u,test}$ is the ultimate axial load in the experimental test (TEST) and $P_{u,NS}$ is the ultimate axial load in the numerical simulation (NS).

Table 2 presents the maximum load and the corresponding displacements of both the numerical and experimental cases. Also the error (defined in the previous section) is computed. From this table (but with the carefulness of not having yet all the cases carried out) it can be inferred that for a ratio e/D = 0.2 the 1-D finite element model is on the safe side, but for a ratio e/D = 0.5 the model does not predict well the column behavior.

From figures of horizontal displacement at midspan versus the vertical load applied (not presented here) and Table 2 it can be concluded that the numerical model is more rigid than the experimental one. It needs to be improved.

5 CONCLUSIONS

In this paper a nonlinear numerical model for the analysis of concrete filled tubular columns using the finite element method is presented. The novelty of the model is focussed in the numerical integration of the cross section using the Gauss-legendre quadrature. Also an

267

Table 2. Comparison of numerical and experimental tests.

						Numerical		Experim.		Error ξ	
N°	D (mm)	t (mm)	L (mm)	(fck)	e (mm)	e/D	Pmax (kN)	dy (mm)	Pmax (kN)	dy (mm)	Pm (kN)	dy (mm)
1	100	3	2135	17	20	0.20	173.6	15	182	18	**1.04**	**1.19**
2	100	3	2135	32	50	0.50	127.7	22	117	28.92	**0.92**	**1.18**
3	100	3	2135	**76.93**	20	0.20	250.7	19	249	23.41	**0.99**	**1.23**
4	100	3	2135	**82.42**	50	0.50	146.7	23	152	32.14	**1.04**	**1.40**
5	100	3	2135	**106.20**	20	0.20	287.1	22	271	25.76	**0.94**	**1.17**
6	100	3	2135	**103.6**	50	0.50	158.3	23	154	34.35	**0.97**	**1.49**
13	100	5	2135	**41.47**	20	0.20	273.0	16	270	22.9	**0.99**	**1.43**
14	100	5	2135	**30.25**	50	0.50	175.4	23	161	35.20	**0.92**	**1.53**
15	100	5	2135	**80.96**	20	0.20	318.7	18	313	27.28	**0.98**	**1.52**
16	100	5	2135	**76.00**	50	0.50	198.3	23	183	38.98	**0.92**	**1.69**
17	101.6	5	2135	**106.00**	20	0.20	353.8	21	330	27.37	**0.93**	**1.30**
18	101.6	5	2135	**92.35**	50	0.49	212.6	25	213	34.42	**1.00**	**1.38**

Figure 2. Test setup.

experimental program and the design of the setup are presented. The numerical model is more rigid and needs to be improved for the cases of higher eccentricities, where the displacement corresponding to the maximum load is not well predicted.

This issue is very important to create new buckling curves for high strength concrete.

ACKNOWLEDGEMENTS

The authors wish to express their sincere gratitude to the Spanish "Generalitat Valenciana" for help provided through project GV04/11/2004, and to the " ICT, Instituto de Construccion Tubular", partner of CIDECT, for their advice.

REFERENCES

Aboutaba R.S., & Machado R.I. 1999, Seismic resistance of steel-tubed high-strength reinforced-concrete columns, J STRUCT ENG-ASCE 125(5): 485–494.
Bergmann, R. 1994. Load introduction in composite columns filled with High strength concrete, Proceedings of the 6th Int Symposium on Tubular Structures, Monash University, Melbourne, Australia.
Bonet, J.L. 2001. Método simplificado de cálculo de soportes esbeltos de hormigón armado de sección rectangular sometidos a compresión y flexión biaxial, PhD Thesis, Civil Engineering Dept., Technical University of Valencia.
CIDECT. 1979. Monograph n°1: Concrete filled hollow section steel columns design manual. British edition.
CIDECT. 1979. Monograph n°5: Calcules Poteaux en Proliles Creux remplis de Beton.
CIDECT. 1998. Guía de Diseño para columnas de perfiles tubulares rellenos de hormigón bajo cargas cíclicas estáticas y dinámicas. TUV-Verlag.
Comité Euro-internacional du beton. 1991. CEB-FIB Model Code 1990. C.E.B. Bulletin N° 203–204 and 205.
Comité Euro-internacional du beton. 1995. High Performance Concrete. Recommended extensions to the Model Code 90 research needs. C.E.B.. Bulletin N° 228.
Crisfield M.A. 1981. A fast incremental/iterative solution procedure that handles "snap-through". Computers & Structures 13(1–3): 55–62.
Eurocode 4. 1992. Proyecto de Estructuras Mixtas de Hormigón y Acero, Parte 1–1: reglas generales y reglas para edificación. (in Spanish).

Gourley B.C., Tort C., Hajjar J.F. & Schiller P.H. 2001. A Synopsis of Studies of the Monotonic and Cyclic Behavior of Concrete-Filled Steel Tube Beam-Columns. Structural Engineering Report No. ST-01-4. Department of Civil Engineering. University of Minnesota, Minneapolis, Minnesota.

Grauers M. 1993. Composite columns of hollow sections filled with high strength concrete. Research report. Chalmers University of Technology, Goteborg.

Gutiérrez G. & Sanmartin, A. 1983. Influencia de las imperfecciones en la carga crítica de estructuras de entramados planos. Hormigón y Acero 147: 85–100.

Hu H.T., Huang C.S., Wu M.H., et al. 2003. Nonlinear analysis of axially loaded concrete-filled tube columns with confinement effect, J STRUCT ENG-ASCE 129(10): 1322–1329

Johansson M. & Gylltoft K. 2001. Structural behavior of slender circular steel- concrete composite columns under various means of load application. Steel and Composite Structures 1(4): 393–410.

Kilpatrick A.E. & Rangan B.V. 1999. Tests on High-Strength Concrete-Filled Steel Tubular Columns. ACI STRUCT J 96(2): 268–275.

Liu D.L., Gho W.M. & Yuan H. 2003. Ultimate capacity of high-strength rectangular concrete-filled steel hollow section stub columns. J CONSTR STEEL RES 59(12): 1499–1515.

Lu X.L., Yu Y., Kiyoshi T., et al. 2000. Nonlinear analysis on concrete-filled rectangular tubular composite columns. STRUCT ENG MECH 10(6): 577–587.

Neogi P.K., Sen H.K. & Chapman J.C. 1969. Concrete-Filled Tubular Steel Columns under Eccentric Loading. Structural Engineer 47(5): 187–195.

Rangan B. & Joyce M. 1992. Strength Of Eccentrically Loaded Slender Steel Tubular Columns Filled With High-Strength Concrete, ACI STRUCT J 89(6): 676–681.

Romero M.L., Bonet J.L., Ivorra S. & Hospitaler A. 2005. A Numerical Study Of Concrete Filled Tubular Columns With High Strength Concrete, Proceedings of the Tenth International Conference on Civil, Structural and Environmental Engineering Computing, B.H.V. Topping, (Editor), Civil-Comp Press, Stirling, United Kingdom, paper 45.

Shams M. & Saadeghvaziri M.A. 1999. Nonlinear response of concrete-filled steel tubular columns under axial loading. ACI STRUCT J 96(6): 1009–1017.

Srinivasan C.N. 2003. Discussion of "Experimental behavior of high strength square concrete-filled steel tube beam-columns" by Amit H. Varma, James M. Ricles, Richard Sause, and Le-Wu Lu, J STRUCT ENG-ASCE 129(9): 1285–1286.

Varma A.H., Ricles J.M., Sause R., et al. 2002. Experimental behavior of high strength square concrete-filled steel tube beam-columns, J STRUCT ENG-ASCE 128(3): 309–318.

Varma A.H., Sause R., Ricles J.M., et al. 2005. Development and validation of fiber model for high-strength square concrete-filled steel tube beam-columns. ACI STRUCTURAL JOURNAL 102(1): 73–84.

Zeghiche J. & Chaoui K. 2005. An experimental behaviour of concrete-filled steel tubular columns. JOURNAL OF CONSTRUCTIONAL STEEL RESEARCH 61(1): 53–66.

Steel and Composite Structures – Wang & Choi (eds)
© 2007 Taylor & Francis Group, London, ISBN 978-0-415-45141-3

Behavior of concrete-filled RHS steel tubes subjected to axial compression combined with bi-axial bending moment

Lanhui Guo & Sumei Zhang

School of Civil Engineering, Harbin Institute of Technology, Harbin, China

ABSTRACT: This paper presents an experimental research on 9 concrete-filled Rectangular Hollow Section (RHS) specimens subjected to the axial compression combined with bending moment. The main parameters for the experimental research are loading angle and depth-to-breadth ratio of the cross-section. At the same time, a numerical analytical method was developed to predict the behavior of the rectangular composite columns in which the rotation of neutral axis was taken into account when the specimen are subjected to axial load and biaxial bending. The validity of the analytical method is established by comparing the results with the corresponding experiments. Both the analytical and experimental results show that, for the rectangular composite columns subjected to axial load combined with biaxial bending, the rotation of neutral axis decreases the bearing capacity and the rigidity along minor axis. With the increase of depth-to-breadth ratio, the decrease of bearing capacity caused by the rotation of neutral axis will become bigger. When the loading angle varies from 0 degree to 90 degree, the decrease amplitude of bearing capacity is the largest when the loading angle is 45 degree.

1 GENERAL INSTRUCTIONS

The use of concrete-filled steel tubes in the construction is becoming popular. It provides not only an increase in the load carrying capacity but also economy and rapid construction, and thus additional cost saving. Their use in multistory buildings has increased in recent years owing to the benefit of increases load carrying capacity for a reduced cross section. This type of column can be constructed using either circular or rectangular steel hollow columns filled with either normal or high strength concrete. Recently, the use of concrete-filled rectangular hollow section (RHS) tubes has increasingly become popular in civil engineering structures. Because it not only has the common virtues of concrete-filled circular tube, but also has its own strong characters, for instance, easier construction joint details and convenience for decoration (Zeghiche, 2005).

Compared to circular composite columns, the research on rectangular composite columns is relatively late and most of them are about stub columns or slender columns subjected to axial load or axial load combined with uni-axial bending. The main research works on concrete-filled RHS steel tubes are as follows. Shakir-Khalil and J. Zeghiche (1989) studied seven concrete-filled rolled rectangular columns with H/t and B/t ratios of 16 and 24, respectively. Three of them subjected to axial load, two of them subjected to axial load combined with uniaxial bending and the

rest subjected to axial load and biaxial bending. Their experimental results indicated that the failure mode was an overall buckling, with no sign of local buckling for steel tube, owing to smaller B/t or H/t ratios. Thereafter, more researches were conducted to investigate intensively the behavior of concrete-filled RHS steel tubes.

Shakir-Khalil & Mouli (1990) investigated nine concrete-filled rolled rectangular composite columns. Seven of them subjected to axial load and biaxial bending and the rest subjected to axial load or uniaxial eccentric load. The results showed that the squash load decreased when the specimen height increased, but there was no guidance in the standard regarding the depth-to-breadth ratio of stub column. Wang Y.C (1997) described eight rectangular composite specimens, both with end eccentricities producing moments other than single curvature bending. Han L.H (2002) tested four square and twenty rectangular short composite columns subjected to axial load. Zhang Sumei (2005) studied 50 rectangular composite columns filled with high strength concrete, and all of them subjected to axial load, and a model is offered to analyze the behavior of rectangular composite columns. Liu Dalin et al. (2003) studied the behavior of 22 rectangular stub composite columns with high strength steel and high strength concrete, which was subjected concentric load. Tian Hua (2003) studied two square composite columns filled with high strength concrete subjected to axial load and biaxial bending.

Liu Dalin (2004) tested 12 high strength rectangular concrete-filled steel hollow section columns subjected to eccentric loading. Considering the horizontal loads are randomly applied by earthquake or wind, the rectangular columns is often under the state of axial load combined with bi-axial bending moment. From the above, it can be seen that the research on the behavior of rectangular composite columns is relatively less. Hence, this paper provides an experimental study to supplement the test data in this particular area so that design rules can be well developed for the design of rectangular composite columns under axial load and biaxial bending. At the same time, a nonlinear program was made to analyze the behavior of rectangular columns subjected to axial load and biaxial bending moment, the rotation of neutral axis is considered in the analysis.

2 EXPERIMENT

A total of 9 specimens were tested. The main parameters are load angle and depth-to-breadth ratio. The specimen is subjected to axial load combined with biaxial bending moment. The load angles varied from 0 degree to 90 degree. The depth-to-breadth ratios were 1.0 and 2.0, respectively. The dimensions of specimen are categorized in Table 1.

2.1 Fabrication

All the steel tubes were manufactured from mild steel plates. These plates were initially tack welded, and an internal bracing was provided to minimize geometric imperfections. The tubes were penetration weld and the internal bracing was removed at the end of welding operation. The steel tubes were filled with plain concrete in the vertical position. Progressive vibration method was employed in order to eliminate air pockets in the concrete and also to give a homogeneous

mix. Prior to the test, the top of the specimens surface was ground smooth and flat using a grinding wheel with diamond cutters. This was to ensure that the load was applied evenly across the cross-section and simultaneously to both steel and concrete.

2.2 Material testing

To determine the steel material properties, three tension coupons were cut from a randomly selected steel sheet. The dimension was decided in accordance with the Chinese standard GBJ2975 (1982). The specimens were tested in tension. The average yield strengths were shown in Table 1.

For each batch of concrete mix, three 100 mm cubes were also cast and cured in conditions similar to those for the related specimens. The test procedure was according to the Chinese standard GBJ81-85 (1985). The strength of concrete cube strength is shown in Table 1. The average compressive cube strength (f_{cu}^{10}) during the testing was 55 N/mm^2.

2.3 Testing procedure

Experiments were tested to ascertain the behavior of column combined effects of bending and compression.

A total of 24 strain gauges were bonded on each face of specimen at the middle point and quarter points of the measured section, which were arranged in 90° angle in longitudinal and transverse direction, as shown in Figure 1.

A total of 12 displacement transducers were also installed to measure the deformation of column. Three of them were arranged to measure the deflection along the minor axis (y axis) and four of them were installed to measure the longitudinal deformation of column. Five of them were used to measure the deformation along the major axis (x axis) and the relative rotation of specimens, two of them were installed at the quarter point of cross section on the mid-height and two of them were installed at the quarter point of cross-section

Table 1. Dimensions and material properties of the specimens.

Se.	B mm	D mm	t mm	L mm	f_{cu}^{10} MPa	f_y MPa
S-K3-1	149.3	149.3	3.64	1350	55.0	289.0
S-K3-2	149.6	149.6	3.66	1350	55.0	289.0
S-K3-3	149.6	149.6	3.66	1350	55.0	289.0
R-K3-1	99.6	200.6	3.56	1800	55.0	289.0
R-K3-2	100.5	200.7	3.62	1800	55.0	289.0
R-K3-3	99.8	198.5	3.63	1800	55.0	289.0
R-K3-4	100.6	200.4	3.60	1800	55.0	289.0
R-K3-5	99.4	200.4	3.64	1800	55.0	289.0
R-K3-6	99.6	200.5	3.60	1800	55.0	289.0

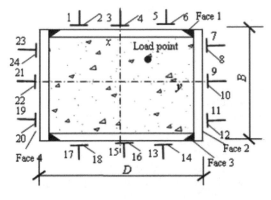

Figure 1. Distribution of strain gauges on the cross section.

272

on one end part of specimen, another displacement transducer is installed on the center of cross section at the other end part.

Spherical hinges were designed and installed at the two ends to ensure that pinned ends were achieved in all direction for the specimen in the test setup. The load transducer was used to measure the load in the experiment. Figure 2 shows the experimental setup for specimen subjected to axial load and biaxial bending.

3 EXPERIMENTAL PHENOMENA

For the specimen subjected to axial load and biaxial bending moment, there was evident deflection along the two axes at the initial loading. The failure position focused on the mid-height of column with an overall

(a) Experimental setup at the end of specimen

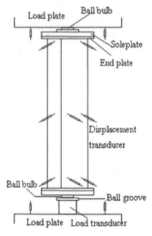

(b) Diagrammatic testing setup

Figure 2. Experimental setup of specimen under axial compression and biaxial bending moment.

buckling failure mode. On the corner between face 2 and face 3, the steel tube had evident local buckling, owning to the compressive stress at this part being the largest. After talking off the steel tube, it was found that the core concrete in the corner was crushed seriously. On the opposite corner, there was evident horizontal cracks appeared. The experimental phenomenon was shown in Figure 3.

4 THEORETICAL ANALYSIS

This part is concerned with a semi-analytical method to prepare the elastic and ultimate load behavior of concrete-filled rectangular steel columns. The nonlinear analysis technique is used in the analysis and the incremental equilibrium equation of a slender composite column is formed. The generalized displacement control method is then applied to solve the incremental equation. The accuracy of the proposed method is assessed by comparing the analytical values with the corresponding experimental results.

For rectangular composite columns subjected to biaxial bending, the asymmetry of cross section about the axis linking the point of application of the load and the center of cross section produces a condition in which the shear center does not fall within this axis. Due to this, the neutral axis is expected to rotate

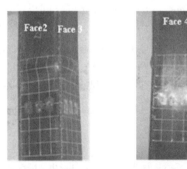

(a) Experimental phenomenon after the test

(b) After taking off the steel tube

Figure 3. Experimental phenomenon of specimen RK3-6.

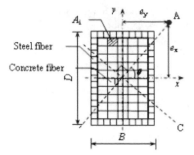

Figure 4. Divisions of cross-section.

and this phenomenon has been observed from the experimental results.

The following assumptions were made in the analysis. (1) Constitutive relationship for concrete and steel are known, which from reference (Zhang, 2005); (2) The deformation shape of the columns is assumed to be a semi-sine wave and consequently the central deflection is related in a simple manner to the curvature at the mid-height section of the column along the two axes; (3) The cross section of columns remains plane during the loading; (4) The tensile strength of concrete is ignored; (5) Shear deformations are small and not considered.

In the analysis, the cross section is firstly divided into many small rectangular elements. The elements are sufficiently refined so that increasing the elements density does not result in any significant change in the results. Figure 4 shows the division of the rectangular cross section. Based on the above assumptions, the numerical analysis proceeds as follows:

(1) Giving a deflection Δ_x (along y axis) at the mid-height, from the assumption (2), the curvature at mid-height can be calculated from equation (1).

$$\phi_x = \pi^2 \Delta_x / l^2 \qquad (1)$$

(2) Based on the curvature ϕ_x, assume the curvature along y axis is that $\phi_y = \phi_x / \tan \theta$, where θ is the load angle, $\theta = \arctan(e_x/e_y)$;

(3) Assume the strain at the center of cross-section is ε_0, the strain ε_i at any strip element center is given in the following equation (2).

$$\varepsilon_i = \varepsilon_0 + \phi_x y_i + \phi_y x_i \qquad (2)$$

x_i y_i are the distance of rectangular element center along y axis and x axis, respectively.

(4) After calculating the strain of element, the concrete element stress and steel element stress can be determined. Then the internal axial load and bending moment are computed by following three equations.

$$N_{in} = \sum_{i=1}^{n} \sigma_i A_i \qquad (3)$$

Table 2. Comparison on the bearing capacity.

Se.	e_x mm	e_y mm	N_e kN	N_c kN	N_e/N_c
S-K3-1	0.0	0.0	1552	1508	1.029
S-K3-2	21.2	21.2	974	957	1.018
S-K3-3	12.3	38.3	902	965	0.935
R-K3-1	0.0	0.0	1461	1313	1.113
R-K3-2	40.0	0.0	983	915	1.074
R-K3-3	30.0	30.0	632	609	1.038
R-K3-4	40.0	20.0	692	691	1.001
R-K3-5	0.0	30.0	655	623	1.051
R-K3-6	12.5	32.5	637	656	0.971

*e_x and e_y is the eccentric distance, which are shown in Figure 4. N_e and N_c are experimental result and calculated result, respectively.

$$M_{ix} = \sum_{i=1}^{n} \sigma_i A_i y_i \qquad (4)$$

$$M_{iy} = \sum_{i=1}^{n} \sigma_i A_i x_i \qquad (5)$$

Where n is the number of element; N_{in} is the internal axial load; M_{ix} is the internal moment along x axis; M_{iy} is the internal moment along y axis.

(5) According to the internal moment M_{ix}, the axial load N_x can be gotten.

$$N_x = M_{ix} / (e_x + \Delta_x + \delta) \qquad (6)$$

If $|(N_x - N_{in})/N_{in}| > 0.005$, then go to step (3) and change the curvature ε_0 until the internal load is equilibrium with the axial load calculated by internal moment along x axis.

(6) According to the internal moment M_{iy} along y axis, the axial load N_y can be gotten.

$$N_y = M_{iy} / (e_y + \Delta_y + \delta) \qquad (7)$$

If $|(N_y - N_{in})/N_{in}| > 0.005$, then go to step (2) and change the ϕ_y until the difference between N_y and N_{in} is within an acceptable tolerance.

(7) After the curvature ϕ_y is known, the deflection Δ_y along x axis can be gotten by the following equation.

$$\Delta_y = \phi_y l^2 / \pi^2 \qquad (8)$$

Then give an increment of deflection $\delta\Delta_x$ along y axis, the deflection along y axis will be $\Delta_x = \Delta_x + \delta\Delta_x$, and go to step (1). For the specimen subjected to axial load and uniaxial bending, the step (6) is not necessary for the calculation.

The program can be used to calculate the longitudinal load-deflection curves at the mid-height of column. The theoretical results were compared with the experimental results, which are shown in Figure 5. And the calculated results are also compared with some experimental results from other references. It can be seen that they agreed well with each.

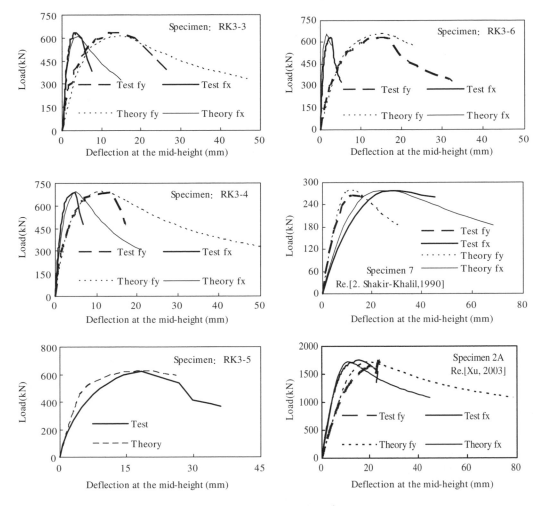

Figure 5. Comparison between calculated results and results from other references.

4.1 *Influence of depth-to-thickness ratio*

Figure 6 shows the influence of neutral axis rotation on the bearing capacity of rectangular composite columns with depth-to-breadth ratio varying from 1.0 to 2.5. During the analysis, the area of steel and core concrete is same and the eccentric distance is equal for rectangular column with different depth-to-breadth ratios. The load angle is 45 degree.

From Figure 6, it can be seen that there is no influence on the bearing capacity for the square composite column. With the increase of depth-to-breadth ratio, the influence of neutral axis rotation on the bearing capacity will become larger. If not considering the neutral axis rotation, the bearing capacity will be larger with the increase of depth-to-breadth ratio. Or else the bearing capacity will decrease with increasing of depth-to-breadth ratio. Hence, it can be concluded that the neutral axis rotation will decrease the bearing

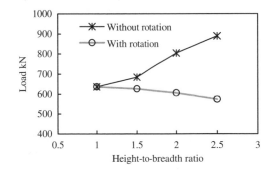

Figure 6. Influence of depth-to-breadth ratio.

capacity of rectangular composite evidently. The influence will rise with the increase of depth-to-breadth ratio. When the depth-to-breadth ratio is 2.0, the decrease ratio of bearing capacity is about 24%.

275

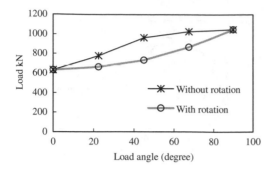

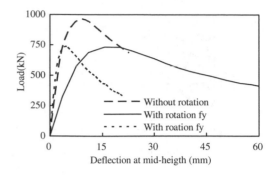

Figure 7. Influence of rotation of load angles.

Figure 8. Influence of rotation on the load-deflection curves.

4.2 Influence of load angle

In this part, the rectangular composite with depth-to-breadth ratio 2.0 are analyzed to study the influence of neutral axis rotation with different load angles. Figure 7 shows the calculated results. The load angle θ is the angle between the x axis and the line connecting the loading point and the center of cross section, as shown in Fig. 4. It can be seen that the specimen with load angle 0 degree or 90 degree was under the state of axial load and uniaxial bending. There was no neutral axis rotation, so the bearing capacity is the same. While for the load angle varying from 0 degree to 90 degree, the rotation of neutral axis will decrease the bearing capacity. The decrease ratio will become larger with the load angle varying between 0 degree and 45 degree. However, the decrease ratio of bearing capacity will be lower with increase of load angle when it changes from 45 degree to 90 degree. The decrease ratio with load angle 45 degree is the largest.

Figure 8 shows the influence of rotation on the load-deflection curves at mid-height. It can be seen that the load-deflection curves is the same if the neutral axis rotation is not considered. While the rotation of neutral axis is considered, the rigidity along minor axis decreases evidently and the rigidity along major axis is near to the rigidity without considering rotation. It also illustrates that the rotation will make the deflection at the mid-height larger evidently, thus the bearing capacity decreases fast.

5 CONCLUSIONS

This paper has presented an experimental research on the behavior of rectangular composite column subjected eccentric load. At the same time, a nonlinear program is produced to analyze the behavior of specimen subjected to axial load and biaxial bending. The influence of neutral axis rotation is also considered. The calculated results agreed well with the experimental results, and the experimental results are also compared with calculated results with different design codes. The following main conclusions can be drawn.

(1) For rectangular composite column under the state of axial load combined with biaxial bending, the neutral axis rotation will decrease the bearing capacity and rigidity along minor axis. When the depth-to-breadth ratio is 1.0, the influence of neutral axis rotation on the bearing capacity is not evident. With the increase of depth-to-breadth ratio, the decrease ratio of bearing capacity will become larger. So the neutral axis rotation should be considered in the analysis.

(2) The influence of neutral axis on the bearing capacity is different with the changing of load angles. The decrease ratio will become larger with the load angle varying from 0 degree to 45 degree. However, the decrease ratio of bearing capacity will be lower with the increase of load angle when it changes from 45 degree to 90 degree. The decrease ratio with load angle 45 degree is the largest.

REFERENCES

GBJ2975. 1982, Test standard for steel material. Beijing: Chinese Planning Press [in Chinese].
GBJ81-85. 198, Test standard for concrete material. Beijing: Chinese Planning Press [in Chinese].
Han Linhai. 2002. Tests on stub columns of concrete-filled RHS sections. *Journal of Constructional Steel Research* 58(3): 354–372.
Liu Dalin, Gho Wie-Min and Yuan Jie. 2003a. Ultimate capacity of high-strength rectangular concrete-filled steel hollow section stub columns. *Journal of Constructional Steel Research* (59)12:1499–1515.
Liu Dalin. 2004b. Behaviour of high strength rectangular concrete-filled steel hollow section columns under eccentric loading. *Thin walled structures* 42(12): 1631–1644.
Shakir-Khalil, H. and Zeghiche, J. 1989. Experimental behavior of concrete-filled rolled rectangular hollow-section columns. *The Structural Engineer* 67(9):346–353.
Shakir-Khalil, H. 1990. Further tests on concrete-filled rectangular hollow-section columns.*The Structural Engineer* (68) 20:405–413.

Tian Hua. 2003. Static behavior of high strength concrete – filled SHS and RHS steel tubular beam-columns. *Dissertation for the master degree in engineering of Harbin Institute of Technology.*

Wang, YC, and Moore, DB. 1997. A design method for concrete-filled hollow section, composite columns. *The Structural Engineer* 75:368–373.

Xu Zheng. 2003. Experimental research of concrete-filled rectangular thin-walled steel tubular beam-columns applied in Hongjiadu hydroelectric station. *Dissertation for the master degree in engineering of Harbin Institute of Technology.*

Zeghiche, J. and Chaoui, K. 2005. An experimental behaviour of concrete-filled steel tubular columns. *Journal of constructional steel research* 61(1):53–66.

ZHANG Sumei, GUO Lanhui, et al. 2005. Behavior of Steel Tube and Confined High Strength Concrete for Concrete-Filled RHS Tubes. *Advances in Structural Engineering* 3(2):101–116.

Steel and Composite Structures – Wang & Choi (eds)
© 2007 Taylor & Francis Group, London, ISBN 978-0-415-45141-3

Comparison between Chinese and three worldwide codes for circular concrete-filled steel tube members

Sumei Zhang & Xinbo Ma
Harbin Institute of Technology, Harbin, China

C.D. Goode
The University of Manchester, Manchester, UK

ABSTRACT: The characteristics and approaches of the design methods for circular concrete-filled steel tube (CCFST) members in three Chinese codes (JCJ 01–89 code, CECS code, DL/T 5085–1999 code) and three typical worldwide codes (American LRFD (99) code, European Eurocode 4 (94), Japanese AIJ (97) code), are introduced and compared. The limitations prescribed by each code are summarized and discussed. The load-carrying capacity calculated by each code is analyzed and compared with 1060 tests (including 775 compressive columns and 285 beam columns collected from a wide range of references), both excluding and including the limitations prescribed by each code, and the results discussed.

1 INTRODUCTION

Circular concrete-filled steel tubes (CCFST) have many advantages over other forms of construction for columns and other compression members. The steel confines the concrete protecting it and enhancing its load carrying capacity. Other advantages are its toughness and plasticity, its good fire resistance, its convenience of construction, all of which promote economical building (Zhong 2003). As a new structural form CCFST are being used more widely in tall buildings, large span bridges and other complex structures. To facilitate its use design codes have been developed in many countries, based on research and the design philosophy used in each country.

In China, three codes for designing CCFST structures are currently used, which were issued by the Chinese National Structural Materials and Industry Bureau (JCJ 01–89), the Association of Chinese Engineering Construction Standardization (CECS 28:90 and CECS 104:99) and the Chinese State Economic and Trade Commission (DL/T 5085–1999) respectively. These three codes are denoted as "CHN-JCJ 01–89", "CHN-CECS" and "CHN-DL/T 5085" in sequence in this paper. In addition, other three worldwide codes are presently used in USA, Europe and Japan, which were issued by the American Institute of Steel Construction (AISC-LRFD (1999)), European Committee for Standardization (Eurocode 4 (1994)) and the Architectural Institute of Japan (AIJ (1997)) respectively. They are denoted as "AISC-LRFD (99)", "Eurocode 4 (94)" and "JAN-AIJ (97)" in this paper.

2 DESIGN METHODS FOR CCFST COMPRESSIVE COLUMNS IN DIFFERENT CODES

In CHN-JCJ 01–89 and CHN-CECS codes, the ultimate strength of a short column (corresponding to the "peak point" of stress-strain curve derived from the experiment) is defined as the load-carrying capacity of the cross section. In CHN-DL/T 5085 code, the load-carrying capacity of a cross section is defined by limiting the longitudinal deformation, which often equals $3000 \, \mu\varepsilon$ in the stress-strain curve. The AISC-LRFD (99) code adopts the plastic strength of the whole cross section without allowing for confinement. Eurocode 4 (94) code also adopts the plastic strength of the whole cross section modified by 'enhancement factors (η)' obtained from experimental results in order to include the confining effect between steel tube and core concrete. JAN-AIJ (97) code adopts a simple superposed strength after considering the confining effect.

Different approaches are used in the processes of building up the formulae for calculating the strength of the cross section. In CHN-JCJ 01–89 and CHN-CECS codes, the formulae were built up by superposing the separate cross sectional strength of steel tube and core concrete based on the "confining theory" and experiments. In CHN-DL/T 5085 code, the whole CCFST cross section was regarded as a kind of new composite material, based on this 'composite material' the formulae were built up using numerical analysis and experiments from which the 'composite material properties' were derived. This method is innovative and is called the "unified theory" in China. The AISC-LRFD (99) code's formulae were built up by simply superposing the plastic cross sectional strength of the two separate parts. In Eurocode 4 (94) the simple superposed formulae for cross sectional strength, after being amended according to the experimental results to allow for confinement, were adopted. The formulae in JAN-AIJ(97) code were called "Superposed method" based on numerical analysis and experiments.

Thus five codes all use the simple superposed form of steel and core concrete parts, with or without enhancement factors for the effect of confinement. So the formula for calculating the load-carrying capacity of a cross section "N_0" can all be written in the form of $N_0 = f_s A_s + k f_c A_c$, where f_s, f_c are strengths of steel tube and concrete respectively and A_s, A_c the cross-sectional areas of steel tube and concrete respectively. The coefficient "k" is called the "enhanced coefficient of concrete strength". From this formula, it can be seen that in most of the codes, the confining effect is considered by increasing the concrete strength based on the assumption that the steel tube will not reduce its stress when it reaches its yield point during the loading process. In the CHN-DL/T 5085 code such form cannot be used because the cross section was regarded as a unified composite material in building up the formulae although the confining effect was considered in deciding the stress-strain relations of this 'composite material'. In addition, CHN-CECS and CHN-DL/T 5085 codes use a "confining coefficient (ξ)", defined as $\xi = (1.5 A_s f_y)/(A_c f_{cuk})$ (where f_{cuk} is the 150 mm cube compressive strength of concrete), as an important parameter for defining the characteristics of materials and geometry.

Overall buckling will reduce the load-carrying capacity of slender columns. The general method for considering the influence of slenderness, as in CHN-JCJ 01–89, CHN-CECS, CHN-DL/T 5085 and Eurocode 4 (94) codes, is by multiplying the cross sectional strength by a 'buckling coefficient', less 1.0, which can be obtained from buckling curves. There are differences in prescribing the buckling curves in the different codes. In the three Chinese codes the buckling curves are obtained from theoretical analysis and experiments. In the European Eurocode 4 (94) the buckling curves are those which have been prescribed for steel columns in Eurocode 3. In the American AISC-LRFD (99) code CCFST composite members are regarded as pure steel members with the formulae for calculating CCFST slender columns using the relative formulae for pure steel columns given in the same code. For the geometrical characteristics of the composite cross section, such as the moment of inertia and area, only the external steel tube is taken into account. However, for aspects of rigidity and the mechanical characteristics, such as strength and elastic modulus

of the composite cross section, the influence of the core concrete must be considered. In the Japanese code JAN-AIJ(97), the columns with L/D greater than 4.0 and less than 12 are classified as "medium length columns", while the columns with L/D greater than 12 are classified as "slender columns". For slender columns, the load-carrying capacity is calculated by superimposing the bearing capacities of the steel part and the concrete part, the influence of slenderness must be considered during the computing of the strength of each part. For "medium length" columns, the formula is a linear equation between the strength when the length-to-diameter ratio L/D equals 4 and when it is 12.

3 DESIGN METHODS FOR CCFST BEAM COLUMNS IN DIFFERENT CODES

In CHN-JCJ 01–89 code, Eurocode 4 (94) and JAN-AIJ (97) code for short and medium length beam columns ($L/D \leq 12$), the load-carrying capacity of CCFST under pure bending adopts the plastic bending strength of a cross section. The shape of internal stress distributions is assumed as rectangular while the concrete strength in the tensile zone are neglected. In CHN-CECS code, the design formula for load-carrying capacity of CCFST under pure bending is an empirical equation derived from two beam specimens which were loaded by two concentrated loads on the one-third points along the span of each beam. The load-carrying capacity of CCFST under pure bending is defined by restricting the mid-span deformation of beam specimen not to exceed 1/50 the length of beam span. In CHN-DL/T 5085 code, the whole CCFST cross section is regarded as a kind of new composite material according to the 'unified theory'. The load-carrying capacity of CCFST under pure bending in this code is defined as the moment whose maximum strain of outer fiber reached 10000 $\mu\varepsilon$ according to the moment-curvature relations derived from numerical analysis and experiments. In AISC-LRFD (99) code and JAN-AIJ (97) code for slender columns ($L/D > 12$), only the bending strength of the outer steel tube is defined as the load-carrying capacity of CCFST under pure bending. The only difference between these two codes is that the former adopts the elastic strength while the latter adopts the plastic strength. So this method for the AISC-LRFD (99) code must be more conservative. In addition, it should be noted that the load-carrying capacities of CCFST under pure bending in CHN-CECS code and JAN-AIJ (97) code are influenced by the length-diameter ratio (L/D).

For a CCFST beam column, a typical real moment-compression ($M-N$) interaction curve is a protruding curve. Different ways for simplifying the M-N interaction curves provide different design methods for

Table 1. Comparison of the limitations prescribed in different codes.

Item	CHN-JCJ 01–89	CHN-CECS	CHN-DL/T 5085	AISC-LRFD(99)	Euro-code4 (94)	JAN-AIJ(97)
f_{cuk} (MPa)	30~50	30~80	30~80	26~65	25~60	≤ 66.5
f_y (MPa)	235~345	235~420	235~390	≤ 415	235~355	235~353
D/t	–	$20{\sim}90\sqrt{(235/f_y)}$	20~100	$\leq \sqrt{(8E/f_y)}$	$\leq 90\sqrt{(235/f_y)}$	$\leq 35280/f_y$
α_s	0.04~0.16	–	0.04~0.20	≥ 0.04	–	–
ξ	–	0.03~3.0	–	–	–	–

* Where f_{cuk} is the 150 mm cube compressive strength of concrete; f_y is the yield strength of steel tube; A_s, A_c are areas of steel tube and concrete respectively; α_s is the steel ratio defined as $\alpha_s = A_s/A_c$ in CHN-CECS and CHN-DL/T 5085 codes but $\alpha_s = A_s/(A_c + A_s)$ in CHN-JCJ 01–89 and AISC-LRFD(99) codes; E is the elastic modulus of steel tube. ξ is the confining coefficient defined as $\xi = 1.5A_s f_y/(A_c f_{cuk})$ in CHN-CECS and CHN-DL/T 5085 codes.

cross-sectional strength of CCFST beam columns in different codes.

In CHN-JCJ 01–89 code and for the short beam columns in JAN-AIJ (97) code, the shape of the simplified cross sectional interaction curve calculated by the parametric equations is similar to the real protruding interaction curve. The shape of the cross sectional interaction curve simplified by CHN-CECS code, CHN-DL/T 5085 code and AISC-LRFD (99) code is two straight lines, which does not exhibit the real protruding interaction curve. It should be noted that the design formulae for the pure steel beam columns were adopted directly for the CCFST beam columns by AISC-LRFD (99) code. In Eurocode 4 (94), the real cross sectional interaction curve can be used or simplified as four straight lines linked between five 'characteristic points'. The values of M and N at these points are easy to be obtained according to the assumed initial stress distributions on the cross section. The shape of the simplified cross sectional interaction curves in Eurocode 4 (94) can exhibit the characters of a protruding curve. For the CCFST slender beam columns in JAN-AIJ (97) code, the simplified cross sectional interaction curve is obtained by superposing the $M - N$ cross sectional interaction curves for the steel tube and core concrete. The shape of interaction curve for the steel tube is a straight line while for concrete it is a protruding curve thus the shape of the simplified cross sectional interaction curves by this method can also exhibit the characters of the real protruding curve. For the design formulae in above codes the more similar to the real protruding curve for the simplified interaction curve is the more complex the design formulae. The cross sectional $M - N$ interaction curve calculated by each code can be referred to Figure 2(a).

For slender CCFST beam columns, the same approach is used in all six codes to consider the influences of slenderness ratio and 'P-δ' secondary effect in building up the formulae for buckling strength. The influence of slenderness is allowed for by substituting the cross-sectional compressive strength by buckling compressive strength. The influence of the 'P-δ' secondary effect is considered by multiplying the applied bending moment by an 'amplification factor', in which the distribution of moment along the beam column is also considered, although values are different in each code.

4 COMPARISON OF THE LIMITATIONS IN DIFFERENT CODES

Different limitations on the compressive strength of concrete, steel yield strength, diameter-to-thickness ratio, steel ratio and confining coefficient are prescribed by the different codes. These limitations are compared and summarized in Table 1. In this Table the concrete strength in the different codes has been converted to 150 mm cube strength f_{cuk} and all units have been converted to SI units because each country uses the compressive strength of concrete based on a specimen with different dimensions. The Chinese codes use $150 \times 150 \times 300$ mm 'prism strength' (f_{ck}); the European and American codes use 150×300 mm 'cylinder strength' (f_{cyl}); the Japanese code uses 100×200 mm 'cylinder strength' (f_{cylj}). The relationship between 'cube', 'prism' and 'cylinder' strengths have been taken, in Table 1, as: $f_{ck} = 0.67 f_{cuk}$ and $f_{cyl} = 0.8 f_{cuk}$, the relationship between f_{cylj} and f_{cyl} have been taken as: $f_{cylj} = 1.04 f_{cyl}$, referring to (Sun & Sakino 2000).

5 COMPARISON OF THE RESULTS CALCULATED BY DIFFERENT CODES

5.1 Comparison of CCFST compressive columns

Figure 1 shows a comparison of the load-carrying capacities N_u calculated by different codes compared with changes of the length-diameter ratio (L/D). The parameters chosen for calculation satisfy the limitations prescribed by each code (listed in Table 1).

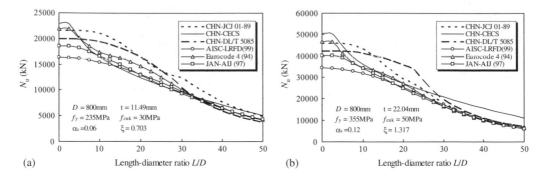

Figure 1. Comparison of N_u calculated by the different codes for two specified columns.

It can be seen from Figure 1 that the trend of N_u-L/D curves calculated by different codes is similar; N_u decreases with the increase of L/D. Although the calculated results have obvious differences for short columns the differences decrease with the increase of L/D until, for slender columns, the results calculated by the different codes are very similar. The main reason for this is that overall buckling plays the most important role on the slender columns. When L/D is less than 12, the results calculated by CHN-JCJ 01–89 code are generally the largest and this phenomenon is more obvious in the elastoplastic region. When L/D is greater than 12 but less than 25 the results calculated by CHN-DL/T 5085 code are usually the largest and this phenomenon is more obvious in the elasto-plastic region. The greatest difference in the strength calculated by each code for short columns ($L/D < 4$) is reflecting the different methods of allowing for confinement in the codes.

5.2 Comparison of CCFST beam columns

Figure 2 shows a comparison of the interaction curves of axial load 'N_u' and bending moment 'M_u' calculated by different codes under the variations of length-diameter ratio L/D. The parameters of the CCFST beam columns in this comparison are: $D = 800$ mm, $t = 22$ mm, $f_y = 345$ MPa, $f_{cuk} = 40$ MPa, $\alpha_s = 0.12$ and $\xi = 1.548$. All these parameters satisfy the limitations prescribed by each code listed in Table 1.

Except for the CHN-CECS code which is still composed of two obvious straight lines, the shapes of the N_u-M_u interaction curves calculated by the other five codes changed from non-linear curves to linear curves with the increase of L/D. This phenomenon indicates that the problem of buckling becomes the main factor for the very slender columns.

For short beam columns ($L/D \leq 4$) the results calculated by CHN-CECS code are the largest with the American AISC-LRFD (99) code well below the others because it does not consider the confining effect of the steel tube to enhance the strength. For medium length beam columns ($4 < L/D \leq 12$) there are still large differences between the codes in the axial load permitted at the same applied moment. The differences between the codes decrease as they become more slender, $L/D > 15$. It should be noted that the CHN-CECS code requires a large reduction in moment capacity as slenderness increases whereas the other codes show no reduction in pure moment capacity with slenderness.

6 COMPARISON BETWEEN RESULTS CALCULATED BY THE DIFFERENT CODES AND THE EXPERIMENTAL DATA

A database contains 1060 experimental tests on CCFST members, which includes 775 experimental tests on CCFST compressive columns from 40 worldwide references and 285 experimental tests on CCFST beam columns from 9 worldwide references, was built up by the author (referring to Ma 2005 and Ma 2006). Another database of over one thousand tests on both circular and rectangular CCFST columns, with and without moment, can be viewed on the website: *http://web.ukonline.co.uk/asccs2* was established by Dr. Goode (Goode 2006), where the test results are also compared with Eurocode 4 (94) predictions. During the process of comparison the different reduced coefficients of material used, which are called 'material partial safety factors' in some codes, were taken as unity. The material data use the given experimental data directly. The limitations required by the codes (Table 1) have not been considered and all tests have been included.

6.1 Comparison of CCFST compressive columns

Table 2 shows the statistical results of the N_u/N_{test} ratio calculated by each code for CCFST compressive column tests, where N_u is the load-carrying capacity calculated by the code and N_{test} is the corresponding test data from the references. The table gives the average value (μ) and standard deviation (σ). It should

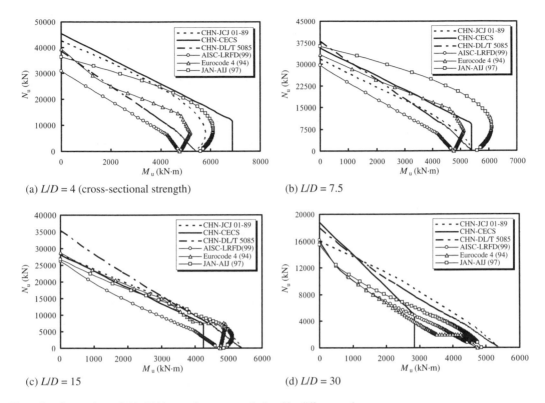

(a) *L/D* = 4 (cross-sectional strength)

(b) *L/D* = 7.5

(c) *L/D* = 15

(d) *L/D* = 30

Figure 2.   Comparison of 'N_u-M_u' interaction curves calculated by different codes.

Table 2.   Statistical results of Nu /Ntest ratio calculated by each code for CCFST compressive tests.

	CHN-JCJ 01–89		CHN-CECS		CHN-DL/T5085		AISC-LRFD(99)		Eurocode 4 (94)		JAN-AIJ(97)	
Item	μ	σ	μ	σ	μ	σ	μ	σ	μ	σ	μ	σ
All tests	No. = 696		No. = 775		No. = 775		No. = 775		No. = 775		No. = 775	
	0.982	0.176	0.952	0.153	0.952	0.190	0.767	0.127	0.926	0.152	0.852	0.136
Tests with	No. = 214		No. = 456		No. = 453		No. = 469		No. = 357		No. = 417	
limitations	1.015	0.169	0.975	0.146	0.958	0.168	0.760	0.123	0.910	0.150	0.843	0.132

* Where μ is the average value and σ is the standard deviation, of the ratio N_u/N_{test}. No. is the number of tests for calculation.

be noted that the number of tests recorded under the CHN-JCJ 01–89 code is smaller than the other codes because the formulae provided by this code are not applicable when the steel ratio is larger than 0.20.

For all 775 compressive column tests the average μ values for each code are less than 1.0. The results calculated by CHN-JCJ 01–89 code are closest to the experimental values with $\mu = 0.982$, followed by CHN-CECS, CHN-DL/T 5085, Eurocode 4 (94) and JAN-AIJ (97) codes in sequence. The results calculated by AI3C-LRFD (99) code are still much lower than the experimental values with $\mu = 0.767$. For tests which satisfy the limitations (listed in Table 1)

incorporated into each code. It will be seen that these limitations exclude many tests yet the average N_u/N_{test} and standard deviation of this ratio differ only slightly from these values when all tests are considered. This would imply that the restrictions imposed by the codes should be reconsidered by code committees and could be extended to cover a wider range of the parameters.

6.2   Comparison of CCFST beam columns

Table 3 shows the statistical results of the N_u/N_{test} ratio calculated by each code for CCFST compressive tests,

Table 3. Statistical results of Nu/Ntest ratio calculated by each code for CCFST beam columns.

| Item | CHN-JCJ 01–89 | | CHN-CECS | | CHN-DL/T5085 | | AISC-LRFD(99) | | Eurocode 4 (94) | | JAN-AIJ(97) | |
	μ	σ	μ	σ	μ	σ	μ	σ	μ	σ	μ	σ
All tests	No. = 285		No. = 285		No. = 285		No. = 285		No. = 285		No. = 285	
	0.866	0.148	0.865	0.132	0.900	0.152	0.735	0.116	0.870	0.190	0.919	0.148
Tests with	No. = 120		No. = 232		No. = 238		No. = 250		No. = 249		No. = 250	
limitations	0.898	0.151	0.872	0.126	0.909	0.153	0.739	0.112	0.857	0.121	0.913	0.130

* Where μ is the average value and σ is the standard deviation, of the ratio N_u/N_{test}. No. is the number of tests for calculation.

where N_u is the eccentric load-carrying capacity calculated by the code and N_{test} is the corresponding test data from the references. The table also gives the average value (μ) and standard deviation (σ).

For all 285 beam column tests the average μ values for each code are less than 1.0. The results calculated by JAN-AIJ (97) code are closest to the experimental values with $\mu = 0.919$, followed by CHN-DL/T 5085, Eurocode 4 (94), CHN-JCJ 01–89, and CHN-CECS codes in sequence. The results calculated by AISC-LRFD (99) code are much lower than the experimental values with $\mu = 0.735$. For tests which satisfy the limitations (listed in Table 1) incorporated into each code. It will be seen that there is also very little difference between the results when all tests are considered and the results when only those tests which satisfy the code limitations (Table 3) are considered. Only in Eurocode 4 (94) and JAN-AIJ (97) code does the average move very slightly down, that is in a safer direction, when these limitations are imposed; thus it might be possible to ease the code limitations.

7 CONCLUSIONS

This paper covered three Chinese codes and three typical worldwide codes for circular concrete-filled steel tube (CCFST) compressive columns and beam columns, compared the design methods the strength calculated by them with 1060 tests which include 775 compressive columns and 285 beam columns. Based on this the following conclusions were drawn:

1. The characteristics and methods for the design of CCFST members are, in many aspects, different in the different codes. The main factor that leads to differences for calculating the load-carrying capacity of compressive column by each code is attributed to how the confining effect of the steel tube was taken into account. On the other hand the main factor that leads to differences for calculating the load-carrying capacity of beam column by each code is attributed to how the pure bending strength was defined and how the real $M - N$ interaction curves from numerical analysis or experiments were simplified. The limitations prescribed by different codes are also different.

2. For short and medium length CCFST compressive columns, whose length-diameter ratios are less than 12, the load-carrying capacities calculated by CHN-JCJ 01–89 code are generally the largest. For slender compressive columns, whose length-diameter ratios are more than 12, the load-carrying capacities calculated by CHN-DL/T 5085 code are generally the largest. For short CCFST beam columns ($L/D \leq 4$) the results calculated by CHN-CECS code are the largest with the American AISC-LRFD (99) code well below the others. For medium length beam columns ($4 < L/D \leq 12$) there are still large differences between the codes but these decrease as they become more slender.

3. For CCFST compressive columns, comparing the load-carrying capacities calculated by each of the six codes with 775 experimental data the results calculated by the CHN-JCJ 01–89 code were closest to the experimental data and they were also slightly (2%) conservative. All the other codes were slightly more conservative with the AISC-LRFD (99) code the most conservative predicting strengths. For CCFST beam columns, comparing the load-carrying capacities calculated by each of the six codes with 285 experimental data, the JAN-AIJ (97) code gave closest agreement with tests while the AISC-LRFD (99) code also gave very safe results and may be too conservative.

4. For all codes there was very little difference between the results when all tests were considered and the results when tests were omitted which did not satisfy the limitations imposed by a code. Thus it might be possible to relax the code limitations.

REFERENCES

AIJ (1997). *Recommendations for design and construction of concrete filled steel tubular structures*. Architectural Institute of Japan.

AISC-LRFD (1999). *Load and Resistance Factor Design specification for structural steel building*. American Institute of Steel Construction.

Chinese Engineering Construction Standardization standard (CECS 28:90). *Specification for design and construction of concrete-filled steel tubular structures*. Beijing: Chinese Planning Press, Beijing. (in Chinese)

Chinese Engineering Construction Standardization standard (CECS 104:99). *Specification for design and construction of high-strength concrete structures*. Beijing: Chinese Planning Press. (in Chinese)

Chinese Electric Power Industry standard (DL/T 5085–1999). *Code for design of steel-concrete composite structure*. Beijing: Chinese Electric Power Press. (in Chinese)

Chinese National Structural Materials and Industry Bureau standard (JCJ 01–89). *Specification for design and construction of concrete-filled steel tubular structures*. Shanghai: Press of Tongji University. (in Chinese)

Eurocode 4 (1994). *Design of composite steel and concrete structures, Part 1.1: General rules and rules for buildings*. Brussels: European Committee for Standardization.

Goode, C. D., 2006. A review and analysis of over one thousand tests on concrete filled steel tube columns. *Proceedings of 8th international conference on steel-concrete composite and hybrid structures, Harbin, August 2006*: 17–23.

Ma, X.B. & Zhang, S.M. 2005. Comparison of design methods of load-carrying capacity of compressive members of CFST in typical codes worldwide. *Journal of Harbin Institute of Technology, SP2*: 96–99,136. (in Chinese)

Ma, X.B., Zhang, S.M. & Goode, C.D. 2006. Comparison of design methods for circular concrete filled steel tube beam columns in different codes. *Proceedings of 8th international conference on steel-concrete composite and hybrid structures, Harbin, August 2006*: 30–37.

Sun, Y.P. & Sakino, K.J. 2000. Simplified design method for ultimate capacities of circularly confined high-strength concrete columns. *The ACI special publication, SP193, September*: 561–578.

Zhong, S.T. 2003. *The concrete-filled steel tubular structures* (Third edition). Beijing: Press of Tingshua University. (in Chinese)

Steel and Composite Structures – Wang & Choi (eds)
© *2007 Taylor & Francis Group, London, ISBN 978-0-415-45141-3*

Modified elastic approach for stability design of in-plane frame columns

Hetao Hou
School of Civil Engineering, Shandong University, Jinan, Shandong, China

Guoqiang Li
School of Civil Engineering, Tongji University, Shanghai, China

ABSTRACT: In order to overcome the limitations of the current Chinese Code, one modified elastic approach for analysis and check for ultimate loading-carrying capacity of steel frames is outlined. This paper provides a simple criterion of frame classification, and describes the notional lateral load approach to account for the initial geometric imperfections within a second-order elastic analysis. The second-order elastic moments can be easily determined according to the first-order elastic analysis of the steel frame. Case studies are drawn to show the detailed design procedures of the modified elastic approach without calculating column effective length factors. Comparisons are made to the results from rigorous plastic zone analyses. In conclusion, the modified elastic approach presented is an efficient, reliable, practical method and therefore to be recommended for general design practice.

1 INTRODUCTION

The present approach to the analysis and design of a frame structures in Chinese Code for Design of Steel Structures(GB50017-2003) has been to conduct two, essentially separate operations, i.e. two-stage process in design: elastic analysis is conducted to determine the distribution of internal moments and forces of steel frame; inelastic analysis and design is carried out to determine the strength of each member treated as an isolated component by using the previously determined set of moments and forces. The interaction between the structural system and its members is represented by the effective length factor.

The effective length method generally provides a good method for the design of framed structures. However, the approach has major limitations. The first of these is that it does not give an accurate indication of factor against failure. The second and perhaps the most serious limitation is probably the rationale of the current two-stage process in design. There is no verification of the compatibility between the isolated member and the member as part of a frame. The effective length method is not user-friendly for a computer-based design. The other limitations of the effective length method include the difficulty of computing a K factor, and the inability of the method to predict the actual strength of framed member, among many others. To this end, there is an increasing awareness of the need for practical analysis/design methods that can account for the compatibility between the member and system without the use of K-factors (Nethercot 2000, Nethercot & Gardner 2005, Chen 2000, Chan & Zhou 2000, Liew et al. 2000).

The purpose of this paper is to present a practical design method based on the second-order elastic analysis for consideration of stability of the in-plane steel frame columns, which is expected to take the place of the effective length method.

2 SECOND-ORDER ELASTIC ANALYSIS METHOD

2.1 GB method

In GB50017-2003(GB), the internal forces and moments may be determined using first-order elastic analysis in all cases; or second-order elastic analysis in the case of $\rho = (\sum N \times \Delta u)/(\sum H \times h) > 0.1$, where $\rho =$ non dimensional parameter, $\sum N =$ the sum of design axial forces of all columns in a story, $\Delta u =$ the lateral inter-story deflection determined from the first-order elastic analysis due to design story shear $\sum H$, $h =$ the story height. GB method does not require any consideration of initial geometric imperfections or distributed plasticity within the first-order elastic analysis, but accounts for these effects through the column strength equations. However, when performing the second-order elastic analysis, notional lateral load of H_{ni} are applied at each story level in computing

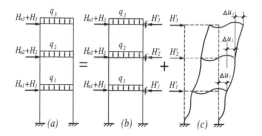

Figure 1. Approximate second-order elastic analysis of unbraced steel frame using GB method.

the moments to account for those mentioned effects. Notional lateral load of H_{ni} is expressed as follow

$$H_{ni} = \frac{\alpha N_i}{250}\sqrt{0.2 + \frac{1}{n_s}} \qquad (1)$$

where N_i = the sum of column design axial forces at story i, n_s = the number of stories, α = the factor to take into account the effect of the steel strength.

For unbraced steel frames in GB method, geometrical nonlinearity is accounted for indirectly by using moment magnification factor (B_2) in lieu of a second-order elastic analysis. In using the moment magnification approach, two first-order elastic analyses are performed on the frame, as shown in Figure 1. In the first analysis, the frame is artificially prevented from sway (by providing fictitious supports at each story level) and analyzed for a given load combination with the national lateral loads of H_{ni}, the reaction forces of H'_i of the fictitious supports are obtained. In the second analysis, the frame is allowed to sway (by removing fictitious supports) and analyzed for applied reaction forces of H'_i in a reverse direction. The maximum design moments obtained for each member from these two analyses, denoted as M_{Ib} and M_{Is}, respectively, are then combined to obtain the design values for the member using the equation

$$M_{II}=M_{Ib}+B_2M_{Is} \qquad (2)$$

where the factor B_2 is given by

$$B_2 = 1/(1-\rho) \qquad (3)$$

2.2 Modified elastic analysis approach

As mentioned above, one of the disadvantages of GB method is that the imperfections are not accounted for in the first-order elastic analysis, but those effects are accounted for by the use of the notional lateral load in

the second-order elastic analysis. Another major disadvantage is that two different frame models (braced and unbraced frame) are utilized to calculate the approximate second-order elastic moment of the unbraced frame from Equation 2. In order to overcome those limitations, this paper provides one modified second-order elastic analysis approach, which is made up of four parts: classification of frame; criterion of first- or second-order elastic analysis; approximate second-order elastic analysis procedure; model of the initial geometric imperfections.

Frame classification is a formalization of the engineering decision as to how the distribution of internal forces within a structure necessary for the selection of suitable members should be determined (Dario et al. 1996). In GB method, frames are classified into unbraced and braced cases which rely on whether there are braced bays or shear walls or not. The latter is then classified into partially-braced frame or fully braced frame. The difference between these two braced frames is that the minimum stiffness of the lateral bracing of the latter satisfies the following equation

$$S_b \geq 3(1.2\textstyle\sum N_{b\,i} -\textstyle\sum N_{0\,i}) \qquad (4)$$

where S_b = the stiffness of the lateral bracing, $\sum N_{bi}$ and $\sum N_{0i}$ = the sum of the nominal compressive strength of all columns at story i, determined in accordance with the effective length factors of the fully braced and unbraced columns, respectively. Its difficulty is that the effective length factors of all the columns must be calculated and the design process is tedious.

For convenience, this paper provides an improved criterion of frame classification to check whether a frame may be regarded as braced. If the lateral bracing have a total stiffness (T, whose unit is in force/displacement) of the frame at least five times the sum of the stiffness of all the columns within the story, i.e. $T = 5\sum 12i_c/L^2$ (or $K_T = TL^2/i_c = 60$), the frame achieves non-sway buckling mode, the frame can be defined as braced frame; otherwise, then defined as partially braced frame (Hou 2005), where L = the column length, $i_c = EI_c/L$, in which E = the modulus of elasticity, I_c = the moment of inertia, K_T = the non dimensional stiffness parameter, Following the improved criterion, the design process of frame columns therefore becomes simple and accurate.

Second-order effects shall be considered if they increase the action effects significantly or modify significantly the structural behavior. First-order elastic analysis may be used for the structure, if the increase of the relevant internal forces or moments or any other change of structural behavior caused by deformations can be neglected. In the modified second-order elastic analysis approach, the second-order elastic moments can be replaced by those obtained from the first-order

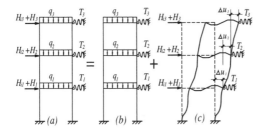

Figure 2. Approximate second-order elastic analysis of partially braced steel frame using modified elastic approach.

elastic analysis if the following criterion is satisfied (Hou 2005):

For $K_T \leq 5$, $\rho < 0.055$ (5a)

For $K_T > 5$, $\rho < 0.1$ (5b)

the terms are as defined as before.

As discussed above, GB method uses the moment magnification approach to calculate the second-order elastic moments, which is only suitable for unbraced columns and frames. One modified elastic analysis method based on GB method can be extended to partially braced frames, the second-order elastic moments, preferably determined from a second-order analysis, may be accounted for alternatively by amplifying the moments obtained from a first-order elastic analysis by the factor βB_2, under lateral loads, where β = the reduction factor to account for the effects of both the stiffness of lateral bracing and slenderness ratio of the frame column. In using the modified elastic analysis approach, two first-order elastic analyses are also performed on the frame, as shown in Figure 2. In the first analysis, the frame is analyzed under vertical loads. A second analysis is then performed with the frame subject to both the real and notional lateral loads. The maximum elastic design moments obtained for each column from these two analyses, denoted as M_{Iq} and M_{IH}, respectively, are then combined to obtain the design values for the column using the equation

$$M_{II} = M_{Iq} + \beta B_2 M_{IH} \qquad (6)$$

where the reduction factor β is given by

for $\lambda \leq 60$ or $\rho \leq 0.32$, $\beta = 1 - \eta \cdot \rho$ (7a)

for $\lambda > 60$ or $\rho > 0.32$, $\beta = 1 - 0.25\eta \cdot \rho$ (7b)

in which λ is the in-plane slenderness ratio of the column, and equals to KL/i, where K = the effective length factor obtained from Equation 8 (Li et al. 2006), L = the column length, i = the radius of gyration; η = the reduction factor due to the lateral bracing, given by Equation 9 (Hou 2005).

$$K = K_0 / (\sqrt{1 + (K_0^2/K_b^2 - 1)}(K_T/60)^{0.5} \qquad (8)$$

$$\eta = 0.1376 + 0.042 K_T \qquad (9)$$

where K_0 and K_b = effective length factors of unbraced and braced columns, respectively, which may be determined from clause 5.5.5 and Annex D of GB.

Initial geometric imperfections in the form of member out-of-straightness, frame out-of-plumbness are always present in real frameworks, which are currently assumed in many design codes are on the conservative side so that imperfection in the Perry-Robertson formula always results in a lower permissible load. The notional lateral load approach is currently used to account for the effect of frame out-of-plumbness by applying a set of notional lateral loads whose magnitudes are expressed as a friction of the gravity loads acting on the frame at each story level in the AS4100(1998), BS5950(2000) and EC3(2003). As mentioned above, the national lateral load approach is recommended to consider the effects of the geometric imperfections and residual stresses within the second-order elastic analysis in GB method. For the partially braced frame, those design codes do not provide the national lateral load approach to take into account of the effect of the member out-of-straightness, while this effect can be accounted for implicitly in the column equations.

Referring to the Chinese Code for Construction of Steel Structure (GB50205-2001), a fabrication tolerance of $L/1000$ for member out-of-straigtness, and $h/1000$ for frame out-of-plumbness are recommended in this study, where L = the column length, h = the height of the story. This paper presents a modified notional lateral load approach to account for both the member out-of-straightness and the frame out-of-plumbness (Hou 2005). This approach is especially suitable for calculating the elastic moments accurately in the partially braced frame. The modified notional lateral load applied at each story level and roof level is given by

$$H_i = \frac{N_i}{300}\sqrt{0.2 + \frac{1}{n_s}} \qquad (10)$$

where N_i = the sum of column design axial forces at story i, n_s = the number of stories.

3 MODIFIED ELASTIC ANALYSIS AND DESIGN PROCEDURES OF FRAME COLUMNS

The present study is limited to two-dimensional steel frames, the spatial behavior of frames is not considered, and the lateral torsional buckling of members is assumed to be prevented by adequate lateral braces. The frames are subjected to static loads, not earthquake or cyclic loads. The modified elastic analysis and design approach is based on the limit-state approach to strength. Analysis and design procedures in using the modified elastic approach are summarized as follows.

1. Calculate the dead load, live load and the load combinations based on the Load Code for the Design of Building Structures (GB50009-2001), the member sizes of frames are determined from an appropriate combination of factored loads.
2. Calculate the sum of the stiffness of all the columns within each story, $\sum 12i_c/L^2$; and the stiffness of the lateral bracing of each story, T and K_T.
3. If $K_T < 60$ is satisfying, the frame is defined as partially braced frame.
4. Firstly, one first-order elastic analysis is performed on the frame to obtain the internal forces (including the design axial forces N_q and the design moments M_{Iq}) under the factored vertical loads. Secondly, another first-order elastic analysis is then performed with the frame subject to both the real and notional lateral loads to obtain the internal forces (including the design axial forces N_H and the design moments M_{IH}) and the lateral inter-story deflection,Δu. The total design axial forces of the column is the combination of the design axial forces obtained from these two first-order elastic analysis $(N = N_q + N_H)$.
5. If both the expressions $K_T \le 5$ and $\rho > 0.055$, or $K_T > 5$ and $\rho > 0.1$ are satisfying, the second-order elastic should be performed on the frame, the second-order moments M_{II} can be calculated using Equation 6; otherwise, the second-order elastic moments can be easily replaced by those obtained from the first-order elastic analysis $(M_{II} = M_{Iq} + M_{IH})$, since the second-order effects may be neglected.
6. Calculate the resistance factor φ_x about x-axis for axial compressive columns according to the slenderness ratio λ_x based on $K = 1$, the steel strength and the type of cross-section. The reduced Euler buckling load about x-axis is calculated using the equation, $N'_{Ex} = \pi^2 EA/(1.1 \; \lambda_x)^2$, where $E =$ the modulus of elasticity, $A =$ the area of the cross-section.
7. Calculate the equivalent uniform moment factor relating to in-plane x-axis bending, $\beta_{mx} = 0.65 + 0.35M_2/M_1$, in which M_2/M_1 is the ratio of the smaller to larger moments at the ends of the column under consideration, and is taken as positive when the column is bent in single curvature, negative when bent in reverse curvature. The value may be determined in accordance with the clause 5.2.2 of GB. γ_x is the coefficient to account for the plastic distribution, defined in Table 5.2.1 of GB.
8. The checks for ultimate loading-carrying capacity of in-plane steel frame columns are carried out, respectively.

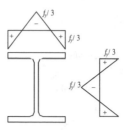

Figure 3. Residual stress.

The check for the section capacity is that

$$\frac{N}{Af} + \frac{M_x}{\gamma_x W_x f} \le 1 \tag{11}$$

The check for in-plane member capacity is

$$\frac{N}{\varphi_x Af} + \frac{\beta_{mx}M_x}{\gamma_x W_x f(1-0.8N/N'_{Ex})} \le 1 \tag{12}$$

where $N =$ the design axial force, $M_x =$ the design second-order elastic moment about x-axis, $W_x =$ the section modulus of the outside fiber of the compression flange about x-axis, $f =$ design axial steel strength.

For the case of braced frame, i.e. $K_T \ge 60$, except the design moment $M_{II} = M_{Iq} + M_{IH}$, the other analysis and design procedures are same as those for the case of partially braced frame.

4 CASE STUDIES

A partially braced portal frame and single-bay, three-story partially braced frame are selected for the present case studies. The design procedures of these frames follow analysis and design guidelines described previously. The case studies are carried out by comprising the results of the proposed modified elastic approach with those of rigorous distributed plasticity of analyses and of the GB method. The stress-strain relationship for all frame members is assumed to be elastic-perfectly plastic with 235 MPa yield stress (f_y) and 200000 MPa elastic modulus. The maximum residual stress is taken as $f_y/3$, shown in Figure 3.

A fabrication tolerance of $L/1000$ for member out-of-straightness, and $h/1000$ for frame out-of-plumbness are recommended in this study. ANSYS, one of the mostly widely used and accepted commercial finite element analysis program is used in plastic zone analysis. The fixed boundary is modeled at the base of the frame columns. All frame members are bent with respect to the major axis (x-axis), and the beam-column joints are all rigid. A compact hot-rolled H-section for all frame members is assumed so

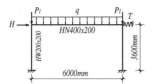

Figure 4. Configuration of portal frame.

Table 1. Elastic moments of rigorous analysis, GB and modified elastic approach for portal frame.

	(a) Rigorous analysis	(b) GB method	(b)/(a)	(c) Modified elastic approach	(c)/(a)
M kN.m	116.98	115.7	0.989	117.23	1.002

that sections can develop full plastic moment capacity without local buckling. The vertical loads are first applied to the frame, and then the lateral loads are applied at the each floor level and roof level until the frame could not resist any more loads.

The ultimate load-carrying capacity indicates the maximum load that the steel frame can sustain. The ultimate loads obtained from the plastic-zone analysis are applied to the frame to carry out first- or second-order elastic analysis to check the section capacity and in-plane stability of the column by using the GB method and the modified elastic approach. The errors in the checks for the section capacity and in-plane stability are calculated as Equations 13 and 14, respectively.

$$\varepsilon_1 = \frac{N}{Af} + \frac{M_x}{\gamma_x W_x f} - 1 \qquad (13)$$

$$\varepsilon_2 = \frac{N}{\varphi_x Af} + \frac{\beta_{mx} M_x}{\gamma_x W_x f (1 - 0.8 N / N'_{Ex})} - 1 \qquad (14)$$

Unconservative error is represented by a negative value (unsafe result, $\varepsilon_1 < 0$ or $\varepsilon_2 < 0$), and conservative error by a positive value; as ε_1 or ε_2 equals to 0, the result of check of the member capacity represents the real ultimate load-carrying capacity of the frame.

4.1 Single-bay partially braced portal frame

Figure 4 shows the partially braced portal frame subjected to a non-proportional loading. The members are HN400 × 200 × 8 × 13 ($A = 8412$ mm², $i_x = 168$ mm, $I_x = 2.37 \times 10^8$ mm⁴, $W_x = 1.19 \times 10^6$ mm³) for the beam and HW200 × 200 × 8 × 12 ($A = 6428$ mm², $i_x = 86.1$ mm, $I_x = 4.77 \times 10^7$ mm⁴, $W_x = 4.77 \times 10^5$ mm³) for the columns. Factored distributed beam force of $q = 59.4$ kN/m, and concentrated gravity loads of $P_1 = 287$ kN are applied to the portal frame. The stiffness of lateral bracing at the top of the column is assumed to be the sum of the stiffness of the columns within the story, i.e. $T = \sum 12 i_c / L^2$ (or $K_T = 12$). The ultimate load-carrying capacity H_u of the frame is calculated to be 131.08 kN using the plastic zone analysis.

The portal frame is subjected to the combined action of gravity loads (q and P_1) and lateral load (H_u). The GB method is based on first- or second-order elastic

analysis of the perfect structure (i.e. no notional lateral loads or modeled imperfections are included). For the modified elastic approach, notional lateral load of H_i (Equation 10) is applied in the load combination of gravity loads and lateral load. The portal frame is also analyzed by a rigorous second-order analysis to determine an accurate distribution of the moments. These moments of the right column are used for comparison to those established by the GB method and the modified elastic approach. The results of analysis and comparison are summarized in Table 1.

Table 1 indicates that the result of the GB method is approximately 1.01% unconservative compared to the rigorous second-order analysis results. This difference is attributed to the fact that the GB method does not consider the second-order effects and the initial geometric imperfections. Whereas the analysis result of the modified elastic approach is only 0.2% conservative compared that of the rigorous analysis.

The errors in the checks for ultimate loading-carrying capacity of in-plane steel frame column determined from Equations 13 and 14 by using the GB method and the modified elastic approach are provided in Table 2. Table 2 shows that the error ε_2 determined by the GB method is more conservative than ε_1, so the in-plane stability capacity governs the ultimate loading-carrying capacity of the frame. Conversely, the section capacity in the modified elastic approach controls the ultimate loading-carrying capacity of the frame. In addition, the error ε_2 in GB method is more conservative than ε_1 in the modified elastic approach. The modified elastic approach gives a more liberal accurate estimate of ultimate loading-carrying capacity of in-plane steel frame column. It can be observed that in the modified elastic approach the section capacity check is needed and the in-plane stability check is not required. The conservative error of the section capacity check in the modified elastic approach is attributed to the fact that the modified elastic approach does not consider the inelastic moment redistribution but the plastic zone analysis includes inelastic redistribution effect.

4.2 Three-story partially braced frame

A single bay, three story frame is selected to demonstrate the accuracy and validity of the modified elastic

291

Table 2. Errors of member check by using GB method and modified elastic approach for portal frame.

	GB method	Modified elastic approach
ε_1 (section capacity)	47.52%	48.03%
ε_2 (in-plane member capacity)	56.20%	−6.47%

Table 3. Elastic moments of rigorous analysis, GB and modified elastic approach for three-story frame.

	(a) Rigorous analysis	(b) GB method	(b)/(a)	(c) Modified elastic approach	(c)/(a)
M kN.m	86.93	82.2	0.945	88.32	1.016

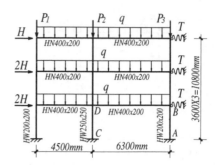

Figure 5. Configuration of three-story frame.

Table 4. Errors of member check by using GB method and modified elastic approach for three-story frame.

	GB method	Modified elastic approach
ε_1 (section capacity)	25.89%	30.71%
ε_2 (in-plane member capacity)	34.48%	−20.50%

approach for evaluation of the loading-carrying capacity of steel frame columns. The configuration of the frame is shown in Figure 5.

The members are HN400 × 200 × 8 × 13 for the beams and HW200 × 200 × 8 × 12 for the exterior columns, HW250 × 250 × 9 × 14 ($A = 9218\,\mathrm{mm}^2$, $i_x = 108\,\mathrm{mm}$, $I_x = 1.08 \times 10^8\,\mathrm{mm}^4$, $W_x = 8.67 \times 10^5\,\mathrm{mm}^3$) for the interior columns. The beams are subjected to factored distributed member forces of $q = 70.3\,\mathrm{kN/m}$, and the concentrated gravity loads of $P_1 = 158.18\,\mathrm{kN}$, $P_2 = 379.62\,\mathrm{kN}$, $P_3 = 221.45\,\mathrm{kN}$ are simultaneously applied to the top of the columns of the frame, respectively. The stiffness of lateral bracing at each floor level is assumed to be the sum of the stiffness of the columns within the story, i.e. $T = \sum 12i_c/L^2$ (or $K_T = 12$). The ultimate load-carrying capacity H_u of the frame is calculated to be 142.86 kN using the plastic zone analysis.

The elastic analysis results of the right exterior column AB of the three-story frame using the rigorous second-order analysis, the GB method and the modified elastic approach are presented in Table 3. Similar to the portal frame, the analysis result of the modified elastic approach is only 1.6% conservative compared that of the rigorous analysis, but the unconservtive error in the GB method is up to 5.5% due to ignoring the geometric nonlinearity and the initial geometric imperfections.

Table 4 presents the errors in the checks for ultimate loading-carrying capacity of in-plane steel frame column determined from Equations 13 and 14 by using the GB method and the modified elastic approach. As previously discussed about the portal frame, Table 4 shows that the GB method is also controlled by the check for the in-plane capacity of the three-story frame; and the modified elastic approach does not rely on

effective length factors to assess the in-plane capacity, its design check is based on the member section capacity. The error ε_2 in the GB method is more conservative than ε_1 in the modified elastic approach. Consequently, the section capacity check in the modified elastic approach is needed only. The modified elastic approach provides the more accurate prediction of the ultimate loading-carrying capacity of steel frame, and also produces an economical design than the GB method.

5 CONCLUSIONS

A modified elastic approach has been developed for analysis and check for ultimate loading-carrying of steel frames. The conclusions of this study are as follows.

1. Compared to GB method, the modified elastic approach provides an improved criterion of frame classification.
2. The modified elastic approach presents a criterion of first- or second-order elastic analysis.
3. By using the modified elastic approach, the second-order elastic moments in a steel frame can be easily determined according to the first-order elastic analysis of the steel frame under vertical loads and lateral loads respectively. The case studies shows that such second-order elastic moments are more accurate than those determined from the GB method.

4. The modified elastic approach proposes a modified notional lateral load applied at each story level to account for the effects of the initial geometric imperfection (including both the member out-of-straightness and the frame out-of-plumbness) of the partially-braced frame.

5. The case studies show that the section capacity in the modified elastic approach controls the ultimate loading-carrying capacity of the frame, whereas the in-plane stability capacity in the GB method governs the ultimate loading-carrying capacity of the frame. The check errors of the section capacity in the modified elastic approach are less than that of the in-plane stability capacity in the GB method, as a result, the modified elastic approach captures well the ultimate load-carrying capacity of the steel frame including its individual members.

6. The modified elastic approach is time-effective in design process because it completely eliminates tedious and often confused member capacity checks including the calculation of K-factors in the GB method.

7. The modified elastic approach presented is an efficient, reliable, practical method and therefore to be recommended for general design practice.

REFERENCES

AS-4100, 1998, Australia Standard for Steel Structures, Sydney.

BS5950, 2000, British Standards Institution. Structural Use of Steel in Building, Part 1, U.K.

CEN.(2003). prEN 1993-1-1:2003 Eurocode 3: Design of steel structures, Part 1-1: General rules and rules for buildings. European Committee for Standardization, Brussels.

Chen, W.F., 2000, Structural stability : from theory to practice, *Engineering Structures* 22: 116–122.

Dario, J. & Aristizabal-Ochoa, 1996, Braced, partially braced and unbraced columns: complete set of classical stability equations, *Structural Engineering and Mechanics*, Vol.4, No.4: 365–381.

GB50009-2001, 2001, Load Code for the Design of Building Structures, Ministry of Construction, China.

GB50017-2003, 2003, Code for Design of Steel Structures, Ministry of Construction, China.

GB50205-2001, 2001, Code for Construction of Steel Structure, Ministry of Construction, China.

Guoqiang Li, Yushu Liu & Xin Zhao, 2006, *Advanced analysis and design of reliability of the steel structure frame*, Beijing, Press of Chinese Building Industry, China.

Hetao Hou, 2005, The research on check method for limit state strengths of steel frame columns, Ph.D. Thesis, Tongji University, China.

Liew, J.Y.Richard, Chen, W.F. & Chen, H., 2000, Advanced inelastic analysis of frame structures, *Journal of Constructional Steel Research* 55: 245–265.

Nethercot, D.A., 2000, Frame structures: global performance, static and stability behavior general report, *Journal of Constructional Steel Research* 55: 109–124.

Nethercot, D.A. & L.Gardner, 2005, The EC3 approach to the design of columns, beams and beam-columns, *Steel and Composite Structures*, Vol.5, No.2–3: 127–140.

Siu-Lai Chan & Zhi-Hua Zhou, 2000, Non-linear integrated design and analysis of skeletal structures by 1 element per member, *Engineering Structures* 22: 246–257.

Steel and Composite Structures – Wang & Choi (eds)
© *2007 Taylor & Francis Group, London, ISBN 978-0-415-45141-3*

Static and linear buckling analysis of changed boundary beam

Yongfeng Luo, Rui Yu & Xue Li
Building Engineering Department, Tongji University, ShangHai, China

Wei Yang
East China Electric Power Design Institute, ShangHai, China

ABSTRACT: A structure with different boundary conditions in construction stage and service stage is a new construction idea used in long span structures. The structural internal forces change with boundary condition, which is a key point in our study. A single beam, which is subjected to uniformly distributed lateral load, is introduced in this paper as an example. Firstly, static analysis is conducted and the results are compared on the beams with three different boundary conditions (simply supported, fixed supported and changed boundaries). Then, linear buckling behaviors of the beam with changed boundary are investigated, and the equilibrium equation is developed subsequently for commutating the critical load. This provides a valuable reference in later analyzing work for similar structures.

1 INTRODUCTION

A structure with different boundary conditions in construction stage and service stage, i.e. changed boundary structure is a new construction idea in long span structures. The structural internal forces do not remain constant while boundary condition changes. During the whole construction process, the period of boundary changing may be so short that the internal forces may change in a few minutes. So, changed boundary structures are treated to be time-dependant. This new form of structure cannot be solved satisfactorily with traditional computation method, and time-varying mechanics should be used.

The background of this paper is The National Grand Theatre of China, a characteristic large spatial structure that has been paid special attention to both national and international. Changed boundary method is employed in this structure. Not only the materials are used more efficiently, but also adopting the new method lowers the cost of whole structure. The system was supposed to be hinge joint under gravity load of shells and roofs. After the construction, each curved truss is connected with concrete ring beam at the bottom. This promises that the whole system is fixed constrained in normal service stage. Currently, few structures have adopted changed boundary method, and thus the study on mechanical characteristics as well as relevant theory is limited.

A beam with changed boundary is adopted in this paper as an example. A two-stage method is employed in the analysis. At Stage 1, the beam subjected to uniformly distributed lateral load q1 is simply supported, showing that the beam is under construction. And at Stage 2, the beam subjected to uniformly distributed lateral load q_2 is fixed supported, as the service stage. The theoretical equations of this kind of beams are deduced on static and linear buckling analysis. This provides a valuable theoretical basis in later analysis work for similar structures.

2 STATIC ANALYSIS OF CHANGED BOUNDARY BEAM

In order to tell the differences between changed boundary beam and unchanged boundary ones, two beams, simply supported and fixed supported respectively, are employed.

All beams analyzed below are supposed to be with the same length of 1 and same cross-sectional moment of inertia. The final loads on three beams are equal. Static analysis of the beams with three different boundary conditions: simply supported, fixed supported and changed boundary are executed respectively, and the results are compared as shown in Figure 1.

The parameters can be obtained consequently and shown as the Equation. (1)~(5).

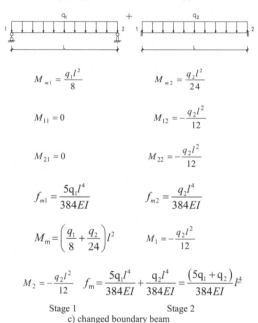

$$M'_m = \frac{(q_1+q_2)l^2}{8}$$

$$M''_m = \frac{(q_1+q_2)l^2}{24}$$

$$M'_1 = 0$$

$$M''_1 = -\frac{(q_1+q_2)l^2}{12}$$

$$M'_2 = 0$$

$$M''_2 = -\frac{(q_1+q_2)l^2}{12}$$

$$f'_m = \frac{5(q_1+q_2)l^4}{384EI}$$

$$f''_m = \frac{(q_1+q_2)l^4}{384EI}$$

a) simply supported beam   b) fixed supported beam

$$M_{m1} = \frac{q_1 l^2}{8}$$

$$M_{m2} = \frac{q_2 l^2}{24}$$

$$M_{11} = 0$$

$$M_{12} = -\frac{q_2 l^2}{12}$$

$$M_{21} = 0$$

$$M_{22} = -\frac{q_2 l^2}{12}$$

$$f_{m1} = \frac{5q_1 l^4}{384EI}$$

$$f_{m2} = \frac{q_2 l^4}{384EI}$$

$$M_m = \left(\frac{q_1}{8}+\frac{q_2}{24}\right)l^2$$

$$M_1 = -\frac{q_2 l^2}{12}$$

$$M_2 = -\frac{q_2 l^2}{12} \quad f_m = \frac{5q_1 l^4}{384EI}+\frac{q_2 l^4}{384EI}=\frac{(5q_1+q_2)}{384EI}l^4$$

Stage 1   Stage 2
c) changed boundary beam

Figure 1.   Bending Moments and Deflections of the Beams with Different Boundary Conditions.

$$\eta'_1 = \frac{M'_m - M_m}{M'_m} = \frac{\dfrac{q_2 l^2}{12}}{\dfrac{(q_1+q_2)l^2}{8}} = \frac{2}{3\left(\dfrac{q_1}{q_2}+1\right)} \tag{1}$$

$$\eta''_1 = \frac{M''_m - M_m}{M''_m} = \frac{\left(\dfrac{1}{24}-\dfrac{1}{8}\right)q_1}{\dfrac{(q_1+q_2)}{24}} = -2\frac{\dfrac{q_1}{q_2}}{\dfrac{q_1}{q_2}+1} \tag{2}$$

$$\eta''_2 = \frac{M''_1 - M_1}{M''_1} = \frac{-\dfrac{1}{12}q_1}{\dfrac{(q_1+q_2)}{12}} = \frac{\dfrac{q_1}{q_2}}{\dfrac{q_1}{q_2}+1} \tag{3}$$

$$\eta'_3 = \frac{f'_m - f_m}{f'_m} = \frac{4q_2}{5(q_1+q_2)} = \frac{4}{5\left(\dfrac{q_1}{q_2}+1\right)} \tag{4}$$

$$\eta''_3 = \frac{f''_m - f_m}{f''_m} = -\frac{4q_1}{(q_1+q_2)} = -\frac{4\dfrac{q_1}{q_2}}{\left(\dfrac{q_1}{q_2}+1\right)} \tag{5}$$

Where $q_1 =$ the uniformly distributed lateral load at Stage 1 on changed boundary beam;
$q_2 =$ the uniformly distributed lateral load at Stage 2 on changed boundary beam;
$q =$ the uniformly distributed load on simply and fixed supported beam $(q = q_1 + q_2)$
$l =$ the length of beam;
$M'_m, M'_1, M'_2; M''_m, M''_1, M''_2 =$ bending moments at midspan and node1, 2 of unchanged boundary beam, respectively;
$M_{m1}, M_{11}, M_{21}; M_{m2}, M_{12}, M_{22} =$ bending moments at mid-span and node 1, 2 of the changed boundary beam at two stages, respectively;
$M_m, M_1, M_2 =$ overall bending moments at mid-span and node 1, 2 of the changed boundary beam, respectively;
$f'_m; f''_m =$ mid-span deflections of simply and fixed supported beam, respectively;
$f_{m1}; f_{m2} =$ mid-span deflections of changed boundary beam at two stages respectively;
$f_m =$ overall deflection at mid-span point of changed boundary beam;
$E =$ modulus of elasticity;
$I =$ cross-sectional moment of inertia;
$\eta'_1 =$ relative deviation of mid-span moment of simply supported beam and changed boundary beam;
$\eta''_1 =$ relative deviation of mid-span moment of fixed supported beam and changed boundary beam;
$\eta''_2 =$ relative deviation of moment at node 1 of fixed supported beam and changed boundary beam;
$\eta'_3 =$ relative deviation of mid-span deflection of simply supported and changed boundary beam;
$\eta''_3 =$ relative deviation of the mid-span deflection of fixed supported and changed boundary beam.
According to the value of q_1/q_2., relative deviations are shown in Table 1.

It is observed in Table 1 that the bending moment and deflection at mid-span point of changed boundary beam are less than those of simply supported beams. Another conclusion can be drawn that the difference of mid-span moment between changed boundary beam

Table 1. Relative Deviations.

q_1/q_2	0	0.5	1	2	∞
η_1'	2/3	4/9	1/3	2/9	0
η_1''	0	−2/3	−1	−4/3	−2
η_2''	0	1/3	1/2	2/3	1
η_3'	4/5	8/15	2/5	4/15	0
η_3''	0	−4/3	−2	−8/3	−4

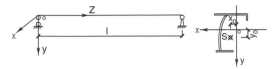

Figure 2. Coordinate System of Beam.

and fixed supported beam is determined by the value of q_1/q_2. However, the deflection of changed boundary beam is larger than that of fixed supported beam, while it is the opposite with the end moment.

3 LINEAR BUCKLING ANALYSIS OF CHANGED BOUNDARY BEAM

In this part, study is focused on a changed boundary beam with I-section only. Several assumptions are included in order to simplify the model.

(1) All members are elastic.
(2) No deformation will occur on the cross-section when subjected to lateral bending moment and torque.
(3) Lateral-torsional deformation is small.
(4) All members are with uniform section and undefected.
(5) Residual stress is out of consideration.
(6) The in-plan rigidity of all members is considered to be large enough that the influence that pre-buckling deformation brings to flexural-torsional buckling can be neglected.

Firstly, load q_1 is applied on the beam with simply support, and the member is stable in plane. Then load q_2 is applied on it with fixed support and the member may lose stability in this stage. It can be easily understood that the displacement of Stage 1 is considered to be the initial displacement of Stage 2. The total internal forces consist of those in two stages. In this paper, the functions are established in the condition of fixed boundary. u is the offset between shear center S and shape center O in X direction, and ϕ is element rotation to sheer center S, shown in Figure. 2.

With the assumption that the buckling deformation distributed according to half sine wave

$$u = C_1(1 - \cos\frac{2\pi z}{l}) \tag{6}$$

$$\varphi = C_1(1 - \cos\frac{2\pi z}{l}) \tag{7}$$

and the boundary condition of the beam which is subjected to q_2 is

$$u(0) = u(l) = 0 \tag{8}$$

$$u'(0) = u'(l) = 0 \tag{9}$$

$$\varphi(0) = \varphi(l) = 0 \tag{10}$$

$$\varphi'(0) = \varphi'(l) = 0 \tag{11}$$

Total potential energy of uniformly distributed beam can be expressed as

$$\Pi = \frac{1}{2}\int_0^l \left[EI_y u'^2 + EI_\omega \varphi'^2 + GI_k \varphi'^2 + 2\beta_x M_x \varphi'^2 + 2M_x u' \varphi - qa\varphi^2 \right] dz \tag{12}$$

where

E = modulus of elasticity
G = shear modulus;
A = cross-sectional area;
I_y = cross-sectional moment of inertia about y axis;
I_w = cross-sectional warping moment of inertia ;
I_k = free torsional constant of cross section;
M_x = bending moment about x axis;
q = uniformly distributed lateral load;
M_ω = restraint torque;
a = loading position.

$$\beta_x = \frac{\int_A y(x^2 + y^2)dA}{2I_x} - y_0$$

The bending moment of an arbitrary section can be obtained by linearly superposition of the moment at two stages:

$$M_x = M_{x1} + M_{x2} = \frac{q_1 lz}{2} - q_1 z\frac{z}{2} + \frac{q_2 l}{2}z - \frac{q_2 l^2}{12} - q_2 z\frac{z}{2} = \frac{1}{2}z(l-z)(q_1+q_2) - \frac{1}{12}q_2 l^2 \tag{13}$$

Putting Equation. 6, 7, 8, 9, 10, 11, 13 to Equation. 12, total potential energy of the member can be expressed as:

$$\Pi = \frac{1}{2} \int_0^l \left\{ EI_y C_1^2 \frac{16\pi^4}{l^4} \cos^2 \frac{2\pi z}{l} + EI_\omega C_2^2 \frac{16\pi^4}{l^4} \cos^2 \frac{2\pi z}{l} + GI_k C_2^2 \frac{4\pi^2}{l^2} \sin^2 \frac{2\pi z}{l} \right.$$
$$+ 2\beta_x \left[\frac{1}{2} z(l-z)(q_1+q_2) - \frac{1}{12} q_2 l^2 \right] C_2^2 \frac{4\pi^2}{l^2} \sin^2 \frac{2\pi z}{l} +$$
$$2 \left[\frac{1}{2} z(l-z)(q_1+q_2) - \frac{1}{12} q_2 l^2 \right] C_1 C_2 \frac{4\pi^2}{l^2} \cos \frac{2\pi z}{l} C_2 \left(1 - \cos \frac{2\pi z}{l} \right)$$
$$\left. - (q_1+q_2) a C_2^2 \left(1 - \cos \frac{2\pi z}{l} \right)^2 \right\} dz$$

$$= \frac{1}{2} \int_0^l \left\{ \left(EI_y C_1^2 + EI_\omega C_2^2 \right) \frac{16\pi^4}{l^4} \cos^2 \frac{2\pi z}{l} \right.$$
$$+ \left[GI_k + \beta_x z(l-z)(q_1+q_2) - \frac{1}{6} \beta_x q_2 l^2 \right] C_2^2 \frac{4\pi^2}{l^2} \times$$
$$\sin^2 \frac{2\pi z}{l} \left[z(l-z)(q_1+q_2) - \frac{1}{6} q_2 l^2 \right] C_1 C_2 \frac{4\pi^2}{l^2} \cos \frac{2\pi z}{l}$$
$$- \left[z(l-z)(q_1+q_2) - \frac{1}{6} q_2 l^2 \right] C_1 C_2 \frac{4\pi^2}{l^2} \cos^2 \frac{2\pi z}{l}$$
$$\left. - (q_1+q_2) a C_2^2 \frac{4\pi^2}{l^2} + 2(q_1+q_2) a C_2^2 \cos \frac{2\pi z}{l} - (q_1+q_2) a C_2^2 \cos^2 \frac{2\pi z}{l} \right\} dz$$

$$(14)$$

where

$$\int_0^l \cos^2 \frac{2\pi z}{l} dz = \frac{l}{2} \qquad \int_0^l \sin^2 \frac{2\pi z}{l} dz = \frac{l}{2} \qquad \int_0^l \cos \frac{2\pi z}{l} dz = 0$$

$$\int_0^l (zl - z^2) \cos \frac{2\pi z}{l} dz = -\frac{l^3}{2\pi^2}$$

$$\int_0^l (zl - z^2) \cos^2 \frac{2\pi z}{l} dz = \frac{(4\pi^2 - 3) l^3}{48\pi^2}$$

$$\int_0^l (zl - z^2) \sin^2 \frac{2\pi z}{l} dz = \frac{(4\pi^2 + 3) l^3}{48\pi^2}$$

Substituting the above integral equations to Equation. 14, then

$$\Pi = \frac{1}{2} \left[\frac{8\pi^4}{l^3} \left(EI_y C_1^2 + EI_\omega C_2^2 \right) + \frac{2\pi^2}{l} C_2^2 \left(GI_k - \frac{1}{6} \beta_x q_2 l^2 \right) + \frac{4\pi^2 + 3}{48\pi^2} \beta_x C_2^2 4\pi^2 l (q_1+q_2) \right.$$
$$\left. - C_1 C_2 2l(q_1+q_2) \frac{4\pi^2-3}{48\pi^2} C_1 C_2 \cdot 4\pi^2 l(q_1+q_2) + C_1 C_2 \cdot \frac{\pi^2}{3} l q_2 - \frac{3a C_2^2}{2} l(q_1+q_2) \right]$$

$$= \frac{4\pi^4}{l^3} EI_y C_1^2 + \frac{1}{2} \left[\frac{8\pi^4}{l^3} EI_\omega + \frac{2\pi^2}{l} \left(GI_k - \frac{1}{6} \beta_x q_2 l^2 \right) + \frac{4\pi^2+3}{48\pi^2} \beta_x \cdot 4\pi^2 l(q_1+q_2) \right.$$
$$\left. - \frac{3a}{2} l(q_1+q_2) \right] C_2^2 + \frac{1}{2} \left[\frac{\pi^2}{3} q_2 l - 2(q_1+q_2) l - \frac{4\pi^2-3}{48\pi^2} \cdot 4\pi^2 (q_1+q_2) l \right] C_1 C_2 \quad (15)$$

According to principle of stationary potential energy, when the Equation. 15 is partial differentiated to C_1 and C_2 respectively, they both equal to zero, i.e.

$$\frac{\partial \Pi}{\partial C_1} = 0, \quad \frac{\partial \Pi}{\partial C_2} = 0$$

$$\begin{cases} \left[\frac{8\pi^4}{l^3} EI_y C_1 + \frac{1}{2} \left[\frac{\pi^2}{3} q_2 l - 2(q_1+q_2) l - \frac{4\pi^2-3}{12} (q_1+q_2) l \right] C_2 = 0 \\ \frac{1}{2} \left[\frac{\pi^2}{3} q_2 l - 2(q_1+q_2) l - \frac{4\pi^2-3}{12} (q_1+q_2) l \right] C_1 \\ + \left[\frac{8\pi^4}{l^3} EI_\omega + \frac{2\pi^2}{l} \left(GI_k - \frac{1}{6} \beta_x q_2 l^2 \right) + \frac{4\pi^2+3}{12} \beta_x l(q_1+q_2) - \frac{3a}{2} l(q_1+q_2) \right] C_2 = 0 \end{cases}$$

$$(16)$$

Hence, the critical buckling condition is:

$$\begin{vmatrix} \frac{8\pi^4}{l^3} EI_y & \frac{4\pi^2+21}{24} q_2 l - \frac{7}{8} q_2 l \\ \frac{4\pi^2+21}{24} q_2 l - \frac{7}{8} q_2 l & \frac{8\pi^4}{l^3} EI_\omega + \frac{2\pi^2}{l} GI_k + \left(\frac{4\pi^2+3}{12} \beta_x l - \frac{3}{2} a \right) q_1 + \left(\frac{1}{4} \beta_x l - \frac{3}{2} a \right) q_2 \end{vmatrix} = 0$$

$$(17)$$

By solving Equation. 17, the critical load q_2 can be expressed by q_1 as follows:

$$q_2 = \frac{-(Bq_1+C) \pm \sqrt{(Bq_1+C)^2 - 4A(Dq_1^2 - Fq_1 - H)}}{2A}$$

$$(18)$$

where

$$A = \frac{49}{64} l^2 \qquad B = \frac{7}{4} \left(\frac{4\pi^2+21}{24} \right) l^2 \qquad C = \frac{8\pi^4}{l^3} EI_y (1.5al - 0.25\beta_x l)$$

$$D = \left(\frac{4\pi^2+21}{24} \right)^2 l^2 \qquad F = \frac{8\pi^4}{l^3} EI_y \left[\left(\frac{4\pi^2+3}{12} \right) \beta_x l - 1.5al \right]$$

$$H = \frac{8\pi^4}{l^3} EI_y \left(\frac{8\pi^4 EI_\omega}{l^3} + \frac{2\pi^2}{l} GI_k \right)$$

If $Dq_1^2 - Fq_1 - H \geq 0$,

then $\sqrt{(Bq_1 + C)^2 - 4A(Dq_1^2 - Fq_1 - H)} \leq Bq_1 + C$,

$q_2^{①} \leq q_2^{②} \leq 0$, which does not have any realistic meaning. ($q_2^{①}, q_2^{②}$ are two answers respectively.) Otherwise, if $Dq_1^2 - Fq_1 - H < 0$,

then $\sqrt{(Bq_1 + C)^2 - 4A(Dq_1^2 - Fq_1 - H)} > Bq_1 + C$

$q_2^{①} < 0 < q_2^{②}$ then $q_2^{②}$ can be solved and $q_2^{①}$ is abandoned.

Although the buckling analysis methods of changed boundary beam and traditional beam are basically the same, the influence of displacement in Stage 1 to that in Stage 2 is taken account in changed boundary beam while analyzing total potential energy, which distinguished the new method from the traditional ones. The eigen values are no longer the same.

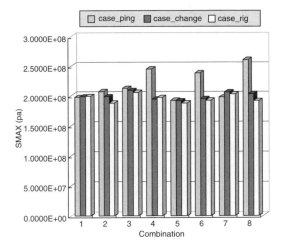

Figure 3.   Comparison of the maximum stresses of members under load combinations.

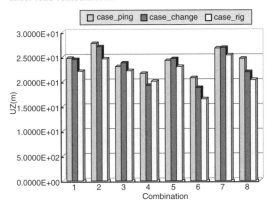

Figure 4.   Comparison of the maximum deflections (Z direction) of joints under load combinations.

4   THE LOADING BEHAVIOR OF THE ROOF STRUCTURE OF THE NATIONAL GRAND THEATRE OF CHINA

The roof of The National Grand Theatre is an ellipsoidal shell. The middle surface of the roof is covered with transparent glass and others with aluminum panels. The total area of the roof is 35790 m². A new construction method is used in the shell construction. The shell is hinged on the ground reinforced ring beam during the construction of the shell structure and the roof cover. After finishing of the shell structure and the roof cover, the structure supports are all fixed on the ring beam. On the other words, the shell supports are hinged under the application of the structure self weight and the roof weight, and are fixed when later live loads are applied. Total 8 load combinations are considered in the computation of the shell. Figure 3

and Figure 4 show the comparison of the maximum stresses of members and the maximum deflections (Z direction) of joints under all 8 load-combinations in 3 models. It is found from the tables and figures that the stress distribution of the structure with changed boundaries is well proportioned. The stress differences $\Delta\sigma$ in the structure are 68.55 MPa for hinged, 17.97 MPa for fixed and 17.37 MPa for changed respectively. The absolute maximum member stress of the structure with changed boundaries is between the stresses of the hinged structure and the fixed structure. The stress distribution is much reasonable and the members are taken full advantages in the structure with changed boundaries.

5   CONCLUSIONS

According to static and linear buckling analysis of changed boundary beam above, several conclusions are drawn.

As to static analysis of changed boundary beam:

(1) The total internal forces change according to the constraints and load q1, q2.
(2) The maximum bending moment in changed boundary beam is less than those of unchanged boundary ones. While designing a structure, the cross section can be chosen much smaller since the most disadvantageous internal force of the beam with changed boundary is smaller. Also, the distribution of internal force is much flatter. Hence, the changed boundary beam is more economic.

As to linear buckling analysis of changed boundary beam:

The linear buckling eigen value of changed boundary beam is different from that obtained from traditional beam. That is because the internal forces and displacement of the beam under load q1 have influence on that of the beam under load q2. To what extent this influence may reach is determined by the properties of the cross section.

REFERENCES

Timoshenko. S.P. & Gere. J.M. 1963. Theory of Elastic Stability. Seconded. McGraw-Hill. New York.
TRAHAIR. N.S. 1993. Flexural-Torsional Buckling of Structures. E & FN SPON.
F.Mohri, et al. 2003. Theoretical and numerical stability analyses of unrestrained mono-symmetric thin-walled beams. Journal of Constructional Steel Research. 59: 63–90
Barsoum. R.S. & Gallagher. R.H. 1970. Finite element analysis of torsional and torsional-flexural stability problems. International Journal for Numerical Methods in Engineering. 2: 335–352

Steel and Composite Structures – Wang & Choi (eds)
© *2007 Taylor & Francis Group, London, ISBN 978-0-415-45141-3*

The form-finding method for the lower-chord pull-cable of bi-directional beam string structure based on balanced load

Renjie Shang, Zhuanqin Wu, Jingliang Liu, Xisheng Gong & Peixun Li
Central Research Institute of Building and Construction, MMC Group Beijing, China

ABSTRACT: The partial differential equations of the curved surface of lower-chord pull-cable of bi-directional string structure is deduced in the paper based on the relationship between the balanced load and the curved surface of the cable and the force of tension and pull of the cable. The FEM and program to calculate the curved form of lower-chord cable-net is provided for the first time based on the equilibrium of the lower-chord nodes. The FEM process and method to determine the curved surface is introduced by 2 examples of curved roof. And stress analysis is made for the string structure determined after form-finding, and the correctness of the results for the form-finding is validated. The computation method given in the paper features definite physical meaning, convenient computation process and accurate computation result.

1 FORWARD

The pre-stressed string structure is one kind of structural form for the pre-stressed steel with big span formed for the pre-stress technology with high efficiency in the application of steel structure. This kind of structure has the advantages of big span ability and saving steel material etc. The structure of roof cover of the National Indoor Stadium for 2008 with the construction under way, using the bi-directional string structure with span up to 114 m × 144.5 m has now become the bi-directional string structure with the biggest span at home and in the world. The form-finding for the curved surface of the lower-chord curved surface in the bi-directional string structure has become the key problem in the design of the bi-directional string structure. Most string structure constructed in the past is uniaxial stress, such as the level projection span for the airport building of the already built Pudong Airport at home up to 82.6 m(Wang,1999), the span of Guangdong Convention and Exhibition Center up to 126.6 m(Chen,2002), the span of Harbin Convention and Exhibition Center up to 128 m(Fan,2002), the span of Shenzhen Convention and Exhibition up to 126 m, and the span for the dome of Anô of Kitakyushu up to 61.8 m(Sailtoh,1999). All these structures use the parallel configuration for the beam string or the string truss with same dimension for the multi-brace structure to from the roof structure. Most lower-chord for the uniaxial string structure is configured in accordance with parabola, and it can provide the upwards balanced load of basic evenness, so as to make the

stress of the truss or beam for the upper chord even. The selection of the suitable pull and the form of curve can ensure basic elimination of the production of bending in the upper and down chords with evenly distributed load. The lower chord of the bi-directional string structure is located at the curved surface. The cable in both directions produces the upwards balanced load. The cable in both directions transfers the balanced load to the upper chord structure through the stay bar. How to decide the curved surface where the lower chord is located, to make the balanced load between the lower chord and the upper chord of the bi-directional string structure keep even is the target for the form-finding in the curved surface of the lower chord for the bi-directional string structure. All the past study in form-finding is to study the deformation of down beam-string under the action of the pre-stress after the determination of the height of the stay bar(Qi,2004;Ma,2004), and to determine the state of zero stress for the beam string, without searching the best form of the lower chord. This article presents the method to establish the FE equation and to compute the best form for the curved surface of the lower-chord cable based on the size of the balanced load and the size of the cable tension. The correctness of the presented method of form-finding based on FE is validated through the computing case of uniaxial bending roof. The general FE software ADINA is used to compute and analyze the internal and deformation under the specified load for the bi-directional string structure gained by form-finding: the bending of the upper chord and the deformation in vertical direction

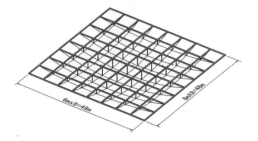

Figure 1. Schematic diagram for bi-directional string beam.

are almost zero, so as to completely reaching the target of form-finding. The method is characteristic of simple computation and less error, able to provide use for reference in the form-finding for the curved surface of the lower chord cable–net of the bi-directional string structure.

2 DEDUCTION OF THE BALANCE EQUATION OF CONTINUOUS LOWER CHORD CABLE-NET

If the coordinate of the plane projection for the bi-directional string structure is plane x-y, the area of cable in x- direction in the unit width of y-direction is Ax, the area of cable in y-direction in the unit width of x-direction is Ay, and the modulus of elasticity of the cable is E, see Fig-1. $\bar{H}_x(y)$ and $\bar{H}_y(x)$ are respectively the pull of the unit width for the cable in x-direction and y-direction, able to change along with y and x. The curved surface of the lower chord cable-net can be expressed by $z(x, y)$,then:

the balanced load produced in x-direction is:

$$q_x(x, y) = \bar{H}_x(y)\frac{\partial^2 z(x, y)}{\partial x^2} \tag{1}$$

the balanced load produced in y-direction is:

$$q_y(x, y) = \bar{H}_y(x)\frac{\partial^2 z(x, y)}{\partial y^2} \tag{2}$$

In order to ensure the smaller bending moment of the upper-chord bending structure for the bi-directional string structure under the vertical load, the sum of the vertical load to counteract the upper–chord arch (if the upper–chord is of curved surface) and the bi-directional cable for the lower chord cable should be identical to the external load. As the lower chord is not ideal curve, but the multi-sectioned fold-line with transition at the position of stay-bar, the balanced load of the lower chord cable-net is embodied by the concentrated force at the position of stay bar too. If all the load of the upper-chord structure acts at the position

of stay-bar, and the balanced load of cable-net for each stay-bar is equal to the concentrated force of external load at the node, then the string structure with the upper chord as plane is located at the state under pressure in bi-direction and without bending moment.

In order to facilitate the computation, the lower-chord cable is represented in accordance with the curved surface $z(x, y)$ with continuous change for the lower chord, if the balance needed by the bi-directional cable is $q(x, y)$ in accordance with (1)(2), then:

$$\bar{H}_x(y)\frac{\partial^2 z(x, y)}{\partial x^2} + \bar{H}_y(x)\frac{\partial^2 z(x, y)}{\partial y^2} = q(x, y) \tag{3}$$

and the boundary conditions:

when $\quad f(x, y) = 0$, $g(x, y, z) = 0 \tag{4}$

the partial differential equation (3) can have the analytical solution, under some special boundary conditions, especially when the both directions of the cable force are not equal, even when $\bar{H}_x(y)$ and $\bar{H}_y(x)$ change along with y and x, it is impossible to gain the analytical solution.

3 FE METHOD-BASED FORM-FINDING METHOD

The form-finding of the cable-net is rather mature, such as force density method (Wang, 2003), support displacement method (Zhang, 2002) and dynamic relaxation method etc. However the form-finding for the cable-net is to find the self-balance position for the cable-net without external load, and the form-finding for the lower chord of the bi-direction string structure is an optimization process to find a best stress form, namely to be able to keep the upper-chord bending-less, so as to reach the purpose to reduce the section of the upper-chord to the utmost. It is very difficult to get the analytical solution for the partial differential equation (3) and the boundary conditions (4), it is possible to make the solution by use of the numerical method, but the solution method is comparatively complex, this article deduces the rigidity matrix and vector of the FE analysis, in accordance with the network of plane configuration for the lower chord cable-net, and uses the FORTRAN language to formulate the FE procedure to find out the ideal form of the curved surface of the lower-chord cable, and to determine the heights of various stay bars.

3.1 The vertical rigidity when the level component of force is certain

All the units are the ones with pulled bar, with only axial pull, and the nodes of unit e are i and j, and the level component of the axial pull T_{i-j}, kept as the

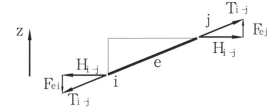

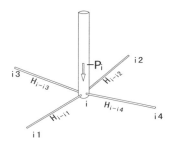

Figure 2. Stress of unit e.

Figure 3. Comprehensive vertical load for node i.

designed pull value without change. The coordinate for node i in the initial state is (x_i, y_i, z_i), and the coordinate of the node j is (x_j, y_j, z_j). As the level projection position for the node is fixed in design, therefore there are only the vertical deformation and the vertical rigidity. The level projection for the length of unit e is: $L_{i-j} = \sqrt{(x_j - x_i)^2 + (y_j - y_i)^2}$, while the vertical component of force produced in node i of unit e is:

$$F_{ei} = \frac{z_j - z_i}{L_{i-j}} \times H_{i-j} \tag{5}$$

And the displacement w produced in the vertical direction for node i of unit e is:

$$F_{ei} + \Delta F_{ei} = \frac{z_j - z_i - \Delta z_i}{L_{i-j}} \times H_{i-j} \tag{6}$$

Therefore we can get: $\Delta F_{ei} = -\frac{\Delta z_i}{L_{i-j}} \times H_{i-j}$, the rigidity of unit e at node i: $k_{ii} = \frac{\Delta F_{ei}}{\Delta z_i} = -\frac{H_{i-j}}{L_{i-j}}$; Similarly we can get $k_{ij} = k_{ji} = \frac{H_{i-j}}{L_{i-j}}$; $k_{jj} = -\frac{H_{i-j}}{L_{i-j}}$, and the rigidity matrix of the unit is:

$$[k^e] = \begin{bmatrix} k_{ii} & k_{ij} \\ k_{ji} & k_{jj} \end{bmatrix} = \begin{bmatrix} -\dfrac{H_{i-j}}{L_{i-j}} & \dfrac{H_{i-j}}{L_{i-j}} \\ \dfrac{H_{i-j}}{L_{i-j}} & -\dfrac{H_{i-j}}{L_{i-j}} \end{bmatrix} \tag{7}$$

where: H_{i-j} is the level component of the pull for the unit, L_{i-j} is the level projection length of the unit.

3.2 Load matrix

The coordinate of the node input at the earlier beginning should ensure the accord between its x,y and the projection position of the stay bar for the bi-directional string structure, namely the position of the stay bar is already determined in beginning the design, and the position of the section for the lower-chord cable is already fixed, but the vertical coordinate z can be input at will. The vertical position under the pull of the cable can not certainly ensure the balance of the nodes. Any node of the bi-directional string structure has four units to connect with the four nodes. The coordinate of node i is $(x_{0,i}, y_{0,i}, z_{0,i})$, the external load is P_i, and the n codes connecting with I are i1 $(x_{0,i1}, y_{0,i1}, z_{0,i1})$ ...in$(x_{0,in}, y_{0,in}, z_{0,in})$. Then the comprehensive load formed by the external load at node i and force at the vertical unbalanced node of node i is:

$$F_i = P_i + \sum_{k=1}^{k=n} \frac{z_{0,ik} - z_{0,i}}{L_{i-ik}} \times H_{i-ik} \tag{8}$$

3.3 The constraint conditions:

The position of the node for the lower-chord cable at the boundary is determined, unable to have deformation, therefore it is possible to suppose that the rigidity of the node is a very large number to simulate the infinity of the rigidity. If node i is the fixed point of the cable boundary, the following result is taken in the program:

$$k_{ii} = 1 \times 10^{19} (N / mm) \tag{9}$$

3.4 FE equation

In accordance with (7)(8)and (9), it is possible to get the FE equation:

$$[K]\{w\} = \{F\} \tag{10}$$

Where, the rigidity matrix $[K]$ is formed by the various units through formula (7) and taking the constraint conditions (9) into consideration: And the load vector of the load $\{F\}$ is formed by the formula (10) for the comprehensive load of various nodes. In accordance with (10), it is very easy to resolve the vertical displacements $\{w\}$ for various nodes, and to finish the computation of the vertical coordination after the completion of form-finding:

$$\{Z_1\} = \{z\} + \{w\} \tag{11}$$

3.5 The computing steps for form-finding in the bi-directional string structure

Step 1: The vertical coordinates $\{Z_1\}$ for various points are computed out on the basis of the preliminarily

303

determined force of tension and pull $H_{1,i-j}$, and the node force $P_{1,id}$ to be balanced at the lower chord. When the upper chord is of the curved surface, the pull of the cable will let the arch-shaped beam of the upper chord bear the pressure, and the upwards balanced load will be produced at the roof. The corresponding node of node i is iu. The balanced load at the node of the upper chord iu is determined, in accordance with the size of level component of force in the upper chord:

$$Q_{iu} = \sum_{k=1}^{k=n} \frac{z_{iu} - z_{iku}}{L_{i-ik}} \times H_{i-ik}$$

where z_{iu} is the vertical coordinate for the corresponding node at the upper chord. z_{iku} is the vertical coordinate for node ku connected with the node iu at the upper chord. And n is the node number adjacent to node iu. In general, n is taken as 4. The load of the node needed to balance at the lower chord is

$$P_{1,id} = P_i - \sum_{k=1}^{k=n} \frac{z_{iku} - z_{iu}}{L_{i-ik}} \times H_{i-ik}$$

The form coordinate $\{Z_1\}$ for the lower chord can be calculated out in considering the upper chord as the curved surface, by $P_{1,id}$ to replace P_i in formula (8), if the upper chord is plane, $P_{1,id} = P_i$.

Step 2: In accordance with the coordinate of the control point and the height of the designed vector, to regulate the pull of the cable, and to make form-finding again, it is possible to reach the form and forces of tension and pull meeting the requirements on construction. The general construction design has determined the coordinate of the control point. If the construction has determined the coordinate for the upper chord as $\bar{Z}_{ku}$ for some control point ku, and the vertical coordinate for the node of the lower chord as $\bar{Z}_{ku}$, the vertical coordinate for the node at the lower chord as $\bar{Z}_{kd}$, then, the force of tension and pull for the cable:

$$H_{i-j} = H_{1,i-j} \times \frac{Z_{1,kd} - Z_{kd}}{\bar{Z}_{ku} - Z_{kd}} \quad \text{is regulated.}$$

The load to be balanced $P_{2,id}$ for the node of lower chord is computed again. The computed form on the basis of the balanced load between the new forces of tension and pull and the node is $\{Z_2\}$, as the form of the lower chord cable-net, without bending moment for the ideal upper chord and without vertical deformation.

4 THE TENSION AND PULL CONTROL FOR THE STRUCTURE AFTER FORM-FINDING

The area of the lower-chord pull-cable is A_d, the elastic modulus is E_d, the preliminary strain of the lower-chord pull-cable is: $\varepsilon_0 = \frac{H}{A_d E_d}$.

As the lower-chord pull-cable is of the configuration of curve line and each section of cable is subject to the vertical load at the position of stay bar, therefore, each section of level component of force upon each piece of cable is not changed, and the vertical pull is different along with the difference of the angle in vertical direction. Thus, the initial strain should be different in the form of being proportional with the pull for each section in time of tension and pull as well as checking computations. This kind of control and computation for the tension and pull are burden-some. In order to simplify the computation, it is possible to take each piece of cable into consideration in accordance with the same initial strain. In this way, there will be sufficient accuracy, and convenient computation. The initial strain can be revised in accordance with the length of cable L and the length of the level projection:

$$\varepsilon = \frac{L}{L_0} \varepsilon_0 = \frac{FL}{A_d E_d L_0}$$

In addition, in the process of tension and pull, the lower chord has the horizontal pull H, and meanwhile the upper chord will be subject to the axial horizontal pressure H, to produce compression. The strain of the pull-cable should be revised again:

$$\Delta\varepsilon \cong \frac{HL_0}{A_u E_u}$$

Therefore, the computation and the control strain of tension and pull should become:

$$\varepsilon \cong \frac{HL}{A_d E_d L_0} + \frac{H}{A_u E_u} \qquad (12)$$

5 EXAMPLES

Bi-directional beam string for the roof of unidirectional curved surface

The bi-directional beam string shown as Fig. 4, with the plane projection shown in Fig. 5, in the form of the square of 48 m × 48 m, and the network of the upper-chord beam at 6 m × 6 m, with the upper chord to be the square steel pipes of 200 × 400 × 14, and the stay bar at $\phi 120 \times 6$, the cable arranged under the upper chord beam, the stay bar set up at the section, all load at the roof up to 1.333 kN/m², the load acting on the cross-nodes in the form of the force of the node, therefore $P_i = 1333 \times 6 \times 6 = 48000\,(N)$. In terms of construction, the unidirectional bending for the upper chord is required to be the parabolic. The highest point of the parabolic is z = 2000 mm, and the lowest point of the lower chord is z = −3000 mm. Namely, the height of the stay-bar at the most middle part is 5000 mm.

The lower chord selects the same specification, with the same horizontal pull. The form of the curved surface for the lower-chord pull-cable of the bi-directional beam string is determined, to make the bending of the upper-chord to be minimum.

First, let us to suppose all the cable pull to be $H = 400\,\text{kN}$, the balanced load at each stay bar is computed out:

$$Q = \frac{8 \times 400000 \times 2}{48 \times 48} \times 6 = 16667 \ (\text{N}).$$

The external load for the node is 48000 N. the load to be balanced at the node of the lower chord is $48000 - 16667 = 31333\,\text{N}$. via computation, the central point $Z = -1353.799$ is obtained. The regulated pull of the cable is:

$$H = 400000 \times \frac{2000 + 1353.8}{2000 + 3000} = 268304 \ (\text{N}).$$

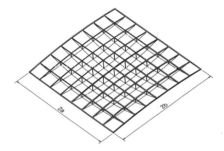

Figure 4.   Side drawing for the bi-directional string.

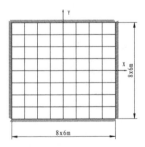

Figure 5.   Plane projection of the bi-directional beam string.

The balanced load for the node of the upper chord is

$$Q = \frac{8 \times 268304 \times 2}{48 \times 48} \times 6 = 11179.3 \ (\text{N}).$$

The load to be balanced at the node of the lower chord is $48000 - 11179.3 = 36820.7\,(\text{N})$. The computation for form-finding is made again. For the result, refer to Table-2. For the central point $Z = -3000.041\,\text{mm}$, the cable of the most outside in x-direction is horizontal without arrangement, to counteract the horizontal component of force for the upper chord of the arch. After form-finding, ADINA is used for computational analysis into the bi-directional string structure. For the computational result, refer to Figs 5~8. Under the above-mentioned form, force of node and the pre-stress, the biggest displacement in

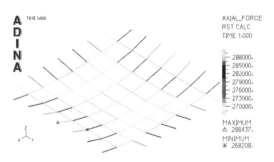

Figure 6.   The biggest deformation in vertical direction up to 0.104 mm.

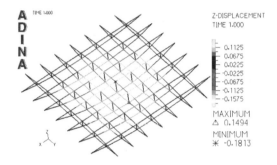

Figure 7.   The horizontal component of force up to 268208 N.

Table 1.   The coordinates after form-finding (mm).

	$x = 0$	$x = \pm6000$	$x = \pm12000$	$x = \pm18000$	$x = \pm24000$
$y = 0$	−3000.041	−2731.688	−1891.113	−371.763	2000.000
$y = \pm6000$	−2856.688	−2606.094	−1818.795	−386.265	1875.000
$y = \pm12000$	−2391.113	−2193.795	−1568.296	−406.093	1500.00.
$y = \pm18000$	−1496.763	−1386.265	−1031.093	−346.399	875.000
$y = \pm24000$	0	0	0	0	0

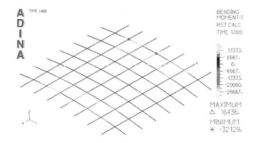

Figure 8. The bending moment for the upper chord.

Figure 9. The pressure of stay bar up.

vertical direction for the upper chord is 0.18 mm; the bending moment $-0.044\sim0.021$ kN-m, almost up to zero; The pressure of the stay bar is $36816\sim36828$ N. the biggest difference is 0.02%, in comparison with load 36821 N; The horizontal component of force for the pull of cable is 268208 N, with a difference up to 0.04%, in comparison with the theoretical 268304 N. thus the form-finding reaches the very ideal effect.

6 CONCLUSIONS

The article presents the relation between the balanced load of the lower-chord pull-cable for the bi-directional string structure and the shape of the curved surface, namely the differential equation:

The unit rigidity and the unbalanced force of the node are deduced on the basis of the balance relation between the balanced load of the lower-chord node and the cable when the horizontal component of force is unchanged;

The computation formula for the size of the initial strain of the cable is given, and thus it is able to apply to the computation analysis into the beam string and the control of tension and pull;

The formulated FE programs are used for the computation in form-finding for the case of unidirectional bending roof. And the stress analysis of the beam string structure for the structure after form-finding and the correctness is validated for the method of form-finding. Under the specified load, the bending moment for the upper chord of the beam string structure is almost zero, reaching the aim of optimization design. The method features simplicity, practicality, less error in the computation results. Thus it completely the requirements on engineering design, and can provide reference for the computation analysis and the structure design in the bi-directional string structure.

REFERENCES

Chen Rongyi, Dong Shilin, Design on the steel dome for the exhibition hall of Guangzhou International Convention and Exhibition Center. Space Structure. 2002 (3)

Fan Feng, zhi Jiudong, and Shen Shizhao, The design on the steel structure with big span for the main hall of Heilongjiang International Convention and Exhibition Gym Center. Proceedings of the 10th academic exchange meeting on space structure. China Construction Material Industry Press, Beijing, 2002 (7) p806–811

Ma Meilin, Study on the form-finding for the beam string structure and the stress performance, Zhejiang University, Dissertatation on Master's Degree, 2004.2

Qi Yongsheng, zhou Hong, and Su Kang, The methods using APDL language to solve the problem of form-finding in the beam string structure, Shanxi Construction, 2004.2

Sailtoh, M. and Okada, A. The Role of String in Hybrid String Structure. Eng. Struct. 1999,21:756–769

Wang Dashui, Zhang Fulin etc. Study and design on steel structure of the airport building of Shanghai Pudong International airport (Term 1 engineering) Construction Struction Transaction.1999, 20 (2)

Wang Chunjiang, Dong Shilin, Wang Renpeng, and Qian Ruojun, Form-finding analysis method for force density and its computer realization, Mechanics Quaterly, No.3, 2003.

Zhang Hua,chuan Jian, The form-finding analysis based on the D.R. method for pre-stressed cable-membrane structure, Engineering Mechanics, No.2, 2002

Steel and Composite Structures – Wang & Choi (eds)
© *2007 Taylor & Francis Group, London, ISBN 978-0-415-45141-3*

Global stability of telescopic box boom with cylinder support

Bing Luo, Nianli Lu & Peng Lan
School of Mechatronics Engineering, Harbin Institute of Technology, Harbin, China

ABSTRACT: Cylinders of telescopic boom of crane bear nearly all the axial compressive component of forces, while telescopic boom bears the other biaxial bending component completely. The stability of telescopic boom is studied, considering the effect of cylinders which offers a non-conservative force, and the friction between booms is not taken account. The critical force of telescopic boom is calculated with three different methods, including differential equation method, elastic support method and finite element method. Effect of cylinders to global stability is discussed subsequently. Moreover, the paper introduces to calculate stability of three segments boom with precise finite element method in detail. The result is compared with that of methods which did not take account of the effect of cylinders. It shows that the effective length coefficient from finite element method is more accurate in theory.

Keywords: axial compressive component, biaxial bending component, effective length, precise finite element method.

1 INSTRUCTIONS

Telescopic boom of crane is made of several box booms with variable sections. Axial direction among booms can slide oppositely. Telescopic mechanism of boom is various and hydraulic drive is often adopted. Actuator of the drive system uses hydrocylinder. Relative movement between cylinder body and piston rod can push telescoping of next boom. Piston rod of cylinder hinges with basic jib and cylinder body with the second boom. Ordinal acting valves are installed in the top of cylinder. When the second boom is expanded, switch of ordinal acting valve pushes cylinder of the second boom. Generally n segment booms correspond to n-1 cylinder body-piston groups. The maximal merit of telescopic boom by hydraulic drive rests with achieving infinitive telescoping and load telescoping at different degrees. So using function of crane can be expanded in complex conditions.

The stability theories of beam column have been developed perfectly[1], and studies on some difficult matters such as taper beam, thin curve beam have also made great progress in recent years[2–5]. Cylinders are important components of hydraulic drive system in engineering machines. So it is necessary that cylinder should be checked on stability[6]. Telescopic boom doesn't bear axial force straightly ignoring friction and the force is born by cylinders collocated in the boom. So calculation of critical axial force is different from that of simple ladder column with variable section, and the critical force is also different. Stability analysis for telescopic boom ignored the function of cylinders and telescopic boom was calculated as ladder column with variable section in the past[7–10]. The result wasn't reasonable obviously. Some papers considered the influence of cylinders, but derivate process was complex and simplification wasn't proper[11]. Finite element method is used usually for stability of beam column in projects. The paper focuses on confirming critical force considering influence of cylinders, and introducing to calculate stability by finite element method in detail.

2 CONFIRMING CRITICAL FORCE BY DIFFERENTIAL EQUATION METHOD

To explain the problem, we analyze constant section inertia moment of two segments telescopic boom at first

Fig.1 (a) is mechanical model of telescopic boom. Fig.1 (b) shows that telescopic boom bears moment only and axial force is born by cylinder. The section inertia moment of telescopic boom is *I*. So the differential equation of deflection is:

$$EIy'' = P(\delta - y) - P(\frac{\delta x}{l} - y)$$
$$- P\delta(1 - \frac{x}{l})$$

(1)

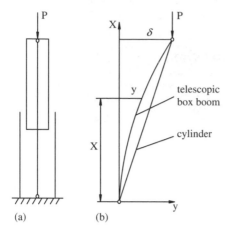

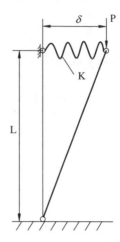

Figure 1.

Figure 2.

The solution is: $y = \dfrac{P}{EI}\delta\left(\dfrac{x^2}{2} - \dfrac{x^3}{6l}\right) + Cx + D$

Substituting the boundary conditions: $\begin{cases} x = 0, y = 0 \\ x = 0, y' = 0 \\ x = l, y = \delta \end{cases}$

into the former formula gives:

$$\begin{cases} C = D = 0 \\ \delta = \dfrac{\delta P}{EI}\left(\dfrac{l^2}{2} - \dfrac{l^3}{6l}\right) = \dfrac{Pl^2\delta}{3EI} \end{cases}$$

So critical force is solved:

$$P_{cr} = \dfrac{3EI}{l^2} = \dfrac{\pi^2 EI}{\left(1.8138l\right)^2} \tag{2}$$

According to the critical force solved by formula (2), we can obtain:

(1) Critical force of telescopic boom is improved due to supporting of cylinder. Effective length coefficient $\mu = 1.8138$ is less than that $\mu = 2$ of cantilever column bearing axial pressure.

(2) In the stable condition of cylinder, critical force of telescopic boom is independent of inertia moment of cylinder. So enhancing inertia moment of cylinder can't improve resistant buckling capability of boom.

(3) In telescopic boom of constant section, the steady condition of cylinder in advance is:

$$P_{cr} = \dfrac{3EI}{l^2} < \dfrac{\pi^2 EI'}{l^2}$$

that is, section inertia moment of cylinder

$$I' \geq \dfrac{3}{\pi^2} I \quad.$$

3 CONFIRMING CRITICAL FORCE BY ELASTIC SUPPORT METHOD

The model of Fig.1 (a) can be simplified as mechanical model shown in Fig. 2. The telescopic boom bearing moment and lateral force supplies elastic support for cylinder. The stiffness of elastic support is K and balanced equation of stability is:

$$P\delta = \delta Kl \tag{3}$$

so critical force is:

$$P_{cr} = Kl \tag{4}$$

When constant section lateral stiffness $K = \dfrac{3EI}{l^3}$ of cantilever beam is substituted in the former formula, we can obtain:

$$P_{cr} = \dfrac{3EI}{l^2}$$

The result is the same as formula (2). Obviously, self-stability of cylinder is similar to the former paragraph.

4 SOLVING CRITICAL FORCE WITH FINITE ELEMENT METHOD

From the model in Fig. 3(a), we can build three elements system in Fig. 3(b). The stiffness matrix of the system is $[K]$. From the condition that determinant of stiffness $[K]$ is zero that is $\det(K) = 0$, critical force P_{cr} can be confirmed.

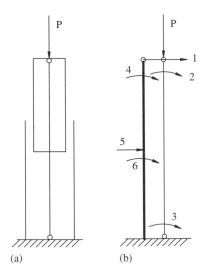

Table 2. Effective length coefficient μ_2 for variable section of two segments boom (Fig. 4).

$\beta_2 = I_1/I_2$	1.0	1.3	1.6	1.9	2.2	2.5
μ_2		0.966	0.984	1.00	1.017	1.035	1.05

(a)   (b)

Figure 3.

Table 1. Effective length coefficient μ and μ_2 of two segments boom (Fig. 3).

$\beta_2 = I_1/I_2$	1.0	1.3	1.6	1.9	2.2	2.5
μ	1.8138	1.83	1.848	1.865	1.882	1.89
μ_2	0.907	0.915	0.924	0.932	0.941	0.949

In the table, μ is totally effective length coefficient of two segments telescopic boom, μ_2 is length coefficient of variable section considering root supporting condition of cantilever beam in design rules for cranes[7], and $\mu_2 = \mu/2$.

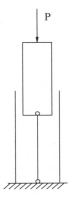

Figure 4.

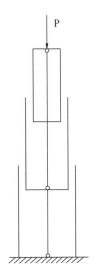

Figure 5.

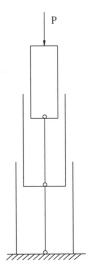

Figure 6.

According to section parameters and sizes of two segments boom in Table J-4 of design rules for cranes[7], we can conclude table of effective length coefficient calculated by precise finite element method.

The table is shown in Table 1 and corresponds to effective length calculated from stability of telescopic boom by cylinder supporting in Fig. 3.

For the other supporting condition that cylinder supports on root of two segments boom as Fig. 4, result of effective length coefficient for variable section is shown in Table 2.

For three segments boom of different cylinder supporting in Fig. 5 and Fig. 6, we can also get effective

Table 3. Effective length coefficient μ_2 for variable section of three segments boom (Fig. 5).

$\beta_2 = I_1/I_2$	1.3		1.6		1.9		2.2		2.5	
$\beta_3 = I_2/I_3$	1.3	2.5	1.3	2.5	1.3	2.5	1.3	2.5	1.3	2.5
μ_2	1.027	1.053	1.068	1.10	1.108	1.144	1.148	1.188	1.186	1.23

Table 4. Effective length coefficient μ_2 for variable section of three segments boom (Fig. 6).

$\beta_2 = I_1/I_2$	1.3		1.6		1.9		2.2		2.5	
$\beta_3 = I_2/I_3$	1.3	2.5	1.3	2.5	1.3	2.5	1.3	2.5	1.3	2.5
μ_2	1.04	1.077	1.086	1.132	1.131	1.186	1.175	1.238	1.218	1.29

length coefficient μ_2 according to section and length parameters in Table J-4 of design rules for cranes[7]. Comparing former data in each table, we obtain μ_2 considering effect of cylinder is less than that of Table J-4 in design rules for cranes[7].

5 CONCLUSIONS

(1) In the original design rules for cranes, effective length of telescopic boom is confirmed by energy method without considering effect of cylinder. In the new revised design rules for cranes[7], Table J-4 adopts precise finite element method without considering effect of cylinder. The result is more precise in theory.

(2) Stability loading capability of telescopic boom is improved slightly considering effect of cylinder. Calculation is inclined to safety without considering effect of cylinder.

(3) In the precondition of stability for cylinder, loading capability of telescopic boom depends on section inertia moment of boom and supporting form. The capability is irrespective with inertia moment of cylinder.

(4) As different placement and supporting form of cylinder for various telescopic booms, effective length coefficients are different too. Slide block in the place of telescopic boom overlapping can pass definite friction force. Some booms use wire rope with pulley system and cylinder at the same time. So it is difficult to calculate length coefficient with unified method. We suggest using the data of Table J-4, because the data is inclined to safety as a whole and most errors are small.

REFERENCES

Timoshenko SP, Gere JM. Theory of elastic stability [M]. 2nd ed. New York: McGraw-Hill, 1961.

Sung-Bo Kim, Moon-Young Kim. Improved formulation for spatial stability and free vibration of thin-walled tapered beams and space frames [J]. Engineering Structures. 2000, 22 (5): 446–458.

Guo-Qiang Li, Jin-Jun Li. A tapered Timoshenko–Euler beam element for analysis of steel portal frames [J]. Journal of Constructional Steel Research, 2002, 58(12): 1531–1544.

S. Naguleswaran. Transverse vibration and stability of an Euler–Bernoulli beam with step change in cross-section and in axial force [J]. Journal of Sound and Vibration, 2004, 270 (5): 1045–1055.

Hasan Öztürk, Mustafa Sabuncu. Stability analysis of a cantilever composite beam on elastic supports. Composites Science and Technology 65 (2005): 1982–1995.

LAN Peng, LU Nianli, LIU Manlan. Stability calculating model of slim hydraulic cylinder. Construction Machinery. 2004(6). (in Chinese)

Design rules for cranes GB3811–83 national standards of People's Republic of China. Standards Press of China. 1983. (in Chinese)

LU Nianli, LAN Peng, LI Liang. Application of FEM of Beam Element with Theory of II Order. Journal of Harbin University of Civil Engineering and Architecture. 1998(4). (in Chinese)

LU Nianli, LAN Peng, BAI Hua. Precise stability analysis of telescopic boom. Journal of Harbin University of Civil Engineering and Architecture. 2000(2). (in Chinese)

JI Aimin, ZHANG Peiqiang, PENG Duo, LUO Yanling. Finite element analysis for local stability of telescopic boom of truck crane. Transactions of The Chinese Society of Agricultural Machinery. 2004(6). (in Chinese)

WANG Shuozhe, GU Diming. Influence of telescopic cylinder to box boom about bearing force condition. Hoisting and Conveying Machinery. 1983 (11): 34–37. (in Chinese)

Steel and Composite Structures – Wang & Choi (eds)
© 2007 Taylor & Francis Group, London, ISBN 978-0-415-45141-3

Equivalent inertia moment method to stability analyze of lattice type beam

Jia Wang, Nianli Lu & Bing Luo
Harbin Institute of Technology, China

ABSTRACT: An inertia moment of lattice type beam equivalent to solid web bending beam is introduced while calculating the stability of lattice type beam. Based on the fact that the equivalent solid web frame has the equal stiffness as the original lattice type beam, the formula to calculate the equivalent inertia moment of lattice type beam is given. It is not only suitable for the stability analysis of single lattice type beam but also suitable for the overall stability analysis of composite structure which is composed by multiple lattice type beams. In this sense, it is better than the effective length method which changes the calculated length of each beam. Eight kinds of lattice type beams are analyzed by the equivalent inertia moment method, and by also finite element method as a compare. The results show that this method is both simple and accurate for engineering use.

Keywords: Stability, equivalent inertia moment, lattice type

1 INSTRUCTIONS

Lattice type beam is a beam system which is composed by thousands of single frames. The stability analysis of this system is very difficult[1]. Both finite element method (FEM) and equivalent method are used to deal with this problem[2,3]. A high accurate result can be obtained by FEM while it is very complex and not suitable for engineering use without well education[4–7]. In many papers, the stability problem is solved by equivalent method, which equals the lattice type beam to a solid web frame. The effective length from lattice type beam to solid web frame is calculated in traditional approach[8]. The method to calculate the effective length is given. As we can see from Fig.1, when we analyze the stability of the lattice type beam,

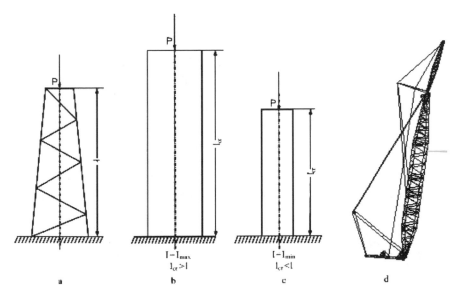

Figure 1. Model of traditional method.

we can use either the maximum or the minimum inertia moment to analyze the stability of the lattice type beam while the length is changed to effective length l_{cr}. Because of this, when we analyze composite system which is made up by more than one lattice type beam such as the lattice jib of crane with main and vice jib structure (fig.1(d)), this method is out of use as the whole structure is changed. In this paper, Equivalent inertia moment method to analyze the stability of lattice type beam is introduced, and the formula to calculate the equivalent inertia moment of lattice type beam is given.

2 SIMPLE FORMULA TO CALCULATE THE LATTICE DISPLACEMENT OF LATTICE TYPE BEAM

Cantilever lattice beam is shown in Fig.2. H, l and a are the Total height, Monolayer height and Span of chords respectively. The lateral displacement of the cantilever under horizontal load Q can be calculated simply by formula [9](1).

$$\Delta = Q\frac{H^3}{3EI}\beta \qquad (1)$$

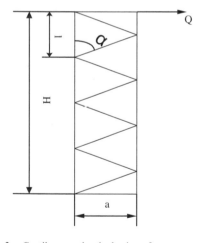

Figure 2.   Cantilever under the horizon force.

Where: β- influence coefficient of web members (table 1)

The influence coefficient β of different lattice type beams with different placement of web member is listed in table1. Common placement of web member is illustrated in Fig.3.

3 CALCULATION OF EQUIVALENT INERTIA MOMENT OF THE LATTICE TYPE BEAM WITH TWO CHORDS

Calculation of the lateral displacement for cantilever lattice beam which is under horizontal force at the top end is performed by formula(1). Alternatively, for a simple solid web beam with the same constraint condition and load, the lateral displacement can be calculated as follow:

$$\Delta_1 = Q\frac{H^3}{3EI_{eq}} \qquad (2)$$

Let $\Delta = \Delta_1$, so the formula to calculate the equivalent inertia moment of the lattice type beam with two chords is obtained:

$$I_{eq} = I/\beta \qquad (3)$$

After calculating the equivalent inertia moment by formula(3), we can analyze the stability of the lattice beam with two chords easily by EULER equation.

$$P_{cr} = \frac{\pi^2 EI_{eq}}{l^2} \qquad (4)$$

Lots of examples have been calculated by SAP program to certify the accuracy of the method which is mentioned in this paper. The details which model the lattice type beam by SAP and calculate the EULER critical force P_{cr} by FEM program are carried out in this work. Afterwards, the accurate inertia moment I_{FEM} of the real model is calculated by Euler's formula inversely. We can see the accuracy of the method

Table 1.   Value of influence coefficient of different web members β.

Type	A, B	C	D	E, F
β	$1 + \dfrac{3}{2H^2}\dfrac{\lambda_2}{\lambda_1}$	$1 + \dfrac{3}{2H^2}\dfrac{\lambda_2 + \lambda_3}{\lambda_1}$	$1 - \dfrac{3}{2n} + \dfrac{3}{4H^2}\dfrac{\lambda_2 + \lambda_3}{\lambda_1}$	$1 + \dfrac{3}{4H^2}\dfrac{\lambda_2}{\lambda_1}$

In the table $\lambda_1 = \dfrac{1}{EA_x a^2}$, $\lambda_2 = \dfrac{1}{EA_f \cos\alpha \sin^2\alpha}$, $\lambda_3 = \dfrac{1}{EA_h}\tan\alpha$; A_x area of cross section of chord, A_f area of cross section of obliquely web member, A_h area of cross section of web member.

mentioned in this paper by comparison between I_{eq} and I_{FEM}. I_{FEM} is calculated as follow:

$$I_{FEM} = \frac{(\mu l)^2 P_{cr}}{\pi^2 E} \qquad (5)$$

The equivalent error is

$$Error = \frac{I_{eq} - I_{FEM}}{I_{FEM}}\% \qquad (6)$$

Geometry and material parameters of the model:

Chord: **sqare steel** 6×6 cm, web member: sqare steel 5×5 cm, elastic ratio: $2.05E5\,MP_a$, monolayer height: $50\,cm$, span of chord: $50\,cm$. The Equivalent Errors of the model with different web member are shown in table 2.

The results indicate that: equivalent error is reduced while the slenderness ratio increased. When the slenderness ratio is less than 30, the error is great. It is true because even Euler's equation is out of use when the slenderness ratio of the beam is too small. When the slenderness ratio is greater than 30, the error becomes very small, about less than 5%. The lattice type beam that is commonly used in engineering has a greater slenderness ratio than 30, the equivalent method mentioned above is suitable for engineering use.

4 CALCULATION OF THE EQUIVALENT INERTIA MOMENT OF THE LATTICE TYPE BEAM WITH FOUR CHORDS

Load model of lattice type beam with four chords is presented in Fig.4. Seen from fig.4, the lateral

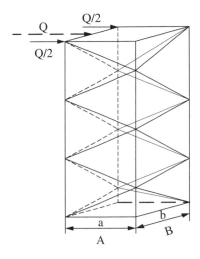

Figure 4. Model of loads decomposition.

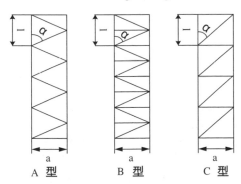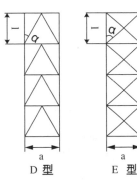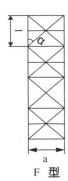

Figure 3. Common dispose from of web members.

Table 2. Equivalent Error of different type.

Level number	Slenderness ratio	Equivalent Errors					
		type A	Type B	Type C	Type D	Type E	Type F
4	16	71.52%	97.51%	74.62%	16.07%	90.24%	60.22%
6	24	−1.31%	9.03%	−4.19%	3.75%	−1.93%	−2.45%
8	32	−1.50%	−1.16%	−2.38%	3.46%	−1.74%	−1.74%
10	40	−1.26%	−1.01%	−1.77%	2.26%	−1.38%	−1.39%
12	48	−1.09%	−0.91%	−1.42%	2.43%	−1.16%	−1.18%
16	64	−0.88%	−0.79%	−1.06%	1.65%	−0.94%	−0.95%
20	00	−0.77%	−0.72%	−0.88%	1.12%	−0.84%	−0.85%
24	96	−0.69%	−0.68%	−0.78%	0.74%	−0.78%	0.79%

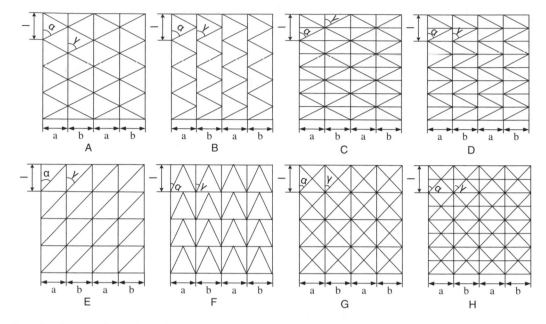

Figure 5. Common dispose from of web members.

Table 3. Value of influence coefficient of different web members β.

Type/ Coefficient	A, B	C, D	E	F	G, H
β	$1 + \dfrac{3\lambda_2}{2H^2\lambda_1}$	$1 + \dfrac{3\lambda_2}{2H^2\lambda_1}$	$1 + \dfrac{3(\lambda_2 + \lambda_3)}{2H^2\lambda_1}$	$1 - \dfrac{3l}{2H} + \dfrac{3(\lambda_2 + \lambda_3)}{4H^2\lambda_1}$	$1 + \dfrac{3\lambda_2}{4H^2\lambda_1}$

displacement of lattice type beam with four chords under horizontal force is equal to lateral displacement of lattice type beam with two chords such as plane A. The influence of plane B and it's opposite can be neglected. So the lateral displacement of lattice type beam with four chords under Q can be calculated as by formula (7):

$$\Delta_1 = Q\frac{H^3}{3EI}\beta \tag{7}$$

Where: $I = A_x a^2$ or $I = A_x b^2$ (use the smaller one of a and b).

Let the left side of formula (1) and (7) equal, the formula to calculate the equivalent inertia moment of lattice type beam with four chords can be deduced:

$$I_{eq} = I / \beta \tag{8}$$

The lattice type beam with four chords that is commonly used in engineering is shown in Fig.5. Different

type has the different value of β. The different value of β is shown in table 3.

In the same way, we can certify the accuracy using FEM program. Geometry and material parameters are as follow: Chord: sqare steel 6×6 cm, web member: sqare steel 5×5 cm, elastic ratio: 2.05E5 MP$_a$, mono-layer height: 50 cm, span of chord: 50 cm (a = b), The result of the model with different web members is listed in table 4.

The result indicates that: equivalent error is reduced while the slenderness ratio increased. When the slenderness ratio is less than 30, the error is great. When the slenderness ratio is greater than 30, the error of type A to F becomes very small, about less than 5%. The lattice type beam that commonly used in engineering has a greater slenderness ratio than 30, so the equivalent method mention in this paper can be used in engineering. But for the type of G and H, the stiffness of the side face is great and can't be ignored. The result in table 4 is too conservative. We can update it by a correctional factor. The correctional result is shown in table 5 when the correction factor is 1.16.

Table 4. Equivalent Error of different types.

Level number	Slenderness ratio	Equivalent Errors							
		Type A	Type B	Type C	Type D	Type E	Type F	Type G	Type H
4	16	167.42%	35.99%	−0.38%	−1.09%	−5.11%	5.00%	3.83%	−16.54%
6	24	48.33%	−3.03%	−2.13%	−2.19%	−3.11%	2.36%	−15.95%	−16.42%
8	32	−2.33%	−2.26%	−2.25%	−2.25%	−2.23%	0.88%	−15.60%	−16.29%
10	40	−1.85%	−1.81%	−2.26%	−2.25%	−1.79%	−0.19%	−15.37%	−16.20%
12	48	−1.56%	−1.54%	−2.25%	−2.23%	−1.54%	−0.96%	−15.23%	−16.14%
16	64	−1.26%	−1.26%	−2.22%	−2.21%	−1.29%	−1.98%	−15.05%	−16.06%
20	80	−1.12%	−1.11%	−2.20%	−2.20%	−1.17%	−2.60%	−14.97%	−16.01%
24	96	−1.03%	−1.03%	−2.20%	−2.19%	−1.10%	−3.03%	−14.89%	−15.97%

Table 5. Corrected result of type G and H.

Level number	4	6	8	10	12	16	20	24
Slenderness ratio	16	24	32	40	48	64	80	96
Type G	20.44%	−2.51%	−2.10%	−1.83%	−1.66%	−1.46%	−1.36%	−1.27%
Type H	−3.20%	−3.05%	−2.90%	−2.79%	−2.72%	−2.62%	−2.57%	−2.52%

5 CONCLUSIONS

(1) Analyzing the stability of lattice type beam by equivalent inertia moment method, the equivalence from lattice type beam to solid web frame is implemented without dimensional change. The theoretical basis is established for the stability analyze of multiplexed system which is made up by more than one lattice type beam.

(2) Calculating the equivalent inertia moment of the lattice type beam by the comparison of displacement is very simple and accuracy. It is suitable for engineering calculation.

(3) The formula to calculate the equivalent inertia moment of lattice type beam is given in this paper. The accuracy of the method mention here is certified by lots of examples. The results indicate when the slenderness ratio of the frame member is greater than 30, the error is very small and it is suitable for engineering calculation. For the method to calculate the lattice type beam with variable cross-section, please reference to paper [10].

REFERENCES

GU Dimin. Engineer crane [M]. China architecture & building press, 1986.(in Chinese)

Timoshenko SP, Gere JM. Theory of elastic stability [M]. 2nd ed. New York: McGraw-Hill, 1961.

Liu Guangdong, Luo Hanquan. Stability of Skeletal Structures[M]. China Communication Press, 1998. (in Chinese)

Klaus-Jürgen Bathe. Finite Element procedures. Prentice Hall, Inc. 1996.

Guo-Qiang Li, Jin-Jun Li. A tapered Timoshenko–Euler beam element for analysis of steel portal frames [J]. Journal of Constructional Steel Research, 2002, 58(12): 1531–1544

Sun Huanchun, Wang Yuefang. Comment on classical theory of stability analysis for truss structures. Chinese Journal of Computational Mechanics, 2005, 22(3): 316–319

Leipholz H. Stability theory. New York: Academic Press; 1970.

Design rules for cranes GB3811-83 national standards of People's Republic of China. Standards Press of China. 1983.

XUE Yuan. An efficient method for calculating of tower mast stiffness of tower crane [D]. Master's thesis of Harbin institute of technology. (in Chinese)

WANG Jia. Research on the overall stability of crane's lattice jib with main and vice-jib structure. Master's thesis of Harbin institute of technology. (in Chinese)

Steel and Composite Structures – Wang & Choi (eds)
© *2007 Taylor & Francis Group, London, ISBN 978-0-415-45141-3*

Developments in structural steel construction in Southern Africa

M. Dundu
University of Johannesburg, Department of Civil Engineering,
Auckland Park, South Africa

ABSTRACT: This paper describes the developments in steel as a construction material in Southern Africa. Case studies of steel structures are presented and discussed. The discussion entails the extent to which the design of steel structures facilitates the ease of construction. It is recognized that steel has been with us for some time and that most of the world's spectacular bridges have taken advantage of the relative lightness of the material for long spans. More recently, high-rise steel-framed buildings have further demonstrated the versatility of steel. This is happening despite the fact that throughout its history, the steel industry has been dogged by high producer costs, almost inevitably resulting in heavily subsidized steel.

1 INTRODUCTION

The dominant construction material in Southern Africa has been concrete. This position as the construction industry's favourite material has been slowly eroded in recent years by steel. The major factor influencing developers and designers has been the speed of erection offered by steel, as the framework is not susceptible to drying-out movement or delays due to slow strength gain. Steel construction is lightweight, particularly in comparison with traditional concrete construction and the elements of the framework are prefabricated and manufactured under controlled conditions to established quality procedures. Fabrication methods have improved since the 1970s and are now partially or wholly automated, thus reducing costs and increasing the rate of throughput. The developments of high capacity transport and lift equipment, such as cranes, has led to the manufacturing of larger and heavier weldable elements in the workshop that can easily be transported and erected. High strength fine grain steel qualities with yield stress up to $450 \, \text{N/mm}^2$ and adequate fatigue strength have improved economy and design of long span structures. Steel codes (SANS 10162 Part 1 and 2 1993), a design handbook (Southern African Steel Construction Handbook 2005) and a detailing guide (Southern African Structural Steelwork Detailing Manual 1994) have been developed by the Southern African Institute of Steel Construction (SAISC) in collaboration with the steel producers and experts to advance the knowledge of steel in the region.

2 MULTI-STOREY CONSTRUCTION

South Africa has over the years evolved to the use of reinforced concrete as the main building frame for most commercial buildings. Multi-storey steel construction got a major boost in South Africa when the decision to erect the Calton Hotel in Johannesburg from a steel skeleton frame was mooted in 1904 (Steel Construction 1999b). The steel girders of the frame came from Germany, since steel was not manufactured in South Africa during that time. The use of steel in multi-storey buildings has increased dramatically in South Africa, especially in the past 25 years, particularly for multi-deck car parks, large-scale office blocks and large shopping centres.

In Johannesburg, a 20-storey composite office building (Project Mutual Building) was completed in 1989 (Steel Construction 1989b). The supporting structure of the building consists of a concrete core and a steel frame. The reinforced concrete core was constructed ahead of the steel frame whilst the shear walls were cast progressively behind the steel frame. The purpose of the reinforced concrete core is to provide the lateral stability of the structure, as well supporting part of the vertical loads. The advantages of having this type of bracing is that the beam-column connections are simple connections (easy to fabricate and erect), the beams are designed as simply supported members, the shear walls are rigid and act as compartment walls for fire. This simplified the design of the steelwork, as the steelwork would support vertical loads only. During the same period, a smaller seven-storey office

Figure 1. Composite concrete-steel slab.

building, Sanburn Building in Benoni, Johannesburg, was constructed within a short time (Steel Construction 1989c). The developer requested that structural steel be used as the frame and subsequent to awarding the contract the developer changed the lift core and stairwell from conventional reinforced concrete to structural steel so as to achieve a rapid construction programme. Simple connections were also used in this structure because of the ease in which the members can be connected.

The popularity of multi-steel construction has largely been due to the availability of steel decking. The slabs of Project Mutual and Sanburn Buildings were constructed from concrete and the Bond-deck profile, and designed to act compositely with the steel framing members. This profile was developed by Brownbuilt for use in steel framed buildings and in other fast-track decking applications. The Bond-dek profile has 75 mm deep troughs and transverse crenellations (see Figure 1). It is roll-formed from specially graded, structural quality steel with guaranteed yield and tensile properties (ASTM 446 Grade C). This allows the product to have an un-propped span of 3.5 m under the weight of concrete – quite incredible, considering that the sheet has a thickness of 0.8 mm. The bond-dek profile has a major advantage in its 900 mm cover width which, together with its interlocking side lap, makes it the fastest erecting steel decking system available. No bolting of the profiles is necessary because of the clip-in system. This facilitates fast, efficient erection of permanent formwork and reinforcing, without the need to engage skilled workers. The boxed-rib profile allows for further savings in concrete volumes, provide flexible space for electrical lines and other services and can efficiently accommodate shear studs.

Bond-dek has undergone stringent testing by the Department of Civil Engineering, University of the Witwatersrand. It has been fire tested by the Council of Scientific and Industrial Research (CSIR), South Africa and complies with a two hour fire rating, when used to construct a 170 mm thick composite concrete slab. Corrosion of the sheeting is prevented by galvanizing the profiles. For normal applications of Bond-dek steel floors, no additional reinforcing other than a light mesh for shrinkage control is required; typically 193 mesh. However, for "Fire Application" of Bond-dek floors, welded steel mesh reinforcement of 8 mm diameter steel bars at 200 mm spacing in each

direction is required, with minimum top cover along supports (typically on top of shear studs). Other steel profiles for use in composite slabs are Bond-lock (also developed by Brownbuilt) and Flexdeck (developed by Macsteel).

In the same city, the developers of a nine storey-Migporov Building (Metal Steel industries Pension Fund), had to switch from concrete to a steel frame for speed reasons (Steel Construction 1989a). In all these projects it was found that with proper design and planning, multi-storey structural steel framing can save the developers more than 25% on construction time, simplify project management and quality control, secure confidence on project delivery dates and save 15% on project costs. Later modification to a building can be achieved relatively easily by unbolting a connection; with traditional concrete construction such modifications would be expensive, more extensive and disruptive.

3 COMPOSITE BRIDGE CONSTRUCTION

In July 2003, Johannesburg opened the largest cable-stayed bridge in South Africa (Nelson Mandela Bridge). The 284 m long bridge crosses over 42 operational railway lines in linking Braamfontein Newtown in the central business district of Johannesburg (Steel Construction 2002a, 2003a). The unsymmetrical dual-pylon cable–stay bridge consists of a 66 m-north span, a 176 m-main span and a 42 m-south span. The southern and northern steel tubular pylons are 27 m and 43 m high above the deck, respectively. To increase the buckling capacity of the pylons, they were filled with concrete. The main span was built as light as possible using structural steel with a composite deck, while the heavier side spans were built from reinforced concrete to counterbalance the long main span. The choice of steel for the mid-span construction had the benefit of allowing off site manufacturing of the steel box and plate girders, thus maximising quality control and minimising construction impact on train disruptions. Box girders were used as the primary beams because of their large torsional resistance. The manufactured components (some of them of abnormal size) were transported to site by rail and erected using 250 and 400 ton lattice boom cranes over 42 railway lines which were in continuous operation.

An interesting study was carried out by Damoulakis (2000) on the problems that can arise from constructing a concrete bridge on an existing highway (New Road Bridge over the Ben Schoeman Highway in Johannesburg, South Africa). The scaffolding to support the bridge under construction narrowed the road causing long traffic jams and accidents. Damoulakis conservatively quantified the monetary impact of the loss of productive time due to delays, increase in accidents and waste of fuel to be R77million. Based on Damoulakis' study one can conclude that the bridge was heavily subsidizing by the public. During the same period Grinaker got a contract to widen the Griffiths Road Bridge near Johannesburg International Airport. Two trapezoidal steel box sections were used as the main girders spanning 2–30 m highway lanes. Pre-cast, post-tensioned concrete elements were placed on the steel box girder. Composite action was achieved by welding shear studs on the steel girder through the holes located in the pre-cast elements. The two lanes of the highway were closed from 8a.m to 3p.m and during this period motorists used existing off and on ramps as a bypass, whilst the four sections of bridge was being erected, using a mobile crane. This reduced the inconvenience experience by motorists to a minimum. A 1.8 m deep plate girder bridge with a 10 m wide composite concrete deck was constructed to create a new crossing across Dwambazi river in Malawi after the existing bridge was washed away during heavy floods in 1998 (Steel Construction 2002b). Most bridgework on existing highways and rivers are likely to be carried out under these conditions in the future, thus putting structural steel at the forefront among other construction materials. The success of these projects was achieved through the speed at which the steel bridges were put into place. It is no coincidence that this resulted in economic structures.

The construction of the Lesotho Highlands Water project (Mohale and Katse dams and the supply tunnel linking the two dams) led to the construction of steel bridges across Senqunyane and Bokong rivers as a replacement to the flooded routes and footpaths previously used by the communities (de Clerq 2002). The Senqunyane and Bokong bridges consist of a 10-span continuous deck (with a total length of 448,8 m) and a 7-span continuous deck (with a total length of 308,55 m), respectively. The corresponding height of the road above the river is 89 m and 71,9 m respectively. To simplify design the same maximum span length was used so as to have the same deck section for both bridges. The superstructure consists of a trapezoidal steel truss girder with a concrete slab on top. The side and bottom members of the steel truss girder are 12 mm thick plates. These members are 16 mm thick at the supports so as to resist the maximum shear forces at the supports. The top member is $100 \times 75 \times 10$ double angle and the cross-section of the girder was braced using $100 \times 75 \times 10$ mm double angle K-bracing to prevent distortion of the section.

Composite action was achieved by fixing pre-cast slabs (1 m wide) on top of the girder by means of channel shear connectors that were welded to the top flange in the workshop. The pre-cast slabs were spaced at 1,3 m centres and had two openings through which the shear connectors protruded. These openings were filled with a non-shrink grout after the pre-cast slabs had been placed in position, thus forming an effective connection between the slabs and the steel girder. The spacing of the pre-cast slabs created a 300 mm wide gap between them, which had to be filled with in-situ concrete infill strips to ensure that the deck formed a continuous slab. Every infill strip was also fixed to the steel girder with a channel shear connector. Construction of the deck at site started with the welding of a number of 10 m long steel box girder sections together in a closed tent at one abutment. An independent testing authority conducted all testing on site.

4 DOME STEEL STRUCTURES

Domes are now slowly becoming popular in the region because of their capability to have large spans. The three-dimensional structures are constructed either from individual elements or prefabricated modules and posses high strength-to-weight ratio and stiffness. Two types of domes (ribbed and lamella domes) have particularly been found to be popular. In both structures, the frames have double layers (lattice frames) as opposed to single layer grids so that the members are subjected to axial or compressive forces only. Single grid domes develop high flexural stresses and are suitable for lesser spans (usually up to 15 m) than double layer grids (can span in excess of 100 m).

The Caesars Gauteng Complex, situated close to Johannesburg International Airport is one of the many examples where ribbed domes were constructed (Steel Construction 1999a, 2000b). The largest dome has a diameter of 80 m and the five smaller domes have diameters ranging from 8 m to 20 m. The domes created a good architectural design to the surroundings. The overall roof loading of 250 kg/m^2 (excluding the weight of trusses and girders) applied to the main dome was found to be about 3 times greater than the load normally applied to standard steel structures. This load equates to 300 tonnes of steelwork supporting a 1100 tonne load. Structatube 300 steel was specified for the members in the domes to take advantage of their efficiency in compression and the ease in which they can be curved to match the various radii of the different domes.

Structatube 300 is Grade 300WA weldable steel tube that complies with SABS 1431. This product was introduced to the South African market in 1997 by Robor Tube, together with other members of the Association of Steel Tube and Pipe manufacturers. Erection procedures were complex, requiring tandem lifts by mobile cranes to initiate erection of the dome.

Another double layer ribbed dome (93 m diameter by 30 m high), constructed at Skorpion Zinc mine in Namibia showed how a steel structure can be part of an innovative solution of preventing stockpile dust from contaminating the surroundings (Steel Construction 2003b, Frisby 2003). Each frame consists of 10 segments of steel lattice girders that are connected together. Spans of the girders were selected so as to minimize splicing, and to fall within the transport and material supply limitations. This was necessitated by the remoteness of the site and the logistics of assembling the structure on site. The dome was designed as a self-supporting structure once the last structural elements have been put in place. Temporary support was required during erection. This was provided by the circular stacker/reclaimer centre column and tower. The steelwork was trial assembled in the workshop prior to transportation to site so as to ensure that the erection progressed without a hitch.

A 2 m deep double layer spherical lamella dome (161 m diameter and 34 apex height) with a plan area of 20360 m^2 was constructed at the Northgate Shopping Centre in Johannesburg (O'Connell 1997). The 14 equidistant ring-dome incorporates a structural step at rings 7 and 8 to ensure that the angles between members are wide enough to make the fabrication of the connections easy. The frame which consists of 1260 nodes and 6015 elements was analysed using the PROKON software and detailed using STRUCAD. Large forces were found in the rings and lower forces in the connecting truss members. Only the first 4 rings experience tension forces whilst the rest experience compression forces. The largest compression member (a 152 mm × 4 mm circular section) resist a force of 430 kN and the largest tension member (a 273 mm × 6 mm) resist a force of 660 kN.

Erection started from the buttress columns and progressed towards the centre, ring by ring. The triangular modules (made up the prefabricated trusses) were pre-assembled at ground level. Hoisting into position was done by three precisely positioned tower cranes. All truss connections are bolted using M24 high strength friction grip bolts, whereas inter-truss connections are welded. To ensure that the spherical shape was maintained after the completion of a ring, the structure was propped.

5 INDUSTRIAL BUILDING

The traditional uses of structural steel have been in factory and warehouse construction since spans in these structures are usually larger than can be achieved economically with timber or reinforced concrete. Industrial buildings such as the Cold Storage Commission (CSC) factories in Zimbabwe's three cities (Harare, Bulawayo and Masvingo), demonstrated the importance of using a lattice steel girder in constructing large span roofs.

The Swaziland Sugar Association's new bulk raw sugar store at Simunye, in northern Swaziland (see Figure 2) has clearly demonstrates the advantages of

Figure 2. Raw Sugar Storage Facility in Swaziland.

using structural steel as the main construction material (Steel Construction 1988). The building is 30 m high, spans more than 60 m and has a capacity of approximately 60 000 tonnes of sugar. An investigation of various construction materials to determine the most suitable structural medium, established that a parabolic arch of structural steelwork and clad with steel sheeting would almost halve capital expenditure on the project. A further important factor in favour of structural steel was the ease with which secondary steel structures (conveyors) could be suspended from the main structure. Steel plates (30 mm thick) were used to fabricate the 14 I-shaped arches of the sugar terminal, each having a mass of 28 tonnes. In steel construction, designers and clients tend to place an excessive amount of emphasis on reducing the mass of a given frame. Too little attention is given to the reduction of the fabrication and erection costs. Raw material costs account for about 35% of the total costs of each tonne of steel erected, and fabrication and erection costs amount to 40–50% of the total costs.

A major contribution to time saving of the sugar storage structure was the unique erection procedure developed by the steelwork contractor. Assembling a complete arch on the ground and lifting it on to the foundations was not ideal because the space required for such an operation was not available, hiring two 75 tonnes capacity mobile cranes to lift the arch was costly and the deflection of the complete arch about its weak axis was excessive. It was decided to assemble pairs of arches of not more than 3 m high on the ground, complete with all bracing and purlins. These were lifted into position and hinged to the concrete bases (Figure 2). Two 19 m long hinged erection poles were fitted under each segment and as each pair was lifted by crane, the pole slid to a vertical position. Jacks were fitted under each pole and the arches erected 40 mm higher than required to enable the center section to be fitted with a slight clearance. On releasing the jacks, the entire arch settled to its true shape. This erection procedure permitted the crane capacity to be reduced to 25 tonnes, which resulted in a considerable cost saving.

Tor Structures and Dorbyl Structural Engineering are one of the leading manufacturers of steel module industrial buildings in Zimbabwe and South Africa respectively. Tor's present range of units offer clear spans from 7 to 38 m while Dorbyl's Superframe spans from 9 to 63 m. Ready-to-use structural steel modules reduce design time, speed up erection and reduce labour costs. Provision can also be made for expansion at a later date with an extension simply bolted on to the original building. The ease, with which these frames can be transported, brings the whole country within reach of operations. Tor Structure's skilled erection crews can complete buildings of up to 25 m clear span without the assistance of cranes; this is a tremendous advantage in remote areas. Besides their normal range of modular structures, they also undertake special designs for one-off projects.

Dundu and Kemp (2006) developed lightweight, cold-formed section portal frame structures for small span building construction. The rafter and column members are formed from single channel sections, which are bolted back to back at the eaves and apex joints and are connected to the foundation through angle cleats. The spans of the frames investigated range from 10 m–18 m, with a constant eaves height of 3 m and a roof pitch of 10°. It is proposed that the frame of the structure should be provided in a kit form and erected in "meccano" fashion. The portal frame will be delivered to site in sections cut to length and with connection holes pre-punched at the factory as part of the manufacturing process. Pre-galvanised cold-formed steel sections may be used for the column and rafter members so as to avoid site painting. It is envisaged that the structure will be erected mainly by unskilled labour, using only site aids such as lightweight scaffolding, ladders and spanners. Such a simple structure will offer the owners the opportunity to construct their buildings with minimum professional help.

6 STEEL FRAMED HOUSING

Significant demand for steel framed houses exists in Southern Africa. Numerous structures have already been successfully built in South Africa and house a great variety of enterprises. These include the "Mobikaya" (Steel Construction 1991), "Symodule" (Steel Construction 1991), "Kit of parts concept" (Steel Construction 1992) and the "Balaton Building System" (Steel Construction 1995). The fact that these systems have been used successfully in low-cost housing and other structures confirms the acceptance of light weight steel-framed buildings. Rural housing construction can be accelerated in many cases by the use of steel-framed classrooms, offices and houses. Hollow concrete blocks can be made on site and used to fill in an existing steel frame. This method was used extensively in Zimbabwe just after independence to expedite the education programme.

7 CONCLUSION

The discussion in this paper illustrates that steel structures are highly competitive and can be much more versatile than concrete structures in meeting the variable construction conditions at each site. With its more efficient load-carrying capacity and comparable lighter weight than concrete, steel members present fewer transportation and erection problems. Aesthetically, the finished steel structure is slimmer and has cleaner lines than its bulkier and heavier competitor. There is considerable potential, especially with anticipated construction resurgence in the region, to specify and use more structural steel in our civil engineering and building construction. Today steel is the world's most recycled material; every piece of steel used in construction can be recycled again at the end of its life. Even steel construction scrap brings value when it is recycled rather than running up costs for disposal. This minimizes the depletion of the natural resource.

REFERENCES

Damoulakis, M. 2000. Bridge over Troubled Travellers. *J. Southern African Institute of Steel Construction* 24(6): 4–5.

de Clerq, H. 2002. New Steel Bridges for Lesotho Highlands Communities. *J. Southern African Institute of Steel Construction* 26(6): 10–11.

Dundu, M. & Kemp, A.R. 2006b. Strength Requirements of Single Cold-Formed Channels Connected Back-to-Back. *J. Construct. Steel Res.* 62(3): 250–261.

Frisby, C. 2003. Skorpion Zinc Mine Project - Skorpion Material Handling System and Stacker/Reclaimer. *J. Southern African Institute of Steel Construction* 27(5): 20–21.

O'Connell, I. 1997. The Dome – A First for Southern Hemishere. *J. South African Institute of Steel Construction* 24(2), 14–19.

Steel Construction 1988. Raw Sugar Storage Facility in Swaziland. *J. South African Institute of Steel Construction* 12(4): 3–5.

Steel Construction 1989a. Migporov Building (Metal Steel Industries Pension Fund) – Johannesburg. *J. South African Institute of Steel Construction* 13(2): 7–8.

Steel Construction 1989b. Project Mutual Building – Johannesburg. *J. South African Institute of Steel Construction* 13(2): 7.

Steel Construction 1989c. Sanburn Building in Benoni – Johannesburg. *J. South African Institute of Steel Construction* 13(2): 5.

Steel Construction 1991. Steel Solutions for Low-Cost Housing. *J. South African Institute of Steel Construction* 11(2): 11–13.

Steel Construction 1992. Steel Plays a Role in Solving SA's Housing Needs. *J. South African Institute of Steel Construction.* 14(1): 20–21.

Steel Construction 1995. International Award for SA system. *J. South African Institute of Steel Construction.* 19(1): 7–9.

Steel Construction 1999a. The Caesars Gauteng Complex – Johannesburg. *J. Southern African Institute of Steel Construction* 24(6): 8–9.

Steel Construction 1999b. Carlton Hotel – Johannesburg. *J. Southern African Institute of Steel Construction* 24(6): 4.

Steel Construction 2000a. BROstruct 300 – An Engineering Solution for Cost-effective Construction. *J. Southern African Institute of Steel Construction* 24(6): 36.

Steel Construction 2000b. Caesar's Gauteng – Johannesburg. *J. Southern African Institute of Steel Construction* 24(6): 15–16.

Steel Construction 2002a. Madiba's Cable Stayed Bridge, Feature – Bridges; SADC Countries. *J. Southern African Institute of Steel Construction.* 26(6): 4.

Steel Construction 2002b. Plate Girder Bridge for Dwambazi River, Feature – Bridges; SADC Countries. *J. Southern African Institute of Steel Construction* 26(6): 18.

Steel Construction 2003a. Mandela bridge – SAISC Steel Awards. *J. Southern African Institute of Steel Construction* 27(4): 7–8.

Steel Construction 2003b. Skorpion Zinc Mine dome. *J. Southern African Institute of Steel Construction* 27(4): 17–18.

Steel and Composite Structures – Wang & Choi (eds)
© *2007 Taylor & Francis Group, London, ISBN 978-0-415-45141-3*

Time-dependent behavior of continuous composite beams

Sandeep Chaudhary
Malaviya National Institute of Technology, Jaipur, India

Umesh Pendharkar
Government Engineering College, Ujjain, India

A.K. Nagpal
Indian Institute of Technology Delhi, New Delhi, India

ABSTRACT: Studies are reported for the time-dependent behavior of continuous composite beams subjected to service load. Magnitude of load, tension stiffening, age of loading, grade of concrete and relative humidity are the parameters whose effects have been studied on the bending moments at the supports and midspan deflections. It is found that the time-dependent changes in bending moments are primarily due to shrinkage, whereas, the time-dependent changes in deflections result from both creep and shrinkage. Relative humidity is found to be the most influential parameter affecting the time-dependent behavior. Magnitude of loading and grade of concrete have a small effect on time-dependent change in bending moments but a siginificant effect on the changes in the mid-span deflection. Age of loading and tension stiffening are found to have a small effect on the changes in both, the bending moments and the mid-span deflections.

1 INTRODUCTION

The steel-concrete composite beam is one of the economical forms of construction. In continuous composite beams subjected to service load, in addition to instantaneous cracking, the time-dependent effects of creep and shrinkage in concrete can lead to the progressive cracking of concrete deck near interior supports and result in considerable moment redistribution along with increase in deflections. It is therefore desirable to study the time-dependent behavior of continuous composite beams taking into account the progressive cracking of concrete along with the creep and shrinkage.

Extensive literature is available on the instantaneous and time-dependent behavior of the continuous composite beams. Dezi and Tarantino (1993) carried out parametric studies on two span continuous composite beams. The effect of three parameters i.e. compressive strength of concrete (f_{ck}), relative humidity (RH) and age of concrete at loading (t_1) was studied on the time-dependent behavior. The concrete was assumed to be uncracked and further the effect of shrinkage was not considered in these studies. Virtuoso and Vieira (2004) have studied the time-dependent behavior of two span

continuous composite beams with flexible connection and concluded that the effect of creep simultaneously with shrinkage is more important than the effect of creep alone. The individual effect of various rheological factors has not been studied. Ranzi (2006) has presented the studies on the behavior of simply supported composite beams stiffened by longitudinal plates, thereby, neglecting the cracking. The effect of relative humidity and compressive strength has been studied and relative humidity has been found to be the major factor influencing the time-dependent behavior. Some other studies (Kwak and Seo 2000; Mari et al. 2003) have also been reported for the time-dependent behavior of continuous composite beams. These studies are mainly related to the effect of construction sequence and in these, the overall effect of creep and shrinkage but not of the individual rheological parameters is presented.

In the regions, where the tensile stress exceeds the tensile strength of the concrete, cracking occurs. The length of the cracked portion, along with the other factors, depends on the magnitude of load, w. Owing to the distributed nature of cracking, the effective rigidity of the sections is higher than the rigidity of the cracked sections (tension stiffening effect). The displacements

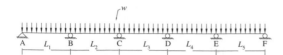

Figure 1. **Example Beam EB**.

may be overestimated if this effect is neglected (Ghali et al. 2002).

In none of the above studies, the effect of all the individual parameters i.e t_1, w, f_{ck}, RH and tension stiffening is reported. The present paper therefore reports systematic studies for the service load behavior of continuous composite beams. The parameters whose effects have been studied on the bending moments at the supports and midspan deflections are t_1, w, f_{ck}, RH and tension stiffening.

The studies are carried out using the hybrid procedure (Chaudhary et al. 2007) developed by the authors for the service load analysis of composite beams and frames. A brief description of the procedure is given in Appendix A.

2 PARAMETRIC STUDY

CEB-FIP MC90 (1993) is used for predicting the properties of the concrete as well creep and shrinkage coefficients. Cement is assumed to be of normal type and mean temperature is assumed to be 20 degree Celsius. The total time-duration of 20,000 days is divided into 20 time intervals. In each time interval, an equal amount of shrinkage occurs. At the time of application of load, instantaneous bending moments, $M^{it}(t_1)$, and instantaneous mid-span deflections, $d_m^{it}(t_1)$ are obtained considering cracking. For time-dependent analysis total bending moments, $M^t(t)$ and total midspan deflections, $d_m^t(t)$ at time t are obtained considering the cracked state at t_1 and subsequent progressive cracking.

A five span continuous composite beam of equal spans designated as example beam EB (Figure 1, $L_1 = L_2 = L_3 = L_4 = L_5 = 7.0\,m$) is considered. All the spans of the beam are assumed to be subjected to uniformly distributed load of same intensity. The cross-section is assumed to consist of a steel section (254 × 146 UB 43) and a concrete slab of dimensions 1000 mm × 75 mm with a reinforcement of area 113 mm² placed at a distance of 15 mm from the top fiber. It may be noted that the increase in steel size would reduce the effects of creep and shrinkage. The shear connectors are assumed to be rigid and at sufficiently close spacing. The slip in such case would be negligible and is therefore neglected.

Studies, for the four parameters (t_1, w, f_{ck} and RH) are carried out by varying each of the parameters in

Table 1. Effect of magnitude of load on instantaneous and total bending moments.

			Bending moment (kN-m)	
			$M^t(t_{21})$	
Support	Load (kN/m)	$M^{it}(t_1)$	Creep	Creep and shrinkage
(1)	(2)	(3)	(4)	(5)
B	2.31	11.89	11.89	50.11
	9.00	45.39	45.80	80.43
	15.00	68.52	71.80	102.61
C	2.31	8.92	8.92	37.58
	9.00	35.01	34.94	64.29
	15.00	56.43	57.10	84.77

turn and keeping other three parameters constant. Tension stiffening is included in these studies. The effect of tension stiffening is studied by keeping the other parameters constant.

Noting the symmetry of the beam, values of $M^{it}(t_1)$ and $M^t(t_{21})$ are reported for only two supports (supports B and C) whereas values of $d_m^{it}(t_1)$ and $d_m^t(t_{21})$ are reported only for three spans (spans AB, BC and CD).

2.1 *Effect of magnitude of load*

Three load cases are considered. In case 1, spans are subjected only to self-weight ($w = 2.31\,kN/m$), whereas in cases 2 and 3 spans are subjected to $w = 9.00\,kN/m$ and $15.00\,kN/m$ (these loads being consisting of dead load and that portion of live load which is permanent in nature) respectively. It may be noted that, at time of application of load, cracking anywhere in beam is initiated at a load of 7.38 kN/m. The other data chosen is: $t_1 = 14$ days, $f_{ck} = 30\,N/mm^2$ and relative humidity (R.H.) = 70%.

The effect of magnitude of load on $M^{it}(t_1)$ and $M^t(t_{21})$ at supports B and C is shown in Table 1. It may be noted from Table 1 that change, due to creep and shrinkage, in $M^t(t)$ is much smaller when only creep is considered. No change is observed when only self-weight is acting since there is no cracking in this case and symmetrical change in curvature along the beam length does not result in additional bending moment. Significant variations occur for all the cases when shrinkage is also considered.

It is also seen that the increases, due to creep and shrinkage, in $M^{it}(t_1)$ at a support [column (5) minus column (3)] are of the same order for different loadings as the main contribution in the changes is from shrinkage, which is independent of loading. The increase, due

Table 2. Effect of magnitude of load on instantaneous and total midspan deflections.

			Bending moment (kN-m)	
			$d_m^t(t_{21})$	
Span	Load (kN/m)	$d_m^{it}(t_1)$	Creep	Creep and shrinkage
(1)	(2)	(3)	(4)	(5)
AB	2.31	1.01	1.40	4.12
	9.00	4.02	5.52	8.64
	15.00	7.25	9.68	13.22
BC	2.31	0.24	0.33	−0.45
	9.00	0.99	1.33	0.87
	15.00	2.32	2.79	2.87
CD	2.31	0.49	0.68	1.07
	9.00	1.89	2.64	2.87
	15.00	3.44	4.63	5.22

Table 3. Effect of tension stiffening (TS) on instantaneous and total bending moments.

		Midspan def. (mm)	
Supp.	Tension stiffening	$M^{it}(t_1)$	$M^t(t_{21})$
(1)	(2)	(3)	(4)
B	Not considered	63.28	96.99
	Considered	68.52	102.61
C	Not considered	53.26	80.28
	Considered	56.43	84.77

Table 4. Effect of tension stiffening (TS) on instantaneous and total midspan deflections.

		Midspan def. (mm)	
Span	Tension stiffening	$d_m^{it}(t_1)$	$d_m^t(t_{21})$
(1)	(2)	(3)	(4)
AB	Not considered	7.68	13.86
	Considered	7.25	13.22
BC	Not considered	2.99	4.02
	Considered	2.32	2.87
CD	Not considered	3.95	6.21
	Considered	3.44	5.22

Table 5. Effect of age of loading on instantaneous and total bending moments

		Bending moment (kNm)	
Supp.	Age of loading (Days)	$M^{it}(t_1)$	$M^t(t_{21})$
(1)	(2)	(3)	(4)
B	7 days	44.40	80.17
	14 days	45.39	80.43
	21 days	45.79	80.32
C	7 days	35.17	63.00
	14 days	35.01	64.29
	21 days	34.94	65.15

and C is shown in Table 3. Both creep and shrinkage are considered in the analysis. In the present case, $M^t(t_{21})$ is less when tension stiffening is not considered. The maximum error in $M^t(t_{21})$ is 5.55% (support C).

The effect of tension stiffening on $d_m^{it}(t_1)$ and $d_m^t(t_{21})$ for spans is shown in Table 4. The values of $d_m^t(t_{21})$ are more in the example considered, when tension stiffening is not considered or accounted for. For example, consider span with the highest $d_m^{it}(t_1)$ and therefore of design significance (span AB). The error in prediction of $d_m^t(t_{21})$ is 4.84%. The percentage error in other spans [with lesser $d_m^{it}(t_1)$] can be much higher. For example, in span BC, the error is 40.07%.

2.3 Effect of age of loading

Three ages of loading ($t_1 = 7$ days, 14 days and 21 days) are considered. The other data chosen is: $w = 9.00$ kN/m, $f_{ck} = 30$ N/mm² and R.H. = 70%. The effect of age of loading on $M^{it}(t_1)$ and $M^t(t_{21})$ is shown in Table 5. Time-dependent analysis is carried out considering both creep and shrinkage. It is

to creep and shrinkage, in $M^{it}(t_1)$ at a support (support B, $w = 2.31$ kN/m) is up to 321.45%.

Table 2 shows the effect of magnitude of load on $d_m^{it}(t_1)$ and $d_m^t(t_{21})$. It is observed that, due to creep and shrinkage, $d_m^{it}(t_1)$ may increase (span AB, CD) or decrease (span BC, $w = 2.31$ and 9.00 kN/m). Differing magnitudes of crack lengths and the influence of adjacent spans result in this increase or decrease. The increase, due to creep and shrinkage, in $d_m^{it}(t_1)$ [column (5) – column (3)] is more for greater loading although the percentage increase is higher for smaller loading. It is observed from Table 2 that the percentage increase, due to creep and shrinkage, in $d_m^{it}(t_1)$ of a span is up to 307.92% (span AB, $w = 2.31$ kN/m).

2.2 Effect of tension stiffening

The data chosen is: $w = 15.00$ kN/m, $t_1 = 14$ days, $f_{ck} = 30$ N/mm² and R.H. = 70%. The effect of tension stiffening on $M^{it}(t_1)$ and $M^t(t_{21})$ at supports B

Table 6. Effect of age of loading on instantaneous and total midspan deflections.

Span	Age of loading (Days)	Midspan deflection (mm)	
		$d_m^{it}(t_1)$	$d_m^t(t_{21})$
(1)	(2)	(3)	(4)
AB	7	4.17	8.69
	14	4.02	8.64
	21	3.96	8.68
BC	7	1.07	0.86
	14	0.99	0.87
	21	0.96	0.91
CD	7	1.89	3.02
	14	1.89	2.87
	21	1.89	2.78

Table 7. Effect of grade of concrete on instantaneous and total bending moments.

Supp.	Concrete grade	Bending moment (kNm)	
		$M^{it}(t_1)$	$M^t(t_{21})$
(1)	(2)	(3)	(4)
B	M20	44.01	75.90
	M30	45.39	80.40
	M40	46.22	83.07
C	M20	35.14	62.19
	M30	35.01	64.30
	M40	34.85	63.90

Table 8. Effect of grade of concrete on instantaneous and total midspan deflections.

Span	Concrete grade	Midspan deflection (mm)	
		$d_m^{it}(t_1)$	$d_m^t(t_{21})$
(1)	(2)	(3)	(4)
AB	M20	4.22	9.54
	M30	4.02	8.64
	M40	3.88	7.78
BC	M20	1.11	1.18
	M30	0.99	0.87
	M40	0.92	0.60
CD	M20	1.91	3.11
	M30	1.89	2.87
	M40	1.89	2.78

observed from the Table that there is a minor effect of age of loading on $M^t(t_{21})$. The minor effect owes to the fact that the time-dependent changes, in $M^{it}(t_1)$ are generated primarily due to shrinkage which is almost independent of the age of loading. The maximum change, due to age of loading, in $M^t(t_{21})$ is 3.30% (support C).

The effect of age of loading on $d_m^{it}(t_1)$ and $d_m^t(t_{21})$ is shown in Table 6. The effect on $d_m^t(t_{21})$ is small since shrinkage is almost independent of age of loading and the final creep coefficients (2.50, 2.36 and 2.26) for the ages of loading considered do not differ much. The value of $d_m^t(t_{21})$ for span with highest $d_m^{it}(t_1)$ (span AB) changes only by 0.11%, when age of loading is changed from 7 days to 21 days.

2.4 Effect of grade of concrete

Three grades of concrete, M20, M30 and M40 ($f_{ck} = 20$ N/mm², 30 N/mm² 40 N/mm² respectively) are considered. The other data chosen is: $w = 9.00$ kN/m, $t_1 = 14$ days and R.H. $= 70\%$. The time-dependent analysis is carried out considering both creep and shrinkage. The effect of grade of concrete on $M^{it}(t_1)$ and $M^t(t_{21})$ at supports B and C is shown in Table 7. It is observed from the table that, as expected, $M^{it}(t_1)$ differ with grade of concrete owing to different crack lengths resulting from different tensile strengths. It is also observed that $M^t(t_{21})$ may be higher or lower for higher grade concrete. The maximum difference in $M^t(t_{21})$ for different grades of concrete is 9.49% (support B). Although, creep and shrinkage differ significantly with grade of concrete, the corresponding difference is not observed in the magnitudes of total bending moments, since it depends additionally on crack lengths (as explained earlier) and on modulus of elasticity of concrete.

The effect of grade of concrete $d_m^{it}(t_1)$ and $d_m^t(t_{21})$ is shown in Table 8. As expected, $d_m^{it}(t_1)$ and $d_m^t(t_{21})$ are higher for lower grade of concrete. The value of $d_m^t(t_{21})$ for span with highest $d_m^{it}(t_1)$ (span AB) is up to 1.22 times when grade of concrete changes from M40 to M20.

2.5 Effect of relative humidity

Three values of relative humidity (50%, 70% and 90%) are considered. The other data chosen is: $w = 9.00$ kN/m, t_1 and $f_{ck} = 30$ N/mm². The effect of relative humidity on $M^t(t_{21})$ at supports is shown in Table 9. Time-dependent analysis is carried out considering both creep and shrinkage. As expected, $M^t(t_{21})$ increases with decrease in relative humidity. The increase, due to creep and shrinkage, in $M^{it}(t_1)$ changes from 38.82% to 99.14% (support C), when relative humidity decreases from 90% to 50%.

The effect of relative humidity on $d_m^t(t_{21})$ is shown in Table 10. It is observed from the table that, as expected,

Table 9. Effect of relative humidity on instantaneous and total bending moments.

Supp.	Relative humidity	Bending moment (kNm)	
		$M^{it}(t_1)$	$M^t(t_{21})$
(1)	(2)	(3)	(4)
B	50%	45.39	85.56
	70%	45.39	80.43
	90%	45.39	63.33
C	50%	35.01	69.72
	70%	35.01	64.29
	90%	35.01	48.60

Table 10. Effect of relative humidity on instantaneous and total midspan deflections.

Span	Relative humidity	Midspan deflection (mm)	
		$d_m^{it}(t_1)$	$d_m^t(t_{21})$
(1)	(2)	(3)	(4)
AB	50%	4.02	9.86
	70%	4.02	8.64
	90%	4.02	6.44
BC	50%	0.99	0.91
	70%	0.99	0.87
	90%	0.99	0.96
CD	50%	1.89	3.02
	70%	1.89	2.87
	90%	1.89	2.63

$d_m^t(t_{21})$ is more for less relative humidity. The increase, due to creep and shrinkage, in $d_m^{it}(t_1)$ changes from 60.20% to 145.27% (span AB), when relative humidity decreases from 90% to 50%.

3 CONCLUSIONS

Systematic studies have been carried out for a five span continuous composite beam. From the studies, following conclusions are drawn:

(a) The time-dependent changes in $M^{it}(t_1)$ can be up to about 321%. Further, these changes are primarily due to shrinkage.
(b) The percentage errors on neglecting the tension stiffening are up to about 6% for $M^t(t_{21})$ in the example considered.
(c) The increase, due to creep and shrinkage, in $M^{it}(t_1)$ changes from 39% to 99%, when relative

humidity decreases from 90% to 50%. Similarly, the increase, due to creep and shrinkage, in $d_m^{it}(t_1)$ changes from 60% to 145%, when relative humidity decreases from 90% to 50%.
(d) Age of loading has a small effect on $M^t(t_{21})$ and $d_m^t(t_{21})$.
(e) The maximum difference in value of $M^t(t_{21})$ for different grades of concrete is about 9% for the range (M 20 to M 40) considered in the example.
(f) The grade of concrete has a significant effect on $d_m^t(t_{21})$. The value of $d_m^t(t_{21})$ is found to be up to 1.20 times for span with highest $d_m^{it}(t_1)$ if the grade of concrete is lowered from M 40 to M 20.

REFERENCES

Chaudhary, S., Pendharkar, U. & Nagpal, A.K. 2007. A hybrid procedure for cracking and time-dependent effects in composite frames at service load. *Journal of Structural Engineering, ASCE* 133(2):166–175.
Comite' Euro International du Beton-Fe'de'ration International de la Pre'contrainte. 1993. *CEB-FIP Model Code 1990 for concrete structures, Bulletin d' information No. 213/214*. Laussane (Switzerland).
Dezi, L. & Tarantino, A.M. 1993. Creep in composite continuous beams. II: Parametric study. *Journal of Structural Engineering, ASCE* 119(1993):2112–33.
Ghali, A., Favre, R. & Elbadry, M. 2002. *Concrete Structures: Stresses and Deformations*. London: Spon Press
Kwak, H.G. & Seo, Y.J. 2000. Long-term behavior of composite girder bridges. *Computers and Structures* 74(2000):583–99.
Mari, A., Mirambell, E. & Estrada, I.2003. Effect of construction process and slab prestressing on the serviceability behavior of composite bridges. *Journal of Constructional Steel Research* 59(2003):135–63.
Ranzi, G.. 2006. Short- and long-term analyses of composite beams with partial interaction stiffened by a longitudinal plate, *Steel and Composite Structures* 3:237–55.
Virtuoso, F. & Vieira, R. 2004 Time dependent behavior of continuous composite beams with flexible connection. *Journal of Constructional Steel Research* 60(2004): 451–63.

APPENDIX A. HYBRID PROCEDURE

Recently, a hybrid analytical-numerical procedure (Chaudhary et al. 2007) has been developed by the authors for the service load analysis of composite frames and beams. The procedure is highly computationally efficient and takes into account the nonlinear effects of concrete cracking and time-dependent effects of creep and shrinkage in concrete portion of the composite beams. The procedure is analytical at the elemental level and numerical at the structural level. A cracked span length beam element consisting of an

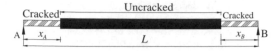

Figure A.1

Figure A.2

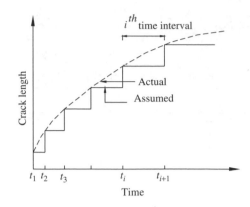

Figure A.3

uncracked zone in the middle and cracked zones at the ends (Figure A.1) has been used in the procedure. The degrees of freedom consdiered for the beam element are shown in Figure A.2.

The analysis in the hybrid procedure is carried out in two parts. In the first part, instantaneous analysis is carried out using an iterative method. In the second part, time-dependent analysis is carried out by dividing the time into a number of time intervals to take into account the progressive nature of cracking of concrete (Figure A.3). In the time-dependent analysis, crack length is assumed to be constant and equal to that at the beginning of the time-interval, as shown in Figure A.3.

Connections

Steel and Composite Structures – Wang & Choi (eds)
© *2007 Taylor & Francis Group, London, ISBN 978-0-415-45141-3*

A proposed lower-bound wind moment method for steel frames having semi-rigid joints

T.C. Cosgrove & C.M. King
The Steel Construction Institute, UK

J.B.P. Lim
University of Strathclyde, UK

ABSTRACT: A simple method that can be used to conservatively estimate the plastic collapse load of a hot-rolled steel portal frame, using only first order linear-elastic frame analysis, is to load the frame until the peak bending moment is equal to the plastic moment capacity of its member, and then to continue the elastic loading using another frame model in which a pin is modeled at the position where the plastic hinge is calculated to have formed. Such a method was popular up to the advent of frame analysis software that incorporated first-order plastic analysis. This paper describes a study that explores whether or not a similar method can be used for the analysis of steel or composite frames having semi-rigid joints. For the frame arrangement considered in the study, two frames are required to be analysed using first-order elastic analysis: Frames A and B. Frame A incorporates semi-rigid joints and column bases and is loaded until the bending moment of the joints reaches the moment capacity of the joints. Frame B is identical to Frame A except that the pre-determined joints are replaced by pins. In both frames, second-order effects are taken into account through amplification of the bending moments in accordance with the British Standards. The ultimate load that the frame carries, calculated through summing the load applied to Frames A and B, is compared against that determined using second-order analysis with non-linear moment-rotation curves modeled for the joints. For the frame and load combinations considered, it is demonstrated that the proposed method results in a lower-bound solution to that obtained using second-order analysis.

1 INTRODUCTION

In a previous paper, the Authors described a parametric study of unbraced steel frames having semi-rigid joints in which the number of bays and storeys and the joint and column base stiffnesses were key variables (Lim *et al* (2007). The frames were analysed using second-order analysis and designed in accordance with the current British Standards (BS5950 (2000)). The semi-rigidity of the joints was taken into account through the definition of non-linear moment-rotation curves representing their behavior. The section sizes used for beams and columns and the strength and stiffness of the joints were practical and easily achievable in practice. It was demonstrated that bay spacings between adjacent primary frames could be designed that would be of a similar order to that of the span of the beam. With respect to joint stiffness, a preliminary conclusion was reached that it was not the strength and stiffness of the joints that was important, but the strength and stiffness of the column bases.

The design of such frames, however, requires the designer to have access to second-order analysis software in which non-linear moment rotation curves

can be defined representing the strength and stiffness of the joints. It is recognized that not all designers have access to such software.

For portal frames, a simple method that can be used to conservatively estimate the plastic collapse load, using only first order linear-elastic frame analysis, is to load the frame until the peak bending moment is equal to the plastic moment capacity of its member, and then to continue the elastic loading using another frame model in which a pin is modeled at the position where the plastic hinge is calculated to have formed. Such a method was popular up to the advent of frame analysis software that incorporated first-order plastic analysis.

King proposed that a similar approach could be applied to unbraced steel frames, to be referred to in this paper as the proposed lower-bound Wind Moment Method. This method would not be limited by the geometrical limitations of the Wind Moment Method described by Salter *et al* (1999) and Hensman and Nethercot (2000).

In this paper, a study of a typical frame is described that explores whether or not the proposed method can be used for the analysis of steel frames having semi-rigid joints.

Using the proposed method, for the frame arrangement, two frames are required to be analysed using first-order elastic analysis: Frames A and B. Frame A incorporates semi-rigid joints and column bases and is loaded until the bending moment of a single joint reaches its moment capacity. Frame B is identical to Frame A except that the pre-determined joints are replaced by pins. In both frames, second-order effects are taken into account through amplification of the bending moments in accordance with BS5950 (2000).

The ultimate load that the frame carries, calculated through summing the load applied to Frames A and B, is compared against that determined using second-order analysis with non-linear moment-rotation curves modeled for the joints. For the frame considered in this paper, it is demonstrated that the proposed method results in a lower-bound design to that obtained using second-order analysis.

2 FRAME DESCRIPTION

2.1 *Frame geometry*

Figure 1 shows the geometry used for the frame considered in the parametric study. As can be seen, the frame is of 6 storeys and 6 bays. The bay spacing, defined as the distance between adjacent bays, is 5.85 m.

2.2 *Section sizes*

The sections capacities adopted for the beams and column members of the frame are shown in Table 1. As an indication of section sizes, the section properties of the column corresponds approximately to that of a $254 \times 254 \times 132$ UC S355.

2.3 *Joint properties*

Figure 2 shows the moment rotation curve assumed for the joints. As can be seen, the moment capacity of the joints is 82 kNm and the initial rotational stiffness is 2.17×10^4 kNm/rad. For the section capacities, such a moment capacity and initial rotational stiffness is achievable in practice using a simple fin plate connection (Van Keulen *et al* (2003) and Lim *et al* (2007)).

Eurocode 3 (2005) defines a non-dimensional joint rotational stiffness as follows:

$$K_j = \frac{S_j}{EI/L} \qquad (1)$$

Under the EC3 classification system, for an unbraced frame, a rigid joint is assumed to have a value of K_j greater than 25, while a semi-rigid joint has a value of K_j between 0.5 and 25. For the frame and joints under consideration, the above value of S_j corresponds to a value of K_j of 3.2 (using a beam length of 7.5 m).

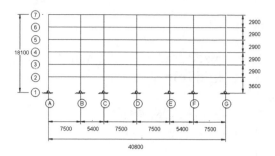

Figure 1.   Details of typical frame used in study.

Table 1.   Section properties and capacities of beam and column members.

Member	Area (cm^2)	Second moment of area Major axis (cm^4)	Moment capacity Major axis (kNm)	Minor axis (kNm)	Axial capacity (kN)
Beam	158	22500	614	270	5609
Column	168	23500	664	312	5964

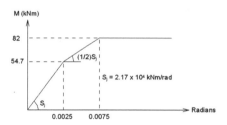

Figure 2.   Tri-linear moment rotation curve for joints.

2.4 *Column base stiffness*

The rotational stiffness of the column base is assumed to be equal to that of the column-end rotational stiffness. Such a rotational stiffness is five times that recommended in BS5950: 2000 for a nominally pinned column base (Lim *et al* (2007)).

3 FRAME LOADS

The loads applied to the frame in the parametric study are as follows:

Dead load (DL) $= 4.1$ kN/m^2
Live load (LL) $= 3.5$ kN/m^2 on floors
$\qquad\qquad\quad\ = 1.0$ kN/m^2 on roof
Wind load (WL) $= 1.0$ kN/m^2

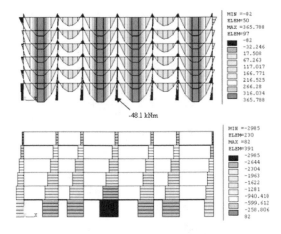

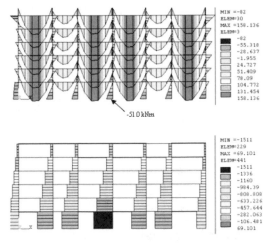

-48.1 kNm

-51.0 kNm

Figure 3. Bending moment and axial force diagram from second order analysis for load combination 1.

Figure 4. Bending moment and axial force diagram from second order analysis for load combination 2.

4 LOAD COMBINATIONS

The following two ultimate limit state load combinations are considered:

LC1: 1.4DL + 1.6LL + NHF
LC2: 1.4DL + 1.4WL

5 DESIGN USING SECOND ORDER ANALYSIS

For each load combination, the frame is analysed using second-order analysis. The joints are modeled using the idealized moment rotation curve shown in Figure 2. For both load combinations, the load that the frame can sustain will be determined. This load is defined, with reference to BS5950 (2000), as the load at which any of the columns at the base of the frame fail through either the local capacity check or a member buckling check.

5.1 Load combination 1

Figure 3 shows the bending moment and axial force diagram resulting from load combination 1. The results of the second-order analysis show that the factored design loads can be increased by a factor, λ_2, of 1.35 before a failure at the column base.

5.2 Load combination 2

Figure 4 shows the bending moment and axial force diagram resulting from load combination 1. The results of the second-order analysis show that the factored loads can be increased by a factor, λ_2, of 2.42 before a plastic hinge forms at the column base.

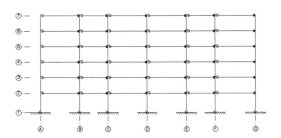

Figure 5. Frame A used for proposed lower-bound WMM.

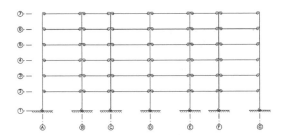

Figure 6. Frame B used for proposed lower-bound WMM.

6 DESIGN USING PROPOSED LOWER-BOUND WIND MOMENT METHOD

Design using the proposed lower-bound WMM requires the first-order analysis of two frames: Frame A and Frame B. For each frame, the factored design loads are applied.

Figure 5 and Figure 6 show Frame A and Frame B, respectively. As can be seen, in the case of Frame A, the joints on both sides of each beam are modeled. The

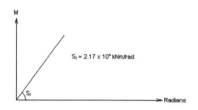

Figure 7. Rotational joint stiffness used for proposed lower-bound WMM.

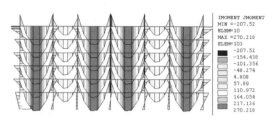

Figure 8. Bending moment diagram for Frame A to load combination 1.

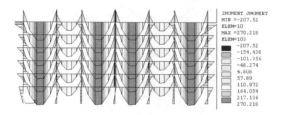

Figure 9. Bending moment diagram for Frame B to load combination 1.

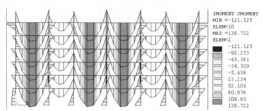

Figure 10. Bending moment diagram for Frame A to load combination 2.

rotational stiffness assumed for the joints is linear and is shown in Figure 7.

In the case of Frame B, only the rotational stiffness of the joint on the left hand side of each beam is modeled. The joint on the right hand side of each beam is pinned.

6.1 Load combination 1

6.1.1 Frame A

Figure 8 shows the bending moment diagram of Frame A under the factored design load. The values of λ_{crit} and k_{amp} were 4.57 and 1.28, respectively.

As can be seen, the maximum bending moment at the joints is -207.5 kNm (hogging moments are taken as negative) and occurs at the joint at level 7 to the left hand side of the column located at gridline D. The load factor, λ_A, at the formation of the first hinge, is calculated as follows:

$$\lambda_A = \text{(hinge resistance moment)}/$$
$$\qquad \text{(max hinge moment from analysis)}$$
$$\qquad = 82/207.5$$
$$\qquad = 0.395$$

6.1.2 Frame B

Figure 9 shows the bending moment diagram of Frame B under the factored design load. The values of λ_{crit} and k_{amp} were 2.45 and 1.69, respectively.

After considering all the joints, the joint at level 6 to the right of the column located at gridline B was identified as being the critical joint.

For Frame A, the bending moment at this joint is -100.7 kNm. Similarly, for Frame B, the bending moment at the same joint is 202.0 kNm. The factor, λ_B,

by which the design load needs to be reduced before a hinge forms, may be calculated from the following expression:

$$82 = (0.395 \times (-100.7)) + (\lambda_B \times 202.0)$$

From which:

$$\lambda_B = 0.602$$

On the formation of hinges on both sides of the joints, the value of λ_{cr} of the frame will be less than one and the frame will no longer be able to carry any additional load. Therefore, in accordance with design using the King-Arthur method, the factor, λ_{KA}, by which the design load can be increased before failure occurs, may be calculated as follows:

$$\lambda_{KA} = \lambda_A + \lambda_B = 0.997$$

6.2 Load combination 2

6.2.1 Frame A

Figure 10 shows the bending moment diagram of Frame A under the factored design load. The values of λ_{crit} and k_{amp} were 9.33 and 1.12, respectively.

As can be seen, the maximum bending moment at the joints is -121.1 kNm (hogging moments are taken as negative) and occurs at the joint at level 7 to the left hand side of the column located at gridline D. The load factor, λ_A, at the formation of the first hinge, is calculated as follows:

$$\lambda_A = \text{(hinge resistance moment)}/$$
$$\qquad \text{(max hinge moment from analysis)}$$
$$\qquad = 82/121.1$$
$$\qquad = 0.677$$

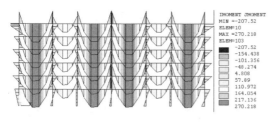

| IMOMENT JMOMENT |
| MIN =-207.52 |
| ELEM=10 |
| MAX =270.218 |
| ELEM=103 |

■	-207.52
■	-154.438
□	-101.356
□	-48.274
□	4.808
□	57.89
□	110.972
□	164.054
▨	217.136
▨	270.218

Figure 11. Bending moment diagram for Frame B to load combination 2.

6.2.2 *Frame B*

Figure 11 shows the bending moment diagram of Frame B under the factored design load. The values of λ_{crit} and k_{amp} were 4.85 and 1.26, respectively.

After considering all the joints, the joint at level 4 to the right of the column located at gridline B was identified as being the critical joint.

For Frame A, the bending moment at this joint is -36.2 kNm. Similarly, for Frame B, the bending moment at the same joint is 81.7 kNm. The factor, λ_B, by which the design load needs to be reduced before a hinge forms, may be calculated from the following expression:

$$82 = (0.677 \times (-36.2)) + (\lambda_B \times 81.7)$$

From which:

$$\lambda_B = 1.303$$

On the formation of hinges on both sides of the joints, the value of λ_{cr} of the frame will be less than one and the frame will no longer be able to carry any additional load. Therefore, in accordance with design using the King-Arthur method, the factor, λ_{KA}, by which the design load can be increased before failure occurs, is calculated as follows:

$$\lambda_{KA} = \lambda_A + \lambda_B = 1.980$$

7 COMPARISON OF DESIGN USING SECOND ORDER ANALYSIS PROPOSED LOWER-BOUND WIND MOMENT METHOD

Table 2 compares the results of the design of the frame using both second-order analysis and the proposed

Table 2. Comparison of design using second-order analysis and proposed lower-bound WMM.

Load combination	λ_A	λ_B	λ_{KA}	λ_2	λ_2/λ_{KA}
1	0.395	0.602	0.997	1.35	1.35
2	0.677	1.303	1.980	2.42	1.22

lower-bound WMM. It can be seen that for both load combinations, design in accordance with the proposed lower-bound WMM is conservative between 22% and 35%, and so will result in an acceptable design.

8 CONCLUSION

It has been demonstrated that the proposed lower-bound Wind Moment Method has produced conservative results for two load combinations for a typical six storeys by six bay frame having practical joint strength and stiffness. Application of this method could potentially allow the safe design of all frames using only first-order analysis software. Further work is still required to assess the method for other load combinations and different frame types.

REFERENCES

BS5950. 2000. Part 1: Code of practice for design in simple and continuous construction: hot rolled sections, London, British Standards Institution

Eurocode 3. 2005. Design of steel structures Part 1.1: General rules and rules for buildings, European Pre-standard, ENV 1993–1–1: 1992, CEN Brussels, Belgium

Hensman. J.S., Nethercot, D.A. 2000. Design of unbraced composite frames using the wind-moment method, The Structural Engineer, 79(11), p28

Lim, J.B.P., Cosgrove, T.C., King, C.M., Nethercot, D.A. 2007. Effect of column base rotational stiffness on the strength of unbraced semi-rigid steel frames, ICSCS07 Manchester.

Salter, P.F., Couchman, G.H., Anderson, D. 1999. Wind moment design of low rise frames, The Steel Construction Institute, Publication 263, Ascot

Van Keulen, D.C., Nethercot, D.A., Snijder, H.H., Bakker, M.C.M. 2003. Frame analysis incorporating semi-rigid joint action: Applicability of the half initial Secant stiffness approach, J.Constr. Steel Res., 59, p1083

Steel and Composite Structures – Wang & Choi (eds)
© 2007 Taylor & Francis Group, London, ISBN 978-0-415-45141-3

Effect of column base rotational stiffness on the strength of unbraced semi-rigid steel frames

J.B.P. Lim
University of Strathclyde, UK

T.C. Cosgrove & C.M. King
The Steel Construction Institute, UK

D.A. Nethercot
Imperial College, UK

ABSTRACT: Steel frames in the UK are traditionally designed as braced and analysed assuming only a notional column base rotational stiffness. One of the limitations of such braced frames is that, unless additional structure is introduced to provide lateral stability against wind loads (or notional horizontal loads), such a design does not easily permit full fenestration. It is well known, however, that the joints of steel frames are semi-rigid and that the rotational stiffness of the joints will resist some lateral load. In addition, it is also known that the actual rotational stiffness of the column bases used in steel frames may be substantially higher than the notional column base rotation stiffness suggested by the British Standards. This paper considers whether or not a combination of the rotational stiffness of the steel joints and that of the column bases can resist sufficient lateral load to enable frames to be designed without bracing. A parametric study is described of frames in which the number of bays, storeys, joints and column base stiffnesses are key variables. The frames are analysed using second-order analysis and designed in accordance with the British Standards, allowing for both vertical loads as well as wind loads; checks are conducted against both ultimate and serviceability limit states. It is demonstrated that the strength of an unbraced steel frame is not very sensitive to the strength and stiffness of the joints; of more importance is the strength and stiffness of the column base. A preliminary conclusion is that the rotational stiffness of the column base should be at least four times that of the column-end rotational stiffness; increasing the rotational stiffness above this value is not of substantial benefit to the design of the frame.

1 INTRODUCTION

Designers normally employ one of two basic approaches for providing adequate lateral stiffness to steel frame structures: bracing or portal action. The former relies on additional structure e.g. diagonal bracing, cores, lift shafts etc and is generally associated with the very popular "simple construction" design method. The latter requires that joints possess rotational stiffness and moment capacity; its simplest form ensures that their moment capacity is equal or greater to that of the surrounding members and that they function as rigid connections. This leads to the alternative "continuous construction" approach to design. Thus both approaches are directly linked to the basic simplifying assumptions for connections; either they are assumed to act as pins or they are assumed to act as rigid.

Research conducted over the past few decades has shown that all practical forms of steel beam to column connection actually function in an intermediate fashion i.e. they are semi-rigid and/or partial strength. Some types operate sufficiently close to one of the basic forms that the design approaches of simple construction or continuous construction are reasonable. Eurocode 3 (2005) leads to a third alternative of semi-continuous construction in which explicit allowance for the stiffness and/or moment capacity of the joints should be made.

The paper considers a class of frame that does not, by inspection, satisfy the requirements for continuous construction i.e. the joints do not have sufficient rotational stiffness and/or moment capacity. However, it is shown that provided a suitable level of rotational stiffness is provided in the nominally pinned column bases then the combination of some beam to column joint

337

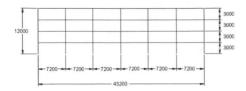

Figure 1. Details of typical frame.

stiffness and some column base stiffness may well be sufficient to satisfy design requirements for both the ultimate and serviceability limit states.

The approach therefore has some relationship to the Wind Moment Method (WMM), in which design under gravity loading assumes pin joints, design under lateral wind loading assumes rigid joints and superposition is used to cover the combined loading case. However, the basic WMM relies on the use of fixed base columns to provide sufficient lateral stiffness that serviceability drift limits be met. It is also restricted to that class of frames for which more rigorous checks have shown that it produces safe designs (Salter et al (1999)). Recently Hensman and Nethercot (2001), in investigation the applicability of the WMM to composite frames, have included the benefits of partially fixed column bases. Readily achievable levels of base fixity, derived from a consideration of all available test data, were shown to ensure satisfactory designs without the complexity and expense of rigid supports; pin bases were unable to deliver frames with sufficient stiffness to meet drift limits.

2 FRAME DESCRIPTION

2.1 Frame geometry

For all the frames considered, the span of each bay is taken as 7.2 m and the storey height as 3 m, such an arrangement being representative of that used by designers in practice. Figure 1 shows an example of such a frame comprising 6 bays and 3 storeys.

2.2 Section sizes

The sections adopted for the frames are $457 \times 191 \times 82$ UB S355 for the beams and $305 \times 305 \times 97$ UC S355 for the columns. Table 1 gives the member resistances of these sections. These section sizes have been chosen with reference to the Structural Engineer's Manual for braced frames (Cobb (2003)).

2.3 Joint properties

Van Keulen et al (2003) describe a parametric study of frames having one of four bolted flush endplate joints,

Table 1. Section capacity of beam and column members calculated in accordance with BS5950–1: 2000.

| | Moment capacity | | |
| | Major axis (kNm) | Minor axis (kNm) | Axial capacity (kN) |
Section			
$457 \times 191 \times 82$ UB S355	572	69.6	3690
$305 \times 305 \times 97$ UC S355	513	170	4370

Table 2. Section capacity of beam and column members calculated in accordance with BS5950–1: 2000.

Characteristic	Initial stiffness (S_j) (kNm/rad)	Moment capacity (M_j) (kNm)	Kj*
Joint 1	19892	111.5	3.5
Joint 2	30455	157.2	5.5
Joint 3	40594	185.8	7.2
Joint 4	90700	266.5	16

*Value of K_j calculated for $457 \times 191 \times 82$ UB and span of 7.2 m

Table 3. Comparison of section properties of beam and column members.

	Section	Area cm^2	I_{major} cm^4	Depth (mm)	Width (mm)
Beam	IPE360	72.8	16270	360	170
	$457 \times 191 \times 82$	104	37100	460	191.3
Colum	HEA260	86.8	10450	250	260
	$305 \times 305 \times 97$	123	22300	307.9	305.3

the joint characteristics being calculated in accordance with Annex J of EC3 (2005). The beam and column members of these frames are IPE 360 and HEA 260, respectively. The joints are referred to as Joints 1 to 4, with Joint 1 having the lowest strength and stiffness and Joint 4 having the highest strength and stiffness. Table 2 summarises the joints characteristics used by Van Keulen et al.

In the absence of calculations to EC3 for joints using the section sizes to be used in this paper, the joint properties of Table 2 will be adopted. Table 3 compares the section sizes of beam and columns used by Van Keulen et al with those used in this paper. As can be seen, the section sizes used by Van Keulen et al are smaller and so can be expected to possess lower rotational stiffness and moment capacity. Adopting joint properties of Van Keulen et al will therefore be conservative but

338

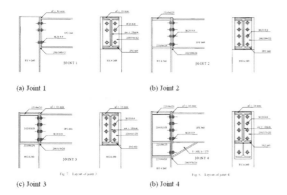

(a) Joint 1

(b) Joint 2

(c) Joint 3

(b) Joint 4

Figure 2.   Details of joints.

indicative of the strengths and stiffnesses that may be achieved in practice.

2.4   *Column base stiffness*

BS5950–1: 2000 Clause 5.1.3.4 states that for a nominal semi-rigid base, a stiffness of up to 20% of the stiffness of the column may be assumed:

$$S_{j,b} = \frac{EI_c}{5} \tag{1}$$

Adopting the non-dimensional rotational stiffness of EC3, for a column height of 3 m, the equivalent value of $K_{j,b}$ may be calculated as follows:

$$K_{j,b} = \frac{S_{j,b}}{(EI_c / H)} = \frac{3}{5} = 0.6 \tag{2}$$

Under the EC3 classification system, a rigid joint is assumed to have a value of K_j greater than 25, while a semi-rigid joint has a value of K_j between 0.5 and 25. The BS5950–1: 2000 definition for a nominal semi-rigid base is therefore consistent with that of EC3.

For the frames described in this paper, it can be presumed that the value of $K_{j,b}$ will be at least 0.5; in the parametric study, the effect on frame strength of increasing the column base rotational stiffness from 0.5 to 25 will be investigated.

3   FRAME LOADS

The loads applied to the frame in the parametric study are as follows:
Dead load (DL) $= 4.1 \, \text{kN/m}^2$
Live load (LL) $= 3.5 \, \text{kN/m}^2$ on floors
$= 1.0 \, \text{kN/m}^2$ on roof
Wind load (WL) $= 1.0 \, \text{kN/m}^2$

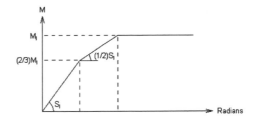

Figure 3.   Idealised tri-linear moment rotation curve for joints.

4   LOAD COMBINATIONS

The following three ultimate limit state load combinations are considered:
LC1: 1.4DL + 1.6LL + NHF
LC2: 1.4DL + 1.4WL
LC3: 1.2DL + 1.2LL + 1.2WL
In addition, four serviceability limit state load combinations are also considered:
LC4: 1.0DL + 1.0WL
LC5: 1.0DL + 0.8LL + 0.8WL
LC6: 1.0DL + NHF
LC7: 1.0DL + 0.8LL + NHF

5   FRAME IDEALISATION

The frames are modelled using the finite element program ANSYS. The beams and columns are modelled using beam elements (BEAM3), while the joints and column bases are modelled using a non-linear rotational spring element (COMBIN39).

Figure 3 presents the idealised EC3 tri-linear moment rotation curve for the joints. This idealised curve can be used in the absence of the actual non-linear design moment-rotation characteristics of the joints. As can be seen, the initial stiffness of the joint Sj is used up to an applied moment equal to two thirds of the moment capacity of the joint. After this value, half the initial stiffness is assumed.

The EC3 non-dimensional joint rotational stiffness for the joints is calculated as follows:

$$K_j = \frac{S_j}{EI / L} \tag{3}$$

For a 457 × 191 × 82 UB of span 7.2 m, Table 1 also shows the values of K_j for each joint of Van Keulen et al, but calculated using the frame geometry and section size adopted in this paper. As described previously, in accordance with the EC3 classification system, for an unbraced frame, a rigid joint is assumed to have a

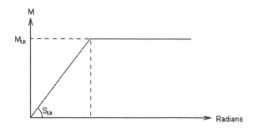

Figure 4. Idealised bi-linear moment rotation curve for column base.

value of K_j greater than 25, while a semi-rigid joint has a value of K_j between 0.5 and 25.

Figure 4 shows the moment rotation curve assumed for the column base. As can be seen, the curve is bi-linear. In the parametric study, the effect on frame strength of different values of $K_{j,b}$ will be investigated. The value of M_j used corresponds to that of the moment capacity of the column.

6 FRAME ANALYSIS

The frames are analysed using second-order analysis. For each frame, only the elastic properties for the beam and column members are defined. However, as discussed in Section 5, the full tri-linear moment-rotation curve of the joints is defined for the rotational spring representing the joints. For the case of the column base, a bi-linear moment-rotation curve is defined.

For each frame, the load that the frame can sustain under each ultimate limit state load combination will be determined. This load is defined, with reference to BS5950–1: 2000, as the load at which any of the columns at the base of the frame fail through either the local capacity check or a member buckling check. Having determined the failure load for each of the ultimate limit state load combinations, the lowest value is taken as the ULS failure load.

For each frame, the working load at which the SLS load is reached is also determined. This load is defined as the load at which the relative drift at the external columns between any two floors is greater than h/300. Again, having determined the load at which the serviceability check fails for each of the serviceability limit state load combinations, the lowest value is taken as the SLS failure load.

The load that the frame can be designed to sustain is then defined as the lowest of the ULS and SLS failure loads calculated. A better physical interpretation of the failure loads is the bay spacing between adjacent frames allowed under the relevant load combination. This bay spacing is the main output of the parametric study.

7 PARAMETRIC STUDY

7.1 Scope of parametric study

The parametric study considers frames having between two and six bays and between two to eight storeys. Each frame is designed using each of the joint properties shown in Table 1. In addition, three rotational stiffnesses for the column bases will be used, corresponding to values of $K_{b,j}$ of 0.5, 4 and 25.

The load that can be sustained by each frame under each load combination has been determined and is converted into a bay spacing (m). The critical bay spacing for both ULS and SLS has been determined and the critical load combination identified.

In practice, the usual bay spacing for a grid in which the beam spans are 7.2 m is also 7.2 m. Nevertheless, comparisons of bay spacing are more useful than comparisons of load, allowing results for ultimate limit states and serviceability limit states to be compared directly. While the results are only valid for the particular arrangements used herein, general trends can be observed.

However, practical designs can also be interpreted. For example, for the member sizes chosen, it can be seen that a 2 bay frame having 4 storeys and base stiffness of $K_{b,j}$ of 4 is feasible (i.e. having a bay spacing 7.2 m), but that frames requiring more storeys are not. As another example, a 6 bay frame having 6 storeys and a base stiffness of $K_{b,j}$ of 4 is only slightly unfeasible.

7.2 Effect of joint strength and stiffness

Figure 5 shows the effect of the different Joint types on the critical bay spacing for 2 bay and 6 bay frames. While the results are shown with K_j on the x-axis, plots with M_j on the x-axis would have been equally valid.

As can be seen, aside from the 2 storey frames, where impractical bay spacings of the order of 20 m are shown, the critical bay spacing does not appear to be particularly sensitive to the joint strength and stiffness. For example, for the 6 bay frame having 8 storeys, the benefit in changing the Joint type from 1 to 4 is only 10%.

As expected, for the 4, 6 and 8 storey frames, the 2 bay frames are more sensitive to changes in joint strength and stiffness than the 6 bay frames. However, it is interesting to note that there is very little difference between the critical bay spacing for the 2 bay frame and 6 bay frames. For example, for the 2 bay frame having 8 storeys, the critical bay spacing is 4.6 m. On the other hand, for the 6 bay frame having 8 storeys, the critical bay spacing has only increased to 5.4 m.

7.3 Effect of column base stiffness

Figure 6 shows the effect of $K_{j,b}$ on SLS bay spacing for two frames: 2 bays by 8 storeys and 6 bays

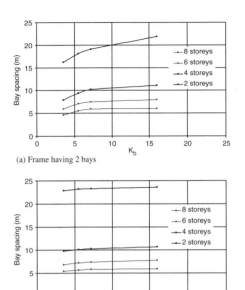

(a) Frame having 2 bays

(b) Frame having 6 bays

Figure 5. Effect of K_j on critical bay spacing.

(a) Frame having 2 bays by 8 storeys

(b) Frame having 6 bays by 4 storeys

Figure 6. Effect of $K_{j,b}$ on SLS bay spacing for different joint types.

by 4 storeys. As can be seen, the results have been non-dimensionalised with respect to that for a value of $K_{j,b}$ corresponding to 25, representing a rigid column base.

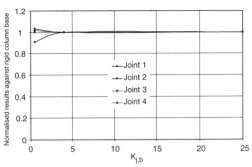

(a) Frame having 2 bays by 8 storeys

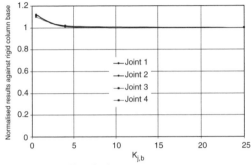

(b) Frame having 6 bays by 4 storeys

Figure 7. Effect of $K_{j,b}$ on ULS bay spacing for different joint types.

For the 2 bays by 8 storeys frame, the effect of increasing the value of $K_{j,b}$ from 0.5 to 4 can be seen to result in an increase in bay spacing of around 35% for all four joints. On the other hand, for the 6 bays by 8 storeys frame, the effect of increasing the value of $K_{j,b}$ from 0.5 to 4 can be seen to result in an increase in bay spacing of around 15% for all four joints. For both frames, however, increasing the value of $K_{j,b}$ further results in only a 5% and 3% increase in bay spacing, respectively.

Figure 7 shows the effect of $K_{j,b}$ on ULS bay spacing for the same two frames. In the case of the 2 bays by 8 storeys frame (Figure 7a), there is a negligible benefit in increasing the value of $K_{j,b}$ any further than 4. As expected, in the case of Joint 1, which has a lower rotational stiffness all the other Joints, there is a benefit in increasing the value of $K_{j,b}$. However, in the case of Joints 2, 3 and 4, the effect of increasing the value of $K_{j,b}$ results in an unexpected reduction in ULS bay spacing. This result can be attributed to the fact that a different load combination controls the ULS bay spacing. In addition, the increased rotational stiffness of the Joints 2, 3 and 4 results in moment being attracted to the joint, which changes the ratio of bending moment at the joint to that at the column base.

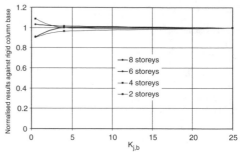

(a) Frame having 2 bays

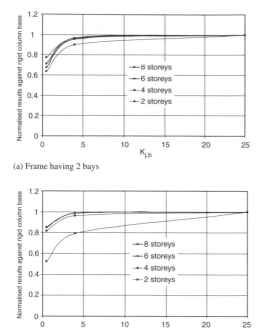

(b) Frame having 6 bays

Figure 8. Effect of $K_{j,b}$ on SLS bay spacing for frames using Joint 1.

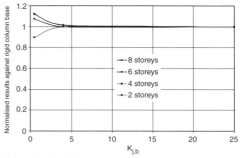

(a) Frame having 2 bays

(b) Frame having 6 bays

Figure 9. Effect of $K_{j,b}$ on ULS bay spacing for frames using Joint 1.

In the case of the 6 bays by 4 storeys frame (Figure 7b), a small reduction in bay spacing is observed for all four joints as the value of $K_{j,b}$ increases. The fact that all four joints exhibit this behaviour may be attributed to the fact that the 6 bay by 4 storeys frame is more laterally stable than the 2 bays by 8 storeys frame.

Figure 8 shows the effect of $K_{j,b}$ on SLS bay spacing for different frames using Joint 1. As can be seen, apart from the frame having 6 bays and 2 storeys, there is little benefit in increasing the value of $K_{j,b}$ above 4. Figure 9 shows the effect of $K_{j,b}$ on ULS bay spacing for different frames using Joint 1. Again, there is little effect in increasing the value of $K_{j,b}$ above 4.

8 CONCLUDING REMARKS

This paper has described a parametric study of frames in which the number of bays and storeys and the joint and column base stiffnesses are key variables. The frames were analysed using second-order analysis and designed in accordance with the current British Standards. It has been demonstrated that the design of an unbraced frame, expressed in terms of bay spacing between adjacent frames, is not very sensitive to the strength and stiffness of the joints. Of more importance is the strength and stiffness of the column base.

A preliminary conclusion is the rotational stiffness of the column base should be at least four times that of the column-end rotational stiffness. Further work will focus on determining whether or not such a rotational stiffness can be achieved in practice through an assessment that takes into account realistic soil conditions and bearing pressures.

REFERENCES

BS5950. 2000. Part 1: Code of practice for design in simple and continuous construction: hot rolled sections, London, British Standards Institution

Cobb, F. 2003. Structural Engineer's Pocket Book, Elsevier, London

Eurocode 3. 2005. Design of steel structures Part 1.1: General rules and rules for buildings, European Pre-standard, ENV 1993-1-1: 1992, CEN Brussels, Belgium

Hensman. J.S., Nethercot, D.A. 2000. Design of unbraced composite frames using the wind-moment method, The Structural Engineer, 79(11), p28

Salter, P.F., Couchman, G.H., Anderson, D. 1999. Wind moment design of low rise frames, The Steel Construction Institute, Publication 263, Ascot

Van Keulen, D.C., Nethercot, D.A., Snijder, H.H., Bakker, M.C.M. 2003. Frame analysis incorporating semi-rigid joint action: Applicability of the half initial Secant stiffness approach, J. Constr. Steel Res., 59, p1083

Steel and Composite Structures – Wang & Choi (eds)
© *2007 Taylor & Francis Group, London, ISBN 978-0-415-45141-3*

Alternative designs for internal and external semi-rigid composite joints

B. Gil & E. Bayo

Department of Structural Analysis and Design. School of Architecture
University of Navarra, Pamplona, Spain

ABSTRACT: An alternative design for semi-rigid composite joints is proposed and tested. The new proposal consists of inserting the central reinforcement bars through the column flanges in order to improve the behaviour. In addition, for external joints, the cantilever needed to ensure sufficient anchorage length for reinforcement can be dispensed with. An additional aim is to determine the effective reinforcement length needed to calculate the stiffness properties. This is achieved by means of a parametric study developed with robust and reliable finite element models, previously validated by the tests results. This parametric study allows us to verify the reliability of the component method and to establish the modifications that account for the proposed alternative design.

1 INTRODUCTION

Experimental investigation on composite joints began in the seventies. Since then the most important works on the subject until 1989 were revised by Zandonini 1989 and summarized by Simoes da Silva et al. 2001. Simoes da Silva also revised the experimental results carried out between 1990 and 2001.

From the more relevant experimental results which have been carried out since 2001 until the present day, only a few of these ones have focused on the study of external joints (Demonceau & Jaspart 2004, Ferreira et al. 1999, Byfield et al. 2004). In fact, it is suggested in publications such as SCI 1998 that the external joints should not be composite as positioning the reinforcement on the necessary full anchorage length becomes a problem. However, the rotational stiffness which they supply in non-braced frames can be beneficial for frame stability.

The use of external composite and semi-rigid joints implies that sufficient anchorage length is provided for reinforcement in order to obtain the required stiffness and strength. In turn, and as a consequence of this, the column needs to be surrounded by a certain percentage of the reinforcement, and therefore the concrete slab has to include a small cantilever, which in many cases may condition the building design. Therefore, one of the aims of this experimental study is to consider the possibility of eliminating the cantilever on semi-rigid external joints. This will be achieved by fitting the reinforcement through the column flanges, and attaching it to the opposite flange either by creating a thread at the end of these rebars or by placing a head to act as a buffer. Using the same system as with the external joints, the aim is to study internal joints

in order to improve their behaviour; in this case passing the rebars through the column flanges without the need for thread.

An additional aim is to determine the effective reinforcement length needed to calculate stiffness properties of the connection. This will be achieved by means of a parametric study developed with robust and reliable finite element models, which have been validated by the tests results.

In turn, the parametric study will allow us to verify the reliability of the component method and to establish the necessary modifications that account for the proposed alternative design.

2 PROTOTYPE DESCRIPTION

The experimental program includes three tests on flush end plate semi-rigid composite joints which have been subjected to static loads. The choice of geometry and materials has been made in accordance with Eurocode 2 (EC2, CEN 2002), Eurocode 3 (EC3, CEN 2003) and Eurocode 4 (EC4, CEN 2003).

A test is carried out on an internal joint T1 (Fig. 1), and two on external joints T2 and T3 (Fig. 2). In all three cases the joints are between steel columns and composite beams with a solid concrete slab.

The T2 and T3 specimens show the proposed alternative external column joint solution (Fig. 2). The conventional solution for this type of joint consists in extending the concrete slab with a small cantilever on the opposite side of the joint in order to provide sufficient anchorage length for the reinforcement. This is achieved by surrounding the column with the slab reinforcement. The alternative solution proposed herein

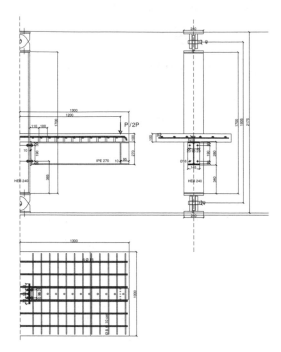

Figure 1. Configuration and geometry measurement of test T1.

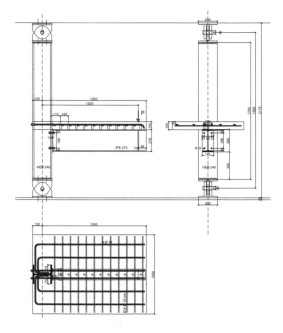

Figure 2. Configuration and geometry measurement of tests T2 and T3.

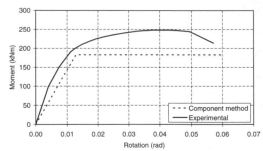

Figure 3. Moment-rotation curve for the left connection of T1.

eliminates the cantilever which acts, from an architectural point of view, as a very limiting factor in building design. The reinforcement therefore is anchored in the following way: The ends of the central rebars in the T2 are threaded and this allows the rebars to be bolted up to the column on the flange opposite to the joint. The remaining reinforcement is folded and lodged in the slab, in the cavity between the column flanges. This results in sufficient anchorage length without the need to extend the slab. Thus, the cantilever is eliminated. The T3 differs from the previous case in that instead of machining the reinforcement that goes through the column by means of a bolt and thread, a head is added in order to act as a buffer.

The loading pattern and test configuration are shown in Figures 1 and 2. The column has pinned ends, and the joint is firmly attached to the laboratory floor and the ceiling, which absorb the vertical and horizontal reactions.

3 EXPERIMENTAL RESULTS

3.1 Test results for T1

In the T1 specimen, the load applied on one of the sides is double the load on the other. The purpose of this load pattern is to study the interaction of the web panel shear with the rest of the components that constitutes the joint. Furthermore, the aim is to verify how the effective length of the reinforcement, which needs to be considered when calculating the joint stiffness, is affected by the different bending moments at each side of the column.

The specimen has behaved just as predicted by the component method. According to this, the reinforcement is the weakest component in this joint, being the first to yield followed by local buckling on the lower beam flange.

The moment-rotation curve of the joint bearing the greatest load is compared in Figure 3 with the bilinear curve obtained using the component method.

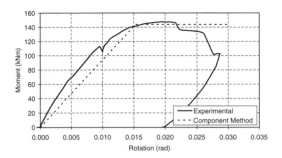

Figure 4. Moment-rotation curve of T2.

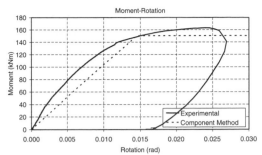

Figure 5. Moment-rotation curve of T3.

The joint secant stiffness and resistance are being underestimated by the component method.

3.2 Test results for T2

The T2 specimen shows an alternative external column joint. It can be observed that the reinforcement yields first and, furthermore, is responsible for the collapse of the joint as the thread area on one of the rebars breaks. This is due to the fact that by machining the rebar a sudden reduction of the cross section has taken place and the diameter has passed from measuring 16 mm to measuring 12 mm. As it is a local break, tension redistribution does not arise and the joint behaviour is no longer ductile. Therefore the rotational capacity expected is not achieved.

The moment-rotation curve is compared with the bilinear curve obtained from the component method in Figure 4. The maximum bending moment reached is nearly 150 kNm and the rotation for that moment is only 22 mrad. It is observed that the component method slightly overestimates the strength.

3.3 Test results for T3

This specimen presents the same configuration as T2 except for the central reinforcement. In this test a welded head, similar to a bolt head, is added to the central reinforcement to act as a buffer.

Figure 5 illustrates the moment-rotation curve for the tested joint. The maximum moment reached is approximately 163 kNm. The component method (EC4) predicted column web shear failure and yet it remains in an elastic state throughout the entire test. The bilinear curve in the graph also shows that the strength is slightly overestimated by the component method.

In general, the solution proposed for external joints can be said to have shown very satisfactory behaviour when compared with conventional external joints, but with the big advantage of avoiding the external cantilever.

4 FINITE ELEMENT MODEL

The three-dimensional finite element model has the configuration and dimensions of the tested specimens. With the objective of achieving better computational efficiency, half joints have been modelled using symmetry conditions for both internal and external joints.

The concrete slab, the slab reinforcement, the steel beams and the column and the bolts are modelled with 8-node solid elements (C3D8). The shear connectors have been modelled with non-linear springs with normal and tangential stiffness located in the beam-slab interface in its real position.

The interactions between the different materials are modelled in the following way:

The reinforcement is embedded in the slab. This technique eliminates the translational degrees of freedom of the embedded nodes and makes them correspond with those of the host element.

Surface to surface contact interactions are defined between the end plate and the column flange, as well as between the shaft and the nut of the bolts and the steel profiles.

5 PROPOSED DESIGN COMPARED TO THE CONVENTIONAL DESIGN

With the finite element model already calibrated, helped by the experimental results, several simulations are carried out both on external and internal joints. The objective is to establish comparisons in the behaviour of the proposed alternative design which consists of crossing the column flanges with the central reinforcement rebars, compared to the conventional design.

5.1 Internal joint

The graphs which are presented below show the moment-rotation curves for joints in which the characteristics of the sections, bolts, concrete and reinforcement are the same as in the T1 sample.

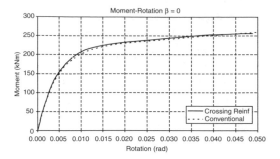

Figure 6. Moment-rotation curve for internal joint with symmetrical load.

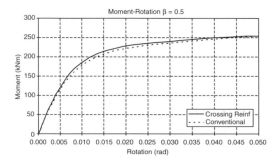

Figure 7. Moment-rotation curve for internal joint with loads P and 0.5 P. More loaded joint.

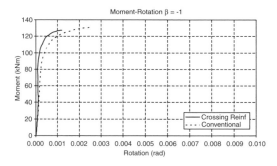

Figure 8. Moment-rotation curve for internal joint with loads P and 0.5 P. Less loaded joint.

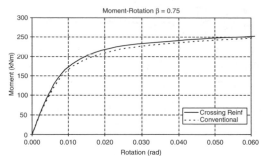

Figure 9. Moment-rotation curve for internal joint with loads P and 0.25 P. More loaded joint.

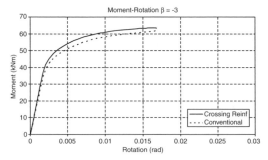

Figure 10. Moment-Rotation curve for internal joint with loads P and 0.25P. Less loaded joint.

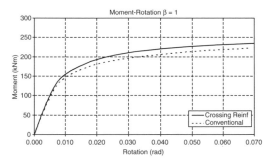

Figure 11. Moment-Rotation curve for internal joint with loads P and 0.01 P. More loaded joint.

Two types of joints are simulated: with through-flange reinforcement and with reinforcement which does not pass through the column flange. Both configurations are simulated for several load conditions, so that the interaction parameter β (EC3-1.8-5.3) takes values from 0 to 1. In this way, it may be seen how the column web in shear is affected in each case.

Figure 6 presents the internal joint curves for β = 0, meaning the loads on both sides of the column are equal. For the two cases studied, the curves are very

similar; their strength and stiffness can barely be differentiated. The difference starts to be appreciated, in Figure 7, when β = 0.5 (the load on one side is double the other). On the side that receives a smaller load (Fig. 8), the difference in the strength and the stiffness is even greater, always in favour of the solution with through-flange reinforcement proposed in this paper.

In the following curves (Figure 9 to 12) one may observe that with relatively high values for β, the

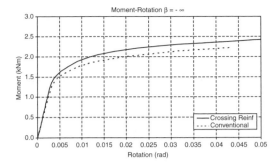

Figure 12. Moment-rotation curve for internal joint with loads P and 0.01P. Less loaded joint.

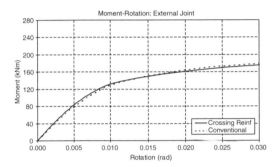

Figure 13. Moment-rotation curve for external joint with load P.

difference in the joint behaviour becomes increasingly noticeable and in all cases favourable for the proposed joint where the reinforcement crosses through the column flanges.

It can be affirmed, therefore, that an improvement is achieved in the joint behaviour with this design, which is more visible when the moments are increasingly unbalanced. The through-flange rebars add stiffness and strength to the whole joint, and in addition they delay the formation of mechanism 1 (see Eurocode 8 Annex C, CEN 2003) (concrete rushing against the column) as the compression strength of the through-flange reinforcement is added to the concrete compression strength in this area.

By means of the simulations, it has been also discovered that, as opposed to what is established in EC4, the effective length of the reinforcement does not differ much for the different load conditions. The EC4 takes $h_c/2$ as its effective length when $\beta = 0$, a value which is less than what is observed in the finite element model, even when the joint configuration is conventional. Therefore the joint stiffness remains overestimated. In the rest of the cases studied (when $\beta = 0.5, 0.75$ and 1) the effective length which comes from applying the EC4 formulas is greater than what is obtained using the simulations, as proven in the following section.

5.2 External joint

The external joints simulated also respond to the tested configuration. For a semi-rigid external joint following the usual configuration, the reinforcement is fitted surrounding the column and thereby extending the concrete slab backwards. Figure 13 shows that both moment-rotation curves are very close together. This shows that the objectives were achieved, meaning that the same stiffness and strength were reached without needing the cantilever on the back of the column.

6 EFFECTIVE LENGTH OF THE REINFORCEMENT

The components method calculates stiffness and strength using the characteristics of each component considered individually. The component which most contributes to the joint stiffness is the reinforcement (Amadio & Fragiacomo 2003). In order to calculate its stiffness, it is necessary to consider the effective length of the reinforcement (EC4, Annex A). The first versions of EC4 proposed an effective length, l_r, equal to 0.5 times the depth of the column, for any joint, either internal or external, and subjected to equal or different moments. However, the latest version of EC4 substitutes this value by several expressions which depend on the interaction parameter β (EC4, A.2.1.1 and Table A.1).

Using the finite element method, the aim is to establish an expression to determine this effective length. For this reason, a parametric study was carried out with different load situations, profiles, bolts and amount of reinforcement.

After studying the simulation results, an identical length is chosen for all the loading conditions, since it has been observed that the difference between them is very small. This makes the effective length independent of β, which is congruent with the fact that the reinforcement axial stiffness should be a characteristic of the joint, which must only depend on its configuration and characteristics and not on the internal forces acting on it.

Following this reasoning it is established a relationship between the length which yield a similar stiffness to that on the model (previously deducting the value corresponding to half the column's depth) and the lever arm of the composite beam. A value close to 0.8 is obtained for internal joints. Therefore, an expression may be established to determine the effective length of the reinforcement in internal joints, as follows:

$$l_r = \left(h_c/2\right) + 0.8z \leq L_{(-)} \tag{1}$$

where $L_{(-)}$ is the length of the beam subjected to negative moment, z the distance from the centre of the beam bottom flange to the centroid of the reinforcement and h_c the depth of the column.

With the proposed effective length, the stiffness comes very close to what is obtained in the simulations, previously calibrated with the experimental results. The reflected error oscillates between 0.3 and 17.4%, smaller than those registered following EC4, which even get 60% in some cases.

Regarding external joints, equation from EC4, gives a stiffness which is quite well adjusted to the values obtained in the simulations. The maximum error is 17.2% in the unsafe side. Following the same reasoning as for the external joints, the effective length will depend on the depth of the column and the lever arm of the beam. The length proposed for the external joints with the through-flange reinforcement is:

$$l_r = 2(h_c + z) \leq L_{(-)} \tag{2}$$

Using this proposed effective length, the stiffness remains in the safe side and the maximum error decreases to 11.4% which compares very favourably with EC4.

7 CONCLUSIONS

In order to study the behaviour of composite semi-rigid joints in more detail, various full-scale tests have been carried out. In these tests, an alternative design is proposed to that normally used. The central rebars of the reinforcement are fitted crossing the column flanges through some holes previously drilled in the workshop. In the case of external joints, there is also a head fitted as a sort of buffer to anchor the reinforcement. In this way, it is not necessary to extend the slab behind the column to give the reinforcement sufficient anchorage length, which gives a great advantage from an architectural and structural point of view. Also passing the rebars through the column flanges improves the behaviour of the internal joints.

By means of the tests carried out, it can be seen that the design for internal joints shows a better behaviour than the prediction obtained through the component method for conventional internal joints. As for the rotation capacity, the tested internal joint shows a large value, way above the required minimum. However, the external joints tested show a smaller rotation capacity, which could be improved by increasing the diameter of the central reinforcement. In the case of strength and stiffness, it can be said that practically the same values are obtained with this design, which eliminates the cantilever used with the conventional design.

The internal joint stiffness and resistance obtained by means of the EC4 component method appear to be very conservative and imprecise, due to the limitations imposed by the interaction parameter β. This suggests the use of formulations that avoid the β parameter such as that proposed by Bayo et al. 2006. On the other hand, the EC4 stiffness and resistance obtained for external joints prove to be more precise, although they need to be improved by some proposed modifications.

The experimental results are also used to validate the finite element models with which to compare conventional joints with the proposed alternative joints. Internal joints showed that the alternative design exhibits an improved behaviour compared with the conventional configuration. In the external joints, on the other hand, barely any difference is seen between both designs; the same values are obtained without requiring cantilever.

In addition, a parametric study, using these results, is carried out with the proposed alternative joints in order to adapt the component method of this new design. The proposed modifications on the model mainly affect the effective length of the rebars under tension, which is made dependent only on the depth of the column and the composite beam, and not on the transformation parameter β, as proposed in EC4. This feature adds efficiency and simplicity at the time of performing the global analysis.

ACKNOWLEDGEMENTS

The financial support provided by the Arcelor Chair of the University of Navarra, as well as the Government of Navarra is greatly acknowledged.

REFERENCES

Amadio, C. & Fragiacomo, M. 2003. Analysis of rigid and semi-rigid steel-concrete composite joints under monotonic loading. Part i: Finite element modelling and validation. *Steel & Composite Structures* **3**(5): 349–369.

Amadio, C. & Fragiacomo, M. 2003. Analysis of rigid and semi-rigid steel-concrete composite joints under monotonic loading. Part ii: Parametric study and comparison with the Eurocode 4 proposal. *Steel & Composite Structures* **3**(5): 371–382.

Bayo, E., Cabrero, J. M. & Gil, B. 2006. An effective component-based method to model semi-rigid connections for the global analysis of steel and composite structures. *Engineering Structures. Elsevier* **28**(1): 97–108.

Byfield, M. P., Dhanalakshmi, M. & Goyder, H. G. D. 2004. Modelling of unpropped semi-continuous composite beams. *Journal of Constructional Steel Research* **60**: 1353–1367.

CEN *Eurocode 2: Design of concrete structures. Part 1: General rules and rules for buildings*, CEN.

CEN 2003. *Eurocode 3: Design of steel structures. Part 1.8: Design of joints (pren 1993-1-8:2003), stage 49 draft edition*, CEN.

CEN 2003. *Eurocode 4: Design of composite steel and concrete structures. Part 1.1: General rules and rules for buildings (pren 1994-1-1:2003)*, CEN.

CEN 2003. *Eurocode 8: Design of structures for earthquake resistance- part 1: General rules, seismic actions and rules for buildings*, CEN.

Demonceau, J. F. & Jaspart, J. P. 2004. Experimental and analytical investigations on single-sided composite joint configurations. *5th International PhD Simposium in Civil Engineering*, Delft (The Netherlands).

Ferreira, L. T. S., de Andrade, S. A. L. & Vellasco, P. C. G. D. 1999. Composite semi-rigid connections for edge and corner columns. *Eurosteel second european conference on steel structures*, Prague.

SCI 1998. *Joints in steel construction: Composite connections*. Ascot, London, Steel Construction Institute. British Constructional Steelwork Association Limited.

Simoes da Silva, L., Simoes, R. D. & Cruz, P. J. S. 2001. Experimental behaviour of end-plate beam-to-column composite joints under monotonical loading. *Engineering Structures. Elsevier* **23**(11): 1383–1409.

Zandonini, R. 1989. Semi-rigid composite joints. *Structural connections: Stability and strength*. Narayanan, R. London, Elsevier: 63–120.

Steel and Composite Structures – Wang & Choi (eds)
© 2007 Taylor & Francis Group, London, ISBN 978-0-415-45141-3

Parametric study of semi-rigid composite joint with precast hollowcore slabs

D. Lam & J. Ye
School of Civil Engineering, University of Leeds, Leeds, UK

F. Fu
Waterman Structure, London, UK

ABSTRACT: This paper presents the parametric studies carried out to investigate the structural behaviour of composite beam – column joint with steel beams and precast hollowcore slabs. A finite element model to simulate the structural behaviour of the composite beam was described and was used to study the structural behaviour of composite joints especially the moment-rotation characteristic. Parametric studies with various parameters such as beam sizes, thickness of the endplate, thickness of column web, depth of precast hollowcore slabs and stud spacing were carried out and results from the studies are presented.

1 INTRODUCTION

In composite construction, extensive researches have been carried out on semi-rigid connections design since it was first proposed by Barnard (1970). It showed these form of connections when used in design will lead to reduction in beam sizes, which in turn will reduce the beam depth, the overall building height and cladding cost, etc. Moment rotation characteristic of the semi-rigid composite connections was first investigated by Johnson and Hope-Gill (1972); they found that neither simple nor rigid beam-column connections are ideal. Simple joints are too unpredictable while rigid joints are often too stiff in relation to their strength and are expensive; therefore, the semi-rigid joint with a large rotation capacity and a predictable flexural strength that does not require site welding or accurate fitting is needed. Numerous researches have been carried out on semi-rigid composite connections mainly with solid reinforced concrete slabs (Bernuzzi *et al*, 1991) and profiled metal deck floors (Li *et al*, 1996). Semi-rigid composite joint incorporating steel beams and precast hollowcore slabs is relatively new and little research has been carried out. The most important properties of this type of connections are moment capacity, rotational stiffness and rotational capacity. These parameters can be defined using the moment-rotation response (M-θ). The best way to capture the M-θ response is by conducting full-scale tests. Series of full-scale testing on composite joints with precast hollowcore slabs have been carried out by Fu & Lam (2006) and further work on this area is currently on-going. In addition to the joint tests, push tests were carried out to determine the structural behaviour of the shear connection in precast hollowcore slabs by Lam (2006). Due to the complexity of the problem and limitation of the test results, non-linear finite elements (FE) method is an attractive tool for modelling large number of variables and complement the experimental studies. Lam *et al* (1998) were the first to simulate the behaviour of composite girders with precast hollowcore slabs; a 2-D finite element model was built using ABAQUS (2005). A 3-D FE model of the headed studs in steel-precast composite beams was built by El-Lobody and Lam (2002) to model the behaviour of the headed stud in precast hollowcore slabs. The model was validated against the test results and good agreement is obtained.

From the available literatures, it is noted that although there are some research works toward the modelling of composite construction, most of the works were focussed on the simulation of the composite beams. Fewer works have been done to model the composite joints, especially using a 3-D finite element model.

2 FINITE ELEMENT MODEL

Using the general-purpose finite element package ABAQUS, a 3-D finite element model was built to simulate the behaviour of semi-rigid composite connection with precast hollowcore slabs. As shown in Figure 1, the model uses three-dimensional solid

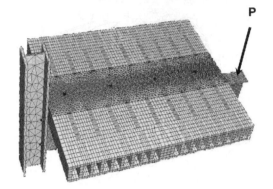

Figure 1.   3-D FE model of the composite joint.

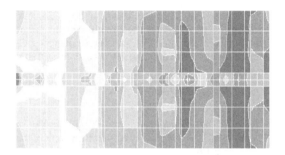

Figure 2.   Effective breadth of the composite slab.

elements to replicate the composite joint of the actual full scale test (Fu & Lam, 2006). The boundary condition and method of loading adopted in the finite element analysis followed closely to those used in the tests. The load was applied at the end of the beam. Material nonlinearity was included in the finite element model by specifying the stress-strain curves of the material taken from the test specimens.

Different mesh sizes have been examined to determine a reasonable mesh that provides both accurate results and reasonable computational time. The results shown in Figure 2 showed that the effective breadth around the joint is confined to the in-situ infill concrete portion of the slabs. Therefore, the model was simplified as shown in Figure 3.

The longitudinal shear force is transferred to the concrete by dowel action of the shear connectors. Fu and Lam (2006) showed that the front of the stud is in compression while the back of the stud is in tension and is detached from the concrete slabs. This mechanism is shown in Figure 4. Therefore in the simulation, only the nodes of the studs at the front side are connected to the nodes of the concrete slab with rigid connection using *TIE options and the other nodes of the studs are detached to the surrounding concrete nodes.

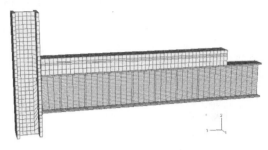

Figure 3.   Simplified FE model.

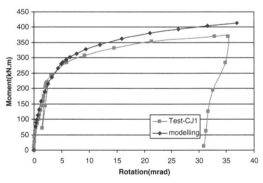

Figure 4.   Model of the shear connector.

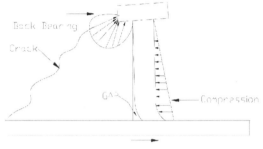

Figure 5.   Comparison of M-ϕ curves for test vs. FE model.

2.1   Validation of the FE model

To validate the accuracy of the model, the FE analysis results are compared with the experimental studies. The comparison results are shown in Figure 5 and 6. It can be seen that the numerical results have good agreement with the tests results.

3   PARAMETRIC STUDIES

Parametric studies were carried out to investigate the structural behaviour of the joint. Table 1 shows the different parameters selected for FE analysis. Only one variable was changed at a time so that its effect can be

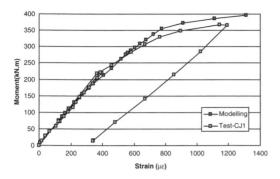

Figure 6. Comparison of steel bottom flange strain.

Table 1. Variables for the parametric studies.

Parameter	Range of variable selected
Position of first stud	235 mm, 535 mm, 835 mm, 1135 mm
End-plate thickness	5 mm, 10 mm, 15 mm, 20 mm
Size of steel beam	457 × 191 × 89 UB, 406 × 140 × 39UB, 254 × 146 × 31UB
Column's web	3.6 mm, 7.2 mm, 14.4 mm,
Slab thickness	150 mm, 250 mm, 400 mm

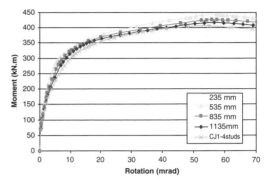

Figure 7. M-ϕ curves with various position of the first studs.

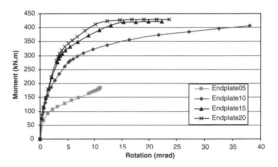

Figure 8. M-ϕ curves with various endplate thickness.

easily identified. They are considered to be the most influential factors for the composite connections.

3.1 Effect of position of first stud

Figure 7 shows the moment rotation capacity with variation of the position of the first stud. It can be seen that there is little difference in the moment capacity at failure. This is because that as long as the shear stud can sustain the longitudinal shear force and be able to fully mobilise the longitudinal rebars, the tensile capacity of the longitudinal rebars will remain the same and therefore there is little difference in the moment capacity. However, the rotation capacity increased as the first stud spacing increased. With the first stud position close to the column face, fewer cracks were formed with a main crack opening near the column faces and eventually led to the fracture of the longitudinal reinforcement. The formation of cracks would appear to be related to the position of the first stud. As the first stud position is placed further away from the face of the column, after the formation of the first crack, additional cracks were formed between the column face and the position of the first stud. Cracks between the column face and the first stud distributed evenly rather than concentrated at a single crack around the column face. This led to lesser demand on the percentage of elongation required by the longitudinal reinforcement

at the position of the cracks and led to less strain concentration in the rebars.

Similar results on metal decking composite flooring were reported by Helmut and Hans (1996) & Schafer and Kuhlmann (2004). They showed that placing the studs further away from the column flange will result in larger rotation capacity. However, when the first stud spacing is increased to some limit, i.e. over 800 mm, the stresses in the longitudinal steel bar became non-uniform within the range from the column flange to the second stud, the high stresses are concentrated near the first stud, the modelling result also shows that there is a slightly reduction of moment capacity.

3.2 Effect of endplate thickness

Four endplate thicknesses of 5, 10, 15 and 20 mm were modelled and their results were shown in Figure 8. It was observed with 5 mm endplate, low moment and rotation capacity were recorded. The mode of failure for the model endplate05 is due to the yielding of the endplate as shown in Figure 9. As the thickness increased from 5 mm to 20 mm, an increases in moment capacity is recorded.

However, as the endplate thickness was increased beyond 15 mm, little difference in behaviour was observed from the model. The reason for this is that

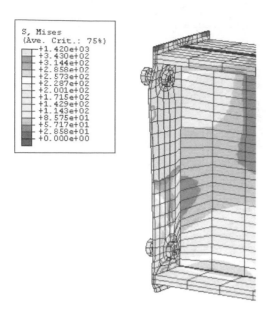

Figure 9. Failure of endplate for FE model endplate05.

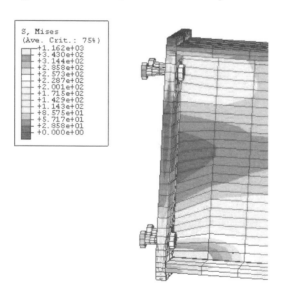

Figure 10. Failure of bolts for FE model endplate20.

as the thickness of the endplate increases beyond a certain thickness, the strength of the bolt becomes the influential factor. As shown in Figure 10 for the model endplate20, most area of the endplate remains elastic and the failure occurred at the bolts. As the thickness of the endplate increased, the mode of failure changed from complete yielding of the flange (Model endplate05) to bolt fracture with flange yielding (Model endplate10) and to bolt fracture (Model endplate20).

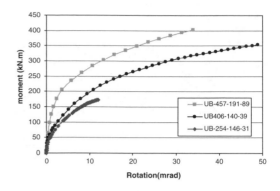

Figure 11. M-ϕ curves with various beam sizes.

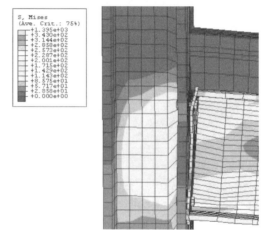

Figure 12. Failure of Model UB-457-191-89.

3.3 Effect of beam sizes

Figure 11 shows the modelling results with variation of beam sizes. The result showed that the maximum stresses in the bottom flange for the model UB-457-191-89 were only 213 N/mm^2 at failure, i.e. still remains elastic. The mode of failure was fracture to the longitudinal rebars. However, with the model UB-254-146-31, yielding of the bottom flange as well as the web was observed. Compared with the model UB-457-191-89, the moment capacity is reduced as the beam size reduced.

3.4 Effect of column web thickness

To study the effect of the column web thickness to the behaviour of the composite connection, a small column size of 203 × 203 × 46 UC was chosen in order to investigate the effect of column web thickness. The test results are shown in Figure 14. In these results, the rotation of the column is included to show the effect. It is found from the modelling results that as the thickness

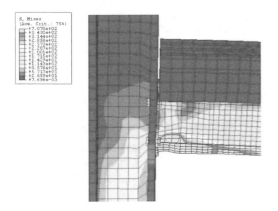

Figure 13. Failure of Model UB-254-146-31.

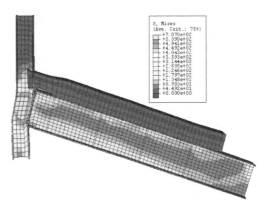

Figure 15. Failure of Model Web36.

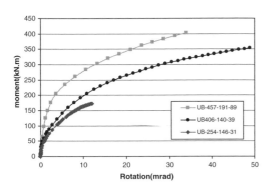

Figure 14. M-φ curves with variation of column web.

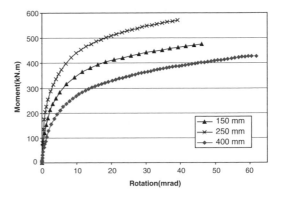

Figure 16. M-φ curves with variation of slab thickness.

of the web decreases, the moment capacity decreased, the initial stiffness decreased and the rotation capacity increased. From Figure 14, it can be seen that model web144 (web thickness of 14.4 mm) and web72 (web thickness of 7.2 mm), there is no obvious difference in moment rotation capacity, but as the web thickness being further reduced, i.e. model web36, yielding of the column's web occurred leading to large rotation and formation of plastic hinges in the column. The reduction of the web thickness reduced the compression capacity of the column web and hence reduced the compressive force at the beam–column interface. The large deformation of the column greatly influences the rotation behaviour of the connection.

3.5 Effect of slab thickness

To study the effect of slab thickness. Slab thickness of 150, 250 and 400 mm were modelled. The results are shown in Figure 16. When the thickness increased, the moment capacity increased with the increases in slab thickness. This is to be expected as the increases in slab thickness would raise the neutral axis of the composite beam and hence increasing the lever arm of

the section. With the other parameters remaining the same, the failure mode of the models is the fracture of the longitudinal rebars.

4 CONCLUSIONS

Parametric studies were carried out to investigate the various parameters that would affect the behaviour of the composite joints with precast hollowcore slabs. Results show that the proposed FE model can accurately represent all the main features of the behaviour of composite joints, it offers a reliable and very cost-effective alternative to laboratory testing as a way of generating results. Different variables have been studied on their influence on the structural behaviour of the joints and following conclusions can be drawn below:

1. Position of the first stud plays an important role to rotation capacity and serviceability of the composite joints. The stud spacing has little effect on the moment capacity of the connection.
2. Increased thickness of the endplate increases the tensile capacity of the steel connection which

355

increases the moment capacity. However, increase in endplate thickness caused brittle failure of the bolts with low rotation capacity of the connection. Therefore, it is necessary to limit the thickness of the endplate in order to avoid brittle failure of the bolts and to ensure that ductile failure by yielding of the endplate.

3. Moment capacity is not only controlled by the amount of longitudinal reinforcement but also limited by the compression capacity of the beam flange, beam web and column web. Very high reinforcement ratio could lead to low rotation capacity of the composite joint.

4. Column web thickness limits the maximum compression force that can be transferred and hence controls the moment capacity of the joints.

5. Increase in slab thickness will lead to increase in moment capacity of the connection due to the increase in lever arm.

ACKNOWLEDGEMENT

The authors would like to acknowledge the financial support from International Precast Hollowcore Association (IPHA) and Overseas Research Scholarship (ORS), the support provided by Severfield–Reeve Structures Ltd. for supplying the steel specimens and Bison Concrete Products Ltd. for supplying the precast hollowcore slabs. The skilled assistance provided by the technical staff in the School of Civil Engineering at Leeds University is also appreciated.

REFERENCES

ABAQUS (2005), Version 6.4, Hibbitt, Karlson and Sorensen, Inc.

Barnard, P.R. (1970), 'Innovations of composite floor systems', Canadian Structural Engineering Conference, Canadian Steel Industries Construction Council, 1970, pp. 13–21.

Bernuzzi, C., Salvatore, N., & Zandonini, R. (1991), 'Semi-rigid composite joints: Experimental studies', Connections in Steel Structures II: Behaviour, strength and design conference, Pittsburgh Pennsylvania, USA.

El-Lobody, E. & Lam, D. (2002), 'Modelling of headed stud in steel – precast composite beams', Steel & Composite Structures, Vol. 2, No. 5, pp. 355–378.

Fu, F. & Lam, D. (2006), 'Full scale tests on Semi-Rigid Composite Connection with Steel Beams and Precast Hollowcore Slabs', Journal of Constructional Steel Research, Vol. 62 (8), pp. 771–782.

Helmut B & Hans J.K (1996), 'Behaviour of composite joints and their influence on semi-continuous composite beams', Composite construction in steel and concrete III Proc., Engineering Foundation Conf., ASCE, New York

Johnson, R.P & Hope-Gill, M. (1972), 'Semi-rigid joints in composite frames', International Association for Bridge and Structural Engineering, Ninth Congress, Amsterdam, May, pp. 133–44.

Lam, D (2006), 'Capacities of Headed Stud Shear Connectors in Composite Steel Beams with Precast Hollowcore Slabs', Journal of Constructional Steel Research (available online).

Lam, D., Elliott, K. S. & Nethercot, D. A. (2000), 'Parametric study on composite steel beams with precast concrete hollow core floor slabs', Journal of Constructional Steel Research, Vol. 54 (2), pp. 283–304.

Li, T.Q., Moore, D.B., Nethercot, D.A. & Choo, B.S. (1996), 'The experiment behaviour of a full-scale, semi-rigid connected composite frame: Overall consideration', Journal of constructional steel research, Vol. 39 (3), 1996, pp. 167–191.

Schafer M. & Kuhlmann U. (2004), 'Innovative sway frames with partial-strength composite joints', 5th international PhD Symposium in Civil Engineering, Walraven, Blaauwendraad, Scarpas & Snijder (eds.), Taylor & Francis Group, London.

Steel and Composite Structures – Wang & Choi (eds)
© 2007 Taylor & Francis Group, London, ISBN 978-0-415-45141-3

Experimental test of semi-rigid minor axis composite seat and web side plate joints

A. Kozłowski & L. Ślęczka

Rzeszów University of Technology, Rzeszów, Poland

ABSTRACT: Composite steel-concrete structures are very effective and attractive to designers because of their greater stiffness and resistance capacity compared to non-composite ones. This enables to reach less depth of used beams and to reduce the height of floor structure. Further decrease of composite beam section can be obtained by appropriate design of beam to column connections. For framed structures, the key behaviour indicator for a joint is its moment-rotation characteristic (M-ϕ curve). The best way to obtain this characteristic is experimental test. A large number of steel and composite joints have been tested during the past years, but most of them were conducted for bending of the column about its major axis. In this paper the results of a four, full-scale tests on minor axis composite beam-to-column connections are reported. The steelwork part of the connection consisted of a seating Tee section bracket, bolted to beam flange and fin plate welded to column web and bolted to beam web. The main variable investigated was the amount of reinforcement. For comparison, three tests on pure steel connections of the same type were conducted. Results of tests show that properly designed composite joint posses appropriate moment resistance, stiffness and rotation capacity. Comparison of bare steel and composite joints behaviour shows that presence of the concrete slab increases moment resistance from 2,5 to 3,8 times. Increase of initial stiffness of composite joints ranged from 3 to 4 times of pure steel joint stiffness, while rotation capacity is of the same range.

1 INTRODUCTION

Composite steel-concrete construction is very effective and attractive to designers because of its greater stiffness and resistance capacity compared to non-composite construction. This enables to reach less depth of used beams and to reduce the height of floor structure (Nethercot 2003) . Further decrease in composite beam section can be obtained by appropriate design of beam to column connections. In so-called "composite connection", resistance to hogging moment is provided by properly anchored tension reinforcement, placed in concrete slab, together with steel part of beam-to-column joints (Zandonini 1989). Efficiency of such composite joints is specially high for joints, where steelwork details is customary associated with "simple" construction, e.g. web cleats with seating cleats, partial depth end plates and so on.

For framed structures, the key behaviour indicator for a joint is its moment-rotation characteristic (M-ϕ curve). The best way to obtain this characteristic is experimental test. A large number of steel and composite joints have been tested during the past years (Anderson 1996, Li et al 1996, Liew et al 2000, Brown and Anderson 2001, Silva et al 2001, Green et al 2004,

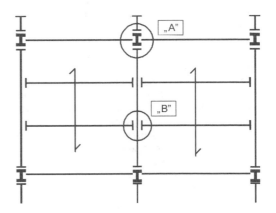

Figure 1. Example of composite floor arrangement.

Loh et al 2006), but most of them were conducted for bending of the column about its major axis.

In flooring system, where the composite beams are used, there is often the need to join the secondary beam to column in weak axis plane (joint "A") or to main beam (joint "B"). Example of such floor layout is shown in Figure 1. By appropriate ratio of main beams

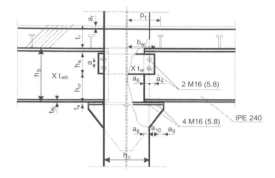

Figure 2.   Details of composite joint specimen.

Table 1.   Details of the specimens.

Specimen	Column	Beam	Reinforcement	No. of specimens
CP-1	HEB200	IPE240	6ϕ10; $\rho = 0,5\%$	1
CP-2.1	HEB200	IPE240	10ϕ10; $\rho = 0,8\%$	1
CP-2.2*	HEB200	IPE240	10ϕ10; $\rho = 0,8\%$	1
CP-3	HEB200	IPE240	14ϕ10; $\rho = 1,1\%$	1

* specimens CP-2.1 and CP-2.2 were nominally identical, but specimen CP-2.2 was loaded non-symmetrical.

Table 2.   Measured cross-sectional dimensions of specimens.

Specimen	CP-1	CP-2.1	CP-2.2	CP-3
t_r [mm]	105,0	106,0	104,0	101,0
t_w [mm]	6,1	6,1	6,1	6,1
b_w [mm]	120,4	120,9	120,5	121,0
h_w [mm]	88,4	88,2	88,8	88,9
h_b [mm]	241,2	241,3	241,3	240,9
t_{fb} [mm]	10,1	9,8	9,9	9,5
t_{wb} [mm]	6,5	6,6	6,59	6,5
b_b [mm]	119,7	120,0	120,1	119,8
t_s [mm]	10,0	10,0	9,9	9,8
reinforcement	6ϕ10	10ϕ10	10ϕ10	14ϕ10
A_r [mm^2]	$\rho = 0,5\%$	$\rho = 0,8\%$	$\rho = 0,8\%$	$\rho = 1,1\%$
$a_4 = a_8$ [mm]	24,0	24,0	24,0	24,0
h_c [mm]	201,1	200,5	200,4	201,2
h_{vr} [mm]	126,0	125,0	125,5	124,8

to secondary beams spans, it is possible to get the same depth of both beams, what gives lower floor structure depth and smaller global storey height.

Shaping of joints connecting beams to column in the weak axis plane can make some difficulties. One of the joint used in such situation is a joint which steelwork part consisted of a seating Tee section bracket bolted to lower beam flange and fin plate welded to column web and bolted to beam web, (Fig. 2).

Such joint has some advantages. In the construction stage, joint produces certain rotation restraint, what results in lower beam size and smaller beam deflection. In the working phase, by introduction of a reinforced concrete slab over the steel beam, the lever arm is increased, reinforcement bars take tension force and bolts connecting lower beam flange to the bracket transmit compression. Amount of reinforcement and the number of bolts in lower flange beam connection are the main variable allowing designers to get joint of required capacity and stiffness. The use of described joints does not require holes to be drilled in the column and can lead to an increased construction speed. Next advantage is presence of some tolerance in the bolt holes, what makes that requirements for beam length are not so strict as for other joints.

The purpose of this paper is to present the test results of minor axis composite joint, shown in Figure 2. Seven tests were conducted, in which three of them were a pure steel connections and the remaining four were composite connections.

The main aims of conducted test were as follows:

– observation of whole joints and their particular components behaviour under loading to find the main source of deformation and failure modes,
– investigation of reinforcement bars number influence on the joint moment capacity, stiffness and rotation capacity,
– investigation of the effect of the slip in the beam to concrete slab plane,
– finding the influence of non-symmetrical loading on joint behaviour,

– prediction of the main joint parameters: moment resistance capacity, initial stiffness and rotation capacity, and finally whole M-ϕ curve.

The proposed design method to predict the moment capacity, initial stiffness and rotation capacity was presented in (Kozłowski 2000).

2   TEST SPECIMENS AND MATERIAL PROPERTIES

All specimens were of the cruciform arrangement as shown in Figure 2. The cantilever beams are made of IPE 240 and columns of HEB 200 sections. Collection of tested composite specimens is shown in Table 1.

Grade S235 steel was used for all beams, columns and plates. Before the tests, all parts of joints were measured to obtain their actual geometric dimensions. Measured dimensions of specimen components are collected in Table 2.

Tensile test were carried out on coupon samples of structural steel used for beams, columns and plates, in

Table 3. Mechanical properties of steel components.

Element	Yield strength [MPa]	Ultimate strength [MPa]	Elongation [%]
IPE 240, beam flange	328	495	28
IPE 240, beam web	342	513	30
HEB 200, column flange	298	441	30
HEB 200, column web	336	472	23
Plate $t_w = 6\,\text{mm}$	293	401	26
Plate $t_s = 10\,\text{mm}$	339	479	34
Rebar $\phi10$	385	598	15

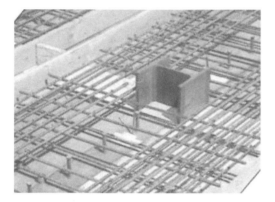

Figure 3. Details of the reinforcement.

accordance with standard methods. The steel coupon test results are summarized in Table 3.

Connection of lower beam flange to seat was made with the use of four bolts and connection of beam web to side plate by two bolts. M16 grade 5.8 bolts were used in all connections. Bolts were only hand tightened. From tensile testing of bolts it was obtained that $f_{yb} = 601\,\text{MPa}$, $f_{ub} = 762\,\text{MPa}$.

Longitudinal reinforcement was made of $\phi10$ bars. The number of bars was varied from 6 to 14 providing reinforcement varying from 0,5% to 1,1%. Transverse reinforcement in the form of 10 mm diameter bars with a spacing equal to 150 mm was supplied. Results of tensile tests of reinforcement bars were included in Table 3.

Headed studs of 12 mm diameter, spacing 188 mm, were used as shear connections. They were fixed to the beam flange at the fabrication shop, using stud-welding equipment.

The steel components of the tested joint specimens were assembled in the laboratory, then they were shuttered and the reinforcement fixed, (Fig. 3). Concrete

Figure 4. View of test rig and specimen CP-2.1 after failure.

for the slab was provided by the local concrete fabricator. The normal weight concrete of the slab was designed to be C20/25 grade. Compression cube tests conducted after 28 days gave the mean value of concrete strength of 31 MPa. After 28 days, the completed specimens were carefully lifted and fastened into the test rig.

3 TEST RIG AND INSTRUMENTATION

A general arrangement of the test rig is shown in Figure 4. The specimen was connected to heavy test rig beam by four M20 10.9 grade bolts. Loads were applied by two independent jacks at 1000 mm from the face of the column section, through the upper transverse box section beams, two steel bars of 30 mm diameter and lower transverse beams. Forces were measured by the load cells placed between the hydraulic jacks and the lower transverse beam.

The instrumentation was designed to measure the following:

- rotations of each side of joint, by the use of two dial gauges facing a plate welded to the column section, and additionally by one dial gauge facing steel bar bolted to lower flange of the beam,
- loads, by two 20 t load cells,
- slips, by dial gauge at the end of each slab (Fig. 5),
- strains in lower and upper beam flange and reinforcement bars, by the tensometer strain gauges,
- crack widths on the top surface of the concrete slab, by a microscope.

4 TESTING PROCEDURES

The specimens were initially loaded to 10 kN on each side and then unloaded in order to remove the slackening in the test rig and to ascertain proper functioning of instrumentation. Specimens CP-1, CP-2.1 and

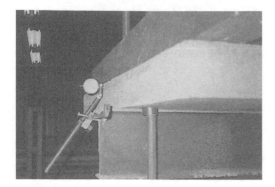

Figure 5. Measurement of slip in the slab-beam flange interface.

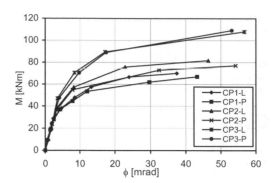

Figure 6. Collection of M-ϕ curves of all specimens.

CP-3 were symmetrical loaded by two balanced loads. In general, loads were subsequently applied in steps of $0,2P_u$ (predicted ultimate resistance of the connections) up to $0,6P_u$. Then, the specimens were unloaded to monitor their unloading stiffness. The specimens were then loaded to failure.

Specimen CP-2.2 had the same geometrical sizes and nominal material properties as CP-2.1, but was loaded non-symmetrical, according to the following:

– $P_L = 0$; P_R increased steeply to $0,6P_u$,
– $P_R = $ const; P_L increased to P_R,
– unloading,
– $P_L = P_R$; increased to failure,

where: P_L, P_R – left and right loads applied to specimen.

All the connection tests were terminated because of excessive deflection; rotations of joints reached 38 mrad and more.

5 COMPOSITE JOINT TEST RESULTS

The main test results are presented in Figure 6, in the form of the connection moment M against joint rotations ϕ. The connection moments were calculated as the product of the real applied load obtained from load cells and the length of the arm, which was the distance from the centre of the applied load to the surface of the column web. The connection rotations were calculated on the base of dial gauges readings.

Table 4 summarizes the main results of the tests: maximum moment M_R obtained in tests, joint stiffness $S_{j.ini}$ and available rotation capacity ϕ_u, for left (L) and right (R) connection of the specimens.

Joint stiffness was established as unloading stiffness of the connection. The moment-rotation curves, in common with previous research, show the unloading stiffness to be approximately equal to the initial tangent stiffness $S_{j.ini}$ of the connection M-ϕ curves for tested joints.

Table 4. Results of composite joint tests.

Specimen	Maximum moment M_R [kNm]	Stiffness $S_{j.ini.}$ [kNm/mrad]	Rotation capacity ϕ_u [mrad]
CP- 1L	69,89	12,9	37,4
CP- 1R	66,77	13,5	43,2
CP- 2.1L	81,78	13,9	46,4
CP- 2.1R	77,12	14,5	54,2
CP-2.2L	83,99	14,2	49,2
CP-2.2R	78,82	15,1	55,9
CP- 3L	108,00	14,1	56,7
CP- 3R	109,08	15,3	53,1

Table 5. Cracks width in specimens.

Specimen	First observed cracks	Cracks width 0,3 mm	Maximum crack width
CP-1	$M_j = 28$ kNm	$M_j = 55$ kNm	1,2 mm
CP-2.1	$M_j = 42$ kNm	$M_j = 81$ kNm	1,1 mm
CP-2.2	$M_j = 38$ kNm	$M_j = 78$ kNm	0,9 mm
CP-3	$M_j = 70$ kNm	$M_j = 89$ kNm	1,0 mm

From the comparison of the curves between specimen CP-1, CP-2 and CP-3 it can be observed that increase in amount of reinforcement leads to higher joint moment capacity, while stiffness and rotation capacity changes much less.

The formation of cracks in the concrete slab was traced as loading progressed. Transverse cracks generally initiated in the vicinity of the connection, followed by new ones further from the column as load increased. Developments of crack widths are presented in Table 5.

A typical distribution of the crack pattern on the top of the concrete slab is presented in Figure 7.

The detailed test results for each joint specimen can be found in (Research project KBN, 1998). Only test results of CP-1L specimen are presented in this paper, but other joint test results are very similar in the nature.

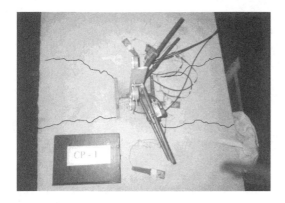

Figure 7. Typical crack pattern.

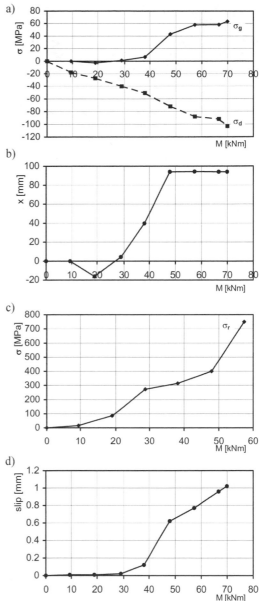

Figure 8. CP-1L specimen test results: a) top (σ_g) and bottom (σ_d) beam flange stresses, b) range of tension zone in steel beam, c) average reinforcement bar stresses, d) slip in the interface of steel beam and slab.

One-dimensional stresses for the beam flanges and reinforcement are computed on the base of the measured strains according to Hook's law. The steel beam flange stresses are shown plotted against the connection moments in Figure 8a. The top flange stresses σ_g were initially compressive, what means that neutral axis of the joint was located in the concrete slab, and gradually changed to tensile, after cracking of the slab. The top flange stresses remained relatively small because tensile forces in the connection were mainly transferred by the reinforcement. Bottom flange stresses σ_d were compressive and increased nearly linearly as loading increased. This was expected, since most of the compression force in the beam had to be carried by the bottom flange and then transferred to the column by bolted connection of flange to seat.

Figure 8b shows the relationship between the range of tension zone in the beam x, calculated as $\left(\dfrac{\sigma_g h_b}{\sigma_d + \sigma_g}\right)$, against connection moment. It is seen that initially, when concrete slab remain uncracked, this value is negative, what means that whole beam section is in compression. Then x value becomes positive and finally reaches 2/5 of the beam section height.

The average stresses in the reinforcement are presented against the connection moment in Figure 8c. It should be mentioned that, because Hooke's law is not applicable beyond the yield limit, stresses shown in Figure 8c are not the real values once the rebar yield stress has been exceeded. At the beginning of loading, these stresses are relatively small and become larger after the concrete slab cracked. It is because the tensile force transmitted previously by the concrete slab had to be redistributed to the rebars. The rebar stresses exceeded the yield point before the ultimate connection moment was reached, what means that failure mode of the joint is yielding of the reinforcement. Any other failure of bolts, welds as well as beam instability has not been observed.

Figure 8d shows the values of the measured slip in the interface of concrete slab and beam flange in relation to the connection moment. This slip is nearly zero until concrete slab cracked, and then becomes larger values of about 1 mm at the end of loading.

361

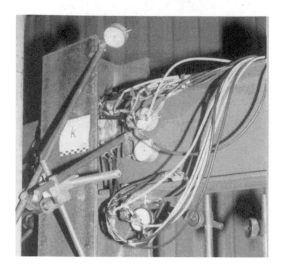

Figure 9. Bare steel joint test.

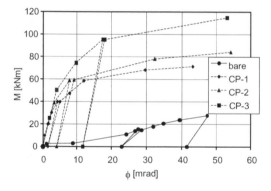

Figure 10. Comparison of bare steel and composite joint behaviour.

6 BARE STEEL JOINT TESTS

In order to compare the behaviour of pure steel and composite joints, it was necessary to test the beam-to-column connection without the presence of the concrete slab. Bare steel joints tests were conducted on the same test rig, using the same loading and measuring equipment (Fig. 9).

Specimens for steel joint tests are nominally identical that steelwork part of the composite joints. Tests were terminated when excessive deformation (i.e. rotation in excess of 60 mrad) was reached. In Figure 10 it is shown a comparison M-ϕ curves of steel and composite joints. For steel joints M-ϕ curve is an average of six (three left and three right) test results. It can be observed that the presence of a concrete slab in composite joints provides significant increase in joint moment capacity, from 2,5 to 3,8 times, compared to bare steel joints. Increase of initial stiffness of composite joints ranged from 3 to 4 times of pure steel joint stiffness, rotation capacity is of the same range.

7 CONCLUSIONS

On the base of presented test results the following conclusions can be drawn:

– Few phases can be distinguished in the composite joints behaviour. During the initial stage of loading, tension force in the upper part of joint is transmitted by the concrete slab, neutral axis is placed in concrete slab and whole steel beam is in compressive zone. After appearing of the first cracks in concrete slab, tension force started to be redistributed to reinforcement bars and bolted connection of beam web to the side plate. Stresses in top beam flange altered to be in tension zone. This is due to reduction of slab stiffness caused by cracking, resulting in the neutral axis moving down. At the failure, range of tension zone in the steel beam reaches 2/5 of beam height.
– Increase in the amount of reinforcement increases significantly the joint moment resistance. Initial stiffness and rotation capacity change much less.
– Investigated composite joints posses relatively large moment resistance, stiffness and rotation capacity. Comparison of bare steel and composite joints behaviour shows that presence of the concrete slab increases moment resistance from 2,5 to 3,8 times. Increase of initial stiffness of composite joints ranged from 3 to 4 times of pure steel joint stiffness, rotation capacity is of the same range.
– The effect of non-symmetrical loading on the joint behaviour is not significant.

REFERENCES

Anderson, D. 1996. *Composite Steel-Concrete Joints in Braced Frames for Buildings.* Brussels, Luxembourg.

Brown, N.D., Anderson, D. 2001. Structural properties of composite major axis end plate connections. *Journal of Constructional Steel Research*, 57 (3): 327–349.

Green, T. P., Leon, R. T., Rassati G., A. 2004. Bidirectional Tests on Partially Restrained, Composite Beam-to-Column Connections. *Journal of Structural Engineering.* 130, 320.

Kozłowski, A. 2000. Analytical model of semi-rigid composite joint. *Inżynieria i Budownictwo* (in Polish). Nr 8: 444–448.

Li, T.Q., Nethercot, D.A, Choo, B.S. 1996. Behaviour of Flush End-plate Composite Connections with Unbalanced Moment and Variable Shear/Moment Ratios - I. Experimental Behaviour. *Journal of Constructional Steel Research*, 38 (2):125–164.

Liew, J.Y.R., Teo, T.H., Shanmugam, N.E., Yu, C.H. 2000. Testing of steel–concrete composite connections and appraisal of results. *Journal of Constructional Steel Research*, 56 (2): 117–150.

Loh, H.Y., Uy, B., Bradford, M.A. 2006. The effects of partial shear connection in composite flush end plate joints Part I — experimental study. *Journal of Constructional Steel Research*. 62 (4): 378–390.

Nethercot, D.A. 2003. *Composite construction.* Spon Press, London.

Research project KBN nr 7T07E06910. 1998. Analysis of the stiffness, stability and carrying capacity of steel structures with semi-rigid joints (in Polish). Rzeszów University of Technology.

Silva, L.S, Simoes, R.D, Cruz, J.S. 2001. Experimental behaviour of end-plate beam-to-column composite joints under monotonical loading. *Engineering Structures*. 23: 1383–1409.

Zandonini R. 1989. Semi-Rigid Composite Joints. In: *Structural Connection. Stability and Strength.* Ed. R. Narayanan. Elsevier. London.

Steel and Composite Structures – Wang & Choi (eds)
© *2007 Taylor & Francis Group, London, ISBN 978-0-415-45141-3*

Finite element analysis of steel and composite top-and-seat and web angle connection

Z. Yuan, S.K. Ting & K.H. Tan
Department of Civil and Environmental Engineering
Nanyang Technological University, Singapore

ABSTRACT: This paper presents a three-dimensional (3-D) analysis of steel and composite top-and-seat and web (TSW) angle connections to simulate structural responses so that their moment-rotation characteristics can be predicted. The reported research work is part of an ongoing project aimed at understanding connection behaviour under fire conditions. A general purpose finite element software MSC.MARC was used to model the connections. Beam, column and angles were modelled using 4-noded shell elements and the concrete slab with 3D brick elements. The bolts and shear studs were also considered in the modelling. Due to a lack of elevated-temperature test data, the predictions from using the proposed 3-D model are compared with the ambient connection experimental data. The proposed 3-D model can predict accurate moment-rotation characteristics of connections which can be incorporated in analysis of global frame behaviour. The predicted connection response can also enrich the database of connection behaviour at ambient temperatures.

1 INTRODUCTION

Bolted connections are widely used in steel frames and usually designed as either simply-supported or fully-rigid type. However, actual non-linear behaviour of bolted connections falls between these two extreme situations. Moreover, it was found that bolted connections will significantly influence structural behaviour of steel frames. A practical solution to this problem was found by adopting a semi-rigid connection design technique. This technique gave designers greater freedom in proposing connection configurations and in analysing connection behaviour not restricted to only 'pinned' or 'fully-rigid' type. Use of semi-rigid connections had been adopted in many countries in the design of steel frames. It had shown to be economical. Hence, it is necessary to understand the joint response and how it affects the frame behaviour for practical usage of semi-rigid design.

Nonlinear moment-rotation behaviour of connections was recognized in the early 1930s. Mathematical functions had been developed from basic bilinear models to more sophisticated models based on polynomial or power functions, which came from cure-fitting. Sherbourne & Bahaari (1997) had compiled an excellent review of these functions. However, the scope of application of these functions is limited to the types of connections which had been tested. Therefore, the usage of mathematical functions is limited.

Although many full-scale experimental studies had been conducted to date, we are still far from fully understanding the non-linear behaviour of semi-rigid connection. This is partly due to a large number of geometrical and mechanical parameters of connections. In many cases experiments are very expensive and too difficult to conduct. The lack of a large database does not give designers confidence to generate reliable moment-rotation behaviour of semi-rigid connections for analysis and design purposes. Therefore, it is rational to use finite element (hereafter referred as 'FE') models to supplement design process by simulating the complex behaviour of connections.

Steel and composite top-and-seat and web (hereafter referred as 'TSW') angle connections are commonly used for its large rotation capacity and ease of construction. However, numerical research on TSW angle connections was rather limited. As part of research program on TSW angle connections, this paper presented a 3-D finite element model by means of shell and brick elements to simulate the non-linear behaviour of steel and composite TSW angle connections. Assumptions and simplifications were made to avoid detailed modelling and to reduce computational effort. The study showed the response of the steel and composite TSW angle connections can be well simulated by a commercially available finite element software MSC.MARC (MSC.MARC, 2005).

2 PREVIOUS EXPERIMENTAL WORKS

The test results from two extensive and detailed experimental studies by Azizinamini & Radziminski (1989) and Ammermann & Leon (1987) were selected to compare with results of finite element models. These test results had been chosen because they are among the best documented and most cited works on study of steel and composite TSW angle connections.

Azizinamini & Radziminski (1989) carried out an extensive experimental study of steel TSW angle connections and angle pull behaviour. The test specimen of Azizinamini is shown in Figure 1. Two identical beams were connected to a column stub with top and seat angles bolted to the flanges of beams and column. The double web angles were bolted to the beam web and the column flange. High strength ASTM A325 bolts and nuts were used with A325 hardened washers. The

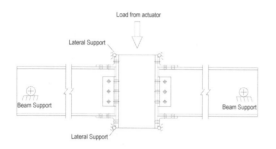

Figure 1. Test specimen of Azizinamini & Radziminski (1989).

ends of beams are supported with rollers with the column restrained and loaded to move vertically through an actuator.

The purpose of these tests was to study the behaviour of steel TSW angle connections which are representative of those used for wind moment design requirements in low rise building structures. Different geometrical parameters were investigated, including angle thickness, angle length, bolt size, spacing and gauge length of bolts on column leg of connecting angles. Two configurations are used: W8 × 21 beam with W12 × 58 column and W14 × 38 beam with W12 × 96 column, respectively. The test specimen details are listed in Tables 1 and 2.

Ammermann & Leon (1987) conducted an extensive experimental study of the composite TSW angle connection in order to compare the behaviour of semi-rigid connection with and without a composite slab. A layer of reinforced concrete slab was added to each angle connection with similar configurations with those in Azizinamini study. The test results showed great improvement in terms of rotational stiffness and moment capacity of the composite TSW angle connections compared to those of the steel TSW angle connections.

The specimen details are shown in Figure 2. A connection similar to 14S1 by Azizinamini was chosen since those moment-rotation characteristics of the non-composite connection were well known. The test specimen consisted of a pair of W14 × 38 beams and a W14 × 99 column. A 1.52-m wide reinforced slab of 100 mm thickness embedded with 8 nos. of Ø12.7 mm rebar was connected to the beams by shear studs to

Table 1. Details of tests specimen of Azizinamini & Radziminski (1989), Part 1 of 2.

Specimen Number	Beam Section	Bolt diameter (mm)	Web Angle	
			Angle (mm)	Length L_c (mm)
8S1	W8 × 21	19.05	2 × 101.6 × 88.9 × 6.35	139.7
8S2	W8 × 21	19.05	2 × 101.6 × 88.9 × 6.35	139.7
8S3	W8 × 21	19.05	2 × 101.6 × 88.9 × 6.35	139.7
8S4	W8 × 21	19.05	2 × 101.6 × 88.9 × 6.35	139.7
8S5	W8 × 21	19.05	2 × 101.6 × 88.9 × 6.35	139.7
8S6	W8 × 21	19.05	2 × 101.6 × 88.9 × 6.35	139.7
8S7	W8 × 21	19.05	2 × 101.6 × 88.9 × 6.35	139.7
8S8	W8 × 21	22.23	2 × 101.6 × 88.9 × 6.35	139.7
8S9	W8 × 21	22.23	2 × 101.6 × 88.9 × 6.35	139.7
8S10	W8 × 21	22.23	2 × 101.6 × 88.9 × 6.35	139.7
14S1	W14 × 38	19.05	2 × 101.6 × 88.9 × 6.35	215.9
14S2	W14 × 38	19.05	2 × 101.6 × 88.9 × 6.35	215.9
14S3	W14 × 38	19.05	2 × 101.6 × 88.9 × 6.35	139.7
14S4	W14 × 38	19.05	2 × 101.6 × 88.9 × 6.35	215.9
14S5	W14 × 38	22.23	2 × 101.6 × 88.9 × 6.35	215.9
14S6	W14 × 38	22.23	2 × 101.6 × 88.9 × 6.35	215.9
14S8	W14 × 38	22.23	2 × 101.6 × 88.9 × 6.35	215.9
14S9	W14 × 38	22.23	2 × 101.6 × 88.9 × 6.35	215.9

Table 2. Details of tests specimens of Azizinamini & Radziminski (1989), Part 2 of 2.

Specimen Number	Top and Seat Flange Angles			
	Angle (mm)	Length L (mm)	Gauge in leg on column flange (mm)	Bolt spacing in leg on column (mm)
8S1	152.4 × 88.9 × 7.94	152.4	50.8	88.9
8S2	152.4 × 88.9 × 9.53	152.4	50.8	88.9
8S3	152.4 × 88.9 × 7.94	203.2	50.8	88.9
8S4	152.4 × 88.9 × 9.53	152.4	114.3	88.9
8S5	152.4 × 88.9 × 9.53	203.2	63.5	139.7
8S6	152.4 × 101.6 × 7.94	152.4	63.5	88.9
8S7	152.4 × 88.9 × 9.53	152.4	63.5	88.9
8S8	152.4 × 88.9 × 7.94	152.4	50.8	88.9
8S9	152.4 × 88.9 × 9.53	152.4	50.8	88.9
8S10	152.4 × 88.9 × 12.7	152.4	50.8	88.9
14S1	152.4 × 101.6 × 7.94	203.2	63.5	139.7
14S2	152.4 × 101.6 × 12.7	203.2	63.5	139.7
14S3	152.4 × 101.6 × 7.94	203.2	63.5	139.7
14S4	152.4 × 101.6 × 7.94	203.2	63.5	139.7
14S5	152.4 × 101.6 × 7.94	203.2	63.5	139.7
14S6	152.4 × 101.6 × 12.7	203.2	63.5	139.7
14S8	152.4 × 101.6 × 15.88	203.2	63.5	139.7
14S9	152.4 × 101.6 × 12.7	203.2	63.5	139.7

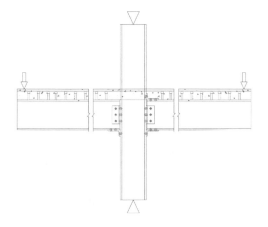

Figure 2. Details of test specimens of Ammermann & Leon (1987).

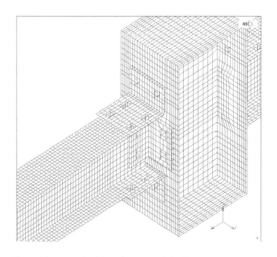

Figure 3. Local view of FE model of steel TSW angle connection.

simulate a continuous floor slab across an interior column. It was expected the heavy section column will behave almost as a fixed end for connected beams under gravity load, so different connection details were used on either side of the column. On the right side a connection was similar to 14S1 labelled by Azizinamini. However, the connection on the left side did not have an upper angle, but the rest of details were similar to the right one. Two numbers of composite TSW angle connections were modelled by FE models in this study. They were, namely, SRCCML & SRCCMR.

3 FINITE ELEMENT MODELS

3-D FE models were proposed to simulate non-linear behaviour of steel and composite TSW angle connections by using MSC.MARC (MSC.MARC, 2005). Member properties and geometries from the experimental studies (Azizinamini & Radziminski 1989, Ammermann & Leon 1987) were adopted in the models. A typical 3-D FE model of steel/composite TSW angle connection is shown in Figure 3 and 4, respectively.

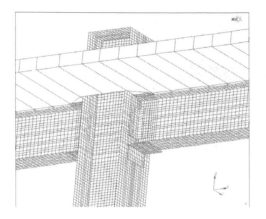

Figure 4.   Local view of FE model of composite TSW angle connection.

Bolts were modelled using beam elements with corresponding properties. For bolts, the action of connecting with angles, beams and columns were modelled with RBE2 element to transfer the necessary forces and moments. The influence due to bolt size could also be modelled with RBE2 element by controlling the distance between bolt centre to connecting structural element (e.g. angle, beam or column). As the shear studs are not the source of structural failure of connections, each shear stud was modelled using three springs to simulate its axial and shear stiffness and strength. Beam, column and angles were modelled by 4-noded thick shell elements, which would not affect the accuracy of results with dense meshes. Dense meshes of shell element were used to capture localised failure in combination with global connection failure. 3-D brick element was chosen to model the concrete slab. Special rebar element was selected to model the embedded rebar using smeared layer approach. The arc-length iterative procedure had been employed to properly trace the non-linear behaviour of connections. Another type of spring with specific force-displacement relationship was used to simulate contact links between beams and columns and angles. A tri-linear elliptic stress-strain steel material model was assumed and adopted in the FE simulation, which incorporates strain-hardening as specified in Eurocode 3 Part 1.1 (prEN1993-1-1).

3.1   Validation of finite element model

Before the model was used to investigate the full-range non-linear moment-rotation response of connections, it was validated against theoretical predictions of moment capacity by adopting rigid-plastic stress-strain steel property. The predicted results by Faella et al. (1996, 2000) had been used to validate the

Table 3.   Comparison of plastic moment capacity of steel TSW connection by FE models and by mechanical models.

Test No.	M_{FEM} (kNm)	M_p (kNm)	$M_{(FEM)}/M_p$
8S1	31.76	33.25	0.96
8S2	39.33	37.93	1.04
8S3	38.96	41.84	0.93
8S4	15.81	16.33	0.97
8S5	45.10	36.25	1.24
8S6	22.20	23.33	0.95
8S7	28.26	28.96	0.98
8S8	34.99	44.21	0.79
8S9	44.50	47.47	0.94
8S10	50.23	50.67	0.99
14S1	63.77	65.91	0.97
14S2	90.63	86.75	1.04
14S3	59.38	61.61	0.96
14S4	92.80	94.96	0.98
14S5	69.20	74.42	0.93
14S6	96.66	96.43	1.00
14S8	110.70	112.88	0.98
14S9	96.66	96.43	1.00
		Avg	0.981
		Std	0.085

finite element model. The work by Faella et al. (1996, 2000) had been chosen since it is among the best documented theoretical work on the study of steel TSW angle connections. A mechanical model was proposed (Faella et al. 1996, 2000) for steel TSW angle connections to calculate the initial rotational stiffness and moment capacity. A total of 18 test specimens were modelled and the comparison is shown in Table 3. Rigid-plastic steel properties were used to estimate the plastic moment capacity of connection. It was illustrated that good agreement between the predicted moment capacities using FE models and the theoretical moment capacities using mechanical models. This showed that the FE models could be used with confidence to predict the moment capacities of steel TSW angle connections. For composite TSW angle connections, it were shown later that the FE model could predict accurately the moment-rotation responses.

The discrepancies between the predicted moment capacity and theoretical values may be due to rigid-plastic analysis used in mechanical approach and the slight differences in support and material properties. The rigid plastic analysis tended to underestimate the moment capacity as the connections behaved more plastically as steel material approaches yield. Moreover, detailed steel stress-strain curves of specimens were not available from test reports. So the mean value or nominal values of material properties and assumptions were adopted in the FE simulation. Nonetheless, the mean value of predicted moment capacity was quite close to the theoretical predicted values.

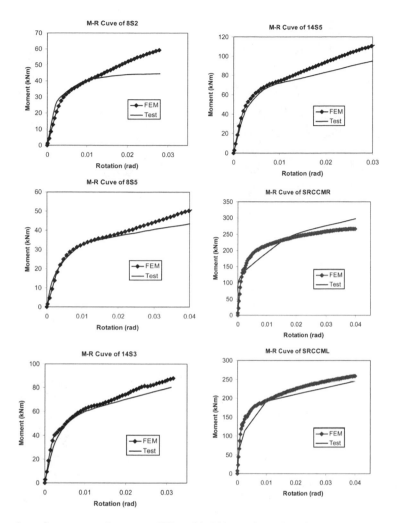

Figure 5. Comparison of moment-rotation curve of FE models with experimental results.

3.2 FE simulation of steel and composite TSW angle connections

With this validation completed, the next step was to input actual member properties to simulate the moment-rotation behaviour of connections. As detailed stress-strain properties of each specimen were not available from test reports, only the mean values of materials were reported. So the mean values of material properties were adopted in the FE simulation. A tri-linear elliptic stress-strain steel material model was assumed and adopted in the FE simulation, which also incorporated strain-hardening as specified in Eurocode 3 Part 1.1 (prEN1993-1-1).

17 nos. of steel angle connections and 2 nos. of composite angle connections were analysed using FE models. Some moment rotation curves from tests and

FE models were also shown in Figure 5. It was noted that for test 14S2 slippage of bolts occurred resulting in a drop in moment until bolt bearing was achieved against bolt holes. For 8S2 an anomaly in the initial rotational stiffness and strength occurred, as noted by Azizinamini & Radziminski (1989). It is shown that the initial stiffness by FE models in general is in good agreement with experimental results. For steel TSW angle connections, it could be seen that for 14S configuration tests the predictions were more consistent than the smaller 8S ones. The reason was that irregularities in fabrication quality and material properties would have smaller influence to connections with bigger beams than those with smaller beams. For composite TSW angel connections, the moment-rotation curves by FE model followed closely with the experimental

369

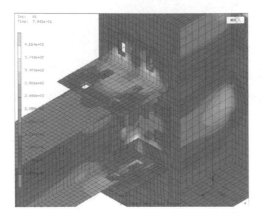

Figure 6. Stress contour of steel TSW angle connection.

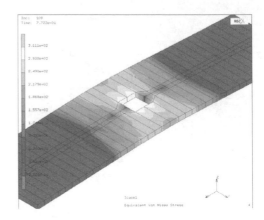

Figure 7. Stress contour of rebar in composite TSW angle connection.

results. The highly non-linear behaviour of the connections was due to material properties (strain hardening) and connection configurations. A reasonably good agreement was also observed between the results of FE models and actual experimental results.

Generally, the stiffer post-yield behaviour of FE models might be due to incomplete material properties and simplification made in modelling of bolts. The detailed investigation is being carried out.

3.3 Failure modes of steel and composite TSW angle connections

The numerical analyses were also used to identify the failure modes of connections. Several representative specimens were presented in this paper.

For steel TSW angle connections, it was observed that local yielding occurred and plastic hinge formed at each toe of the fillet in the angle attached to the tension flange of beam. Another plastic hinge developed in the vicinity of bolt line on the leg of the flange angle attached to the column. Progressive plastic hinge line also occurred in the outstanding legs of the web angle (Azizinamini & Radziminski, 1989).

For composite TSW angle connections, progressive yielding of rebar, bolt yielding, yielding of column web in compression were among the primary contributions to overall structural failure (Ammerman & Leon, 1987).

It should be noted that the FE models could predict the location of plastic hinges compared with the experimental tests. Figure 6 shows, for steel TSW angles, the plastic hinges were formed at the toe of the fillet of the top angle attached to the tension flange of beam.

Similarly, for composite TSW angle connections, the rebar had yielded at the maximum moment and column web in compression had also reached plastic strength with noticeable deformation, as shown in Figure 7 & 8.

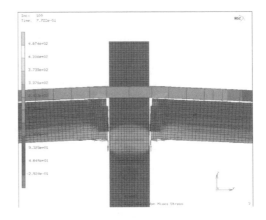

Figure 8. Stress contour of part of composite TSW angle connection.

4 CONCLUSION

Although current design codes had included the concept and analysis of semi-rigid connection, we were still far from sufficiently understanding the non-linear behaviour of semi-rigid connections.

Therefore, it is logical to use FE models to supplement the design process. A supplementary numerical analysis had been conducted to simulate moment-rotation response of steel and composite TSW angle connections by using commercial FE software MSC.MARC. It was observed that the overall non-linear behaviour of the connections could be well predicted with the proper construction of FE models. The results would be helpful in overall frame analysis and could enrich the current database of connection behaviour at ambient temperature. Once we understand the behaviour mechanism of the connection under different loads, we can have the confidence to use FE models to analyse the semi-rigid connections.

REFERENCES

Azizinamini, A. & Radziminski, J. B. 1989. Static and cyclic performance of semirigid steel beam-to-column connections.*Journal of Structural Engineering*, 115(2): 2979–2999.

Ammerman, D. J. & Leon, R. T. 1987. Behaviour of semi-rigid composite connections. Engineering Journal, 2nd quarter: 53–61.

Faella, C., Piluso, V. & Rizzano, G. 1996. Prediction of the flexural resistance of bolted connections with angles. *IABSE, International colloquium on semi-rigid structural connections. September 1996.* Istanbul: Turkey.

Faella, C., Piluso, V. & Rizzano, G. 2000. *Structural steel semirigid connections : theory, design and software.* Florida, CRC press LLC.

Eurocode 3 1992. ENV 1993-1-1, Design of steel structures, Part 1.1: general rules and rules for building. CEN. Brussels.

prENV 1993-1-1/A2 1994. New revised Annex J: joints in building frames. CEN. Brussels.

LRFD 1986. AISC, Manual of Steel Construction – Load and Resistance Factor Design, 1st edn. American Institute of Steel Construction (AISC), Chicago, IL.

MSC.MARC User Manual, Version 2005. MSC Software Corporation, 2 MacArthur Place, Santa Ana, CA 92707. USA.

Sherbourne, A. N. & Baharri, M. R. 1994. 3D simulation of end-plate bolted connection response. *Journal of Structural Engineering*, 120(11): 3122–3136.

Steel and Composite Structures – Wang & Choi (eds)
© 2007 Taylor & Francis Group, London, ISBN 978-0-415-45141-3

Development of a component mechanical model for composite top-and-seat and web angle connection

Z. Yuan, S.K. Ting & K.H. Tan

Department of Civil and Environmental Engineering Nanyang Technological University, Singapore

ABSTRACT: This paper describes the development of a component-based mechanical model for composite top-and-seat and web (hereafter referred as 'TSW') angle connections. It also describes the proposal of new component characteristics of reinforced concrete (hereafter referred as 'RC') slab and shear stud. The reported research work is part of an ongoing project aimed at understanding connection behaviour under fire conditions. This paper summarises the derivation of this mechanical model, based on spring elements, for this type of connections. Due to lack of elevated-temperature test data, predictions from using the proposed mechanical model were compared with connection experimental data conducted at the ambient temperature. Good agreement is observed between mechanical model predictions and test results at ambient temperature. The proposed model is general and can be applied to other connection configurations with modification of geometrical and material parameters. It can also provide a quick and reliable assessment of influence of each important parameter on connection behaviour.

1 INTRODUCTION

It was found that the behaviour of beam-to-column connection had significant influence on the frame structure. However, in practice due to a large number of combinations of geometrical and mechanical parameter of connections, typical beam-to-column connection was generally designed as perfectly pinned or completely fixed. In fact, the behaviour of beam-to-column connection falls between these two extreme cases and is called "semi-rigid".

Beam-to-column connections are complicated structures, which consist of steel members, bolts and welds. This results in complex interactions among structural elements. Because of non-linear element behaviour and a large number of possible variables in element details within a beam-to-column connection, a general approach for calculating rotational and axial stiffness and moment-rotation characteristics of connections is necessary.

The methods available for predicting the behaviour of beam-to-column connection include four categories. They are empirical curves, mathematical formulations, finite element models and component mechanical models. Mathematical formula focuses on analytical prediction of joint rotational stiffness and moment resistance, while modelling of moment-rotation behaviour still depends on curve-fitting. However, component mechanical model can predict the

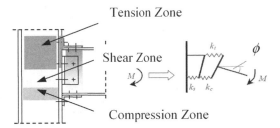

Figure 1. Component of a steel top-and-seat and web angle connection (Faella et al. 1995, 1996, 2000).

moment-rotation behaviour by predicting progressive yielding of each joint component without knowing the sequence of failure of each joint component.

The first component mechanical model was proposed by Tschemmernegg et al. (1988) and later adopted into EC 3 Part 1.8 (2002). This model had the original feature of simulating the behaviour of a joint using a set of individual rigid and flexible components.

The component-based mechanical model for steel top-and-seat and web (hereafter referred as 'TSW') angle connection had been proposed by Faella et al. (1995, 1996, 2000) based on EC3 Annex J (1994), which is shown in Figure 1. The joint was divided into three major zones (tension, shear and compression), and then each zone was divided into components, represented by spring elements. It was found that the

Mechanical Model of Composite TSW angle connection

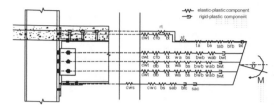

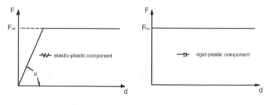

Figure 2. The proposed mechanical model for composite top-and-seat and web angle connection.

mechanical model could describe steel TSW angle connection behaviour accurately and consistently.

However, not much study had been conducted on composite TSW angle connections to date. When a layer of concrete slab is added onto a steel TSW angle connection, the composite action of concrete slab and bare-steel connection will greatly enhance the initial rotational stiffness and moment-capacity of the connection. Composite TSW connections are popular for structures in earthquake-prone zones for large rotation capacity, ease of fabrication and good energy absorption characteristics. Since not much study had been devoted to this type of connection, so far there was no component mechanical model for composite TSW angle connection.

In this paper the authors proposed a component mechanical model for composite TSW angle connection. The authors also identified several key components with new proposal of structural characteristics.

2 THE PROPOSED COMPONENT-MECHANICAL MODEL FOR COMPOSITE TSW ANGLE CONNECTION

The proposed component mechanical model is shown in Figure 2. The joint components include reinforced concrete (hereafter referred as 'RC') slab in tension (rt), shear stud in shear (st), column web in tension (cwt), column flange in bending (cfb), bolt in tension (bt), top angle in bending (ta), bolt in shear (bs), top angle in bending (tab), beam flange in bearing (bfb), top angle in tension (tat), web angle in bending (wa), beam web in bearing (bwb), web angle in bending (wab), beam web in tension (bwt), column web in shear (cws), column web in compression (cwc), seat angle in bearing (sab), beam flange in compression (bfc), and seat angle in compression (sac).

2.1 Initial rotational stiffness

The calculation of the initial rotational stiffness of the joint requires the input of stiffness of each component and the position of each component in the joint itself.

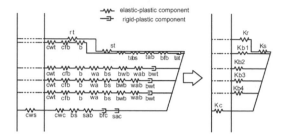

Figure 3. The force-displacement relationship of elastic-plastic and rigid-plastic component.

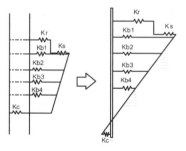

Figure 4. Simplification process of component mechanical models into spring models.

Figure 5. The simplified spring model for composite top-and-seat and web angle connection.

The behaviour of each component can be defined as either elastic-plastic or rigid-plastic, as shown in the Figure 3 (Ahmed & Nethercot, 1997).

The component mechanical model can be simplified by the procedure shown in Figure 4.

The simplified component mechanical model is illustrated in Figure 5.

Using the proposed spring model shown in Figure 5 the force equilibrium conditions give rise to Equation 1:

$$F_r = F_s ; \quad F_r + F_{b1} + F_{b2} + F_{b3} + F_{b4} = F_c \qquad (1)$$

where F_r is the tension force in RC slab and F_{bi} (i = 1,2,3,4) is the tension force in other spring row, respectively. F_s is the shear force in shear stud and F_c is the compressive force in the column web in compression.

From Hook's law the elongation of each row of spring is calculated as

$$F_r = K_r \Delta_r ; F_{bi} = K_{bi} \Delta_{bi}, i = (1,2,3,4) ; F_s = K_s \Delta_s \qquad (2)$$

where K and Δ are the stiffness and elongation of the respective spring row.

The compatibility equilibrium gives:

$$\frac{\Delta_r + \Delta_s + \Delta_c}{H_b} = \frac{\Delta_{b,i} + \Delta_c}{H_{b,i}} = \phi \qquad (3)$$

where H_b is the beam depth and H_{bi} (i = 1,2,3,4) is the distance of each spring row from the seat angle centreline.

After combining and substituting the above equations, a set of equations can be obtained as follows.

$$F_r\left(\frac{1}{K_r}+\frac{1}{K_s}+\frac{1}{K_c}\right)+\frac{F_{b1}+F_{b2}+F_{b3}+F_{b4}}{K_c}-H_b\phi = 0$$

$$\frac{F_r}{K_c}+F_{b1}\left(\frac{1}{K_{b1}}+\frac{1}{K_c}\right)+\frac{F_{b2}+F_{b3}+F_{b4}}{K_c}-H_{b1}\phi = 0$$

$$\frac{F_r}{K_c}+F_{b2}\left(\frac{1}{K_{b2}}+\frac{1}{K_c}\right)+\frac{F_{b1}+F_{b3}+F_{b4}}{K_c}-H_{b2}\phi = 0 \quad (4)$$

$$\frac{F_r}{K_c}+F_{b3}\left(\frac{1}{K_{b3}}+\frac{1}{K_c}\right)+\frac{F_{b1}+F_{b2}+F_{b4}}{K_c}-H_{b3}\phi = 0$$

$$\frac{F_r}{K_c}+F_{b4}\left(\frac{1}{K_{b4}}+\frac{1}{K_c}\right)+\frac{F_{b1}+F_{b2}+F_{b3}}{K_c}-H_{b4}\phi = 0$$

After solving the above set of equations, the forces in each spring row can be obtained. Since at initial stage of loading the internal tensile forces are low, all the components' behaviour is elastic. Therefore, the moment can be calculated using rebar force and spring row forces in this case. The moment is:

$$M = F_r H_r + F_{b1} H_{b1} + F_{b2} H_{b2} + F_{b3} H_{b3} + F_{b4} H_{b4} \quad (5)$$

where M is the moment and H_r is the distance from rebar centre to seat angle centreline. Hence, the initial rotational stiffness $K_{\phi,i}$ can be determined from:

$$K_{\phi,i} = M / \phi \qquad (6)$$

2.2 Moment capacity

The connection moment capacity $M_{j,Rd}$ can be calculated by summing up the product of tension or compression capacity of each component row and the distance between each row of component to the centreline of seat angle (Ahmed & Nethercot, 1997).

$$M_{j,Rd} = F_{r,Rd}H_r + F_{b1,Rd}H_{b1} + F_{b2,Rd}H_{b2} + F_{b3,Rd}H_{b3} + F_{b4,Rd}H_{b4} \qquad (7)$$

where $F_{r,Rd}$ and $F_{bi,Rd}$ (i = 1,2,3,4) are the tension or compression capacity of rebar and each bolt row, respectively.

Force equilibrium shall be maintained when calculating the force distribution in the spring rows. The RC slab is located farthest from the centre of compression and is assumed to reach it tension capacity. For other spring rows, the tension/compression capacity is analysed successively from the farthest spring row to the nearer spring row, towards the centre of compression. The tension/compression capacity of each spring row is limited by the minimum strength of its constituent components, and individual row capacity when considering the failure mechanism involving a group of spring rows. Moreover, it is obvious that the tension/compression capacity of any combination of spring rows shall not exceed the total capacity of all the spring rows (Ahmed & Nethercot, 1997).

2.3 Moment-rotation (M-φ) curve including influence of straining hardening

For advanced analysis of steel frame structures, the accurate prediction of full-range moment-rotation behaviour is indispensable for a successful prediction of overall structural behaviour.

The general concept is to superimpose the moment-rotation relationship of each row of components to get the overall moment-rotation behaviour of the joint. The rotation of a joint for a certain moments is calculated on the principles of equilibrium, compatibility and components force-displacement relationship to be satisfied at all rows of components. The complexities of the calculation process result from the interactions among the constitutive relationships and the non-linearity of each row of components.

It should be noted that the evaluation of initial rotational stiffness and moment capacity according to EC 3 Part 1.8 (2003) neglected the influence of strain hardening. Consequently, it was observed that, for semi-rigid connections, the moment capacity was always less than the ultimate moment capacity. However, strain-hardening behaviour of each component after reaching plastic capacity is important for correctly predicting the full-range moment-rotation curve of the connection and ultimate moment capacity.

To account for strain-hardening, Faella et al. (1995, 1996) has proposed a simple method without a significant increase of the computational effort. Figure 6 shows the typical quadric-linear moment-rotation relation of a component. This approach was adopted in the proposed mechanical model.

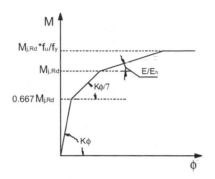

Figure 6. The typical quadric-linear moment-rotation relation of a component row (Faella et al. (1995, 1996).

3 SELECTION OF KEY PARAMETERS

The successful application of the proposed model depends on the accurate description of the behaviour of key individual components. MSC.MARC (MSC. MARC, 2005), a general-purpose finite element software, had been extensively used by the authors to simulate the behaviour of composite angle connections. These studies provided valuable data on the detailed aspects of connection behaviour. The authors found that the RC slab and shear stud behaviour have the most influence. Column web in compression is less important, while other parameters are even less significant.

3.1 The force-displacement relationship of RC slab

To obtain the accurate force-displacement relationship of RC slab, the stress-strain relationship of concrete and embedded rebar should be determined as the first step, which shall be derived from the experimental tests and analytical model. Equally important was the effective length of RC slab.

3.1.1 The stress – strain relationship of concrete and embedded rebar

RC slab of composite connection will be subjected to tension and will start cracking under increasing hogging moment. Previous studies (Ahmed & Nethercot, 1997, Anderson & Najafi, 1994, Leston Jones, 1997, Al-Jabri, 1999) had assumed that the nonlinear behaviour of RC slab depended only on the bare rebar properties. Only rebar's stiffness and strength were considered in predicting initial rotational stiffness and moment capacity of composite connection. Moreover, the effective length of rebar for calculating initial rebar stiffness was assumed to match the experimental results. These assumptions were then made due to limited knowledge on the behaviour of RC members. Hence, the important influence of concrete on behaviour of RC members had been neglected.

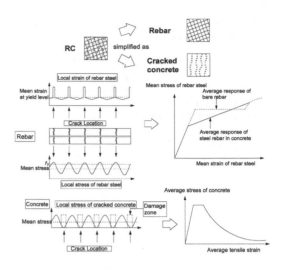

Figure 7. Constitute laws and averaging scheme (Maekawa, et al. 2003).

In this paper, the authors proposed a new component model to account for the behaviour of RC (embedded rebar + concrete) slab, based on the research work and analytical model conducted by Maekawa et al. (2003). Non-linear mechanical properties of RC slab can be determined by constitutive laws within the frame work of continuum or quasi-continuum mechanics. Constitutive models are mathematically expressed force-deformation relations. The behaviour of RC slab under tension, which includes concrete slab and rebar, can be simulated by incorporating the constitutive models of cracked concrete slab and embedded rebar on an averaged basis with respect to a finite control volume containing several primary cracks. Here the control volume is defined as the whole reinforced concrete slab (Maekawa et al. 2003).

The nonlinearity of a RC slab results primarily from cracking, reinforcement plasticity and bond action between concrete and reinforcement. Although the local stress and strain distribution within cracked RC slab is not uniform, such locality need not be explicitly considered in analysis through space averaging (Maekawa et al. 2003). Figure 7 shows the structural modelling of an in-plane RC element under tension. The total stress carried by reinforced concrete is the sum of the averaged stresses of cracked concrete and rebar at equal average strain. Here the concrete stress is assumed to comprise the stress normal to cracks only, as is the case in reinforced slab under pure tension. In the same manner, the average stress of a rebar is derived from its average stress developed in a reinforced concrete slab. Figure 7 shows schematically how this idea can be incorporated into mechanical model. This model has been adopted to define the tension force carried by RC slab at various strains of concrete and rebar.

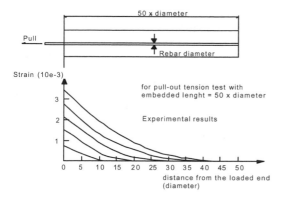

Figure 8. Strain distributions in pull-out tension specimen (Maekawa, et al. 2003).

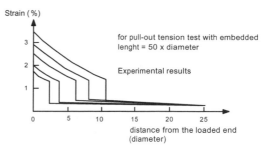

Figure 9. Strain distributions in pull-out tension specimen after first major crack (Maekawa, et al. 2003).

3.1.2 The effective length of embedded rebar and concrete

After we have obtained the constitutive stress-strain model of embedded rebar and cracked concrete slab under tension, it is also critical to decide how to choose the effective length to calculate the tension stiffness of RC slab.

First, let us look at the behaviour of a long embedded rebar when subjected to a pull-out force, as shown in Figure 8.

The strain at the loaded end will increase with increasing stress and the rebar strain at other locations will increase correspondingly. At the opposite end, the strain and stress are always zero. It has to be noted that when stress is low, i.e. at initial stage of loading, the length of rebar where there is visible strain is limited to only about 5 diameters of rebar. Therefore, the authors have suggested the effective length to calculate the initial tensile stiffness of RC slab to be 6 times the diameter of rebar, on the conservative side.

With increasing stress at the loaded end, concrete slab will crack through the thickness of RC slab at certain stage of loading. Rebar may also reach yielding under increasing external load. Generally, after the concrete slab has shown the first sudden major crack, the rebar will soon start yielding locally at the crack so that the tensile force carried by uncracked concrete slab is transferred to embedded rebar after the major crack has occurred. This will cause a jump in strain in the embedded rebar. Although in the post-yield range of rebar behaviour, the strain distribution is not smooth in the region where rebar yields, this discontinuity does not affect the definition of effective length of rebar. This is because the locality in strain discontinuity has already been incorporated in the average stress-strain relationship of rebar and concrete in RC member.

In Figure 9 the length of rebar that has visible strain can increase up to the whole length, assuming that the behaviour is still elastic.

Therefore, the authors suggest that, after the concrete cracks or the rebar yields, the whole length of rebar will be used to calculate the stiffness of RC slab.

Similarly, the same approach is adopted for concrete in RC slab to calculate the effective length.

3.2 The force-displacement relationship of shear studs

Although previous studies (Ahmed & Nethercot, 1997, Anderson & Najafi, 1994) recommended adopting an assumed stiffness of shear studs, the authors had proposed another component model to account for the behaviour of shear studs.

The research work done by Loh et al. (2004a, 2004b) had suggested that all the shear studs had interface slip along the composite beam under hogging moment and therefore, participated in the deformation of whole composite beam. Hence, this finding supports the proposal to include all the shear studs in the calculation of stiffness and strength.

Gattesco & Giuriani (1996) conducted direct shear tests, with boundary conditions to impose direct shear on the shear connectors. This arrangement was perceived to simulate the actual behaviour of connectors. The connector load–slip curve was represented by the following equation with $\alpha = 0.97$, $\beta = 1.3$ mm^{-1} and $\gamma = 0.0045$ mm^{-1}. This model has been adopted in the proposed mechanical model.

$$F_j = F_{s.Rd} \left[\alpha \left(1 - e^{-\beta s/\alpha} \right)^{0.5} + \gamma s \right] \qquad (8)$$

3.3 Strength and stiffness of column web in compression

The column web in the compression zone can have a significant influence on the overall joint behaviour especially in the case of unstiffened joints. Spyrou et al. (2004) conducted ambient and elevated-temperature tests and proposed a semi-analytical model which not only provide the ultimate capacity of the column web but also determine the stiffness of the column web in the elastic and plastic regions. This model has been adopted in the proposed mechanical model.

4 COMPARISON OF PREDICTIONS BETWEEN THE PROPOSED MECHANICAL MODEL AND EXPERIMENTAL RESULTS

The comparison between the predicted behaviour by proposed mechanical model and test data was made with experimental data of Ammerman & Leon (1987),

Table 1. Comparison between the predicted values of the initial rotational stiffness K_ϕ and the experimental results.

Test				
No.	Test No.	$K_{\phi,exp}$ (kNm/rad)	$K_{\phi(predicted)}$ (kNm/rad)	$K_{\phi(predict)}/$ $K_{\phi,exp}$
1	SRCC1ML	225969	241229	1.068
2	SRCC1MR	255346	257834	1.010

Davison et al. (1990), and Altmann et al. (1990). The results of the comparison were shown in Tables 1 and 2.

Since the test results on initial rotational stiffness were limited, only two numbers of tests were used for comparison. It could be observed that the proposed model could predict the initial rotational stiffness closely with experimental results.

It was important to stress that this model adopts a generalised philosophy that leads to a good prediction of composite joints.

A total of 28 experimental results (Ammerman & Leon. 1987, Davison et al. 1990, Altmann. et al. 1990) had been compared with predicted ultimate moment capacities in Table 2. A good degree of accuracy was observed from the average of $M_{u,Rd}/M_{u,Exp}$.

The accurate determination of component force-displacement relationship would help to determine the full-range moment-rotation curve of composite top-and-seat and web angle connection. Figure 10 shows, for some specimens, the moment-rotation curve was compared with that from the proposed model. When

Table 2. Comparison between the predicted values of the ultimate flexural resistance $M_{u,Rd}$ and the experimental results $M_{u,Exp}$.

Test		$M_{u,Exp}$ (kNm)	$M_{u,Rd}$ (kNm)	$M_{u,Rd}$ $M_{u,Exp}$
No.	Details			
1	Ammerman–SRCC1ML	245	235	0.959
2	Ammerman–SRCC1MR	298	298	1.000
3	Davison–C7	140	143	1.021
4	Davison–C9	160	167	1.044
5	Altmann $36 \times 3c \times 10$ mm web $\times$ T10	285	288	1.011
6	Altmann $36 \times 3c \times 10$ mm web $\times$ T14	333	335	1.006
7	Altmann $36 \times 3c \times 10$ mm web $\times$ T18	370	372	1.005
8	Altmann $36 \times 3c \times 13$ mm web $\times$ T10	295	305	1.034
9	Altmann $36 \times 3c \times 13$ mm web $\times$ T14	354	352	0.994
10	Altmann $36 \times 3c \times 13$ mm web $\times$ T18	393	375	0.954
11	Altmann $36 \times 2c \times 10$ mm web $\times$ T10	250	248	0.992
12	Altmann $36 \times 2c \times 10$ mm web $\times$ T14	338	332	0.982
13	Altmann $36 \times 2c \times 10$ mm web $\times$ T18	366	358	0.978
14	Altmann $36 \times 2c \times 13$ mm web $\times$ T10	254	255	1.004
15	Altmann $36 \times 2c \times 13$ mm web $\times$ T14	344	338	0.983
16	Altmann $36 \times 2c \times 13$ mm web $\times$ T18	363	337	0.928
17	Altmann $30 \times 3c \times 10$ mm web $\times$ T10	252	258	1.024
18	Altmann $30 \times 3c \times 10$ mm web $\times$ T14	312	281	0.901
19	Altmann $30 \times 3c \times 10$ mm web $\times$ T18	286	289	1.010
20	Altmann $30 \times 3c \times 13$ mm web $\times$ T10	292	274	0.938
21	Altmann $30 \times 3c \times 13$ mm web $\times$ T14	282	287	1.018
22	Altmann $30 \times 3c \times 13$ mm web $\times$ T18	329	321	0.976
23	Altmann $30 \times 3c \times 10$ mm web $\times$ T10	206	214	1.039
24	Altmann $30 \times 3c \times 10$ mm web $\times$ T14	264	281	1.064
25	Altmann $30 \times 3c \times 10$ mm web $\times$ T18	323	327	1.012
26	Altmann $30 \times 3c \times 13$ mm web $\times$ T10	208	207	0.995
27	Altmann $30 \times 3c \times 13$ mm web $\times$ T14	292	285	0.976
28	Altmann $30 \times 3c \times 13$ mm web $\times$ T18	324	309	0.954
			Average	0.993
			Std	0.037

percentage of steel rebar in the slab varies between two extreme values 0.5% (e.g. Davison – C7) to 2.1% (e.g. Altmann 36 × 2c × 10 mm web × T18), the proposed mechanical model can give close approximation of the actual behaviour because the number of deformable components considered and the sources of deformation are captured in this model. A shift in rotation due to bolt slip was also considered for some test specimens. The magnitude of shift in rotation angle can be estimated by dividing bolt hole allowance (usually 1 mm for bolts smaller than M24 or 2 mm for bolts bigger than M24) by the beam depth.

5 CONCLUSION

From the basic mechanism of force transfer within the components of composite connection, a component mechanical model was proposed to predict the initial rotational stiffness, moment capacity and moment-rotation characteristics for composite top-and-seat and web angle joints. Comparisons against test data had shown that the model was able to describe the behaviour of composite top-and-web and web angle connection with very good accuracy and consistency. If the effect of temperature and other

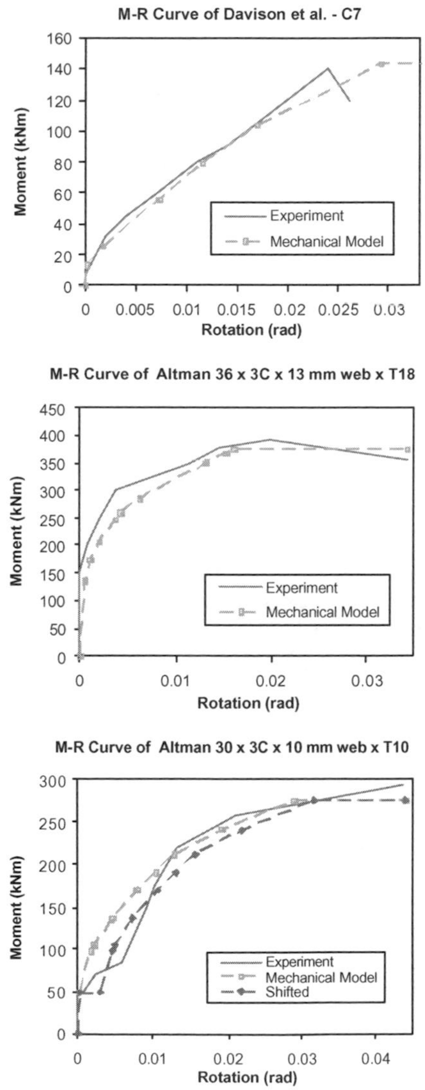

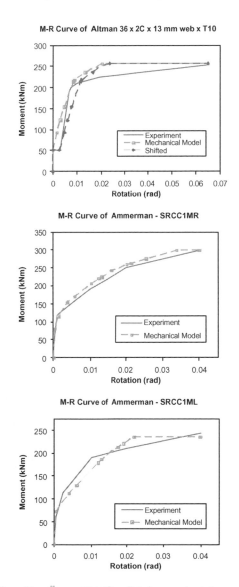

Figure 10. Comparison of predicted moment-rotation curve with experimental results (Part 1 of 2).

Figure 11. Comparison of predicted moment-rotation curve with experimental results (Part 2 of 2).

fire-induced member forces were considered, the elevated-temperature behaviour of this type of joint could be modelled with good accuracy. Remaining study would focus on the elevated-temperature TSW angle connection behaviour and the results be reported on this aspect.

REFERENCES

Ahmed, B. & Nethercot, D. A. 1997. Prediction of initial stiffness and available rotation capacity of major axis composite flush endplate connections. *Journal of constructional steel research*. 41(1): 31–60.

Al-jabri, K.S. 1999. *The behaviour of steel and composite beam-to-column connections in fire*. PhD thesis. Department of civil and structural engineering. University of Sheffield.

Altmann, R., Maquoi, R. & Jaspart, J. P. 1990. Experimental study of the non-linear behaviour of beam-to-column composite joints. *Proceedings of International Colloquium on Stability of Steel Structures*. Hungary, Budapest.

Ammerman, D. J. & Leon, R. T. 1987. Behaviour of semi-rigid composite connections. Engineering Journal, 2nd quarter: 53–61.

Anderson, D. & Najafi, A. A. 1994. Performance of composite connections: major axis end plate joints. *Journal of constructional steel research*. 31(1): 31–57.

Bailey C.G. & Moore D.B. The influence of local and global forces on column design. *Final report for DETR*. Partners in Technology contract no. CC1494, September 1999.

BS5950 : Part 1: structural use of steelwork in building: code of practice for design: rolled and welded sections. London. BSI; 1990.

Davison, J. B., Lam, D. & Nethercot, D. A. 1990. Semi-rigid action of composite joints. The Structural Engineer. 68(24): 489–499.

European Committee for Standardization (CEN) 1995, "Eurocode 3: Design of Steel Structures, Part 1.2: General Rules – Structural Fire Design", ENV 1993, Brussels, Belgium.

European Committee for Standardization (CEN) 2002, "Eurocode 3: Design of Steel Structures, Part 1.8: Design of Joints", prEN1993-1-8, Brussels, Belgium.

European Committee for Standardization (CEN) 1994, "Eurocode 3: Design of Steel Structures, Part 1.1: Revised Annex J: Joints and Building Frames", (Draft), Document CEN/TC250/SC3 N419E, Brussels, Belgium.

Faella, C., Piluso, V. & Rizzano, G. 1995. Modelling of the moment-rotation curve of welded connections : proposals to improve Eurocode 3 Annex J, C.T.A., *Italian conf. on steel construction. October 1995*. Riva del Garda: Italy.

Faella, C., Piluso, V. & Rizzano, G. 1996. Prediction of the flexural resistance of bolted connections with angles. *IABSE, International colloquium on semi-rigid structural connections. September 1996*. Istanbul, Turkey.

Faella, C., Piluso, V. & Rizzano, G. 2000. *Structural steel semirigid connections: theory, design and software*. Florida, CRC press LLC.

Gattesco N. & Giuriani E. Experimental study on stud shear connectors subjected to cyclic loading. *J Construct Steel Res* 38(1): pp 1–21.

Leston-Jones, L.C. 1997. *The influence of semi-rigid connections on the performance of steel framed structures in fire*. PhD thesis. Department of civil and structural engineering. University of Sheffield.

Loh, H. Y., Uy, B. & Bradford, M. A. 2004a. The effects of partial shear connection in the hogging moment regions of composite beams Part I – Experimental study. *Journal of constructional steel research*. 60 : 897–919.

Loh, H. Y., Uy, B. & Bradford, M. A. 2004b. The effects of partial shear connection in the hogging moment regions of composite beams Part II – Analytical study. *Journal of constructional steel research*. 60 : 921–962.

Maekawa, K., Pimanmas, A. & Okumura, H. 2003. Nonlinear mechanics of reinforced concrete. Chapter 2, 8, 9.

MSC.MARC User Manual, Version 2005. MSC Software Corporation, 2 MacArthur Place, Santa Ana, CA 92707. USA.

Ollgaard J.G., Slutter R.G. & Fisher J. W. 1971, Shear strength of stud connectors in lightweight and normal weight concrete. AISC *Engineering Journal* 8(2): 55–64.

Spyrou S., Davison J.B., Burgess I.W. & Plank R.J. 2004. Experimental and analytical investigation of the compression zone' component within a steel joint at elevated temperatures, *Journal of Constructional Steel Research* 60 : 841–865.

Tschemmernegg F. 1988. On the nonlinear behaviour of joints in steel frames. *Connections in steel structures : behaviour, strength and design*. R. Bjorhovde et al. (eds). Elservier applied science, London.

Steel and Composite Structures – Wang & Choi (eds)
© *2007 Taylor & Francis Group, London, ISBN 978-0-415-45141-3*

Developments in semi-rigid joint moment versus rotation curves to incorporate the axial versus moment interaction

A.A. Del Savio
Imperial College London, London, UK & PUC-Rio, Rio de Janeiro, RJ, Brazil

D.A. Nethercot
Imperial College London, London, UK

P.C.G.S. Vellasco
UERJ – State University of Rio de Janeiro, Rio de Janeiro, RJ, Brazil

S.A.L. Andrade & L.F. Martha
PUC-Rio – Pontifical Catholic University of Rio de Janeiro, Rio de Janeiro, RJ, Brazil

ABSTRACT: Under certain circumstances, beam-to-column joints can be subjected to the simultaneous action of bending moments and axial forces. Although, the axial force transferred from the beam is usually low, it may, in some situations attain values that significantly reduce the joint flexural capacity. Few experimental tests are available and are usually described by their associated moment-rotation curves. An interesting question is how to incorporate these curves into a structural analysis, for the various required axial force levels. The main aim of the present paper is the development of a consistent and simple approach to determine moment versus rotation curves for any level of axial force using a limited set of experiments including the axial versus bending moment interaction.

1 INTRODUCTION

1.1 *Generalities*

Under certain circumstances, beam-to-column joints can be subjected to the simultaneous action of bending moments and axial forces. Although, the axial force transferred from the beam is usually low, it may, in some situations attain values that significantly reduce the joint flexural capacity. These conditions may be found in: Vierendeel girder systems (widely used in building construction because they take advantage of the member flexural and compression resistances eliminating the need for extra diagonal members); regular sway frames under significant horizontal loading (seismic or extreme wind); irregular frames (especially with incomplete storeys) under gravity/horizontal loading; and pitched-roof frames.

On the other hand, with the recent escalation of terrorist attacks on buildings, the study of progressive collapse of steel framed building has been highlighted, as can be seen in Vlassis et al. (2006). Examples of these exceptional conditions are the cases where structural elements, such as central and/or peripheral columns and/or main beams, are suddenly removed, sharply increasing the joint axial forces. In these situations the structural system, mainly the connections, should be sufficiently robust to prevent premature failure modes that may lead to progressive structural collapse.

Unfortunately, few experiments considering the bending moment versus axial force interactions have been reported. Additionally, the available experiments are associated with a small number of axial force levels and associated bending moment versus rotation curves, $M - \phi$. Nevertheless, a question still remains on how to incorporate these effects into a structural analysis. There is a need for $M - \phi$ curves, associated with numerous axial force levels, which accurately represent the joint rotational stiffness.

This has led to the development of an approach to incorporate any moment versus rotation curve from tests including the axial versus bending moment interaction, as well as its evaluation and validation against experiments. This approach is not restricted to the use of experiments, but can be applied to results obtained analytically, empirically, mechanically, and numerically.

As this approach is exclusively based on the use of $M - \phi$ curves, it can be easily incorporated into a nonlinear semi-rigid joint finite element formulation because the moment versus axial force interaction is associated with a specific $M - \phi$ curve. The nonlinear joint finite element formulation does not change. It only requires a rotational stiffness update procedure. This approach has been used to improve the joint finite element model proposed by Del Savio (2004) and Del Savio et al. (2004, 2005), which was initially based on the semi-rigid joint force independence.

1.2 *Component method*

The component method consists of relatively simple joint mechanical models, based on a set of rigid links and spring components. The component method, introduced in Eurocode 3 (2003), can be used to determine the joint's resistance and initial stiffness. Its application requires the identification of active components; evaluation of the force-deformation response of each component; and the subsequent assembly of the active components for the evaluation of the joint moment versus rotation response.

The Eurocode 3 (2003) component method permits the evaluation of the semi-rigid joint's rotational stiffness and moment capacity when subjected to pure bending. However, this component method is still not able to calculate these properties when, in addition to the applied moment, an axial force is also present. Eurocode 3 (2003) suggests that the axial load may be disregarded in the analysis when its value is less than 10% of the beam's axial plastic resistance, but provides no information for cases involving larger axial forces. Even though, the Eurocode 3 (2003) component method has not considered the axial force, its general principles could be used to cover this situation, since it is based on the use of a series of force versus displacement relationships, which only depend on the axial force level, to characterize any component behaviour.

1.3 *Background: experimental and theoretical models*

The study of the semi-rigid characteristics of beam to column connections and their effects on frame behaviour can be traced back to the 1930s, Li et al. (1995). Since then, a large amount of experimental and theoretical work has been conducted both on the behaviour of the connections and on their effects on complete frame performance.

Despite the large number of experiments, they do not cover all possible connection ranges. As an alternative to tests, different methods have been proposed by researchers to predict bending moment versus rotation curves. These methods are usually classified as: empirical, analytical, mechanical (component-based approaches) and numerical (finite element).

Recently, several researchers have paid special attention to joint behaviour under combined bending moment and axial force. The investigators concluded that the presence of the axial force in the joints modifies their structural response and, therefore, should be considered. A number of experimental works deserve mention:

– Guisse et al. (1997) performed tests on six prototypes of column bases with extended endplates with bolts placed outside of the beam height and six tests on flush endplates with bolts inside the beam height. In these tests, the compressive axial force was first applied and kept constant during the test while the bending moment was subsequently increased up to failure.
– Wald et al. (2000) conducted two tests on beam-to-beam and beam-to-column joints. The loading system adopted a proportional increase of axial force and bending moment. However, a test without axial forces was not performed, making it difficult to assess the axial force influence on the joint response.
– Lima et al. (2004) and Simões da Silva et al. (2004) performed tests on fifteen prototypes, i.e. eight flush and seven extended endplate joints. All the tests adopted a loading strategy consisting of an initial application of the total axial force (tension or compression), held constant during the entire test, and the subsequent incremental application of the bending moment.

Regarding the theoretical models recently developed to predict the behaviour of beam-to-column joints under bending moment and axial force, it is possible to mention:

– Jaspart (1997, 2000), Finet (1994) and Cerfontaine (2001, 2004) have applied the principles of the component method to establish design predictions of the M-N interaction curves and initial stiffness.
– Simões da Silva & Coelho (2001), using the same general principles, have proposed analytical expressions for the full non-linear response of a beam-to-column joint under combined bending and axial forces.
– Sokol et al. (2002) proposed an analytical model to predict the behaviour of joints subjected to bending moment and axial force for proportional loading.

Table 1 presents a summary of recent studies carried out to investigate joint behaviour when subjected to bending moment and axial force.

2 CORRECTION FACTOR

2.1 *Concepts of the correction factor*

The Correction Factor has initially been proposed by Del Savio et al. (2006) to consider the bending

Table 1. Summary of studies of joints subjected to bending and axial force, Lima et al. (2004).

Authors	Analysis type
Finet (1994)	AM*
Jaspart (1997, 2000)	AM* and ET**
Cerfontaine (2001, 2004)	AM*
Simões da Silva & Coelho (2001)	AM*
Simões da Silva et al. (2001)	AM*
Lima (2003)	ET**
Wald & Svarc (2001)	AM*
Sokol et al. (2002)	AM*

*AM = Analytical model.
**ET = Experimental tests.

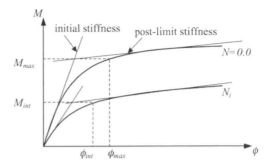

Figure 1. Correction factors parameters for the moment and rotation axis.

moment versus axial force interaction, by scaling original moment values present in the moment versus rotation curves (disregarding the axial force effect). This strategy shifts this curve up or down depending on the axial force level. However, as it only modifies the bending moment axis, it is not able to fully describe the bending moment versus rotation associated with different axial force levels. This fact is highlighted when the joint is subject to a tensile axial force, where there is a significant difference, principally, in terms of initial stiffness.

Aiming to improve the Correction Factor's basic idea, the Correction Factor was divided into two parts: one for the moment axis and another for the rotation axis. Both corrections are in principle independent, and do not depend on the moment versus axial force interaction diagram, as was the case for the initial idea presented by Del Savio et al. (2006). It is now only a function of the moment versus rotation curves for different axial force levels.

2.2 Evaluation of the correction factor

As previously noted, there are two corrections, one to the moment axis and another to the rotation axis.

As a general approach, the Correction Factor for the moment axis is evaluated in terms of the bending moment versus rotation curves considering the axial force effect. Using the design bending moment ratio and considering the axial force effect, the Correction Factor for the moment axis, CF_M, can be evaluated by:

$$CF_M = \frac{M_{int}}{M_{max}} \qquad \begin{aligned} M_{int} &= f\left(Mx\phi(N_i)\right) \\ M_{max} &= f\left(Mx\phi(0.0)\right) \end{aligned} \tag{1}$$

where $Mx\phi$ = bending moment versus rotation curve; M_{int} = design bending moment for the $M - \phi(N_i)$ considering the axial force; M_{max} = design bending moment for cases without axial forces; and N_i = axial force present in the i interaction. M_{int} and M_{max} can be

determined according to Eurocode 3 part 1.8 (2003), through the intersection between two straight lines, one parallel to the initial stiffness and another parallel to the $M - \phi$ curve post-limit stiffness, Figure 1.

Similarly, the rotation axis Correction Factor, CF_ϕ, is evaluated using the design rotation ratio, i.e.:

$$CF_\phi = \frac{\phi_{int}}{\phi_{max}} \qquad \begin{aligned} \phi_{int} &= f\left(Mx\phi(N_i)\right) \\ \phi_{max} &= f\left(Mx\phi(0.0)\right) \end{aligned} \tag{2}$$

where ϕ_{int} = design rotation related to M_{int}; and ϕ_{max} = rotation related to M_{max}. Both design rotations are found by tracing a horizontal straight line at the design moment level until it reaches the $M - \phi$ curve. At this point a vertical straight line is drawn until it intersects the rotation axis, Figure 1.

With the Correction Factors evaluated for both the moment (Equation 1) and rotation (Equation 2) axes, they are incorporated into the joint structural response considering the moment versus axial force interaction, modifying the $M - \phi$ curve for the zero axial force case, i.e.:

$$Mx\phi(N=0) \rightarrow Mx\phi(N_i)$$
$$Mx\phi(N_i) = Mx\phi(M_{N=0} \times CF_M, \phi_{N=0} \times CF_\phi) \tag{3}$$

Basically, the $M - \phi$ curve point coordinates, $M_{N=0}$ and $\phi_{N=0}$, for the case without axial forces, are multiplied by the Correction Factors, where CF_M and CF_ϕ, multiply the moment and the rotation axis coordinates, respectively.

However, using only a pair of Correction Factors throughout the whole $M - \phi$ curve, for the case without axial forces, does not provide a good approximation to $M - \phi$ curve considering the axial force, because it is very sensitive to the adopted initial stiffness and post-limit stiffness angles.

This motivated the division of the $M - \phi$ curve into three segments with different pairs of Correction Factors. This division was made for two-third, one, and 1.1 times the design moment, as shown in Figure 2.

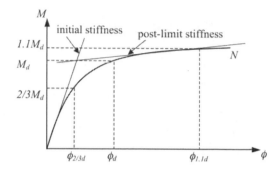

Figure 2. Correction factor strategy approach using a three segment division of the $M - \phi$ curve.

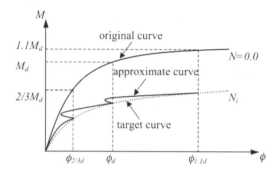

Figure 3. Approximate $M - \phi$ curve using three correction factor pairs.

With this division, the Correction Factors cannot be applied as presented in Equation 3. This is justified, in fact, because they would provoke two abrupt variations of stiffness throughout the approximate $M - \phi$ curve at around the point of intersection of the approximate curve with the vertical lines at the points $\phi_{2/3d}$ and ϕ_d, Figure 3. This is due to the use of three different pairs of Correction Factors.

To avoid the problem of abrupt alterations of stiffness presented in Figure 3, it is proposed, in Figure 4, to use a tri-linear representation of the $M - \phi$ curve.

With this tri-linear approach, Figure 4, the moment levels of the required $M - \phi$ curve, associated with a certain axial force level (N_i), can be evaluated by:

$$M_p = \left(M_{N,p} - M_{0,p}\right)\frac{N_i}{N} + M_{0,p}$$

$$p = 2/3M_d;M_d;1.1M_d$$

$$0 < N_i \leq N \rightarrow tensile\ axial\ force \qquad (4)$$

$$N \leq N_i < 0 \rightarrow compressive\ axial\ force$$

where M_p = evaluated moment for the new $M - \phi$ curve; $M_{N,p}$ = moment on the reference $M - \phi$ curve considering the axial force; $M_{0,p}$ = moment on the

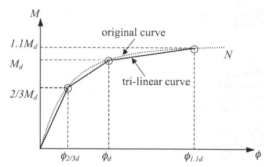

Figure 4. Tri-linear representation of the $M - \phi$ curve approach.

reference $M - \phi$ curve without axial forces; and N = axial force load level associated with the reference $M - \phi$ curve.

Likewise, the rotations of the evaluated $M - \phi$ curve, for the associated N_i, can be calculated by:

$$\phi_p = \left(\phi_{N,p} - \phi_{0,p}\right)\frac{N_i}{N} + \phi_{0,p}$$

$$p = 2/3M_d;M_d;1.1M_d \qquad (5)$$

$$0 < N_i \leq N \rightarrow tensile\ axial\ force$$

$$N \leq N_i < 0 \rightarrow compressive\ axial\ force$$

where ϕ_p = evaluated rotation for the new $M - \phi$ curve; $\phi_{N,p}$ = rotation on the reference $M - \phi$ curve considering the axial force; and $\phi_{0,p}$ = rotation on the reference $M - \phi$ curve without axial force effects.

3 APPLICATION OF THE CORRECTION FACTOR

3.1 Uses and input of the correction factor

The main focus of the proposed Correction Factor was to determine $M - \phi$ curves for any axial force level from the $M - \phi$ curve for zero axial force. The quality of the approximations obtained will depend on the quality of $M - \phi$ used as input to the method. This method requires three $M - \phi$ curves, one disregarding the axial force effect and two considering the compressive and tensile axial force effects.

3.2 Example of application and validation

The goal of this section is to demonstrate how to use the Correction Factor method to obtain $M - \phi$ curves for any axial force level, as well as to validate it, using experimental tests carried out by Lima et al. (2004) and Simões da Silva et al. (2004), on eight flush endplate joints. The geometric properties of the flush endplate

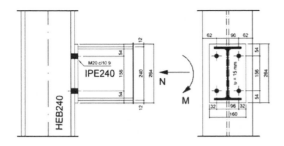

Figure 5. Flush endplate joint layout, Simões da Silva et al. (2004).

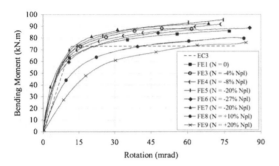

Figure 6. Experimental moment versus rotation curves, Simões da Silva et al. (2004).

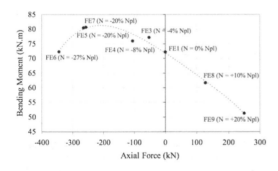

Figure 7. Flush endplate bending moment versus axial force interaction diagram, Simões da Silva et al. (2004).

analysed, the $M - \phi$ curves describing the experimental behaviour of each test, and the bending moment versus axial force interaction diagram are shown in Figures 5–7, respectively.

The experimental data, Figure 6, provides the necessary input for the Correction Factor method. The minimum input is composed of two $M - \phi$ curves, disregarding and considering either the tensile or compressive axial force. The flush endplate joint, tested by Simões da Silva et al. (2006), exhibited a decrease in the moment resistance for the tensile axial forces whilst achieving the highest moment resistance for the compressive axial force of 20% of the beam's

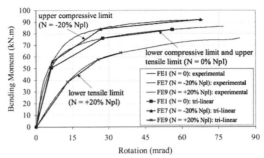

Figure 8. Tri-linear approach for the experimental $M - \phi$ curves.

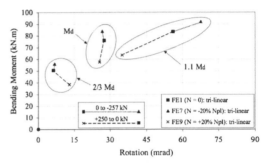

Figure 9. Paths used to define the way to find any $M - \phi$ curve contained within these limits imposed on the correction factor.

axial plastic resistance (see Figure 7, FE7). Therefore, three $M - \phi$ curves: FE1 ($N = 0$); FE7 ($N = -257\,\text{kN}$, -20% Npl), and FE9 ($N = 250\,\text{kN}$, $+20\%$ Npl), where Npl is the beam's axial plastic resistance, have been used. These three experimental curves and their tri-linear approximations are shown in Figure 8.

Tri-linear $M - \phi$ curves, Figure 8, are used to define paths between each curve at points $2/3M_d$, M_d and $1.1M_d$, Figure 9. These paths will be used to guide the Correction Factor throughout the range of axial force levels to determine the required set of $M - \phi$ curves.

Subsequently, Figures 10–12 show the results obtained using the Correction Factor approach for three experimental $M - \phi$ curves: FE8, FE3 and FE4. The evaluated $M - \phi$ curve requires two $M - \phi$ reference curves defining the maximum and minimum limit for the associated axial force level. Figure 13 gives the complete results of the Correction Factor approach.

4 RESULTS AND DISCUSSION

Three experimental $M - \phi$ curves, by Lima et al. (2004) and Simões da Silva et al. (2004), were subjected to the Correction Factor concept, as can be seen in Figures 10–13.

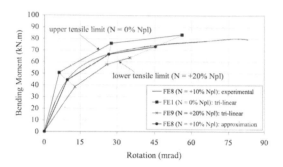

Figure 10. FE8 $M - \phi$ curve approximation, considering a tensile force of 10% of the beam's axial plastic resistance.

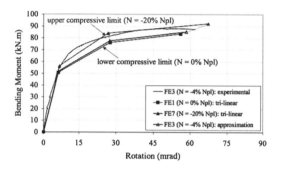

Figure 11. FE3 $M - \phi$ curve approximation, considering a compressive force of 4% of the beam's axial plastic resistance.

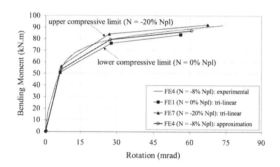

Figure 12. FE4 $M - \phi$ curve approximation, considering a compressive force of 8% of the beam's axial plastic resistance.

Figure 10 illustrates an approximation for FE8 $M - \phi$ curve that considers a tensile force of 10% of the beam's axial plastic resistance. This approximation was obtained from two tri-linear $M - \phi$ curves, disregarding and considering a tensile force of 20% of the beam's axial plastic resistance. This approximation was very close to FE8 $M - \phi$ test curve.

Figures 11 and 12, present approximations for FE3 and FE4 $M - \phi$ curves that consider compressive forces of 4% and 8% of the beam's axial plastic resistance, respectively. These approximations were

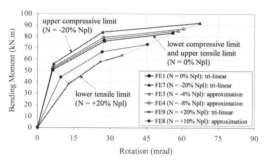

Figure 13. Final correction factor approach curves.

obtained from two tri-linear $M - \phi$ curves, disregarding and considering a compressive force of 20% of the beam's axial plastic resistance. The approximation acquired for FE4 $M - \phi$ curve, Figure 12, was relatively close to the experimental curve. However, for FE3 $M - \phi$ curve, Figure 11, the obtained estimation was not as good. This was due to the differentiable behaviour of this experimental curve when compared to the others. It is possible to observe in Figure 7 that there is an increase in the flush endplate joint moment capacity from FE1 $M - \phi$ curve ($N = 0\% Npl$) to FE7 $M - \phi$ curve ($N = -20\% Npl$). However, within this range, with 4% beam's compressive plastic resistance the final moment is larger than the maximum moment obtained with the 8% test. A possible reason for that could be related to problems with this specific experimental test, i.e., measuring errors, eccentricities in the construction and assembly of this test, etc.

5 CONCLUSIONS

The main goal of this work was to present an approach to determine any moment versus rotation curve from experimental tests, including the axial versus bending moment interaction. It can also be applied to results obtained analytically, empirically, mechanically, and numerically. Due to its simplicity and the fact that its basis is $M - \phi$ curves that already consider the moment versus axial force interaction, it can be easily incorporated into a nonlinear semi-rigid joint finite element formulation. The use of the proposed approach does not change the basic formulation of the non-linear joint finite element, only requiring a rotational stiffness update procedure.

This approach is a simple and accurate way of introducing semi-rigid joint experimental test data into structural analysis, through of $M - \phi$ curves.

Application and validation of this approach to obtain $M - \phi$ curves for three axial force level, using experimental tests carried out by Lima et al. (2004)

and Simões da Silva et al. (2004), on eight flush end-plate joints, were performed with results close to the experiments.

ACKNOWLEDGEMENTS

The authors gratefully acknowledge the financial support provided by the Brazilian National and State Scientific and Technological Developing Agencies: CNPq, CAPES and FAPERJ.

REFERENCES

Cerfontaine, F. 2001. Etude analytique de l'interaction entre moment de flexion et effort normal dans les assemblages boulonnés. *Construction Métalique*, 4: 1–25 (in French)

Cerfontaine, F. 2004. Etude de l'interaction entre moment de flexion et effort normal dans les assemblages boulonnés. *Thèse de Docteur en Sciences Appliquées, Faculté des Sciences Appliquées, University of Liège, Belgium.* (in French)

Del Savio, A.A. 2004. Computer Modelling of Steel Structures with Semi-rigid Connections. *MSc. Dissertation,* Civil Eng. Depart. – PUC-Rio, Brazil, (in Portuguese), 152p.

Del Savio, A.A., Andrade, S.A. de, Vellasco, P.C.G.S. & Martha, L.F. 2004. A Non-Linear System for Semi-Rigid Steel Portal Frame Analysis. *7th Int. Conf. Comp. Struct. Tech., CST2004,* 1: 1–12.

Del Savio, A.A., Andrade, S.A. de, Vellasco, P.C.G.S. & Martha, L.F. 2005. Structural Evaluation of Semi-Rigid Steel Portal Frames. *Proceedings of the Eurosteel 2005, Forth European Conference on Steel and Composite Structures, Maastricht.* vol. A.: 4.49–4.56.

Del Savio, A.A., Andrade, S.A. de, Vellasco, P.C.G.S., Martha, L.F. & Lima, L.R.O. de. 2006. Semi-Rigid Portal Frame Finite Element Modelling Including the Axial Versus Bending Moment Interaction in the Structural Joints. *International Colloquia on Stability and Ductility of Steel Structures – SDSS'06, Lisboa,* vol. 1: 1–8.

Finet, L. 1994. Influence de l'effort normal sur le calcul des assemblages semi-rigides. *CUST, Clermont-Ferrand. Travail de Fin d'Etudes, CRIF, Liege.* (in French)

Guisse, S., Vandegans, D. & Jaspart, J.-P. 1997. Application of the component method to column bases. Experimentation and development of a mechanical model for characterization. *Rapport CRIF, MT 295, CRIF, Liege.*

Jaspart J.-P. 1997. Recent advances in the field of steel joints. Column bases and further configurations for beam-to-column joints and beam splices. *Department MSM, University of Liege.*

Jaspart J.-P. 2000. General report: session on connections. *Journal of Constructional Steel Research,* vol. 55, n.1–3: 69–89.

Li, T.Q., Choo, B.S. & Nethercot, D.A. 1995. Connection Element Method for the Analysis of Semi-rigid Frames. *Journal Constructional Steel Research*, 32: 143–171.

Lima, L.R.O. de. 2003. Behaviour of endplate beam-to-column joints under bending and axial force. *Ph.D. Thesis. PUC-Rio, Pontifical Catholic University, Civil Engineering Department, Rio de Janeiro, Brazil.* [in Portuguese]

Lima, L.R.O. de, Simões da Silva, L., Vellasco, P.C.G.S. & Andrade, S.A. de. 2004. Experimental evaluation of extended end-plate beam-to-column joints subjected to bending and axial force. *Engineering Structures.* Vol. 26, No. 10: 1333–1347.

Simões da Silva, L. & Coelho, A.G. 2001. An analytical evaluation of the response of steel joints under bending and axial force. *Computers & Structures,* 79: 873–881.

Simões da Silva, L., Lima, L.R.O. de, Vellasco, P.C.G.S. & Andrade, S.A. de. 2001. Experimental and numerical assessment of beam-to-column joints under bending and axial force. *First International Conference on Steel and Composite Structures, Pusan, Seoul: Techno Press,* vol. 1: 715–722.

Simões da Silva, L., Lima, L.R.O. de, Vellasco, P.C.G.S. & Andrade, S.A. de. 2004. Behaviour of flush end-plate beam-to-column joints under bending and axial force. *Steel and Composite Structures,* vol. 4, n. 2: 77–94.

Sokol, Z., Wald, F., Delabre, V., Muzeau, J.P. & Svarc, M. 2002. Design of Endplate Joints Subject to Moment and Normal Force. *Third European Conference on Steel Structures – Eurosteel 2002, Coimbra, Cmm Press*: 1219–1228.

Vlassis, A.G., Izzuddin, B.A., Elghazouli, A.Y. & Nethercot, D.A. 2006. Design Oriented Approach for Progressive Collapse Assessment of Steel Framed Buildings. *Structural Engineering International* (Report), SEI Editorial Board: 129–136.

Wald, F., Pertold, J. & Xiao, R.Y. 2000. Embedded steel column bases. I. Experiments and numerical simulation. *Journal Constructional Steel Research*, 56: 253–270.

Wald, F. & Svarc, M. 2001. Experiments with endplate joints subject to moment and normal force. Contributions to experimental investigation of engineering materials and structures. *CTU Reports No.:2-3, Prague*: 1–13.

Steel and Composite Structures – Wang & Choi (eds)
© *2007 Taylor & Francis Group, London, ISBN 978-0-415-45141-3*

Numerical analysis and experimental studies concerning the behaviour of steel and steel-concrete composite joints under symmetrical and asymmetrical loads

D. Dan, V. Stoian & T. Nagy - Gyorgy

"Politehnica" University of Timisoara, Romania

ABSTRACT: At the design of composite structures located in seismic areas the designer must conceive the failure mechanism of the structure, thus some inelastic deformations must be concentrated in predefined zones, in order to have an efficient dissipation of the seismic energy. A theoretical approach and an experimental testing program were developed at the "Politehnica" University of Timisoara for a specific steel and composite (steel-concrete) joint, used at construction work of an administrative building in Timisoara. The numerical analysis was performed in the elastic and post-elastic range, via a numerical method. The elastic analysis was realised used dedicated software SAP2000, for the steel joint. Dedicated nonlinear software was used for post-elastic analysis of the composite steel-concrete joint. The failure modes were studied for steel and composite joint. Two loading hypotheses were considered: symmetrical vertical eccentric loads and asymmetrical vertical eccentric loads. The paper presents a comparative study between numerical analysis (elastic and post-elastic range) for the steel and steel-concrete composite joints symmetrical loads and asymmetrical loads. The results were also compared with the recorded data of the experimental test made on steel and composite joints, under symmetrical loads.

1   INTRODUCTION

In the case of the steel and the steel-concrete composite structures the designer must conceive special details in accordance with the specific codes: EC3 and EC4.

An experimental test program for a specific steel-concrete composite joint, used at construction work of an administrative building, was developed at the "Politehnica" University of Timisoara.

Starting with the joint type used, four joints were considered as experimental specimens. The joints were initially analysed together with their connections – the beams and the columns in order to determine the dimensions of the joint components, thus satisfying the desired collapse mechanism at the joint zone.

As is known in reality the plasting hinges must appear outside of the joints into the beams. The aim of the study was to obtain the collapse in the joint panel rather than outside the joint, in order to compare the bending moment resistant of steel joint with the bending moment resistant of composite joint. Were studied also the failure mechanism of the steel joint and compared with the composite one. Initially, the joint was design by using the EC4 code. Then a numerical study was performed in the elastic and post elastic range. Finally the experiment was performed by using special testing equipment, and the international recommended testing procedures.

2   STRUCTURAL JOINT TYPE

Due to the technological process, a composite structure is initially a steel structure. After placing the reinforcement and the concrete casting the structure becomes a composite one.

The type of studied joint was used at a multi-storey building in Timişoara, consisting of 9 and partially 12 storeys. A general view of building is represented in Figure 1.

The entire structure is realised as a steel-concrete composite construction. The structure is realised as space skeleton bar structure using plane frames placed on two orthogonal directions, being connected through the floor slabs. The structural solution is justified by the span width with unexaggerated cross sectional dimensions for the columns, adequate lateral stiffness and cost effective fire protection due to the presence of the concrete.

The detail of the structural steel for composite joint, during construction is represented in Figure 2.

As it is known the structure becomes a composite one after the longitudinal reinforcement and the transversal reinforcement are placed on site and the concrete are cast in to the column mould.

Some aspects concerning the details of composite column and joint during the constructional work is presented in Figure 2 and 3.

Figure 1. Front view – Multi-storey building in Timişoara.

Figure 2. Details of structural steel for composite joint.

3 NUMERICAL ANALYSIS OF STEEL AND STEEL - CONCRETE COMPOSITE JOINT CALIBRATION OF THE MODEL

3.1 Numerical analysis and calibration of steel joint model

In order to evaluate the stress state in the joint and the behaviour study of the dimensioning element, on the geometrical dimensions basis, some numerical analyses have been done using the finite element method. In

Figure 3. Details of reinforcements for composite joint.

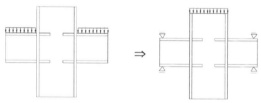

Figure 4. Schematization of the real load case and experimental testing.

the first stage the SAP 2000 numerical analysis programme was used, the modelling been done by type SHELL finite elements, for the structural steel in the joint.

3.1.1 Symmetrical loads

To have a clearer view of the stress state in the joint, the model analysed was done according to the testing mode. Taking into consideration the possibility of making an experimental test, we drew the conclusion that the instruments that we had at our disposal allow the loading of the column and the mounting of some joint supports at the extremity of the two beams that concur in the structural joint. In fact this loading type simulates the real situation when the loading is actually on the beams. The schematization of the testing mode and of the actual loading is presented in Figure 4.

The results of the numerical analysis on the structural steel of the experimental element, obtained initially as a result of the design, are presented in Figure 5.

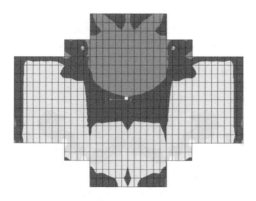

Figure 5. Stresses σ_{max} for steel joint (mid-plane view).

By the analysis of the diagrams of the efforts in the joint, we can observe that there is a concentration of stresses in the vertical stiffeners which connect the beam to the column. The value of the maximum stresses in the stiffeners is of 450 N/mm^2, and in the web of the column or in the beam of $\sim$300 N/mm^2.

These observations lead to the idea that in the case of an experimental testing on a model, made up as above, there is the possibility of tearing outside the joint, starting from the vertical stiffeners and continuing with the beam.

Because the purpose of the testing was the study of the joint failure mode and the checking of its bearing capacity, in the situation presented above, it was suggested the increase of the beam bearing capacity by increasing its flange width and its web as well, and by maintaining the column section and the height of the beam respectively.

The purpose of the testing being to obtain information on the stress state inside the joint and thus to cause the failure in the joint, the decision was to eliminate the vertical stiffeners, which became useless in this case. The vertical stiffeners are useful in real structures because they increase the bearing capacity in the joint zone, the plastic hinge taking place in the beam and not in the joint.

In Figure 6 we present the issostresses, obtained after the elimination of the stiffeners and after remaking the numerical analysis.

3.1.2 Asymmetrical loads

To have a general view of the stress state in the joint, for asymmetrical loads the model analysed was done according to the testing mode. The schematization of the testing mode for asymmetrical loading is presented in Figure 7.

Starting with the geometrical dimensions obtained at the previous analysis for asymmetrical load cases the new numerical analysis on the structural steel of the experimental element was performed. The results obtained are presented in Figure 8.

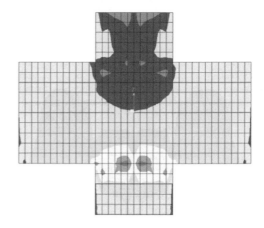

Figure 6. Isostresses σ_{max} for the steel joint.

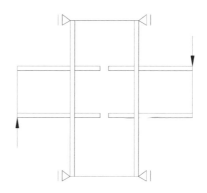

Figure 7. Schematization of the experimental test asymmetrical loads.

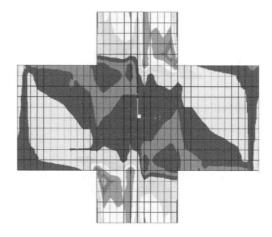

Figure 8. Isostresses σ_{max} for the steel joint asymmetrical load (mid-plane view),

391

3.2 Numerical analysis of composite steel-concrete joint

The evaluation of the stress state in the composite joint elements has been done after several numerical analyses in the post elastic range using nonlinear analysis software. A vertical section was considered in the mid-plane of the joint, practically in the middle of the joint panel axis. In this case it was assumed that the joint is in a plane stress state.

3.2.1 Symmetrical loads

The finite element mesh of composite steel concrete joint was done at the level of mid-plane of joint in vertical direction. All the constitutive elements (structural steel, longitudinal reinforcements, stirrups, stiffeners etc) were considered in this analysis.

The analyses made on the composite joint showed a similar state of distribution of the efforts inside the joint, as the distribution obtained for steel joint, but the stress values in composite being slightly lower. This can be explained if we take into account the fact that at the analysis of the composite joint all the components were considered, i.e. the structural steel, the concrete, the transversal and the longitudinal reinforcements.

The crack evolution and distribution for symmetrical load is presented in Figure 9.

3.2.2 Asymmetrical loads

Another nonlinear analysis was performed for the vertical section and asymmetrical load case. As it can be seen in the Figure 10, the crack distribution is similar to those of a typical reinforced concrete joint.

4 EXPERIMENTAL TESTS UNDER SYMMETRICAL LOAD

All the experimental tests were performed using the procedure indicated by ECCS. The load was applied at the top of column for each tested element. The tests were controlled using a displacement device of a hydraulic actuator.

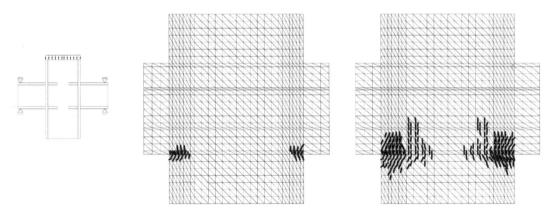

Figure 9.   Evolution and distribution of cracks – symmetrical load.

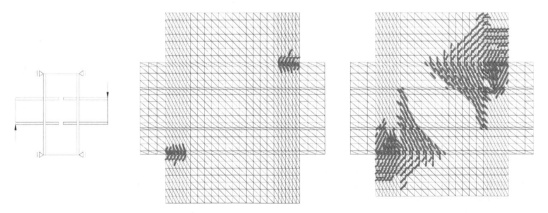

Figure 10.   Evolution and distribution of cracks – asymmetrical load.

Four specimens were tested in the research program. Two of the specimens were steel joints (structural steel) and the others two were composite steel-concrete joins. Two tests were performed as classical push over increase displacement tests in order to evaluate the conventional limit for the steel joint and for the composite joint respectively. The other two tests were cyclic with increase displacement.

4.1 Monotonic displacement increase test

In order to record the behaviour of the tested joints a basic instrumentation was used for both elements. The instrumentation consisted in displacement transducers, inclinometers and strain gauges (Fig. 11). Using recorded data from the monotonic displacement increase tests made on the steel joint (SJ1) and the composite joint (CJ1) there were evaluated the limit of the elastic range F(kN) and the corresponding displacement e_y (mm). The elastic limit was calculated using the push over load-displacement diagram.

The behaviour of the joints was analysed by comparing the characteristic diagram moment rotation recorded at the exterior face of the joint (Fig. 12).

4.2 Cyclic tests

The history of the cyclic tests was established using the elastic limit obtained in the monotonic tests for both elements, and the ECCS procedure.

In the photos represented in Figures 13 and 14 can be observed a comparative study between the failure mechanism of steel joints and steel-concrete composite joints.

The comparative study between the experimental elements is based on moment rotation characteristic diagram recorded at the lateral face of joints, represented in Figure 15. The behaviour parameters are presented in Table 1.

5 CONCLUSIONS

Using the results from the experimental tests the following conclusions were formulated for the composite joints:

– the behaviour of steel and composite joint under symmetrical loads is similar;
– for the tested elements the stiffness of composite joint increase with 30% due to presence of concrete;

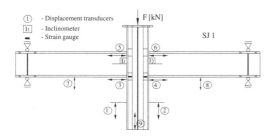

Figure 11.   Basic instrumentation used in monotonic test of the steel joint and composite joint (SJ1, CJ1).

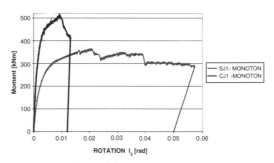

Figure 12.   Moment – rotation diagram (SJ1/CJ1).

a. General view after testing

b. Tear of vertical stiffener

c. Large vertical crack in joint panel

Figure 13.   Failure mechanism of steel joint.

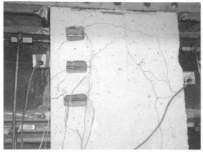

a. Front view of cracks distribution   b. Tear of vertical stiffener   c. Small vertical crack in joint panel

Figure 14. Failure mechanism of composite joint.

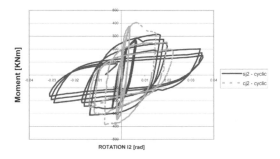

Figure 15. Moment – rotation diagram for steel and composite joints.

Table 1. Basic parameters for cyclic tests.

	Steel joint (SJ2)/ Cyclic test		Composite joint (CJ2)/Cyclic test	
Maximum bending moment [kNm]	+315.5	−310.39	+405.6	−382.8
Maximum displacement [mm]	+41.17	−36.8	+27.36	−25.51
Ultimate rotation [mrad]	+35.2	−29.3	+22.3	−13.8
Elastic limit e_y [mm]	6.18		6.38	
Experimental bending moment (elastic limit) [kNm]	+201.4	−215.8	+273.4	−284.9

- the simplified tendency to take into account only the structural steel is inadequate;
- in the composite joint a redistribution of the stresses occurs between the concrete, reinforcement and structural steel;
- the connection between the structural steel flanges and the web is in a zone where the stress distribution must take into account the presence of the reinforcement and the concrete and therefore the stress state is far from a pure steel stress state;
- the buckling of joint panel and vertical stiffeners in compression zone at the composite joint is avoided due to presence of concrete and transversal reinforcement (stirrups) in the joint;
- the presence of the concrete in the joint has the effect of increasing the load bearing capacity of the joint;
- the tested joints are dissipative, with similar values of energy dissipation;
- the cracks distributions on numerical analysis are similar with the experimental distribution of cracks;
- the concrete inside the composite joint after testing is crushed although at the exterior face the concrete has only smeared cracks;
- due to the technology of welding the vertical stiffeners cannot be weld by complete penetration; it is considered that the vertical stiffeners play a significant part to the increase of the joint bearing capacity, the weak point being the welding at the column flanges;
- the results of numerical analysis were confirmed during the testing process, the failure mechanism consisting in tearing of vertical stiffeners between column and beam flanges, tearing of joint panel, cracking of tensioned zone and crushing of compressed zone of concrete;
- for the composite joint the crack distribution of concrete is similar to those of a typical reinforced concrete joint.

The observations made during the tests and the joint failure mode lead to certain considerations which can improve the joint simplified calculus mode.

REFERENCES

Eurocode 3 – Design of steel structure –EN 1993.1.1
Eurocode 4 – Design of composite steel and concrete structures – EN 1994.1.1.

ECCS – Recommended testing procedure for assessing the behaviour of structural steel elements under cyclic loads.

Dan D. – Contribution to the design of the composite steel-concrete elements, Ph. D. Thesis, Politehnica University of Timişoara, 2002.

Stoian V., Dan. D., Monotonic and cyclic behaviour of steel-concrete composite joints, STESSA 2003 Behaviour of steel structures in seismic areas, Italy, Napoli 2003.

Sung-Mo Choi., Yen-Sang Yun, Jin-Ho Him – Experimental study on seismic performance of concrete filled tubular square column-to-beam connections with combined cross diaphragm, Journal Steel and Composite Structures, Vol. 6, No. 4, 2006.

Dan D., Stoian V., Nagy T., Theoretical and experimental studies concerning the load bearing capacity of steel and composite joints, Brasov, Romania, 2006.

Steel and Composite Structures – Wang & Choi (eds)
© *2007 Taylor & Francis Group, London, ISBN 978-0-415-45141-3*

Comparison of shear connection design methods in continuous steel-concrete composite beams

A. Dall'Asta, G. Leoni & A. Zona

Dipartimento di Progettazione e Costruzione dell'Ambiente, University of Camerino, Ascoli Piceno, Italy

ABSTRACT: The behavior of continuous steel-concrete composite beams designed for the same flexural bearing capacity but with different shear connection distributions is analyzed by means of a nonlinear finite element model including partial interaction. The numerically simulated response results permit to study and quantify the effect of shear connection distributions on the global and local behavior both under service loads and at collapse. The analysis focuses its attention on the ductility requirements for the shear connectors when varying the connection design approach and distribution.

1 INTRODUCTION

Continuous beams are widely adopted in steel-concrete composite construction for both building floors and bridge decks, owing to the many advantages arising from combining the two materials (Oehlers & Bradford 2000). Many aspects related to the service state and ultimate state behavior of continuous composite beams have thoroughly been investigated in the past, by using both experimental tests and linear and nonlinear numerical modeling (see for example Viest et al. 1997, Oehlers & Bradford 2000, Spacone & El-Tawil 2004). Some other aspects deserve more attention since reduced information is available. Among them is the evaluation of the effects of various design methods and distributions of shear connectors between the reinforced concrete slab and the steel beam.

In this work the behavior of continuous steel-concrete composite beams designed for the same flexural bearing capacity but with different shear connection distributions is analyzed. For this purpose a finite element code (Dall'Asta & Zona 2002) specifically developed by the authors for the nonlinear analysis of steel-concrete composite beams is employed. The adopted finite element numerical model includes material nonlinearity of concrete, beam steel and reinforcement steel, as well as slab-beam nonlinear partial interaction due to the deformable shear connection. The inclusion of the partial interaction in the composite beam model provides information not only on the slab-beam interface slip and shear force but also enables to model the failure of shear connectors. In this way it is possible to analyze and quantify the effect of shear connector distributions on the global and local response

of continuous steel-concrete composite beams, both under service load levels and at collapse. Particular attention is focused on the ductility requirements on the shear connectors when varying the connection design approach and distribution.

2 SHEAR CONNECTION DESIGN METHODS

2.1 *Basic concepts and assumptions*

The shear connection in steel-concrete composite beams provides composite action by transferring the longitudinal shear force between the reinforced concrete slab and the steel beam. Full or partial shear connections are possible. A span of a beam has full shear connection when increase in the number of shear connectors would not increase the design bending resistance of the member. Otherwise, the shear connection is partial. In this work design methods for full shear connection only are studied. No effects of local and global instabilities are included, thus materials can fully develop inelastic deformations up to their ultimate strain. In the numerical application presented symmetric two-span continuous beams under uniform distributed load are used as testbed structural configuration. Due to the problem symmetry only the first span is analyzed. Propped construction only is considered.

In two-span continuous beams a sagging region of length $2L_1$ and a hogging region of length L_3 can be identified (Figure 1) once that the distribution of the bending moment is known. Various design methods for the shear connection can be employed. In the sequel two approaches are adopted, namely: (i) the approach proposed in Eurocode 4 Part 1 (CEN 2004) for beams in which plastic theory is used for resistance

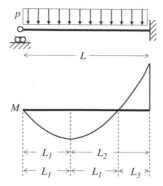

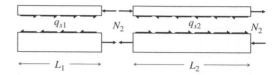

Figure 2. Balance of axial forces and longitudinal shear forces in the EC4-1 design approach.

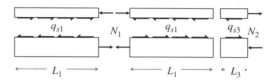

Figure 3. Balance of axial forces and longitudinal shear forces in the alternative design approach.

Figure 1. Structural configuration and bending moment distribution under uniform distributed load (first span of a symmetric and symmetrically loaded two-span continuous beam).

of cross sections; (ii) an alternative approach based on alike considerations except for a different choice of the tracts in which longitudinal forces are balanced.

2.2 Eurocode 4 Part 1 design approach

In the Eurocode 4 Part 1 design approach two tracts are considered: (i) from the external support to the maximum sagging bending moment (tract of length L_1); (ii) from the maximum sagging bending moment to the internal support (tract of length L_2).

In the sagging cross sections the maximum longitudinal force N_1 that the shear connection may possibly be required to transfer between the slab and the steel beam is given by:

$$N_1 = \min\{N_a; N_c\} \tag{1}$$

where N_a and N_c are the ultimate plastic axial forces of the steel beam in tension and of the reinforced concrete slab in compression respectively.

In the hogging cross sections the maximum longitudinal force N_2 that the shear connection may possibly be required to transfer is given by:

$$N_2 = \min\{N_a; N_s\} \tag{2}$$

where N_s is the ultimate plastic axial force of the reinforcement of the slab in tension.

A full shear connection must be able to develop shear forces q_{s1} and q_{s2} so to satisfy the longitudinal balance conditions (Figure 2):

$$\int_0^{L_1} q_{s1}(z)\mathrm{d}z = N_1 \tag{3}$$

$$\int_{L_1}^{L_1+L_2} q_{s2}(z)\mathrm{d}z = N_1 + N_2 \tag{4}$$

with z the abscissa along the beam axis of a reference system having its origin in the external support.

2.3 Alternative design approach

An alternative design approach is studied for comparative purposes. Three tracts are considered: (i) from the external support to the maximum sagging bending moment (tract of length L_1); (ii) from the maximum sagging bending moment to the point of null bending moment (tract of length L_1); (iii) from the point of null bending moment to the internal support (tract of length L_3).

As in the previous case N_1 and N_2 are the ultimate plastic axial forces that the shear connection may possibly be required to transfer between the slab and the steel beam in the sagging and hogging cross sections respectively. A full shear connection must be able to develop a shear force q_{s1} in the first two tracts and a shear force q_{s3} in the third tract (Figure 3) satisfying the longitudinal balance condition in the sagging tract given by Equation (3) and the balance condition in the hogging tract:

$$\int_{2L_1}^{2L_1+L_3} q_{s3}(z)\mathrm{d}z = N_2 \tag{5}$$

3 NUMERICAL SIMULATIONS

3.1 Tested beams

The two approaches considered in this study are adopted for the design of different shear connection distributions in a continuous beam having the cross section shown in Figure 4 and kept constant along the two spans both of length $L = 40\,\mathrm{m}$. The $5000 \times 300\,\mathrm{mm}^2$ concrete slab is reinforced with rebars of resisting area $A_{s,top} = A_{s,bot} = 7797\,\mathrm{mm}^2$ in the top and bottom layers respectively with a concrete cover of 30 mm. The steel beam has a $1770 \times 22\,\mathrm{mm}^2$

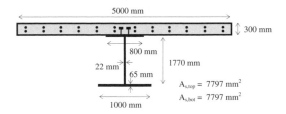

Figure 4.   Cross section of the tested beams.

Table 1.   Designed connection strengths.

Approach	f_{s1} (kN/m)	f_{s2} (kN/m)	f_{s3} (kN/m)
EC4	3072.94	2610.30	–
AA	3072.94	–	1299.50

web, a 800×65 mm^2 top flange and a 1000×65 mm^2 bottom flange.

The following material properties are considered: concrete compressive strength $f_c = 35$ MPa; beam steel yield stress $f_{ya} = 335$ MPa, ultimate stress $f_{ta} = 490$ MPa, elastic modulus $E_a = 200$ GPa, hardening strain $\varepsilon_{ha} = 0.04$, ultimate strain $\varepsilon_{ulta} = 0.11$; reinforcement yield stress $f_{ys} = 500$ MPa, collapse stress $f_{ts} = 540$ MPa, elastic modulus $E_s = 200$ GPa, hardening strain $\varepsilon_{hs} = 0.02$, ultimate strain $\varepsilon_{ults} = 0.10$.

The shear connection in all beams considered in this study has an assumed ultimate slip $s_{ult} = 6$ mm and thus it can be considered as ductile connection (CEN 2004). According to Eurocode 4 (CEN 2004) ductile connectors may be spaced uniformly over each tract. Considering the bending moment distribution at collapse as computed from plastic analysis (i.e., sagging and hogging moment hinges develop the full positive and negative plastic moments respectively), the tracts defined in Figure 1 are $L_1 = 17$ m, $L_3 = 6$ m, and $L_2 = L_1 + L_3 = 23$ m. The designed connection strengths f_{s1}, f_{s2} and f_{s3} for uniform connection distributions in the tracts of length L_1, L_2 and L_3 respectively are given in Table 1. These strengths are computed from the assumption that the connection is fully plasticized (i.e., uniform distributions of q_{s1}, q_{s2} and q_{s3}) thus $f_{s1} = N_1/L_1, f_{s2} = (N_1 + N_2)/L_2, f_{s3} = N_2/L_3$. The relevant shear connection distributions for the two design approaches considered in this study are indicated as EC4u and AAu and depicted in Figures 5 and 6 respectively.

Because of its ductility, an arbitrary connection distribution for each tract may be employed, provided that limits in maximum and minimum spacing of connectors are satisfied (CEN 2004). Feasible connection distributions other that the uniform distribution for each tract are considered for each design approach, as shown in Figure 5 for the EC4-1 approach and in Figure 6 for the alternative approach. Distributions

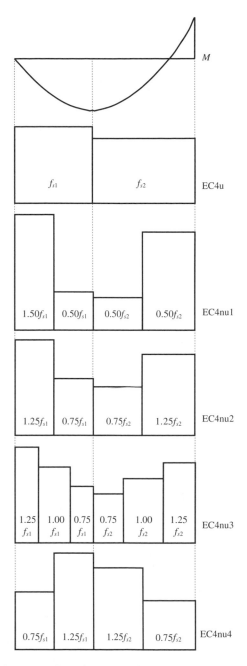

Figure 5.   EC4-1 shear connection design: distributions of connection considered in the numerical study.

indicated as EC4nu1, EC4nu2 and EC4nu3 have a stronger connection at beam ends where the vertical shear is larger (Figure 5). On the other hand the distribution indicated as EC4nu4 has a weaker connection at beam ends while the connection is stronger in the region where the sagging bending moment is larger.

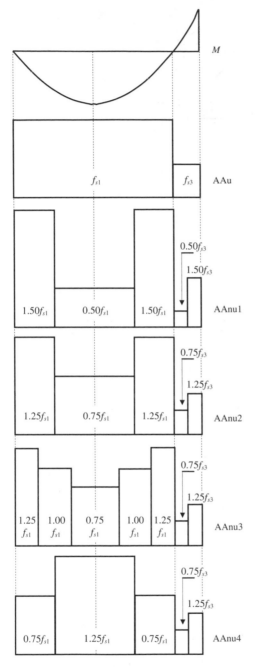

Figure 6. Alternative approach shear connection design: distributions of connection considered in the numerical study.

Similar considerations apply to the four non uniform distributions of Figure 6 with the difference that the connections in the hogging regions are always stronger near the support where both vertical shear and bending moment are larger.

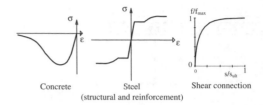

Figure 7. Uniaxial constitutive laws adopted in the finite element nonlinear analysis.

Table 2. Load, mid-span deflection and maximum absolute slip at collapse.

Beam	p_{ult} (kN/m)	v_{ult} (mm)	$|s|_{max}$ (mm)
EC4u	43.37	580	1.61
EC4nu1	43.14	555	3.37
EC4nu2	43.28	585	2.32
EC4nu3	43.28	575	2.08
EC4nu4	43.45	596	3.52
AAu	43.42	585	3.82
AAnu1	43.21	585	2.34
AAnu2	43.35	585	2.85
AAnu3	43.28	575	2.87
AAnu4	43.46	595	4.78

The various distributions are designed to transfer the same longitudinal shear force of the uniform distribution in each tract. Thus the total amount of shear connection for all shear connection distributions considered in this study is the same.

3.2 Response simulation results

The beams with different shear connection distributions described in the previous section were analyzed by using a finite element model (Dall'Asta & Zona 2002) previously validated by comparisons with experimental tests (Dall'Asta & Zona 2004). Realistic uniaxial nonlinear constitutive models are adopted for the materials and the shear connection (Figure 7). See (Dall'Asta & Zona 2002) for details on constitutive laws, numerical integration and solution procedures.

The ultimate load, ultimate mid-span deflection and maximum absolute value of the slip between slab and steel beam are given in Table 2 for each beam, as defined in Figures 5 and 6. It is observed that the flexural bearing capacity is almost the same for all beams, with differences between and lower and largest collapse loads smaller than 1%. Difference are more noticeable for the mid-span deflection at collapse, with differences between the lower and largest deflection values of about 7%. It is observed that the beams with stronger connections where the sagging bending moment is larger (EC4nu4 and AAnu4) have the largest ultimate loads and ultimate deflections. Indeed the more significant differences in the tested

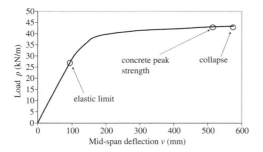

Figure 8.   Load – deflection curve for beam EC4nu3.

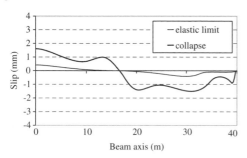

Figure 9.   Slip distribution for beam EC4u.

beams are observed in the slip demand, with differences of almost 3 times between the beam with lower slip demand (EC4u) and the beam with larger slip demand (AAnu4).

The load–deflection curves for the tested beams are virtually superimposed apart in the very last branch where small differences can be observed. Due to space limitations only the load – deflection curve for one of the beams is shown (Figure 8).

The distributions of the slip between the slab and the steel beam are shown for each beam in Figures 9 through 18. Two load levels are considered: (i) beam at the elastic limit, i.e., condition representative of the maximum slip requirements at service state; (ii) beam at collapse. It can be observed that in all tested beams the ultimate slip of the shear connectors ($s_{ult} = 6$ mm) is not reached, thus the connection is not responsible for beam failure.

Figure 9 shows that in beam EC4u the slip is less than 0.5 mm at the elastic limit and is less than 2 mm at collapse. If the connection is not distributed uniformly in each of the two tracts L_1 and L_2 and a stronger connection is adopted at beam ends (and consequently a weaker connection in the span central region being the total connection the same), then there is a reduction of the slip at beam ends alongside a significant increase of the slip in the span central region. This increase is particularly evident in beam EC4nu1 (Figure 10).

A more gradual transition from stronger to weaker connection from beam ends to the span central region reduce the slip demand in the central part of the

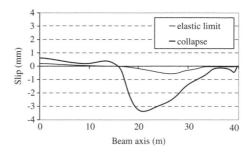

Figure 10.   Slip distribution for beam EC4nu1.

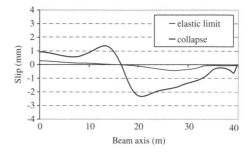

Figure 11.   Slip distribution for beam EC4nu2.

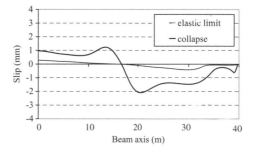

Figure 12.   Slip distribution for beam EC4nu3.

beam (Figures 11 and 12) but still the higher slip requirements are near mid-span and larger that the slip requirements when the connection is uniform in the two tracts L_1 and L_2. On the other hand, if the connection is stronger in the central part (and thus weaker at beam ends) then a significant increase in slip can be observed at beam ends (Figure 13) together with an increase of global ductility of the beam and a slight increase of the load carrying capacity, as previously observed (Table 2).

A discussion similar to the one made for beams with shear connection designed according to EC4-1 can be repeated for beams with shear connection designed with the presented alternative approach. The main difference in this case is a larger requirement in slip demand in the hogging region (Figures 14 through 18) due to the reduced amount of shear connection in tract L_3 given by the alternative approach.

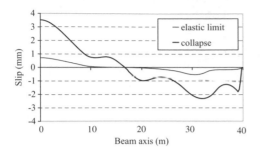

Figure 13. Slip distribution for beam EC4nu4.

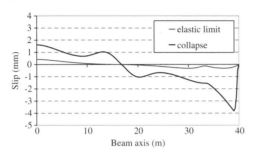

Figure 14. Slip distribution for beam AAu.

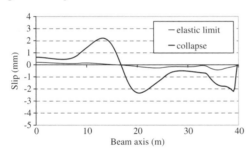

Figure 15. Slip distribution for beam AAnu1.

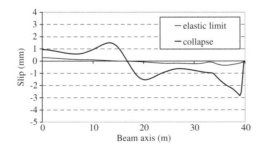

Figure 16. Slip distribution for beam AAnu2.

4 CONCLUSIONS

The numerical simulated response analyses presented in this study for two-span simply supported steel-concrete composite beams designed for the same

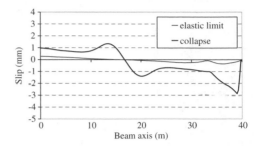

Figure 17. Slip distribution for beam AAnu3.

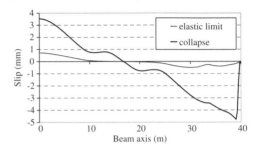

Figure 18. Slip distribution for beam AAnu4.

flexural bearing capacity but with different shear connection distributions, show large differences in the slip demand (ductility) of the shear connectors. However the differences at the local level have a limited influence on the beam global behavior. It is worth observing that shear connection distributions proportional to the vertical shear do not appear to permit an enhanced structural behavior compared to uniform distributions in each tract of the beam between a support and the adjacent location of the maximum bending moment.

REFERENCES

CEN Comité Européen de Normalization. 2004. *EN 1994-1-1 Eurocode 4: Design of composite steel and concrete structures – Part 1-1: General rules and rules for buildings.* Brussels.

Dall'Asta, A. & Zona, A. 2002. Non-linear analysis of composite beams by a displacement approach. *Computers and Structures*, 80(27–30): 2217–2228.

Dall'Asta, A. & Zona, A. 2004. Comparison and validation of displacement and mixed elements for the non-linear analysis of continuous composite beams. *Computers and Structures* 82(23–26): 2117–2130.

Oehlers D.J. & Bradford M.A. 2000. *Elementary Behaviour of Composite Steel and Concrete Structural Members.* Butterworth-Heinemann, London.

Spacone, E. & El-Tawil, S. 2004. Nonlinear analysis of steel-concrete composite structures: state-of-the-art. *Journal of Structural Engineering ASCE* 130(2): 159–168.

Viest, I.M., Colaco, J.P., Furlong, R.W., Griffis, L.G., Leon, R.T. & Wyllie, L.A. 1997. *Composite Construction Design for Buildings.* ASCE McGraw-Hill, New York, NY.

402

Steel and Composite Structures – Wang & Choi (eds)
© 2007 Taylor & Francis Group, London, ISBN 978-0-415-45141-3

Shear connectors in steel fibre reinforced ultra high performance concrete

J. Hegger & S. Rauscher

Institute of Structural Concrete, RWTH Aachen University, Aachen, Germany

ABSTRACT: The paper summarizes the results of an experimental research program involving the testing of push-out specimens with innovative continuous shear connectors in ultra-high performance concrete. In the push-out tests the initial stiffness, the load carrying capacity and the ductility of the shear connectors are determined and compared for various test series. The parameters included the steel fibre content, the concrete cover between the shear connector and the concrete surface, the thickness of the puzzle strip, the transverse reinforcement ratio and the geometry. The test results indicate that the steel fibre reinforcement ratio has only a minor influence on the load carrying capacity whereas the transverse reinforcement ratio is more vital.

1 INTRODUCTION

Composite structures combine the favourable features of structural steel and concrete. Taking into account the mechanical properties the steel carries the tensile forces and the concrete is arranged in the compression zone. Due to the composite action a significant increase in load carrying capacity and stiffness of the beam is achieved, resulting in savings in the dead load, construction depth and construction time. So far composite structures made of high strength steel and high performance concrete have been established (Hegger et al. 2001, Hechler et al. 2006). Within a collaborative research project (SPP 1182) ultra high performance concrete (UHPC) with steel fibres is applied for hybrid structures. Due to its high compressive strength of up to 200 MPa without thermal treatment even more slender and attractive structures are feasible. In addition, ultra high performance concrete features a high tensile strength under considerable tensile strains which is primarily caused by the steel fibres. Fibres are added to the concrete not only for the strength but also for ductility reasons.

The increase in bearing capacity of a composite beam also leads to higher stresses in the composite joint. So far the shear connection is commonly accomplished by headed studs. Due to their high load carrying capacity and high initial stiffness continuous shear connectors, e.g. the puzzle strip described in (Hegger et al. 2006), are very appropriate. In Figure 1 the puzzle strip and the saw tooth are presented with transverse reinforcement. These two types of shear connectors were tested in push-out tests. The shear connectors are expected to withstand high loads with sufficient ductility.

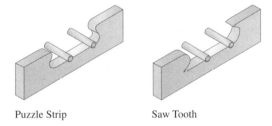

Puzzle Strip   Saw Tooth

Figure 1. Continuous shear connectors.

In the paper the results from push-out tests are presented. In order to determine the load carrying capacity and the ductility of the composite joint between high strength steel and ultra high performance concrete, several push-out tests with different amounts of steel fibres and transverse reinforcement were performed.

2 ULTRA HIGH PERFORMANCE CONCRETE

The composition of ultra high performance concrete (UHPC) differs fundamentally from conventional concrete such as normal strength concrete (NSC) or high strength concrete (HSC). The main characteristics of UHPC are a high compressive and tensile strength as well as a high durability compared to NSC and HSC. The UHPC features a compressive strength of up to 200 MPa without thermal treatment. The high performance of UHPC is based on the following vital factors: a low water-cementitious materials ratio (w/cm) of 0.19 and a high density which is achieved by a high content of cement, fine sands (mainly quartz) and silica fume. This is essential to fill the cavities between

Table 1. Characteristics of UHPC.

	Unit	–
Cement CEMI 52.5 R	kg/m^3	650
Silica Fume	kg/m^3	177
Quartz Powder	kg/m^3	456
Coarse Aggregates	kg/m^3	951
Steel Fibres	% per volume	0.9–2.5
w/cm – ratio	–	0.19

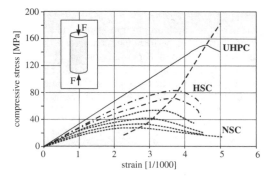

Figure 2. Stress strain diagram.

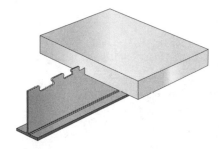

Figure 3. Slender composite beam without upper flange.

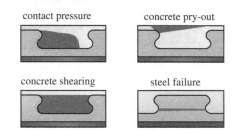

Figure 4. Failure modes.

the solid particles. The silica fume functions as a super-plasticizer. Table 1 shows the components of UHPC which has been employed for the tests.

To ensure a sufficient ductility thin steel fibres are added. The fibre content was varied between 0.9 and 2.5% of the concrete volume. Due to the steel fibres UHPC not only features a high compressive strength but also a linear-elastic behaviour until about 90% of its compressive strength which is achieved at a strain rate of about 4.5‰ (Figure 2). Conventional concrete (NSC and HSC) shows a distinctive nonlinear behaviour due to the micro-cracks which develop at a stress level of about 40% of the compressive strength.

Besides the ductility the tensile strength of the UHPC is also increased because of the steel fibres. Their function in the UHPC is described in the following section.

3 CONTINUOUS SHEAR CONNECTORS

Continuous shear connectors have been used for about 20 years. The most common one is the perfobond strip (Leonhardt et. al. 1987) where the shear forces between the steel beam and the concrete slab are transferred by vertical steel plates with holes. In (Schmitt et al, 2005, Hegger et al. 2006) the so-called puzzle-strip is introduced. Its main advantage is the symmetrical geometry. This way it is possible to receive two shear connector strips with one cut and no material is wasted. If the cut is performed in the web of a steel I-beam, two composite beams can be produced (Figure 3). Since

the neutral axis of a composite beam under positive bending moments is in the region of the upper flange the strains are usually very low. Thus there is little loss in load carrying capacity compared to a conventional composite beam with equal height. However, there are significant savings in production costs.

The load bearing behaviour of continuous shear connectors has been investigated by several researchers (Hegger et al. 2001, 2006, Wurzer 1997). Four failure modes have to be considered (Figure 4): local concrete failure in front of the shear connector, concrete pry-out failure, shear failure of the concrete and steel failure due to the moment stress of the puzzle profile which is described in (Hegger et al. 2006).

According to the model of (Wurzer 1997) there are lateral tensile forces between the steel teeth resulting from the concentrated load introduction of the steel profile (Figure 5).

In the early state of testing fine micro-cracks begin to form in the concrete. The steel fibres are capable to bridge the micro-cracks and thus the formation of macro-cracks is deferred (Figure 6, left). However, as the micro-cracks grow the tensile forces have to be sustained by additional reinforcement and the fibres fail due to the missing anchorage (Figure 6, right).

4 TEST PROGRAM

4.1 *Push-Out Tests*

The specimens were designed for both, concrete and steel failure. Concrete failure is mainly governed by

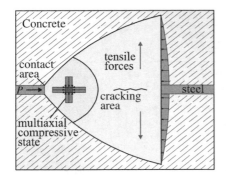

Figure 5. Stresses in the concrete (outline).

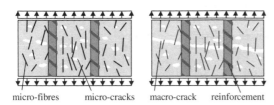

micro-fibres   micro-cracks   macro-crack   reinforcement

Figure 6. Principle of steel fibre reinforced concrete.

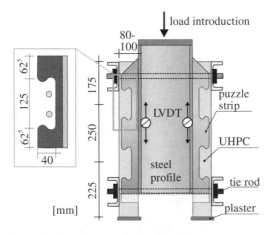

[mm]

Figure 7. Push-Out Standard Test POST.

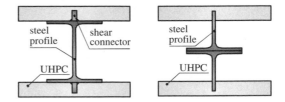

Figure 8. Push-Out Test with (left) and without (right) flange.

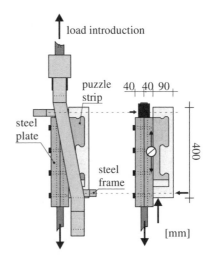

Figure 9. Single Push-Out Test SPOT.

the clearance between shear connector and concrete surface and the degree of transverse reinforcement by means of steel fibres and steel rebar, respectively.

Five test series were performed where two types of shear connectors were tested: the puzzle-strip and the saw-tooth. The tests were performed using two different set-ups, the Push-Out Standard test and the Single Push-Out Test.

The Push-Out Standard Test (POST) according to (EC 4, 2004) simulates the shear transfer in the composite joint of composite girders (Figure 7).

The standard specimen consists of a steel I-beam where the shear connectors are welded on the outer side of the flange. In slender composite beams as shown in Figure 3 there is no upper flange and thus there is less confinement of the concrete in this area. For this reason push-out tests were performed on steel profiles where the geometry is cut in the web (Figure 8). In this case the thickness of the shear connector equals the thickness of the web.

The dimensioning of the POST specimen is matched to standard strength concrete. For higher concrete quality grades, as it is the case with UHPC, the dimensions can be reduced to a slab thickness of 8 to 10 cm. Due to the eccentricity of the action lines of the forces the studs are not equally stressed. This leads to a horizontal uplift and thus to horizontal forces in the shear connectors. To minimize the influence, tie rods were attached. LVDTs were applied to measure the slip between the steel profile and the concrete slab.

The Single Push-Out Test (SPOT) consists of a steel frame embracing the UHPC block. Due to its compactness it is an appropriate test set-up to pre-select shear connectors effectively. In the SPOT a single shear connector is tested under nearly pure shear load.

Table 2. Test program.

Series	Number of Tests	Test Set-up	Description	
A1	2	POST	Puzzle	2.5 Vol.-%, $c_o = 10$ mm
A2	3	POST	Puzzle	0.9 Vol.-%, $c_o = 10$ mm
A3	2	POST	Puzzle	0.9 Vol.-%, $c_o = 20$ mm
B1	1	POST	Puzzle	Reinf. 2 Ø 12
B2	2	POST	Puzzle	Reinf. 2 Ø 12 + Ø 8/5
C1	2	SPOT	Puzzle	$t_w = 10$ mm
C2	2	SPOT	Puzzle	$t_w = 15$ mm
C3	2	SPOT	Puzzle	$t_w = 20$ mm
D1	3	POST	Puzzle	No flange, $c_o = 30$ mm, Reinf. 2 Ø 12 + Ø 10/10
E1	2	SPOT	Saw Tooth	Direction 1
E2	2	SPOT	Saw Tooth	Reverse direction 2

4.2 Test parameters

The parameters investigated within five test series are summarized in Table 2. As a reference, the puzzle strip with a thickness of 20 mm was tested (Series A) in which the effects of the amount of steel fibres and the concrete cover c_o between the shear connector and the concrete surface was determined. For the following tests the concrete cover c_o is 20 mm and the fibre content 0.9% unless stated otherwise. The influence of the transverse reinforcement was investigated in Series B. In Series C the thickness of the shear connector t_w was varied. As a basis for further beam tests push-out tests without a steel flange underneath the concrete slab were performed (Series D). Furthermore, in Series E a new geometry, the saw tooth profile was tested.

5 TEST RESULTS

5.1 General

For each test series the load-slip diagrams were evaluated according to the procedure of EC 4. The characteristic slip δ_{uk} describes the deformation measured in the tests during the plastic period when the characteristic load P_{Rk} was maintained.

The concrete compressive strength for a 100-mm-cube $f_{c,cube100}$, the yield strength of the shear connector, the mean maximum loads $P_{max,mean}$ the characteristic loads for one recess of the shear connector and corresponding slip δ_{uk} are presented in Table 3.

In almost all tests failure of the concrete due to pry-out occurred. Only in the specimens with thin shear connectors (Series C1 and D1) the yielding of the steel determined the ultimate load. In the following the results are discussed in detail.

5.2 Load-slip behaviour

As a result of the push-out tests the load-slip diagrams are plotted for each series. In the evaluation

Table 3. Compressive strength and test results.

Series	$f_{c,cube100}$ N/mm^2	$f_{y,connector}$ N/mm^2	$P_{max,mean}$ kN	P_{Rk} kN	δ_{uk} mm
A1	191.0	499	426.0	372.4	2.2
A2	177.9	499	389.1	340.7	1.4
A3	194.4	499	432.9	369.5	4.4
B1	177.9	499	492.7	443.4	5.1
B2	179.5	499	562.5	493.5	7.1
C1	182.6	441	326.2	282.1	15.1
C2	178.0	521	375.2	327.0	11.8
C3	179.5	499	412.6	330.5	5.2
D1	179.1	472	587.0	495.0	8.4
E1	176.1	499	457.5	396.8	6.5
E2	182.6	499	418.4	354.5	1.2

not only the ultimate load but also the initial stiffness and the ductility, which is the slip after reaching the characteristic load P_{Rk}, were examined.

5.2.1 Effect of steel fibres
The load-slip diagram in Figure 10 shows that the amount of steel fibres added to the concrete influences the load carrying capacity of the puzzle-strip. With 2.5% of steel fibres the load carrying capacity was increased by approximately 10% compared to a 0.9% fibre content. In addition the steel fibres had a positive effect on the descending branch after reaching the maximum load. However, an increase in the maximum load of 10% with about 2.7 times more steel fibres cannot be recommended for practical use in terms of cost effectiveness since the steel fibres are the most cost-intensive factor in UHPC.

5.2.2 Effect of concrete cover
Due to the larger concrete cover the ultimate load could be increased by 11% with equal initial stiffness. The ductility after reaching the maximum load

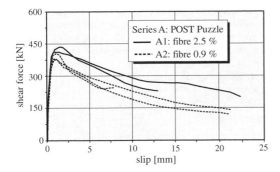

Figure 10.   Load-slip diagram – series A.

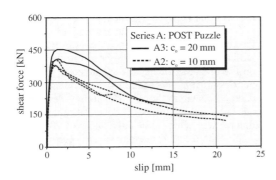

Figure 11.   Load-slip diagram – series A.

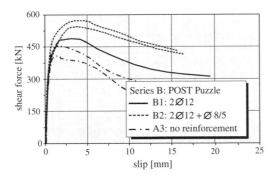

Figure 12.   Load-slip diagram – series B.

was improved significantly due to the larger concrete cover.

5.2.3   *Effect of transverse reinforcement*

In Series B the effect of transverse reinforcement in the puzzle recesses as well as above the shear connector was investigated. The transverse reinforcement increased the confinement of the concrete in front of the puzzle and thus led to higher ultimate loads and an increase in ductility in case of concrete failure.

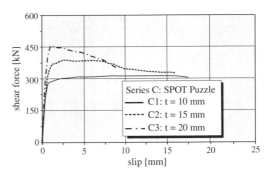

Figure 13.   Load-slip diagram – series C.

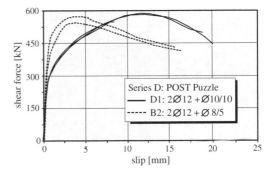

Figure 14.   Load-slip diagram – series D.

5.2.4   *Effect of the shear connector thickness*

The influence of the shear connector's thickness was investigated in the SPOT. There is a slight difference in initial stiffness. With increased thickness of the shear connector higher ultimate loads were achieved (Figure 13). The specimens of Series C1 with a thickness of 10 mm failed due to the yielding of the puzzle. After the test the puzzles were deformed noticeably and a horizontal crack was observed. In Series C2 with a connector thickness of 15 mm a combined steel and concrete pry-out failure was observed. Due to the larger thickness pure concrete pry-out failure occurred in Series C3 with no deformation in the steel. With increasing thickness and predominating concrete failure the ductility was significantly reduced compared to pure steel failure. However, even for concrete failure the behaviour is not brittle due to the steel fibres.

5.2.5   *Effect of the confinement through the flange*

So far the push-out tests have been performed with a flange as shown in Figure 7. The flange which holds the shear connector increases the confinement of the concrete. The diagram in Figure 14 shows the load-slip curve of the puzzle-strip without a flange compared to the puzzle-strip with flange.

The thickness of the shear connector with flange (Series B2) is 20 mm. The shear connector in Series

Direction 1   Direction 2

Figure 15. Load directions for the saw tooth.

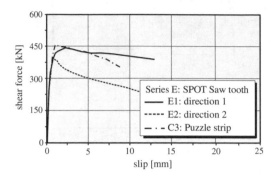

Figure 16. Load-slip diagram – series E.

D1 results from a cut in the web of an IPE 600 pro-file with a web thickness of 12 mm. Even though the thickness of the web was only 60% of the one in series B there was no difference in the ultimate load. How-ever, the thinner shear connector was more flexible and its ultimate load was reached after 10 mm. The failure mode in Series D was steel failure with a horizontal crack in the puzzle strip. On the outer surface there were signs of concrete pry-out failure, however, on the inside of the concrete surface no cracks were visible.

5.2.6 *Effect of the geometry*

With the saw tooth it was intended to increase the con-fined area in front of the shear connector and thus to improve the load carrying behaviour of the connector. Since the saw tooth has no symmetrical shape it was tested in the SPOT in two directions (Figure 15).

In Figure 16 the results of the SPOT with both saw tooth-directions are compared to the results with the puzzle strip. The initial stiffness was equally high inde-pendent of the load direction and geometry. The saw tooth direction 1 and the puzzle show no noticeable differences in ultimate load. However, when the saw tooth was stressed in reverse direction 2, both, the load carrying capacity and the ductility were significantly reduced.

Since in most structures and especially in bridges the direction of the shear forces between steel profile and concrete slab changes depending on the load state, the symmetrical puzzle should be preferred.

6 SUMMARY AND CONCLUSIONS

Experiments with Push-Out standard test specimens (POST) and single Push-Out test specimens with ultra-high performance concrete were carried out. Two types of shear connectors made of high strength steel S460 were tested: the puzzle strip and the saw tooth. The parameters of the tests were the concrete cover, the transverse reinforcement, the thickness of the shear connector, the confinement by a flange and the geometry.

The results from the push-out tests can be summa-rized as follows:

– The influence of the amount of steel fibres added to the UHPC is of minor importance. With almost 3 times more steel fibres only an increase of 10% in ultimate load could be achieved. In all cases concrete failure occurred.
– Arranging transverse reinforcement in the puzzle recesses and between the shear connector and the concrete surface leads to an increase in ultimate load of up to 30%. Also the ductility was improved.
– Different failure modes were observed depending on the thickness of the shear connector. For thin puzzles (10 mm) steel failure occurred with a very ductile behaviour showed. Concrete failure (pry-out failure as well as local concrete compression failure) was achieved at a thickness of 20 mm. A combined steel and pry-out failure occurred at a thickness of 15 mm.
– Even without a confinement by a flange, the ulti-mate load of the puzzle-strip which is directly cut into the web of a steel profile IPE 600 is compara-ble to the load carrying capacity of a puzzle-strip with flange. However, the behaviour is more elastic due to the thinner thickness of the shear connector. A combined steel and concrete failure was noticed.
– The saw tooth loaded in direction 1 shows the same load carrying behaviour as the puzzle. However, loaded into reverse direction 2, the load carrying capacity and the ductility are significantly reduced. For this reason the puzzle strip is more appropriate for structures where the shear load across the com-posite joint changes its direction, e.g. in composite bridges.

The research program is continued. The perfor-mance of UHPC in composite beams in combination with high strength steel will be investigated in future tests.

ACKNOLWEDGEMENT

This research project is part of the SPP 1182 which is founded by the DFG. The authors express their thanks to the DFG and to Arcelor S.A. for their support.

REFERENCES

Hegger, J., Sedlacek, G., Döinghaus, P., Trumpf, H. 2001. Testing of Shear Connectors in High Strength Concrete.

Proc. RILEM – Symposium on *Connections between Steel and Concrete, Stuttgart, Germany, Sept. 9–12, 2001.*

Hechler, O., Feldmann, M., Rauscher, S., Hegger, J. 2006. Use of shear connectors in high performance concrete, Proc. intern. symp. *Stability and Ductility of Steel Structures,* D. Camotim et. al (ed.), *Lisbon, Sept. 6–8, 2006.*

SPP 1182. Priority program SPP 1182: Nachhaltig Bauen mit UHPC, German research foundation, 2007.

Hegger, J., Feldmann, M., Rauscher, S., Hechler, O. 2006. Load-Deformation Behavior of Shear Connectors in High Strength Concrete subjected to Static and Fatigue Loading. Proc. *IABSE Symposium on Responding to tomorrow's challenges in structural engineering, Budapest, Hungary, Sept 13–15, 2006.*

Leonhardt, F., Andrä, W., Harre, W. 1987. Neues, vorteilhaftes Verbundmittel für Stahlverbund-Tragwerke mit hoher Dauerhaftigkeit. *Beton- und Stahlbetonbau, 82(12),* pp. 325–331.

Schmitt, V., Seidl, G., Hever, M. 2005. Composite Bridges with VFT-WIB-Construction Method. Proc. *Eurosteel 2005, Maastricht, Netherlands, June 8–10, 2005.*

Wurzer, O. 1997. *Zur Tragfähigkeit von Betondübel*, PhD thesis. Institut für Konstruktiven Ingenieurbau, Universität der Bundeswehr, München, Juni 1997.

EC 4, prEN 1994-1-1, 2004. Design of composite steel and concrete structures Part 1.1 – General rules and rules for buildings. Brussels.

Steel and Composite Structures – Wang & Choi (eds)
© *2007 Taylor & Francis Group, London, ISBN 978-0-415-45141-3*

Effect of steel fibre reinforcement on the shear connection of composite steel-concrete beams

O. Mirza & B. Uy
University of Western Sydney, Australia

ABSTRACT: Continuous composite steel-concrete beams are becoming increasingly popular in multistorey buildings due to their higher span/depth ratio, reduced deflections and higher stiffness value. However, their performance is greatly dependent on the load-slip characteristics of shear connectors. Nowadays, the trapezoidal slab is becoming popular for high rise buildings compare to solid slabs because it can cover a large span with little or no propping required with less concrete. This paper describes an accurate nonlinear finite element model using ABAQUS to study the behaviour of shear connectors for both solid and profiled steel sheeting slab. In addition to analysing the influence of shear connectors on the structural performance, steel fibres are introduced to further augment the ductility and strength of the shear connection region in the slab. The results obtained from the finite element analysis were verified against experimental results and indicate that strength and load-slip behaviour are greatly influenced by steel fibres

1 INTRODUCTIONS

Martin (2003) stated that currently, composite structures have been commonly used in construction in order to save time of construction and reduce the cost. Composite beam in designs are greatly affected by the behaviour of the shear connection. The main factors affecting shear connection are the strength of the shear connectors and the strength of concrete. One improvement to the existing concept is the application of steel fibres. Steel fibres are preferred to steel reinforcement/mesh because it reduces the load transferred to the footings and at the same time improves the moment capacity, improves fire resistance and controls cracking. Hence, experimental push tests from other experiments that were used to evaluate the shear connector capacity and load versus slip relationship of the connector were compared against finite element software packages known as ABAQUS. The main objectives of this paper are to develop a three dimensional finite element model to simulate the shear connector behaviour for both solid slabs and profiled slabs with the addition of steel fibres into the concrete slab for the push test and to determine the advantages related to finite element analysis of push tests.

2 PREVIOUS EXPERIMENTAL INVESTIGATION

2.1 *Description of push off test specimen*

Two types of experimental investigations were studied to compare with finite element model. They are Becher (2005) for the solid slabs and Wu (2006) for the profiled slabs. All experimental tests are based on Eurocode 4 (2004) where push test specimen consist a steel beam connected to two concrete slabs either a solid slab or profiled slab through shear connectors. The concrete slabs are bedded in mortar or gypsum onto the floor and a point load is applied onto the upper end of the web of the steel beam. However, some compromise has been made due to time limits, equipment and budget restriction. The modified push test as shown in Figure 1 and Figure 2 involved a standard push test as described with a roller support at one end to eliminate any horizontal resistance being imposed on the slab. The slab dimensions were 600 mm × 600 mm × 120 mm; 50 mm shorter and 30 mm thinner than Eurocode 4 to save material and cost. In order to give identical lateral and longitudinal reinforcement strength, 4 × 4 reinforcing mesh bar were used as a replacement for 5 × 4 reinforcing mesh which was outlined in Eurocode 4.

The steel beam used was a 200UB 25. Each specimen was initially loaded to 40% of the expected failure load then cycled 25 times between 5% and 40% of the expected failure load. Finally, each specimen was loaded until failure occurred.

Twelve push tests were carried out to determine the load-slip behaviour of the shear connectors in both solid slab and profiled slab. Three specimens of different steel fibres with two push tests each were tested. See table 1 for the tests.

3 FINITE ELEMENT MODEL

3.1 General

The finite element program ABAQUS was used to quantify the behaviour of the shear connection in composite beams. The main components affecting the behaviour of the shear connection in composite beams are the concrete slab, steel beam, profiled steel sheeting, reinforcement bars and shear connectors. These components must be accurately modeled to obtain an accurate result from the finite element analysis.

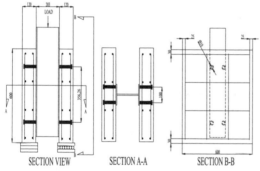

4 x 4 reinforcing mesh are provided both ways.

Figure 1.   Details of push test specimen for solid slab.

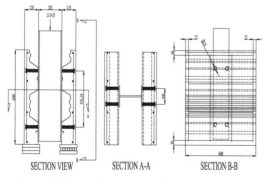

4 x 4 reinforcing mesh are provided both ways.

Figure 2.   Details of push test specimen for profiled slab.

A three dimensional finite element model has been developed to simulate the geometric and material non-linear behaviour of composite beams. The accuracy of the analysis depends greatly on the constitutive laws involved to define the mechanical behaviour.

3.2 Concrete

Plain concrete was recommended by Carreira & Chu (1985) where the linear elastic stress in compression is up to $0.4f'_c$. Beyond this point, it is in the plastic region.

$$\sigma_c = \frac{f'_c \, \gamma(\varepsilon_c / \varepsilon'_c)}{\gamma - 1 + (\varepsilon_c / \varepsilon'_c)} \tag{1}$$

where $\gamma = \left| \frac{f'_c}{32.4} \right|^3 + 1.55$ and $\varepsilon'_c = 0.002$

As for concrete in tension, the tensile stress is assumed to increase linearly with tensile strain until the concrete cracks. After the concrete cracks, the tensile stresses decrease linearly to zero. The value of strain at zero stress is usually taken as 10 times the strain at failure which can be shown in Figure 3.

For concrete with steel fibres Lok & Xiao (1999) described the stress-strain relationship as:

$$\sigma = f_c \left[2 \left(\frac{\varepsilon}{\varepsilon_{co}} \right) - \left(\frac{\varepsilon}{\varepsilon_{co}} \right)^2 \right] \Rightarrow, \varepsilon \leq \varepsilon \tag{2}$$

$$\sigma = f_c \Rightarrow (\varepsilon_0 \leq \varepsilon \leq \varepsilon_{cu}) \tag{3}$$

where $f_c = 0.85f'_c$ and $e_{cu} = 0.0038$ for 0.5 % to 2 % of steel fibres.

For concrete in tension, it can be expressed by function shown below:

$$\sigma = f_t \left[2 \left(\frac{\varepsilon}{\varepsilon_{co}} \right) - \left(\frac{\varepsilon}{\varepsilon_{co}} \right)^2 \right] \Rightarrow, 0 \leq \varepsilon \leq \varepsilon_{co} \tag{4}$$

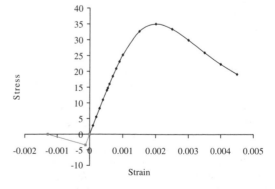

Figure 3.   Stress-strain relationship for concrete with steel fibres (Lok & Xiao, 1999).

412

$$\sigma = f_t \left[1 - \left(1 - \frac{f_{tu}}{f_t} \right) \left(\frac{\varepsilon - \varepsilon_{t0}}{\varepsilon_{t1} - \varepsilon_{t0}} \right) \right] \Rightarrow, \varepsilon_{t0} \leq \varepsilon \leq \varepsilon_{t1} \tag{5}$$

$$\sigma = f_{tu} \Rightarrow, \varepsilon_{t1} \leq \varepsilon \leq \varepsilon_{tu} \tag{6}$$

where f_t = the ultimate tensile strength which can be determined through a direct tensile test, ε_{t0} = the ultimate strain, $f_{tu} = \eta V_f \tau_d \frac{L}{d}$, $\varepsilon_{t1} = \tau_d \frac{L}{d} \frac{1}{E_s}$, V_f = the fibre volume fraction, τ_d = the bond stress, $\frac{L}{d}$ = the fibre aspect ratio, E_s = the elastic modulus of steel fibres, η = the fibre orientation in three dimensional random distribution = 0.405. The stress-strain relationship can be shown in Figure 4.

3.3 Shear connectors, structural steel beam, profiled steel sheeting and steel reinforcing

Shear connectors, structural steel beam, profiled steel sheeting and steel reinforcing were modelled in non-linear analysis as an elastic-plastic material with strain hardening. The model is according to Loh et al. (2003) shown in Figure 5.

3.4 Finite element mesh, boundary condition and load application

Three dimensional solid elements were used to model the push off test specimens. These are a three dimensional eight node element (C3D8R) for both concrete slab and structural steel beam, a three dimensional thirty node quadratic brick element (C3D20R) for shear connectors, a four node doubly curved thin shell element (S4R) for profiled sheeting and a two node linear three dimensional truss element (T3D2) for steel reinforcing. Figure 6 shows the finite element mesh used to represent half a stud of push test to reduce the simulation cost. For the boundary condition, surface

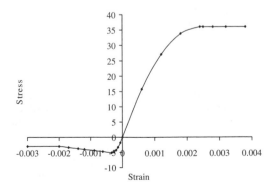

Figure 4. Stress-strain relationship for concrete with steel fibres (Lok & Xiao, 1999).

1,2 and 3 are restricted to move in the x, y and z direction respectively. As for the load application, a static concentrated load is applied to the center of the web using a modified RIKS method which can obtained a series of iteration for each increment for nonlinear structure.

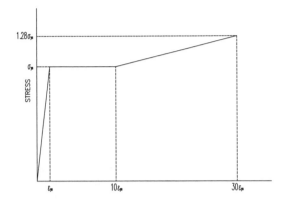

Figure 5. Stress-strain relationship for structural steel beam, shear connectors, profiled steel sheeting and steel reinforcing (Loh et al., 2003).

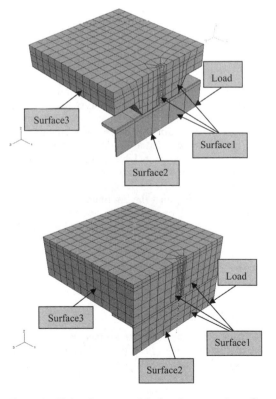

Figure 6. Finite element model of push test specimen for profiled slab and solid slab.

413

Table 1. Beam tests parameter.

Test No.	Slab profile	Percentage of steel fibres
1	Solid slab	0.0
2	Solid slab	0.3
3	Solid slab	0.6
4	Lysaght W-Dek Profiled slab	0.0
5	Lysaght W-Dek Profiled slab	0.3
6	Lysaght W-Dek Profiled slab	0.6

Table 2. Comparison of experimental result and finite element model.

Test No	Concrete strength (N/mm^2)	Exp. result (kN)	FEM result (kN)	Result Discrepancy (Exp./FEM)
1	31.71	104.77	113.65	0.92
2	34.89	118.26	135.00	0.89
3	30.58	106.37	113.13	0.94
4	33.93	49.93	52.01	0.96
5	36.77	51.23	53.27	0.96
6	33.30	47.45	50.18	0.95

4 FINITE ELEMENT ANALYSIS AND DISCUSSION

4.1 General

In order to determine the effect of steel fibres on composite steel-concrete beam, push tests were carried out to verify the load-slip and shear capacity behaviour. Six specimens of different steel fibres quantity and slab profiles were tested to validate the present finite element model shown in Table 1. Table 2 shows the ultimate load of the six tests compared with the finite element analysis.

Table 2 verifies that the optimum dosage of steel fibres is F = 0.3 because an increment of F = 0.6 proved that no visible advantage is gained in concrete strength or ductility. The ultimate load discrepancy is in good agreement with experimental. Generally it can be observed that the ultimate load for solid slab is higher than profiled slab.

4.2 Solid slab

The solid push tests show that no cracks were found on the concrete slab until the ultimate load was reached. Cracks only found around the stud after solid slab failure. All three tests for solid slab showed that the failure mode was stud yielding where shear connectors sheared off near the weld collar. The failure mode of the shear connector is similar as mentioned by Yam (1981)

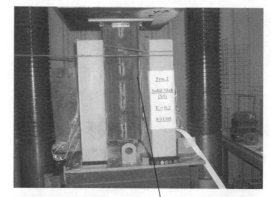

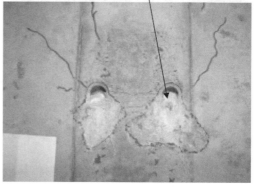

Figure 7. Concrete damage at the base of shear connectors.

and Lam (2005). Figure 7 and Figure 9 proved that the shear connectors experienced significant deformation around their base. In the experiment, yielding of the stud element was discerned near the shear connector collar followed by maximum compressive reached by concrete elements around shear connector. It was observed that the cracking is less when the fibre content is higher. In finite element model stress contour showed a higher stress values before failure occur with the increased of fibre content. In Figure 8, higher fibre content gives higher stiffness and ductility. One very obvious observation was that the deformation decrease with the increase of fibre content.

4.3 Profiled slab

The profiled slab shows obvious first cracking occur in the middle of the slab along the trough of W-Dek which is caused by concrete failure shown in Figure 10 and 12. As shown in Figure 12, the concrete element in the trough of W-Dek reached maximum stresses before the shear connector element. All three tests for profiled slab showed that the failure mode was concrete failure where concrete cracked before shear connectors sheared off near the weld collar. The failure mode observed is similar as mentioned by El-Lobody (2006).

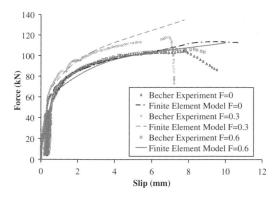

Figure 8. Comparison load versus slip relationship with different fibre content.

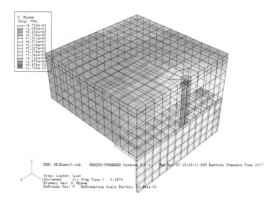

Figure 9. Stress contour of push test for solid slab.

It was observed that the profiled slab had less deformation and less stress in the concrete and shear connector because of the contribution of the profiled steel W-Dek. It was observed that the cracking through the trough is less when the fibre content is higher which proved that the steel fibres increase the stiffness and ductility of concrete. Figure 11 showed that higher fibre content gives higher stiffness and ductility.

5 CONCLUSION

This paper discusses three key issues in push test of the shear studs. An accurate finite element model has been developed to investigate the behaviour of shear connection in composite steel-concrete beam for both solid and profiled slab. Based on the comparisons between the result obtained from finite element models and available experimental result, it was observed that they are in good correlation. All the failure modes were accurately predicted by finite element model. The maximum discrepancy is 11%.

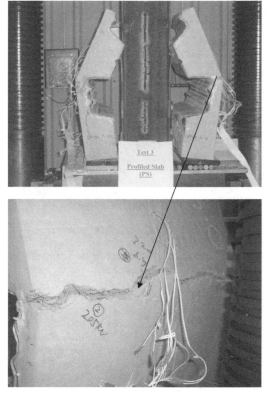

Figure 10. Concrete cracking along the trough of W-Dek.

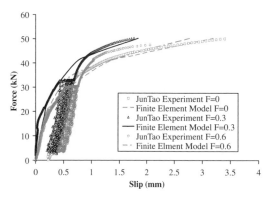

Figure 11. Comparison push test for profiled slab with different fibre content.

One primary issue that has been solved when steel fibres were included in composite steel-concrete beam was the improved stiffness and ductility for both solid and profiled slab. Even though, steel fibres did not show any major gain in concrete strength, the effectiveness of steel fibres only shows when the concrete begins cracking. From both experimental test and finite element model showed that Inclusion of steel

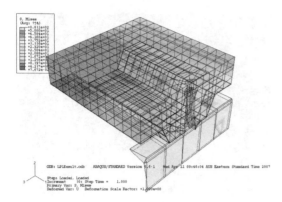

Figure 12. Stress contours of push test for profiled slab.

fibres improved the cracking load of concrete. This is a major advantage especially for continuous beams in hogging moment region. Moreover, the ultimate load of the push tests also increased with the inclusion of steel fibres for both solid and profiled slab.

There are three differences between profiled sheeting and solid slab. One of them is the failure mode. For the solid slab, the failure model is shear connection failure where else the profiled slab failure is concrete failure as discussed in Section 4. Stresses in the shear connector and concrete are lower compared to solid slab due to addition of the steel profiled W-Dek. It also can be observed that the solid slab generally have higher ultimate load compared with profiled sheeting.

6 RECOMMENDATION FOR FURTHER RESEARCH

Composite steel-concrete beams with the inclusion of steel fibres showed the improvement in cracking load. Studies considering continuous beam in hogging moment region with the inclusion of steel fibres to look at the cracking behaviour and ductility will be subjected for future research in this project.

ACKNOWLEDGEMENTS

The authors would like to thank BlueScope Lysaght ARC Linkage Project in Sydney for providing funding to this project and University of Wollongong for their assistance in the preparation of the experimental work described herein.

REFERENCES

Becher, L. 2005. *Behaviour and design of composite beams using fibre reinforced composite slabs*. Faculty of Engineering. Wollongong, University of Wollongong. Bachelor of Engineering: 75.

Carreira, D. and Chu, K. 1985. Stress-strain relationship for plain concrete in compression. *Journal of ACI Strcutral* 82(11): 797–804.

El-Lobody, E. and Young, B. 2006. Performance of shear connection in composite beams with profiled steel sheeting. *Journal of Constructional Steel Research.* 62(7): 682–694.

Eurocode 4, ENV 1994-1-1. 2004. *Design of composite steel and concrete structures, Part1-1. General rules and rules for building.* British Standard Institutions, London.

Lam, D. and El-Lobody, E. 2005. Behavior of headed stud shear connectors in composite beam. *Journal of Structural Engineering-ASCE.* 131(1): 96–107.

Loh, H. Y., Uy, B. and Bradford, M.A. 2003. The Effects of Partial Shear Connection in the Hogging Moment Region of Composite Beams Part II - Analytical Study. *Journal of Constructional Steel Research.* 60: 921–962.

Lok, T. and Xiao, J. 1999. Flexural strength assessment of steel fiber reinforced concrete. *Journal of Material in Civil Engineering.* 11(3): 188–196.

Martin, D. A. 2003. Steel-fibre-reinforced concrete floors on composite metal decking. *Concrete (London).* 37(8): 31–32.

Pawtucket, R. I. 2006. *ABAQUS Theory Manual Version 6.5.* Hibbitt, Karlsson and Sonrensen, Inc.

Pawtucket, R. I. 2006. *ABAQUS User's Manual Version 6.5.* Hibbitt, Karlsson and Sonrensen, Inc.

Pawtucket, R. I. 2006. *ABAQUS Analysis User's Manual Version 6.5.* Hibbitt, Karlsson and Sonrensen, Inc.

Wu, J. 2006. *Behaviour and design of composite beams using fibre reinfroced composite slabs*. Faculty of Engineering. Wollongong, University of Wollongong. Master of Engineering Practice: 191.

Yam, L. C. P. 1981. *Design of composite steel-concrete structures.* London, Surrey University Press.

Steel and Composite Structures – Wang & Choi (eds)
© 2007 Taylor & Francis Group, London, ISBN 978-0-415-45141-3

Quasi-static tests of rigid joint between steel beam and concrete wall in high-rise mixed structures

Jun Huang
Design & Research Institute of Wuhan University of Technology, Wuhan, China,
School of Civil Engineering of Wuhan University, Wuhan, China

Shaobin Dai
Hubei Key Laboratory of Roadway Bridge & Structure Engineering, Wuhan, China

ABSTRACT: Through quasi-static tests, the failure mechanism and seismic behaviour of three rigid joints between steel beam and concrete wall in steel frame-concrete wall mixed structure were studied. The results of the experiments show that the quality of the welding seam between the flange girth of steel beam and the seal plate is very important to the performance of the joint; the high-strength bolts connection between the web plate of steel beam and connecting plate can affect the performance of the joint in certain degree. At the same time, the relevant design suggestions based on the results of the tests are presented.

Keywords: rigid joint of steel beam and concrete wall; high strength bolt connection; quasi-static test.

1 GENERAL INSTRUCTIONS

In steel frame-concrete wall mixed structures concrete wall and steel frame are connected with hinged joints or rigid joints. As the main resistance structure, the concrete wall with a large lateral stiffness resists most horizontal load, and high strength steel frame resists most vertical load. Steel frame has the merits of high strength, weight light and can be made greater span structure; concrete wall has the merits of large lateral stiffness, and can resist horizontal load effectively. So in the structure system, the merits of steel frame and concrete wall have been combined organically. In some degree it reduces the dead load of the structure and accelerates the construction progress. So it has superiority in high-rise civil building. In the structure system the connection joint of steel beam and concrete wall connects steel frame and concrete wall, which are two different structure systems, so the performance of the joint will affect the function of the whole structure system.

In previous design, the joint is usually designed according to the connection way of the embedded steel piece. But in fact because the construction errors are big in steel structure and concrete structure, it is impossible to align bolt holes of embedded steel piece in concrete wall with bolt holes of the beam of steel frame accurately. In order to solve this problem, local formed steel frame is set in concrete wall in many practical engineering, and then steel beam of external steel frame is connected directly with formed steel beam in concrete wall. The connecting way is rigid connection in our standards.

But, because of the complexity of this connection joint, at present the study of the performances of the connection joint, especially earthquake resistant capability, is very few in China or other countries. The failure mechanism of the connection joint under seismic action is not clear. The design method of the connection joint is not clear in current codes and standards too.

In order to study the practical working performance of rigid joint between steel beam and concrete wall under seismic action, in this paper quasi-static tests are done in three rigid connection joints, and then the characteristics of failure form, ultimate bearing capacity and so on are studied under cyclic loading.

2 EXPERIMENT SCHEME

2.1 *Specimen design*

The specimen design considers the requirements of the codes (Design code of steel structure(GB50017-2003).2003, Design code of combined steel and concrete structure(YB9082–97).1998, Design standard of high-rise steel-concrete mixed structure.2003) comprehensively. Figure 1 is the diagram of the specimens. The material of steel structure is Q235. The name of constituents, specification, and connection way of

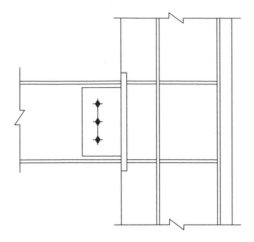

Figure 1. Model of the specimen.

Table 1. Information of steel specimens.

Serial number	Name	Specification
1	Formed steel column	HW150 × 150 × 7 × 10
2	Hidden corbel	HN250 × 125 × 6 × 9
3	Sealing plate	−290 × 165 × 16
4	Connecting plate	−200 × 125 × 12/16
5	Formed steel beam	HN250 × 125 × 6 × 9
6	Stiffening rib of column	−130 × 71.5 × 8
7	High strength bolt	10.9 grade M16/M20

*1. The connecting plate thickness of the two specimens, GJ1 and GJ2 is 12 mm; the thickness of GJ3 is 16 mm.
2. GJ1 and GJ3 adopt M16 high strength bolt; GJ2 adopts M20 High strength bolt.

structural members are shown in Table 1 and Figure 2. In the specimens all bolts are 10.9 grade high strength friction-type bolts; the diameter of bolt hole is the diameter of bolt pulsing 1.5 mm; the friction surfaces are dealt with by sandblast; the anti-slipping coefficient is above 0.3. High strength bolts adopt the construction method of torque. The pre-tensile forces of the two kinds of bolts (M16 and M20) are 100 kN and 155 kN respectively. When the specimens are being made, three standard specimens are selected from the same group of hot rolled H sectional steel, and carried through the test of material performance. The test indicates that the mean ultimate strength of hot rolled H sectional steel is 378.5 MPa; the mean percentage of elongation is 28.3%.

Figure 3 is the constitution of concrete wall. The strength grade of concrete is C30. The concrete is cast in site. During the process, three group including nine standard cubic specimens are made, then using press machine, the compression test is carried through. The

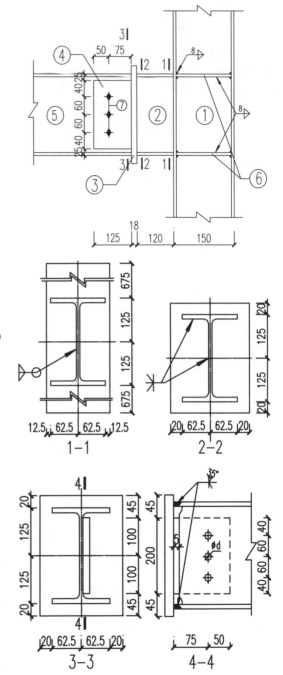

Figure 2. Details of steel specimens.

test indicates that the mean compression strength is 37.5 MPa. The type of distributing reinforcing bar in the concrete wall is HRB335, and the type of tie bar is HPB235.

418

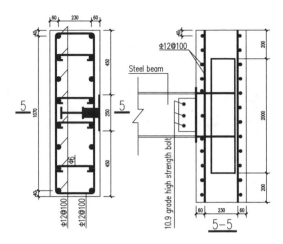

Figure 3. Details of concrete.

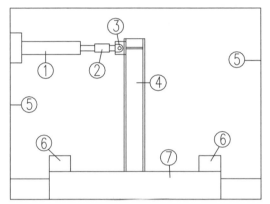

1. jack 2.fiber sensor 3.connecting member 4.steel beam
5.reaction frame 6.steel beam 7.concrete wall

Figure 4. Details of test apparatus.

(a)   (b)

Figure 5. Fiber sensor.

2.2 Test apparatus and loading scheme

The test is finished in the structure laboratory of Wuhan University of technology. Figure 4 is the diagram of test apparatus.

In the test, the concrete wall is inverted on the rigid ground, and then horizontal load is applied to the steel beam. In order to guarantee that the specimens don't move in the process of loading, some restriction measures have been adopted in horizontal and vertical direction. Two jacks are set between the end of concrete wall and reaction frame respectively in the horizontal direction. When the two jacks are working, they restrict the displacement of concrete wall in horizontal direction. It can also partially simulate the ratio of axial compression stress to strength of concrete wall in high rise building. At the end of concrete wall, two steel beams are set respectively in vertical direction. On the one hand it can compact the concrete wall and rigid ground tightly; on the other hand it can prevent the concrete wall buckling in the process of loading.

The loading device is double-acting hydraulic jack which is fixed in the reaction frame. The load is controlled by tensile and pressure sensing device, which is installed at the end of hydraulic jack. Function recorder reflects the data measured by sensing device directly. In order to prevent steel beam buckling in the outside plane, the lateral bracings, which are paralleling to the direction of loading along the both sides of steel beam, are installed. This experiment is a quasi-static test, and the method and the process of loading comes from the quasi-static test scheme (Guoqiang Li et al. 2003). The process of loading is controlled by load and deformation. Before the joint yields, it is controlled by gradation loading. Every grade load is 5 kN, reciprocally circulates a turnaround. After the joints yield, it is controlled by deformation. The value of deformation is the maximum value when the joints yield, and then take the multiple of the displacement value as gradation to control loading. Every grade of load reciprocally circulates two times.

2.3 Content of test

The content of test includes four aspects. The first is the stress distribution of flange girth, web plate and connecting plate of steel beam in the open part of the joint. The second is the displacement and deformation of steel beam. The third is the stress distribution of flange girth and web plate of the steel hidden corbel. The fourth is the stress distribution of concrete wall at the bilateral sides of steel hidden corbel.

The stress is tested mainly through strain gauge pasted in the surface of steel members. The displacement of steel beam is observed mainly through resistance type displacement device. At the same time the stress of concrete wall is measured with stain gauge and fiber sensor, which are buried in concrete wall (Figure 5a). In addition, a surface bended fiber sensor and a strain gauge are set at the flange girth of steel

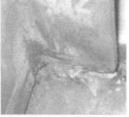

Figure 6. Fracture of welding seam.

Figure 8. Buckling of flange girth of steel.

Figure 7. Deformation of bolt holes.

beam in GJ2 (Figure 5b), and then the two results are contrasted in order to check the measuring effect of the fiber sensor.

3 ANALYSIS OF THE TEST RESULTS

3.1 *Main damage phenomenon of the test*

In the test, joint areas of the three specimens have obvious damage. According to the observed phenomenon, damages mainly occur in three positions: the welding seam between the flange girth and the sealing plate of the steel beam, the bolt hole in the web plate of steel beam, and the web plate at the end of steel beam near the sealing plate. The process of test may simply summarize as follows: at the beginning, the moment of the end of steel beam is commonly undertaken by the flange girth of steel beam, the welding seam and the friction force between the web plate of steel beam and connecting plate. When the load increases to a certain degree, the function of high strength bolts will be gradually out of work, and the slipping between the flange girth of steel beam and connecting plate will occur, then the moment of the end of steel beam is mainly undertaken by the flange girth of steel beam and the welding seam. When the load is close to yield load, the flange girth of steel beam starts buckling (Figure 8), and the tensile force undertaken by the welding seam increases continuously until the welding seam fails (Figure 6), then the specimen damages finally.

From theoretical aspects, the strength of the welding seam should be higher than the strength of steel beam, so the final damage is the buckling of the flange girth

of steel beam. But because the welding seams are all made by hand in site, it is difficult to guarantee the quality of the welding seam. It results in the tensile failure of the welding seam, when the buckling of the flange girth of steel beam occurs. In addition, after the test there is obvious slipping near the bolt hole in the web plate of steel beam, and the two sides of the bolt hole appear some deformation along the longitudinal direction under squeezing force (Figure 7). It indicates that high strength bolts undertake a part of load, and it also explains that the specimens can still provide a part of bearing force, when the welding seam is broken and the specimen appears great rotation.

3.2 *Results and analysis of bearing capacity*

The results of bearing capacity of three specimens are listed in Table 2. *Fy* is the yield load; *Fu* is the ultimate load in the table. According to the results of bearing capacity, it indicates that although the thickness of connecting plates, GJ1 and GJ2 which have same bolts, differs in 4 mm, the yield and ultimate bearing capacity are almost same. Although the connecting plate of GJ2, in which the diameter of the high strength bolt is bigger, is not the thickest, the bearing capacity of GJ2, is higher than the other two specimens evidently. It indicates that when the size of steel beam is same, the specification of the high strength bolts and the friction force between the web plate and connecting plate of steel beam are the main factors that affect the bearing capacity of the connection joint. So if the coefficient of friction surface and the prtightening force of high strength bolt do not change, the increasing of the thickness of connecting plate will not improve the bearing capacity of the connection joint.

The end moments of connection joints and connecting plates are listed in Table 3. The end moments of connection joints are theoretical moment value calculated by cantilever model. The moments of connecting plates are calculated by the stress value of connecting plate and the formula $\sigma = M/W$.

The data listed in Table 3 indicate that a part of the end moments of connection joints are delivered

Table 2. Experimental results of yield load, ultimate load and ductility coefficients.

Serial number	Positive load				Reverse load				Mean of positive and reverse load				
	F_y	F_u	u_y	u_d	F_y	F_u	u_y	u_d	F_y	F_u	u_y	u_d	u
GJ1	55.3	56.3	7.7	9.1	54.4	55.2	5.6	6.9	54.9	55.8	6.6	8.0	1.21
GJ2	71.0	78.8	14.3	36.3	64.5	66.3	10.1	12.1	67.8	72.6	12.2	24.2	2.00
GJ3	50.0	54.6	6.7	10.7	45.1	46.6	5.7	6.7	47.6	50.6	6.2	8.7	1.40

Table 3. End moment of joints and connecting plates.

Serial number	Load/kN	End total Moment(M)/ kN.m	End Moment connecting of plate(M1)/ kN.m	M1/M
GJ1	50	78.0	11.5	14.7%
GJ2	65	98.5	21.0	21.3%
GJ3	45	70.2	11.1	15.8%

through the high strength bolts before the specimens yield. If the rigidity of high strength bolts is bigger, they can deliver more moment. Thus the moment undertaken by the welding seam between the flange girth of steel beam and the seal plate is less, and the bearing capacity of the connection joint is improved.

3.3 Hysteresis and ductility performance analysis

In Table 2 there are the horizontal displacements and corresponding ductility coefficients of three specimens under yield load and ultimate load. In the table F_y is the yield load, u_y is the corresponding horizontal displacement; and F_u is the ultimate load, u_d is the corresponding horizontal displacement. u is the ductility coefficient. Figure 9(a) ~ Figure 9(c) is the hysteretic curves of load-displacement. From the hysteretic curves, some results are showed as follows:

(1) The shape of hysteretic curves and ductility coefficients of GJ1 and GJ3 have small differences. This indicates that the performances of the two specimens are almost same under reciprocal load.
(2) The hysteretic curve of GJ2 is fuller than that of GJ1 and GJ3, the middle retrenchment phenomenon of GJ2 is not more evident than the other two specimens. The final performance of deformation skeleton curve of GJ2 excels the other two specimens. It indicates that dissipative capacity and deformability of GJ2 have some improvement.
(3) Although the ultimate displacements of the three specimens are large, the ultimate displacement comparing with the yield displacement increases very few. So the ductility coefficient, which is calculated by the ultimate displacement, is not very

big. It indicates that the three specimens have some dissipative capacity, but the deformation capacity is not very ideal.

3.4 Stress analysis of hidden corbel

In the connection joints, the function of hidden corbel is to deliver the moment, which is delivered from the sealing plate, to the steel column and concrete wall. Because the steel column is buried in the concrete wall, the stress status is different from the simple cantilever members.

In the test, the stress status of the web plate and the flange girth in the middle of the hidden corbel is measured under reciprocal load. The results show that the stress status of bending section is the same as cantilever beam under reciprocal load.

Figure 10 is the load-stress curve of the flange girth in the middle of the hidden corbel of the specimen GJ1. In the figure, series 1 and series 3 are the theoretical stress values of simple cantilever beam respectively. Series 2 and series 4 are the actual measuring value. It indicates that no matter what the flange girth is under tension or compression status, the actual measuring value has no difference with the theoretical value (Bahram M et al. 2004).

3.5 The stress status of the concrete wall on both sides of the hidden corbel

When the hidden corbel is under bending, the concrete wall on both sides of the hidden corbel will be squeezed. In order to monitor the local squeezing, the imbedded strain gauge and imbedded fiber sensor are used in the test. The data indicate that even in GJ2, which has the highest, bearing capacity, the maximum stress of the concrete wall is only 5.8 MPa. It indicates that the function of the hidden corbel is not evident for the compression of concrete wall. This is also the main reason why the stress status of the hidden corbel is the same as the cantilever beam.

4 CONCLUSIONS

According to the observed phenomenon and test results, there are some conclusions and design suggestions.

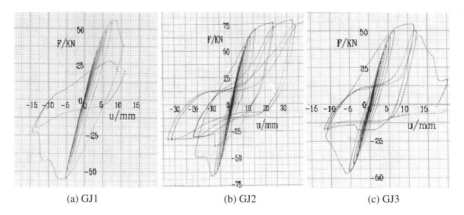

(a) GJ1   (b) GJ2   (c) GJ3

Figure 9.   Hysteresis curves of specimens.

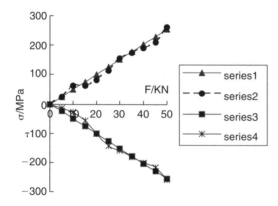

Figure 10.   Load – stress curves of flange girth of the hidden corbel.

(1) The connection joint of steel beam and concrete wall has enough bearing capacity and better dissipative performance. If the quality of welding seam between the flange girth and sealing plate of steel beam is guaranteed, it is proper to use the connecting type of the joint in the steel-concrete mixed structure.

(2) Because the flange girth and sealing plate of the steel beam must be welded in site, the quality of the welding seam can not be better guaranteed. The problem must be considered in design. It is suggested to reduce the designing strength of welding seam or adopt dog-bone connection joint in the flange girth of steel beam in order to guarantee the steel beam buckle when the damage occurs, and then improve the ductility capacity and dissipative performance.

(3) The high strength bolt connection of the web plate and connecting plate can bear a part of load, and it is also a kind of stable dissipative mechanism. It can reduce the earthquake energy input to the welding seam between the flange girth and sealing plate of steel beam, delay the brittle damage. So high strength bolt connection can affect the performance of steel beam and concrete wall connection joint greatly. In design, it is suggested to adopt high strength bolt of bigger specification when the tectonic requirement can be satisfied. So under the condition that the section of the steel beam do not augment, and the number of bolts do not increase, the bearing capacity and seismic performance of the connection joint are effectively improved.

(4) In the connection joint of steel beam and concrete wall, because the connecting plate is much thicker than the web plate of steel beam, after the friction of high strength bolt is out of work, the squeezing damage at the edge of the bolt hole must appear at the web plate of steel beam. Therefore the thickness of connecting plate has no influence to the performance of the connection joint. In design, it is only to satisfy the strength of the connecting plate in the plane and the stability out of the plane.

REFERENCES

Design code of steel structure (GB50017-2003)2003. Beijing: Press of china construction industry.
Design code of combined steel and concrete structure (YB9082–97)1998. Beijing: Press of metallurgy industry.
Design standard of high-rise steel-concrete mixed structure 2003. Shanghai: Engineering construction code of Shanghai City.
Guoqiang Li, Bing Qu, Feifei Sun etc 2003. Cyclic Loading Tests Of Steel Beam To Concrete Wall Joints In Steel Concrete Mixed Structures. Journal of Building Structure, 24(4):1–7.
Bahram M. Shahrooz, Jeremy T. Deason and Gokhan Tunc 2004. Outrigger Beam-Wall Connections. I : Component Testing and Development of Design Model. Journal of Structural Engineering. 130(2):253–261.

Steel and Composite Structures – Wang & Choi (eds)
© 2007 Taylor & Francis Group, London, ISBN 978-0-415-45141-3

Headed studs surrounded with cylinders steel behavior in HSC

M.R. Badr & A.E. Morshed

National Research Center for Housing and Building, Cairo, Egypt

ABSTRACT: This paper presents the main results of an experimental investigation of headed studs surrounded by steel cylinders welded to the flange of steel member and embedded in high strength concrete (HSC). This modification is suggested to increase the load carrying capacity of the welded headed studs embedded in HSC. The specific push out test according to Eurocode 4 1994 was applied on three samples of headed stud connectors embedded in HSC (f_{cu} average $= 76.4$ MPa). The test program included headed studs with 100 mm long and a diameter of 20 mm surrounded by cylinders 50 mm diameter, 50 mm height and 2 mm thickness. The obtained results were compared with previous work in literature and gave higher results. The experimental results indicated that the use of the proposed technique improved the load carrying capacity of headed studs by 1.65 and the ductility of composite member by twice.

1 INTRODUCTION

The headed studs are the most common shear connector in composite construction. Therefore, several modifications have been applied to improve the load shearing capacity of headed studs, especially when embedded in high strength concrete (HSC). Recently, numerous tests performed indicated that the behavior of headed studs in high strength concrete is different from that in case of normal strength concrete (Hegger J. et al. 2005).

In NSC where $f_{ck} < 50$ MPa, the load carrying capacity of the headed stud is governed by the concrete strength. The failure is initiated by concrete crushing around the shank of the stud. Due to the lower stiffness of the normal strength concrete, the stud shank is deformed along its height and a concrete crushing surrounded the shank of the stud. As shown in Figure 1 the headed stud is subjected to shearing force P_w at the base, shearing force P_F & P_S and tensile axial force P_T. These forces cause plastic deformation of the shank, shearing stress at the base and tensile force in axis of the shank. In this case, the shear studs fails immediately above the collar.

In HSC where $f_{ck} > 50$ MPa, the high strength material increases the load carrying capacity of structural members. Thus, the shear connectors should be developed to resist the expected high stresses.

The shear connectors have to be capable of transferring high shear forces. In the composite joint, the shank is tightly held in the surrounding concrete while

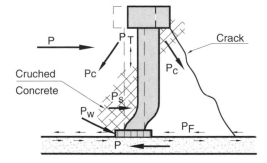

Figure 1. The behavior of headed studs in normal strength concrete (Hegger J. 2000).

the headed stud deformation is concentrated on the stud base directly above the welded collar. This leads to a high ultimate load but at the same time to significantly reduced ductility. Three research projects (ECSC 2000) has been investigated the load carrying behavior of shear connectors in HSC. They indicated that due to the high strength of materials, the headed studs is subjected to pure shearing force and fails directly above the welded collar (Hegger et al 2005).

Dabaon 2005 conducted an experimental and theoretical headed study to investigate the behavior of headed stud embedded in both NSC and HSC. The experimental tests indicated that there were two modes of failure for headed studs embedded in HSC. The

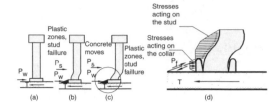

Figure 2. Mechanism of headed studs in HSC (Hegger J. 2005).

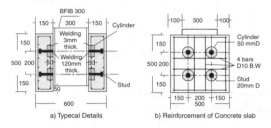

a) Typecal Details   b) Reinforcement of Concrete slab

Figure 3. Typical details of the specimen.

first mode was the failure of weld-collar in case of fillet welds but the second mode was the shearing off of the stud above the weld collar shank in case of K-shape butt welds. Hegger 2005 el al. suggested a new technique to improve the shearing resistance of the headed studs. They covered the headed studs by ultra – high performance concrete (UHPC). This modification increased the shearing resistance of the headed studs and hence, increased the load carrying capacity as well as improved the ductility of the headed studs.

From the previous investigation, it can be notice that the traditional headed studs failed under low levels of stress and it cannot be carry the higher stresses from HSC. Therefore, the headed stud shear connectors should be developed to be capable the higher shearing forces in the composite member in case of HSC and achieve the economic of using of the HSC. The aim of this research is to apply a modification of headed studs embedded in HSC to increase the load carrying capacity of the headed stud and examine the efficiency of this modification.

2   BEHAVIOR OF THE HEADED STUDS IN HSC

As shown in Figure 2. Doinghaus P. 2001 described the failure mechanism of headed studs in HSC as follows. At the beginning, the shearing force P translated from the steel flange to the concrete slab through Pw. The compression forces Pw are concentrated directly in front of the welded collar as shown in Figure 2-a. When the load increases the compressive forces concentrate within a concrete wedge. As shown in Figure 2-b, the main part of the load is carried by the welded collar. Subsequently the force in front of the welded collar moves upwards and the stud's shank starts to deform directly above the welded collar since it is subjected to shear force Ps. The concrete body slides over the concrete wedge as shown in Figure 2-c. Due to this uplift there is no full three –axial stress condition in compressive wedge starts to plasticize at approximately 150% of its uni-axial compressive strength. Friction forces Pf are generated in the interface between the concrete wedge and the concrete body as shown in Figure 2-d which lead to further compression of the concrete

wedge. Finally, the stud plasticizes directly above the welded collar and fails due to shearing.

The idea of welding cylinder steel around the heated stud is filling the space in between with HSC to perform a composite block of steel and concrete which produces additional resisting force to the direct shearing force between steel and concrete parts. Then, the direct shearing forces Ps&Pw are resisted by the total area of the composite block and the cross section of the studs. This leads to increase the carrying capacity of the headed studs.

3   EXPERIMENTAL INVESTIGATIONS

3.1   Details of the specimen

Three specimens were prepared to carry out this study. The dimensions of specimens were performed according to Eurocode 4. 1994 requirements. The headed studs were 20 mm diameter and 100 mm long. As shown in the Figures 3–5 the test specimen has the following features: Two concrete slabs 500 × 500 mm with thickness 150 mm, the steel member is standard BFIB 300 of height 500mm, each side of the steel member is provided with 4 headed studs. Each slab of the specimens is reinforced by 4 bars 10 mm diameter in two layers and in both direction of the slab.

3.2   Materials

The concrete mixture was designed to achieve high strength concrete. The mixture consisted of coarse aggregate (sand stone – dolomite), fine aggregate (sand) and silica fume. The workability was improved by using super plasticizer (P.V.F). The mixture was designed by the weight with ratio 1 : 1.13 : 2.9 cement: sand: dolomite. The water cement ratio was 0.3. The silica fume ratio was 0.16 and the super plasticizer (P.V.F) ratio was 0.04 from the cement content by weight.

The slabs of all the specimens were cast horizontally. All the first three slabs of the specimens were cast and after 5 days the other slabs of the specimens were cast. All the specimens were tested after 60 days to minimize the difference of concrete strength in the two slabs of the same specimen. 12 cubes for each

Figure 4. Details of headed studs and the welded cylinder around it.

Figure 5. Details of reinforcement of the concrete slab.

mixture were prepared for each time of cast. Half of those were tested after 28 days and remain cubs were tested at the same time of the specimen tests. The average strength of the standard cubes after 60 days was 77.2 MPa for the first specimen, 77 MPa for the second specimen and 75 Mpa for the third specimen.

The steel of the headed stud and the cylinder was mild steel 37 type. Where the ultimate strength is 360 MPa and the yield strength is 24 MPa . The conventional welding system was used for welding the headed studs and the cylinders by a professional welder and the welding was tested.

3.3 Loading

After 60 days from the casting of concrete the push out test of the specimens were tested and at the same time the standard cubes were examined. Each specimen was placed in the testing machine and was plumbed and leveled. A layer of grout was placed beneath the specimen to provide uniform support. Both vertical displacements on either side were monitored using LVDT at both sides of the specimen. All tests were carried out using 500 ton capacity machine which it is available in the concrete research laboratory, National Research Center for Housing and Building, Cairo.

The load was applied to the exposed steel member at the top of the specimen, and the reaction was provided through the concrete at the base. Firstly, the load was applied incrementally 5 ton per minute until 40% of the expected ultimate load. After this the load was released and started to increase incrementally with constant rate of 5 ton per minute for 0.001 mm slip. The load was increased by this rate until the failure load of each specimen.

4 EXPERIMENTAL RESULTS

4.1 Test results

During the test, the load-longitudinal slip readings were recorded by data acquisition system for each test specimen. Figure 6 shows the load – longitudinal slip curve for the three specimens. The maximum loads were 213.0, 224.22, 206.05 ton for the specimens number 1, 2 & 3 respectively.

Table 1 indicates the maximum average shear capacity of each headed stud which calculated by dividing the failure load of the specimen by the total number of the headed studs embedded within the concrete slabs. Table 2 indicates the results of Daboan 2005 for comparison the results with the obtained results. It is clear to notice the difference between the two results in spit of the difference of the concrete strength.

4.2 The failure mode

The observation of the failure mode indicated that all the specimens failed above the welds due to fracture

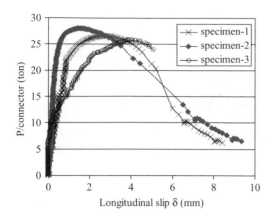

Figure 6. The load–longitudinal slip relationship of the tested specimens.

Table 1. The obtained results from the experimental tests.

No	Description	Fcu (MPa)	Pmax for each stud (ton)	δ at Pmax (mm)
1	20 mm Diameter headed studs +	77.2	26.62	2.816
2	50 mm diameter cylinder, 2 mm	77	28.03	1.691
3	thickness & 50 mm height	75	25.75	4.072

Table 2. The results of the previous test (Daboan 2005).

No	Description	Fcu (MPa)	Pmax for each stud (ton)	δ at Pmax (mm)
1	20 mm Diameter	43	11.27	4.4
2	headed studs	54	11.3	1.92
3		65	16.25	1.39

Figure 7. Typical failure mode of headed studs in the steel member.

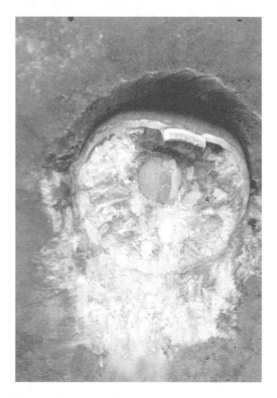

Figure 8. The failure of headed studs and the cylinder in the concrete slab.

of the shanks of the headed studs and the around cylinders accompanied with local failure of concrete around and inside the cylinder. No significant cracks in the concrete appeared until the failure of the specimens. As shown in Figure 7. the welding parts of the headed studs and the around cylinders attach to the steel member while the main parts of the headed studs buried in the concrete slab as shown in Figure 8.

Also, Figure 8. indicates gab in the concrete slab above the cylinder due to amount of deformation of the cylinder at the interface prior the failure accompanied with crushing of the concrete inside the cylinder.

4.3 Evaluation of results of push out tests

The investigation of the results in Figure 6. indicates that:

- The actual slip started at load of 30-60% of the maximum load.
- The slip at the maximum load is between 1.7 and 4.0 mm
- The failure load occurred suddenly after the maximum load.

426

From the comparison between the results in Tables 1 & 2 it is obvious that:

- The ultimate load of the same headed studs 20 mm diameter and nearly concrete strength increased by 1.65 due to using of the proposed technique.
- The slip at the maximum load is improved by twice and the ductility of the proposed technique is approximately twice the ordinary system.

4.4 *The failure load calculation*

Although, the calculation of the load capacity of the headed studs in this case is not easy, but according to the plasticity theory, the strength of headed studs depends on the local concrete pressure and shearing strength of the shear connector. Where, the general equation for the headed studs is:

$$Prd = k1 \ d^2 \ (E_c.f_{ck})^{1/2} < k2 \ F_u . As \qquad (1)$$

The first part of the equation controls the failure of concrete while the second part controls the failure of shank cross section.

Where, d is the diameter of the shank; E_c, f_{ck} are the modulus of elasticity and compressive strength of concrete; F_u, As are the ultimate strength and area of the steel of the headed studs; and k1 & k2 are constant. k1 = 0.29 and k2 = 0.8 according to EC-4 1994 for normal concrete.

By applying this equation in the proposed shear connector, according to the fracture failure of steel and putting Prd=, average Pmax of the tests, As = As of the shank + As of the cylinder and Fu = 360 MPa, then: k2 average = 1.12 this value is bigger than the shearing capacity of the steel sections, while k2 = 0.8 for normal concrete. So, try to calculate the first part of the equation 1. if the modulus of elasticity of HSC is calculated according to Nawy E.G. 2000:

$$Ec = (3.32\sqrt{f_{ck}} + 6895) \ (wc/2320)^{1.5} \qquad (2)$$

Where, Ec in MPa and wc is the unit weight of concrete (k/m3), using the first part of equation 1 and d = Diameter of the cylinder then the average of k1 equals 0.15 while k1 = 0.29 for normal concrete.

This indicates that the failure occurred at level over than the capacity of the shear strength of the steel section and lower than the strength of the concrete.

5 CONCLUSION

An innovative technique of headed shear studs was developed and tested. The proposed technique included headed studs with 20 mm diameter and 100 mm long surrounded by welded cylinder steel 50 mm in diameter, 50 mm height and 2 mm thickness. The space in between the stud shanks and the steel cylinders were filled by concrete during the casting of the slabs performing composite blocks. These blocks capable to resist the high shearing force which subjected to the shear connectors embedded in HSC. The push out test according to Eurocode 4 1994 was carried out on three specimens. The load – longitudinal slip relationship were established for the specimens and the mode failure was observed. The following conclusions are obtained:

- By using the welded cylinder steel around the headed studs, the load capacity of the headed studs increased by 1.65 and the ductility improved by twice approximately.
- The failure of all the tests specimens was a fracture of the stud shanks and surrounded cylinders steel above the welds directly. No significant cracks in concrete slab appeared until the failure of the specimens. A crushing of concrete occurred inside the steel cylinder.

6 RECOMMENDATION

The improvement of the carrying capacity load of headed studs as well as the ductility encourage to make further investigations on the proposed technique to achieve the optimum dimension of the cylinder dimensions and let the failure occurred in concrete not in the shank of the studs.

REFERENCES

Dabaon M. & Fahmi M. 2005 . Behavior of shear connectors using normal and high strength concrete. *11the ICSGE*. Ain Shams University. Cairo. Egypt: P E05st27-1-12.
Doinghaus P. 2001. Zum Zusammenwirken hochfester Baustoffe in Verbundtragen. *Doctoral thesis*. RWTH Aachen University. Germany.
ECSC . 2000. Use of high strength steel 5460. *Research project No. 7210-SA/129,130,131,325,523*. RWTH Aachen. Germany.
Eurocode 4 Part 1.1 1994: Design of Composite Steel and Concrete Structures.
Hegger, J., Raucher, S. & Goralski . C. 2005 . Push out Tests on Headed Studs Covered with UHPC . *4th European Conference on Steel and Composite Structures*. Maastricht: P4.2-49-55.
Nawy E.G. 2000. Prestressed concrete: A fundamental approach. 3ed edition (upper saddle river . NJ: practice-Hall: P.38.

Steel and Composite Structures – Wang & Choi (eds)
© 2007 Taylor & Francis Group, London, ISBN 978-0-415-45141-3

Behavior of strengthened channel connectors in HSC

M.R. Badr & A.E. Morshed
National Research Center for Housing and Building, Cairo, Egypt

ABSTRACT: The paper presents a development of the channel shear connector to increase its capacity load especially when embedded in high strength concrete (HSC). The channel was provided by middle stiffener welded back perpendicular to the channel. The back stiffener strengthens the stiffness of the channel connectors as well as increases its welding to resist high stresses due to using HSC. The specific push out test according to Eurocode 4 was applied on three samples of channel connectors embedded in HSC (f_{cu} average = 73.6 MPa). The development included standard channels UPN 100 with length 100 mm strengthened by plate 80 × 80 mm in dimension and 8 mm thickness which welded back to the channel and top of the flange of the steel member. The obtained results were compared with the previous work in literature and gave improvement. The experimental results indicated that the using of the proposed technique improve the load carrying capacity of the channel connectors in HSC by 3 times as well as the ductility by 3 to 4 times.

1 RESARCH PROBLEM AND BACKGROUND

The channel connectors have been used in composite structure due to higher load carrying capacity. As results, one channel connector may be replaced number of headed studs. The channel connectors can be used in case of convention welding. Therefore, several modifications have been applied to investigate their shear capacity. Badr 2000 produced an experimental study on channel shear connectors embedded in solid normal concrete (NSC) slab using a specific push-out test based on Eurocode 4 1994. The specific push-out test was applied on three different types of channel shear connectors. He found that the factor of safety in the available codes for designing the channel shear connectors is conservative. Pashan 2006 investigated the behavior of channel shear connectors embedded in NSC. He studied the effect of number of parameters on the shear capacity of channel shear connectors. Depending on the results, he suggested a new equation for predicting the shear strength of the channel connectors rather than the CAN/CSA equation. The failure modes of the channel connectors embedded in NSC were two different modes (Badr 2000). The first mode was deformation of channels inside the concrete slab as shown in Figure 1. The other mode was fracture of web above the welded flange of the channel which attached with steel member while remain part of the channel buried in the concrete slab.

The use of channel connectors in HSC is different. High strength materials increase the load carrying

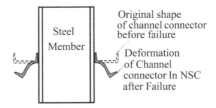

Figure 1. The failure of channel connectors in NSC, first mode of failure (Badr 2000).

capacity of structural members. Thus, the shear connectors have to be capable of transferring high shear forces in the composite joint. Dabaon 2005 conducted experimental and theoretical study to investigate the behavior of channel connector embedded in both NSC and HSC. The experimental tests indicated that there were two modes of failure for channel connectors embedded in HSC. The first mode was the failure of welding due to the fillet weld while the second mode was fracture of web at the welded flange of the channel with the steel member. The problem of using the traditional channel shear connectors in HSC is not applicability to carry the higher stresses.

The objective of this study is to strengthen the normal channel connectors embedded in HSC by providing perpendicular stiffener in the back of the channel to increase the load carrying capacity of the connectors. This modification is increased the stiffness

BFIB 300

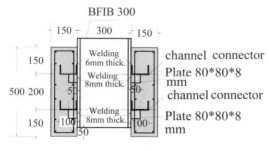

channel connector
Plate 80*80*8 mm
channel connector
Plate 80*80*8 mm

a. Typical Details of the specimens

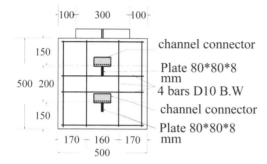

channel connector
Plate 80*80*8 mm
4 bars D10 B.W
channel connector
Plate 80*80*8 mm

b. Reinforcement of concrete slab

Figure 2. Typical details of the specimens.

Figure 3.   Welding of the channel and the back stiffeners.

Figure 4. Typical reinforcement of the slabs of the specimens.

of the channel as well as the welding length of the connectors to prevent the fracture of the welds.

2 EXPERIMENTAL PROGRAM

2.1 *Specimens*

Three specimens were prepared to carry out this connectory. The dimensions of specimens were performed according to Eurocode 4 1994 requirements. The channels were standard UPN 100 with 100 mm length. The stiffeners were 80 × 80 with thickness 8 mm. As shown in the Figures 2–4 the test specimen has the following features: two concrete slab 500 × 500 mm with thickness of 150 mm; the steel member was standard BFIB 300 of height 500 mm; each side of steel member was provided with 2 channel connectors; the slabs was reinforced by 4 bars 10 mm in both ways in two layers as shown in Figure 2.

2.2 *Materials*

The concrete mixture was designed to achieve high strength concrete. The mixture consisted of coarse aggregate (sand stone-dolomite), fine aggregate (sand) and silica fume. The workability was improved by

using super plasticizer (P.V.F). The mixture was designed by the weight with ratio 1 : 1.13 : 2.9 cement: sand: gravel. The water cement ratio was 0.3. The silica fume ratio was 0.16 and the super plasticizer (P.V.F) ratio was 0.04 from the cement content by weight.

The slabs of all the specimens were cast horizontally. All the first three slabs of the specimens were cast and after 5 days the other slabs of the specimens were cast. All the specimens were tested after 60 days to minimize the difference of concrete strength of the two slabs of specimens. 12 cubes for each mixture were prepared for each time of cast. Half of them were tested after 28 days and remain cubes were tested at the same time of the specimens test. The average strengths of the standard cubes were 74.8, 72.5 and 73.8 MPa for the first, second and third specimen respectively. The steel connector and the stiffener were mild steel 37 type. Where the ultimate strength is 360 MPa and the yield strength is 24 MPa. The conventional weld system was used for welding the channels and the stiffeners to the steel member. The welds were performed by a professional welder as well as the weld was tested by the x-ray test.

2.3 Test progress

All the specimens were tested after 60 days from the casting of concrete and in the same time the standard cubes were examined. Each specimen was placed in the testing machine. A layer of grout was placed beneath the specimen to provide uniform support. Both vertical displacements on either side of the specimen were monitored by using LVDT at both sides of the specimen and the data were collected by data acquisition system connected by personal computer.

All tests were carried out using the machine 500 ton capacity which is available in the concrete research laboratory, National Research Center for Housing and Building, Cairo.

After adjustment of the specimen, load was applied to the exposed steel at the top of the specimen, and the reaction was provided through the concrete at the base. The load was applied incrementally with constant rate of 5 ton per minute until 40% of the expected failure load according to Eurocode 4. Then, the specimen was unloaded and the load was increased incrementally with constant rate of mm 5 ton per minute for 0.001 mm slip until the failure of the specimen.

3 TEST RESULTS

3.1 Push out test results

Each specimen was loaded until failure of the specimen. The load-longitudinal slip readings are recorded by data acquisition system and stored by PC. Figure 5. shows the load–longitudinal slip curve for the three specimens. The max loads were 240.73, 220.85 and 231.89 ton for the specimens number 1,2&3 respectively.

The maximum average shear capacity of each channel connector was calculated by dividing the maximum load failure by the total number of the connectors embedded within the concrete. Table 1 indicates the maximum average shear capacity of each connector and the longitudinal slip at maximum load. Table 2 indicates the results of Daboan 2005 for comparison with the obtained results. It is clear to notice the difference between the two results in spit of the difference of the concrete strength.

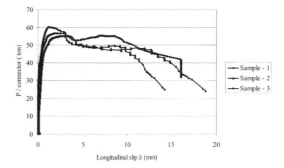

Figure 5. The load-longitudinal slip relationship of strengthened channel connectors in HSC.

Table 1. The results of experimental tests.

No	Description	Fcu (MPa)	Pmax for each connector (ton)	δ at Pmax (mm)
1	channel UPN 100 mm	74.6	60.18	1.25
2	& 80*80 mm	72.5	55.21	4.049
3	stiffener plate 8 mm thickness	73.8	56.47	3.074

Table 2. The results of the previous tests (Daboan 2005)

No	Description	Fcu (MPa)	Pmax for each connector (ton)	δ at Pmax (mm)
P2	channel connector	56	18.5	0.99
P3	UPN 100	85	18.6	0.93

3.2 Test failure

The failure in the first specimen occurred due to fracture of the web of the channel accompanied with failure in the welds of the stiffener as shown in Figure 6 & 7. The fracture of welds of the stiffener followed by fracture of the channel web near the fillet welds of the flange. But the observation of Figure 5. indicates that this may be occurred at last stage of the failure. Figure 6. indicates that the flange of the channel attached to the steel member while the remain part of the channel buried in the concrete stab as shown in Figure 7.

The failure of second and third specimens was crushing of concrete slabs with some deformation of the channel web and flanges as shown in Figures 8–9. It is clear to notice that the failure in concrete slab was a shear failure where it is in diagonal shape. After the removal of concrete slab, it observed that the channel

431

Figure 6. Indicates the attached part of the flange of the channels and the welds of the stiffener in the steel member.

Figure 8. The failure of the concrete slab.

Figure 7. Indicates part of the channel and stiffener which buried in the concrete slab.

Figure 9. The crushing part of the concrete slab at the failure load.

had undergone considerable deformations but remain attached to the steel member.

3.3 Analysis of the results

The investigation of the results in Figures 5–8 indicates that:

- The relationship between the load- slip started linear until 60% of the ultimate load. The cracks

developed in the slab and the plastic slip was started due to the increasing of load until the failure.
- The slip at the maximum load is between 3&4 mm.

From the comparison between the results in Tables 1 & 2 it is obvious that:

- The ultimate load of the channel connectors is increased due to using of the proposed technique by 3 times.
- The max slip is increased due to the proposed technique by 3–4 times.

432

The available equation in case of channel connectors in NSC is the equation of the CAN/CSA-S16 2001. If applied this equation here:

$$Q = 36.5 \, \phi \, (\, t + 0.5 \, w \,) \, Lc \, (fc)^{0.5} \qquad (1)$$

Where, ϕ = resistance factor; t = flange thickness of the channel in mm; w = web thickness of channel in mm; Lc = length of the channel in mm; fc = specified cylinder compressive strength of concrete MPa.

If $\phi = 1$, t = 8.5 mm, w = 6 mm, Lc 100 mm, fc = 58.88 MPa.

Then, Q = 32.2 ton, but the obtained failure load is approximately twice time of this value. Then, this equation is not valid for using of HSC.

4 CONCLUSION

A new technique of strengthening the channel shear connectors was applied to increase its shearing load capacity in composite structure especially when embedded in HSC. The channel was strengthened by perpendicular stiffener welded back to the channel to increase channel stiffness as well as the welding length of channel in the steel member. The push out test according to Eurocode-4 1994 was carried out on three specimens. The following conclusions are obtained:

– By using the suggested technique, the shearing load capacity of the channel connectors was increased by 3 times and the ductility was increased by 3–4 times

– The failure mode of the specimens was due to the crushing of the HSC concrete slabs which indicates that this type of shear connectors is suitable to use in case of HSC.

It is recommended to make further investigation on this technique of channel shear connectors for different heights, lengths, stiffener size to develop general roles which leads to predict the shear capacity of this type of shear connectors.

REFERENCES

Badr M.R. 2000 . Experimental study on channel shear connectors embedded in sold concrete slab. *ASCE-Egypt.*

Dabaon M. & Fahmi M. 2005 . Behavior of shear connectors using normal and high strength concrete. *11the ICSGE.* Ain Shams University. Cairo. Egypt : P E05st27-1-12

CAN/CSA –S16 2001. Limit states design of steel structures. Canadian Standard Association. Rexdale. Ontario.

ECSC. 2000 . Use of high strength steel 5460. *Research project No. 7210-SA/129,130,131,325,523, RWTH Aachen.* Germany.

Eurocode 4 Part 1.1 1994 . Design of Composite Steel and Concrete Structures.

Pashan A. 2006. Behavior of channel shear connectors: Push out tests. Thesis for degree of master of science. *University of Saskatchewan.* Canada.

Steel and Composite Structures – Wang & Choi (eds)
© 2007 Taylor & Francis Group, London, ISBN 978-0-415-45141-3

Hot spot stress of concrete-filled circular hollow section T-joints under axial loading

Lewei Tong, Ke Wang, Weizhou Shi & Yiyi Chen
Tongji University, Shanghai, China

Bin Shen & Chubo Liu
Shanghai Baoye Construction Co. Ltd, Shanghai, China

ABSTRACT: Welded trusses made of concrete filled circular hollow sections (CFCHS) have come into increasing use in large span highway bridges in China. It is worth while to pay attention to the fatigue problem of welded CFCHS joints under daily vehicle loads. Hot spot stress approach can be used for their fatigue assessment, like welded circular hollow section (CHS) joints. Static experiments on ten welded CFCHS T-joints subjected to axial loads in the braces were carried out in the paper. Different non-dimensional geometric parameters were considered for the ten test specimens. The measuring method and results of hot spot stress concentration factors (SCF) are presented. The effects of geometric parameters and concrete strength on the SCF around the CFCHS joint are discussed. It is found that the concrete filled in the tubular chord significantly increases the stiffness, and then decreases the SCF around the CFCHS joint and makes the position of maximum SCF changed, compared with the welded joint of made circular hollow sections (CHS). Furthermore, the CFCHS joint shows lower SCF values under axial tension than under axial compression in the brace.

1 GENERAL INSTRUCTIONS

Due to their excellent structural performance, concrete-filled circular hollow sections (CFCHS) have been increasingly used in large span highway bridges such as arch bridges and cable stayed bridges in recent years in China. The arch can be made of CFCHS trusses for a arch bridge and the main girder can be made of CFCHS reticulated structures for a cable stayed bridge as shown in Fig. 1 The larges CFCHS truss arch bridge in China comes up to 460 meters (Zhou and Chen 2003). For these bridges, circular hollow section (CHS) chord members and web members are connected by cutting and welding techniques, and then concrete is filled into chords, but webs are keeping empty for the convenience of the construction. It is realized that such CFCHS trusses have better structural behaviour than CHS trusses in both the whole structures and the connections between chords and webs.

As we know, welded joints of CHS trusses are prone to fatigue failure under a long period of repeated loads, because of very high stress concentrations and welding defects at the joints. A considerable amount of research over more than twenty years has resulted in good understanding of fatigue behaviour of welded

CHS joints and relevant design guidelines (Zhao et al. 2000 and 2001). The fatigue assessment based on the advanced concept of hot spot stress is adopted in current guideline, in which hot spot stress concentration factors (SCF) along the welded joints are key important data.

The welded joints of CFCSH arch truss bridges under daily vehicle loads may face fatigue problem, but their research is quite shortage currently (Udomworarat et al. 2000). It is estimated that the welded CFCHS joints have the similar fatigue behaviour to the welded CHS joints in some aspects and different in other aspects. The existence of concrete in the CHS chords certainly changes the original fatigue behaviour of welded CHS joints. It is a new research area how concrete exerts its effect on CHS joints. It is considered that the hot spot stress approach used in the fatigue assessment of the CHS joints is still suitable for that of the CFCHS joints.

A research project on the fatigue behaviour of welded CFCHS T-joints is performing in Tongji University, which deals with experiments and numerical analysis of hot spot stresses, fatigue strength and crack propagation life prediction. The paper presents the experimental results of our first research period on stress concentration factors.

(a) Arch bridge

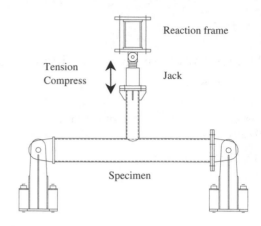

Figure 2. Test specimen and setup.

(b) Cable stayed bridge

Figure 1. Applications of concrete-filled circular hollow sections in bridges of China.

2 EXPERIMENTAL SETUP

2.1 Test specimens

Ten test specimens of welded CFCHS T-joints shown in Fig. 2 were designed for the static tests in the paper and the fatigue tests in the future. For each specimen, the CHS chord member was filled with concrete, but the CHS brace was kept empty. Table 1 gives the details of these specimens, in which different non-dimensional geometric parameters and concrete grades were arranged to present their effect on joint behaviour. Both the CHS braces and chords were made of steel Q345. The mechanical test of the steel showed its yield stress was 324 MPa.

In order to make the stress distribution along the joint reasonable, not being influenced by the member end conditions, the lengths of the chord and brace for each specimen were taken to be about $6d_0$ and $5d_1$ respectively (Wingerde, 1992). The brace was connected to the chord by manual arc welding with full penetration welds according to AWS D1 (AWS 2000).

Each specimen was simply supported at the two ends of the chord member. An axial force was applied to the end of the brace member through a hydraulic jack that can push and pull. Static tests under both the axial compressive and tensile forces were carried out for each specimen to see if there would be any difference in joint behaviour. Each specimen was loaded under the condition of linear elasticity.

2.2 Determination Method of SCF on CFCHS T-joints

The hot spot stress (also called geometric stress) approach is a dominant method for the fatigue assessment of welded joints made of circular and rectangular hollow sections (CHS and RHS). The hot spot stress can be obtained from stress concentration factor (SCF). SCF is ratio between the hot spot stress at a joint and the nominal stress in the member due to a basic member load which causes this hot spot stress. It is considered that the method is also suitable for the fatigue assessment of welded CFCHS joints, although their fatigue behaviour is more or less different from that of welded CHS joints.

Like the method used for a CHS joint (Zhao et al. 2000), four lines of measurement of hot spot stresses at the CFCHS joint of the weld toe were arranged respectively on both the chord and the brace, namely two lines for crowns and the other for saddles as shown in Fig. 3. Besides, three supplementary lines, 22.5 degrees apart each other in a quarter quadrant of intersection, were added respectively onto the chord and brace in order to observe whether there would be any other higher hot spot stress between the crown and the saddle. A set of four strain gauges manufactured specially for the tests was used in each line of measurement, in which the four measuring points was set within the extrapolation region in Fig. 3 and Fig. 4. The linear extrapolation method was adopted.

Table 1. Details of test specimens.

Specimens	$d_0 \times t_0$ (mm × mm)	$d_1 \times t_1$ (mm × mm)	$\beta = d_1/d_0$	$\gamma = d_0/2t_0$	$\tau = t_1/t_0$	Steel Grade	Concrete Grade
CFCHS-1	245 × 8	133 × 8	0.54	15.31	1.00	Q345	C50
CFCHS-2	180 × 6	133 × 6	0.74	15.00	1.00	Q345	C50
CFCHS-3	133 × 4.5	133 × 4.5	1.00	14.78	1.00	Q345	C50
CFCHS-4	245 × 8	133 × 6	0.54	15.31	0.75	Q345	C50
CFCHS-5	245 × 8	133 × 4.5	0.54	15.31	0.56	Q345	C50
CFCHS-6	245 × 8	133 × 8	0.54	15.31	1.00	Q345	C20
CFCHS-7	245 × 8	133 × 8	0.54	15.31	1.00	Q345	C70
CFCHS-8	203 × 8	140 × 8	0.69	12.69	1.00	Q345	C50
CFCHS-9	203 × 10	140 × 10	0.69	10.15	1.00	Q345	C50
CFCHS-10	203 × 12	140 × 12	0.69	8.46	1.00	Q345	C50

Note: d_0 and t_0 = diameter and thickness of the chord respectively. d_1 and t_1 = diameter and thickness of the brace respectively.

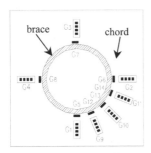

(a) Measurement lines

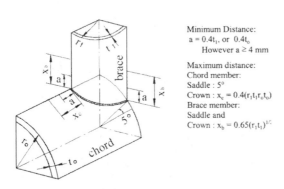

Minimum Distance:
a = 0.4t_1, or 0.4t_0
However a ≥ 4 mm

Maximum distance:
Chord member:
Saddle : 5°
Crown : x_c = 0.4$(r_1t_1r_0t_0)$
Brace member:
Saddle and
Crown : x_b = 0.65$(r_1t_1)^{1/2}$

(b) Boundaries of extrapolation region

Figure 3. Measurement of hot spot stress for CFCHS joint.

Figure 4. A set of four strain gauges on CFCHS T-joint.

2.3 Experimental results and discussion

Table 2 shows the experimental SCFs of CFCHS T-joints under both the axial compressive and tensile forces on the braces, which were obtained from the conversion of the original strain concentration factors (SNCFs) measured using strain gauges, namely SCFs = 1.2 SNCFs (Zhao X.L. *et al.* 2001). Only the hot spot stresses at the saddle and crown positions are listed in the table 3, because they are larger than the hot spot stresses between the saddle and the crown in a chord or a brace based on the test results of measurement lines shown in Fig. 3. It is worth while to compare the SCFs of CFCHS joints with those of CHS joints under the same conditions such as the joint type, load and geometric parameters of tubes. So the CFCHS T-joints investigated in the paper can be regards as CHS T-joints without any concrete inside the chords. Their SCFs can be calculated using the existing SCF formula (Zhao X.L. *et al.* 2001) and listed in Table 2.

It can be seen from the table 2 that the maximum SCF along the intersection of the CFCHS joint is always smaller than the maximum SCF of the CHS joint, no matter what axial load is applied on the brace of CFCHS joint, compression or tension. Furthermore, the maximum SCF of CFCHS joint almost occurs at the chord crown, whereas the maximum SCF of CHS joint does at the chord saddle. It indicates that the concrete filled in the chord exerts an important influence on the behaviour of CFCHS joint. The

Table 2. Experimental SCFs of CFCHS joints and comparison with calculational SCFs of CHS joints.

	Experimental values under compression (1)				Experimental values under tension (2)				Calculational values as CHS joints (3)						
	Chord		Brace		Chord		Brace		Chord		Brace				
Specimens	S	C	S	C	S	C	S	C	S	C	S	C	(1)/(2)	(1)/(3)	(2)/(3)
CFCHS-1	2.35	**4.86**	3.55	3.82	4.20	**6.79**	5.21	2.39	**15.28**	7.10	8.20	3.34	0.71	0.32	0.44
CFCHS-2	0.94	**5.11**	0.22	4.06	1.50	**8.86**	0.58	3.36	**13.90**	8.22	7.07	3.76	0.58	0.37	0.64
CFCHS-3	0.14	**8.00**	1.42	6.26	0.78	**12.50**	3.55	6.08	6.40	**11.66**	3.56	4.93	0.64	0.69	1.07
CFCHS-4	1.52	**2.72**	1.55	1.34	1.99	**3.98**	2.10	1.52	**11.15**	5.32	7.24	3.07	0.68	0.24	0.36
CFCHS-5	1.37	2.33	2.24	**3.10**	1.43	**4.08**	2.72	2.33	**8.14**	4.00	6.41	2.87	0.76	0.38	0.5
CFCHS-6	2.99	**5.34**	2.83	5.02	3.30	**7.42**	2.95	3.16	**15.28**	7.10	8.20	3.34	0.72	0.35	0.49
CFCHS-7	2.40	**4.06**	2.16	4.12	3.66	**6.54**	3.29	2.59	**15.28**	7.10	8.20	3.34	0.62	0.27	0.43
CFCHS-8	2.08	**6.84**	1.82	4.00	2.16	**7.98**	2.42	2.66	**10.01**	7.00	5.41	3.72	0.85	0.68	0.8
CFCHS-9	1.46	**7.78**	0.79	5.68	2.44	**8.77**	1.32	2.51	**12.18**	7.18	6.44	3.56	0.89	0.64	0.72
CFCHS-10	4.52	**8.72**	2.99	2.11	3.52	**8.65**	2.86	1.76	**8.56**	6.84	4.73	3.82	1.01	1.02	1.01

Note: (1) S – Saddle position; C – Crown position (2) The SCF highlighted is the maximum SCF among the four positions of the chord saddle, chord crown, brace saddle and brace crown.

concrete increases greatly the stiffness of the whole joint, restrains ovalisation of the chord and makes the stiffness distribution around the intersection evener so that the hot spot stresses in the joint become much smaller and the position of the maximum SCF changes from the chord saddle of the joint without concrete to the chord crown of the joint with concrete.

It can be found from the table 2 that the SCF of the CFCHS joint under axial compression on the brace is quite different from that under axial tension. The maximum SCF under compression is about 10% ~ 40% less than that under tension, but their positions do not change, both being at the chord crown. The phenomenon can be explained that the wall of the chord in the vicinity of intersection tends to separate from the concrete inside the chord due to low cohesive force between steel and concrete when the brace is subjected to an axial tension, which makes the ability of collaboration between the chord and concrete weaker at the local area. As a result, the bending deflection of the chord wall around the intersection grows larger and the hot spot stress becomes higher.

Figure 5 presents the curves of maximum SCFs versus non-dimensional geometric parameters (β, τ, γ) and concrete grade based on the relative data in the table 1 and the table 2. The following observations can be made from Fig. 5 based on the non-dimensional geometric parameters (β, τ, γ) investigated in the paper:

(1) The effect of β on SCFmax of the CFCHS joint is significant. SCFmax increases with increasing β value. However, the effect of β is opposite for the CHS joint, that is to say, SCFmax decreases with increasing β value. It indicates that small β value is good for decreasing the hot spot stress.

(2) Both τ and γ exert no important effect on SCFmax of the CFCHS joint. This behaviour is quite different from that of the CHS joint. SCFmax of the CHS joint increases obviously with increasing τ and γ respectively. τ is the ratio of wall thickness (t_1/t_0) and γ is the ratio of half chord diameter to thickness ($d_0/(2t_0)$). Both are the function of wall thickness of the chord. The thickness of tubular chord seems to have no meaning after concrete is filled, because the tubular chord becomes solid and its diameter can be regarded as the whole thickness of the chord. Compared with the diameter of the tubular chord, the wall thickness of tube is so small that it can not play important role in the CFCHS chord.

(3) The concrete grade has not much effect on SCF$_{max}$ of the CFCHS joint within the period of elasticity, which indicates the main function of concrete is to enhance the stiffness of the joint, but the strength itself, no matter what grade is, has almost nothing to do with the SCFmax of the joint.

3 FINITE ELEMENT MODELLING AND VALIDATION

The FE models were generated using the software package ANSYS 9.0. An higher order solid element SOLID95 was used to simulate the steel tube and the filled concrete. The element is defined by 20 nodes having three degrees of freedom per node: translations in the nodal x, y, and z directions. The element may have plasticity, creep, stress stiffening,

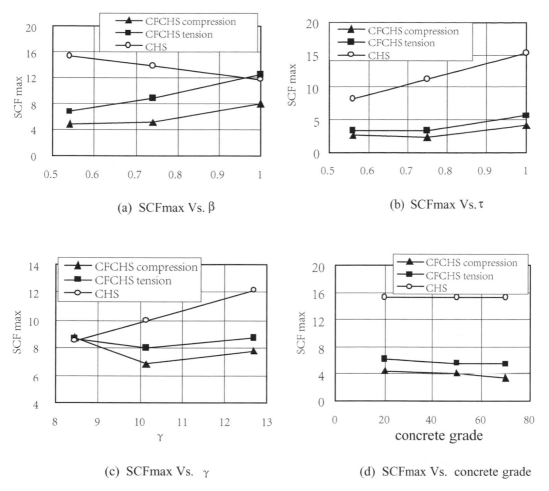

(a) SCFmax Vs. β

(b) SCFmax Vs. τ

(c) SCFmax Vs. γ

(d) SCFmax Vs. concrete grade

Figure 5.   Effect of geometric parameter and concrete grade on SCF_{max}.

large deflection, and large strain capabilities. The integration older chosen for the present work is $2, \times 2 \times 2$. And the surface-to-surface contact elements (targe 170 and conta 174)were employed to simulate the flexible-flexible contact between the steel and concrete surfaces. The The Young's Modulus $E = 200$ KN/mm^2 and the Poisson ratio $\nu = 0.3$ were used.

The weld geometry at the intersection of two tubular members is very complicated and difficult to model accurately. An approximation to the actual weld profile can be obtained by using concentric cylinders to model the weld volume. And it is improtant to ensure that the weld profile at the expected hot-spot stress site conforms to design specification (AWS , 2000).

As a general rule, the mesh density required for a stress analysis is significantly higher than that required for a strength analysis. In order to get the accurate stresses, which are perpendicular to the weld toe, element dimensions of $0.5t_1$ to $0.5t_0$ can be used inside

the joint intersection. The element size in regions of less interest can be increased to save computing time. A typical finite element mesh is shown in Figure 6.

4   CONCLUSIONS

The hot spot stress and the stress concentration factor of welded T-joints made of concrete filled circular hollow sections (CFCHS) were investigated experimentally. The test results were compared with the existing research results of welded T-joints made of circular hollow sections (CHS) with the same geometric parameters as the CFCHS joints. The following conclusions can be drawn:

(1) The CFCHS joint has much better static behaviour than the CHS joint. The maximum SCF around the CFCHS joint is much smaller than that around the CHS joint and their positions is different. The

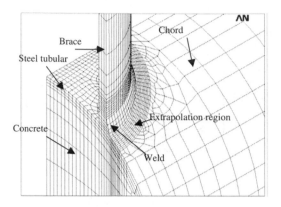

Figure 6.   The finite element mesh of the joint intersection.

former is located at the chord crown, whereas the latter at the chord saddle.

(2) The CFCHS joint has quite different SCFs when its brace is loaded in compression and in tension. The maximum SCF under compression is smaller than that under tension, although their positions are the same. The wall of the tubular chord tends to separate from concrete for the brace in tension, and then the deflection around the joint becomes large.

(3) The maximum SCF around the CFCHS joint increases with increasing β within the range of 0.5 to 1.0, which is opposite to the behaviour of the CHS joint.

(4) The effects of τ and γ on the SCF around the CFCHS joint are so slight that they can be neglected, which are different from behaviour of the CHS joint.

(5) The Concrete filled in the CHS joint improves significantly the stiffness and strength of the joint, but its grade, namely the strength of concrete itself, does not exert much influence on the SCF around the joint.

ACKNOWLEDGEMENTS

The authors wish to thank the Natural Science Foundation of China for financially supporting the research in the paper through the grant No. 50478108.

REFERENCES

Zhou S. and Chen S. 2003, "Rapid Development of CFST arch bridges in China" Advances in Structures, Vol. 2, Proceedings of the International Conference on Advances in Structures: Steel, Concrete, Composite and Aluminium, ASSCCA'03, Editors: Hancock G.J., Bradford M.A., Wilkinson T.J., Uy. B and Rasmussen K.J.R., 23–25 June 2003, Sydney, Australia, pp. 915–920.
Van Wingerde et al., 1992, The Fatigue Behaviour of T and X Joints made of Square Hollow Sections, Herion, The Netherlands, pp. 1–180.
Zhao X.L. et al. (2001): Design Guide for Circular and Rectangular Hollow Section Welded Joints under Fatigue Loading, CIDECT and TUV-Verlag, German.
Zhao X.L. and Packer, et al.(2000), Fatigue Design Procedures for Welded Hollow Section Joints, IIW, Abington Publishing, Cambridge, UK.
Udomworarat P., et al. (2000), "Fatigue and ultimate strengths of concrete filled tubular K-joints on truss girder", Journal of Structural Engineering, Vol. 46A, March 2000, pp. 1627–1635, Japan Society of Civil Engineers.

Steel and Composite Structures – Wang & Choi (eds)
© 2007 Taylor & Francis Group, London, ISBN 978-0-415-45141-3

Pull-out strength of interacting bond-type anchors

M. Obata & Y. Goto

Department of Civil Engineering, Nagoya Institute of Technology, Nagoya, Japan

ABSTRACT: A bond-type anchor is a typical post-installed type anchor and resists pull-out force with enhanced bond stress on a bolt surface. This type of anchor has been used primarily for retrofits of concrete and masonry structures. As this type of anchor exhibits a complex failure mode involving cone and bond failure at the same time, its design method is still to be investigated in depth. As for the pull-out strength of a single anchor bolt, the authors have already proposed a rational formula without any particular assumptions on a failure mode. The objective of this manuscript is to investigate the strength and failure mode of the multiple bond-type anchors under interaction both analytically and experimentally. In analytical formulation, a semi-analytical formula of the 1.5 power law for cone failure mode is employed. The experimental results agree with the analytical estimation well.

1 INTRODUCTION

Anchors that fasten structures and objects on concrete slabs or walls are classified into two categories. One is a cast-in-place type, or a pre-installed type. The other is so-called a post-installed type that is placed after construction. A post-installed type has an advantage to pre-installed type because it can be used in various location, size and time with great flexibility. However, a post-installed type carries less load than a pre-installed type and its application had long been limited to fastening small and light objects. After development of high performance anchor bolts and adhesive, they have been widely used in such important cases like retrofitting of structures and diaphragm walls in Japan.

A post-installed type anchor transfers loads to base material by friction (an expansion anchor), mechanical interlock (an undercut anchor) or bond stress (a bonded anchor) (CEB (1994)). Among these, a bonded anchor using chemical compound has higher load carrying capacity than usual expansion or undercut anchors.

In addition, the use of a bonded anchor, unlike an expansion anchor, is less restrictive when an anchor is located near free edges. Because of these advantages, the numbers of application of bonded anchors have recently increased. At present, the design of bonded anchors is based on the code for various composite concrete and steel structures (e.g., AIJ (1985)) in Japan. Specifically, the design of bonded anchors conforms to those of reinforced bar anchor. The modes of pull out failure of a single bonded anchor are classified as those in Fig. 1. One is bond failure (Fig. 1(a)) and another is so called a cone failure where an anchor pulls out with cone shaped mass of concrete (Fig 1(c)). The other is the mixture of these previous two (Fig. 1(b)). This partial cone failure mode appears when embedded length is large. However, in these design codes, the mixture mode failure is excluded without proper reasoning. Since it is not specifically shown when the mixed mode appears and governs pull out strength, ignorance of mixed-mode is not justified. In fact, with proper material and geometrical parameters, the

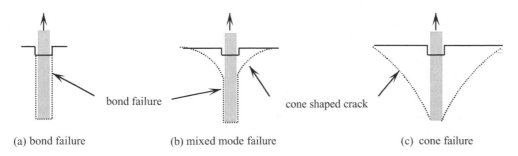

(a) bond failure   (b) mixed mode failure   (c) cone failure

Figure 1. Failure modes of a bonded anchor.

maximum pull-out load does appear with mixed mode failure (Maeno et al. (1992a,b)).

Regarding this sort of mixed mode failure, Ozaka et al. (1991) studied the decrease in embedded length of reinforced bars caused by partial cone failure and estimate the pull out strength involving mixed mode failure. Although they consider the mixed mode failure, their estimation of the location of crack initiation is yet to be examined in detail. On the other hand, Goto et al. (1992) proposed the method to estimate the pull out strength of a bonded anchor using an equilibrium equation. The method predicts all three different failure modes in a consistent way. They also explained the 1.5 power law of pull out strength of a concrete cone using fracture mechanics analysis (Obata et al. (1998)) and showed that the equilibrium method with the 1.5 power law predicted the pull out strengths of a single bonded anchor near free edge very well.

Thus, the objective of this work is to investigate the group effect on the pull-out strength of multiple bonded anchors that interact among them. The equilibrium method and the 1.5 power law are modified to apply the problem of multiple bonded anchors.

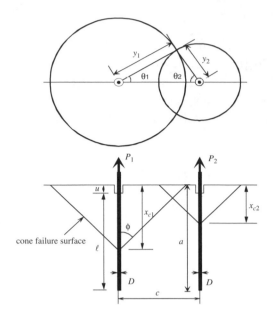

Figure 2. Interaction of two bonded anchors.

2 PULL OUT STRENGTH OF TWO BONDED ANCHORS

2.1 *Equilibrium method*

In this section, the frame work of equilibrium method is briefly explained in application to a case of two bonded anchors subjected to pull out force P_1 and P_2 (Fig. 2). These anchors fail in three distinct modes depending on embedded length, strength of base concrete, bond strength, and the distance between anchors bolts.

Since it is easy to estimate the pull out strength in the case of bond failure (Fig. 1(a)), in what follows, the formulation is focused on the case when cone shaped cracks involve. In the current design method, a cone shaped crack initiates from the bottom of an anchor bolt irrespective of embedded length apparently because there is no established method to deal with what we call mixed-mode failure. However, as far as a cone shaped crack occurs for a deeply embedded anchor bolts, it is not reasonable to exclude the case when a cone shaped crack emanates from a point other than the bottom. Such assumption apparently overestimates pull out strength of a bond type anchor.

Suppose that either cone or mixed mode failure occurs at the depths of x_{1c} and x_{2c} respectively, the equilibrium of bond and cone shaped crack strength must hold in the part upper than cracked surfaces. In addition, as pull out force increase, a bond force curve approaches to a cone strength curve from lower side (Fig. 3). The equilibrium equation requires that

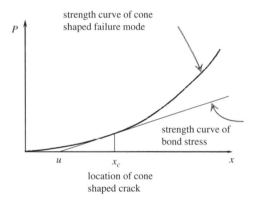

Figure 3. Location of mixed-mode failure.

mixed-mode failure takes place when these two curves contact. Then, we have the following equations.

$$P_c\left(x_{1c}, x_{2c}\right) = P_b\left(x_{1c}, x_{2c}\right) \tag{1}$$

$$\left.\frac{\partial P_c}{\partial x_1}\right|_{x_1 = x_{1c}} = \left.\frac{\partial P_b}{\partial x_1}\right|_{x_1 = x_{1c}} \tag{2}$$

$$\left.\frac{\partial P_c}{\partial x_2}\right|_{x_2 = x_{2c}} = \left.\frac{\partial P_b}{\partial x_2}\right|_{x_2 = x_{2c}} \tag{3}$$

where P_c and P_b are pull out strength of cone failure and bond failure up to the depths of x_{1c} and x_{2c}

respectively. The interactive effects between multiple anchors can be included implicitly in the expression of $P_c(x_1, x_2)$ and $P_b(x_1, x_2)$. There are several possibilities for specific expressions of $P_c(x_1, x_2)$ and $P_b(x_1, x_2)$. As for the distribution of bond stress on an anchor bolt, it is nearly uniform unless embedded length is very long. This can be confirmed by numerical simulation using a distributed spring model. In practical applications, it is safely assumed that the bond stress distribution is uniform and we have

$$P_b\left(x_1, x_2\right) = P_1 \frac{x_1 - u}{\ell} + P_2 \frac{x_2 - u}{\ell}$$

$$= \frac{P_1}{\ell} \left\{\left(x_1 - u\right) + \alpha\left(x_2 - u\right)\right\} \tag{4a}$$

$$\alpha = P_2 / P_1. \tag{4b}$$

On the other hand, $P_c(x_1, x_2)$ is often given by assuming uniform tensile failure stress acts on the cone shaped crack (ACI (1985), AIJ (1985)). Among the effects of interacting bonded anchors on pull out strength, the most important is reduction of effective resistant area due to overlapping of cone shaped crack surface (Fig. 2). If pull out strength is proportional to projected cone failure area, for a pair of anchors, it is given by

$$P_c\left(x_{1c}, x_{2c}\right) = \sigma_{cone} \left\{y_1^2\left(\pi - 2\theta_1\right) + y_2^2\left(\pi - 2\theta_2\right)\right.$$

$$\left. + \frac{1}{2}y_1^2 \sin 2\theta_1 + \frac{1}{2}y_2^2 \sin 2\theta_2 - \frac{\pi}{2}D^2\right\}, \tag{5a}$$

$$y_i = x_i \tan \phi + \frac{D}{2}, \quad (i = 1, 2), \tag{5b}$$

$$y_1 \cos \theta_1 + y_2 \cos \theta_2 = c, \quad y_1 \sin \theta_1 = y_2 \sin \theta_2, \tag{5c,d}$$

where σ_{cone} is a strength of cone shaped cracks and usually equal to tensile strength. Despite its simplicity, the formula is known to overestimate the pull out strength of cone shaped failure when embedded length is large. In fact, Bode and Roik (1987) gives the 1.5 power law based on the regression analysis of many experimental results. The authors also theoretically derived the 1.5 power law on the basis of linear fracture mechanics and experimentally confirmed the applicability to a bonded anchor near a free edge (Obata et al. (1998)). According to the theoretical basis of the 1.5 power law, the governing factor of pull out strength is not a projected area of a failure surface but the length of front of a cone shaped crack because fracture strength is critically controlled by fracture process in the immediate neighborhood of a crack tip. Thus, the following equation was proposed instead of Eq. (5) (Obata et al. (1998))

$$P_c\left(x_1, x_2\right) = C\left\{\left[\left(1 - \frac{\theta_1}{\pi}\right)y_1\right]^{1.5} + \left[\left(1 - \frac{\theta_2}{\pi}\right)y_2\right]^{1.5}\right\}. \tag{6}$$

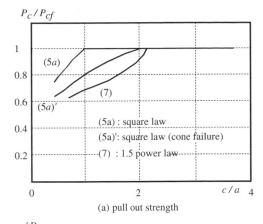

(a) pull out strength

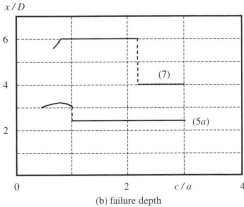

(b) failure depth

Figure 4. Effect of distance of anchor bolts.

However, as Eq. (6a) does not yield a consistent result when $\theta_1 \approx \theta_2 \approx \pi/2$, in this work, Eq. (6) is further modified to the following expression.

$$P_c\left(x_1, x_2\right) = C\left\{\left(1 - \frac{\theta_1}{\pi}\right)y_1^{1.5} + \left(1 - \frac{\theta_2}{\pi}\right)y_2^{1.5}\right\}, \tag{7}$$

where is a coefficient that theoretically depends on fracture toughness and Young's modulus of concrete. ϕ is a half apex angle of a failed concrete cone and usually set to 45 degree. Pull out strength and depth where a cone shaped crack initiates are given as solutions of nonlinear simultaneous equations of Eqs. (1), (2), (3), (4) and (7).

2.2 Numerical example

Figure 4 shows how pull out strength and failure depth depends on the distance of two bolts when $a/D = 6$, $u/D = 1$, $P_1 = P_2$. a and u are the anchor length and the unbonded length respectively (Fig. 2) The unbonded

443

part is necessary to avoid unfavorable cracking at low level load (Doerr and Klingner (1989)). When bond strength is large enough and the 1.5 power law is adopted, the first mixed mode failure occurs at the following depth,

$$x_c = 3u + D. \tag{8}$$

Note that $\phi = 45°$. Therefore, the interactive effect prevails when the distance between the bolts becomes close to $2x_c$. However, in the equilibrium formulation, depending on the shape of cone strength curve, the effect could appear at farther distance. In fact, Fig. 4 shows that the interactive effect occurs at $c/(a + D/2) \approx 2$. On the other hand, with the adoption of the square rule, interactive effects does not prevails until anchor bolts are located as close as $c/a \approx 1$. Employment of the square law leads to considerable underestimation of interactive effects than the use of the 1.5 power law. At the same time, the pull out strength is much higher than in the case of the 1.5 power law. The locations of conical crack initiation are also very different. A crack initiates at the bottom of bolts under the 1.5 power law and $x/D \approx 3$ under the square law. In Fig. 4(a), for comparisons, the result is also given when cone shaped failure is assumed to occur at the bottom of anchor bolts and the other two failure modes are excluded like in the current design code. The last result is closer to the one by the 1.5 power law. As explained in the above, however, as far as the exclusion of mixed mode failure has no rational basis, the validity of the result is not known.

3 PULL OUT EXPERIMENT OF A PAIR OR BONDED ANCHORS

Pull out experiments of a pair of bonded anchors were performed to verify the theoretical formulation in the preceding section. The dimension and configuration of a specimen is illustrated in Fig. 5. The lengths of embedded part and unbonded part are respectively 5D and D with $D(= 12\,\text{mm})$ being a diameter of an anchor bolt. Base mass concrete has no reinforcement to exclude effects on cone shaped cracks. The dimension of the specimen is given so that it has enough strength against bending fracture and the effect of free edge can be ignored. There are 6 specimens with different distance between the anchor bolts (Table 1). If the 1.5 power law is assumed, the first mixed mode fracture takes place at the depth of $4D$ for a single anchor bolt according to Eq. (8). However, Eq. (8) does not necessarily assure that interactive effects does not take place when the distance is larger than $8D$ or $c/a = 1.5$. In fact, as seen in the numerical results in Fig. 4, interactive effect can occur when $c/a \approx 2.0$. Thus, the distance between the bolts is set to $c/a \approx 3.5$

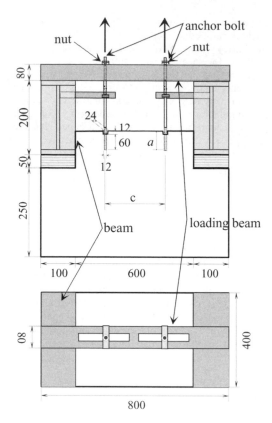

Figure 5.   Specimen.

Table 1.   Specimens.

Specimen	Distance (mm)	c/a
FC	250	3.46
F2.0	144	2.0
F1.7	122	1.7
F1.4	100	1.4
F1.0	72	1.0
F0.6	44	0.6

Table 2.   Material properties.

Material	Property
Anchor bolt	F10T M12, $P_y = 5.15 \times 10^4\,\text{N}$
Concrete	$\sigma'_c = 33.9\,\text{MPa}$
Adhesive	Epoxy polymer based compound

in specimen FC where no interaction is supposed to involve. Specimen FC is used as reference. The material properties of anchor bolts and base concrete are summarized in Table 2.

444

Table 3. Pull out strength.

Specimen	Pull out strength (per bolt)(kN)
FC	31.8
F2.0	27.2
F1.7	28.2
F1.4	27.0
F1.0	20.6
F0.6	18.5

Anchor bolt is a high tension bolt of 12 mm diameter. The embedded part is threaded to enhance bond strength. These anchor bolts are installed in a usual way. That is, after drilling a hole of 16 mm diameter by a hammer drill, a bolt is fixed by epoxy polymer based compounds. The diameter of unbonded part is 24 mm.

Pull out force is applied to the bolts by torqueing each nut statically until pull out failure occurs(Fig. 5). The drift of the force ratio is kept within 10% throughout experiments by monitoring strain of each bolt. Numerical analysis based on the equilibrium method shows that the results are insensitive to force ratio α. Pull out force, pull out displacement and cracking pattern are observed. Internal cracks are also examined after experiments.

4   EXPERIMENTAL RESULTS AND DISCUSSION

The averaged bond strength of the bolts was found to be more than 177 MPa by an experiment with a similar specimen as shown in Fig. 5. With this high strength, bond failure (Fig. 1(a)) did not dominate in either specimen. It is confirmed that mixed mode or cone shaped failure actually occurred in all cases by the observation of external or internal cracks. Pull out strengths are summarized in Table 3. In this table, the pull out strength is the average of two anchor bolts'. Figure 6 illustrates the crack patterns of specimens FC, F1.7 and F1.0. These patterns were observed at the load maximum points. The maximum load occurs when the first major cracking takes place. Figure 6(a) suggests that the interactive effect is negligible in specimen FC because independent conical cracks appear. On the other hand, quite different type of crack pattern appeared in the other specimens. For example, Fig. 6(c) clearly shows that a single large crack embraces two anchor bolts. To examine the crack patterns in further details, internal cracking pattern after the experiments are shown in Figs. 7. In these figures, it can be seen that cone shaped cracks initiates from the bottom of anchor bolts in all cases. In the case of specimen FC,

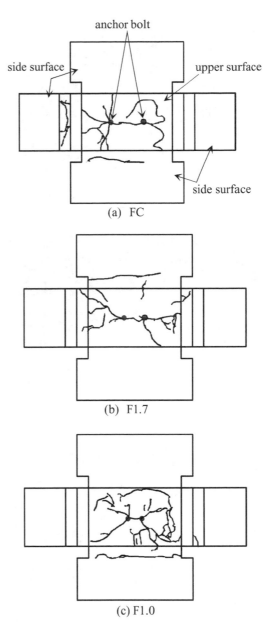

(a) FC

(b) F1.7

(c) F1.0

Figure 6.   External crack pattern.

however, this results does not agree with the numerical prediction shown in Fig. 4(b). This difference is partly due to the fact that the diameter of a drilled hole is not 12 mm but 16 mm. As bond strength is large enough, the diameter of the bolt is effectively larger than 12 mm and a conical crack initiates at deeper location. These results show the good agreement to the prediction by the 1.5 power law and the equilibrium method (Fig. 4(b)). The apex angles of cone shaped

(a) FC

(b) F1.7

(b) F1.0

Figure 7. Internal crack pattern.

cracks are larger than 45° as observed in many cases. In case of specimen FC, it is about 50° and gets larger when interactive effect predominates.

Figure 8 illustrates the comparisons between numerical and experimental results with respect to pull out strength. In this figure, pull out load is normalized

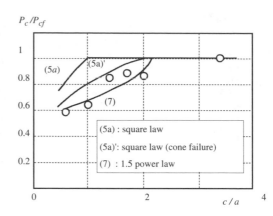

Figure 8. Pull out strength and bolt distance.

by the result of FC that is supposed to be the one without interactive effect. Figure 8 clearly shows that the prediction based on the 1.5 power law of Eq. (7) gives the best agreement with the experimental results. The square law yields only poor prediction as observed in the case of the bonded anchor near a free edge (Obata et al.(1998)). It is not acceptable to assume that pull out strength depends on the projected area of a cone shaped crack by ignoring the fracture process. In Fig. 8, the results of Eq. (5a)' corresponds to those by the current design method, where a cone shaped crack is supposed to occur at the bottom of a bolt. As in the present experiments, cone shaped cracks happen to initiate at the bottom of bolts in these cases, the cracks govern the pull out strength and the square law without the equilibrium consideration lead to better prediction. It should be noted, however, that the validity of the assumption is not known with respect to the location of cone shaped crack initiation. The present design method does not have a sound theoretical basis on which factors like material strength, bond strength, embedded length are accounted in a consistent way. For example, the current design method is ambiguous when pull out force ratio is not equal to 1.

5 SUMMARY

The interactive effects on the pull out strength of bonded anchors are theoretically examined using the 1.5 power law and the equilibrium method and experimentally confirmed. The reduction of cone shaped failure area accounts for the interactive effect. In this work, the overlapping is represented by total length of crack front by considering linear fracture mechanics. The experimental results showed that the 1.5 power law yields good agreement with experimental results even under the interactive effect of multiple bonded anchors. The specific dependence of the coefficient C on material properties is yet to be examined.

REFERENCES

Architectural Institute of Japan (AIJ) 1985. The design code of composite structures.

American Concrete Institute (ACI) 1985. Code requirements for nuclear safety related structures.

Bode, H. & Roik, K. (1987). Headed studs embedded in concrete and loaded in tension, Anchorage to Concrete, SP103, pp.61–89, ACI.

Comite Euro-International du Beton (CEB) 1994. Fastenings of to concrete and masonry structures, state of the art report.

Goto, Y., Obata, M., Maeno, H., & Kobayashi, Y. 1992. Failure mechanicsm of new bond-type anchor bolt subject to tension, J. Struct. Engrg., ASCE, 119, pp.1168–1187.

Maeno, H., Goto, Y., Obata, M., & Matsuura, S. 1992a. Bonding characteristics of a very large deformed bar with headed studs, Proceedings of JSCE, I-18, pp.87–96 (in Japanese).

Maeno, H, Goto, Y., Obata, M., & Matsuura, S. 1992b. Ultimate strength of bond-type anchorage under tensile force, Proc. JSCE., I-18, pp. 185–192 (in Japanese)

Obata, M., Inoue, M., & Goto, Y. 1998. The failure mechanism and the pull-out strength of a bond-type anchor near a free edge, Mechanics of Materials, 28, pp. 113–122.

Ozaka, Y., Otsuka, K., Maki, Y., & Kobayashi, S. 1991. Anchorage failure and group effect of deformed bars embedded in massive concrete, Proceeding of JSCE, V-17, pp. 77–96.

Doerr, G.T. & Klingner, R.E. 1989. Adhesion anchors: behavior and spacing requirements. Center fro Transportation Researtch, University of Texas at Austin, Austin, Research Report 1126–2.

Steel and Composite Structures – Wang & Choi (eds)
© 2007 Taylor & Francis Group, London, ISBN 978-0-415-45141-3

Load-carrying capacity of post-installed steel anchors to concrete, subjected to shear

M. Karmazínová, J. Melcher & Z. Kala
Brno University of Technology, Brno, Czech Republic

ABSTRACT: This paper deals with the problems of the fastening using post-installed anchoring elements. The paper is oriented to the presentation of some results of the investigation of the actual behaviour of steel expansion anchors to concrete under shear loading due to the transverse force. Namely, this research is directed to the analysis of strain mechanisms in the process of loading and failure mechanisms and corresponding reached loadings. For the obtaining of information and knowledge on the actual behaviour of expansion anchors in concrete, the results of experimental verification are used as the basis for the subsequent statistic and, eventually, probability and reliability elaboration. The experimental analysis and accompanying theoretical analysis are aimed at the determination of ultimate load-carrying capacities for all important failure mechanisms, both for steel failure (anchor element) both for concrete failure (base material). The problems mentioned above were investigated on the large set of test results for the test specimens with various dimensions and various parameters of both materials, steel and concrete.

1 INTRODUCTION

The effectiveness and accurate placement together with the new easy technologies and techniques are the most important advantages of the post-installed fastening systems. For the fastening to concrete in the new constructions as well as in repair works the steel expansion anchors are very suitable. Their behaviour can be rather complicated considering the influence of concentrated loading, their different directions and, especially, the failure mode depending on the way of the load transfer from anchor body into the concrete base. Thus the experimental verification together with statistical analysis of appropriate test results should be an authority for the theoretical modelling and practical design of the fastening systems.

In this paper the brief information on the main results (failure mechanisms and adequate load carrying capacities) of the loading tests of steel expansion anchors under shear loading is presented.

2 BASIC INFORMATION OF EXPERIMENTAL RESEARCH PROGRAMME

During several last years in our testing laboratory the loading tests of torque-controlled expansion anchors to concrete produced in our country have been conducted (Karmazínová 2005b). Together 368 specimens under static loading actions as well as under dynamic

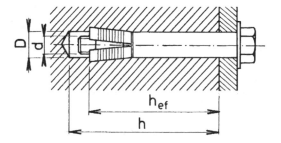

Figure 1. Scheme of applied type of expansion anchor.

loading actions have been tested so far. From all tests 277 specimens under static and also dynamic tension loading (Karmazínová 2005, Melcher & Karmazínová 2001, Karmazínová 2001) and 91 specimens under static shear loading have been tested.

For the test specimens under shear loading the expansion torque-controlled anchors – see Fig. 1, with following parameters have been used: For the anchor bolts the steel grade of 8.8 – with nominal values of the ultimate tensile strength $f_{ub} = 800$ MPa, and diameters of $d = 10$, 12 and 16 mm have been used. The external diameters of anchor sleeve were $D = 14$, 18 and 24 mm. The cube concrete strength f_{cc} of the specimen bodies was in the range of 19 MPa to 37 MPa and the effective anchor embedment depth h_{ef} was in the range of 50 mm to 80 mm.

Figure 2.   Lateral concrete-cone failure mode.

Figure 3.   Concrete spalling failure mode.

3   ANALYSIS OF EXPANSION ANCHORS
  STRENGTH PARAMETERS

During the test process the failure mechanism and
load-carrying capacity of the tested specimens have
been verified. Depending on basic parameters of fastening arrangement different types of failure mode can
be established. For the behaviour of anchor under shear
loading the edge distance c_1 and concrete strength f_{cc}
together with the bolt dimensions d, D and its strength
f_{ub} are decisive, especially. Also the embedment depth
h_{ef} can influence the fastening strength.

Depending on the edge distance and embedment
depth, both steel failure both concrete fracture, can
occur. Steel failure by yielding and fracture of the
anchor bolt is characteristic for deep embedment, low
steel strength and large edge distance. Concrete failure is characterized by one of the following possible
modes: lateral concrete-cone failure – see Fig. 2, for
the anchors close to edge, and concrete spalling – see
Fig. 3 or concrete crushing – see Fig. 4, respectively,
for the large edge distance. The failure by concrete
spalling can occur for deeper embedment, concrete
crushing combined with subsequent anchor pull-out

Figure 4.   Concrete crushing failure mode (with subsequent
anchor pull-out).

may occur if the anchoring embedment depth is small
($h_{ef} \leq 4D$ to $6D$).

Within the frame of this information the results
covering the set of lateral concrete-cone failure and
concrete spalling are presented (altogether 43 and 23
tests, respectively). Additionally, in 38 cases other
failure modes occurred (bolt failure in 29 cases and
concrete crushing in 9 cases, respectively).

For the primary evaluation of the test results
the basic calculation models according to published
results (Eligehausen 2006, Eligehausen 2001, Bulletin
d'Information 1994) were used. For the subsequent
analysis directed to the determination of characteristic and design values of fastener strength the method
according to the normative documents (EN 1990) for
design assisted by testing can be used.

3.1   Lateral concrete-cone failure mode

Based on Concrete Cone Method (Eligehausen 2006,
Eligehausen 2001, Bulletin d'Information 1994) the
shear breakout capacity can be expressed by the
general basic format of

$$V_u = A_c \cdot f_{ct} \tag{1}$$

where A_c is projected area of the lateral cone and
f_{ct} is the concrete tensile strength. According to the
45-Degree Cone Method considering the 45° lateral
failure cone – see Fig. 5, and for the edge distance of e
and the tensile strength expressed by the cube strength
as $f_{ct} = k \cdot f_{cc}^{0,5}$, we can write

$$V_u = \frac{\pi \cdot e^2}{2} \cdot k \cdot f_{cc}^{0,5}, \tag{2}$$

where k is units dependent. For the ACI Method – for
example (Bulletin d'Information 1994), the format of
(2) comes to the format of

$$V_{u,m} = 0.137 \cdot \pi \cdot e^2 \cdot f_{cc}^{0.5} \tag{3}$$

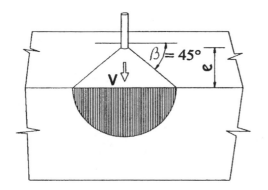

Figure 5.   Principle of the 45-Degree Cone Method.

Based on Concrete Capacity Design Method (CC Method) (Eligehausen 2006, Eligehausen 2001, Bulletin d'Information 1994) the corresponding general expression can be presented in the basic form of

$$V_u = k_c \cdot e^{1.5} \cdot f_{cc}^{0.5},\qquad(4)$$

where the factor of k_c is usually derived from the test results using the regression analysis (usually, for example, $k_c \approx 5$ to 6). For the ψ-method – for example (Bulletin d'Information 1994), the format of (4) comes to the format of

$$V_{u,m} = 4.68 \cdot e^{1.5} \cdot f_{cc}^{0.5}.\qquad(5)$$

Comparing our experimental results (Karmazínová 2005b) with theoretical values based on (3) the mean value $V_{u,m}$ of shear breakout capacity can be given by

$$V_{u,m} = 0.24 \cdot \pi \cdot e^2 \cdot f_{cc}^{0.5}\qquad(6)$$

and the appropriate characteristic value $V_{u,k}$ and design value V_d considering the number of the tests is

$$V_{u,k} = 0.11 \cdot \pi \cdot e^2 \cdot f_{cc}^{0.5},\qquad(7)$$

Similarly for the format based on (5) the corresponding mean and characteristic values of shear breakout capacity are

$$V_{u,m} = 7.30 \cdot e^{1.5} \cdot f_{cc}^{0.5},\qquad(8)$$

$$V_{u,k} = 4.41 \cdot e^{1.5} \cdot f_{cc}^{0.5}.\qquad(9)$$

In Figs 6, 7 the relationship between experimental strengths $V_{u,exp}$ and mean values $V_{u,m}$ according to (6) and (8) derived by regression analysis are drawn.

Based on (3) and using the procedure for design assisted by testing the design value V_d of shear breakout capacity determined with respect to the number of

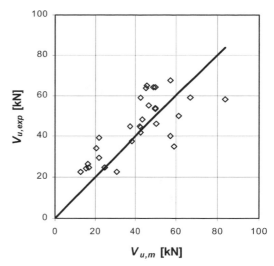

Figure 6.   Relationship between $V_{u,exp}$ and $V_{u,m}$ based on (3).

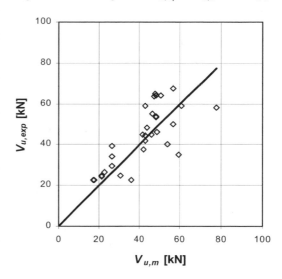

Figure 7.   Relationship between $V_{u,exp}$ and $V_{u,m}$ based on (5).

the tests (Karmazínová 2005b), can be given by the format of

$$V_d = 0.065 \cdot \pi \cdot e^2 \cdot f_{cc}^{0.5}.\qquad(10)$$

Based on (5) and using the procedure for design assisted by testing the design value V_d of shear breakout capacity determined with respect to the number of the tests, can be given by the format of

$$V_d = 2.87 \cdot e^{1.5} \cdot f_{cc}^{0.5}.\qquad(11)$$

Recommended characteristic and design values are based on following. the characteristic value $V_{u,k}$ is

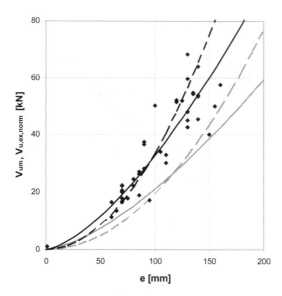

e [mm]

Figure 8. Comparison of mean values and test values. Mean values according to: (3) – – – , (5) ——— , (6) – – – , (8) ———.

determined as 5% fractile, the design value V_d is determined so that the probability of observing the lower value is about 0.1%.

The test results compared to the course of the mean values of shear breakout capacity are plotted as a function of the edge distance e, for the relationships going out of (3) and of (5) in Fig. 8. The experimental values are normalized for the concrete cube strength of $f_{cc} = 20$ MPa according to

$$V_{u,norm} = V_{u,exp} \cdot \left(\frac{20}{f_{cc}} \right)^{0.5}. \qquad (12)$$

From the evaluation above mentioned and from Fig. 8 it implies that the equation in the format of (5) is more apposite to this type of failure comparing with the format of (3).

3.2 Concrete spalling failure mode

The shear capacity for the failure mode by concrete spalling is influenced by the concrete tensile strength, flexural stiffness and diameter of the anchor shaft, embedment depth and by concrete E_c – modulus. The mean value can be given according to (Eligehausen 2001, Bulletin d'Information 1994) as

$$V_{u,m} = 0.5 \cdot A \cdot \sqrt{f_c \cdot E_c} = 0.39 \cdot D^2 \cdot \sqrt{f_c \cdot E_c}, \qquad (13)$$

where the cylinder strength of concrete f_c can be expressed using the cube strength f_{cc} [1]

$$f_c \cong 0.8 \cdot f_{cc}. \qquad (14)$$

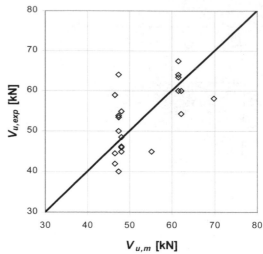

$V_{u,m}$ [kN]

Figure 9. Relationship between $V_{u,exp}$ and $V_{u,m}$ based on (13).

Based on the format (13) and using the regression analysis we can obtain the verification of the *mean value* for our test results as

$$V_{u,m} = 0.226 \cdot D^2 \cdot \sqrt{f_c \cdot E_c}. \qquad (15)$$

The corresponding characteristic and design values (Karmazínová 2005b) determined using design assisted by testing regarding the number of the tests then can be written

$$V_{u,k} = 0.174 \cdot D^2 \cdot \sqrt{f_c \cdot E_c}, \qquad (16)$$

$$V_d = 0.117 \cdot D^2 \cdot \sqrt{f_c \cdot E_c}. \qquad (17)$$

The relationship between experimental strengths $V_{u,exp}$ and mean values $V_{u,m}$ according to (13) and (15) derived by regression analysis are drawn on the graph in Fig. 9.

4 CONCLUSIONS

The partial results above mentioned can be presented as the example of application of experimental methods for the verification of actual behaviour of expansion anchors under shear loading. The ultimate load-carrying capacities implied from the tests can be taken as the basic information for the verification of existing known theoretical models for their calculation. The determination of characteristic and design values described here indicates the possibility of using design assisted by testing. Recently this experimental research programme of the investigation of post-installed fastening systems is continuing.

452

ACKNOWLEDGEMENTS

This paper has been elaborated under the support of the Research Centre Project of MSMT No. 1M0579 and of the grant projects of GA CR (Czech Grants Foundation) No. 103/06/1107, No. 103/05/0085, No. 103/05/2059 and No. 103/07/0628.

REFERENCES

Karmazínová, M. 2005a. Resistance of the fastening to concrete under repeated loading, In *Proc. of the 4-th European Conference on Steel and Composite Structures EUROSTEEL 2005, Maastricht 2005*, Vol. C, pp. 33–40. Rotterdam: Balkema. ISBN 3-86130-812-6.

Karmazínová, M. 2005b. *To the Problems of the Design Philosophy and the Experimental Verification of the Steel Expansion Anchors*, Assoc. Prof. Theses, Brno University of Technology, Faculty of Civil Engineering, 2005, 154 pp.

Melcher, J. & Karmazínová, M. 2001. The analysis of fastener strength using the limit state approach, In *Proceedings of the International Symposium "Connections between Steel and Concrete 2001"*, University of Stuttgart, pp. 212–219. Stuttgart: RILEM. ISBN 2-912143-25-X.

Karmazínová, M. & Melcher, J. 2006. Fastening systems load-carrying capacity based on the reliability approach applied for the test results evaluation, In *Proc. of the International Conference VSU*, Sofia 2006, Vol. I, pp. 219–224. Sofia: VSU. ISBN 954-331-009-2.

Eligehausen, R. et al. 2006. *Anchorage in Concrete Construction*, Ernst & Sohn, GmbH, Berlin 2006. ISBN 3-433-01143-5.

Eligehausen, R. 2001. *Proceedings of the International Symposium "Connections between Steel and Concrete"*, University of Stuttgart, RILEM 2001. ISBN 2-912143-25-X.

Bulletin d'Information 1994. *Fastening to Reinforced Concrete and Masonry Structures, State-of-art-report*, CEB, Thomas Telford Services Ltd., 1994.

Steel and Composite Structures – Wang & Choi (eds)
© 2007 Taylor & Francis Group, London, ISBN 978-0-415-45141-3

An experimental investigation on stud and perfobond connectors with push-out test

Suhaib Yahya Kasim Al-Darzi, Ai Rong Chen & Yu Qing Liu
Department of Bridge Engineering, Tongji University, Shanghai, China

ABSTRACT: In the present work, investigations of both stud and perfobond shear connector are conducted using push out test. A total of 24 specimens are prepared and tested using the two types of connectors in order to investigate the shear resistance of stud and perfobond connector and the effect of several parameters, such as diameter, concrete placement method and availability of transverse reinforcement in perfobond connector rib holes, on connector shear resistance. The results are compared with some available provisions stated by several specifications and the available numerical expressions suggested by researchers. The comparison results show that some expression gives unconservative results. Also shown that, some numerical expressions give misleading results. Suggestions for future studies are also presented.

1 INTRODUCTION

The investigation of connection between steel beam section and concrete deck slab is one of the main important subjects adopted to improve steel-concrete composite bridges. The connection is usually performed through several types of shear connector; the common types of shear connector adopted in codes are the stud and channel connectors. (Salmon, C.G., & Johanson, J.E., 1990 and Liu, Y. Q., 2004).

Several stud strength prediction provisions have been developed since the 1970s. Some of these methods are part of the specifications used in China, in US, and Canada. These are the Chinese code, the AASHTO, the AISC specification and the Canadian Standards Association CSA. (GBJ17-88, 1988, AASHTO, 2004, AISC, 2005 & CSA, 1994) The push-out test is usually used to investigate shear resistance of connector. (Roddenberry, M.D.R., 2002) The perfobond connector is a new types of connector established and developed as an alternative shear connector. The applicability of such new type of shear connector required extensive investigation on behaviour, shape of failure and shear resistance. Since a few researches were conducted on perfobond connector aiming to develop suitable numerical model used to calculate shear resistance. (Oguejiofor, E.C., & Hosain, M.U., 1996) In the present work, push-out test is attempted to be used to investigate the shear resistance of both stud and perfobond connectors and choosing the optimum one to be used in constructing Naning bridge in China, which is designed with a new shape of

steel-concrete composite arch using shear connector connecting concrete and steel sections.

An experimental works are achieved using push-out test adopting both stud and perfobond connector specimens with different types of concrete casting to investigate:

(1) The effect of concrete casting method on failure of both stud and perfobond connectors. Whereas, different positions of steel and concrete necessitate different methods of concrete placement.
(2) The shear resisting capacity for both stud and perfobond shear connectors.
(3) The effect of transverse reinforcement on resisting capacity of perfobond connector.
(4) To compare the predicted experimental results with provisions available in codes.

2 THE STRENGTH PREDICTION EXPRESSIONS

2.1 *Resistance capacity of stud shear connector*

Several stud strength prediction provisions are available as a part of specifications used in China, in US and in Canada as stated in the following.

2.1.1 *Chinese code*

The strength of stud shear connector Q_n is given as (N_v^c) in Chinese code (GBJ17-88, 1988) given as.

$$N_v^c = 0.43 A_s \sqrt{E_c f_{cc}} \leq 0.74 A_s f \tag{1}$$

where: N_v^c is design value of shear capacity of a connector, As cross sectional area of stem of the stud, Ec elastic modulus of concrete, fcc design value of axially compressive strength of concrete, f design value of tensile strength of the stud.

2.1.2 AASHTO specification

According to AASHTO–specification, 2004 the factored shear resistance Q_r of a single shear connector limit state shall be taken as:

$$Q_r = \phi \ Q_n \qquad (2)$$

where ϕ is resistance factor for shear connector specified by AASHTO, 2004 given as 0.85, and Q_n is nominal shear resistance of one stud shear connector embedded in concrete deck given as:

$$Q_n = 0.5 A_{SC} \sqrt{f_C' E_C} \le A_{SC} F_U \qquad (3)$$

where: A_{SC} is cross sectional area of stud (mm^2), F_U specified minimum tensile strength of stud shear connector (MPa), fc' a 28 day compressive strength of concrete, (MPa), and Ec modulus of elasticity of deck concrete (MPa), for concrete densities between 1440 and 2500 kg/m^3 the modulus of elasticity is given as: ($E_C = 0.043 \gamma_c^{1.5} \sqrt{fc'}$) where, γ_c is unit weight of concrete in (kg/m^3) and fc' in (MPa). (AASHTO, 2004).

2.1.3 AISC – Specification

According to AISC-(LRFD) specification, 2005 nominal strength of one stud shear connector embedded in solid concrete or in composite slab is:

$$Q_n = 0.5 A_{SC} \sqrt{f_C' E_C} \le R_g R_p A_{SC} F_U \qquad (4)$$

where: Q_n is nominal strength of one stud, A_{SC} cross sectional area of stud shear connector (mm^2), fc' a 28 day compressive strength of concrete, (MPa), Ec modulus of elasticity of concrete (MPa) ($E_C = 0.043 w_c^{1.5} \sqrt{fc'}$) where, w_c weight of concrete per unit volume ($1500 \le w_c \le 2500$ kg/m^3) and fc' in (MPa), R_g, R_P are factors given by AISC-specification depending on the number of stud, deck orientation and welding method and F_U the specified minimum tensile strength of stud shear connector (MPa). (AISC, 2005).

2.1.4 CSA provisions

The Canadian Standards Association (CSA) gives equations used for strength prediction of a stud in a solid slab.

$$Q_{SOL} = 0.5 A_S \sqrt{f_C' E_C} \le 1.0 A_s F_u \qquad (5)$$

where Q_{SOL} is strength of stud in solid slab, A_S is cross sectional area of stud, f_c' compressive strength of concrete, E_C modulus of elasticity of concrete and F_u tensile strength of stud.

2.2 Resistance capacity of perfobond connector

2.2.1 Expressions suggested by Oguejiofor, E.C. & Hosain, M.U.

The resistance capacity of perfobond shear connectors suggested by Oguejiofor, E.C., & Hosain, M.U., (1996) based on regression analysis of normal weight concrete specimens' results, with different connector geometries and reinforcement distribution is stated below:

$$q_u = 0.590 A_c \sqrt{f_c'} + 1.233 A_{tr} f_y + 2.871 n d^2 \sqrt{f_c'} \quad (6)$$

where f_c' is the concrete compressive strength, f_y the steel yield strength, A_c the concrete shear area, A_{tr} the area of transversal reinforcement that pass through holes, d the diameter of perfobond rib holes and n is the number of Perfobond rib holes. The same author conducted more tests and established expression given as:

$$q_u = 4.5 h t f_c' + 0.91 A_{tr} f_y + 3.31 n d^2 \sqrt{f_c'} \qquad (7)$$

where h is the rib height, t the thickness, A_{tr} the total area of transversal reinforcement and the other parameters are the same as in expression (6). (Oguejiofor, E.C., & Hosain, M.U., 1996 & Valente,I., & Cruz, P.J.S., 2004).

3 PUSH – OUT TEST

3.1 Classification and establishing push-out test

As stated previously, push-out test for both stud connector and perfobond connector specimens are conducted with four groups of each connector, having three specimens for each group prepared and tested to failure. The load-relative displacement curves for each specimen are plotted for each single case, using the predicted results to choose the group give an optimum connector shear resistance.

Two types of connector associated with two groups containing a total of 24 specimens are used and classified as: NS-groups referring to stud connector, containing a total of 12-specimens, and NP-groups referring to perfobond connector, containing a total of 12-specimens, dividing each group into four subgroups. Three different methods of concrete placement are used for each connector's specimens types, Figure 1, these are: TCC top concrete casting where steel section lies at bottom and concrete cast from the top,

Figure 2a, SCC side concrete casting where steel section is standing and concrete cast at sides, Figure 2b, and BCC bottom concrete casting where steel section lies at top and concrete cast at bottom Figure 2c. (Liu, Y. Q., 2004).

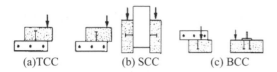

(a)TCC   (b) SCC   (c) BCC

Figure 1.   Direction of concrete casting.

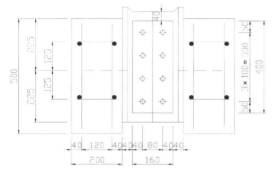

a. Stud shear connector

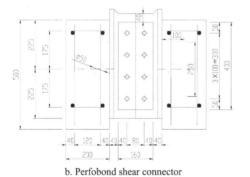

b. Perfobond shear connector

Figure 2.   Push-out specimen dimensions. (Unit: mm).

Four connectors are used for each specimen. Both 19 mm and 22 mm stud diameters are used in stud specimens. Using reinforcing pass through perfobond connector hole and the case with no reinforcement is also investigated in perfobond connector specimen, as listed in Table 1. Two types of stud connectors adopted in test, namely ϕ 19 mm × 100 mm, and ϕ 22 mm × 100 mm having yielding stress of $fy_{stud}, = 320$ MPa. The results predicted from test are used to decide the optimum choice, recommending the types of connectors predicting best resisting capacity and the best concrete casting method. Yielding stress of steel section, perfobond rib connector, and steel reinforcement used are $fy_s = 345$ MPa. The C60 concrete grade is attempted to be used. Three standard concrete cubes of $150 \times 150 \times 150$ mm dimensions were used to test cube compressive strength of concrete at 28-days given values of 52.65 MPa, 52.72 MPa and 58.52 MPa having average value of $f_{cu} = 54.6$ MPa.

3.2   Preparing push-out specimens

Dimensions of both welded stud and perfobond connectors are shown in Figure 2. Preparing the welded stud connector and perfobond connector specimens were conducted for TCC and BCC samples by preparing two halves of steel section for each specimen separately, Figure 3, casting and treating concrete parts then assembling by guest plates and bolts, Figure 4. Meanwhile, the steel parts of SCC samples for both welded stud and perfobond shear connector specimens are made as one I-beam steel section, Figure 4. Planks of 10 mm at the top of steel section web are made for all specimens, with corners of 20 mm, ensuring that the load applied on flanges.

3.3   Conducting test

To investigate the failure in connections between connector and concrete, the vertical shear force V applied slowly in small steps to failure through steel flanges using hydraulic jack supported on thick steel plate, Figure 5. Dial gauges are used to record relative

Table 1.   Classification of specimens' groups.

Group	No.	Connector type	Casting	Connectors/ specimen	Details
NS-1	3	weld stud	TCC	4	Stud ϕ 22 mm
NS-2	3	weld stud	TCC	4	Stud ϕ19 mm
NS-3	3	weld stud	SCC	4	Stud ϕ19 mm
NS-4	3	weld stud	BCC	4	stud ϕ19 mm
NP-1	3	Perfobond	TCC	4	no reinforcing
NP-2	3	Perfobond	TCC	4	with reinforcing
NP-3	3	Perfobond	SCC	4	with reinforcing
NP-4	3	Perfobond	BCC	4	with reinforcing

displacement between concrete and steel through fixing four gauges for each specimen at same level to record the displacement for each load step. The average of four values for each specimen is then used to plot

(a) Welded stud steel part

(b) Perfobond steel part

Figure 3.   Steel parts of for TCC and BCC specimens.

(a) Stud TCC, BCC and SCC

(b) Perfobond TCC, BCC and SCC

Figure 4.   Steel parts of TCC, BCC and SCC specimens.

a. Stud connector

b. Perfobond connector

Figure 5.   Push-out test specimens during test.

load–relative displacement curve for each specimen. The tests are used to obtain the ultimate shear load and largest relative displacement, the shape of failure for each specimen is also investigated. The test's records of load and relative displacement for each specimen to failure stage are then used to plot he load-relative displacement curve, locating yielding points and ultimate points. Defining V_u as ultimate strength obtained from test and V_y as yield.

3.4   Experimental tests results

3.4.1   Welded Stud connector specimens' results
The load and relative displacement of stud connector specimens are listed in Table 2 and Table 3 respectively.

The shear resistance of stud specimens is stated in Table 2, including yielding and ultimate shear resistance for each specimen and the average values for each group. As well as, relative displacement at yielding and ultimate points for each specimen and average values for each group are listed in Table 3. It can be seen that, result of specimen NS3-2 is not included because of technical problem occurred after the beginning of test, therefore, the yielding and ultimate values are note available. The shear failure of stud connectors are the most shape of failure noticed, as shown in Figure 6a and 6b. The results show the differences between each group and the effects of the considered parameters. The difference between the largest ultimate value for both NS1 and NS2 groups is about 47%. However, the effect of concrete casting direction can be shown clearly from the comparison shown in Figure 7 for ultimate shear resisting.

3.4.2   Perfobond connector specimens' results
The load and relative displacement curves for perfobond connector specimens are shown in Table 4 and Table 5 respectively. The shear resistance of stud specimens is listed in Table 4, including yielding

Table 2.   Welded stud specimens' shear resisting.

Test group		Diam mm	Cast	Yield force Vy (kN)	Ave. (kN)	Ult. force V_u (kN)	Ave. (kN)
NS1	1	22	TCC	106.6	115.3	153.5	175.1
	2			117.7		173.6	
	3			121.5		198.2	
NS2	1	19	TCC	87.0	92.4	124.7	128.4
	2			92.7		125.8	
	3			97.6		134.8	
NS3	1	19	SCC	75.6	79.9	128.0	136.9
	2			–		–	
	3			84.2		145.9	
NS4	1	19	BCC	88.8	89.8	129.1	118.4
	2			87.7		103.1	
	3			93.0		123.0	

Table 3. Welded stud specimens' relative displacement.

Test group		diam. mm	Cast	Yield mm	Ave. mm	Ultimate mm	Ave. mm
NS1	1	22	TCC	0.38	0.45	2.29	3.89
	2			0.47		3.55	
	3			0.49		5.83	
NS2	1	19	TCC	0.41	0.42	2.19	2.57
	2			0.43		2.21	
	3			0.42		3.33	
NS3	1	19	SCC	0.58	0.51	5.63	4.81
	2			–		–	
	3			0.43		3.98	
NS4	1	19	BCC	0.45	0.42	3.00	3.45
	2			0.41		4.78	
	3			0.38		2.58	

(a) stud failure remarks   (b) stud connectors after failure

Figure 6.   Shape of stud connector failure.

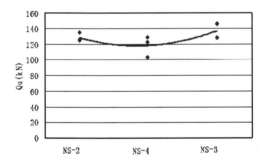

Figure 7.   Influence of casting on stud connector shear resistance.

and ultimate shear resistance for each specimen and average values for each group.

As well as, relative displacement at yielding and ultimate points for each specimen and average values for each group are listed in Table 5. Results show differences and effects of the considered parameters. The difference between ultimate value of NP1 and NP2 groups is 80% with a noticeable ductility. As well as, the effect of concrete casting direction on shear resisting is shown clearly from comparison shown in Figure 8.

The failure of perfobond connectors' specimens are noticed in concrete, Figure 9, cracks in concrete external faces at bottom of concrete slab and shear failure in concrete dowels is noticed which seems to be the main shape of failure in perfobond connector specimens.

4   DISCUSSION AND CONCLUSIONS

4.1   Stud with perfobond connectors

Results of eight groups tested are listed in Table 6. The shear resistance and the displacement predicted from both stud and perfobond specimens are listed in Table 7 and Table 8. It can be seen obviously that, welded stud connector produce low shear resistance comparing with perfobond connector. The differences illustrated in Table 7 shows that shear resistance from perfobond specimen with different casting directions are about 220% to 300% at yielding stage and about 260% to 300% at ultimate stage. Table 8 also shows that more rigid connection produced using perfobond connector, since ratio of stud displacement to perfobond displacement at yielding stage shows that displacements occurred in stud connector specimen is more than that in perfobond one of about 120% to 130%, as well as at ultimate stage is about 160% to 310%.

Another comparison between shear resistance and relative displacements of 22 mm and 19 mm diameters stud specimens is also illustrated in Table 9. It can be shown that 22 mm stud diameter specimen have more shear resistance and less displacement comparing with 19 mm diameter.

Where V_{sy} is the stud specimen shear resistance at yielding stage, V_{py} is perfobond specimen shear resistance at yielding stage, V_{su} is stud specimen shear resistance at ultimate stage, and V_{pu} is perfobond specimen shear resistance at ultimate stage. Hence,

Table 4. Perfobond connector specimens' shear resisting.

Test group		Reinf.	Cast	Yield force Vy (kN)	Ave. (kN)	Ult. force Vu (kN)	Ave. (kN)
NP1	1	No	TCC	–	–	204.7	203.1
	2			–		167.2	
	3			–		237.5	
NP2	1	Yes	TCC	275.5	300.1	340.7	392.1
	2			297.4		401.5	
	3			327.4		434.0	
NP3	1	Yes	SCC	213.9	219.2	362.8	359.3
	2			243.0		375.9	
	3			200.7		339.2	
NP4	1	Yes	BCC	256.0	248.4	363.0	358.7
	2			248.0		351.0	
	3			241.2		362.2	

Table 5. Perfobond connector specimens' relative displacement.

Test group		Reinf.	Cast	Yield mm	Ave. mm	Ultimate mm	Ave. mm
NP1	1	No	TCC	–	–	0.49	0.54
	2			–		0.63	
	3			-		0.48	
NP2	1	Yes	TCC	0.37	0.34	1.69	1.62
	2			0.29		1.44	
	3			0.34		1.74	
NP3	1	Yes	SCC	0.34	0.38	1.66	1.52
	2			0.37		1.34	
	3			0.42		1.56	
NP4	1	Yes	BCC	0.37	0.34	1.46	1.71
	2			0.34		1.88	
	3			0.32		1.79	

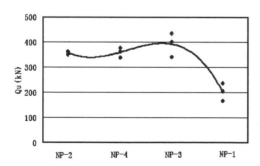

Figure 8. Influence of casting on perfobond shear resistance.

d_{sy} is stud specimen displacement at yielding stage, d_{py} is perfobond specimen displacement at yielding stage, d_{su} is stud specimen displacement at ultimate stage, and d_{pu} is perfobond specimen displacement at ultimate stage.

(a)Without reinforcement   (b) With reinforcement

Figure 9. Shape of perfobond connectors after failure.

4.2 Experiment and code results

4.2.1 Stud connector results
Results predicted from the above specifications for stud shear connector are listed in Table 10.

4.2.2 Perfobond connector results
Results from numerical expression for perfobond shear connector are listed in Table 11.

Push-out test results in this study for stud shear connector specimen confirm that most of these methods specified by codes give unconservative predictions for the strength of studs shear connector. The results of perfobond shear connector specimen also illustrated that the predicted equations given by Oguejiofor, E.C., & Hosain, M.U., (1996) may produce some misleading results. Since the expected shear resistance value from equation are greater than the experiment of 32%.

However, depending on the test results, it is concluded that using perfobond connector would give better connection between steel beam and concrete slab than stud connector. The perfobond connector show greater shear resistance with more rigid connection comparing with stud connector. Comparing with codes show that, the available code's models give unconservative value for stud shear strength but such value would be more safely. The available numerical model used for calculating perfobond shear connector strength is not accurate enough to used for design therefore the test is required to check the shear resistance of the connector.

5 RECOMMENDATIONS

Several recommendations can be stated as; (1) More researches need to investigate true behaviour and the shear resistance of the perfobond connector. (2) Modeling push-out test using the finite element method may give a basic idea on the behaviour of the connector and can be used to investigate the shear resistance of connector. (3) An optimization can be used to choose the optimum properties produce the required strength and relative slip.

Table 6. Stud and perfobond experiment groups results.

Group	Properties	Cast	Yield Vy kN	Ult. Vu kN	Ult. displ. mm	Ult. displ. mm
NS1	22 mm stud	TCC	115.3	175.1	0.45	3.89
NS2	19 mm stud	TCC	92.4	128.4	0.42	2.57
NS3	19 mm stud	SCC	79.9	136.9	0.51	4.81
NS4	19 mm stud	BCC	89.8	118.4	0.42	3.45
NP1	No reinf.	TCC	–	203.1	–	0.54
NP2	With reinf.	TCC	300.1	392.1	0.34	1.62
NP3	With reinf.	SCC	219.2	359.3	0.38	1.52
NP4	With reinf.	BCC	248.4	358.7	0.34	1.71

Table 7. Shear resistance of stud and perfobond connectors.

Cast	Vsy kN	Vpy kN	Vsy/Vpy	Vsu kN	Vpu kN	Vsu/Vpy
TCC	92.4	300.1	3.25	128.4	392.1	3.05
SCC	79.9	219.2	2.74	136.9	359.3	2.62
BCC	89.8	248.4	2.77	118.4	358.7	3.03

Table 8. Displacement of stud and perfobond connectors.

Group	dsy mm	dpy mm	dsy/dpy	dsu mm	dpu mm	dsu/dpy
TCC	0.42	0.34	1.24	2.57	1.62	1.59
SCC	0.51	0.38	1.34	4.81	1.52	3.16
BCC	0.42	0.34	1.24	3.45	1.71	2.02

Table 10. Experiment and specification stud shear resistance kN.

Group	Experiments		Chinese Code	AASHTO	AISC	CSA
	Vy	Vu				
NS1	115.3	175.1	85.15	103.39	121.64	121.64
NS2	92.4	128.4	63.51	77.12	90.73	90.73
NS3	79.9	136.9				
NS4	89.8	118.4				

Table 11. Experiments and numerical perfobond shear resistance kN.

Group	Experiments		Expression 6	Expression 7
	Vy	Vu		
NS1	–	203.1	61.94381	437.2942
NS2	300.1	392.1	195.5824473	535.9245228
NS3	219.2	359.3		
NS4	248.4	358.7		

Table 9. Displacement of 22 mm and 19 mm stud specimens.

Welded stud connector		(1)22mm	(2)19 mm	Welded stud connector		(1)22 mm	(2)19 mm
Shear resistance	Vsy kN	115.3	92.4	Displacement	dsy mm	0.45	0.42
	(1)/(2)	1.25			(1)/(2)	1.07	
	Vsu kN	175.1	128.4		dsu mm	3.89	2.57
	(1)/(2)	1.36			(1)/(2)	1.51	

ACKNOWLEDGEMENT

This research was supported by Chinese scholarship Council, Tongji University Bridge Engineering Department and Naning Bridge project.

REFERENCES

AASHTO. 2004. *AASHTO-LRFD Bridge Design Specification*. Washington DC: American Association of State Highway and Transportation Officials, 3rd ed, USA.

AISC Specification. 2005. *Specification for Structural Steel Buildings, ANSI/AISC 360-05*, American Institute Of Steel Construction, Inc., One East Wacker Drive, Suite 700, Chicago, Illinois 60601-1802, USA.

CSA–Canadian Standards Association. 1994. *Steel Structures for Buildings Limit States Design*, CAN3-S16.1-M84, Rexdale, Ontario.

GBJ 17-88. 1988. *Code for Design of Steel Structures*, National Standard of the People's Republic of China, GBJ 17–88, Beijing, China.

Liu, Y. Q. 2004. *Composite Bridge*, Renmin Jiaotong Publisher, Shanghai, China. (Chinese).

Oguejiofor, E.C. & Hosain, M.U. 1996. *Numerical analysis of Push-Out specimens with Perfobond rib connectors*, Journal of Computer and Structure, Vol. 62 (4), pp. 617–24.

Roddenberry, M.D.R. 2002. *Behaviour And Strength Of Welded Stud Shear Connectors*, PhD Dissertation, Civil Engineering, Virginia Polytechnic Institute and State University, Blacksburg, Virginia.

Salmon, C.G. & Johanson, J.E. 1990. *Steel Structures, Design and Behaviour*, 3rd Ed, Harper and Row Publishers, New York, USA.

Valente, I. & Cruz, P.J.S. 2004. *Experimental analysis of Perfobond shear connection between steel and lightweight concrete*, Journal of Constructional Steel Research, Vol. 60, pp. 465–479.

Steel and Composite Structures – Wang & Choi (eds)
© 2007 Taylor & Francis Group, London, ISBN 978-0-415-45141-3

The numerical simulation of experiment process of high strength bolt joint of steel beam and steel framed concrete wall

Shaobin Dai
Hubei Key Laboratory of Roadway Bridge & Structure Engineering, Wuhan, China

Jun Huang
Design & Research Institute of Wuhan University of Technology, Wuhan, China
School of Civil Engineering of Wuhan University, Wuhan, China

ABSTRACT: Simulating the experiment process by using the finite element program, contrastively analysing the experiment data and simulating data, and confirming the application range of the finite element program in the structure experiment. Further more, this paper analyses the wall's stress distributing state by the finite element program. Results indicate that the maximum stress area of wall lies in the contact position between the wall and anchor plate, and the double-shear-plate connection could better the bearing performance of the experimental specimens.

Keywords: High strength bolt; bearing performance; numerical simulation

1 GENERAL INSTRUCTIONS

As the development of the numerical technology, finite element software is more and more popular in civil engineering field. Using finite element software to simulate structure experiment can not only break through the limitations of experiment condition, solve a lot of difficult problems which can not be solved in laboratory and get relative analytical data, but also save a great deal of money, manpower and time. So it is very worthy to use in civil engineering field. Based on the experiment, the paper uses the ANSYS finite element software to simulate the experiment process and compares the calculation results with the experiment data in order to ascertain the effective range of numerical analysis. Otherwise, the status of strain distribution of the concrete wall, which is difficult to express through experiment data, is analyzed in the effective range of numerical simulation using finite element software (Huan Yu. 2006).

2 MODEL OF SPECIMEN

2.1 *Dimensions of model*

According to the specification, the quantity, the way of layout, the thickness of web plate and the way of connection of high strength bolts, six specimens, numbering SJ-1 to SJ-6, are designed in the experiment. Steel beams and concrete wall, which are connected with the six specimens, are same. The length of steel beam is 1.7 m; the dimension of the section is HN250 × 125 × 6 × 9 mm. The dimension of concrete wall, in which formed steel is imbedded, is 2000 × 1100 × 350 mm. The length of formed steel column, which is imbedded in the concrete wall, is 1.6 m; the dimension of the section is HM200 × 150 × 6 × 9 mm. The length of steel corbel is 120 mm; the dimension of the section is same as steel beam. They are connected with formed steel column by butt weld. The reinforced bar which is distributed in the concrete wall is first-level steel bar, its diameter is 12 mm, the spacing is 200 mm. Table 1 and Figure 1 show the general situation of the six specimens and the connection way of joints separately. (In order to explain the structure conveniently, the concrete wall is left out in Figure 1.)

2.2 *Characteristics of material*

The material of beams and columns is Q235. The strength grade of concrete is C30. In addition, all bolts are 10.9 grade high strength friction-type bolts, the friction surfaces are dealt with by sandblast. In the experiment of material characteristics, Steel and concrete specimens come from the same collection and the same curing condition. Then we can get the two constitutive relations of the two materials (Figure 2 and Figure 3) The constitutive relation

Table 1. General situation of specimens.

Serial number	Way of connection	Type and thickness of web plate	Specification and type of bolts	Quantity of bolts
SJ-1	Onc row, three bolts	Single shearing plate (12 mm)	M16 (conventional type)	Three
SJ-2	One row, three bolts	Single shearing plate (12 mm)	M20 (conventional type)	Three
SJ-3	One row, three bolts	Single shearing plate (16 mm)	M16 (conventional type)	Three
SJ-4	Two rows, four bolts	Single shearing plate (16 mm)	M22 (torsion shear type)	Four
SJ-5	Two rows, six bolts	Single shearing plate (16 mm)	M20 (torsion shear type)	Six
SJ-6	One row, three bolts	Double shearing plate (12 mm)	M20 (torsion shear type)	Three

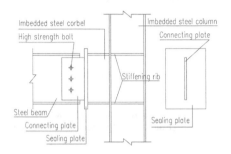

SJ-1~SJ-3 Diagram of connection ioint

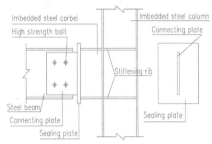

SJ-4 Diagram of connection ioint

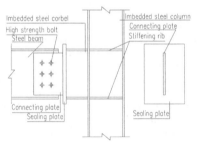

SJ-5Diagram of connection ioint

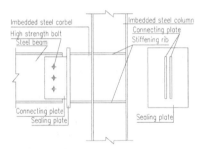

SJ-6 Diagram of connection ioint

Figure 1. Diagram of connection joint.

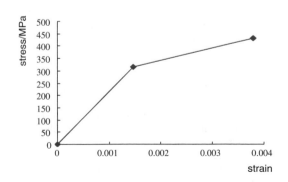

Figure 2. Constitutive relations curve of steel.

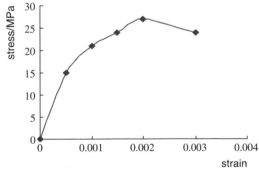

Figure 3. Constitutive relations curve of concrete.

curve of concrete is a multi-thread type follow-up strengthening curve. The constitutive relation curve of steel is a double-thread type follow-up strengthening curve, its yield strength is 317.37N/mm^2, ultimate strength is 430.77 N/mm^2, and Young's modulus is 2.16×105 N/mm^2.

2.3 *Experiment of anti-slipping coefficient of friction*

Experiment of anti-slipping coefficient of friction is used to measure the anti-slipping coefficient of web plates with high strength bolt. It can provide the parameter for the contact problem in finite element analysis. According to the requirements and methods of design, construction, acceptance specification (JGJ82-91) and criteria for test and evaluation of steel structure engineering (GB50221-95) in high strength bolt connection, three tensile specimens, numbering MS-1, MS-2 and MS-3, are made. They have double friction surfaces and two bolts connection joints. According to the requirement of specifications, the pretightening force is applied in order to measure the anti-slipping coefficient of specimens (Qicai Li et al. 2003). The experiment results are showed in Table 2.

3 MODEL BUILDING OF FINITE ELEMENT

3.1 *Type of element*

According to material characteristics and structure bearing features, in ANSYS software, SOLID45 element, which has three-dimensional eight nodes, is used to simulate formed steel and bolts; SOLID65 element, which has three-dimensional eight nodes and has considered the distributing steel bars uniquely, is used to simulate concrete wall; LINK10 element is used to simulate the vertical steel bars in concrete wall; 3D target170 and pt-to-surf175 contact element are used to simulate the contact of web plates of high strength bolts. The coefficient of friction comes from the result of anti-slipping coefficient experiment. Pre179 element is used to simulate the pretightening force of high strength bolts. The value of the pretightening force comes from the standards according to the specifications of bolts (Xianlei Ye et al. 2003).

3.2 *Boundary condition and mesh dividing*

Finite element model analysis adopts one-way step-by-step direct loading. The deforming process and stress changing rules of the structure are studied during the bearing process. All degrees of freedom are restrained at the up-down end of the concrete wall in the model, and then the boundary condition is the same as experiment condition basically. In order to make the calculation of model accurate, hexahedral mapping mesh dividing element is adopted. Figure 4 is the finite element model.

3.3 *Solution control*

The solution adopts complete Newton-Lapusen asymmetric iterative method. Considering big deformation effects of structure, the linear searching is used, and predicting apparatus option is opened. By these methods it can strengthen the convergence of analysis results and guarantee the stability of iterative calculation. Time step is opened automatically in order to adjust time step at any moment, then computing time is saved, and the good balance between precision and cost is got.

4 RESULTS ANALYSIS OF FINITE ELEMENT

4.1 *Contrastive analysis of skeleton curve*

The skeleton curve of load-displacement can describe the whole one-way loading process of the experiment. From the result of finite element analysis, six skeleton curves of load-displacement of steel beam are drawn. Through the contrastive analysis of skeleton curves between the experimental results and finite element analytical results, it can verify the reasonability of finite element models and the accuracy of finite element method which is used to simulate the experiments. With the help of finite element simulation technology, it can help us to reasonable evaluate the experimental results, then reduce the experiment errors because of the experiment condition and so on. It can also help us complete the unfinished research contents of experiment using reasonable finite element method. Figure 5 ∼ Figure 10 are the contrastive figures of skeleton curves of six specimens under one-way load.

From the contrastive figures of skeleton curves, they indicate that before the slipping damage of the six specimens appears, the results of finite element analysis coincide with the results of experiment, the skeleton curves coincide basically, and the inflexion point positions of the skeleton curves are uniform basically. So the results of finite element analysis can accurately predict the displacement and load when the slipping damage appears, then verify the accuracy of the finite element models. But when the displacement increases, the errors between the results of finite element analysis and the experiment results increase after the slipping damage appears.

The skeleton curves of finite element analysis have not descending branch of load. The final damage situation of the specimens can not be predicted, because the present finite element software does not consider the actual situation that the pretightening force will

Table 2. The results of anti-slipping coefficient experiments.

Serial number	Specification of bolts	Treatment of friction surface	Thickness of plate (t1)	Thickness of plate (t2)	Width of plate(b)	Anti-slipping coefficient
MS-1	M16	sandblast	6 mm	12 mm	60 mm	0.414
MS-2	M20	sandblast	6 mm	12 mm	75 mm	0.406
MS-3	M22	sandblast	6 mm	12 mm	80 mm	0.433

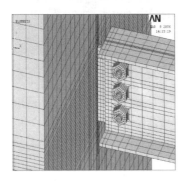 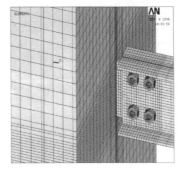 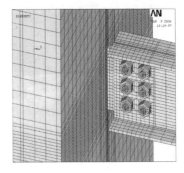

Figure 4. Mesh dividing of finite element model.

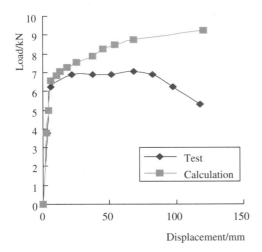

Figure 5. Contrastive figures of skeleton curves of SJ-1.

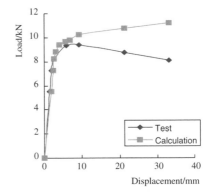

Figure 6. Contrastive figures of skeleton curves of SJ-2.

gradually come down, and the coefficient friction will gradually reduce after the slipping damage appears.

4.2 Stress diagram of finite element

The stress distribution of the concrete wall is one of the problems what the experiment wants to solve. However, because the formed steel is buried in the concrete wall, the bearing situation is very complex.

The visual experiment results are difficult to get by present experiment methods, so it needs finite element method to solve this kind of problems. Figure 11 is the stress distribution diagrams of the six concrete walls under normal working condition.

From the diagrams, the stress distribution status of the six concrete walls has the same characteristic. Under load two stress concentration areas appear at the contact area of the concrete walls and the top and bottom edges of the anchor slab, thereinto the upper area is the maximum tensile stress area, the lower area is the maximum compressive stress area. When the two stress areas expend all around, the stress of the concrete walls will reduce continuously, then the sector stress loops are formed. This indicates that when the load increases gradually, the first crack place will appear at the top and bottom edges of the anchor slab,

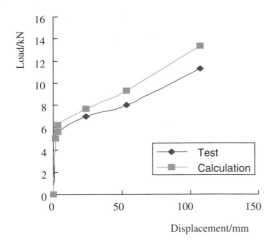

Figure 7. Contrastive figures of skeleton curves of SJ-3.

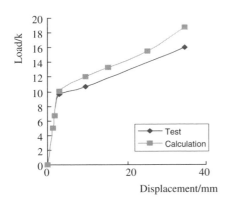

Figure 8. Contrastive figures of skeleton curves of SJ-4.

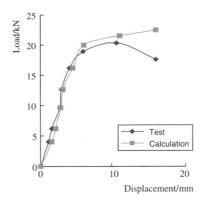

Figure 9. Contrastive figures of skeleton curves of SJ-5.

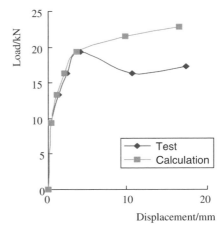

Figure 10. Contrastive figures of skeleton curves of SJ-6.

and then continuously expend to outside radically. So strengthening the reinforcement of the concrete walls at the edge of the anchor slab can delay the occurrence of cracks, prevent the development of cracks effectively, and then improve the anti-crack capability and the integral bearing capacity.

In addition, from the deformation shape of steel beam, the evident torsion of steel beams has happened in SJ-1~SJ-5 except SJ-6. From the stress distribution diagram of the concrete walls, the stress distribution diagram of SJ-6 is more bilaterally symmetrical than others. This indicates that the specimen of single-shear plate can easily result in the torsion of steel beam; the moment of torsion can be transferred to concrete wall through anchor slab and buried formed steel, then result in more complicate stress of concrete wall; it is bad to the bearing capacity of concrete wall and steel beam. But double-shear plate is symmetrical by itself; it can reduce the action of torsion and the stress of other members. So double-shear plate can make the

bearing state more reasonable, and then improve the bearing capacity of specimens (Bahram M et al. 2004, Guoqiang Li. 2003).

5 CONCLUSIONS

(1) Before the slipping damage of the six specimens appears, the results of finite element analysis coincide with the results of experiment, the skeleton curves coincide basically. But when the displacement increases, the errors between the results of finite element analysis and the experiment results increase after the slipping damage appears. It indicates that before the slipping damage appears, the application of the finite element program in the structure experiment is accurate in the period of loading.

(2) Under load, the maximum stress area of the concrete wall appears the contact area between the

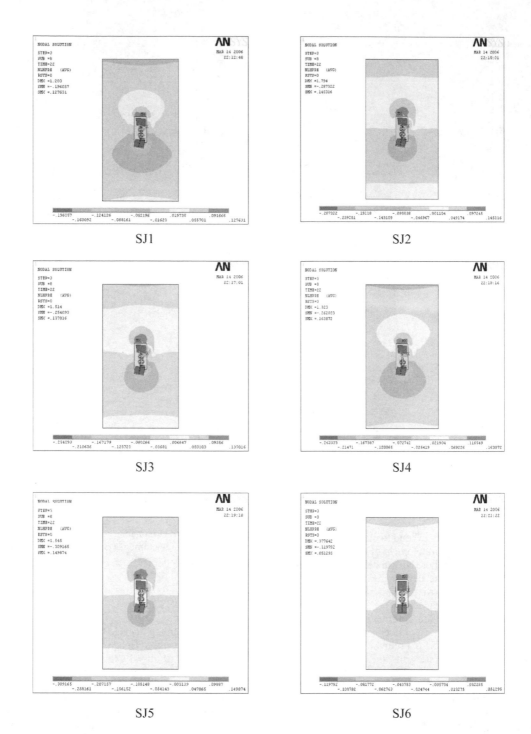

Figure 11. Stress distribution diagram of concrete.

concrete wall and the edge of anchor slab. It indicates that strengthening the reinforcement of the concrete walls at the edge of the anchor slab can delay the occurrence of cracks, prevent the development of cracks effectively, and then improve the anti-crack capability and the integral bearing capacity.

(3) Under load, the evident torsion of steel beams has happened in SJ-1 ~ SJ-5 except SJ-6. This indicates that because double-shear plate is symmetrical by itself; it can reduce the action of torsion and make the bearing state more reasonable, then improve the bearing capacity of specimens.

REFERENCES

Xianlei Ye, Yajie Shi 2003. *Application examples of ANSYS engineering analysis software.* Beijing: Press of Tsinghua University.

Huan Yu 2006. *The research of high strength bolt connection joints of steel beam and buried formed steel concrete wall.* Wuhan: Master Dissertation of Wuhan University.

Qicai Li, Mingzhou Su, Qiang Gu etc 2003. Experimental study of steel beam to column connection with cantilever beam splicing under cyclic loading. *Journal of Building Structure*, 24(4):54–59.

Bahram M. Shahrooz, Jeremy T. Deason and Gokhan Tunc 2004. Outrigger beam-wall connections.component testing and development of design model. *Journal of Structural Engineering.* 130(2):253–261.

Guoqiang Li, Bing Qu, Feifei Sun etc 2003. Cyclic loading tests of steel beam to concrete wall joints in steel concrete mixed structures. *Journal of Building Structure*, 24(4):1–7.

Bridges

Steel and Composite Structures – Wang & Choi (eds)
© *2007 Taylor & Francis Group, London, ISBN 978-0-415-45141-3*

Load-consistent effective widths for composite steel-concrete decks

L. Dezi & F. Gara
DACS, Università Politecnica delle Marche, Ancona, Italy

G. Leoni
PROCAM, Università di Camerino, Ascoli Piceno, Italy

ABSTRACT: This paper proposes new effective widths for the analysis of twin-girder and single-box girder steel-concrete composite bridge decks. The method proposed catches the effects of a single load layout, as well as the maximum effects obtained by considering the envelopes of bending moments, support settlements and shrinkage effects. The method permits evaluating the normal longitudinal stress on the slab by means of a cross sectional analysis starting from the results of the structural analysis performed by considering the geometric width of the slab. Various comparisons between the slab stresses of a realistic continuous twin-girder composite bridge deck calculated by means of the method proposed, a refined shell finite element model, are shown demonstrating the simplicity and effectiveness of the method.

1 INTRODUCTION

In steel-concrete composite continuous decks with wide slab (Figure 1), the usual assumption of bending theory, according to which the plane cross sections remain plane after loading, is not realistic. Due to the interaction with the steel beams, the concrete slab undergoes a significant warping which is responsible of a non-uniform stress distribution (shear-lag effect). This happens on the whole deck when vertical loads are applied inducing non-uniform shear and, locally, when longitudinal forces (shrinkage, thermal action, prestressing) act on the slab or on the steel beams as a result of beam-slab interaction.

In the design of such decks, the main codes of practice, e.g. the Eurocode 4 (prEN 1994-2, 2003), recommend taking into account the effects of shear-lag by suitably reducing the slab width in order to catch the stress peaks by assuming the preservation of the plane cross section.

The effective widths proposed are valid for external vertical loads (Sedlacek & Bild 1993) whereas no suggestion is given for other actions even if they must be considered in usual bridge design.

This paper proposes a simplified method for the verification of twin-girder and single-box girder steel-concrete composite decks based on the definition of new effective widths consistently with the load case (uniformly distributed loads, traffic loads, support settlements and shrinkage). The method permits evaluating the normal longitudinal stress on the slab for each load condition by means of a cross sectional analysis

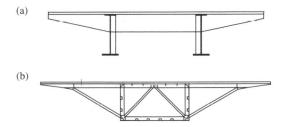

(a)

(b)

Figure 1. Composite decks with very wide slab: (a) twin-girder deck with cantilevered transverse beams, (b) box girder with lateral struts.

starting from the results of the structural analysis performed by considering the geometric width of the slab.

2 MODEL RECALLS

The analytical model used in the research proposed by this paper, was presented firstly by the authors in (Dezi et al. 2001) and then in (Dezi et al. 2005) where a new beam finite element was proposed. The twin-girder deck, or the single-box girder deck, is schematised as a beam for which the concrete slab may undergo warping.

The bridge deck cross section is assumed to be rigid in its own plane so that, by considering actions applied so as to avoid twisting rotation and transverse horizontal displacements, its generic deformed configuration is characterized by vertical deflections and

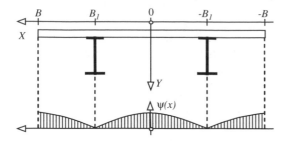

Figure 2. Composite cross section.

longitudinal axial displacements only. The beam-slab interface slip due to the longitudinal flexibility of the shear connection is considered while the slab uplift is assumed to be fully prevented (Newmark et al. 1951). For the steel beam, the Euler-Bernoulli kinematical model is assumed, whereas the assumption of preservation of the plane cross section is removed for the concrete slab. The non-uniform out-of-plane displacements, due to the shear strains, are taken into account by separating variables and considering a known warping function on the cross section multiplied by an intensity measure that is a function of the beam axis abscissa (Reissner 1946). The warping function is obtained according to Dezi et al. (2003). Due to the small slab thickness, it is considered to be constant on the slab depth and, in order to catch the planarity loss of the whole slab taking into account the actual beam spacing. In the case of a twin-girder deck it is described by three parabolic branches (Figure 2).

The steel beam and the slab reinforcements are assumed to be linear elastic while the concrete is considered to be visco-elastic and uncracked even under traction. The shear connectors are considered to be spread along the beam and a linear elastic relationship is assumed between the longitudinal shear flow and the beam-slab slip.

The problem is defined in a time-space domain and the numerical solution must be sought accordingly by introducing a double discretization, the first for the time domain and the second for the beam axis. The problem is then solved thanks to a step-by-step procedure used in conjunction with the finite difference method (Dezi et al. 2001) or with the finite element method (Dezi et al. 2005). The procedure was validated by comparing the results obtained with those given by refined shell finite element models (Dezi et al. 2005).

The analytical method proposed permits calculating the variation in time of displacements, stress resultants, and stress distribution in the concrete slab and in the steel beam, and is particularly straightforward in the calculation of the effective slab width B_{eff}. In fact, if σ_c^{max} and N_c are the peak and the resultant of the stresses in the concrete part of the cross section, B_{eff} is defined by the following expression obtained by

equating N_c to the resultant of uniformly distributed stresses σ_c^{max} on a reduced area of the slab:

$$\frac{B_{eff}}{B} = \frac{N_c}{A_c \sigma_c^{max}} \qquad (1)$$

3 EFFECTIVE WIDTH METHOD

Even if Von Kármán (1924) introduced the effective width as a parameter describing the stress concentration phenomenon due to shear-lag, the main codes of practice suggest considering a priori an effective width for the slab in order to catch the shear-lag effects.

This praxis, that reverses the early meaning of effective width, permits simplifying the structural analyses that may be performed by considering the preservation of the plane cross section without taking into account explicitly the cross section loss of planarity. Practice demonstrated that this approach leads usually to errors acceptable for routine verifications.

Eurocode 4 (prEN 1994-2, 2003) proposes a method whose philosophy is common to other codes of practice. For continuous bridge decks, the Eurocode method consists in the definition of two effective widths: one for the global structural analysis, defined for each span, and one for the cross section analysis, variable along the span. The effective width is calculated on the basis of equivalent lengths L_e defined as the distance between points of zero bending moment. In the case of continuous composite decks where moment envelope from various load arrangements governs the design, the equivalent lengths are conventionally defined only on the basis of the geometry (span lengths, slab width and beam spacing). Even if not specified, the formulas suggested should be valid only for transverse loads (Sedlacek and Bild, 1993). For other kinds of actions introducing longitudinal forces (e.g., concrete shrinkage, thermal action and prestressing), no specific suggestions are given even if they are very important in the common bridge design. This may induce some designers to believe that the effective slab width only depends on the deck geometry – and not on the load case – and to adopt the method suggested for continuous decks, valid for transverse loads, for the verifications. In some cases, this may lead to non conservative results as for prestressing actions (Dezi et al. 2006).

In this section a simplified method for the analysis of shear-lag effect, based on the adoption of new effective widths defined according to load schemes, is proposed.

3.1 Parametrical analysis

In order to calibrate the formulas of the effective slab widths for continuous steel-concrete composite decks, a wide parametrical analysis was performed by using the analytical model previously described. Both simply supported and continuous beams were considered under the actions of interest in the bridge design, namely vertical concentrated and uniformly distributed loads, shrinkage of concrete and thermal actions, jacking of supports, and prestressing with internal or external cables. Each of these was considered to act alone in order to understand how shear-lag depends on the kind of action. Under the assumption of linear behaviour of the materials, the superposition principle ensures the validity of this kind of investigation.

All the geometrical parameters that define the deck were varied in the ranges of interest for realistic bridges, namely twin-girder spacing, beam geometry, slab thickness and width, reinforcement geometrical ratio and shear connection stiffness; furthermore different span lengths were considered in order to investigate a wide range of the width-to-length ratio. All the analyses were performed taking into account the creep behaviour of the concrete in order to investigate the possible variation in time of the effective width. Due to lack of space, the results of the parametrical analysis cannot be shown here and only qualitative remarks will be made.

Analyses confirm that the effective width strongly depends on the kind of action (Dezi et al. 2003, Dezi et al. 2006); furthermore the span width-to-length ratio as well as the ratio between beam spacing and slab width proved to be the most important among the parameters investigated.

The other parameters are not significant with two exceptions: the reinforcement geometrical ratio whose influence on the effective width is however not of practical importance in the range of interest for slab cracking control (1–2%), and the shear connection stiffness whose influence is however insignificant for the values of stiffness usually adopted in bridge design. Finally, concrete creep induces weak variations of the effective width only in the case of longitudinal prestressing while it is not important for the other actions.

3.2 Effective width formulas

Based on the parametric analysis previously summarised, new effective widths are proposed for the analysis of continuous composite decks. The formulas are deduced by considering simplified static schemes (simply supported beam and two-span continuous beam) obtained by extracting deck sections from the continuous girder. They are obtained by linear and second order polynomial regressions by performing the least squares fit of data obtained from the parametrical analysis. The method proposed is valid for the SLS and the elastic ULS and is based on the concept that

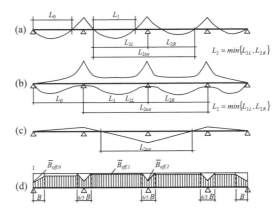

Figure 3. Definition of the geometrical quantities appearing in Equations 2–4: (a) uniformly distributed loads; (b) traffic load envelope; (c) settlement of support; (d) longitudinal variation of the non dimensional effective width B_{eff}/B.

an effective width, different for each kind of action, must be considered when the stress state of a given cross section is evaluated by means of a cross sectional analysis.

More specifically, the method is articulated according to the following steps:

1. Definition of the effective widths, different for each load cases, for the deck cross sections at the deck ends ($B_{eff,0}$), at each sagging region ($B_{eff,1}$), and at each internal support ($B_{eff,2}$) with the equations

$$\overline{B}_{eff,0} = \frac{B_{eff,0}}{B} = \left(C_1 \frac{B}{L_0} + C_2 \right) \left[C_3 \left(\frac{B_1}{B} \right)^2 + C_4 \frac{B_1}{B} + C_5 \right] \quad (2)$$

$$\overline{B}_{eff,1} = \frac{B_{eff,1}}{B} = \left(D_1 \frac{B}{L_1} + D_2 \right) \left[D_3 \left(\frac{B_1}{B} \right)^2 + D_4 \frac{B_1}{B} + D_5 \right] \quad (3)$$

$$\overline{B}_{eff,2} = \frac{B_{eff,2}}{B} = \left[E_1 \left(\frac{B}{L_{2tot}} \right)^2 + E_2 \frac{B}{L_{2tot}} + E_3 \right] \cdot$$
$$\cdot \left[E_4 \left(\frac{B_1}{B} \right)^2 + E_5 \frac{B_1}{B} + E_6 \right] \left(E_7 \frac{L_2}{L_{2tot}} + E_8 \right) \quad (4)$$

where the geometrical quantities are defined by Figures 3a–3c and the constants are reported by Table 1.

2. Approximation of the variation of the effective width along the deck axis by considering constant values at the sagging regions and linear variations at the hogging regions and at the deck ends (Fig. 3d); in particular, the linear variation at the deck ends is obtained by interpolating between the values $B_{eff,0}$ and $B_{eff,1}$ in a deck section of length B; in the hogging regions, the linear variation is obtained by interpolating between $B_{eff,2}$ and the geometric value D in a deck section of length $3/5\ B$.

475

Table 1. Coefficients for the definition of the effective widths.

External supports	C_1	C_2	C_3	C_4	C_5			
UDL	−0.75	0.97	−1.45	1.20	0.76			
TLE	−0.75	0.87	−1.45	1.20	0.76			
SS	0	1	0	0	1			
SH	0	1	−2.5	1.4	0.425			
Span sections	D_1	D_2	D_3	D_4	D_5			
UDL	−0.67	1.05	−0.66	0.72	0.81			
TLE	−0.67	0.95	−0.66	0.72	0.81			
SS	0	1	0	0	1			
SH	0	1	0	0	1			
Internal supports	E_1	E_2	E_3	E_4	E_5	E_6	E_7	E_8
UDL	6	−3.75	0.95	−2.81	2.07	0.67	−0.35	1.17
TLE	6	−3.75	0.95	−2.81	2.07	0.67	−0.35	1.17
SS	0	−0.83	0.97	−1.24	1	0.81	0	1
SH	0	0	1	0	0	1	0	1

UDL = uniformly distributed load; TLE = traffic load envelope; SS = support settlement; SH = shrinkage.

3. Calculation of the stress state by the usual theory of bending for each load case and superposition of the results.

For the concrete shrinkage action (or thermal action), the slab width has to be reduced only at the deck ends where the effective width is given by Equation 2. The stresses calculated at the end decks are assumed to be constant over a deck section of length B.

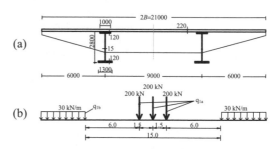

Figure 4. (a) Composite cross section; (b) traffic loads according to DM (1999).

4 NUMERICAL COMPARISONS

Four numerical applications on a twin-girder composite continuous deck with the cross section of Figure 4a are reported to demonstrate the potential of the proposed method. In order to underline the dependence of the shear-lag effect on the different load cases, uniformly distributed load, traffic loads of the Italian Code of Practice (1990) (Fig. 4b), settlement of a support, and uniform thermal action on the slab are considered separately. The diagrams of effective widths and maximum longitudinal normal stresses at the slab middle plane, obtained with the proposed method, are compared with those obtained with the analytical model proposed in Dezi et al. (2005). The results obtained with the Eurocode 4 method are also given for the case of traffic load, by using explicit formula given by the code, and, for the cases of uniformly distributed load and settlement of a support,

by considering the equivalent lengths equal to the distances between points of zero bending moment.

In the case of uniformly distributed load, the upper part of Figure 5 shows that the proposed method accurately catches the effective width obtained with the analytical method at the span and support sections.

Discrepancies are only in the sections where the bending moment is nearly zero. The method proposed leads to errors that are not significant in the calculation of the stress state. In Figure 5 the results obtained by applying rules of Eurocode 4 are also reported both considering the effective width suggested for moment envelopes (dashed curve) and by defining the equivalent lengths equal to the distance between zero bending moments (spots). The Eurocode method overestimates the effective width at span sections and especially at

476

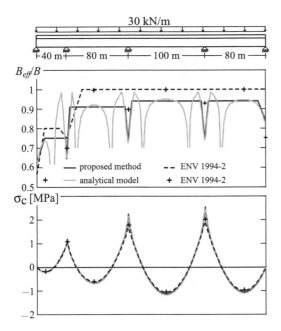

Figure 5. Effective width diagram and stress at the middle plane of the slab: uniformly distributed load.

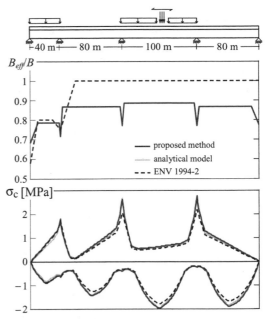

Figure 6. Effective width diagram and stress at the middle plane of the slab: traffic loads (envelope).

the internal supports in both the cases. The lower part of the Figure 5 shows that the diagrams of the maximum longitudinal normal stress at the slab middle plane obtained with the proposed simplified method and with the analytical method are practically superimposed while the Eurocode method gives some errors at the span and at the internal supports even if the method of the *zero bending moment points* is more accurate at supports.

Figure 6 shows the results of the deck under the traffic load. The comparison between the maximum and the minimum longitudinal stresses on the slab middle plane shows good agreement between the results of the method proposed and the analytical method, while the Eurocode underestimates the stresses value by about 10% at the span and 25% at the internal support.

Figure 7 shows the results of the deck subjected to a settlement of an internal support. The method proposed catches both the effective width and the stress state given by the analytical method. In this case only the *zero bending moment points* method suggested by the Eurocode 4 is applied for the calculus of the effective width at internal supports.

At the span the effective width is assumed to be equal to the geometrical width. The Eurocode method underestimates the value of the slab stresses at the internal supports. This is conservative when the support settlement is imposed to pre-stress the deck.

In Figure 8 shows the case of a deck subjected to a thermal action on the slab. The effective width diagram of the proposed method is very close to that given by

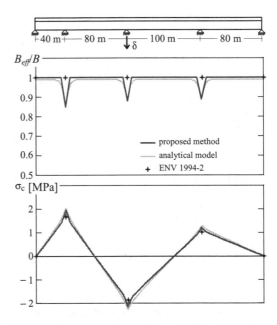

Figure 7. Effective width diagram and stress at the slab middle plane: settlement of an internal support ($\delta = 0.20$ m).

the analytical method and, consequently, the relevant slab stress diagrams are practically superimposed. No comparison with Eurocode 4 is done because this case is not considered in the code.

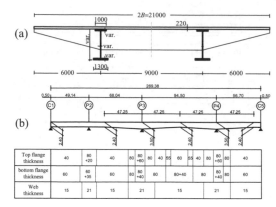

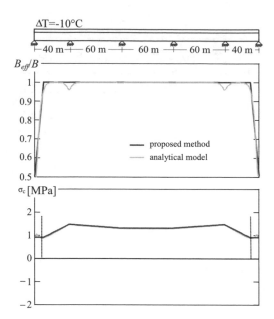

Figure 8. Effective width diagram and stress at the middle plane of the slab: thermal action on the concrete slab.

Figure 9. (a) cross section; (b) I-beam geometry.

4.1 Application to a real bridge

The method has been checked by comparing the results with those obtained by refined analyses performed by using shell finite elements (SAP 2000 2004). The results of an application performed with reference to a real four-span continuous bridge having I-beams with variable geometry (Kretz et al. 2002) are shown in Figure 9. Both the results of a uniformly distributed dead-load and the traffic loads considered by the Italian Code of Practice (1990) are reported.

The results obtained with the Eurocode method are also shown: with dashed line those obtained with the formula of the effective widths for the composite beams whose design is governed by a moment envelope; with spots, the results obtained in the case of uniformly distributed load, from the *zero bending moment points* method. In both the cases, the effective width values obtained with the Eurocode are greater than those given by the proposed method.

In the case of the uniformly dead-load an acceptable approximation of the stress values in the spans is obtained with both the methods while greater errors are obtained at the inner supports, especially with the Eurocode method. For the traffic load case, the method proposed furnishes better estimation of the stresses both at sagging and hogging regions.

5 FINAL REMARKS

The proposed method is useful for the verification of twin-girder and single-box girder steel-concrete

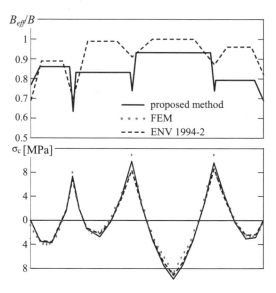

Figure 10. Effective width diagram and stress at the mean layer of the slab in the case of uniformly distributed self-weight.

composite decks subjected to uniformly distributed loads, traffic loads, support settlements and shrinkage. Since the effective widths depend on the load case, different structural models should be used in the design. To this purpose new effective widths are defined consistently with the load cases. The method is applied similarly to that suggested by prEN 1994-2 (2003) but the effective widths are considered only for the sectional analysis whereas the global structural analysis is performed with the real geometry of the deck. For

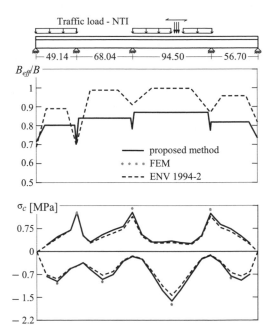

Figure 11. Effective width diagram and stress at the mean layer of the slab in the case of traffic loads (envelope).

this reason the method proposed is suited for bridge analyses.

REFERENCES

D. Min. LL.PP. 4 maggio 1990. Aggiornamento delle norme tecniche per la progettazione, l'esecuzione e il collaudo dei ponti stradali. *Italian Code of Practice.*

Dezi, L., Gara, F., and Leoni, G. 2003. Shear-lag effect in twin-girder composite decks. *Steel and Composite Structures*, 3(2), 111–122.

Dezi, L., Gara, F., and Leoni, G. 2005. A beam finite element for the shear-lag analysis of steel-concrete composite decks. *Proceedings of ICASS'05, Shanghai, China*, 735–740.

Dezi, L., Gara, F., and Leoni, G. 2006. Effective slab width in prestressed twin-girder composite decks. *J. Struct. Engrg., ASCE*, 132(9), 1358–1370.

Dezi, L., Gara, F., Leoni, G., and Tarantino, A.M. 2001. Time dependent analysis of shear-lag effect in composite beams. *J. Mech. Engrg., ASCE*, 127(1), 71–79.

prEN 1994-2, 2003. *EUROCODE 4: Design of composite steel and concrete structures - Part 2: Rules for bridges.* European Committee for Standardization.

Kretz, T., Michotey, J.L, et Svetchine, M. 2002. Le cinquième pont sur la Nive. *OTUA Bulletin Ponts métallique*, 15, 59–72.

Newmark, N. M., Siess, C. P., and Viest, I. M. 1951. Test and analysis of composite beams with incomplete interaction. *Proc. Soc. Exp. Stress Anal.*, 9(1), 75–92.

Reissner, E. 1946. Analysis of shear lag in box beams by the principle of the minimum potential energy. *Quarterly of Appl. Math.*, 4(3), 268–278.

SAP 2000, 2004. CSI analysis reference manual. *Computer and Structures, Inc.*, Berkeley, California, U.S.A..

Sedlacek, G., and Bild, S., 1993. A simplified method for the determination of the effective width due to shear lag effects. *J. Construct. Steel Research*, 24, 155–182.

v. Kármán, Th. 1924. *Die mittragende poreite*. Berlin: Springer.

Steel and Composite Structures – Wang & Choi (eds)
© 2007 Taylor & Francis Group, London, ISBN 978-0-415-45141-3

Local buckling of class 4 cross-section – application to a steel-concrete continuous beam of bridge at real scale

S. Guezouli

INSA, Rennes, France

ABSTRACT: The present paper is concerned with the local buckling of class 4 cross-section on hogging zone of steel-concrete continuous beams of bridges. The first investigation concern a 3D composite panel supposed on hogging zone and loaded until buckling. Several numerical simulations show the significant influence of the ratio (length/height) of the panel on the moment-rotation (M-θ) curve. A simplified model representing this curve is proposed in the aim to include it in the user-friendly finite element program "Pontmixte". The second investigation concern an application to a continuous beam at real scale in order to show the influence of local buckling on the moment redistribution percentage when non-linearity of materials, concrete creep and shrinkage, tension stiffening and temperature difference effects are also taken into account.

1 INTRODUCTION

Recent research concerning the moment redistribution percentage (Guezouli & et al., 2004) in steel-concrete composite continuum beams shows that in general case, if an uncracked elastic global analysis is used to design the beam, 10% moment redistribution is allowed from intermediate support to mid-span. Nevertheless, this value becomes questionable if the length ratio (shorter/longer) becomes too small. The moment redistribution is due to material non-linearities combined to some phenomena like: creep and shrinkage of the concrete slab, the tension stiffening, the temperature range... In this work, the influence of local buckling is pointed out especially for class 4 cross-section (Eurocode classification). In this aim, the first investigation will concern a panel of beam supposed on hogging zone. Knowing that the class of the cross-section is the maximum between the one of the web and the other of the compressed bottom flange of the girder, the moment-rotation (M-θ) curves will surely depend on the combination of both classes. In addition, the influence of the stiffeners as well as the influence of the panel length must be taken into account. A 3-D model is developed on CASTEM finite element code. The finite element code PONTMIXTE (Guezouli & et al., 2006) is used to investigate on the influence of local buckling on the moment redistribution percentage.

The symmetrical 3-spans beam under investigation is presented on Figure 1. The effective width of the slab is about 5500 mm and its height is equal to 230 mm; the longitudinal cross-section of the bars

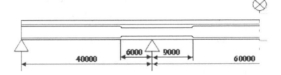

Figure 1. The beam under investigation.

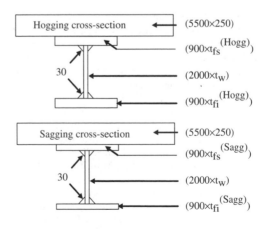

Figure 2. Hogging and sagging cross-sections.

corresponds to 1% of the slab area (Fig. 2). The beam is fully connected with two studs by cross-section. The hogging zone is supposed covering 15% of the span length at each side of the intermediate

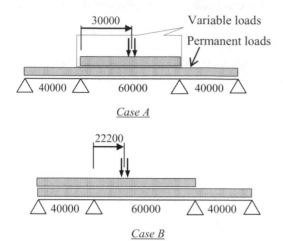

Case A

Case B

Figure 3.   Design results.

Table 1.   Numerical investigation.

Cross-section	t_w (mm)	Web class	t_{fi} (mm)	Flange class	Cross-section class
1	20	3	50	3	3
2	15	4	60	1	4
3	15	4	55	2	4
4	15	4	50	3	4
5	15	4	35	4	4

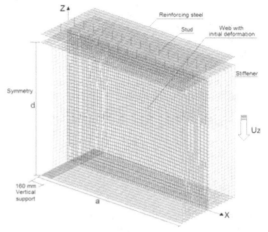

Figure 4.   3D model (a/d = 1).

supports. At the beginning, the continuous beam must be pre-designed using uncracked elastic global analysis, keeping t_w equal to 20 mm and searching appropriate values for $t_{fi}^{(Hogg)}$ and $t_{fi}^{(Sagg)}$ ($t_{fs}^{(Hogg)} = 0.9t_{fi}^{(Hogg)}$ and $t_{fs}^{(Sagg)} = 0.9t_{fi}^{(Sagg)}$). The unknown values of bottom flange thicknesses could be found by an iterative process balanced between two critical loading cases (More details could be found in Guezouli, 2001):

– Critical loading case in sagging zone:
   $M_{Sd} = M^+pl, Rd$.
– Critical loading case on hogging zone:
   $M_{Sd} = M^-_{el,Rd}$.

The pre-design results are shown on figures 3. Permanent loads correspond to the steel girder self weight, the concrete slab and the superstructure. The variable loads correspond to traffic loads applied to the bridge (here a twin-steel girder bridge) and have values in accordance with the Model 1 of EN 1991–2 Part 2 (CEN, 2003), namely a "U.D.L." of 9 kN/m² for the lane 1 and 2.5 kN/m² for the lanes 2 and 3, and a Tandem System with 2 axle loads each equal to 300 kN. In the transverse direction, the traffic loads are distributed according to an influence line. Finally, the numerical characteristic values of traffic loads for the most loaded lane are: q = 31.2 kN/m and Q = 406 kN (with the distance of 1200 mm). Elastic material characteristics are as follows: *Concrete*: $E_{cm} = 35000$ MPa and $f_{ck} = 40$ MPa ; *Steel girder*: $E^{(a)} = 210000$ MPa and $f_y^{(a)} = 355$ MPa; *Reinforcing steel*: $E^{(s)} = 200000$ MPa and $f_y^{(s)} = 400$ MPa; *Studs*: h = 125 mm, $\phi = 22$ mm, $P_{Rd} = 121,6$ kN, Y-spacing = 225 mm and X-spacing ≈140 mm.

The flange thicknesses obtained are:
$t_{fi}^{(Hogg)} = 50$ mm, $t_{fi}^{(Sagg)} = 16$ mm.

2   FIRST INVESTIGATION – (M-Θ) CURVES

Several numerical investigations are presented in table 1. It can be seen that the hogging cross-section obtained by the pre-design is not interesting our investigation (Class 3); so, some arbitrary modifications are proposed for the web (to be always of class 4) and for the bottom flange (to be of class 1 to class 4) giving always a girder cross-section of class 4.

2.1   The 3D-Model

It is supposed that on hogging zone the concrete slab is totally cracked so it does not need to be modellized. This simplification requires common mesh nodes between the studs and the reinforcing bars. The model will use the symmetry of the panel and will be loaded by applying displacement at its end to avoid possible geometrical element distortion that could appear if the model is loaded by concentrated force. The steel girder as well as the stiffeners is meshed by 4-nodes shells, the studs are meshed using 3D beams to ensure the displacements continuity with the shells (same degrees of freedom) and the reinforcing bars are replaced by equivalent shells supposed at the top

482

Figure 5. 3D model (a/d = 2, 3 and 4).

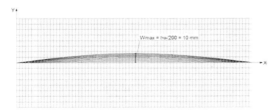

Figure 6. Initial deformation of the web – Top view.

of the studs as shown on figures 4 and 5 corresponding respectively to (a/d) ratio equal to 1, 2, 3 and 4.

The weld between the web and the flanges being not modellized, the distance "d" (Fig. 4) is supposed equal to h_w (web height). The stiffeners thickness is equal to 20 mm. The vertical displacement Uz equal to 200 mm is applied by steps of 20 mm. Initial deformation is applied to the web according to Eurocode specifications (Fig. 6) with: $w_{max} = h_w/200$. In the aim to obtain zero deformation at the four borders of the web, its mesh must use the following formula for node positions in Y direction, (1/2 sine wave in both X and Z directions):

$$y = w_{max} \sin\left(\frac{\pi x}{a}\right) \sin\left(\frac{\pi z}{h_w}\right) \quad (1)$$

2.2 (M-θ) curves

(M-θ) curves are plotted for different cross-sections (Tab. 1) on figures 7 to 10. The rotation capacity of the cross-section decreases if the ratio (a/d) increases whatever the bottom flange class. The maximum point of each curve represents the beginning of the local buckling (M$_v$, θ_v). This couple of values increases if the ratio (a/d) increases whatever the bottom flange class. According to the equations given on each figure, in general case:

$$\begin{cases} M_v = A_1 (a/d)^3 + B_1 (a/d)^2 + C_1 (a/d) + D_1 \\ \theta_v = A_2 (a/d)^3 + B_2 (a/d)^2 + C_2 (a/d) + D_2 \end{cases} \quad (2)$$

it is easy to obtain on tables 2 and 3 the parameters values depending on the bottom flange class. These

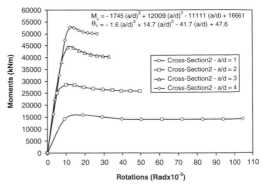

Figure 7. (M-θ) curve – Cross-section2.

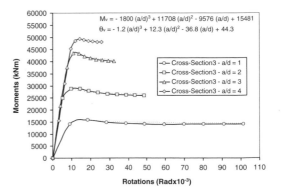

Figure 8. (M-θ) curve – Cross-section3.

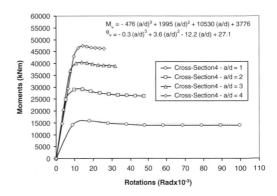

Figure 9. (M-θ) curve – Cross-section4.

values could be used while the web class remains of class 4. Figures 7 to 10 show also that it is possible to define a horizontal asymptotic line: $M = M_0$ (Fig. 11) depending on the ratio (a/d) and the bottom flange

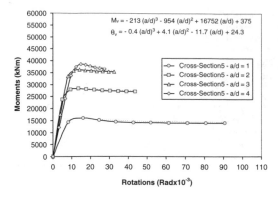

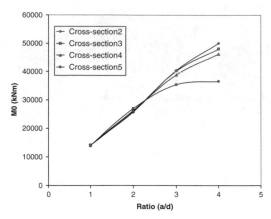

Figure 10. (M-θ) curve – Cross-section5.

Figure 11. Horizontal asymptotic line.

Table 2. Variation of M_v parameters.

Bottom flange class	A1	B1	C1	D1
1	−1745	12009	−11111	16661
2	−1800	11708	−9576	15481
3	−476	1995	10530	4776
4	−213	−954	16752	375

Table 3. Variation of θ_v parameters.

Bottom flange class	A2	B2	C2	D2
1	−1745	12009	−11111	16661
2	−1800	11708	−9576	15481
3	−476	1995	10530	4776
4	−213	−954	16752	375

class. Finally, it can be proposed a hyperbolic model for (M-θ) curve after buckling (Fig. 12):

$$M = M_0 + \left(M_v - M_0\right)\frac{\theta_v}{\theta}, \quad \forall \theta \geq \theta_v \qquad (3)$$

This model mainly describes local buckling behaviour alone (cases: a/d = 1 and 2), but it could sometimes include lateral buckling (cases: a/d = 3 and 4). The isovalues of the web Uy displacements (Figs. 13) show this phenomenon.

On figures 7 to 10 are also plotted the prediction curves (dotted lines) obtained by the proposed model equations 2 and 3 and figure 11. These curves begin from the buckling point (M_v, θ_v) and follow a similar curve to the symbolic one given on figure 12. The predictions appear very close to the numerical simulations. Nevertheless, this model could be improved (for example by including an inflection point not far from the buckling point) but this modification should

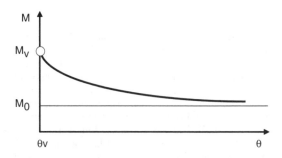

Figure 12. Local buckling model.

add another parameter and complicates the calibration procedure.

The curve given on figure 12 must be generalized whatever the composite cross-section we have but always in the case of full connection including reinforcing bars about 1% of the slab area and stiffeners. The elastic M_{el}^- and plastic M_{pl}^- resistant moments of each cross-section as well as the elastic rotation θ_{el}^- can be easily calculated (Tab. 4). The rotation θ_{el}^- is calculated with the following formula:

$$\theta_{el}^- = \frac{M_{el}^-}{E^{(a)}I_2}a_0 \qquad (4)$$

I_2 is the moment of inertia of the composite cross-section neglecting the concrete slab (supposed cracked) and a_0 is the longitudinal length on witch the rotation could be assumed to be constant (mostly used: $h_a/2$, with h_a the total height of the steel girder). Comparing the elastic values (moment and rotation – Tab. 4) to the buckling values (M_v, θ_v), it can be seen the influence of the reinforcing bars connected to the top flange and the stiffeners to make the local buckling late (over the elastic limit) especially for height values of the ratio (a/d).

484

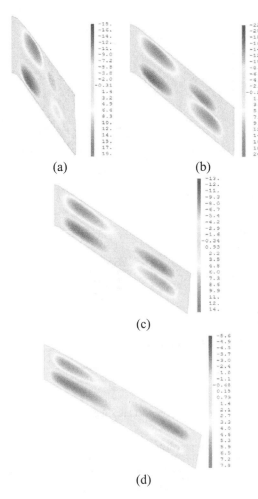

(a)   (b)

(c)

(d)

Figure 13. Final Uy displacements of the web – Cross-section2.

Table 4. Elastic and plastic resistant moments.

Cross-section	M_{el}^- (kNm)	θ_{el}^- (Rad $\times 10^{-3}$)	M_{pl}^- (kNm)
2	38291	3.06	49232
3	35668	3.04	45866
4	33034	3.02	42519
5	25053	2.94	32593

Now, the equation 3 could be generalized using the following formula:

$$\frac{\tilde{M}}{\tilde{M}_{el}} = \frac{M_0}{M_{el}} + \left(\frac{M_v - M_0}{M_{el}}\right)\frac{\left(\theta_v / \theta_{el}^-\right)}{\left(\tilde{\theta} / \tilde{\theta}_{el}\right)}, \forall\, \tilde{\theta}/\tilde{\theta}_{el}^- \geq \theta_v / \theta_{el}^- \quad (5)$$

with: $\tilde{M}$, $\tilde{\theta}$, $\tilde{M}_{el}^-$ and $\tilde{\theta}_{el}^-$ are moments and rotations of other dimensions of cross-section but in the case of full connection including reinforcing bars about 1% of the slab area and stiffeners.

3 SECOND INVESTIGATION – LOCAL BUCKLING INFLUNECE ON MOMENT REDISTRIBUTION PERCENTAGE

The following formula is used to calculate the moment redistribution percentage:

$$P^{support}(\%) = \frac{\left|M_{Ed}^{-(uncr,comp)}\right| - \left|M_{el,Rd}^{-(num,comp)}\right|}{\left|M_{Ed}^{-(uncr,comp)}\right|} \times 100 \quad (6)$$

where the exponent (uncr,comp) is related to the hogging bending moment obtained from the uncracked elastic global analysis, (num,comp) is the one related to the numerical value obtained from the non linear calculation and both of them concern the composite cross-section. Four steps loading are considered:

Step 1:   girder self-weight,
Step 2:   concrete slab still wet so not resisting yet,
Step 3:   superstructure (the composite cross-section is now resisting),
Step 4:   distributed and concentrated variable loads.

The bending moment corresponding to the loading steps 1 and 2 (M_{1-2}) is previously subtracted from the total values before using the equation (6). (M-θ) curve should be integrated as soon as the hogging moment reaches the buckling value M_v. The beam under investigation (Fig. 1) will be supposed on hogging zone with a cross-section of type 3 (web of class 4 and compressed bottom flange of class 2). The finite element code "PONTMIXTE" is used to run a non-linear calculation with following options:

– Material non-linearities,
– Tension stiffening effect,
– Concrete creep and shrinkage,
– Temperature difference effect ($\pm 5°C$).

The behaviour of the steel girder and the reinforcing bars is perfect elastic-plastic. The concrete slab and the stud behaviours are given on figure 14.

Concrete slab: $E_{cm} = 35000$ MPa, $f_{ck} = 40$ MPa, $f_{cm} = 48$ MPa, $f_{sc} = 3.5$ MPa, $\varepsilon_m = 0.0025$ and $\varepsilon_r = 0.0035$.

Stud: $Q_u = 152$ MPa, $C_1 = 0.85$, $C_2 = 0.51$ and $\gamma_{max} = 6$ mm.

The distance between the stiffeners "a" is supposed equal to "2d" (case a/d = 2). The corresponding (M-θ) curve is plotted on figure 8. For reminder, figure 15 shows this curve over the buckling point with the corresponding elastic resistant moment and the asymptotic

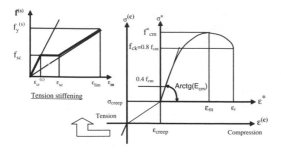

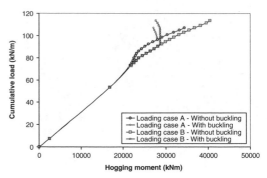

Figure 16. (Load-Moment) curves.

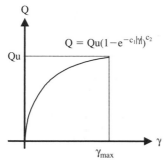

Figure 14. Materials behaviour.

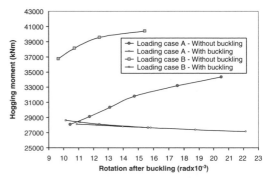

Figure 17. (M-θ) curves after buckling.

Table 5. Moment redistribution percentage.

Calculation	$\left|M_{Ed}^{-(uncr,comp)}\right|$ (kNm)	$\left|M_{cl,Rd}^{-(num,comp)}\right|$ (kNm)	$P^{support}$ (%)
Case A Without buckling	18342	17548	4%
Case B Without buckling	25571	23605	8%

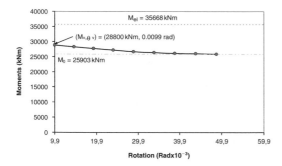

Figure 15. (M-θ) curve after buckling values.

line $M = M_0$. Two numerical calculations will be compared, with and without taking into account the local buckling phenomenon.

Both sagging and hogging critical loading cases (Figs. 3) have to be investigated with new appropriate critical position of the concentrated loads. The buckling curve (Fig. 15) remains available if the rotation is greater than the buckling value ($\theta \geq \theta_v$); for the beam under investigation, figure 16 shows that, for the loading case of type A, as soon as the hogging moment becomes greater than the buckling value (M_v), $\theta \geq \theta_v$; is always satisfied (big rotations at intermediate support – Fig. 17). For the loading case of type B, it appears that, for first values of hogging moments over

the buckling value, the corresponding rotations are till less than the buckling value (small rotations at intermediate support - Fig. 17); for these points we adjust $\theta = \theta_v$; that gives $M = M_v$ (Fig. 16).

The equation 6 is used now to calculate the moment redistribution percentage only if the local buckling is not taken into account. Actually, this formula is not available if the local buckling occurs before reaching the elastic resistant moment. The numerical calculations without considering the local buckling (Tab. 5) were stopped for following reasons: Case A (Plastic resistant moment in sagging zone), Case B (Elastic resistant moment on hogging). The hogging moment to be subtracted is constant $M_{1-2} = 16817$ kNm.

4 CONCLUSIONS

The present work proposes a simplified approach to the local buckling phenomenon for cross-section of class 4. The beam-example shows that, in despite of the hogging cross-section class (class 4 due to the web), a compressed bottom flange of class 2, the material non-linearity, the concrete creep and shrinkage, the tension stiffening and the temperature difference effects seams leading to a moment redistribution percentage about 4% for case A and 8% for case B; these values could not be used during uncracked global analysis because they are reduced to zero when the local buckling phenomenon is taken into account. More numerical investigations should follow this work in the aim to propose more generalized buckling models (partial connection, reinforcing bars percentage different than 1% and with or without stiffeners). Nevertheless, the algorithm included in "Pontmixte" remains available whatever the computed (M-θ) curve; for example the rule of 10% redistribution adopted by the Eurocodes could be verified for more resistant cross-sections.

REFERENCES

Guezouli, S. & Aribert, J.M. 2004. Numerical Investigation of moment redistribution in continuous beams of composite bridges. *Engineering Conferences International – Composite Construction V : 18 to 23 July, 2004*. Sth Africa.

Guezouli, S. & Yabuki, T. 2006. "Pontmixte" a User Friendly Program for Continuous Beams of Composite Bridges. *International Colloquium on Stability and Ductility of Steel Structures (SDSS'06). Lisbon, September 6–8 2006.* Portugal.

Guezouli, S. & Aribert, J.M. 2001. Approche aux éléments finis pour l'étude du comportement des poutres de ponts mixtes à l'échelle réelle. *XVème Congrès Français de Mécanique, Nancy, 3–7 septembre 2001.* France.

CEN (European Committee for Standardisation), prEN 2003 – Part 2. Design of composite steel and concrete structures – Rules for bridges. Stage 34 draft revised, Brussels, August 2003.

Yabuki, T., Arizumi, Y., Aribert, J.M. & Guezouli, S. 2001. Instability testing of steel plate girders with holded webs. *3rd International Conference on Thin-Walled Structures. Poland 5th to 7th june 2001.* Poland.

Yabuki, T., Arizumi, Y., Matsushita, H. & Guezouli, S. 2006. Buckling strength of welded stainless steel girder webs in shear. *International Colloquium on Stability and Ductility of Steel Structures (SDSS'06). Lisbon, September 6–8 2006.* Portugal.

Steel and Composite Structures – Wang & Choi (eds)
© 2007 Taylor & Francis Group, London, ISBN 978-0-415-45141-3

Analysis on structural system and mechanical behavior of box girder bridge with corrugated steel webs

Jun He, Ai-rong Chen & Yu-qing Liu
Department of Bridge Engineering, Tongji University, Shanghai, China

ABSTRACT: This paper describes in details the structure system for the box girder bridge with corrugated steel webs, including structural layout, determination values of web size and parameters of section of this type bridge. The relationship between different parameters (sectional height, height and thickness of corrugated web) with the maximum span is given to guide the design. Based on the actual bridge- Kuan Dian interchange, the design outline and main material of this bridge were presented, and the mechanics behavior of box girder with corrugated steel web has been introduced and analyzed to guide the design and construction, including the deflection, the normal and shear stress, all the calculate results meet the requirements of the code.

1 INSTRUCTIONS

Hybrid concrete box-girder bridges that include pre-stressed slabs and corrugated steel webs provide a major improvement over traditional pre-stressed concrete box-girder bridges. To reduce the self-weight, high strength concrete is used for the top and bottom slabs and corrugated steel webs are employed for the webs. The way to substitute corrugated steel webs for concrete webs of a box girder bridge will result in no restraint among the upper deck slab, lower deck slab and webs of the bridge, alleviate influences due to concrete creep, drying shrinkage and temperature differences.

Pre-stressed concrete (PC) box girders with corrugated steel webs are one of the promising concrete–steel hybrid structures applied to highway bridges.

Some advantages of using the corrugated steel web are summarized as follows:

(1) The decreased dead weight of corrugated steel web, compared to concrete web, leads to reduced seismic forces and smaller substructures, which will result in a lower construction cost of the bridge.
(2) The corrugated steel webs have a higher shear-buckling strength than flat plate steel webs.
(3) The corrugated steel webs are more easily fabricated and constructed than concrete webs.
(4) Pre-stress can be efficiently introduced into the top and bottom concrete flanges due to the so-called "accordion effect" of corrugated webs.

(5) The external post-tensioning is used for PC box girders with corrugated steel webs, which has many advantages over internal bonded tendons.

The Cognac Bridge is the first pre-stressed hybrid structure with corrugated steel webs in France in 1986, (Combault1988). This kind bridge is also built in Germany, Japan, USA, China and other countries. Based on the promising new type hybrid structure, their flexural (Elgaaly et al. 1997), shear (Elgaaly et al.1996), and partial compressive edge loading (Elgaaly & Seshadri 1997) behaviors were experimentally studied. Corrugated steel web fabrication (Johnson & Cafolla 1997a), as well as local (Johnson & Cafolla 1997b), and global (Johnson & Cafolla 1997c) buckling, were recently investigated. The torsional behavior and design of hybrid concrete box girder with corrugated web have recently been studied (Mo et al. 2006). The behavior of bridge girders with corrugated webs under monotonic and cyclic loading were studied (Sherif et al.2005). The fatigue life of steel bridge I-girders with trapezoidal web corrugations was studied and large-scale test girders were experimented (Sause et al.2006). The fatigue performance of the web-flange weld of steel girders with trapezoidal corrugated webs was examined experimentally and analytically, using large-scale girder specimens (Anami et al.2005). This paper describes in details the structure system for the box girder bridge with corrugated steel webs, including structural layout, web size values and parameters of section. Then an engineering example- Kuan Dian interchange was introduced, the design outline, main

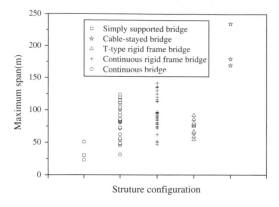

Figure 1. Relationship between the maximum span and structural configuration.

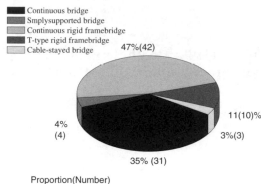

Figure 2. Number and proportion for each structural configuration.

material of the bridge and the FEM simulation analysis were presented subsequently.

2 STRUCTURE SYSTEM FEATURE

2.1 *Outline of configuration*

Hybrid concrete box-girder bridges include concrete slabs, corrugated steel webs, transverse diaphragm, internal and external tendons. Pre-stress can be efficiently introduced into the top and bottom concrete flanges due to the replacement of corrugated webs. Comparing with the pre-stressed concrete box girder, the transverse and the torsional rigidity for the PC box girders with corrugated steel webs reduce at different degree. So the requirements for the layout of spans, internal and external tendons, and transverse diaphragm are different from traditional PC bridge in part.

Fig. 1 shows the relationship between the maximum span and structural configuration, The maximum span of the simply supported bridge, continuous bridge, T-type rigid frame bridge, continuous rigid frame bridge and cable-stayed bridge are 51 m, 124 m, 92 m, 142 m and 235 m respectively. Fig. 2 shows the number and proportion for each structural configuration. Fig. 3 shows the relationship between the heights in the middle span/end support to the maximum span (LIU Yu-qing 2005). There are some characteristic of the ratio (height to span) for composite bridge with corrugated web: (1) The ratio is larger for T-type rigid frame girder bridge than other kind girder bridge or rigid frame bridge; (2) The ratio in the support point is similar to the concrete box girder bridge; (3) The ratio is larger in middle span than the concrete box girder bridge, to increase the eccentricity of external tendon.

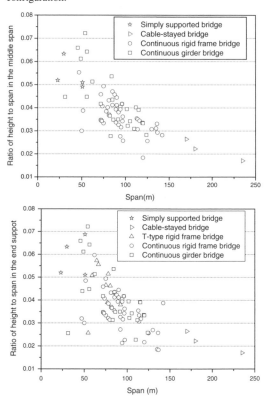

Figure 3. The relationship between the heights in the middle span/end support to the maximum span.

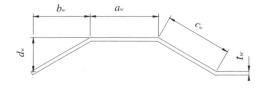

Figure 4. Configuration of corrugated web.

490

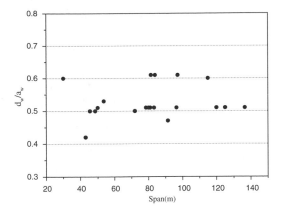

Figure 5. Relationship between d_w/a_w and maximum span.

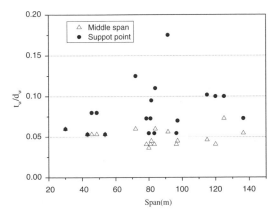

Figure 6. Relationship between t_w/d_w and maximum span.

2.2 Parameter of corrugated web section

The configuration of the corrugated web is determined by the condition that shear buckling does not occur before shear yielding under the ultimate load. The parameters of a_w and c_w indicate the width of the straight plate and oblique plate, b_w and d_w indicate the projected length in longitudinal direction and the height of the oblique plate, t_w indicate the thickness respectively, show in Fig. 4. Comparing most established PC box girder bridges with corrugated web, the width of the straight plate and oblique plate almost is the same, and the ratio of the height to the straight plate or oblique plate for the corrugated web is between 0.4 to 0.6, shows in Fig. 5. The thickness of the web should meet the minimum demand of the steel plate, and generally the thickness for the minimum demand is 8 mm or 9 mm. Fig. 6 indicate the relationship between the thickness of corrugated web and the maximum span of the bridge. The ratio of the thickness to the height for the web (t_w/d_w) is about 0.05 in the middle span, but the value in the support point change a lot.

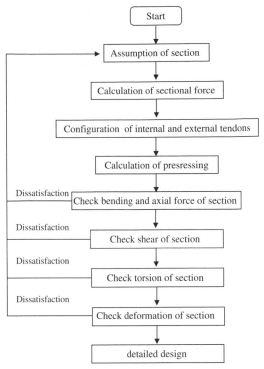

Figure 7. Flow chart of the design for a composite bridge with corrugated web.

3 OUTLINE OF DESIGN

Due to so many similarities comparing with concrete box girder bridges in design, some specialties for composite bridge with corrugated web in design are introduced, such as the calculation of the sectional stiffness, checking the bending, shear, torsion and deformation and so on. Fig. 7 is the flow chart of the design for a composite bridge with corrugated web. The details of each part can be referred to related literature due to limited space.

4 ENGINEERING EXAMPLE

4.1 Design outline

4.1.1 General layout
Kuan Dian interchange is located at the highway from Liaojijie to Dandong in Liaoning province. The bridge is an external pre-stressed continuous composite box girder bridge with corrugated steel web. The total length of the bridge is 117.40 m, and the length of the three spans is 30 m, 48 m and 30 m respectively, show in Fig. 8.

4.1.2 Cross section
Base on the requirements of Bridge design Code and the data of composite box girder bridges with

Figure 8. Outline of Kuan Dian bridge (side view) (Unit: cm).

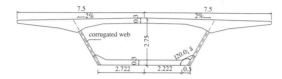

Figure 9. Cross section (Unit: m).

corrugated web, the design parameters as following: The composite bridge adopts single box girder with the identical height of 2.75 m. The transverse gradient of the upper slab is 2%. The width of the upper slab is 15 m, but the lower slab change from 5.1 to 5.44 along with the corrugated web. Both flange plate of the up slab is 3.572 m long. The thickness of the up and lower slab is 0.3 m. The angle of the web to the lower slab is 120°. The parameters of the corrugated web a_w, b_w, c_w and d_w is 300 mm, 260 mm 300 mm, 150 mm, The thickness of the corrugated steel web (t_w) is 16 mm. The connection between the straight line and the oblique line of the web adopt circular curve whose bend radius is 180 mm to reduce the stress concentration. Fig. 9 shows the detail of cross section.

4.1.3 Transverse diaphragm and deviation

To improve the torsional rigidity of the box girder, restrict the torsional deflection, control the distortion and warping, transverse diaphragm is set in the supporting point, and the thickness is 2 m. One deviation in the side span and two deviations in the middle span are set to change the direction of the external tendon, and the thickness of all the deviations is 0.9 m.

4.1.4 Configuration of prestressing

There are many advantages for external prestressing, such as:

(1) Dimensions of the concrete cross section, especially the webs, can be reduced due to the partial or full elimination of internal tendons, it is suitable for corrugated steel web;
(2) Grouting is improved and it is better protect and easily inspect the tendon;
(3) External tendons can be removed and replaced if necessary;
(4) Friction losses are reduced because external tendons are linked to the structure only at deviation and anchorage zones.

Alternately, external prestressing has some disadvantages:

(1) External tendons are more easily accessible than internal ones;
(2) At ultimate limit states, the contribution of external tendons to flexural strength is reduced compared to internal grouted tendons, and failure with little warning because of insufficient ductility is a major problem for external prestressed structure.

Combination of both internal and external tendons can be cover the shortage and exert the advantages respectively. The combination of internal and external tendons is adopted in the Kuan Dian Bridge.

The external tendons are anchored on the transverse diaphragm in the end support point, and divert on the deviation with large distance in the middle span.

4.1.5 Outline of connection

4.1.5.1 Connection of the corrugated steel web with the concrete slab

The connection of the corrugated steel web with the concrete slab is the most important part to make sure the longitudinal horizontal shear can be transferred effectively, and ensure all the parts of the cross section undertake load integrally. The stud on the upper and lower steel flange welded to corrugated web connect the steel web to the concrete slab. Fig. 10 shows the connection of the corrugated steel web with the upper slab and lower slab respectively.

4.1.5.2 Connection between corrugated steel web

The corrugated steel web is divided into several parts to meet the requirement of manufacture, transportation and construction. The connection of the corrugated web generally is welded or bolted. The corrugated web is fillet welded for Kuan Dian bridge, and the thickness of two side fillet welding should more than 12 mm.

4.1.5.3 Connection of the corrugated steel web with transverse diaphragm

The connection of the corrugated steel web with transverse diaphragm is also the important connect part to make sure the shear force transfer to substructure effectively from the corrugated web. The corrugated steel web is continuous along longitudinal direction, and the corrugated web is connected with transverse diaphragm by the stud and the internal concrete, show in Fig. 11.

4.2 Main material

4.2.1 Concrete

The upper and lower concrete slab, transverse diaphragm and deviation adopt C50 concrete, the substructure adopt C40 and C30 concrete.

4.2.2 Steel

The under relaxation steel strand of ASTM270 ϕ^j 15.24 are used for internal and external tendons.

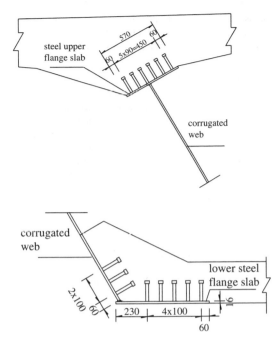

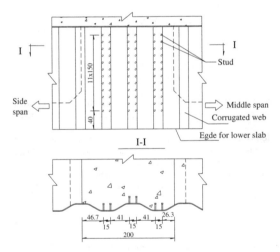

Figure 10. The connection of the corrugated steel web with the upper/lower slab (mm).

Figure 11. Connection of the corrugated steel web with transverse diaphragm(mm).

The nominal diameter is 15.2 mm, and the nominal area is 140.0 mm². Internal tendons compose of 12 binds, while external 27 binds. The control stress for prestressing of internal and external tendons are 1395 *MPa* and 1209 *MPa* respectively.

Weathering steel of Q345E is adopted for corrugated steel web, upper and lower steel flange slab.

Table 1. Material list

Component	The concrete slab transverse diaphragm deviation	Corrugated web	Tendons
Material	C50	Q345	Φj15.24
Elastic modulus (MPa)	3.45×10^4	2.06×10^5	1.95×10^5
Density(g/m³)	2600	7850	7850
Design strength (MPa)			
Tensile	1.89	310	1260
Compressive	22.4	310	390
Shear	–	180	–

4.3 FEM simulation analysis

4.3.1 Assumption

To analyze the performance of prestressing composite continuous bridge with corrugated steel web, some assumptions in perfect condition are presented.

(1) The connections between the steel web and the concrete slab are stable, slip and shear failure do not occur.
(2) Corrugated steel web have enough buckling strength, buckling failure does not occur.
(3) The internal tendon and concrete bond together completely. The external tendon connects with the structure in the anchor and deviation area, the slip of external tendon in deviation area is not considered.
(4) The prestressing loss of internal and external tendon is not taken account of.

4.3.2 Computation model

4.3.2.1 Element selection

The rationality of the FEM model affects the efficiency and precision. ANSYS is adopted to analyze the performance of composite box girder with corrugated web. Corrugated steel web is simulated by 8 nodes shell element considering shearing deformation, the upper and lower concrete slab, transverse diaphragm and deviation are simulated by 20 nodes solid element, internal and external tendons are simulated by cable element, and set real constant to apply prestressing.

4.3.2.2 Boundary condition and loading state

According to the actual support condition of bridge, one of the middle bearings is fixed and the others are hinged. Loading state only consider the dead load.

4.3.3 Calculating result

4.3.3.1 Deflection

The deflections of the middle position in both middle and side span are 0.02 m and 0.002 m respectively under the dead load. All the calculated values are less than the specified value L/600 (0.08 m and 0.05 m).

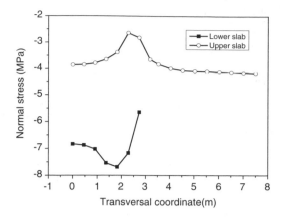

Figure 12. Normal stress of concrete slab in the middle of side span.

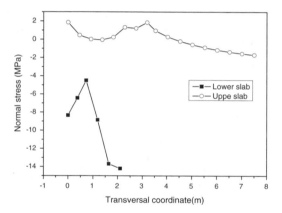

Figure 13. Normal stress of concrete slab in the middle bearing.

4.3.3.2   Stress analysis

The normal stress of concrete slab was calculated to analyze the stress distribution. Fig. 12 and Fig. 13 show that the stress distribution is not uniform in the upper and lower slab along the transversal direction of the bridge, the stress concentration appear at the position of the connection between concrete and steel web, the connection between the concrete and tendon, so these positions should be taken measures to reduce the stress concentration. The trend of the normal stress distribution in the middle span and the side bearing is the same as Fig. 12 and Fig. 13, only the value is different, so they are not presented due to the limit space. The maximum value of the tensile and compressive stress is 1.85 Mpa and 14.2 Mpa respectively, meet the design strength in Tabel.1. The maximum value of shear stress for the corrugated steel web is 30 Mpa in the connect position of the middle support point, and is much less than the shear design value (180 Mpa). The Von Mises stress of the corrugated steel web is 128 Mpa in the connect position of the middle support point, and is much less than the tensile design value (310 Mpa).

5   CONCLUSION

1 The structure system for the box girder bridge with corrugated steel webs, including structural layout, determination of web size values and parameter of section is presented. The relationship between different parameters (sectional height, height and thickness of corrugated web) with the maximum span is given to guide the design.
2 The flow chart of the design for a composite bridge with corrugated web is introduced to guide the design.
3 An engineering example (Kuan Dian interchange) is presented, including the design outline, main material and FEM analysis of the bridge. All the calculate results meet the requirement of the code.

REFERENCES

Combault J. 1998. The Maupre Viaduct near Charolles. *Proc., 1988. National Steel Construction Conf.*, American Institute of Steel Construction, Chicago.
Elgaaly. M., Hamilton, R. W., & Seshadri, A. 1996. Shear strength of beams with corrugated webs. *J. Struct. Eng.*, 122(4): 390–398.
Elgaaly, M., & Seshadri, A. 1997. Girders with corrugated webs under partial compressive edge loading. *J. Struct. Eng.*, 123(6): 783–791.
Elgaaly, M., Seshadri, A., & Hamilton, R. W. 1997. Bending strength of steel beams with corrugated webs. *J. Struct. Eng.*, 123(6): 772–782.
Johnson, R. P., & Cafolla, J. 1997a Fabrication of steel bridge girders with corrugated webs. *Struct. Eng.*, 75(8): 133–135.
Johnson, R. P., & Cafolla, J. 1997b. Local flange buckling in plate girders with corrugated webs. *Proc. Inst. Civ. Eng., Struct. Build.*,123(May): 148–156.
Johnson, R. P., & Cafolla, J. 1997c. Corrugated webs in plate girders for bridges. *Proc. Inst. Civ. Eng, Struct. Build.*, 123(May): 157–164.
Kengo Anamia, Richard Sauseb, & Hassan H. Abbasb. 2005. Fatigue of web-flange weld of corrugated web girders: 1. Influence of web corrugation geometry and flange geometry. *International Journal of Fatigue* 27: 373–381.
LIU Yu-qing. 2005. Composite Bridge. China Communications Press.
Richard Sause. 2006. Fatigue Life of Girders with Trapezoidal Corrugated Webs. *J. Struct. Eng*, 132(7), 1070–1078.
Sherif A. Ibrahima, Wael W. El-Dakhakhni, Mohamed Elgaaly. 2005. Behavior of bridge girders with corrugated webs under monotonic and cyclic loading. *Engineering Structures* (Sept.)
Y. L. Mo, Yu-Lin Fan. 2006. Torsional Design of Hybrid Concrete Box Girders. *Journal of Bridge Engineering*, (May/June): 329–339.

Steel and Composite Structures – Wang & Choi (eds)
© 2007 Taylor & Francis Group, London, ISBN 978-0-415-45141-3

A new-type prestressed composite girder bridge

P.G. Lee
Civil Engineering Research Department, RIST, Hwasung, Korea

C.E. Kim
SAMHYUN P.F Co., Ltd., Seoul, Korea

C.S. Shim
Department of Civil Engineering, Chung-Ang University, Ansung, Korea

ABSTRACT: There has been a strong demand on more economic and lower depth girder bridges for short and medium span range in Korea. A new type steel-concrete composite girder has been developed to realize a more economic bridge system with a lower depth girder. In the girder bridge, a steel plate girder is simply supported and then concrete form is hung to the girder. Thus, the self-weight of the concrete is loaded to the steel girder. To increase the resistance of concrete in the lower casing against tensile stress, compressive force is introduced by prestressing tendons without preflexion processes as the main processes for PREFLEX girders. Conventional prestressed concrete girders cannot be transported to a separate worksite from where they are cast due to the concrete's self weight unless they are prestressed. However, the new method in this paper enables easier transport and piling since the concrete in the lower casing is maintained in a non-stress state. Likewise, losses of compressive stress in the lower casing concrete due to its long-term behavior can be minimized, since prestressing work is carried out immediately before placing the girders on the piers. To evaluate the manufacturability and performances of the completed bridge, three 15-m girders and a bridge specimen with two 20-m girders were constructed. The camber during the construction and introduction of an appropriate compressive force was evaluated. Static loading test was also conducted to examine cracking and to evaluate the decrease in stiffness and failure behavior under extreme conditions.

1 INTRODUCTION

1.1 Background

There has been a strong demand on more economic and lower depth girder bridges for short and medium span range in Korea. A prestressed composite girder bridge uses prestressing tendons instead of preflexion of a steel beam for the introduction of compression to concrete. Preflex girder bridges are mainly utilized for the design condition of extremely limited deflection criterion even though the cost is much higher than normal composite girders. Serviceability limit states such as deflection and vibration are crucial in design of railway bridges.

Although high steel ratio of the composite girder is helpful for the flexural stiffness, it is unfavorable for the effectiveness of the prestressing at early age of concrete. A new-type prestressed composite girder is proposed to reduce the steel ratio and to minimize the prestressing losses by allowing flexible prestressing time. This concept allows us to utilize steel and

concrete effectively. In this paper, the design concept of the girder is introduced and experiments for the verification of the concept were performed.

1.2 Concept of the new composite girder

A new type steel-concrete composite girder has been developed to realize a more economic bridge system with a lower depth girder. In the girder bridge, a steel plate girder is simply supported and then concrete form is hung to the girder. Thus, the self-weight of the concrete is loaded to the steel girder. To increase the resistance of concrete in the lower casing against tensile stress, compressive force is introduced by prestressing tendons without preflexion processes as the main processes for PREFLEX girders. Conventional prestressed concrete girders cannot be transported to a separate worksite from where they are cast due to the concrete's self weight unless they are prestressed. However, the new method in this paper enables easier transport and piling since the concrete in the lower casing is maintained in a non-stress state. Likewise, losses of compressive stress in the lower casing concrete due to its long-term behavior can be minimized, since prestressing work is carried out immediately before placing the girders on the piers.

2 FABRICATION PROCEDURE

2.1 Fabrication

An ordinary PREFLEX girder (Evans, 1966) has the following fabrication procedures;

a) placement of a steel girder with a precamber on end bearings;
b) preflexion of the steel girder by application of two concentrated loads at 1/4 and 3/4 of the span;
c) casting of the first phase high strength concrete at the level of the bottom flange of the steel girder while keeping the loads of the preflexion phase on the girder;
d) X days after casting of the concrete, removal of preflexion loads: the beam goes up, the precamber becomes smaller than the original precamber and the concrete is submitted to compression;
e) casting of the second phase concrete on site.

The first and second phases concretes increase significantly the stiffness of the beam by comparison with the stiffness of the steel girder alone. Since the concrete of the bottom flange is submitted to compression before the application of the service loads, this concrete becomes useful in this part of the beam and the requirements of no cracking of the concrete are satisfied.

Table 1 summarizes the fabrication procedures. A steel girder is fabricated considering precamber and is simply supported at both ends. Concrete formworks and reinforcements for the casing concrete are hung to the girder. After casting concrete, formworks are removed when the concrete reaches the design strength. At this step, a little compression is introduced to the concrete. This compression is very useful for the control of shrinkage cracks. Prestressing by internal tendons is introduced just before the erection of the composite girder on sites for the reduction of the prestressing losses. Usually, the interval between casting concrete and prestressing is longer than 3 months (even 1 year) resulting in relatively smaller losses than normal PREFLEX girders.

2.2 Design issues

A prestressed composite girder is considered as non-economical alternatives except for the limited design conditions. The new composite girder has smaller steel area than the PREFLEX beam and uses the internal prestressing tendons. Main issues for the verification

Table 1. Fabrication procedure.

Precambered steel beam	
Placing the reinforcements	
Hang the formwork to the steel beam	
Casting concrete & removing formworks	
Lower casing concrete is under non-stress or small compression Prestressing just before erection of the girders on sites	

of the behavior are flexural stiffness, crack control and ultimate strength of the girder.

3 EXPERIMENTS

3.1 Test specimens

Three girder models (CB1 ~ CB3)and a bridge model (BM) were fabricated as in Table 2. Three models have different prestressing forces and one girder has smaller cover depth than the other two. The minimum prestressing force was calculated to obtain no tensile stress under service loads. Cover depth means the distance from the bottom surface of the casing concrete to the center of the prestressing tendons.

Table 2. Test specimens.

Specimen		Prestressing Force (kN)	Cover Depth (mm)
Girder	CB1	8,000 (Maximum)	175
Model	CB2	7,600	175
	CB3	6,600 (No Tensile Stress Under Service Load)	110
Bridge Model	BM	9,200/girder	200

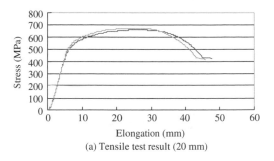

(a) Tensile test result (20 mm)

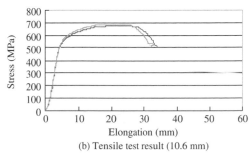

(b) Tensile test result (10.6 mm)

Figure 2. Tensile test results of steel.

Table 3. Concrete properties.

Concrete	Time	Design strength (MPa)	At Prestressing (MPa)		At testing Time (MPa)	
CB	Lower casing	40	5days	41.8	30days	43.8
	Slab	27	–		6days	23.3
BM	Lower casing	40	11days	45.6	38days	50.2
	Slab	35	–		12days	37.0

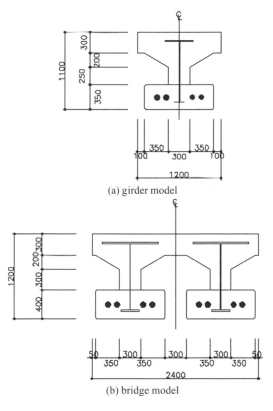

(a) girder model

(b) bridge model

Figure 1. Test specimens.

Figure 1 shows section dimensions of the girder model and the bridge model. The bridge model has two girders and total width was 2400 mm. Four 15.2 mm diameter prestressing tendons were used for a girder and two tendons were jacked simultaneously. At each step, 50% of the initial prestress force was introduced sequentially. Initial prestressing losses were measured and showed good agreement with calculated values.

Dimensions were determined by designing the girder for the simply supported railway bridge with 15 m and 20 m span length. Lower steel flange has smaller dimensions than upper flange for the effectiveness of prestressing. Block shear connectors were welded to the flanges for the composite action between steel and concrete. Constructability and camber were verified through the fabrication procedures.

3.2 Material properties

Steel beams for the girder were SM570TMC steel and tensile test results from coupon tests were presented in Figure 2. Concrete for the lower casing was designed to have 40 MPa and 27 MPa for the upper concrete slab. Table 3 summarizes the material properties of the concrete. Compressive strength of concrete was obtained twice at prestressing time and testing time.

3.3 Measurements and test setup

In order to measure the prestressing losses, a load cell was installed for each girder. Local stresses of the

497

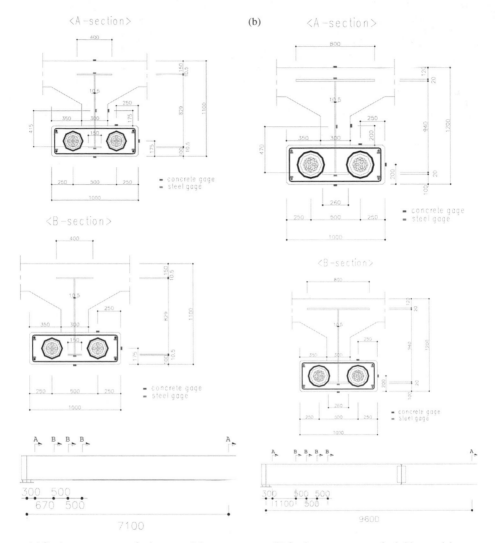

(a) Strain measurements for beam models

(b) Strain measurements for bridge models

Figure 3.   Measurements.

lower casing concrete at the prestressing time were evaluated by measuring concrete strain at 300 mm from the end of the supports in the longitudinal and transverse direction. Strain distribution of the mid-span section was also estimated by strain gauges as in Figure 3. Displacements were also measured to evaluate the flexural stiffness of the composite girder.

A concentrated load was applied at the center of the span by a closed loop electro-hydraulic testing system of 10,000 kN capacity. Figure 4 shows the test setup. Four steps of static tests were conducted to evaluate elastic, cracking and ultimate behavior of the girder. Final failure test was done by displacement control.

Figure 4.   Test setup.

Figure 5. Finite element model.

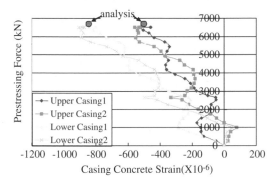

Figure 6. Concrete stain at prestressing.

Table 4. Test results.

Specimen				
Result	CB1	CB2	CB3	BM
Cracking (kN)	1,020	998	1,180	2,744
Ultimate (kN)	2,400	2,375	2,650	6,314
(Calculation)	(2,060)	(2,060)	(2,200)	(5,194)

3.4 Test results

When the prestress was introduced, the effectiveness of prestressing was verified through finite element analysis and strain measurements. Figure 5 shows the finite element model using DIANA (TNO, 1998). Material properties for the F.E. model were the tested values as shown in Table 3. Figure 6 shows the measured concrete strain and analyzed values. The graph showed good agreement and compressive stress to the casing concrete was ensured.

Table 4 summarizes the test results in terms of cracking load and ultimate load. Ultimate capacity of the composite section was calculated by considering the ordinary rigid plastic concept. Failure of the composite section initiated from the yielding of the steel beam and ended with crushing of concrete. A new composite girder showed ductile behavior. Test specimens

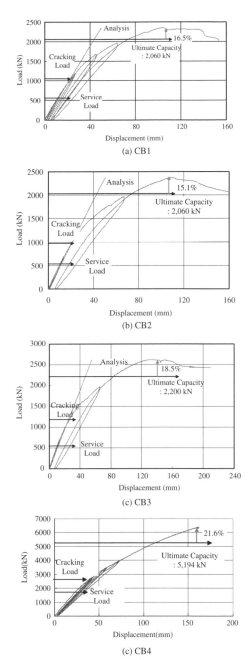

Figure 7. Load-displacement curves.

showed 17 ~ 20% higher strength than the calculation. The difference is from the steel hardening.

As shown in Figure 7, ultimate strength of the composite section is 6 times higher than service load and more than 2 times higher than cracking load. After cracks were observed at bottom concrete, the flexural

499

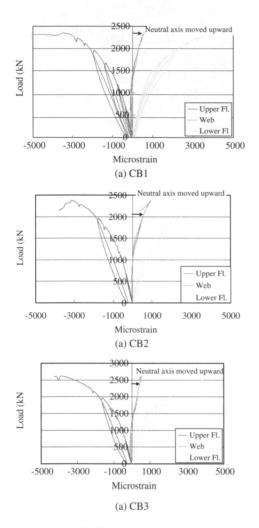

Figure 8.   Strain distribution of the steel section.

stiffness was not reduced until the load reached about 3 times of service load. In terms of the ultimate limit states, the new composite girder is quite conservative because the critical design criterion is the flexural stiffness.

Figure 8 shows strain distribution of steel girder according to load increment. Three strain gauges were attached at the girder and a strain gauge at web section was located at elastic neutral axis. Strain distribution showed perfect composite action at elastic range. Until the applied load reached 1500 kN, the neutral axis

was not changed. If the concrete section under tension is considered ineffective, the flexural stiffness of the composite section will be too underestimated. When the crack width is controlled properly, the composite section can have the similar flexural stiffness of the uncracked section. Lower flange and upper flange showed yielding at ultimate load.

4   CONCLUSIONS

A new type steel-concrete composite girder has been developed to realize a more economic bridge system with a lower depth girder. In this paper, the design concept of the girder is introduced and experiments for the verification of the concept were performed. From the tests, the following conclusions were derived.

Ultimate capacity of the composite section was calculated by considering the ordinary rigid plastic concept. Failure of the composite section initiated from the yielding of the steel beam and ended with crushing of concrete. A new composite girder showed ductile behavior. Test specimens showed 17~20% higher strength than the calculation.

If the concrete section under tension is considered ineffective, the flexural stiffness of the composite section will be too underestimated. When the crack width is controlled properly, the composite section can have the similar flexural stiffness of the uncracked section.

Ultimate strength of the composite section is 6 times higher than service load and more than 2 times higher than cracking load. After cracks were observed at bottom concrete, the flexural stiffness was not reduced until the load reached about 3 times of service load. In terms of the ultimate limit states, the new composite girder is quite conservative because the critical design criterion is the flexural stiffness.

REFERENCES

Evans, R.H., White, A.D., Ford, J., Lash, S.D., Nicholas, R.J., Lipsky, A., Leung, K.W., Bruggeling, A.S.G., Blake, R.J., Haseltine, B.A. 1966 Discussion; Characteristics of Preflexed Prestressed Concrete Beams, ICE Proceedings, Vol. 34, pp. 266–284.
Thewalt, A., Tröger, A., Rusch, A., and Wanzek, T. 2005, Hermann-Liebmann Bridge in Leipzig – Construction method with Spannverbund-Girder, Stahlbau 74 , Heft 9, pp. 664–672.
Rigot, G. 2003, Preflexed & prestressed = flexstress the synthesis of the techniques, Proceedings of IABSE.
TNO Building and Construction Research 1998. DIANA User's Manual release 7

Steel and Composite Structures – Wang & Choi (eds)
© *2007 Taylor & Francis Group, London, ISBN 978-0-415-45141-3*

Prototype of Structural Health Monitoring System (SHMS) for the Nanpu bridge in Shanghai, China

R. Wang
Department of Building Engineering, Tongji University, Shanghai, China

W. Huang
Intelligent Transportation Systems Research Centre, Southeast University, Nanjing, China

X. Meng
Institute of Engineering Surveying and Space Geodesy (IESSG), The University of Nottingham, Nottingham, UK

Y. Luo
Department of Building Engineering, Tongji University, Shanghai, China

L. Yao
Department of Surveying and Geo-Informatics, Tongji University, Shanghai, China

ABSTRACT: In this paper, we introduce a prototype of structural health monitoring system (SHMS) which is developed for the Nanpu Bridge in Shanghai, China. The Nanpu Bridge is the first and also the longest cable-stayed suspension bridge in China which was designed independently by the Chinese engineers in the beginning of 1990s and it is playing an important role in the transport system of Shanghai. We employ the Global Position System (GPS) technique to acquire in-situ ambient dynamic responses of the Nanpu Bridge in September 2006. Several signal processing methods are applied to the GPS data and some preliminary results are presented in the paper. A finite element (FE) model of the bridge is created and serves as a baseline model for the further analysis of the deformation of the Nanpu Bridge.

1 INTRODUCTION

Monitoring the condition and performance of structures is a research focus within the structure engineering community in last decade. Online monitoring of the condition of an operational structure provides a procedure to keep tracking of the variations of a structure under normal service loads. Research demonstrates that structural health monitoring system (SHMS) that utilized non-destructive evaluation (NDE) techniques is a feasible tool in the direct identification of the global structural damage or degradation.

In this paper, we develop a prototype of structural health monitoring system (SHMS) for the Nanpu Bridge which is the first cable-supported long-span suspension bridge designed independently by the Chinese engineers and is playing an important role in the transport system in Shanghai, China. An ideal SHMS may be composed of two components: the data acquisition system, and the damage identification and assessment process. Global Navigation Satellite System (GNSS), particularly real-time kinematical Global Positioning System (RTK GPS) has been used for monitoring structural displacement for more than a decade. Many studies demonstrate that GPS has its feasibility and obvious advantages over other more traditional monitoring systems and is in deed a very promising alternative for structural health monitoring. We employ the Global Position System (GPS) technique to acquire in-situ ambient dynamic responses of large civil structures. Several signal processing methods are applied to the analysis of GPS data for structural dynamics (Meng 2002, Wang et al. 2004).

Cable-stayed bridges are one of the most important structure types selected by the engineers for long-span bridges. Although these structures have many advantages comparing with other types of bridge structures, these structures have relatively softer stiffness, so they are normally sensitive to dynamic loadings such as earthquakes, wind and vehicle loadings. The health monitoring and condition assessment of large span

Figure 1. The Nanpu Bridge, Shanghai, China.

cable-stayed bridges become a crucial issue to ensure safety during the bridge service life (Ren and Peng 2005).

The generation of a finite element (FE) model for an in-service structure is the core in the development of a practical and reliable monitoring system. The created FE model can serve as a baseline model to describe the structure before its damage. This is a start point of many identification and updating methods and a model updating technique can be utilised to evaluate the condition of the structure from time to time. The focus of this paper is the introduction of an FE model which acts as a baseline model in the establishment of a structural health monitoring system (SHMS) for the Nanpu Bridge.

2 THE NANPU BRIDGE AND TRIALS WITH GPS POSITIONING TECHNIQUE

The Nanpu cable-stayed bridge is the first steel and concrete composite girder cable-stayed bridge built in China and is playing an important role in transport system of Shanghai. Figure 1 shows part of the Nanpu Bridge.

The bridge has a composite-deck system consisting of seven spans with an overall length of 846 m (41.5 m + 76.5 m + 423 m + 76.5 m + 40.5 m). The bridge was completed in November 1991 and was designed by the Shanghai Municipal Engineering Research Institute in cooperation with Tongji University. It was constructed by the Shanghai Huangpujiang Bridge Engineering Construction Co. Figure 2 shows an elevation view of the Nanpu Bridge.

On September 21, 22, 23 and 24, 2006, a four-day bridge monitoring trial was carried out on the Nanpu Bridge. The instrument setup for the trial can be seen in Figure 3. The total numbers of the GPS receivers are 14, which consisted of 3 Leica 530, 1 Trimble 5700, 4 Thales Z-max, 6 Ashtech Z-xtreme GPS receivers.

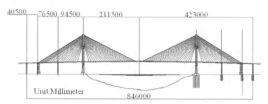

Figure 2. An elevation view of the Nanpu Bridge.

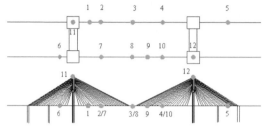

Figure 3. The layout of the GPS receivers on the bridge.

Figure 4. The layout of two reference stations atop a nearby building, a monitoring station antenna clamped to the bridge handrail and the tower top GPS antenna.

They are all dual frequency GPS receivers. Among these GPS receiver, one Leica 530 GPS receiver was installed on top of nearby building as a reference station, two Leica 530 GPS receivers located at point 3 and point 4 were used as two monitoring stations, one Trimble 5700 was also used as a reference station at another point on the top of the same building, four Thales Z-max were employed as the monitoring stations and were located at point 1, 2, 8, 9, respectively. Other six Ashtech Z-xtreme GPS receivers were installed at points 5, 6, 7, 10, 11, and 12 as the monitoring stations as indicated in Figure 3.

All GPS receivers with their antennas that were situated at the monitoring stations were clamped firmly to the bridge handrail using specially designed U-shaped clamps which is shown in Figure 4.

According to the Nyquist sampling theorem and the preliminary analysis of the predictions of a finite element model of the bridge structure, all the sampling rate of the GPS receivers was set to 10 Hz which will satisfy the extraction of all vibration frequencies of this bridge. The GPS raw measurements were recorded using two types of compact flashcards with memory sizes of 64 Mb and 128 Mb. To overcome the limitation of GPS receiver's internal memory card while sampling at a high data rate, the data acquisitions had to be divided into three observation sessions on each day. Three observation sessions of each day are set as 9:00–11:00, 16:00–18:00 and 21:00–1:00, respectively. GPS raw measurements were post-processed in a kinematic manner to obtain each epoch positioning solution for each monitoring site on the bridge.

3 GPS DATA ANALYSIS

Since a huge amount of GPS data was collected during these four-day trials, we can only present some representative data set to demonstrate the data processing and analysis procedure. The coordinates of the monitoring stations have been computed from GPS raw measurements through special GPS algorithms. The coordinates have been converted into those in a dedicated Bridge Coordinate System (BCS) with one of its axes along the longitudinal direction of the bridge main axis and the other two to form a complete Cartesian coordinate system.

Figure 5 shows a representative coordinate time series of point 11 in the BCS and the data sets were acquired on September 24, 2006. The points 11, 12 were located on the two top sites of the bridge towers and the points 3, 8 were set at two midspan points of the main span of the bridge.

Because of data size collected in the trials, only the data acquired at point 11 on September 24 2006 is selected as the representative data set for the following analysis. We applied Fast Fourier transforms (FFTs) to the point 11 data which is shown in Figure 6. From the results, we found that the frequencies between 0.0–1.0 Hz are intensive and overlapped together.

We also applied various averaged power spectrum algorithms to the point 11 data and the results are shown in Figure 7. We further applied MUltiple SIgnal Classifier (MUSIC) method which was introduced by Mathworks (2006) to the point 11 data and the results are shown in Figure 8. It needs to be pointed out that for reducing the computation burden when MUSIC method is utilised only a small part of point11 data is analysed.

4 FINE FINITE ELEMENT MODELING OF THE NANPU BRIDGE

Modern large span cable-stayed bridges have complex geometry that involves a variety of decks, towers and stay cables that are connected together in different ways. Traditionally simplified three-dimensional finite element models have developed for the initial design of the bridge structure, such as the single-girder (spine) model. The simplified finite element model assumes a simple one-dimension type element

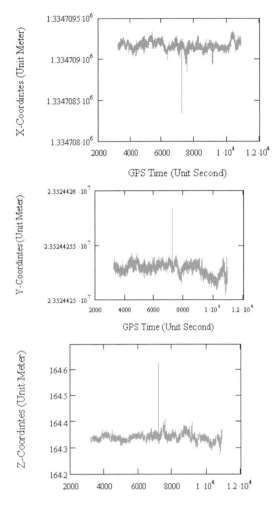

Figure 5. Point 11 coordinates.

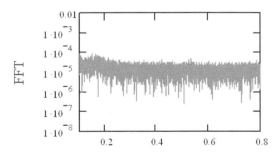

Figure 6. Fast Fourier transform for point 11 data.

to simulate the main members of the bridge structure, such as using elastic beam elements to model the towers and deck, and truss elements to model the cables. The single-girder (spine) model is probably the earliest three-dimensional finite element model

503

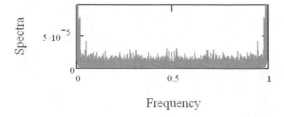

Figure 7. Averaged power spectrum for point 11 data.

of cable-stayed bridges for the concept design. These simplified finite element models may not fully capture the complex mechanical properties, especially the very complex dynamic characters of large span cable-stayed bridges. Since the simplified finite element model uses beam elements to model the deck and tower, they cannot well simulate the lateral and torsional vibrations. The analysis of data obtained from GPS in-situ monitoring of the Nanpu Bridge has shown the complex dynamic nature of the bridge structure.

For full simulation of the dynamics of the Nanpu Bridge and the identification of complex dynamic parameters, a fine three-dimensional finite element model must be created using various types of elements such as beam elements, truss, elements, shell elements, solid elements and link elements to well model different components of the cable-stayed bridges.

The MUSIC method can also be used to extract several important dynamic parameters.

Aimed at establishing a fine finite element model for the long-term health monitoring of the Nanpu Bridge, we adopt the following model techniques. We first build the almost exact geometry of the main members of the towers, girders, floor beams, concrete slab using a modern CAD system, Solidworks based on the design blueprints of the bridge. The real 3-D solid model of entire bridge structure can be assembled together that forms a detailed geometry of the whole bridge structure (Solidworks 2006). Figure 9 shows the 3-D solid model of entire bridge structure, the main members and some important geometry connections. The assembled members have well described the physical connections of the members such as the slabs and towers that serve as the boundary condition of the finite element model.

Secondly, we import the detailed geometry model to the MSC.Marc program for the numeric simulation of the dynamic properties of the Nanpu Bridge (MSC.Marc 2006). MSC.Marc can carry out the most intense nonlinear and dynamic simulation for the structures efficiently and reliably, so we choose the MSC.Marc to simulate complex dynamic properties of the bridge in its service durations.

Quad and line types of MSC.Marc are employed to create a finite element model. We employ several techniques to mesh the very complex geometry model in

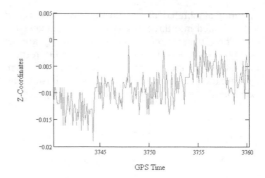

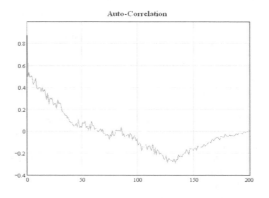

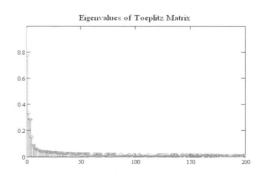

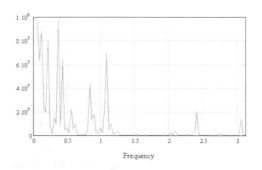

Figure 8. MUSIC method for point 11 data.

504

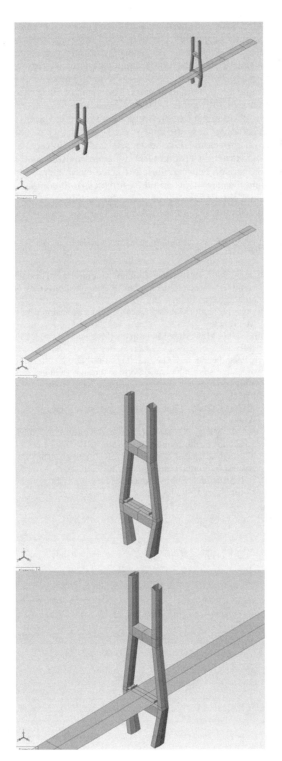

Figure 9. 3-D solid model of entire bridge structure, the main members and their connections.

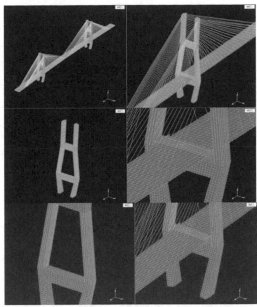

Figure 10. A finite element model of entire bridge structure.

order to obtain the fine finite element model. The final finite element model includes about 20000 elements. Figure 10 shows the created finite element model of entire bridge structure. The boundary conditions of the finite element model have a high impact on the simulation result. The contact boundary conditions have be set to simulate the connections of slabs, girders and towers.

For obtaining the most reliable simulation, we have set up almost the nonlinear options in MSC Marc that suitable for dynamic computation.

The fine finite element model of the Nanpu Bridge expresses very complex dynamics behavior. In a lower frequency band, the Nanpu Bridge holds very dense lower frequencies. Figure 11 shows the first 20 lower frequencies of the Nanpu Bridge. The dense low frequency behaviors make the identification of dynamics parameter very difficulty. The complex dynamic behaviors are very different from the behaviors obtained from a controlled field test by engineers. The dynamic frequency simulation shows that the complex dynamic behavior of a service bridge.

5 CONCLUSIONS

Construction of a successful Structural Health Monitoring System (SHS) encounters huge difficulty and maybe takes a long time to obtain a practical SHMS. GPS technology, model validation and damage identification form the main components of SHMS. In this

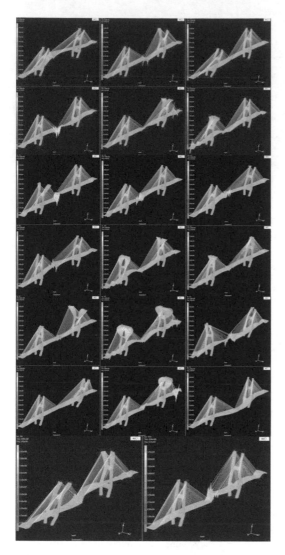

Figure 11. 20 simulated low frequencies of entire bridge structure.

paper, we establish a fine finite element model for the Nanpu Bridge and analyse a huge amount of measurements acquired from GPS receivers mounted on the Nanpu Bridge. Our practice shows that a finite element model for SHMS should contain fine geometry of the bridge structure in order to capture the complex dynamic properties. The number of the extracted frequencies from the GPS data maybe only represents partial of the vibration frequencies and the extraction of other parameters such as dynamic modeshapes can be a much difficult work. The damage identification will fully reply on these extractions. Therefore, the development of the reliable and applied SHMS must solve these difficult tasks first. Further results from processing GPS data sets and the interpretation/extraction of structural dynamics as well as the comparison between these extracted parameters from field measurements and the estimates from computational simulation will be presented in future research contributions.

REFERENCES

Meng, X. 2002. Real-time deformation monitoring of bridges using GPS/accelerometers. PhD thesis, The University of Nottingham, Nottingham, UK.

Mathworks. 2006 Mathcad User's manual (revision 13.1). Mathworks.

MSC. 2006. MSC.Marc User's manual (revision 2006). MSC software.

Ren, W. and Peng, X. 2005. Baseline finite element modeling of a large span cable-stayed bridge through field ambient vibration tests, Computers and Structures 83 (2005) 536–550.

SolidWorks. 2006. User's manual (revision 2006). Solid-Works.

Wang, R. Meng, X., Roberts, G. and Dodson, A. 2004. Structural health monitoring system (SHMS) for bridge with hybrid sensor system. 1st FIG International Symposium on Engineering Surveys for Construction Works and Structural Engineering, 28 June – 1 July 2004, Nottingham, United Kingdom.

Steel and Composite Structures – Wang & Choi (eds)
© 2007 Taylor & Francis Group, London, ISBN 978-0-415-45141-3

Stability behavior and new construction process for a five-span CFST arch bridge

Yuyin Wang, Sumei Zhang, Changyong Liu & Yue Geng
Harbin Institute of Technology, Harbin, China

ABSTRACT: This paper developed a finite element model to simulate and analyze the mechanical behavior of a five-span tied rigid-frame concrete filled steel tubular arch bridge with ANSYS. The stability behavior of the main span was predicted by means of four methods and the differences of the predicted results were compared and discussed. Some parameters were investigated on the bridge's stability and the dominating factors among them were finally outlined. In addition, it is quite time-consuming to build such a continuous CFST arch bridge in the traditional construction method, so a new construction scheme was brought forward in this paper in order to raise the constructional efficiency. The analysis results proved that with the new construction scheme, all the stress, deformation and displacement can meet the Chinese design requirements. Thus, this new construction scheme was reasonable and reliable.

1 GENERAL INSTRUCTIONS

Concrete-filled steel tubes (CFST) are widely used in arch bridges due to their high compressive strength, efficiency and simplicity in construction. Two problems (heavy self-weight and complexity in construction) impeding the arch bridges with a large span can be solved by CFST arches. This kind of arch bridges has some advantages over the traditional ones in architecture, economy, functionality, service and flexibility. The back to back type CFST arch bridge is a new kind of bridge avoiding using of lateral braces. To raise the transverse rigidity, a side arch rib is set. The vertical arch rib is mainly used to carry the vertical loads, and the side arch rib to increase the stability. They build a stable structure together (Zhang & Li 2006).

The Han River North Bridge is located in Chaozhou City of Guangdong Province, China. The bridge is a five-span tied rigid-frame connected by V-shaped components. The continuous spans are $85 + 114 + 160 + 114 + 85$ m, whose main span is the longest back to back type concrete-filled steel tubular arch bridge in the world (Fig. 1).

2 FINITE ELEMENT MODEL

The pivot and difficult point for the finite element model is how to simulate the arch, because both of the 'Plane Cross-Section Assumption' of the vertical arch rib and the entirety of the arch should be concerned. Four methods were built in reference (Wang et al 2005), i.e.: a.Model of double equivalent bars; b. Model of using equivalent steel web; c. Model of using latticed web members; d. Model of restraining equations (Fig. 2). The Model of double equivalent bars are simplified by means of the equivalence of the connecting bars; 'Model of using equivalent steel web' and

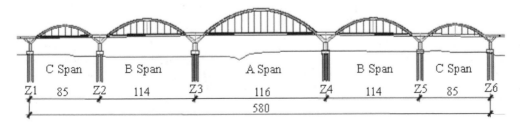

Figure 1. Elevation view of the Han River North Bridge (Unit: m).

a) Model of double equivalent bars

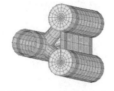

b) Model of using equivalent steel web

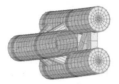

c) Model of using latticed web members

d) Model of restraining equations

Figure 2. Four finite element models of the arch.

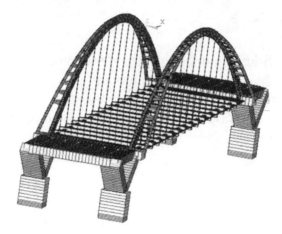

Figure 3. Finite element model of 'A' span.

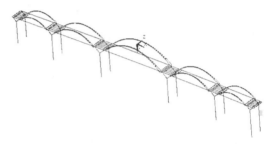

Figure 4. Finite element model of the whole bridge.

'Model of using latticed web members' are simplified by means of the equivalence of the vertical rigidity between the vertical arch in the two models and the one built by dumbbell-shape cross section.

Each model has its merits and drawbacks: the first three models certainly will weaken the rigidity of the arch because of the simplification; the fourth model could simulate the arch accurately, however, the restraining equations is forbidden using in large deformation in ANSYS, therefore, it is only used as a criterion to check the first three models in small deformation. Reference (Wang et al 2005) gave a general appraise on above methods and found that, 'Model of double equivalent bars' can be not only used in large deformation, but also the most approximate to 'Model of restraining equations' on lateral rigidity and stability; in addition, the vertical arch rib in 'Model of double equivalent bars' was built by dumbbell-shape cross section, making it convenient for simulation analysis on construction process. Consequently, 'Model of double equivalent bars' is adopted as the final finite element model.

During stable analysis of the Han River North Bridge, the effects of approach spans are neglected since the rigidity of the V-shaped components is rather big. Thus, the 'A' span arch bridge was taken out in stable analysis for it is the longest span.

For simulation analysis on construction process, the stiffness contribution of the floor system is neglected as a simplification, since it has little effect on the stiffness of the whole bridge in construction stages. Its self-weight is considered as the action of the suspenders and is applied on the vertical arch ribs in terms of concentrated forces. The birth and death technique is adopted to simulate series of changes in configuration of the structure, such as the closure of arch ribs, the pumping of core concrete, and the generation of

the stiffness in core concrete. The bridge models are shown in Figures 3, 4.

3 STABILITY OF 'A' SPAN

Reference (Wang et al 2006) has researched on the stability behavior of 'A' span subjected to five different loading combinations (Tab.1) by the eigenvalue buckling analysis, geometric nonlinear analysis, material nonlinear analysis and dual-nonlinear analysis. Refer to Reference (Wang et al 2006) for the detailed characteristic values of the permanent loads, lane loads, crowd loads and wind loads. The safety factors are listed in Table 2

It is shown in Table 2 that the eigenvalues are ranging from 7.068 to 8.901, and the lowest value is in load case I, while the highest one in load case V. The safety factors using geometric nonlinear buckling method vary from 6.200 (in load case III) to 7.859 (in load case V). And the average value of these coefficients using geometric nonlinear buckling method decrease 12 percent compared with those using the eigenvalue buckling method. Take stock of the stress of the arch

Table 1. Loading combination cases.

No.	Limit state	Load combinations
I	Bearing capacity limit state	Permanent loads + lane loads + crowd loads
II&III		Permanent loads + lane loads + crowd loads + wind
IV	Serviceability limit state (short-term loads)	Permanent loads + lane loads + crowd loads + wind
V	Service stability limit state (long-term loads)	Permanent loads + lane loads + crowd loads + wind

Table 2. Safety factors of Han River North Bridge.

Case No.	I	II	III	IV	V
Eigenvalue buckling	7.068	7.083	7.208	8.662	8.901
Geometric nonlinear	6.647	6.556	6.200	7.556	7.859
Material nonlinear	3.743	2.961	2.664	3.677	3.740
Dual-nonlinear	3.636	2.807	2.216	3.303	3.395

rib under ultimate load, it can be found that the critical stress in steel tube and core-concrete has been in excess of the yield stress (for steel) or compressive strength (for concrete), so the material nonlinearity was employed: the steel tube is assumed to be ideal elasto-plasticity, while the material property of the core-concrete is described by the stress-strain curve of the plain concrete. The safety factors range from 2.806 to 3.743, which are distinctly smaller than those calculated by the geometric nonlinear buckling analysis. From the analysis above, it can be seen that both the geometric nonlinearity and the material nonlinearity affect the stability of the bridge, and the latter influence is more distinctive. Consequently, a dual-nonlinear buckling analysis was carried out in the five different load cases. The safety factors range from 2.216 to 3.636, among which the lowest occurs in load case III. Contrast the safety factors considering material nonlinearity and those taking dual-nonlinearity into account, it is easily found that there is little change between them, which approves that the material nonlinearity affects the stability of the bridge further more than the geometric nonlinearity.

Some parameters were investigated on the bridge's stability in reference (Wang et al 2006), it was shown that the safety factor of the structure under distributed wind forces is 2.216, and will fall to 2.081 when the wind load is simplified to the equivalent concentrated forces, by a decrease of 6%. Changing the value of the pretension of the tied bars and the loads transferred from the adjacent spans will result in little alteration (less than 1%) to the safety factors. The stability of the single dumbbell-shaped arches and that of the back to

back shaped arches were compared, the former value is 1.5 times larger than the latter one, which shows that the side ribs have played an important role on increasing the stability of the structure.

4 MECHANICAL BEHAVIOR AND FAILURE MODES OF THE ARCH UNDER ULTIMATE LOADS

Based on reference (Wang et al 2006), reference (Zhang et al 2007) analyzed the mechanical behavior and the failure modes of the Han River North Bridge. This article defined that Time $= P/P_k$, where P means the loads applied on the bridge, and P_k means the sum of dead loads and life loads. It was shown that the two arches had different failure modes under ultimate loads:

Mode I: The springings on both side arch ribs and vertical ones yielding → The plastic zone developing → Over-buckling (Fig. 5);

Mode II: The springings on side arch ribs yielding → The plastic zone developing → Over-buckling;

Two factors result in such different failure modes: first, the Han River North Bridge is a back to back type arch bridge without lateral braces, leading to poor entirety for the two arches; second, the side arch ribs cause the two arches dissymmetrical when the wind acts on them. Take out the two arches from the 'A' span model as following: the effect of suspenders is neglected; the restraint of V-shaped components is simplified as fixed ends; the load on deck is applied to the nodes on which the suspenders connect with the vertical arch ribs; the wind is applied respectively windward and leeward to arch. The windward arch rib is named as positive arch, and the leeward one as negative arch (Fig. 6).

The inner forces in the springings of the two arches are listed in Tab. 3. From Table 3, it can be seen that the inner forces in the arches are the superposition of the forces generated by vertical loads and horizontal loads. Under the vertical loads, the springings of side arch ribs and vertical arch ribs mainly bear compressive forces; under the horizontal loads, they bear compressive forces and tensile forces respectively, which generates a moment to resist the 'turnover moment' created by the horizontal loads.

It can be deduced from Table 3 that the inner forces in the springings of the positive side arch are smaller than those in the negative one, because of the counteracting of the tensile forces and the compressive ones. At the same time, the springings of positive vertical arch rib are in an unfavorable situation for the two compressive forces, so it will yield simultaneously with those on side arch rib under ultimate loads. For negative arch rib, the springings will yield earlier than those

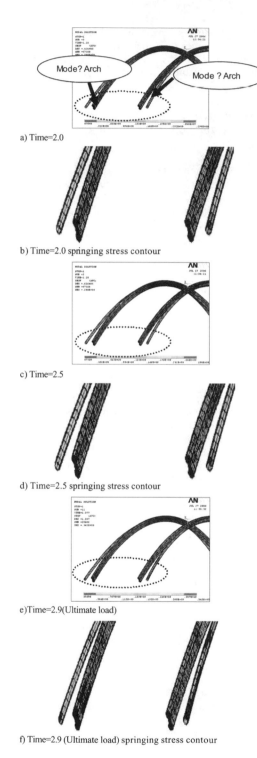

a) Time=2.0

b) Time=2.0 springing stress contour

c) Time=2.5

d) Time=2.5 springing stress contour

e)Time=2.9(Ultimate load)

f) Time=2.9 (Ultimate load) springing stress contour

Figure 5.   Stress contours in two arches of 'A' span.

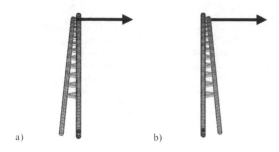

Figure 6.   Two cases for different failure modes: a) Positive arch (Failure mode I) b) Negative arch (Failure mode II).

on vertical arch rib for they have to bear the two compressive forces. According to calculating results, the safety factor for positive arch rib is 3.065, and 2.854 for the negative one. It shows that the stability of negative arch rib is more unreliable. Therefore, though both of the positive arch rib and negative one yield partially, the negative arch rib, or rather, the springings on side arch rib of the negative arch rib determine the ultimate bearing capacity of the bridge. Consequently, the overbuckling of the A span under ultimate loads is induced by the weakness of the springings on the negative side arch rib.

5   NEW CONSTRUCTION PROCESS

The initial construction process of the Han River North Bridge is: the steel hollow tubes of the 'C', 'B' and 'A' spans would be erected successively after all the V-shaped support piers and box girders were built, and when all the tubes were closed, the subscquent construction procedure would start in the same order. A T-shape beam would be hoisted when one procedure was finished on 'C' span. However, the V-shaped support piers and box girders supporting the arches of the 'A' spans can not be completed in time. If the original construction sequence is still adopted, the bridge will not be finished on schedule. Furthermore, the original hoisting scheme of the T-shape beams requires a larger construction site and the utilization ratio of the equipment is low. In this case, a new construction scheme characterized by the relatively separate construction process of the 'A', 'B' and 'C' spans was brought forward (Wang et al 2007). On one hand, the erection of the steel arch ribs of 'B' spans will not begin until the transverse and longitudinal girders of 'C' spans have been hoisted. On the other hand, while the construction on 'B' spans is proceeding, the 'C' spans continue their own subsequent working procedures. 'A' span will be constructed in the same way. The hoisting of T-shape beams and the subsequent procedures in bridge approach will be performed at the end of the construction. Such new construction scheme can

Table 3. The inner forces in springings.

Position	Positive arch rib		Negative arch rib	
	Horizontal load	Vertical load	Horizontal load	Vertical load
Vertical arch rib	Compressive force	Compressive force	Tensile force	Compressive force
Side arch rib	Tensile force	Compressive force	Compressive force	Compressive force

increase the constructing speed and raise the efficiency and benefit, though it is not the best one in minimizing the stress and deformation in components.

To make sure the security and applicability of the structure, a proper finite element model was developed to simulate this five-span bridge with composite arch ribs (Wang et al 2007). The variation of the displacement and the stress statement in V-shaped support piers and arch ribs during construction were summed up, and the feasibility of the new construction scheme was analyzed.

5.1 Variation of stress state and displacement during the construction

The feasibility of the new construction scheme rests with the stress and deformation of the structure. Due to the specialty in the deformation and stress state of the five-span continuous composite arch ribs without wind braces, the out-of-plane displacement of arch ribs and the maximum compressive stress in steel tubes are analyzed. Besides, considering the mechanical properties of the concrete, the variations of the maximum tensile stress in V-shaped support piers and core concrete in arch ribs are studied in reference (Wang et al 2007).

V-shaped support piers are the support components of the whole superstructure. Thus, the stress in such components should be monitored in each construction stage in order to ensure that the stress vary within the allowable range all the time. It is found that the maximum tensile stresses in these two components of Z1 are much higher than those in Z2 and Z3 and that they keep increasing before the T-shape beams are hoisted. But as soon as the T-shape beams are hoisted, both of them dropped down dramatically. This is mainly because that Z1 is under unilateral action before the T-shape beams are hoisted. As a result, the imbalanced moment in Z1 is much larger than those in Z2 and Z3. Further more, Z1 has the lightest self-weight, which results in the lowest axial press in Z1. Thus, the tensile stress in Z1 should be monitored carefully before the T-shape beams are hoisted. Hoisting T-shape beams can reduce the imbalanced moment in Z1 greatly, which is effective in lessening the tensile stress in Z1.

The following conclusions can be drawn through the analysis of the stress statement in arch ribs. Firstly, all the arch ribs will in total-cross-section compression after the pavage of the pre-casting concrete hollow slabs, which indicates that the arch ribs will mainly under axial compression force after the construction of the hollow slabs and the risk of core concrete cracking will not exist during the following construction. Secondly, influence between off spans is little enough to be neglected during construction. Thirdly, owing to the large stiffness of the V-shaped support piers, construction in approach spans effects little on the stress state of arch ribs. The maximum variation of the compressive stress in steel tubes is 9.0 MPa, accounting for only 3% of the design value of its compressive strength (according to GB 50017-2003). The maximum variations of the tensile stress in core concrete before and after the hoisting of the hollow slabs are 0.85 MPa and 1.10 MPa respectively. But due to the low tensile strength of concrete (according to GB 50010–2002), the maximum variations are approximately 40% and 50% of the standard value of tensile strength for concrete. Thus, the core concrete should also be monitored in approach spans for fear of its cracking.

Commonly, in-plane camber is considered in both design and construction to prevent the excess deflection of arch ribs under vertical loads. However, the Han River North Bridge has composite arch ribs, which will lead to their centers of rigidity off the axes of vertical arch ribs due to the existence of the side ones. This will cause an additional torque making the arch ribs leaning inward. It was concluded in reference (Wang et al 2007) that the variation of out-of-plane displacement on the arch crown decline sharply when their own spans are under construction, while they remain relatively aclinic when the ties are stretched or other spans are under construction. So the out-of-plane displacement of the arch ribs only depends on the vertical loads on the arch ribs. The larger the vertical loads are, the more dramatic the inward deflection is.

5.2 Feasibility of the new construction scheme

To realize the emulation analysis on construction process of the whole bridge, the V-shaped support piers

Table 4. Maximum stresses in arch ribs during construction (MPa).

| Components | A span | | | | B span | | | | C span | | | |
| | Duration | | Completion | | Duration | | Completion | | Duration | | Completion | |
	max	min	max	min	max	min	max	min	max	min	max	min
Steel tubes	4.1	−93	−40.8	−93.1	5.7	−92.9	−28.4	−91	7.6	−86.6	−12.5	−86
Core concrete in upper chords	1	−11	−3.8	−11	0.7	−9.4	−3.3	−9.4	0.5	−11.7	−2.7	−8.7
Core concrete in bottom chords	0.6	−8.7	−4	−8.5	0.6	−11.3	−3.7	−11.1	0.7	−11.2	−2.1	−11

Table 5. Displacements of control points on arch ribs (mm)*.

| Positions | Left spring | | 1/4 span | | | Crown | | | 3/4 span | | | Right spring | |
	Ux	Uy	Ux	Uy	Uz	Ux	Uy	Uz	Ux	Uy	Uz	Ux	Uy
A span	−5.9	9.4	−10.4	38.2	−48.9	0.3	37.4	−95.3	10.4	38.2	−48.9	5.9	9.4
B span	−4.1	7.1	−7.4	29.7	−26.1	−4.3	31.6	−45.8	−2.5	15.5	−23.6	−3	−0.5
C span	−2	4.3	1.6	16.1	−18	−2.2	22.9	−28.5	2.7	15.2	−15.3	−2	0.1

*The positive direction of X axis directs from Z1 to Z6, while the positive direction of Y axis directs downwardly. (same infra)

and box girders are simulated with BEAM4 for a simplified calculation. Such a simplification will not affect the whole structure results. And, when the stress state of the V-shaped support piers is to be regarded, a further analysis on these components can be performed with solid models. Based on the solid model analysis, the maximum tensile stresses in V-shaped components and support piers during the whole construction are 5.1 MPa and 3.2 MPa respectively, while the compressive prestress effect in the corresponding components can reach to 4.5 MPa and 2.3 MPa. Thus, the maximum tensile stresses in V-shaped support piers are within the designing prescribed tolerance and will not induce concrete cracking. It also can be seen from the calculated results that, though concrete cracking will not happen in V-shaped support piers during construction, the tensile stress in Z1 will stay in a high level for a long time. Consequently, it is suggested that the T-shape beams be hoisted a little earlier during construction, so that the maximum tensile stress in Z1 can be reduced and the economy, efficiency as well as rationality can be ensured simultaneously.

The maximum stresses in arch ribs of each span during and after construction are listed in Table 4. From this table we can see that arch ribs of each span are mainly under compression. The maximum compressive stresses of steel tubes and core concrete are about 30% and 50% of their design values respectively. The maximum tensile stress of core concrete accounts for 36% of its standard value of tensile strength. Therefore,

Table 6. Variational range of displacements and final deflections at the top of support piers (mm).

| Pier number | Z1 | | Z2 | | Z3 | |
	Ux	Uy	Ux	Uy	Ux	Uy
Variational range	−3.2	0.4	−3.2	0.4	−2.9	0.4
	3.2	0.6	3.3	0.7	5	0.8
Final deflection	0.1	0.6	0	0.7	−0.1	0.8

the stress in arch ribs can satisfy the engineering design requirements. The compressive prestress in core concrete is large enough to keep it from cracking during operation. The stress distribution in each component is relatively even, contributing to a full material use. Displacement of each control point on arch ribs at the completion of the construction is presented in Tab.5. From Table 5 we can see that the final lateral displacements at arch crowns are larger than the vertical displacements. In A span, the lateral displacement (95 mm) is almost three times as large as the vertical one (37 mm), which indicates the necessity of the out-of-plane camber. Meanwhile, the displacements on springings are within the allowable range. So by means of proper in-plane and out-of-plane cambers, the desired final configuration of arch ribs can be achieved.

Tab.6 lists the variational range of the displacements and the final deflections at the top of support piers. It can be seen that a proper stretching force in ties can ensure that the amplitude of right-and-left swing at each top point of support pier is relatively equal and that all these points will return to the designing position at the completion of the construction.

6 CONCLUSIONS

With above study, some conclusions can be drawn:

'Model of double equivalent bars' can be not only used in large deformation, but also the most approximate to 'Model of restraining equations' on lateral rigidity and stability; in addition, the vertical arch rib in 'Model of double equivalent bars' was built by dumbbell-shape cross section, making it convenient for simulation analysis on construction process. Consequently, 'Model of double equivalent bars' is adopted as the final finite element model.

The material nonlinearity influences the safety factor of Han River North Bridge greatly. And the dual-nonlinear buckling analysis which considers the nonlinearity of both the material and the geometry is the most accurate method.

The weak bearing capacity of the springings on side arch rib in the negative arch is the reason why the bridge fails under ultimate loads.

The new construction scheme is feasible for the Han River North Bridge, since the stress and displacement in V-shaped support piers and arch ribs can satisfy the engineering design requirements in each construction stage.

REFERENCES

Wang, Y.Y., Geng, Y., Zhang, S.M., Hui, Z.H. 2007. New Construction Process of a Five-span Tied Rigid-frame Concrete Filled Steel Tubular Arch Bridge. *Pacific Structral Steel Conference, Wairakei, New Zealand,* (accepted).

Wang, Y.Y., Zhang, S.M., Geng, Y., Liu, C.Y. 2006. Nonlinear Finite Element Analysis on the Stability of a Concrete Filled Steel Tubular Arch Bridge. *The Eighth Conference in the Computational Structures Technology, Las Palmas de Gran Canaria, Spain:* 101–105.

Wang, Y.Y., Zhang, S.M., Geng, Y., Ranzi G. 2005. Numerical simulation and research on the Han River North Bridge. *ASCCS' 8th International Conference on Steel-Concrete Composite and Hybrid Structures, Harbin, China:* 227–233.

Zhang, S.M., Liu, C.Y., Wang, Y.Y. 2007. Mechanical Behavior and Failure Modes of Han River North Bridge under Ultimate Loads, *Pacific Structral Steel Conference 2007, Wairakei, New Zealand,* (accepted).

Zhang, T.H., Li, Q.F. 2006. The Design and Stability Analysis of Continuously Multi – span Slanting Arched Bridge, *Journal of Zhengzhou University* 27: 107–109.

Steel and Composite Structures – Wang & Choi (eds)
© *2007 Taylor & Francis Group, London, ISBN 978-0-415-45141-3*

Local deformation analysis for the box girder of a cable-stayed bridge during erection

J.Y. Xu & D.J. Han

School of Architecture and Civil Engineering, South China University of Technology, Guangzhou, Guangdong, China

Q.S. Yan & W.F. Wang

School of Traffic and Communications, South China University of Technology, Guangzhou, Guangdong, China

ABSTRACT: Steel box girder with truss-type diaphragms and internal webs is adopted in Zhanjiang Bay Bridge. The local deformation of the box girder is obviously different between the girder segment to be hoisted and the cantilever system supporting the walking crane during the cantilever erecting due to different applied loads. 3D finite element models for girder segments are developed. The local deformations of the box girder are computed. The relative deformation between the adjacent segments and some critical parameters to the local deformation are studied. Analysis results show that it is feasible to adopt the steel box girder with truss-type diaphragms and internal webs in long span cable-stayed bridges, and the local deformation is governed by the transverse location of the crane acting point and the global stiffness of the steel box girder.

1 INTRODUCTION

Zhanjiang Bay Bridge is located on the Zhanjiang City, Guangdong Province in the South of China. The main structure of the bridge is a hybrid girder cable-stayed bridge with five continuous spans. The total length of the main bridge is 840 m; the central span is 480 m long and the side spans are 120 m long and 60 m long, respectively. Two 155-meter high torch-shaped concrete pylons with varying hollow cross sections hold up the two slightly inclined planes of cable stays, which are anchored along the outsides of the deck. Figure 1 shows the elevation view of the bridge and the transverse view of the pylon. The main girder is composed of 720 m long steel close-box girder between the two

intermediate piers and two 60 m long prestressed concrete (PC) girders. The steel girder is a flat hollow streamlined box girder with a curve bottom plate, and the width and height of the box girder is 28.5 m and 3.0 m, respectively. The box girder has two truss-type internal webs. Truss-type diaphragms are setting every 3.2 m at lengthways. Figure 2 and Figure 3 shows the cross-sectional view of the box girder and the longitudinal view of the internal web, respectively. The total steel girder is divided into 51 girder segments, and the standard segment is 16 m in length and is about 200 t in weight. They are connected by welding and high strength bolts (the internal webs and the longitudinal stiffeners are connected by high strength bolts other part connected by welding). The PC girder is a three

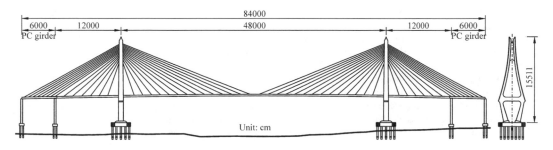

Figure 1. General view of Zhanjiang Bay Bridge.

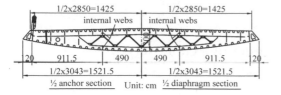

Figure 2. Cross-sectional view of standard segment.

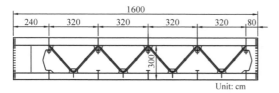

Figure 3. Longitudinal view of internal web.

cells box girder and has the same shape, width and height with the steel box girder. There are 56 pairs of cable stays in the bridge with the spacing of 8 m and 16 m along the PC girder deck and the steel girder deck, respectively. Also, all of these cables are galvanized steel wires, 7 mm in diameter and with high strength and low relaxation. The cable stays are prefabricated in the factory. Zhang & Yang (2002) describe details.

Except for the 0# girder segments near the pylon and the JH segment that connecting the PC girder and the steel girder all steel girder segments are erected by walking cranes on the erected girder segments deck. Li (2006) describes details.

The steel girder segments of the bridge are erected with the free cantilever method. Each steel girder segment is hoisted up by the walking crane, when it reaches the design hoisting level then the following working procedure such as matching, connecting the high strength bolts and welding will be done. Figure 4 shows the structure that hoisting girder segment. It can be found that the girder segment being hoisted bears crane pulling force and weight, the cantilever system supporting the crane (the erected segment) bears not only weight but also the cable forces and the crane acting forces during cantilever hoisting, as shown in Figure 5. The deformation is obviously different in the splice due to two different applied load series. Matching will be difficult, and the welding quality can not be guaranteed if the relative deformation is large enough. On the other hand, errors will be enlarged because this kind of local deformation is not taken into account in the model of traditional erection control analysis that composes of beam and truss elements.

There are two types of diaphragm and internal web for steel box girder (Zhang & Zhou 2005): solid-type and truss-type. The former can provide more

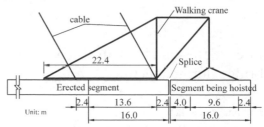

Figure 4. Structural drawing of hoisting erection.

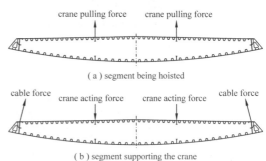

(a) segment being hoisted

(b) segment supporting the crane

Figure 5. Sketch for the forces applied on the box girder during hoisting erection.

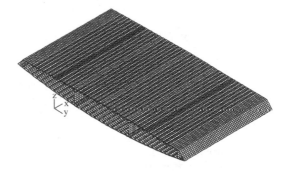

Figure 6. Finite element model of standard segment.

stiffness for box girder, the latter is more economical and is convenient for erection and maintenance. Truss-type internal web and solid-type diaphragm are widely used in the steel box girder of long span cable-stayed bridges, while truss-type diaphragm is rarely used (Huang et al. 2001, Liu et al. 2004, Zhou & Wu 2006, Hao & Qiu 2002). Zhou & Wu (2006) studied the deformation of box girder during cantilever erecting; Hao & Qiu (2002) also studied the relative deformation of box girder during cantilever erecting and influence parameters. However, none of these studies aims at box girders with truss-type diaphragms, so it is necessary to study the deformation of box girder with truss-from internal webs and diaphragms during cantilever erecting to make up the shortage of traditional

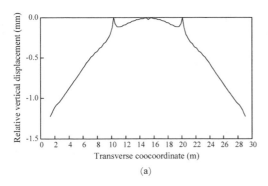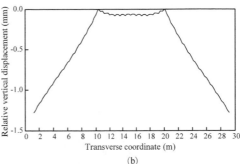

| (a) | (b) |

Figure 7. Relative vertical displacement of splice of girder segment being hoisted (a-top deck, b-bottom plate).

erection control analysis. In addition, the study provides advice for the box girder erection of Zhanjiang Bay Bridge.

2 ANALYSIS MODEL

All components must be taken into account in the finite element method (FEM) model for all of them deform together when the girder segment is hoisted. Standard box girder segment is meshed into mixed FEM model composes of beam elements and shell elements. Top deck, bottom plate, external webs and the stiffeners are meshed with shell element; trusses of internal webs and diaphragms are meshed with beam element. In order to obtain accurate displacements and a high computing speed, four nodes shell elements are used in the model and elements' shape are restricted to rectangle. A standard segment is meshed into 37600 elements, as shown in Figure 6.

Two independent FEM models are developed: one for the girder segment being hoisted and the other for the cantilever system supporting the crane. The former contains one standard segment. There is only dead weight of the box girder itself applies on it, and fix at the four hoisting points. The latter is a cantilever system. According to the Saint-Venant principle, enough length must be kept between the fixed end and the section that concerned about so as to reduce the influence of the fixed end on the deformation of the section, so three erected segments are employed in the model. There are the cable forces and the crane acting forces apply on it besides the dead weight of the box girder itself.

3 ANALYSIS RESULTS

With the developed FEM model in the previous section, the deformation of the girder segment being hoisted and the cantilever system supporting the crane

are computed under the applied loads. The deformation of the girder segment being hoisted and the cantilever system supporting the crane under the applied load consist of global deformation and local deformation, the latter is the part that are concerned about by the bridge engineers. The intersection points of internal webs with the top deck and bottom plate are select as the reference point because aligning the internal webs of two segments is the first procedure of matching during cantilever erecting. The displacements of the box girder shown in the following figures and listed in the analysis results are relative values with the reference point.

The results show that the relative vertical displacements of the box girder segment being hoisted are quite small. The maximum value at the top deck is −1.2 mm and the maximum value at the bottom plate is −1.2 mm. The top deck and the bottom plate has the same displacement curves, as shown in Figure 7, and the displacement curves are similar to that of a cantilever beam with two supports.

The relative vertical displacements of the cantilever system supporting the crane are quite large under the applied loads. The maximum value at the top deck is 23.7 mm; the maximum value at the bottom plate is 23.9 mm. The displacement curves between the two internal webs of top deck and bottom plate are discrepant. The displacement curves are similar to that of a continuous beam with two forces applying on it, as shown in Figure 8.

It can be seen that the relative vertical displacements are opposite in the splice of two segments. The relative deformation is mostly caused by the segments supporting the crane. The relative deformation curves can be draw by choosing the girder segment being hoisting as the reference, as shown in Figure 9. It shows that the maximum relative deformation at the top deck is 24.9 mm and the maximum relative deformation at the bottom plate is 25.2 mm. Matching and welding quality can be guaranteed by taking assisting construction methods with jacks.

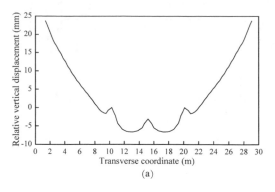
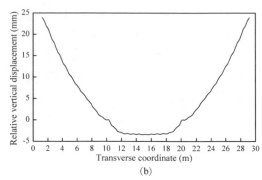

Figure 8. Relative vertical displacement of splice of girder segment supporting the crane (a-top deck, b-bottom plate)

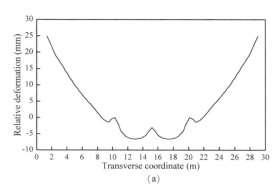
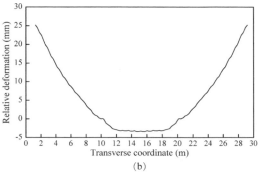

Figure 9. Relative deformation of segment splice (a-top deck, b-bottom plate).

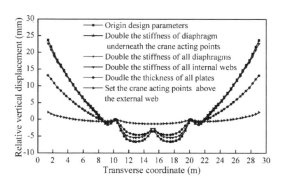

Figure 10. Relative vertical displacement of splice of segment supporting the crane under different parameters.

4 INFLUENCE PARAMETERS STUDY

It is necessary to study the influence parameters such as the truss stiffness, the transverse location of the crane acting point and the total stiffness of the girder segments supporting the crane for it is the major factor that caused the relative deformation in the splice between two segments. Figure 10 shows the compute results. It shows that the truss stiffness of the webs

and the diaphragms have little influence to the relative displacement, however, the maximum relative displacement reduce from 23.9 mm to 13.1 mm after doubling the thickness of all plates; the maximum relative displacement reduce from 23.9 mm to 2.1 mm after setting the crane acting points above the external webs.

5 CONCLUSIONS

With the above computation and study, some conclusions can be drawn:

(a) It is feasible to adopt the steel box girder with truss-type diaphragms and internal webs in the long span cable stayed bridges. Matching and welding quality can be guaranteed by taking some assisting construction methods for the Zhanjiang Bay Bridge.

(b) The deformation of the cantilever system supporting the crane is obvious large than that of the girder segment to be hoisted, and it is the major factor that caused the relative deformation in the splice between tow segments.

(c) The transverse location of the crane acting point and the total stiffness of the steel box girder are the chief influence parameters to the relative deformation in the splice between two segments, which should be taken into account carefully during the box girder design phase. Excessive large relative deformation that affects matching and welding quality can be avoided by choosing proper total stiffness of the steel box girder and the transverse location of the crane acting point.

REFERENCES

Hao, Chao & Qiu, Songding 2002. The relative deformation study of flat steel box-shaped girder of long-span cable-stayed bridge during cantilever construction. *Steel Construction* 17(58):34–37

Huang, Jielin & Wang, lin & Rao, Changxue 2001. Design, manufacture and installation of steel box girder of Wuhan Junshan Changjiang highway bridge. *Sci. & Tech. information of Water Transportation* 184:18–23

Li, Yong 2006. Introduction to construction of steel box girder of Zhanjiang Bay Bridge and experience. *Guangdong Highway Communications* 94:33–36.

Liu, Liping & Wang, Yingliang 2004. New analysis methods of Nanjing Yangtze River Second Bridge streamline flat thin-walled box girder. *Journal of Highway and Transportation Research and Development* 21(7):51–53

Zhang, Qiang & Yang, Jin 2002. General design of main spans of Zhanjiang Bay Bridge. *Bridge Construction* 2002(6):31–34

Zhang, Taike & Zhou, Xiaorong 2005. Nutshell of steel box girder design factor of long span bridge. *Journal of China & Foreign Highway* 25(4):139–141

Zhou, Jianlin & Wu, Chong 2006. Analysis of section deformation of flat steel box girder of long span cable-stayed bridge during cantilever installation. *Bridge Construction* 2006(1):29–35

Steel and Composite Structures – Wang & Choi (eds)
© 2007 Taylor & Francis Group, London, ISBN 978-0-415-45141-3

Investigation of the Compressive Membrane Action in composite steel-concrete bridge decks

Y. Zheng, D. Robinson, S.E. Taylor & D. Cleland

SPACE, Queen's University, Belfast

ABSTRACT: Over the past 50 years, a high proportion of bridges built are composite structures with a deck slab of reinforced concrete supported by longitudinal steel beams. The longitudinal or ladder deck steel supporting beams connect to the slab through shear studs. The steel beams and the surrounding concrete edge slab provide restraint against lateral expansion to the transverse deck slab. As a result, a compressive membrane thrust is developed which ultimately enhances the load carrying capacity of the slab. In order to investigate compressive membrane action (CMA) in this type of bridge, experimental and numerical studies were conducted. A series of one third scale steel-concrete composite bridge models with varied concrete compressive strength, reinforcement percentage and the size of steel supporting beam, were tested. The effect of these variables on the degree of arching action was investigated. Of particular interest was the influence, of the size of supporting steel beam on the loading carrying capacity of the slab as this is generally ignored in the local design of the slab. It was shown that increasing the size of the supporting beam significantly enhanced the ultimate capacity of the slab. Numerical modelling was used to simulate the behaviour of composite bridge deck structure using the non-linear finite element package—ABAQUS. A quasi-static analysis method was adopted. In order to predict the ultimate load capacity, a failure criterion based on the energy condition and satisfactory equilibrium equations was used. The numerical predictions of the ultimate capacity showed a good correlation with the experimental tests.

Keywords: Composite steel-concrete bridge, compressive membrane action (CMA), Non-linear finite element analysis

1 INTRODUCTION

1.1 *Background*

The durability of bridge structures has been of concern since the 1990's. In the 1960's structures were built without a full appreciation of long-term durability. Bridge deck slabs are an important structural component and their behaviour is crucial to the whole structure. However, in the past 30 years, there have been many occurrences of corrosion of the reinforcement in concrete bridge deck slabs by the penetration of chloride ions form de-icing salts or a marine environment. A reduction in the loading capacity of the bridge deck slab influences the integrity of the whole structure. However, bridge deck slabs in typical beam-and-slab or ladder deck type bridge have inherent strength due to the in-plane forces set-up as a result of restraint provided by the slabs boundary conditions, such as beams, diaphragms and slab continuity. This is known as compressive membrane action (CMA) or aching action.

Although the effect of arching action in concrete bridge deck slabs has been recognised for some time,

it only recently that there has been acceptance of rational treatment of arching action in concrete slabs and some design and assessment codes now acknowledge the benefits of CMA. These include the Department of Regional Development (NI), Design Specification for Bridge Decks[1]; The Canadian Bridge Design Code[2] and the UK Highways Agency Standard BD81/02[3] which came about as a direct result of research at Queen's University Belfast.

The ultimate strength of a reinforced concrete slab is affected by its end conditions and particularly by the stiffness of the external lateral restraint. Arching action then occurs due to the difference between tensile and compressive strengths of concrete. On the application of loads, a crack occurs and, the neutral axis ascends towards the compressive surface. If the ends of the slab are restrained by a stiff boundary, an arching thrust develops and ultimate this enhances the flexural capacity of the slabs[4]. Previous research at Queen's University, Belfast by Rankin and Long[5] resulted in a method for predicting the enhanced strength, due to CMA, in reinforced concrete slab and column specimens. Concurrently, tests by Kirkpatrick et al[6] were

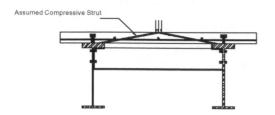

Assumed Compressive Strut

Figure 1. Compressive membrane action in Composite Steel-Concrete Bridge Deck Slabs.

able to establish substantial reductions in the reinforcement required in the slabs for M-beam bridge decks due to the beneficial influence of arching action.

The flexural strength of simply supported concrete slabs is generally governed by the amount and strength of the reinforcement and only minimally by the concrete strength. In contrast, laterally restrained slabs strips generally fail by crushing of the concrete and so the capacity is significantly influenced by the concrete compressive strength. Tests by Taylor et al[4] demonstrated that the arching action was particularly significant in high performance strength concrete slabs. The research also led to a prediction method with incorporated an assessment of the magnitude of the lateral restraint stiffness inherent in bridge deck type structures[7].

Composite steel-concrete bridges are one of the most common forms of bridge. The presence of supporting steel beams together with shear stud connectors provide restraint against expansion of the deck slab. As a result, compressive membrane forces are developed (see Fig. 1). Previous researchers[8] agree that reinforced concrete bridge deck girder-slab type bridge decks have a load carrying capacity far greater than the ultimate static load predicted by flexural theory.

1.2   *Research significance*

Many of the current design codes used in the strength assessment of reinforced concrete slabs underestimate the ultimate capacities of the slab, because compressive membrane action is neglected. Its existence can result in the slab having a higher strength and stiffness than that determined by standard flexural theory or punching shear predictions. Furthermore, in the past research on arching action in composite concrete-steel bridge decks, the influence of steel supporting beam was always not considered[9,10] or regarded as a constant in the estimation of restraint stiffness[11].

Past research7 on concrete bridge deck slabs has concluded that the loading capacity was enhanced with the increasing size of edge beams. In this study, emphasis was put on the effects of varying the size of steel supporting beam and their effect on the load

capacity of the slab. This paper presents and analyses the response of steel reinforced-concrete composite bridge decks under static concentrated load using both experimental test and non-linear finite element analysis (NLFEA).

1.3   *Objectives*

(1) To study the structural behaviour of bridge decks under service load and at ultimate load conditions.
(2) To investigate the influencing factors on compressive membrane action through variations in the structural parameters.
(3) To assess, experimentally and analytically, the significance of CMA in a realistic bridge deck model.
(4) To present the numerical prediction and compare these with standard design methods.

2   EXPERIMENTAL STUDY

2.1   *Experimental parameters*

The test models were designed to represent an external bay of a typical composite steel-concrete bridge at one third scale. The continuity of slabs in the central bays of a typical bridge structure provides additional restraint and therefore enhances the compressive membrane action. An external bay represented the area with lowest lateral restraint and hence the most conservative ultimate capacity for the deck slab. The tests models indicated the enhancement which would occur in a bridge deck slab and verified the presence of arching action under realistic loading conditions. Additionally, it provides information on the most effective restraint system, thereby enabling the most efficient improvement in the load carrying capacity.

A summary of the experimental details is presented in Table 1. As shown as in Table 1, the name of the model includes all of its structural variables. For example, for Model M36SB05, 36 is the concrete compressive strength, SB means the support beam is small, and 05 is the reinforcement percentage (0.5%).

Tests were carried out on six composite bridge decks models of one-third scale. The dimensions were as shown as Fig. 2. A typical model consisted of a one-way spanning concrete slab of 50-mm thickness supported by two steel I-beams connected at the end by channel diaphragms. Shear studs were simulated as $25 \times 25 \times 3$ steel equal angles with a spacing of 150 mm and provide the equivalent shear area as typical studs. In all of the test models, the steel reinforcement was positioned at the mid-depth of concrete slabs. Two concrete compressive strengths, normal and high strength (NSC and HSC), were used. The other variables were the support beam size and the reinforcement percentage.

Table 1. Nominal variables in experimental model.

Model	h (mm)	ρ(%)	f_{cu} (N/mm^2)	f_t (N/mm^2)	Support Beam Size	P_t (kN)	Ultimate Deflection/ Thickness
M36SB05	50	0.5	35.8	3.83	305 × 102 × 25 kg	58	0.4
M77SB05	50	0.5	77	4.8	305 × 102 × 25 kg	78	0.4
M38BB05	50	0.5	37.7	3.51	305 × 165 × 54 kg	79	0.2
M69BB05	50	0.5	68.8	4.4	305 × 165 × 54 kg	99	0.3
M33SB10	50	1	32.8	3.77	305 × 102 × 25 kg	63	0.2
M34BB10	50	1	33.8	3.43	305 × 165 × 54 kg	95	0.2

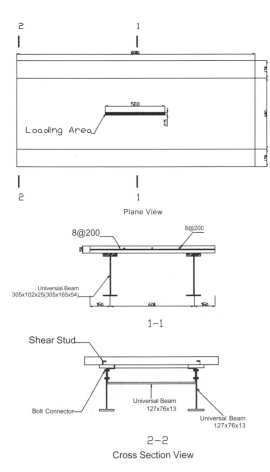

Figure 2. Test Arrangement.

2.2 Instrumentation and data acquisition

The instrumentation consisted of displacement transducers and strain gauges. The arrangement of transducers and gauges is shown in Fig. 2 The monitoring was by a computer data acquisition system. The deflection of the concrete bridge deck slab was measured along the slab midspan parallel to the steel beams. All of the transducers were placed directly under the

slab mid span with one exception; one was set on top of the loading beam. Concrete strains were measured with vibrating wire strain gauges. The top and bottom surface strains were recorded in an attempt to identify strain due to in-plane or membrane forces. To observe the deformation and load transfer through the reinforcement, epoxy-bonded electrical resistance strain (ERS) gauges were used to measure axial strains in the reinforcement, steel supporting beams and diaphragms.

2.3 Test setup and procedure

For testing, the model was supported at each corner. The experimental model was loaded via a 500 mm × 25 mm knife edge load applied at the mid-span of the slab and equivalent to two wheel loads. It was applied via an accurately calibrated 600 kN hydraulic actuator (see Fig. 2). Each model was loaded incrementally up to service loads of 10–15 kN for NSC and 20–25 kN for HSC. The service loads were to ensure bedding-in of the test model (see Fig.4). The model was then loaded incrementally to the failure load. The crack pattern at the bottom and top surface was observed continuously through the tests.

3 ANALYSIS OF EXPERIMENTAL RESULTS

3.1 Observed behaviour

The cracking load corresponds to the point on the load-deflection curve where the curve changes slope and consequently a permanent deformations starts accumulating. The cracking load on the bottom surface was about 15–20 kN for all test, which corresponds to ~20% of the ultimate load value. The ratio between cracking load on the bottom surface and ultimate loads for the bridge deck models is not significantly influenced by the reinforcement percentage. The cracking load level on the top surface was influenced by the size of supporting beams. Pcrt/Pu was 0.6 and 0.5 for the low restraint stiffness and high restraint stiffness respectively. It was noted that the ductility of the structure reduced with an increase of restraint stiffness.

The yield load corresponds to the level of applied load that causes yielding (about 2000 με) in the

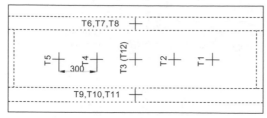

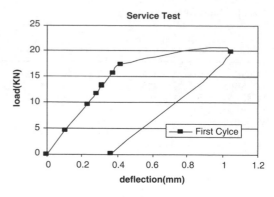

displacement transducer arrangement

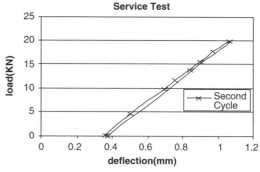

Figure 4. Service load test (M69BB05).

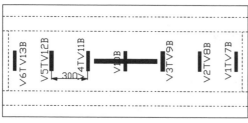

vibrate gauge arrangement

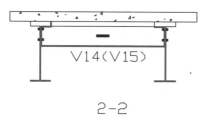

2-2

Figure 3. Measurement arrangement.

transverse flexural reinforcement under the load area. It was found that this yield load value was not influenced by the other factors.

The bridge deck slab spans predominately one-way between the steel I-beams and was subjected to the concentrated load at midspan, the mode of failure can be described as following:

1. When the applied load exceeded the cracking load, cracking first occurred on the underside of the slab directly below the load point in the longitudinal direction and parallel to the steel support beams.
2. As the load increased, transverse cracks occurred along the bottom surface in the diagonal direction and the cracks radiated from the end of loaded area towards the slab corners.
3. With further loading, the cracking was associated with yielding of the reinforcement and a decrease in the depth of the compression zone.
4. When the loading was close to the ultimate capacity, cracks formed on the top surface adjacent to and parallel to the edges of the supporting beams.
5. Failure then occurred in the highly stressed compression zone and may have been accompanied by a sudden punching of concrete immediately under the concentrated load.

The failure mode of all the deck specimens is presented in Table 2. All the models except M36SB05 punched through although the tests also validated with the other type of failure mode. Model M36SB05 exhibited a more ductile behaviour, as shown by its load-deflection response, and it exhibitted a more flexural behaviour. Crushing of the concrete in the compression zone was the primary failure mode for all the experimental models.

Typical crack patterns at failure for different loading level are present the Fig. 5. An increase in the restraint stiffness or concrete strength reduced the ductility of the structural behaviour. Punching effect also becomes more pronounced with an increase in

Table 2. Summary of failure load result.

Model	Pu (kN)	Pcrb (kN)	Pcrt (kN)	Py (kN)	Pcrb/Pu	Pcrt/Pu	Py/Pu
M36SB05	58	12	40	30	0.2	0.68	0.51
M77SB05	78	18	50	30	0.23	0.64	0.38
M38BB05	79	15	45	29	0.19	0.57	0.37
M68BB05	99	20	43	29	0.2	0.43	0.29
M33SB10	63	12	40	27	0.19	0.63	0.43
M34BB10	95	12	45	40	0.13	0.47	0.44

Model	(I_y) (cm^4)	fcu (N/mm^2)	Pu (kN)	Failure Cracking Pattern (Bottom)	Primary Failure Mode
M36SB05	123	36	59	LF + R	Flexure
M77SB05	123	77	78	LF + R + CR	Flexure + Punching
M38BB05	1063	38	79	LF + R + CR	Flexure + Punching
M68BB05	1063	68	99	LF + R + CR	Flexure + Punching
M33SB10	123	33	63	LF + R + CR	Flexure + Punching
M34BB10	1063	34	95	LF + R + CR	Flexure + Punching

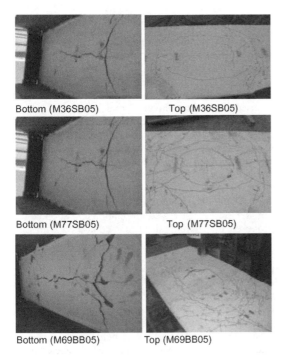

Bottom (M36SB05) Top (M36SB05)

Bottom (M77SB05) Top (M77SB05)

Bottom (M69BB05) Top (M69BB05)

Figure 5. Crack pattern in Model 36SB05 M77SB05 M69BB05.

these variables. Except for model M36SB05, a circular crack developed on the bottom surface when the applied load increased to ultimate strength. The failure in models with this type of cracking occurred suddenly, especially Model M68BB05. In all of the tests, the crack patterns on the top surface developed in a circular style surrounding the rectangular load area. At the occurrence of failure, a crack occurred in the top surface close to the load edge.

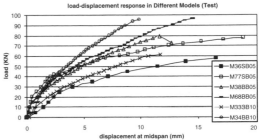

Figure 6. Load-deflection response in validated experimental test.

3.2 Model deflections

The applied load vs. deflection in the mid-span is shown in Fig. 7. As expected, the static ultimate strength of the concrete bridge deck slab with the stiffer steel girders and high strength concrete slab had the highest ultimate capacity. Furthermore, the tests showed that the ultimate strength and hence the membrane action, which was the major load-carrying mechanism under a static load, were more sensitive to an increase in the slab external restraint stiffness than an increase in the steel reinforcement ratio. Also, with the increase of the restraint stiffness, the ductility of the structure reduced, which was reflected in the decrease at the mid-span deflection. The failure mode transferred from predominantly flexural to flexural punching modes with increasing ultimate capacity.

Fig. 7 shows the deflected shape at the mid-span of the models M77SB05 and M69BB05 at ultimate loads. The deflections have been magnified by 5 with respect to the cross-sectional dimensions. Moreover, the rotation point of the steel beam was assumed to occur in the mid-depth of the beam. Firstly, it can be seen

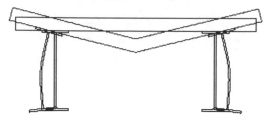

M77SB05: Beam size=305x102x25kg

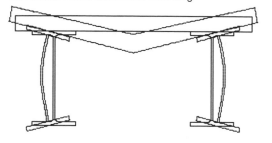

M69B05: Beam size=305x165x54kg

*Note: Deflectionx5 (Assuming linear interpolation between measured deflections)

Figure 7. Idealised Deflected shapes at Ultimate Loads based on Deflection Measurements.

that the vertical deflection in the mid-span reduced as the size of steel supporting beams increased. The deflected shapes clearly show the horizontal deflections in the supporting beams, which is the results of membrane actions. It was found that the horizontal deflection at the top of the steel beams in two models were similar. In contrast, the horizontal deflections measured in the bottom of the Model with big beam were smaller than those measured in the small beam. At the same time, it could be noted that the larger rotation angles were obtained in model M69BB05. This highlights the effect of varying the degree of external edge restraint stiffness. Thus an increase in the external lateral restraint stiffness reduced the vertical deflections in the slabs.

4 NONLINEAR FINITE ELEMENT ANALYSIS

4.1 Finite element model

This section discusses the characteristics of the FEA modelling including the constitutive material properties for structural steel and alternative concrete models. The numerical model used the ABAQUS[12] software package. An eight-node shell element with reduced integration was used to model the concrete slab, steel beams and end diaphragm while the reinforcement was modelled using a one-dimensional rebar elements (see Fig. 8). The shear connection between the reinforced

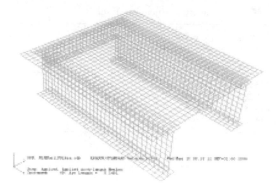

Figure 8. Finite Element Model.

concrete slab and the steel beam was represented by multiple-point constraint equations.

4.2 Material modelling of concrete

4.2.1 Steel and reinforcement
The Von Mises yield criteria, with associated plastic flow rule was used to represent the steel and the reinforcement. The reinforcement was modelled by adding layers of reinforcement at specified depths and angle of orientation within the virtual thickness of the shell element. These layers provide the equivalent of rod elements and hence apply the appropriate stiffness needed to represent the behaviour of the reinforcement bars. However provision for the slippage between steel and concrete was not directly catered for in the model.

4.2.2 Concrete constitutive model
Fig. 9 shows the stress-strain relationship adopted as proposed by Thorenfeldt et al[13] combined with the Hognestad's[14] relationship for the elastic modulus of the concrete. This applies a factor to increase the post-peak decay, which reflects the behaviour of high strength concrete. This relationship is based on the concrete cylinder strength. The tensile property of the reinforced concrete was modelled using a simple tension stiffening model. A linear softening model was used to represent the post failure behaviour in tension. The softening rate depends on the size of the elements in the crack region[15], Fig. 9. A linear relationship was assumed for simplicity. In Fig. 10, an ultimate strain value of 2.5E-3 was recommended[16], but the results using other values of ultimate strain (2.0-3, 3.0E-3) were similar.

4.2.3 Numerical concrete model—concrete plasticity
Isotropic elasticity in combination with isotropic tensile and compressive plasticity was used in the concrete model to represent the inelastic behaviour of concrete more accurately. This concrete model uses the

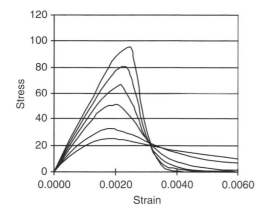

Figure 9. Stress-strain relationship of concrete.

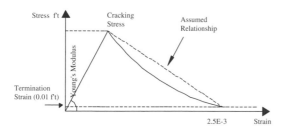

Figure 10. Stress-strain relationship for reinforced concrete in tension.

yield function proposed by Lubliner et al[17] with the modification suggested by Lee and Fenves[18] to consider the evolution of strength characteristics under tension and compression. The evolution of the yield surface is defined by tabular values giving an inelastic stress-strain relationship. This concrete model employs non-associate plastic flow potential using the Drucker-Prager hyperbolic function for the flow potential.

4.3 Instability problems

When the composite steel-concrete bridge deck slabs are subjected concentrated patch loads, the failure modes are usually punching modes with strong unstable effects.

From previous research[19], the punching failure can be regarded as a type of snap-through instability. The instability in the bridge deck resulted in a post-buckling behaviour with a sudden energy release and the failure mode is very sudden and brittle. The researchers simplified the bridge deck subjected concentrated loads as a truss model with a laterally restrained three-hinge compressive strut mechanism. The snap-through instability response of the truss model under a vertical concentrated load at its apex

was able to describe the instability characteristics of a sudden brittle ultimate response of a bridge deck in the transverse direction.

In the concrete slab or beam analysis, a softening is usually provided in the material model. Softening materials are known to induce 'strain-localization', in which a local region softens (or cracks) while the adjoining materials unloads elastically. These localizations maybe accompanied dynamic 'snap-through'. In the research by Crisfield[20], it was found that the strain softening can lead to local maxima that, under load control, would lead to dynamic snap-through instability.

Modelling unstable models as a static analysis with ABAQUS could result in convergence difficulties. This resulted in that several of ABAQUS models using the Static General or the Static Riks solution method terminated or softening before the maximum load was accomplished. In ABAQUS/Explicit, it is possible to treat the unstable problem dynamically, thus modelling the response with inertia effects included as structure snaps. This method is accompanied by ABAQUS, by terminating the static solution procedure and switching to a dynamic procedure—Quasi-Static analysis in ABAQUS/Explicit, when the static solution becomes unstable.

4.4 Explicit analysis—Quasi-static analysis

The explicit dynamic method was originally developed for high-speed dynamic events in which inertia plays a dominant role in the solution. In order to prevent the numerical instability, the time increment should be limited to be lower than a critical velocity, which can be calculated from the equation:

$$\Delta t = \frac{2}{\omega_{max}} \qquad (1)$$

where ω_{max} is the highest frequency of the system. If the time increment is larger than the maximum time increment, the increment is said to have exceeded the stability limit, this will cause the unbounded solution. However, it is difficult to determine the stability limit exactly. ABAQUS provides a conservative estimate of this critical time step using the formula:

$$\Delta t = \min\left(\frac{Le}{c_d}\right) \qquad (2)$$

where Le is the characteristics element dimension and c_d is the wave speed of the material. C_d can be expressed as:

$$c_d = \sqrt{\frac{E}{\rho}} \qquad (3)$$

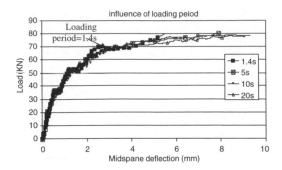

Figure 11. Influence on different loading rates (Model M77SB05).

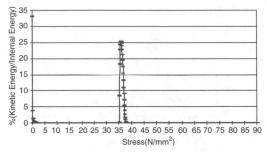

Figure 12. Kinetic Energy vs Internal Energy (Compressive Test by Smith[21], $f_c'' = 35.17 \, \text{N/mm}^2$).

where E is the Young's modulus and ρ is the mass density.

The main advantage of the explicit method in limit load analysis is the absence of convergence problem. A further advantage of dynamic analysis is that, in the vicinity of a critical point, the inertia forces stabilize the system motion even in the post-critical range where the load the system can carry decreases. Thus, the character of the post-critical behaviour can be studied.

In this type of analysis, applying a dynamic procedure to static problems requires some special considerations, such as the loading rate. It can be seen from Fig. 11, that fast loading rates cause a structural response with significant inertia effects. In a static analysis, the lowest mode of the structure usually dominates the response. Therefore the time required to obtain the proper static response can be estimated from the period of the lowest mode. It is highly recommended by ABAQUS to increase the loading time to ten times the period of the lowest mode to be certain that the solution is truly quasi-static. The natural period was obtained by performing eigenvalue frequency analysis in ABAQUS/Standard. In these numerical models, the obtained period for natural frequency is 0.14 seconds. When the loading period is adopted as 10 times of this period, it was found that the inertia effects were still too strong for quasi-static analysis. Therefore, loading rates should be the critical parameter to obtain a true quasi-static analysis. When the applied load is fixed, the time for loading period should be selected carefully. Although the longer loading period can reduced the dynamic effects, it may reduce the analysis efficiency significantly. To obtain an economical solution, a compromise has to be reached. The loading rate has to slow enough to avoid inertia effect but not too slow so as to increase the time taken to complete the analysis. A suitable loading period was obtained after several trials in the optimum analysis. ABAQUS/Explicit successfully uses quasi-static simulations in problems which is related to local instability and post-buckling during the processing.

However, the analysis procedure will not terminate, or soften, after the ultimate strength is reached. Therefore, it was necessary to establish failure criteria for identifying the ultimate load capacity.

4.5 Failure check in quasi-static analysis

When the structure starts to collapse, it is a mechanism and there is an increase of kinetic energy. Therefore it is possible to use the step change in the kinetic energy to identify the onset of failure. A normalised ratio of kinetic energy and internal energy was used.

Failure check based on the energy condition
In dynamic analysis, if a simulation is quasi-static, the work applied by the external forces is nearly equal to the internal energy of the system. The viscously dissipated energy is generally small unless viscoelastic materials, discrete dashpots, or material damping are used. It has been established that the inertial forces are negligible in a quasi-static analysis because the acceleration of the material in the model is very small. The corollary to both of these conditions is that the kinetic energy is also small. As a general rule the kinetic energy of the deforming material should not exceed a small fraction (typically 5% to 10%) 5 of its internal energy throughout most of the process. Therefore, in the quasi-static analysis, when the kinetic energy surpasses 5%–10% of internal energy or the difference between external energy and internal energy is larger than 5%–10% of the external energy, failure is deemed to have occurred. As shown as Fig. 12, in a uniaxial compressive concrete cylinder test, when the applied load reaches the compressive strength of the concrete material, the crushing occurs and the ratio between Kinetic Energy and Internal Energy surpasses 5%.

Failure check based on mechanism equilibrium
However, in some cases, it is difficult to capture the real failure loads by the failure criterion based on energy condition. In the analysis of highly redundant structure, when redistribution of forces close to failure load

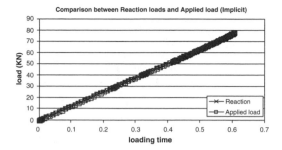

Figure 13. Comparison between reaction forces and applied load in M77SB05 in implicit static analysis.

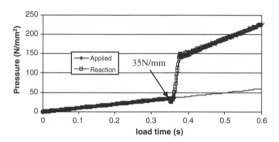

Figure 14. Comparison the Reaction Forces and Applied Forces in Quasi-Static Analysis (Compressive Test by Smith[21], $f_c' = 35.17$ N/mm^2).

occurs, the critical cut-off point for the energy condition becomes blurred. Therefore, the static equilibrium should be checked. In the implicit static analysis, it was found that the reaction forces are always equal to the applied loads as shown as Fig. 13. However, in the quasi-static analysis, it is possible for the non-equilibrium to occur in the analysis due to the dynamic effects. The same validation models for the failure check based on energy condition were adopted. As shown as Fig. 14, it can be seen that non-equilibrium occurs after the applied load reaches the concrete compressive strength.

The proposed method for predicting the failure load in a quasi-static analysis needs to be validated against the experimental tests.

5 RESULTS AND DISCUSSIONS FOR NLEFA IN PREDICTIONS OF COLLAPASE LOADS

FEA modelling of the Authors' experimental models was performed using the procedure discussed above and a typical mesh of these composite steel-concrete bridge decks is shown in Fig. 9.

In order to increase to computing efficiency and keep the practical aspect ratio to be close to one (element size is 25 mm and loaded are is 25 mm × 500 mm), half-symmetric finite element

model was established. Constitutive laws shown in Fig.9 and Fig.10 are used in concrete material along with the concrete plasticity model. The quasi-static analysis in ABAQUS/Explicit is employed in this section.

5.1 Comparisons of results

Table 3 compares the experimental ultimate loads in experimental tests, predictions using the current design codes[22,23] and NLFEA. From the comparison between NLFEA and experimental test, the results from NLEFA show a good correlation with the experimental test.

In the explicit analysis, the two types of failure determinations are used. It can be seen that during the initial load increment **kinetic energy/internal energy** has a peak but within the next increment it had reduced drastically shown as the Fig 15 (a). The ratio at the peak is due to the initial discontinuous loading and further exaggerated value by the small numbers at the starting. However, this initial inertial instability was quickly recovered and the energy system stabilised showing that the energy ratio less than 5% was reached after the applied load increased more than 10 kN. In order to make it easier to identify the point of failure, it is better to discard the initial peak and hence Fig. 15(b) shows the energy ratio without the initial peak. In Fig.15 (b), it is presented that when the ratio of kinetic energy and internal energy increases more than 5%, the applied load become close to the ultimate strength of the experimental tests. However, the critical point of the limit load is not particularly clear. Fig. 16 shows the reaction forces and applied load against the loading increment times. Before the applied load reaches the ultimate loads, the response is the same as the implicit static analysis (see Fig. 13). However, as soon as the applied surpass the ultimate value, a state of non-equilibrium occurred with an obvious dynamic effect.

5.2 Analysis of load-deflection response

In a comparison between the load-displacement responses in the experimental test and NLFEA as shown as Fig.17 and Fig.18, it can be seen that the adopted numerical model had an excellent ability to predict the structural responses of composite bridge deck when the adopted failure criteria was used. The trends of the load-displacement responses in experimental test and numerical analysis are similar.

5.3 Influence of varying the structural parameters on loading capacity

Table 4 shows that an increase in the variables; reinforcement percentages, concrete compressive strength and steel supporting beam size, resulted in an increase in the ultimate capacity of the concrete bridge deck

Table 3. Comparison between Design Standard, Numerical Analysis and Experimental Results.

Model	P_{test}	BS5400[22] (Flexural Capacity)	BS5400[22] (Shear Capacity)	ACI318-05[23] (Flexural Capacity)	ACI318-05[23] (Shear Capacity)	Pexplicit (energy 5%KI/IE)	Pexplicit (equilibrium)	Pt/Pexplicit (KI/IE)	Pt/Pexplicit (equilibrium)
M36SB05	58	12	45	12	51	60	59	1.03	1.02
M77SB05	78	13	58	13	75	73	75	0.94	1.04
M38BB05	79	12	46	12	53	85	81	1.08	1.03
M69BB05	99	13	56	13	71	107	103	1.08	1.04
M33SB10	64	22	56	22	48	68	67	1.06	1.05
M34BB10	95	22	57	22	50	99	90	1.04	0.95
Average								1.04	1.01
St.deviation								0.05	0.04
Coeff.Var.								0.05	0.04

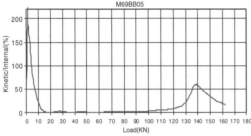

(a) Original Curves

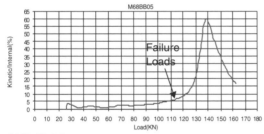

(b) Modified Curves

Figure 15. Applied Load vs. Kinetic and Internal Energy Ratio (M69BB05).

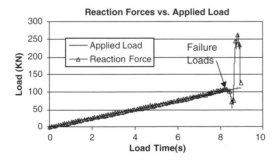

Figure 16. Comparison between Reaction Forces and Applied Loads (M69BB05).

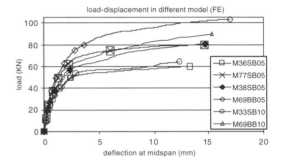

Figure 17. Load-deflection response in nonlinear finite element analysis (explicit analysis).

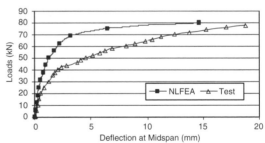

Figure 18. Load-deflection Response Comparison.

slab. The concrete compressive strength and the steel beam size had a more significant effect on the ultimate loading capacity compared to the reinforcement percentage which made only a marginal difference. A 50% increase in the reinforcement percentage equated to an average increase in loading capacity of only ~10% in both the actual test and numerical analysis predictions. However, a 50% increase in the concrete compressive strength provided nearly 25% enhancements in the ultimate capacity. In previous research the restraint stiffness of the supporting steel I-beam (in composite

Table 4. Influence on the structural parameters on loading capacity.

Reinforcement

Model	Pt (KN)	Ratio of Increase	PFEA (explicit) (KN)	Ratio of Increase
M36SB05	58	0.07	59	0.14
M33SB10	62		67	
M38BB05	79	0.2	81	0.11
M34BB10	95		90	

Concrete Strength

Model	Pt (KN)	Ratio of Increase	PFEA (explicit) (KN)	Ratio of Increase
M36SB05	58	0.34	59	0.79
M77SB05	78		75	
M38BB05	79	0.25	81	0.27
M69BB05	99		103	

Steel Beam Size

Model	Pt (KN)	Ratio of Increase	PFEA (explicit) (KN)	Ratio of Increase
M36SB05	58	0.36	59	0.37
M38BB05	79		81	
M77SB05	77	0.29	75	0.73
M69BB05	99		103	
M33SB10	62	0.53	67	0.34
M34BB10	95		90	

steel-concrete bridge decks) has been regarded as a constant. In this study, it was found that increasing the lateral stiffness of the supporting beam can significantly increase the ultimate load capacity of the deck slab. The load carrying capacity of the slab was enhanced by nearly 30% when the supporting beam was increased (a ten times increase of Iy-y value resulting from an increase of 305 × 102 × 25 run of beam to 305 × 165 × 54 run of beam). This phenomenon is neglected in current design methods (Table 3). In the comparisons, it also shown that the influence of the concrete compressive strength is less sensitive when the external lateral restraint stiffness is high compared to when the external lateral restraint is low.

6 CONCLUSIONS

(1) Current design codes predict the ultimate loads conservatively for composite steel-concrete bridge decks structures, because it does not take account of the existing of internal arching mechanism;
(2) The increase of lateral/rotational stiffness and concrete strength enhanced the ultimate strength. The sensitivity to restraint stiffness and concrete strength are more than that from changes of reinforcement percentage;
(3) The ductility of the structure was lowered by the increase of compressive membrane action, which was reflected by the punching failure modes in most of the experimental models;

(4) The ultimate loads predicted by explicit method are reliable, because the failure criteria are based on the global structural behaviour. In the explicit analysis, it was found that it can be difficult to define a precise failure check based on energy condition. Therefore, it can be concluded that the failure criterion based on the check of equilibrium between the reaction forces and applied loads in explicit analysis is the most suitable analysis method for determination of the loading capacity;
(5) The sensitivity to restraint stiffness and concrete strength are more than that from changes of reinforcement percentage. According to the accurate prediction in the ultimate loads, a compressive membrane effect can be accurately reflected by the NLFEA.
(6) Based on the lower cost and lower time requirement compared with the experimental tests, the proposed NLFEA models can be used by design engineers for the analysis of composite steel-concrete bridge decks, assessment of the loading capacity and parametric studies.

REFERENCES

1 Department of Regional Development for Northern Ireland (formerly Department of the Environment or DOE) Design of M-beam bridge decks – Amendment No.3 to Bridge Design Code N.I. Roads Service Headquarters, 1986, 11.1–11.5.
2 Canadian Standards Association, Canadian Highway Bridge Design Code CAN/CSA-S6-00 (new version available for comment 2005 CSA-Technical Committee), Canada.
3 UK Highways Agency, BD 81/02: Use of compressive membrane action in bridge decks, Design manual for Roads and Bridges, Volume 3, Section 4, Part 20, August 2002.
4 Taylor, S.E., G.I.B. Rankin and D.J. Cleland, 'Arching action in high-strength concrete slabs', Structures & Building ICE-Proceeding, Vol. 146, ISSUE 4, pp 353–362.
5 Rankin, G.I.B. and Long, A.E., Arching action strength enhancement in laterally restrained slab strips, ICE Proceedings—Structures and Buildings, No. 122 Nov. 1997, pp 461–467.
6 Kirkpatrick, J., Rankin, G.I.B and Long, A.E., Strength of evaluation of M-beam bridge deck slabs, Structural Engineer, Vol. 62b, No. 3, Sep. 1984, pp 60–68.
7 Taylor, S.E., Rankin, G.I.B. and Cleland, D.J., (2002) Guide to compressive membrane action in bridge deck slabs, UK Concrete Bridge Development Group/British Cement Association, Technical Paper 3, June 2002.
8 O. Shervan Khanna, Aftab A. Mufti, and Baidar Bakht (2000), 'Experimental investigation of the role of reinforcement in strength of concrete deck slabs', Canadian Journal of Civil Engineering, Vol. 27, 2000, pp 475–480.
9 Barrington deV. Batchelor, Brain E. Hewitt , P. Casgoly and M. Holowka (1978), ' An investigation

of the ultimate strength of deck slabs of composite steel/concrete bridges', Transportation Research Record No.664, Transportation Research Board, 1978 pp 162–170.

10 Aftab A. Muft and John P. Newhook (1998), 'Punching shear strength of restrained concrete bridge deck slabs', ACI Structural Journal Vol. 95, No. 4 July–August 1998.

11 Yogush M. Desai, Aftab A. Mufti, Gamil Tadros, 'Finite element analysis of steel-free decks—User Manual for FEM Punch Version (2.0)', July 2002.

12 ABAQUS User's Manuals, Version 6.5 Hibbit, Karlsson and Sorensen, Inc. USA 2005.

13 Thorenfeldt, E., Tomaszemicz, A. and Jensen, J.J.(1987), "Mechanical properties of high-strength concrete Application in Design", Proceedings of the symposium utilization of high strength concrete, Tapir Trondheim.

14 Mattock, A.H. , Kriz,l.B. and Hognestad, E. (1961), "Rectangular concrete stress distribution in ultimate strength design", Proceeding of the ACI, Vol.57, No. 8, Feb 1961, pp 875–928.

15 Gilbert, R. I. and Warner, R. F. (1978), "Tension stiffening in reinforced concrete slabs", Proceedings of the American Society of Civil Engineers, Vol.104, No. S12, December, 1978, pp 1885–1900.

16 Report of a Concrete Society Working Party (2004), "Influence of tension stiffening on deflection of reinforced concrete structures", Technical Report No. 59, UK, 2004, pp 35

17 Lubliner, J., J. Oliver, S. Oller, and E. Oñate, "A Plastic-Damage Model for Concrete," International Journal of Solids and Structures, vol. 25, pp 299–329, 1989.

18 Lee, J., and G. L. Fenves (1998), "Plastic-Damage Model for Cyclic Loading of Concrete Structures,"Journal of Engineering Mechanics, vol. 124, no. 8, pp 892–900

19 Petrou, M.F. and Perdikaris, P.C. (1996), 'Punching shear failure in concrete decks as a snap-through stability loads', ASCE Journal of Structural Engineering, Vol.122, No.9, Sep 1996, pp 998–1005.

20 Crisfield, M. A., "A Fast Incremental/Iterative Solution Procedure that Handles 'Snap-Through'," Computers and Structures, vol. 13, pp 55–62, 1981.

21 S.H. Smith. (1987), 'One fundamental aspects of concrete behaviours', Master thesis, University of Colorado at Boulder.

22 British Standards Institute, "BS 5400: Parts 2 & 4, British Standard for the design of steel, concrete and composite bridges", London, 1978 and 1990.

23 ACI Committee 318 (ACI318-05), "Building code requirement for reinforced concrete and commentary" American Concrete Institute, Detroit, USA, 2005.

Fire resistance

Steel and Composite Structures – Wang & Choi (eds)
© *2007 Taylor & Francis Group, London, ISBN 978-0-415-45141-3*

Temperature distributions in unprotected steel connections in fire

X.H. Dai, Y.C. Wang & C.G. Bailey
School of Mechanical, Aerospace and Civil Engineering, University of Manchester, UK

ABSTRACT: Four unloaded unprotected steel beam to column connections including a composite slab at the top of the steelwork have recently been fire tested at the University of Manchester as part of a research project to investigate the behaviour and robustness of connections in steel framed structures in fire. The connection types were flush endplates, flexible endplates, fin plates and web cleats. This paper presents an analysis of temperatures in the steel members and joint components, using the FE package ABAQUS and the simple design calculation method in EN 1993-1-2 (2005). The objectives of this analysis are (1) to check whether the design calculation method is appropriate and (2) to propose appropriate heat transfer coefficients and section factors to be used in the simple design calculation method.

1 INTRODUCTION

Connections are critical members in steel framed structures. Despite extensive researches on steel-framed structures in fire in the past, large gaps exist in understanding of connection behaviour in fire. Under fire conditions, the behaviour of a steel structure is complex with forces in different members changing during the entire course of fire exposure. These forces are transmitted from one connected member to another, mainly dependent on the behaviour and performance of the connections, making understanding connection behaviours in fire a key factor in structural fire design. To fill some of the knowledge gaps, a comprehensive research project on robustness of joints in steel-framed structures at high temperatures is under way at the University of Manchester in collaboration with the University of Sheffield, funded by the UK's EPSRC. The overall objective of this project is to investigate the structural behaviour and robustness of connections in steel-framed structures in fire. As part of this research project, this paper will present results related to temperatures in unprotected connections.

one column, four beams (two bolted to the column flanges through the aforementioned different types of connection and two bolted to the column web via fin plates) and a concrete composite slab with profiled steel decking and reinforcements connected to the steel beams via shear connectors to capture realistic temperature distribution in composite connections. The column section is UC254 × 254 × 89 and its length is 1000 mm. The beam section is UB305 × 165 × 40. The length of the steel beams connected to the flanges of the column is 605 mm and the length of the steel beams connected to the web of the column is 485 mm. The dimension of the fin plates welded to the column web is 100 × 200 × 10 mm. The dimensions of the connection components connecting the beams to the column flanges are: flush endplates 200 × 314 × 10 mm; flexible endplates 150 × 200 × 8 mm; fin plates 150 × 200 × 10 on one side and 100 × 200 × 10 on the other side; web cleats 90 × 150 × 10 mm (depth 200 mm) on one side and 90 × 90 × 10 mm (depth 200 mm) on the other side. The dimension of the composite slab is 1000 × 1000 mm with an overall depth of 130 mm.

2 FIRE TESTS

2.1 *Description of connection assemblies*

Four specially designed unprotected connections, including flush endplates, flexible endplates, fin plates and web cleats have recently been tested. Figure 1 shows the 3D configurations of these connection assemblies. Each connection specimen consists of

2.2 *Description of test arrangements*

Fire tests were conducted in the large furnace with internal dimensions of 3.5 m × 3.5 m × 2.5 m in the fire laboratory in the University of Manchester. The interior faces of the furnace were lined with ceramic fibre materials of thickness 200 mm that efficiently transfer heat to the specimen. Two gas burners and

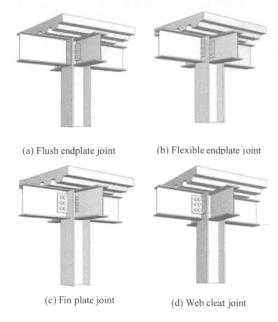

(a) Flush endplate joint   (b) Flexible endplate joint

(c) Fin plate joint   (d) Web cleat joint

Figure 1. Composite connections with various joint components.

two exhausts were connected to the furnace. The firing and control equipments were installed with the gas burners. The furnace temperatures were recorded by six conventional bead thermocouples. To ensure that the concrete slab surface was exposed to ambient temperature air, the test specimen was rotated by 90° and hung inside the furnace via three protected steel ropes, as shown in Figure 2. Fire exposure was according to the standard fire condition (ISO 834).

2.3 Test observations

Figure 3 and 4 present typical distributions of recorded temperatures, obtained from the test specimen with flush end plates and fin plates connecting to the column flanges. The following observations can be noticed:

- The temperatures in the joint components were much lower than the temperatures in the connected beams remote from the connection region.
- For temperature distributions in connection components in the vertical direction, the closer the monitored point to the slab bottom surface, the lower the measured temperature. In the horizontal direction, the connection temperatures increase as the monitoring points move away from the connection zone.
- Temperature differences in vertical direction in connections with small vertical dimensions (flexible endplate $150 \times 200 \times 10$ mm, fin plate

(a) View the specimen inside furnace

(b) View the specimen outside furnace

Figure 2. Set-up of connection model test.

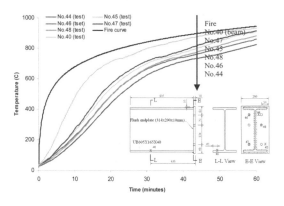

Figure 3. Temperatures in the beam and flush endplate.

$100 \times 200 \times 10$ mm and web cleat $90 \times 90 \times 10$ mm with depth 200 mm) are moderate with the maximum being less than 60 °C as shown in Figure 5. However, for connection components with larger vertical dimension, such as flush endplate ($200 \times 314 \times 10$ mm), the temperature differences in the connection components may be over 80 °C (Figure 5). Here the temperature differences in the vertical direction were calculated between the bolt

536

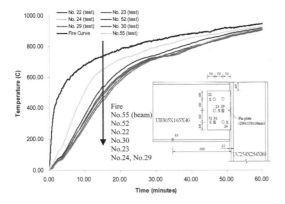

Figure 4. Temperatures in the beam and fin plate.

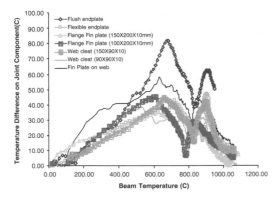

Figure 5. Maximum vertical temperature differences in all joint components.

monitoring points that were the remotest and closest to the slab bottom surface. For example, the temperature difference for flush endplate was calculated using temperatures at points No.46 and 47 in Figure 3, and that for the larger fin plate was calculated using temperatures at points No.22 and 52 in Figure 4.

3 ANALYSIS OF TEST RESULTS

3.1 Numerical simulations and empirical calculations

Comprehensive FE models using ABAQUS-6.5 (HKS 2005) were built to predict temperatures of the test connection specimens. In the ABAQUS FE models, 8-noded heat transfer solid elements were used. The maximum element dimension was less than 20 mm for the steel members and less than 15 mm for the composite slab. The actual bolts and nuts were not modelled exactly but replaced by cylinder-shaped blocks (radius 17 mm and height 15 mm) that were tied

to the surfaces of connected members. Also the welds were not modelled but the welded member surfaces were tied together to allow heat transfer between the two connected members, but the mass of the weld was ignored. For boundary conditions, it was assumed that the bottom surface (except the part in contact with the beam flanges) of the composite slab and all the steelwork surfaces (except the top flanges in contact with the concrete slab) were exposed to a uniform fire temperature field. The top surface of the concrete slab was exposed to the ambient air environment. The fire temperature used in the ABAQUS simulations was the average furnace temperature recorded by the six monitoring thermocouples. The thermal properties of steel and concrete were taken from EN 1993-1-2 (2005) and EN 1994-1-2 (2005). The water content of concrete was measured at 3% by weight.

Calculations were also performed using the empirical formulae provided in EN 1993-1-2 (2005) for unprotected steelwork. It is possible to use this to account for radiation heat exchange between the cooler slab and the joint components.

3.2 Determination of key parameters

In order to accurately calculate the steel temperatures, accurate data of boundary condition appropriate to the fire test furnace has to be determined. The two boundary condition quantities are convective heat transfer coefficient "h" and resultant emmissivity ε. To obtain the necessary data, temperatures of the beam cross-section remote from the joint were predicted by using different sets of boundary condition data. The beam cross-section remote from the joint was selected for the reason that calculation methods for plain steel sections are well established. Figure 6 presents comparisons of temperatures for two sets of boundary condition data. It is clear that the predicted and measured beam temperatures are in excellent agreement by using a resultant emmissivity of 0.5 and convective heat transfer coefficient of 25 W/m^2K. It is also clear that the unprotected steel temperature calculation equation in EN 1993-1-2 (2005) gives excellent results. For calculations using EN 1993-1-2(2005), the section factors of the steel beam web and flanges were calculated separately, being 333 and 205 m^{-1} respectively.

In ABAQUS simulations, it was extremely difficult to introduce thermal radiation exchange between the concrete slab and steel connection components to account for the recorded vertical temperature distributions indicated in Figure 3 to Figure 5. For example, figure 7 shows a set of ABAQUS prediction results, which are almost identical in the same connection component. What figure 7 does indicate is that one single section factor may be used to represent connection components in the same region, which is the basis of the connection temperature calculation method in EN

537

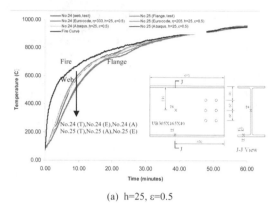

(a) h=25, ε=0.5

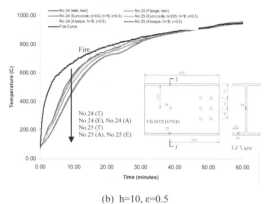

(b) h=10, ε=0.5

Figure 6. Temperatures in beam cross-section remote from joint.

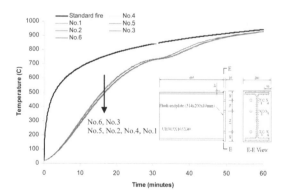

Figure 7. Temperatures in joint component predicted by ABAQUS.

1993-1-2 (2005). By comparing the prediction results from ABAQUS and EN 1993-1-2(2005) for the same boundary conditions in both models, it is possible to derive appropriate section factors for different connection regions for different types of connections. Table 1 summarises the section factors that may be used for

Table 1. Section factors for various connection components.

Component		Section factor: η (1/m)
Beam	Flange	205
	Web	333
Column	Flange	121
	Web	194
Fin on web		145
Flush endplate		88
Flexible endplate		100
Fin on flange	$150 \times 200 \times 10$ mm	138
	$100 \times 200 \times 10$ mm	145
Web cleat (depth:	$150 \times 90 \times 10$ mm	88
200 mm)	$90 \times 90 \times 10$ mm	93

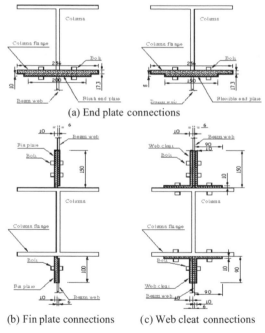

(a) End plate connections

(b) Fin plate connections   (c) Web cleat connections

Figure 8. Methods of calculating section factors for joints, shaded area indicating steel being heated, bold lines indicating exposed surfaces.

different types of connections as well as the beam and column cross-sections remote from joints. Figure 8 shows the method of calculation of section factors for connection components.

4 TEMPERATURES IN JOINT COMPONENTS

For connection components, Figure 3 to Figure 5 suggest that radiation exchange between the concrete slab and connection components would influence temperature distributions in the vertical direction. Also due to obstacles, gas flow in the joint region would be slower than around the steel beam and column cross-sections remote from the joint region. Therefore, it is expected that the convective heat transfer coefficient in the joint region would be lower than in regions remote from the joint. In order to obtain appropriate quantities to describe these two effects, calculation results using EN 1993-1-2 (2005) have been compared against test results. Joint section factors are the same as those given in Table 1. The EN 1993-1-2 method was used due to its accuracy and its ability to easily account for radiation exchange between the concrete slab and joint components. To use the EN 1993-1-2 method to account for radiation exchange, the heat flux to a location in the joint is given as:

$$h_{net\,r} = (1-\Phi)\varepsilon_r\sigma[(\theta_r + 273)^4 - (\theta_m + 273)^4] + \Phi\varepsilon_r\sigma[(\theta_c + 273)^4 - (\theta_m + 273)^4]$$

where θ_m is the steel temperature (°C), Φ the configuration factor, ε_r the resultant emmissivity, σ the Stephan Boltzmann constant, $5.67 \times 10{-}8\,(W/m^2K^4)$, θ_r the effective radiation temperature of the fire environment (°C) which may be taken as the measured gas temperature and θ_c the temperature of the fire exposed concrete surface (°C).

4.1 Temperatures in fin plates welded to column web

The same connection was used in all four tests, which recorded very similar temperatures.

Figure 9 shows a typical comparison between recorded and predicted temperatures in the fin plate. It can be seen that by using the different configuration factors relevant to the different joint temperature monitoring locations, the differences in the recorded temperatures in these monitoring locations may be reproduced. Nevertheless, for the fin plate connection, the recorded temperature differences are small so it may be acceptable to predict the connection temperatures without taking into account the radiation exchange between the slab and the connection. Figure 9 also indicates that it is more appropriate to use a convective heat transfer coefficient of $10W/m^2K$ than $25W/m^2K$.

4.2 Temperatures in joint components connected to column flanges

Similar calculations as in the previous section were performed for joint components connected to column flanges and the conclusions are similar. Figure 10 presents a typical example of the comparison results.

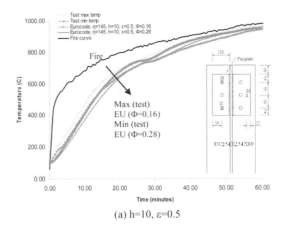

(a) h=10, ε=0.5

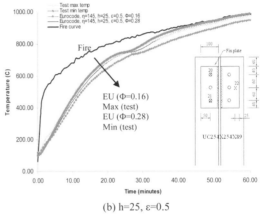

(b) h=25, ε=0.5

Figure 9. Comparison between measured and predicted temperatures in fin plate welded to the column web.

The large differences in measured temperatures were very accurately predicted by using different appropriate configuration factors for the different temperature measuring locations. However, it must be pointed out that the average measured slab bottom service temperatures were used as input data in the predictions.

5 CONCLUSIONS

This paper has presented measured and predicted steel temperature results of four fire tests on unloaded steel beam to column connections with a concrete slab on top. The following conclusions may be drawn:

- The finite element package ABAQUS and the empirical method provided in EN 1993-1-2 (2005) can be used to predict temperatures of unprotected connections.
- For each connection component, a single section factor may be used in the EN 1993-1-2 equation to calculate the connection component temperature.

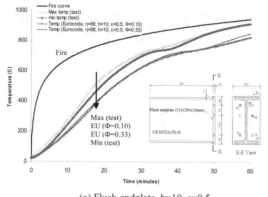

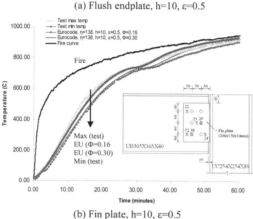

(a) Flush endplate, h=10, ε=0.5

(b) Fin plate, h=10, ε=0.5

Figure 10. Comparison between measured and predicted temperatures in joint components connected to column flanges.

For the four types of connections tested in this study, Figure 8 shows how the single section factors may be calculated.

• The convective heat transfer coefficient for beam and column cross-sections remote from the connection region may be taken as $25\ W/m^2K$, which is the value recommended in EN 1991-1-2 (2002). To calculate temperatures in the joints, a convective heat transfer coefficient of $10\ W/m^2K$ may be used to account for the slow gas field in the connection region.

• Radiation exchange between the cooler concrete slab surface and connection components may produce noticeable differences in the same connection component in the direction perpendicular to the concrete slab. Although this difference may be adequately accounted for in EN 1993-1-2 calculations, by introducing an appropriate configuration factors, the calculations will be more involved than not considering this radiation exchange effect and will depend on availability of temperature data for the fire exposed concrete surface. Not accounting for this radiation exchange effect will lead to overestimation of connection component temperatures, which will be on the safe side when evaluating connection structural performance, which may be acceptable in practical design calculations.

ACKNOWLEDGEMENTS

This research is funded by a research grant from the UK's Engineering and Physical Science Research Council (EP/C003004/1). The authors would like to thank Mr. Jim Gorst and Mr. Jim Gee for assistance with the fire tests.

REFERENCES

CEN (2002), BS EN 1991-1-2:2002, Eurocode 1: Actions on structures – Part 1–2: General actions – Actions on structures exposed to fire, British Standards Institution, London.

CEN (2005a), BS EN 1993-1-2:2005, Eurocode 3: Design of steel structures – Part 1–2: General rules – structural fire design, British Standards Institution, London.

CEN (2005b), BS EN 1994-1-2:2005, Eurocode 4: Design of composite steel and concrete structures – Part 1–2: General rules – Structural fire design, British Standards Institution, London.

HKS (2005) ABAQUS 6.5/Standard User's Manuals.

Steel and Composite Structures – Wang & Choi (eds)
© 2007 Taylor & Francis Group, London, ISBN 978-0-415-45141-3

Experimental investigation of the behaviour of fin plate connections in fire

Hongxia Yu, I.W. Burgess & J.B. Davison
Department of Civil and Structural Engineering, The University of Sheffield, Sheffield, UK

R.J. Plank
School of Architecture, The University of Sheffield, Sheffield, UK

ABSTRACT: Fire hazards and full-scale structural tests have indicated that steel connections could be subjected to large deformations and fractural failure in fire. This is not currently considered in design approaches because the connections are assumed to heat up more slowly than the structural frame members and therefore retain more relative strength. A project at the Universities of Sheffield and Manchester is currently investigating the robustness of common types of steel connections when subjected to fire. This paper reports part of the test results on fin plate connections. The test results indicate that bolts are vulnerable to shear fracture and failure is controlled by bolt shear rather than plate bearing. Furthermore, the rate of reduction of load resistance with temperature is greater than predicted by the steel strength reduction factors in Eurocode 3 Part 1.2

1   INTRODUCTION

Current design codes generally consider that steel connections will be heated more slowly than beams or columns in fire situations, and are therefore less likely to be the critical components in fire safety design. However, evidence from the collapse of the WTC buildings (FEMA, 2002) and full-scale fire tests at Cardington (Newman et al., 2004) indicates that connections may often be the weakest link in a structural frame in fire conditions. This is because at ambient temperature connections are designed to transfer shear and/or moment, whereas in fire they can be subjected to additional compressive or tensile forces due to restraint to thermal expansion or catenary action arising from large deflections. At very high temperatures, beams lose most of their bending capacity, and develop axial tensile forces which, in combination with large deflections, may support the lateral loads by second-order effects. In consequence, the connections may eventually be subjected to large rotations and significant tensile forces. Under such conditions there is a clear possibility of connection fracture, which may lead either to fire spread to upper floors, or to progressive collapse of the building.

In the past, connections have been extensively investigated to determine moment-rotation behaviour. However, the importance of connections in providing tying resistance to hold the whole structure together should not be overlooked (DSI, 2001).

Previous researches (Yu and Liew, 2005) have shown that in a fire situation connections can be subjected to significant tying force when the beams are heated and deformed to extensive deflection so that they work like cables to hold the upper floor loads. When this behaviour was proposed to be used in design to enhance the fire resistance of structural steel frames, it was implicitly assumed that the connections that hold the beams to the columns had sufficient rotation ductility and were able to transfer any catenary force up to the tensile capacity of the beam section at elevated temperatures. However, the real behaviour of steel connections under such circumstances has never been investigated. The Universities of Sheffield and Manchester are conducting a joint research programme with the aim to investigate the behaviour of steel connections at elevated temperatures, especially when the beam is in the catenary phase. The University of Sheffield is investigating the behaviour of single joints And the University of Manchester is studying the behaviour of structural sub-frames, taking into consideration the interaction between the connection and the framed members. The two researches complement each other in that the sub-frame behaviour provides information on the level of loads and deformations the connections are subjected to and understanding of the connection helps interpret the sub-frame behaviour.

The behaviour of end plate connections in fire have been successfully modeled with a component-based approach (Block, 2006) The method divides the joint

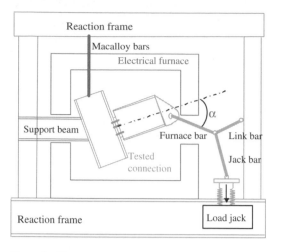

Figure 1. The test setup.

into several independently working components. As the behaviour of each component can be described by a spring, the overall behaviour of the whole joint can be calculated from an assembly of springs. It is believed that this method provides a practical solution to semi-rigid analysis of steel and composite framed structures in fire. This current research project aims to develop component-based models capable of predicting behaviour up to failure for the common types of steel connections and incorporate them into non-linear structural analysis programs.

A total of four types of connections will be studied. They are the flush endplate flexible endplate, fin plate and web cleat connection. An estimated 16 fire tests will be performed for each type of connection. This paper reports the test results to date on fin plate connections.

2 TEST SETUP

2.1 Furnace and specimen setup

The electric furnace at the University of Sheffield has an internal volume $1.0\,m^3$ and takes about two hours to heat a specimen to 700°C. This dictates that the fire test is a steady-state test, that is, the specimen is heated to a specified temperature and tested at the constant temperature. The furnace has one 300 mm diameter hole on each side. During the designing of the tests, several factors were considered: (i) connections should be full size and represent practical designs; (ii) the connection should be loaded by a combination of shear force and tensile force; (iii) the test should allow the connection to deform until fracture. The final design is shown in Figure 1. The beam-to-column connection is placed in the middle of the furnace. A support beam

of UC203 × 86 extends from the left hole into the furnace and is connected to the flange of the column. Two Φ25 mm 1030 grade Macalloy bars are used to hang the column top to the reaction frame from the top. To allow very large rotation to the beam, a complicated load system is designed to allow large deformation to the connection through the 300 mm hole. The load is applied through three Φ26.5 mm 1030 grade Macalloy bars that are all connected to a central pin. One bar goes into the furnace (hereafter referred to as the furnace bar) and is pin-connected to the beam end. The ultimate strength of the material is around $1030\,N/mm^2$ and the 0.1% proof resistance of these bars is 460 kN. One link bar is pin-connected to the reaction frame. The third bar, the jack bar, is connected to the head of the displacement-controlled load jack. When the jack moves downward, it brings the central pin downward, thus apply an inclined tensile force to the beam end through the furnace bar. The angle between the furnace bar and the axis of the beam determines the ratio of shear and tensile forces applied to the connection. This value is pre-determined. To allow free movement of the furnace bar through the furnace hole, the whole specimen is tilted by 25° in the furnace.

Except for tests at normal temperature, the support cable and the support beam as well as its connection to the column flange are wrapped in thermal blankets. The loaded end of the beam plus the end connector are also wrapped in thermal blankets. Both holes are filled with thermal blankets to prevent the heat from escaping. The connection, the column and a significant length of the beam are unprotected. A thermal test was performed before loading tests, in which 25 thermal couples were installed to monitor the temperature distribution of the connection. From the test, differences in temperature at various locations of the connection were less than 5° at a stable furnace temperature of 700°.

2.2 Measurements

Measurement of the forces is achieved with strain-gauges attached to the three load bars. At elevated temperatures, the strain-gauge readings could be affected by temperature change and possibly damaged by high temperature. A thermal test was therefore performed to record the temperatures of these three bars. The furnace was heated to 700°C and kept constant at 700°C for two hours. The temperature of the furnace bar reached 120°C, but the other two bars remained at normal temperature when a fan was provided. It was decided that at elevated temperatures, the forces in the link and jack bars should be recorded and used to determine the force in the furnace bar by force equilibrium. One test at normal temperature, in which, all three bar forces were monitored, was conducted to validate this approach. To calculate the force equilibrium, a camera

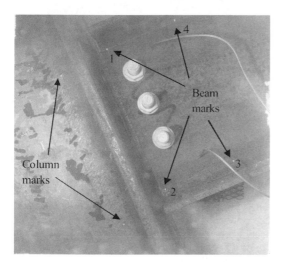

Figure 2. Camera measurement of the specimen deformation.

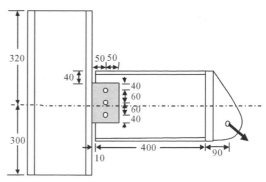

Figure 3. The geometry of the test specimen.

Table 1. List of test results.

Test No	Temperature (°C)	Initial α (Degree)	Ending α (Degree)	Force (kN)	Rotation (Degree)
1	20	53.85	32.41	145.95	8.107
2	450	51.47	41.37	70.48	6.093
3	550	53.44	42.68	34.81	6.558
4	650	53.09	44.02	17.99	6.255
5	20	33.80	34.06	185.11	7.805
6	450	39.04	33.52	84.47	6.237
7	550	40.94	31.51	37.46	7.121
8	650	40.50	30.60	19.30	7.367

was placed facing the central pin to measure change of the angles between the three bars during the whole test. Tests at normal temperature have shown that force equilibrium can be achieved with a maximum of 5% discrepancy.

Measurement of specimen deformations in fire tests is not easy because tests are generally performed in furnaces that are well enclosed and insulated. The electric furnace has an observation hole in the front 100 mm high by 200 mm wide. The specimen deformations are measured with a camera placed in front of the observation hole. To ensure high-quality digital images, extra light was provided through half of the observation hole into the furnace transparent thermal insulation was installed in front of the hole to prevent the camera from becoming over-heated. Figure 2 shows a photo of the specimen taken by the camera through the observation hole. Two marks are made on the web of the column and 4 marks are made on the web of the beam. Rotations and displacements were calculated by processing the pixels between the marks.

2.3 Specimens

To fit into the internal space of the furnace, a UC254 × 89 is used for the column and a UB305 × 165 × 40 is used for the beam. The geometry of the tested fin plate connections is shown in Figure 3. The thickness of the fin plate is 8 mm. A custom made connector was bolted to the end of the beam and the load was applied to the connector through a pin hole. The effective distance of the load to the connection is 490 mm. All bolts used are M20 Grade 8.8 and all steel used is S275.

3 TEST RESULTS

At the time of writing, eight tests have been conducted on the fin plate connection. The connection was tested subjected to two different combinations of shear and tying forces and at four temperatures. Various combinations of shear and tying forces correspond to different angles α in Figure 1. Two initial angles, 55° and 35° were chosen. During the tests, the angle α is slightly different from the chosen value depending on the assembled position of the load system and the angle changes during the loading process. The values of α at the beginning and end of the test are listed in Table 1. Also listed in Table 1 are the maximum resistance and the connection rotation at maximum resistance.

All the tests failed by shearing the bolts. At 20°C, the bolts deformed the bolt holes on the beam web significantly before shear fracture. At elevated temperatures, the bolts sheared, causing little deformation to the bolt holes. A comparison of the beam web deformation after ambient temperature test and fire test is shown in Figure 4. This appears to show that the bolts become relatively weaker compared to the connected steel plates at elevated temperatures. Reductions of the maximum connection resistance relative to their resistance at ambient temperature are shown in Figure 5 for each load angle. Figure 4 also shows the strength

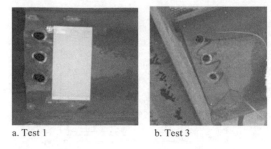

a. Test 1   b. Test 3

Figure 4.   Deformation of the beam web bolt holes.

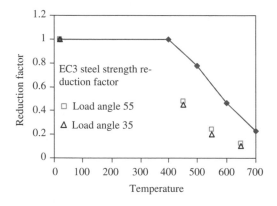

Figure 5.   Reduction of the connection resistance at elevated temperatures.

reduction curve for normal steel according to EC3: Part 1.2 (CEN, 2005a). The resistances of the connection decreased much faster than normal steel with the increase of temperature. The maximum resistance reduced to less than half at 450°C, and at 650°, the connection resistance is only 10% of the connection resistance at normal temperature. Since in all the tests, the failures were controlled by bolt shear fracture, this indicates that bolts lost their strength at elevated temperatures faster than the strcutural steel. Considering that the temperatures of steel beams may be above 700°C when they enter the catenary action, the use of fin plate connection in fire engineered designs is not recommended.

The rotation of the connection is calculated as the rotation of the beam minus the rotation of the column. Plots of the total force versus the connection rotation for each test are shown in Figure 3. Each sub-figure shows the force-rotation for each temperature at two different angles. Except for the first one, the others show similar responses for different load angles. The maximum resistance at a smaller load angle is bigger because a smaller load angle leads to a low proportion of shear force in the total force and therefore a lower bending moment generated at the bolt locations. Test 1 has a longer phase of initial slip because it was a pilot

test. The specimen was tested up to approximately half its capacity, unloaded and re-tested. Before re-testing the specimen had about 1 mm bearing deformation to the bolt holes on the beam web.

Special attention was given to the assembly of each specimen so that the bolts were in the middle of the bolt hole and the beam was initially perpendicular to the column. Then the bolts were hand-tightened. The long flattened initial phase in the response curves is probably to overcome the friction forces generated between the fin plate and beam web. The connection starts to pick up strength when the bolts are in contact with the bolt holes. The resistances are further enhanced when the bottom flange of the beam touches the column flange. Fracture of the top bolt at the peak force happened suddenly at ambient temperature and was accompanied by a loud noise. It can be seen from Figure 6a that each bolt failure is followed by rapid drop of the curve. When the load angle was 35, the test was terminated after one bolt failure. When the load angle was 55, two bolts were sheared; the bottom bolt underwent large rotation at a constant load of around 50 kN. At higher temperatures, the bolt became softer and bolt fracture was not distinguishable. Take the load angle 35 at 550°C as an example, it was observed after the test that two upper bolts were completely sheared and the third bolt was subjected to significant shear deformation as shown in Figure 7. The gradual decrease of the resistance at the last phase of the force-rotation curve must accompany the shearing deformation of the bottom bolt.

4   INVESTIGATION OF THE FAILURE MECHANISM

The tested specimens were designed according to UK design guide (SCI & BCSA, 2002). Bolt shear fracture is an unfavorable failure due to its lack of ductility. It is implicitly assumed to be avoided by limiting the ratio of the bearing plate thickness to the nominal bolt diameter to ≤ 0.5 for S275 steel. However, bolt shear failure was still widely observed in this series of test. If the capacity of this connection is calculated according to EC3: Part 1.8 (CEN, 2005b), the maximum bearing resistance of the beam web is 83.6 kN and the maximum shear resistance of single bolt to be 107.4 kN suggesting plate bearing to be the governing failure mode in the connection. Reasons for the bolt shear failure are investigated below and its consequences are discussed.

4.1   Double shear test of the bolt

One double shear test was performed on the bolt used in the connection tests to find out its shear resistance at normal temperature. The test arrangement is shown

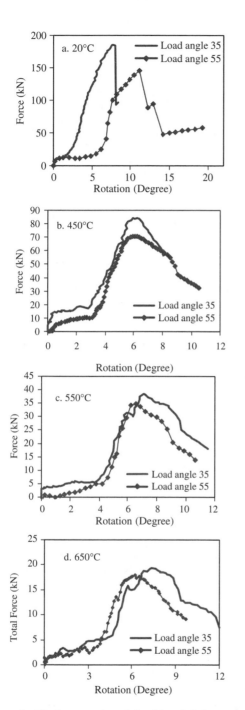

Figure 6. The force-rotation relationships of all the tests.

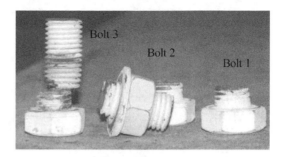

Figure 7. Deformed shape of the bolts.

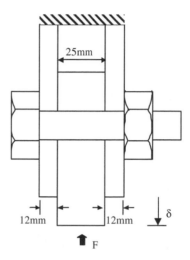

Figure 8. Arrangement for double shear bolt test.

including the bolt deformation and the plate deformation. The plates used were not strong enough to resist damage completely and deformations of around 2 mm were observed around the bolt holes from all three plates. Bending of the bolt may have contributed to the overall deformation as well. A measurement to the deformed shape of the bolt showed that the pure shear deformation is about 2.5 mm.

The maximum shear resistance of the bolt is about 155 kN which is well above the design value of 107.4 kN given by EC3: Part 1.8. According to the test result, the bolts used in the tests are not underperforming or substandard.

A simple analysis was performed to determine the maximum forces working on the single bolts in Test 1 and Test 5. At the peak point, the maximum force for test 1 is 145.95 kN and the load angle is 44.13°. The tensile force, shear force and the bending moment on the connection are respectively:

in Figure 8 and the test result is shown in Figure 9. The measured force is the total force for two shear planes, which is halved to give the shear resistance of the bolt. The measured displacement is the total displacement

$$
\begin{aligned}
V: & \quad 145.95 \times \sin\left(44.13^\circ\right) = 101.62 \text{ kN} \\
T: & \quad 145.95 \times \sin\left(44.13^\circ\right) = 104.76 \text{ kN} \\
M: & \quad 145.95 \times \sin\left(44.13^\circ\right) \times 0.45 = 45.73 \text{ kN.m}
\end{aligned} \tag{1}
$$

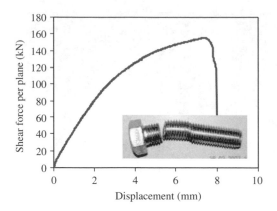

Figure 9. Response of the bolt in shear.

It can be seen from Figure 3 that every bolt has reached its plastic resistance. Using plastic design, the plastic modulus of the bolt group is

$$S = 65 + 125 + 185 = 375 \, \text{mm}^3 \qquad (2)$$

Assuming tensile force and shear force are uniformly distributed to three bolts, the force taken by each bolt should be

$$F = \sqrt{\left(\frac{101.62}{3}\right)^2 + \left(\frac{104.76}{3} + \frac{45.73}{375}\right)^2} = 166 \, \text{kN} \qquad (3)$$

The maximum force for test 5 is 185.11 and the load angle is 25.26°. Using the same method, the force taken by each bolt is calculated to be 155.5 kN. They are both very close to the double shear test results. Therefore, shear fracture of the bolts is reasonable. The behaviour of the connection is different from initial expectation or design code suggestions probably because the bearing resistance of the steel plates is extremely high. It was suggested by Owens and Cheal (1989) that "real bearing failure will not occur until about 3x ultimate plate tensile strength".

4.2 Influence of bolt shear

Both the steel sections and the bolts used in the tests are representative of the common industrial practice. Therefore, the possibility of bolt shear is very high in real steel constructions, especially in fire situation. This has significant influences to the design practice. At ambient temperature, although it has been proven in the previous section that the connection has achieved its plastic resistance, this failure mode still affects the ductility. At high temperatures, as bolts become weaker compared to the steel plates, the upper bolt may have failed before all three bolts reach their plastic resistance so that plastic design of the connection is no longer valid.

4.3 Future work

This series of connections has shown that fin plate connections have little resistances at elevated temperatures and are unlikely to be suitable to resist the tying force in catenary action. Future studies will examine if simple adjustments can be made to the connection, such as enlarging the bolt size or increasing the bolt grade, to change the failure mode to beam web bearing and significantly enhance the connection ductility and resistance. The shear resistance of the bolt will also be tested for the complete temperature range.

5 CONCLUSIONS

Utilization of catenary action to enhance the fire resistance of structural steel beams call for investigations into the capacity of steel connections to resist the tying forces required to support the beams. This paper reports eight test results on typical fin plate connections subjected to a combination of shear and tension forces. The test results show that the resistances of fin plate connections are significantly affected by temperatures. At 650°C, the residual resistance is only around 10% of its ambient temperature resistance.

Bolt shear failure was observed in all tests. At ambient temperature, bolt shear occurred after significant bearing deformations. However, at elevated temperatures, little deformation was observed around the bolt holes. This indicates that the bolt strength reduced faster than strcutural strength steel with the increase of temperature.

ACKNOWLEDGEMENT

The authors gratefully acknowledge the support of the Engineering and Physics Sciences Research Council of Great Britain. Contributions of the steel sections by Corus and fabrication of the steel sections by Billington Structures Ltd are also acknowledged.

REFERENCES

FEMA. 2002. *World Trade Center, Building Performance Study: Data Collection, Preliminary Observations, and Recommendations.* Federal Emergency Management Agency, Washington, DC and New York, USA.

Newman G.M. Robinson J.T. Bailey C.G. *Fire Safety Design: A New Approach to Multi-Storey Steel-Framed Buildings.* The Steel Construction Institute, 2004.

BSI. 2001. *BS5950 Structural Use of Steelwork in Building-Part 1: Code of practice for design-Rolled and welded sections.* British Standard Institution.

Yu H.X. and Liew J.Y. Richard. 2005. *Considering Catenary Action in Designing End-restrained Steel Beams in Fire.*

Advances in Structural Engineering, V8, No 3. pp309–324.

Block Florian. 2006. *Development of a component-based finite element for steel beam-to-column connections at elevated temperatures*. Ph.D thesis, University of Sheffield.

European Committee for Standardization (CEN). 2005a. *BS EN 1993-1-2, Eurocode 3: design of steel structures, Part 1.2:, General rules- structural fire design*. British Standards Institution, UK.

European Committee for Standardization (CEN). 2005b. *BS EN 1993-1-8, Eurocode 3: design of steel structures, Part 1.8: Design of Joints*. British Standards Institution, UK.

SCI & BCSA. 2002. *Joints in steel connection, Simple connections*. The Steel Construction Institute and The British Constructional Steelwork Association Limited, UK.

Owens G.W. Cheal B.D. 1989. Structural Steelwork Connections, Butterworth & Co. Ltd. pp54.

Steel and Composite Structures – Wang & Choi (eds)
© 2007 Taylor & Francis Group, London, ISBN 978-0-415-45141-3

Development of a component model approach to fin-plate connections in fire

M. Sarraj, I.W. Burgess & J.B. Davison
University of Sheffield, Department of Civil & Structural Engineering, Sheffield, UK

R.J. Plank
University of Sheffield, School of Architectural Studies, Sheffield, UK

ABSTRACT: The properties and behaviour of semi-rigid joints in both steel and composite structures have been widely studied for some time. The focus has recently been on improving the design of structural frames by taking advantage of realistic connection moment-rotation response. This has necessitated the development of an effective and practicable methodology to describe steel connection behaviour, despite its inherent complexity. Over a number of years the Component Method has been developed to describe the moment-rotation characteristics of end-plate connections, and the method is now included in Eurocode 3. To date, most of the research conducted on steel connections using the Component Method has focused on relatively stiff and strong connections – flush end-plates and extended end-plates. The modelling of more flexible ("pinned") connections using the Component Method has not received much attention, since the benefits arising from consideration of their behaviour in overall frame response are usually modest. However, in fire conditions connections are subject to complex force combinations of moment and tying forces, as well as vertical shear forces, and the real behaviour even of nominally pinned connections can have a significant effect on the overall response of the frame. The Component Method offers an opportunity to model the often complicated behaviour of joints in fire. Steel fin-plate shear connections, which are assumed to act as pins in normal service conditions, are economic to fabricate and simple to use in erection. An intensive investigation has been conducted to develop a representation of this connection type via a simplified component model, enabling prediction of the connection response at both ambient and elevated temperatures.

The three main components of a fin-plate connection have been identified as plate bearing, bolt shearing and web-to-plate friction. These components have been described in detail via intensive parametric FE analyses, leading to a simplified component model of a fin-plate connection which has been evaluated against FE models of complete fin-plate joints. The component method presented in this paper can be incorporated into frame analysis for fire conditions.

1   GENERAL INTRODUCTION

Inclusion of the real behaviour of connections in structural design leads to reduction in both beam bending moments and beam deflections, and may result in an overall improvement in the design efficiency. However, in order to integrate the semi-rigid connection approach in the design of a steel structure, it is necessary to identify the joint characteristics, (rotational stiffness, moment resistance, shear capacity, ductility). These joint characteristics can be established by experimental testing or mathematical modelling according to the geometrical and mechanical properties of the joint. Full-scale experimental testing is naturally the most reliable method to describe the

rotational behaviour of structural joints. However, this is time consuming, expensive, and can not be considered as a normal design tool. Existing joint test data is limited in scope and tends to be restricted to certain connection details; therefore these results cannot be extended to different joint configurations. However, many mathematical modelling approaches[1] can be applied to represent the connection behaviour; these include:

i) Curve fitting to test results.
ii) Numerical models.
iii) Component models.

Due to the nonlinear interaction of joint loads and the large number of possible variables in the design of a bolted joint, the component method is believed to be the most efficient procedure to predict the complex behaviour of bolted joints with reasonable accuracy. The Component Method was initially developed by Tschemmernegg et al.[2,3] and later introduced into Eurocode 3 Part 1.8[4].

In the last decade the component method approach has gradually gained great popularity, simply because of the advantages it offers over the other analytical methods. For instance, it may apply to any joint type, different connection configurations and different loading conditions, provided that the description of the load and deflection of each component is properly characterized. The method consists of three basic steps[5]:

i) Identification of the active components for a given joint.
ii) Characterization of the load-deflection response of each individual component.
iii) Assembly of a mechanical model of the joint which is made up of an assembly of extensional springs and rigid links.

The active joint components for a fin-plate connection with three bolt rows are: the fin-plate in bearing, the beam web in bearing, bolts in single shear, the beam web-to-fin-plate interface in friction and the weld in tension. Also, if a large rotation occurs, then the beam flange in compression is another component to be considered.

2 PLATE BEARING COMPONENT

Through an intensive FE parametric study, factors that influence the plate bearing behaviour (Figure 1) were investigated. The resulting graphs were used to generate general expressions to describe the force-deflection behaviour associated with plate bearing. After several cycles of curve fitting procedures using non-linear equations, it was found that the Richard equation[6,7] (Eqn. 1) gave the best fit to the plate bearing force-deflection behaviour (Figure 2).

The process distinguishes between two cases of bearing. The first is bearing on a narrow edge clearance ($e_2 \leq 2d_b$) in tension, and the second is bearing onto a wide edge distance ($e_2 \geq 3\ d_b$), in compression. The temperature effect was considered by a applying the reduction factors of Eurocode 3 part 1–2[8] to material ultimate stress (f_u) and varying the Ω value (Eqn. 5) with the plate temperature.

$$\frac{F}{F_{b,rd}} = \frac{\psi\ \overline{\Delta}}{\left(1 + \overline{\Delta}^{0.5}\right)^2} - \Phi\ \overline{\Delta} \quad (1)$$

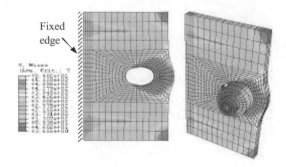

Figure 1. Von Mises stress contour of FE model for bolt bearing onto different plate thicknesses at 20°C.

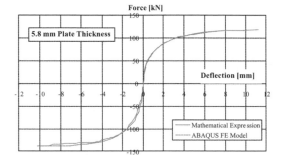

Figure 2. Force-deflection comparison between the proposed expression and ABAQUS FEM.

$$\overline{\Delta} = \Delta\ \beta\ K_i \big/ F_{b,rd} \quad (2)$$

$$F_{b,Rd} = \frac{e_2}{d_b} \times f_u \times d_b \times t \quad (3)$$

where F is the applied force [N]; $F_{b,Rd}$ is the nominal plate strength [N]; $\overline{\Delta}$ is the normalized deformation ($\overline{\Delta} = \Delta\beta\ K_i/F_b$); Δ is the hole elongation [mm]; β is a steel correction factor ($\beta = 30\%/\%$Elongation), for typical steels taken as unity; and K_i is the initial stiffness [N/mm].

$$K_i = \cfrac{1}{\cfrac{1}{K_{br}} + \cfrac{1}{K_b} + \cfrac{1}{K_V}} \quad (4)$$

Bearing stiffness $\quad K_{br} = \Omega\ t\ F_y (d_b/25.4)^{0.8} \quad (5)$

Bending stiffness $\quad K_b = 32\ E\ t\ (e_2/d_b - 0.5)^3 \quad (6)$

Shearing stiffness $\quad K_V = 6.67\ G\ t\ (e_2/d_b - 0.5) \quad (7)$

Table 1. Curve fitting parameters corresponding to different temperatures for all bolt sizes (tension).

$f_{u,\theta}$	$f_{u,\theta}$	T [°C]	Ω	ψ	Φ
$1.0 \times f_u$	445	20	145	2.1	0.012
$1.25 \times f_y$	343.75	100	180	2	0.008
$1.25 \times f_y$	343.75	200	180	2	0.008
$1.25 \times f_y$	343.75	300	180	2	0.008
$1.0 \times f_y$	275	400	170	2	0.008
$0.78 \times f_y$	214.5	500	130	2	0.008
$0.47 \times f_y$	129.25	600	80	2	0.008
$0.23 \times f_y$	63.25	700	45	2	0.008
$0.11 \times f_y$	30.25	800	20	1.8	0.008

Table 2. Curve fitting parameters corresponding to different temperatures for all bolt sizes (compression).

$f_{u,\theta}$	$f_{u,\theta}$	T [°C]	Ω	ψ	Φ
$1.0 \times f_u$	445	20	250	1.7	0.008
$1.25 \times f_y$	343.75	100	220	1.7	0.008
$1.25 \times f_y$	343.75	200	220	1.7	0.008
$1.25 \times f_y$	343.75	300	220	1.7	0.008
$1.0 \times f_y$	275	400	200	1.7	0.008
$0.78 \times f_y$	214.5	500	170	1.7	0.008
$0.47 \times f_y$	129.25	600	110	1.7	0.008
$0.23 \times f_y$	63.25	700	40	1.7	0.007
$0.11 \times f_y$	30.25	800	20	1.7	0.007

2.1 Bearing in tension on narrow edge distance

The key parameters for tension given a narrow end distance ($e_2 \leq 2d_b$) for all sizes of bolt are given in Table 1.

2.2 Bearing in compression on large end distance

$$F_{b, Rd} = 0.92 \times \frac{e_2}{d_b} \times f_u \times d_b \times t \qquad (8)$$

where the edge distance $e_2 \geq 3 \times d_b$.

3 SINGLE-SHEAR BOLT COMPONENT

There is very little literature reporting investigations of bolts in single shear, at elevated temperatures or even at ambient temperature, although Eurocode 3 Part-1.8[4], has presented an equation for bolt shear strength (Eqn. 9), and another equation (the basis of which is unclear) for the initial stiffness of bolt shearing at ambient temperature (Eqn. 10). This stiffness was actually proposed in order to calculate connection rotational stiffness (Eqn. 11). The term (z) in this

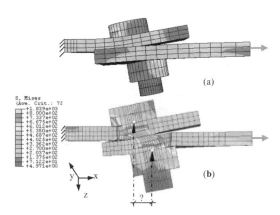

Figure 3. FE model of lap joint composed of two 10 mm thickness plates and M20 8.8 bolts at 20°C.

equation represents the lever arm between bolts in tension and the others in compression within a moment connection.

$$F_{v, rd} = 0.6 \times f_{ub} \times A \qquad (9)$$

$$K_{11} = \frac{16 \, d_b^2 \, f_u \, n_b}{E \, M_{16}} \qquad (10)$$

$$S_{ij} = \frac{E \, z^2}{K_{11}} \qquad (11)$$

In this research programme the bolt shearing component was investigated at both ambient and elevated temperatures. In order to conduct the investigation, FE models were created and evaluated[9] of two lapped S275 steel plates, each one having a thickness of $0.5d_b$ (Figure 3). One of these plates was clamped on its edge and the other plate was constrained to move axially by applying displacement boundary conditions to its far edge. Analyses were conducted on bolt shearing, varying the diameters of the 8.8 high strength bolts in the range 12, 16, 20 and 24 mm. The proposed FE model for an M20 bolt was studied at elevated temperatures ranging from 100°C to 900°C. The resultant shear force is plotted against bolt relative deflection for each FE model in Figure 4. It is important to highlight here that the bolt deflections Δ (Figure 3) were taken at certain nodes on the bolt centre line and relative to the opposite node during the entire bolt shearing analysis.

In order to represent the bolt shear load-deflection data at elevated temperature in a form suitable for incorporation in component modelling, the load–deflection–temperature data obtained from the FE

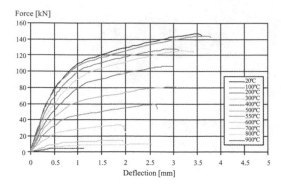

Figure 4. Load-deflection curves of (M20) 8.8 bolts shearing under varied elevated temperatures from FE analyses.

modelling were fitted using a modified Ramberg–Osgood expression [10,11] of the following form:

$$\Delta = \frac{F}{k_{v,b}} + \eta \left(\frac{F}{F_{v,Rd}} \right)^n \tag{12}$$

in which

Δ is the relative bolt deflection [mm];
F is the corresponding level of shear force [N];
$K_{v,b}$ is the temperature-dependent bolt shearing stiffness [N/mm];
$F_{v,Rd}$ is the temperature-dependent bolt shearing strength [N];

$$F_{v,Rd} = R_{f,v,b} \times f_{u,b} \times A \tag{13}$$

$R_{f,v,b}$ is the strength reduction factor for bolts in shear;
$n(=6)$ is a parameter defining the curve sharpness,
η is a temperature-dependent parameter for curve fitting.

With regard to the bolt shear stiffness, this can be represented by the following temperature-dependent expressions:

$$K_v = \frac{k \, G A}{d_b} \tag{14}$$

G is the shear modulus $G = \dfrac{E_\theta}{2\,(1+\nu)}$ (15)

In addition, E_θ is the temperature-dependent Young's modulus, k is a shear correction factor introduced to account for the error in the shear strain energy caused by assuming a constant shear strain through the bolt section, as opposed to the classical parabolic distribution. The shear correction factor depends upon the cross-sectional shape and material properties[12,13]. It was found that a value of $k = 0.15$ is suitable for the bolt shearing analyses.

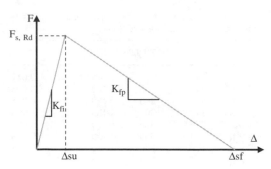

Figure 5. Simplified friction load-deflection description and the corresponding parameters.

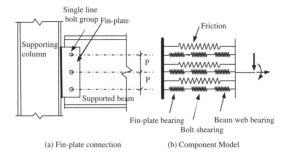

(a) Fin-plate connection   (b) Component Model

Figure 6. Fin-plate component model.

4 FRICTION COMPONENT

A surface friction component was investigated through analysing two FE model lap joints, one with a friction coefficient of 0.25 between the surfaces, and the other model with an almost zcro friction coefficient value. The second force-deflection plot was subtracted from the first to produce the final graph in Figure 5.

This graph represents the friction behaviour history throughout the analysis of a lap joint model. To generalize this graph as a mathematical expression, firstly it has been simplified into two straight lines in Figure 5. The five main parameters describing the force-deflection relationship were then formulated into an appropriate expression.

5 COMPONENT MODEL ASSEMBLY

It has been shown through extensive analysis in this research that a fin-plate connection under tying force can be represented by a series of lap joints attached to each other in parallel. Similarly, the component-based mechanical model of the complete 3-bolt fin-plate connection presented in Figure 6(a) has been assembled as a series of lap joint component models as shown in Figure 6(b).

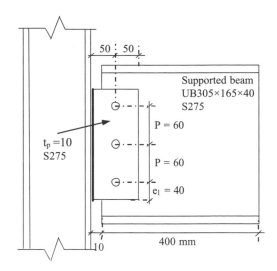

Figure 7. FEM geometric details.

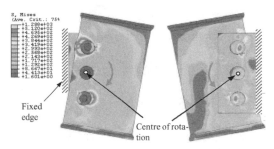

Fixed edge

Centre of rotation

Figure 8. 3-bolt FE model under pure rotation.

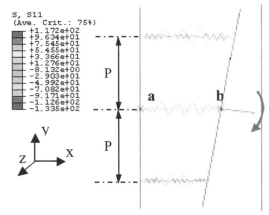

Figure 9. 3-bolt component model under pure rotation.

Each spring group contains four springs; three connected in series (plate bearing, bolt shearing and web bearing) and a friction spring connected in parallel with the other three springs. This model has two basic assumptions. Firstly, the weld forms a strong link between the column flange and the fin-plate, and does not tear under any fire conditions. This is usually a reasonable assumption, as demonstrated by tests at Cardington [14] and the Czech Republic. Secondly, beam end rotation are not large before the beam bottom flange comes into contact with the column flange.

6 COMPONENT MODEL EVALUATIONS

The proposed component model for fin-plate connections was evaluated through intensive FE analyses by which the shear, tying and rotation response were examined. The evaluation for rotation at ambient temperature and tying force resistance at elevated temperature was only included herein. A three bolt fin-plate connection shown in Figure 7 was chosen to be analysed by a FE model and the corresponding component model under pure rotation at ambient temperature.

The centre of rotation in the FE model (Figure 8) was chosen at a node between the top and bottom bolts and between the fin-plate and the beam web.

The corresponding component model (Figure 9) consisted of two rigid bars with a reference node each (a, b). Node (a) was clamped in all degrees of freedom, while node (b) was restrained in the z and y directions but permitted to rotate about the z-axis. Three series of spring groups link rigid bars separated by the pitch distance (P). The comparison of moment-rotation behaviour for the FEM and the component

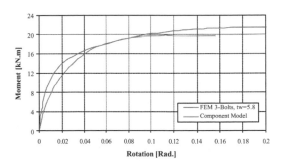

Figure 10. Moment-rotation comparison for the 3-bolt FEM and the equivalent component model.

models demonstrates good agreement between them (Figure 10).

The proposed component model for fin-plate connections was also evaluated at elevated temperatures via FE modelling. In this paper only the evaluation of tying resistance is included.

In order to investigate the component response for tying force at elevated temperatures the connection detail proposed in Figure 7 and its FE model were analysed under a 45° tying force. The corresponding

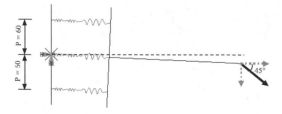

Figure 11. Component model under 45° tying force at 550°C–750°C.

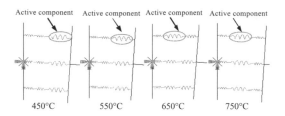

Figure 12. Comparison of FEM and the Component model under 45° tying force at 450°C–750°C.

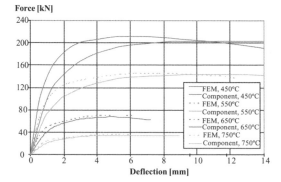

Figure 13. Fin-plate component model.

component model (Figure 11) was applied following the component model principle. The component model's capability to predict response to tying force at elevated temperature was assessed through a tying force inclined at 45° applied at temperatures of 450°C, 550°C, 650°C and 750°C.

The load-deflection responses for both FE and the component model (Figure 12) demonstrate good correlation. The component model shows a consistently weaker response than the FE model which makes it safer for design purposes.

Observation of the component model at each elevated temperature (Figure 13) confirms the weakest component in the model, which is the beam web bearing up to 550°C and the bolt shearing at temperatures above this.

7 CONCLUSION

In this paper an effective computational connection model for fin-plate joints based on component principles has been proposed for use in global structural analysis of steel frames at ambient and elevated temperature. The fin-plate shear connection was successfully modelled via a component method approach to represent, with reasonable accuracy, the actual non-linear connection behaviour. Three main fin-plate connection components (plate bearing, bolt single shear and slippage) were identified. Each component was investigated in detail via intensive parametric FE studies. Furthermore, the load-deflection characteristics for each individual component were described and generalized via mathematical curve fitting procedures. Multi-linear, temperature-dependent elastic-plastic expressions have been used to describe the bolt single shear and plate bearing components. In addition, several important conclusions can be drawn from the component simulations performed on this connection type:

- Fin-plate connections can be modelled by the component method approach.
- A bearing component has been introduced and justified. Also the bolt single shear component has been developed.
- The component model has the capacity to predict the vertical shearing, horizontal tying and rotation capacity of fin-plate connections with reasonable accuracy.
- The Eurocode 3 equation for calculating the bolt stiffness in single shear does not appear to have appropriate justification.
- The component model identifies clearly the weakest component under the analysed actions and temperatures.

REFERENCES

1. Nethercot, D. A., and Zandonini R. "Methods of prediction of joint behaviour: beam-to-column connections", Chapter 2 in Structural connections, stability and strength (Ed.: R. Narayanan). Elsevier Applied Science, London, UK; 23–62, (1989).
2. Tschemmernegg, F., Tautschnig, A., Klein, H., Braun, Ch. and Humer, Ch., "Zur Nachgiebigkeit von Rahmenknoten (Semi-rigid joints of frame structures)", Stahlbau **56**, 299–306, (1987).
3. Tschemmernegg F. "The Design of Structural Steel Frames under Consideration of the Nonlinear Behaviour of Joints" *J. Constructional Steel Research*, **11**, pp. 73–103, (1988).
4. European Committee for Standardization (CEN). "Eurocode 3: Design of steel structures, Part 1.8: Design of joints", BS EN 1993–1–8, British Standard Institution, London, (2005).

5. Jaspart, J. P., "General report: session on connections" *J. Constructional Steel Research*, Vol. **55**, pp. 69–89, (2000).

6. Rex, C. O., and Easterling, S. W. "Behaviour and modeling of a bolt bearing on a single plate", *Journal of Structural Engineering*, ASCE, **129**, (6), pp. 792–800, June (2003).

7. Richard, R. M., and Elsalti, M. K. (1991). "PRCONN, Moment-Rotation Curves for Partially Restrained Connections", Users manual for program developed at The Univ. of Arizona, Dept. of Civil Engineering and Engineering Mechanics, Tucson, Arizona.

8. European Committee for Standardization (CEN), "Eurocode 3: Design of steel structures, Part 1.2: General rules - Structural fire design", BS EN 1993-1-2, British Standards Institution, London, (2005).

9. Sarraj, M., Burgess, I.W., Davison, J.B. and Plank, R.J., "Finite Element Modelling of Fin Plate Steel Connections in Fire", Structures in Fire Conference, SiF'06, University of Aveiro, Portugal, 10–12 May, 2006.

10. Ramberg W., and Osgood WR., "Description of stress–strain curves by 3 parameters", Technical Report 902, National Advisory Committee for Aeronautics, (1943).

11. Jones, S. W, Kirby, P. A., and Nethercot, D. A. "The analysis of frames with semi-rigid connections a state-of-the-art report". *J. Constructional Steel Research*, **2**, pp. 2–13, (1983).

12. Hayes, M. D. "Structural analysis of a pultruded composite beam: shear stiffness determination and strength and fatigue life predictions", PhD thesis in Engineering Mechanics, Faculty of the Virginia Polytechnic Institute and State University, (2003).

13. Madabhusi-Raman, P., and J.F. Davalos, Static shear correction factor for laminated rectangular beams. *Composites Part B: Engineering*, 1996. **27** (3–4): p. 285–293.

14. Newman, G. M., Robinson, J. T. and Bailey, C. G., SCI Publication P288, "Fire Safety Design a New Approach to Multi-Storey Steel Framed Buildings", The Steel Construction Institute, pp.49, (2000).

Steel and Composite Structures – Wang & Choi (eds)
© 2007 Taylor & Francis Group, London, ISBN 978-0-415-45141-3

Experimental studies and numerical analysis of the shear behavior of fin plates to tubular columns at ambient and elevated temperatures

M.H. Jones & Y.C. Wang

School of MACE, University of Manchester, Manchester, UK

ABSTRACT: Following recent events such as the World Trade Center building collapse and the Cardington large scale structural fire research program, the fire behavior of connections has become a prominent research subject. This paper reports the results of recent experimental studies into the behavior of welded fin-plate connections to both hollow and concrete filled tubular (CFT) columns under shear and bending force. Experiments have been performed at both ambient and elevated temperatures with the aid of an electric kiln. A finite element (F.E.) model, developed using ABAQUS, is presented and validated against the experimental results in order that extensive parametric tests may be subsequently performed.

1 INTRODUCTION

Following recent events such as the World Trade Center building collapse and the Cardington large scale structural fire research program, the fire behavior of connections has become a prominent research subject. Recent research has moved away from the 'prescriptive' approach of assuming that fire protection of structures is required towards a 'performance-based' approach. This approach endeavors to account for more specific circumstances to which a structure may be exposed in fire. The European standard, Eurocode 3 for steel structures (CEN 2003), incorporates some of these computation methods for a variety of welded and bolted connections between steel I-beam and H-columns.

There is, however, a comparative lack of research into the behavior of simple welded connections to hollow or concrete-filled tubular (CFT) columns either at ambient or elevated temperatures. CFTs are increasingly used in tall, multi-storey buildings and, as well as pleasing aesthetic properties, they possess structural advantages such as allowing for a comparatively reduced column cross-sectional area and inherent fire-resistance properties due to the insulative properties of the concrete in-fill.

The Eurocode guidance does not encompass welded fin-plate connections but guidance is given as to the design shear resistance of a plate under combined bending and shear force. This guidance has been summarized in previous work by Jones and Wang (2006). The Eurocode guidance does not, however, take account of the effect of tubular column properties

upon the behavior of such a fin-plate under shear load when connection to a tubular column. The research presented in this paper is part of an attempt to supply this deficiency in the research field.

There have been several experimental investigations concerning the interaction between fin-plate and steel hollow section (SHS) under shear load at ambient temperatures. White and Fang (1966) conducted research into the behavior, including shear performance, of five types of welded connections to SHS columns. The closed nature of the column cross-section led White and Fang to conclude that welding was the only practical method for fastening the connection to the tube. The research highlighted the significance of the ratio of width of tube wall to tube thickness. It was noted that as this ratio increases, connections fastened directly to the tube wall tended to become more flexible. Simple plate connections, fillet welded to the tube wall, were observed to produce distortion of the tube wall as the plate rotated under load. When such connections were loaded directly in shear White and Fang encountered several failure modes in different specimens including local buckling of the tube, weld tearing and web crippling of the connected beam. Sherman (1995) conducted a large series of tests upon realistically loaded framing connections between SHS columns and wide flange beams under predominant shear load. These tests expanded upon those conducted by White & Fang. One limit state was identified for the column face under such load. This limit state was punching shear failure due to the attached beam end rotation when the shear tab connection was thick relative to the SHS wall. The criterion to avoid this failure mode, as described by Packer & Henderson (1997), is to ensure that the tension resistance of the shear tab under axial load (per unit length) is less than the shear resistance of the SHS wall along two planes (per unit length). Thus the thickness of the plate, t_p, is limited with respect to the tube wall thickness, t_O, by the following relationship.

$$t_p < \left(\frac{F_{uO}}{F_{yp}} \right) t_O \qquad (1)$$

To move from this prescriptive approach and gain a deeper understanding of the behavior of welded fin-plate to CFT column connections an analytical performance-based approach is required in addition to an experimental results base. More recently, analytical research has been undertaken into tensile behavior of similar connections using the yield-line method to develop design equations. In particular, Yamamoto, Inaoka & Morita (1994) developed a yield line mechanism for transverse branch plate to circular CFT connections under axial load. It was noted that concrete in-fill significantly increased the strength of such connections. A yield line analysis of such connections to rectangular hollow section (RHS) columns has also been developed by Cao, Packer & Yang (1998) and this work, together with further observations by Kostecki & Packer (2003) appears in the current CIDECT design guide for structural hollow section column connections (Kurobane, Packer, Wardenier & Yeomans 2004). Lu (1997) developed yield line mechanisms for several types of connection including I-beam to RHS columns under in-plane bending moments. The findings of this research are also in the CIDECT guide. The analyses developed by Kostecki & Packer and Lu define the connection strength in terms of a serviceability deformation limit of 1% of the column breadth, b_0, and an ultimate deformation of 3% of b_0. The actual ultimate failure load is not calculated by their analyses.

There is currently a gap in the research regarding the shear behavior of fin-plate to CFT connections, particularly the column behavior of such connections. The aim of this research program is to contribute to the understanding of the shear behavior of simple connections to CFTs by developing a mathematical method based upon extensive parametric studies. These parametric studies are based on models created using finite element modeling software with the models validated against test results. This paper presents the results of such a validation.

2 TEST PROGRAM

A series of tests has been performed in order to investigate the interaction between welded connections and both hollow and CFT columns when a shear force is present in the connection. The purpose of these tests

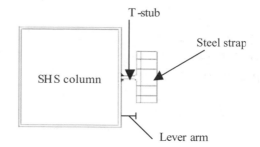

Figure 1. Cross-sectional overview of shear test specimen.

is also to provide data from which a finite element model for use in extensive parametric studies may be developed and validated against.

2.1 Test set-up

The configuration of the test rig, specimens and instrumentation is described below.

2.1.1 Test rig

The test specimens consist of T-stubs machined from UB or UC sections fillet welded at the web to the CFT column prior to concrete filling. For the specimens with fin plate thickness, t_f, of 6 mm and 10 mm, a UB section of $203 \times 102 \times 23$ and a UC section of $203 \times 203 \times 71$ respectively were used to manufacture the fin-plate connection. All column lengths were 450 mm and each connection was welded halfway along the column length and, in the case of SHS sections, breadth. The machined web of the T-stub section thus simulates a fin-plate and these connections are referred to as such subsequently in this paper. A typical specimen cross-section is depicted in Figure 1.

The test rig, pictured in Figure 2, consists of a frame housing a reaction beam. This beam is pinned at one end with the load being applied via a hydraulic jack at the opposite end. The top of the test specimen column is anchored to the reaction beam and the T-stub flange of the connection is bolted to a 50 mm thick steel hinged strap. This strap is in turn bolted to an assembly forming part of the base of the test rig which is housed on rollers. The roller base and the hinged strap allow rotation of the connection, thereby allowing shear and bending force to be applied to the connection. For the elevated temperature tests an electric kiln was installed with holes cut into the lid and base to allow the specimen to be lowered into the kiln and enclosed. The kiln position is shown as the shaded area in Fig. 2.

2.1.2 Specimen Details

The dimensions of the test specimens are detailed in Table 1 in which ambient and elevated temperature tests are prefixed with A and T respectively. The column type has been categorized by the four

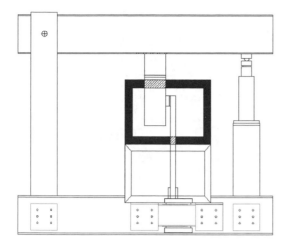

Figure 2. Test rig with specimen mounted inside kiln.

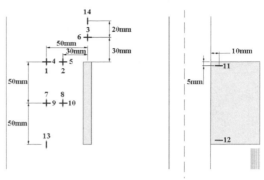

Figure 3. Strain gauge positions on column face and fin-plate.

Table 1. Test specimen dimensions.

Test No.	Column Type	Column Thickness, t_c (mm)	Fin-plate thickness, t_f (mm)	Lever arm (mm)
A1	SFS	5	6	60
A2	CFS	5	10	60
A3	CHS	5	6	60
A4	CHS	5	10	60
A5	SHS	5	6	60
T1	CFS	5	6	30
T2	CFS	12.5	6	30
T3	CFS	5	6	30
T4	CFS	12.5	6	30
T5	SFS	5	6	30
T6	SFS	5	6	30
T7	CFS	5	10	60
T8	CFS	5	10	60

fundamental variants of the columns tested. These variants are denoted as square hollow section (SHS), square filled section (SFS), circular hollow section (CHS) and circular filled section (CFS). Each circular column section tested has an external diameter of 193.7 mm. Each square section has an external cross-section of 200×200 mm.

2.1.3 *Instrumentation*

A number of strain gauges were attached to each ambient test specimen in order to provide data against which to calibrate the F.E. model. Each specimen had five bi-directional and one longitudinal gauge attached to the column face around the connection and two uni-directional gauges attached to the fin-plate perpendicular to the column face. A total of thirteen

gauges, assigned numerical positions, were attached to Tests A2 through A5. An additional uni-directional gauge was attached to specimen A1 in position 14. Figure 3 depicts the placing and numbering of the gauges.

External thermocouples were attached to each elevated temperature test specimen at central positions on the rear column face opposite the connection, on one side face of the column and on the fin-plate itself. Additional internal thermocouples at positions of 25%, 50% and 75% of the column breadth or radius from the connection were placed within specimen T3. The external thermocouples were used as a guide in order that the test load could be applied a the desired specimen temperature. Displacement gauges were also placed both at a fixed point along the reaction beam and against the roller base in order to give reference points by which to calculate the vertical displacement with load at the connection.

3 TEST RESULTS DISCUSSION

The main results of the test program are summarized in Table 2. Two principle failure modes were identified;

(a) Failure of the fin-plate due to bending and shear force
(b) Tear-out failure of the column face around the connection weld

Test specimens A1 and A5 are similar except for the presence of concrete in-fill. Tests A3 and A4 are also comparable in this way. It may be observed from the results that the presence of concrete in-fill increases the strength of the connection compared to the empty tube case. The increase in strength for the two sets of comparable cases is of the order of 30%. In addition, the failure mode observed for A1 is different to that of A5 indicating that the concrete in-fill can be influential in determining the failure mode. The

Table 2. Test program results summary.

Test No.	Column Type	Test Temperature (°C)	Failure Load (kN)	Failure Mode
A1	SFS	ambient	94.6	(a)
A2	CFS	ambient	163.7	(b)
A3	CHS	ambient	92.5	(a)
A4	CHS	ambient	127.4	(b)
A5	SHS	ambient	69.7	(b)
T1	CFS	400	130.7	(a)
T2	CFS	600	51.1	(a)
T3	CFS	600	49.6	(a)
T4	CFS	500	85.5	(a)
T5	SFS	400	133.8	(a)
T6	SFS	600	53.3	(a)
T7	CFS	600	67.4	(b)
T8	CFS	400	133.4	(b)

Figure 4. Failure mode (a) – failure of the fin-plate due to shear and bending force (Test A3).

elevated temperature tests were all performed on concrete-filled specimens and the failure load is observed to fall with increased temperature for similar specimen configurations. The test temperatures were all in the range 400°C to 600°C.

For longer lever arm lengths of 60 mm the failure mode observed was generally that of tear-out of the column face around the weld. However, for concrete-filled specimens A1 and A3 with the lower t_f/t_c case of 1.2 the failure mode was shear failure in the fin-plate. This was also exclusively the failure mode for specimens with the smaller lever arm of 30 mm. The two failure modes are depicted in Figures 4 and 5.

4 FINITE ELEMENT MODEL

ABAQUS (2004) is used to create the F.E. model which has been validated against the experimental results in

Figure 5. Failure mode (b) – tear-out of the column face around the weld (Test A2).

order that it may be used in parametric studies with confidence. Four basic models with slight variations were created in order to take account of the test specimen column variations. In addition, the fin-plate was modified to take account of the different specimen lever arm lengths. In each model the T-stub flange is included to provide a region to which the load, as a body force, may be applied.

4.1 Element type

The steel components of each model are constructed using liner shell elements. Shell elements are used because the shell thickness may be easily modified at the input file level, a characteristic which is useful in large-scale parametric studies. They are also less costly in terms of computational resources than solid elements. Modified second order linear elements, with four nodes per side, were also be used for contact analyses as convergence problems concerning mid-face nodes in contact arise with the higher order eight-noded quadratic elements. The Riks method, an iterative procedure, was used to obtain the solution.

4.2 Circular column model characteristics

The geometry of the model developed for the circular column was dictated by the complexities of solving the 'contact' equations between two surfaces which come together during an analysis. In the particular case of a CFS column, the shell elements modeling the steel column will come into contact with the solid elements representing the concrete core. During analyses conducted in which the nodes of the steel column inner surface and the core outer surface were co-incident in the model space it was noted that 'chattering', where nodes continually open and close in contact without convergence, occurred frequently, often preventing a solution from being reached.

Figure 6. F.E. model mesh and partition detail.

To solve this problem only half of the column cross-section was modeled with the cutting plane restrained by boundary conditions. The column ends were restrained with the Encastre boundary condition and the CHS edges and concrete core rear face along the cutting plane were restrained symmetrically as would be the case if the entire cross-section was modeled. The half-model was compared with full models for several CHS cases and it was found that this did not affect the results.

The half-model core for the CFS case was then offset from the column elements by a distance of 0.0001 mm in order that the nodes did not share coincident space at the beginning of the analysis step, thus preventing chattering.

Adopting a similar method to Yamamoto et al (1994) and Lu (1997) the model was partitioned around the connection, both in the fin-plate and the column face in order to simulate the weld footprint. This footprint was assigned a slightly thicker thickness than the surrounding material to simulate the physical presence of the weld. The mesh and partition detail for the CFS model are depicted in Figure 6.

4.3 Square column model characteristics

The square column models were constructed in the same way as the circular sections with the exception that one column face only, that with the connection, was modeled for the SFS case. This was to prevent chattering of the other surfaces of the column with the core. A pinned boundary condition was applied around the edges of this face. The core was modeled in full and restrained at the ends by Encastre boundary conditions.

4.3.1 Material properties

Tensile tests were performed on three coupons from each UB or UC section web and three coupons from each of the column types. The engineering stress-strain data, $\sigma_e - \varepsilon_e$, obtained was converted to true stress-strain, $\sigma_t - \varepsilon_t$, by the following relationships,

$$\sigma_t = \sigma_e(\varepsilon_e + 1) \qquad (2)$$

$$\varepsilon_t = \ln(\varepsilon_e + 1) \qquad (3)$$

Table 3. Comparison of experimental and numerical failure loads for ambient tests.

Test no.	Test failure load, T_L (kN)	F.E. failure load, A_L (kN)	T_L/A_L
A1	94.6	99.4	0.952
A2	163.7	165	0.992
A3	92.5	94.1	0.983
A4	127.4	128	0.995
A5	69.7	67.3	1.036

For each CFT model the solid core elements were assigned elastic properties equivalent to those of normal strength concrete and fracture of the encased concrete core was not considered at this stage. These simplifications were also practiced by Lu (1997).

5 F.E. MODEL VALIDATION

The F.E. models have been validated against the test results using the criteria of agreement with failure load, agreement with failure mode and agreement with experimental strain gauge readings. The results are presented separately for ambient and elevated temperature tests.

5.1 Ambient test model validation

The experimental and numerical results for failure load of the ambient tests are depicted in Table 3.

The numerical results for failure load compare excellently with the experimental results with a maximum difference of 4.8%. As a further check upon the validity of the model, strain data was extracted and compared with the data obtained from the experimental gauges.

5.2 Strain gauge validation for ambient tests

As at least thirteen strain gauges are attached to each of the five ambient tests only a selection of graphs charting strain against load for both the experimental and numerical results have been included in this paper. Examples are included for each ambient test.

The numerical results showed good agreement with the test results considering that numerical results may only be extracted from discrete nodes which may not fall in the exact position of the experimental gauge due to human error when attaching the gauges. In regions of high stress the strain was observed to vary significantly over small distances. Despite this, the numerical results agree well with the experiments across the range of gauge positions analyzed.

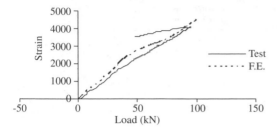

Figure 7. Load vs. strain comparison between experimental and numerical results for gauge position 1, Test A1.

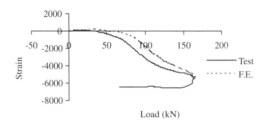

Figure 8. Load vs. strain comparison between experimental and numerical results for gauge position 3, Test A2.

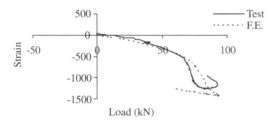

Figure 9. Load vs. strain comparison between experimental and numerical results for gauge position 1, Test A3.

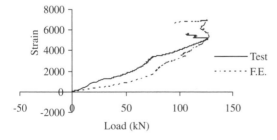

Figure 10. Load vs. strain comparison between experimental and numerical results for gauge position 6, Test A4.

5.3 Elevated temperature test model validation

When modeling steel the elevated temperature tests it is necessary to take account of the reduction in strength of the steel at the higher temperatures. Eurocode 3

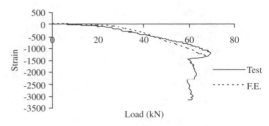

Figure 11. Load vs. strain comparison between experimental and numerical results for gauge position 9, Test A5.

Table 4. Comparison of experimental and numerical failure loads for elevated temperature tests using Eurocode reduction factors.

Test no.	Test failure load, T_L (kN)	F.E. failure load, A_L (kN)	T_L/A_L
T1	130.7	128	1.021
T2	51.1	64	0.798
T3	49.6	60.2	0.824
T4	85.5	106	0.807
T5	133.8	134	0.998
T6	53.3	62.1	0.858
T7	67.4	67.3	1.001
T8	133.4	140	0.953

contains guidance for this through a table of reduction factors applying to the proportional yield strength, the ultimate yield strength and the elastic modulus of the steel for a range of temperatures from ambient to 1100°C. Table 4 depicts the comparison between experimental and numerical results when the Eurocode reduction factors are applied to the F.E. material model.

It may be observed from Table 4 that the correlation between experimental and F.E. results when applying the Eurocode reduction factors for steel at elevated temperatures is erratic. Whilst good agreement is achieved for tests T1, T5, T7 and T8 the agreement for the other tests is inferior. Tests T2, T3, T4 and T6 all failed at loads some 80–85% of the F.E. failure load.

Alternative investigations to Eurocode into the reduction in strength of steel at elevated temperatures may be found in published literature. Li, Jiang, Yin, Chen & Li (2003) performed tests upon Chinese grade 16 Mn steel (equivalent to European grade S355J2) at elevated temperatures and reported results which indicate greater reductions in steel strength at elevated temperatures than Eurocode. Table 5 shows the comparison between Eurocode and the findings of Li et al for the reduction factors $k_{y,\theta}$ and $k_{E,\theta}$ where:

$$k_{y,\theta} = f_{y,\theta}/f_y \qquad (4)$$

Table 5. Comparison of reduction factors for steel at elevated temperatures.

Temperature (°C)	Eurocode		Li *et al* (2003)	
	$k_{y,\theta}$	$k_{E,\theta}$	$k_{y,\theta}$	$k_{E,\theta}$
20	1	1	1	1
400	1	0.7	0.85	0.89
500	0.78	0.6	0.62	0.80
600	0.47	0.31	0.38	0.64

Table 6. Comparison of experimental and numerical failure loads for elevated temperature tests using Li et al (2003) reduction factors.

Test no.	Test failure load, T_L (kN)	F.E. failure load, A_L (kN)	T_L/A_L
T1	130.7	119	1.099
T2	51.1	56.1	0.911
T3	49.6	0.7	0.979
T4	85.5	89.4	0.956
T5	133.8	129	1.038
T6	53.3	49.5	1.077
T7	67.4	56.6	1.190
T8	133.4	127	1.050

$$k_{E,\theta} = E_{,\theta}/E \qquad (5)$$

in which the subscript, θ, denotes the value of effective yield strength, f_y, or elastic modulus, E, at elevated temperature.

Table 5 shows that Li *et al* observed a reduction in the steel ultimate strength, f_y, relative to Eurocode and a relative increase in elastic modulus, E.

The results presented in Table 6 indicate that incorporating the reduction factors observed by Li *et al* into the F.E. model gives good agreement with the observed experimental failure load for seven of the eight test specimens. The agreement is more consistent than when using the Eurocode reduction factors.

6 SUMMARY

This paper has presented the results of an experimental study of the shear behavior of fin-plates welded to steel tubes at ambient and elevated temperatures. A numerical model based on the finite element method is also presented and has been validated against the test data. Based on the results of this study the following main conclusions may be drawn:

- Concrete in-fill significantly increases the strength of welded fin-plate to CFT connections.
- Concrete in-fill has been observed to influence the failure mode for otherwise similar connections.
- The numerical model developed for each ambient specimen configuration has been shown to agree well with the experimental failure load and strain profiles.
- The numerical results agree well with experimental results at elevated temperatures when applying reduction factors similar to those observed by Li *et al* (2003). The Eurocode reduction factors appear, in the main, to be less than conservative.

Based on the above observations the authors conclude that the numerical model developed against experimental results is appropriate for use in parametric studies.

ACKNOWLEDGEMENTS

This work was made possible by the sponsorship of an EPSRC industrial CASE award to the first author in which the industrial partner is Corus Plc. The authors would like to thank Mr. Andrew Orton of Corus for his interest in the research. Thanks are also due to Mr. Jim Gee and the other technicians of the Structures Division, University of Manchester, for fabricating the test specimens and assisting with the tests.

REFERENCES

Cao, J.J., Packer, J.A. & Yang, G.J. 1998. Yield line analysis of RHS connections with axial loads. *Journal of constructional steel research*, Vol. 48 (1998) pp. 1–25.
CEN. 2001. Eurocode 3: Design of steel structures, Part 1.2: General rules – structural fire design. Draft for development, European committee for standardization. Document DD ENV 1993-1-2: 2001.
CEN. 2003. Eurocode 3: Design of steel structures, Part 1.8: Design of joints. European committee for standardization. Document prEN 1993-1-8: 2003.
Hibbitt, Karlsson & Sorensen, inc. 2004. ABAQUS manuals version 6.4. Providence RI, USA.
Jones, M.H. & Wang, Y.C. 2006. Shear behavior of fin plates to tubular columns at ambient and elevated temperatures. *Proc. 11th International Symposium on Tubular Structures, Tubular Structures X1, Packer & Willibald (eds), Quebec City, Canada.* 2006. pp 425–432.
Kostecki, N. & Packer, J.A. 2003. Longitudinal plate and through plate-to-hollow structural section welded connections. *Journal of structural engineering*, American Society of Civil Engineers, 10.1061/(ASCE)0733–9445(2003)129:4 (478).
Kurobane, Y., Packer, J.A., Wardenier, J. & Yeomans, N. 2004. Design guide for structural hollow section column connections. CIDECT Design Guide 9 for Construction with Hollow Steel Sections. 2004 TÜV-Verlag.

Li, G-Q., Jiang, S-C., Yin, Y-Z., Chen, K. & Li, M-F. 2003. Experimental studies on the properties of constructional steel at elevated temperatures. *Journal of structural engineering*, American Society of Civil Engineers, 10.1061/(ASCE)0733–9445(2003)129:12 (1717).

Lu, L.H. 1997. The static strength of I-beam to rectangular hollow section column connections. PhD thesis, University of Delft, Delft University Press, Netherlands.

Packer, J.J. & Henderson. J.E. 1997. *Hollow structural section connections and trusses: a design guide, 2nd Ed.*, Canadian Institute of Steel Construction.

Sherman, D.R. 1995. Simple framing connections to HSS columns. *Proc. American Institute of Steel Construction, National steel construction conference, San Antonio, Texas, USA*. 30.1-30.16.

White, R.N. & Fang, P.J. 1966. Framing connections for square structural tubing. *Journal of the structural division*, American Society of Civil Engineers 92(ST2): 175–194.

Yamamoto, N., Inaoka, S. & Morita, K. 1994. Strength of unstiffened connection between beams and concrete filled tubular column. *Proc. 6th International Symposium on Tubular Structures, Tubular Structures VI, Grundy, Holgate & Wong (eds), Rotterdam, Netherlands*. 1994. pp 365–372.

Steel and Composite Structures – Wang & Choi (eds)
© 2007 Taylor & Francis Group, London, ISBN 978-0-415-45141-3

Experimental investigation of steel beam-to-column joints at elevated temperatures

K.H. Tan & Z.H. Qian
School of Civil and Environmental Engineering, Nanyang Technological University, Singapore

I.W. Burgess
Department of Civil and Structural Engineering, University of Sheffield, UK

ABSTRACT: This paper presents the results of an experimental investigation of typical extended end-plate steel beam-to-column joints at elevated temperatures. Six beam-to-column joints were tested. These included three tests conducted at 400°C, 550°C and 700°C, and another three tests on specimens at 700°C with different axial compression forces applied to the beams. Moment-rotation-temperature characteristics are summarized in order to investigate the degradation of the behaviour of this type of steel joint at elevated temperatures.

1 INTRODUCTION

In conventional steel structural analysis at ambient temperature, a beam-to-column joint is defined as either "pinned" (no moment resistance) or "rigid" (full moment resistance). However, the actual joint behaviour lies somewhere between these two extremes (Jaspart, 2000), hence being known as a "semi-rigid" joint. The flexural behaviour of a semi-rigid joint is commonly defined by the relationship between the end moment and the relative rotation at the joint. The gradients of the moment-rotation curves from different experimental results represent different joint stiffnesses. Until relatively recently, the provision of adequate fire protection for structural steel has tended to cost from £6 to £30 per square metre for different fire resistance ratings or, in effect, around 50% of the steel price (Alan 1996). This represents a significant addition to the construction cost, and places steel structures at a disadvantage compared with concrete structures. Therefore, minimizing the cost of fire protection has provided a strong impetus to research work on the effects of fire on steel structures and their components (Cooke *et al.* 1987, Ali *et al.* 1998, Tang *et al.* 2001, Liu *et al.* 2002, Huang *et al.* 2004, 2006, Tan *et al.* 2006, etc.). Because steel members are assembled using joints, there is a pressing need to understand better the joint behaviour and its effect on the overall behaviour of steel framed structures under fire conditions. However, to date, there have only been limited numbers of experimental investigations conducted, on a few types of joints, at elevated temperatures.

The first experimental work on steel joints was conducted by Wilson and Moore (1917) to determine the joint rigidity in steel structures at normal temperature. It was only in 1990 that Lawson carried out an experimental investigation of eight beam-to-column joints under fire conditions. In 1997, with the aim of developing full moment-rotation-temperature characteristics for typical bare-steel and composite joints, Leston-Jones *et al.* (1997a, 1997b) reported on two series of tests conducted at the Building Research Establishment on bare steel and composite flush end-plate joints at ambient and elevated temperatures. Another series of transient tests was conducted on beam-to-column flush and flexible end-plate joints by Al-Jabri (1998). However, because of the complexity and diversity of steel joints, and especially due to the high cost of testing, it is impractical to conduct numerous high temperature tests over a wide range of joint types and assemblies. It has also been realized that individual active joint components and their mechanical descriptions can be collected and assembled together for overall analysis of beam-to-column joints. This method for the determination of the mechanical properties of the joint has become known as the "Component Method". Adequate analytical descriptions have been developed by several researchers (Spyrou *et al.* 2002, Vimonsatit *et al.* 2006, Tan *et al.* 2006) for the main components, and these form the basis for overall mechanical modelling of steel beam-to-column joints. The study described here focuses on establishing elevated-temperature rotational characteristics for an extended end-plate beam-to-column steel joint in the presence of axial compression. This has been achieved through a series of tests conducted on full-scale beam-to-column joints under three different isothermal conditions, including tests which

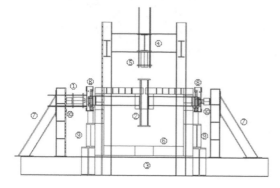

Figure 1. Elevation of test set-up.

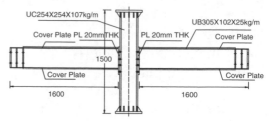

Figure 2. Bare steel extended end-plate assembly detail.

model the effect of thermal restraint induced by adjacent unheated structure, as well as those under pure moment.

2 TEST SET-UP

In total, six extended end-plate beam-to-column joints were tested as "cruciform" assemblies in two groups. In the first group, three cruciform specimens (CR1, CR2 and CR3) were tested at 700°C with axial compressive force maintained at 0%, 2.5% and 4% of the plastic squash capacity of the beam section at ambient temperature. In the second group, three more cruciform assemblies (CR4, CR5 and CR6) were tested at 400°C, 550°C and 700°C respectively, this time without thermal restraint. These tests were conducted so as to obtain the moment-rotation-temperature characteristics under the three isothermal conditions. These two groups of assemblies all used the same beam and column sections for comparison.

2.1 Test arrangement

The whole loading system is shown in Figure 1.

The connection moment was produced by a hydraulic jack⑤, applying vertical load at the top of a cruciform specimen, and the twin reaction frames⑦ at the ends of the cantilever beams. A horizontally-mounted hydraulic jack① provided an active axial force onto these cantilevers. In addition, the two forked supporting systems⑧ prevent out-of-plane rotations at the beam ends. For the second group of tests (CR4 to CR6), in order to prevent lateral-torsional buckling and out-of-plane global buckling of the beam sections, an additional lateral restraint system was fabricated and attached to the external reaction frame. This consisted of three pairs of supporting plates providing lateral restraint to the upright column and beam sections. The roller plate system allowed the column and the attached beam sections to deflect vertically.

2.2 Cruciform specimen specifications

A typical cruciform specimen, consisting of two UB305 × 102 × 25 kg/m beams, symmetrically framing into the flanges of a UC254 × 254 × 107 kg/m section, is shown in Figure 2. An extended end-plate joint was chosen for investigation, as one of the more conventional steel joints in construction practice. This investigation focused particularly on the shear component of the beam-to-column joint. Thus, it seemed prudent to select a joint which would be designed as rigid rather than as hinged. Prior to the assembly testing, material properties at ambient temperature and the geometry had been measured. Material properties at elevated temperatures were assumed to follow the steel material strength and stiffness reduction factors specified in Eurocode 3 part 1.2 (CEN, 2005).

2.3 Deflection measurement

In order to obtain a good description of the development of in-plane deflections, ten sets of LVDTs were utilized to monitor the beam's vertical deflections for CR1 to CR3. These were situated at five locations along the centre-line of the top flange of each beam. For the unrestrained cruciform tests, CR4 to CR6, in order to obtain explicit rotation measurements of the beam sections, two sets of LVDTs were placed at each end of the beam sections outside the furnace. Based on experience from the first group of specimens, column deflections were monitored by two LVDTs at each side of the column-top stiffener.

2.4 Temperature acquisition

During the elevated-temperature tests, eight K-type thermocouple wires were attached to the surfaces of the web-plate, and the top and bottom flanges. Eight additional thermocouples were used to monitor the temperatures of the steel bolts. For the first group of specimens, parts of the beam and column sections remote from the joint zone were thermally protected using ceramic fibre blanket, to reduce the possibility of failure and to reduce heat loss to the external air and the horizontally-mounted hydraulic jack. The top flanges and webs of the beam sections within about

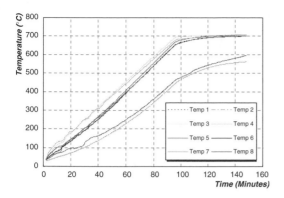

Figure 3. Typical Temperature Measurement for CR1.

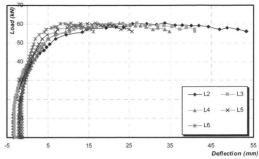

Figure 4. Vertical Deflection Measurements for CR2.

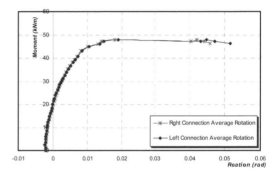

Figure 5. Rotational behaviour of CR2.

600 mm from either face of the joint, and the column section within the steel joint, were exposed to radiation by the furnace. In contrast, for the second group of specimens, CR4 to CR6, remote parts of the beam sections were not thermally insulated since no hydraulic jack was attached.

3 RESTRAINED CRUCIFORM TEST MEASUREMENTS

3.1 Temperature measurements

A steady-state heating method was used in this experimental programme. This was because the main objective of this study was to obtain moment-rotation-temperature characteristics of bare steel end-plate joints in an isothermal condition. Temperature distributions in the restrained cruciform specimens CR1 to CR3 were recorded by thermocouples. Figure 3 shows a typical temperature development, in this case for CR1.

Figure 3 shows the temperature distributions for the left beam. Due to thermal insulation, the bottom flange temperatures increase slowly compared to those for the beam web-panels and the top flange. Temperature differences between the bottom flange and web-panel (which have high exposure) and the top flange of the beam section increase to about 30°C after 100 minutes of heating, when the joint area reaches an average temperature of 700°C. The temperature difference between the insulated area and the exposed area approaches approximately 120°C after CR1 reaches its desired temperature at 700°C, as shown in Figure 3.

3.2 Deflection characteristics

From experimental measurements it is possible to obtain the moment-rotation curves of joints by vertical deflection measurements at different locations along the cantilever beams. For a typical restrained cruciform test (CR 2), 2.5% of the compressive capacity of the beam section at normal temperature was applied to the beam sections before vertical load was applied. Recorded deflections and rotations are presented in Figure 4.

From the vertical beam deflection measurements in Figure 4, negative deflection measurements can be found for both left- and right-hand beams prior to loading. This upward movement was due to axial compressive force and to thermal bowing of the beams at elevated temperature. It can also be seen that both the left- and right-hand beams show linear load-deflection characteristics before the load reached 30 kN. When the external load reached 40 kN the beams started to sag downwards. Applied compression can induce a significant increase of beam deflections due to the P-δ effect.

Figure 5 shows the joint rotations plotted against applied moment. It can be seen that there is good agreement between the left- and the right-hand joints. Both of these joints can sustain moments up to 30 kNm without any significant rotation, after which, joint rotations start to increase progressively due to spread of plasticity in the joints.

3.3 Beam out-of-plane deflection measurements

From the first group of steel joint tests (CR1 to CR3), all three specimens showed some out-of-plane deflections for the beam web panels adjacent to the steel joints. In order to obtain accurate deflection behaviour, cruciform specimens were dismantled, and the beam sections were measured by a special scanning system. These three specimens were measured after the tests to ascertain whether shear buckling had occurred in the web. From these measurements, both the left- and right-hand beam webs displayed out-of-plane buckling near the steel joint. The maximum out-of-plane deflection measurements were 6 mm and 8 mm, respectively. Although these out-of-plane deflections are insignificant in comparison with the vertical deflections of the tested cruciforms, it is clear that, for this group of specimens, shear buckling had occurred in the beam web near to the joint.

4 UNRESTRAINED CRUCIFORM TEST MEASUREMENTS

4.1 Initial imperfection measurements

From the first group of tests, it was realized that the initial out-of-plane imperfection has a considerable influence on beam web shear behaviour. Thus, for the second group of specimens, CR4 to CR6, prior to assembly with the columns, initial imperfections of the beam web panels were measured by the scanning system. It was noticed that the initial imperfections along the three reference lines indicate similar deflection trends along the beam length. These measured initial imperfections have been introduced into the numerical models at discrete points (Qian *et al.* 2006).

4.2 Temperature measurements

For the second group of cruciform tests, three specimens, CR4 to CR6, were tested at 400°C, 550°C and 700°C. Typical temperature developments for CR4 test are shown in Figure 6.

In contrast to the first group of tests, the beam sections were not insulated at their outer ends, since there was no axial load applied at these beam ends. From the temperature development shown in Figure 6 it can be seen that, disregarding fairly small discrepancies, the beam sections eventually reached approximately uniform temperature distributions. This is because the furnace was normally allowed to continue heating for an additional 200 minutes after its control temperature had reached the target level. For this group of tests, the observation windows of the furnace were blocked by

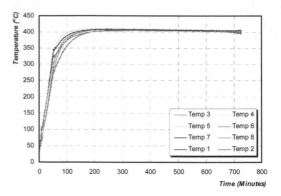

Figure 6. Typical Temperature Measurement for CR4.

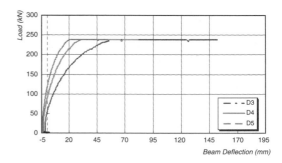

Figure 7. Vertical deflection measurements for left beam of CR5.

the lateral restraint beams. This reduced heat radiation losses.

4.3 Deflection characteristics

Specimen CR 5 was loaded up to failure at a constant 550°C. The vertical deflections recorded by the three transducers, at different distances along the left-hand beam, are shown in Figure 7.

Initial negative deflections could be seen for both left- and right-hand beams due to the slightly higher test temperatures of the compression flanges. Both beams showed a smooth deflection response until the vertical load reached 150 kN. Above this limit, there was a rapid increase in vertical deflection, indicating run-away failure of the joints.

Beam rotations obtained from the displacement measurements from different LVDT combinations are compared in Figure 8. It can be seen that the indicated rotations derived from different pairs of vertical deflections show slightly different flexibilities, as would be expected since the curvature of the cantilever beams is not taken into account. However, the agreement is generally acceptable.

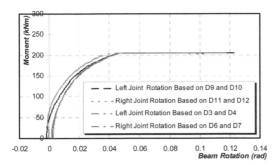

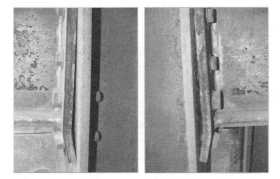

Figure 8.   Rotation Measurements for CR5.

Figure 9.   Typical Failure Mode for CR1.

5   CRUCIFORM TEST FAILURE MODE AND SUMMARY

The first group of specimens, CR1 to CR3, was tested at 700°C, with axial load ratios of 0, 2.5% and 4% of the beam's plastic squash capacity. In addition to out-of-plane deflection of the beam web panels, significant bending deformation was observed at the end-plate adjacent to the beam's tension flange. CR1 shows the greatest end-plate bending deflection, due to its having the highest tensile force transferred to the column from the beam tension flange (Figure 9).

The beam sections were subjected to axial compressive forces in tests CR2 and CR3, hence the gap between the end-plate and column flange decreased with increasing axial restraint. However, these applied axial forces induce significant beam deflections due to P-δ effects.

For the second group of cruciform specimens, the test results show that both the left- and the right-hand beam web panels developed out-of-plane deflections adjacent to the end-plates. A typical overall failure mode is shown for CR4 in Figure 10.

This specimen is selected as it shows the typical behaviour of a cruciform specimen under fire conditions. It also had a more uniform temperature field

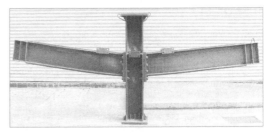

Figure 10.   Failure mode for CR4.

than the first three specimens and its behaviour was quite similar to that of CR5 and CR6. It can be seen that tensile yield zones are formed in the beam web-panels adjacent to the column flanges. Out-of-plane deformations of the left- and right-hand beam webs are in opposite directions, the anti-symmetry possibly being induced by small rotations of the column. Clearly, shear deformation is significant even in universal rolled sections, and should be taken into account in the joint modelling when using a component-based method.

Moment capacities for both restrained and unrestrained joint tests have been compared with finite element analysis and analytical predictions (Qian et al. 2006). In general, the experimental capacities M_{test} and the FE predictions M_{FEM} compare well, with a mean ratio and standard deviation of 1.005 and 0.035 respectively. Discrepancies between the finite element predictions for the unrestrained tests and the test results seem primarily due to friction at locations of lateral restraint during the tests. In addition, finite element analyses slightly over-predict the moment capacities for CR1 and CR2. This can mainly be attributed to differences in the assumed and actual temperature fields. However, this discrepancy becomes negligible for CR5 and CR6, since there were more uniform temperature distributions in these specimens.

6   CONCLUSION

In this research, six steel cruciform assemblies have been tested at elevated temperatures, both with and without thermal restraint effects. Moment-rotation-temperature characteristics of both the restrained and unrestrained steel beam-to-column joints have been obtained. It is anticipated that these results will be beneficial in validating component-based models for steel joints under fire conditions. Such models offer the most practical and economical way of introducing a predictive capability for connection behaviour into whole-frame modelling in fire, since it is impractical for designers to use either complex finite element analysis or expensive furnace testing.

REFERENCES

Alan, T. 1996, '*Steel or Concrete; the Economics of Commercial Buildings*', British Steel Corporation.

Al-Jabri, K. S., Lennon, T., Burgess, I. W. and Plank, R. J. 1998, 'Behaviour of steel and composite beam-column connections in fire', *J. Construct. Steel Research*, **46** (1–3), Paper No. 180.

Cooke, G. M. E., and Latham, D. J. 1987, 'The Inherent fire resistance of a loaded steel framework', *Steel Construction Today*, **1**, pp49–58.

CEN 2005, '*EN 1993-1-2: Design of steel structures: Part1.2 General rules-structural fire design*', European Committee for Standardization, Brussels, Belgium.

Ali, F. A., Shepherd, P., Randall, M., Simms, W. I., O'Connor, D. J. and Burgess, I. W. 1998, 'Effect of axial restraint on the fire resistance of steel columns', *J. Construct. Steel Research*, **46** (1-3), pp305–306.

Huang, Z. F. and Tan, K. H. 2004, 'Effects of external bending moments and heating schemes on the responses of thermally restrained steel columns', *Engineering Structures*, **26** (6), pp769–780.

Huang, Z. F., Tan, K. H. and Phng, G. H. 2006, 'Axial restraint effects on the fire resistance of composite columns encasing I-section steel', *J. Construct. Steel Research*, (In press).

Jaspart, J. P. 2000, 'General report: session on connections', *J. Construct. Steel Research*, **55**, pp69–89.

Lawson, R. M. 1990, 'Behaviour of steel beam-to-column connections in fire', *The Structural Engineer*, **68** (14), pp263–271.

Leston-Jones, L. C. 1997a, 'The influence of semi-rigid connections on the performance of steel framed structures in fire', PhD thesis, Department of Civil and Structural Engineering, University of Sheffield.

Leston-Jones, L. C., Burgess, I. W., Lennon, T. and Plank, R. J. 1997b, 'Elevated-temperature moment-rotation tests on steelwork connections', *Proc. Institution of Civil Engineers, Structures and Buildings*, **122** (4), pp410–419.

Liu, T.C.H., Fahad, M. K. and Davies, J. M. 2002, 'Experimental investigation of behaviour of axially restrained steel beams in fire', *J. Construct. Steel Research*, **58**, pp1211–1230.

Qian, Z. H., Tan, K. H. and Burgess, I. W. 2006, 'Behaviour of steel beam-to-column joints at elevated temperature: I. Experimental investigation', *J. Structural Engineering, ASCE* (under review).

Spyrou, S. 2002, 'Development of a component-based model of steel beam-to-column joint at elevated temperatures', PhD Thesis, Department of Civil and Structural Engineering, University of Sheffield, Sheffield, UK.

Tang, C. Y., Tan, K. H., Ting, S. K. 2001, 'Basis and application of a simple interaction formula for steel columns under fire conditions', *J. Structural Engineering, ASCE*, **127** (10), pp1206–1213.

Tan, K. H., and Qian, Z. H. 2006, 'Experimental behaviour of thermally restrained plate girder loaded in shear at elevated temperature', *J. Construct. Steel Research*, (under review)

Vimonsatit, V., Tan, K. H., and Qian, Z. H. 2006, 'Testing of plate girder web panel loaded in shear at elevated temperature', *J. Structural Engineering, ASCE*, (in press).

Wilson, W. H. and Moore, H. F. 1917, 'Tests to determine the rigidity of riveted joints in steel structures', *University of Illinois, Engineering Experimentation Station, Bulletin No. 104*, Urbana, USA.

Steel and Composite Structures – Wang & Choi (eds)
© 2007 Taylor & Francis Group, London, ISBN 978-0-415-45141-3

Effects of high temperatures on the stiffness of concrete-filled steel tubular column-to-steel beam connections

J.F. Wang & L.H. Han
Department of Civil Engineering, Tsinghua University, Beijing, China

ABSTRACT: This paper studies the effects of high temperatures on the stiffness of concrete-filled steel tube column to steel beam connections by ABAQUS software. The finite element (FE) modeling is proposed and validated by the experimental results. Based on the FE models of connections, the effects of seven parameters, such as width of stiffening ring, steel ratio, cross-sectional dimension, beam-to-column stiffness ratio, beam-to-column strength ratio, axial load level and column slenderness ratio, on the moment-rotation relation and the stiffness of beam-to-column connection are studied and evaluated. It was found that this kind of composite connections at room temperature exhibit rigid and have high load-carrying capacities and stiffness. However, the load-carrying capacities and stiffness of the beam-to-column connections with an increasing of temperature decrease significantly and they behave semi-rigid. When temperature is 300°C, 500°C, 700°C, and 900°C, the moment capacity of beam-to-column connection respectively decreases 17%, 33%, 54%, and 92% moment capacity of beam-to-column connection at room temperature, while the initial rotation stiffness of beam-to-column connection respectively decreases 12%, 28%, 78%, and 96% initial rotation stiffness of beam-to-column connection at room temperature.

1 INSTRUCTIONS

Concrete-filled steel tubes (CFST) have several structural and constructional benefits, such as high strength and fire resistance, large stiffness and ductility, restraint on local buckling of the steel tube provided by the infill concrete core, omission of formwork and thus reducing the construction cost and time (Han, 2004). These advantages have been recognized and have led to the increased use of CFST columns in some of the recent tall buildings. However, the behavior of the beam-to-column connections for CFST is complicated and their design has not been sufficiently verified, thus leading to their application limited.

Due to composite action between the steel tube and its core concrete under fire, CFST columns have high fire resistance. In the past, there have been a large number of research studies on the realistic performance of concrete filled steel tubular columns under fire conditions by experimental and theoretic methods, such as Lie and Chabot (1990), Lie (1994), Wang (2000), Kodur (1999), Han et al.(2001, 2003); however, there is very little research work on the behavior of the beam-to-column connection for CFST at high temperature. The only report findings is an experimental study of the fire performance of eight concrete-filled steel tubular column assemblies using extended

end plate connections by Wang and Davies (2003), who investigated the effects of rotational restraint on column bending moments and column effective lengths.

In engineering practice, the concrete-filled steel tube column to steel beam connections with stiffening rings are popularly used in the high-rise building structures. At room temperature, the type beam-to-column connections exhibit rigid and have high load capacities and stiffness. However, at high temperature, the load capacities and stiffness of beam-to-column connections decrease and they behave semi-rigid. In addition, the behavior of connections at high temperatures is research foundation of the performance-based fame design.

This paper studies the effects of high temperatures on the stiffness of concrete-filled steel tube column to steel beam connections by ABAQUS software. Determination methods for the constitutive models of steel and concrete, interface model between the steel tubes and their core concrete are presented. The proposed FE model is validated by the experimental results. Based on the FE models of connections, the effects of seven parameters, i.e. width of stiffening ring, steel ratio, cross-sectional dimension, beam-to-column stiffness ratio, beam-to-column strength ratio, axial load level and column slenderness ratio, on the moment-rotation relation and the stiffness of the beam-to-column connection are studied and evaluated.

2 MATERIAL PROPERTIES

2.1 Steel

In this paper, the yield stress and the Young's modulus of steel material proposed by Li et al. (2006) are used in terms of temperature. The Poisson's ratio for steel material is assumed to be constant, taken as 0.3. For the creep deformation of steel at elevated temperatures, the model proposed by Skowroński (1993) was adopted.

The stress-strain relation for steel material is taken from data provided by Lie and Chabot (1990) and Lie (1994) as follow:

$$
\sigma_s = \begin{cases} \dfrac{f(T,0.001)}{0.001}\varepsilon_s & \varepsilon_s \le \varepsilon_p \\ \dfrac{f(T,0.001)}{0.001}\varepsilon_p + f\left[T,(\varepsilon_s - \varepsilon_p + 0.001)\right] - f(T,0.001) & \varepsilon_s > \varepsilon_p \end{cases}
$$

(1)

where $\varepsilon_p = 4 \times 10^{-6} f_y$;

$$f(T,0.001) = (50 - 0.04T) \times \left(1 - e^{\left[(-30 + 0.03T)\sqrt{0.001}\right]}\right) \times 6.9 \, ;$$

$$f\left[T,(\varepsilon_s - \varepsilon_p + 0.001)\right] = (50 - 0.04T) \times \left(1 - e^{\left[(-30 + 0.03T)\sqrt{\varepsilon_s - \varepsilon_p + 0.001}\right]}\right) \times 6.9$$

Fig. 1 shows a set of relationships between the stress and the strain of steel material with yield stress $f_y = 345$ MPa at different temperature, which illustrate how the stress-strain response varies with temperature.

2.2 Concrete

The cube compressive strength of concrete is one of most basic and important material parameters, because it influences the ultimate strength, Young's modulus and maximum strain etc. The cube compressive strength and the cube tension strength of concrete provided by Guo and Li (1991) are adopted in this paper. The Young's modulus of concrete at high temperature suggested by Lu (1999) is used. The Poisson's ratio for concrete is assumed to be constant, taken as 0.2.

The stress-strain relations of concrete are provided by Lin (2006) based on the results of work by Han (2004) and Han et al. (2007). The equations describing these relations are as follows:

$$
y = \begin{cases} 2x - x^2 & x \le 1 \\ \dfrac{x}{\beta_0 \cdot (x-1)^\eta + x} & x > 1 \end{cases}
$$

(2)

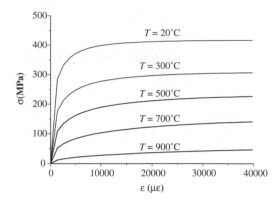

Figure 1. Typical stress-strain curves of steel.

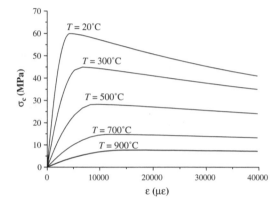

Figure 2. Typical stress-strain curves of concrete.

where $x = \dfrac{\varepsilon}{\varepsilon_0}$; $y = \dfrac{\sigma}{\sigma_0}$; $\sigma_0 = \dfrac{f_c}{1 + 2.4 \times (T - 20)^6 \times 10^{-17}}$;

$\varepsilon_0 = \varepsilon_c + 0.0008 \times \xi^{0.2}$;

$\varepsilon_c = (1300 + 12.5 f_c) \times 10^{-6} \times \left[1 + 0.0015T + 5 \times 10^{-6} \times T^2\right]$;

$$\eta = \begin{cases} 2 & \text{circular} \\ 1.6 + 1.5/x & \text{square or rectangle} \end{cases} \; ;$$

$$\beta = \begin{cases} \left(1.18 \times 10^{-5}\right)^{\left[0.25 + (\xi - 0.5)^7\right]} \cdot f_c^{0.5} \ge 0.12 & \text{circular} \\ \dfrac{f_c^{0.1}}{1.2\sqrt{1 + \xi}} & \text{square or rectangle} \end{cases}$$

in which f_c is the circular compressive strength of concrete at room temperature.

Composite action between the steel tube and its concrete core had been considered by use of constraining factor $\xi = (A_s f_y)/(A_c f_{ck})$ in Eq.(2). A_s and A_c are the cross-section area of the steel tube and the concrete, respectively; f_y is the yield stress of steel; and $f_{ck} =$ the compression strength of concrete at room temperature.

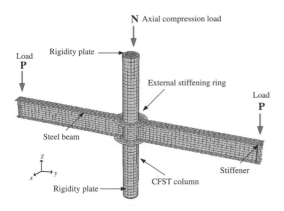

Figure 3. Analysis model of connection.

Figure 4. Deformed failure of connection.

Typical stress-strain curves of confined concrete of the concrete-filled steel tube with a cube compressive strength $f_{cu} = 60$ MPa and constraining factor $\xi = 0.84$ at high temperatures are shown in Fig. 2.

3 FINITE ELEMENT MODEL OF CONNECTION

The commercial finite element software ABAQUS was used to simulate the behavior of concrete-filled steel tube column to steel beam connections with external stiffening rings at high temperature. The following assumptions were made:

(1) The stress-strain relationship for steel at high temperature given in Eq. (1) is adopted for both tension and compression.
(2) The stress-strain relationship for concrete at high temperature given in Eq. (2) is adopted for compression only.
(3) The tension softening model is modified by considering the effect of high temperature and given in in Eq. (3).
(4) The interaction friction coefficient between the steel tube and the core concrete is assumed to 0.35.

Fig. 3 shows a 3-D Finite Element (FE) model of this type composite connection. The FE model was to estimate the vertical deflection of the beam subjected to an initial constant concentrated force applied downward at the free end of the beam section. The connection consists of placing external stiffening ring between concrete-filled steel tube column and steel beam connected together by welding. Fig. 4 shows the predicted deformed failure of composite connection.

A C3D8R element with reduced integration (1 Gauss point) has been chosen for the simulation of the core concrete in the model as it can provide more accurate results. The efficiency of this type of element

has been verified by Han et al. (2007). The structural steel components such as steel beam, steel tube and external stiffening ring are modeled by using S4R element with reduced integration. For accurate computation, 9 Gauss point Simpson integral are used in the shell element thickness. The analyses account for material non-linearities through classical metal plasticity theory based on the Von Mises yield criterion. With *ELASTIC and *PLASTIC options, the yield and ultimate tensile strength of the structural steel components are converted into the true stress and plastic strain with appropriate input format for ABAQUS. Different mesh sizes have been examined as well to determine a reasonable mesh that provides both accurate results with less computational time. The mesh selected is shown in Fig. 3.

The concrete damaged plasticity model is used to simulate for elastic-plastic behavior and nonlinear behavior of concrete. The model adopts non-correlation flowage principle and its plastic potential energy equation uses Drucker-Drager double curve. The plastic behavior of concrete is expressed by stress-plastic strain of the core concrete. The energy breakage criterion is simulated for tension softening behavior of concrete. The tension softening model is modified by considering the effect of high temperature, shown Fig. 5. In Fig. 5, $\sigma_{t0,T}$ is the maximum tension stress of concrete at temperature T and can be expressed as:

$$\sigma_{t0,T} = 0.26 \times (1.25 \times f_c)^{0.67} \times (1 - 0.001 \times T) \qquad (3)$$

A surface-based interaction, with a contact pressure model in the normal direction, and a Coulomb friction modeling in the tangential direction between surfaces of steel tube and the core concrete was used in the FE modeling proposed by Han et al.(2007). The classical isotropic Coulomb friction model is shown in Fig. 6. It assumes that no relative motion occurs if the average bond stress is less than the critical stress, which can be expressed as follows:

$$\tau_{crit} = \mu \cdot p \geq \tau_{bond} \qquad (4)$$

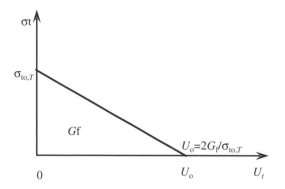

Figure 5.   Soft tension model of concrete.

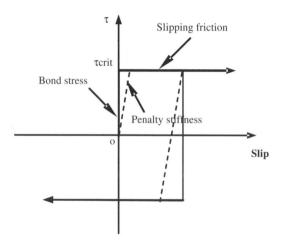

Figure 6.   Coulomb friction model.

where μ = the friction coefficient; P = the contact pressure.

The average bond stress τ_{bond} for circular CFST column provided by Roeder (1999) is expressed as

$$\tau_{bond} = 2.314 - 0.0195 \cdot (D/t) \qquad (5)$$

Where D = the diameter of the core concrete; t = the thickness of steel tube.

A kind of "gap element" with high stiffness in the normal direction was adopted to simulate the contact and separation between surfaces of steel tube and core concrete.

The * CONTACT PAIR, SMALL SLIDING and * SURFACE BEHAVIOR, SEPAPATION options are used to define contact surfaces between steel tube and concrete in ABAQUS. In section 3, parameter analysis of CFST column at high temperatures shows that it is reasonable for the interaction friction coefficient assumed 0.25–0.5. In this paper the friction coefficient is taken as 0.35.

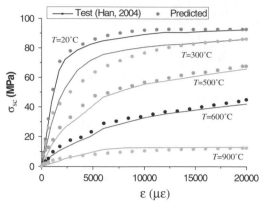

Figure 7.   Axial stress (σ_{sc}) versus axial stress (ε) curves of CFST column at high temperatures.

In order to simulate loading plate and pinned constraining condition of column bottom side, each rigidity plate is respectively placed to the top and bottom sides of column, which is modeled by C3D8R element. The nodes of the outer layer element on both the rigidity plates and the corresponding concrete surface were paired up and constrained using the "TIE" option provided by ABAQUS. The nodes of the outer layer element on both the rigidity plates and the corresponding steel tube surface were paired up and constrained using the "SHELL TO SOLID COUPING" option.

The structure was initially subjected to a predefined concentrated axial compression load on the rigidity plate at the top side of column, then the other concentrated forces at a distance from the face of the column flange which generates the required moment about the connection. The beam was allowed to deflect downward only, while horizontal movement was restrained to prevent any possibility of premature failure of the beam by lateral torsional bucking. The whole loading is controlled by displacement. Newton-Raphson method and line search are used to effectively solve the nonlinear problem of composite connections, so the convergence is well.

4   VALIDATION OF PROPOSED FE MODEL

To validate the proposed FE model described as above, the predicted results using the FE model are compared with previously test results in Han (2004), Schneider and Alostaz (1998) and Wang et al. (2005). Fig. 7 shows the predicted axial stress (σ_{sc}) versus axial stress (ε) curves of CFST column at high temperatures are compared with the measured cures (plotted in dashed lines). Fig. 8 shows the predicted M_u/M_p versus θ_r curves of circular CFST column to steel beam connections are compared with the measured cures. Fig. 9

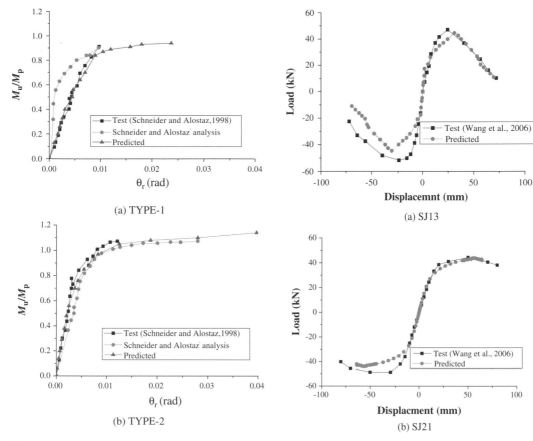

(a) TYPE-1

(b) TYPE-2

Figure 8. M_u/M_p versus θ_r curves of circular CFST column connection.
Note: M_u is ultimate moment capacity of connection; M_p is ultimate moment capacity of beam.

(a) SJ13

(b) SJ21

Figure 9. Load versus lateral displacement curves of square CFST column connection.

shows the predicted load (P) versus later displacement (Δ) curves of square CFST column to steel beam connections are compared with the measured cures. It can be found that, generally good agreement is obtained between the predicted and the test results.

5 FINITE ELEMENT MODEL OF CONNECTION PARAMETRIC STUDIES

It is expected that the moment-rotation relation of a composite connection may be influenced by a number of parameters, such as width of stiffening ring (b), steel ratio of CFST cross-section (α), dimension of CFST cross-section (D), axial load level (n), and slenderness ratio of CFST column (λ). The limited number of tests cannot be used to examine the influence of the aforementioned factors and this section presents a parametric study using the FE model.

The parametric analyses were conducted without taking the strength of the materials into account because the parametric analysis results in Han (2004) showed that the strength of the materials hardly had any influence on the relative strength of the concrete-filled steel tube column. The studied parameters include the width of stiffening ring b, the steel ratio α, the cross-sectional dimension D, the beam-to-column stiffness ratio $k(k = (K_s H)/(K_e L)$, K_s and K_e are the sectional stiffness of steel beam and CFST column respectively), the beam-to-column strength ratio k_m ($k_m = M_{ub,T}/M_{uc,T}$, $M_{ub,T}$ and $M_{uc,T}$ are ultimate bending moment capacity of steel beam and CFST column at high temperature respectively), the level of axial load n ($n = N_0/N_{u,T}$, N_0 is the axially compression load applied on CFST column and $N_{u,T}$ is the ultimate compression resistance of composite cross-section at high temperature), and the slenderness ratio λ ($\lambda = 4H/D$ for CFST columns with circular).

The influences of the studied parameters on the moment-rotation relations are shown in Fig. 10. The basic calculating conditions are: CFST column $D \times t = 400 \times 12\,\text{mm}$, $f_y = 345\,\text{MPa}$, $f_{cu} = 60\,\text{MPa}$, $\alpha = 0.132$, $\xi = 1.11$, $H_c = 3.6\,\text{m}$, $\lambda = 36$; steel beam $h \times b_f \times t_w \times t_f = 450 \times 250 \times 12 \times 12\,\text{mm}$, $L_b = 3.6\,\text{m}$; $k_m = 0.8$; $k_m = 4.9$; $n = 0.4$; stiffening ring $b \times t_1 = 200 \times 12\,\text{mm}$. The thickness of stiffening ring is assumed same with the thickness of steel beam flange.

Fig. 10a show the influence of the width of stiffening ring b on the moment-rotation relations of the connection at 500°C. It can be found that the elastic stiffness and the ultimate moment capacity of the composite connection increase significantly with an increase in b. The increasing of b means the connected area between CFST column and steel beam. The steel beam constraints the CFST column strongly and the forces can be transferred reliably, thus leading to a decreasing strength and stiffness of the composite connection. However, the reasonable size of b is related with both force transfer and architecture fitment, so the effective size of b still needs to be studied homely.

Fig. 10b shows the influence of steel ratio α of the composite column on the moment-rotation relations of the connection at 500°C. From this figure, it can be found that, the elastic stiffness and the ultimate moment capacity of the composite connection increase significantly with an increase in steel ratio. Because a higher α means higher strength and stiffness of the CFST column, and thus it gives a higher strength and elastic stiffness of the composite connection.

Fig. 10c shows the influence of CFST column cross-sectional dimensions D on the moment-rotation relations of the connection at 500°C. It can be found that, with decreasing D, the elastic stiffness and the ultimate moment capacity of the composite connection degrease significantly. Under the same temperature, the smaller the sectional size is, the higher is the temperature in the core concrete, which means a greater loss of the strength and stiffness of the composite columns, thus leading to a decreasing strength and stiffness of the composite connection.

Fig. 10d shows the influence of beam-to-column stiffness ratio k on the moment-rotation relations of the connection at 500°C. It can be found that the elastic stiffness and the ultimate moment capacity of the composite connection increase remarkably with an increase in k. This comes about because a higher k means stronger restraint of the steel beam against the CFST column, which would result in a smaller effective length of the composite column, and thus leading to a higher stiffness and ultimate strength of the composite connection.

Fig. 10e shows the influence of beam-to-column strength ratio k_m on the moment-rotation relations of the connection at 500°C. The change in k_m can be achieved by changing the cross-sectional dimensions of the columns gradually, while the span of steel beam should be adjusted to keep the beam-to-column

stiffness ratio constraint $k = 0.49$. It can be found that, the ultimate moment capacity increases with an increasing in k_m. The reason is that a higher k_m means higher strength of the steel beam, which would result in a higher ultimate strength of the composite connection. However, k_m has moderate influence on the elastic stiffness. Because the increase in the yield strength of the steel beam doesn't change its elastic modulus, the stiffness of the composite connection remains constant.

Fig. 10f shows the influence of axial load level n in CFST column on the moment-rotation relations of the connections at 500°C. It can be found that, the elastic stiffness and the ultimate moment capacity of the composite connection degrease little with an increase in axial load level.

Fig. 10g shows the influence of column slenderness ratio λ on the moment-rotation relations of the connection at 500°C. The height of CFST column H and the span of steel beam L was adjusted to keep the beam-column stiffness ratio constant, i.e. $k = 0.49$. It can be found from Fig. 10 g that, as it was expected, the ultimate moment capacity of the composite connection decreases with increasing λ.

It should be mentioned that, from Fig. 11, at room temperature most of composite connections exhibit rigid and have high load capacities and stiffness. However, with an increasing of temperature, the Yong's modulus and the strength of steel and concrete reduce and the axially ultimate compressive strength decreases. The restraint of the beam on the CFST column weakens, thus leading to the load capacities and stiffness of beam-to-column connections decreasing gradually and they behave semi-rigid. From Fig. 11, It was also found that the ultimate moment capacity decreases gradually with an increasing of temperature. In addition, when the temperature is less than 500°C, the increment of the initial connection stiffness is small; but the initial connection stiffness decreases significantly beyond 500°C.

Fig. 12 shows the moment capacity (M_u) versus temperature (T) relation with various widths of stiffening ring. It is found that when temperature is 300°C, 500°C, 700°C, and 900°C, the moment capacity of beam-to-column connection decreases 17%, 33%, 54%, and 92% moment capacity of beam-to-column connection at room temperature, respectively.

Fig. 13 shows the initial rotation stiffness (K_o) versus temperature (T) relation with various widths of stiffening ring. The initial rotation stiffness (K_o) of connection is defined as the secant stiffness corresponding to a bending moment of $0.2 M_u$. It is found that when temperature is 300°C, 500°C, 700°C, and 900°C, the initial rotation stiffness of beam-to-column connection decreases 12%, 28%, 78%, and 96% initial rotation stiffness of beam-to-column connection at room temperature, respectively.

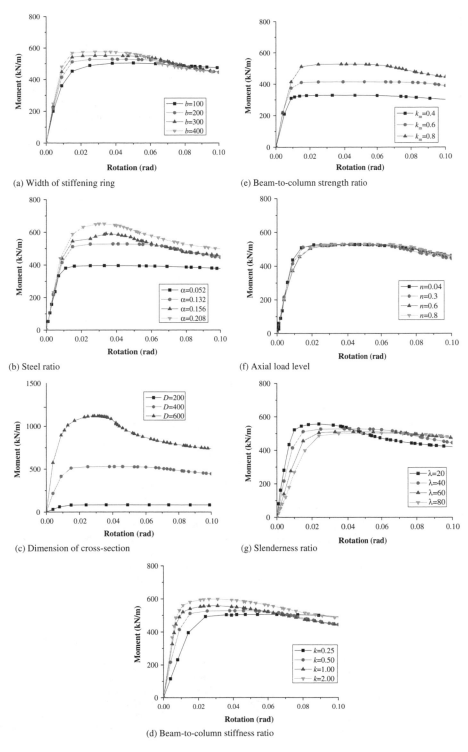

Figure 10.   Moment-rotation relations of connections at 500°C.

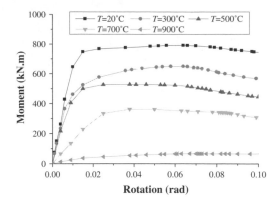

Figure 11. Moment-rotation relations of connections with various temperatures.

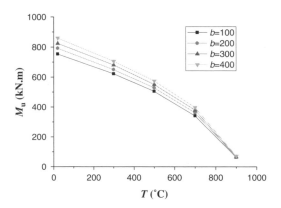

Figure 12. Moment capacity versus temperature relations.

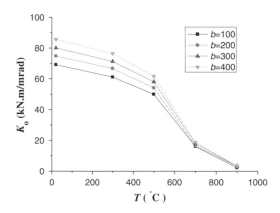

Figure 13. Initial rotation stiffness versus temperature relations.

6 CONCLUSIONS

To understand the effects of high temperatures on the stiffness of concrete-filled steel tubular column-to-steel beam connections, the 3-D nonlinear finite element analysis are carried out with seven parameters. From this study the following conclusions are made:

(1) The validity of the proposed FE models is verified by comparing $\sigma_{sc} - \varepsilon$ curves of CFST column at high temperatures, $M_u/M_p - \theta_r$ curves of circular CFST column connection, $P - \Delta$ curves of square CFST column connection derived from the tests.

(2) The effects of seven parameters, such as the width of stiffening ring, the steel ratio, the cross-sectional dimension, the beam-to-column stiffness ratio, the beam-to-column strength ratio, the axial compression level, and the slenderness ratio, on the moment-rotation relations of the composite connections were analyzed and the results are presented in this paper.

(3) At room temperature most of composite connections exhibit rigid and have high load capacities and stiffness. However, with an increasing of temperature, the load capacities and stiffness of beam-to-column connections decrease gradually so that they behave semi-rigid.

(4) When temperature is 300°C, 500°C, 700°C, and 900°C, the moment capacity of beam-to-column connection respectively decreases 17%, 33%, 54%, and 92% moment capacity of beam-to-column connection at room temperature, while the initial rotation stiffness of beam-to-column connection respectively decreases 12%, 28%, 78%, and 96% initial rotation stiffness of beam-to-column connection at room temperature.

ACKNOWLEDGEMENTS

The research reported in the paper is part of Project 50425823 supported by National Natural Science Foundation of China, and the project supported by Start-Up Fund for Outstanding Incoming Researchers of Tsinghua University. Their financial supports are highly appreciated.

REFERENCES

Han, L.H. 2004. *Concrete filled steel tubular structures-theory and practice.* Beijing: Science Press (in Chinese).

Lie, T.T. & Chabot, M. 1990. A Method to predict the fire resistance of circular concrete filled hollow steel columns. *Journal of Fire Engineering* 2(4): 111–126.

Lie, T.T. & Stringer, D.C. 1994. Calculation of the fire resistance of steel hollow structural section columns filled with plain concrete. *Canadian Journal of Civil Engineering* 21(3): 382–385.

Wang, Y.C. 2000. A simple method for calculating the fire resistance of concrete-filled CHS columns. *Journal of Constructional Steel Research* 54(3):365–386.

Kodur, V.K.R. 1999. Performance-based fire resistance design of concrete-filled steel column. *Journal of Constructional Steel Research*, 51(1):21–26.

Han, L.H. 2001. Fire performance of concrete filled steel tubular beam-columns, *Journal of Constructional Steel Research* 57(6):695–709.

Han, L.H., Zhao, X.L., Yang, Y.F. & Feng, J.B. 2003. Experimental study and calculation of fire resistance of concrete-filled hollow steel columns, *Journal of Structural Engineering*, ASCE, 129(3):346–356.

Wang, Y.C. & Davies, J.M. 2003. An experimental study of the fire performance of non-sway loaded concrete-filled steel tubular column assemblies with extended end-plate connections. *Journal of Constructional Steel Research* 59(7):819–838.

Li, G.Q., Han, L.H., Lou, G.B., Jiang, S.C. 2006. *Fire design of steel structures and steel-composite composite structures*, Beijing: Building Industry Press of China (in Chinese).

Skowroński, W. 1993. Bucking fire endurance of steel columns. *Journal of Structural Engineering* 119(6):1712–1732.

Guo, Z.H. Shi, X.D. 2002. *Behavior of reinforced concrete at elevated temperature and its calculation*, Beijing: Tsinghua University Press (in Chinese).

Lu, Z.D. 1999. *Fire behavior of reinforce concrete beam*, Doctoral's Degree dissertation, Tongji University. (in Chinese).

Schneider, S.P. 1998. Axially loaded concrete-filled steel tubes. *Journal of Structural Engineering*, ASCE, 124(10):1125–1138.

Susantha, K.A.S., Ge, B.H. & Usami, T. 2001. Confinement evaluation of concrete-filled box-shaped steel columns. *Steel and Composite Structures*, 1(3):313–328.

Hu, H.T., Huang, C.S & Wu, M.H. 2003. Nonlinear analysis of axially loaded concrete-filled tube columns with confinement effect. *Journal of Structural Engineering* 129(10):1322–1329.

Roeder, C.W., Cameon, B. & Brown, C.B. 1999. Composite action in concrete filled tubes. *Journal of Structural Engineering*, ASCE, 125(5):477–484.

Schneider, S.P. & Alostaz, Y.M. 1998. Experimental behavior of connections to concrete-filled steel tubes. *Journal of Constructional Steel Research* 45(3): 321–352.

Wang, W. D., Han, L.H. & You, J.T. 2006. Experimental studies on hysteretic behavior s of steel beam to concrete filled SHS column connections with stiffening ring, *China Civil Engineering Journal* 39(9): 17–25, 61. (in Chinese).

Lin, X.K. 2006. Cyclic performance of concrete-filled steel tubular columns after exposure to fire. Doctoral Degree dissertation. Fuzhou University (in Chinese).

Han, L.H., Yao, G.H. & Tao, Z. 2007. Performance of concrete-filled thin-walled steel tubes under pure torsion. *Thin-Walled Structures* (accepted for publication).

Steel and Composite Structures – Wang & Choi (eds)
© *2007 Taylor & Francis Group, London, ISBN 978-0-415-45141-3*

Modelling of bolted T-stub assemblies at elevated temperatures

A. Heidarpour & M.A. Bradford

*Centre for Infrastructure Engineering & Safety, School of Civil & Environmental Engineering, The University of
New South Wales, Sydney, Australia*

ABSTRACT: This paper describes an analytical model to investigate the behaviour of steel T-stub assemblies
in steel frame connections at elevated temperatures. Unlike similar proposed analytical models in which the
T-stub is attached to an infinitely stiff support stratum, the present paper considers the physical interaction of
the end plate and the column flange. The theoretical approach for evaluating the elastic and plastic behaviour
of flexible bolted T-stub assemblies herein is based on a premise that the possible failure modes can be derived
from simple beam theory using a knowledge of the ratio of the flexural stiffness of the end plate (which is in
series with the column flange) and the axial stiffness of the bolts. The analytical model is validated against test
results published elsewhere, and it is demonstrated that the proposed analytical model can predict accurately the
behaviour of the T-stub component in a steel beam-to-column frame connection in fire.

1 INTRODUCTION

It is commonly accepted that the so-called component
method affords a rational procedure for joint or connection design in steel frames and it has been embraced
in detail in the Eurocode 3 (2006). A useful part of the
component approach is the concept of a T-stub, that
appears to have been first introduced by Zoetemeijer
(1974), and it can model specific components such as
a column flange in bending bolted to a beam end-plate
in bending.

Predictions of the force-displacement response of
bolted T-stubs at ambient temperature have been
developed by several researchers, including Agerskov
(1976, 1977) who developed an analytical model for
the tension zone of a connection which he assumed
reached its yield capacity when the end plate (the T-
stub flange) developed its yield moment at the toe of
the fillet weld; this model assuming the end plate to be
attached to an infinitely rigid stratum comprising of the
column flange. This model was extended by Shi *et al.*
(1996) to consider a flexible stratum, and other contributions have been provided by Piluso *et al.* (2001a,b),
Swanson & Leon (2000, 2001) and others.

The behaviour of T-stub assemblies at elevated temperatures, as caused by fire attack, has been studied by
relatively few researchers. Spyrou & Davison (2001)
and Spyrou *et al.* (2004a,b) reported both analytical
modelling and test results for bolted T-stubs in work
undertaken at The University of Sheffield. Their work
assumed the end plate to be founded on a rigid medium
(i.e. the column flange was rigid), but in reality the
bolt forces are influenced significantly by the flexible

column flange, and some failure modes associated
with a flexible column flange (Shi *et al.* 1996) are
potentially excluded under the rigid-stratum assumption. The type of T-stub assembly considered in the
Sheffield work is depicted in Figure 1.

The intention of the present paper is to extend
the Sheffield work by incorporating a more thorough
interaction of the end plate and the column flange.
By incorporating empirical formulae for the degradation of the mechanical properties of the components
of the T-stub assembly with temperature, it is possible to elucidate the 'ductile', 'semi-ductile' or 'brittle'
nature of a T-stub. It is shown that increased temperatures can render an assembly that is ductile at ambient
temperature as being brittle at elevated temperatures.

2 ANALYTICAL MODEL

2.1 *General*

The analytical modelling herein is founded on simple
beam theory that incorporates both the flexural stiffness of the end plate and of the column flange, as well
as the axial stiffness of the bolts. Theoretical predictions of the response of T-stub assemblies at elevated
temperature are clearly complex, and so it is assumed
that:

- 3-D effects including geometric nonlinearity can be
  disregarded.
- Bolt bending is neglected, so that only axial bolt
  deformations are considered,

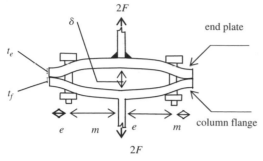

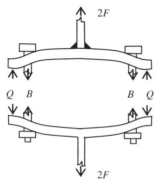

(a) T-stub assembly

(b) Free body diagram of T-stub assembly

Figure 1. T-stub model.

- Prying action is incorporated by assuming the actions are localised to the edges of the end plate and column flange assembly.
- No prying force develops at plastic hinges at the line of the bolts or when the bolts have yielded.
- The end plate and column flange thicknesses are significantly small to preclude through-depth partial plasticisation.

By invoking classical beam theory (Timoshenko & Gere 1985), if the tension force acting on the T-stub assembly is $2F$ (Figure 1), then compatibility between the deflection of the T-stub at the bolt line and the elongation of the bolt at temperature T requires that

$$\left(\frac{1}{K_{cT}} + \frac{1}{K_{eT}}\right)\left[F\left(\frac{\lambda}{8} - \frac{\lambda^3}{6}\right) - B\left(\frac{\lambda^2}{2} - \frac{2\lambda^3}{3}\right)\right] = \frac{B}{n_b K_{bT}} \quad (1)$$

where $\ell = e/L$, n_b is the number of bolts at each bolt line and K_{cT}, K_{eT} and K_{bT} are the flexural stiffness of the column flange and end plate and axial stiffness of the bolt at the elevated temperature T, given by

$$K_{cT} = \frac{E_{sT}I_c}{L^3}; \quad K_{eT} = \frac{E_{sT}I_e}{L^3}; \quad K_{bT} = \frac{A_{sb}E_{bT}}{\lambda_b} \quad (2)$$

in which ℓ_b is the length of the bolt, and E_{sT} and E_{bT} are the elastic moduli of the structural steel and bolt material at temperature T respectively; related to their ambient temperature values E_{s0} and E_{b0} by

$$E_{sT} = \eta_{Es} \cdot E_{s0} \quad \text{and} \quad E_{bT} = \eta_{Eb} \cdot E_{b0}. \quad (3)$$

Using these, Equation 1 can be written as

$$\frac{1}{\eta_{Es}\overline{K}_0}[\alpha F - \beta B] = \frac{B}{n_b \eta_{Eb}K_{b0}}, \quad (4)$$

where

$$\alpha = \frac{\lambda}{8} - \frac{\lambda^3}{6}; \quad \beta = \frac{\lambda^2}{2} - \frac{2\lambda^3}{3}; \quad \frac{1}{\overline{K}_0} = \frac{1}{K_{c0}} + \frac{1}{K_{e0}}. \quad (5)$$

Prying forces (Bahaari & Sherbourne 1996) develop to maintain equilibrium when the relative stiffnesses of the bolt and end plate and column flange cause changes in curvature along the free edges of the tension region, and these forces can be obtained from Equation 4 as

$$B = \frac{\alpha}{\dfrac{\eta_{Es}\overline{K}_0}{n_b \eta_{Eb}K_{b0}} + \beta} \cdot F \quad \text{and} \quad Q = B - F. \quad (6)$$

For prying action to develop, $Q > 0$ and so the relative stiffnesses of the bolt and the column flange and end plate assembly must be limited by

$$\overline{K}_0/K_{b0} \le n_b(\alpha - \beta)\eta_{Eb}/\eta_{Es}. \quad (7)$$

If prying forces are present, the bending moment developed at the fillet weld on the end plate and the bolt line in either the end plate or column flanges are

$$M_f = Bm - Q(e + m) \quad \text{and} \quad M_b = Qe, \quad (8)$$

and if $Q = 0$ then $M_f = Bm$ and $M_b = 0$. A plastic hinge develops at the fillet weld location, or at the bolt line, when

$$M_f \text{ or } M_b = \min(M_{epT}, M_{cpT}), \quad (9)$$

where

$$M_{epT} = \eta_{Ys}f_{ye0}S_e \quad \text{and} \quad M_{cpT} = \eta_{Ys}f_{yc0}S_c \quad (10)$$

are the plastic moments of resistance of the end plate and column flange respectively at temperature T, f_{ye0} and f_{yc0} are the respective yield stresses, and

$$\eta_{Ys} = f_{ysT}/f_{ys0} \quad (11)$$

describes the temperature-dependent degradation of the yield strength of structural steel.

582

2.2 Formulation of first yield point

First yield is characterised by the force F reaching the value

$$^1F_y = \begin{cases} \min(F_f, F_{bl}, F_b) & Q > 0 \\ \min(F_f, F_b) & Q = 0 \end{cases} \qquad (12)$$

where F_f and F_{bl} are defined as the values of F applied to the T-stub to produce the first plastic hinge at the weld location in the end plate and the bolt line in the end plate (or in the column flange) and F_b is the value of F causing tensile yielding of the bolts. In the presence of prying forces ($Q > 0$) these forces are given by

$$F_f = \frac{\eta_{Ys}\left(\dfrac{\eta_{Es}}{n_b \eta_{Eb}} + \beta \dfrac{K_{b0}}{\overline{K}_0}\right) \cdot \min(M_{ep0}, M_{cp0})}{(e+m)\dfrac{\eta_{Es}}{n_b \eta_{Eb}} + e(\beta - \alpha)\dfrac{K_{b0}}{\overline{K}_0} + m\beta \dfrac{K_{b0}}{\overline{K}_0}} \qquad (13)$$

$$F_{bl} = \frac{\eta_{Ys}\left(\dfrac{\eta_{Es}}{n_b \eta_{Eb}} + \beta \dfrac{K_{b0}}{\overline{K}_0}\right) \cdot \min(M_{ep0}, M_{cp0})}{e(\alpha - \beta)\dfrac{K_{b0}}{\overline{K}_0} - \dfrac{e\eta_{Es}}{n_b \eta_{Eb}}} \qquad (14)$$

$$F_b = \eta_{Yb}\left(\frac{1}{\alpha}\frac{\eta_{Es}}{\eta_{Eb}}\frac{\overline{K}_0}{K_{b0}} + \frac{n_b\beta}{\alpha}\right)B_{y0}, \qquad (15)$$

while in the absence of prying forces ($Q = 0$)

$$F_f = \eta_{Ys} \cdot \min\left(\frac{M_{ep0}}{m}, \frac{M_{cp0}}{m}\right) \text{ and } F_b = n_b\eta_{Yp}B_{y0}, \qquad (16)$$

where $\eta_{Yb} = f_{ybT}/f_{yb0}$, with f_{ybT} and f_{yb0} being the yield strength of the bolt material at elevated and ambient temperatures respectively. The total central deflection of the T-stub at first yield (Figure 1) can be obtained from

$$^1\delta_y = \begin{cases} \dfrac{^1F_y}{24\eta_{Es}\overline{K}_0}\left(1 - \dfrac{24\alpha^2}{\dfrac{\eta_{Es}\overline{K}_0}{n_b\eta_{Eb}K_{b0}} + \beta}\right) & Q > 0 \\[4mm] ^1F_y\left(\dfrac{1}{n_b\eta_{Eb}K_{b0}} + \dfrac{m^3}{3L^3\eta_{Es}\overline{K}_0}\right) & Q = 0 \end{cases} \qquad (17)$$

2.3 Formation of subsequent yield points

As the temperature increases and the force F increases above its first yield point value ^1F_y, other yielding locations develop that depend on the location of the first hinge.

For case (I) which is defined as that for which the first hinge occurs at the fillet weld on the end plate (or in the column flange), the bending moment at this location equals M_{epT} (or M_{cpT}). A second yield point can form at the bolt line of the T-stub assembly, or in the bolts themselves, characterised by the force F reaching the value ^2F_y, where

$$^2F_y = {}^1F_y + {}^1\Delta F \qquad (18)$$

with

$$^1\Delta F = \min(\Delta F_{bl}, \Delta F_b) \qquad (19)$$

in which

$$\Delta F_{bl} = \frac{\min(M_{epT}, M_{cpT}) - e \, {}^1Q_y}{m} \qquad (20)$$

$$\Delta F_b = \frac{e(n_b\eta_{Yb}B_{y0} - {}^1B_y)}{m+e} \qquad (21)$$

with ^1Q_y and ^1B_y being determined by substituting ^1F_y from Equation 12 into Equations 6. Applying the condition $M = M_{epT}$ (or $M = M_{cpT}$) in the model based on classical beam theory results in the central deflection of the T-stub assembly being given by

$$^2\delta_y = {}^1\delta_y + {}^1\Delta F \cdot \left[\frac{2m^2(e+m)}{3L^3\eta_{Es}\overline{K}_0} + \left(1 + \frac{m}{e}\right)^2 \frac{1}{n_b\eta_{Eb}K_{b0}}\right] \qquad (22)$$

It should be noted that when $^1\Delta F = \Delta F_{bl}$, the second and third plastic hinges form simultaneously owing to symmetry when $M_{epT} = M_{cpT}$, and a mechanism will develop, the mechanism being identified as Mode I in Figure 2a and with ultimate values $F_u = {}^2F_y$ and $\delta_u = {}^2\delta_y$. However, if $M_{cpT} < M_{epT}$ when $^1\Delta F = \Delta F_{bl}$, the force can be increased until the bolts yield, and then fracture, as identified by Mode VII in Figure 2b. For this, yielding of the bolts occurs when

$$^2\Delta F = n_b\eta_{Yb}B_{y0} - {}^2B_y \qquad (23)$$

and

$$^3\delta_y = {}^2\delta_y + {}^2\Delta F \cdot \left(\frac{1}{n_b\eta_{Eb}K_{b0}} + \frac{m^3}{3L^3\eta_{Ets}\overline{K}_{t0}}\right) \qquad (24)$$

where

$$^2B_y = {}^1B_y + \Delta F_{bl}\left(\frac{m}{e} + 1\right) \text{ and}$$

$$\frac{1}{\overline{K}_{t0}} = \frac{L^3}{E_{ts0}}\left(\frac{1}{I_e} + \frac{\eta_{Ets}E_{ts0}}{\eta_{Es}F_{s0}} \cdot \frac{1}{I_e}\right) \qquad (25)$$

583

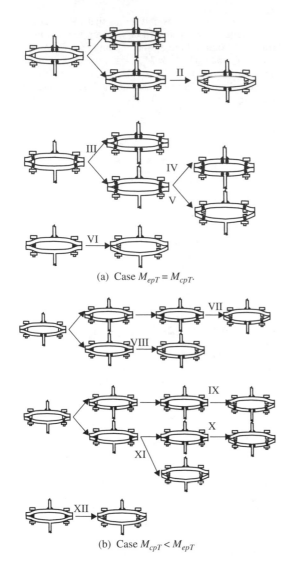

(a) Case $M_{epT} = M_{cpT}$.

(b) Case $M_{cpT} < M_{epT}$

Figure 2. T-stub failure modes.

and $\eta_{Ets} = E_{tsT}/E_{ts0}$ defines the degradation of the tangent modulus of structural steel. Finally, to reach bolt fracture in this case,

$$^{3}\Delta F = n_{b}(\eta_{Ub}B_{u0} - \eta_{Yb}B_{y0})$$ (26)

and so the ultimate values are

$$F_{u} = {}^{1}F_{y} + {}^{1}\Delta F + {}^{2}\Delta F + {}^{3}\Delta F$$ (27)

$$\delta_{u} = {}^{3}\delta_{y} + {}^{3}\Delta F\left(\frac{1}{n_{b}\eta_{Etb}K_{tb0}} + \frac{m^{3}}{3L^{3}\eta_{Ets}\overline{K}_{t0}}\right)$$ (28)

where

$$K_{tb0} = \frac{E_{tb0}A_{s}}{\lambda_{b}}$$ (29)

and $\eta_{Ub} = f_{ubT}/f_{ub0}$ defines the degradation of the ultimate tensile strength of the bolt material and $\eta_{ETb} = E_{tbT}/E_{tb0}$ defines the degradation of its tangent modulus at elevated temperature.

If $^{1}\Delta F = \Delta F_{b}$ in Equation 19, the bolts yield and the applied load and deflections can be increased until they fracture (failure modes II of figure 2a and VIII of figure 2b) and the corresponding increment load ($^{2}\Delta F$) can be given by the right hand side of Equation 26; however the ultimate values of deflection and lode are then obtained from:

$$F_{u} = {}^{1}F_{y} + {}^{1}\Delta F + {}^{2}\Delta F$$ (30)

$$\delta_{u} = {}^{2}\delta_{y} + {}^{2}\Delta F\left(\frac{1}{n_{b}\eta_{Etb}K_{tb0}} + \frac{m^{3}}{3L^{3}\eta_{Es}\overline{K}_{0}}\right)$$ (31)

For case (II) which is that when first yielding occurs at the bolt line, the prying force cannot be increased further (Spyrou & Davison 2001, Spyrou et al. 2004a,b) and Equation 12 can be rewritten as

$$^{1}\Delta F = \min\left(\Delta F_{f}, \Delta F_{b}\right)$$ (32)

where

$$\Delta F_{f} = \min\left(\frac{M_{epT}}{m}, \frac{M_{cpT}}{m}\right), \quad \Delta F_{b} = n_{b}\eta_{Yb}B_{y0} - {}^{1}B_{y},$$ (33)

and the central separation of the T-stub becomes

$$^{2}\delta_{y} = {}^{1}\delta_{y} + {}^{1}\Delta F \cdot \left(\frac{1}{n_{b}\eta_{Eb}K_{b0}} + \frac{m^{3}}{3L^{3}\eta_{Es}\overline{K}_{0}}\right).$$ (34)

Following a similar procedure to the previous case, when $^{1}\Delta F = \Delta F_{f}$ and $M_{epT} = M_{cpT}$, the T-stub assembly develops a mechanics (Figure 2a, mode III) and the ultimate values are $F_{u} = {}^{1}F_{y} + {}^{1}\Delta F$ and $\delta_{u} = {}^{2}\delta_{y}$. When $M_{cpT} < M_{epT}$, the force can be increased further until the bolts yield and then fracture (failure mode IX of Figure 2b), and the corresponding load and deflection increments can be determined using Equations 23 and 24, and Equations 26 to 28, in which

$$^{2}B_{y} = {}^{1}B_{y} + \Delta F_{f}.$$ (35)

On the other hand, if $^{1}\Delta F = \Delta F_{b}$ in Equation 32, a third plastic hinge may form at the fillet weld, or the bolts may fracture before the formation of this hinge at the fillet. In the former case, when $M_{epT} = M_{cpT}$ the

T-stub develops the final mechanism shown in Figure 2a (failure mode IV), whilst when $M_{cpT} < M_{epT}$ the applied load and deflections can be increased (Equations 26 to 28) until the bolts fracture (failure mode X of Figure 2b). In the latter case where bolt fracture occurs first (Figure 2a, mode V and Figure 2b, mode XI), the second load increment is given by

$$^{2}\Delta F = ^{2}\Delta F_{b} \qquad (36)$$

in which $^{2}\Delta F_{b}$ is given by the right hand side of Equation 26. The corresponding ultimate values of F_{u} and δ_{u} are then obtained from Equations 30 and 31.

For case (III) for which the bolts yield first, the T-stub is not subjected to prying forces and any increase of load is carried by the bolts. In this case, the load may increase until the bolts fracture (failure modes VI of Figure 2a and XII of Figure 2b). Equation 19 may therefore be written as

$$^{1}\Delta F = ^{1}\Delta F_{b} \qquad (37)$$

where $^{1}\Delta F_{b}$ is the same as the right hand side of Equation 26. The ultimate separation and tensile resistance of the T-stub assembly are then

$$^{2}\delta_{y} = ^{1}\delta_{y} + ^{1}\Delta F \cdot \left(\frac{1}{n_{b}\eta_{Etb}K_{tb0}} + \frac{m^{3}}{3L^{3}\eta_{Es}\overline{K}_{0}} \right) \quad \text{and} \qquad (38)$$

$$F_{u} = ^{1}F_{y} + ^{1}\Delta F . \qquad (39)$$

3 MODEL VALIDATION

Spyrou *et al.* (2004a,b) reported an experimental investigation of the behaviour of T-stub assemblies, in which separate results were given for the end plate and column flange separately. This is consistent with their modelling of the column flange or plate as a rigid stratum, whilst the present modelling accounts for both components being flexible. Because of this, the values of the appropriate stiffnesses in Equations 2 were chosen such that K_{cT} or $K_{eT} \rightarrow \infty$.

Tables 1 and 2 show the geometrical and mechanical properties of typical specimens from Spyrou's experimental results used herein. The mechanical properties of the bolt material was taken from Theodorou (2001) which was assumed to exhibit a bilinear stress-strain relationship, and the temperature-dependent properties of structural steel were taken from the Australian AS4100 standard (1998). For all tests, the Young's modulus of the bolts at ambient temperature was taken as 215,000 MPa, and the yield and ultimate tensile strengths of the bolts at ambient temperature were 811 MPa and 886 MPa.

Figures 3 to 6 show a comparison of the test data of Spyrou with the predictions of the current model. It

Table 1. Geometric properties of test specimens used (mm).

Test	e	m	t_e	t_f	d_b
1	48.55	32.03	20.10	∞	20
2	83.22	28.03	∞	21.12	20
3	82.46	28.51	∞	21.14	20
4	31.85	35.895	∞	9.20	20

Table 2. Mechanical properties of test specimens.

Test	T_{T-stub} (°C)	T_{bolt} (°C)	$f_{y,T-stub}$ (MPa)	E_{T-stub} (MPa)
1	505	505	284	192000
2	615	618	288	189000
3	700	703	288	189000
4	705	705	285	198000

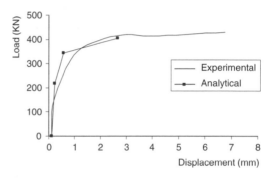

Figure 3. Comparison of experiments and theory for Test 1.

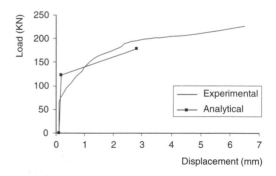

Figure 4. Comparison of experiments and theory for Test 2.

can be seen that the agreement is generally very good, with any differences being believed to be attributable to strain hardening and the post yield strength which were not accounted for in the present model.

585

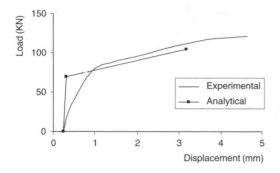

Figure 5. Comparison of experiments and theory for Test 3.

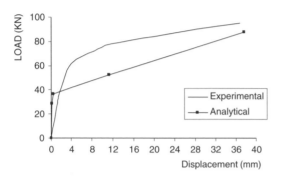

Figure 6. Comparison of experiments and theory for Test 4.

Nevertheless, any discrepancy is considered acceptable for providing practical design guidance.

4 CONCLUSIONS

This paper has considered the behaviour of the beam-to-column connection in a steel frame when subjected to elevated temperatures. The study has been undertaken in the context of the component method of joint design that is used widely in the Eurocode and elsewhere. Unlike previous research, the present application has incorporated the flexibility of the end plate and the column flange, as well as the temperature degradation of the bolt material and structural steel. The analytical results have been compared with tests undertaken at University of Sheffield that are reported in the literature, where acceptable correlation has been seen to be achievable. The analytical model herein is simple, and the good correlation shows that the component method is appropriate for elevated temperatures, and that acceptable design procedures can be formulated. It forms a useful foundation for steel joint design in a structural fire engineering approach.

ACKNOWLEDGEMENT

The work reported in this paper was supported by the Australian Research Council through a Federation Fellowship awarded to the second author.

REFERENCES

Agerskov, H. 1976. High-strength bolted connections subject to prying. *Journal of the Structural Division, ASCE* 102(ST1): 162–175,

Agerskov, H. 1977. Analysis of bolted connections subject to prying. *Journal of the Structural Division, ASCE* 103(ST11): 2145–2163.

Bahaari, M.R. & Sherbourne, A.N. 1996. Structural behavior of end-plate bolted connections to stiffened columns. *Journal of Structural Engineering, ASCE* 122(8): 926–935.

British Standards Institute 2006. *Eurocode 3: Design of Steel Structures – Part 1.8: Design of Joints*. London: BSI.

Piluso, V., Faella, C. & Rizanno, G. 2001a. Ultimate behavior of bolted T-stubs. I: Theoretical model. *Journal of Structural Engineering, ASCE* 127(6): 686–693.

Piluso, V., Faella, C. & Rizanno, G. 2001b. Ultimate behavior of bolted T-stubs. I: Model validation. *Journal of Structural Engineering, ASCE* 127(6): 694–704.

Shi, Y.J., Chan, S.L. & Wong Y.L. 1996. Modeling for moment-rotation characteristics for end plate connections. *Journal of Structural Engineering, ASCE* 122(11): 1300–1306.

Spyrou, S. & Davison J.B. 2001. Displacement measurement studies in steel T-stub connections. *Journal of Constructional Steel Research* 57(6): 647–659.

Spyrou, S., Davison, J.B., Burgess, I.W. & Plank, R.J. 2004a. Experiments and analytical investigation of the 'tension zone' components within a steel joint at elevated temperatures. *Journal of Constructional Steel Research* 60(6): 867-896.

Spyrou, S., Davison, J.B., Burgess, I.W. & Plank, R.J. 2004b Experimental and analytical studies of steel joint components at elevated temperatures. *Fire and Materials* 28: 83–94.

Standards Australia 1998. *Steel Structures*. Sydney:SA.

Swanson J.A. & Leon, R.T. 2000. Bolted steel connections: Tests on T-stub components. *Journal of Structural Engineering ASCE* 126(1): 50–56.

Swanson, J.A. & Leon, R.T. 2001. Stiffness modeling of bolted T-stub connection components. *Journal of Structural Engineering, ASCE* 127(5): 498–505.

Theodorou, Y. 2001. Mechanical properties of Grade 8.8 bolts at elevated temperatures. MSc Thesis, The University of Sheffield.

Timoshenko, S.P. & Gere, J.M. 1977. *Mechanics of Materials*. New York: Van Nostrand Reinhold Company.

Zoetemeijer, P. 1974. A design method for the tension side of statically loaded bolted beam-to-column connections. *Heron* 20(1): 1–59.

Steel and Composite Structures – Wang & Choi (eds)
© *2007 Taylor & Francis Group, London, ISBN 978-0-415-45141-3*

Comparative study of the behaviour of BS 4190 and BS EN ISO 4014 bolts in fire

Y. Hu, J.B. Davison & I.W. Burgess
Department of Civil & Structural Engineering, Sheffield, UK

R.J. Plank
School of Architectural Studies, Sheffield, UK

ABSTRACT: A comparative study on bolt assemblies to BS 4190: 2001 and BS EN ISO 4014: 2001 has been conducted recently at the University of Sheffield. The intention of this study was to observe the performance of these two types of bolts in tension at both ambient and elevated temperatures. In addition, the project attempted to identify an approach to eliminate premature failure due to thread stripping at elevated temperatures, as had been seen in earlier connection tests, and thereby to prevent such failures occurring in a subsequent series of connection tests.

Experimental results have shown the pattern of failure of Grade 8.8 bolts with Grade 10 nuts to differ from those of Grade 8.8 bolts with Grade 8 nuts at normal temperatures. Furthermore, the bolts specified to British and European Standards performed slightly differently in the tensile tests. At elevated temperatures, the study focused on bolt assemblies: using Grade 8.8 bolts with Grade 10 nuts. The bolts performed differently in the elevated-temperature tensile tests, possibly reflecting different specifications in these two standards, or variations in the manufacturing (quenching and tempering) processes used by different fastener manufacturers.

1 GENERAL INTRODUCTION

Bolted connections are widely used in steel structures as they enable prefabricated beams and columns to be erected quickly on site without the complication and expense of on-site welding. The available deformation (or ductility) for bolts is relatively small, and a "brittle" failure of bolts could adversely affect the rotation capacity of joints, and possibly the behaviour of complete structures, unless the necessary ductility is provided through plate deformation capacity.

A thorough understanding of the performance of bolts at both ambient and elevated temperatures is important in studies of connection behaviour. This programme of research was undertaken to study the performance of Grade 8.8 bolts to the new British Standard BS 4190 (2001) and the European Standard BS EN ISO 4014 (2001) at ambient and elevated temperatures.

2 BACKGROUND REVIEW

2.1 Bolt standards

In the UK, non-preloaded bolts have been specified to both BS 3692: 1967 and BS 4190: 1967 for many years.

With the introduction of BS EN standards, the British Standards for bolts have been declared obsolete. However, the introduction of European standards has not been fully accepted by the UK construction industry, and bolts are still ordered to the familiar British Standards. Therefore, BS 3692 and BS 4190 were revised and re-published in 2001. These new British Standards specify mechanical properties of steel in line with BS EN ISO 898-1 (1999), and BS 4190: 2001 now covers the full range of bolt grades from 4.6 to 10.9 for non-preloaded bolts. In consequence, the National Structural Steelwork Specification for Building Construction (2007) recommends that bolts used in steel or composite structures should be ordered to either the European Standard BS EN ISO 4014 or the British Standard BS 4190.

2.2 Previous bolt tests

Godley and Needham (1983) conducted a series of comparative tests on Grade 8.8 bolts and HSFG bolts in pure tension. The results displayed two typical failure modes for bolts: tensile bolt breakage and thread stripping. In 1994, a group of tensile tests was completed in the Materials Research Laboratory, Australia. The failure mechanism of thread stripping was presented as a three-stage process: threads elastically deformed, threads plastically deformed and threads sheared off (Mouritz, 1994).

In order to understand the performance of bolts in fire, Kirby (1995) carried out a series of tests on Grade 8.8 bolts. Interestingly, he found that thread stripping happened not only at ambient temperature but also at elevated temperatures. A remarkably high loss of strength of bolts occurred in the temperature range 300°C to 700°C for both tension and shear. To avoid thread stripping, Kirby highlighted the importance of the interaction between the threads of bolts and nuts (the degree of fit between these two components). The Tolerance Class between bolts and nuts affects the mechanism of failure to some extent. In addition, the Strength Grade of nuts used in combination with bolts is an important factor affecting the failure mechanism of the assembly. This was noted by both Barber (2003) and by Godley and Needham (1983).

However, a number of comments should be borne in mind in the context of this review. Firstly, before the publication of new bolt standards, bolts were always ordered to either BS 4190:1967 or BS 3692: 1967; the difference between the two bolt standards is that the requirement for the Tolerance Class of Grade 8.8 bolts is not the same. To BS 3692:1967 the tolerance class is 6 H/6 g, but to BS 4190: 1967 it is 7 H/8 g. Secondly, the new European Standard (BS EN ISO 4014) accepts the value of 6 H/6 g as a Tolerance Class for bolts, while British Standards (BS 4190) adopts Tolerance Class 7 H/8 g. Finally, both Kirby and Godley and Needham used bolts ordered to BS 3692: 1967, rather than BS 4190: 1967. These old British Standards have now been declared obsolete, and are superseded by BS 4190: 2001 and BS EN ISO 4014. A knowledge gap on bolts ordered to the new standards therefore needs to be covered. The focus of this research is on this point, and on attempting to understand the behaviour of Grade 8.8 bolts specified to different standards at normal and elevated temperatures.

3 THE OBJECTIVES FOR BOLT TESTS

Grade 8.8 20 mm diameter bolts were obtained to both British Standards (BS 4190) and European Standards (BS EN ISO 4014), partly-threaded and 100 mm long. The bolts were supplied from two manufacturers A and B. The nuts were supplied according to BS 4190 and BS EN ISO 4032. The bright-finish nuts were of Property Class 8 and black nuts were of Property Class 10. Throughout these tests, the bolt assemblies were divided into four classes as illustrated in Table 1:

The tolerance class for bolts and nuts can be considered as the degree of fit between these two components. Kirby (1995) suggests that premature failure due to thread stripping could be controlled by the degree

Table 1. Details of bolt assemblies in the experiments.

Bolt sets	Bolts Standards	Nuts Standards	—Tolerance (nut/bolt)
Group C	Grade 8.8 (BS 4190, Br*)	Grade 8 (BS 4190, Br*)	7 H/8 g
Group A	Grade 8.8 (BS 4190, Br*)	Grade 10 (BS 4190, Ba*)	7 H/8 g
Group D	Grade 8.8 (ISO 4014, Ba*)	Grade 8 (ISO 4032, Br*)	6 H/6 g
Group B	Grade 8.8 (ISO 4014, Ba*)	Grade 10 (ISO 4032, Ba*)	6 H/6 g

Br* = Bright finish (zinc plated) , Ba* = Black finish

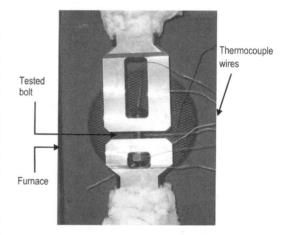

Figure 1. The arrangement of bolt tests.

of fit between bolt and nut threads. Therefore, the first objective in this research was to identify an approach which would eliminate premature failure due to thread stripping at elevated temperatures. The second was to observe the performance of bolts ordered to BS 4190 and to BS EN ISO 4014 in tension at both ambient and elevated temperatures.

4 TEST ARRANGEMENT

A test rig used for all the bolts was designed to fit a compression/tension testing machine, shown in Figure 1.

Since the tests were carried out under displacement control, a nominal strain rate of 0.003 mm/min was adopted in both the elastic and post-elastic regions, which is consistent with the recommendation by Kirby (1995). Each bolt assembly was heated to the desired temperature, maintained for a period of 15 min for stabilization to establish a uniform temperature distribution, and then tested under displacement control. For

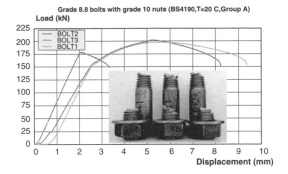

Figure 2. BS 4190 bolts with Property Class 10 nuts (T = 20°C, Group A).

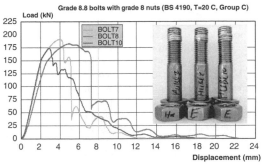

Figure 4. BS 4190 bolts with Property Class 8 nuts (T = 20°C, Group C).

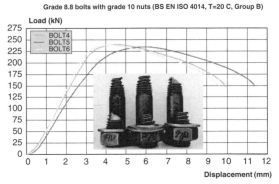

Figure 3. ISO 4014 bolts with Property Class 10 nuts (T = 20°C, Group B).

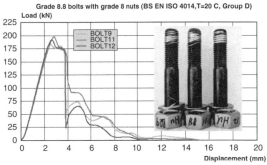

Figure 5. ISO 4014 bolts with Property Class 8 nuts (T = 20°C, Group D)

the heating rate, a value of 2–2.5°C /min was adopted throughout this investigation. This is in accordance with Kirby's (1995) statement that a prolonged "soaking" period and a slow heating rate had little influence on the ultimate capacity of bolt assemblies. A total of 42 tests was carried out with bolts of diameter 20 mm, in addition to a few pilot tests to check the equipment and procedures.

5 TEST RESULTS AT AMBIENT TEMPERATURE

Figures 2 to 5 display a typical set of failed specimens with the corresponding load-deformation curves for each tested group. From the figures, it can be seen that the bolts with Property Class 10 nuts failed by bolt breakage, and for the bolts with Property Class 8 nuts, the failure mode was always the stripping-off of the threads of the nuts. The Tolerance Class (the degree of fit) did not appear to influence the final failure mode of the bolt assemblies.

Table 2. Ultimate capacities of bolts at ambient temperature.

Bolt sets	Bolt mark	Recorded failure loads	Nuts' grade
Group A (BS 4190)	BOLT1	197.8 kN	grade 10
	BOLT2	202.3 kN	grade 10
	BOLT3	178.5 kN	grade 10
Group B (ISO 4014)	BOLT4	239.7 kN	grade 10
	BOLT5	234.4 kN	grade 10
	BOLT6	238.4 kN	grade 10
Group C (BS 4190)	BOLT7	191.1 kN	grade 8
	BOLT8	182.7 kN	grade 8
	BOLT10	173.2 kN	grade 8
Group D (ISO 4014)	BOLT9	183.8 kN	grade 8
	BOLT11	197.6 kN	grade 8
	BOLT12	191.4 kN	grade 8

Table 2 summarises the maximum failure loads recorded for each bolt test. The minimum ultimate load is 192.3 kN (= 785 N/mm^2 × 245 mm^2).

Comparing the recorded bolt failure loads to 192.3 kN, Grade 8.8 bolts with Property Class 10 nuts (Groups A and B) can guarantee the minimum ultimate tensile value specified previously, except for one questionable bolt BOLT3. However, the bolts (Groups C and D) with the premature failure due to thread stripping cannot meet the specified minimum ultimate value. In addition, comparing the average loads between Groups A and C (BS 4190 bolts) the reduction in capacity using grade 8 nuts compared with grade 10, was 5.5% (excluding BOLT3) and for ISO 4014 bolts (Groups B and D) the corresponding reduction was 19.6% in their ultimate capacity. Premature failure happened in the bolts, leading to a reduction of the full capacity of nut and bolt assemblies, which could affect the performance of bolted steel connections at both ambient and high temperatures.

Two sets of M20 nuts were used in this test programme. The Property Class 10 nuts were supplied with a black finished surface and the Property Class 8 nuts were bright finished (Zinc plated or galvanized). Kirby (1995) states that the process of manufacturing for black nuts involves hot forging from steel bar, quenching from 870°C and tempering back at around 540°C. By contrast, the bright finished nuts are produced by cold forging steel bar. In Kirby's test programme, the bolts with black nuts always failed by bolt breakage. In contrast, the failure of the same bolt assemblies using the bright nuts occurred entirely by thread-stripping. Therefore, the test results and failure modes of bolts suggest that the black nuts may have a slightly higher hardness value and proof load than bright finished nuts. Kirby also believes that the threads of the bright nuts are slightly weaker than those of the black finished nuts, which is caused by the threads of bright finished nuts commonly being 'over-tapped' in order to accommodate the additional thickness of surface protective coatings. Similarly, in the bolt test programme reported here, the bolts in combination with the black nuts failed by ductile necking in the threaded portion; the bolts using the bright nuts failed by thread-stripping. From this discussion, it can be concluded that the failure patterns of bolt assemblies can be controlled by the type of nuts which are used.

Grade 8.8 bolts are manufactured by either hot or cold forging of steel, followed by a quenching and tempering process to develop the required mechanical properties. The processing method for bolts and nuts varies between manufacturers in the market, and is controlled by the chemical composition of the feedstock, the size range and the demand for a particular bolt, all of which are a function of producing a competitively priced product (Kirby, 1995). At ambient temperature, the bolts are made to meet the specifications of national or international standards, but the performance of bolts both during and after a fire will vary depending on how they have been manufactured. Therefore, it is important for the high temperature bolt tests to focus on the performance of two different types of bolts: BS 4190 bolts and ISO 4014 bolts.

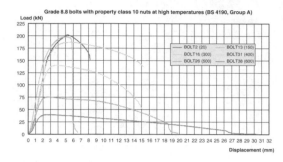

Figure 6. BS 4190 bolts with property class 10 nuts (T = 20°C–600°C, Group A).

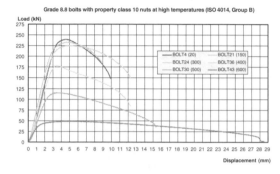

Figure 7. ISO 4014 bolts with property class 10 nuts (T = 20°C–600°C, Group B).

6 ELEVATED-TEMPERATURE TEST RESULTS

The test programme at high temperatures used the Grade 8.8 bolts with Property Class 10 nuts in order to prevent thread-stripping. A total of 36 bolt tests were carried out in the temperature range from 20°C to 600°C; two bolt sets were involved:

- Group A: Grade 8.8 bolts with Grade 10 nuts (BS 4190)
- Group B: Grade 8.8 bolts with Grade 10 nuts (BS EN ISO 4014)

Figures 6 and 7 display the load-displacement performance of bolts at different temperatures. In the high temperature programme 34 bolts with black nuts failed by bolt breakage across the threaded part at all temperatures, and just 2 bolt assemblies (BOLT18 and BOLT19) failed by thread stripping at the temperature of 150°C. It is evident from the these two graphs that the ultimate tensile load decreased for each bolt as the furnace temperatures increased, and elongation increased with temperature.

Figure 8 summarises the ultimate capacities at different temperatures and Figure 9 displays the strength reduction factors indicated for both BS 4190 and ISO 4014 bolts, compared to the factors given by Kirby

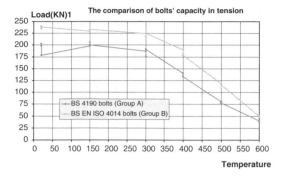

Figure 8. Comparison of bolt ultimate tensile loads at various temperatures.

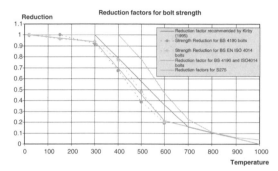

Figure 9. Comparison of strength reduction factors indicated for BS 4190 and ISO 4014 bolts, and by Kirby (1995).

(1995). The tests in tension using Grade 8.8 bolts and black nuts have highlighted a marked loss in ultimate capacity and bolt strength between 300°C and 600°C, whereas only a slight loss happened in the temperature range 20°C to 300°C. The test results for ISO 4014 bolts in combination with black nuts are quite close to those for Bolt Set A with Nut A in Kirby's test programme.

For grade 8.8 bolts acting in either shear or tension, Kirby (1995) presents a strength reduction factor to describe the ultimate capacity at the fire limit state, which is defined by a tri-linear relationship as:

$$
\begin{aligned}
SRF &= 1.0 & (T \le 300\ ^\circ C) \\
&= 1 - 0.2128\,(T - 300) \times 10^{-2} & (300\ ^\circ C < T \le 680\ ^\circ C) \\
&= 0.17 - 0.5312\,(T - 680) \times 10^{-3} & (680\ ^\circ C < T \le 1000\ ^\circ C)
\end{aligned}
$$

where T = temperature.

The mechanical properties for Kirby's Grade 8.8 bolts meet the requirements of BS 3692 design rules. However, the bolts used in the reported test programme are specified to either BS 4190 or BS EN ISO 4014 standards. According to the test results, it would be

appropriate to amend the Kirby's reduction factors to take into account the possible loss in bolt strength from 20 to 300°C and the slight change in reduction gradient for bolts between 300 and 600°C. For temperatures above 600°C it is reasonable to assume that the reduction curve for bolt strength is similar to Kirby's.

The Strength Reduction Factors for BS 4190 and ISO 4014 bolts may be defined by an amended tri-linear relationship (Figure 9):

$$
\begin{aligned}
SRF &= 1 - (0.2275\,T - 4.55) \times 10^{-3} & (20\ ^\circ C \le T \le 300\ ^\circ C) \\
&= 0.9363 - 0.24\,(T - 300) \times 10^{-2} & (300\ ^\circ C < T \le 600\ ^\circ C) \\
&= 0.5407\,(1 - T/1000) & (600\ ^\circ C < T \le 1000\ ^\circ C)
\end{aligned}
$$

where T = temperature.

Bolts tested in tension commonly experience the problem of premature failure by thread-stripping at both normal and high temperatures. Kirby (1995) argues that this type of failure mode may be controlled by improving the interaction of the threads between the nut and bolt. In order to achieve the full capacity of the bolts in tension, it is therefore recommended that bolts with the tolerance class given in BS EN ISO 4014 (6 H/6 g) be used, preferably together with nuts supplied to the higher strength Grade 10, specified in BS EN ISO 4032 (Barber, 2003). Moreover, it seems that the final stage in manufacture of bright nuts is commonly galvanizing, and to accommodate this the nut threads are commonly 'over-tapped' by 0.4 mm to accommodate the coating thickness (Thomas William Lench, 2007). The protective coatings such as zinc are relatively soft compared to steel, and zinc melts at 420°C. In consequence, this process results both in a reduction of cross-sectional area and a reduction in the tolerance in the nut threads. The evidence, from both the current series of bolt tests and Kirby's results, that the threads of bright nuts stripped off in all the tensile tests, supports this statement. Consequently, selecting black nuts instead of bright nuts is an effective measure in achieving improved performance of bolts in tension at both ambient and elevated temperatures.

7 CONCLUSIONS AND RECOMMENDATIONS

This research has investigated Grade 8.8 bolts, which are specified to the standards BS 4190 and BS EN ISO 4014, at both ambient and elevated temperatures. From the test results, the ISO 4014 bolts with Property Class 10 nuts had a better performance than BS 4190 bolts in all tests. In addition, their avoidance of thread stripping has been illustrated at both ambient and elevated temperatures in this study. Finally, the amended reduction factors for BS 4190 and ISO 4014 bolts may be used to predict the strength reduction for bolts in a temperature-varying environment.

In this study, the bolts have been tested in pure tension at both ambient and elevated temperatures. It may be worth putting some research effort into testing BS 4190 and ISO 4014 bolts for performance in shear, especially in elevated-temperature conditions.

ACKNOWLEDGEMENT

The research work described in this paper is part of a project funded under Grant EP/C510984/1 by the Engineering and Physical Sciences Research Council of the United Kingdom. This support is gratefully acknowledged by the authors. In this project, the authors would also like to thank the technical staff for their assistance and brilliant work.

REFERENCES

Barber, H. (2003) Chapter 23 Bolts, In J.B. Davison and Owens, G.W.(Eds), *Steel Designers' Manual* : 671–684 London and Cambridge: Blackwell Science.

BS 3692 (2001), "BS 3692: ISO metric precision hexagon bolts, screws and nuts – Specification" British Standards Institution, London.

BS 4190 (2001), "BS 4190: ISO metric black hexagon bolts, screws and nuts – Specification", British Standards Institution, London.

BS EN ISO 4014 (2001), "BS EN ISO 4014: Hexagon head bolts–Products grades A and B", European Committee for Standardisation, Brussels.

BS EN ISO 4032 (2001), "BS EN ISO 4032 Hexagon nuts, style 1–Products grades A and B", European Committee for Standardisation, Brussels.

BS EN ISO 898-1 (1999), "BS EN ISO 898–1 Mechanical properties of fasteners made of carbon steel and alloy steel – Part 1: Bolts, screws and studs", European Committee for Standardisation, Brussels.

Godley, M. H. R. and Needham, F. H. (1983), "Comparative tests on 8.8 and HSFG bolts in tension and shear", *The Structural Engineer* **60** (3), pp 21–26.

Kirby, B.R. (1995), "The Behaviour of High-strength Grade 8.8 Bolts in Fire", *J. Construct. Steel Research* **33** (1), pp3–38.

Mouritz, A.P. (1994), "Failure Mechanisms of Mild Steel Bolts under Different Tensile Loading Rates", *Int. J. Impact Engng* **15** (3), pp311–324.

National Structural Steelwork Specification for Building Construction 5th Ed(2007), http://www.steelconstruction.org/

Thomas William Lench (2007), http://www.thomas-william-lench.co.uk/structural.htm

Steel and Composite Structures – Wang & Choi (eds)
© *2007 Taylor & Francis Group, London, ISBN 978-0-415-45141-3*

Experimental study of the fire behavior of high strength bolt connections in steel frames

J.Q. He, Y.F. Luo & R.P. Wang
College of Civil Engineering, Tongji University, Shanghai, P. R. China

ABSTRACT: Few researches on the connection behavior of steel frames during fire and after fire have been found till now. It is of great importance to investigate the performance of steel connections after fire. Whether the steel frame can be reutilized and whether it is safe for rescuers to enter the damaged steel building as soon as possible after fire are based on the loading behavior of steel members and connections after fire. According to an actual fire state, a fire test for simulating fire effects on joints of a steel frame is conducted in this paper. The joint used for fire test is composed of steel plates and high-strength bolts. The change of the behavior of high-strength bolts after fire is investigated. The experimental results show that the friction coefficient of high-strength bolt connections decreased after fire. A significant relaxation of the pretension of high-strength bolts is caused by the fire. Finally, fire effects result in the weakness of the joint stiffness of the steel frame. It is a valuable reference for future similar researches and actual engineering.

1 INTRODUCTION

High strength bolt joints which have become main construction methods of steel joints for its high load-bearing capacity and convenience for construction have gradually replaced some traditional steel joints such as rival ones and partially seamed ones. Pre-torsion coefficient and friction coefficient have important influence on the bearing capacity of high strength bolt joints. Although significant developments have been made in analyzing the loading behavior of high strength bolts under natural condition, there have been few researches on the post-fire behavior of such kind of joints up till now. The main objectives of this paper are to investigate the change of the pre-torsion coefficient and slip-resistance force of the high strength bolt after fire through an experiment of post-fire behavior of high-strength bolt joints.

2 TEST SPECIMENS

High-strength bolt joints with the steel plate of Q235 and the bolt of 10.9S which are referenced to Chinese code for design of steel structures were used to simulate web connections of a beam from an actual project; There are 12 high-strength bolts in each specimen; see Fig. 1 and Fig. 2.

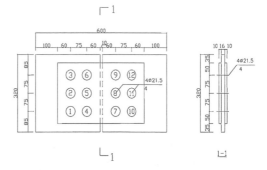

Figure 1. Specimen design details.

Figure 2. The actual specimen.

Table 1. Fire lasting time and cooling method of specimens.

Specimen No.	Fire lasting time	Cooling method
1	60 min	Water spray
2	30 min	Natural cooling
3	60 min	Natural cooling
4	–	–

* Specimen No.4 is tested under natural condition.

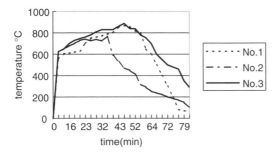

Figure 3. Temperature-time relation.

Four specimens which had no fire protection (three specimens under fire and one under natural condition) were tested and compared. The fire lasting time and the cooling methods are listed in table 1.

3 TEST ARRANGEMENT AND MEASUREMENTS

Each specimen without loading was set on the outdoor test shelf. The experiment was conducted under natural fire. A certain quantity of kerosene is taken as fuel for fire source.

The measured parameters are listed as below: (1) the temperature of the specimens under fire and after fire; (2) the pre-torsion of the high strength bolts before fire and after fire; (3) the pre-tension of the high strength bolts before fire and after fire; (4) the slip resistance forces of specimens before fire and after fire.

An infrared-ray temperature measurement equipment is used to measure the surface temperature of specimens every 2 minutes. Through the thermacam software, the temperature of the specimen at a specified time was collected; see Fig. 3. After specimens were cooled (the cooling method was listed in table 1), the pre-torsion, pre-tension and slip-resistance force of high strength bolts after fire were measured.

4 EXPERIMENTAL DATA AND DISCUSSION

The pre-torsion, the pretension, the pre-torsion coefficients of high strength bolt joints of three specimens before fire and after fire are listed in table2–4.

The pre-torsion coefficients of high strength bolt joints can be described as follows:

$$K_i = \frac{T}{P \times d} \qquad (1)$$

where K_i = pre-torsion coefficient; T = pre-torsion; d = diameter of bolt, and P = pre-tension.

A summary of the temperatures in bolts of specimen No.2 is presented in Fig 6. The pre-torsion of high strength bolts which reached the highest temperature in the heating phrase, reduced greatest after cooling, see Table 2.

The maximum reduction of the pre-torsion of specimen No.1 is bolt 4, a decrease of 94% because of the highest temperature; the minimum reduction is bolt 6 with a decrease of 33.73%. The reduction of the pre-torsion of the other bolts is all over 60%. Furthermore, the pre-tensions of bolts reduced greater. The reduction of pre-tension of bolt 4 reached 96.11%, and that of bolt 6 reached 74.99%. The pre-torsion coefficient of bolts of specimen No.1 increased after fire and the dispersion of the pre-torsion coefficient is greater than that before fire. These changes have resulted in the reduction of the load-bearing capacity of the high strength bolt connection.

To specimen No.2, the greatest reduction of the pre-torsion coefficient is bolt 1 reaching 91%. The least reduction appears in bolt 12, a reduction of 9.7%. The greatest reduction of the pre-torsion is bolt 7 reaching 95.8%. The least reduction of the pre-torsion is bolt 12 reaching 72.2%. The dispersion of pre-torsion coefficients of high strength of specimen No.2 is greater than that of specimen No.1. There some drastic reductions of a number of bolts in figure 4, such as the No. 5 bolt and No.9 bolt in specimen No.3, No.6 bolt in Specimen No.1 and No.12 bolt in Specimen No.2. The drastic reduction is caused by the good contact between steel plates, so some pre-torsions of high strength bolts reduced few after fire.

Similar phenomena appear in specimen No.3 (listed in table 4).

The reduction of pre-torsion of high strength bolts before fire and after fire is shown in figure 4 and the reduction of pre-tension of high strength bolts before fire and after fire is listed in figure 5. From figure 4 and figure 5, it is concluded that the reduction of the natural cooling is greater than that of the water-cooling. The loading performance of joints after natural cooling is worse than that after water spray.

In order to have a better understanding of the change of the friction surface of the high-strength blot joint,

Table 2. The pre-torsion coefficients and the comparison of specimen No.1 before fire and after fire.

Condition Item No.	Before fire				After fire			
	Pre-torsion (N·m)	Pretension (kN)	Pre-torsion coefficient	Average value	Pre-torsion (N·m)	Pretension (kN)	Pre-torsion coefficient	Standard deviation
12	501	168.1	0.149	0.149	200	36.8	0.272	0.098
9	504	169.1	0.149		118	12.3	0.480	
6	510	171.1	0.149		338	42.8	0.395	
3	507	170.1	0.149		106	12.7	0.417	
11	505	169.5	0.149		103	14.1	0.365	
8	501	168.1	0.149		102	10.7	0.477	
5	516	173.2	0.149		76	9.4	0.404	
2	502	168.5	0.149		119	12.9	0.461	
10	502	168.5	0.149		50	10.8	0.231	
7	501	168.1	0.149		198	20.9	0.474	
4	506	169.8	0.149		28	6.6	0.212	
1	514	172.5	0.149		71	11.8	0.301	

Table 3. The pre-torsion coefficients and the comparison of specimen No.2 before fire and after fire.

Condition Item No.	Before fire				After fire			
	Pre-torsion (N·m)	Pretension (kN)	Pre-torsion coefficient	Average value	Pre-torsion (N·m)	Pretension (kN)	Pretension coefficient	Standard deviation
1	503	168.8	0.149	0.149	45	9.9	0.227	0.109
4	510	171.1	0.149		126	17.8	0.354	
7	522	175.2	0.149		61	7.3	0.418	
10	506	169.8	0.149		162	30.2	0.268	
2	576	193.3	0.149		56	10.8	0.259	
5	506	169.8	0.149		69	8.3	0.416	
8	521	174.8	0.149		56	9.4	0.298	
11	506	169.8	0.149		194	17.4	0.557	
3	522	175.2	0.149		72	12.5	0.288	
6	503	168.8	0.149		158	15.2	0.520	
9	502	168.5	0.149		77	11.7	0.329	
12	504	169.1	0.149		455	47	0.484	

Table 4. The pre-torsion coefficients and the comparison of specimen No.3 before fire and after fire.

Condition Item No.	Before fire				After fire			
	Pre-torsion (N·m)	Pretension (kN)	Pre-torsion coefficient	Average value	Pre-torsion (N·m)	Pretension (kN)	Pretension coefficient	Standard deviation
1	515	172.8	0.149	0.149	102	13.4	0.381	0.146
4	530	177.9	0.149		71	17	0.209	
7	517	173.5	0.149		140	35.4	0.198	
10	506	169.8	0.149		220	49.6	0.222	
2	508	170.5	0.149		151	23.4	0.323	
5	558	187.2	0.149		443	40.2	0.551	
8	511	171.5	0.149		49	8.5	0.288	
11	502	168.5	0.149		98	12.9	0.380	
3	500	167.8	0.149		39	3.2	0.609	
6	506	169.8	0.149		57	6.3	0.452	
9	506	169.8	0.149		343	29.8	0.576	
12	505	169.5	0.149		77	8	0.481	

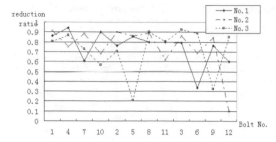

Figure 4. Reduction ratio of pre-torsion of high strength bolt
* Reduction ratio = (value under natural condition-value after fire)/value under natural condition.

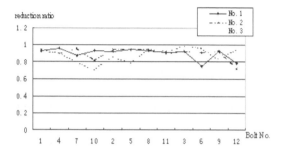

Figure 5. Reduction ratio of pre-tension of high strength bolt
* Reduction ratio = (value under natural condition-value after fire)/value under natural condition.

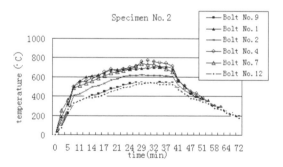

Figure 6. Differential relative temperature variation of specimen No.2.

the slip resistance experiment is conducted. Each specimen is cut into three parts from top to the bottom. Each part has four bolts in a line. The experimental results are shown in table 5. From table 5, it can be found that the slip-resistance force of specimen No.1, No.2 and No.3 has reduced obviously. The reduction of specimen 1 reached 83.96%, that of specimen No.2 reached

Table 5. Slip-resistance force

Bolt	Slip resistance force (KN)			
	No.1	No.2	No.3	No.4
1~10	34	18	32	212
2~11	34	32	12	202
3~12	22	64	152	214

*1~10 represents the slip resistance experimental specimen consisting of bolt No.1, No.4, No.7 and No.10; 2~11 represents the slip resistance experimental specimen consisting of bolt No.2, No.5, No.8 and No.11; 3~12 represents the slip resistance experimental specimen consisting of bolt No.3, No.6, No.9 and No.12.

91.5% and that of specimen No.3 reached 94.06%. Among the three specimens, the reduction of the slip-resistance force of Specimen No.1 is most uniform and that of the specimen No.3 is most un-uniform.

5 CONCLUSION

Through the fire experiment of high strength bolt connection , the conclusion is reached as below:

(1) The slip-resistance force, pre-torsion and the pre-tension of the high strength bolt joints after fire reduced apparently.
(2) The pre-torsion coefficient of high strength bolt joints after fire increased.
(3) The dispersion of the pre-torsion coefficients of high strength bolts after natural cooling is greater than that after water cooling.
(4) Fire results in the weakness of the joints stiffness, which may become the weak place of a structural after fire. Therefore it is of great importance to conduct fire protection of the joints.

REFERENCES

Yu, Zhiwu. Ding, Faxing &Wen Hailin 2005. Experimental research on the mechanical properties of steel after high temperature, Advanced in steel structures, Vol II:1071–1082.
W.Y.Wang & S.L.Dong. Experimental research of steel welded flange bolted web under fire, Hebei Architecture and science Academy, Vol.23.No.2.Jun.2006.
G, Shi, Y.J.Shi & Y.Q.Wang. Experimental research on the tightening of sequences and strain relaxation of high strength bolts in end-plate connections, Advanced in steel structures, Vol I:866–860.
Chinese specification, 1992, JGJ82–91.

Steel and Composite Structures – Wang & Choi (eds)
© *2007 Taylor & Francis Group, London, ISBN 978-0-415-45141-3*

Fire behavior of steel-concrete composite members with austenitic stainless steel

C. Renaud & B. Zhao

Fire Research and Engineering Section, CTICM, St-Rémy-lès-Cheuvreuse, France

ABSTRACT: Within the scope of a European research project, a part of the work was focused on both experimental and numerical investigation of the fire performance of steel-concrete composite members with austenetic stainless steel. For the experimental investigation, an important number of fire tests have been carried out with different composite members. Based on these fire tests, numerical analysis has been made using an advanced numerical model developed by the authors for simulating the mechanical behaviour of steel-concrete composite structures exposed to fire.

1 INTRODUCTION

Although, many European research projects have already shown its great fire resistance, the use of high-strength austenitic stainless steel as steel-concrete composite members remains not very common in practice due to the lack of enough knowledge on fire behavior of such type of structural members. With its good behavior at elevated temperature, stainless steel could become a practical alternative solution to conventional structural carbon steel with for example reduced cross-section size of steel profiles or lower ratio of additional reinforcing bars which are often needed with carbon steel to achieve the required fire resistance. So, in order to investigate the fire behavior of composite members with austenitic stainless steel, a total of nine fire tests have been carried out with both RHS columns filled with concrete and partially protected slim-floor beams with exposed part in stainless steel and concrete protected part in carbon steel. All test members were grade EN 1.4404 stainless steel currently used in construction. Based on fire tests, corresponding numerical analysis have be made using an advanced finite element model specifically developed by the authors for simulating the mechanical behavior of composite structures exposed to fire (C. Renaud, 2003) (B. Zhao, 1994). It is noticed that the model was already proved to be in good agreement with several fire tests performed on composite members with conventional structural carbon steel.

The actual paper is then devoted to present test results as well as corresponding numerical simulations.

2 EXPERIMENTAL PROGRAMME

Fire tests on seven columns and two beams were conducted in France (Fire station of CTICM in "Maizières-Les-Metz") for the purpose of obtaining experimental evidence about the structural behavior of stainless steel-concrete composite members subjected to fire.

2.1 Test specimens

The main structural properties of tested composite columns are summarized in Table 1. All columns were square hollow steel sections with cross-section sizes ranging from 150×8 to 300×8 mm. The column length was 4000 mm. Steel tubes were filled with either reinforced or non-reinforced concrete core. Additional reinforcement, if used, was in 4 identical longitudinal bars, with a diameter chosen to achieve a ratio of reinforcement $A_s/(A_s + A_c)$ of approximately 2% and an axis distance of reinforcing bars $u_s = 30$ mm. The stainless steel grade was EN 1.4401. All columns were tested under eccentric load. For centrically loaded columns, a small loading eccentricity of 5 mm was applied to both column ends in order to induce an overall flexural buckling mode of failure under the fire condition.

The main structural properties of tested beams are reported in figure 1. Beams were simply supported hybrid I-section (stainless steel lower flange, carbon steel web and top flange) of 5 m span. Two different I cross-sections have been tested. The first beam is composed of 1/2 HEA 450 and 15 mm thick × 500 mm

Table 1. Structural details of composite columns with hollow steel section.

Cross-section	Reinforcement		Loading	
	Diameter (mm)	cover* (mm)	Load (kN)	Eccentricity (mm)
150 × 8 mm	none	–	400	5 mm
200 × 8 mm	none	–	240	0.25 × b**
200 × 8 mm	4φ14	30	630	5 mm
200 × 8 mm	4φ14	30	240	0.25 × b
300 × 8 mm	none	–	750	0.5 × b
300 × 8 mm	4φ14	30	1000	0.125 × b
300 × 8 mm	4φ14	30	800	0.25 × b

* distance between the axis of longitudinal reinforcements and the border of concrete core
** external side of hollow steel section

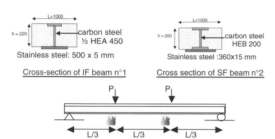

Figure 1. Structural details of tested beams.

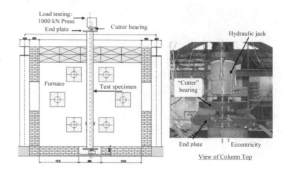

Figure 2. Test set-up of columns.

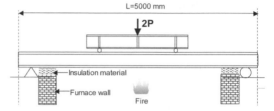

Figure 3. Test set up of simply supported beam.

stainless wide plate. The second beam is composed of carbon steel HEB 200 and 15 mm thick × 360 mm stainless wide plate. Beams were partially encased with concrete. The loading was applied in two points so that uniform bending moment was present in mid-span area of beams. The load P applied on beams was 100 kN and 75 kN respectively.

2.2 *Test arrangement of columns*

Test set-up of column is described in Figure 2. Each column was located at the centre of the furnace. The load was applied before fire test and kept constant during the test until the failure. Columns were exposed to heating condition according to ISO-834 standard fire curve. Columns were tested with both ends hinged. For that, support conditions at the top end and the bottom end of columns were built-up from additional end plates and cutter bearings (see figure 2). Moreover, to obtain eccentric load, cutter bearings were shifted in comparison to the gravity center of the cross-section. Both ends of the specimens were free to rotate about the axis perpendicular to cutter bearings but restrained to rotate about the other axis. Loading was applied by a hydraulic jack of capacity one hundred tons located outside and above the furnace chamber. The constant

load was controlled manually and measured using pressure transducers.

During all the tests, the furnace temperature will be continuously measured with twelve plate thermometers on four sides of the specimen at 100 mm from the surface of the specimen. Thermocouples were also installed on the hollow steel section and the reinforcing bars as well as in the concrete core: Three cross-sections were equipped with thermocouples along the column length in order to measure the temperature field. In fact, only three-quarter of the columns were heated because the top of the column should be outside the furnace to allow its loading. Axial deformations of the test specimen were determined by measuring the displacement of the top of the column (outside the furnace) using transducers. The rotations of the lower supporting end plate along two axes were also measured during the test by two rotation sensors. Failure time measured during all the test corresponds to the condition when each column could not bear the applied load any more.

Specimens from steel profiles, reinforcing bars and concrete were used to determine the real mechanical properties (yield and ultimate tensile strengths of steel and compressive strength of concrete).

2.3 *Test arrangement of beams*

The experimental set-up for the beam is shown in figure 3. The ISO-834 standard fire curve was followed inside the furnace to heat the test beam. The load was applied to the beam at least 15 minutes before

Table 2. Failure times of columns.

Cross-section	Load ratio*	Failure time (min)
150 × 8 mm	0.42	42.0
200 × 8 mm	0.22	59.0
200 × 8 mm	0.31	56.0
200 × 8 mm	0.20	71.0
300 × 8 mm	0.46	38.0
300 × 8 mm	0.29	70.5
300 × 8 mm	0.29	62.0

* The load ratio is defined as the ratio between the test load and the buckling resistance of column (according to the load eccentricity) calculated using numerical model.

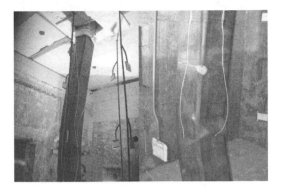

Figure 4. View of some composite column after tests.

Table 3. Failure times of beams.

Beam	Fire duration (min)	Max. defection* (mm)	Failure time (min)
1	90	440	79
2	86	515	76

* Maximum deflection measured at end of fire test.

the heating period and was maintained until the beam failed.

During the tests, the heating of the beams was measured on five sections, uniformly distributed along the heated part of the beams. Temperatures were recorded by means of thermocouples located at several points over the cross-section (stainless steel plate, carbon steel profile and concrete). Furnace temperatures were recorded using 5 plate thermometers located at the level of the five preceding sections. Two linear displacement transducers were positioned above the mid-span of the beam to measure the central deflection of the beam during the test.

2.4 Test results

Failure times measured during the tests are reported in Table 2 which correspond to the condition when columns could not bear the applied load any more. All experimental failure times were higher than expected fire ratings (R30 or R60). The main reason is that initial design of columns was made using nominal values of the mechanical properties of materials and assuming a uniform temperature distribution along full column height.

Column failures were caused by simple global flexural buckling or flexural buckling combined with local buckling (see figure 4). The observation after the test shows that the maximum deflection of the column was located either close to the bottom (lower part of the column) or near the mid-height of the specimen. For specimens with the larger cross-section (300x8mm), maximum deflection was observed on the lower part of the columns (due to the non-uniform heating of columns during fire tests). The thermocouple recordings on the other RHS columns showed that columns can be considered to be uniformly heated. In this case, the maximum deflection was observed near the mid-height of the specimen.

It is observed also that the local buckling had occurred in hollow steel section of tested columns

with large cross-section sizes (200 × 8 mm and 300 × 8 mm). This local buckling may be explained by the fact that these columns have excessive wall slenderness. No local buckling has been observed in the column with small cross-section sizes (150 × 8 mm).

Elapsed time of the fire tests carried out on beams and their failure times are reported in table 3. The beams are deemed to have failed when they can no longer supports the test load. The failure criteria is considered to be reached as a deflection of L/20 is exceeded (where L is the span of the specimen). The first beam reached the limiting deflection just after 79 minutes. Just before the imposed load was removed (at 90 minutes) the rate of deflection reached a maximum value of 15 mm/min. The second beam reached the limiting deflection just after 76 minutes. Just before the imposed load was removed (at 86 minutes) the rate of deflection reached a maximum value of 10 mm/min.

Photograph of the first beam after the fire test is given in figure 5.

3 NUMERICAL SIMULATIONS

The mechanical behavior of tested composite members has been simulated using the FEM model SISMEF. Temperature distributions in members have been obtained separately, either from 2D heat transfer analyse (based one finite difference or finite element method) or from test data. These temperatures have been used as input data for SISMEF.

Figure 5.   View of first beam during and after fire test.

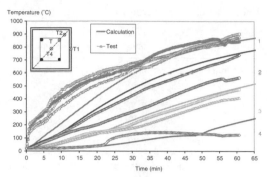

Figure 6.   Calculated and measured temperatures in the test column n°3

3.1   Assumptions for numerical simulations

In addition to the loading, boundary and heating conditions described in previous paragraph, the fire behaviour of composite members has been analysed with following assumptions:

The thermal and mechanical material properties of concrete and reinforcing steel bars as a function of temperature are in accordance with EN 1994-1-2. Material models for stainless steel are from EN 1993-1-2. It should be underlined that the creep strains of steel and concrete are implicitly included in the stress-strain relationships at elevated temperatures. Moreover, the effect of residual stresses is neglected.

Temperature distributions have been assumed to be uniform over the column height, except at the top of the column where a temperature gradient has been taken into account. The reason is that the top of the column was outside the furnace during the test and was not heated directly by the fire. Temperature distribution over the cross-section of column has been computed separately from 2D heat transfer analysis. To calculate the heat flow transmitted to the surface of hollow steel section during the fire exposure, it is necessary to introduce into the model the values of the convection heat transfer coefficient (h_c), the emissivity of fire (ε_f) and the emissivity of steel (ε_m). In practice, whatever the nature of materials, the convection coefficient inside the furnaces is taken equal to $h_c = 25\,W/m^2K$ under ISO fire. In EN 1991-1-2 and EN 1993-1-2, the emissivity of the fire is taken in general as unity. In the present work, the emissivity of the fire is also assumed as unity. The surface emissivity of the column is applied in accordance with EN 1993-1-2, namely $\varepsilon_m = 0.4$. The influence of moisture is considered in a simplified way in the calculation of the temperature filed of columns by assuming that all moisture evaporates, without any moisture mass transfer, at the temperature of 100°C or another temperature within a narrow range with the heat of evaporation giving a corresponding change in the enthalpy-temperature

curve. Direct heat transfer was assumed between stainless steel hollow section and concrete core (no gap due to differential thermal elongation of materials).

For beams, temperature distribution has been assumed uniform along their length. The temperature development of the two beam tests was modelled numerically with the computer code ANSYS, using the same parameters as those adopted for columns and an emissivity of 0.7 for concrete slab.

The mechanical interaction between the hollow steel section and the concrete core is neglected, that is slipping has been assumed to occur without significant bond between the steel wall and the concrete core.

Contribution of the concrete slab on the mechanical fire resistance of beams is neglected.

3.2   Presentation of some results

3.2.1   Thermal response

The measured temperature rises on composite members are systemically compared to the predicted temperature rises.

As example, figure 6 shows comparisons for reinforced column n°3.

The temperature rises predicted for the hollow steel sections are in good agreement with the measured ones. However, figure 3 shows fairly large discrepancies between the predicted temperatures and the measured temperatures, particularly for the first 20 minutes. In fact, predicted temperatures rise more slowly. The difference is less than 150°C during the early stage of the tests and decreases quickly with the fire duration. The faster rise in temperature of the hollow steel section might be explained by the role of "heat shield" played by the gap which occurs usually between the hollow section and the concrete core of heated composite columns. This gap is due to the differential thermal elongation of materials (steel and concrete) in the radial direction. It interrupts direct

600

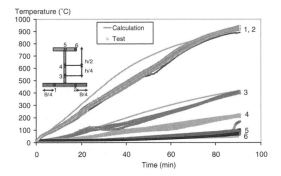

Figure 7. Calculated and measured temperatures in the test beam n°1.

heat conduction between the steel wall and the concrete core. The concrete core is mainly heated then by the thermal radiation from the heated hollow steel section.

The temperatures in the longitudinal reinforcing steel bars are evaluated satisfactorily between 0 to 100°C. Once the temperature of 100°C is reached close to the reinforcement, calculated temperatures become more important than those measured (the maximum difference is about 200°C). Globally, the predicted temperature rise rate beyond 100°C is analogous with that observed in test, but somewhat translated towards lower time instants. This translation between the curves is due, on the one hand to the delaying effect as a result of the gap between the hollow steel section and the concrete core (not taken into account in the thermal analyses) and on the other hand to the time necessary to the vaporisation of water really enclosed in the concrete.

With regard to the point inside the concrete core, the agreement is not so good, but can nerveless be considered as satisfactory. The difference with calculated curves is of no significant consequence: for low temperatures, the concrete mechanical properties are not affected and for higher temperatures the calculated curve is on the safe side.

Finally, assuming the thermal parameters recommended in EN 1993-1-2 for stainless steel, the temperature rises predicted for the hollow steel sections are in good agreement with the measured ones. Globally, all calculated temperatures remain overall on the side safe.

Figure 7 compares the measured temperature rises with the predicted curves for beam n°1.

It can be noted that the temperatures of stainless steel plates obtained from the numerical model are higher than the test data between 10 to 70 minutes. Then, they become quite close to the measured temperatures. The largest temperature difference in this case is about 100°C at 45 minutes. One reason for

this difference might be that the surface properties of the stainless steel plate undergo some changes, which could affect the value of emissivity during the fire exposure. Globally, temperatures calculated for the carbon steel profile are quite close to the measured values.

3.2.2 Mechanical response

The many temperature rises recorded during the test allow introducing temperature fields with enough accuracy in the mechanical simulations. So, calculated as well as experimental temperature fields have been systematically used to check the mechanical analyses conducted with SISMEF.

As example, axial displacement calculated at the top of the non-reinforced column n°2 is shown in figure 8. These displacements are compared with the test values. Same comparison is given in figure 6 for reinforced columns n°10.

From these figures, it can be noted that there is a reasonably good agreement between measured and calculated displacements. The numerical model SIS-MEF predicts the deformation behavior of the tested composite columns at a satisfactory level of accuracy. During the early stages of the fire exposure, axial displacement of column increases rapidly due to the quick heating of the external unprotected hollow steel section. As the steel column expands more rapidly than the concrete core, it carries alone the applied load. With temperature increase high enough, the load becomes critical due to the decrease of steel strength at elevated temperatures. Then the steel (as well as the column) suddenly contracts with eventual local buckling. At this time the concrete core is loaded almost suddenly and then supports progressively more and more the load with temperature rises. The concrete core, due to lower thermal conductivity and higher heat capacity of concrete, loses its strength more slowly than steel. It provides the fire resistance of the column at these later stages of fire exposure. If the concrete core is reinforced, the composite column can remain stable and the axial displacement decrease more slowly (see figure 9). The strength of the concrete also decreases with time and ultimately, when the concrete core can no longer support the load, failure occurs by buckling. When concrete core are non-reinforced, the vertical displacement increases approximately linearly and reaches a maximum just before failure occurs, whereupon it reduces very rapidly as the column buckles (see figure 8).

Comparison result between experience and calculation for beam n°2 is given in figure 10. It can be seen that there is a good correlation between the predicted and the measured curves. The agreement is quite good during the first stage of test and some differences are observed at the end of test.

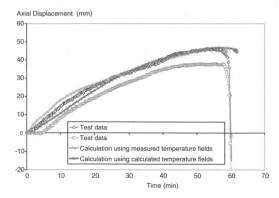

Figure 8. Comparison between test and calculation for column n°2.

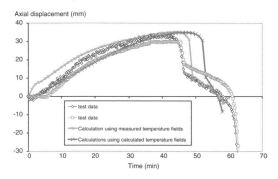

Figure 9. Comparison between test and calculation for column n°7.

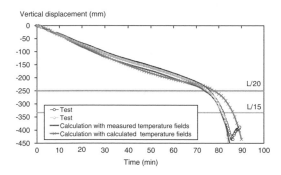

Figure 10. Comparison between test and calculation for beam n°2.

3.3 Synthesis of results

All calculated failure times of the composite members are reported in the figure 11. They are quite close to the experimental failure times. Globally, the comparison between failure times ascertained either numerically or experimentally shows a divergence less than 10%,

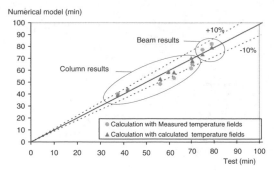

Figure 11. Comparison of fire resistances between numerical model and test.

what is reasonable considering uncertainties inherent to tests data, such as the uniformly heated length of members, the degree of rotational restraint at the ends, unintentional eccentricity of load and the initial out-of straightness.

4 CONCLUSION

In order to investigate the fire resistance behaviour of steel and concrete composite members, a series of fire tests are conducted within the scope of a European project. The corresponding experimental results are presented in detail in this paper. In parallel, using the advanced finite element model SISMEF, the fire performance of several composite members with stainless austenitic steel have been evaluated and compared with fire test results. These comparisons have shown that the model is capable of predicting the behaviour of composite columns and partially protected Slim-floor beams (failure time, displacements...) with a good accuracy.

As a consequence, parametric studies with the purpose of developing a simple calculation model providing a practical rule for daily design will be performed by varying sensitive parameters susceptible of affecting the fire resistance of composite members under standard fire conditions, such as cross section size, load conditions, eccentricity of loading, buckling length, etc.

REFERENCES

CEN, prEN 1994-1-2, Eurocode 4 "Design of Composite Steel and Concrete Structures: Structural Fire Design", June 2004.
CEN, EN 1993-1-2 – Eurocode 3 "Design of steel structures" – Part 1–2: General rules – Structural fire design, CEN, April 2005.

CEN, EN 1991-1-2 – Eurocode 1 "Actions on structures" – Part 1–2: General Actions – Action on structures exposed to fire, CEN, 2005.

Renaud, C. 2003. Modélisation numérique, expérimentation et dimensionnement pratique des poteaux mixtes avec profil creux exposés à l'incendie, Thèse de Docteur en génie civil, INSA de Rennes, France.

Renaud, C., Aribert, J.M., and Zhao, B. 2003. Advanced numerical model for the fire resistance of composite columns with hollow steel section, *Steel and composite structures,* Vol. 3, 75–95, No. 2 .

Zhao, B. and Aribert, J.M. 1996. Finite Element Method For Steel-Concrete Composite Frames Taking Account of Slip and Large Displacements, *European Journal of Finite Element,* Vol. 5, n°2, 221–249.

Zhao B. 1994. Modélisation numérique des poutres et portiques mixtes acier-béton avec glissements et grands déplacements, Thèse de docteur en Génie Civil, INSA de RENNES, France.

Steel and Composite Structures – Wang & Choi (eds)
© *2007 Taylor & Francis Group, London, ISBN 978-0-415-45141-3*

Simulation of composite columns in a global structural approach

P. Schaumann
Institute for Steel Construction, Leibniz University Hannover, Germany

F. Kettner
oemig + partner, consulting engineers, Kiel, Germany

ABSTRACT: Global structural analysis becomes more and more a standard design procedure as an answer to the demand for performance based fire design. Furthermore, composite columns are an economical solution for high performance and slender vertical load bearing elements with particular advantages in the fire situation. This paper presents the development of a numerical model to incorporate composite columns in the analysis of global structures. The cross section of the composite column is analysed in pre-processing routines. The obtained stiffness properties are applied to simple beam elements and hence are included in ABAQUS routines. For this contribution the developed model is tested at ambient temperatures. Load bearing capacities of several composite columns are calculated and compared to the results of the design Eurocode model. In the next step of upcoming investigations the elevated temperature will be included in the model.

1 INTRODUCTION

The demand for the application of performance based design methods drives forward the development and improvement of numerical methods for global structural fire design. This contribution presents a calculation method for the application of composite columns in global structural fire design. This is an intermediate report of the development process.

2 COMPOSITE COLUMNS FOR FIRE DESIGN

Through the last decades steel and concrete composite elements have become an important possibility for an economic and effective design of multi storey buildings. The high load bearing capacity of composite columns allows the design of slender vertical load bearing elements in open and light building structures.

Composite columns combine the advantages of the two structural materials steel and concrete. Profiled steel sections allow a high degree of prefabrication with fast and easy erection on the building site. Furthermore, concrete supports the high load bearing capacity of steel and prevents slender steel section from buckling.

But the main advantage of concrete is its thermal isolation behaviour as buildings have to fit fire safety requirements. Normally, no application of additional fire protection material is necessary.

Due to the high load bearing capacity even at elevated temperatures composite columns are economical and architectural appealing elements, particularly where fire safety requirements have to be met.

3 RECENT METHODS FOR STRUCTURAL FIRE DESIGN

3.1 *Prescriptive rules and performance based design*

The most common way for fire design in practice is the application of prescriptive rules. Regarding load carrying elements, this means that single structural members will be classified by standard fire tests. To avoid expensive fire tests, codes like the national German DIN 4102 or the Eurocodes present tables and simple calculation methods which can be used for the verification of structural members. These methods are easy to use and give quick results. However, the methods are constricted to single elements and exclusive cross sections. Eurocode 4-1-2 for example, covering the fire design of composite elements, offers tabulated data (method A) and simple calculation models (method B) only for cross sections given in Table 1. Furthermore, the methods are only valid for columns in braced frames.

Using prescriptive rules means using simplified rules that cannot consider the individual performance of the structure. In some cases this might be sufficient but in other, a performance based design, considering the individual conditions of the building, leads to a much more economic and attractive design.

Table 1. Composite column cross section considered by Eurocode 4-1-2

		Method	
	Cross section	A	B
1	Totally encased I-section	X	–
2	Partially encased I-section	X	X
3	Concrete filled hollow section	X	X

Figure 1. Slender composite columns in the German Technical Museum, Berlin: crossed I-sections with concrete encasement.

A request for modern fire design methods demands provision of appropriate performance based standards and codes, as it was demanded in recent reports by the National Institute of Standards and Technology (NIST) (see Gann (2005)) or the Institution of Structural Engineers (IStructE (2002)).

The structural Eurocodes provide primary approaches for performance based design as they offer the possibility of using "general calculation methods".

3.2 Global structural design

There has been an impressive research progress during the last years concerning the design of load bearing structures. The Cardington Fire tests (e.g. Newman, Robinson & Bailey (2000)) demonstrated that the whole structure showed significant differences in load bearing behaviour when exposed to fire compared to a single structural element in standard fire tests: Under certain conditions beams and slabs are acting together and loads are distributed to colder parts of the ceiling. Large deflections occur but the structure does not fail, although secondary beams are left unprotected. Slab and secondary beams develop membrane action and endure the conflagration.

Figure 2. Cardington Fire tests: large deflections after a compartment fire. No failure occurred.

From the results of Cardington a simple calculation method has been derived by Bailey (2000), which allows the design of concrete or composite slabs with unprotected secondary beams exposed to fire. The reduction of fire protection material leads to a much more economic solution for the structure.

Finally, the Cardington Fire tests showed the economical benefit of a global structural design compared to single member verification with standard fire conditions.

In view of these effects and with the high capability of modern computers it is now possible to simulate even complex structures exposed to fire, not only for research work but as well as standard design process in practice. Numerical software like ABAQUS, ANSYS or SAFIR can be used for those calculations. However, the simulations demand high expertise and know-how, which only can be provided by few engineering offices.

3.3 Modelling composite columns

It is a typical method in practice to model slabs and steel beams and columns with shell elements. A consistent approach for composite columns would be an approach with 3D-volume elements (see Figure 3) because composite columns are stocky members and temperature gradients all through the member have to be considered. However, such modelling is difficult because the existing material models are not sufficient for three dimensional modelling at elevated temperatures. Finally the modelling fails due to the size of the model. Just the 3D-model of a single composite column requires more than 10^5 degrees of freedom. This effort is tolerable for single column analysis or research work, but not for practical fire design performing global structural analysis.

Thus the aim was to develop an adequate method that allows the application of composite columns even for global structural analysis at elevated temperatures.

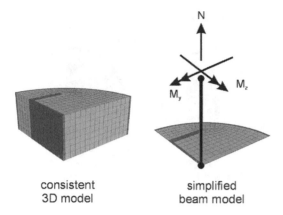

consistent
3D model

simplified
beam model

Figure 3. Different modelling of composite columns.

4 REDUCED BEAM MODEL FOR COLUMNS

4.1 General

The reduced model is based on beam elements. This adds the fewest number of additional degrees of freedom to the structural model. Stiffness properties at the nodes of the beam element are needed. They are determined in pre-processing routines by cross sectional analysis. Therefore the cross section is divided in finite elements (see Figure 3) and stresses will be integrated at the nodes of the mesh.

This contribution deals with the determination of the inner forces of a cross section. Temperature variation of the cross section due to fire exposure is neglected at this stage of the investigations and will be considered in future. However, the meshed cross section can be used as basis for 2D-temperature calculations as described for example in Schaumann (1985) and Schaumann & Hothan & Kettner (2005).

4.2 Material properties according to Eurocodes

Eurocode 2, 3 and 4 provide stress strain relationships for steel and concrete for normal temperature design and for fire design as well in the respective part 1-2. An evaluation of the given curves for structural steel and normal weight concrete at elevated temperatures is plotted in Figure 4 and Figure 5. The stresses are related to the yield strength and the compressive strength at 20°C. The characteristic strength properties for normal temperature design are equivalent to the fire design curves at 20°C.

4.3 Cross section analysis

The inner forces of a cross section can be calculated by integration of the normal stresses according to Equations 1a to c.

$$N = \int \sigma \, dA \qquad (1a)$$

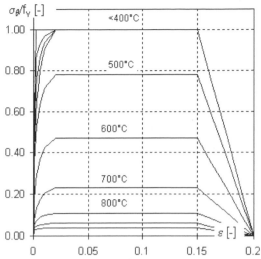

Figure 4. Stress strain relation for structural steel at elevated temperatures according to Eurocode 3-1-2.

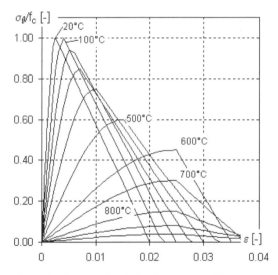

Figure 5. Stress strain relation for normal weight concrete at elevated temperatures according to Eurocode 2-1-2.

$$M_y = \int \sigma \cdot z \, dA \qquad (1b)$$

$$M_z = \int \sigma \cdot y \, dA \qquad (1c)$$

Shear forces play a secondary role for the load bearing capacity of columns and are neglected for these investigations.

Following Bernoulli hypotheses of plain cross section, strains at point i of the cross section can be calculated according Equation 2:

$$\varepsilon_i = \varepsilon_0 + \kappa_y \cdot z_i + \kappa_z \cdot y_i \qquad (2)$$

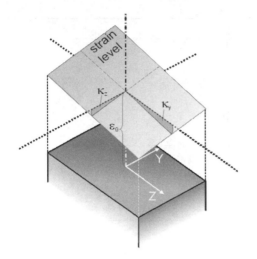

Figure 6. 3D-strain distribution over the cross section according to Bernoulli-hypothesis.

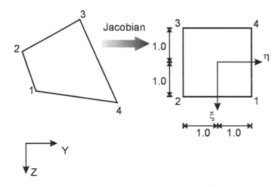

Figure 7. Transformation of general 4 node elements on a standard element.

A graphical demonstration of the strain distribution is in Figure 6. Stresses at each point i of the cross section can be determined with the given stress strain relations of section 4.2. Considering Bernoulli hypothesis according to Equation 2 the stresses are a function of the strain level:

$$\sigma_i = f(\varepsilon_i) = f(\varepsilon_0, \kappa_y, \kappa_z) \qquad (3)$$

The inner forces of the cross section are determined by numerical integration. The stress distribution over the element is approximated by bilinear shape functions according to Equation 4 and a transformation of stresses and coordinates to a standard element as it is shown in Figure 7.

$$N_i(\eta, \xi) = \frac{1}{4}(1 + \eta\eta_i)(1 + \xi\xi_i) \qquad (4)$$

With Equation 4 the stress distribution $\sigma(\eta, \xi)$ over the element can be expressed by the stress ordinates σ_i at

Table 2. Investigated cross section.

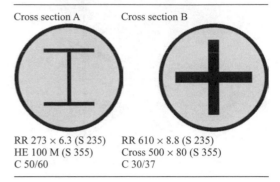

Cross section A	Cross section B
RR 273 × 6.3 (S 235)	RR 610 × 8.8 (S 235)
HE 100 M (S 355)	Cross 500 × 80 (S 355)
C 50/60	C 30/37

the nodes i:

$$\sigma(\eta, \xi) = \sum_i N_i(\eta, \xi) \cdot \sigma_i \qquad (5)$$

The same circumstances apply for the transformation of the coordinates.

The integration of the stresses over the element is done by Gauß-point evaluation. A summation over all elements of the cross section Ω_e transforms Equations 1a to:

$$N = \iint \sigma \, dy \, dz = \iint \sigma(\eta, \xi) \cdot \det J \, d\eta \, d\xi$$
$$= \sum_{\Omega_e} \sum_{GP} w_{GP} \cdot \det J \cdot \mathbf{N} \cdot \boldsymbol{\sigma} \qquad (6)$$

where $\det J$ = Jacobian, GP = Gauß points, $\mathbf{N}$ = vector with shape functions, $\boldsymbol{\sigma}$ = vector with stress values at nodes.

The evaluation of the inner bending moments M_y and M_z are similar to Equation 6.

The result of this procedure is that one strain distribution defined by ε_0, κ_y and κ_z leads to a triple of inner forces N, M_y and M_z. This means that for example the inner axial force N is not only a function of the plain strain ε_0 but changes with varying curvatures as well due to the nonlinear material behaviour (see section 4.2). However, it was the intention of the authors to make each single inner force depend on the other two inner forces, e.g. $N = f(\varepsilon_0, M_y, M_z)$, i.e. that one $N - \varepsilon_0$-curve belongs to a constant pair of bending moments M_y and M_z. Each pair of $N - \varepsilon_0$-values is obtained by Newton iteration until the integrated bending moments coincide with the predefined values for M_y and M_z.

This procedure can be devolved to the $M_y - \kappa_y$ and $M_z - \kappa_z$ relation, respectively.

4.4 Results of the cross section analysis

Results of the procedure described above are presented below for two exemplarily selected cross sections. Both sections are circular hollow steel section filled with concrete and an embedded stocky steel section (see Table 2). The advantage of this type of cross

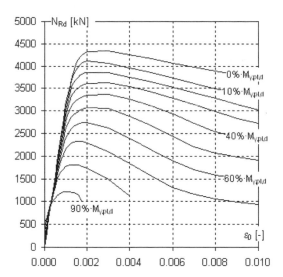

Figure 8. $N - \varepsilon_0$-curves for cross section A in dependency of different values for M_y.

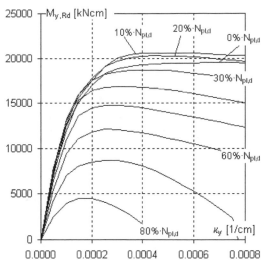

Figure 9. $M_y - \kappa_y$-curves for cross section A in dependency of different values for N.

section is to be able to leave out additional reinforcement. A typical small and large tube diameter was chosen to analyse the results of the calculation method for slender and non slender columns. Cross section B has about half the slenderness of cross section A.

The results of the integration process (Equation 6) can be seen in Figure 8 for cross section A. The $N - \varepsilon_0$-curves have been evaluated for uniaxial bending around the strong axis and different ratios of $M_{y,pl,d}$. For better illustration the variation of the curvature κ_z and hence the bending moment M_z was neglected.

Obviously the obtained axial force N_{Rd} decreases with increasing bending moments M_{Rd}, which complies with the assumptions of plasticity theory. The peak value of the curve for $M_{y,Rd} = 0$ is the value for plastic resistance of the composite cross section to compressive normal force $N_{pl,Rd}$. Investigations of Schaumann & Kettner & Hothan (2005) had shown that this coherence applies no longer for elevated temperatures.

$M_y - \kappa_y$-curves for cross section A are shown in Figure 9. Over all the bending moment capacity $M_{y,Rd}$ decreases with increasing axial compression N. However, an increase of the bending moment capacity can be observed for small acting axial forces. The reason is that the increasing axial forces turn concrete parts of the cross section which have been in tension before to compression mode. As tension strength of concrete is much smaller than compression strength the integration of the stresses finally results in a higher bending moment capacity. This phenomenon is typical for concrete and composite members and can be identified in the concave shape of design interaction curves of the standards.

4.5 Adaptation to ABAQUS

The cross section behaviour in terms of $N - \varepsilon_0$-curves and $M_{y(z)} - \kappa_{y(z)}$-curves is implemented in ABAQUS as element stiffness property at the nodes. Finally ABAQUS is used to solve the geometrical non-linear system equation.

The stiffness properties are implemented as general nonlinear beam section. However, the dependency on other inner forces (e.g. $N = f(M_y)$) has to be defined with field variables. The field variables have to be updated in every load increment in dependency of the inner forces, which can be obtained from the solution of the system equation of the previous increment. This is done using user subroutines.

5 LOAD BEARING CAPACITIY OF COMPOSITE COLUMNS FOR NORMAL TEMPERATURE DESIGN

Results of ABAQUS calculations are compared to results of the design procedure of ENV 1994-1-1 to analyse the quality of the simplified model. Calculations are performed for the two cross sections presented in Table 2. The static system is a single hinged column with typical storey heights varying between 2.35 m and 4.00 m. The load was applied either axial or eccentrically with 1/5 of the tube diameter. An imperfection of 1/1000 was applied on the system to consider the structural imperfections and the influence of the residual stresses. Figure 10 shows a comparison of absolute load bearing capacity design values. A good accordance between the design procedure of ENV 1994-1-1 and the simplified ABAQUS model can be observed.

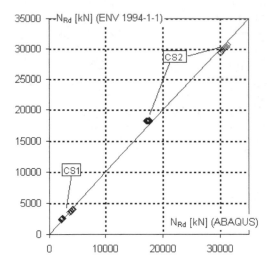

Figure 10. Comparison of normal temperature load bearing capacities between simplified ABAQUS and design model of ENV 1994-1-1.

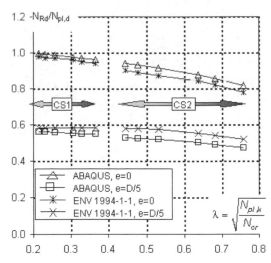

Figure 11. Comparison of normal temperature load bearing capacities between simplified ABAQUS and design model of ENV 1994-1-1.

Figure 11 shows the ratio of load bearing capacity to plastic resistance of the composite section to compressive normal force against the slenderness λ of the analysed system. The compliance of Figure 10 can be approved by the results. Furthermore, the results from the ABAQUS model for axial loading are marginal more conservative than the Eurocode results. However, the ABAQUS calculation leads to higher load bearing capacities when considering regular bending. The reason for this discrepancy has to be analysed in further investigations.

6 CONCLUSIONS

Global structural analysis becomes more and more a practical procedure for modern and economic fire design following the demand for performance based design. Furthermore, composite columns combine the advantages of steel and concrete to high performance and slender vertical load bearing element.

Thus the authors have developed a substructure model to establish the possibility of using composite columns in global structural analysis but the computational effort should be in economical dimensions. The model consists of beam elements and the stiffness properties at the nodes are developed from pre-processing routines, which calculate the inner forces as $N - \varepsilon_0$ or $M - \kappa$-curves by numerical integration. The cross section properties are implemented in ABAQUS to solve the system equation.

The model is tested with two different cross sections in combination with different static systems. The load bearing capacities are compared to the design procedure of ENV 1994-1-1 and showed good accordance.

In the next step of the investigations the temperature development has to be included in the procedure. A 2D-temperature calculation will be connected to the numerical integration process. Furthermore it will be important to consider the thermal stresses and the influence on the stiffness properties.

The results of the simplified model have to be compared to test results at normal and elevated temperatures.

REFERENCES

Bailey, C.G. 2000. Design of Steel Structures with Composite Slabs at the Fire State. Final Report. Prepared for: Department of the Environment, Transport and the Regions and The Steel Construction Institute. Garston Watford: Building Research Establishment.

Gann, R.G. 2005. ed., "Final Report of the National Construction Safety Team on the Collapse of the Word Trade Center Twin Towers," NIST NCSTAR 1, National Institute of Standards and Technology, Gaithersburg, MD.

IStructE 2002. Safety in tall buildings, The Institution of Structural Engineers (IStructE), London, UK.

Newman, G. M. & Robinson, J. T. & Bailey, C. G. 2000. Fire Safe design: A New Approach to Multi-Storey Steel-Framed Buildings, SCI Publication P288, Berkshire: The Steel Construction Institute.

Schaumann P. 1984. Calculation of steel elements and frame structures exposed to fire, Technisch-wissenschaftliche Mitteilungen Nr. 84-4, Institut für konstruktiven Ingenieurbau, Ruhr-Universität Bochum, Dissertation, Bochum.

Schaumann, P. & Hothan, S. & Kettner F. 2005 Recent improvements on numerical methods in structural fire safety, ICASS '05, Advances in Steel Structures, Vol. II, Z.Y. Shen et. al. (Eds.), ©2005 Elsevier Ltd., Proceedings, Page 925–932.

Steel and Composite Structures – Wang & Choi (eds)
© *2007 Taylor & Francis Group, London, ISBN 978-0-415-45141-3*

Numerical analysis of composite steel and concrete columns under fire conditions

R.B. Caldas & R.H. Fakury
Dept. Structural Engineering, Universidade Federal de Minas Gerais, Belo Horizonte, MG, Brazil

J.B.M. Sousa Jr.
Dept. Civil Engineering, Universidade Federal de Ouro Preto, Ouro Preto, MG, Brazil

ABSTRACT: This paper describes a finite element model for the thermomechanical analysis of steel-concrete composite beam-columns, aiming at the simulation of these elements under fire conditions. Three-dimensional nonlinear elements, based on the corotational technique and displacement based interpolation, are the basis for the mechanical model. Thermal analysis is performed at the cross section or integration point level using a local finite element scheme, and material property degradation is taken into account using a fiber model. At the global level, thermal strains are included in the analysis using the effective strain concept. Generic boundary fire conditions may be applied. The work focuses on the numerical simulation of composite concrete encase and concrete filled columns. Some examples are analyzed by the model and the results are compared with other numerical and experimental results.

1 INTRODUCTION

The use of composite columns of steel and concrete, either concrete filled tubes or concrete encased hollow sections has grown considerably worldwide. Filling hollow steel columns with concrete increases their load bearing capacity and a higher fire performance can be achieved in comparison with bare steel tubular columns. Concrete filled hollow steel sections under fire conditions also have better endurance characteristics than reinforced concrete columns as the steel casting prevents concrete spalling. Another benefit is the reduction or elimination of formwork. For concrete encased steel columns under fire, the concrete cover delays the rise of temperature in encased steel, due to its much lower conductivity.

1.1 Literature overview

Several research works on concrete filled hollow steel sections (HSS) have been published recently. Kodur (1999, 2000) presented experimental results and simplified design equations for evaluating the fire resistance of composite columns with concrete filled HSS. Composite columns with plain concrete, bar reinforced and fiber reinforced concrete were studied. Experimental results and solutions for enhancing the fire resistance of HSS filled with high strength concrete (HSC) were also presented by Kodur (2006). Bar

reinforcement and steel fibers are presented as good solutions for enhancing the behaviour of HSC filling under fire conditions.

A simple method for calculating the squash load and rigidity of composite columns with circular hollow sections under fire, based on Eurocode 4, was presented by Wang (2000). In the proposed method there is no need to evaluate the temperature distribution in the column.

Han (2000, 2001) applied the finite element method for the calculation of temperature distribution of concrete filled HSS under fire and proposed a theoretical model to evaluate deformations and strength in fire.

Al-Kalleefi *et al.* (2002) presented a functional relationship between the fire resistance of concrete filled steel columns and the parameters which influence the fire resistance using an artificial neural network. Structural, materials and loading conditions were taken into account to develop the neural network.

An investigation on the behaviour of composite columns with circular hollow sections (CHS) under fire using the finite element method (FEM) was carried out by Zha (2003). The results agreed well with current design codes. Eight-node solid elements were employed and both steel and concrete were assumed to be isotropic.

Yin *et al.* (2006) investigated the fire behaviour of concrete filled CHS and RHS (rectangular hollow sections). A strength analysis is used and the authors concluded that the circular column has slightly better fire resistance than the rectangular column.

Wang and Tan (2006a) presented a numerical approach, based on analytical Green's function solution, to predict the temperature distribution in the composite column with CHS. A computational model is developed based on Duhamel's principle incorporating time-varying fire conditions.

Although most of the research focused on concrete filled HSS, some works present studies on the behaviour of composite columns with concrete encased steel sections. Huang et al. (2007) presented an experimental study of the axial restraint effect on fire resistance of unprotected encased I-section columns. Axial restraints were applied to simulate thermal restraints from the adjoining cool structures onto a heated composite column in a compartment. The results show that axial restraint significantly reduces the column fire resistance.

Experimental and numerical methods were employed by Yu et al. (2007) to investigate the behaviour of steel concrete encased steel columns subjected to fire. A combined finite element/finite difference scheme was adopted to analyze the temperature distribution and the behaviour of these columns.

Wang and Tan (2006b) proposed a residual area method to calculate the equivalent thickness of concrete for temperature analysis of composite columns with concrete encased steel I section in fire. The equivalent concrete thickness is based on a regression analysis which fits experimental results. The proposed method was applied to composite columns exposed to hydrocarbon, external and parametric fire curves.

The purpose of this work is to present and test a FE model for general composite steel and concrete columns under fire action.

The model consists on 3D beam column elements, with thermal analysis performed at the cross section level, employing an explicit finite element scheme with four-noded isoparametric elements. Different stress-strain relations for normal strength concrete found in the literature are verified and compared against the numerical and experimental results.

2 NUMERICAL MODEL

The numerical model is based on a thermomechanical analysis, characterized by the influence of temperature increase on the strain state and material property degradation under fire. The effects of temperature variation are taken into account at the cross section (integration point) level.

The finite element employed is a displacement-based generic cross section beam column element (Sousa Jr. and Caldas, 2005) combined with the corotational formulation to account for the large displacements and rotations.

2.1 Sectional thermomechanical analysis

A uniaxial stress state is considered for the mechanical evaluations, which are carried out employing the effective strain concept, by which $\varepsilon_{ef} = \varepsilon_x - \varepsilon_{th}$, where ε_x is the total strain and ε_{th} are the thermal strains. Total strains are evaluated by $\varepsilon_x = \varepsilon_0 + k_y z - k_z y$, where ε_0 is the strain at a reference point (usually the cross section centroid) and k_y, k_z are the curvatures of the section, supposed to lie in the $y - z$ plane.

The section is then subdivided in 4-noded isoparametric elements. The resistant forces and tangent moduli (derivatives of the forces with respect to the deformational variables) of the cross section are obtained as a sum in the elements of the section, $i = 1...n$, with area A:

$$N_x = \sum_{i=1}^n (\sigma_x A)_i \, , \; M_y = \sum_{i=1}^n (\sigma_x z A)_i$$

$$(1)$$

and $M_z = -\sum_{i=1}^n (\sigma_x y A)_i \, .$

The tangent moduli are obtained by differentiation of Eq. 1 with respect to the generalized strains ε_0, k_y and k_z. The transient heat transfer analysis which permits the evaluation of the thermal strains is performed employing an explicit time integration finite element scheme (Bathe, 1996). In order to ensure stability, the nodal temperatures vector ΔT of the cross section is evaluated for a limited time increment Δt as

$$\Delta T = M^{-1}(Q - KT)\Delta t \, . \quad (2)$$

where K is the conductivity matrix, Q is the heat flux vector due to convection/radiation for nodes along the cross section boundary and M is the "lumped" (diagonalized) mass matrix (Bathe, 1996). Concrete, steel and fire protection material properties were implemented, enabling a versatile representation of fire situations. Concrete spalling is not taken into account, and moisture is considered by modifying the thermal properties, as specified by the Eurocode 2 (2004) and Eurocode 4 (2005).

2.2 Beam-column finite element

The finite element employed in the present work must be able to undergo large displacements and rotations, but small strains. Basic assumptions include: (a) plane sections remain plane; (b) there is perfect bond at the interface of the materials; (c) each element that subdivide the cross section is subjected to a uniaxial state of

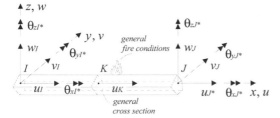

Figure 1. Finite element degrees of freedom.

stress; (d) local buckling and cross section warping are not taken into account. The corotational formulation is employed for the treatment of large displacements and rotations.

The present work follows the procedure developed by Crisfield (1997). In this formulation, geometrical transformations eliminate rigid-body movements from the overall element formulation. In that sense one needs just to provide a "local" finite element which not necessarily accounts for large displacement effects. In the expression of the element stiffness matrix of the 3D beam corotational beam formulation the only requirement on the "local" element formulation is to have the same degrees of freedom shown in Figure 1 (indicated by the symbol "*").

The finite element employed by Sousa Jr. and Caldas (2005) for the numerical analysis of composite columns at ambient temperature is used here, with proper elimination of degrees of freedom and linear torsional stiffness included. The element is based on a total lagrangian formulation, suitable for large displacements and moderate rotations. Cubic hermitian functions (collected in vectors $\mathbf{N}_v$ and $\mathbf{N}_w$) are used for local transverse displacement (v,w) interpolations. For the axial displacement (u) quadratic functions are employed ($\mathbf{N}_u$), with a hierarchical axial displacement degree of freedom. The additional d.o.f. is dealt with by means of static condensation at the incremental level. The local nodal displacements are grouped in vector $\mathbf{q}$.

The Principle of Virtual Work, associated with the strain-displacement relation usually considered for beams

$$\varepsilon_x = u' + 1/2[(v')^2 + (w')^2] - yv'' - zw'', \qquad (3)$$

where the prime indicates differentiation with respect to x, allows one to obtain the internal force vector

$$\mathbf{p}_m = \int_{\ell_m} \begin{bmatrix} N_x\mathbf{N}'_u \\ N_xv'\mathbf{N}'_v + M_z\mathbf{N}''_v \\ N_xw'\mathbf{N}'_w - M_y\mathbf{N}''_w \end{bmatrix} dx. \qquad (4)$$

The "local" element tangent stiffness is obtained after differentiation of the internal force with respect to the displacements:

$$\mathbf{k} = \int_{\ell_m} \begin{bmatrix} \phi_u\left\{\dfrac{\partial N_x}{\partial \mathbf{q}}\right\}^T \\ \mathbf{N}_v\left(v'\left\{\dfrac{\partial N_x}{\partial \mathbf{q}}\right\}^T + N_x\lfloor \mathbf{0}_u \ \ \mathbf{N}_v \ \ \mathbf{0}_w \rfloor\right) + \mathbf{N}'_v\left\{\dfrac{\partial M_z}{\partial \mathbf{q}}\right\}^T \\ \mathbf{N}_w\left(w'\left\{\dfrac{\partial N_x}{\partial \mathbf{q}}\right\}^T + N_x\lfloor \mathbf{0}_u \ \ \mathbf{0}_v \ \ \mathbf{N}_w \rfloor\right) - \mathbf{N}'_w\left\{\dfrac{\partial M_y}{\partial \mathbf{q}}\right\}^T \end{bmatrix} dx. \qquad (5)$$

Eliminating lines and columns from the previous expressions, and including a linear torsional stiffness term the "local" element is obtained. All the thermal effects and material nonlinearities are carried out within this "local" element, with the corotational transformations taking care of the large displacement and rotation effects.

In the Newton-Raphson scheme for fire analysis, load increments are applied prior to temperature increments due to fire exposure. After a typical temperature increment, a heat transfer analysis is performed on each integration point. With the new thermal strains and material properties, the internal force is no longer is in equilibrium with the applied loads, and an iterative Newton-Raphson scheme is carried out to recover equilibrium. The stresses are evaluated in terms of the effective strains. Structural collapse is identified when the numerical scheme is not able to reach an equilibrium point. Different stress-strain relations with distinct property degradation laws may be employed.

2.3 Concrete stress-strain relations

Two different stress-strain relations for the concrete in compression are employed in the present work. The first one is taken from the Eurocodes 4 (2005) and 2 (2004), see Figure 2.

Han (2000, 2001) and Han et al. (2003) present a concrete model that considers the composite action between steel and concrete. This stress-strain relation is based on test results of concrete filled steel tubular axial short columns under constant high temperature. Composite action between the steel tube and its concrete core is considered by a constraining factor

$$\xi = \frac{A_s f_{y,T}}{A_c f_c}, \qquad (6)$$

in which A_s is the cross sectional area of the steel tube, A_c is the cross-sectional area of the concrete, f_{yT} is the yield stress of steel at temperature T and f_c is the compression strength of concrete at ambient temperature.

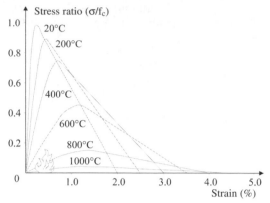

Figure 2. Eurocode stress-strain relations for concrete.

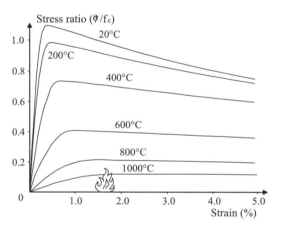

Figure 3. Example of Han's model for CIIS.

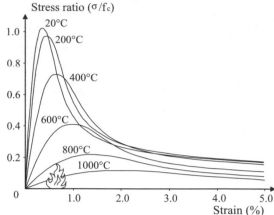

Figure 4. Example of Han's model for RHS.

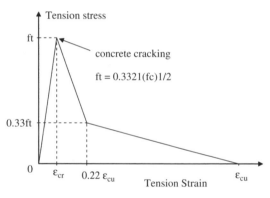

Figure 5. tensile stress strain relation for concrete.

The expressions for stress-strain for concrete filled steel tubular circular, rectangular and square section used in this work are lengthy and will not be shown here. Figures 3 and 4 show typical stress-strain curves for confined concrete of the concrete filled steel tube with 41 MPa concrete compressive strength and 0.29 constraining factor.

The concrete tensile stress is considered as in Huang *et al.* (2003), see Figure 5, with $\varepsilon_{cu} = 15\varepsilon_{cr}$. The tension resistance reduction factor is taken, for all the stress-strain relations, as the cube of its compressive counterpart. This approximation agrees well with the Eurocode 2 (2004) prescriptions.

3 EXAMPLES

The two main types of composite columns were analyzed using the proposed formulation. The results are compared with numerical and especially with experimental results from the literature.

3.1 *Composite column with encased steel section*

Huang *et al.* (2007) present experimental and numerical results of the analysis of a composite column with steel I section encased in concrete subjected to fire. Concrete encasing a UC 152 × 152 × 37 steel section has 300 × 300 mm and there are four T13 bars with 30 mm of concrete cover. The column has 3540 mm length, concrete compression strength of 43 MPa and steel and reinforcement yield stress of 460 MPa. The gas temperature curve has two ascending phases at 5°C/min and 8°C/min, respectively.

In this work, the material properties were considered in accordance with the Eurocode 4 (2005) with a moisture content of 8% of the concrete weight. The upper limit of thermal conductivity was used. Four beam elements with four Gauss points each were adopted and the section was subdivided in 1198

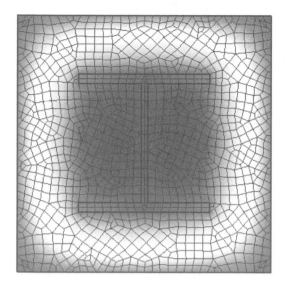

Temperature (°C) Step (210)
426   587   748

Figure 6.   Cross section temperature distribution (420 min).

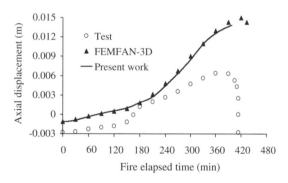

Figure 7.   Column axial displacement.

four-noded elements, Figure 6. The axial load applied was 1106 kN. A sinusoidal initial imperfection with maximum amplitude of 0.1mm was considered in the analysis.

Figure 7 shows that the calculated axial displacement agrees well with numerical analysis performed by Huang *et al.* (2007) using the program FEMFAN-3D and the calculated time of exposure to fire is about 95% of the time measured in the test. In the final stages, however, the predicted column contraction is lower than measured in the test. This also happened when modeling concrete filled steel tubes, as will be seen in the next example.

3.2   Concrete filled CHS columns

Several research works presented results of tests for concrete-filled CHS columns. Kodur (2006) presents the results of 17 fire tests on composite columns with concrete filled hollow circular, square and rectangular steel sections, subjected to concentric loads and eccentric loads, with and without reinforcement, and with protection material.

The first test (Test 1) involved a 273 mm diameter 6.35 thickness CHS, filled with 38.2 MPa concrete and 712 kN applied load. The second test (Test 2) was of a 219 mm diameter 4.78 mm thickness CHS, filled with 42.7 MPa concrete and 560 kN applied load. Both columns were 3810 mm long.

Four beam elements with four Gauss points each were used to analyze the column exposed to standard fire (ISO834, 1975; ASTM E119, 1990). The cross section was subdivided in 1400 four-noded elements and a sinusoidal initial imperfection with 0.1 mm of maximum amplitude was considered in the analysis. Moisture content was taken as 10% of concrete weight. Material properties are taken from the Eurocode 4 (2005) and Eurocode 2 (2004) for calcareous concrete, and the upper limit of the conductivity was considered.

For Test 1, Figure 8 displays the typical axial displacement at the top of the unprotected composite columns with concrete filling. During the first stages of heating the steel section resists most of the loading due to its larger expansion relative to the concrete component. Subsequent heating causes loss of resistance of the steel section and its yielding, with contraction taking place at an exposure time usually between 15 and 30 minutes. Due to its low thermal conductivity, the concrete resistance degradation is slower, resulting in additional fire resistance. Further increases in temperature then cause the loss of strength of the concrete component, which collapses under compression or instability. It may be seen from Figure 8 that the additional strength provided by the concrete component leads to significant fire performance way after the yielding of the steel section.

The prediction of the behaviour employing the proposed formulation is accurate in the first stages of fire exposure and also on the value of the critical time. However, close to the collapse the predicted contraction of the column is lower than the measured in the test.

The fire resistances predicted by the numerical formulation of Test 1 were 138.5 minutes for the Eurocode concrete model and 142.5 minutes for the model by Han *et al.* (2003), while the test resistance was 144 min. For Test 2, the values are 77.5, 82 and 108 minutes for the Eurocode, Han's model and the test results respectively.

Tests 3 to 10 were carried out on concrete-filled rectangular hollow sections, and Tests 11–13 on square hollow sections (Han, 2003). Tests 14 and 15 employed

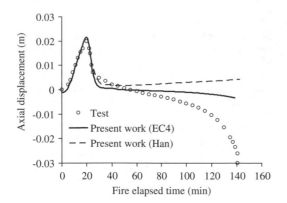

Figure 8. Axial displacement.

Table 1. Fire resistance for Tests 3-17 (minutes).

Test	EC4	Han	Test	EC4 Test	Han Test
3	18.5	25.5	21	0.88	1.21
4	10	9.5	24	0.42	0.40
5	24.5	24.5	16	1.53	1.53
6	22	20.5	20	1.10	1.03
7	60	69	104	0.58	0.66
8	95	110.5	146	0.65	0.76
9	50.5	53	78	0.78	0.81
10	109.5	114	122	0.90	0.93
11	88	108.5	169	0.52	0.64
12	165.5	188	140	1.18	1.34
13	129	108	109	1.18	1.17
14	133	136.5	188	0.71	0.73
15	79	75	96	0.82	0.78
16	94.5	104.5	150	0.63	0.70
17	107.5	102.5	113	0.95	0.91
			Average:	0.85	0.90
		Standard	Deviation:	0.28	0.29

circular hollow sections (Lie, 1994), and Tests 16 and 17 square hollow sections (Lie & Irwin, 1995). The comparison between the numerical and experimental analyses, using the Eurocode 4 Model and Han's Model (Han et al., 2003) from Tests 3 to 17 are shown in Table 1.

4 CONCLUSIONS

This paper presented the formulation and application of a three-dimensional finite element to the simulation of composite steel-concrete columns under fire action. The numerical results were compared with test results in the case of concrete encased steel profiles and concrete filled circular hollow sections.

The results displayed a good agreement in terms of the values of fire time exposure until collapse, although the displacements at the final stages, when the column experiences significant contraction, were not captured by the model.

For the tests carried out in the present work, the Eurocode concrete model, as well as the model by Han et al. (2003) produced safe results in terms of resistance.It must be emphasized, however, that the behaviour of concrete when subjected to fire is very complex, and simple uniaxial models as the one presented here may not be capable of taking into account every aspect of this complexity. Further numerical and experimental research must be carried out in order to develop reliable one-dimensional models.

ACKNOWLEDGEMENTS

The authors wish to thank V&M Tubes do Brasil and CNPq, Brazilian Development Agency.

REFERENCES

Al-Kalleefi, A., Terro, M.J. Alex, A.P., Wang, Y. 2002. Prediction of the Fire Resistance of Concrete Filled Tubular Steel Columns Using Neural Networks. Fire Safety J., 37, 339–352.
ASTM E119 1990. American Society for Testing and Materials. Standard Methods of Fire Tests on Building Construction and Materials.
Bathe, K.J. 1996. Finite Element Procedures. Prentice-Hall.
Crisfield, M.A. 1997. Nonlinear Finite Element Analysis of Solids and Structures. Volume 2: Advanced Topics. John Wiley & Sons.
Eurocode 2: Design of Concrete Structures, Part 1.2: General Rules, Structural Fire Design. BS EN 1992-1-2:2004.
Eurocode 4: Design of Composite Steel and Concrete Structures, Part 1-2: General Rules, Structural Fire Design. BS EN 1994-1-2:2005.
Han, L.L. 2000. Fire Resistance of Concrete Filled Steel Tubular Beam-Columns in China-State of the Art. Proc. Conf. Composite Construction in Steel and Concrete IV.
Han, L.H. 2001. Fire Performance of Concrete Filled Steel Tubular Beam-Columns. J. Constr. Steel Res., 57, 695–709.
Huang, Z.F., Tan, K.H. and Phng, G.H. 2007. Axial Restraint Effects on the Fire Resistance of Composite Columns Encasing I-Section Steel. J. Constr. Steel Res., 63, 437–447.
Han, L.H., Yang, Y.F. and Xu, L. 2003. An Experimental Study and Calculation on the Fire Resistance of Concrete-Filled SHS and RHS Columns. J. Constr. Steel Res., 59, 427–452.
Huang, Z., Burgess, I.W. and Plank, R.J. 2003. A Non-Linear Beam-Column Element for 3D Modelling of General Cross-Sections in Fire. University of Sheffield, Res. Report DCSE/03/F/1.
ISO-834 Fire Resistance Tests-Elements-Elements of Building Construction. International Standard, 1975.
Kodur, V.K.R. 1999. Performance-Based Fire Resistance Design of Concrete-Filled Steel Columns. J. Constr. Steel Res., 51, 21–36.

Kodur, V.K.R. and Mackinnon, D. H. 2000. Design of Concrete-Filled Hollow Structural Steel Columns for Fire Endurance. AISC, Engineering Journal, 13–24.

Kodur, V.K.R. 2006. Solutions for Enhancing the Fire Endurance of HSS Columns Filled with High-Strength Concrete. AISC, Engineering Journal, 1–7.

Sousa Jr., J.B.M., Caldas, R.B. 2005. Numerical Analysis of Composite Steel-Concrete Columns of Arbitrary Cross Section. J. Struct. Eng., 131(11), 1721–1730.

Wang, Y.C. 2000. A Simple Method for Calculation the Fire Resistance of Concrete-Filled CHS Columns. J. Constr. Steel Res., 54, 365–386.

Wang, Z.H., Tan, K.H. 2006a. Green's Function Solution for Transient Heat Conduction in Concrete-Filled CHS Subjected to Fire. Eng. Struct., 28, 1574–1585.

Wang, Z.H., Tan, K.H. 2006b. Residual Area Method for Heat Transfer Analysis of Concrete-Encased I-Section in Fire. Eng. Struct., 28, 411–422.

Zha, X.X. (2003). FE Analysis of Fire Resistance of Concrete Filled CHS Columns. J. Constr. Steel Res., 59, 769–779.

Yin, J., Zha, X.X., Li, L.Y. 2006. Fire Resistance of Axially Loaded Concrete Filled Steel Tube Columns. J. Constr. Steel Res., 62, 723–729.

Yu, J.T., Lu, Z.D. Xie, Q. 2007. Nonlinear Analysis of SRC Columns Subjected to Fire. Fire Safety J., 42, 1–10.

Steel and Composite Structures – Wang & Choi (eds)
© *2007 Taylor & Francis Group, London, ISBN 978-0-415-45141-3*

Design of corrugated sheets exposed to fire

Z. Sokol, F. Wald & P. Kallerová
Faculty of Civil Engineering, Czech Technical University, Prague, Czech Republic

ABSTRACT: Fire tests of corrugated sheets used as load bearing structure of roofs of industrial buildings are presented. Additional tests of bolted sheet connections to the supporting structure at ambient and elevated temperatures are described. Three connection types were tested and their resistance, stiffness and deformation capacity was evaluated. Finite element simulation of the corrugated sheet based on the experimental observations are briefly described and design models are presented.

1 INTRODUCTION

Corrugated sheets are designed for roofs of span up to 9 metres. Their fire resistance is usually evaluated by experiments. The load bearing criterion R, the integrity criterion E and thermal insulation criterion I are observed for roof structures.

The mechanical load during the test corresponds to accidental design situation and the temperature follows nominal temperature curve. The section factor of corrugated sheets A_m/V (m^{-1}) exceeds $1\,000\,\mathrm{m}^{-1}$, therefore the temperature of the steel sheet θ_a can be approximately taken equal to the gas temperature θ_g. The change of the mechanical properties of steel of thin-walled elements at elevated temperature can be found in literature (Outinen & Mäkeläinen, 2002) and was introduced into European standard (EN 1993-1-2).

The reliability of the roof structure at fire is highly influenced by connection of the sheets to the supporting structure. The connection is loaded by forces induced by thermal expansion and contraction and its resistance is influenced by the temperature.

2 EXPERIMENTS

Experimental programme was carried out in the past years in fire test laboratories PAVUS (Veselí nad Lužnicí, Czech Republic) and FIRES (Batizovce, Slovak Republic) in cooperation with Czech company Kovové profily.

The test programme was focussed on experiments where the corrugated sheets were used as the load bearing structure and the thermal insulation of the roof was made from mineral wool or polystyrene, see Table 1 ((Hůzl, 2002).

Table 1. Summary of test results with corrugated sheets exposed to fire.

Test	Span L, mm	Load q, kN/m^2	Deflection δ_1, mm	δ_2, mm	Fire resistance
C1	6000	0,83	735	805	R 14, E 14, I 12
C2	6000	0,72	380	390	R 21, E 21, I 21
C3	6000	0,74	709	829	R 20, E 20, I 20
S1	6000	1,00	196	326	R 28, E 27, I 27
C4	6000	0,75	610	591	R 60, E 60, I 60
F1*	4800	0,98	–	442	R 22, E 30, I 30
F2**	4800	0,90	364	469	R 57, E 60, I 60
F3	4800	0,79	429	397	R 42, E 45, I 45
F4	4800	0,79	489	418	R 23, E 30, I 30

Corrugated sheet TR 150/280/0,75 was used for all tests, except
* TR 200/375/0,88, 4 bolts in rib,
** TR 160/250/0,75.

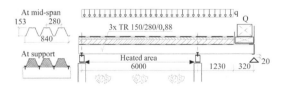

Figure 1. Test geometry of the fire resistance test C3 (Hůzl, 2002).

One experiment was carried out on simple beam (S1). For the other tests, beam with cantilever end simulating two-span continuous beam was used, see Figure 1. Only the span of the beam was heated during the test, the cantilever end and supports were located outside the furnace. The maximal deflection of the

Figure 2. Fire resistance test C3, load introduced by lightweight concrete block.

Figure 3. Fire resistance test C3, the specimen after the test.

cantilever end was limited to 20 mm to avoid collapse of the structure when a plastic hinge forms at the cantilever support.

The gas temperature in the furnace θ_g followed the standard temperature curve given by

$$\theta_g = 20 + 345 \, \log(8\,t + 1) \qquad (1)$$

where t is time in minutes.

The structure was supported on rigid frame made from two channels U 200. Two self-tapping bolts E-VS BOHR 5-5,5 × 38 in every rib of the sheet were used to connect the sheets to the frame. The sheets were connected along the longitudinal joints by self-drilling bolts 4,8 × 20 mm spaced at 500 mm.

Material tests were performed at 20°C. The yield limit of the steel was $f_y = 374$ MPa, the ultimate strength $f_u = 461$ MPa and the ductility = 18,9%.

The mechanical load representing snow load and technology (air conditioning, etc.) was introduced by lightweight concrete block and steel plates, see Figure 2.

Figure 4. Fire resistance test F2, the specimen after the test.

Figure 5. Fire resistance test F2, the specimen after the test.

The fire resistance criterion R was evaluated from the experiment using the following conditions which should be satisfied simultaneously. The maximal deflection (mm) should not exceed the limit

$$\delta_{lim} = \frac{L}{400\,r} \qquad (2)$$

where r is the lever arm, and the rate of deformation (mm/min) should no exceed

$$\frac{d\delta_{lim}}{dt} = \frac{L^2}{9\,000\,r} \qquad (3)$$

after the deflection of $L/30$ is reached.

The experimental results are summarized in Table 1 and plotter on Figure 6. All the experiments were stopped when the resistance criteria listed above were exceeded. The collapse of the structure was not observed. Local buckling on the bottom surface of the sheet was observed as a result of clamped support, see Figure 4. Plastic strains and deformation of the cross section were observed at mid-span of the sheets, see Figure 5.

Temperature of the connections was measured during the test C3, see Figure 7. The ribs at the end support were filled with mineral wool to achieve

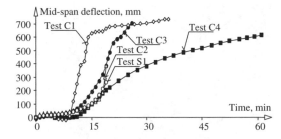

Figure 6. Mid-span deflection of specimens C1–C4 and S1.

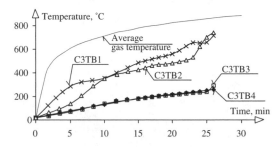

Figure 7. Temperature of the connections of the test C3, measured temperature (Hùzl, 2002).

thermal insulation of the bolts (measured by thermocouples C3TB1 and C3TB2). No thermal insulation was applied at the cantilever support (thermocouples C3TB3 a C3TB4). The temperature at time $t = 15$ min reached 135°C at the end support (insulated bolts) and 352°C (non-insulated bolts).

The connections were placed outside the furnace. It is expected, the temperature of the connections of real structure exposed to fire will be higher than those obtained during the laboratory tests.

3 BEHAVIOUR OF THE CONNECTIONS

The connection of the sheet to the support was subjected to experimental study at the laboratory of Czech Technical University in Prague. The tests were performed on steel sheets thickness 0,75 mm bolted to 10 mm steel plate. Three connection types were used for the tests:

- bolt E-VS BOHR 5-5,5 × 38 with sealed washer Ø19 mm,
- bolt E-VS BOHR 5-5,5 × 38 with steel washer Ø29 mm,
- bolt SD8-H15-5,5 × 25 without washer.

The tests were carried at constant temperatures 20, 200, 300, 400, 500, 600 and 700°C, see Figures 9, 10 and 11. The resistance, stiffness, deformation capacity and collapse mode of the connection were observed.

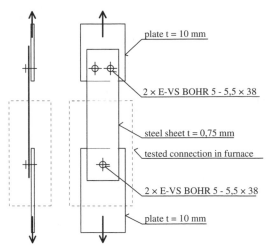

Figure 8. Test set-up of bolted connection at elevated temperature.

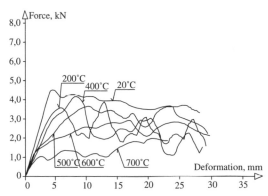

Figure 9. Connection behaviour at elevated temperature, bolts E-VS BOHR 5-5,5 × 38 with sealed washer Ø19 mm (Sokol, 2005).

The resistance of the connection with the sealed washer was limited by bearing resistance of the sheet. The sealant burns at higher temperatures and the flexible washer does not influence the bearing resistance and the stiffness of the connection. For the typical collapse mode of the connection see Figure 13.

When the steel washers were used, the stiffness of the connection increased and the resistance was almost doubled compare to the connection with sealed washers, see Figure 10. The benefit of using the steel washers is related to the collapse mode, see Figure 14, where the thin sheet was deformed and pushed in front of the washer. However, at temperatures exceeding 500°C the shear failure of the bolt was observed. This failure mode is accompanied by reduced deformation capacity and therefore should be avoided. This can be achieved by increasing number of the bolts.

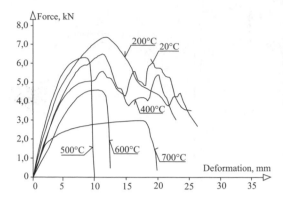

Figure 10. Connection behaviour at elevated temperature, bolts E-VS BOHR 5-5,5 × 38 with steel washer Ø29 mm (Sokol, 2005).

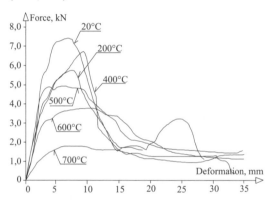

Figure 11. Connection behaviour at elevated temperature, bolts SD8-H15-5,5 × 25 without washer (Sokol & Kallerová, 2006).

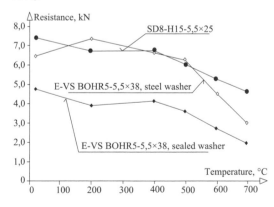

Figure 12. Resistance of the connections at elevated temperatures.

4 FEM MODELLING

Finite element code ANSYS was used for numerical simulation. Plastic beam element with three degrees

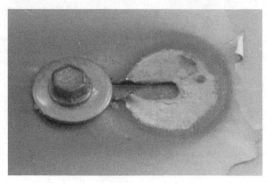

Figure 13. The collapse mode of bolted connection at 400°C, bolt E-VS BOHR 5-5,5 × 38 with sealed washer Ø19 mm.

Figure 14. The collapse mode of bolted connection at 200°C, bolt E-VS BOHR 5-5,5 × 38 with steel washer Ø29 mm.

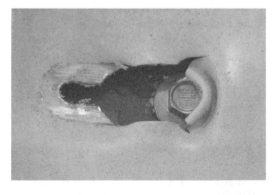

Figure 15. The collapse mode of bolted connection at 200°C, bolt SD8-H15-5,5 × 25 without washer.

of freedom at each node (BEAM 23 in the ANSYS element library) was used for the trapezoidal sheet, see Figure 16.

Temperature dependent multi-linear isotropic material stress strain relationship ($\sigma - \varepsilon$) based on the

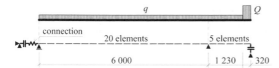

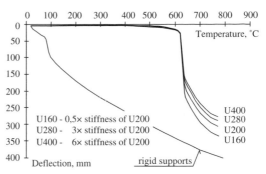

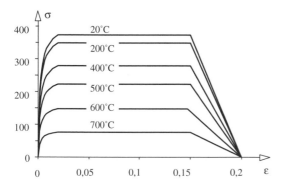

Figure 16. The FEM model of the corrugated sheet.

Figure 17. Temperature dependent stress-strain diagram.

Figure 18. Mid-span deflection of the sheet depending on stiffness of the support.

can be seen that the membrane effect can be observed for wide range of the stiffness of the support.

rules in EN 1993-1-2 was introduced, see Figure 17. The influence of the connections (stiffness and resistance) was modelled by non-linear translational spring elements NONLIN 39. The force-deformation relationship of the bolted connection corresponds to the experiments. Contact element representing the auxiliary support at the cantilever end was used to limit its deformation.

Non-linear analysis with large strains and large deformations was performed. The load was introduced in two steps, the first representing the mechanical and the second the thermal load.

The numerical model was used to perform a numerical study to find influence of various parameters on the behaviour at fire.

Stiffness of the supports has significant influence on the overall response. When restrained supports are assumed the thermal expansion is prevented which leads to large deflection of the structure. At higher temperatures membrane effect develops which contribute to the resistance of the structure because the bending moment resistance is significantly reduced by the temperature.

In reality, the supports do not need to be rigid but certain stiffness is required. The behaviour is quite different in this case: deformation of the supports allows for the thermal expansion and does not lead to large deflection at the early phase of the fire. Large deformation is observed at higher temperature when the reduced bending moment resistance is equal to the applied bending moment. However, collapse of the structure is not observed because the membrane effect will also contribute to the resistance, see Figure 18. It

5 DESIGN MODELS

The design model of thin-walled elements at ambient temperature is based on effective cross section taking into account local buckling. The bending moment resistance is calculated using effective moment of inertia $I_{a,eff}$ (Rondal, 2003). In addition, the resistance of the web to concentrated load at the supports needs to be checked.

The resistance check at elevated temperature is based on similar assumptions. The effective cross section can be derived for the steel temperature using reduced yield limit and modulus of elasticity (Ranby, 1998). As a simplification, the effective cross section derived at ambient temperature can be used for check at elevated temperatures.

Assuming unrestrained supports of the corrugated sheet, the bending moment resistance is given by

$$M_{\theta,Rd} = W_{a,eff,\theta} \, k_{y,\theta} \, f_{y,p}, \qquad (4)$$

where $W_{a,eff,\theta}$ is effective section modulus at temperature θ_a, $k_{y,\theta}$ is reduction factor of yield stress at temperature θ_a and $f_{y,p}$ is yield stress.

When restrain at the supports can be maintained during the fire, membrane effect will contribute to the resistance. As a simplification, it is assumed the bending moment resistance can be neglected and the total load is transferred by membrane action of the sheets.

Axial force in the sheets needs to be calculated which depends on deflection of the trapezoidal sheet. The extension of the sheet consists of three parts:

• thermal extension of the material,
• slip in the bolted connections at the support and
• extension caused by membrane force N.

The thermal expansion of the corrugated sheet is equal to

$$\Delta_p = \alpha\, L \left(\theta_a - 20\right) \qquad (5)$$

where α is coefficient of thermal expansion of steel and L is span of the sheet. The influence of slip of the connection Δ_b at the support can be introduced

$$\Delta_b = \frac{N_H}{n\, K_{b,\theta}} \qquad (6)$$

where n is number of bolts per unit width of the sheet, $K_{b,\theta}$ is stiffness of the bolted connection in N/mm at temperature θ_a and N_H is the horizontal component of the axial force N, see Figure 19.

The extension of the sheet caused by axial load N is given by

$$\Delta_N = \frac{N\, L}{E_\theta\, A}. \qquad (7)$$

The horizontal component of the axial force N_H, see Figure 19, is obtained from the catenary equation

$$N_H = \frac{q\, L^2}{8\, \delta} \qquad (8)$$

and the deflection of the trapezoidal sheet δ is

$$\delta = \frac{\sqrt{6\, L\, \Delta L}}{4} = \frac{\sqrt{6\, L \left(\Delta_p + 2\,\Delta_b + \Delta_H\right)}}{4}. \qquad (9)$$

The rotation α at the support, see Figure 19, can be evaluated from

$$tg\, \alpha = \frac{4\delta}{L} \qquad (10)$$

and the axial force N is equal to

$$N = \frac{N_H}{\cos\,\alpha}. \qquad (11)$$

Iterative procedure is necessary to solve the deflection, because the slip in the connection and extension of the sheet depend on the axial force N. Initial estimation of the force and several cycles of the calculation is necessary for accurate results.

The resistance of the bolted connections loaded by shear force N_H should be checked. Design model for the connection at elevated temperature is not available at present therefore the design resistance of the connection needs to be evaluated by testing.

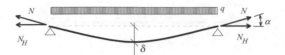

Figure 19. Design model of corrugated sheet with restrained supports at elevated temperature.

6 CONCLUSION

The resistance of roof structure made from corrugated sheets can reach the fire resistance R 60. It is achieved by low stress in the corrugated sheets at ambient temperature which is the reason of design to meet the SLS criteria (deflection).

Two design models are available depending on the support of the sheets. For unrestrained support, the model is based on bending resistance of the effective section. When thermal elongation/contraction is restrained, membrane model is used to predict the resistance.

The numerical study shows that even small stiffness of the supports is sufficient for the membrane effect to develop. The resistance and stiffness of the bolted connections and of the supporting structure should be checked in this case.

ACKNOWLEDGEMENT

The research work was supported by grant MSM 6840770001 Reliability, optimization and durability of building materials and constructions.

REFERENCES

EN 1363 Fire resistance tests. Part 1. General requirements. CEN: Brussels.
EN 1993-1-2. 2005 Eurocode 3. Design of steel structures. Part 1–2. General rules. Structural fire design. CEN: Brussels.
Hřebíková P. 2003. Experimenty s přípoji trapézových plechů, Část 1 Zkoušky za pokojové teploty. Výzkumná zpráva (Experiments with connections of corrugated sheets. Part 1. Tests at ambient temperature. In Czech). Praha: ČVUT.
Hůzl J. 2002. Vodorovná nosná konstrukce, skládaný střešní plášt. Protokol o zkoušce požární odolnosti Pr-02-02.014, 02.043. (Research report on fire test No. Pr-02-02.014, 02.043. In Czech). Veselí n. L.: PAVUS.
Lawson R.M. Burgan B. & Newman G.M. 1993. Building design using cold formed steel section, fire protection. Publication No. P129, SCI: London.
Outinen J. Kaitila O. & Mäkeläinen P. 2001. High-temperature testing of structural steel and modelling of structures at fire temperatures. Helsinki: HUT.
Outinen, J. & Mäkeläinen, P. 2002. High-temperature strength of structural steel and residual strength after cooling down. In: Ivanyi, M. (ed), Stability and ductility of steel structures: 751–760. Budapest: Akadémiai Kiadó.

Ranby A. 1998. Structural fire design of thin walled steel sections. *Journal of Constructional Steel Research*, 46. 1–3.

Rondal J. 2003. Lesson 2. Design of Cold–Formed Steel Sections. In: *CISM Course*. CISM: Udine.

Sokol Z. & Vácha J. 2002. *Numerická simulace trapézového plechu za zvýšených teplot. Výzkumná zpráva. (Numerical simulation of corrugated sheets at elevated temperature. Research report. In Czech)*. Praha: Kovové profily.

Sokol Z. 2005. *Experimenty s přípoji trapézových plechů, Část 2 Zkoušky za zvýšené teploty. Výzkumná zpráva (Experiments with connections of corrugated sheets. Part 2. Tests at elevated temperature. In Czech)*. Praha: ČVUT.

Sokol Z & Kallerová P. 2006. *Experimenty s přípoji kazetové stěny. Zkoušky za zvýšené teploty. Výzkumná zpráva (Experiments with connections of thin-walled sheets. Tests at ambient and elevated temperatures. In Czech)*. Praha: ČVUT.

Steel and Composite Structures – Wang & Choi (eds)
© 2007 Taylor & Francis Group, London, ISBN 978-0-415-45141-3

3D thermal and structural analyses of Bi-Steel composite panels in fire

C. Yu, Z. Huang & I.W. Burgess
Department of Civil and Structural Engineering, University of Sheffield, Sheffield, UK

R.J. Plank
School of Architectural Studies, University of Sheffield, Sheffield, UK

ABSTRACT: This paper presents some results on the performance in fire of Bi-Steel panels used as predominantly compressive structural elements such as building core-walls. In the three-dimensional heat transfer analysis, results generated from ABAQUS were compared with the temperature values provided by the 'Bi-Steel Design and Construction Guide' (1999). The structural response to elevated temperatures was simulated by modelling the structural arrangement using an assembly of 3D brick elements, which have been developed by the principal author. In the formulation of these elements, geometric nonlinearity, material nonlinearity and material degradation at elevated temperatures have been taken into account. Parametric studies were carried out with different boundary conditions, load intensities and temperature distributions. From the results of these analyses, some useful conclusions can be drawn.

1 INTRODUCTION

The Bi-Steel panel, which is composed of two steel facing plates connected by an array of transverse friction-welded shear connectors, and is filled with concrete, was developed from the older steel-concrete-steel double-skin composite construction method. In a Bi-Steel panel, the steel faceplates provide resistance to both in-plane and bending forces. The steel bar connectors carry longitudinal shear force between the faceplates and the concrete core, and prevent buckling of the faceplates by providing a tie between the front and rear plates. The concrete core provides resistance to compressive and shear forces. Because of its structural and constructional advantages, such as enhanced blast and fire resistance, thickness reduction in core walls, leak resistance, and optimization of site work and site time, Bi-Steel could be a particularly economic construction solution in applications where high security against blast and fire is required. However, since their structural characteristics are complicated, few studies have been reported of the behaviour in fire of Bi-Steel structures.

In the building fire resistance context, it is necessary to do some detailed research in this field. In this research, the performance in fire of Bi-Steel panels used as predominantly compressive structural elements such as building core-walls has been studied. In a fire situation, walls are generally subjected to heating on one face, with the other face exposed

to ambient condition. Therefore, a thermal gradient is introduced through the thickness of the wall (Yu et al. 2006). Owing to the differential thermal expansion of the material, large thermal stresses can be introduced into the wall, and it can bow thermally towards the fire if it is unrestrained (Nadjai 2004). Hence, its structural response can be very complex.

For Bi-Steel panels, the calculation of fire resistance involves the determination of temperature distribution, deformation and stress under various types of loading.

2 THERMAL ANALYSIS

A Bi-Steel panel without any additional fire protection, exposed to the fire, is presented for thermal analysis. The finite element analysis software ABAQUS was used to generate temperature information. Because of the inherent symmetry of the case, only a cuboid of Bi-Steel was considered here (see Fig. 1). The properties are listed in Table 1. For analysis by the finite element method, the model required a mesh of nodes and elements. 3D diffusive heat-transfer solid elements were chosen to represent both steel and concrete materials in ABAQUS simulations. In order to determine the appropriate mesh density, some sensitivity analyses were undertaken to calibrate the accuracy of computation. On the basis of the results of sensitivity analysis, the Bi-Steel model was divided into 729 DC3D8

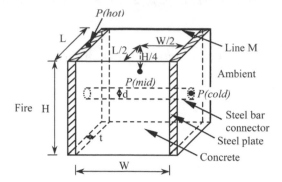

Figure 1. 3D Bi-Steel thermal model.

Table 1. Properties used in thermal analysis.

W/H/L mm	t mm	d mm	Steel Density kg/m³	Concrete Density kg/m³	Concrete Moisture content
200	15	25	7850	2400	5%

elements and 1000 nodes. For convenience, the steel bar connector was also represented by a column of 8-noded solid elements, with cross-section equal to the real area.

The chosen analysis type was the uncoupled heat transfer analysis within ABAQUS. In order to work out the heat transfer characteristic of Bi-Steel completely, the Standard ISO 834 and Hydrocarbon fire curves in Eurocode 1 Part 1.2 (2002) were set as the boundary fire conditions. Corresponding to these two fires, the coefficients of heat transfer by convection were respectively 25 W/m²K and 50 W/m²K. The emissivity of the fire was taken as 0.8. Other thermal properties, such as the specific heats and the thermal conductivities of both steel and concrete, are given in Eurocode 4 Part 1.2 (2003).

Results generated by ABAQUS simulations were compared with the temperature values provided by 'Bi-Steel Design and Construction Guide' (The Bi-Steel Guide) (1999). The compared key positions are shown in Figure 2. Results from the two models under the two fire conditions are plotted in Figures 2 to 4. At the point *P(hot)*, the histories of steel temperatures obtained from ABAQUS and The Bi-Steel Guide were quite similar (Fig. 2). At the position *P(mid)*, concrete temperatures predicted by the two models were slightly different, especially at the later heating stage (Fig. 3). Nevertheless, at the position *P(cold)*, there were significant (up to 39.8%) differences in temperature between ABAQUS and The Bi-Steel Guide (Fig. 4). This inconsistency might be induced by the different values of

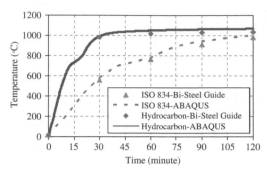

Figure 2. Temperatures at *P(hot)*.

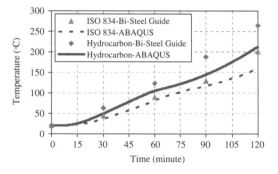

Figure 3. Temperatures at *P(mid)*.

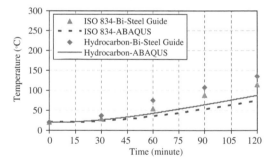

Figure 4. Temperatures at *P(cold)*.

thermal properties implied in ABAQUS analysis and The Bi-Steel Guide. In the latter case their values were not specified, even though they were based on Eurocode 4 Part 1.2.

Moreover, from the results it can be noticed that the difference between the temperatures predicted by the two models subjected to the Hydrocarbon fire were more obvious than those subjected to the ISO 834 fire. This is because the Hydrocarbon fire gives a much faster temperature rise than the ISO 834 fire.

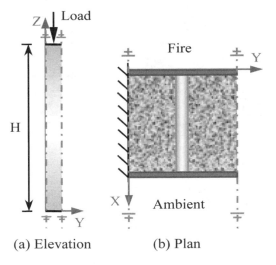

(a) Elevation   (b) Plan

Figure 5. Bi-Steel structural model.

Table 2. Properties of 3D Bi-Steel model.

Steel faceplate		Steel bar connector		$E_s^* \times 10^5$ N/mm^2	Concrete Strength N/mm^2
Material grade	Strength N/mm^2	Material grade	Strength N/mm^2		
S355JR	355	070M20	370	2.05	32

* Young's modulus of steel

Table 3. Specification of the Bi-Steel study.

Case name	End boundary condition	Load (kN)	Fire condition
Case 1	Both pinned	780	ISO 834
Case 2	Both pinned	1300	ISO 834
Case 3	Both pinned	780	Hydrocarbon
Case 4	Both pinned	1300	Hydrocarbon
Case 5	Both fixed	780	ISO 834
Case 6	Both fixed	1300	ISO 834

Additionally, in order to analyse the thermal behaviour of the Bi-Steel structure properly, some parametric studies were conducted (Yu et al. 2006).

3 STRUCTURAL ANALYSIS

After obtaining the temperature information, the Bi-Steel panel, used as a compressive building core-wall, was considered in the structural analysis. The in-house finite element program *Vulcan* was used to model the structural behaviour of Bi-Steel panels. A three-dimensional eight-noded brick element had been developed and implemented into *Vulcan* by the principal author (Yu et al. 2006). In the formulation of these elements, geometric nonlinearity, material nonlinearity and material degradation at elevated temperatures have been taken into account.

In finite element analysis, a compressive core-wall presents a huge-order problem. It is not efficient or necessary to model the whole structure by thousands of the relatively tiny 3D brick elements. However, due to the symmetry of the case, the structural model can be reduced to a strip of the shear wall (see Fig. 5). In this paper, only the results from the models with one cross-sectional boundary condition, which represents the symmetric boundary condition within a short wall, are presented. Most of the dimensions of the 3D model were the same as those applied in the Section 2, except H and t. Here, their values are 3140 mm and 10 mm. Corresponding to these dimensions, some heat transfer analyses were again conducted on the cuboid model (see Fig. 1). The material properties of the model are listed in Table 2. In analysis, Poisson's ratios of steel and concrete were taken as 0.3 and 0.2 respectively.

In addition, steel and concrete were assumed to be completely bonded.

In structural analysis, for efficiency of calculation, it is not necessary for the mesh to be identical with that in thermal analysis, as long as they share common nodes. Thus, a couple of mesh sensitivity studies were conducted on the model at ambient and high temperatures respectively. The mesh density determined was then used in the following structural analyses.

As a compressive structure in fire, the study focused on the effects of different boundary conditions, load intensities and temperature distributions on its fire resistance and structural performance. A series of parametric studies were carried out under the considerations listed in Table 3.

In analysis, the amount of load imposed concentrically on the model was proportional to its ultimate strength (2600 kN), which was calculated from its ambient-temperature model. As shown in Table 3, 780 kN and 1300 kN were taken as the values of imposed load. They are equivalent to 0.3 and 0.5 load intensities, since practical values are always between these levels. To ensure that the steel faceplates and the concrete core were loaded simultaneously, a stiffened steel plate was attached to each end of the strip model. Its thickness was 25 mm. For the model with both ends pinned, due to the symmetry of the structural layout and boundary conditions, only half of the length of the Bi-Steel strip was considered. Including the end plate, the model was divided into 2744 3D brick elements and with 3648 nodes. However, for the model with both ends fixed, because of the limitation on the freedom of rotations within the 3D brick elements, a

629

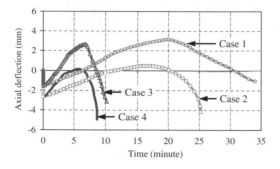

Figure 6. Axial deflections of the models plotted against time.

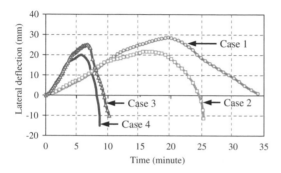

Figure 7. Lateral deflections of the models plotted against time.

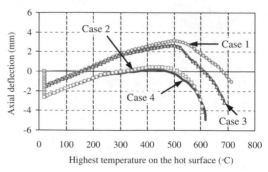

Figure 8. Axial deflections of the models plotted against the highest temperature.

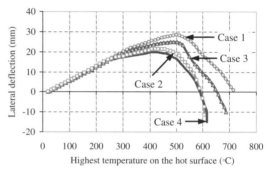

Figure 9. Lateral deflections of the models plotted against the highest temperature.

full-length strip wall, which was meshed with 5439 3D brick elements and 7168 nodes, had to be considered.

Firstly, the results of the Bi-Steel models, with the pin-ended boundary conditions but under different load intensities and fire conditions, were compared with each other. In Figures 6 and 7, the axial and lateral deflections are plotted against heating time. It is evident that the fire resistances of Cases 3 and 4 were far less than those of Cases 1 and 2. This is because of the inherent characteristics of the two fire curves.

In order to understand the essential influences of the different fires on the structural responses of the Bi-Steel model, deflections produced by the Standard ISO 834 and Hydrocarbon fires are plotted against the highest temperature on the hot surface of the model (see Figs. 8 and 9). It can be seen that the differences between the results of the cases with the ISO 834 and Hydrocarbon fire conditions were not significant, particularly to Case 2 and Case 4.

To explain this phenomenon, the thermal gradients within Model B under the Hydrocarbon fire curve were compared with those within Model A under the ISO 834 fire curve at different heating times (see Fig. 10). In other words, the temperature distributions along Line M from the hot steel faceplate to the cold steel faceplate (see Fig. 1) are plotted using the same

temperatures on the hot surface in these two thermal models. From Figure 10, it can be observed that the thermal gradients within Model B were slightly more extreme than those of Mode A, especially in the field between 20 mm to 100 mm from the surface exposed to fire in the later stages of heating. Their differences are shown in Figure 11. The differences between the temperatures of Model A and Model B increased with heating time, but their trends were quite similar. Therefore, the differences between the deflections calculated from the models under the ISO 834 and Hydrocarbon fire conditions were insignificant in the early heating stage. The considerable discrepancy between Case 1 and Case 3 was because of the combination of a low load intensity and the effects of thermal gradients and thermal expansion.

In addition, from Figures 7 and 9, it is found that the directions of the central lateral deflections reversed and the model bowed away from the fire until the analyses stopped. This reverse bowing was caused by the degradation of the material properties at high temperatures on the fire-exposed side, and it effectively created a load eccentricity towards the fire side.

Consequently, the Bi-Steel models with different end boundary conditions in the Standard ISO 834 fire

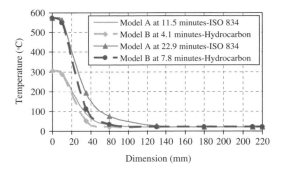

Figure 10. Temperature distributions from the models in two different fires.

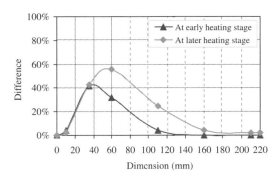

Figure 11. Differences between the temperatures of the models in two different fires.

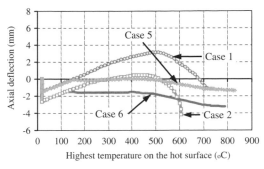

Figure 12. Axial deflections of the models plotted against time.

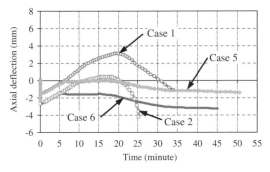

Figure 13. Axial deflections of the models plotted against the highest temperature.

Table 4. Summary of fire resistances.

Case name	Highest temperature reached °C	Fire resistance minute
Case 1	715.0	34.03
Case 2	607.5	25.27
Case 3	687.1	10.18
Case 4	616.6	8.69
Case 5	832.1	50.63
Case 6	788.1	45.13

were studied. Since the lateral deflections of Case 5 and Case 6 were negligible, only the axial deflections of the four cases were compared (see Figs 12 and 13). According to the results, it can be determined that the failure modes of the models with both-ends-pinned boundary conditions were overall buckling; the failure modes of the models with both-ends-fixed boundary conditions were compression. Furthermore, the influences of the different load intensities on the deformations of the models are obvious. For the cases with low load levels, the deformation caused by the thermal expansion was much larger than the deformation induced by the stress within the structure.

Additionally, it can be noticed that the fire resistance periods of Cases 5 and 6, especially Case 6, were longer than those of Cases 1 and 2. This is because the fixed-end boundaries helped the models to reduce the instability induced by buckling. Moreover, in Case 5, the load intensity was lower than that in Case 6. Thus, when the model was heated, the thermal expansion in Case 5 was more obvious compared to Case 6. This thermal behaviour, combined with the higher restraint from the boundaries, reduced the impact brought about by the fixed ends. The fire resistances of all six cases are listed in Table 4.

On the deflection curves of Cases 5 and 6, evident kinks can be observed. According to the results, it was found that the stresses at a number of Gauss points of the steel elements were beyond the linear-elastic limit at the time of the kink. At ambient or low temperatures, the imposed load on the Bi-Steel strip model was mostly carried by the steel faceplates. In the linear- elastic stage, the thermal expansion dominates the deformation behaviour of the model. However, as soon as the stress in the steel plate reached the transitional elliptical stage, which is nonlinearly elastic-plastic (Purkiss 1996), its tangent modulus was reduced and then its mechanical strain was increased. Thus, the thermal expansion was not as pronounced as it was under the linear-elastic phase. The occurrence of the kink was also partially due to the assumption that steel and concrete are completely bonded. The different coefficients of thermal expansion of steel and concrete introduced some interactive stresses into them. Nevertheless, in Cases 1, 2, 3 and 4, owing to the less restrained boundaries, the kinks were not apparent. In other words, since the structure could relatively freely expand, the concentrations of the stresses at steel plates were not as obvious as those in Cases 5 and 6.

4 CONCLUSIONS

From the results of the thermal and structural analyses on Bi-Steel panels used as compressive core-walls, a series of conclusions can be drawn:

(1) The temperatures of the 3D Bi-Steel model subjected to the Standard ISO 834 and Hydrocarbon fires, obtained from ABAQUS, were reasonable, though there were some differences in the values provided by The Bi-Steel Guide. This is due to the different values of thermal properties implied in the two models.

(2) The analyses indicated that there was a large temperature gradient from the heated steel faceplate to the cold steel faceplate.

(3) The difference between the deflections of the Bi-Steel models subjected to the Standard ISO 834 and Hydrocarbon fires were not significant in terms of the hot-surface temperature, although in terms of time the hotter Hydrocarbon fire achieves these temperatures more quickly.

(4) The load intensities had an influence on the fire resistance of the model. In general, low load intensities showed high fire resistance; high load intensities showed reduced fire resistance, if the models studied were based on the same boundary conditions and temperature distributions. In addition, in the cases with low load intensity, the effect of thermal expansion was more pronounced.

(5) The effects of different boundary conditions on the fire resistances and deformation behaviour of the models were significant. In the above analyses, the failure modes of the models with pinned-end boundaries were overall buckling; the failure modes of the models with fixed-end boundaries were compression.

These results of the models and numerical analyses could benefit the further development and study of Bi-Steel panels as a structural system, and underpin a theoretical basis for calculating their fire resistance.

REFERENCES

British Steel Ltd. 1999. *Bi-Steel Design and Construction Guide*, Volumes 1 and 2. Scunthorpe: British Steel Ltd.

European Committee for Standardization. 2002. *Actions on structures. Part 1.2: General actions-Actions on structures exposed to fire. Eurocode 1*. Brussels: CEN.

European Committee for Standardization. 2003. *Design of composite steel and concrete structures. Part 1.2: General rules-Structural fire design. Eurocode 4*. Brussels: CEN.

Nadjai, A. 2004. Behaviour of masonry compartment walls in fire situations. *Proceedings of the Third International Workshop Structures in Fire*, Ottawa, Canada: 99–118.

Purkiss, J. A. 1996. *Fire safety engineering, Design of structures*. Butterworth-Heinemann, Linacre House, Jordon Hill, Oxford.

Yu, C., Huang, Z., Burgess, I.W. & Plank, R.J. 2006. 3D modelling of Bi-Steel structures subject to fire. *Proceedings of the Fourth International Workshop Structures in Fire*, Aveiro, Portugal: 393–404.

Steel and Composite Structures – Wang & Choi (eds)
© 2007 Taylor & Francis Group, London, ISBN 978-0-415-45141-3

Behaviour of fire-exposed CFST columns repaired with CFRP wraps

Zhong Tao & Zhi-Bin Wang
College of Civil Engineering, Fuzhou University, Fuzhou, Fujian Province, P.R. China

Lin-Hai Han
Department of Civil Engineering, Tsinghua University, Beijing, P.R. China

ABSTRACT: This paper presents the test results of fire-exposed concrete-filled steel tubular (CFST) beam columns repaired by unidirectional carbon fibre reinforced polymer (CFRP) composites. The test results showed that the load-bearing capacity of the fire-exposed CFST beam-columns was enhanced by the CFRP jackets to some extent, while the influence of CFRP repair on stiffness was not apparent. The strength enhancement from CFRP confinement decreased with the increasing of eccentricity or slenderness ratio. Ductility enhancement was also observed except those axially loaded shorter specimens with rupture of CFRP jackets at the mid-height, occurred near the peak loads. However, the strengths of all repaired specimens have not been fully restored due to the long exposure time of them in fire. From the test results, it is recommended that, in repairing severely fire-damaged CFST members, slender members or those members subjected to comparatively large bending moments, other appropriate repair measures should be taken.

1 INTRODUCTION

It is well-known that fire is always one of the most serious potential risks to modern high-rise buildings. With the increasing use of concrete-filled steel tubes as structural members, there is a growing need to understand the effect of fire on them and to provide measures for postfire repair of such kind of composite members (Tao & Yu, 2006). In the past, many studies have been performed on the fire resistance of concrete filled steel tubes (CFST) columns, as well as their residual strengths after exposure to fire (Han, 2004; Han & Lin, 2004). However, very limited research has been conducted so far concerning the repair of fire-damaged CFST members (Tao et al., 2007).

Fibre reinforced polymer (FRP) is composed of fibres embedded in a resin matrix, and is characterized by high strength-to-weight ratio, high corrosion resistance and ease of installation. Over the past few decades, FRP material has gained its popularity as a jacketing material in retrofitting/repairing existing concrete structures or steel structures (Teng et al., 2002). More recently, some research results have been reported on the strengthening of CFST stub columns, which have preliminarily demonstrated the effectiveness of the carbon fibre reinforced polymer (CFRP) wraps in increasing the axial loading capacity of CFST stub columns (Xiao et al., 2005; Tao & Yu, 2006).

In the past, section enlargement method has been proved suitable to be used in rehabilitation of reinforced concrete columns (Han & Yang, 2004). It is expected that this method will also be effective in repairing fire-damaged CFST columns. However, it will result in a significant increase in the column cross-section and a comparatively long construction time. FRP jacketing is a more recent method and is particularly attractive in that it does not significantly increase the section size and is easy to install.

In order to evaluate the effectiveness of the above two repair methods and identify method gaps, the authors have been engaged in research studies on this topic recently. Test results of 25 CFST specimens, in which nine of them were fire-damaged and then strengthened by wrapping the original columns by concrete and a thin walled steel tube, have been reported in Han et al. (2006). These specimens were tested under a constant axial load and a cyclically increasing flexural load. The test results indicate that the strength and the stiffness of the strengthened columns can be restored over the previous level of the specimens.

As far as the wrapping FRP method is concerned, Tao et al. (2007) have carried out preliminary tests on fire-exposed CFST stub columns and beams repaired with CFRP wraps. It demonstrated that FRP composites were effective in enhancing the load-carrying capacity of stub columns, while the strengthening effect for beams was quite limited if only unidirectional FRP was used to confine the beams. The research results reported in this paper are part of a wider study concerning the repair of fire-exposed CFST beam columns.

New test data concerning concrete-filled steel tubular beam columns, that have been exposed to and damaged by the standard ISO-834 fire and subsequently repaired are provided in this paper. Fourteen beam columns with circular cross-sections were tested. The fire-exposed CFST specimens were repaired by wrapping CFRP as shown in Fig. 1. The repair effects of CFRP wraps on strength, stiffness and ductility are investigated in this paper.

2 EXPERIMENTAL PROGRAM

2.1 General

A total of 14 beam column specimens were tested. The specimens can be classified as three categories: specimens without fire exposure and no repairing, specimens with fire exposure but without repairing, and

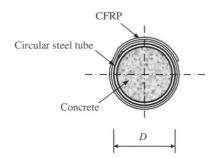

Figure 1. Cross-section of CFRP-repaired concrete-filled steel tubes.

specimens with fire exposure and repairing. Repair of specimens was achieved by the external wrapping of one or two layers of unidirectional carbon fibre sheets with the fibres oriented in the lateral direction. Since no design method is available at this time for the repair of fire-damaged CFST members and the main purpose of this paper is just to validate the repair effectiveness of wrapping FRP method, no repair design is done to determine the required number of FRP layers. Specimen details of beam columns are provided in Table 1, where the first letter C in specimen designations refers to specimens with circular cross-sections. In order to designate the specimens with fire exposure, an additional letter F is used to label them. The last number of 0, 1 or 2 in the labels of these fire-exposed specimens indicates the wrapped number of CFRP layers. Two kinds of slenderness ratio (λ), with values of 25 and 50 were selected. The slenderness ratio (λ) herein is defined as:

$$\lambda = \frac{L}{i} \qquad (1)$$

where L is the effective length of a column, which is the same as the physical length of the column (L) with pin-ended supports; $i = \sqrt{I_{sc}/A_{sc}}$, is the section radius of gyration, I_{sc} and A_{sc} are the second moment of area and area of a composite cross section, respectively. In calculating I_{sc} and A_{sc}, it should be noted that the contributions from FRP jackets were ignored.

2.2 Material properties

Cold-formed steel tubes with wall thicknesses (t_s) of 3.0 mm were used in the construction of the specimens. The measured yield strengths for these tubes are 356. All the specimens were cast with one batch of self-consolidating concrete. The average cube strength (f_{cu})

Table 1. Specimen labels and member capacities.

No.	Specimen label	$D \times t_s$ (mm × mm)	f_y (MPa)	f_{cu} (MPa)	L (mm)	λ	e (mm)	t (min)	n_{frp}	N_{ue} (kN)	RSI (%)	SEI (%)	K_e (kN/mm)	DI
1	C-1	150 × 3.0	356	75	940	25	0	0	0	1815	–	–	4033	1.91
2	CF-1-0	150 × 3.0	356	75	940	25	0	180	0	550	69.7	–	244	3.62
3	CF-1-2	150 × 3.0	356	75	940	25	0	180	2	1110	–	101.8	247	1.39
4	C-2a	150 × 3.0	356	75	940	25	50	0	0	735	–	–	267	3.96
5	C-2b	150 × 3.0	356	75	940	25	50	0	0	735	–	–	237	3.02
6	CF-2-0	150 × 3.0	356	75	940	25	50	180	0	290	60.5	–	112	9.61
7	CF-2-1	150 × 3.0	356	75	940	25	50	180	1	347	–	19.7	110	10.2
8	CF-2-2	150 × 3.0	356	75	940	25	50	180	2	390	–	34.5	113	11.0
9	C-3	150 × 3.0	356	75	1875	50	0	0	0	1430	–	–	1300	1.69
10	CF-3-2	150 × 3.0	356	75	1875	50	0	180	2	493	–	–	720	2.36
11	C-4a	150 × 3.0	356	75	1875	50	50	0	0	530	–	–	61	4.0
12	C-4b	150 × 3.0	356	75	1875	50	50	0	0	570	–	–	62	3.75
13	CF-4-0	150 × 3.0	356	75	1875	50	50	180	0	222	59.6	–	36	9.32
14	CF-4-2	150 × 3.0	356	75	1875	50	50	180	2	235	–	5.9	42	9.71

at the time of tests was 75 MPa. The tensile strength and elastic modulus of the cured CFRP determined from tensile tests of flat coupons are 3950 MPa and 247 GPa respectively, which were calculated on the basis of the nominal thickness of 0.17 mm.

2.3 Specimen preparations

In preparing the CFST specimens, empty tubes were accurately machined to the required length. One end of each tube was tack welded by a steel plate with a thickness of 12 mm. The specimens were cast in a vertical position. Concrete mix was filled in layers and was vibrated by a poker vibrator. These specimens were then naturally cured in the indoor climate of laboratory. Before testing, the top surfaces of the CFST specimens were ground smooth and flat using a grinding wheel with diamond cutters. A steel plate with a thickness of 12 mm was then welded to the top of each of those specimens.

For specimens to be heated with a slenderness ratio of 25, four circular holes with 10 mm in diameter, located at quarter points of the specimens, were drilled in each section wall. Besides holes drilled in the above-mentioned locations, two additional holes were also drilled in the mid-height for specimens with a slenderness ratio of 50. These holes were provided as vent holes for the water vapour pressure produced during fire exposure, and then repaired after fire exposure by welding them shut with steel plates 9 mm in diameter.

At 28 days after the concrete was filled, the related specimens were heated by exposing them to heat in a furnace specially built for testing building structures in Tianjin, China. The ambient temperature at the start of the test was about 20°C. The furnace heating was controlled as closely as possible to the ISO-834 standard fire curve (ISO-834, 1975). The fire duration time (t) for all relevant specimens were set to be 180 min.

It should be noted that because of the limitation of the furnace faculty, the beam columns were not loaded during the fire exposure, which does not reflect the actual fire damage situation. However, the test data should provide the basis for further theoretical studies of the postfire repair of CFST beam columns.

The fire-damaged specimens were then transported to the laboratory of Fuzhou University and repaired by jacketing with CFRP, where carbon fibre sheets were applied using a hand lay-up method. The finishing end of a sheet overlapped the starting end by 150 mm. The wrapped specimens were left to cure in the laboratory environment at room temperature for about 3 months before testing.

2.4 Test setup and instrumentation

The beam columns were tested as pin-ended supported and subjected to single curvature bending.

Axial loading was applied through V-shape edges to each specimen, which were installed with steel caps at both ends prior to loading. Grooves of 6 mm in depth were machined on each steel cap to receive the V-shape loading edge so that the load eccentricity (e) could be precisely controlled. The effective lengths of all beam-columns (L) shown in Table 1 were defined as the distance between the tips of the V-shape edges.

For each specimen, eight strain gauges mounted on the specimen surface were installed to measure the axial and transverse strains at the mid-height of the beam-column. Displacement transducers were used along the specimens span to monitor the deflections. Two other transducers, at the end of the specimen, were used to monitor the axial shortening.

A load interval of less than one tenth of the estimated load capacity was used. Each load interval was maintained for about 2 to 3 minutes. At each load increment the strain readings, the deflection and axial shortening measurements were recorded. To study the softening response of the specimens, the loads were applied at closer intervals near the ultimate load.

3 EXPERIMENTAL RESULTS AND DISCUSSIONS

3.1 Test observations and failure modes

For unrepaired fire-exposed specimens, the oxide layers formed during the fire exposure began to peel when the load attained 40–60% of the ultimate load in the pre-peak stage. Almost all the oxide layers spalled off when the load attained the ultimate load. For repaired specimens, flexural cracks of CFRP were observed on the tension surfaces at a load level of about 20–30% of the ultimate loads. Despite this, the FRP jackets worked well together with the infilled steel tubes generally.

No local buckling was observed before their ultimate strengths had been developed for most beam columns except several short undamaged specimens. Marked points were used in Figs. 4 and 5 to identify the approximate location of observed local buckling of the steel tubes. It is interesting to note that, the steel tubes of most unrepaired fire-damaged specimens buckled a little earlier than those undamaged ones at ambient temperature. However, the phenomenon observed above is not applicable to specimen C-2b. This is due to the fact that specimen C-2b failed by cracking of the top end plate to steel tube weld, thus induced an abrupt failure different from that of the companion specimen C-2a. The buckling of steel tubes for repaired specimens was postponed to some extent with the increasing number of CFRP layers (n_{frp}). It seems that it was the confining effect of the CFRP jackets that delayed the local buckling of the steel tubes (Xiao et al., 2005).

Figure 2. A general view of typical repaired specimens after testing.

Figure 3. Typical failure modes of specimens.

All beam columns, except specimens C-1 and C-2b have shown an overall buckling failure mode. Fig. 2 provides a general view of typical repaired specimens after testing. During the tests, the deflection curves of these specimens were approximately in the shape of a half sine wave.

For specimen C-1, which were axially loaded undamaged specimens with a smaller slenderness ratio, a local buckling failure mode was observed (see Fig. 3). Generally, the local buckling occurred near the top ends due to the effect of end conditions. It seems overall imperfection had moderate influence on such kinds of short specimens.

For axially loaded short specimen CF-1-2, rupture of CFRP jackets at the mid-height was observed near the peak load (shown in Fig. 3). The jacket rupture is due to hoop tension which is the same failure mode that is often observed for FRP-confined concrete cylinders or short RC columns (Teng et al., 2002). No CFRP

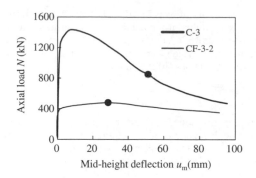

Figure 4. Effect of fire exposure on N-um relations.

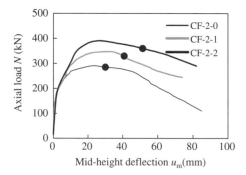

Figure 5. Effect of CFRP repair on N-u_m relations.

rupture was observed for other specimens whose hoop strains were less than the rupture strain of CFRP. This is attributed to the fact the compression zone of the column decreases and the FRP will experience smaller hoop strain as the column becomes slender or the load eccentricity increases.

3.2 Test results

The maximum loads (N_{ue}) obtained in the test are summarized in Table 1. Axial load (N) versus mid-height lateral deflection (u_m) curves of all specimens are shown in Figs. 4 and 5.

Fig. 4 shows the effect of fire exposure on $N - u_m$ curves for circular specimens. It can be seen from Fig. 4 that the exposure of specimens to fire can result in significant loss of strength and stiffness. However, the post peak curves of the undamaged specimens are much steeper compared to those of the damaged specimens.

Fig. 5 illustrates the effect of CFRP repair on $N - u_m$ curves for typical specimens. It seems that the load-carrying capacity of CFRP wrapped specimens increase with the increasing number of CFRP layers. The repair effects of CFRP wraps on stiffness and ductility of the composite beam columns will be discussed in subsequent sections.

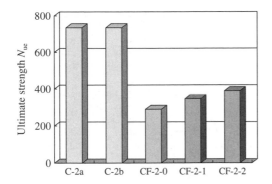

Figure 6. Tested ultimate loads for typical beam columns.

3.3 Effect of FRP repair on load-bearing capacity

As mentioned above, the exposure of specimens to fire have resulted in significant loss of strength, and on the other hand, strength enhancement was observed when CFRP wraps were applied. Fig. 6 shows the effects of fire exposure and FRP repair on load-bearing capacity for typical specimens.

For convenience of analysis, residual strength index (RSI) defined in Han & Huo (2003) is used to quantify the strength of the CFST beam columns after exposure to ISO-834 standard fire. It is expressed as

$$RSI = (N_u - N_{u,r})/N_u \qquad (2)$$

where N_u is maximum strength for composite specimens at ambient temperature; $N_{u,r}$ is residual strength of the fire-exposed composite columns.

Similarly, in order to gauge the enhancements in the load-bearing capacity of repaired CFST columns due to FRP wrapping, a strength enhancement ratio is defined as the percentage increase in the ultimate load:

$$SEI = (N_{eS} - N_{eU})/N_{eU} \qquad (3)$$

where N_{eS} and N_{eU} are maximum loads for repaired and unrepaired specimens after exposure to fire.

The values of RSI and SEI determined from Eqs. (1) and (2) are listed in Table 1. It can be seen that the values of RSI range from 59.6% to 69.7% and those for SEI range from 5.9% to 101.8%.

From the comparisons, it can be found that the load eccentricity (e) and the slenderness ratio (λ) have very moderate influence on the strength loss ratio (RSI) in the current tests.

As can be seen from Table 1, the SEI ratio reduces as the column becomes more slender or the load eccentricity increases, as a result of bending becoming more dominant and confinement becoming less effective at the time of failure. The SEI ratio increases with increasing number of CFRP layers.

It should also be noted that the strength of all damaged specimens has not been fully restored. This is due to the fact that the loss of strength in the current tests was considerably significant for the long exposure time of specimens in fire. Thus, it is not reasonable to expect the strength of the damaged specimens to be restored to a great extent while wrapped with only one or two layers of CFRP. Moreover, less strength enhancement from CFRP confinement can be obtained when the eccentricity or slenderness ratio increase. Therefore, it is recommended that, in repairing severely fire-damaged CFST members, slender members or those members subjected to comparatively large bending moments, other appropriate repair measures such as section enlargement method should be taken.

3.4 Effect of FRP repair on stiffness

Stiffness (K_e) in the present context is determined from load versus mid-height deflection curves as shown in Figs. 4 and 5. In this paper, K_e is defined as secant modulus corresponding to axial load of $0.6N_{ue}$ in the pre-peak stage. The values of K_e so determined are summarized in Table 1. Since no apparent deflection was detected for those axially loaded specimens, especially undamaged ones (such as specimens C-1 and S-1), the determined values of K_e for these specimens were not considered completely valid to reflect actual stiffness. However, it is useful to examine the effects of fire exposure and FRP repair on stiffness. It can be seen from Table 1, the exposure of specimens to fire results in considerable loss of stiffness. However, no apparent stiffness increase is observed after wrapping with CFRP jackets. This is due to the fact that the confinement from CFRP wraps is moderate when the CFST beam columns are remain in the elastic stage.

3.5 Effect of FRP repair on ductility

In order to quantify the effect of CFRP jackets on member ductility, a ductility index (DI) is defined herein:

$$DI = \frac{u_{85\%}}{u_y} \qquad (4)$$

where $u_{85\%}$ is the mid-height deflection when the load falls to 85% of the ultimate load, u_y is equal to $u_{75\%}/0.75$, and $u_{75\%}$ is the mid-height deflection when the load attains 75% of the ultimate load in the pre-peak stage.

Since no apparent deflection was detected for those axially loaded specimens, a ductility index (DI) defined as follows is used:

$$DI = \frac{\varepsilon_{85\%}}{\varepsilon_y} \qquad (5)$$

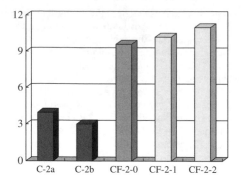

Figure 7. Ductility indexes for typical beam columns.

where $\varepsilon_{85\%}$ is the nominal axial shortening (Δ/L) when the load falls to 85% of the ultimate load, ε_y is equal to $\varepsilon_{75\%}/0.75$, and $\varepsilon_{75\%}$ is the nominal axial shortening when the load attains 75% of the ultimate load in the pre-peak stage.

The ductility indexes (DI) so determined are given in Table 1. Fig. 7 shows the tested ductility indexes DI for typical specimens. It is not surprising that the DI index increases as the column becomes more slender or the load eccentricity increases, which is a common phenomenon for beam columns. It can also be found that the DI index of a fire-exposed specimen is larger than that of its counterpart at ambient temperature. However, it is interesting to note that the DI index decreases for specimen CF-1-2 since rupture of CFRP jackets at the mid-height was observed near the peak loads. On the other hand, ductility enhancement was achieved for other CFRP-wrapped specimens without rupture of CFRP jackets are occurred during the loading process.

4 CONCLUSIONS

This paper provides new test data concerning concrete-filled steel tubular beam columns that have been exposed to and damaged by the standard ISO-834 fire and subsequently CFRP-repaired. The following conclusions can be drawn based on the experimental results of the study:

(1) The load-bearing capacity of the fire-exposed CFST beam columns was enhanced by the CFRP jackets to some extent. The strength enhancement from CFRP confinement decreased with increasing of eccentricity or slenderness ratio. However, the strength of all damaged beam columns has not been fully restored due to the long exposure time of them in fire.

(2) The influence of CFRP repair on stiffness was not apparent due to the fact that the confinement from CFRP wraps is moderate when the CFST beam columns are remain in an elastic stage.

(3) To some extent, ductility enhancement was observed except those axially loaded shorter specimens with rupture of CFRP jackets at the mid-height occurred near peak loads.

(4) It is recommended that, in repairing severely fire-damaged CFST members, slender members or those members subjected to comparatively large bending moments, other appropriate repair measures such as section enlargement method should be taken.

ACKNOWLEDGEMENTS

The research work reported herein was made possible by grants from the National Natural Science Foundation of China (No. 50425823 and No. 50608019), a key Grant of Chinese Ministry of Education (No. 205083), and assistance from the Project of Fujian Province Start-Up Fund for Outstanding Incoming Researchers; their financial support is highly appreciated. The authors also express special thanks to Mr. Zhuang Jin-Ping and Miss Wang Ling-Ling for their assistance in the experiments.

REFERENCES

Han L.H. 2004. *Concrete filled steel tubular columns – theory and practice*. Beijing: Science Press (in Chinese).

Han L.H., Huo J.S. 2003. Concrete-filled hollow structural steel columns after exposure to ISO-834 fire standard. *Journal of Structural Engineering*, ASCE 129(1): 68–78.

Han L.H. & Lin X.K. 2004. Tests on cyclic behavior of concrete-filled hollow structural steel columns after exposure to the ISO-834 standard fire. *Journal of Structural Engineering*, ASCE 130(11): 1807–1819.

Han L.H., Lin X.K. & Wang Y.C. 2006. Cyclic performance of repaired concrete-filled steel tubular columns after exposure to fire. *Thin-Walled Structures* 44(10): 1063–1076.

Han L.H. & Yang Y.F. 2004. Modern concrete filled steel tubular structures. Beijing: China Architecture & Building Press (in Chinese).

Tao Z., Han L.H. & Wang L.L. 2007. Compressive and flexural behaviour of CFRP repaired concrete-filled steel tubes after exposure to fire. *Journal of Constructional Steel Research* (in press).

Tao Z. & Yu Q. 2006. *New types of composite columns – experiments, theory and methodology*, Beijing: Science Press (in Chinese).

Teng J.G., Chen J.F., Smith S.T. & Lam L. 2002. *FRP-strengthened RC structures*. John Wiley & Sons, Ltd.

Xiao Y., He W.H. & Choi K.K. 2005. Confined concrete-filled tubular columns. *Journal of Structural Engineering*, ASCE 131(3): 488–497.

Steel and Composite Structures – Wang & Choi (eds)
© 2007 Taylor & Francis Group, London, ISBN 978-0-415-45141-3

Implications of transient thermal strain for the buckling of slender concrete-filled columns in fire

S. Huang, I.W. Burgess & Z. Huang
Department of Civil and Structural Engineering, University of Sheffield, Sheffield, UK

R.J. Plank
School of Architecture, University of Sheffield, Sheffield, UK

ABSTRACT: Pre-compressed concrete has been observed to acquire a large amount of irreversible strain (called Transient Thermal Strain) when it is heated. This effect seems not to occur when heating precedes the application of compressive stress. The objective of the research reported in this paper is to assess how this phenomenon affects the buckling resistance of concrete-filled hollow-section columns in fire. A self-coded programme has been established based on a Shanley-like model. Preliminary analyses indicate that TTS does have a significant effect on the buckling of slender concrete columns subjected to fire.

1 INTRODUCTION

The temperature and loading history of concrete prior to thermal exposure can have a significant influence on its behaviour (Khoury 1996). In particular, pre-compressed concrete experiences a considerable increase in strain during first heating compared to concrete heated without pre-loading. This additional strain is called Transient Thermal Strain (TTS) (Anderberg 1976). It has been noticed that TTS is absent for concrete which is unloaded prior to heating. Its large magnitude, which may be up to 45–50% of the total strain, makes it the dominant compressive strain component during first heating, and causes significant relaxation to the thermal expansion. TTS is also irrecoverable; it increases with temperature rise and loading but never decreases. Moreover, its occurrence is limited only to first heating, but is not observed during cooling or reheating (Khoury 1996).

When structures are subjected to accidental fires they are usually already carrying in-service loading. In multi-storey frames the columns in particular continue to carry these preloads throughout the fire, although the forces in the elements of flooring systems may change radically due to the thermo-mechanical effects of the fire as it progresses. Therefore, the occurrence of transient thermal strain in concrete columns subjected to fire is definite. However, this property is not normally taken into account in structural analysis, which may lead to results that are not representative of reality.

Concrete-filled hollow-section columns have become increasingly popular recently, due to their considerably enhanced compressive strength and assumed

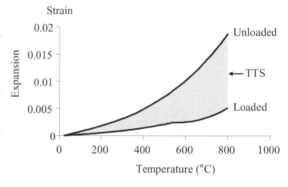

Figure 1. TTS as difference between total strain of preloaded and unloaded concrete heated for the first time.

inherent fire-resistance which enables the size of the cross section to be minimised, and to the possibility of reducing, or even eliminating, passive fire protection.

Although research has been done on TTS, the published experimental and theoretical studies focus solely on stocky specimens, generally cylinders (Anderberg 1976, Khoury et al. 1985a, b, Khoury 1996, Li & Purkiss 2005). For concrete columns, especially when the slenderness ratio is high, there is no current knowledge on how this large amount of transient strain may affect the buckling resistance. This is probably because reinforced concrete columns usually have modest slenderness and hence buckling is not very significant for them. But because the cross-sections of concrete-filled columns can be much smaller than those of RC columns of the same ultimate capacity, buckling may become a more significant issue for concrete-filled columns than for the equivalent RC columns. Therefore, the effect of TTS within the concrete infill on their buckling resistance is worth investigating.

The objective of the research reported in this paper is to assess how transient thermal strain, in combination with other temperature-dependent material properties of concrete, affects the buckling resistance of concrete-filled hollow-section columns in fire. This involves an initial study of the mechanics of inelastic buckling followed by elevated-temperature analyses taking into account TTS.

2 INITIAL STUDY OF THE MECHANICS OF INELASTIC BUCKLING

If imperfection effects are set aside, the strength of a column depends on its geometry (slenderness ratio) and its material properties (stiffness and strength). Very slender columns fail by buckling when the material is still linear-elastic (elastic buckling), and the classic Euler formula is applicable in the determination of the critical buckling load. In contrast, very stocky columns fail by yielding and crushing of the material, and hence their strength depends solely on the ultimate compressive strength of the material; no consideration of buckling or stability is necessary. Between these extremes, for columns with intermediate slenderness, buckling would occur after the material becomes plastic but before it crushes, which is known as inelastic buckling (Gere & Timoshenko 1997). In this case, the simple elastic buckling solution is no longer valid, and the inelastic behaviour of material must be taken into account.

For normal design of concrete-filled columns, their slenderness usually lies in the 'intermediate' range, and so their global failure mode would usually be inelastic buckling. Therefore, for the purpose of this research, an insight into the mechanics of inelastic buckling is a good start at this initial stage.

2.1 Inelastic buckling theories

Various inelastic buckling theories have been published since the late 1880s, including the tangent-modulus theory, the reduced-modulus theory and Shanley's theory (Shanley 1947, Bleich 1952, Bazant & Cedolin 1991). Both the tangent-modulus theory and the reduced-modulus theory assume that inelastic buckling has characteristics analogous to elastic buckling. It is assumed that a neutral equilibrium exists, and the column starts to bend and finally fails by buckling at a constant load. This critical load is determined by generalizing Euler's formula with varying Young's modulus.

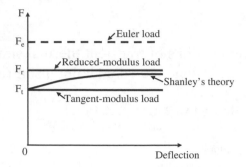

Figure 2. Load-deflection curves given by various column buckling theories.

In the tangent-modulus theory, Young's modulus E is replaced by a tangent modulus E_t, which is the slope of the compressive stress-strain curve at the critical stress (Gere & Timoshenko 1997). This theory over-simplifies inelastic buckling by using only one tangent modulus. In reality, the tangent modulus depends on the stresses, which vary through the cross-section and may even unload elastically on the convex side due to bending.

The reduced-modulus theory attempts to mitigate the error of neglecting unloading on the convex side of deflected columns in the tangent-modulus theory. It considers both loading with tangent modulus E_t on the concave side, and unloading with Young's modulus E on the convex side, when buckling occurs. An effective modulus (known as the reduced modulus E_r), which lies between E and E_t, is introduced to replace E_t in the tangent-modulus formula (Bazant & Cedolin 1991).

The three basic column formulas may be written as follows (assuming pin ends and zero eccentricity):

$$\text{(Euler)} \qquad F_e = \pi^2 EI/L^2 \qquad (1)$$

$$\text{(Tangent - modulus)} \qquad F_t = \pi^2 E_t I/L^2 \qquad (2)$$

$$\text{(Reduced - modulus)} \qquad F_r = \pi^2 E_r I/L^2 \qquad (3)$$

where F = critical load; I = second moment of area of the column cross section; L = column length; E = Young's modulus; E_t = tangent modulus; E_r = reduced modulus.

However, whether a perfect column will remain straight and suddenly bifurcate at a certain critical load (shown as the straight horizontal lines in Figure 2) in the inelastic range is questionable. Shanley (1947) showed by tests and mathematical analysis that, unlike elastic buckling, inelastic buckling does not have a unique critical load. The column starts to bend at the tangent-modulus critical load, and the bending proceeds simultaneously with a further increase of axial

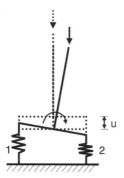

Figure 3. Shanley's model.

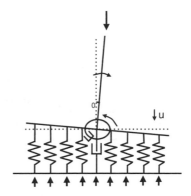

Figure 4. Multi-spring model.

load, but the load does not exceed the reduced-modulus load, shown as the rising curve in Figure 2.

Shanley's theory also indicates that the manner of inelastic buckling significantly depends on the rates of vertical loading and bending. When a column starts to deflect laterally the stresses and strains through a cross-section due to both vertical compression and bending are superposed. If the axial strain is imposed more rapidly than the bending strain then it is possible for buckling to occur without strain reversal, as described in the tangent-modulus theory. On the contrary, if the column deflects so rapidly that only the reversal itself can keep the load constant, then what is described in the reduced-modulus theory would be true. This suggests why the tangent-modulus and reduced-modulus theories give the two practical extremes of inelastic buckling. Between these two extremes, the combinations of compression and bending are infinitely variable, explaining why no unique critical load exists for inelastic buckling.

2.2 Numerical analysis with Shanley-like column model

Although in practice engineers tend to use the over-conservative tangent-modulus theory to obtain simple and safe solutions to inelastic buckling problems, theoretically only Shanley's theory correctly describes the mechanics of inelastic buckling. Shanley demonstrates his theory with a simplified column model consisting of two rigid legs and an elastic-plastic hinge composed of two axial elements (Shanley 1947). A modification of Shanley's model has been used as a basic model in this research.

2.2.1 Geometry

Shanley's model may be expressed in a modified form as a rigid cantilever supported by two identical springs, as shown in Figure 3. It has two degrees of freedom:

– Vertical movement u, which is the mean vertical movement of both springs;

– Rotation θ, which is proportional to the differential displacement of the two springs.

Shanley's model is good for a general investigation of buckling, but is over-simplified for a rational numerical analysis of the problems addressed previously. Therefore, this basic model has been modified and extended as shown in Figure 4.

Two dampers, one vertical and one rotational, were added to the basic model. They respectively control the rates of the two degrees of freedom u and θ. They damp the movement of the model in a controlled manner, which enables it to obtain the full buckling load-deflection path rather than a sudden bifurcation when the column fails by buckling. In addition, by changing the values of the two damping coefficients the two extreme situations of inelastic buckling (bifurcation at the tangent-modulus or reduced-modulus loads) can be simulated.

The two-spring model was extended to a multi-spring model, taking into account the material continuity through the cross-section. The axial deformation of each spring is consistent with the linear strain-gradient assumption, and hence the mean and differential displacements of each pair of springs at the equivalent location on both sides are still functions of the two DoFs u and θ. The springs are identical, having the same stress-strain curve and representing the same material, but the stress level of each spring can differ from the others at any given time, depending on the global deformation and the force equilibrium of the column. For instance, some springs may already be in the plastic range when the others are still linear-elastic, and some may have already started unloading while others continue to load. A bilinear stress-strain curve with elastic strain reversal was used in this analysis.

In Shanley's mathematical analysis, the model is initially perfectly symmetric, with the load perfectly central. It is then assumed to start to bend at the tangent-modulus buckling load, and the axial load continues to increase thereafter. Shanley makes these

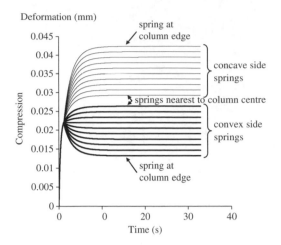

Figure 5. Development of the deformation of each spring over time under constant load as equilibrium is reached.

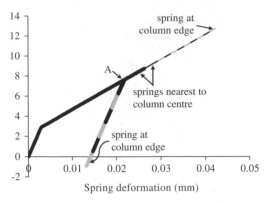

Figure 6. Compressive load-deformation curves of the springs.

assumptions because the purpose of his analysis is only to support his previous test results. However, for the purpose of this research, these assumptions were considered as over-simplified and hence inappropriate. Without assuming the column to start to bend at a certain point, the introduction of an initial imperfection as a rotation θ_0 was necessary to enable the analysis to guarantee numerical solutions. This also pre-defined the direction of deflection.

2.2.2 Dynamic approach

Dynamic numerical analysis was conducted with a self-coded program based on the multi-spring model. The equations of motion were written for the static and dynamic forces and moments caused by the imposed load and the reaction forces on the springs and dampers.

In this preliminary investigation, constant loading was imposed on the model. This simulated the application of a weight on top of the column, in one step but very slowly, so that no initial velocity or acceleration was induced. In the initial time step, the unbalanced external and internal forces (whose difference is identical to the damping force) induce velocity, causing the model to move. The model continued to deform gradually through the time steps until a new static equilibrium was reached, and this equilibrium position was recorded. The same procedure was repeated for successively higher loads until the rotation of the model was seen to diverge, indicating its final failure by buckling. Plotting all the loads against the corresponding rotations θ recorded at equilibrium gives the full equilibrium path.

2.3 Results & discussion

The results of this initial study of the mechanics of inelastic buckling are presented in Figures 5–8. It

should be noted that the results presented in this section are all from an example model consisting of 20 springs.

Figures 5 and 6 show the structural responses of the model under constant loads until a new equilibrium state is reached.

Figure 5 shows the development of the deformation of each spring through time under a constant load. Initially, the compressive deformations of the springs all increase almost identically from zero, indicating that very little rotation occurs, and the springs are compressed almost uniformly. At about 1 second, the deformations of the springs start to differ more markedly from each other. The deformations of the springs on the concave side (the thin lines) continue to increase rapidly while the deformations of the springs on the convex side (the thick lines) increase rather slowly; some even start to decrease. This indicates that the column starts to rotate. In addition, due to the linear strain-gradient assumption the differential deformation of the pair of springs at the column edges is larger than that of the pair of springs nearest to the centre for a certain global angle of rotation θ. At the end of this load step, the movement of the model gradually stabilizes as shown in Figure 5 and the deformation of each spring reaches a constant value.

Figure 6 plots the compressive load-deformation relationships of these pairs of springs. The thick lines are for the springs on the convex side while the thin lines are for those on the concave side. At the beginning of this load step, because almost no rotation occurs, the forces and deformations of the springs are nearly identical, shown as the single line from zero to point A. Soon after the column starts to rotate, some of the springs on the convex side start to unload while the others continue to load; this is also shown in Figure 5. Moreover, the springs at the column edges deform more, either in loading or unloading, due to the linear strain-gradient assumption, which is consistent with Figure 5.

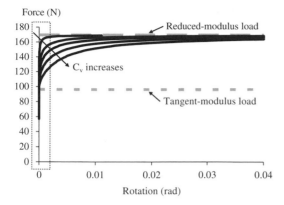

Figure 7. Buckling load paths of the model with various damping ratios.

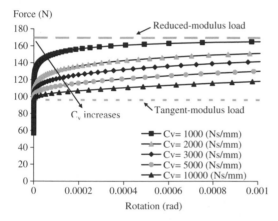

Figure 8. Magnification of the framed section of Figure 7.

Figure 7 shows the buckling load paths of the model with various damping ratios. Figure 8 is a magnification of the framed part of Figure 7. The full buckling load path is obtained by plotting the applied loads against the angle of rotation achieved for each load step. These are then validated against the theoretical buckling loads: the tangent-modulus load (the short-dashed line in Figures 7 and 8) and the reduced-modulus load (the long-dashed line in Figures 7 and 8). Irrespective of the variation of damping, the rotation always starts to increase significantly at the tangent-modulus load and then continues to increase as the force approaches the reduced-modulus load, but the force never goes beyond this upper bound. This indicates that the column fails by buckling in the exact manner described by Shanley.

The effects of damping have been investigated by varying the vertical damping coefficient C_v while the rotational damping remains constant. Figures 7 and 8 show that with larger vertical damping the load-deflection curve is closer to the tangent-modulus bifurcation line. It would be expected that, if C_v is large enough, the buckling curve would be very much like the tangent-modulus line, because its approach towards the reduced-modulus line would require so much rotation that the column could be considered as having failed long before reaching this upper bound. On the other hand, if C_v is very small, the curve would resemble the reduced-modulus line. This phenomenon corresponds to Shanley's statement that the relationship between the rates of loading and bending controls the manner of inelastic buckling. By increasing C_v the vertical movement is damped more and decelerated. This results in a decrease of the vertical damping force, which weakens the resistance to the vertical movement. In other words, with larger C_v the vertical movement becomes slower but easier (It is similar to how a bicycle tyre pump works, the slower it is pushed the easier it moves downwards.). In this case vertical movement dominates over rotation; therefore, it is possible for the column to buckle without strain reversal (unloading), as described in the tangent-modulus theory.

3 ELEVATED-TEMPERATURE ANALYSIS

After the program was evaluated at ambient temperature the temperature-dependent material properties of concrete, including transient thermal strain, were included in the spring model.

3.1 Material model of concrete at elevated temperatures

Anderberg's (1976) mathematical model of the material properties of concrete at elevated temperatures was adopted in this analysis. The total strain of uniaxially compressed concrete subjected to elevated temperature consists of four components:

$$\varepsilon = \varepsilon_{th}(T) + \varepsilon_\sigma(\sigma) + \varepsilon_{tr}(T,\sigma) + \varepsilon_{cr}(T,\sigma,t) \qquad (4)$$

In which ε = total strain; ε_{th} = thermal strain; ε_σ = instantaneous stress-induced strain; ε_{cr} = basic creep; ε_{tr} = transient thermal strain; T = temperature; σ = stress; and t = time.

Thermal strain ε_{th} is the strain measured on unloaded concrete subject to a uniform temperature increment. It is a unique function of temperature.

Instantaneous stress-induced strain ε_σ is the mechanical strain derived from the stress-strain curve. For any given temperature, ε_σ is only stress-dependent but it should be noted that the stress-strain curve may vary with temperature.

Transient thermal strain ε_{tr} is found to be reasonably linear with stress. It is also a nonlinear function of temperature and approximately proportional to ε_{th} :

$$\varepsilon_{tr} = -k_{tr} \frac{\sigma}{\sigma_{u0}} \varepsilon_{th} \qquad (5)$$

where $k_{tr} = $ a constant from 1.8 to 2.35; and $\sigma_{u0} = $ the strength at 20°C.

As mentioned by Anderberg, the basic creep ε_{cr} is often small compared to the other strain components, due to the relatively short period of fire, and therefore it was neglected in this analysis.

3.2 Loading

For the elevated-temperature analysis, the loading procedure was upgraded from the constant loading used in the ambient-temperature analysis to step loading. The load was no longer applied at once to the column; it was increased step by step. Within each load step the load remained constant until a new equilibrium position was reached, then moved on to the next load step. The step loading was applied at constant temperature until buckling occurred, in order to assess the buckling resistance of the column at this temperature. By incorporating the transient thermal strain property, pre-loading prior to heating was taken into account. This loading procedure represented a pre-loaded column which was heated under this load to a stable temperature; it was then unloaded and reloaded from zero at this temperature until failure.

3.3 Results & discussion

The results of the elevated-temperature analysis on an example model consisting of four springs are briefly presented in this section. Figure 9 shows the buckling resistance of the model, with and without transient thermal strain, at various temperatures and their comparison with the theoretical buckling loads. The two solid curves show the ultimate buckling loads of the model at various temperatures, with and without the consideration of transient thermal strain (as marked). Their comparison shows that TTS causes a remarkable reduction of the buckling resistance of slender concrete columns subjected to fire. This result is revealing because the effect of TTS on buckling has never previously been considered in structural analysis.

On the other hand, this is less surprising if the stress-dependence of TTS is highlighted. The study of the mechanics of inelastic buckling reveals that when the column starts to bend, the bending causes differential stresses between the concave and convex sides. At elevated temperatures this differential between compressive stresses will cause differential TTS, which further increases the differential total strain between

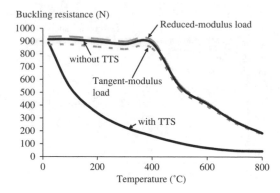

Figure 9. Buckling resistance of the model with and without TTS comparing with theoretical buckling loads at elevated temperatures.

the two sides, leading to further bending. Due to the rather large magnitude of TTS this interactive effect will not be insignificant. Figure 9 also shows that the result of the analysis without TTS lies between the two theoretical buckling loads, which is consistent with Shanley's theory.

4 CONCLUSIONS & FUTURE WORK

The analyses have examined the buckling of slender concrete columns subjected to temperature rise, including the effect of transient thermal strain. A self-coded program has been evaluated at both ambient and elevated temperatures. It has shown that transient thermal strain may cause significant reduction of the buckling resistance of slender concrete columns at elevated temperatures, which indicates the inappropriateness of ignoring it. It is accepted that uniform temperature conditions will not happen in realistic cases. For future work, the thermal gradient through the cross-section and the consequent redistribution of stresses will be taken into account. In addition, the current 2D model will be extended to a 3D finite element model, so that the column can be represented in much more detail.

ACKNOWLEDGEMENT

The first author is a Dorothy Hodgkin Scholar, funded by the Engineering and Physical Sciences Research Council and Corus Ltd.

REFERENCES

Anderberg, Y. & Thelanderson, S. 1976. *Stress and deformation characteristics of concrete at high temperatures*

2. Experimental investigation and material behaviour model. Bulletin 54, Lund.

Bazant, Z. P. & Cedolin, L. 1991. *Stability of structures.* Oxford: Oxford University Press.

Bleich, F. 1952. *Buckling strength of metal structures.* New York: McGraw-Hill.

Gere, J. M. & Timoshenko, S. P. 1997. *Mechanics of materials 4th edition.* Boston: PWS Publishing Company.

Khoury, G. A. 1996. *Performance of heated concrete mechanical properties.* London.

Khoury, G. A., Grainger, B. N. & Sullivan, P. J. E. 1985a. Transient thermal strain of concrete: literature review, conditions within specimen and behaviour of individual constituents. *Magazine of Concrete Research* **37**(132):131–144.

Khoury, G. A., Grainger, B. N. & Sullivan, P. J. E. 1985b. Strain of concrete during first heating to 600°C under load. *Magazine of Concrete Research* **37**(133): 195–215.

Li, L. & Purkiss, J. 2005. Stress–strain constitutive equations of concrete material at elevated temperatures. *Fire Safety Journal* **40**(7): 669–686.

Shanley, F. R. 1947. Inelastic column theory. *Journal of the Aeronautical Sciences* **14**(5): 261–268.

Steel and Composite Structures – Wang & Choi (eds)
© *2007 Taylor & Francis Group, London, ISBN 978-0-415-45141-3*

Effects of slab panel vertical support on tensile membrane action

A.K. Abu & I.W. Burgess
Department of Civil & Structural Engineering, University of Sheffield, Sheffield, UK

R.J. Plank
School of Architecture, University of Sheffield, Sheffield, UK

ABSTRACT: A recently-developed design method predicts composite slab capacities in fire, incorporating the effects of tensile membrane action. The method designs rectangular slab panels including unprotected beams within the panels, and are supported on edges that resist vertical deflection. In practice, vertical support is achieved by protecting the perimeter beams. Generic fire protection ensures that beam temperatures stay below critical temperatures of 550°C or 620°C within the required fire resistance time. However, large vertical displacements of the protected edge beams may cause a loss of the membrane mechanism, inducing single-curvature bending, which may lead to a catastrophic failure of the structure. A finite element investigation into the provision of adequate vertical support along slab panel boundaries has been conducted. The study has examined various degrees of protection relative to the development of the membrane mechanism. Comparisons are made with the membrane action design method and various acceptance criteria.

1 INTRODUCTION

Developments in structural fire engineering over the past 20 years have led to an improved understanding of structural behaviour under fire conditions. More recently, the influence of tensile membrane action in sustaining steel-framed buildings with composite floors, well beyond their traditional failure times, has emphasised the need to incorporate performance-based approaches into the design of structures in fire. Tensile membrane action is a mechanism producing increased load-bearing capacity in thin slabs undergoing large vertical displacements, in which stretched central areas of the slab induce an equilibrating peripheral ring of compression. This large-deflection mechanism relies on two-way bending and the availability of vertical support along the edges of the slab, and occurs with or without horizontal restraint along the edges of the slab. Use of the mechanism in the design of composite floors in fire may help to reduce building costs, as a large number of floor beams can often be left unprotected (Bailey 2004). Also, repair of fire-damaged structures may only involve replacing a few compartments, instead of rebuilding the entire structure.

In the design of composite slabs for tensile membrane action, a floor plate is divided into several rectangular slab panels of low aspect ratio. The beams on the perimeter of each slab panel are protected to provide vertical support, while those in the interior region are left unprotected, as shown in Figure 1. On exposure to fire, the unprotected beams lose strength, and their loads are progressively borne by the slab, under large vertical deflections. The slab's capacity increases as its vertical deflections increase.

Tensile membrane action can be modelled effectively by sophisticated finite element software, such as *Vulcan* (Huang et al. 2002, 2003a, b). Finite element processes are time-consuming, and simpler performance-based methods, such as the Building Research Establishment's membrane action method, are preferred alternatives for routine design.

2 BRE MEMRANE ACTION METHOD

The design method developed by Bailey is based on rigid-plastic theory subject to large change of geometry. It calculates the enhancement that tensile membrane action adds to the traditional (Yield-Line) flexural capacity of the slab. The membrane action method divides a floor plate into several rectangular slab panels supported on edges that primarily resist vertical deflection, and which incorporate unprotected composite beams (Bailey 2004) as shown in Fig. 1. The panels are horizontally unrestrained; the unprotected beams and slabs are simply supported, with their capacities determined by lower-bound limit analyses (Bailey & Moore 2000). The method conservatively

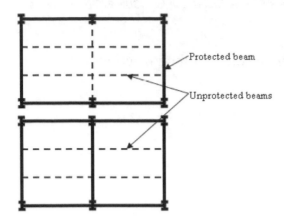

Figure 1. Typical rectangular slab panels for tensile membrane action.

ignores any contribution of the tensile strength of concrete to the capacity of the slab, and does not provide any information on the state of the protected boundary beams apart from the need to keep them vertically supported. Failure is based on the formation of a full-depth crack across the shorter span of the slab (Bailey 2001).

2.1 Slab panel capacity

After a floor slab has been divided into a number of panels, and the required fire resistance of the structure has been established, the residual resistance of each panel is calculated by considering the dimensions of the panel, the resistance of the internal beams and the reinforcement mesh size. This is then compared with the applied load at various times within the required design time. If the residual capacity of the panel is found to be less than the applied load at the fire limit state, then either the capacity of the internal beams or the reinforcement mesh size is increased (Newman et al. 2006).

At first, all floor loads are carried by the unprotected composite beams. As these lose strength, large vertical displacements are induced, which generate self-equilibrating in-plane tensile and compressive forces capable of producing enhancements to the theoretical yield-line capacity of the slab. The method, initially developed for isotropic reinforcement (Bailey & Moore 2000, Bailey 2001), has been extended to include orthotropic reinforcement (Bailey 2003). Recently, the change of in-plane stress distributions and compressive failure have been added (Bailey & Toh 2006).

To facilitate use of the tensile membrane mechanism and the BRE method, a design guide known as P-288 (Newman et al. 2006) has been produced by the Steel Construction Institute. This provides tables

of slab panel arrangements and reinforcement sizes that should be used, based on the type of concrete, the geometry of the slab cross-section and the load ratio of the secondary beams. It also provides information on the use of free software (TSLAB) for quick analyses of these slab panels.

2.2 Panel vertical support

As vertical support along the slab panel boundaries is realised in practice with edge beam protection, the assumption of continuous vertical restraint at all times during the fire cannot hold true. Numerical analyses have shown that the protected perimeter beams lose strength and stiffness, at some point, allowing the formation of a single-curvature slab-bending mechanism, which will pull on its connections, leading to a catastrophic failure of the structure if these connections are not adequately designed against such forces. This potential failure type has led to the series of finite element studies reported here, into the adequacy of vertical support along the slab panel boundaries.

3 FINITE ELEMENT ANALYSES

The finite element analyses were performed with *Vulcan* (Huang et al. 2000, 2003a, b), a geometrically nonlinear finite element program that includes the effects of nonlinear material behaviour at elevated temperatures. The program models reinforced concrete slabs with 9-noded nonlinear layered rectangular elements, which represent temperature distributions through the depth of the slab by assigning different, but uniform, temperatures to each layer of the element. The slab element can adequately represent bending and membrane action at large displacements. Reinforcement is smeared across the area of the slab element. The program uses a biaxial failure surface for concrete, and so it can adequately represent failure in tension or compression. Beams are modelled with 3-noded nonlinear beam elements.

3.1 Loading

Using the SCI P-288 document, a slab panel of dimensions 9.0 m × 7.5 m (Figure 2) was selected, to be designed for a 60-minute fire resistance. The panel had two unprotected beams spanning in the short direction. Following the guidance in the document for office-type buildings, the unprotected beams could be loaded to their full capacity, and an A193 mesh was sufficient if the slab profile of Figure 3 was used, with normal weight concrete and the loads listed in Table 1.

Ambient and elevated-temperature design of the floor beams, using BS5950-3 and BS5950-8, resulted

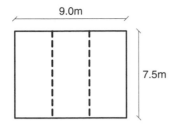

Figure 2. Slab panel geometry.

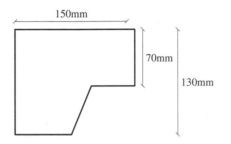

Figure 3. Concrete slab cross-section profile.

Table 1. Design Loading.

Permanent Load	kN/m²
Slab selfweight	2.40
Beam selfweight	0.20
Mesh (A193)	0.03
Imposed Load	
Variable load	3.5
Ceilings/ Services	1.7

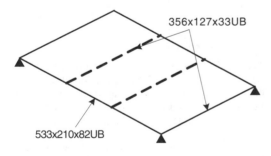

Figure 4. Slab panel, showing protected and unprotected beams.

in $356 \times 127 \times 33$UB as secondary beams and $533 \times 210 \times 82$UB as primary beams (Fig. 4).

The beam design factor of the secondary beams was 0.74 (0.44 load ratio in fire with increased loading). The critical temperatures of the protected secondary and primary beams were 631°C and 646°C respectively. A generic protection scheme

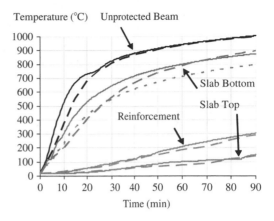

Figure 5. Unprotected beam and slab bottom surface, top and reinforcement temperatures of the slab panel system.

was adopted with lightweight fire resisting gypsum boards (density = 800 kg/m³, specific heat capacity = 1700 Jkg/K, conductivity = 0.2 W/mK), so that their temperatures were limited to 550°C at 60 minutes.

3.2 Thermal analysis

An average slab thickness of 100 mm (100 mm = (70 mm + 130 mm)/2) was chosen for the structural analysis. A thermal analysis was performed to ascertain the temperature distributions through the unprotected beams and the concrete slab. Temperature distributions calculated to the simplified process from Eurocode 3 Part 1.2 (CEN 2005) were used for the beams, while a one-dimensional thermal analysis was performed to generate the temperature distributions across 13 layers of concrete and reinforcement, using a thermal analysis program (FPRCBC-T), developed by Huang et al. (1996). A comparison of the temperature distributions in the *Vulcan* model and TSLAB, for a 90-minute exposure to the standard temperature-time curve is shown in Figure 5.

TSLAB results are shown as dashed lines while the *Vulcan* temperature distributions are shown as continuous lines. For the bottom of the slab, the dotted line shows the average temperature of the lowest layer of the *Vulcan* slab element while the solid line shows the temperature of the bottom surface. It is observed from the figure above that there was generally good correlation between the thermal distributions in the *Vulcan* and TSLAB models.

3.3 Structural analyses

The primary *Vulcan* model was a horizontally-unrestrained slab panel, with vertical restraint only available at the corners (Fig. 4), providing vertical edge support by using protected beams around the

Table 2. *Vulcan* Analyses and Parameters.

Condition	*Vulcan* Analyses								
	1	2	3	4	5	6	7	8	9
Generic Protection	✓	✓	✓			✓	✓	✓	
2X Generic Protection									✓
Cold Perimeter Beams				✓	✓				
Corner Vertical Restraint	✓	✓	✓	✓	✓	✓	✓	✓	✓
Edge Vertical Restraint		✓		✓	✓				
Concrete Tensile Strength ignored					✓				
Rotational Restraint on 2 edges								✓	
Rotational Restraint on 3 edges						✓	✓		
Rotational Restraint on 4 edges			✓						

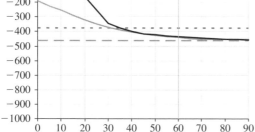

Figure 6. Absolute maximum vertical displacements of the slab panel model given by the BRE Method.

perimeter. Other variations of this model (see Table 2) were investigated to evaluate the influence of several parameters on tensile membrane action.

The *Vulcan* analyses were compared with the TSLAB limiting deflection curve, the BRE allowable vertical deflection limit (for fire resistance times up to 90 minutes), the required vertical deflection (from the generic BRE approach) and a limiting deflection of span/20 (750 mm/20 = 375 mm). Figure 6 shows the vertical displacements used in the comparison. The black continuous line shows the required vertical displacements obtained by the generic BRE Method with an A193 mesh.

The faint continuous curve shows the limiting vertical displacements generated by the TSLAB program, obtained from Equation 1 below, at each time step with T_2 and T_1 as the bottom and top slab temperatures respectively:

$$V = \frac{\alpha(T_2 - T_1)l^2}{19.2h} + \sqrt{\left(\frac{0.5f_y}{E}\right)_{\text{Re inf } 20°C} \times \frac{3L^2}{8}} \quad (1)$$

in which α is the coefficient of thermal expansion; l is the length of the shorter side of the slab panel; L is

the length of the longer side of the slab panel; h is the thickness of the concrete slab; f_y is the yield stress of the steel reinforcement; and E is the Young's modulus of the reinforcement.

The faint dashed line in Fig. 6 is the allowable absolute vertical deflection limit, as defined in the BRE method. The value of this deflection limit is obtained when $T_2 - T_1 = 770°C$ in Equation 1. The line with short dashes represents the deflection at span/20 (375 mm). A faint line has been placed on each of the graphs to show that the slab panels were designed for 60 minutes fire resistance.

Figure 6 shows the adequacy of the A193 mesh for the chosen slab panel size and the design criteria of the simplified design method.

4 RESULTS & DISCUSSION

The results of the various *Vulcan* analyses are presented in Figures 7–9 and 11–12. It should be noted that, unless otherwise stated, the results presented in this section all show absolute maximum vertical displacements of the middle of the slab panel system.

4.1 *Vertical restraint*

Figure 7 shows plots from *Vulcan* analyses which were the same except for V2 being supported vertically along its edges. In contrast, V1 suffered collapse of the protected secondary beams, caused by their loss of strength and stiffness as they approached the design temperature of 550°C. This is emphasized in Figure 8 which shows vertical displacements of the middle of the slab panel relative to the vertical displacements of the mid-span of the protected primary and secondary beams.

From this figure it is observed that, as the secondary beams reached 550°C, they began to fold, losing the tensile membrane mechanism and allowing the formation of a single-curvature mechanism that produced excessively large displacements. The reduction in difference between the deflections of the centre of the slab and the mid-span of the protected secondary

Vertical Displacement (mm)

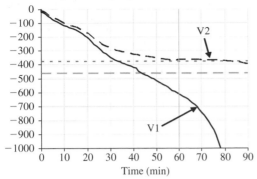

Figure 7. Absolute vertical displacements from *Vulcan* analyses with generic protection and vertical support at corners and along the slab panel edges.

Vertical Displacement (mm)

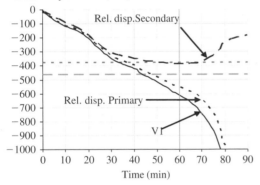

Figure 8. Absolute and relative vertical displacements of *Vulcan* analysis with generic protection and vertical restraints at corners only.

beams shows how rapidly they deflected while the primary beams effectively stayed vertical.

4.2 *Influence of the tensile strength of concrete*

The results (Figure 9) show that neglecting the tensile strength of concrete requires higher vertical deflections to generate the required slab capacity. The *Vulcan* model which included the tensile strength of concrete compared very well with the generic BRE required deflection, and just crosses the allowable vertical displacement limit at about 55 minutes. Part of the large deflections recorded here was due to the buckling of the slab, caused by its expansion against the cold perimeter beams.

4.3 *Effect of continuity at the panel boundary*

The effect of continuity of the slab panels was also investigated. The analyses examined 4 scenarios: one

Vertical Displacement (mm)

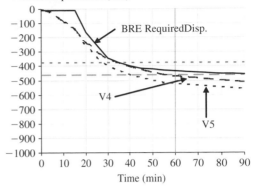

Figure 9. Absolute vertical displacements of *Vulcan* analyses with cold perimeter beams (20°C) and vertical support along all slab panel edges.

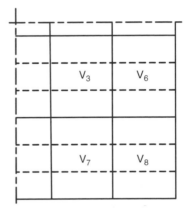

Figure 10. Layout of continuous slab panels in a typical floor plate. Labels correspond to analyses shown in Figure 11.

with continuity on 2 adjacent sides, two with continuity on 3 sides and one with continuity on all 4 sides. The analyses were performed on the typical floor layout shown in Figure 10. Figure 11 shows the absolute vertical displacements of the centre of each of the slab panels in Figure 10.

The results confirm the dependence of the survival of each slab panel on the stiffness of the protected beams, the positive influence adjacent slabs have on the stiffness of the fire-exposed slab.

4.4 *Thickness of protection material*

A final analysis of Type V1 was conducted. This time, twice the thickness of fire protection was applied to the perimeter beams. Figure 12 shows that the absolute vertical displacements were within the required TSLAB limit. This suggests that the requirement for vertical restraint at the boundaries of the slab panel

Vertical Displacement (mm)

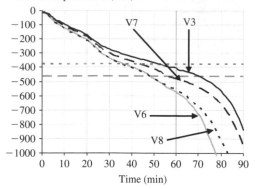

Figure 11. Absolute vertical displacements of *Vulcan* analyses with generic protection, vertical support at corners and rotational restraints along the slab panel edges.

Vertical Displacement (mm)

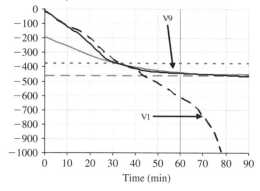

Figure 12. Maximum absolute vertical displacements of *Vulcan* analyses with 2 different thicknesses of protection.

may require the use of heavier sections or thicker protection, which may have the effect of increasing construction costs.

5 CONCLUSION

The analyses have examined three protection schemes, and several support conditions. The results suggest that, on condition that the slab panel edges stay vertical throughout the fire, the BRE simplified method and its failure criteria provide good estimates of composite slab behaviour in tensile membrane action. This degree of vertical support is, however, not practical as protected perimeter beams deflect, even at ambient temperature.

In practice, the protected beams would have been designed against attaining their critical temperatures of 631°C and 646°C. The analyses have shown that, even for a design temperature of 550°C, the protected secondary beams lost their necessary degree of support, although they had been designed for the additional loads to be expected due to tensile membrane action.

Restraint from adjacent slabs is clearly beneficial, but for slab panels verging on the façade of a building, increasing the level of protection seems a viable option, although this could potentially be an expensive thing to do.

REFERENCES

Bailey, C. G. & Moore, D. B. (2000), *Design of Steel Structures with Composite Slabs at the Fire Limit State*. Final Report No. 81415, prepared for DETR and SCI, The Building Research Establishment, Garston.

Bailey, C. G. (2001), Membrane action of unrestrained lightly reinforced concrete slabs at large displacements, *Engineering Structures*, 23, pp470–483.

Bailey C. G. (2003), Efficient arrangement of reinforcement for membrane behaviour of composite floors in fire conditions. *J. Construct. Steel Research,* 59, pp931–949.

Bailey C. G. (2004), Membrane action of slab/beam composite floor systems in fire. *Engineering Structures*, 26, pp1691–1703.

Bailey, C. G. & Toh, W. S. (2006), Experimental Behaviour of Concrete Floor Slabs at Ambient and Elevated temperatures. Proc. Fourth International Workshop on Structures in Fire, Aveiro, Portugal, Vol 2, pp709–720.

CEN (2005), Eurocode 3: Design of Steel Structures – Part 1.2: General Rules – Structural Fire Design, European Committee for Standardization.

Huang, Z, Burgess, I. W. & Plank, R. J. (2002), Modelling of Six Full-Scale Fire Tests on a Composite Building, *The Structural Engineer*, 80 (19), pp30–37.

Huang, Z., Burgess, I. W. & Plank, R. J. (2003a), Modelling Membrane Action of Concrete Slabs in Composite Buildings in Fire. I: Theoretical Development, *ASCE Journal of Structural Engineering*, 129 (8), pp1093–1102.

Huang, Z., Burgess, I. W. & Plank R. J. (2003b), Modelling Membrane Action of Concrete Slabs in Composite Buildings in Fire. II: Validations, *ASCE Journal of Structural Engineering*, 129 (8), pp1103–1112.

Huang, Z., Platten, A. & Roberts, J. (1996), Non-linear Finite Element Model to Predict Temperature Histories within Reinforced Concrete in Fires, *Building and Environment*, 31, (2), pp109–118.

Newman, G. M., Robinson, J. T. & Bailey, C. G. (2006), *Fire Safe Design: A New Approach to Multi-Storey Steel-Framed Buildings (Second Edition)*, SCI Publication P288.

Steel and Composite Structures – Wang & Choi (eds)
© *2007 Taylor & Francis Group, London, ISBN 978-0-415-45141-3*

Realistic modelling of Concrete Filled Tubular (CFT) columns in fire

J. Ding
now at Bodycote Warringtonfire, Warrington, Cheshire, UK

Y.C. Wang
School of Mechanical, Aerospace and Civil Engineering (MACE), The University of Manchester, Manchester, UK

ABSTRACT: Concrete filled tubular (CFT) columns are widely used all over the world, due to their significant advantages in construction speed and high fire resistance. This paper employs the commercial finite element analysis package ANSYS to model the behaviour of isolated CFT columns in fire. The finite element model is validated by comparing the simulation results with 34 fire test results of various CFT columns with different boundary conditions and under different loading. A numerical parametric study is then performed to investigate the sensitivity of simulation results to different assumptions introduced in the finite element model. The results of these numerical studies show that whether or not including slip between the steel tube and concrete core in the numerical model has little influence on the calculated column fire resistance time. However, including slip gives better prediction of column deflection behaviour. Introducing an air gap between the steel tube and the concrete core can improve the accuracy of predictions for both structural temperatures and structural performance. Using different tension strength or tangent stiffness of concrete has minor effect on the calculated column fire resistance. Different amounts of column initial deflection have some influence on column fire resistance times. Nevertheless, the influence is relatively small so that it is acceptable to use a maximum initial deflection of L/1000 as commonly assumed by other researchers.

1   INSTRUCTIONS

The advantages of concrete filled tubular (CFT) columns are numerous, including attractive appearance, structural efficiency, reduced column footing, fast construction and high fire resistance without external fire protection. In fire, the behaviour of a structure is much more complex than at ambient temperature. Changes in the material properties and thermal movements will cause the structural behaviour to become highly nonlinear and inelastic. The easy accessibility of high-speed computers has made it possible to quickly perform elaborate modelling in which the temperature dependence of material properties can be taken into account. Consequently, a considerable amount of research has been carried out by numerous researchers to investigate fire behaviour of CFT columns (Ding & Wang 2005, Wang & Davies 2003, Lie & Chabot 1994, Kimura et al. 1990).

This paper employs the commercial finite element analysis package ANSYS to model the behaviour of isolated CFT columns in fire. The temperature distribution within a CFT column cross-section is simulated by using a 2-D model. And the fire resistance of the CFT column is calculated by using a 3-D finite element model. A series of numerical parametric studies is then

performed using the validated model. The objectives of the numerical studies are: (1) to verify applications of the ANSYS software to CFT columns in fire; (2) to examine the effects of a number of factors that have, in general, not been considered in other numerical simulations. These factors include: (i) thermal resistance of a possible air gap between the steel tube and the concrete core, which can affect both the temperature results and structural behaviour; (ii) slip between the steel tube and the concrete core, which affects the structural behaviour of CFT columns; (iii) stress-strain model of concrete in tension at high temperatures.

2   THERMAL ANALYSIS

In general transient distribution within a structural member is described by a differential equation with 3 variables for space coordinates and 1 variable for the time. For practical analysis, the column is treated as uniformly heated lengthwise in the analyses. Thus, the temperature distribution within a composite column cross section can be analyzed by using a 2-D model. Because of symmetry of the geometry and the boundary conditions, only one quadrant is analyzed.

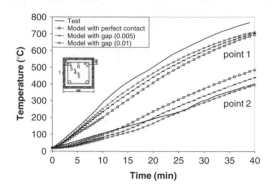

Figure 1. Comparison between calculated and measured temperatures for SHS 200 × 200 × 12.5.

Non-linearity due to the temperature dependent material properties and boundary conditions (convection and radiation) are taken into account. This study adopted the material thermal properties for steel and concrete recommended in EN 1994-1-2 (CEN 2005) and the recommended heat transfer coefficients given in EN 1991-1-2. The moisture content in the concrete is chosen as 4%.

When a CFT column is exposed to fire attack, an air gap may develop at the interface between the steel tube and the concrete core. This comes about because (1) the thermal expansion coefficient of steel is higher than that of concrete, leading to a higher radial expansion of the steel tube than the concrete core; (2) the difference in thermal expansion and strains in the longitudinal direction may overcome the bond between the steel tube and the concrete at the interface, which causes separation.

An air gap at the interface directly affects heat conduction between the steel tube and the concrete core. The result of not considering this air gap is higher temperatures in the concrete core. The effect of an air gap may be represented by its thermal resistance, which is defined as L/λ, where L is the air gap thickness and λ is the effective thermal conductivity of the air gap. It is difficult to obtain exact values for both quantities. Therefore, in this study, a nominal air gap thickness of 1mm is assumed. The effective air gap thermal conductivity is determined to suit the experimental results, but a sensitivity analysis will be carried out to check its appropriateness.

Figure 1 presents comparisons between predicted and measured temperatures for SHS 200 × 200 × 12.5 (test 1/3) from the CIDECT Report C1/C2 (Kordina and Klingsch 1983). Temperature predictions were carried out for both cases with or without an air gap between the steel tube and concrete core. Two values of the air gap thermal resistance were used, being 0.01 $[K \cdot m^2 \cdot W^{-1}]$ and 0.005 $[K \cdot m^2 \cdot W^{-1}]$. It can be seen that introducing an air gap in prediction produces quite

noticeable difference in temperatures from those without an air gap. Using different air gap resistance values gives different temperature results. The figure shows that introducing an air gap improves the correlation between the prediction and test results. Particularly, Figure 1 indicates that by introducing an air gap, the temperature difference between the steel tube and the out layer of the concrete core (points 1 and 2) becomes much greater (as shown by the measured temperature results) than without an air gap, thereby improving accuracy of predictions for both points.

3 STRUCTURAL ANALYSIS

3-D models are used to predict the structural behaviour of columns with non-uniform temperature distribution. The temperature results were input into the structural model as a body load. In order to read the temperature data of each node, the mesh on the cross-section of the CFT column in the structural model was kept the same as the mesh in the thermal model. By taking advantage of symmetry, only half of the column is analyzed. The 3-D structural solid element SOLID45 was used to mesh CFT columns. The effect of slip between the steel tube and concrete core was simulated by 3-D Surface-to-Surface Contact. The temperature dependent material properties recommended in EN 1994-1-2 are adopted in the 3-D structural analysis.

3.1 Column deflections

Figures 2(a) & (b) show comparison between predicted and measured column deflection curves for test 1/23 from the CIDECT Report C1/C2. Although there are differences between the measured and calculated results, the figures show very good agreement between the measured and calculated displacements. Particularly, the four different stages (thermal extension, contraction due to steel tube local yielding, plateau and accelerated shortening) of axial deformation of CFT column behaviour are clearly captured by the simulation results. A possible explanation of the slight difference between the measured and calculated axial deformations may be a result of the column not being uniformly heated over its whole length in the test (thus the measured column elongation being lower), as well as possible inaccuracy in thermal analysis and coefficient of thermal expansion of concrete (being according to EN 1994-1-2 in this study). The column has some applied bending moments and Figure 2(b) indicates that the transverse deflections are very accurately calculated. The column fire resistance time is also closely calculated.

3.2 Fire resistance time

The finite element model has been used to predict the fire resistance times of a large number of CFT columns

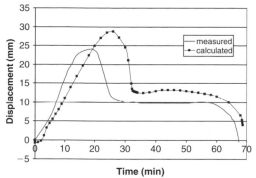

(a) Vertical displacement at the top of column

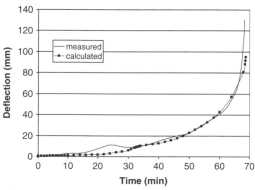

(b) Transverse deflection at mid-height of column

Figure 2. Comparison between calculated and measured deflections for test 1/23 from the CIDECT Report C1/C2.

Table 1. Summary of CFT test parameters and comparison of fire resistance times.

No.	Dimensions (mm)	Rebar (mm)	Length (mm)	Load (KN)	Failure time (min) Test	Calculated
1	SHS200 × 6.3	4φ18	4200	432	63	61.3
2	SHS200 × 6.3	4φ18	4200	318	58	60.4
3	SHS200 × 12.5	4φ18	4200	612	39	43.1
4	SHS200 × 12.5	4φ18	4200	456	34	36.2
5	SHS200 × 6.3	4φ18	4200	537	61	58.1
6	SHS200 × 6.3	4φ18	4200	213	79	70.9
7	SHS260 × 7.1	4φ18	4200	1237	37	40.3
8	SHS300 × 7.0	4φ18	4200	1000	90	87.2
9	SHS200 × 6.3	4φ18	3700	649	39	46.6
10	SHS300 × 7.0	4φ18	5200	636	92	88.8
11	CHS273 × 5.0	4φ18	4200	695	56	57.9
12	SHS200 × 6.3	4φ10	4200	551	23	27.5
13	SHS200 × 6.3	-/-	4200	400	22	29.9
14	SHS200 × 6.3	4φ18	3700	649	56	55.4
15	SHS200 × 6.3	4φ18	4200	550	59	58.7
16	SHS200 × 6.3	4φ18	3700	294	82	78.2
17	SHS220 × 6.3	4φ18	4200	375	68	68.7
18	SHS220 × 6.3	6φ20	4200	421	88	86.5
19	SHS260 × 7.1	6φ22	4200	869	64	66.4
20	SHS300 × 7.0	6φ25	4200	1507	56	55.1
21	SHS150 × 5.0	4φ12	3810	140	82	78.9
22	SHS300 × 8.0	4φ32	3810	1400	58	57.2
23	CHS273.1 × 6.4	4φ20	3810	1050	188	150.2
24	CHS273.1 × 6.4	4φ20	3810	1900	96	90.8
25	SHS203 × 6.4	4φ16	3810	930	105	99.8
26	SHS254 × 6.4	4φ20	3810	1440	103	98.4
27	SHS254 × 6.4	4φ20	3810	2200	70	59.5
28	CHS168.3 × 4.8	-/-	3810	150	76	73.0
29	CHS168.3 × 4.8	-/-	3810	150	81	75.0
30	CHS219.1 × 4.8	-/-	3810	492	80	66.8
31	CHS219.1 × 4.8	-/-	3810	384	102	85.9
32	CHS273.1 × 5.6	-/-	3810	574	112	109.7
33	CHS273.1 × 5.6	-/-	3810	525	133	121.3
34	CHS355.6 × 12.7	-/-	3810	1050	170	147.9

that have been tested by others. Table 1 shows comparison of measured and calculated fire resistance times for 34 test results, of which 20 column tests (No. 1–20 in Table 1) are from CIDECT Report C1/C2 tested at the University of Braunschweig, Germany and 14 tests (No. 21–34 in Table 1) from National Research Council of Canada (Lie and Caron 1988a, 1988b, Lie and Chabot 1994). Some details of these test specimens are also included in Table 1. In most cases, a moisture content of 4% and a thermal resistance of 0.005 [K·m²·W⁻¹] for the air gap have been introduced in the thermal analysis. For a few tests, these values were adjusted so as to give as good fit as possible between the measured and simulated steel and concrete temperatures.

Table 1 shows that the model can provide a good estimation of the column fire resistance times. Considering existence of various uncertainties in the fire tests (such as the heating condition along the column height, the stiffness of rotational restraint at the column ends, the initial imperfection of the column, the unintentional eccentricity of axial load, etc) as well as the difficulty of obtaining various material properties,

the agreement shown in Table 1 can be considered excellent and acceptable.

4 PARAMETRIC STUDIES

In previous thermal and structural analyses of CFT columns in fire, a number of assumptions were introduced, including slip between the steel tube and concrete core, air gap at the interface of the steel tube and concrete core, column initial imperfections and concrete tensile properties. This section will present the results of a numerical parametric study investigate the sensitivity of simulation results to different assumptions.

4.1 Slip between steel tube and concrete core

For simulation results in earlier sections, a friction coefficient of 0.2 was assumed. Other different values

Table 2. Failure time of test 1/1 calculated using different friction coefficients and maximum bond strengths.

Friction coefficient	Maximum bond strength (MPa)	Failure time (min)
0.2	0.5	61.348
0.47	0.5	61.351
0.8	0.5	61.348
0.2	0.2	61.348
0.2	0.8	61.348

Table 3. Column failure times calculated with or without slip.

No	Dimension (mm)	End	Load (kN)	e (mm)	Failure time (min) with slip	Failure time (min) without slip
1	200 × 100 × 5	p-f	268	0	21.18	21.62
2	200 × 100 × 5	p-p	268	0	18.72	15.84
3	200 × 100 × 5	p-f	199	10	23.27	23.67
4	200 × 100 × 5	p-p	199	10	20.56	18.12
5	200 × 100 × 5	p-f	626	0	7.08	8.04
6	200 × 100 × 5	p-p	626	0	5.91	5.43
7	200 × 100 × 5	p-f	465	10	11.87	11.84
8	200 × 100 × 5	p-p	465	10	7.96	6.40
9	200 × 6.3	p-f	660	0	38.87	39.24
10	200 × 6.3	p-p	660	0	25.40	26.86
11	200 × 6.3	p-f	508	20	38.15	39.03
12	200 × 6.3	p-p	508	20	26.36	27.19
13	200 × 6.3	p-f	1540	0	18.58	17.73
14	200 × 6.3	p-p	1540	0	7.65	7.48
15	200 × 6.3	p-f	1185	20	17.68	18.27
16	200 × 6.3	p-p	1185	20	10.77	10.56
17	200 × 12.5	p-f	969	0	33.98	34.00
18	200 × 12.5	p-p	969	0	27.72	28.16
19	200 × 12.5	p-f	740	20	33.92	35.06
20	200 × 12.5	p-p	740	20	29.52	29.82
21	200 × 12.5	p-f	2261	0	19.72	18.44
22	200 × 12.5	p-p	2261	0	8.97	9.45
23	200 × 12.5	p-f	1728	20	21.42	21.42
24	200 × 12.5	p-p	1728	20	13.68	13.55

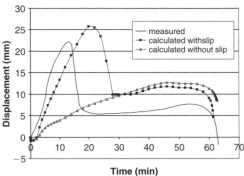

(a) Vertical displacement at the top of column

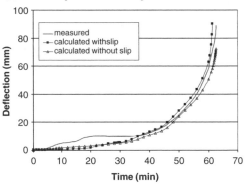

(b) Transverse deflection at mid-height of column

Figure 3. Comparison of measured and predicted deflections with and without slip, test 1/1 from the CIDECT Report C1/C2.

have been suggested by some researchers (Wium 1992, Roeder *et al.* 1999). The results in Table 2 using different values of friction coefficient and maximum bond strength for test 1/1 of CIDEC Report C1/C2 indicate that there is virtually no difference in the simulation results. This is not surprising because the steel tube and the concrete core would have been separated when exposed to fire, due to the different thermal expansions in both the longitudinal and radial directions.

To further confirm that the steel tube and concrete core would have been separated during the fire tests, Figures 3(a) & (b) compare the column deflections between simulations with and without slip and between them and test results. Figure 3(a) indicates that without incorporating slip in the simulation model, it would not be possible to simulate the CFT column shortening stage when the applied load was entirely resisted by the extended steel tube causing it to shorten when local yielding occurs.

Table 3 summarizes results of a parametric study comparing the effects of with and without slip on fire resistance times. The column sizes are RHS 200 × 100 × 5 mm, SHS 200 × 200 × 6.3 mm and SHS 200 × 200 × 12.5 mm. The columns are 3 m long, made from S275 steel and filled with plain C30 concrete. The column with cross-section size RHS 200 × 100 × 5 mm was designed to fail about the minor axis. Different loads, eccentricities and rotational restraints were considered in this parametric study. The different loads represent 0.3 and 0.7 times of the column load carrying capacities at ambient temperature.

The results in Table 3 suggest that the finite element models with or without slip between the steel tube and the concrete core give very similar column fire resistance times. In general, including slip gives slightly higher fire resistance times than without slip. However, as previously demonstrated, including slip in the numerical model gives more realistic column behaviour and displacement results.

Table 4. Column failure times calculated with various initial eccentricities.

Column No. in Table 2	Test (min)	Initial eccentricity			
		$L/2000$	$L/1000$	$2L/10003$	$L/1000$
1	63	61.8	61.3	60.4	59.6
8	90	88.3	87.2	86.1	85.1
16	82	78.9	78.2	77.5	76.8
18	88	87.2	86.5	85.8	85.2
21	82	80.4	78.9	77.3	75.9
28	76	75.1	73.0	70.7	68.7
32	112	113.0	109.7	106.3	103.0
34	170	152.0	147.9	143.7	139.6

Table 5. Column failure times calculated with different air gap thermal resistance.

No	Dimension (mm)	Length (mm)	Load (kN)	Thermal resistance ($K \cdot m^2 \cdot W^{-1}$)	Failure time (min)
1	SHS200 × 5	3000	659	0	30.8
2	SHS200 × 5	3000	659	0.005	33.8
3	SHS200 × 5	3000	659	0.01	38.6
4	SHS200 × 5	3000	1098	0	20.6
5	SHS200 × 5	3000	1098	0.005	21.1
6	SHS200 × 5	3000	1098	0.01	21.7
7	SH200 × 12.5	3000	1019	0	29.1
8	SH200 × 12.5	3000	1019	0.005	28.2
9	SH200 × 12.5	3000	1019	0.01	27.6
10	SH200 × 12.5	3000	1698	0	21.6
11	SH200 × 12.5	3000	1698	0.005	20.5
12	SH200 × 12.5	3000	1698	0.01	19.3
13	SHS400 × 10	5000	2747	0	70.0
14	SHS400 × 10	5000	2747	0.005	76.0
15	SHS400 × 10	5000	2747	0.01	84.0
16	SHS400 × 10	5000	4579	0	36.2
17	SHS400 × 10	5000	4579	0.005	37.6
18	SHS400 × 10	5000	4579	0.01	39.5
19	SHS400 × 20	5000	3745	0	51.5
20	SHS400 × 20	5000	3745	0.005	56.7
21	SHS400 × 20	5000	3745	0.01	61.3
22	SHS400 × 20	5000	6241	0	35.6
23	SHS400 × 20	5000	6241	0.005	35.5
24	SHS400 × 20	5000	6241	0.01	35.4

4.2 Initial eccentricity

To predict overall buckling of the columns, an initial eccentricity is introduced in the model. 8 columns in Table 1 have been analyzed using four different values of initial eccentricity. Table 4 presents the simulation results. As expected, the predicted fire resistance of CFT columns is higher when the initial displacement is smaller. Also initial eccentricity has more effect on slender columns (No.21) or columns without reinforcement (No. 28, 32 and 34). However, the calculation results indicate that the effect of using a large range different eccentricity values is relatively small. The value of $L/1000$ has commonly been used by various researchers and is acceptable to model CFT column behaviour in fire.

4.3 Thermal resistance between steel tube and concrete core

Table 5 compares the simulated fire resistance times of a range of CFT columns with no air gap between the steel tube and the concrete core (thermal resistance = 0) and with an air gap of two different values of thermal resistance (0.005 and 0.01 [$K \cdot m^2 \cdot W^{-1}$]). The columns are made from S275 steel and filled with plain C40 concrete. They are pinned at both ends and loaded axially. The load ratios of the columns range from 0.3 to 0.5. In all simulations, slip between the steel tube and the concrete core is included.

It can be seen from the results of numerical analyses in Table 5 that including an air gap in the calculation model makes noticeable difference to the predicted column fire resistance times. As discussed in section 2 on thermal analysis, introducing an air gap slightly increases the steel tube temperature and has larger influence on the predicted concrete core temperatures. Since the applied load on a CFT column is mainly sustained by the concrete core, including an air gap generally gives higher column fire resistance times. Therefore, not including the air gap in the numerical simulation model would generally give results that are on the safe side. For CFT columns that use thick steel tubes, introducing an air gap gives slightly lower fire resistance times than without the air gap, therefore not including an air gap in the model may give unsafe results. Fortunately, the differences are very small so that the predicted results without including an air gap are still acceptable. Based on the validation exercise in section 3.3, an air gap thermal resistance value of 0.005 [$K \cdot m^2 \cdot W^{-1}$] may be used.

4.4 Concrete tension properties

Modelling concrete behaviour at elevated temperatures has presented a particular challenge, particularly when dealing with the negative tensile stiffness due to tension softening after the concrete stress has exceeded its tensile strength. Table 6 presents a few calculation results of column fire resistance times for tests 1/1, 1/2 and 1/5 from the CIDECT Report C1/C2 by using different concrete tensile stiffness (E_θ^-) after the concrete stress has exceeded its tension strength $f_{c,\theta}^t$. It can be seen that provided the value of E_θ^- is non-negative, using different tension strength $f_{c,\theta}^t$ or tangent modulus E_θ^- of concrete has relatively small effect on the column fire resistance time. However, using a negative value of E_θ^- can give very low values. The tensile stress-strain curve of a positive tensile strength and a

Table 6. Column failure times calculated with different concrete tension properties.

No.	$f_{c,\theta}^t: 0$ $E_\theta^-: 0$	0 $E_\theta/100$	$f_{c,\theta}^c/10$ -880	$f_{c,\theta}^c/10$ 0	$f_{c,\theta}^c/10$ $E_\theta/100$
1/1	60.2	61.0	34.5	61.3	61.3
1/2	59.4	60.0	33.7	60.4	60.4
1/5	57.3	57.8	31.2	58.1	58.1

negative stiffness is within the two extremes of stress-strain curve of (1) zero tensile strength and (2) positive tensile strength with zero or positive tensile stiffness. Since the predicted CFT column fire resistance times of these two extreme sets of stress-strain curve are very close, it is acceptable to conclude that these low fire resistance times associated with a negative tensile stiffness are not real CFT fire resistance times but are consequences of the numerical convergence problem associated with negative stiffness in the material model. They should be discarded. Instead, it is acceptable to use a small or zero tensile stiffness in modelling to avoid numerical convergence difficulties.

5 CONCLUSIONS

This paper has presented a finite element analysis of the behaviour of CFT columns in fire. The ability of the model is verified by comparing the simulation results with a number of fire test results for temperatures in the CFT columns, column deflections and column fire resistance times. In particular, this study has incorporated a few assumptions that have rarely been considered by other researchers, these being an air gap between the steel tube and the concrete core and allowing for slip between the steel tube and concrete core. Incorporating these assumptions has been shown to improve the correlation between the calculated results and measured results for a large number of CFT columns previously tested by others. In addition, this paper has presented a limited number of sensitivity studies to investigate the sensitivity of CFT column behaviour and fire resistance times to concrete tensile properties and initial column deflections.

From the results of the various numerical parametric studies, the following conclusions may be drawn:

(1) Whether or not including slip between the steel tube and concrete core in the numerical model has little influence on the calculated column fire resistance time. However, including slip gives better prediction of column deflection behaviour. As long as slip is allowed, the predicted CFT column behaviour is practically identical regardless of the properties of the slip interface.

(2) Introducing an air gap between the steel tube and the concrete core can improve the accuracy of predictions for both structural temperatures and structural performance. Compared to the model without an air gap, introducing an air gap gives a slightly higher steel tube temperature, but noticeably lower concrete core temperatures. Therefore, in general, including an air gap will give higher column fire resistance times than without an air gap. For columns with very thick steel tubes so that the contribution of the steel tubes to column fire resistance is high, introducing an air gap may lead to predictions of lower column fire resistance times, suggesting that the predicted column fire resistance times without an air gap may not be safe. Fortunately under this circumstance, the differences in predicted CFT column fire resistance times are very small.

(3) Using different tension strength $f_{c,\theta}^t$ or tangent stiffness E_θ^- of concrete has minor effect on the calculated column fire resistance, unless the value of E_θ^- is negative. The predicted column fire resistance with a negative value of E_θ^- is a consequence of lack of numerical convergence and should be discarded.

(4) Using different amounts of column initial deflection has some influence on column fire resistance times. Nevertheless, the influence is relatively small so that it is acceptable to use a maximum initial deflection of L/1000 as commonly assumed by other researchers.

ACKNOWLEDGEMENT

Funding for this research was provided by an Overseas Research Scholarship to the first author.

REFERENCES

Ding, J. & Wang, Y.C. 2005. Finite element analysis of concrete filled steel columns in fire, *Fourth international conference on advances in steel structures, June, 2005.* Shanghai, China.

European Committee for Standardisation (CEN). 2005. Eurocode 4: Design of Composite Steel and Concrete Structures, *Part 1.2: Structural Fire Design.* London: British Standards Institution.

Kimura, M., Ohta, H., Kaneko, H., Kodaira, A. & Fujinaka H. 1990. *Fire resistance of concrete-filled square steel tubular columns subjected to combined loads.* Takenaka Technical Research Report No.43: 47–54, Japan.

Kordina, K. & Klingsch, W. 1983, *Fire Resistance of Composite Columns of Concrete Filled Hollow Sections*, Research report, CIDECT 15 C1/C2 −83/27, Germany.

Lie, T. T. & Caron, S. E. 1988a. *Fire Resistance of Circular Hollow Steel Columns Filled with Siliceous Aggregate Concrete*, Test results, Internal report No.570, Institute for Research in Construction, National Research Council of Canada, Ottawa, Canada.

Lie, T. T. & Caron, S. E. 1988b. *Fire Resistance of Circular Hollow Steel Columns Filled with Carbonate Aggregate Concrete*, Test results, Internal report No.573, Institute for Research in Construction, National Research Council of Canada, Ottawa, Canada.

Lie, T. T. & Chabot, M. 1994. *Fire Resistance Tests of Square Hollow Steel Columns Filled with Reinforced Concrete*, Test results, Internal report No.673, Institute for Research in Construction, National Research Council of Canada, Ottawa, Canada.

Roeder, C. W., Cameron, B. & Brown, C. B. (1999), Composite action in concrete filled tubes, *Journal of Structural Engineering*, **125**(5): 477–484.

Wang, Y.C. & Davies, J.M. 2003. An experimental study of the fire performance of non-sway loaded concrete-filled steel tubular column assemblies with extended end plate connections. *Journal of Constructional Steel Research* 59: 819–838.

Wium, J. A. 1992, *A Composite Column: Force Transfer from Steel Section to Concrete Encasement*, Ph.D thesis, Swiss Federal Inst. of Tech., Lausanne, Switzerland.

Steel and Composite Structures – Wang & Choi (eds)
© 2007 Taylor & Francis Group, London, ISBN 978-0-415-45141-3

Analytical inelastic buckling loads of restrained steel columns under longitudinal non-uniform temperature distribution

W.F. Yuan & K.H. Tan

School of Civil & Environmental Engineering, Nanyang Technological University, Singapore

ABSTRACT: Columns under natural fire conditions are usually exposed to non-uniform temperature distribution in the longitudinal direction. The difference in temperature between the top and bottom ends of a column can be quite significant, particularly prior to flashover condition. In this paper, the inelastic stability of a pin-ended steel column under non-uniform temperature distribution is studied analytically. Across a column section, the temperature is assumed to be uniform. Two linear elastic springs connected to the column ends simulate axial restraints from adjoining unheated structures.

1 INTRODUCTION

Thermal restraint from adjoining unheated structure plays a key role in the stability of steel columns. The structural response depends largely on the temperature distribution in the cross-sectional and longitudinal direction due to i) thermal load will change the material properties of steel and ii) under elevated temperature, the magnitude of thermal induced compressive stress arising from thermal restraint is of the same order as initial applied stress at ambient temperature. The effect of temperature variations in cross-sectional direction has been studied by Ossenbruggen et al. [1]. However, this paper focuses on the analytical derivations of column stability subjected to longitudinal temperature variations since there has not been any significant theoretical development on this aspect.

At present, numerical approach is very popular in the study of structures under fire because finite element program offers a wide range of flexibility. A review of recent literature shows that the effects of axial restraint have been investigated numerically by Neves [2], and Shepherd et al. [3]. Besides, the effects of rotational restraint have also been studied numerically by Franssen et al. [4], Wang [5] and Valente et al. [6]. It was found that the critical temperatures of columns will be reduced by axial restraint but enhanced by rotational restraint. Huang and Tan et al. [7] presented a series of numerical studies conducted on thermally-restrained steel columns subjected to predominantly axial loads. A finite element program FEMFAN3D was developed for the fire resistance analysis and creep strain has been explicitly considered. With incorporation of creep strain, the rate of

increase in temperature on mechanical response can be modeled.

However, for design purpose, theoretical analysis that can be performed manually is much needed as it enables engineers to quickly ascertain the column buckling loads, particularly under local fire scenarios. Culver et al. [8] proposed an approach to determine the buckling loads for pin-ended steel columns subjected to a uniform temperature increase along the member length. In their work, the effects of residual stress and the influence of temperature on the buckling strength in both the elastic and inelastic range were considered but thermal restraints were not taken into account. On the other hand, it is acknowledged that the behavior of a steel column in fire is mostly affected by the restraints of its adjoining structure [9]. Ali et al. [10] reported that axial restraint reduced the fire resistance of columns after a series of test on 37 axially-restrained steel columns subjected to quasi-standard fire. Tang et al. [11] proposed a simple approach based on Rankine interaction formula to obtain a realistic estimate of column fire resistance. However, in that paper, columns are subjected to uniform temperature distribution. Huang and Tan [12, 13] considered the axial restraint on an isolated heated column using a linear spring attached to the column top end. They extended the traditional Rankine formula to predict the critical temperature of an axially-restrained steel column. The proposed Rankine approach incorporating both the axial restraint and creep strain yields very good agreement with finite element predictions.

Although the stability of axially-loaded columns at elevated temperature and subjected to elastic restraints has been studied by some researchers, the assumption about uniform temperature distribution may give overly conservative prediction since the temperature distribution in the longitudinal direction under fire conditions is usually non-uniform. This is because through the convective process, the hottest layer of air will rise up to the top with a relatively cooler layer at the bottom. Thus, based on uniform temperature assumption, for conservatism, engineers usually ascribe the hottest temperature at the column top as the uniform column design temperature. In 1972, Culver [14] analyzed the stability of wide-flanged steel columns subject to elevated temperature using a finite difference approach. The buckling loads were determined by solving the governing differential equation based on finite difference method. Various cases of non-uniform temperature distribution along the member length were considered, but the influence of end restraints was not investigated. In this paper, as shown in Fig. 1, the temperature in the longitudinal direction (x − axis) is assumed to be non-uniform. Two linear springs attached to the column ends simulate the linear restraints from adjoining unheated structure. To simplify the ensuing derivations, the two elastic springs can be replaced by one equivalent spring (k_e) at the top end of the column. The critical load is derived analytically using Galerkin method. Tan and Yuan [15] studied the stability of a pin-ended steel column subjected to varying longitudinal temperature distribution but the analysis was only conducted for elastic material model. In this paper, the inelastic buckling load is derived analytically.

2 STABILITY

2.1 Stress in the column

For a pin-pin ended column under non-uniform temperature distribution, Young's modulus of steel is no longer constant since temperature varies over the column height. Adopting the coordinate system shown in Figure 1, Young's modulus can be expressed by (1):

$$E = E(T) = E(x) \tag{1}$$

Due to force equilibrium and geometry compatibility, one obtains:

$$\int_0^L \varepsilon_{ep} dx - \int_0^L \beta(x) dx = \frac{P_0 - A\sigma}{k_e} \tag{2}$$

where $\varepsilon_{ep} = \dfrac{\mu - (\mu - 1)H}{\mu E(x)} \cdot \sigma + \dfrac{(\mu - 1)H}{\mu E(x)} \cdot \sigma_Y$ is the mechanical elasto-plastic strain and ε_T is the thermal expansion ratio.

In the expression of ε_{ep}, the term σ is the compressive axial stress in the column and σ_Y is the yield

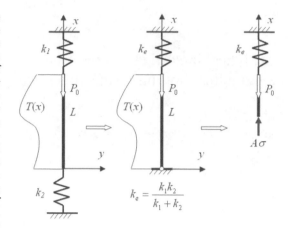

Figure 1.   Column Member under Compressive Load.

strength of steel under temperature. E is Young's modulus in elastic phase, μ is a material constant and μE is the tangent value of stress-strain curve when material is in plastic phase. H is the Heaviside function.

Denoting the shortening of the equivalent elastic spring by ΔL, the external applied force P_0 is resisted by the equivalent spring and the column itself:

$$P_0 = k_e \Delta L + A\sigma \tag{3}$$

Hence, the total internal compressive axial force acting on the cross section is:

$$P_c = A\sigma = P_T + \varphi_0 P_0 \tag{4}$$

where $\quad \varphi_0 = \dfrac{1}{\dfrac{k_e}{A\mu} \int_0^L \left[\dfrac{\mu + (\xi - 1)(\mu - 1)H}{E(x)} \right] dx + 1}$

and $P_T = k_e \varphi_0 \int_0^L \beta(x) dx$. The term ξ is defined as $\xi = \dfrac{\sigma_Y}{\sigma}$. Obviously, P_T is the force induced by temperature and restraint effect.

2.2 Critical load

Consider the equivalent column in Figure 1 subjected to external load P_0 and buckling, the following classical Euler equation differential equation can be used to describe the column deflection:

$$E(x)I_{ep} \frac{d^2 y}{dx^2} + P_c y = 0 \tag{5}$$

where P_c is the total internal compression force calculated from (4). In this paper, P_{c-cr}, the critical buckling load is obtained by using Galerkin method:

$$P_{c-cr} = \frac{-K_B \pm \sqrt{K_B^2 - 4K_A K_C}}{2K_A} \tag{6}$$

The terms K_A, K_B and K_C are given in Appendix.

During the evaluation of P_{c-cr}, the trial function $\hat{y}$ for (5) is selected as follows:

$$\hat{y} = a_1 N_1 + a_2 N_2 = a_1 x(L-x) + a_2 x^2(L-x) \qquad (7)$$

where N_1 and N_2 are the shape functions at both ends of the column, and a_1 and a_2 are arbitrary constants.

To simplify the derivation, Eq. (6) is rewritten as:

$$P_{c-cr} = \eta \cdot P_E \qquad (8)$$

where the term $P_E = \pi^2 E_0^{20} I / L^2$. is the Euler buckling load of the column under ambient temperature. η is a factor which varies with temperature.

The expression in (8) gives the conventional column buckling load. Indicating the critical situation by subscript 'cr', one can obtain the critical external load from (4):

$$P_{0-cr} = \frac{P_{c-cr} - P_T}{\varphi_0} = \frac{\eta \cdot P_E - P_T}{\varphi_0} \qquad (9)$$

2.3 Material model

For steel under fire conditions, several material models from design codes are available. However, based on the ECCS data [16], this paper proposes that elastic modulus ratio is given by a continuous function of t:

$$E/E_0^{20} = c_4 t^3 + c_3 t^2 + c_2 t + c_1 \qquad (10)$$

with $t = T/100$, $c_1 = 0.96483$, $c_2 = 0.07922$, $c_3 = -0.03622$ and $c_4 = 0.00192$.

In this paper, thermal expansion is described by Harmathy model [17]:

$$\beta(T) = \frac{\Delta l}{l} = a + bT + cT^2 \qquad (11)$$

with $a = -2.6601 \times 10^{-4}$ and $b = 1.0923 \times 10^{-5}$ $c = 5.3006 \times 10^{-9}$.

2.4 Temperature distribution

The temperature distribution on the column is assumed to follow a piecewise function given by:

$$T(x) = \begin{cases} 100 a_{1T} & 0 \le x \le L_1 \\ 100 a_{2T} & L_1 \le x \le L_2 \end{cases} \qquad (12)$$

where a_T and b_T are two constants.

For step temperature distribution, the critical axial force P_{c-cr} can be calculated from (6) by considering Eqs. (10) and (11). In this case, P_T is given as:

$$P_T = k_e \varphi_0 SL \qquad (13)$$

with $S = a + b(T_1 \alpha + T_2 \vartheta) + c(T_1^2 \alpha + T_2^2 \vartheta)$ and

$$\varphi_0 = \varphi_{0-I} H(\sigma_{Y-2} - \sigma_I) + \varphi_{0-III} H(\sigma_{III} - \sigma_{Y-1})$$
$$+ \varphi_{0-II} H(\sigma_{II} - \sigma_{Y-2}) H(\sigma_{Y-1} - \sigma_{II})$$

where the terms σ_I, σ_{II}, σ_{III}, φ_{0-I}, φ_{0-II} and φ_{0-III} can be found in Appendix. σ_{Y-1} and σ_{Y-2} are the respective yield stresses at segment temperature T_1 and T_2.

For convenience, Eq. (8) is further written to the following form:

$$P_{c-cr} = \eta \cdot P_E = \eta \cdot \sigma_e A \qquad (14)$$

where $\sigma_e = \dfrac{P_E}{A}$ is defined as the critical internal stress of the column at ambient temperature. The formulae used to calculate η are given in Appendix.

3 EXAMPLE

3.1 Geometry

Considering a typical steel column made of UB178 × 102 × 19, it has the following dimensions and material properties: $I_z = 136.71\,cm^4$, $A = 24.26\,cm^2$, $L = 300\,cm$, $E_0^{20} = 2.1 \times 10^5\,MPa$, $\sigma_Y^{20} = 3.0 \times 10^2\,MPa$, $\mu = 0.01$.

The stiffnesses of the springs are assumed to be $k_1 = 0.1 \dfrac{E_0^{20} A}{L}$ and $k_2 = 3.0 \dfrac{E_0^{20} A}{L}$, which result in an equivalent stiffness $k_e = 0.097 \dfrac{E_0^{20} A}{L}$. A series of studies are conducted for different values of L (column length), α (segment length ratio) and μ (Young's modulus softening ratio).

3.2 Result

In Figure 2, the hotter segment is only one quarter of the whole column. In the figures, $P_{Y-1} = A\sigma_{Y-1}$ and $P_{Y-2} = A\sigma_{Y-2}$. The figure shows that P_{0-cr} is always greater than zero. That means, the column will not buckle for whatever T_2 unless external load is applied. It can also be seen that P_{c-cr} is greater than P_{Y-2} when $T_2 \geq 675°C$. Theoretically, the column may not buckle even the hotter segment is at plastic stage. One finds that when $T_2 \geq 675°C$, P_{c-cr} is nearly constant while P_T decreases as T_2 grows because the hotter segment is in plastic stage. In this case, P_{0-cr}, which represents

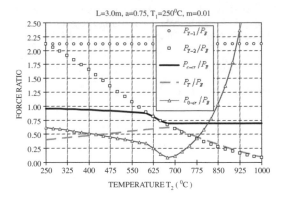

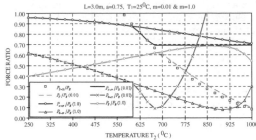

Figure 3. Comparison between results obtained by elastic and inelastic assumptions.

Figure 2. Forces vs temperature ($\alpha = 0.75$).

the column capacity to undertake external compressive load has a positive growth when T_2 increases. The reason is: i) The column cannot afford higher internal compression since the steel is plastic, but ii) the increase of external load is mostly resisted by the equivalent spring at the top column end. However, it must be realized that a higher P_{0-cr} does not mean a higher failure load because a large axial plastic deformation will take place. This is also regarded as one kind of structural failure before the column buckles.

The comparison between the results obtained by elastic ($\mu = 1.0$) and inelastic ($\mu = 0.01$) assumptions for the case $\alpha = 0.75$ is shown in Figure 3. It is obvious that the difference is significant. Based on inelastic assumption, P_T will decrease even though T_2 goes up because the hotter segment is plastic ($T_2 > 675°C$). However, if plasticity is not considered, P_T will continuously increase until $T_2 > 850°C$. As for critical axial load, the elastic assumption will result in a higher P_{c-cr} than inelastic assumption when $T_2 > 600°C$. In other words, elastic assumption will overestimate the bearing capacity of the particular column. In Figure 3 it can be observed that the elastic critical external load P_{0-cr} is lower than the plastic P_{0-cr} if $T_2 > 740°C$. However, as discussed above, large plastic axial deformation, which is also one kind of structural failure will occur at this stage. The figure also shows that if $T_2 > 740°C$, the value of P_{0-cr} for $\mu = 0.01$ grows as T_2 increases because the increase of external load is mostly resisted by the equivalent line elastic spring connected to the top column end.

4 CONCLUSION

The stability of a pin-ended steel column at elevated temperature is studied analytically. In this model, two elastic springs connected to the ends of the member are considered to simulate the linear restraints at the column ends. The temperature distribution along the column is assumed to be non-uniform and bilinear material model is employed. Comparing the results based on elastic and inelastic assumptions, it can be concluded that the effect on critical load caused by plasticity is considerable especially at high temperature. The explicit formulation for step temperature distribution can be used conveniently to evaluate the critical load. The results can also be combined with fire modeling. For instance, in zone modeling, the compartment contains fire is normally divided into two layers. Hence, the fire modeling results can be directly used as input in this study.

It must be noted that the proposed results are not valid if the temperature rises very slowly because Tan et al [18] suggested that transient analysis should be carried out to consider the creep effect. Furumura et al [19] and Anderberg [20] also reached similar conclusions. In the future study, the results of heat transfer should be adopted and the effect of rotation restraints will also be investigated.

REFERENCES

[1] Ossenbruggen P. J., Aggarwal V. and Culver C. G. "Steel column failure under thermal gradients by member", Journal of the Structural Division, April 1973; Vol. 99, No.4: 727–739.

[2] Neves I. Cabrita. "The critical temperature of steel columns with restrained thermal elongation", J. Construct. Steel Res. 1995; Vol. 24: 211–227.

[3] Shepherd P., Burgess IW., Plank RJ. and O'Connor DJ. "The performance in fire of restrained steel columns in multi-storey constructions", 4th Kerensky International Conference, Hong Kong, 1997; 333–342.

[4] Franssen JM. and Dotreppe JC. "Fire resistance of columns in steel frames", Fire Safety Journal (UK), 1992; 19, 22: 159–175.

[5] Wang Y. C. "The effects of frame continuity on the behavior of steel columns under fire conditions and fire resistant design proposals", J. Construct. Steel Res. 1997; 41: 93–111.

[6] Valente Joaquim C. and Neves I. Cabrita. "Fire resistance of steel columns with elastically restrained axial elongation and bending", J. Construct. Steel Res. 1999; Vol.52: 319–331.

[7] Huang Z. F., Tan K. H. and Ting S. K. "Heating Rate and Boundary Restraint Effects on Fire Resistance of Steel Columns with Creep", Engineering Structure, 2005; Vol. 28(6): 805–817.

[8] Culver C., Aggarwal V. and Ossenbruggen, P. "Buckling of steel columns at elevated temperatures", Journal of the Structural Division. 1973; Vol. 99, no. 4: 715–726.

[9] Rodrigues JPC., Neves IC & Valente JC. "Experimental research on the critical temperature of compressed steel elements with restrained thermal elongation", Fire Safety Journal, 2000; 35: 77–98.

[10] Ali F., Shepherd P., Randall M., Simms IW., O'Connor DJ. and Burgess IW. "The effect of axial restraint on the fire resistance of steel columns", J. Construct. Steel Res. 1998; 46(1–3): Paper 177.

[11] Tang C. Y., Tan, K. H., Ting, S. K. "Basis and application of a simple interaction formula for steel columns under fire conditions", ASCE Journal of Structural Engineering (United States). 2001; Vol. 127, No.10: 1206–1213.

[12] Huang Z. F. and Tan K. H. "Analytical fire resistance of axially-restrained steel columns", J. of Struct. Engng. 2003; ASCE Vol. 129(11): 1531–1537.

[13] Huang Z. F. and Tan, K. H. "Rankine approach for fire resistance of axially-and-flexurally restrained steel columns", J. of Construct. Steel Res. (United Kingdom). 2003; Vol.59(12): 1553–1571.

[14] Culver C. "Steel column buckling under thermal gradients", Journal of the Structural Division. Aug. 1972; Vol. 92, no. 8: 1853–1865.

[15] Tan K. H., Yuan W. F. "Stability of Elastically Restrained Steel Columns under Longitudinal Non-Uniform Temperature Distribution", Structural Engineering and Mechanics. Accepted, 2007

[16] ECCS (European Convention for Constructional Steelwork). "European recommendation for the fire safety of steel structures", 1983; Elsevier, Amsterdam.

[17] Harmathy, T. Z. "Creep deflection of metal beams in transient heating processes, with particular reference to fire", Can. J. Civ. Eng., 1976; Vol.3(2): 219–228.

[18] Tan K. H., Ting S. K. and Huang Z. F. "Visco-elasto-plastic analysis of steel frames in fire. J. Struct. Engng", ASCE, 2002, 128(1): 105–114.

[19] Furumura F., Ave T., Kim WJ. and Okabe T. "Nonlinear elasto-plastic creep behaviour of structural steel under continuously varying stress and temperature", J. Struct. Constr. Engng., 1985; 353: 92–100.

[20] Anderberg Y. "Modelling steel behaviour", Fire Safety Journal, 1988; 13: 17–26.

APPENDIX

$$C_1 = \int_0^L [x(L-x)F_1(x)]dx \tag{1-1}$$

$$C_2 = \int_0^L [x^2(L-x)^2]dx = \frac{1}{30}L^5 \tag{1-2}$$

$$C_3 = \int_0^L [x(L-x)F_2(x)]dx \tag{1-3}$$

$$C_4 = \int_0^L [x^3(L-x)^2]dx = \frac{1}{60}L^6 \tag{1-4}$$

$$D_1 = \int_0^L [x^2(L-x)F_1(x)]dx \tag{1-5}$$

$$D_2 = \int_0^L [x^3(L-x)^2]dx = \frac{1}{60}L^6 \tag{1-6}$$

$$D_3 = \int_0^L [x^2(L-x)F_2(x)]dx \tag{1-7}$$

$$D_4 = \int_0^L [x^4(L-x)^2]dx = \frac{1}{105}L^7 \tag{1-8}$$

$$K_A = C_2D_4 - C_4D_2, \quad K_C = C_1D_3 - C_3D_1 \tag{1-9}$$

$$K_B = C_1D_4 - C_4D_1 + C_2D_3 - C_3D_2 \tag{1-10}$$

with $F_1(x) = -2E(x)I_{ep}$, $F_2(x) = (2L - 6x)E(x)I_{ep}$.

$$\sigma_I = \frac{S}{\left(\dfrac{\alpha}{E_1} + \dfrac{\vartheta}{E_2}\right) + \dfrac{A}{k_e}} \tag{1-11}$$

$$\sigma_{II} = \frac{S + \dfrac{(1-\mu)}{\mu}\dfrac{\vartheta\sigma_{Y-2}}{E_2}}{\left(\dfrac{\alpha}{E_1} + \dfrac{\vartheta}{\mu E_2}\right) + \dfrac{A}{k_e}} \tag{1-12}$$

$$\sigma_{III} = \frac{S + \dfrac{(1-\mu)}{\mu}\left(\dfrac{\alpha\sigma_{Y-1}}{E_1} + \dfrac{\vartheta\sigma_{Y-2}}{E_2}\right)}{\dfrac{1}{\mu}\left(\dfrac{\alpha}{E_1} + \dfrac{\vartheta}{E_2}\right) + \dfrac{A}{k_e}} \tag{1-13}$$

$$\varphi_{0-I} = \frac{1}{\dfrac{k_e L}{A}\left(\dfrac{\alpha}{E_1} + \dfrac{\vartheta}{E_2}\right) + 1} \tag{1-14}$$

$$\varphi_{0-II} = \frac{1}{\dfrac{k_e L}{A}\left(\dfrac{\alpha}{E_1} + \dfrac{\vartheta}{E_2}\dfrac{\xi_{II}\mu - \xi_{II} + 1}{\mu}\right) + 1} \tag{1-15}$$

$$\xi_{II} = \frac{\sigma_{Y-2}}{\sigma_{II}}$$

$$\varphi_{0-III} = \frac{1}{\dfrac{k_e L}{A}\dfrac{(\xi_{III}\mu - \xi_{III} + 1)}{\mu}\left(\dfrac{\alpha}{E_1} + \dfrac{\vartheta}{E_2}\right) + 1} \tag{1-16}$$

$$\xi_{III} = \frac{\sigma_{Y-1}}{\sigma_{III}}$$

$$\eta = \eta_1 + \eta_2 + \eta_3 \tag{1-17}$$

$$\eta_1 = \eta_I H(\sigma_{Y-2} - \eta_I \sigma_e) \tag{1-18}$$

$$\eta_2 = \begin{bmatrix} \eta_{II} H(\eta_{II}\sigma_e - \sigma_{Y-2}) H(\sigma_{Y-1} - \eta_{II}\sigma_e) \\ + \dfrac{\sigma_{Y-2}}{\sigma_e} H(\sigma_{Y-2} - \eta_{II}\sigma_e) \end{bmatrix} \\ \cdot H(\eta_I \sigma_e - \sigma_{Y-2}) \tag{1-19}$$

$$\eta_3 = \begin{bmatrix} \eta_{III} H(\eta_{III}\sigma_e - \sigma_{Y-1}) \\ + \dfrac{\sigma_{Y-1}}{\sigma_e} H(\sigma_{Y-1} - \eta_{III}\sigma_e) \end{bmatrix} \\ \cdot H(\eta_I \sigma_e - \sigma_{Y-2}) H(\eta_{II}\sigma_e - \sigma_{Y-1}) \tag{1-20}$$

$$\eta_I = \frac{2}{\pi^2}\begin{pmatrix} G_{1\alpha}a_{1E} + G_{2\alpha}a_{2E} \\ -\sqrt{G_{3\alpha}a_{1E}^2 + G_{4\alpha}a_{2E}^2 + G_{5\alpha}a_{1E}a_{2E}} \end{pmatrix} \tag{1-21}$$

$$\eta_{II} = \frac{2}{\pi^2}\begin{pmatrix} G_{1\alpha}a_{1E} + \mu G_{2\alpha}a_{2E} \\ -\sqrt{G_{3\alpha}a_{1E}^2 + \mu^2 G_{4\alpha}a_{2E}^2 + \mu G_{5\alpha}a_{1E}a_{2E}} \end{pmatrix} \tag{1-22}$$

$$\eta_{III} = \mu \eta_I \tag{1-23}$$

with $\sigma_e = \dfrac{P_E}{A}$, $L_{d1} = L_1$, $L_{d2} = L - L_1$, $\alpha = \dfrac{L_{d1}}{L}$ and $\vartheta = 1 - \alpha$.

Table 1. Values of $G_{i\alpha}$

α	$G_{1\alpha}$	$G_{2\alpha}$	$G_{3\alpha}$	$G_{4\alpha}$	$G_{5\alpha}$
0.00	0.0000	13.0000	0.0000	64.0000	0.0000
0.10	1.1805	11.8195	1.3926	48.7449	13.8625
0.20	3.2874	9.7126	10.7549	30.4771	22.7680
0.30	5.0306	7.9694	24.7968	25.5865	13.6167
0.40	6.0275	6.9725	33.8881	29.6078	0.5040
0.50	6.5000	6.5000	34.4570	34.4570	−4.9141
0.60	6.9725	6.0275	29.6078	33.8881	0.5040
0.70	7.9694	5.0306	25.5865	24.7968	13.6167
0.75	8.7539	4.2461	26.4076	17.8451	19.7474
0.80	9.7126	3.2874	30.4771	10.7549	22.7680
0.85	10.7764	2.2236	38.2175	4.9343	20.8482
0.90	11.8195	1.1805	48.7449	1.3926	13.8625
1.00	13.0000	0.0000	64.0000	0.0000	0.0000

Table 2. Values of a_{iE}

$T_i(°C)$	100	200	300	400	500
a_{iE}	1.000	0.994	0.928	0.825	0.695
$T_i(°C)$	600	700	800	900	1000
a_{iE}	0.551	0.403	0.264	0.144	0.055

Steel and Composite Structures – Wang & Choi (eds)
© *2007 Taylor & Francis Group, London, ISBN 978-0-415-45141-3*

Tracking the progressive collapse of axially-restrained steel beams under fire conditions

K.H. Tan & Z.F. Huang
School of CEE, Nanyang Technological University, Singapore

G.L. England
Department of Civil Engineering, Imperial College, UK

ABSTRACT: This paper investigates structural responses of three axially-restrained steel beams under mono-tonically rising temperature. Only uniform temperature distribution within a beam is considered. An elastic spring attached at an end of a beam is adopted to simulate axial restraint from the adjoining structure. Three different axial restraint ratios are investigated. Progressive collapse processes of all beams throughout heating are examined, whilst the formulation of plastic hinge at the mid-span is emphasised. Besides, developments of beam internal forces, including axial force and mid-span bending moment, are discussed in detail. Based on the definitions of limiting temperature and collapse temperature adopted in this study, it is found that (1) Increasing axial restraint stiffness will reduce limiting temperature; and (2) beams subjected to greater axial restraint will have longer duration before collapse.

1 INTRODUCTION

Unlike under cool environment, a steel beam within a compartment fire received noticeable axial restraint (Fig. 1) from its adjoining cool structure. There exist remarkable differences between the behavior of a heated beam and the behaviour of the same beam under normal service condition. For instance, a heated beam tends to receive strong axial restraint at its ends which is normally neglected at room temperature. To date, there are some published works examining the axial restraint effect on the response of steel beams in fire. The following outlines recent research works.

In UK, Liu et al (2002) conducted a series of fire tests to study axial restraint effect on unprotected steel beams. Three axial restraint ratios were investigated and three different load levels were applied onto the beams. Both flush end-plate beam-to-column joint and web-cleat joint were examined. It is observed that catenary action prevented the beams from failing mechanically. Catenary action, being mobolised at large deflection, was observed in beams with high axial restraint and subjected to low load levels.

In the analytical aspect, Yin & Wang (2005) proposed a simple analytical model to compute the catenary forces in steel beams at elevated temperature. Both axial and flexural restraints at the beam ends were considered. Validation of the proposed method showed that normally the predicted catenary force was greater than the associated numerical results.

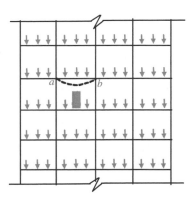

Figure 1. Typical response of a steel beam subject to a compartment fire in a multi-storey steel frame.

In Australia, Wong (2005) proposed a formula to calculate limiting temperature of axially-restrained steel beams. He also presented a simple technique to quantify the axial restraint exerted onto a heated beam from its neighbouring members. The axial force induced by thermal expansion was incorporated through axial-bending interaction of a critical cross section. The catenary action of beams was also considered.

Bradford (2006) derived a generic modelling of axially and flexurally-restrained steel beams at elevated temperature. The analysis was elastic and therefore

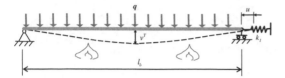

Figure 2. Isolated axially-restrained beam model.

yielding and catenary action were ignored. Based on theorem of virtual work, governing differential equations were derived and solved for specific cases of restraint stiffness.

To make use of catenary action in the fire resistant design of steel beams that may substantially decrease fire protections to the beams, the plastification process of an axially-restrained steel beam at elevated temperature should be understood fundamentally. This becomes the objective of this study.

This study presents numerical simulation for examining the structural responses of three steel beams of square cross section at elevated temperatures. The choice of square cross section is incidental; the observations can equally apply to other types of cross section. Three beams are associated with there different axial restraint ratios.

2 ISOLATED AXIALLY-RESTRAINED BEAM MODEL AND SCOOP OF ANALYSES

To study the axial restraint effect on a heated beam, an isolated beam model is proposed (Fig. 2). The linear spring is used to approximate axial restraint at beam ends. For consistency, in all the following case studies a square steel section of $100 \times 100 \, mm^2$ is used. All three beams are of an identical slenderness ratio of 40. They differ only in their axial restraint ratios β_l, which takes 0, 0.5 and ∞, respectively. Here, ratio β_l is defined as the ratio of linear spring stiffness to beam axial stiffness:

$$\beta_l = \frac{k_l}{E_0^{20} A_b / l_b} \tag{1}$$

in which the denominator $E_0^{20} A_b / l_b$ represents the beam axial stiffness at 20° C, where E_0^{20} is elastic modulus, A_b is the cross sectional area and l_b is beam span.

All beams are subjected to a typical external load level $R = 0.5$. Only uniformly distributed lateral load (hereafter, UDL) is accounted for. The load level R is defined as the ratio of mid-span bending moment for a pinned-rollered beam to its cross-sectional plastic moment capacity M_p^{20} in the absence of axial force. That is,

$$R = \frac{q l_b^2}{8 M_p^{20}} = \frac{q l_b^2}{8 f_y^{20} W_p} \tag{2}$$

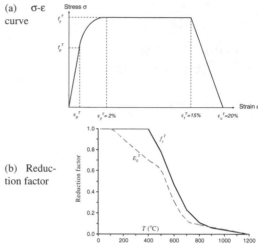

(a) σ-ε curve

(b) Reduction factor

Figure 3. EC3 steel material model at elevated temperature.

where q is UDL value, and W_p is plastic modulus of a section.

3 ASSUMPTIONS

Assumptions adopted in the numerical simulation are listed below:

- Beams are straight and without residual stress.
- Beams are restrained transversely, thus neither local nor lateral-torsional buckling is considered.
- Linear springs are assumed to be elastic throughout a heating, that is, its stiffness k_l remains unchanged.
- Temperature across a beam section as well as along its length is uniformly distributed.
- Steel temperature rises monotonically.
- EC3 steel model (CEN 2000) is employed in this study. Fig. 3a shows the adopted bilinear-elliptical stress-strain relationship at elevated temperature, while Fig. 3b shows the reduction factors of yield strength f_y^T and elastic modulus E_0^T.

In FE analysis, each beam is divided into ten elements of equal length. Program FEMFAN, being capable of elasto-plastic and creep analysis of steel plane frames at elevated temperatures, is used in this study (Tan et al. 2002). Both material and geometrical nonlinearities are considered.

4 FORMING OF A PLASTIC HINGE

Fig. 4 shows that at the instant when a plastic hinge forms at the beam mid-span, there are mainly 5 types of cross-sectional stress distribution along the beam length.

668

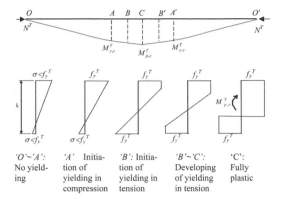

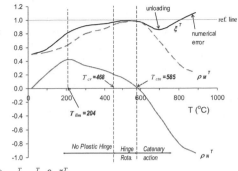

(a) ρ_N^T, ρ_M^T & ξ^T

Figure 4. Cross-sectional stress distributions along a beam when mid-span is fully plastified.

Fig. 4 shows that within any cross section between point 'O' and 'A', neither extreme fibers in compression zone nor those in the tension zone yield. At point 'A', the top fiber in compression just yields. From point 'A' to 'C', with increasing applied moment M_c^T (subscript c denotes presence of compression force) and subjected to constant axial force, yielding gradually spreads from the compression to tension zone. Finally and right at point 'C', the whole section becomes fully plastic.

With the presence of internal axial force N^T, the *reduced plastic moment capacity* $M_{p,c}^T$ for the square section can be expressed as (Horne & Morris 1981):

$$\xi^T = (\rho_N^T)^2 + \rho_M^T = 1.0 \qquad (3)$$

where $\rho_N^T = N^T/N_y^T$, $\rho_M^T = M_{p,c}^T/M_p^T$. and ξ^T denotes the plasticity index taking account of interaction between axial force N^T and bending moment $M_{p,c}^T$.

With a plastic hinge forming at point 'C', any attempt to increase the moment further will cause the member to rotate at point 'C'. The beam will survive under the pulling force from the boundary restraint, this is termed as *catenary action*.

In the numerical analysis, Eq. (3) is adopted as a criterion that indicates the formation of plastic hinge. Due to potential numerical error, this study assumes a plastic hinge forms if $\xi^T \geq 0.98$.

5 COLLAPSE PROCESS OF A GENERAL BEAM

5.1 *Overall behaviour of $\lambda = 40$ beam with $\beta_l = 0.5$*

Firstly, a stocky beam of $\lambda = 40$ is chosen for study. This beam receives with $\beta_l = 0.5$ is subjected to a moderate load level of 0.5. The numerical results are shown in Fig. 5, in which Fig. 5(a) shows the development of

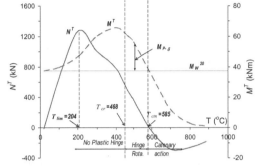

(b) N^T & M^T

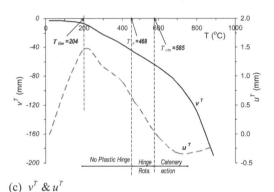

(c) v^T & u^T

Figure 5. Developments of internal forces and deflection in $\lambda = 40$ beam with $\beta_l = 0.5$ (unloading considered).

three indices ρ_N^T, ρ_M^T & ξ^T at the mid-span during the heating, Fig. 5(b) shows the internal axial force N^T and mid-span moment M^T and Fig. 5(c) the mid-span deflection v^T and right-end expansion u^T.

Generally, the collapse behaviour of the heated beam can be divided into three stages, namely, no plastic hinge stage, hinge formation and rotation stage, and catenary action stage. The details are shown below.

During the first stage, axial index ρ_N^T increases nearly linearly to around 204°C, after which ρ_N^T

decreases monotonically (Fig. 5a). Beyond 585°C, catenary action begins with ρ_N^T becoming negative. At the same time, moment index ρ_M^T increases from 0.50 to 0.9442 at 468°C, while the plastification index ξ^T progresses from 0.50 to 0.98 which signifies a plastic hinge forms at the mid-span. On the other hand, the axial force N^T linearly increases to 1274 kN at 204°C and then decreases to 482 kN at 468°C (Fig. 5b). During the first stage, the mid-span moment M^T increases slightly from 37.5 kNm to 47.1 kNm. Theoretically, at a particular temperature M^T comprises two parts as shown below:

$$M^T = M_V^T + M_{P-\delta}^T \qquad (4)$$

in which

$$M_V^T = \frac{q\left(l_b^T\right)^2}{8} = \frac{ql_b^2}{8}(1+\alpha \cdot \Delta T)^2 \approx \frac{q\left(l_b\right)^2}{8} = M_V^{20} \qquad (5)$$

$$M_{P-\delta}^T = N^T \cdot v^T \qquad (6)$$

Clearly, M_V^T is generated by external loading while $M_{P-\delta}^T$ is from $P-\delta$ effect. Fig. 5c exhibits that v^T increases at an accelerating rate during the first stage due to increasing $P-\delta$ effect resulting from compressive force N^T. Fig. 5b shows that the axial expansion u^T develops in the same manner as N^T.

In this study, 204°C is defined as the beam *limiting temperature* $T_{\lim}$, at which both N^T and longitudinal expansion u^T attain their respective maximum values. This implies that under actual fires, a heated steel beam will push away its adjoining columns to the greatest distance at $T_{\lim}$. In the other words, the columns may experience the greatest $P-\Delta$ effect at that moment.

On the other hand, the temperature associated with the formation of a plastic hinge forms is defined as beam *critical temperature* T_{cr}. When temperature rises beyond T_{cr}, the beam enters into its second stage, during which ξ^T remains at unity while ρ_N^T decreases steadily to zero and ρ_M^T slightly increases up to unity. Fig. 5b shows that during the second stage both N^T and M^T keep on decreasing while the beam experiences plastic hinge rotation. Due to a decreasing N^T, the $P-\delta$ effect is diminishing. In terms of deformation, the beam continues the behaviour during the later period of Stage-1 — that is, although the beam continues to deflect, the beam end also continues to contract at a steady rate (see u^T and v^T in Fig. 5c). This leads to a net increase in M^T, and consequently in ratio ρ_M^T. Fig. 5 shows that axial force N^T, ratio ρ_N^T and horizontal displacement u^T reduce steadily while ξ^T is maintained at unity during this stage.

At the end of stage-2, with N^T finally reduces to zero, stage-3 begins with mobilisation of catenary action. In this study, the temperature at which N^T

decreases to 0 is defined as *catenary temperature* T_{ctn}. Obviously, beams with different axial restraint ratios but subjected to the same load level R have a typical T_{ctn}.

The mid-span bending moment at T_{ctn} is computed as:

$$M_V^{T_{ctn}} = \frac{q\left(l_b^T\right)^2}{8} = \frac{ql_b^2}{8}(1+\alpha \cdot \Delta T)^2 \approx \frac{q\left(l_b\right)^2}{8} = M_V^{20} \quad (7)$$

$$M^{T_{ctn}} = M_p^{T_{ctn}} = f_y^{T_{ctn}} W_p = \left(\psi_{f_y}^{T_{ctn}} f_y^{20}\right) \cdot W_p \qquad (8)$$

in which W_p is section plastic modulus and $\psi_{f_y}^{T_{ctn}}$ is reduction factor for steel yield strength at T_{ctn}.

By equating Eq. (7) and Eq. (8), there is

$$\psi_{f_y}^{T_{ctn}} = \frac{ql_b^2}{8f_y^{20}W_p} = R \qquad (9)$$

Eq. (9) implies that T_{ctn} is the temperature at which the reduction factor $\psi_{f_y}^{T_{ctn}}$ becomes equal to R. For current case, the predicted T_{ctn} is 585°C, very close to 590°C at which $\psi_{f_y}^{T_{ctn}} = R = 0.5$. The negligible difference of 5°C is attributable to stress unloading at the top half of the section, where tension effect from N^T becomes the dominant one compared to compression effect from M^T.

A beam attains its T_{ctn} when either one of the following criteria is met:

- Internal axial force N^T reduces to zero, and so do parameters ρ_N^T and $M_{p-\delta}$;
- Mid-span moment M^T resorts to its initial value M_V^{20} when heating begins;
- Ratio ρ_M^T increases up to 1.0.

During Stage-3, ρ_N^T becomes negative (note: tension force) and its value keeps on reducing. Moment index ρ_M^T also reduces steadily while ξ^T decreases first and then increases above 1.0 beyond 800°C. The decrement of ξ^T below unity is due to stress unloading at the top half part within a cross section nearly mid-span of the beam. This can be confirmed by the following examination. Without consideration of unloading effect (i.e., a nonlinear-elastic stress unloading is assumed), the beam exhibits a slightly different behaviour during Stage-3. Fig. 6 which shows that instead of reducing below 1.0, ξ^T remains at unity during Stage-3 before 800°C. Nonetheless, ξ^T rises slightly above 1.0 beyond 800°C — again, this is due to numerical error in computing very small M^T (Fig. 5b).

Fig. 5b shows that beyond T_{ctn}, the tensile axial force N^T increases first and then decreases while bending moment M^T decreases to zero until the beam fails. The beam continues to deflect steadily while it begins

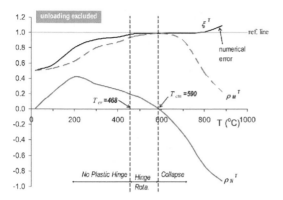

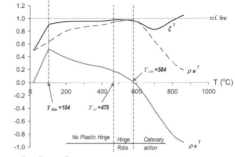

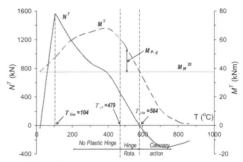

(a) ρ_N^T, ρ_M^T & ξ^T

Figure 6. Developments of plastification indices for $\lambda = 40$ beam with $\beta_l = 0.5$ under a rising temperature (unloading excluded).

to contract axially (Fig. 5c). During this stage, from the free body diagram the tension force in the beam can be approximated as:

$$N^T = \left(M_V^{20} - M^T\right)/v^T \qquad (10)$$

where M_V^{20} is mid-span bending moment before heating starts (Eq. (5)). That is, there is beneficial effect from tension action to counteract the initial imposed M_V^{20}.

At the end of heating, M^T approaches zero and thus

$$N^T \approx \frac{M_V^{20}}{v^T} \qquad (11)$$

Eq. (11) explains why beyond 700°C with deflection v^T keeps on increasing, N^T decreases accordingly.

5.2 Axial restraint effects

Entire collapse process of a $\lambda = 40$ beam with limited axial restraint of $\beta_l = 0.5$ has been examined in Sec. 5.1. For the completeness of study, it is useful to conduct FE analyses on the axial restraint effect. Two extreme cases are chosen, viz., a beam with full restraint and the other one without. Load level is remained at 0.5. The response of the fully-restrained beam and non-restrained beam is shown in Fig. 7 and Fig. 8, respectively.

Firstly, the fully-restrained beam is taken for study. Fig. 7 shows that this beam responds in a similar manner to the previous beam with $\beta_l = 0.5$ (Fig. 5). They start to experience catenary action nearly at the same temperature (484°C vs 485°C), which was defined as the catenary temperature T_{ctn}. Nevertheless, although the axial restraint ratio β_l has negligible effect on T_{ctn}, it remarkable reduces the limiting temperature T_{lim} at which N^T achieves its maximum value in

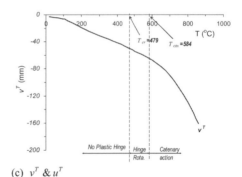

(b) N^T & M^T

(c) v^T & u^T

Figure 7. Developments of internal forces and deflection in $\lambda = 40$ beam with fully axial restraint ($\beta_l = \infty$).

compression. T_{lim} of the current beam attains 104°C, a significant drop from 204°C attained by the previous beam with $\beta_l = 0.5$ (cf. Fig. 5a). Furthermore, this study adopts a concept of *warning time* t_{wn}, which is the time difference between the *catenary time* t_{ctn} and the *limiting time* t_{lim} associated respectively with T_{ctn} and T_{lim}, respectively. Clearly, by increasing β_l from 0.5 to infinity, t_{wn} is substantially increased since T_{lim} occurs much earlier (compare Fig. 5b & Fig. 7b).

In addition, comparison between Fig. 5b and Fig. 7b illustrates that under the fully restrained condition, the heated beam experiences much more $P - \delta$ effect, represented by an increase in $M_{P-\delta}$. This is directly

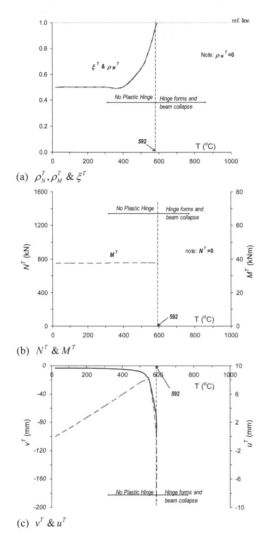

(a) $\rho_N^T, \rho_M^T \ \& \ \xi^T$

(b) $N^T \ \& \ M^T$

(c) $v^T \ \& \ u^T$

Figure 8. Developments of internal forces and deflection in $\lambda = 40$ beam without axial restraint ($\beta_l = 0$).

owing to the greater axial force N^T attained by the fully-restrained beam.

With regard to mid-span deflection v^T, there is little difference between the two beams.

After studying the fully-restrained beam, it is time to examine the beam without restraint which shows a very different behaviour at elevated temperature. Without axial restraint, the beam collapses as soon as a plastic hinge forms at the mid-span (at 592°C). Neither $P - \delta$ effect nor catenary action is experienced. At the end of heating, the mid-span deflection v^T experiences run-away failure (Fig. 8c). It should be highlighted that there is no warning time t_{wn} in this beam. That is, this beam will fail in brittle manner which should be avoided in structural fire resistance design.

6 CONCLUSIONS AND DISCUSSIONS

This paper presents a FE study on the progressive collapse process of three axially-restrained steel beams under a rising temperature while subjected to the same external load level of 0.5. The axial restraint ratios of the beams are 0, 0.5 and ∞, respectively. Beam slenderness ratio is fixed at 40. Only uniform temperature distribution is studied.

Forming and developing a plastic hinge at beam mid-span is examined. The response of a restrained beam can be divided into three stages: no plastic hinge, hinge forming and rotating, and catenary stages. This study defines the temperature when internal axial compression force within a heated beam reaches its ultimate value as limiting temperature $T_{\lim}$. Besides, the temperature when catenary action starts is defined as catenary temperature T_{ctn}. FEA show that:

- Increasing axial restraint will reduce $T_{\lim}$ due to $P - \delta$ effect;
- A beam receiving stronger restraint has a longer warning time t_{wn} as defined in Sec.5.2.

ACKNOWLEDGEMENT

The work represents part of an investigation on "Mitigation of Progressive Collapse of Tall Buildings", sponsored by both Nanyang Technological University research account RGM 28/03 and the Ministry of Education of Singapore grant ARC 5/03. The authors would like to express their appreciation to the university and the ministry.

REFERENCES

Bradford M.A. 2006. Elastic analysis of straight members at elevated temperatures, *Adv Struct. Engng*, 9(5):611–618.
European Committee for Standardization (CEN) 2000. "Design of Steel Structures: Part 1.2: General Rules Structural Fire Design (*EC3 Pt.1.2*)", *Eurocode 3*, British Standards Institution, London, U.K.
Horne M.R. & Morris L.J. 1981. *Plastic Design of Low-rise Frames*, Granada, London, U.K.
Liu T.C.H., Fahad M.K. & Davis J.M. 2002. Experimental investigation of behaviour of axially restrained steel beams in fire, *J. Construct. Steel Res.*; 58(9):1211–1230.
Tan K.H, Ting S.K. & Huang Z.F. 2002. Visco-Elasto-plastic analysis of steel frames in fire, *J. Struct. Engng, ASCE*; 128(1):105–114.
Yin Y.Z. & Wang Y.C. 2005. Analysis of catenary action in steel beams using a simplified hand calculation method, Part 1: theory and validation for uniform temperature distribution, *J. Construct. Steel Res.*; 61:183–211.
Wong M.B. 2005. Modelling of axial restraints for limiting temperature calculation of steel members in fire, *J. Construct. Steel Res.*, 61:675–687.

Steel and Composite Structures – Wang & Choi (eds)
© *2007 Taylor & Francis Group, London, ISBN 978-0-415-45141-3*

Experimental and numerical investigation on composite floor cellular steel beams in fire

A. Nadjai, N. Goodfellow, D. Talamona & F. Ali
University of Ulster, Belfast, UK

C.G. Bailey & B.M. Siamak
University of Manchester, Manchester, UK

ABSTRACT: This paper describes an experimental study at elevated temperatures on the behaviour of full scale composite floor cellular steel beams. A total of four specimens, comprising two different steel geometries, applied load ratios and different temperature time curves. All beams were designed for a full shear connection between the steel beam and the concrete flange using shear studs. The beams were designed to fail by web buckling, which was observed in all the tests. A finite element model is then established with both material and geometrical non-linearity using shell elements to compare the experimental results. The comparison between the finite element prediction and actual tests results are quite good in terms of failure modes, load deflection behaviour and ultimate loads.

1 INTRODUCTION

Cellular beams (CBs) are currently being widely used in multi-storey buildings where, as well as reducing the total weight of the steelwork, they help decrease the depth of floors by accommodating pipes, conduits, and ducting. They are also used in commercial and industrial buildings, warehouses, and portal frames. CBs produced by modern automated fabrication processes can be competitive for the construction of both floor and roof systems. The widespread use of CBs as structural members has prompted several investigations into their structural behaviour. Early studies concentrated on in-plane response in both the elastic (Report 1957, Gibson & Jenkins 1957) and plastic (Hosain & Speirs 1973, Sherbourne 1966) ranges. Extensive measurements were made of the stress distributions across the cross-section, and these were compared with the predictions of various theoretical studies employing a Vierendeel analogy (Mandel et al. 1971), finite difference techniques (Mateesco & Mercea, 1981, Redwood, 1978, Shoukry 1965), various finite element schemes (Srimani, & Das 1978), and a complex variable analytical method (Gotoh 1976). As a result of various series of tests a number of different failure modes have been observed (Kerdal & Nethercot 1984, Nethercot & Rockey 1971, Nethercot & Traihair 1976, Okubo & Nethercot 1985). The main failure modes are a Vierendeel collapse mechanism in which plastic hinges form at the section touching the four reentrant corners of a castellation, buckling of a web-post, and web weld failure. Several collapse mechanisms have been proposed (Dougherty 1993, Knowles 1991, Zaarour & Redwood 1996) and the lateral buckling of the web posts has been analysed; however only limited investigations of composite floors using cellular steel beams at ambient temperature have been conducted (Megharief & Redwood 1997). These beams have been used widely in roof and composite construction without having been rigorously investigated under fire conditions.

A composite concrete floor-slab has the effect of significantly increasing the flexural resistance of a steel section; however its effect on shear resistance is more complex. Investigation of the behaviour of composite beams with isolated web openings in otherwise solid webs has shown that the slab significantly increases the shear-carrying capacity beyond that of the steel beam alone. This is due to the enhanced flexural and shear capacity of the upper part of the beam across an opening, although an unsupported web-post is more susceptible to buckling. In fire, the temperature distribution across a composite member is non-uniform, since the web and bottom flange have thin cross-sections and a greater exposed perimeter than the top flange. The deterioration of the material properties of the web will therefore become an important effect on the overall performance of the member in the event of fire.

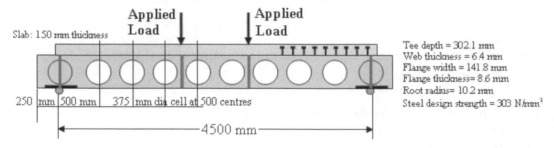

Figure 1. Symmetrical composite Cellular Beam.

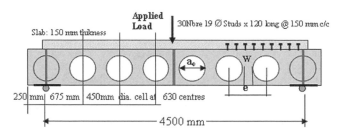

TOP TEE

Tee depth = 287.4 mm; Flange width = 141.8 mm
Web thickness = 6.4 mm; Flange thickness 8.6 mm
Root radius 10.2 mm

BTM TEE

Tee depth = 328.0 mm; Flange width = 152.4 mm
Web thickness = 7.6 mm; Flange thickness= 10.9 mm
Root radius= 10.2 mm; Steel design strength = 319 N/mm²

Figure 2. Detailing of Asymmetric Composite Cellular Beam: 406 × 140UB39 Top & 457 × 152UB52 Bottom Tee.

The fire resistance of CBs has been very controversial in recent years, with most of the debate being concerned with their requirements for intumescent protection. A rather illogical prescriptive "rule" (ASFP 1992) for beams with web openings, requiring 20% extra coating thickness compared with solid-web beams, has recently been subjected to much criticism on the basis of thermal tests on beams coated with specific intumescent products. The previous mentioned document (ASFP 1992) was followed by a most recent publications (RT1006 2004, Bailey 2004, Newman et al. 2006) justifying the traditional prescriptive fire protection rules, based on understanding of the mechanics of structural behaviour in fire.

The experimental and numerical aspects of this project have the potential to provide essential data in several areas currently lacking systematic research. It will underpin the current lead in expertise, which is held by European fire engineering designers, and will assist European-based fabricators who have made cellular beams the most popular long-span system in current construction. The Purpose of these tests is not to evaluate the intumescent performance but to provide data on the structural web post failure temperature. Figures 1 & 2 summarizes the test beams configurations.

2 EXPERIMENTAL TEST PROGRAM

The tests were carried out on six full-scale composite cellular steel beams using span lengths of 4500 mm

subjected to one point load and two-point loading using two types of beams:

a- Symmetrical composite beams (Test A1): The beam for Test A1 has been produced on the basis of UB 406X140X39 as a top tee section and of UB 406X140X39 as a bottom tee section having a finished depth of 575 × 140 CUB 39 kg/m

b- Asymmetric Composite beams (Test B1): The beam for Test B1 has been produced on the basis of UB 406X140X39 as a top tee section and of UB 457X152X52 as a bottom tee section having a finished depth of 630 × 140/152 ACUB 46 kg/m

The first two composite beams were tested initially at ambient temperature, to evaluate and calibrate the systems and equipment. The loading on the other specimens was applied at the start, and maintained during the fire test until failure occurred. Two different fire temperatures-time curves were adopted. Load ratios, 0.3 and 0.5 of the failure load capacity obtained from the ambient experimental investigation was used in the fire tests.

The concrete slabs were all nominally 150 mm thick and 1200 mm wide using normal-weight concrete (Grade 35 N/mm²). The slab reinforcement consisted of welded wire mesh reinforcement A142 having yield strength of 460 N/mm². Full interaction between the slab and beam was ensured in all specimens by the use of a high density of shear connectors of 19 mm diameter studs at height 120 mm. The shear studs have been equally distributed in one row with a distance

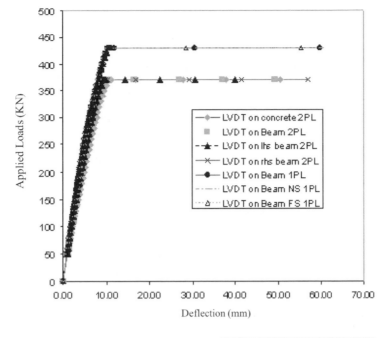

Post web buckling mode failure A1

Post web buckling mode failure B1

Figure 3. Load vs. deflection for Tests A1 and B1 at ambient temperature.

of 150 mm over the beam length. A Holorib sheets HR 51/150 with a thickness of 1.25 mm was used as sheeting. The measured yield stress from a tensile test was $F_y = 327 \, N/mm^2$. Concrete compressive strength was determined at different stages of time: after 2 weeks, 28 days and during the testing days giving an average of 35 N/mm² using a compressive strength calibrated machine at the University of Ulster.

3 TESTING PROCEDURE FOR THE AMBIENT TESTS

Both tests 1A and 1B were conducted using portal frames with a capacity of 160 Tones loading system. The composite beam specimen was simply supported at both ends. A 600 KN hydraulic jack was used to apply the monotonic load. The load was applied to the top concrete flange through distribution beams, as shown in Figures 1 and 2, for two points and one point loading.

For both test A1 and B1 load cycles at a load level of 20% and 60% of the pre-design load were applied to avoid slippage of the load introduction and supports as well as friction in the shear joint and the structure. All the load cells used for the experimental work were calibrated before testing procedure took place. Twenty percent of the pre-design load used by Westok was maintained for one-hour before the load was increased. Each load step with a value of 10 kN/step was kept for 3 minutes intervals Ultimate failure of both specimens was associated with web-post buckling, and study of various measures of buckling load led to the conclusion that this maximum load represented the web-buckling load. Variation of central deflection with load is shown in Figure 3. In each case the buckling mode comprised double curvature bending of the post (see Figure 3). Before this occurred high strains had developed following tensile yield of the lower part of the steel beam. Tensile strains developed above the opening indicating that the neutral axis was close to, or in, the slab.

After the post web buckling took place the beam was then followed by hinges around the openings causing the webs to buckle in an S form of shape. BS 5950: Part 1 specifies that the maximum deflection under factored load for a beam of this type should not exceed span/200. The maximum deflection in both test beams just before failure was 10 mm, which is well within British Standards recommendations. Diagrams of the measurements as well as photographs of the failure mechanism are given in the following figures.

4 EXPERIMENTAL RESULTS TESTS AT ELEVATED TEMPERATURES

The composite cellular steel beam tests A2 and B2 were tested under the same fire slow curve shown in Figure 4. This fire curve was set up in order to produce lower peak temperatures but of longer duration sufficient to permit significant heat conduction, which may produce a large build up of vapour pressure and the creation of significant thermal expansion producing a restraint force coming from the concrete slab. The applied loads for tests A2 and B2 are calculated as 180 KN and 210 KN respectively from 0.5 × failure load obtained from the cold tests A1 and B1. Both beams A2 and B2 were kept loaded to its respective applied load for an hour before the furnace started functioning.

Whereas, the last two composite beams A3 and B3 were tested under the ISO 834 temperature-time curve subjected to 0.3 × failure load obtained from the cold tests given respectively as 108 kN and 126 kN. The positions of the thermocouples were located at each web post along its depth of the section and also around the openings.

Figures 5 and 6 demonstrated that a linear elastic response with almost equal deflections in both beams can be seen in the time deflection curves at 30 minutes time when the temperature is equal to 320°C. Beam A2 starts to show loss in stiffness at time 40 min when the temperature is 432°C while beam B2 still shows no sign of weakness. At 60 min time the deflection is increasing in both beams, and stiffness decreasing at a temperature equal to 634°C. Between 60 and 70 min both beams deflection are becoming larger with less stiffness and strength in the member when the temperature achieved 712°C. After that time the rate of deflection of the beams reduce for a short time due to the presence the reinforced concrete slab on the top

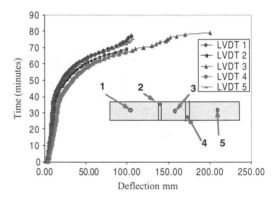

Figure 5. Time versus deflection of test A2.

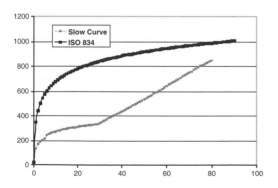

Figure 4. Furnace temperature time curve used in the experimental study.

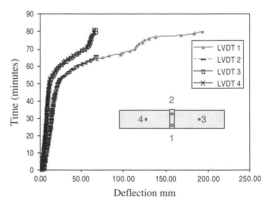

Figure 6. Time versus deflection of test B2.

of the steel members. Final instability of the beams was due to the rapid loss of web post capacity relative to the loss of capacities in the Tee-sections. This failure mechanism is known as web post buckling and a typical form can be seen in Figure 9.

The stiffness in beam A3 (Figure 7) is linear elastic up to 13 minutes (T$_{fur}$ = 717°C) with a reading deflection of 23 mm, whereas B3 (Figure 8) stayed linear up

to 16 minutes (T$_{fur}$ = 748°C) with a deflection equal to 26 mm. At 30 minutes of time (T$_{fur}$ = 842°C) both beams A3 and B3 failed with a recording different respectively 179 mm and 235 mm. The instability of the beams and the rapid loss of stiffness of web post capacity occur faster than the beams A2 and B2 and this is due to the severity of the ISO834 temperature time curve.

Figure 10 demonstrates that the stiffness and strength in beams A2 and B2 are expected to fail at temperatures given respectively 540°C and 600°C. In comparison with the experimental results the beams failed beyond the expected time at temperature equal to 650°C while the furnace temperature was around 740°C. Similarly it can be seen that in beams A3 and B3 the temperatures expectation should fail at 600°C (stiffness) and 660°C (strength) whereas the experimental failure recorded is 770°C as demonstrated in Figure 10. We may conclude that from Figure 10 that the Young modulus decreases quicker than the steel strength limit, which resulted in web-post buckling. The difference between the temperature of the top

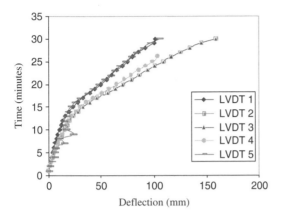

Figure 7. Time versus deflection of test A3.

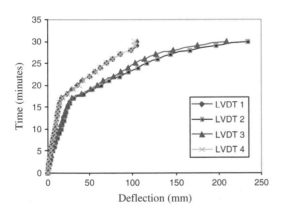

Figure 8. Time versus deflection of test B3.

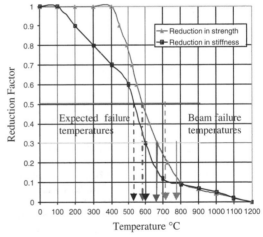

Figure 10. Temperature versus reduction factor.

Figure 9. Web post buckling failure mechanism.

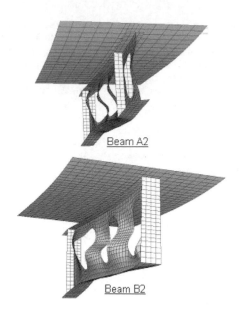

Figure 11. Post web buckling in beams A2 and B2.

and the bottom tee is not relevant for the web-post buckling.

5 FINITE ELEMENT MODEL AND COMPARISONS

The two fire tests on Beam A2 and B2 were modelled using the commercial finite element software ABAQUS (Hibbitt et al. 2004) at the University of Manchester. Shell elements, with the ability to handle large strains, large deformations, and plasticity were used to model the cellular steel beam. A composite shell element, incorporating a smeared crack approach for the concrete, was used to model the composite slab. Both the steel deck, as a bottom layer, and the reinforcing mesh as a layer within the concrete was included with the composite shell element.

Due to the high density of the shear connectors used in the test, full interaction between the beam and supporting composite slab was assumed. This assumption is also justified from test observations (Nadjai 2005) which confirmed that no stud failure occurred before web-post buckling of the beam.

The measured yield and ultimate tensile strength of the steel was used in the simulation (Figure 11).The beam geometry was based on measured dimensions. The stress-strain curve at elevated temperatures for steel given by EN1994-1-2 (ENV 2003) was adopted. For the stress-strain of the concrete in compression, at elevated temperatures, the relationship given in EN1994-1-2 (ENV 2003) was adopted. For concrete in tension, at both ambient and elevated temperatures,

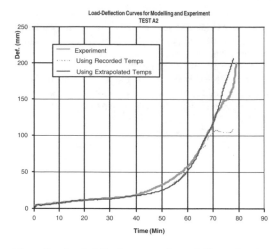

Figure 12. Load-deflection curves for modelling and equipment test A2.

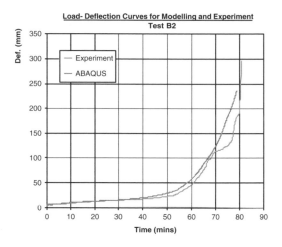

Figure 13. Load-deflection curves for modelling and equipment test B2.

the fracture energy concept (Telford 1990) is used to estimate the ultimate tensile strain of the concrete, as shown in Figure 11. Imperfections were introduced, based on an eigenvalue buckling analysis, with the amplitude of the imperfections being governed by the thickness of the steel plate for local bucking and the overall length of the section for global buckling (Schafar & Pekoz 1998). An implicit analysis was conducted in two steps, where the load was applied in the first step and the temperature was applied in the second step. Figures 12 and 13 show the load-deflection comparison between the test and FEA at elevated temperatures. A very good correlation was obtained between the predictions and test results. In both cases (Figure 11) web post buckling is clearly observed in the modelling corresponding to the failure mode witnessed in the tests.

6 CONCLUSION

This paper describes an experimental and analytical study of the behaviour of composite floor cellular steel beams in fire conditions conducted at the FireSERT, University of Ulster. The study suggests the following:

The two beams failed due to web post buckling and the instability resulted in sudden loss of stiffness and strength in the beams.

The experimental data has compared well with the results from the Finite Element Modelling, giving confidence that it can be used for further parametric studies.

The numerical model is capable to simulate the mechanical behaviour of composite Cellular beam sections in both cold and at elevated temperature conditions with a relatively high accuracy.

ACKNOWLEDGEMENT

The principal author wishes to acknowledge Mr. Mike Hawes (Westok LTD, Wakefield, England WF4 5ER) for the support funding of this research project.

REFERENCES

ASFP/SCI/FTSG 1992. "Fire Protection for Structural Steel in Buildings." Second Edition.

Bailey, C.G. 2004. Indicative Fire Tests to Investigate the Behaviour of Cellular Beams Protected with Intumescent Coatings. Fire Safety Journal. 39, pp 689–709.

Bazile et Texier 1968. Essais de poutres ajourées, revue Construction métallique n°3-1968, CTICM, p.12–25.

Bitar, D. et al. 2005. Steel and non composite cellular beams – Novel approach for design based on experimental studies and numerical investigations, brochure EUROSTEEL.

Delesques, 1968. Stabilité des montants de pouters ajourées, revue Construction métallique n°3-1968, CTICM, p. 26–33.

Dougherty, J. 1993. South African Institution of Civil Engineers, 35 (2), pp 12–20.

ECSC contract 7210-PR-315Large web Openings for service integration in composite floor. 2004.

ENV 1993-1-2:200. 2003. Design of steel and composite structures, Part 1–2: Structural fire design.

Gibson & Jenkins 1957. The Structural Engineer, 35 (12), pp 467–479.

Gotoh 1976. Trans. JSCE, 7, pp 37–38.

Hibbitt et al. 2004 ABAQUS/ Standard Users Manual, Version 6.5.

Hosain & Speirs 1973. J., American Welding Society, Welding Res. Supp., 52 (8), pp 329–342.

Kerdal & Nethercot 1984. J. Construct. Steel Research, 4, pp 295–315.

Knowles 1991. Proc. Instn. Civ. Engrs., Part 1, 90, pp 521–536.

Mandel et al. 1971. J. Structural Division, ASCE, 97 (ST7), pp 1947–1967.

Mateesco, D. et Mercea, G. 1981. Un nouveau type de poutres ajourées, revue Construction Métallique, n°3-1981, CTICM, p.3–14

Megharief & Redwood 1997. Proc. Annual conference, Canadian Society for Civil Engineering, pp 239–248.

Nadjai A. 2005. Performance of Cellular Composite Floor beams at ambient temperatures, Report Nbre: FRC 734/part1.

Nethercot & Rockey 1971. The Structural Engineer, 49 (7), pp 312–330.

Nethercot & Traihair 1976. The Structural Engineer, 54 (6), pp 197–204.

Newman G.M. et al. 2006. Fire Safe design: A New Approach to Multi-Storey Steel-Framed Buildings (Second Edition). SCI Publication P288. The Steel Construction Institute, Ascot.

Okubo & Nethercot 1985. Proc. Instn. Civ. Engrs., Part 2, (79), pp 533–557.

Redwood, R. G. 1978. Analyse et dimensionnement des poutres ayant des ouvertures dans les âmes, revue Construction métallique n°3-1978, CTICM, p.15–27.

Report D.GE. 71/262 1957. "Properties and strength of castellated beams. Consideration of previous tests." The United Steel Companies Ltd, Swinden Laboratories, Rotherham.

RT1006 Version 02 2004. "Fire Design of Cellular Beams with Slender Web Posts." SCI, Ascot.

Schafar B.W. & Pekoz T.1998. Computational modelling of cold-formed steel: characterizing geometric imperfections and residual stresses. Journal of Constructional Steel Research.

Sherbourne 1966. Proc. 2nd Commonwealth Welding Conference, Institute of Welding, London, C2 pp 1–5.

Shoukry, J. 1965. American Welding Society, Welding Res. Supp., 44 (5), pp 231–240.

Srimani, & Das 1978. Computers and Structures, 9, pp 169–174.

Telford T. 1990. CEB-FIP

Zaarour & Redwood 1996. Journal of Structural Engineering, pp 860–866.

Steel and Composite Structures – Wang & Choi (eds)
© 2007 Taylor & Francis Group, London, ISBN 978-0-415-45141-3

Experimental investigation of the behaviour of a steel sub-frame under a natural fire

A. Santiago & L. Simões da Silva
Civil Engineering Department, University of Coimbra, Coimbra, Portugal

P.M.M. Vila Real
Civil Engineering Department, University of Aveiro, Aveiro, Portugal

ABSTRACT: This paper describes an experimental programme under development at the Department of Civil Engineering of the University of Coimbra. The principal aim of this experimental programme is to study the behaviour of a system consisting of steel sub-frame beam-to-column under a natural fire, and to assess the influence of the connection typology on the beam behaviour. This experimental programme is also concerned with the detailed behaviour of each joint component during the heating and cooling phases. The experimental layout is defined by two thermal insulated HEA300 columns and an unprotected IPE300 beam with 5.68 m span, supporting a composite concrete slab. Beam-to-column connections are representative of the most common joint type used on building: welded joints and extended, flush and partial depth plate. Finally, the available results are presented and discussed: evolution of the steel temperature; development of displacements and local deformations and failure modes on the joints zone.

1 INTRODUCTION

Current design codes for fire resistance of structures are based on isolated member tests subjected to standard fire conditions. Such tests do not reflect the behaviour of a complete building under fire conditions. Many aspects of behaviour occur due to the interaction between members and cannot be predicted or observed in tests of isolated elements. Unlike room temperature conditions, joint behaviour can not be adequately represented by a moment-rotation relationship alone. Large variable axial forces are induced in the connection under fire conditions: thermal expansion during the heating phase and to thermal contraction during cooling phase. This axial force, in combination with bending moment and shear force can cause fracture of the tensile joint components (bolts, welds, end-plate, ...).

This paper describes an experimental programme under development at the Department of Civil Engineering of the University of Coimbra. The principal aim of this experimental programme is: *i*) to study the behaviour of a system consisting of steel sub-frame beam-to-column under a natural fire, and to assess the influence of the connection typology on the beam behaviour; *ii*) to study the detailed behaviour of each joint component during the heating and cooling phases.

Six high temperature tests are being conducted. The test arrangement, instrumentation, testing procedure and the resulting behaviour of the performed tests are described in this paper.

2 TESTING ARRANGEMENT

The experimental layout is defined by two thermal insulated HEA300 cross-section columns and an unprotected IPE300 cross-section beam with 5.68 m free span, supporting a composite concrete slab (Figure 1). These dimensions were largely dictated by the steel profile dimensions of the compartment on the 7th Cardington fire test (Wald et al. 2006). The beam cross-section is class 1 at room temperature as well as at elevated temperatures. The slab construction is of steel deck and light-weight in-situ concrete composite floor and is intended to reproduce the thermal boundary condition in typical composite frames. Beam-to-column connection configurations are representative of the most usual joint types used in building frames: header plate; flush and extended end-plate and welded. Columns are simply supported at the top and hinged at the bottom ends. The end fixtures act as hinged bearings. The hinged bearing on the bottom is fixed to a reinforced concrete footing that is secured in position by Dywidag bars passing through

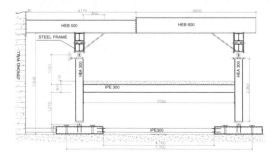

Figure 1. General layout.

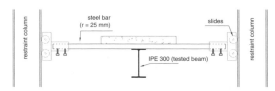

Figure 2. Lateral restraint system.

the laboratory strong floor and fixed horizontally using a steel profile connected between both reinforced concrete footings. To restrain the reaction frames at the top, HEB 600 and HEB 500 cross-section beams were connected to the strong wall. In order to avoid rotation at the top of the reaction frames and consequently mistakes on the measurements, struts were positioned between the top of the reaction frames and the top beam.

Lateral movement of the beam was not allowed. The beam top flange was restrained at three points: i) at mid-span; ii) at 1500 mm to the left side from the mid-span, and iii) at 1500 mm to the right side from the mid-span. The restraint system is able to slide vertically as shown in Figure 2.

3 PROGRAMME AND TEST PROCEDURE

Six high temperature tests are being conducted. An overview of the test programme is presented in Table 1. Tests FJ01, FJ02, EJ01 and HJ01 are already performed and tests FJ03 and WJ01 are under preparation.

These tests are transient temperature tests or nonsteady state tests. The testing programme was characterized in two different and sequential steps: step 1: the mechanical load is applied instantaneously and measurements are recorded. The load is frozen for the remaining part of the test; step 2: after applied the mechanical load, the heating unit is switched on. The mechanical loading is maintained constant and the thermal load is incremented according to the prescribed strategy.

Table 1. Test programme.

Test ID	Joint typology	End-plate dimensions	Bolts/Weld
FJ01	Flush end-plate	$(320 \times 200 \times 10)$; S275	2 bolt row M20, 8.8
FJ02		$(320 \times 200 \times 16)$; S275	2 bolt row M20, 10.9
FJ03		$(320 \times 200 \times 16)$; S275	2 bolt row M20, 8.8
EJ01	Extended end-plate	$(385 \times 200 \times 16)$; S275	3 bolt row M20, 8.8
HJ01	Header plate	$(260 \times 150 \times 8)$; S275	4 bolt row M20, 8.8
WJ01	Welded joint	——————	$a_f = a_w = 8$ mm

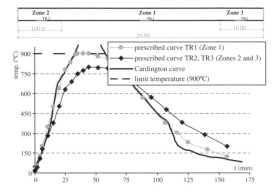

Figure 3. Thermal loading.

4 LOADING DEFINITION

4.1 Mechanical loading

The mechanical loading is applied at two points of the beam top flange, 700 mm to each side from the beam mid-span. Each concentrated load is equal to 20 kN, approximately, which corresponds to a joint hogging bending moment equivalent to 20% of the beam resistance. This mechanical loading was applied using two pairs of concrete blocks.

4.2 Thermal loading and exhaustion system

Thermal loading is time dependent (heating and cooling phases) and is also variable along the beam span. The tested beam is divided into three independent heating zones: zone 2 (central zone) and zones 1 and 3 (end zones) (Fig. 3). The beam temperature-time curves applied at each beam zone reproduce the values measured in a previous full-scale test (Wald et al. 2006). The first 10 min. of the full-scale fire are neglected because the corresponding temperatures are very low and difficult to reproduce;

Figure 4. (a) Heating system; (b) Temperature control system.

Figure 5. General layout, including the exhaustion system.

for safety reasons the maximum temperature applied in the tests is 900°C at the beam bottom flange (35 min < time < 50 min). In order to prevent global structural instability, columns are thermally protected by 30 mm of ceramic wood ($\lambda = 0.06$ W/mK (200°C); $\lambda = 0.27$ W/mK (1000°C)).

To apply the fire load, a special heating system was developed. This heating system is constituted by 11 individual gas burners distributed along each side of the beam length. The burners are fitted by external control continuous valves that control each zone individually and allow specifying the range of flames intensity and thus archiving the thermal load strategy (Fig. 4). At the same time, reference thermocouples at each zone were installed to measure instantaneous temperature inside the steel. The control of the gas delivered to the system is adjusted comparing the thermal load strategy with the instantaneous temperature at each reference thermocouple.

The burners are fed by propane gas through flexible copper pipes (to allow adjustments at the support structure) and are supplied by a battery of gas reservoirs located on the outside of the Laboratory. Propane gas allows a definition of a yellow flame that characterizes urban fires.

In order to allow the exhaustion of the smoke and combustion gases, the system presented in Figure 5 is used. This exhaustion system, fixed on the reaction frame, is constituted by a semi-circular steel sheet around the top of the composite slab and closed on the ends. This semi-circular steel sheet drives the combustion gases through flexible steel pipes to a ventilator that force out these gases to the outside of Laboratory through an opening on the roof. In test situation, the ventilator allow a flows of 10460 m³/h.

Gas temperature control is particularly difficult. In order to reduce the heat losses of at the vicinity of the beam, rock wood panels are fixed vertically from the steel sheet to the floor. The internal face of the rock wood is aluminized to reflect radiation. This way, the beam heats up not only because of direct incidence of flames but also through reflected radiation.

5 INSTRUMENTATION

The results are recorded by means of the following instrumentation: thermocouples, displacement transducers and high strain gauges. Thermocouples of K type with two 0.5 mm wire were attached to the specimen in order to monitor the temperature in the connection elements (end-plate and bolts) and the thermal gradient across the beam and column cross-sections. Displacement transducers are used to describe displacements and deformations of the beam cross sections. They measure i) beam deflection: at mid-span and 300 mm from the joints; ii) horizontal movement of the columns: external column flange at the level of the beam axis, external flange at the top and bottom ends; residual displacements of the support structure. In the beam, measurements are made outside the fire zone via refractory glass, with a very low thermal expansion coefficient, and a sheaves system. A pair of 200 mm displacement transducers located at the mid-span of the beam is used to measure the maximum deflection (2 × 200 mm). High strain gauges are placed at the column ends to measure stresses and, consequently, to assess the moments transmitted by the joints to the columns.

6 RESULTS

6.1 Beam temperature versus prescribed fire curves

Figure 6 compares the prescribed fire curves (control system) with the temperatures measured on the beam reference points (TR1, TR2 e TR3). Good agreement is observed.

6.2 Beam temperature

Due to the size limit of this paper, test EJ01 is selected to represent temperature results of the performed tests. Figure 7 depicts the thermal gradient across the beam mid-span cross section on this test. During the heating phase, the web and bottom flange temperatures are quite similar; on cooling, the web temperature decreases faster than the observed at the bottom flange

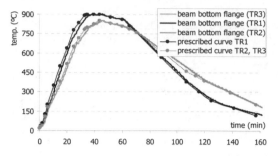

Figure 6. Beam temperature versus prescribed fire curves (test EJ01).

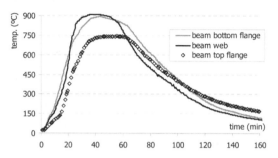

Figure 7. Thermal gradient across the beam mid-span (test EJ01).

Table 2. Temperatures at beam mid-span.

		Temperature			
Test		25 min.	max. temp.	80 min.	150 min.
FJ01	bottom flange	752	900	648	153
	web	773	902	545	119
	top flange	516	762	556	227
FJ02	bottom flange	730	899	647	183
	web	725	889	552	150
	top flange	(a)	(a)	(a)	(a)
EJ01	bottom flange	728	895	604	140
	web	812	909	500	121
	top flange	501	743	567	188
HJ01	bottom flange	753	888	(b)	(b)
	web	753	856	(b)	(b)
	top flange	512	700	(b)	(b)

(a) not measured; (b) beam failure

because during this phase, the flame length decreases, and at a certain instant it only surrounds the bottom flange. The top flange shows the lowest temperature during the heating phase with a maximum temperature at about 743°C; in contrast, cooling down is slower because of shielding by the concrete slab.

Table 2 summarises the thermal gradient across the depth of the beam mid-span. All the tests show similar temperature development during the fire.

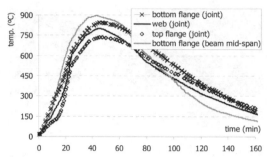

Figure 8. Temperature in the beam near the joint Z3 (test EJ01).

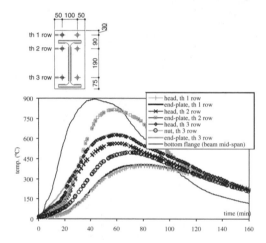

Figure 9. Temperature within the joint Z3 (test EJ01).

6.3 Joint temperature

Figure 8 compares the thermal gradient across the depth of the beam near the joint Z3 with the bottom flange temperature at mid-span (test EJ01). In the heating phase, the joint temperature is significantly lower than the remote bottom flange, which is usually the critical element that defines the limiting temperature of the beam; in contrast, the cooling down in the joint is slower, because of shielding by the adjacent column.

Figure 9 compares the temperature curves in the joint elements (test EJ01). Again, in the heating phase, the joint elements temperature is significantly lower than the remote bottom flange; on the contrary, the cooling down in the joint elements was slower. The first bolt row from the top is significantly cooler than the lower bolts, because of shielding by the adjacent slab and column. The end-plate temperature is quite similar to the bolts head at the same level. Exception is made at the level of the second bolt-row; in this case, the flames engage the plate thermocouple more than the bolt head thermocouple and the end-plate is warmed more than the bolt head. This measurement should not

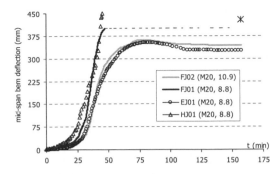

Figure 10. Mid-span deflection of the beams.

Figure 11. Deformed structure after test FJ01.

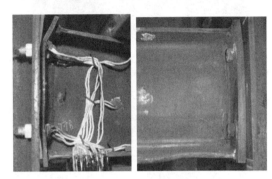

Figure 12. Joint deformation at (a) test FJ01; (b) FJ02.

be considered as representative of the average plate temperature at this level. Temperature gradient along the bolts is also measured; Figure 9 shows that the maximum temperature in the head of the third bolt-row is about 230°C higher than the corresponding nut.

It is also evident the effect of the heat transfer by conduction on the joint element; first the beam reaches the highest temperature while the joint elements only reaches it already during the prescribed cooling phase: first the bolts heads and plate at the level of the second and third rows, afterwards the corresponding nuts and finally, the bolt and plate at the level of the first row.

6.4 Structure deformation

Figure 10 compares the evolution of the beams mid-span deflections during the fire. Most of them are able to sustain the load with reduced deflection up to 10 minutes ($T_{bbf}{}^1 < 150°C$), during this stage, the deflection is mainly due to the mechanical loading and accommodate deformation on the beam. After that, due to the loss of stiffness, the mid-span deflection increases gradually. Beyond 20 minutes, a further rise in temperature ($T_{bbf} > 550°C$) leads to a progressive run-way of beam deflection as the loss of stiffness and strength accelerates. In the case of the FJ02 and EJ01 test beams, a maximum deflection of 350 mm is approximately reached (these values are measured already during the cooling phase). For the PJ01 test, Z3 joint was broken during the heating phase of the fire ($T_{bbf} = 900°C$) as a result of the run-way deflection at high temperatures ($\delta_{beam} = 393$ mm). Once the cooling phase starts, the heated beams begin to recover strength and stiffness from an inelastic state, together with a reduction of thermal strains. This induces tensile axial forces and the reversal of the deflection. Because of the limited range of the displacement transducers (400 mm), FJ01 curve is incomplete; however, a maximum deflection of 428 mm was measured at the end of the fire. Figure 11 shows the deformed structure at the end of the FJ01 test.

---

[1] Temperature of the beam bottom flange.

6.5 Failure modes

The main failure modes observed in the joints of the tested structures are:

FJ01 – Deformation of the end-plate occurred, particularly around the bolts, accompanied by local buckling of the beam bottom flange and shear buckling of the beam web (Fig. 12a). The deformation at the top of the end-plate is mainly observed during the heating phase while deformation at the bottom is developed during the cooling phase, due to the tensile force in this zone. This failure mode is not surprising due to the relative dimensions of the end-plate.

FJ02 – Failure modes are similar to those observed for FJ01 test: end-plate deformation accompanied by local buckling on the beam bottom flange and shear buckling of the beam web (Fig. 12b). The end-plate deformation is smaller than for test FJ01, because the end-plate is thicker.

EJ01 – During the heating phase, due to the end-plate thickness and the connection configuration (three bolt rows), no significant deformation at the connection elements is developed, it is only observed deformations at the beam: local buckling on the beam bottom flange and shear buckling of the beam web; however, during the cooling phase and due to the large tensile forces developed during this phase, localized deformation at the bottom of the end-plate and bolt failure at the bottom bolt-row is observed (Fig. 13).

Figure 13. (a) Joint deformation and (b) nut and bolt stripping – bottom bolt row at test EJ01.

Figure 14. (a) End-plate failure and (b) beam failure at test HJ01.

HJ01 – During the heating phase, some local buckling on the bottom flange is observed; shear failure at the beam web is insignificant. At the maximum joint temperature (850°C), the end-plate was broken along both beam web welds (joint Z3), due to the considerable rotation before the beam and column flange came into contact (Fig. 14a); rapidly the beam suffer a large deflection and shear forces are developed at the joint Z2, leading to beam rupture (Fig. 14b).

Furthermore, others failure modes are observed on the beam and on the concrete slab: shear buckling of the beam web near the load points; large cracks on the concrete slab due to the separation of the shear studs from the concrete slab; major cracks perpendicular to the slab that occurred as a result of the beam and joints deformations. Due to the considerable size of the columns, columns deformations are irrelevant.

7 FINAL REMARKS

This paper describes an experimental programme of fire tests on a steel sub-frame beam-to-column, under development at the University of Coimbra. The experimental programme includes six tests; four of them are already performed and the other two are under development. The testing programme is characterized by the application of the mechanical load at room temperature followed by a heating-cooling curve.

The results show that the heating-system reproduces the prescribed fire curves with high reliability. At the beam mid-span two different situations are observed: during the heating phase, the web and bottom flange temperatures are quite similar; on cooling, the web temperature decreases faster than the observed at the bottom flange. Again, in the heating phase, the joint elements temperature is significantly lower than the remote bottom flange; in contrast, the cooling down is slower. The first bolt row from the top is significantly cooler than the lower bolts. The end-plate temperature is quite similar to the bolts head at the same level.

Measurements also show that the maximum beam mid-span deflection depends of the stiffness and strength of the beam-to-column joint; though this value is about $1/15\ L_{beam}$. These tests demonstrated that the failure modes are squite distinct to those which happened to the same structure at room temperature. Most of these tests produced similar failure modes on the beam near the connections: local buckling on the beam bottom flange and shear buckling of the beam web; however the failure modes on the connection components are largely dependent of it configuration and dimensions. A highlight point evidenced is the developments of high tensile force on the lower zone of the connection due to the thermal contraction that lets to the failure of some joint components which is not stressed on room temperature cases.

ACKNOWLEDGEMENTS

Financial support from the Portuguese Ministry of Science and High Education under FCT research project POCI/ECM/55783/2004 and propane gas from GALP SA are acknowledged.

REFERENCE

Wald, F. & Chladná, M., Moore, D., Santiago, A., Lennon, T. 2006. Temperature distribution in a full-scale steel framed building subject to a natural fire. *International Journal of Steel and Composite Structures*, vol. **6**(2), pp. 159–182.

Steel and Composite Structures – Wang & Choi (eds)
© 2007 Taylor & Francis Group, London, ISBN 978-0-415-45141-3

The behaviour of single-storey industrial steel frames in fire

Y. Song, Z. Huang & I.W. Burgess
Department of Civil and Structural Engineering, University of Sheffield, Sheffield, UK

R.J. Plank
School of Architectural Studies, University of Sheffield, UK

ABSTRACT: In this paper a simplified dynamic procedure for structural analysis of steel portal frame at elevated temperatures is presented. The procedure takes into account both geometric and material nonlinearities. The model is validated against other software using fully dynamic procedures, and reasonable agreements are achieved. A detailed case study is conducted to investigate the failure mechanism of a portal frame subject to different base fixity conditions in fire.

1 INTRODUCTION

The pitched portal frame is a typical single-storey steel structure widely used for industrial applications in the UK. For ultimate limit state design at ambient temperature, such frames are usually designed assuming that column base connections act as frictionless hinges.

In fire conditions, it is imperative that boundary walls stay close to vertical, so that fire is not allowed to spread to adjacent property. A current UK fire design guide (Newman 1990) allows the collapse of the rafters, which are normally unprotected against fire. To ensure the lateral stability of boundary walls, column bases and foundations have to be designed to resist the forces and moments generated by the collapse of the rafters. However, the method makes a number of apparently arbitrary assumptions and does not attempt to model the true behaviour of the frame during fire. Therefore the method can lead to very uneconomical design details.

A fundamental aspect of portal frame rafter collapse is that they often lose stability in a "snap-through" mechanism, which is capable of re-stabilising at high deflection, when the roof has inverted but the columns remain close to vertical. By most static analysis methods only the initial loss of stability is identifiable. In a model-scale test (Wong 2001) on a single-storey single-bay portal frame, all the static modelling stopped at the point where the snap-though occurred. The maximum vertical displacement at the apex in a numerical analysis was about 17% of the roof depth, while the deflection of the apex almost reached the eaves level in the actual fire test, so it was not possible to identify the final equilibrium state or the intermediate column movements.

In this research a new simplified dynamic solution procedure has been developed in which both damping and inertia effects have been added to the force equilibriums and an effective stiffness matrix, which always has positive diagonal values, is generated. The procedure can deal with the instabilities encountered in previous static analyses, and the snap-through behaviour of portal frames at elevated temperatures can be properly investigated.

The main objective of dynamic analysis is tracing the global structural behaviour after a transient loss of stability, or partial failure, occurs. The vibration introduced by the dynamic behaviour can be damped out in a very short time period by super-critical damping, above the critical damping which can be determined by the natural frequency of the structure (see Fig. 1).

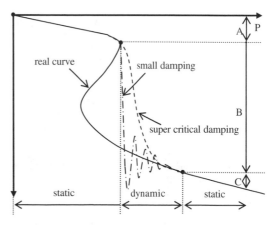

Figure 1. Effect of damping in snap-through process.

The exact nature of the dynamic motion during period B is not very important for this study. The limit point at the end of period A and the behaviour beyond period B are the main considerations in this research, as statically stable regions. It is the ability of the dynamic system to cross the region of temporary instability that is important.

The true static snap-through behaviour should follow the black line in Fig. 1 when nonlinearity is considered. However, in load-controlled numerical analysis this behaviour cannot be traced beyond the initial limit-point because the equilibrium becomes unstable. Instead of tracing the static deformation curve, the dynamic motion can be used to allow the analysis to find the second stable state C, as shown in Fig. 1. Damping is used to ensure that the analysis can go back to a static one when the motion is sufficiently slow to assume that the structure has reached a stable position.

2 DYNAMIC MODEL

In static analysis, the internal forces balances the external forces to achieve equilibrium, so the governing equilibrium equation is:

$$\mathbf{F}^{Int}(t) = \mathbf{F}^{Ext}(t) \tag{1}$$

where $\mathbf{F}^{Int}(t)$ is the internal nodal force vector, and $\mathbf{F}^{Ext}(t)$ is the external nodal force vector. This nonlinear equation is solved by the Newton-Raphson iterative solution method. The iterations are based on Equations (2) and (3):

$$\overline{\mathbf{g}}_{t+\Delta t} = \mathbf{F}^{Int}(t+\Delta t) - \mathbf{F}^{Ext}(t+\Delta t) \tag{2}$$

$$\overline{\mathbf{g}}_{t+\Delta t,n+1} = \overline{\mathbf{g}}_{t+\Delta t,n} + \frac{\partial \overline{\mathbf{g}}}{\partial \mathbf{U}} \delta \mathbf{U}_{n+1} = 0 \tag{3}$$

where $\overline{\mathbf{g}}_{t+\Delta t,n}$ is the unbalanced force at time-step $t + \Delta t$ and the nth iteration. If a snap-through is encountered in an analysis, some negative or null values will appear on the leading diagonal of the stiffness matrix which leads to divergence of the analysis.

In fact, the snap-through is a dynamic behaviour between two critical points. When the snap-through finishes, the whole structure returns to stable static equilibrium provided that the structure is loaded in a low rate.

The Newmark integration scheme for nonlinear dynamic problems (Bathe 1982, Crisfield 1998) is adopted due to its significant advantage in numerical computing efficiency. Because the behaviour of portal frames within Stage B (see Fig. 1) is quasi-static, it is not necessary to carry out a full dynamic solution procedure which needs more computing time but shows the same static results. In this model, if singular values on the diagonal of the stiffness matrix are detected the dynamic analysis is then activated.

For dynamic analysis, the inertia and damping terms are added to the left-hand side of equation (1):

$$\mathbf{F}^{I}(t) + \mathbf{F}^{D}(t) + \mathbf{F}^{Int}(t) = \mathbf{F}^{Ext}(t) \tag{4}$$

The dynamic unbalanced force at step $t + \Delta t$ turns into equation (5):

$$\overline{\mathbf{g}}_{t+\Delta t} = \mathbf{F}^{Int}(t+\Delta t) - \mathbf{F}^{Ext}(t+\Delta t) + \mathbf{F}^{I}(t) + \mathbf{F}^{D}(t) \tag{5}$$

where $\mathbf{F}^{I}(t)$ is the inertial force and $\mathbf{F}^{D}(t)$ is the damping force. When negative values appear on the diagonal of the stiffness matrix, the damping and inertial forces are added, to generate an effective stiffness matrix which always has positive diagonal values. This is very helpful to the Newton-Raphson solver which can still solve for the incremental displacement as the quotient of the dynamic unbalanced force, dividing the effective stiffness when partial failure of the structure happens.

To restrain the vibration of the structure about equilibrium states in fire, a high damping in excess of the system's critical damping is required, so numerical damping, which only gives convergence at relatively small damping values, is not ideal for this model. Because the exact path of the motion is not very important in this case, the mass and damping are assumed to be lumped at the nodes and the critical damping, relative to the lowest natural frequency and mass, is adopted in the model. The procedure discussed here has been incorporated into *Vulcan* for modelling of steel structures in fire.

3 VALIDATIONS

3.1 *Pitched frame*

To test the capability of the dynamic model in dealing with geometrically nonlinear problems, a simplified pitched frame (Fig. 2) with an increasing point load was tested and the results compared with the commercial software *Abaqus*. Purely elastic material properties were assumed in this case, in order that the geometrical nonlinearity of the roof frame could be presented and the integrated snap-through of the roof frame could be traced.

For this ideal symmetrical case, a bifurcation is encountered under completely symmetric loading. This causes a problem, even for dynamic analysis, in breaking the inherent symmetry, and the motion must be given the facility to move in an asymmetric fashion by introducing a small imperfection, which in this case is a very small additional load on one rafter. Beyond the limit-point the frame collapses dynamically until it reaches its almost-inverted position.

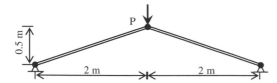

Figure 2. The initial configuration of the pitched roof frame.

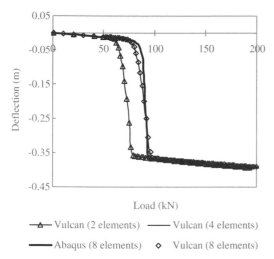

Figure 3. Vertical displacements of the apex.

Because only two Gauss points control integration on each beam element, the accuracy of the result largely relies on the element mesh density. According to the parametric study, it is found that after the deflection jump the load-deflection curves of models with different element mesh density always tend towards the same stable static curve where the stiffness of the frame recovers due to the generation of tension in the members. A four-element (per member) mesh is adequate for convergence of the results, and the position of the limit points is very close to the full dynamic results from *Abaqus*. It is evident that this new model is capable of capturing the new stable position after snap-through of the pitched roof frame in a load-controlled analysis.

3.2 *Double-bay single-storey steel portal frame*

A test was carried out on a double-bay single-storey steel portal frame, as shown in Fig. 4, which was designed by Franssen and Gens (2004). This was modelled using 61 geometrically nonlinear beam-column elements. The two-dimensional analyses of this model have been compared with the results generated from several finite element softwares, such as *Safir Dynamic*, *Abaqus Dynamic* and *Ansys Dynamic* (Vassart *et al.* 2005). Figs. 5(a) to 5(g) show the comparisons. It is evident that very good agreement was

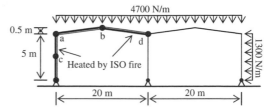

Figure 4. Initial configuration and loading arrangement.

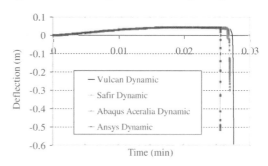

Figure 5(a). Vertical displacement at node a.

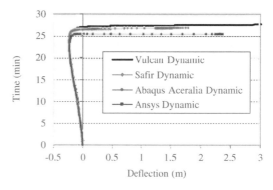

Figure 5(b). Horizontal displacement at node a.

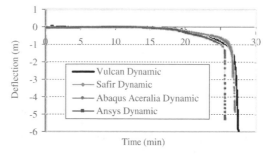

Figure 5(c). Vertical displacement at node b.

achieved between the current model and the other software.

The half-frame on the right kept almost its original shape when the left-hand locally heated part had

689

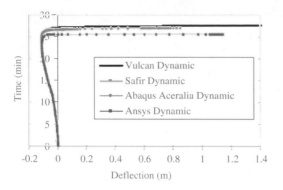

Figure 5(d). Horizontal displacement at node b.

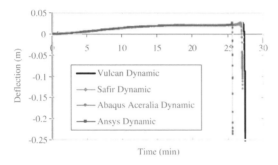

Figure 5(e). Vertical displacement at node c.

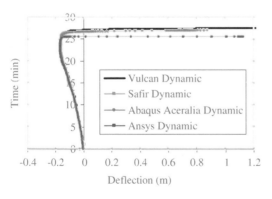

Figure 5(f). Horizontal displacement at node c.

collapsed to column-base level. The collapsed shape of the frame is shown in Fig. 6. This also attests that the current dynamic model can handle partial or local failure conditions in steel structures.

4   CASE STUDY

According to previous research conducted at the University of Sheffield (Wong, 2001), the static analysis

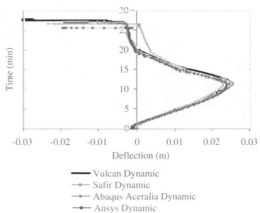

Figure 5(g). Horizontal displacement at node d.

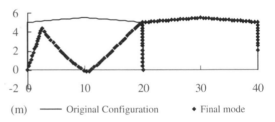

Figure 6. Deflection shape of the frame.

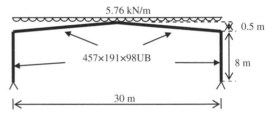

Figure 7. Numerical model of the pitched portal frame.

in *Vulcan* has shown very good accuracy in comparison with experimental fire test results up to the large deflection jump which happens at the limit-load. This study is designed to trace the post-failure behaviour of the portal frame and explore the failure mechanism under fire conditions. A two-dimensional portal frame model was designed as shown in Fig. 7. For convenience in modelling, the whole frame was assumed to have the same section size, and was designed using plastic theory at ambient temperature. The design loadings are listed in Table 1, and these loads were applied to the frame before the heating. It was assumed that the whole frame was exposed in fire, and a uniform temperature distribution across the member sections was used. The distance between centres of adjacent portal frames was 6 m, and a horizontal imperfection force

Table 1. Load combinations.

Load Type	Unfactored load kN/m²	Ambient Load factors	Fire load factors
Dead load	0.66	1.4	1.0
Imposed load	0.60	1.6	0.5

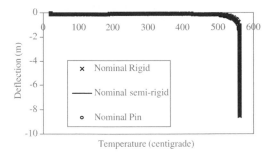

Figure 8(a). Vertical displacement of apex.

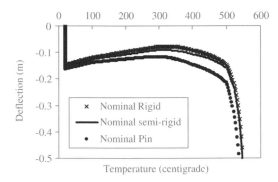

Figure 8(b). Vertical displacement of apex between 0 to 0.5 m.

of 1.7 kN, was added to the left eaves connection. The temperature of the steel used in this test was calculated according to Eurocode 3.

A semi-rigid column base was assumed, and modelled by a spring with a rotational stiffness which was calculated by the rule recommended in the SCI document (Salter *et al.* 2004). For a nominally rigid base the stiffness of this spring is taken to equal $4EI_{column}/L_{column}$, and for a nominally pinned base the stiffness of the spring equals $0.4EI_{column}/L_{column}$, where E is the Young's modulus, I_{column} is the second moment of area of the column section and L_{column} is the length of the column. A mean value of rotational stiffness between the nominally pinned and rigid bases was defined for a semi-rigid base.

The vertical deflection of the apex and the horizontal deflections of the eaves are presented in Figs. 8 to 10. It is clear that the failure modes are very similar, and the difference in failure temperatures between

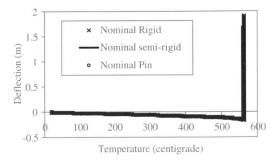

Figure 9. Horizontal displacement of left eave.

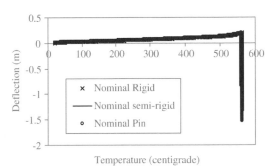

Figure 10. Horizontal displacement of the right eave.

the defined cases was not more than 2°C. The base rotational stiffness clearly does not show a significant influence on the behaviour of the portal frame for this heating profile.

In this case study the frame was loaded at ambient temperature first and then uniformly heated. Due to the thermal expansion of the steel, the apex deflected upward until about 330°C. With the elevated temperature and the original loading, the roof frame began to lose stiffness and pushed the columns outward. The maximum outward deflection of the eaves was reached when the roof frame had fallen to eave level at around 560°C. At this moment, only the two sections near the eaves had a flange in full plasticity. Beyond this point the apex deflected below the eaves level and the column tops were pulled inward. After a further temperature increment of about 2°C, one fire hinge developed at each eaves, which provided more flexibility. Near to the apex, the strains in the lower part of the rafter cross-section exceeded the plastic limit strain when the displacement of the apex was over half the original height of the frame. The plastic region expanded to about one fifth of the length of the rafters when the apex had deflected to base level, and because of the imperfection force at the left eaves, the plastic regions were mainly in the left rafter. It is worth noting that at this stage more than half of the cross-section at the Gauss point closest to the left-hand column base

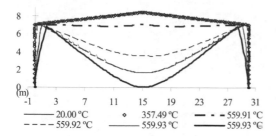

Figure 11. Deflection shape of the nominal pin based single portal frame at different temperatures.

reached its plastic strength. Fig. 11 presents the whole progress of collapse of the frame.

The phenomenon of re-stabilisation after snap-through was not detected during these analyses because, when the plastic hinges were generated at the eaves, the columns were too weak to provide enough restraint to the roof frame in this case. However, in other cases, especially where column sections are stronger and stiffer than the rafters and where the rafters become much hotter in fire than columns, this will undoubtedly happen. In the next phase of this work such cases will be examined.

5 CONCLUSION

The following conclusions can be drawn from this study:

• The simplified dynamic procedure developed for this study is suitable to deal with the temporary instabilities encountered in numerical analysis of structures.
• The current model can be used to predict progressive collapse of portal frames and other steel frames in fire accurately.

• For the simplified portal frame case study presented here, the rotational stiffness and strength of column bases has no significant effect on the structural behaviour when the whole frame is uniformly heated.
• The restraint provided by the columns is important to the snap-through mechanism of the roof frame. Hence, some parametric studies considering different heating profiles and column sizes are being carried out. The results will be reported soon.

REFERENCES

Newman, G. M. 1990, The behaviour of steel portal frames in boundary conditions. *Fire and Steel Construction: The Behaviour of Steel Portal Frames in Boundary Conditions 2nd eds)*. The Steel Construction Institute.

Wong, S. Y. 2001, *The structural response of industrial portal frame structures in fire*. PhD Thesis. University of Sheffield.

Bathe, K. J. 1996, *Finite Element Procedures*. New Jersey: Prentice-Hall.

Crisfield, M. A. 1998, *Non-linear Finite Element Analysis of Solids and Structures, Volume 2: Advanced Topics,* John Wiley & Sons Ltd., Chichester.

Franssen, J. M. & Gens, F., 2004, Dynamic analysis used to cope with partial and temporary failures, *Third International Workshop – Structures in Fire:* Paper S6–3

Vassart, O., Brasseur,M., Cajot, L. G., Obiala, R., Spasov, Y., Friffin, A., Renaud, C., Zhao,B., Arce, C. & De La Quintana, J. 2005, *Final Report: Fire Safety of Industrial Halls and Low-rise Buildings.*

Salter, P. R., Malik, A. S. & King, C. M. 2004, *Design of single-span steel portal frames to BS 5950–1:2000 (2004 Edition)*, SCI Publication P252, Ascot: The Steel Construction Institute.

Steel and Composite Structures – Wang & Choi (eds)
© *2007 Taylor & Francis Group, London, ISBN 978-0-415-45141-3*

Fire tests on industrial building

F. Wald, M. Pultar & J. Chlouba
Czech Technical University, Prague, Czech Republic

ABSTRACT: The paper presents the results of the fire test on an industrial building before demolition in steel plant in city Ostrava and its comparisons to Eurocode 3 design models. Two fire tests were performed; local fire and compartment fire. The local fire was concentrated to heating of a column and to the temperature distribution in close structure. The compartment fire was focussed on the temperature distribution into the connections during heating and during cooling of the structure; on the internal forces in columns due to the elongation and shortening of an unprotected floor; on the behaviour of steel beams; on the heating of the unprotected front beam; and on heating of the external beam outside the fire compartment. The models of two timber panels, two timber and concrete composite elements and two sandwich panels were exposed to natural fire in the fire compartment as well. The prediction of the temperature by parametric fire curve and of the development of the temperature in the beams by step by step method show a good accuracy. The model of the transfer of heat into the external element was found conservative compare to the measured temperatures.

1 INTRODUCTION

The tests of separate structural elements, e.g. beams, columns, and joints in furnaces allowed to prepare the prediction design models of elements, see (Buchanan, 2003). The main aim of the fire test at structure of the Ammoniac Separator II in company Mittal Steel Ostrava, see Figure 1, was to learn more about the connection temperatures and the internal forces into structure. The behaviour of restrained beams during compartment fire at elevated temperatures, the heating of external element as well as column during local fire and the temperature of sandwich panels, light timber based panels and timber concrete element was studied under the heating by natural fire as well.

Simplified design of structure in fire is based on the design of structure at ambient temperature. The advanced design takes into account the structure loaded by a realistic temperature fire curve and the joints are exposed to forces caused by the elongation during the warming and by the contraction during the cooling faze as well which may be evaluated by the natural fire test only, see (Wald, 2006).

Structural elements outside the fire compartment are heated during the fire by radiation from the flames in the openings of the fire compartment only. The amount of heat transferred from the fire is affected by the position against the opening except of the intensity of the fire. The transfer of heat outside the fire compartment may be modelled by FEM or by energy based analytical prediction models on more levels of sophistication, see (Wang, 2002). The analytical

Figure 1. The building before demolition.

prediction is based on the gas temperature in the window, which may be evaluated by fluid dynamic models (CFD), zone models, by parametrical curves, or nominal ones with an asked accuracy. The presented work concludes, that the good prediction of the gas temperature by parametric fire curve enable to estimate the temperature of the external steelwork with asked accuracy, see (Law, 1987).

2 DESCRIPTION OF THE TEST

2.1 *Steel frame*

The structure for the test is three storey industrial building attached to single storey framed building, see Figure 1 (Kallerová & Wald, 2006). The load-bearing

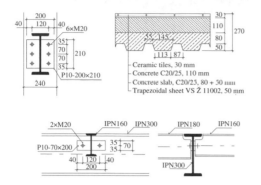

Figure 2. The concrete slab, beam to beam and beam to column connections.

Figure 4. The fire load for the compartment fire test.

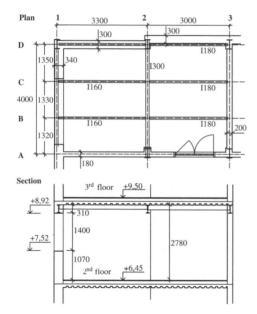

Figure 3. Dimensions of the fire compartment.

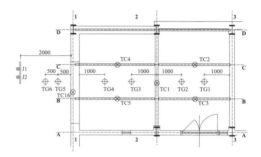

Figure 5. Location of thermocouples on first floor of column D2.

The steel columns were partially encased in the walls and only the flange was exposed to fire. During the fire test, the exposed column flanges were protected by fibre-silicate boards. The beams and connections were not protected during the test.

Mechanical load was introduced on the third floor. The total load, including self weight of the structure, was 5.7 kN/m². Wooden cribs created the fire load, 60 kg/m² e.g. 1060 MJ/m², by the compartment fire test, see Figure 4. 170 kg located on area 1 × 1 m was used for the local fire.

2.3 Measuring devices

Thermocouples, strain gauges and displacement transducers recorded the behaviour of the structure during the fire test. In total there were 42 thermocouples to measure the temperature of the air in the fire compartment, of the structural elements and beam connections, see Figure 5. Vertical deformations were measured by five transducers located on third floor at mid-span of the primary and secondary beams. In addition, the relative horizontal displacements of the columns A2-D2, D1-D2 a D2-D3 were observed by three transducers. Strains in the columns were measured by 16 strain gauges attached to the column flanges on first and third

structure consists of steel columns and beams supporting concrete slab thickness 130 mm, including the ribs. No shear connection between steel beams and the slab was installed. The beam-to-column and beam-to-beam connections are designed as simple end plate connections using two or six bolts M20, see Figure 2.

2.2 Fire compartment

The fire compartment sized 3.80 × 5.95 m height 2.78 m was located on 2nd floor. The walls were made from hollow ceramic bricks except the front wall which was made from light weight concrete. One window of width 2.40 m and height 1.40 m was located in the front wall, see Figure 3.

694

Figure 6. Local fire test in 15 min of fire.

Figure 7. Compartment fire with the element outside the compartment.

floors. The fire and smoke development was recorded by five video cameras and one thermo imaging camera.

2.4 *Fire tests*

Two fire tests were performed in the compartment. The local fire test performed June 15, 2006 was designed to measure the temperature in steel beams and column close to the fire. The fire was located in the middle of the compartment just below the primary beam A2-D2, see Figure 6. A column was erected in the middle of the compartment. It did not support any load but was used only for measurement of the temperature along its length.

The compartment test was designed to obtain the gas temperature, temperature of the structure including the joints, the tie forces and temperature of steel structure in front of the compartment window, see Figure 7.

Figure 8. Beams after the compartment fire test.

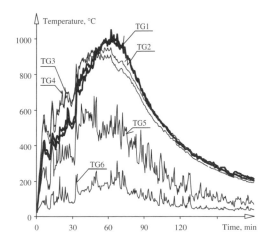

Figure 9. Measured gas temperatures.

No collapse occurred during the tests, however, deformations and lateral-torsional instability of the beams was observed, see Figure 8.

3 GAS AND STEEL TEMPERATURES

3.1 *Measured temperatures*

Figure 9 shows the gas, beam in the mid span and connection temperatures measured during the experiment. On Figure 10 are demonstrated the temperature differences between the beams.

3.2 *Predicted temperatures*

Figure 12 shows the comparison of the predicted gas temperature in the fire compartment by parametric fire curve according to EN 1993-1-2.2005 Annex A, see (Kallerová & Wald, 2006) to the measured values. The parametrical curve is derived for values $b = 1059\,\mathrm{Jm^{-2}s^{-1/2}K^{-1}}$; $O = 0.0398\,\mathrm{m^{1/2}}$; $\Gamma = 1.188$; $q_{fd} = 238\,\mathrm{MJ\,m^{-2}}$; $t^*_{max} = 1.413$ hod.

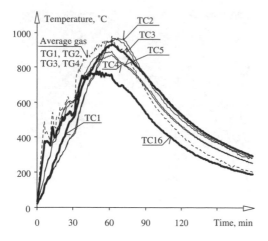

Figure 10. Measured temperatures on the top of the beam lower flanges at mid span.

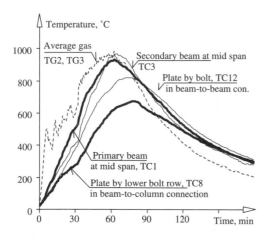

Figure 11. Comparison of the measured temperatures on the beam-to-column and beam-to-beam header plate connection to the gas and beam mid span connection.

In the step by step method, see (EN 1993-1-2. 2005), is the steel temperature calculated by the simple principle, where the heat is brought in/brought out by the member surface and the member volume is heated/cooled. Geometrical characteristic of the section is the section factor A_m/V of the steel parts of which the joint is composed. The section factor indicates the relation between the surface area of the connection A_m per unit of length exposed to the fire and the volume of the connection V per unit length which is being heated. The temperature of the unprotected inner steel structure is given by

$$\Delta\theta_{a,t} = k_{sh}\frac{A_m/V}{c_a\rho_a}\dot{h}_{net}\Delta t \qquad (1)$$

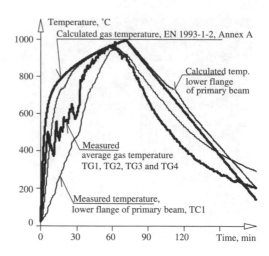

Figure 12. Comparison of the predicted temperature by the parametric fire curve according to EN 1993-1-2.2005 Annex A, see (Kallerová & Wald, 2006) to the measured average gas temperature and primary beam temperature.

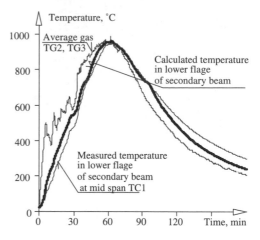

Figure 13. Comparison of the measured temperatures on the header plate connection to the gas and beam mid span connection.

where A_m/V is the section factor in m^{-1}, c_a the specific heat depending on the temperature in $Jkg^{-1}K^{-1}$, ρ_a the volume weight of steel in kgm^{-3}, $\dot{h}_{net}$ the design value of the net heat flux per unit area in Wm^{-2}, Δt the time increment in s, and k_{sh} the correction factor for the shadow effect which is used by the heating using the nominal fire curve. The temperature of the secondary beam was calculated from the measured temperature of the gas, see Figure 13 according to eq. (1).

Figure 14. Location of the external steelwork exposed to the heat during the test.

Figure 15. Position of the plate to the window.

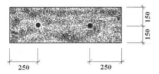

Figure 16. Location of the thermocouples at plate.

4 EXTERNAL STEELWORK

4.1 Measured temperatures

In the experimental fire compartment was one opening of size 2400 × 1400 mm, see Figure 14. In front of the opening was in distance 2000 mm suspended on thermo protected hangers plate 950 × 300 mm of thickness 10 mm. The position of the plate to the window is shown at Figure 15. The steel temperature was measured by two thermocouples placed horizontally in the centre of the plate 250 mm from the edges, see Figures 7 and 16. The comparison of the external plate measured temperatures to the gas measured temperatures during the compartment fire is summarised at Figure 17. Both thermocouples measured similar temperatures. The lower temperatures were recorded by the thermocouple closer to the edge of the opening.

4.2 Predicted temperatures

The fire in the building affects its surrounding by heat of the elements heated by the transfer by radiation.

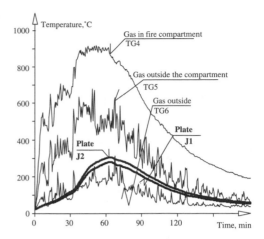

Figure 17. Measured gas and extended steel plate temperature during the experiment.

A distinction should be made between members not engulfed in flame and members engulfed in flame, depending on their locations relative to the openings in the walls of the fire compartment. The heating of the exposed elements depends on the intensity of the heat source, size and position of the openings, in our case the windows only, the position of the element to the source, the emisivity of the elements and its material properties (thermal conductivity) and mass. The description of the effect and basic physical explanation may be found in (Wald at all, 2005). The temperature of the plate exposed to the fire was predicted based on the heat balance, see Annex B in (EN 1993-1-2: 2005), which is for design applied for prediction of the maximal temperature only,

$$\sigma T_m^4 + \alpha T_m = \sum I_z + \sum I_f + 293 \, \alpha \qquad (2)$$

where σ is the Stefan Boltzmann constant, α is the convective heat transfer coefficient, I_z a I_f the radiactive heat flux from the flame and from the corresponding openings respectively. The calculated results are compared to the measured values, see Figure 18.

The heat flux was further conducted in the step by step procedure, see (Pultar & Wald, 2006) based on

$$\phi_{1,2} = \frac{\varphi_{1,2} \, S_1 \, \sigma \left(T_1^4 - T_2^4\right)}{1 + \varphi_{1,2} \left(\frac{1}{\alpha_{e1}} - 1\right) + \varphi_{2,1} \left(\frac{1}{\alpha_{e2}} - 1\right)} +$$

$$+ \lambda \left(T_1^4 - T_2^4\right) \qquad (3)$$

697

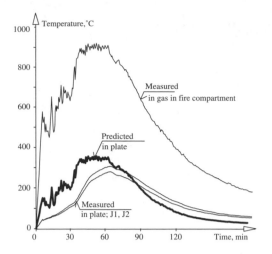

Figure 18. Comparison of the calculated temperature, based on simplified model, eq. (2), to the measured values.

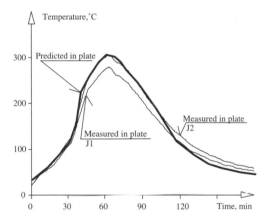

Figure 19. Comparison of the calculated temperature, based on advanced calculation model, see eq. (3), to the measured values.

where index 1 relates to the opening, index 2 to the exposed element, $\phi_{1,2}$ is the resultant radiative heat flux,

$$\varphi_{1,2} = \int_{S_j} \frac{\cos \psi_i \cos \psi_j}{\pi \, r^2} \, dS_j \qquad (4)$$

is the average radiative factor, S_1 is the window area, T_i are the temperatures of the flames in the window and exposed element, α_{ei} is absorptivity of the flames, is λ the heat transfer coefficient under the cooling of the element of temperature T_0. The differential equation derived from eq. (2) for the temperature of the element T_2 was solved numerically.

The final graph, see Figure 19, shows the good accuracy in prediction of the external steelwork temperatures from the gas temperatures by applying the procedure according to eq. (3).

5 CONCLUSION

Basic results according to the gas a steel temperature development during the fire test performed on three storey building are summarised in this paper. The good quality of the prediction of the temperature in the fire compartment and of the beam temperatures was confirmed.

The calculation of the external steelwork temperature from the measured temperature in the fire compartment confirmed the simple calculation model in EN 1993-1-2: 2005. The advanced calculation model gives rather more accurate estimation of the temperature.

ACKNOWLEDGEMENT

The work was prepared with support of Czech Grant Agency GAČR 103/07/1142. This outcome has been achieved with the financial support of the Ministry of Education, Youth and Sports of the Czech Republic, project No. 1M0579, within activities of the CIDEAS research centre.

REFERENCES

Buchanan A. H. 2000. Structural design for fire safety, John Wiley & Sons 2000, ISBN 0-471-89060-X.
EN 1991-1-2: 2002. Eurocode 1, *Basis of design and actions on structures* – Part 2-2: Actions on structures – Actions on structures exposed to fire, CEN, Brussels.
EN 1993-1-2: 2005. Eurocode 3 – *Design of steel structures* – Part 1-2: General Rules – Structural fire design, CEN, Brussels.
Kallerová P. & Wald F. 2006. Ostrava fire test, Czech Technical University, Praha, CIDEAS report No. 3-2-2-4/2, p.18.
Law M., O'Brien T. 1989. *Fire safety of bare external structural steel*, SCI, č. 009, London, 88 s., ISBN 0-86200-026-2.
Pultar M. & Wald F. 2006. *Temperature of external steelwork*, Czech Technical University, Praha, CIDEAS report No. 3-2-2-3/3, p.18.
Wald F. et al. 2005. *Calculation of fire resistance of structures*, in Czech, Czech technical University in Prague, Praha, p. 336, ISBN 80-0103157-8.
Wald, F., Simões da Silva, L., Moore, D.B., Lennon, T., Chladná, M., Santiago, A., Beneš, M. & Borges, L. 2002. Experimental behaviour of a steel structure under natural fire, *Fire Safety Journal*, Volume 41, Issue 7, October 2006, p. 509–522.
Wang Y.C. 2002. *Steel and Composite Structures*, Behaviour and Design for Fire Safety, Spon Press, London, ISBN 0-415-24436-6.

Steel and Composite Structures – Wang & Choi (eds)
© 2007 Taylor & Francis Group, London, ISBN 978-0-415-45141-3

The optimisation of Fire-Engineered Composite Cellular Beams in tall steel structures

P. Toczek

Fabsec Limited, Wakefield, UK

ABSTRACT: This paper presents developments in the design of Fire-Engineered Composite Cellular Beams (FECCBs) using Fabsec's FBEAM®design software and Leighs Paints Firetex®FB120 Intumescent coating. Significant progress has been made in the field of optimisation of FECCBs in tall structures and other major projects. The method of optimisation of multiple beam designs for the ambient and fire condition is explained. Progress in the development of two-way links between FBEAM design software and proprietary analysis and design software is explored. The recent step change that permits optimisation of multiple FECCBs in tall structures is explained. Conclusions relating to the efficient design of FECCBs are also presented with recommendations for future work.

1 INTRODUCTION

Composite design was introduced to the UK from the USA in the 1960s. This method of construction permitted the introduction of much longer beam spans. Cellular beams were first used in the UK in 1988, as a result of manufacturing and design improvements to the castellated method, to create beams with circular openings. Applications for cellular beams have grown rapidly and now include tall buildings, hospitals, car parks, airport terminals, shopping malls and many other structures. Fabsec manufactures cellular beams using fabricated plated sections and now has a dominant position in the UK market. In 2005 a total of c75,000 tonnes of cellular beams were manufactured in the UK with roughly 1-in-4 steel building in the commercial sector constructed using cellular beams – the majority of these being FECCBs

There is considerable pressure upon structural steelwork designers to seek optimum solutions in a limited design time period. Similar pressure exists upon construction periods. As a consequence suppliers of cellular beams have invested heavily in the development of design software for structural engineers, low cost methods of manufacture and the off-site application of fire protection. Development of beam design software has advanced, especially in the creation of links to structural frame analysis and design packages and 3D modelling software. FBEAM is an analysis and design tool (provided free of charge) for the design of Fabsec beams fabricated from plate. The beams normally have web openings (for the passage of services within the web depth) which may be cellular or bespoke. In

Figure 1. Composite construction using FECCBs.

addition, beams can be designed for ambient temperature or the fire condition (for a period of up to 120 minutes' fire resistance). The development of two-way links to proprietary frame design packages now permits the steel designer to easily and swiftly optimise multiple beams in whole structures to reduce both cost and construction period. An extensive program of fire testing has allowed Fabsec to optimise fire protection thicknesses for its range of cellular beams.

2 THE CHOICE OF FIRE ENGINEERED COMPOSITE CELLULAR BEAMS

The UK market has moved considerably in the direction of long span solutions for steel framed buildings. The primary reason for this trend is the adoption of

Figure 2.   FECCBs with varying cell shape.

composite designs which make long spans economical, with typical secondary beam spans in the range of 12 to 24 m. Cellular beams benefit from economical production plus flexible cell geometry and location which permits services to pass through the web. This method of construction reduces the overall floor sandwich depth saving wall and cladding costs.

The FBEAM software is used to design floor and roof beams on a variety of buildings. However, the particular types of building for which optimised FECCBs can provide the greatest efficiency are tall structures, such as Renzo Piano's London Bridge Tower, dubbed the 'Shard of Glass' which will top 306 m and be Europe's tallest building when completed.

3   FIRE TESTING AND ANALYSIS OF TEMPERATURE DATA

3.1   *Test program at warrington fire research centre*

The FBEAM software also permits engineers to design highly efficient FIREBEAMs with a fire rating of 30–120 minutes based on a comprehensive 2-year test programme at Warrington Fire Research Centre which included 7 loaded, coated, composite fire tests. During these tests temperature measurements were taken over time at numerous thermocouple positions in beams with a variety of cross sections and web openings. All the tests exceeded the required period of fire resistance without failure.

The intumescent paint for FIREBEAM is provided by Leighs Paints of Bolton, UK. The product used is FIRETEX FB120, a unique and innovative product that has been specifically formulated and tested for use in FIREBEAM. The application of the coating is undertaken off-site in selected factory locations. This process takes a wet trade off-site and ensures a high level of quality control for the application.

The test programme was devised by the Steel Construction Institute of Ascot, UK, who also analysed the resulting temperature data. Temperature data recorded

Figure 3.   Artist's impression of The Shard of Glass.

Figure 4.   Intumescent coated beam; protective char visible.

throughout the test programme was analysed and incorporated in a data file which is interrogated by the software. As a result the checks that determine a section's capacity at the Fire Limit State, in particular vierendeel bending, web post buckling and shear buckling for webs with high d/t ratios are all based on reduced values of yield strength and Young's Modulus at elevated temperatures.

3.2 Thermal model

The purpose of the thermal model is to enable the temperature of various parts of a beam to be predicted for fire resistance periods of 30, 60, 90 and 120 minutes and for practical thicknesses of Firetex FB 120.

The fire test results were analysed using various methods. The best correlation was found using a regression method from ENV13381-4. See for example Equation 1 below:

$$t = a_0 + a_1 d_i + a_2 \frac{d_i}{A/V} + a_3 \theta_s + a_4 d_i \theta_s + a_5 d_i \frac{\theta_s}{A/V} + a_6 \frac{\theta_s}{A/V} + \frac{a_7}{A/V} \quad (1)$$

where:

t	=	Time in minutes to steel temperature
d_i	=	Thickness of fire protection(mm)
a_0 etc	=	Regression coefficients (determined from tests)
θ_s	=	Steel temperature (°C)

During the tests, the temperature at various positions on the beam was recorded. The bottom flange temperature for one test is shown (Figure 5). The shape of the curve is typical of steel protected by an intumescent coating. Initially the steel heats up rapidly, at about 240°C the intumescent process starts and the rate of temperature rise reduces. The steel then heats at a steady rate until the coating begins to detach or break up. At this point, the rate of rise of temperature increases.

To use the regression technique, the individual tests were analysed to obtain the time when the top flange, web and bottom flange reached temperatures in the range of 350 to 850°C (in steps of 50°C). This data was then analysed using the above regression equation and the coefficients, a_0 etc., were obtained.

For each constituent element of the section, tables of steel temperature were computed for 30 to 120 minutes, for a range of section factors and thicknesses of FB120 from 0.2 to 2.2 mm. The three tables were combined into one. To obtain the temperature of a given steel beam, FBEAM reads this table from a secure data-file.

In addition to the basic top flange, web and bottom flange temperatures, the data file contains information on the temperature distributions around openings.

4 OPTIMISATION OF COMPOSITE FABRICATED CELLULAR BEAMS

4.1 The optimisation process for a single beam

The optimiser in FBEAM automatically finds an optimum design without the need for the user to manually iterate by adjusting the input data to optimise the design. Essentially, the software completes the iteration automatically and hence generates an optimum

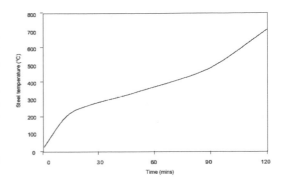

Figure 5. Rise in temperature for a steel beam protected with an intumescent coating.

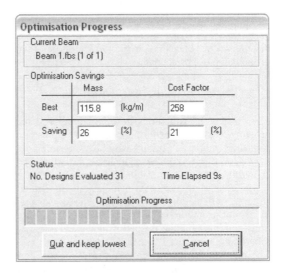

Figure 6. Optimisation progress.

design without the user having to adjust any of the beam parameters.

At the heart of the optimiser is the Beam Wizard® algorithm which solves the problem of iterating every possible beam design. The Beam Wizard® algorithm constantly monitors the entire collection of cost and engineering parameters of the beam under design, and calculates which parameters to change to yield the best result. This process generates the optimum design, with the least number of iterations, in the shortest time.

4.2 The optimisation process for multiple beams

FBEAM 2006 has an enhanced project optimisation facility, allowing the designer to optimise many beams at the click of a single button. Selecting "Optimise" will load the Beam Wizard® with some additional features not available previously. Extra features include an additional tab on the form to allow selection of beams to be optimised. To select a beam for optimisation

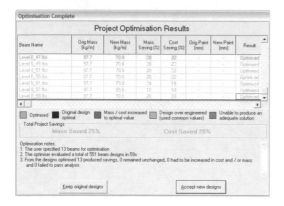

Figure 7.   Project optimisation results.

either click the check box to the left hand side of the beam name, or if the entire group needs to be optimised click the check box to the left hand side of the group name or project name.

5   STRUCTURAL FIRE ENGINEERING

The concept of structural fire engineering and the growing role of the design engineer in specifying fire resistance methods explains why software improvements are supporting the engineer in these new and changing areas of responsibility.

Understanding fire, human behaviour, mitigation strategies (detection, suppression, smoke management) and structural behaviour are all part of this more holistic thinking about the performance and integrity of structures in a fire situation. (Torero 2005)

Structural modelling is a very useful tool for performance prediction. For its part Fabsec's FBEAM software undertakes modelling of the effect of fire upon cellular steel beams using a built-in model of fire performance based on extensive tests at Warrington Fire Research Centre, a UKAS approved laboratory. The model (developed by the Steel Construction Institute for Fabsec) predicts temperatures in the top flange, web, bottom flange and around openings to determine the steel temperature and strength in a fire condition.

At 600°C the yield strength is reduced by over 50% and the elastic modulus by 70%. As a tool for the engineer, the ability to have an automatically fire engineered beam design is a huge benefit. The software takes the concept a stage further by allowing the user to find the optimum fire design, including the ability to balance the cost of the solution, to achieve the most economical combination of steel mass and coating thickness.

5.1   Structural model

The fire design rules are expressed in a step by step manner similar to that followed for normal design. The

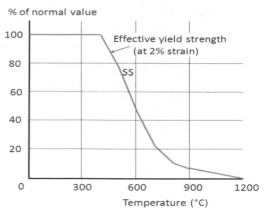

Figure 8.   Relationship between yield strength and temperature.

rules follow the principles of BS5950-8 and EC4-1–2. They apply to Class 1 or 2 sections.

The major checks carried out are as follows:-

– Bending resistance of a non- perforated beam
– Shear resistance of a non-perforated beam
– Shear resistance of beam at an opening
– Bending resistance
– Vierendeel bending
– Horizontal web shear
– Web buckling

6   IMPROVING THE STRUCTURAL PERFORMANCE OF TALL STEEL STRUCTURES IN FIRE

The lessons learned from recent events including the fires at WTC 1 and 2 and Madrid's Torre Windsor suggest that the structural engineer should be more closely involved with all aspects of structural fire engineering. (Lane 2005).

Risk assessment is an important part of building design, particularly for the fire condition of tall buildings. When considering the possibility of extreme events the impact on tall buildings of a blast / impact followed by a fire (which may be cellulose or hydrocarbon based) will need to be assessed. In this situation it is clear that thin coat intumescents (which have proven blast resistance) perform much better than board or sprayed alternatives.

Fabsec's solution for 90 and 120 minute fire resistance for cellular beams utilises Leighs Paints Firetex FB120 which is able to resist a blast and subsequent burn.

Recent advances in thin coat intumescent technology mean that fire resistance periods of 90 and 120 minutes can be economically achieved by quality

Figure 9.   Plasma web plate cutting.

Figure 10.   Factory applied intumescent fire protection.

control off-site application. The thickness required to achieve 90 and 120 minutes fire resistance for a typical tall structure floor beam, subject to office loading, has fallen to as little as 1.0 mm and 1.5 mm respectively – a thickness which can be easily applied in a single coat.

7   INTELLECTUAL PROPERTY RIGHTS AND LICENSED MANUFACTURE

Over the last 5 years of development of the FBEAM software (and the resulting design of optimised FEC-CBs) Fabsec has invested heavily in IPR including Patents, Trade Marks and Copyright.

In the UK, the Fabsec Licensees' automated plate-lines have the capacity to produce some 50,000 tonnes per annum of plated sections using modern profiling and welding plant. The cutting of flange and web plates – including openings for the provision of services – is carried out using the most modern CNC controlled gas and plasma machinery. Plate elements that comprise the section to be fabricated are brought together using automated handling plant and welded by means of a twin-wire submerged arc process. An in-line painting facility, with the capability to apply intumescent coatings, ensures the efficient flow of the FECCBs through the production process.

8   TWO-WAY LINKS BETWEEN FBEAM AND PROPRIETARY ANALYSIS AND DESIGN PACKAGES

8.1   *The interface*

An interface in computer software terms allows one piece of software to directly communicate with another using a clearly defined set of data without resorting to exchanging data files. There are a number of technologies within Microsoft Windows that allow this to happen but the most commonly used mechanisms fall

under the heading Component Object Model (COM). A COM interface not only allows data to be exchanged between two programs while they are both running but also allows one program to "call" or initiate actions within the other.

8.2   *RAM SS, Arup GSA and CSC Fastrak*

Fabsec sought to link its FBEAM software to proprietary analysis and design packages, to make data transfer on a beam-by-beam basis a thing of the past. The primary need for the facility to transfer beam design details to and from packages such as RAM Structural System (RSS), CSC Fastrak and Arup's GSA. The 'Open Link' philosophy followed by RAM made it possible to implement an easy to use link that allows for two-way data exchange between RSS and FBEAM.

FBEAM holds a project of multiple beam designs organised into categories, for example, arranged by floor. The beams required for a Fabsec design can be selected from the RAM model by either a graphical model browser or automatically if the user specifies them with an identifiable designation (for example FAB1234). FBEAM then automatically extracts the beam data including all the load information and creates a full project with all beam data already filled in.

As a result, Fabsec's Beam Wizard® function automatically finds the optimum beam design for all the beams in the project in one operation. Once the optimum solutions are found the data is produced in one of FBEAM's report formats and the optimised section information and web opening data is linked back to RAM offering a true import/export capability via the interface. The whole process takes place automatically without the user having to worry about importing or exporting the data.

The tall building steel designer is now able to link FBEAM to a RAM model and create a project automatically. This permits easy storage and retrieval of

Figure 11.   RAM Model created by Tier Consult.

multiple beam designs which can be stored by project, by floor and by beam type. Multiple beam analysis, design and group editing is now possible.

For the engineer this means improved design efficiency and lower costs. A single beam may have taken 10–15 minutes to design after defining span, position and size of web penetrations, loads and slab information. Now with the link to the RAM model, once the building is modelled in the 3D model generator, a typical structure with hundreds of beams can be optimally designed in minutes rather than hours or days.

The beam data is transferred very swiftly and the FECCBs for the whole project can be optimised using FBEAM's Beam Wizard in a single step. Optimised beams can then be saved back to the RAM model along with web opening penetration details. ·

9   THE BENEFITS OF OPTIMISED FECCBS FOR A GLOBAL MARKET

There are a number of advantages to the steel designer and building owner arising from the inclusion of optimised FECCBs into a building structure which include:

– Much lower weight solutions than alternatives (typical saving 15%–20%)
– Swift multiple beam design optimisation for tall buildings or major projects
– Ease of service integration and reduced building height (typically 300 mm per storey)
– Solutions for 60, 90 and 120 minutes fire resistance
– High quality factory fabrication and off-site application of fire protection
– Reduced structural member piece count and faster erection of the frame
– Reduced construction cost and period of build
– Intumescent coatings that are effective in the extreme case of a blast/burn scenario.

10   CONCLUSIONS AND FUTURE WORK

This paper has described the development of software which permits multiple optimisations of FECCBs in tall structures. FECCBs are now typically used for long spans, commonly 12–24 m. The increased adoption of FECCBs, in the UK and elsewhere, has been driven by the availability of design software that not only facilitates rapid analysis but also generates reductions in weight and cost of the structure. Significant recent advances in optimisation mean that entire projects can be handled on a floor-by-floor basis with or without a link to proprietary analysis packages. The fabrication process is highly automated so the quality of FECCBs is consistently very high. The use of long spans and FECCBs results in fewer structural members and greater repetition in production, the consequence of which is lower cost and shorter construction periods. FECCBs are innovations which are rapidly being adopted in many steel structures around the world.

It should be borne in mind that the use of passive fire protection on beams is only one part of the structural fire engineering of tall structures. The prescriptive approach e.g. 120 minutes' fire protection to all steel elements of the building may be appropriate to many structures; however, the direction laid out in BS9999 is to differentiate between buildings above and below 60 m.

Also, investigations of recent fires and 9/11 suggest that threat and risk assessments and whole frame structural analysis should be employed on tall buildings. (Lane & Lamont 2005)

Fabsec plans to continue the collaboration with RAM and to accelerate integration with (or two-way links to) other structural design and analysis packages. Beam design software should automatically rationalise the number of designs to achieve batch production cost savings. There is the potential to develop an engineering or user interface module which could 'plug in' to most proprietary building analysis and design packages. In addition, Fabsec will explore the potential to link to database modelling packages such as Autodesk's Revit Structure and will add new features including non-composite beam design in the fire condition.

REFERENCES

Lane, B. 2005. Comments on Structural Fire Response and Collapse Analysis, *Session 6, NIST World Trade Centre Conference*, 13–15 September 2005, USA
Lane, B. & Lamont, S. 2005. *Proceedings of New Civil Engineer Seminar, Developing the Role of Fire Engineering*, 12 April 2005, London
Torero, J. 2005. What is Fire Engineering? Where has it come from and where is it going? *Proceedings of New Civil Engineer Seminar, Developing the Role of Fire Engineering*, 12 April 2005, London

Steel and Composite Structures – Wang & Choi (eds)
© 2007 Taylor & Francis Group, London, ISBN 978-0-415-45141-3

Effect of fire on tall buildings: Case study

A. Heise, G. Flint & B. Lane
ArupFire, London, UK

ABSTRACT: The number of tall buildings being designed is steadily increasing around the world. Due to the recent high profile tall building fires, e.g. Torre Windsor in Madrid and the World Trade Center in New York, clients and authorities are expressing increasing interest in designing robust structures for fire loading. Recent increases in computing power and performance allows designers to investigate ever more complex designs in greater detail. Similarly computer modelling provides a greater level of understanding of building response to various load cases, including fire. Greater understanding of structural systems allows for increased safety and efficiency in design.

This paper presents a case study of a tall, composite steel framed office building under severe, multi-storey fires. The aim of this paper is to highlight the structural forms underlying the good response of the structure.

The building consists of a 40 storey vierendeel stress tube supporting a relatively long span composite floor system. A distinguishing feature of the design is the inclusion of 3 storey, local atria over the full height of the building. Although a highly redundant sprinkler system is installed for life safety reasons the lack of compartmentation caused by the atria could lead to a flashed over fire affecting 3 storeys simultaneously.

A study was made of the structure using several severe fire regimes affecting full floor single- and multi-storey structural models. The aim of the study was to investigate the robustness of the structure with respect to stability and compartmentation when subjected to the thermal load of the fire regimes. As a full structural analysis has been conducted reliance on prescriptively specified fire protection is reduced. In this case the inherent fire performance of the structure has a direct bearing on understanding the level of safety of the design. The outcome of the study was an understanding of the strengths and weaknesses of the design and allowed the design team to specify a suitable level of passive fire protection to the structural elements. The conclusion of the study is that a redundant primary frame (i.e. the strong vierendeel tube structure) in conjunction with rolled UB floor beams creates a robust structure for withstanding severe fire scenarios. When such structural systems are provided then a more efficient use of fire protection may be considered.

1 INTRODUCTION

1.1 *Advanced structural fire engineering*

The current state of the art in structural fire engineering is to use advanced analysis techniques to quantify real structural response to fire. These techniques allow designers to quantify the global and local structural mechanisms throughout the duration of a design fire. Therefore the stability and likely performance of compartmentation can be determined. Compartmentation is often required to prevent fire spread from the room of origin and is dependent on the ability of a floor or wall (e.g. to an escape core or fire fighting core) to maintain its integrity and/or insulation for the fire duration. Integrity prevents the passage of smoke or flame, insulation limits temperature rise on the unheated side of the enclosure/floor.

Such advanced analysis techniques can therefore be used to assess a cold temperature structural design, in specific fire events, to determine if there are any particular areas of strength or more importantly weakness in the fire limit state. These techniques can be used in structural steel design because substantial validation work has taken place, such as at the Large Building Test Frame Program at Cardington in the UK (Kirby, 2000).

Fire related responses are very different to the responses of other forms of cold temperature structural design. The response in the fire limit state cannot easily be predicted based on analysis for load cases at ambient. Hence including this form of analysis in the design process increases the understanding of the structure and may allow enhancements to be made to any portions of the structure underperforming in fire.

Advanced analysis techniques also allow the development of refined passive fire protection strategies. In the case of steel structures it allows applied passive fire protection to be installed in areas of the structure which require it, rather than uniform coverage throughout the structure. The traditional uniform coverage approach stems from meeting Building Code requirements, and carries with it associated ongoing maintenance concerns for the life time of a structure. In addition relying on code provisions means the fire resistance design is based on testing of shorter beam or floor spans than used in reality because of the size of the standard fire resistance test furnace (e.g. BS 476, 1987). As client demand for wide, open and long span column free spaces increases, the validity of such measures becomes increasingly of concern. Full scale structural fire modeling however, can provide additional information to improve the reliability of a design.

This is particularly true, if architects include modern design features into buildings, for instance; a number of atria connecting several floors within a high-rise building as it is the case in the presented case study.

1.2 Methodology

It is important that the assessment of the structure using cutting edge tools is accompanied by close consultation with the design team and approving authorities. Other primary stakeholders should also be consulted to ensure that real value is brought to a design in a timely fashion. The input to the advanced analysis needs to be understood and agreed by the relevant parties, within a time frame that permits changes to the cold temperature structural design as needed.

Typically, a structural fire engineering assessment involves a four step process as follows:

(1) determine credible design fires
(2) calculate the heat transfer from these fires to the elements in a representative structural model
(3) quantify the mechanical response of these elements for the duration of the fire
(4) determine strengths and weaknesses of the design and recommend simple enhancements to the structure, to improve performance.

Typically, stability is deemed to be maintained if the columns continue to carry their design load for the duration of the fire, and runaway failure of the beams or slab does not occur (Lamont et al 2006).

For compartmentation, an assessment of strain in the slab is necessary to ensure severe cracking will not cause breach of integrity or insulation. Deformations of structure in the vicinity of walls is also assessed especially around protected escape or fire fighting cores.

1.3 Multi-storey fires

Multi-storey fires are currently not considered as a possible threat when assessing a building according to the UK Building Regulations, or equivalent codes in other countries. In a traditional design it is assumed that existing requirements limit the fire to a fire compartment, which usually comprises one storey within a high-rise building.

The building presented in this case study included a number of multi-storey atria over the whole height of a high-rise office building. This produces a possible fire compartment that includes 3 full storeys challenging current practice and requiring additional considerations beyond existing design experience.

Multi-storey fires comprise an increased threat to the structure. Not only must the maximum temperatures be considered (e.g. a code compliant single element check), but also the complex interaction of structural elements.

In multi-storey fires the columns are heated over a greater height and are therefore more likely to buckle. This is because several floors are on fire and heated columns cannot provide the same rotational restraint as columns at ambient temperature. Which means the fire performance of the structure may be reduced compared to the level the code assumes when prescriptive levels of passive fire protection are applied to the structure.

Furthermore complex redistribution of force occurs in beam-column connections as the load carrying mechanism of the floor changes from bending/shear to tensile membrane action. This acts together with the longer unrestrained length of columns to actively move columns into unstable configurations.

2 THE BUILDING

2.1 Architectural layout and fire fighting facilities

The office building is a high-rise building approximately 200 m tall and consisting of 43 storeys. The building includes sets of multi-storey atria, also called villages. Each village consists of three office floors connected via an open atrium (Figure 1). A typical floor layout, including engineered fire protection layout, may be seen in Figure 3.

The circulation core for the building is on the south side of the building outside the office area. Two fire fighting shafts including fire fighting lifts are provided inside this core.

A glazed façade is provided on all sides of the building.

The high rise building is provided with an automatic detection system, first aid fire fighting equipment and an enhanced, multiple redundant life safety sprinkler system, as well smoke control on every floor.

2.2 The structure

The primary support structure consists of a robust vierendeel stress-tube designed to resist both gravity and wind loads. The stress-tube forms the perimeter of the office floor areas.

Figure 1.   Three storey village.

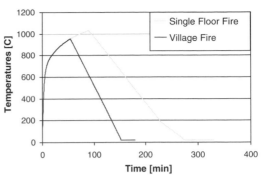

Figure 2.   Design fire temperature-time curves.

Early involvement in the project resulted in a robust structural design for fire and the ambient loads. The following elements of the structural design are considered to improve the robustness of the structure in fire:

- Vierendel structure with capacities to transfer forces
- Universal beam sections and solid fabricated girders (significant openings in beam webs avoided)
- A dovetail metal deck, shear studs staggered or placed side by side to ensure metal deck integrity at join
- Rebars and tie forces concentrated over protected beams

Typical spans for secondary beams range from 12.5 m to 14 m while primary beams are typically 10.5 m long.

2.3   The design fires

Two design fires were assessed for this building: A single storey fire and a multiple storey fire which may be seen in Figure 2. Each fire is representative of a post flash-over condition and it is assumed that all active fire fighting measures fail. This includes failure of the enhanced sprinkler system designed to prevent fire spread over multiple floors.

Assuming sprinkler failure was considered a conservative assumption, which was acknowledged by a reduced fire load density for the three storey fire in comparison to the value taken for the single storey model. The single storey fire is for this reason more severe taking into account the greater probability that such a fire might develop.

The Eurocode 1 parametric fire equation was used to create the design fires. A low amount of ventilation was agreed with the approving authority to provide a suitable level of conservatism. It is considered that the design fires are more severe than 120 minutes of the standard temperature time curve which is the level of fire resistance required by code for this building.

2.4   Fire protection proposal

The aim of the analysis was to assess the structure, identify design weaknesses and strengths and provide fire protection as required. The typical engineered protection layout may be seen in Figure 3.

From previous experience of advanced structural fire engineering and the research performed to date (Kirby 2000, Huang et al 2000, Bailey et al 1999, The University of Edinburgh 2000), a second load bearing mechanism in fire is well known. This is based on tensile membrane action of the slab and to some extent catenary action of beams. This secondary load bearing mechanism allows secondary beams to be unprotected, if all beams attached to internal columns are fire protected. Columns were generally protected to a 90 minute standard using thin film intumescent paint. The model showed that internal columns need to be protected to a limiting temperature of 600°C therefore an appropriate thickness of board protection is being provided to these columns.

3   FINITE ELEMENT MODELS

3.1   Single storey and multiple storey

Structural models of a 3 storey village and a single floor at the upper part of the tower were constructed using the finite element software ABAQUS.

The upper levels of the tower were chosen as this area shows relatively high utilization ratio in the columns under ultimate limit state gravity loading and they are also the most slender column sections.

Typical section sizes may be seen in Table 1.

The columns were modelled on the fire floor, 1.5 storeys below the fire floor and 2.5 storeys above the fire floor. This allowed a minimum number of boundary conditions to be applied while retaining a realistic column response to a multi-storey fire.

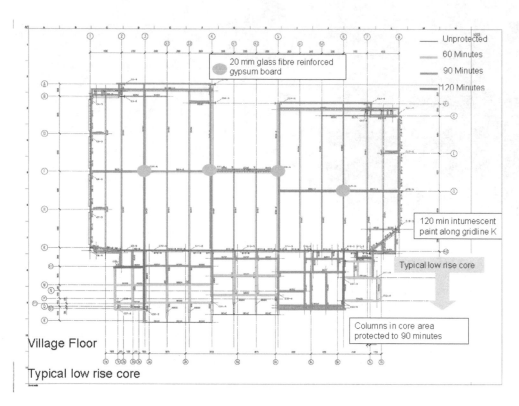

Figure 3. Typical Floor Layout.

Table 1. Typical section sizes.

	Secondary beams	Primary and perimeter beams	Slab	Columns
Building	$457 \times 191 \times$ 98UB	500–1000 mm deep Fabricated Girders	130 mm Reentrant A193 Mesh	Fabricated Girders and SHS
Model assumptions		[–]	120 mm Flat Slab	[–]

As the perimeter stress-tube structure is designed to resist wind loading as well as gravity, wind loading has been included as an additional axial load in the columns. This represents the additional vertical loading as the stress-tube acts in bending.

Four noded shell elements were used to represent the slab. Two noded linear beam elements were used for all beams and columns. Each element was associated with its appropriate section properties and material characteristics.

Connections within the model were assumed to be pinned, except for the perimeter vierendeel stress structure which was assumed to include fully moment resisting connections.

3.2 Material properties

Eurocode values were assumed for material properties of concrete and of structural and reinforcing steel. The material properties all include material and stiffness degradation with increasing temperatures.

3.3 Heat transfer analysis

Heat transfer through the equivalent-thickness floor slab was modeled using ABAQUS. Heat transfer to unprotected members was conducted in accordance with the equations in EC1993.1-2. Temperature variations within the cross section were considered.

All protected members, except for internal columns, were protected by intumescent paint. Internal columns were provided with a prescribed thickness of board protection, taking into account the severity of the applied design fires. The board provided a higher level of protection than strictly required by code provisions.

The applied fire protection layout may be seen in Figure 3.

3.4 Engineered and code compliant models

In addition to the models considering the engineered fire protection layout, a fully code compliant model was assessed. In this case edge beams and all beams attached to internal columns were protected to 120 minutes and all other secondary beams to 90 minutes respectively. The traditional or code compliant fire protection layout was assessed in order to compare certain responses, with the engineered model.

3.5 Failure criteria

Two criteria were suggested as being primary failure mechanisms that required monitoring. The first criterion is the runaway of the structural elements, primarily slab or columns. This is where displacements increase disproportionally with no further increase of load or temperature. This is an instability mechanism.

The second mechanism is failure of compartmentation which could lead to fire spread to previously unaffected parts of the building. Either of these failure mechanisms would mean the structure did not meet the life safety requirements of the Building Regulations.

4 RESULTS

4.1 Models

The results of the following models are presented:

- Multi-storey Model with unprotected secondary beams.
- Single Storey Model with unprotected secondary beams.
- Code Compliant Single Storey Model

The single storey models provide information about the response of the structure with and without all of the code recommended fire protection. The multi-storey model provides an overview of the structural response to fires affecting more than one floor. The code compliant model was used to show a comparable level of safety between the engineered and code compliant solution.

4.2 Overview

Deflections of the slab were monitored and stability was maintained in all 3 models with no indications of runaway.

An analysis of slab strains in all models indicated that integrity was maintained. Despite relatively large maximum displacements (see Figure 4), results from all models indicated that mid-span strains were typically between 5–8% with locally increased strains of up to 16% occurring over some protected members and around internal columns.

An analysis of the curvature of the slab over protected members was conducted using simple engineering calculations. This confirmed that the observed strains in the model were reasonable.

There are currently no specific strain criteria for measuring floor compartmentation failure in fire.

Therefore the material data for rebar presented in EC1992-1-2 was used as a guide.

Tests (data from Certification Authority for Reinforcing Steels) indicate that the ultimate strength of reinforcing steel is typically reached at approximately 8%. This is above the observed midspan strains.

The Eurocode indicates that rebar can maintain maximum stress up to 15% strain and then ruptures at 20% strain. If the rebar strain output from the model is shown to be generally below the 15% limit, as seen in the model presented here, it is considered that rupture will not occur.

In order to increase robustness extra rebar was specified for these areas, which had not been included in the structural model. This was deemed to satisfy the requirements for integrity of the slab.

The increased protection in the Code Compliant model was shown to not substantially reduce slab strains.

4.3 Beams

Beams were monitored for runaway deflections. Due to high temperature and therefore significant thermal expansion, midspan deflections were relatively large but runaway was not observed.

Unprotected beam displacements reach approx. L/8 in the single storey model and L/11 in the multi-storey model. These large deflections were comparable to deflections seen in the Cardington tests. The large deflections help the floor system to activate tensile membrane action, which enables a secondary load bearing mechanism. These large deflections did not destabilize the structure.

More important for global stability is the response of the protected primary beams. Deflections in these members reached a maximum of L/23 in the single storey model with unprotected secondary beams. This was the model with the most severe fire in terms of temperature and fire duration. By comparison the single storey Code Compliant model showed primary member deflections of L/25, which is not significantly less, considering the substantial increase in applied passive protection (120 minute primary beams, 90

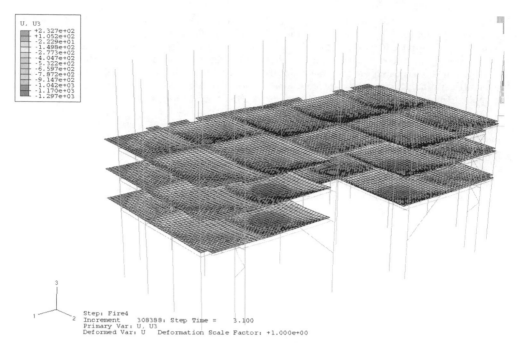

```
U, U3
    +2.327e+02
    +1.052e+02
    -2.229e+01
    -1.498e+02
    -2.773e+02
    -4.047e+02
    -5.322e+02
    -6.597e+02
    -7.872e+02
    -9.147e+02
    -1.042e+03
    -1.170e+03
    -1.297e+03
```

```
3
          Step: Fire4
1      2  Increment   308388: Step Time =    3.100
          Primary Var: U, U3
          Deformed Var: U   Deformation Scale Factor: +1.000e+00
```

Figure 4.   Multi-storey Model Deflection Response.

minute secondary beams versus 90 minute primary beams and unprotected secondary beams).

The response of the Multi-storey model was less extreme due to the less intense (shorter duration) design fire. Unprotected secondary beams attained a maximum deflection of L/11, while primary beams reached a maximum deflection of L/65.

4.4   Columns

Columns were monitored both for stability and for horizontal displacements. With multi-storey fires the stability of columns becomes an issue due to a combination of increased buckling length and the action of sagging floors actively pulling columns inward.

Horizontal displacements of the columns affect both stability and compartmentation.

The maximum horizontal displacements of the columns in the multi-storey model were comparable to those of the single storey model. It would be expected that the multi-storey model would have larger displacements; however it is considered that the less intense fire in the multi-storey case mitigates the movement.

Comparison between Engineered and Code Compliant models indicated that the response on the columns was similar in each case. Deflections were of the same order of magnitude and in some cases the movement of the Code Compliant model columns was larger than that of the Engineered models. This was primarily seen in outward movement of columns at the

interface between the office and circulation core. The lower midspan deflections seen in the floor beams in the Code Compliant model is combined with an overall expansion of the floor slab that is comparable between Code Compliant and Engineered cases. This leads to an increase in outward movement in columns. Thermal expansion is realised in lateral expansion rather than increased midspan deflection.

4.5   Stability

Stability was maintained throughout all of the analyses of the final design solution.

4.6   Compartmentation

Due to the presence of the atrium, floor to floor compartmentation within the village is already breached, leading to the multi-storey fire. However the compartmentation around the whole village fire compartment must be maintained to prevent further spread.

The primary issues of concern in maintaining the compartmentation lines are the midspan deflection of the floor in the boundary slab between villages and the horizontal deflection of columns along the boundary between the office area and the circulation core.

While the Engineered models showed higher midspan deflections compared to the Code Compliant models horizontal deformations were of the same order of magnitude indicating equivalence of engineered and traditional solution with respect to vertical compartmentation. Similarly lateral columns movements, which affect the compartmentation line between the office areas and the circulation core, were comparable.

4.7 Overall structural performance

The excellent performance of the structure is attributed to the holistic approach to the design implemented at an early stage in the project (see Section 2.2.)

The vierendeel frame allows a significant redistribution of forces between columns and allows the frame to act as a whole entity, rather than a collection of parts.

The use of solid section beams over a reasonable span contributes to a response that is relatively easy to analyse and is comparable to existing research data.

The dovetail deck provides a greater amount of material compared to a trapezoidal slab of the same overall thickness. This provides greater insulation both to the rebar and the floor above.

5 CONCLUSION

This paper reports on a case study, where a high rise building was designed to resist multiple storey fires. The aim of this study was to understand the inherent level of safety and robustness of the building in order to provide an appropriate level of passive fire protection to the structure.

It is the conclusion of this analysis that stability and compartmentation, according to the requirements of Part B3 of ADB[1] are maintained using the proposed fire protection layout.

From the comparison of the code compliant fire protection layout with the engineered layout it is concluded that equivalence with respect to compartmentation is achieved. In general, displacements of the primary structure in both cases are comparable.

It should be noted that provision of code compliant fire protection does not prevent significant structural movement during a fire.

The performed analysis required an increase in fire protection level to internal columns (compared to a strict code compliant case), thus increasing the robustness of the structure in a fire. This would not have been quantified in a prescriptive code compliant approach.

Thus it is concluded that an equivalent or increased safety level is obtained as result of this structural analysis.

It is understood that the building presented in this paper is the first to consider a multi-storey fire as part of the design approval process.

REFERENCES

ABAQUS, Inc., Rising Sun Mills, 166 Valley Street, Providence, RI 02909-2499 USA.

Eurocode 1: Actions on structures – Part 1–2: General Actions – Actions on structures exposed to fire. BS EN 1991-1-2:2002, London UK, British Standards Institute, 2002.

Eurocode 2: Design of Concrete Structures - Part 1.2: General Rules – Structural Fire Design. BS EN 1992-1-2:1996. London UK, British Standards Institute, 1995.

Eurocode 3: Design of Steel Structures - Part 1.2: General Rules – Structural Fire Design. BS EN 1993-1-2:1995. London UK, British Standards Institute, 2001.

Eurocode 4 — Design of composite steel and concrete structures — Part 1–2: General rules — Structural fire design BS EN 1994-1-2:2005. London UK, British Standards Institute, 2005.

BS 476 : Part 20 : 1987 Fire tests on building materials and structures. London: BSI, 1987.

BS 5950-8: 2003, Structural use of steelwork in building — Part 8: Code of practice for fire resistant design, 2003.

BS 5950 Part 3 Structural use of steelwork in building design in composite construction code of practice for design of simple and continuous composite beams, 1990.

CARES information taken from http://www.ukcares.co.uk

Lamont, S.; Lane, B., Flint, G., and Usmani, A. S. "Behavior of structures in fire and real design – a case study" Journal of fire protection engineering, Volume 16, Number 1, February 2006.

G. Flint, A.S. Usmani, S. Lamont, J. Torero, B. Lane, ""Effect of Fire on Composite Long Span Truss Floor Systems"", Journal of Constructional Steel Research, Volume 62, Issue 4, April 2006, Pages 303–315

G. Flint, "Fire Induced Collapse of Tall Buildings", Edinburgh University PhD Thesis, 2005 (available at http://www.civ.ed.ac.uk/research/fire/thesis.html)

Bailey C.G., White D.S., and Moore D.B. "The tensile membrane action of unrestrained composite slabs simulated under fire conditions." *Engineering Structures*, 22:1583–1595,2000.

Huang Z., Burgess I.W. and Plank R.J. "Non-linear modeling of three full scale structural fire tests", In First International Conference Structures in Fire, Copenhagen, June 2000.

Kirby B.R. British Steel data on the Cardington fire tests. Technical report, British Steel, 2000.

The University of Edinburgh (2000) Final report of the DETR-PIT project: Behavior of steel framed structures under fire conditions. Technical report. www.civ.ed.ac.uk/research/fire/project/main.html

Steel and Composite Structures – Wang & Choi (eds)
© *2007 Taylor & Francis Group, London, ISBN 978-0-415-45141-3*

Numerical simulation of the 7th Cardington Compartment Fire Test using a full conjugate heat transfer approach

Nuno R.R. Almeida & António M.G. Lopes

Dept. de Eng. Mecânica, Universidade de Coimbra, Portugal

Gilberto C. Vaz

Dept. de Eng. Mecânica, ISEC, Instituto Politécnico de Coimbra, Portugal

Aldina Santiago & Luis S. Silva

Dept. de Eng. Civil, Universidade de Coimbra, Portugal

ABSTRACT: In a previous work, numerical simulation was employed for the calculation of the temperature distribution and its time evolution for the 7th Cardington Compartment Fire Test. At the time, a simplified compartment geometry was considered and the thermal inertia of the walls and ceiling was not taken into account. The present work represents a better approximation, where the conjugate heat transfer at the walls is fully modeled. Additionally, this work shows the influence of a beam placed at the ceiling in the draining and in the distribution of temperatures of the involving fluid. The fire was modeled using a simplified approach, where combustion is simulated as a volumetric heat release. Radiation was modeled with a differential approximation (P1 model), while turbulence effects in the mean gas flow were dealt with the SST turbulence model, integrating the effects of buoyancy on the production/destruction rate of kinetic energy and its dissipation. Results are provided for the temperature and flow field time evolution, thus allowing a direct comparison with experimental data. A detailed analysis on the role of the wall thermal inertia is provided and the influence of the beam on the temperature and air velocity field is described. The computations were performed with commercial code ANSYS CFX 5.7.1.

1 INTRODUCTION

Many works were made in the last years on fire models aimed at the prediction of the most important aspects of structural fire security. Most of these models are based in Parametric Temperature-Time Curves and Zone Models, which represent a simplified approach, but that do not fully predict important aspects of heat transfer mechanisms to and from the walls and structures, as well as fluid flow within compartments.

Computational Fluid Dynamics (CFD) is the most sophisticated approach to simulate transfer phenomena in a fire compartment. Its methodology is based on fundamental conservation laws for mass, momentum and energy. The equations that govern these laws are solved over a mesh that represents the physical compartment under study.

Several CFD commercial codes are available nowadays. In the present work, ANSYS CFX 5.7.1, a well known software, was employed to simulate fluid flow and thermal phenomena for the 7th Cardington fire test.

The present work represents an improvement of the simulations described in Lopes et al. (2005), where the thermal inertia of the solid domains was not taken into account, as well as some geometric details which showed to play an important role, such as the presence of beams.

2 GOVERNING EQUATIONS

Conservation of momentum is described by the Navier-Stokes equations, which time dependent formulation is:

$$\frac{\partial(\rho u_i)}{\partial t} + \frac{\partial}{\partial x_j}(\rho u_j) = -\frac{\partial p}{\partial x_i} + \frac{\partial \tau_{ij}}{\partial x_j} - \frac{2}{3}\frac{\partial(\rho k)}{\partial x_i} + S_i \quad (1)$$

where τ_{ij} represents the stress tensor, k is the turbulence kinetic energy, ρ is the fluid density and S_i is the momentum source due to temperature differences in the fluid:

$$S_i = (\rho - \rho_{ref})g_i \quad (2)$$

The continuity equation formulates the principle of mass conservation:

$$\frac{\partial \rho}{\partial t} + \frac{\partial}{\partial x_j}(\rho u_j) = 0 \tag{3}$$

The energy equation is:

$$\frac{\partial \rho h}{\partial t} - \frac{\partial p}{\partial t} + \vec{\nabla}(\rho \vec{V} h) = \vec{\nabla}(\lambda \vec{\nabla} T) + S_E \tag{4}$$

where the specific enthalpy is given by $dh = c_p\, dT$ and S_E is the energy source term.

For modeling the entire radiation phenomena, a transport equation must be solved for the radiative heat transfer in fluid and gas media. The spectral radiative transfer equation can be written as

$$\frac{dI_\nu(\mathbf{r,s})}{ds} = -(K_{a\nu}+K_{s\nu})I_\nu(\mathbf{r,s}) + K_a I_b(\nu,T) + \frac{K_{s\nu}}{4\pi}\int_{4\pi} dI_\nu(\mathbf{r,s'})\Phi(\mathbf{s\cdot s'})d\Omega' + S \tag{5}$$

where: ν – frequency, $\mathbf{r}$ – position vector, $\mathbf{s}$ – direction vector, s – path length, K_a – absorption coefficient, K_s – scattering coefficient, I_b – blackbody emission intensity, I_ν – spectral radiation intensity, T – local absolute temperature, Ω – solid angle, Φ – in-scattering phase function, S – radiation intensity source term.

The equation (4) is a first order integro-differential equation for I_ν in a fixed direction. To solve this equation within the domain, a boundary condition for I_ν is required. In this study, the walls are considered as diffusely emitting and reflecting opaque boundaries and the openings are considered fully transparent boundaries.

Due to the dependence on 3 spatial coordinates, 2 local direction coordinates, s and ν, the formal solution of the radiative transfer equation (4) is very time consuming, requiring the use of approximate models for the directional and spectral dependencies. For the directional approximation the P1 model is adopted (see Raithby, 1991 for details). This model, also known as the Gibb's model or Spherical Harmonics model, assumes that the radiation intensity is isotropic or direction independent at a given location in space. The spectral radiative heat flux in the diffusion limit for an emitting, absorbing, and linearly scattering medium can be computed as:

$$q_{r\nu} = -\frac{1}{3(K_{a\nu}-K_{s\nu})-AK_{s\nu}}\nabla G_\nu \tag{6}$$

where A is the linear anisotropy coefficient and G_ν is the spectral incident radiation. Then, the equation for the spectral incident radiation is

$$-\nabla\cdot\left(\frac{1}{3(K_{a\nu}-K_{s\nu})-AK_{s\nu}}\nabla G_\nu\right) = K_{a\nu}(E_{b\nu}-G_\nu) \tag{7}$$

where $E_{b\nu}$ is the spectral blackbody emission.

For the wall treatment, assuming that the radiation intensity arriving at and leaving a wall are directionally independent, the boundary condition for (7) at wall is:

$$\mathbf{n}\cdot\mathbf{q}_{r\nu} = -\frac{1}{3(K_{a\nu}-K_{s\nu})-AK_{s\nu}}\frac{\partial G_\nu}{\partial n^+} = \frac{\varepsilon_\nu}{2(2-\varepsilon_\nu)}[E_{b\nu}-G_\nu]|_w \tag{8}$$

where $\mathbf{n}$ is the unit vector outward normal to the wall, n^+ is a distance coordinate in the same direction, ε_ν is the spectral emissivity and w represent the value at the wall.

For the spectral approximation the gray model is adopted. This model assumes that all radiation quantities are nearly uniform throughout the spectrum. Then, the dependency of Eq. (4) on frequency can be dropped. The scattering model adopted is the isotropic model, assuming that in-scattering is uniform in all directions.

For closing the problem, a state equation is necessary for describing the relation between temperature, pressure and density. In the present case, this is achieved considering the gases inside the compartment to behave as a perfect gas with constant specific heat capacity c_p.

Turbulence effects upon the mean flow were simulated with the SST model (Menter 1993, 1994), which is a weighted blending of the k-ω model (Wilcox, 1993) and the standard k-ε of Launder and Spalding (1974), according to the boundary layer region where the solution takes place. The equations are:

$$\frac{\partial}{\partial x_i}(\rho u_i k) = \frac{\partial}{\partial x_i}\left[\left(\mu+\frac{\mu_t}{\sigma_{k3}}\right)\frac{\partial k}{\partial x_i}\right] + P_k - \beta'\rho k\omega \tag{9}$$

$$\frac{\partial}{\partial x_i}(\rho u_i \omega) = \frac{\partial}{\partial x_i}\left[\left(\mu+\frac{\mu_t}{\sigma_{\omega3}}\right)\frac{\partial \omega}{\partial x_i}\right] + \alpha_3\frac{\omega}{k}P_k - \beta_3\rho\omega^2 + (1-F_1)2\rho\sigma_{\omega2}\frac{1}{\omega}\nabla k\nabla\omega \tag{10}$$

where the index 3 coefficients are obtained as a linear combination of the index 1 (corresponding to the k-ε model) and the index 2 (corresponding to the k-ω model) coefficients, as defined next:

$$\beta'=0.09 \quad \alpha_1=5/9 \quad \beta_1=3/40 \quad \sigma_{k1}=2 \quad \sigma_{\omega1}=2$$
$$\alpha_2=0.44 \quad \beta_2=0.0828 \quad \sigma_{k2}=1 \quad \sigma_{\omega2}=0.856$$

As a means of limiting the turbulence kinetic energy production in stagnation regions, its term is computed as follows:

$$\tilde{P}_k = \min(P_k, 10\varepsilon) \tag{11}$$

The weighting factor F_1 is obtained with the following equation:

$$F_1 = \tanh\left(\min\left(\max\left(\frac{\sqrt{k}}{\beta'\omega y}, \frac{500\nu}{y^2\omega}\right), \frac{4\rho\sigma_{\omega2}k}{CD_{k\omega}y^2}\right)\right)^4 \tag{12}$$

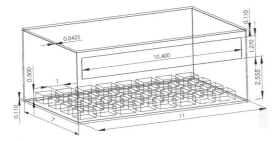

Figure 1. Simplified geometry of the fire compartment.

$$CD_{k\omega} = max\left(2\rho\sigma_{\omega 2}\frac{1}{\omega}\nabla k\nabla\omega, 1e-10\right) \quad (13)$$

Modeling of the shear stress transport is accomplished with a limiting factor in the formulation of the turbulence viscosity:

$$\upsilon_t = \frac{a_1 k}{max\left(a_1\omega, SF_2\right)} \quad (14)$$

where:

$$F_2 = tanh\left(max\left(\frac{2\sqrt{k}}{\beta'\omega y}, \frac{500\upsilon}{y^2\omega}\right)\right)^2 \quad (15)$$

and S quantifies the fluid deformation rate (cf. Menter 1993, 1994 for details).

3 THE PHYSICAL PROBLEM

The Cardington tests were performed in the early 90's in Bedfordshire, UK, in a steel structure built inside a former airship hangar. Construction details of the building structure are given in Bravery (1993). The 7th Cardington tests were performed in the middle of the building in the 4th floor. The plan area of the construction was about $11 \times 7\,m^2$, with a ceiling height of 4.185 m. Ventilation was ensured through an opening with $10.4 \times 1.27\,m^2$, as represented in Fig. 1.

The walls were composed by three plasterboard layers (15 mm + 12.5 mm + 15 mm) with a thermal conductivity k in the range 0.19–0.24 W/mK.

Ceiling and floor were made of concrete (k = 0.7 W/mK) with about 0.11 m thickness. The fire load was provided by a geometrically regular arrangement of 32 wooden cribs made of 10 piles of wooden sticks, representing a fuel load of 40 kg/m² (of floor area), as shown in Fig. 2. Each pile was approximately 0.5 m high.

Temperature distribution in the fire compartment was registered by a set of thermocouples placed 300 mm from the ceiling, with a distribution as depicted in Fig. 2.

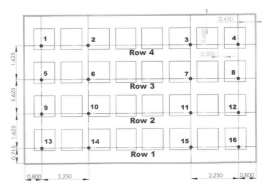

Figure 2. Position of the thermocouples (compartment opening is at the figure bottom).

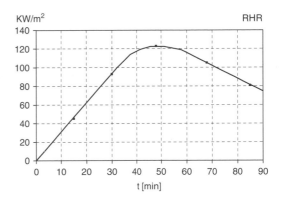

Figure 3. Estimated heat release rate.

4 PHYSICAL MODELING AND NUMERICAL APPROACH

4.1 Rate of heat release

To simulate the rate of heat release (RHR), a simplified approach was adopted, where the fire is modeled as a heat source in the volume initially occupied by the wood. This methodology was necessary due to the lack of information about the RHR history during the fire.

The total energy released may be written as follows:

$$E_{tot} = \eta H \rho_{f,w} A_f \quad (16)$$

where a value $\eta = 0.875$ was assumed for the combustion efficiency, H = 14 MJ/kg is the wood heat content, $\rho_{f,w} = 40\,kg/m^2$ is the fuel load density and $A_f = 77\,m^2$ is the compartment floor area. Substitution of the corresponding values in the previous equation gives a total of 4.312×10^{10} J of available energy.

In the previous work (Lopes et al., 2005), two curves of RHR were considered, both described in Cadorin (2003). The curve adopted (cf. Fig. 3) was obtained in a laboratory fire, referred as NFSC19 in the NFSC database.

4.2 Grid and boundary conditions

As the compartment is symmetric relatively to the middle plane (cf. Fig. 4), the problem was solved only at half the domain, imposing symmetry conditions. Thus, the computational requirements are reduced.

The volume grid is formed by tetrahedral elements and by an inflation layer made of prismatic elements, in order to better resolve the boundary layer gradients. For each geometry (without and with the beams), different regions (solid and fluid) were meshed, as may be observed in Fig. 4.

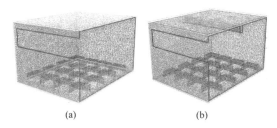

(a)   (b)

Figure 4.   Meshes created. (a) – without beam; (b) – with beam.

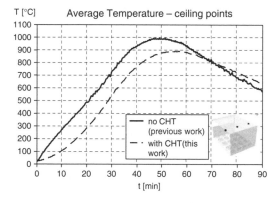

Figure 5.   Average temperature at the points shown.

Simulations for different mesh sizes were compared, in steady state simulations, in order to ensure that results were grid independent.

The energy transfer at the domains interfaces (solid and gas) is described by conservative conditions. Solid surfaces are considered as opaque, with an emissivity of 0.9 and a diffuse fraction of 0.9. An average absorption coefficient of $0.3 \, \mathrm{m}^{-1}$ was considered for the gases inside the compartment.

External solid surfaces were assigned a temperature of 20°C, which is approximately the ambient temperature measured in the day of the experiments.

The compartment opening was taken as an open boundary, with a static relative pressure of 0 Pa.

5   RESULTS

5.1   Simulation 1 (conjugate heat transfer – CHT)

The aim of this simulation was to study the importance of adopting a conjugate heat transfer approach (CHT), by simultaneously solving the gas and the solid domains. Thermal inertia of the walls may, thus, reveal its importance. Results were compared with the simulations described in Lopes et. (2005) (previous work), where a simple heat flux condition was applied at the inner compartment surfaces. Comparisons are presented for temperatures vs. time, averaged at three locations in the ceiling of the compartment (cf. Fig. 5).

Results clearly show that the thermal inertia of the solid domain plays a non-negligible role, mainly at the heating phase.

Moreover, three-dimensional effects taking place near the compartment corners show only when solving the conduction problem inside the solid domain (Figs. 6a, 7a, 8a). These simulations with CHT show the maximum heat transfer occurring at the compartment corners (Fig. 7a). As a conclusion, a CHT approach should be adopted for a correct simulation of this problem.

5.2   Simulation 2 – presence of a beam

The fundamental purpose of this simulation is to make an evaluation of the differences in the draining fluid

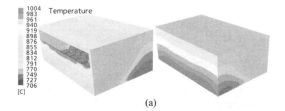

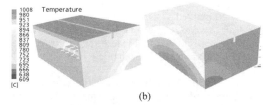

(a)   (b)

Figure 6.   Temperature contours (t = 54 min from fire ignition). (a) – without beam; (b) – with beam.

(a)   (b)

Figure 7. Wall heat flux contours at t = 54 min. (a)–without beam (points in red showing the maximum value); (b)–with beam.

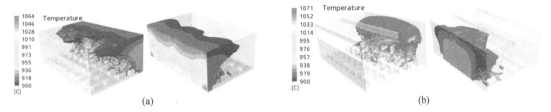

(a)   (b)

Figure 8. Iso-volume of air temperatures in the range 900°C to 1064°C; t = 54 min. (a) – without beam; (b) – with beam.

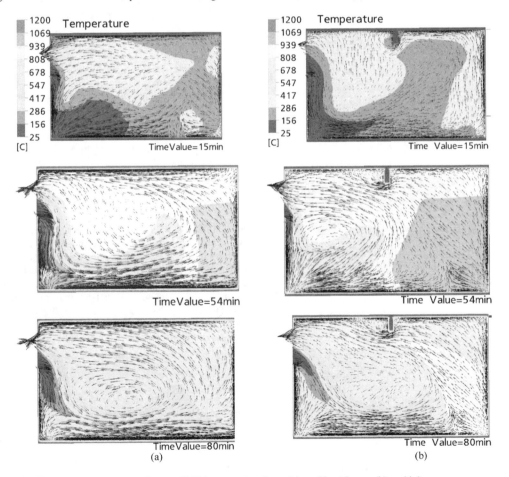

(a)   (b)

Figure 9. Temperature contours and vector field in symmetry plane; (a) – without beam; (b) – with beam.

717

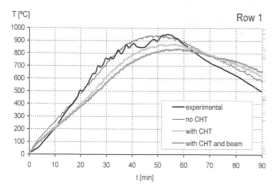

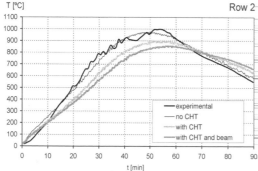

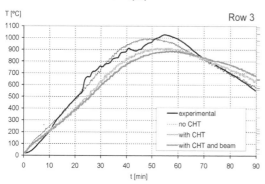

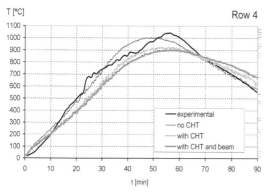

Figure 10. Temperature versus time, averaged at selected points.

and the consequences in the temperature field, due to the presence of a beam with 0.5 m height, located at the ceiling.

Figures 6b, 7b and 8b show the temperature and the wall heat flux field which, when compared with the a) counterpart, lead to the conclusion that the beam affects these results quite significantly. This is due to the manner in which the beam changes the flow pattern. Figures 9 depict the velocity field at the symmetry plane, for both simulations (with and without beam). Analysis of these results reveal that the vortex from the simulation with the beam is reduced in the middle back of the compartment and is directed towards the opening as time proceeds. The presence of the beam shows to have a significant role in the general vortex development, thus affecting the velocity and temperature field.

5.3 *Time temperature curves*

Figure 10 presents the temperature averaged at the punctual locations characterized in Fig. 2. Results are provided for thermocouple experimental data and for simulations of all the cases described previously.

This figure clearly shows that the CHT approach leads to a delay in the temperatures time evolution, which is further enhanced by the presence of the beam. Improvements in the simulation did not, in general, conduct to better agreement with experimental results, which can be explained by the shape of the simplified RHR curve adopted. Further work will be done in studying other curves for the heat release rate and their influence on the results.

6 CONCLUSIONS

Simulation of phenomena in a fire compartment is a very complex subject that must be addressed with some physical simplifications. The present work showed the importance of the conjugate heat transfer on the thermal inertia of the system, as well as the influence that small geometrical detail may have on the results. Further work will include the study of other RHR curves, simulation of the external surrounding environment and detailed analysis of structural components.

REFERENCES

Bravery, P.N.R. (1993), "Cardington Large Building Test Facility – Construction details for the first building". England. Building Research Establishment, Internal paper, Watford.

Launder, B.E. and Spalding, D.B. (1974), "The Numerical Computation of Turbulent Flows", Computer Methods in Applied Mechanics and Engineering, 3, pp. 269–289.

Lopes, A.M.G., Vaz, G.C., Santiago, A. (2005), "Numerical Predictions of the Time-Dependent Temperature Field for the 7th Cardington Compartment Fire Test", Steel and Composite Structures, An International Journal, Vol.5 No.6, pp. 421–441.

Menter, F.R. (1993), "Multiscale Model for Turbulent Flows", 24th Fluid Dynamics Conference, American Institute of Aeronautics and Astronautics.

Menter, F.R. (1994), "Two-equation Eddy Viscosity Turbulence Models for Engineering Applications", AIAA Journal, 32 (8).

Raithby, G.D. (1991), "Equations of motion for reacting, particle-laden flows", Progress Report, Thermal Science Ltd.

Wilcox, D.C. (1993), "Turbulence modeling for CFD", DCW Industries, Inc., La Canada, Califórnia, 460 p.

Steel and Composite Structures – Wang & Choi (eds)
© 2007 Taylor & Francis Group, London, ISBN 978-0-415-45141-3

Fire safety engineering analysis of a single storey framed building based on structural global behaviour

F. Morente & J. de la Quintana
Labein Tecnalia

ABSTRACT: An optimized cold design based steel industrial hall has been chosen to develop a FSE analysis, with the aim of developing cost-effective steel solutions for industrial buildings, taking advantage of a reduced passive protection.

1 INTRODUCTION

A new regulation for fire safety in industrial buildings came into force in Spain in January 2005. This regulation prescribes high periods of fire resistance for structures resulting necessarily in the need to protect steel structures with passive protection systems. This fact rends the steel structures less competitive than concrete for industrial buildings. FSE may be a mean to overwhelm the difficulties posed to steel structures by providing information about structural behaviour in case of fire.

LABEIN-Tecnalia has developed a FSE study for a steel industrial hall producer, PRADO T.M., with the aim of analysing the behaviour of their model structure in case of fire.

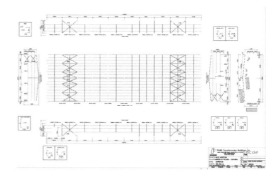

Figure 1. General drawing of the PRADO T.M. industrial hall.

2 CASE STUDY

A real representative steel industrial hall was chosen for the FSE analysis (See figure 1). The hall is a single storey framed building without beams for a bridge crane; the distance between frames is 10 meters. The frames are pined on their connection with the foundations and are composed of class 4 variable sections. The connections of the beam and the columns are welded/built in ones. Secondary structural elements play an important role in the structural stability of the frames (purlins, bracing system and tension members used to connect the purlins with the compressive flange of the beams in the hogging bending moment length).

The structure was conceived to support wind, snow and dead loads. There is no live load to be supported by the steel structure. The wind and snow loads were not considered to verify the structural response in case

of fire as the Eurocodes propose; therefore, the only action considered is the dead load.

Some steps were done to reach the final goals of the project. Firstly an analysis of the structural response under standard exposure (ISO curve – See figure 4) was carried out, and later it was also analyzed the structural response under natural fire conditions. The structural response was analyzed with the non-linear finite element code Abaqus. In both cases, firstly an analysis of a single frame was done to obtain information about potential failure modes and later a simulation of the whole structure (ten frames and secondary elements) to ascertain the global behaviour of the hall.

As the frames are composed of class 4 sections, they were modelled with shell elements. The density of the mesh was not uniform and it was enhanced in the regions of the frame likely to develop hinges due to local buckling failures (See figure 2).

Figure 2.   Modelling of the structural members.

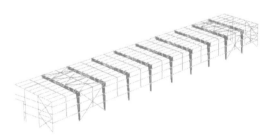

Figure 3.   Modelling of the 3D structure.

The secondary members were modelled with 3D beams and in case of the bracing members with truss elements (See figure 3).

2.1   *Analysis of the structure submitted to the standard exposure curve*

A single frame was analyzed. The temperature rise within the structure was obtained solving the conduction within the steel sections in the shell model of the frame; the standard time – temperature curve was given as a boundary condition.

In a first simulation, they were introduced global and local imperfections, and later one simulation was performed neglecting them and another one considering only half a frame with symmetry boundary condition and without imperfections. As the results did not differ too much, it was decided to neglect in the model the imperfections; their effect is negligible by

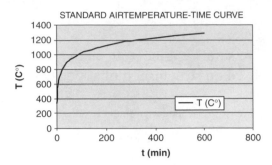

Figure 4.   ISO curve. Temperature – time.

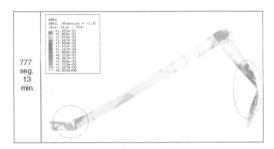

Figure 5.   Failure time and mode. Structure submitted to ISO curve fire.

comparison with the thermal induced actions (See figure 5). This conclusion matches with some references of other FSE studies.

2.2   *Analysis of the structure submitted to natural fire*

The structure was analyzed considering thermal action due to natural fire. Two different fire scenarios were considered depending on the origin of the fire, in one case, fire was ignited far from the façade and in other case fire was developed near the façade, engulfing directly a column. In the case of the fire beginning by the centre of the hall, the zone model Ozone was used to calculate fire evolution.

During the first stage of the fire, it was considered to create a two layer distribution in the hall, a hot upper one representing the smoke accumulated under the ceiling heating beams and upper parts of the column and a cold lower one affecting the lower part of the columns. In case of fire near a column, the temperature evolution within the steel sections was calculated considering localized effect of the flame on the column.

In case of fire beginning in the centre of the hall, the collapse occurred after the onset of flash – over. Two different growing phase velocities were simulated (See figure 6). It was concluded that, although the structure

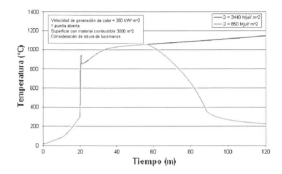

Figure 6. Gas temperature.

Figure 7. 3D structure failure mode. Submitted to natural fire.

Figure 8. Real case fire in a similar industrial hall.

collapses, it happens in safe conditions, when collapse occurs, the environment inside the hall is untenable neither for occupants nor firemen and the failure mode is inwards, preventing firemen outside the building from being trapped by debris (See figure 7).

A real fire in a similar industrial hall occurred accidentally showing exactly this predicted behaviour (See figure 8).

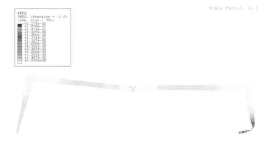

Figure 9. Local collapse of un-protected column.

In case of localized fire in the column, only one frame was simulated. In this case, the collapse of the frame occurred before the onset of flash – over, and it was outwards, endangering the people outside the building (See figure 9). It was recommended to protect columns with passive protection systems to prevent that failure mode because it is a common practice in industrial halls to accumulate significant fire load leaning on the façade.

3   CONCLUSIONS

In conclusion, this analysis shows that the considered steel structure provides an acceptable level of safety on the basis of a performance based approach if localized passive protection is applied rather than the costly protection prescribed by the regulation.

This way the resultant safety of the building has been assessed in a technical and cost-effective way resulting in gaining a good competitive position for the product.

ACKNOWLEDGEMENTS

The authors wish to thank the steel industrial hall producer PRADO T.M.

REFERENCES

1994–1998. Competitive steel buildings through Natural Fire Safety Concept (Contract No.: 7210-SA/522; 01/07/1994–30/06/1998)
1999–2000. Natural fire safety concept – Valorisation project of 7210-SA/522 (Contract No.: 7215-PP/042; 1/07/1999–30/06/2000)

Steel and Composite Structures – Wang & Choi (eds)
© *2007 Taylor & Francis Group, London, ISBN 978-0-415-45141-3*

Fire analysis and design of a composite steel-concrete structure

R. Zaharia, D. Pintea & D. Dubina

The "Politehnica" University of Timisoara, Romania

ABSTRACT: The paper presents the fire analysis for the columns of a multi-storey steel-concrete building which will be the tallest building in Bucharest, Romania. Four different types of steel-concrete sections are analysed: octagonal sections with identical steel profiles, octagonal sections with different steel profiles, double-symmetric rectangular sections and rectangular sections with one axis of symmetry. The analysis, made according to EN1994-1-2, EN1991-1-2 and EN1990, is realised with the SAFIR computer program, a special purpose program for the analysis of structures under elevated temperature conditions. The thermal and structural analysis is carried out considering the standard ISO fire, but also a natural fire scenario.

1 INTRODUCTION

The basic principle in determining the fire resistance of a structural element is that the elevated temperatures produced by the fire reduce the materials strength and stiffness until possible collapse. When the temperatures on the cross-section of a structural element produce the reduction of the element resistance bellow the level of the effect of actions for fire design situation, it is considered that that element lost its load-bearing function under fire action.

The fire resistance of composite steel-concrete structures is calculated according to EN1994-1-2 (EN1994 2005). Three methods are available in order to evaluate the fire resistance: the tabulated data method, the simple calculation models and the advanced calculation models.

The tabulated data method is based on observations resulted from experimental study. It is the easiest to apply, but it is limited by the geometrical conditions imposed to the composite cross-section.

The simple calculation models compute the ultimate load of the element by means of formulas or design charts, established on the basis of experimental data.

The advanced calculation models suppose an advanced numerical analysis of the elements, parts of the structure, or of the entire structure under fire, using specialized software for the mechanical analysis of structures under elevated temperatures.

There are several fire models, accepted by the European Standard EN1991-1-2 (EN1991 2005), which describes the thermal and mechanical actions to be considered for a structure under fire.

The nominal standard temperature-time ISO 854 fire curve does not take into account any physical parameter, and can be far away from reality. From the beginning, the nominal curve supposes that the entire compartment is in the flashover phase and the temperature is increased continuously, without taking into account the cooling phase.

The parametric fire model considers the cooling phase and gives the temperature-time curve function of the fire load density and openings. This model is, however, limited to the surface and the height of the fire compartment considered, and supposes that the temperature is the same on the entire compartment, from the beginning of the fire.

A modern fire model approach is the "Two Zone" model. In this natural fire model, in the pre-flashover phase, the fire compartment is divided in a hot upper zone and a cold inferior one. For each zone, with uniform temperature, mass and energy equations are solved. Complex equations describe the air movement in the fire plume, the radiative exchanges between the zones and the gas movements on the openings and adjacent compartments. After the flashover, the temperature is considered uniform and is determined by solving the equations of mass and energy of the compartment, taking into account the walls and openings. In the frame of the ECSC research "Natural Fire Safety Concept" (CEC 2001) it was considered necessary to develop a computer program for this model. This objective is now reached, a computer program called OZone is available in order to determine the temperature-time curve by means of the "Two Zone" model, and was built at the Liege University, Belgium, in collaboration with the "Politehnica" University of Timisoara (Cadorin et al 2003).

The fire is considered an accidental situation which requires, with some exceptions, only verifications against the ultimate limit state. The combinations of actions for accidental design situations are given in the European Standard for basis of structural design EN1990 (EN1990 2004), by the following formulas:

$$G_k + P_k + \Psi_{1,1}Q_{k,1} + \sum_{i>1} \Psi_{2,i}Q_{k,i} \quad \text{or} \quad (1)$$

$$G_k + P_k + \sum_{i\geq1} \Psi_{2,i}Q_{k,i} \quad (2)$$

in which G_k, P_k, Q_k are the characteristic values of the permanent, variable and prestressing action. According to the European Standard for actions on structures exposed to fire EN1991-1-2 (EN1991 2005), the representative value of the variable action Q_1 may be considered as the quasi-permanent value $\Psi_{2,1}Q_{k,1}$, or as an alternative the frequent value $\Psi_{1,1}Q_{k,1}$.

A structure, substructure or element in fire situation may be assessed in the time domain, where the failure time must be higher than the required fire resistance time. The failure time is the time for which the resistance of the structure (or substructure, or element, as considered) under elevated temperatures reach the effect of actions for the fire design situation, considering the combination of action in fire situation.

2 COMPOSITE STEEL-CONCRETE STRUCTURE ANALYSED

The paper presents the calculation of the fire resistance for the composite columns of "Bucharest Tower Centre" structure, which will be the tallest building in Bucharest for the moment. The building has 3 basements, one ground floor, 21 floors, 3 technical floors at a total height of 106.3 m. The building is 25.5 m by 41.5 m in plan and has a total construction gross area of approximately 24135 m².

The columns are made by cruciform cross sections made of hot rolled European profiles, partially encased in reinforced concrete, in order to increase strength, rigidity and fire resistance. According to Romanian fire regulations, considering the specific and particularities of the building, the columns must have 150 minutes of fire resistance.

Four different cross section types were used, as Figure 1 shows (the values in parathesis represent the maximum load level, i.e. the maximum ratio to the ambient temperature capacity for each set): octagonal sections with identical steel profiles 2HEB500 (0.319), 2HEA800 (0.153), 2HEB800 (0.257), 2HE800x373 (0.303), octagonal sections with different steel profiles HEM800HEM700 (0.310), HEB800HEB700 (0.257), HEA800HEA700 (0.172), double-symmetric rectangular sections HEB1000HEB500 (0.293) and rectangular sections with one axis of symmetry HEB1000HEM500 (0.250), HEB1000 HEB500 (0.201).

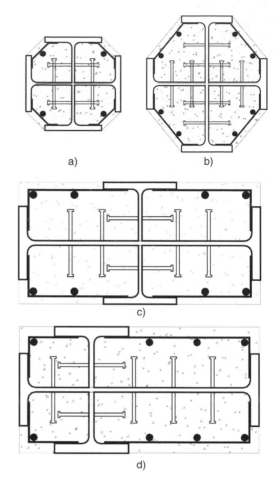

a)   b)

c)

d)

Figure 1. Composite cross-sections.

The rebars have 25 mm diameter and the concrete is C30/37.

For the purpose of this paper, only the 2HEB500, HEM800HEM700, HEB1000HEB500 and HEB1000HEM500 cross-sections will be presented, as they have the lowest fire resistance under ISO fire from each type of cross section types, respectively.

3 THERMAL ANALYSIS UNDER ISO FIRE

For the calculation of the fire resistance of the composite column, the SAFIR computer program was used (Franssen & Kodur & Mason 2004), which is a special purpose program for the analysis of structures under ambient and elevated temperature conditions. The program, developed at the University of Liege, accommodates various elements for different idealization, calculation procedures and various material

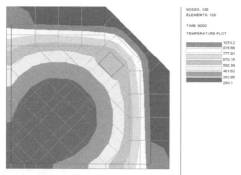

a) 2HEB500 cross-section

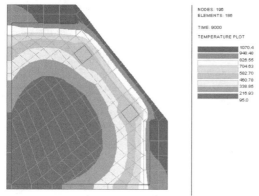

b) HEM800 HEM700 cross-section

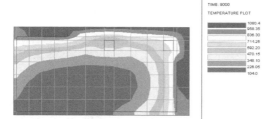

c) HEB1000 HEB500 cross-section

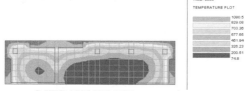

d) HEB 1000 HEM500 cross-section

Figure 2. Temperature distribution at 150 minutes of ISO fire.

models for incorporating stress- strain behaviour under elevated temperatures.

The analysis of a structure exposed to fire consists of two steps. The first step involves predicting the temperature distribution inside the structural members, referred to as "thermal analysis". The second part of the analysis, termed the "structural analysis" is carried out to determine the structural response due to static and thermal loading.

Figure 2 shows the temperature distribution on the cross sections of the considered columns, after 150 minutes of ISO fire. Due to symmetry, only a quarter, or half of the cross-sections was modelled. The round reinforcing bars are represented by quadrilateral elements, with equivalent area. For all cross-sections, after 150 minutes of ISO fire, the steel profiles flanges exhausted practically their load capacity, having temperatures greater than 900°C, while the profiles webs and the reinforcing bars have lower temperatures and there is an important core of concrete with quite low temperatures. Consequently, after 150 minutes of ISO fire, the sections have a reserve of load capacity.

4 MECHANICAL ANALYSIS UNDER ELEVATED TEMPERATURES

The columns, considered as isolated elements, loaded with the axial force and the bending moments on both principal cross-section axes (efforts corresponding to the fire combination of actions), were modelled with 3D beam elements. The buckling length of the columns was considered, conservatively, as the system length. Equivalent imperfections according to EN1994-1-1 (EN1994 2005) were imposed on both directions of the principal cross-section axes.

The horizontal displacement evolutions at the mid height of the columns of ground floor (which have the highest efforts in fire situation) are presented in figure 3. As the characteristic time-displacement demonstrates, excepting for the columns with rectangular cross-section with one axis of symmetry (d), all other columns of the ground floor does not resist to the 150 minutes of ISO fire under the imposed static loads.

Similar analysis was made for the columns of the upper floors. The fire resistance grows with each floor, as the stress level in the columns decrease on the height of the building. Excepting the columns with octagonal cross section and identical profiles 2HEB500, all columns above the ground floor fulfil the 150 minutes requirement. For the 2HEB500 columns, the R150 fire resistance requirement is fulfilled only from the 11th floor forth.

Table 1 gives the corresponding fire resistance times for all considered columns. Fire protection is needed for all the columns on the ground floor, except for the columns with rectangular cross-section with one axis of symmetry, while the 2HEB500 columns need protection up to the 11th floor.

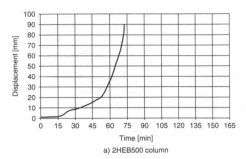

a) 2HEB500 column

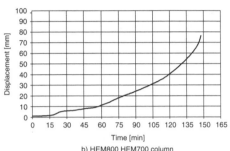

b) HEM800 HEM700 column

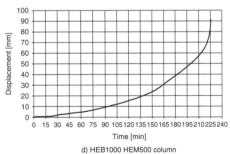

c) HEB1000 HEB500 column

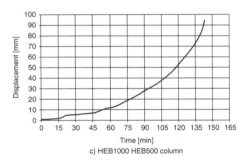

d) HEB1000 HEM500 column

Figure 3. Displacement evolution under ISO fire.

5 THE NATURAL FIRE SOLUTION

Since the ISO fire solution is much too conservative, an approach based on the Natural Fire concept was considered. The same element was subjected to a natural fire curve, obtained using the OZone v2 computer model (Cadorin et al 2003).

Table 1. Fire resistance times [minutes].

Column	Ground floor	Floors 1–10	Floor 11 – forth
2HEB500	70	100–149	>150
HEM800HEM700	146	>150	>150
HEB1000HEB500	143	>150	>150
HEB1000HEM500	>150	>150	>150

Several assumptions must be made when dealing with the natural fire scenario, like the maximum fire area, the fire load, and the surface of the openings.

The most challenging is to establish how the curtain walls will behave in a fire situation. The rate of heat release of fires is limited by the flow of oxygen available to it. In all except very rare circumstances, the flow of oxygen into a room comes largely from open doors and open windows and to a slight extent from any mechanical ventilation systems and from building leakage. Once a fire gets going, however, windows previously closed may crack and break out. The results will often be drastically different, depending on whether the windows break or not. Thus, it becomes of significant interest to be able to predict if, and when, glass may break out.

Here, an important distinction needs to be made. When a window pane of ordinary float glass is first heated, it tends to crack when the glass reaches a temperature of about 150–200°C. The first crack initiates from one of the edges. At that point, there is a crack running through the pane of glass, but there is no effect on the ventilation available to the fire. For the air flows to be affected, the glass must not only crack, but a large piece or pieces must fall out.

Understanding the conditions under which pieces actually fall out has been of considerable interest to fire specialists. Since the fire ventilation openings need to be known in order for fire models to be used, glass breakage has been of special interest to fire design engineers. This has prompted a number of theoretical and simplified studies and a few empirical ones as well.

The earliest guidance to be found in the literature on the question of when glass breaks out in fires comes from the Russian researcher Roytman (Roytman 1975) who notes that a room gas temperature of around 300°C is needed to lead to glass breakage. Hassani, Shields, and Silcock (Hassani et al. 1994/5) conducted a series of experiments in a half-scale fire test room using 0.9 × 1.6 m single-glazed windows where they created a natural top-to-bottom temperature gradient in the room and in the glass. At the time the first crack occurred in 4 or 6 mm thick glass panes, gas temperatures in the upper layer of 323–467°C were recorded. By the end of their 20 min tests, gas temperatures

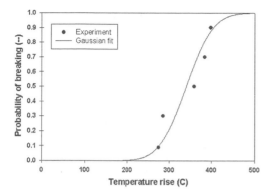

Figure 4. Probability of glass breaking (Tanaka et al. 1998).

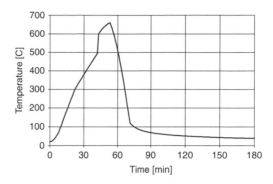

Figure 5. Temperature distribution in the upper layer.

were at ca. 500°C. Yet in only 1 of 6 tests glass fall-out was noticed. Temperature differences between the glass exposed surface and the shielded portion ranged between 125°C to 146°C at the time of crack initiation. These temperatures were about twice that predicted from the no-vertical-gradient theories. The authors do not give the exact room fire temperature at which the glass fall-out began in the one test where this occurred, but this was higher than 431°C (crack initiation) and lower than ca. 450°C (end of test).

One can put these data together, to conclude that at a room gas temperature of around 450°C the probability for glass to break out is 1/6. Shields (Shields et al. 2001) later conducted further tests using a room with three windows glazed with 6 mm thick panes. Glass fell out when the exposed surface temperature reached 415–486°C on the average. But there were quite a lot of variability and individual values ranged from 278 to 615°C at failure. A heat flux of around 35 kW/m² was needed for fall-out to occur. In a follow-on test series (Shields et al. 2002) it was noted that the lowest temperature of the glass at fall-out was 447°C.

The only probabilistically-based results concerning glass exposed to a uniform hot temperature come from the Building Research Institute (BRI) of Japan (Tanaka et al. 1998). In that study, researchers used a large-scale high-temperature door-leakage testing apparatus that resembles a large muffle furnace. Only single-glazed, 3 mm thick window glass was studied. For this type of glass, however, enough tests were run so that a probability graph could be plotted.

The results are presented in terms of a probability of glass breaking out, as a function of temperature rise above ambient (Fig. 4).

The Gaussian fit that can correlate this data corresponds to a mean temperature rise of 340°C, and a standard deviation of 50°C.

The fire compartment can be modelled as a square of 20 m × 20 m, with three walls made of bricks and a fourth wall entirely made of curtain wall. A linear

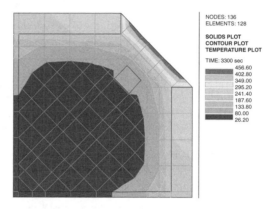

Figure 6. Temperature distribution on the 2HEB500 column under natural fire.

variation of the openings was considered, i.e. the curtain wall. Based on the research mentioned above, at 300°C 30% of the windows were considered broken, while at 500°C all the windows are broken.

The fire load inside the compartment was the one for office occupancy, with a rate of heat release of 250 kW/m². All the other parameters were considered (automatic water extinguishing systems, fire detection, etc). The evolution of the temperature in the upper layer is presented in Figure 5.

The natural fire approach was considered for the 2HEB500 column, which as shown in Table 1 has the lowest fire resistance when subjected to the ISO fire. The temperature distribution on the 2HEB500 column cross section under natural fire at 55 minutes (corresponding to the peak temperature in figure 3) is presented in Figure 6. The horizontal displacement evolution at the mid height of the column 2HEB500 of ground floor is presented in Figure 7. It may be observed that the column is stable; there is no collapse of the element under natural fire.

The same conclusion is valid for all the other columns on the ground floor which did not pass the R150 fire resistance requirement under ISO fire

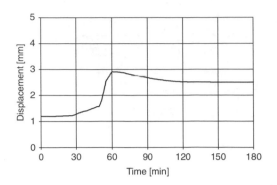

Figure 7. Displacement evolution under natural fire.

6 CONCLUSIONS

The fire resistance of the composite steel-concrete columns of the "Bucharest Tower Centre" structure was determined using advanced computation models, considering both the ISO 834 fire curve and the natural fire approach.

When using ISO fire, the analysis showed that fire protection is needed on some columns of the building, in order to attain the R150 fire resistance. Considering a natural fire scenario, all columns fulfil the R150 requirement, and consequently no protection is needed.

Natural fire solutions must be considered to the fire design of steel and composite steel concrete structures.

Careful consideration of the occupancy of the fire compartment and that of the materials that make out the fire compartment walls lead to a more realistic behaviour of the structural elements under fire.

REFERENCES

Cadorin, J.F, Pintea, D., Dotreppe, J.C, Franssen, J.M, 2003, A tool to desgn steel elements submitted to compartment fires- Ozone V2. Part2: Methodology and application, Fire safety journal, Elsevier, 38, 439–451.

CEC Agreement 7210-PA/PB/PC/PD/PE/PF/PR-060, Natural Fire Safety Concept, Implementation in the Eurocodes and Development of an User friendly Design Tool, 2001.

EN1990 Eurocode: Basis of Design, September 2004, European Comittee for Standardisation, Brussels.

EN1991-1-2: Eurocode 1 – Actions on structures – Part 1–2: General actions – Actions on structures exposed to fire, 2005, European Committee for Standardization, Brussels

EN1994-1-1: Eurocode 4 – Design of composite steel and concrete structures – Part 1-1: General rules and rules for buildings, 2005, European Committee for Standardization, Brussels.

EN1994-1-2: Eurocode 4 – Design of composite steel and concrete structures – Part 1–2: General rules – Structural fire design, 2005, European Committee for Standardization, Brussels.

J. M. Franssen, V. K. R. Kodur, J. Mason, User Manual for SAFIR. A computer program for analysis of structures subbmited to the fire. University of Liege, Department 'Structures du genie civil' Service 'Ponts et charpentes', 2004.

Hassani, S. K., Shields, T. J., and Silcock, G. W., An Experimental Investigation into the Behaviour of Glazing in Enclosure Fire, J. Applied Fire Science 4, 303–323 (1994/5).

Roytman, M. Ya., Principles of Fire Safety Standards for Building Construction, Construction Literature Publishing House, Moscow (1969). English translation (TT 71-580002) from National Technical Information Service (1975).

Shields, T. J., Silcock, G. W. H., and Flood, M. F., Performance of Single Glazing Elements Exposed to Enclosure Corner Fires of Increasing Severity, Fire and Materials 25, 123–152 (2001).

Shields, T. J., Silcock, G. W. H., and Flood, M., Performance of a Single Glazing Assembly Exposed to a Fire in the Centre of an Enclosure, Fire and Materials 26, 61–75 (2002).

Tanaka, T., et al. 1998, Performance-Based Fire Safety Design of a High-rise Office Building.

Steel and Composite Structures – Wang & Choi (eds)
© 2007 Taylor & Francis Group, London, ISBN 978-0-415-45141-3

Mathematical modeling of intumescent coating fire protection for steel structures

J. F. Yuan & Y. C. Wang

School of MACE, The University of Manchester, Manchester, UK

ABSTRACT: This paper presents a mathematical model to calculate temperatures in intumescent coating protected steel under different fire conditions. The mathematical model, based on the finite difference method, considers mass and volume variations due to chemical decomposition in intumescent coating. The dynamic aspects of the intumescence process and the nonlinearity of thermal properties have been included in this model. Parametric studies have been carried out to identify key variables of intumescent coating properties for the model. The results of this paper indicate that the key variables that have been extracted from a small number of small scale fire tests may be used to predict other fire tests under different fire conditions.

1 INTRODUCTION

Intumescent coating is designed to expand and form a thick porous charred layer when exposed to heat in fire. The charred layer insulates the underlying substrate by providing a physical barrier. The applications of intumescent coating are widely ranged and the demands of this material have significantly increased. However, the mechanisms determining the fire-resistant properties of intumescent coating are not well understood yet, due to their highly complex physical and chemical natures. This hampers the application of intumescent coating in performance-based fire engineering. A mathematical model has been established in this study to investigate the fire protection performance of intumescent coatings. In the model, once some basic chemical characteristics and physical properties of the intumescent coating have been quantified, the fire performance of this kind of intumescent coating exposed to different external heat conditions can be determined.

In the mathematical model, the governing equations are based on energy and mass conservation, and the model is solved by using the Finite Difference Method (FDM). Arrhenius equation is used to describe chemical kinetics of coating reaction (typically including three parts: melting, blowing, and charring) [1], which in turn is responsible for intumescence. The main objective of this research is to identify the main fundamental parameters that are responsible for describing the intumescence performance with sufficient accuracy under different fire conditions.

2 MATHEMATICAL MODELLING

Assuming 1-D behaviour and from energy conservation, the change in heat flow conducted through a layer of intumescent coating is equal to the change in internal energy of the coating, giving:

$$
\begin{aligned}
\frac{\partial}{\partial x}(\lambda \frac{\partial T}{\partial x}) &= (m_s C_s + m_g C_g)\frac{\partial T}{\partial t} \\
&+ C_g T \frac{\partial(\varepsilon A \Delta x \rho_g)}{\partial t} + C_s T \frac{\partial m_s}{\partial t} \\
&+ \frac{\partial(\dot{g} T)}{\partial x} C_g + \Delta h(\frac{\partial m_r}{\partial t})
\end{aligned}
\tag{1}
$$

The left hand of equation (1) is the conducted heat to the intumescent coating; the 1st term on the right hand side (RHS) is the heat increase due to change in coating temperature; the 2nd and 3rd terms on RHS are heat increases due to change in gas and solid masses respectively; the 4th term on RHS describes convective heat loss due to gas movement and the last term on RHS is heat release from the coating. To enable quantification of equation (1), it is necessary to derive the various mass change terms, which are determined by the decomposition process. Arrhenius law is used to describe decomposition process, giving:

$$
K_{(j,t)} = A_j \exp(-\frac{E_j}{\Re T}), j = 1,2,3
\tag{2}
$$

The transient solid mass changing rate can be described as:

$$\frac{\partial m_{(j,t)}}{\partial t} = m_{(j,t)} K_{(j,t)} \tag{3}$$

It is assumed that only the blowing agent is responsible for swelling:

$$\frac{\partial x_t}{\partial t} = \frac{1}{\rho_g} \frac{\partial m_{(2,t)}}{\partial t} \qquad \left(x \le E_{max} x_0 \right) \tag{4}$$

The gas density may be obtained from the ideal gas law.

It is necessary to quantify the porosity ε, which will influence both heat capacity and thermal conductivity of intumescent coating.

$$x_s = (1 - \varepsilon_0) x_0 \frac{m_t}{m_0} \tag{5}$$

The total porosity, ε, is:

$$\varepsilon_t = \frac{x_t - x_s}{x_t} \tag{6}$$

The maximum expansion factor E_{max} is probably the most important parameter in this predictive mathematical model. It determines the effectiveness of insulation of the intumescent coating. Further research is now being conducted to establish a reliable model to estimate this quantity. In this research, experimental or nominal values will be used at this stage.

A total mass continuity equation for gas transportation can be written as:

$$\frac{\partial \dot{g}}{\partial x} = m_t K_t - \frac{\partial (\varepsilon x \rho)}{\partial t} \tag{7}$$

The formation of a multi-cellular char leads to a significantly reduced thermal conductivity. Rather than using simplest parallel and serial models, a model employing theory closer to reality has been introduced into this study. For a porous material, Russell [2] estimated the thermal conductivity as:

$$\lambda^* = \lambda_s \frac{\frac{\lambda_g}{\lambda_s} \varepsilon^{\frac{2}{3}} + 1 - \varepsilon^{\frac{2}{3}}}{\frac{\lambda_g}{\lambda_s} (\varepsilon^{\frac{2}{3}} - \varepsilon) + 1 - \varepsilon^{\frac{2}{3}} + \varepsilon} \tag{8}$$

Radiative heat transfer cannot be neglected at high temperature. Although the opaque solid material does not allow radiation, it takes place within the void in the coating. Separating the total heat conduction

coefficient of a gas into that of pure conduction and that due to radiation, one has:

$$\lambda_g = \lambda_{cond} + \lambda_{rad} \tag{9}$$

The thermal conductivity of gas due to pure conduction can be obtained from standard heat transfer textbooks.

For a porous structure with unconnected uniform spherical pores of diameter "d", the radiation contribution to the overall thermal conductivity of the single pore is [3]:

$$\lambda_{rad} = \frac{8}{3} d e \sigma T^3 \tag{10}$$

t Equation 10 clearly shows the importance of bubble diameter. In intumescence process, it is assumed that bubble growth has two distinct stages: initial formation and unifying. Bubbles nucleate and grow rapidly during expansion stage:

$$d_t = d_b \frac{x_t}{E_{max} x_0} \tag{11}$$

Once charring process begins, bubbles start to burst and unity. Bubbles grow slowly but the bubble diameter will increase significantly compared with that in expansion stage. It is described as [4]:

$$d_t = d_b \frac{v_{3,t}}{v_{30}} + d_f (1 - \frac{v_3}{v_{30}}) \tag{12}$$

3 VALIDATION

As part of validation of the mathematical model, figure 1 compares predictions of the model with experiments conducted by Cagliostro et al[5], with an external radiative heat flux of $157\,kW/m^2$. The back face of test sample was thermally insulated. The material properties are listed in Table 1.

In all simulations, the coating has been equally divided into 100 discrete layers. The time increment is 0.01s. Figure 1 indicates good agreement between the simulation and Cagliostro's experimental results.

4 PARAMETRIC STUDIES

It would be difficult to obtain accurate data for all the required properties of intumescent coating. Therefore, it is important that a sensitivity study is carried out to identify key parameters that should be accurately obtained.

The sensitivity studies have been performed by varying the values of input parameters individually, by $\pm 20\%$ from reference case (Table 1). E_{max} has been artificially set to be 20 to represent thin intumescent coatings applied in steel building structures.

Figure 2 shows substrate temperature histories with variable activation energies and pre-exponential

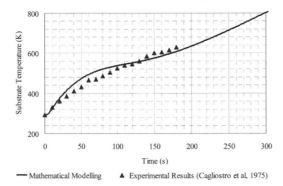

Figure 1. Validation of the model.

Table 1. Input values from Cagliostro [5] (or estimated values "*" from [4])

A_1 (s)	800	C_g(kJ/kg/K)	1.0
E_1 (kJ/mol)	53.384	λ_c(Kw/mK)	0.345×10^{-3}
A_2 (s)	6.9×10^5	λ_s(Kw/mK)	37.68×10^{-3}
E_2 (kJ/mol)	93.035	ρ_c(kg/m^3)	1400.0
A_3 (s)	5.0 (+)	ρ_s(kg/m^3)	7850.0
E_3 (kJ/mol)	63.786	e_c	1.0
H_1(kJ/kg)	-1256(*)	d_b(m)	5.0×10^{-6}(*)
H_2(kJ/kg)	-1256(*)	d_f(m)	325.0×10^{-6}(*)
H_3(kJ/kg)	9789	E_{max}	3.0(*)
v_{10}	0.28	W_g(kg/mol)	30.0×10^{-3}(*)
v_{20}	0.17	h (kW/m^2)	20.0(*)
v_{30}	0.55	Q(kW/m^2)	157.0
C_c(kJ/kg/K)	1.884	l_{c0}(m)	0.2×10^{-2}
C_s(kJ/kg/K)	0.42	l_s(m)	0.15×10^{-2}

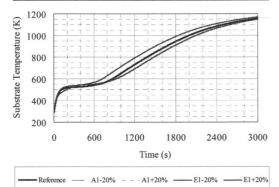

Figure 2. Effect of changing reaction kinetics of inorganic acid source.

factors (E_1 and A_1) of the inorganic acid source. The results indicate that changing these parameters has small effect on overall performance of intumescence. It is not surprising: although this acid release process must precede other reactions, it has little to do with the intumescence process and char structures.

Figure 3 considers the effect of varying the kinetics constants (E_2 and A_2) of the blowing agent. The

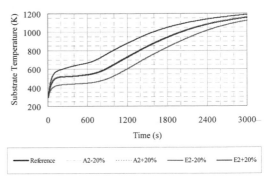

Figure 3. Effect of changing reaction kinetics of blowing agent.

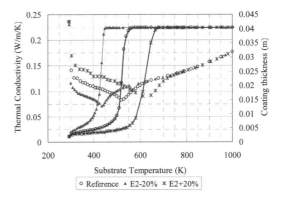

Figure 4. Expansion and thermal conductivity records for different E_2 values (Marks: thermal conductivity; Solid lines: expansion).

substrate temperature histories showed that the kinetic parameters of this process have significant effect on the substrate temperature, with figure 3 indicating a substrate temperature variation of about 100K in the sensitivity study. The variation of intumescent coating performance is directly reflected in the thickness and thermal conductivity of the coating. Figure 4 shows that a lower activation energy E_2 gives a faster expansion of intumescent coating, leading to a quicker drop in thermal conductivity.

Figure 5 indicates substantial effects of varying the charring kinetics. The effects of changing E_3 can be related to decomposition heat and thermal conductivity. Figure 6 shows decomposition heat histories during the intumescence process. It can be seen that the coating with a lower E_3 value has stronger and sharper decomposition heat peak. If this heat cannot be transported away, the excess part will contribute to heat-up of the surrounding materials. Charring process governs bubble growth (equation 12), which in turn influences the thermal conductivity after full expansion (equation 10, figure 7).

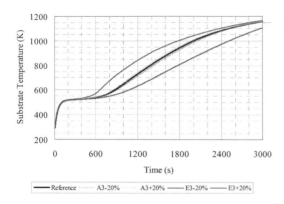

Figure 5. Effect of changing reaction kinetics of charring material.

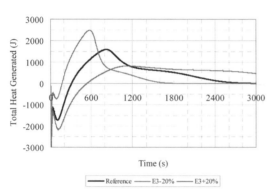

Figure 6. Total decomposition heat histories for different E_3 values.

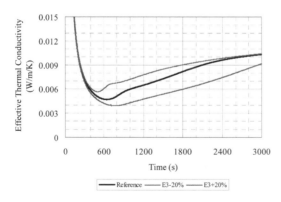

Figure 7. Profiles of effective thermal conductivity for different E_3 values.

Thermal conductivity of the coating plays the most important role in insulation performance of intumescent coating. In the discussions below, thermal conductivity is referred to as the effective thermal conductivity: $\lambda_{eff} = \frac{\lambda_{app} x_0}{x}$.

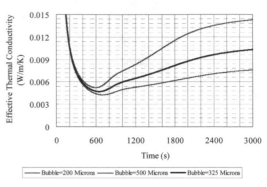

Figure 8. Effect of bubble size variation on effective thermal conductivity.

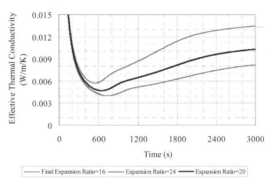

Figure 9. Effect of maximum expansion coefficient on effective thermal conductivity.

As indicated in equation 10, the radiative component of gas thermal conductivity is inversely related to the bubble size. Figure 8 shows the variations of the total thermal conductivity with different bubble diameters of 200, 325, and 500 microns. The expansion ratio is a key factor that affects the effective thermal conductivity. Figure 9 compares the effective thermal conductivity by varying the maximum expansion ratio from 20 by ± 20%.

5 APPLICATIONS

The previous two sections indicate that it is possible to use the model to predict steel temperatures with good accuracy. However, it is necessary to have accurate information on some of the key material properties, which include: activation energies of the blowing agent and charring material, bubble size and maximum expansion coefficient. A possible method is to extract these key parameters from the results of some small scale tests (e.g. cone calorimeter). To test this hypothesis, the following section presents the results of a preliminary study based on cone calorimeter tests carried out by Wang [6].

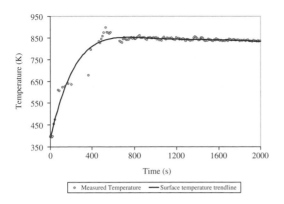

Figure 10.   Surface temperature histories in tests.

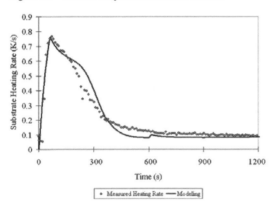

Figure 11.   Substrate heating rate in test B_{12}.

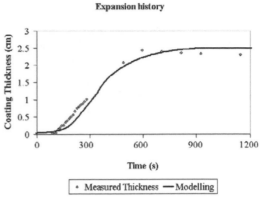

Figure 12.   Expansion histories in B12.

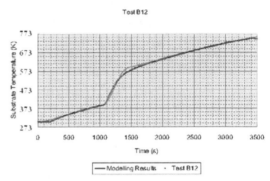

Figure 13.   Agreement between simulation and experiment for test B_{12}.

The test samples were steel plates with 10 and 20mm in thickness. The samples were painted with an intumescent coating (Char 21, supplied by KBS Brandskydd AB) with dry film thicknesses (DFT) of 0.4, 0.8 and 1.2 mm. The radiant heat flux was 50 kW/m^2. The back of the steel plate was attached by a layer of 40 mm- thick mineral wool to minimize heat loss. To minimise error caused due to inaccurate calculation of boundary conditions under cone radiation, the measured coating surface temperatures [7] were used.

Thermogravimetric Analysis (TGA) would normally be used to obtain the reaction kinetics. Unfortunately, TGA tests were not carried out so an alternative method was used in this study. Test B_{12} was chosen for extracting the key variables. In this test, the steel plate was 10mm and the coating was 0.4 mm in thickness. The surface temperature was recorded at different times, and a smoothed trend line was used to define the boundary condition in the simulations, as shown in Figure 10. Without TGA data for this test, an E_2 value of 115 kJ/mol was estimated to give accurate agreement between the measured and predicted substrate heating rate (shown in figure 11) based on the

reasoning that E_2 value governs the intumescent coating expansion process whereby the substrate heating rate before full coating expansion (around 900s in test B_{12}) is determined. The maximum expansion coefficient for test B_{12} was $E_{max} = 62.5$ which was used as input data in the simulations. Figure 12 compares the measured and calculated coating expansion histories, the good agreement of which further confirms the estimated E_2 value. From the calculated effective thermal conductivity of the coating by Wang [6], the bubble size "d" has been estimated as 300 μ m after expansion and 3.5 mm at the end of the experiment after charring. The activation energy of the charring process (E_3) was estimated as 55 kJ/mol to give best agreement with the test results of B_{12} (figure 13). It should be pointed out that with TGA data, E_3 would be obtained first and the bubble size would be obtained by trial and error.

Having obtained the key properties based on test B_{12}, Figures 14 and 15 present comparisons between simulations using the same key properties and experiment results for tests A_3 and A_6, respectively. The steel plate was 20 mm in thickness and the DFT values were 0.8 mm and 0.4 mm for tests A_3 and A_6, respectively.

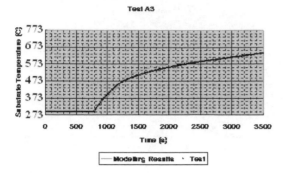

Figure 14.   Prediction for test A_3.

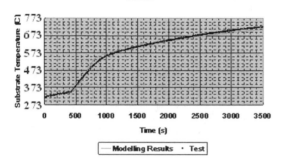

Figure 15.   Prediction for test A_6.

These results indicate the feasibility of using the same set of key parameters for different tests.

6   CONCLUSIONS

A mathematical model has been constructed in this study to model intumescent coating behaviour under different fire conditions. Good agreement has been achieved between modelling and experimental results. Parametric studies have been carried out to evaluate individual parameters. Activation energies of blowing agent and charring material, maximum expansion coefficient and bubble size are among the key parameters. It has been successful to use key variables extracted from one test to predict other tests, which indicates the potential to use this mathematical model to predict intumescent coating behaviour under different fire conditions once these key parameters have been quantified. To obtain the key parameters, TGA test is necessary. It is also necessary to find a reliable method to predict the maximum expansion coefficient and bubble size.

7   NOMENCLATURE

A Pre-exponential factor (s^{-1})
C specific heat campacity $(JK^{-1}kg^{-1})$

d bubble size(m)
e Emissivity
E activation energy $(kJmol^{-1})$
E_{max} maximum expansion coefficient
h convection heat transfer coefficient (Wm^{-2})
H heat of pyrolysis per unit mass of material (Jkg^{-1})
K reaction rate constant
m mass(kg)
P gas pressure(Pa)
Q heat flux(kWm^{-2})
R universal gas constant$(Jmol^{-1}K^{-1})$
t time(s)
T temperature(K)
v mass fraction
x coordinate along coating thickness(m)
$\dot{g}$ mass flow rate of gas per unit area $(kgs^{-1}m^{-2})$
ε Porosity
λ thermal conductivity$(Wm^{-1}K^{-1})$
ρ density(kgm^{-3})
σ Stefan-Boltzmann constant$(Wm^{-2}K^{-4})$

Subscripts:
c Coating
g Gas
s Solid
st Steel plate
r reactive component
$j =$ Inorganic source, blowing agent, charring
$1, 2, 3$ material, respectively

REFERENCES

1. Camino, G., L. Costa, and G. Martinasso "Intumescent Fire-retardant Systems," Poly. Deg. and Stab., 23:359(1989)
2. Russell, H.W., "Principles of heat flow in porous insulators," J. of Am Ceram Soc 18:1(1935)
3. Loeb, A.L., J.Am Ceram. Soc. 37:96 (1954)
4. Di Blasi, C., Branca, C., "Mathematical Model for the Nonsteady Decomposition of Intumescent Coatings" AIChE J. 47:2359 (2001)
5. Cagliostro, D. E., S. R. Riccitello, K. L. Clark, and A. B. Shimizu, "Intumescent Coating Modelling," J. of Fire & Flammability, 6: 205(1975)
6. Wang Y.C., Goransson U., Holmstedt, G and Omrane A., " A model for prediction of temperatures in steel structures protected by intumescent coating, based on tests in the Cone Calorimeter", Proceedings of the 8th International Symposium on Fire Safety Science, Beijing, China, pp. 235–246(2005)
7. Omrane A., Wang Y.C., Goransson, U., Holmstedt G. and Alden M., "Intumescent coating surface temperature measurement in a cone calorimeter using laser-induced phosphorescence", Fire Safety Journal, Vol. 42, pp. 68–74(2007)

Steel and Composite Structures – Wang & Choi (eds)
© *2007 Taylor & Francis Group, London, ISBN 978-0-415-45141-3*

Development of an engineering methodology for thermal analysis of protected structural members in fire

Hong Liang & Stephen Welch
University of Edinburgh, Edinburgh, UK

ABSTRACT: In order to overcome the limitations of existing methodologies for thermal analysis of protected structural members in fire, a novel CFD-based methodology has been developed. This is a generalised quasi-3D approach with computation of a "steel temperature field" parameter in each computational cell. The methodology accommodates both uncertainties in the input parameters and possible variants to the specification by means of parallel calculations. A framework for the inclusion of temperature/time-dependent thermal properties, including the effects of moisture and intumescence, has been established. The method has now been implemented as the GeniSTELA submodel within SOFIE RANS CFD code, with initial validation against results from full-scale fire tests. Model sensitivities have been demonstrated revealing the expected strong dependencies on certain properties of thermal protection materials. The code is verified as a generalised thermal analysis tool, with potential to provide a much more flexible means of assessing the thermal response of structure to fire than has been available hitherto.

1 INTRODUCTION

Increasing interest in assessing the performance of structures in fire is driving the development of an array of modelling methodologies to be used in fire safety engineering design. Whilst traditionally most code-based design has been based on simple calculations, referencing measured fire performance in standard tests, the progressive shift towards performance-based design has opened the door to use of advanced methods based on numerical models. These approaches will not replace standard testing, but they can already be used in a complementary fashion, to extend the application of test data.

Some simplified modelling methods have also been established, such as the protected member equation in Eurocode 3 (EC3) (BSI 2002), but as with all semi-empirical methods the results will tend to be conservative and there are of necessity a number of simplifying assumptions. CFD-based methodologies can in principle provide a much more detailed description of the thermal environment and the effects of localised heating, which could be used in conjunction with thermal analysis models to examine structural performance. In previous work (Kumar et al. 2005), a dedicated fine-mesh thermal modelling tool, known as STELA (Solid ThErmaL Analysis), has been implemented with the RANS CFD code SOFIE (Lewis et al. 1997). However, this research suggests that detailed thermal analysis of structural members in the context

of simulations of full-scale building fires remains problematic. This is partly due to the difference of scale between the mesh which can be afforded for the fire and that required for the thermal analysis of the structure, a particular problem with structured meshes, and also the generally high computational demands for coupled analyses. Moreover, existing approaches are limited to a specific structural arrangement of interest since it is necessary to define all model parameters in advance. Simulations must be repeated if details such as the structural geometry or the thermal properties are changed, a very inefficient procedure.

A more general and flexible methodology has now been proposed, still within the context of a CFD fire simulation, as reported previously (Liang & Welch 2006). This is based on computation of a set of "steel temperature field" parameters within the whole of the calculation domain, accommodating, by means of parallel calculations, both uncertainties in the input parameters and possible variants to the specification. Hence the need for repeat simulations is bypassed. This new generalised methodology is called GeniS-TELA (Generalised Solid ThErmaL Analysis) and also implemented in SOFIE.

This paper addresses the further development of this methodology and its extended application. In particular, the development of a modelling representation for the effects of intumescent performance in fire is described.

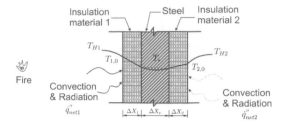

Figure 1. Schematic of heat transfer to protected steel member.

2 METHODOLOGY DEVELOPMENT

2.1 Brief description

When protected steelwork is exposed to fire, heat is transferred to the structure through a layer of insulation. The transient heating response of the member can in principle be described using conventional methods based on numerical heat transfer. However, full 3D analyses impose great computational demands, due to the large numbers of cells required in order to adequately resolve the steep thermal gradients during the initial heating. Even if the computational resource is available, in simple deterministic models there is no direct mechanism to accommodate uncertainties in the thermal properties and member specification. To overcome these problems, with an appropriate balance between accuracy and tractability, a novel quasi-3D analysis methodology has been developed (Liang & Welch 2006). This is achieved by constructing a generalised 1D model and further considering the 2D or 3D effects within the heat transfer processes by appropriate approximations and corrections. The computations are performed in each gas-phase CFD cell in the computational domain.

2.2 Generalised 1D model

The generalised 1D model is constructed through analysing the heat transfer to and within an element in an idealised protected steel member assumed to be exposed to heat on two faces, as shown in Figure 1 below:

This element is supposed to be representative of a slice of a protected steel structure, e.g. a finite section of a flange or a web; two faces are used to allow for situations where the exposure conditions on each side might vary, encompassing also the case of hollow sections with very different exposures on the inside of the structure, though in that case the insulation thickness on the inside is reduced to zero.

The generalised 1D model provides a modelling framework which exploits a simple thermal penetration model for the protection coupled to an essentially lumped parameter representation of the steel heating.

The governing equations for this model are derived by considering the net energy balance together with surface heat transfer boundary conditions (Carslaw & Jaeger 1959), as given below:

Energy balance equation:

$$\frac{\partial E_{system}}{\partial t} = \dot{q}''_{net} \tag{1}$$

i.e.

$$\rho_s \cdot c_{ps} \cdot \frac{\partial T_s}{\partial t} \cdot \Delta x_s + w_{p1} \cdot \rho_1 \cdot c_{p1} \cdot \frac{\partial T_1}{\partial t} \cdot \Delta x_1 + w_{p2} \cdot \rho_2 \cdot c_{p2} \cdot \frac{\partial T_2}{\partial t} \cdot \Delta x_2$$
$$= h_{c1} \times (T_{H1}^{(n)} - T_{1,0}^{(n)}) + \dot{q}''_{r1} - \varepsilon_{m1} \cdot \sigma \cdot T_{1,0}^{(n)4} +$$
$$h_{c2} \times (T_{H2}^{(n)} - T_{2,0}^{(n)}) + \dot{q}''_{r2} - \varepsilon_{m2} \cdot \sigma \cdot T_{2,0}^{(n)4}$$

The terms shown in the expanded equation here represent, respectively, the transient heating of the steel and protection layer on each side, and the convection, radiation and reradiation for each surface of the protected member. A semi-empirical treatment is adopted for transient heating, allowing for spatially- and temporally-varying temperature gradients within the solid. The boundary conditions are supplied from the heat transfer solution for the surfaces, using the following equations:

$$\dot{q}''_{net1} = \frac{k_1}{w_{p1}\Delta x_1} \cdot (T_{1,0}^{(n)} - T_s) \tag{2}$$

i.e. $h_{c1} \times (T_{H1}^{(n)} - T_{1,0}^{(n)}) + \dot{q}''_{r1} - \varepsilon_{m1} \cdot \sigma \cdot T_{1,0}^{(n)4} = \frac{k_1}{w_{p1}\Delta x_1} \cdot (T_{1,0}^{(n)} - T_s)$

$$\dot{q}''_{net2} = \frac{k_2}{w_{p2}\Delta x_2} \cdot (T_{2,0}^{(n)} - T_s) \tag{3}$$

i.e. $h_{c2} \times (T_{H2}^{(n)} - T_{2,0}^{(n)}) + \dot{q}''_{r2} - \varepsilon_{m2} \cdot \sigma \cdot T_{2,0}^{(n)4} = \frac{k_2}{w_{p2}\Delta x_2} \cdot (T_{2,0}^{(n)} - T_s)$

where:

σ is Stefan-Boltzmann constant (5.67×10^{-8} W/m²/K⁴);

$\dot{q}''_{r1}, \dot{q}''_{r2}$ are incident heat fluxes on each side;

$T_{1,0}^{(n)}, T_{2,0}^{(n)}$ are surface temperatures at gas/solid interfaces;

T_s, T_1, T_2 are steel and average protection layer temperatures, respectively;

h_{c1}, h_{c2} are convection coefficients;

$\varepsilon_{m1}, \varepsilon_{m2}$ are emissivities of protection layers;

ρ_s, ρ_1, ρ_2 are densities of steel and protection layers, respectively;

$\Delta x_s, \Delta x_1, \Delta x_2$ are thicknesses of steel and protection layers, respectively;

w_{p1}, w_{p2} are weight factors of protection layers, defined in terms of the thermal penetration depth of the protection, given in the form of Equation 4:

$$w_p = \min\{\frac{A_{actual}}{A_{model}} = \frac{\delta}{\Delta x_p}, 1\} \qquad (4)$$

where

$$\delta = 2 \cdot \left(\frac{k_p \cdot t}{c_p \cdot \rho}\right)^{1/2}$$

c_s, c_{p1}, c_{p2} are specific heat of steel and protection layers, respectively;

k_1, k_2 are thermal conductivity parameters of protection layers.

The temperature/time dependent characteristics, including moisture and intumescence effects, are incorporated in certain parameters for generalisation of the methodology. The treatment for moisture was described in the previous paper (Liang & Welch 2006) whilst intumescence is described in section 3 below.

It is well-known that the above situation is a strongly coupled problem, with the net heat fluxes at the gas-solid interface very much dependent on the surface temperature, but both also related to the transient thermal response of the structure itself. Numerical instabilities might become evident if inadequate solution procedures are used; these are overcome using a Newton-Raphson method to update the surface temperature from the heat transfer boundary condition governing equations and thereafter, with the updated surface temperature as a boundary condition, solving the overall energy balance equation (Equ. 1) with the Runge-Kutta method to obtain the steel temperature. Further details of the solution procedure are provided in the earlier paper (Liang & Welch 2006).

2.3 Quasi-3D model

Use of a fundamentally 1D treatment is essential, considering the costs of doing a full 3D analysis in every computational cell and including a sufficient number of parametric variations. However, adoption of a simple 1D model for thermal analysis could clearly lead to some modelling inaccuracies. These could in principle be in either direction, resulting in either conservative (over-design) or non-conservative (unsafe) results. The former aspect is not a major concern since the method is in any case far more flexible than other simple models, and by using generalised treatments conservatism is already greatly reduced. The latter aspect is a more obvious problem, and in order to overcome it, methods for treating important 2D and 3D effects are needed. A number of corrections factors have been

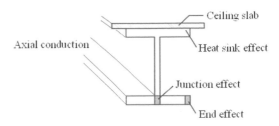

Figure 2. Cross-section of the beam with locations of possible correction effects.

implemented in the model, encompassing the factors indicated in Figure 2, i.e. the junction effect, end effect, heat sink effect and axial temperature gradient effect, as described in the earlier paper (Liang & Welch 2006). It is important to note that these effects are only critical where they negatively impact the performance of the member, i.e. increase the solid temperatures, and in the majority of cases the opposite is true, i.e. the default procedure is a good representation of the "worst" case. Thus, while it is vital to show that these possible corrections have been appropriately considered, their effect on the final results has been found to be fairly limited.

3 MATERIAL PROPERTIES

The aforementioned model might be considered as a reasonable representation of the *fundamental* aspects of the heat transfer phenomena. However, in practice several factors are found to have a great impact on the transient response, in particular the thermal properties of the protection materials, which affect the surface temperature and thus the steel temperature. It is known that these properties are often strongly temperature/time dependant and the use of constant values may result in significant errors in some cases. The methodology developed here aims at generalising the thermal analysis to accommodate all important phenomena; conventional approaches to treatment of moisture effects have already been implemented, referencing modified specific heats and thermal conductivities (Liang & Welch 2006). This is now extended further to include the effects of intumescence, clearly of great practical relevance to the case of protected steelwork. In order to do so, geometrical and density variations must also be explicitly treated.

Intumescent materials are an increasingly popular form of fire protection, due to a number of advantages arising from the fact that they can be applied as thin, aesthetically pleasing, coatings either before or after construction (Goode 2004, Jimenez et al. 2006, Bailey 2006a). When in contact with high temperatures, they will swell and form a layer of carbonaceous char which has much greater thickness than the initial state. The char subsequently acts as a thermal barrier to effectively protect the substrate against increase in temperature. Nevertheless, during the process of intumescence, the material properties are severely changed along with mass transport and endo- and exothermic reactions. These properties include thermal conductivity, specific heat, density and thickness of the intumescent layer.

Several research studies have been carried out to determine the effective intumescent thermal properties by experimental tests, in conjunction with some form of numerical analysis. These include bench-scale cone calorimeter tests and small-scale furnace tests on coated plates (Bartholmai et al. 2003, Bartholmai & Schartel 2007), and furnace tests on cellular beams (Bailey 2006b). The first authors conducted studies on typical water-based and solvent-containing intumescent systems (Bartholmai et al. 2003) and later on a high-performance material, i.e., epoxy resin containing boric acid and phosphate-based flame retardant (Bartholmai & Schartel 2007). The results from the former showed a significant slow down of temperature increase between 200–300°C, due to intumscence, i.e. the formation of an insulating char and other co-acting energy absorbing processes; temperature influences during the latter tests also resolved a damping effect at 150°C due to the endothermic reaction of boric acid, which also releases water. Layer thickness effects were non-linear. Considering first the geometrical expansion, a simple conceptual model would suggest that thermal equivalence to a finite thickness problem can be achieved by simply scaling the thermal conductivity by the layer thickness, $d(=1\,\text{m})$, giving an effective thermal conductivity, k/d. Density is scaled in the same way, and specific heat by the inverse of d, but these parameters always appear as factors of each other so these scalings vanish in the term ρc_p.

The description of the temperature-dependent intumescent thickness, d, can be determined from an expression for the expansion ratio. We postulate that this will fit the general form:

$$R = 1 \qquad\qquad T < T_{lower} \qquad (5)$$

$$R = 1 + \frac{1}{2}\left(R_f - 1\right)\left(\frac{T - T_{lower}}{T_{mid} - T_{lower}}\right)^n \qquad T_{lower} \leq T < T_{mid} \quad (6)$$

$$R = R_f - \frac{1}{2}\left(R_f - 1\right)\left(\frac{T_{upper} - T}{T_{upper} - T_{mid}}\right)^n \qquad T_{mid} \leq T < T_{upper} \quad (7)$$

$$R = R_f \qquad\qquad T \geq T_{upper} \qquad (8)$$

where

R is the time-dependent expansion factor;
R_f is the final expansion factor;

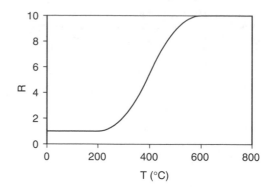

Figure 3. Scaling factor R change with temperature.

$T_{lower}, T_{upper}, T_{mid}$ are the critical temperatures where scaling factor changes;

T is the current averaged intumescent temperature;

n is a shape factor power.

Here, besides the relevant temperatures limits, the critical controlling parameters are the shape factor n and the overall expansion ratio R_f. An approximate calibration has been performed by comparison with test data, including the results of Bartholmai (Bartholmai & Schartel 2007), giving a value of $n=2$. Taking an approximate temperature range from the DTG results of the latter study, and assuming $R_f = 10$, gives the following curve:

In fact, a variety of overall 1D expansion ratios have been reported in the literature (10 – Desanghere & Joyeux 2005; 15–30 – Goode 2004; 50–220 – Bartholmai et al. 2003).

The key parameter for the thermal model is the conductivity, or its scaled value, i.e. k/d. The conductivity itself is affected by fundamental changes in the material as it intumesces (Tan et al. 2004, Bartholmai et al. 2003, Bartholmai & Schartel 2007). Unfortunately, the effect is non-linear and very dependent on initial thickness, and most pronounced at the smaller thicknesses typical of real applications; hence, there would appear to be no substitute for its direct experimental determination. Work is currently underway at Edinburgh to determine thermal properties of intumescents in cone calorimeter tests. In the meantime, the various literature results would suggest an initial increase followed by a fall during intumescence and finally a sharp rise during material degradation. For our initial model, we have fitted values derived from the results of Bartholmai (Bartholmai & Schartel 2007), as shown in Figure 4 below:

4 MODEL IMPLEMENTATION AND VERIFICATION

The above conceptual model has been implemented as a submodel called GeniSTELA within the SOFIE

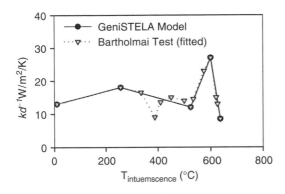

Figure 4. Comparison of thermal conductivity between generalised model and Bartholmai test (Bartholmai & Schartel 2007).

Figure 5. BRE 12×12 m large compartment fire test.

RANS CFD code (Lewis et al. 1997). Representative empirical values are adopted for some terms such as the initial conditions, the dry thermal properties, moisture content, etc., and their influence has been studied by exercising the model with different sets of input parameter values. The performance of the model was assessed by performing sensitivity studies, looking at the effects of a range of numerical and physical parameters. Comparisons were also made with the results from the EC3 protected member equation (BSI 2002).

The case used for verification studies is the protected steel indicative, UC254 × 254/73, in the full-scale tests on a 12 × 12 m compartment undertaken at BRE Cardington (Welch et al. 2007); this member was protected with about 25 mm of Fendolite MII sprayed fibre (baseline $\rho = 680$ kg/m³, $k = 0.19$ W/m/K). Figure 5 shows the test compartment while Figure 6 shows the SOFIE temperature predictions.

In the test a variety of thermal parameter measurements were made, encompassing conditions in the gas phase (temperatures, velocities and heat fluxes) and in the solid phase (steel temperatures in protected beams, columns and indicatives, with and without protection); this study also serves for an initial validation of the model, comparing the model predictions with the measured steel temperatures in the protected indicative.

Figure 6. BRE 12×12 m large compartment fire modelling.

with the fact that the thermal exposures are more severe deeper into the fire (Welch et al. 2007), and the model predictions from GeniSTELA are heavily influenced by the radiative terms, $\dot{q}_r''$, derived directly from the CFD calculation.

Figure 7 shows the temperature predictions for the protected indicative within the compartment. There is a large temperature gradient across the protection.

Figure 8 shows a comparison of the predictions of steel temperature with the test together with EC3 prediction. The latter exceeds the measure temperature reflecting some conservatism in this semi-empirical method, while the prediction from GeniSTELA indicates a sufficient match with the test.

Studies of computational requirements showed that of single instance of the GeniSTELA solver was equivalent to 10 fluid flow steps for a simple localised fire problem (Liang et al. 2007). No significant errors arose when this was increased to a 30 step interval, suggesting that larger parametric case studies will be feasible, especially if the call interval is optimised by automatic selection.

5 RESULTS

5.1 Simulation results

Gas and steel temperatures were computed using SOFIE and the coupled GeniSTELA code. In qualitative terms the results showed the expected differences in steel and gas temperature fields, with relatively higher steel temperatures within the depth of the compartment compared to the openings. This is consistent

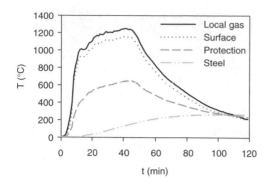

Figure 7. Temperatures at protected indicative, test 8.

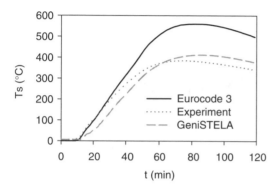

Figure 8. Comparison of steel temperatures.

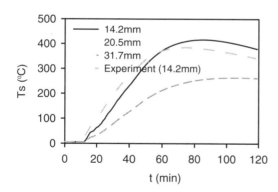

Figure 9. Effect of flange thickness on steel temperature.

5.2 *Sensitivity study results*

Some results from the sensitivity study are shown in Figures 9 & 10 for the effects of changing the steel flange thickness (spanning UC $254 \times 254/73,107,167$) and the protection thickness (12.5 to 50 mm). The results for changing the protection thermal conductivity mirror the latter, and show the expected strong influence of protection properties.

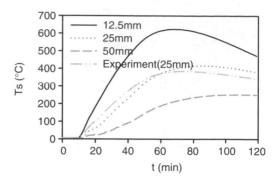

Figure 10. Effect of protection thicknesses on steel temperature.

6 CONCLUSIONS

A generalised methodology for thermal analysis of protected steel structures in fire is described. A framework for the inclusion of treatments for intumescence effects has now been established, with provision for simple calibration in the case of each specific formulation of interest. The GeniSTELA implementation of the method is based on parallel computations spanning the range of cases of interest, providing a generalised methodology. The initial results confirm the sufficiency of the algorithms adopted and comparisons with measurements in a post-flashover compartment fire test are satisfactory. Computational demands are found to be acceptable. Strong dependencies on the thermal properties of the protection materials are observed in the sensitivity studies. These results serve to illustrate the importance of using generalised methodologies in tackling thermal response problems.

ACKNOWLEDGEMENTS

The first author acknowledges the support of her industrial sponsor, the BRE Trust, and the School of Engineering & Electronics. Useful comments on intumescent properties were received from Matthias Bartholmai, and Daniel Joyeux and Mario Fontana via IAFSS email list.

REFERENCES

Bailey, C.G. 2006a. Advances in fire engineering design of steel structures. Proc. ICE, Structures and Buildings 159, issue SB1, pp. 21–35.

Bailey, C.G. 2006b. Indicative fire tests to investigate the behaviour of cellular beams protected with intumescent coatings, Fire Safety Journal, vol. 36, no. 8, pp. 689–700.

Bartholmai, M. & Schartel, B. 2007. Assessing the performance of intumescent coatings using bench-scaled cone calorimeter and finite difference simulations. Fire and Materials, *in press*.

Bartholmai, M., Schriever, R. & Schartel, B. 2003 Influence of external heat flux and coating thickness on the thermal properties of two different intumescent coatings using cone calorimeter and numerical analysis, Fire and Materials, vol. pp. 151–162.

British Standards Institution. 2002. Eurocode 3: Design of steel structures – Part 1–2: General rules – Structure fire design.

Carslaw, H.S. & Jaeger, J.C. 1959. Conduction of Heat in Solids. Oxford University.

Desanghere, S. & Joyeux, D. 2005. Development of design rules for the fire behaviour of external steel structures. ECSC project no. 7210-PR-380, final report.

Goode, M.G. (ed.) 2004 Fire protection of structural steel in high-rise buildings, NIST GCR 04–872.

Jimenez, M., Duquesne, S. & Bourbigot, S. 2006. Characterisation of the performance of an intumescent fire protective coating, Surface and Coatings Technology, vol. 201, issue 3–4, pp. 979–987.

Kumar, S., Welch, S., Miles, S.D., Cajot, L.-G., Haller, M., Ojanguren, M., Barco, J., Hostikka, S., Max, U. & Röhrle, A. 2005. Natural Fire Safety Concept – The development and validation of a CFD-based engineering methodology for evaluating thermal action on steel and composite structures. European Commission Report EUR 21444 EN, 150 pp., ISBN 92-894-9594-4.

Lewis, M.J., Moss, J.B. & Rubini, P.A. 1997. CFD modelling of combustion and heat transfer in compartment fires. Proc. 5th Int. Symp. on Fire Safety Science, pp. 463–474.

Liang, H. & Welch, S. 2006. A novel engineering tool for thermal analysis of structural members in natural fires. Proc. 4th Int. Workshop on "Structures in fire". Aveiro, Portugal.

Liang, H., Welch, S. Stratford, T. & Kinsella, E.V. 2007. Development and validation of a generalised engineering methodology for thermal analysis of structural members in fire. Proc. 5th Int. Seminar Fire & Explosion Hazards. Edinburgh, UK.

Tan, K.H., Wang, Z.H. & Au, S.K. 2004. Heat transfer analysis for steelwork insulated by intumescent paint exposed to standard fire conditions. Proc. 3rd Int. Workshop on "structures in fire". Ottawa, Canada.

Welch, S., Jowsey, A., Deeny, S., Morgan, R. & Torero, J.L. 2007. BRE large compartment fire tests – characterising post-flashover fires for model validation, Fire Safety Journal, *in press*.

Steel and Composite Structures – Wang & Choi (eds)
© *2007 Taylor & Francis Group, London, ISBN 978-0-415-45141-3*

Transient heat transfer model on external steel elements and external hollow steel sections filled with concrete

F. Morente & J.A. Chica
Labein Tecnalia

ABSTRACT: A new model has been developed by LABEIN TECNALIA within the RFCS project "Development of design rules for the fire behaviour of external steel structures", improving the Eurocode's model.

1 INTRODUCTION

It is becoming more usual nowadays to see how some outstanding buildings leave, by architectonic requirements, part of their structure outside the building envelope. This is the case, for instance, of the George Pompidou Centre in Paris or the Arts Hotel in Barcelona among many others. A fire originated inside this kind of buildings can affect in a very different way the outer elements than others which are within the fire enclosures.

The Eurocode 3 : "Design of steel structures" Part 1.2 : "General rules Structural fire design", gathers a stationary model to analyze the thermal response of the external elements exposed to fire. This methodology allows determining a uniform steel temperature in the most heated section of the external members, taking into account the shadow effect in a simplified way based on the use of section factors.

A new model has been developed by LABEIN TEC-NALIA within the RFCS project "Development of design rules for the fire behaviour of external steel structures", improving the Eurocode's model in the following aspects:

- Against the stationary model available currently, the new methodology deals with the transient effects related to a changing exposure to heat from flames and openings during the whole development of the fire.
- The Eurocode's model treats open and hollow sections in the same way, with a section factors based method. The new approach uses an improved definition of configuration factors to take into account the shadow effect. Shadow zones are established in a profile's section and the net heat flux that reaches different points in the section is computed, considering the radiation flux from other parts of the section at different temperatures.

- In open cross sections, due to the different exposure to heat from flames and openings, a thermal gradient is created through the section. Therefore, all the parts do not undergo the same temperature evolution and need to be dealt with separately. The temperatures of web and flanges are calculated for each step of time.
- A new methodology has been implemented to study concrete filled hollow steel sections, calculating heat flux boundary conditions in order to use them with a FEM code to calculate the temperature field across the section.

The model has been implemented in a software (EXTFIRE) developed by LABEIN TECNALIA to make easier its application, and it has been validated with several fire tests carried out by CTICM and TNO during the development of the RFCS project mentioned before, being in a good agreement with test results.

The new methodology provides an approach to the performance based design in which structural fire resistance of the external members could be evaluated in function of time, helping the designer to calculate the time to collapse.

2 DESCRIPTION OF THE MODEL

In open cross sections, due to the transient effects related to different exposition to heat from flames and openings, a thermal gradient is created trough the section. Therefore, all the parts do not undergo same temperature evolution and need to be dealt with separately (web and flanges). Other effects are also accounted for in this kind of sections, such as shadow effect and radiating from other parts of the section at different temperatures. The temperature of each part is calculated for each step of time.

In concrete filled hollow steel sections, heat flux boundary conditions are calculated to use them in other FEM programs to know the temperature distribution in the section.

2.1 Basis

- The fire compartment is assumed to be confined to one storey only. All windows are assumed to be rectangular and with the same height.
- The temperature of the compartment and the parameters of the flames projecting from the openings (dimensions and temperatures) should be calculated apart.
- According to the EC3, a distinction is made between members not engulfed and engulfed in flame, depending on the size of the external flames and the position of the elements. A member that is not engulfed in flame is assumed to receive radiative heat transfer from all the openings and flames. An engulfed member is assumed to receive radiative and convective heat transfer from the engulfing flame and radiative transfer from the opening from which it projects.
- Flames and openings are considered like radiating surfaces with a fixed temperature. An emissivity of 1 is considered for openings and a specific emissivity depending on the thickness for the flames.
- No forced draught condition is assumed.

2.2 Open cross section columns not engulfed in flames

The section is divided in three zones, the two flanges and the web. Each part of the section is going to receive different heat fluxes from openings and flames, so three different temperatures should be considered. Openings and flames are considered like radiating surfaces with a specific temperature and emissivity.

- The emmisivity of the radiating screen is 1 for windows, but for flames is calculated according to the EC:

$$\varepsilon = 1 - e^{-0.3\lambda} \tag{1}$$

where λ is the flame thickness at the top of the window (2h/3).

- The temperature of the radiating screen is the compartment temperature for windows. For flames, the temperature at a distance $\frac{h}{2}$ from the opening measured along the flame axis should be taken (calculated according to EC1).

The contributions of all flames and openings are summed to compute the whole radiative heat flux that arrives at each part of the element, considering that the absortivity of flames is taken as 0 for not engulfed

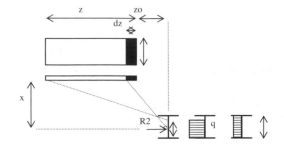

Figure 1. Calculation of the configuration factor for the web.

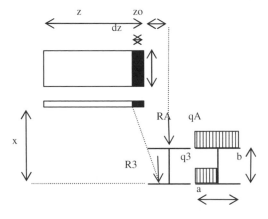

Figure 2. Calculation of the configuration factor for the flanges.

members. The convective heat transfer is neglected for not engulfed elements.

Shadow effect is studied for each radiating screen, in all the surfaces of the section. First the zone in shadow is calculated for each surface. Then, a uniform heat flux is considered in the radiated zone, which value is calculated like the heat flux that arrives at the central point of this radiated zone (calculated according to EC1), considering the dimensions of the effective radiating surface for this point. Finally, the total heat flux that comes into contact with the radiated surface is assumed to be applied in the whole surface (flange or web, see figures 1 and 2).

$$\phi_\perp(x,y,z) = \frac{1}{2\pi}[\tan^{-1}(\tfrac{y}{x}) - \frac{1}{\sqrt{1+(\tfrac{z}{x})^2}}\tan^{-1}(\frac{\tfrac{y}{x}}{\sqrt{1+(\tfrac{z}{x})^2}})] \tag{2}$$

$$\phi_2 = \phi_\perp(x,y,zo+z) - \phi_\perp(x,y,zo+dz) \tag{3}$$

$$R_2 = \phi_2 \cdot \varepsilon \cdot \sigma \cdot T^4 \cdot (1-a_z) \tag{4}$$

$$q_2 = R_2 \cdot c / b \tag{5}$$

$$\phi_\parallel(x,y,z) = \frac{1}{2\pi}[\frac{\tfrac{y}{x}}{\sqrt{1+(\tfrac{y}{x})^2}}\tan^{-1}(\frac{\tfrac{z}{x}}{\sqrt{1+(\tfrac{y}{x})^2}}) + \frac{\tfrac{z}{x}}{\sqrt{1+(\tfrac{z}{x})^2}}\tan^{-1}(\frac{\tfrac{y}{x}}{\sqrt{1+(\tfrac{z}{x})^2}})] \tag{6}$$

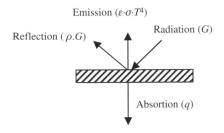

Figure 3. Interaction between surfaces.

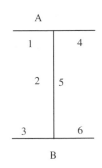

Figure 4. Surfaces in the section.

$$\phi_A = \phi_{//}(x-b, y, zo+z) - \phi_{//}(x-b, y, zo) \quad (7)$$

$$\phi_3 = \phi_{//}(x, y, zo+z) - \phi_{//}(x, y, zo+dz) \quad (8)$$

$$R_A = \phi_A \cdot \varepsilon \cdot \sigma \cdot T^4 \cdot (1-a_z) \quad (9)$$

$$R_3 = \phi_3 \cdot \varepsilon \cdot \sigma \cdot T^4 \cdot (1-a_z) \quad (10)$$

$$q_A = R_A \quad (11)$$

$$q_3 = R_3 \quad (12)$$

The interaction between surfaces is considered to compute the net radiative heat flux that each surface absorbs. The surfaces of the member are considered like gray surfaces ($\varepsilon = \alpha = 1 - \rho$). It is assumed that fictitious surfaces absorb all the radiation they receive (See figure 3).

$$J = \varepsilon \cdot \sigma \cdot T^4 + \rho \cdot G \rightarrow Radiossivity \quad (13)$$

$$q = G - J \quad (14)$$

The radiossivity of each part is the sum of the emission due to the temperature, the reflection of the radiation from the environment and the reflection of the radiation from the other parts of the section.

$$J_i = \varepsilon \cdot \sigma \cdot T^4 + \sigma \cdot T_o^4 \cdot \rho + q_i \cdot \rho + J_j \cdot F_{j,i} \cdot \rho + J_k \cdot F_{k,i} \cdot \rho \quad (15)$$

The temperature of each part of the section used to compute the term of emission is delayed a step with the others ones, because this is just the wanted parameter.

The whole phenomenon of reflections, emissions . . . between all the parts of the section can be studied solving a linear equations system which collects all the heat balances.

$$J = [A]^{-1} \cdot v \quad (16)$$

After that it is possible to solve the heat balance to calculate the temperature in the different zones of the section (See figure 4).

$$Q_i = q_i + \sigma \cdot T_0^4 - J_i + F_{j,i} \cdot J_j + F_{k,i} \cdot J_k \quad (17)$$

The increase of temperature in each step of time is computed according to EC3

$$h_{flange1} = \frac{a}{2} \cdot Q_1 + \frac{a}{2} \cdot Q_4 + a \cdot Q_A \quad (18)$$

$$h_{web} = b \cdot Q_2 + b \cdot Q_5 \quad (19)$$

$$h_{flange2} = \frac{a}{2} \cdot Q_3 + \frac{a}{2} \cdot Q_6 + a \cdot Q_B \quad (20)$$

$$\Delta\theta = \frac{h_s \cdot \Delta t}{\rho \cdot C_p \cdot l \cdot e} \quad (21)$$

When the column is between two openings, two radiating surfaces are considered in each side of the element. These surfaces have a specific temperature and emissivity.

• The emmisivity of the radiating screen is calculated for each side according to the EC

$$\varepsilon_m = 1 - e^{-0.3 \cdot \lambda m} \quad (22)$$

$$\lambda_m = \sum_{i=1}^{m} w_i \quad (23)$$

$$\varepsilon_n = 1 - e^{-0.3 \cdot \lambda n} \quad (24)$$

$$\lambda_n = \sum_{i=1}^{n} w_i \quad (25)$$

where w_i is the width of each opening, m is the number of openings on side m and n is the number of openings on side n.

• The temperature of the radiating screen is the temperature at a distance h/2 from the opening measured along the flame axis should be taken (calculated according to EC1).

The contributions of the two flames are summed to compute the whole radiative heat flux that arrives at each part of the element. The radiative heat transfer

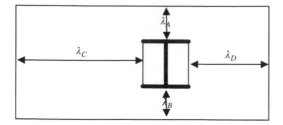

Figure 5. Surfaces in the section.

from the windows and the convective heat transfer are neglected.

Shadow effect is studied for each radiating screen in a similar way to the method presented in part 3.1. In this case there are not zones in shadow, but the effective screen dimensions for each zone must be calculated.

The interaction between surfaces and the calculation of heat fluxes and temperatures are calculated like it has been explained before.

2.3 Open cross section columns engulfed in flames

An engulfed member is assumed to receive radiative and convective heat transfer from the engulfing flame and radiative transfer from the opening from which it projects. Four fictitious radiating surfaces are considered enclosing the section in order to model the engulfing flame. The opening is considered like a radiating surface with a specific temperature and emissivity.

- The emmisivity of the radiating screen is 1 for windows, but for flames is calculated according to the EC (See figure 5):

$$\varepsilon = 1 - e^{-0.3 \cdot \lambda} \qquad (26)$$

where λ is the flame thickness at the top of the window (2h/3).

- The temperature of the radiating screen is the compartment temperature for windows. For flames, the temperature at a distance $\frac{h}{2}$ from the opening measured along the flame axis should be taken (calculated according to EC1).

The absortivity of flames and the convective heat transfer are calculated according to Eurocode:

$$a_z = \frac{\varepsilon_A + \varepsilon_B + \varepsilon_C}{3} \qquad (27)$$

$$\alpha = 4.67 \cdot (\frac{1}{d_{eq}})^4 \cdot (\frac{Q}{A_v})^{0.6} \qquad (28)$$

$$d_{eq} = \frac{a + b}{2} \qquad (29)$$

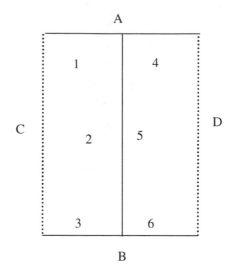

Figure 6. Surfaces in the section.

where A_v is total area of vertical openings on all walls.

Shadow effect is only studied for the window radiating screen in a similar way to the method presented before.

The interaction between surfaces is considered to compute the net radiative heat flux that each surface absorbs. The surfaces of the member are considered like gray surfaces ($\varepsilon = \alpha = 1 - \rho$) (See figure 6).

The radiossivity of each part is the sum of the emission due to the temperature, the reflection of the radiation from the environment and the reflection of the radiation from the other parts of the section.

$$J_i = \varepsilon \cdot \sigma \cdot T^4 + \sigma \cdot T_o^4 \cdot \rho + q_i \cdot \rho + J_j \cdot F_{j,i} \cdot \rho + J_k \cdot F_{k,i} \cdot \rho + J_C \cdot F_{C,i} \cdot \rho \qquad (30)$$

The temperature of each part of the section used to compute the term of emission is delayed a step with the others ones, because this is just the wanted parameter.

The whole phenomenon of reflections, emissions . . . between all the parts of the section can be studied solving a linear equations system which collects all the heat balances.

$$J = [A]^{-1} \cdot \vec{v} \qquad (31)$$

After that it is possible to solve the heat balance to calculate the temperature in the different zones of the section.

The increase of temperature in each step of time is computed according to EC3.

$$h_{flange1} = \frac{a}{2} \cdot Q_1 + \frac{a}{2} \cdot Q_4 + a \cdot Q_A + 2 \cdot a \cdot \alpha \cdot (T_{flame} - T_{flange1}) \qquad (32)$$

$$h_{web} = b \cdot Q_2 + b \cdot Q_5 + 2 \cdot b \cdot \alpha \cdot (T_{flame} - T_{web}) \qquad (33)$$

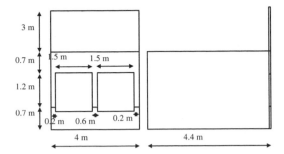

Figure 7. Experimental set up.

$$h_{flange2} = \frac{a}{2} \cdot Q_3 + \frac{a}{2} \cdot Q_6 + a \cdot Q_B + 2 \cdot a \cdot \alpha \cdot (T_{flame} - T_{flange2}) \qquad (34)$$

$$\Delta\theta = \frac{h_s \cdot \Delta t}{\rho \cdot C_p \cdot l \cdot e} \qquad (35)$$

2.4 Close cross section columns

Due to the shape of the section, neither shadow effect nor interactions between surfaces are taken into account.

Heat flux boundary conditions are calculated to use them in other FEM programs to know the temperature distribution in the section.

3 VALIDATION

A fire compartment was built at the TNO Centre for Fire Safety for the purpose of carrying out full-scale experiments of external flaming acting on unloaded steel structures. Several tests were carried out for various compartment and opening dimensions, to study the effect of each parameter on the thermal action and thermal response of the steel structures.

The size of the furnace is 4.4 m of length, 4 m of width and 2.6 m of height. The fire compartment is constructed from cellular concrete blocks of thickness 150 mm.

Two openings are placed symmetrically as shown in figure 7. A promatec panel of height 3 m was attached to the top of the compartment in order to represent the façade of the upper floor of a typical multi-stored building.

Three columns are placed in front of the compartment as shown in figure 21, a HEA 200 and two rectangular concrete filled SHS columns (300 mm × 300 mm) (See figure 8).

The fire source is made up of 42 wooden Europallets stacked on a platform hanging about 0.1 m above the floor of the compartment (1050 Kg of wood). The fire load density is 60 Kg/m².

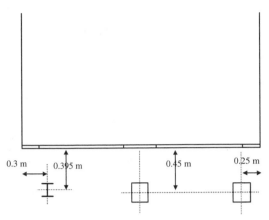

Figure 8. Columns in front of the fire compartmente.

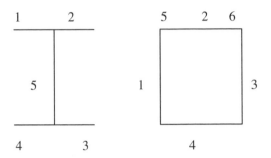

Figure 9. Thermocouples distribution.

Test data are used to simulate the compartment and flame properties:

- The highest measured values are taken for the opening and flames temperatures, and the mean value for the room temperature.
- The convection coefficient is calculated according to the Law model, so the heat release rate is not necessary for columns.
- External flaming occurs from 660 to 3000 sec, as shown in the test.
- The other flame properties not described in the report [2] are calculated according to the Law model.

The steel temperature is measured at 2.1 m above floor level. The measuring section is equipped by thermo-couples distributed as shown in the next figure 9.

4 CONCLUSIONS

There is a very good agreement between test and model results, discrepancies are due to how the thermal action is defined.

The current model of the Eurocode gives results in the unsafe side, because peak temperatures are higher than the steady-state temperature predicted with the current model.

For open sections, the results given by the model are quite close to tests results, lying in the safe side. Discrepancies are not quite large to render the model too conservative.

For composite sections, due to the low conductivity of concrete, large gradients occur, and a finite element calculation is addressed after calculation of heat fluxes on steel perimeter. A simplified 1D model has been checked, and in some cases it is a good estimate of heat transfer within the section. If the peak temperature is the target of the calculation, this approach trends to overestimate the temperature of hottest point of the section because it neglects the conduction to colder parts of steel perimeter.

Otherwise, if a full description of temperature rise is targeted, it trends to underestimate temperature rise of coldest points. This result in an overestimation of thermal gradient within the section due to accidental actions. If the bending moment induced by this thermal gradient sums up to the initial bending moment this result is not unsafe, otherwise, in the first moments of the fire exposure it leads to a reduction of the actual bending moment acting on the member. This effect may be neglected once high temperatures within the section take the role in structural response and accidental induced actions do not play the main role in collapse mechanism.

ACKNOWLEDGEMENTS

The authors wish to thank the partners of the RFCS research project "Development of design rules for the fire behaviour of external steel structures (Contract No.: 7210-PR/380; 1/07/ 2002-21/06/2005)".

REFERENCES

1994. Eurocode 1: Basis of design and actions on structures, Part 2.2: Actions on structures exposed to fire.
1995. Eurocode 3: Design of steel structures, Part 1.2: General rules, Structural fire design.
2002–2005. Development of design rules for the fire behaviour of external steel structures (Contract No.: 7210-PR/380; 1/07/ 2002-21/06/2005)
2004. A. Blanguernon CTICM. External steel structures research. Synthesis of fire test n° 94 – E – 158.
2005. A D Lemaire et al. TNO. Fire test with external flaming on unprotected steel columns included concrete filled columns (Version 1.0).

Dynamics, earthquake and impact

Steel and Composite Structures – Wang & Choi (eds)
© *2007 Taylor & Francis Group, London, ISBN 978-0-415-45141-3*

Numerical modelling of tubular steel braces under cyclic axial loading

K.H. Nip, L. Gardner & A.Y. Elghazouli

Department of Civil and Environmental Engineering, Imperial College London, London, UK

ABSTRACT: Concentrically braced frames are commonly employed as lateral load resisting structures. Their response largely depends on the behaviour of the braces, which are the key energy dissipative zones. Although steel braces under cyclic axial loading have been studied experimentally, there is a need for further assessment focusing on quantifying failure. This paper describes the development of sophisticated finite element models of hollow sections subjected to cyclic axial loading. Also, an accurate, though simple method to predict fracture life was adopted. The effects of initial imperfections and strain hardening were taken into consideration and the models were validated against data from several test programmes. A parametric study was carried out to study the effect of member and cross-section slendernesses on the performance of the braces. The results were compared with the results from other experimental programmes and design specifications.

1 INTRODUCTION

One of the design objectives of concentrically braced frames is to resist severe earthquakes with structural damage but no collapse. This can be achieved by allowing damage in the diagonal braces, which are the key dissipative zones. When the frames are subjected to large cyclic horizontal loading, the braces undergo inelastic deformation. As the braces are the critical members for resisting earthquake loads, it is important to understand their behaviour under cyclic loading.

A number of studies have been performed to investigate the cyclic behaviour of steel structural members. An early study of the hysteretic behaviour of steel braces was conducted by Popov et al. (1979). Circular hollow sections (CHS) with various end conditions and member and cross-section slendernesses were tested. Linearised hysteresis loops for computer analysis were proposed. Jain et al. (1980) tested axially loaded square hollow sections (SHS) and rectangular hollow sections (RHS) together with angle sections. A hysteresis model for computer analysis was proposed and it was concluded that the member slenderness is the most important parameter influencing hysteretic behaviour.

Tremblay (2002) conducted a survey on the cyclic behaviour of steel bracing members. Buckling resistance, post-buckling compressive resistance, maximum tensile resistance and other properties of 50 sections were studied and formulae for predicting these properties were proposed. It was concluded that the fracture life of a bracing member depends more strongly on the member slenderness than the cross-section slenderness.

Elchalakani et al. (2003) confirmed this conclusion from their tests on CHS. It was also found that the influence of the number of repeated cycles at each displacement amplitude is significant on the ductility capacity but has only a minor effect on both the ultimate compressive strength and the corresponding deflection. Shaback and Brown (2003) carried out tests on SHS with gusset plate end connections. Empirical formulae were proposed to predict the fracture life of an SHS brace, as given by Equations 1 and 2:

$$\Delta_f = C_s \frac{(350/F_y)^{-3.5}}{[(b-2t)/t]^{1.2}} \left(\frac{4(b/d)-0.5}{5} \right)^{0.55} (70)^2$$

for $\dfrac{KL}{r} < 70$ (1)

$$\Delta_f = C_s \frac{(350/F_y)^{-3.5}}{[(b-2t)/t]^{1.2}} \left(\frac{4(b/d)-0.5}{5} \right)^{0.55} \left(\frac{KL}{r} \right)^2$$

for $\dfrac{KL}{r} > 70$ (2)

where Δ_f = fracture life; C_s = an empirical constant which was determined to be 0.065; and K = effective length factor.

Recently, Goggins et al. (2006) carried out cyclic axial tests on SHS and RHS. The influence of yield strength and member and cross-section slendernesses on the displacement ductility capacity was studied. Displacement ductility capacity, $\mu\Delta$, is defined as the displacement at fracture normalised by yield displacement. Relationships between ductility and slenderness were obtained:

$$\mu\Delta = 26.2\bar{\lambda} - 0.7 \tag{3}$$

$$\mu\Delta = 29.1 - 1.07(d/t) \tag{4}$$

where $\mu\Delta =$ the displacement ductility capacity; $\bar{\lambda} =$ the non-dimensional member slenderness; and $d/t =$ local width-to-thickness ratio.

Ductility increases with member slenderness, $\bar{\lambda}$, but sections with high cross-section slenderness show lower ductility. Goggins et al. (2006) also agreed that the cross-section slenderness is less influential than member slenderness in terms of the ductility capacity. The findings from the experiments were compared with the recommendations of design codes. Initial bucking capacities predicted by the AISC (1999) standard were found to be close to the experimental results, whilst EC3 (2006) gave more conservative predictions.

Despite these attempts to propose rules to predict the ductility and failure criteria of the braces, there is uncertainty in these rules due to the scatter of test results and the semi-empirical nature of the suggested criteria. Numerical modelling is a good way to understand the behaviour of braces beyond the tested range. This paper presents the development of numerical models of SHS and RHS braces, an approach to predict fracture and the effects of member and cross-section slenderness on the performance of the braces.

2 FINITE ELEMENT MODELS

2.1 General

The general purpose finite element (FE) package ABAQUS (2003) was used to develop models to replicate cyclic loading tests on SHS and RHS braces for cyclic axial loading tests. The analysis of the cyclically loaded members comprised two parts. The first part was an eigenvalue analysis, a linear elastic analysis which determines the eigenmodes. Appropriate eigenmodes were used as the initial geometric imperfections in the subsequent nonlinear analysis. In this non-linear analysis, the model was subjected to cyclic axial displacement with large plastic deformation.

2.2 Element type and mesh density

The 4-node doubly curved general-purpose shell elements with reduced integration, S4R (ABAQUS, 2003), were employed. These elements allow transverse shear deformation and can be used for thick or thin shell applications. They have been used in modelling high strength hollow section columns and gave

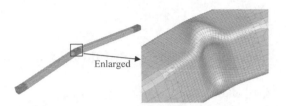

Figure 1. Refined mesh at the middle and ends of member (left), and the profile of local buckling (right).

accurate results for compressive resistance and failure modes (Ellobody & Young, 2005).

A sensitivity study on the mesh density was conducted. It was found that the overall strength of the members was not sensitive to the mesh density, but the prediction of strains in the areas of local buckling was sensitive. The peak strains that the elements experienced increased with mesh density. This is because coarse meshes inhibit local buckling and cannot accurately reflect the actual buckling shape. Also, elements in coarse meshes can only give the average strain over a relatively large area. Therefore, lower peak strains are obtained in models with coarse meshes.

In order to achieve a balance between the accuracy and computational efficiency, refined mesh was used only in areas where local buckling occurs, i.e. at the middle and the ends of the members, since all the models were fixed-ended braces. The extent of the refined mesh was 1.5 times of the dimension of the larger face of the section. This ensured that the refined area was large enough to cover the half wavelength of the buckling shape which is approximately equal to the dimension of the larger face of the section (Figure 1).

2.3 Material modelling

In most literature, the cyclic material properties of the specimens were not given. Instead, monotonic tensile coupon tests were usually carried out and reported. Therefore, stress-strain data obtained from monotonic tensile tests were used to define the material properties in the current FE analysis. ABAQUS calibrates the parameters of the nonlinear kinematic hardening model, by best-fitting the change of backstress of the material model to the stress-strain data from the monotonic tensile tests.

2.4 Boundary conditions and displacement history

All degrees of freedom of nodes at both ends of the members were restrained, except for the axial displacement at the loaded end. The braces were uniformly loaded and displacement-controlled. The displacement history applied was according to the provisions of the ECCS (1986), i.e. one cycle at each level of 0.25, 0.5, 0.75 and 1.0 δ_y, followed by three cycles at each

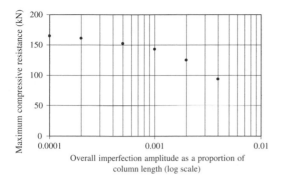

Figure 2. Imperfection sensitivity on maximum compressive resistance for global imperfection amplitudes ranging from L/250 to L/10000.

level of 2, 4, 6, 8 δ_y, etc. where δ_y is the estimated axial yield displacement.

2.5 *Initial imperfections*

All structural members contain initial geometric imperfections. These imperfections trigger out-of-plane deflections before the stresses reach the theoretical critical values. As the specimens in the current study are likely to fail by local or global buckling, or a combination of both, it is important to include the effect of imperfections in the models.

2.5.1 *Global imperfections*

The effect of global imperfection amplitude in tubular columns has been studied numerically by Gardner and Nethercot (2004). Column tests covering specimens with a range of cross-section and member slenderness were carried out. The ultimate loads obtained from the tests and FE models were compared. The results showed that FE models with global imperfection amplitudes of L/2000 agreed most accurately with the tests. Therefore, a global imperfection amplitude of L/2000 was used in the current numerical study, although larger imperfections were required in some cases to overcome numerical problems.

In the current investigation, a parametric study was conducted to examine the effect of initial global imperfection on the behaviour of SHS subjected to cyclic loading. Figure 1 shows the imperfection sensitivity of the maximum compressive resistance.

The results agree with those given by Gardner and Nethercot (2004) and the response is relatively insensitive to the amplitude of the global imperfection when it is below about L/1000.

2.5.2 *Local imperfections*

Eigenvalue buckling analyses were conducted to generate local imperfection modes. The lowest local buckling modes, factored by the amplitudes predicted by

Dawson & Walker (1972), were assumed to be the initial local imperfections. The predicted imperfection amplitude, ω_0, is given by:

$$\omega_0/t = 0.2(\sigma_{0.2}/\sigma_{cr}) \tag{5}$$

where $t =$ plate thickness; $\sigma_{0.2} =$ the material 0.2% proof stress; and $\sigma_{cr} =$ plate critical buckling stress.

3 DAMAGE PREDICTION

Components of machines and structures subjected to cyclic loading suffer microscopic damage due to the cyclic stresses. The microscopic damage accumulates and results in macroscopic damage, such as cracks, even if the cyclic stresses are well below the ultimate strength of the material.

There are three major approaches to analyse fatigue and fracture of metals. The first one is a stress-based approach, which is based on the analysis of the nominal, or average, stress in the region of the component being analysed. The second one is a strain-based approach, which considers the localised plastic deformation where a fatigue crack begins. The third one is a fracture mechanics approach, which deals with the processes of crack initiation and crack growth. In the current study, the braces were subjected to cycles of large displacements which cause localised plastic deformation. They were expected to fail by low cycle fatigue. Since fracture and other kinds of damage were not explicitly incorporated in the current FE models, an indirect approach to assess the damage and predict the occurrence of fracture was employed. It is a strain-based approach as it is most suitable for situations where local yielding is involved and the fatigue life is short.

3.1 *Strain-based approach*

The basic stain-life relationship was firstly proposed by Mason (1960) and it is now commonly referred to as the Coffin-Manson relationship (Equation 6).

$$\frac{\Delta\varepsilon}{2} = \frac{\Delta\varepsilon_e}{2} + \frac{\Delta\varepsilon_p}{2} = \frac{\sigma'_f}{E}(2N_f)^b + \varepsilon'_f(2N_f)^c \tag{6}$$

where $\Delta\varepsilon_e/2 =$ elastic strain amplitude; $\Delta\varepsilon_p/2 =$ plastic strain amplitude; $2N_f =$ number of reversals to failure; $\sigma'_f =$ fatigue strength coefficient; $\varepsilon'_f =$ fatigue ductility coefficient; $b =$ fatigue strength exponent; and $c =$ fatigue ductility exponent.

3.2 *Multiaxial loading*

Many engineering structures and components are subjected to random multiaxial loading. Multiaxial fatigue assessment is important to predict the failure under repeated loading and unloading. One of the approaches is to modify the Coffin-Manson equation. Mowbray (1980) introduced a correction method for biaxial stress using the biaxiality ratio $\lambda = \sigma_2/\sigma_1$ where $\Delta\sigma_1 > \Delta\sigma_2$. The biaxial strain-life relationship is given by:

$$\frac{\Delta\varepsilon}{2} = \frac{\sigma'_f}{E} f(\lambda,v)(2N_f)^b + \left(\frac{3}{3-m}\right)\varepsilon'_f g(\lambda,m)(2N_f)^c \quad (7)$$

where $v = $ Poisson's ratio; and $m = $ material constant.

3.3 *Variable amplitude loading*

The strain-life relationship mentioned above is based on constant-amplitude fatigue test data. For cases of variable-amplitude loading, a cumulative damage rule is required. The simplest form is the linear Palmgren-Miner rule (Miner, 1945).

$$D = \sum_{i=1}^{n} \frac{n_i}{N_{fi}} \quad (8)$$

where $N_{fi} = $ number of cycles to failure at a strain (or stress) level of ε_i (or σ_i); and $n_i = $ number of cycles that particular strain (or stress) level is applied. Fatigue occurs when the damage index, D, reaches unity.

3.4 *Method of fatigue assessment adopted*

In the current study, the correction method proposed by Mowbray (1980) was adopted. The fatigue life and the damage suffered in each cycle were calculated from the strain and stress ranges experienced by the elements. Miner's rule was used to obtain the cumulative damage from cycles of different strain amplitudes.

As mentioned in Section 2.2, the peak strains experienced were sensitive to the mesh density. However, in the sensitivity study, it was found that the average value of element strains over a certain area is relatively insensitive to mesh density. Also, it was observed that the highest strains occurred at the corners and around the middle eighth of the buckling wave in the longitudinal direction. Therefore, average values of strains of an area covering the corner and one-eighth of the buckling wavelength were used to calculate fatigue life and predict damage.

Figure 3 shows the strain distribution around the compressive face of the section at the region of local buckling. The corner areas experience the highest strains. This is explains why fractures begin in the corner areas instead of in the flat faces.

4 VERIFICATION OF FE MODELS

The FE models were verified by the analysis of cyclic axial loading tests on cold-formed steel hollow sections. Three specimens from the tests carried out by Walpole (1996) and Goggins (2006) were modelled and the results of the FE analysis and experiments were compared.

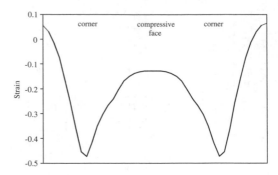

Figure 3. Strain distribution around the compressive face of the section at the region of local buckling.

Table 1. Comparisons between maximum compressive resistance of test and FE results.

Specimen	Test kN	FE kN	Test/FE
$150 \times 100 \times 6$	1047	1049	1.00
$40 \times 40 \times 2.5$	103	109	0.94
$50 \times 25 \times 2.5$	81	103	0.79

Table 2. Comparisons between maximum tensile resistance of test and FE results.

Specimen	Test kN	FE kN	Test/FE
$150 \times 100 \times 6$	1329	1228	1.08
$40 \times 40 \times 2.5$	112	122	0.92
$50 \times 25 \times 2.5$	110	116	0.95

The FE models were able to simulate the important features such as the loss of stiffness under tensile loads, reduction in buckling resistance and low post-buckling strength observed in the tests. Figure 3 shows a comparison of the hysteresis curves of $150 \times 100 \times 6$ SHS tested by Walpole (1996).

Tables 1–3 summarise the comparisons between the test and FE results. The maximum compressive resistance was obtained in the first cycle whilst the maximum tensile resistance occurred during later cycles as the material hardens. The numbers of cycles to failure predicted by the models were lower than those in the tests. The discrepancy in the results is believed to be largely due to insufficient cyclic material properties reported in tests.

5 PARAMETRIC STUDY

A parametric study was carried out to investigate the effect of member and cross-section slenderness on the

Table 3. Comparisons between the number of cycles to failure of test and FE results.

Specimen	Test	FE	Test/FE
150 × 100 × 6	6	5	1.2
40 × 40 × 2.5	14	8	1.75
50 × 25 × 2.5	17	11	1.54

Table 4. Dimensions of FE models.

Specimen	$\overline{\lambda}^*$	$b/t\varepsilon^{**}$
40 × 40 × 1.5 × 1100	0.44	26.6
40 × 40 × 2.5 × 1100	0.45	14.1
40 × 40 × 3.0 × 1100	0.46	11.0
40 × 40 × 4.0 × 2500	1.09	7.0
40 × 40 × 5.0 × 2500	1.11	4.7

* Non-dimensional slenderness defined in EC 3 (2006).
** Cross-section slenderness where $\varepsilon = \sqrt{235/f_y}$.

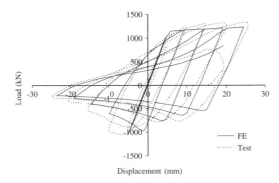

Figure 4. Comparison of the FE and experimental load-displacement response of 150 × 100 × 6 SHS (Walpole, 1996).

strength and ductility of the members. A total number of five models was analysed in the parametric study (Table 4).

5.1 Compressive resistance

The initial buckling load is also the maximum compressive resistance of the member throughout the test. Subsequent buckling resistances are always lower than the initial one. The FE results show that the normalised maximum compressive resistance decreases when the member slenderness increases.

As shown in Figure 4, during the compressive half cycle, the compressive resistance drops rapidly after buckling. This is due to the formation of plastic hinges at the mid-length and near the ends of the members. When the applied displacement is large, the compressive resistance can become as low as 15% of the buckling load. Also, the peak compressive resistance

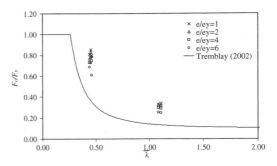

Figure 5. Compressive resistance at cycles of ductility demand of 2.

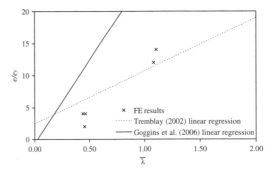

Figure 6. Relationship between ductility at fracture and member slenderness.

continues to decrease during the second and third cycles at a given amplitude. Figure 5 shows the peak compressive resistance of five models during the three cycles at the ductility demand of 2 ($e/e_y = 2$). The prediction given by Tremblay (2002) is also given in the graph.

5.2 Displacement ductility capacity

Ductility of braces is vital to the performance of a frame when it is subjected to large cyclic loading. This is because the larger the number of cycles the braces can resist, the more energy is dissipated, which reduces the damage to other parts of the structure. Therefore, it is desirable to understand how member properties affect the ductility.

Displacement ductility capacity is a measurement of ductility at fracture normalised by yield displacement. It was found that the displacement ductility capacity increases with member slenderness but decreases with cross-section slenderness. During the first few cycles of the displacement history, when the applied displacements were small, the member suffered little damage and retained much of its ductility capacity. The amount of damage increased significantly when local buckling was initiated. Most of the

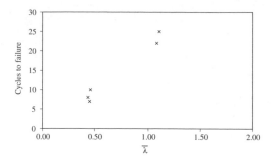

Figure 7. Relationship between cycles to failure and member slenderness.

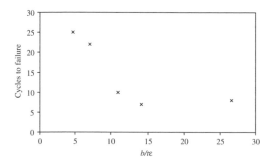

Figure 8. Relationship between ductility at fracture and cross-section slenderness.

damage was caused in the last few cycles before the member fractures. The relationship between the ductility capacity and member slenderness is presented in Figure 6. The FE results are compared with the experimental results of Tremblay (2002) and Goggins et al. (2006).

Besides displacement ductility capacity, the number of cycles to failure can also be used as an indicator of ductility. In Figure 7, it is shown that members with higher member slenderness can sustain more cycles of loading before failure.

As stockier cross-sections are less prone to local buckling, low cross-section slenderness results in a higher number of cycle to failure. Figure 8 shows the relationship between the ductility at fracture and cross-section slenderness.

6 CONCLUSIONS

FE models were built to simulate cyclic axial tests on hollow sections. Results show that the models are capable of capturing key features of the hysteretic behaviour of the braces. In particular, the strengths predicted by the models are close to the results of experiments and those obtained from design codes. A method to predict fracture from strain and stress data was adopted. The predictions of this method

were reasonably close to the results of other experimental programmes. It was found that displacement ductility capacity increases with member slenderness but decreases with cross-section slenderness. Further detailed investigations are needed to obtain a better understanding of the fracture behaviour of steel braces subjected to cyclic loading.

REFERENCES

ABAQUS. 2003. *ABAQUS/Standard user's manual volumes I-III and ABAQUS post amnual. Version 6.4.* Pawtucket, USA: Hibbitt, Karlsson & Sorensen, Inc.

AISC 1999. *Load and Resistance factor Design Specification for Structural Steel Buildings.* Chicago (IL): American Institute of Steel Construction, Inc.

Dawson, R. G., & Walker, A. C. 1972. Post-buckling of geometrically imperfect plates. *Journal of Structural Engineering, ASCE* 98(1): 75–94.

ECCS 1986. *Recommended testing procedure for assessing the behaviour of structural steel elements under cyclic loads.* Brussels.

Elchalakani, M., Zhao, X. L., & Grzebieta, R. 2003. Tests of Cold-Formed Circular Tubular Braces under Cyclic Axial Loading. *Journal of Structural Engineering, ASCE* 129(4): 507–514.

Ellobody, E. & Young, B. 2005. Structural performance of cold-formed high strength stainless steel columns. *Journal of Constructional Steel Research* 61(12): 1631–1649.

EN 1993-1-1 2006. *Eurocode 3: Design of Steel Structures – Part 1-1: General Rules and Rules for Buildings. Part 1.3: General rules – Supplementary Rules for Cold-Formed Thin Gauge Members and Sheeting.* CEN.

Gardner, L. & Nethercot, D. A. 2004. Numerical Modeling of Stainless Steel Structural Components—A Consistent Approach. *Journal of Structural Engineering, ASCE* 130(10): 1586–1601.

Goggins, J. M., Broderick, B. M., Elghazouli, A. Y. & Lucas, A. S. 2006. Behaviour of tubular steel members under cyclic axial loading. *Journal of Constructional Steel Research,* 62(1–2): 121–131.

Jain, A. K., Goel, S. C. & Hanson, R. D. 1980. Hysteretic cycles of axially loaded steel members. *Journal of Structural Engineering, ASCE* 106: 1777–1795.

Manson, S. S. 1960. Thermal stress in design. Part 19, cyclic life of ductile materials. *Machine Design* 32: 139–144.

Mowbray, D. F. 1980. A Hydrostatic Stress-Sensitive Relationship for Fatigue Under Biaxial Stress Conditions. *Journal of Testing and Evaluation* 8(1): 3–8.

Popov, E. P., Zayas, V. A. & Mahin, S. A. 1979. Cycle inelastic buckling of thin tubular columns. *Journal of the Structural Division* 105: 2261–2277.

Shaback, B. & Brown, T. 2003. Behaviour of square hollow structural steel braces with end connections under reversed cyclic axial loading. *Canadian Journal of Civil Engineering* 30: 745–753.

Tremblay, R. 2002. Inelastic seismic response of steel bracing members. *Journal of Constructional Steel Research* 58(5–8): 665–701.

Walpole, W. R. 1996. *Behaviour of cold-formed steel RHS members under cyclic loading.* Christchurch, New Zealand: Department of Civil Engineering, University of Canterbury.

Steel and Composite Structures – Wang & Choi (eds)
© *2007 Taylor & Francis Group, London, ISBN 978-0-415-45141-3*

FBG sensor based dynamic monitoring and analysis on sliding construction of a Prestressed Reticulated Shell-String structure

X. Xu
Space Structures Research Center, Zhejiang University, Hangzhou, Zhejiang Province, China

P. Qiu
Shanghai Institute of Mechanical & Electrical Engineering Co., Ltd., Shanghai, China

Y. Z. Luo*
Space Structures Research Center, Zhejiang University, Hangzhou, Zhejiang Province, China

ABSTRACT: The application of optical fiber sensors in monitoring of civil engineering structures is increasing continuously. One of the most frequently used sensor types is the so-called Fiber Bragg Grating (FBG) sensor which is ideally suited for long-term monitoring purposes due to some inherent advantages. This paper presents a real time monitoring of the sliding construction process of a prestressed reticulated shell-string (PRSS) structure with FBG sensors. Dynamic responses excited by two jumps during the sliding process are recorded by the monitoring system. Time and frequency domain analysis of the responses are conducted to explore the possible structural damage brought by the jump excitations. Dynamic properties of the structure are evaluated based on the data from monitoring. An approximate method for estimation of cable forces of the structure is also proposed. The conclusions of the study may benefit the researchers and practicers interested in construction process monitoring and analysis.

1 INTRODUCTION

The prestressed reticulated shell-string (PRSS) structure is a recently developed style of prestressed space truss structures. It is a self-equilibrium system consisting of grid shell and prestressed cables (Luo et al. 2004). The prestressed cables spanned at the bottom of the barrel vault shell force the structure to reshape itself. And internal forces and displacements opposite to those caused by static loads are introduced into the structure, i.e. the peak values of internal forces, vertical and horizontal displacements of the structure are lowered by the prestressed cables. As a result, PRSS structure has a lower steel consumption compared to traditional grid shells and thus it is more economical. In last five years, this kind of prestressed space trusses are widely used in roof structures for large-scale industrial buildings in China, especially for coal storages of power plants (Dong 2001).

In the construction of these structures which are usually symmetrically supported and have large length to span ratios, a construction method so-called segment cumulation sliding is widely used (Li & Gao 1999). To build a structure using the segment cumulation sliding method, the structure should be divided into several segments in length direction. Only a section of scaffolds as wide as a segment needs to be set up at one end of the structure. At every time one segment of the roof is assembled on the scaffolds. As soon as it is finished, all the finished part of the roof (including all the finished segments) are slid towards the other end of the structure to free the space above the scaffolds. Then the assembly of another segment can be started on the scaffolds. By repeating the above procedure, the whole roof structure will be assembled and slid to its right position finally (Li & Gao 1999). As a result few scaffolds and little work space are needed throughout the whole process of construction, which means low construction cost and little effect to the industrial activities may still carried out under the unfinished roof. While in practices, it is found that the sliding roof may jump forward suddenly due to the roughness or lack of lubrication of the paths (Sun 2005). Structural vibration is induced by the jump. No research focuses on the jump which may do harm to the structure has been reported. To investigate this problem,

---

* Correspondence Author, Luoyz@zju.edu.cn

dynamic response of the structure caused by the jump need to be collected firstly. FBG sensors would be a good choice to achieve this purpose.

Compared to traditional electrical sensors, FBG sensors have many advantages including (i) remarkable dynamic range of measurement (from 1 μm to 10% strain); (ii) high resolution; (iii) non-electrical measurement system (thunder-proof and explosive-proof); (iv) high durability; (v) ability to measure huge numbers of channels, simultaneously and on a long distance (Sumitrol & Wang 2005). In the monitoring or detection of steel structures, the Bragg sensors are usually installed by connectors welded to the structural members. Though this way of installation makes the sensors be reusable, it brings in residual stresses and is time cost which is very unwelcome in the construction. To overcome the above shortcomings, a mountable FGB sensing system has been developed and successfully applied in healthy detection of an in-service space truss structure (Zhai 2005). With the demountable FGB sensing system, the monitoring and detection of space truss structures can be conducted nondestructively, time economically and free of residual stresses.

This paper presents a real time monitoring of the sliding construction process of a PRSS structure with FBG sensors. Dynamic responses excited by two jumps of the sliding roof are recorded by the monitoring system. Time and frequency domain analysis of the responses are conducted to investigate the possible structural damage brought by the jump excitations. Then dynamic properties of the structure is evaluated based the information revealed by the vibrations. An approximate method for cable forces estimation of the structure is also proposed. The conclusions of the study may benefit the researchers and practicers interested in construction process monitoring and analysis.

2 DYNAMIC MONITORING

2.1 Apparatus used in monitoring

There are three different ways, i.e. embedding, sticking and fixing, to attach FBG sensors to a host structure. Accordingly, there are three different types of FBG sensors. In this project, fixing-type sensors were used. Considering that the structural members of space truss structures are usually tubes, a special clamp which can be quickly mounted onto and demounted from circular members is developed for sensor fixing. How the clamp works is shown in Figure 1a. The feasibility of the clamp has been verified both by lab experiments and site tests (Zhai 2005).

The fiber grating sensing analyser FONA-2004A, produced by Shanghai Synetoptics Technology Corporation, is used in the monitoring. It is a four-channel

Figure 1. (a) Sensor fixed to structural member by demountable clamp and (b) data collecting and process system.

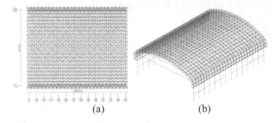

Figure 2. (a) Plan view and (b) axonametric view of the prestressed reticulated shell-string structure.

analyser with a wavelength resolution of 1pm and a scanning frequency of 50 Hz. There is a broadband light source with wavelength range of 1525–1565 nm integrated in it. The shift of the Bragg wavelength could be automatically recorded by a special software compatible with the analyser. Connecting the analyser to a computer installed with necessary softwares constitutes the data collecting and process system of the monitoring system, as shown in Figure 1b.

2.2 Sensors layout

The roof for a coal storage of a power plant is a PRSS structure with a plan dimension of 108 m by 81.8 m and a rise of 22.1 m, as shown in Figure 2. The grids of the structure are 3 m high, with a plan dimension of 4 m by 4 m. there are totally 14 tension cables with a length of 81 m. The lower support part of the structure consists of RPC frame columns with an intercolumniation of 8 m. The elevation of the column tops is 16.5 m. there are connecting beams at the middle and on the top of the columns, respectively. The whole roof is supported on the top beams.

Considering the purpose of the monitoring, some lower chords near to the supports, some upper chords at midspan and all the end nodes of the cables are chosen to be as sampling points. The layout of the sensors and fibers are shown in Figure 3. According to the locations of sampling points, temperature compensation is carried out in three areas, as shown in Figure 4. In each area, there is a temperature compensation sensor responsible for the temperature compensation of the sensors in this area.

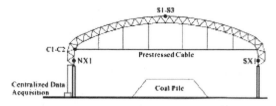

Figure 3. Layout of the sensors and fibers: Si, NXi, SXi and Ci represents midspan upper chord, north side lower chord, south side lower chord and cable of the axis i, respectively.

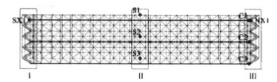

Figure 4. Area division for temperature compensation.

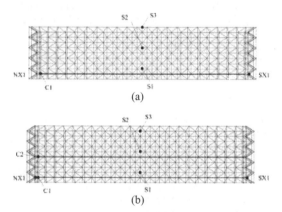

Figure 5. Structure configurations in the two jumps. (a) Structure configuration in the first jump: elements of the first 3 axes were assembled and cable of axis 1 was tensioned. (b) Structure configuration in the second jump: elements of the first 3 and a half axes were assembled and cable of axes 1 and 2 were tensioned.

3 DYNAMIC MONITORING

3.1 *Dynamic responses*

During the whole sliding construction process two dynamic responses induced by two jumps of the sliding roof were monitored. The jumps happened in the first and second slides of the structure. The structure configurations when the jumps happened are shown in Figure 5. Since the two dynamic responses are similar to each other, only the internal force responses recorded in the first jump are given here, see Figure 6.

It is clearly shown that the dynamic responses of the bars are similar in configuration, like response to triangular impulse. The vibration of lower chord NX1 has a

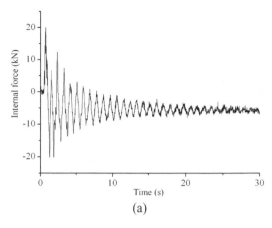

(a)

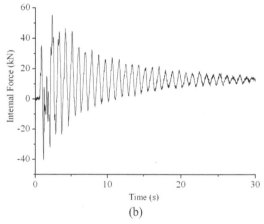

(b)

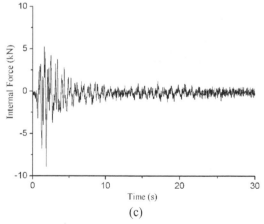

(c)

Figure 6. Internal force spectra for the monitored members in the first jump. (a) Internal force response of lower chord SX1. (b) Internal force response of lower chord NX1. (c) Internal force response of upper chord S1. (d) Internal force response of upper chord S2. (e) Internal force response of upper chord S3. (f) Internal force response of cable C1.

761

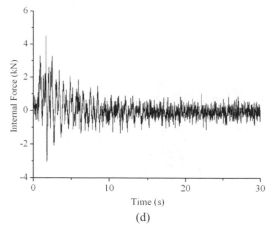

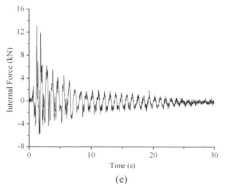

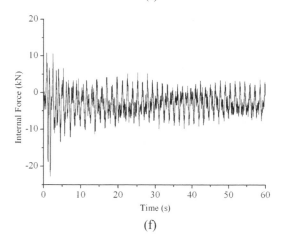

(d)

(e)

(f)

Figure 6.   Continued

Table 1.   Maximum internal force responses of the monitored bars.

Bar No.	SX1	NX1	S1	S2	S3
Maximum internal force response Nd (kN)	20.1	55.3	8.9	4.5	13.1
Static internal Forces N (kN)	−289	−289	−151	−160	−170
Internal force increments ΔN (kN)	5.7	12.4	0.2	0	0
$\|N_d/N\|$	0.07	0.191	0.059	0.03	0.08
$\|\Delta N/N\|$	0.019	0.043	0.001	0	0

maximum internal force responses of the monitored bars are picked out and listed in Table 1, as well as the internal force increments after the vibration.

It is obvious that the dynamic internal forces of the bars are relatively small, with dynamic to static ratios less than 20%. The internal force increments brought by the vibration are also insignificant. All these indicate that the structure keeps in elastic range during the vibration and the jump has little damage to the structure.

3.2   Frequency domain analysis

The internal force responses shown in Figure 6 describe the vibration in the time domain. Another way for vibration description is frequency domain spectrum by which the dynamic properties of the structure can be analyzed conveniently. The time domain spectra can be transformed to frequency domain spectra by discrete Fourier transformation which is given by Equation 1 (Ray & Joseph 1995):

$$X(f) = \sum_{n=-\infty}^{\infty} x(t_n)e^{-j2\pi f t_n}, \, t_n = n\Delta t, (n = 0, 1, 2, \cdots) \quad (1)$$

where X and x are responses in frequency domain and time domain, respectively, and Δt is the interval of discrete time series. The FBG sensors used in this project have a sampling frequency of 60 Hz, therefore $\Delta t = 0.01667$ sec. According to Equation 1, the internal force responses of the structural members can be transformed to frequency domain spectra, as shown in Figure 7.

3.3   Natural frequency of the structure

From frequency spectra, the natural frequencies of the structure can be identified. Peak values in frequency spectra correspond to different mode frequencies of the structure. It is found that the peak frequency values of the bars are at a same level, though they are located at different parts of the structure. A comparison between

relatively significant amplitude, which reflects the fact that the jump is excited at the north support which bar SX1 directly connects to. The vibration of the cable C1 is decayed more slowly and has a longer period than that of bars, as it is softer than bars. To explicitly reflect the affect of the jump to the structure, the

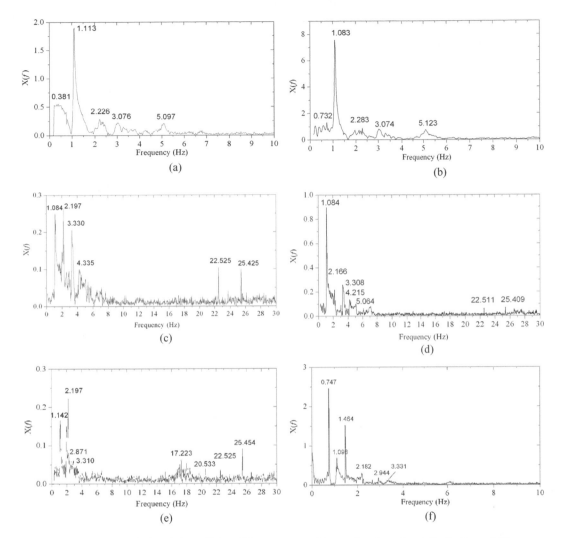

Figure 7. Frequency domain descriptions of internal force responses of the monitored members in the first jump. (a) Frequency spectrum of lower chord SX1. (b) Frequency spectrum of lower chord NX1. (c) Frequency spectrum of lower chord S1. (d) Frequency spectrum of lower chord S2. (e) Frequency spectrum of lower chord S3. (f) Frequency spectrum of cable C1.

the natural frequencies revealed by Figure 7 and those from FEM simulation are carried out in Table 2.

It shows that most bars are sensitive to lower mode frequencies of the structure. the first peak values of the frequency spectra are in well agreement with the first mode frequency calculated by FEM, which indicates that the first peak values of the frequency spectra correspond to the basic frequency of the structure. Due to the high frequency density of space truss structures, some frequencies having close values may correspond to a same peak value in the spectra of the bars. Taking the second peak value in the spectra of bars as an example, it is near to the values of the 2th-4th frequencies of the structure, which indicates that the second

peak value may represent all those frequencies. And some bars are sensitive to certain higher frequencies, such as the bars S1 and S3 whose 7th and 8th peak values correspond to the 24th and 29th frequencies of the structure, respectively. For impulsive excitation, the main problem is that the detected response of it is usually combined with a considerable proportion of noise signals, i.e. its signal-to-noise ratio is relatively low, which affects the accuracy of the analysis result of it. Among all the bars, the peak values of the spectrum of lower chord NX1 tally with the structural frequencies best. Hence the signal-to-noise ratio of the response of it is the highest among all the monitored bars. It reveals the fact that the jump is excited at the north support

Table 2. Frequencies of monitored members.

| Monitored frequencies | | | | | FEM simulation | |
| SX1 | NX1 | S1 | S2 | S3 | Frequencies | Modes |
Hz					Hz	
1.113	1.083	1.084	1.084	1.142	1.129	1
					1.799	2
2.226	2.283	2.197	2.166	2.197	1.972	3
					2.589	4
3.076	3.074	3.330	3.308	2.871	3.098	5
5.097	5.123	4.335	4.215	3.310	5.039	6
		5.038	5.064	17.22	17.74	18
				20.53	20.77	22
		22.53	22.51	22.53	22.81	24
		25.43	25.41	25.45	25.21	29

Table 3. Peak values of frequency spectra of cables.

| | | Frequency modes | | | | | | |
Jumps	Cables	1	2	3	4	5	6	7
First jump	C1	0.747	1.093	1.464	2.182	2.944	3.331	–
Second jump	C1	0.703	1.384	2.387	3.455	5.778	24.78	–
	C2	0.600	1.186	1.774	2.343	2.973	5.067	24.81

again, because the bar nearest to the excitation point should have the highest signal-to-noise ratio.

The peak frequency values of the cables are listed in Table 3. It shows that the first peak frequency values of the cables are smaller than those of bars. Considering that the cable are much softer than the structure, the basic frequencies of them would be smaller than that of the structure. Thus the first frequency peak values of the cables are their own basic frequencies. And the other peak frequency values of the cables reflect the frequencies of the structure as those of the bars do.

3.4 Critical damping ratio of the structure

The damping of a structure is usually described by logarithmic damping ratio or critical damping ratio. For a single-degree-of-freedom system, its mean logarithmic damping ratio γ can be determined from the free damping response spectrum of itself by (Ray & Joseph 1995):

$$\gamma = 2 \times \frac{1}{k} \ln \frac{a_n}{a_{n+k}} \tag{2}$$

where is the number of interval periods, and a_n and a_{n+k} are the values of the nth and $(n+k)$th peak values in the free damping response spectrum. And then

Table 4. Mean critical damping ratios of typical members.

Bar numbers	SX1	NX1	S3
Critical damping ratio	0.018	0.016	0.014

its critical damping ratio ξ can be obtained by (China Ministry of Construction 2003):

$$\xi = \frac{\gamma}{2\pi} \tag{3}$$

For a multi-degree-of-freedom system, if we assume that the normal modes are orthogonal to each other with respect to the damping matrix, the critical damping ratio of each normal modes can be determined by applying Equations 2 and 3 to the corresponding decoupled component of the response spectrum. The decoupling and damping estimation process can be automatically conducted by some commercial packages for random signal analysis.

The responses of bars SX1, NX1 and S3 which have relatively clear damping process are chosen to calculate the critical damping ratios. Only the critical damping ratios corresponding to the basic modes are given, see table 4, as they are concerned most. It

Table 5. Comparison of the estimated values of cable forces and jack readings.

		First mode		Second mode		Third mode		Fourth mode		
		Frequency	tension	Frequency	tension	Frequency	tension	Frequency	tension	Jack readings
Jumps	Cables	Hz	kN	Hz	kN	Hz	kN	Hz	kN	kN
First jump	C1	0.747	211	1.464	207	2.182	205	2.944	208	270
Second jump	C1	0.703	187	1.384	181	2.387	239	3.456	282	220
	C2	0.600	135	1.186	133	1.774	133	2.343	132	170

reveals that the basic modes of the monitored bars are in the range of 0.014~0.018 which is close to 0.02 — the critical damping ratio for steel structures adopted by the design code (China Ministry of Construction 2003).

3.5 Cable forces estimation

According to structural dynamic mechanics, the relationship between the frequencies and the tension force of a cable with fixed ends is (Li & Li 2002):

$$f_n = \frac{n}{2l}\sqrt{\frac{T}{\rho}}, \quad n = 1, 2, 3, \cdots \qquad (4)$$

where f_n is the nth frequency of the cable, n is the corresponding mode number, T is the internal force, l is the length and ρ is the linear density of the cable.

The cables used in this project have $l = 81$ m and $\rho = 14.4$ kg/m (including the weight of PE protector). Substituting the first four frequencies of the cables and the values of l and ρ into Equation 4 gets the tension forces of the cables, see Table 5. The cable forces calculated from frequencies are about 20% lower than the jack readings. The reasons for this may include (i) there is force losing in the cables; (ii) the cable ends are not absolutely rigid, while the calculations are based on the assumption of rigid fixed ends; and (iii) Equation 4 has not taken the slope and deflection of the cable into account. However, the proportionment and change trend of the cable forces calculated from frequencies are agree with those read from jacks. Thus it could be an approximate method for cable forces check or estimation.

4 CONCLUSIONS

This practice proves that the FBG sensors are feasible to the monitoring of dynamic behavior of space truss structures. It is found that the jumps during the sliding construction process do little harm to the PRSS

structure. The maximum dynamic stress is only about 20% of the static one and the residual internal forces brought by the vibrations are also insignificant. The under constructed structure is more sensitive to the lower frequencies. Besides some bars that are sensitive to certain higher order frequencies, the dynamic responses of the structure caused by the jumps are mainly free damping vibrations based on its basic frequency. And the damping ratio calculated from the monitoring results is in the range of 0.014~0.018 which is close to the value adopted by the design code. The proportionment and change trend of the cable forces determined from the monitored frequencies are agree with those read from jacks. Therefore it supplies a simple and approximate way for checking or estimation of cable forces.

REFERENCES

China Ministry of Construction. 2003. *Code for design of steel structures GB50017-2003*. Bcijing: China Plan Press.

Dong, S.L. 2001. State of art of prestressed large-span spatial steel structures. *Spatial Structure* 7(4): 1–14.

Li, G.Q. & Li, J. 2002. *Dynamic Test on Engineering Structures: Theory and Application*. Beijing: China Science Press.

Li, L. & Gao, H. 1999. Application of Sliding Construction Method in Large-span Grid Structures. *Journal of Xi'an University of Architecture and Technology* 31(1): 77–80.

Luo, Y.Z. et al. 2004. Design and analysis on grid structures with prestressed cables. *China Civil Engnieering Journal* 37(2): 52–57.

Ray, W.C. & Joseph, P. 1995. *Dynamics of Structures, 3rd ed.* Berkeley: Computers & Structures, Inc.

Sumitro1, S. & Wang, M.L. 2005. Sustainable structural health monitoring system. *Structure Control and Health Monitoring* 12(3,4): 445–467.

Sun, X.G. 2005. *Monitoring and analysis on the sliding construction process of grid shell structures*. Dissertation for master's degree of Zhejiang University, Hangzhou, China.

Zhai, Z.F. 2005. *Application of Optical Fiber Sensing Technique to Health Detection in Space Trusses*. Dissertation for master's degree of Zhejiang University, Hangzhou, China.

Steel and Composite Structures – Wang & Choi (eds)
© 2007 Taylor & Francis Group, London, ISBN 978-0-415-45141-3

The effect of friction coefficient on the cyclic behaviour of bolted beam-to-RHS column connections

G.J. van der Vegte
Delft University of Technology, Delft, The Netherlands

Y. Makino
Kumamoto University, Kumamoto, Japan

ABSTRACT: The Kobe earthquake in 1995 revealed that the conventional type of beam-to-column connections used in Japan is vulnerable to brittle fracture under seismic loading. As part of a project to develop a new design concept to prevent the occurrence of brittle fracture in beam-to-column connections, Miura et al. (2002) reported test results on three beam-to-RHS-column connections. The large-scale specimens not only included welded but also bolted connections. In order to evaluate the effect of the friction coefficient on the cyclic behaviour of the bolted beam-to-column specimens, this study presents numerical simulations of Miura's specimens, employing the explicit finite element (FE) technique. Prior to the FE analyses, four experiments on simple bolted specimens were conducted to obtain representative values of the friction coefficient. Two types of surface conditions were considered: untreated (i.e. rusted) and shot brushed. The effect of the surface conditions on the friction coefficient and the cyclic behaviour has been made clear, both in the tests on the simple bolted specimens as well as in the FE analyses of the beam-to-column connections tested by Miura.

1 INTRODUCTION

High-rise steel structures in Japan usually consist of an assembly of rectangular hollow section (RHS) columns and wide flange beams. The Kobe earthquake in 1995 revealed that the conventional type of beam-to-column connections is vulnerable to brittle fracture under strong ground motions. In a new design concept developed at Kumamoto University (Kurobane et al. 1998), the beams are no longer welded to the diaphragm plates as is the case in the conventional designs, but are connected to the through-diaphragm plates by bolts some distance away from the column face. After conducting a series of tests on both conventional and bolted connections (with identical beam- and column sections), it was found that the bolted connections did not exhibit fracture as observed in the conventional connections. In addition, the bolted connections showed a much larger energy dissipation capacity than the conventional type.

Miura et al. (2002) reported the results of three large-scale tests on beam-to-column specimens under cyclic loading. The specimens included both welded and bolted connections. Similar to the connections tested by Kurobane et al. (1998), Miura's project was primarily meant to develop and evaluate design concepts different from the conventional type in order to prevent the occurrence of brittle fracture of beam-to-column connections under seismic loading.

As experimental work involves high costs, the number of tests carried out on bolted beam-to-column connections is limited. Since numerical methods offer great flexibility and provide the possibility to study a wider range of parameters than covered by experimental programs, research on bolted beam-to-column connections at Kumamoto University was further extended using the finite element (FE) method.

An essential parameter for the cyclic behaviour of bolted beam-to-column connections is the value of the friction coefficient. In practice, the friction coefficient has to be assumed depending on the type of surface with a certain amount of safety. In order to obtain representative values of the friction coefficient to be used in the FE simulations of actual beam-to-column connections, four experiments on simple, bolted specimens were conducted. Two types of specimens were considered for two different surface conditions: untreated (i.e. rusted) and shot brushed.

Because of the detailed description and relevance of his experiments, Miura's tests were selected to study the effect of the friction coefficient on the hysteresis loops, using the FE method.

With respect to the FE method, two different types of solution strategies can be distinguished: the explicit and the implicit solution procedure. The most important differences between the implicit and explicit approaches can be found in the way in which equilibrium is maintained and the solution technique. The implicit method is based on static equilibrium and is characterized by the simultaneous solution of a set of linear equations. The explicit method is based on dynamic equilibrium. "Explicit" means that the state of the model at the end of an increment is solely a result of the displacements, velocities and accelerations at the beginning of the increment plus the changes in these variables that occur during a very small time interval. Implicit FE packages are not able to analyse highly non-linear problems effectively. In contrast, explicit solvers are well-suited to simulate complex processes such as multiple contact interactions between independent bodies.

As numerical simulations of bolted connections include a range of material and geometric non-linearities and rigid body movements, the explicit solution scheme has been employed in the numerical analyses.

2 EXPERIMENTS ON SIMPLE CONNECTIONS OF HIGH STRENGTH FRICTION BOLTS

2.1 *Outline of experiments*

In general, the value of the friction coefficient not only depends on the type of surface but may also vary under repeated loading. In order to obtain measured values of the friction coefficient, experiments on simple test specimens have been conducted. A series of four bolted specimens has been tested. The configuration of each specimen consists of a centre plate enclosed by a top- and bottom plate. Two types of specimens have been considered for two different surface conditions : untreated (i.e. rusted) and shot brushed.

Two of the four specimens are designed in accordance with the dimensions recommended by the Architectural Institute of Japan (AIJ 2001), describing the standard procedure to determine friction coefficients of connections with high-strength friction bolts. Each of these two specimens has two bolts, placed in series, with the shape of the bolt holes being circular. The specimens are subjected to monotonic, tensile force, causing the centre plate to shear and slip between the bottom- and top plate.

The other two specimens are illustrated in Figure 1. Each specimen contains a single bolt. The bolt holes of the top-and bottom plates are circular, while the hole of the centre plate is slotted. These two specimens are subjected to repeated loading : four cycles of ± 5 mm, ± 10 mm and ± 20 mm. As a result, the total slip length is equal to $4 \times (20 + 40 + 80) = 560$ mm.

The bolts employed in the four tests, M20 bolts with steel grade F10T have been pre-tensioned by a force of

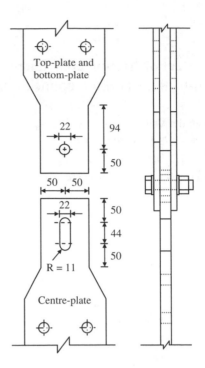

Figure 1. Configuration and dimensions of the single-bolt test specimens (sizes in mm).

181.9 kN. To monitor the strain history (i.e. the axial stress) in the bolt during testing, each bolt has been equipped with a strain gauge inside the shank. A narrow shaft with a diameter of 1 mm has been drilled into the bolt to accommodate the strain gauge.

2.2 *Results*

From the monotonic tests of the two specimens following the standard procedure of the AIJ (2001), a friction coefficient of 0.62 is found for the specimen with the rusted surfaces, while a value of 0.30 is obtained for the specimen with the shot brushed surfaces.

A comparison between the results obtained from the single-bolt specimens under cyclic loading reveals distinct differences between the behaviour of the two specimens.

Figure 2 displays the load applied to the centre plate as a function of the cumulative slip length. Because the applied load is cyclic, the sign of the applied load reverses. Hence, the absolute value is displayed in Figure 2.

For both specimens, the applied load initially drops for the cycles of ± 5 mm. However, unlike the specimen with rusted surfaces for which a gradual decline of load is observed as cycling proceeds, the load applied to the specimen with shot brushed surfaces no longer reduces but increases steadily after the initial cycles. A possible

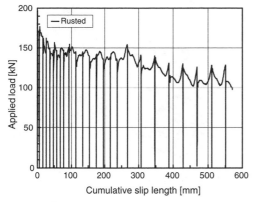

(a) Rusted surfaces

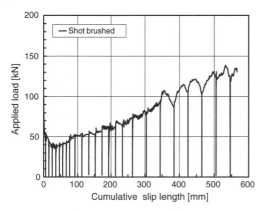

(b) Shot brushed surfaces

Figure 2. Applied load as a function of cumulative slip length.

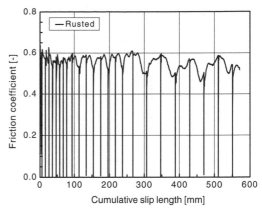

(a) Rusted surfaces

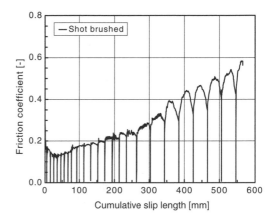

(b) Shot brushed surfaces

Figure 3. Friction coefficient as a function of cumulative slip length.

explanation can be found by examining the contact surfaces of the specimens after the tests were completed. It was found that both specimens exhibit considerable scratches of the surfaces around the slotted hole and the circular bolt holes of the top- and bottom plates. For the rusted specimen, the size of the scratched area as well as the less damaged area where only rust disappeared, is larger than for the shot brushed specimens. However, for the shot brushed surfaces, the scratches penetrate significantly further into the plates than for the rusted specimen. It was further observed that for both surface conditions, the tensile force in the bolt reduces as the total slip length increases.

In Figure 3, the (absolute) value of the friction coefficient is displayed versus the cumulative slip length. From the tests, it is found that for the rusted specimen, the initial value of the friction coefficient is equal to 0.64 and subsequently varies between 0.45 and 0.64 for the entire range of slip length. For the shot brushed specimen, the initial value of the friction coefficient of 0.29 drops to 0.13, but then starts rising fast for increasing values of the cumulative slip length, due to the severe deterioration of the contact surfaces of the plates.

3 FE ANALYSES OF LARGE-SCALE BEAM-TO-COLUMN SPECIMENS

3.1 Experiments by Miura et al. (2002)

In 2002, Miura et al. reported the results of three large-scale tests on bolted beam-to-column connections under cyclic loading. The connection details of the three configurations are shown in Figure 4.

In each of the three specimens, a wide flange beam with nominal dimensions of 500 mm × 200 mm (beam height × flange width) was welded to a cold-formed square hollow section column with nominal dimensions of 400 mm × 400 mm (column width × depth).

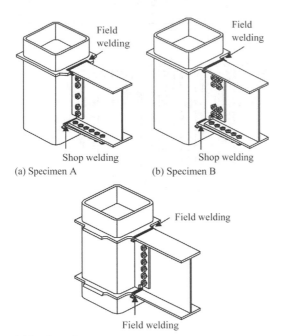

(a) Specimen A   (b) Specimen B

(c) Specimen C

Figure 4. Beam-to-column connections tested by Miura et al. (2002).

Both Specimens A and B contain an internal diaphragm plate inside the column located at the bottom flange of the beam. The beam is welded at the top flange and bolted at the web and bottom flange. The most relevant differences between Specimens A and B are (i) the height of the web plate which for Specimen B has been extended to the top- and bottom diaphragm plates (ii) the presence (Specimen A) and absence (Specimen B) of haunches at the top diaphragm plate and (iii) the number of bolts through the web-plate. Specimen A contains a single row of five web-bolts, while Specimen B has a double row of four web-bolts each.

Specimen C is rather different from Specimens A and B. In Specimen C, both beam flanges are welded to the external diaphragm plates while the beam web is bolted to the web plate by a single row of six bolts.

Although not illustrated in Figure 4, the column and the beam have a length of 3.5 m and 2.4 m respectively. Both ends of the column are fixed against displacements.

3.2 Numerical model

Since the load transfer between the bolted parts takes place solely through contact and friction, a detailed FE model has been developed, taking account of bolt pre-tensioning according to AIJ (2001). The explicit

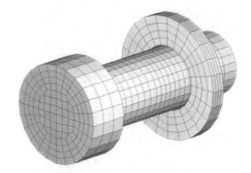

Figure 5. FE mesh generated for bolt and washers.

solver (ABAQUS/Explicit, 2003) has been used for the numerical analyses.

Since the configurations tested by Miura lack symmetry, a full scale FE model is generated for each specimen. As an example, Figure 5 displays the mesh created for the bolts and washers. The FE model primarily consists of eight-noded, linear solid elements. The FE model is in line with the recommendations regarding mesh density made by Van der Vegte et al. (2002). Major features of the numerical models include:

- The following contact interactions for each set of bolt and two washers are considered : (i) contact between bolt head and washer (ii) contact between washer and nut (iii) contact between washer and beam (iv) contact between washer and flange- or web plate and (v) contact between bolt shank and washers, beam and flange- or web-plate. In addition, two more interactions are modelled : (i) contact between the beam web and the web plate and (ii) contact between the beam bottom flange and the flange plate.
- The material properties (yield strength) used for the various components are obtained from tensile tests. Because Miura's tests involve cyclic loading, kinematic hardening is required to describe the material behaviour correctly. A linear hardening model is adopted, assuming a value of E/100 for the hardening modulus, where E is Young's modulus.
- Loading of the specimens is different from the procedure followed in the experiments. In the experiments, the amplitude of the beam rotation was increased as $2\theta_p$, $4\theta_p$, $6\theta_p$ ... up to failure, with θ_p being a theoretical value of beam rotation at full plastic beam moment (Miura et al. 2002). Since simulation of the experimental loading path would require excessive computational time, the numerical parametric study considers fewer cycles, displayed in Figure 6. First, a "small" cycle with the beam rotation θ_m varying from $+\theta_p$ to $-\theta_p$ has been simulated, followed by a "large" cycle from $+6\theta_p$ to $-6\theta_p$ and back to $+6\theta_p$. Beam rotation is

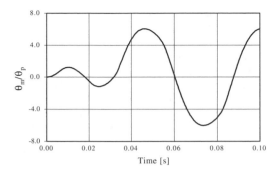

Figure 6. Loading history employed in parametric study.

defined as positive when the beam bottom flange is in tension.

- A comparison between the experimental and numerical hysteresis loops obtained for Specimens A and C shows that the numerical model can predict the behaviour of the specimens reasonably accurate. It is further found that the FE model is able to give reliable assessments of the observed failure modes observed in the experiments (Van der Vegte, unpubl.). Not only do the locations where large strains are observed in the numerical model match well with the locations of crack initiation found in the experiments, buckling of the beam top flanges is also correctly simulated by the explicit FE analyses.

3.3 *Friction coefficient*

Laboratory tests can be used to determine the value of the friction coefficient f. In practice, the friction coefficient has to be assumed depending of the type of surface with a certain amount of safety. An additional problem to choose an appropriate value of the friction coefficient is the fact that the initial value may not be constant. As made clear in the experiments on the single bolted specimens under cyclic loading, the condition of the contacting surfaces may deteriorate, thus changing the value of the friction coefficient.

As mentioned before, from the experiments of the single-bolted specimens, an initial friction coefficient of approximately 0.6 was found for contact interaction between untreated surfaces covered by rust. From the same series of experiments, it was observed that contact between shot brushed surfaces showed an initial friction coefficient of 0.3.

To evaluate the effect of different surface conditions of the beam, flange- or web plates, Specimens A-C have been analysed for the following two values of the friction coefficient (i) f = 0.3 to simulate shot brushed surfaces and (ii) f = 0.6 to model rusted surface conditions. For the beam-to-column connections under investigation, these surfaces refer to the contact surfaces between the beam and the flange- or web

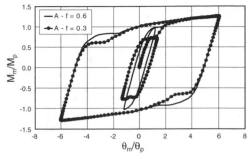

(a) Specimen A

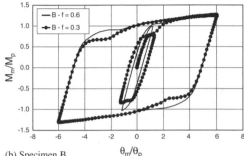

(b) Specimen B

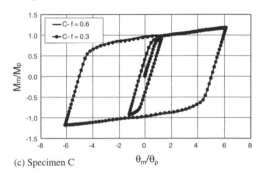

(c) Specimen C

Figure 7. The effect of the friction coefficient f.

plates. For the contact surfaces involving bolts, in both sets of analyses, the friction coefficient is kept at 0.3.

3.4 *Results*

The hysteresis loops obtained for each of the three specimens are shown in Figure 7. The vertical axis plots the moment at the column face M_m and is non-dimensionalized by the full-plastic moment M_p of the beam. The moment is defined as positive when the beam bottom flange is in tension. The horizontal axis displays the rotation of the beam θ_m non-dimensionalized by θ_p.

Because Specimens A and B do not differ significantly, it is evident that the hysteresis loops for Specimens A and B are similar. Both curves display distinct horizontal plateaus after the (initial) elastic

771

stage, primarily caused by slip between the flange plate and the beam bottom flange and to some extent by slip between the web plate and the beam web. A further increase of the beam rotation causes the bolts to contact the adjacent bolt holes, i.e. part of the load transfer will take place through bearing stress and shear stress in the bolts. This will further strengthen the connections as evidenced by the ascent of the hysteresis loops.

Due to the close resemblance of the specimens, the effect of a different value of the friction coefficient on the hysteresis loops is similar for Specimens A and B. For $f = 0.3$ frictional slip between the contact surfaces initiates at significantly smaller bending moments as compared with the corresponding moments for $f = 0.6$. In addition, for $f = 0.3$, the tips of the hysteresis loops for $+\theta_p$ and $-\theta_p$ are further apart in terms of relative beam rotation. In other words : for $f = 0.3$ the height of the loops between $+\theta_p$ and $-\theta_p$ is less, but the width of these loops is larger as compared to the hysteresis loops for $f = 0.6$. On the other hand, for both values of the friction coefficient, the tips of the loops between $-6\theta_p$ and $+6\theta_p$ are identical. In this case, full contact is established between the bolts and the adjacent bolt holes in the beams while no additional slip is possible. For both friction coefficients, larger deformations will cause a further increase in load transfer through shear in the bolts. These observations are confirmed by evaluating contour plots of the shear stress distribution through the bolts. At these deformation levels, ultimate strength is determined by the strength of the beams, independent of the friction coefficient.

The behaviour of Specimen C under cyclic loading is different from the hysteresis loops obtained for Specimens A and B. Since Specimen C does not contain bolts in the flanges, no distinct "slip plateau" is observed. In addition, for Specimen C hardly any difference can be observed between the loops for $f = 0.3$ and 0.6. Because Specimen C has web-bolts only, it is clear that for both the "small" as well as the "large" loops, the contribution of the web and web-bolts to the load transfer is marginal as compared to the load transfer by the bolts in the beam flanges of Specimens A and B. Hence, the effect of the friction coefficient on the cyclic behaviour is almost negligible.

4 CONCLUSIONS

Numerical analyses have been conducted to evaluate the effect of the friction coefficient on the cyclic behaviour of bolted beam-to-column connections employing the explicit solution technique. Prior to the FE analyses, experiments on simple bolted connections have been carried out to determine typical values of the friction coefficient for two types of surface conditions (rusted and shot brushed) under monotonic and cyclic loading. Based on the presented research, the following conclusions can be drawn:

– From the experiments on the simple bolted specimens, it is found that for contact between untreated (i.e. rusted) surfaces, the friction coefficient slightly reduces as the cumulative slip length increases. For reversed loading between shot brushed surfaces, the friction coefficient may vary sharply i.e. may initially decline and increase later as cycling continues due to the severe damage of the contact surfaces. For the surface conditions investigated, the friction coefficient varies between approximately 0.3 and 0.6.

– From the FE analyses of the beam-to-column connections, it is concluded that the effect of the friction coefficient on the hysteresis loops primarily depends on the location of the bolts within the beam-to-column connection. For specimens where the beam bottom flange is bolted to the flange plate (Specimens A and B), a change in value of the friction coefficient significantly alters the hysteresis loops. On the other hand, for a connection with web-bolts only (Specimen C), the hysteresis loops are hardly affected by a variation of the value of the friction coefficient.

– The explicit method is a suitable tool to simulate the behaviour of bolted connections effectively. The numerical simulations of the highly non-linear FE models investigated do not yield any computational difficulty.

ACKNOWLEDGEMENT

The first author would like to express his appreciation to the Japan Society for the Promotion of Science for the opportunity to conduct and present the research reported herein.

REFERENCES

ABAQUS/Explicit 2003. Version 6.3, Hibbitt, Karlsson & Sorensen, USA.

AIJ 2001. Recommendation for design of connections in steel structures. Architectural Institute of Japan (in Japanese).

Kurobane, Y., Ochi, K., Yamashita, Y., Tokutome, Y. & Tanaka, M. 1998. Testing of new bolted moment connections between RHS columns and I-section beams. *Structural Technology*. Vol. 11, No. 115, pp. 37–43 (in Japanese).

Miura, K., Makino, Y., Obukuro, Y., Kurobane, Y., Vegte, G.J. van der, Tanaka, M. & Tokudome, K. 2002. Testing of beam-to-RHS column connections without weld-access holes. *International Journal of Offshore and Polar Engineering*. Vol. 12, No. 3, pp. 229–235.

Vegte, G.J. van der, Makino, Y. & Sakimoto, T. 2002. Numerical research on single-bolted connections using implicit and explicit solution techniques. *Memoirs of the Fac. of Engineering, Kumamoto University*, Vol. 47, No. 1.

Steel and Composite Structures – Wang & Choi (eds)
© *2007 Taylor & Francis Group, London, ISBN 978-0-415-45141-3*

Study on non-linear natural frequency with large deformation for plane prestressed cable-net

Shang Renjie, Wu Zhuanqin, Liu Jingliang, Li Qian & Zhou Jianfeng
Central Research Institute of Building and Construction, MMC Group Beijing, China

ABSTRACT: Taking the 30 m × 70 m plane cable-net of Zhongguancun Culture Building as the engineering background, the differential equation of non-linear vibration is deduced by Ritz method and the solution is given in the paper, considering the influence of nonlinearity produced by large deformation for the rectangular cable-net structure. The method to calculate the frequency of linear vibration, the vibration frequency with large amplitude and the vibration frequency with small amplitude after large deformation is studied. The 3 frequencies given in the example are calculated and analyzed in combination with the cases. The vibration equation and the solution method provided in the paper are simple and accurate.

1 FORWARD

Cable is one kind of flexible unit, with the side rigidity appeared only mainly via the exerting the pre-stress. The cable-net is divided into single-layered and double-layered cable-net(Huo,2002; Zhao,2002). And the cable-net is further divided into the curve-surfaced cable-net (You, 2005; Bobrowski, 1985; Shen, 1990) and plane cable-net (Schlaich, 2005; Guo, 2004). The familiar saddle-shaped cable-net is single-layered cable-net. As the curved surface of cable-net possesses negative Gauss curvature, and can possess strong side rigidity with very small pull. However the plane cable-net is different. As all plane cable-net is arranged within a plane, the side rigidity is provided mainly by the pre-stress. For the plane cable-net structure, its rigidity at side is smaller, and the displacement is bigger under the action of the side load; As the plane cable-net has small structure rigidity, and has large deformation under the load of side wind, therefore in time of considering the vibration of wind, it is necessary to consider the vibration characteristics under the large deformation of the cable-net, and it is also necessary to consider the vibration frequency with small amplitude under the wind load after the occurrence of large deformation. But at present, the study on the vibration of the cable-net structure is most the linear vibration(Yang,1997), and less study is made on the non-linear vibration. As the cable-net structure has good permeable property, with structure to the eyes, therefore, in recent years, in the engineering structure, especially in the glass curtain wall, it is mostly used.

Taking the 30 m × 70 m curtain wall of plane cable-net of Zhongguancun Culture Building as the

engineering background, the differential equation of non-linear vibration is deduced by Ritz method and the solution is given in the paper, considering the influence of geometrical nonlinearity produced by large deformation for the rectangular cable-net structure. The method to calculate the frequencies of linear vibration, the vibration frequency with large amplitude and the vibration frequency with small amplitude after large deformation is studied. The comparison and analysis are made for the 3 frequencies in combination with the cases.

2 DEDUCTION OF THE DIFFERENCE EQUATION FOR NON-LINEAR VIBRATION

Let us suppose the length in x-direction as 2a, and the length in y-direction as 2b for the rectangular cable-net structure, and the area of the cable in x-direction as A_x, and the area of the cable in y-direction as A_y, and the elastic modulus for the cable as E. The deformed curved-surface for the cable-net in the process of vibration can be represented by the following formula:

$$z = w_0 \cos\frac{\pi x}{2a} \cos\frac{\pi y}{2b} \tag{1}$$

then $\frac{\partial z}{\partial x} = -w_0 \frac{\pi}{2a}\sin\frac{\pi x}{2a}\cos\frac{\pi y}{2b} = c\sin\frac{\pi x}{2a}$ (2)

where, $c = -w_0\frac{\pi}{2a}\cos\frac{\pi y}{2b}$, even in large deformation, under normal conditions $\frac{w_0}{2a} < \frac{1}{20}$, therefore

$c^2 < 0.025$ is small quantity, after dislocation, the elongation in x-direction is:

$$\Delta s = 2\int_0^a \sqrt{1+\left(\frac{\partial z}{\partial x}\right)^2}\,dx - 2a = 2\int_0^a \sqrt{1+\left(c\sin\frac{\pi x}{2a}\right)^2}\,dx - 2a$$

$$= 2\int_0^a \sqrt{1+c^2\sin^2\frac{\pi x}{2a}}\,dx - 2a \cong 2\int_0^a \left(1+\frac{c^2}{2}\sin^2\frac{\pi x}{2a}\right)dx - 2a \quad (3)$$

$$= \left(\frac{c^2 x}{2} - \frac{ac^2}{2\pi}\sin\frac{\pi x}{a}\right)\Big|_0^a = \frac{ac^2}{2} = \frac{\pi^2 w_0^2}{8a}\cos^2\frac{\pi y}{2b}$$

The error of formula (3) is not bigger than $(\sqrt{1.025} - (1+0.025/2))/(\sqrt{1.025}-1) = 0.6\%$, the total energy of strain in x-direction is:

$$E_x = 2\int_0^b \left[H_x + \frac{1}{2}EA_x\frac{\Delta s}{2a}\right]\cdot(\Delta s)dy$$

$$= 2\int_0^b \left[H_x + \frac{EA_x\pi^2 w_0^2}{32a^2}\cos^2\frac{\pi y}{2b}\right]\frac{\pi^2 w_0^2}{8a}\cos^2\frac{\pi y}{2b}\,dy$$

$$= \frac{\pi^2 w_0^2}{4a}H_x\int_0^b \cos^2\frac{\pi y}{2b}\,dy + \frac{EA_x\pi^4 w_0^4}{128a^3}\int_0^b \cos^4\frac{\pi y}{2b}\,dy \quad (4)$$

$$= \frac{\pi^2 bH_x}{8a}w_0^2 + \frac{3EA_x\pi^4 b}{1024a^3}w_0^4$$

Similarly, the strain energy for the cable in y-direction is :

y-direction is : $E_y = \frac{\pi^2 aH_y}{8b}w_0^2 + \frac{3EA_y\pi^4 a}{1024b^3}w_0^4 \quad (5)$

The total strain energy of the cable-net is:

$$E_1 = E_x + E_y$$

$$= \left(\frac{\pi^2 bH_x}{8a} + \frac{\pi^2 aH_y}{8b}\right)w_0^2 + \left(\frac{3EA_x\pi^4 b}{1024a^3} + \frac{3EA_y\pi^4 a}{1024b^3}\right)w_0^4 \quad (6)$$

The resilience in the process of vibration is proportional to the acceleration in x-direction, that is the restoring force:

$$q = m\ddot{z} = m\ddot{w}_0\cos\frac{\pi x}{2a}\cos\frac{\pi y}{2b} \quad (7)$$

The potential energy produced by the distribution load $q = mz = m w_0 \cos\dfrac{\pi x}{2a}\cos\dfrac{\pi y}{2b}$ is:

$$E_2 = 4\int_0^b \int_0^a qw\,dx\,dy$$

$$= 4\int_0^b \int_0^a w_0\cos\frac{\pi x}{2a}\cos\frac{\pi y}{2b}m\ddot{w}_0\cos\frac{\pi x}{2a}\cos\frac{\pi y}{2b}\,dx\,dy \quad (8)$$

$$= 4m w_0 \ddot{w}_0 \int_0^b \int_0^a \cos^2\frac{\pi x}{2a}\cos^2\frac{\pi y}{2b}\,dx\,dy$$

$$= abm w_0 \ddot{w}_0$$

The total energy(Qian,1980):

$$E = E_1 + E_2 = \left(\frac{\pi^2 bH_x}{8a} + \frac{\pi^2 aH_y}{8b}\right)w_0^2$$

$$+ \left(\frac{3EA_x\pi^4 b}{1024a^3} + \frac{3EA_y\pi^4 a}{1024b^3}\right)w_0^4 + abm w_0 \ddot{w}_0$$

In accordance with the Ritz method, the energy should the minimum(Qian,1980), then $\delta E = 0$, namely:

$$2\left(\frac{\pi^2 bH_x}{8a} + \frac{\pi^2 aH_y}{8b}\right)w_0 + 4\left(\frac{3EA_x\pi^4 b}{1024a^3} + \frac{3EA_y\pi^4 a}{1024b^3}\right)w_0^3 + abm\ddot{w}_0 = 0 \quad (9)$$

$$\ddot{w}_0 + \left(\frac{\pi^2 H_x}{4ma^2} + \frac{\pi^2 H_x}{4mb^2}\right)w_0 + \left(\frac{3EA_x\pi^4}{256ma^4} + \frac{3EA_y\pi^4}{256mb^4}\right)w_0^3 = 0 \quad (10)$$

Taking $A = \dfrac{\pi^2 H_x}{4ma^2} + \dfrac{\pi^2 H_x}{4mb^2} \quad (11)$

$$B = \frac{3EA_x\pi^4}{256ma^4} + \frac{3EA_y\pi^4}{256mb^4} \quad (12)$$

$$\ddot{w}_0 + Aw_0 + Bw_0^3 = 0 \quad (13)$$

when the non-linear item of large deformation is ignored:

$$\ddot{w}_0 + \left(\frac{\pi^2 H_x}{4ma^2} + \frac{\pi^2 H_x}{4mb^2}\right)w_0 = 0 \quad (14)$$

$\omega^2 = \dfrac{\pi^2 H_x}{4ma^2} + \dfrac{\pi^2 H_x}{4mb^2}$, the same with the other computation results(Shen,2006).

For the convenience in expression, we take the new variable x to substitute w_0 to express the maximum dislocation value in the span of cable-net, then the equation (13) changes to:

$$\ddot{x} + Ax + Bx^3 = 0 \quad (15)$$

In this way, the structure with multiple degree of freedom for the plane cable-net can be regarded as the structure with single degree of freedom, in time to make the computation with large deformation for the main frequency. The displacement at the half span point can be regarded as the generalized displacement x, and x becomes the generalized acceleration or restoring force. Then it is possible to use the single degree of freedom to study the non-linear influence of the basic frequency in time of large deformation.

3 THE SOLUTION METHOD FOR THE DIFFERENTIAL EQUATION OF THE NON-LINEAR VIBRATION

Equation (15) is the second- order differential equation. At present, it has not the analytic solution yet. It is necessary to reduce the order:

Let: $y = \dfrac{dx}{dt}$, to substitute (15) to gain the first- order non-linear differential equation set:

$$\begin{cases} \dfrac{dy}{dt} = -Ax - Bx^3 \end{cases} \tag{16}$$

$$\begin{cases} \dfrac{dx}{dt} = y \end{cases} \tag{17}$$

$$\frac{dy}{dx} = \frac{-Ax - Bx^3}{y} \tag{18}$$

$$ydy = (-Ax - Bx^3)dx \tag{19}$$

Via solving the above-mentioned formula to get:

$$Ax^2 + 0.5Bx^4 + y^2 = C \tag{20}$$

Let us suppose the maximum amplitude as D, then in time of $y = 0$, namely when the speed is 0, x reaches D, therefore we get:
$C = AD^2 + 0.5BD^4$. In accordance with (19), we get:

$$\frac{dx}{dt} = y = \sqrt{AD^2 + 0.5BD^4 - Ax^2 - 0.5Bx^4} \tag{21}$$

$$dt = \frac{dx}{\sqrt{AD^2 + 0.5BD^4 - Ax^2 - 0.5Bx^4}} \tag{22}$$

Cycle $T = 4\displaystyle\int_0^D \frac{dx}{\sqrt{AD^2 + 0.5BD^4 - Ax^2 - 0.5Bx^4}}$ (23)

Via formula (21) or (22), it is easy for us to calculate out via the numerical method. The phase trajectory can be drawn through formula (20).

4 VIBRATION WITH SMALL AMPLITUDE AFTER BIG DEFORMATION

For the cable-net, it is very often that the small vibration with small amplitude near the large deformation can be found after the occurrence of large deformation. Due to the difference of load experienced by the cable-net, the form of curve for the cable-net is different when the same big deformation appears at the half span point. Here the consideration is given nearly in accordance with the

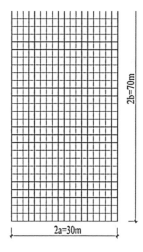

Figure 1. Drawing of plane rectangular cable-net.

above-mentioned $z = mw_0 \cos \dfrac{\pi x}{2a} \cos \dfrac{\pi y}{2b}$, we suppose that the vibration is the small vibration of x, the x in $x + Ax + Bx^3 = 0$ is replaced by w+x, we get:
$x + (A + 3Bw^2)x + 3Bwx^2 + Bx^3 + w + Aw + Bw^3 = 0$

In accordance with the balance near the large deformation, we get: $w + Aw + Bw^3 = 0$, then:

$$\ddot{x} + (A + 3Bw^2)x + 3Bwx^2 = 0 \tag{24}$$

The item x^2 included in formula (24) shows the non-symmetry for the upper and lower vibration at the position of large deformation: That is, when the deformation continues to increase, the rigidity of the hard spring will increase. When the deformation reduces, it can be seen that the rigidity reduces, that is the characteristics of the soft spring. For the case with small amplitude, the item x^2 is ignored, then we get:

$$\omega^2 = A + 3Bw^2 \tag{25}$$

5 VIBRATION PATTERN FOR HIGH ORDER

For the vibration pattern of high order, it is necessary first to judge the form of the vibration pattern, it is O.K. to transform a, b. In the following example, the vibration pattern of second order is upper and lower counter-symmetric, it is O.K. to use b/2 to substitute the original b.

6 ANALYSIS INTO THE ENGINEERING CASE

For the orthogonal cable-net of Zhongguancun Culture Building (Fig. 1). There are very good rigid support around, with height $2b = 70\,\text{m}$ (y-direction), width

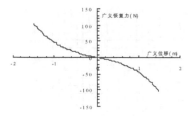

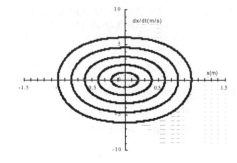

Figure 2. relation between generalized restoring force and generalized displacement.

Figure 3. Phase trajectory for different amplitude.

$2a = 30\,m$ (x-direction), the area of cross-section for the cable-net in x-direction at $672\,mm^2$, the initial force of tension and pull $F_x = 320\,kN$, spacing up to $2\,m$; area of section in y-direction at $336\,mm^2$, the initial force of tension and pull up to $F_y = 120\,kN$, spacing up to $1.67\,m$, the elastic modulus for cable up to $E = 1.7 \times 10^{11}\,N/m^2$, the quality per meter for cable and glass curtain wall $m = 70\,kg/m^2$. The needed basic parameter is:

$$H_x = 320000/2 = 160000(N)$$

$$H_y = 120000/1.67 = 71856(N)$$

$$A_x = 672/2 = 336\,mm^2 = 336 \times 10^{-6}\,m^2$$

$$A_y = 336/1.67 = 201.6\,mm^2 = 201.6 \times 10^{-6}\,m^2$$

$$A = \frac{\pi^2 H_x}{4ma^2} + \frac{\pi^2 H_x}{4mb^2} = \frac{160000\,\pi^2}{4 \times 70 \times 15^2} + \frac{71856\,\pi^2}{4 \times 70 \times 35^2} = 27.133$$

$$\begin{aligned}B &= \frac{3EA_x\pi^4}{256ma^4} + \frac{3EA_x\pi^4}{256mb^4}\\ &= \frac{3 \times 1.7 \times 10^{11} \times 336 \times 10^{-6}\pi^4}{256 \times 70 \times 15^4} + \frac{3 \times 1.7 \times 10^{11} \times 201.6 \times 10^{-6}\pi^4}{256 \times 70 \times 35^4} = 18.772\end{aligned}$$

We get: $\qquad \ddot{x} + 27.133x + 18.772x^3 = 0 \qquad (26)$

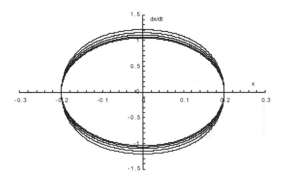

Figure 4. Phase trajectory with different amplitude based on standardization.

6.1 Generalized rigidity

In accordance with (26), we can draw the relation curve drawing (Fig. 1) to show the relation between the generalized displacement x and the generalized restoring force. It can be clearly seen from the drawing that the structure has geometric non-linearity, with the curve of restoring force for the hard spring, to show the increase of rigidity along with the increase in deformation.

6.2 Phase trajectory

The influence of the non-linearity of cable-net on the vibration frequency in time of big deformation is computed and analyzed as follows::

When amplitude $D_1 = 0.2\,m$, the equation of phase trajectory is $27.133x^2 + 9.386x^4 + y^2 = 1.100$;

When amplitude $D_2 = 0.4\,m$, the equation of phase trajectory is $27.133x^2 + 9.386x^4 + y^2 = 4.5816$;

When amplitude $D_3 = 0.6\,m$, the equation of phase trajectory is $27.133x^2 + 9.386x^4 + y^2 = 10.9843$;

When amplitude $D_4 = 0.8\,m$, the equation of phase trajectory is $27.133x^2 + 9.386x^4 + y^2 = 21.2096$;

When amplitude $D_5 = 1.0\,m$, the equation of phase trajectory is $27.133x^2 + 9.386x^4 + y^2 = 36.5190$;

For the phase trajectory of the vibration with different amplitude, refer to Fig. 3, in order to be able to reflect the cyclic characteristics with different amplitudes. X and y with different amplitudes will be respectively standardized into the maximum displacement $x = 0.2$ based on the same proportion. For example, for $D_3 = 0.6\,m$, both x and y are divided by 3, such operation will not influence the cycle of the structure vibration. For the curve of phase trajectory after standardization, refer to Fig. 4, the increase of amplitude from internal to the external. It can be seen clearly from Fig. 4 that along with the augmentation of the amplitude, the speed at the position x of the same displacement increases after the standardization. Therefore, along with the increase of amplitude, the cycle is surely reducing.

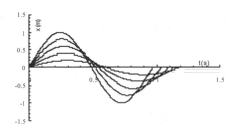

Figure 5.   vibration curve for different amplitude.

6.3   Cycle of vibration with big deformation

If the items of higher order with big deformation for $x + 27.133x + 18.772x^3 = 0$ is ignored, we get $x + 27.133x = 0$, and the computation result for non-linear vibration can be obtained: $\omega = \sqrt{27.133} = 5.209$, $f = 5.209/(2\pi) = 0.829\,(Hz)$, cycle T = 1.206 second;

The manner of numeral integral is used to calculate formula (22), we can respectively get:

when amplitude $D_1 = 0.2$ m, $T_1 = 1.192$ second, $f = 0.839\,(Hz)$, the difference in frequency up to 1.2% in comparison with the linear computation;

When amplitude $D_2 = 0.4$ m, $T_2 = 1.146$ second, $f = 0.873\,(Hz)$, the difference in frequency up to 5.3% in comparison with the linear computation ;

When amplitude $D_3 = 0.6$ m, $T_3 = 1.094$ second, $f = 0.914\,(Hz)$, the difference in frequency up to 10.3% in comparison with the linear computation;

When amplitude $D_4 = 0.8$ m, $T_4 = 1.034$ second, $f = 0.967\,(Hz)$, the difference in frequency up to 16.6% in comparison with the linear computation;

When amplitude $D_5 = 1.0$ m, $T_5 = 0.9692$ second, $f = 1.032\,(Hz)$, the difference in frequency up to 24.5% in comparison with the linear computation;

Similarly, we can use the numerical computation to get the relation curve between x and y, refer to Figure 5. It can be seen from Fig. 5 that the vibration cycle is in reduction along with the augmentation of amplitude.

6.4   The vibration cycle with small amplitude in time of big deformation

Substituting A = 27.133, B = 18.772 into (25), we can get: $\omega^2 = A + 3Bw^2 = 27.133 + 56.316w^2$

When deformation $w_1 = 0.2$ m, $T_1 = 1.159$ second, $f = 0.863\,(Hz)$, a 4.1% increase in frequency in comparison with the computation of plane position;

When deformation $w_2 = 0.4$ m, $T_2 = 1.045$ second, $f = 0.957\,(Hz)$, a 15.5% increase in frequency in comparison with the computation of plane position;

When deformation $w_3 = 0.6$ m, $T_3 = 0.913$ second, $f = 1.096\,(Hz)$, a 32.2% increase in frequency in comparison with the computation of plane position;

When deformation $w_4 = 0.8$ m, $T_4 = 0.791$ second, $f = 1.265\,(Hz)$, a 52.6% increase in frequency in comparison with the computation of plane position;

When deformation $w_5 = 1.0$ m, $T_5 = 0.688$ second, $f = 1.454\,(Hz)$, a 75.4% increase in frequency in comparison with the computation of plane position;

For the engineering, under the action of wind load up to 1 Kn/m^2, the displacement reaches 0.6 m. In accordance with the formula given in the article, under the action of wind load, the vibration frequency with small amplitude is about $f = 1.096\,(Hz)$, a 32% higher than the vibration frequency $f = 0.829\,(Hz)$ at the position without deformation; For the cable-net, the vibration frequency $f = 0.914\,(Hz)$ under the conditions of big amplitude up to 0.6 m about 10.3% higher than $f = 0.829\,(Hz)$ for linear computation frequency; It can be seen that it is necessary to use the non-linear differential equation for the solution.

7   CONCLUSION

This article deduce the non-linear differential equation (15) for the vibration of plane rectangular cable-net, when the geometric non-linearity is considered, and the method and process for the order reduction of the non-linear differential equation and the numeral solution are given; From the non-linear differential equation, the formula (25) on the computation of frequency with vibration of small amplitude made near the big deformation for the plane rectangular cable-net is further deduced; The computation and analysis are made on the formula and computation method presented in the article through the structure of the glass curtain cable-net of Zhongguancun Culture Building. The five different amplitudes with the maximum deformation up to 0.2 m, 0.4 m, 0.6 m, 0.8 m, and 1.0 m are given comparison in computation. It can be seen that when the deformation is 0.6 m , the vibration with big deformation is $f = 0.914\,(Hz)$, about 10.3% higher than the linear computation frequency $f = 0.829\,(Hz)$; The vibration frequency with small amplitude with deformation near 0.6 m is $f = 1.096\,(Hz)$, about 32% higher than the vibration frequency $f = 0.829\,(Hz)$ for the position without deformation. Therefore, in the structure of plane pre-stress cable-net, the influence of non-linearity with big deformation on the vibration can not be ignored.

REFERENCES

Bobrowski, J. Calgary's Olympic Saddle Dome. Int. J. Space Structure. 1985(1)

Guo Shihong, Wu Yue, and Feng Ruoqiang, Analysis into the anti-vibration performance for the single-cable support system for the curtain wall of New Poly Plaza. Dissertation collection of 18th academic meeting on the structure of high-rise construction, 2004, Chongqing.

Huo Wenying, Li Qian, and Shang Renjie, Design and construction for the structure of plane pre-stress cable-net. Industry Construction. March 2002.

Qian Weichang, Variational mehod and finite element, Science Press, 1980

Schlaich, J., Schober, H. and Mosehner, T. Prestressed Cable-net Facades. Structural Engineering International, 2005, 8(1):36–39

Shen Shizhao and Jiang Zhaoji. The combined dome of cable-net for Chaoyang Gym of the Asian Games. Journal of Building Structures. 1990 (3)

Shen Shizhao, Xu Zhongbao, zhao Ren, and Wu Yue Structural design for suspension cable, China Construction Industry Press, January 2006, Beijing, p183–186

Yang Qingshan, and Sun Xuedong. The natural vibration characteristics for the hyperbolic parabolic with elliptic Plan. January of Harbin Construction University, 1997, 30 (4) 35–40

You Guimu, Xiao Qianfeng, and Shang Renjie, The stress analysis for the structure of cable-net with flexible boundary, January of Changzhou Industry College, December 2005, Volume 18 Overall No. 81.

Zhao Xian, The steel structure for the curtain wall of the large public building in our country. China Construction Industry Press, 2002, 17 (6): 1–3.

Steel and Composite Structures – Wang & Choi (eds)
© *2007 Taylor & Francis Group, London, ISBN 978-0-415-45141-3*

Behavior of different connectors under monotonic and cyclic loading

A.L. Ciutina & A. Dogariu
Department of Steel Structures and Structural Mechanics, "Politehnica" University of Timisoara, Romania

ABSTRACT: The investigation of the composite structures with connection between the concrete and the steel structure is of scientific interest nowadays, due to the economical reasons resulted from the practical application of the solution. The paper presents the results of a laboratory test program on standard push specimens, on which various parameters have been monitored, such as the type of connectors (UPN profiles, LL profiles, Φ22 shear studs, Φ16 shear studs, hooks of reinforcing bars), steel profile flange class (class 1, class 2 and class 3 respectively) concrete class (C25/30 and C30/37). Cyclic loading was applied on 5 specimens. The results are commented in terms of resistance, ductility and judged in function of their capabilities to sustain a slip deformation between the concrete slab and steel profile.

1 INTRODUCTION

The composite steel-concrete systems for beams are already used for a considerable time in building design (Oehlers & Bradford, 1995). The behavior of composite beams is governed by the shear connection between the concrete slab and the steel section. For this reason, many types of devices have been conceived and tested by researchers in order to realize an optimum shear connection. The economic considerations continue to motivate the development of new products, while in other cases the researchers try to use new techniques for an economic use of traditional connectors (Hosain & Pashan, 2004). The new generation of connectors, such as the perfobond connectors (Marececk et al., 2005) seems to be a good alternative to the standard headed-stud connectors.

Standard push-out procedure given in the Annex B of Eurocode 4 (2004) provides a good tool for investigation of the shear connectors. The procedure, however, gives information only about the monotonic loading. In case of reversal loading of main beams of composite moment resisting frames under seismic loading, the shearing of the connectors could change from one sense to the other as the bending moments on the beam changes from hogging to sagging. In literature, there are relatively few reports on the behavior of connectors under cyclic loading, and most of them deal with the dowel behavior of standard headed-studs connectors.

Feldmann & Gesella (2004) make a very detailed analysis on the fatigue analysis on the headed studs under non-static loading of headed studs. In low-cycle fatigue, Bursi & Ballerini (1997) analyzed the low-cycle fatigue behavior of welded stud connectors under variable and constant non-sequential phase displacement histories. A comparison between the cyclic and monotonic results led the authors to the conclusion that the strength predicted by the Eurocode 4 appears unsafe when directly applied to seismic design.

Aribert & Lachal (2000), on a research conducted on two different types of connectors, have also proved that the cyclic response lead to a major decrease in the global resistance and ductility, stressing out that in the case of cyclic loading, as is the seismic action, the use of partial-shear connection should be avoided.

The present paper reports the results of eleven push-out shear specimens under monotonic loading and five push-pull specimens under cyclic loading. This represents the first part of an ongoing research activity on the behavior of shear connectors that is under progress in the CEMSIG Laboratory at the Politehnica University of Timisoara, Romania. The experimental results will be completed by the numerical simulation by Finite Element Methods in order to calibrate and extend the study on other parameters. The final purpose of the research activity is to find the behavior of different types of connectors subjected to cyclic loading as in the case of by seismic activity.

2 SPECIMENS, TESTING SET-UP AND INTERPRETATION OF RESULTS

2.1 *General configuration and description of specimens*

The dimensions of specimens for the push-out standard tests (including the metallic section and the reinforcing plan) were initially derived from the paragraph B 2.2

Table 1. Description of push-out specimens.

Specimen	Type of connectors	No. of connectors	Concrete class	Steel profile
PT-16/I-M	8Φ16 (2 rows)	8	C25/30	HEB 260
PT-16/II-M	8Φ16 (2 rows)	8	C25/30	Class 2*
PT-16/III-M	8Φ16 (2 rows)	8	C25/30	Class 3**
PT-16/S-M	8Φ16 (2 rows)	8	C30/37	HEB 260
PT-16/I-C	8Φ16 (2 rows)	8	C30/37	HEB 260
PT-22-M	4Φ22 (1 row)	4	C25/30	HEB 260
PT-22-C	4Φ22 (1 row)	4	C25/30	HEB 260
PT-A-M	Reinf. hooks (Φ 10 mm)	4	C25/30	HEB 260
PT-A-C	Reinf. hooks (Φ 10 mm)	4	C25/30	HEB 260
PT-A/S-M	Reinf. hooks (Φ 10 mm)	4	C30/37	HEB 260
PT-II-M	Perforated steel plate ***	2	C25/30	HEB 260
PT-II-C	Perforated steel plate ***	2	C25/30	HEB 260
PT-LS/II-M	L120 × 80 × 8	4	C25/30	Class 2*
PT-LS/III-M	L120 × 80 × 8	4	C25/30	Class 3**
PT-LS/I-C	L120 × 80 × 8	4	C25/30	HEB 260
PT-US-M	UNP 120	4	C25/30	HEB 260

* 260 × 260 profile $t_f = 10$ mm
** 260 × 260 profile $t_f = 8$ mm
*** longitudinal steel plate ($t = 8$ mm) on each side of steel profile, perforated for the passage of reinforcement

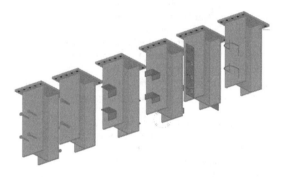

Figure 1. Disposition of connectors on steel profiles.

from the Annex B of Eurocode 4. The concrete and steel profile global dimensions were kept constant for all the tests, as well as the configuration and diameter of the reinforcing bars (Φ10 mm) according to Figure B.1 of Eurocode 4.

The description of push-out specimens it is shown in the Table 1. Four parameters have been taken into consideration in the conception of specimens:

- type of connectors (Φ16 KOCO headed studs on two rows, Φ22 KOCO headed studs on one row, LL120 × 80 × 8 angle profile, UNP 120 channel profile, perforated steel plate and reinforcement anchor hooks of Φ 10 mm,). Figure 1 shows a 3D view of the steel specimens (connectors included);
- concrete strength class (C25/30; C30/37);
- steel profile class (class 1 corresponding to standard HEB 260 profile, class 2 by considering a

10 mm steel flange and class 3, by a flange thickness of 8 mm respectively);
- monotonic (11 specimens) and cyclic loading (applied to 5 specimens).

2.2 Testing arrangement and loading procedure

Figure 2 shows the testing set-up. The load was applied in displacement control through a traction and compression actuator. In case of monotonic loading, the load was first applied in increments up to 40% of the expected failure load and then cycled 25 times between 5% and 40% of the expected failure load, in accordance to the Eurocode 4 stipulations. Subsequent load increments have been imposed up to failure. In case of cyclic loading was applied the ECCS (1986) procedure, with the yielding displacement computed on the results obtained from the monotonic test.

2.3 Tests evaluation

The following basic interpretation of the results was adopted: The connector shear capacity $P_{R,k}$ represents the maximum load capacity reduced by 10% and divided to the number of shear connectors; the connector's slip capacity δ_u is taken from the load-slip deformation curve, as corresponding to the shear capacity $P_{R,k}$ (see Figure 3).

Displacement transducers have been disposed to measure the relative slip between the concrete slabs and the steel profile, but also the separation between the two elements. In total a number 10 displacement transducers have been used for each test – 6 of them

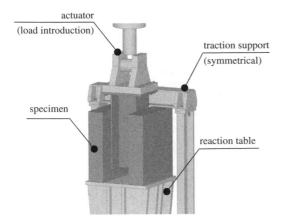

Figure 2. Testing arrangement.

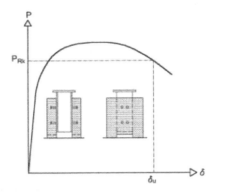

Figure 3. Determination of slip capacity δ_u (Eurocode 4).

for measuring the relative longitudinal slip and respectively 4 for measuring the separation (top and bottom) of concrete slab from steel profile.

For the interpretation of results, the force was considered from the actuator, while the slip was recorded as the mean value between the two displacement transducers disposed at the top of the specimen.

3 EXPERIMENTAL RESULTS

Table 2 summarizes the interpretation of the results derived from the monotonic curves (given in Figure 4) and the envelopes of the cyclic tests, in which:

- F_y represents the yielding force, in the sense of ECCS loading procedure (1986);
- δ_y the corresponding yielding displacement;
- $S_{j,ini}$ the initial stiffness of the F-δ curve;
- F_{max} is the maximum recorded force during testing;
- δ_u is the slip capacity of the connectors;
- $P_{R,k}$ is the connector's shear capacity.

3.1 Results of monotonic tests

The results of the monotonic curves are quite dispersed. The maximum applied load ranges from 590 kN (PT-A-M specimen) to 1256kN (PT-US-M specimen) while the slip capacity ranges from 2.36 mm (PT-LS/II-M) to almost 22 mm (PT-II-M specimen).

A short analysis, taking into account the parameters of the conception of specimens is made.

Influence of type of connectors

For the purpose of our study, six types of different connectors were chosen. All of the six connection typologies have the same steel cross-section and concrete class. Figure 5 presents the behavior curves for these typologies.

In terms of resistance, the specimen having channel connectors (PT-US-M) have shown the greater resisting force, while the specimen with anchoring bars (PT-A-M) had the resistance less than half from the first specimen. In terms of ductility, the PT-II-M specimen have the larger ductility, but it has to be noted the fact that it represents the ductility of reinforcing bars – concrete system (the failure in this case was by splitting and crushing of concrete nearby the perforated steel plate, and bending of reinforcement).

The specimen having LL120 (PT-LS/II-M) connectors behaved rather badly, the failure being in this case very early by puling-out of the angle connectors from the concrete. This is in fact demonstrated by the fast descending of the strength in monotonic curve characteristic of the specimen.

The headed stud connectors proved an expected behavior, by a rather good ductility and a resistance according to their classic design. However, there are quite important differences between the two specimens (PT-16/I and PT-22) although their shear area are almost the same, both in terms of resistance and ductility.

Influence of concrete class

The differences between the characteristics of specimens with different concrete classes is summarized in Table 2 and are shown graphically in Figure 6 (for specimens PT-16/I and PT-A for which have been designed similar specimens with a different concrete class).

Because the observed failure mode of specimens was by shear of the steel connectors and not by concrete crushing, there is no clear evidence in the increase of the specimen's shear resistance, neither of the ductility with the increase of the concrete class. However, the initial stiffness of the curves specific to the specimens with C30/37 concrete class is significantly greater than the usual concrete specimens and seems to be the only influence of the concrete class in the case of monotonic loading.

Table 2. Main test results derived from the interpretation of experimental data.

Specimen	F_y [kN]	δ_y [mm]	$S_{j,ini}$ [kN/mm]	F_{max} [kN]	δ_u [mm]	$P_{R,k}$/conn. [kN]
PT-16/I-M	396.0	0.20	1983.4	801.1	4.91	99.7
PT-16/II-M	514.6	0.37	1349,8	862.6	7.52	100.1
PT-16/III-M	501.2	0.34	1399.6	831.8	6.42	103.1
PT-16/S-M	436.2	0.17	2359.8	844.0	7.18	102.9
PT-16/I-C	430.0	0.29	1454.0	579.0	2.33	60.66
PT-22-M	405.8	0.35	1093.5	737.6	14.43	177.2
PT-22-C	407.0	0.44	871.0	474.0	2.11	95.6
PT-A-M	396.0	0.20	1813.0	590.6	7.84	132.8
PT-A-C	339.0	0.08	4361	544.3	1.98	117.9
PT-A/S-M	356.9	0.08	4331.1	649.6	6.51	146.6
PT-II-M	557.5	0.13	3701.4	1033.0	21.46	465.6
PT-II-C	403.0	0.09	3939	586	0.60	272.5
PT-LS/II-M	586.8	0.21	2677.1	794.4	2.36	179.1
PT-LS/III-M	494.6	0.10	4726.9	790.4	3.09	177.8
PT-LS/I-C	650.0	0.27	2274.0	774.3	0.98	165.9
PT-US-M	804.9	0.41	2052.9	1256.0	9.65	314.0

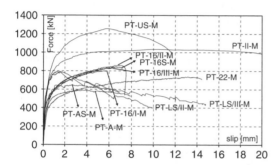

Figure 4. Comparison of the monotonic curves.

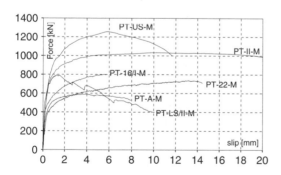

Figure 5. Force-slip curves on different types of connectors.

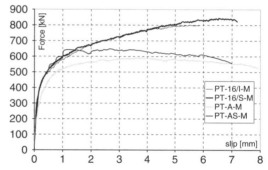

Figure 6. Force-slip curves showing concrete class influence.

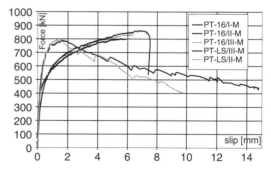

Figure 7. Force-slip curves showing steel flange class influence

Influence of steel flange class

The 16 mm headed stud (PT-16) and the L (PT-L) specimen series have been chose to have different thicknesses of steel flange (17.5 mm for HEB 260 profile representing the class I, 10 mm for class II and 8 mm for class III respectively). The results given in Table 2 and also graphically shown in Figure 7, shows

the fact that there is no major difference among the series specimen's behavior. This is valid for both resistance and ductility. However, the initial stiffness is rather different within the specimens of each series,

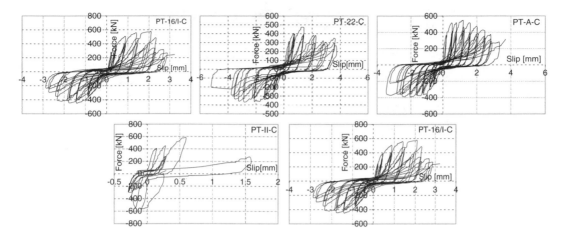

Figure 8. Force-slip curves for cyclic specimens.

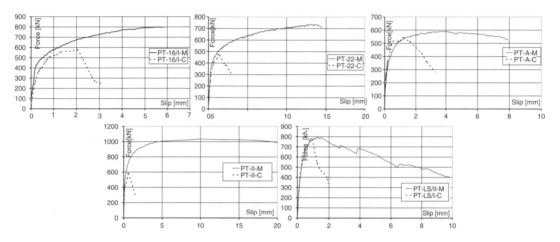

Figure 9. Differences in monotonic and cyclic envelopes.

but this does not follow a certain rule with the change of the thickness of the steel flange.

It seems that the reduction in the steel profile flange thickness should be more important (to about 6 or even 4 mm) in order get a logical increase of the ductility due to local bending of the flange in the vicinity of the connector.

3.2 Results of cyclic tests

Figure 8 displays the cyclic behavior of the five cyclic tests, while Figure 9 shows the comparison of the positive cyclic envelopes to the monotonic corresponding curves. Generally, all the monotonic tests have proved a reduction both in the resistance and ductility capacity. For example, for the specimen PT-16 with 16 mm headed studs, the P_{Rk} resistance reduction is about 40%, while the slip capacity is reduced by more

than 50%, as compared to the monotonic tests. The same conclusion could be stated also for the PT-22 specimens, but with a higher degree of reduction in ductility.

The PT-II specimens with prefobound represent, as proved by the tests the most dangerous situation. Although the monotonic test have shown a very good behavior, with a very good resistance and the highest ductility among the monotonic tests, the cyclic test have proved a very bad cyclic behavior, with about 40% reduction in resistance and an ultimate slip displacement δ_u of 0.6 mm. In fact, the problem in this case appears due to the fact that the shear connection between the steel connector and the concrete is lost in the very first cycles due to the "knife" effect of the prefobound connector. In this way, practically there is no energy dissipation during the cycles and a rapid decrease of its resistance.

Figure 10. Failure modes by shear for headed studs and reinforcement hooks.

However, an interesting situation appears in the case of LL (PT-L) and anchor hook (PT-A) specimens, where despite the important reduction in ductility ($\delta_u < 2$ mm in both cases), the shear capacity reduction remains under 10% for the cyclic tests.

It has to be stressed out that in all the cyclic cases, the slip capacities do not pass the ductility criterion of 6 mm stipulated in Eurocode 4-1, 6.6.1.1.

Although a limited study, the cyclic tests show very clear that the monotonic results do not necessarily conduct to the same cyclic results and in consequence the use of shear connectors in seismic prone areas should be reconsidered.

3.3 *Failure modes*

The Figures 10 to 11 shows the typical failure modes for different types of connectors.

In case of shear studs (Figure 10 a), although the global behavior could be characterized as ductile, the failure was brittle in nature, by the shearing of the headed studs situated on one side of the specimen. It is to be added that there are not evidences of concrete crushing, with the exception of the base of the shear stud.

Failure mode of specimens having hook anchors (Figure 10 b) was by shear of the reinforcement hooks. No crushing of concrete was observed. In nature, this type of failure was considered to be ductile, due to the fact that not all the shear fractures were in the same time.

The LL and UPN specimens, the failure occurred by bending of the shear profiles (Figure 11), but without a real fracture of the steel material. In the case of LL specimens, the flange embedded in concrete was pulled out very rapidly from the concrete after reaching the maximum load. Maybe the use of reinforcing bars through the embedded flange could retard the failure. Besides, the channel connector was very well embedded in concrete, and in this case, the failure mode was combined, by local crushing of the concrete, and bending of the channel web near the welding to the steel flange.

Figure 11. Failure modes by bending for LL and channel profiles.

The CP-II-M specimen represents a special case, due to the fact that the failure was not belonging to the steel connector, but to the concrete slab and reinforcement. Practically, it acted as a "knife" into the body of the concrete slab. The first signs of the failure were by longitudinal cracks of the concrete slabs in the middle of the slabs, and continued by the crushing of concrete at the bottom of the slab. After loading, there was found that the transversal reinforcing bars were bent by the perforated steel plates. In fact, this type of connection gave a ductile behaviour for the monotonic specimen, but led to a very rapid lost of the adherence in case of cyclic specimen.

4 CONCLUSIONS

Within the restrained study presented in the paper, the following conclusions could be drawn:

– the change in the connector type induced practically the dispersion of results in terms of Force-Slip curves for the monotonic tests. A very good monotonic performance in terms of resistance was obtained for the channel connectors and for the prefobound specimen. The worse performance in terms of ductility was obtained for LL specimens, which should be avoided without using additional reinforcing bars passing through the LL flange. The headed stud specimens provided a good response both in terms of resistance and ductility, in accordance to their well-known design characteristics and expected failure. The anchor hook specimens provided a good monotonic ductility but a rather small resistance;

– the influence of the concrete class is not evident in the studied cases, with the exception of the initial stiffness of the Force-Slip curve. The other parameters seem not to be affected by the change of concrete class, unless the failure mode of the specimens is changed (by crushing of concrete for example);

– for the studied specimens, the change in the steel flange class do not affect significantly the monotonic response in terms of Force-Slip deformation curve;

– the cyclic loading introduces for all the specimens a significant reduction in the slip capacity of the connectors, reduction that leads to non-ductile connectors in accordance to Eurocode 4. Also, there is an important reduction in the characteristic resistance P_{Rk}, ranging from 10 to 40 %. The opinion of authors is that the research on connectors under cyclic loading should be continued, in order to give a safer design of composite beams under seismic loadings and proper regulations regarding the connectors in the Eurocode 8.

ACKNOWLEDGEMENT

This work was carried out in the CEMSIG Laboratory of the Department of Steel Structures from the Politehnica University of Timisoara. The financial support from the Romanian Ministry of Education and Research (CEEX-ET no. 3153/13.10.2005) is gratefully acknowledged.

REFERENCES

Aribert, J-M. & Lachal, A., 2000 *Moment Resistant Connections of Steel Frames in Seismic Areas. Design and Reliability* (Editor F.M. Mazzolani), E&FN SPON, London, ISBN 0-415-23577-4 Chapter. 4.3. *Cyclic behaviour of Shear Connectors.*

Bursi, O.& Ballerini M., 1997 *Quasi-static cyclic and pseudo-dynamic tests on composite substructures with softening behaviour* 5th International Colloquium on Stability and Ductility of Steel Structures, Nagoya, Japan.

European Convention for Constructional Steelwork, Technical Committee 1, TWG 1.3 – Seismic Design, No.45, 1986, *Recommended Testing Procedures for Assessing the Behaviour of Structural Elements under Cyclic Loads.*

EN 1994-1-1, 2004. *EUROCODE 4: Part 1.1 – Design of composite steel and concrete structures.* Brussels: CEN, European Committee for Standardisation, Final Version, September 2004.

Feldman, M., Gesella, H. & Leffer, A. 2004 *The Cyclic Force-Slip Behaviour of Headed Studs under Non-Static Service Loads – Experimental Studies and Analytical Description* Composite Constructions in Steel and Concrete V, Ed. Leon, T. and Lange, J., ASCE 2004.

Hosain, M.U. & Pashan, A. 2004 *Channel Shear Connectors in Composite Beams: Push-out Tests* Composite Constructions in Steel and Concrete V, Ed. Leon, T. and Lange, J., ASCE 2004.

Marececk, J., Samec, J. & Studnicka, J. 2005 *Perfobound Shear Connector Behaviour* Proceedings of EUROSTEEL 2005 4th European Conference on Steel and Composite Structures, Maastricht, 2005 Ed. Hoffmeister, B. and Hechler, O.

Oehlers, D.J. & Bradford, M.A., 1995 *Composite Steel and Concrete Structural Members*, ISBN 008041919 Pergamon.

Steel and Composite Structures – Wang & Choi (eds)
© *2007 Taylor & Francis Group, London, ISBN 978-0-415-45141-3*

Evaluating steel structural damping ratio using the Fisher rule and Nonlinear Mapping based on experimental data

Z.Y. Yang
Department of Civil Engineering, Wuhan Univ. of Technology, Wuhan, China

Y.C. Wang
School of MACE, University of Manchester, UK

L.P. Zhang
Wuhan Meteorology Center Observatory, Wuhan, China

ABSTRACT: It is well known that the damping ratio of a steel structural building is affected by many factors such as the structural natural frequency, material used, foundation type and building size. This paper presents a method to evaluate the damping ratio of a building structure based on experimental measurements of other buildings, by applying the Fisher rule and the nonlinear mapping theory (NLM). The experimental data come from observations of real full-scale buildings. Firstly, the Fisher rule is applied to determine the λ value of each influential factor. The factors which possess high values of λ construct an M-dimension data space, which is then mapped to a low (2) dimension space by making use of the Mahalanobis distance. In the lower space, the building damping ratio can be evaluated more easily. An example is provided to show the calculation procedure.

1 INSTRUCTIONS

Damping is a measure of the efficiency of a system in dissipating the energy that is accumulated during the system dynamic action. The dynamic response of a structural system to dynamic loading is fundamentally governed by the amount of damping exhibited by each mode of vibration. Structural damping is governed by many factors. Although so far there is not a good method which can be used to accurately evaluate structural damping[1], over the years, there has been considerable research work in pursuit of descriptions of inherent damping of structures. Several investigations have shown that the damping of steel and concrete structures increases with the amplitude of vibration[2] and the damping ratio at high levels of vibration can be up to 4 times that at low levels of vibration[3][4]. The construction materials used in a building affect its damping as a result of different material internal damping due to different interactions among molecules of the different construction materials. Structural damping of a building is also closely related to testing methods, foundation type, depth of embedment, building weight, the planar dimensions of the building, building construction method. Under different conditions, different factors may affect the structural damping differently. Because of the almost randomness of structural damping and the factors that affect structural damping, it is very difficult to predict the damping ratio of a building structure objectively.

Yet different building codes require that the displacement of a building under normal use should be limited so as not to affect normal functions of the building. This requires that the structural damping value should be estimated at the design stage. Presently, different countries give highly different values of structural damping. For example, in Japan[8], the steel structural damping ratio is given as 0.02 and the reinforced concrete structural damping ratio is 0.03; whilst in the Chinese loading code[9], the damping ratios for steel and concrete structures are taken as 0.01 and 0.05 respectively. There are also some countries whose building design codes separate the damping ratio for free vibration from that for forced oscillation, giving damping ratios of 0.005 & 0.0125 for free and forced vibrations for steel structures and 0.025 & 0.05 for reinforced concrete structures.

Clearly, there is a need to give better estimate of damping ratios of realistic building structures. Based on measurements of damping ratios of a large number of real full-scale structural buildings, this method presents a practical method for estimating damping ratios.

2 BASIC PRINCIPLES

The method used in this research is based on the Fisher rule[6][7]. In this method, the measurements of damping ratios from a large number of full-scale tests are used to estimate the magnitude of influence of each factor, from which the most influential factors are identified. Details of this method are described below.

For simplicity the structural damping values may be considered to be two kinds: either large or small. Naturally, it is hoped that samples with large damping values can be distinctly separated from samples with small damping values. Not only this means the distance between samples with large damping values and those with small damping values is large, but also the distances between samples with either large or small damping ratios are small. In statistics terminology, this means that if factor X is selected as a factor that has high influence on structural damping, the square sum of deviations of observations of the two kinds of samples (large and small damping ratios, β as defined by equation 1) should be as far as possible but the square sum of deviations of observations of samples of the same kind (large or small damping ratios, ω as defined by equation 2) should be as small as possible.

$$\beta(x) = \sum_{g=1}^{G}\sum_{k=1}^{n_g}(\overline{x}_g - \overline{x})^2 = \sum_{g=1}^{G} n_g(\overline{x}_g - \overline{x})^2 \quad (1)$$

$$\omega(x) = \sum_{g=1}^{G}\sum_{k=1}^{n_g}(\overline{x}_g - x_{gk})^2 \quad (2)$$

where: G is the total number of kinds (G = 2 for large and small damping ratios); $\overline{x}_g$ is the mean of g-kind values and $\overline{x}$ is the mean value of all values (large and small damping ratios). Samples whose structural damping ratios are higher than the total mean belong to the first kind (tends to large) and those whose structural damping ratios are lower than the total mean belong to the second kind (tends to small). From definition:

$$\overline{x}_g = \frac{1}{n_g}\sum_{k=1}^{n_g} x_{gk} \quad g = 1,2$$

$$\overline{x} = \frac{1}{N}\sum_{g=1}^{G}\sum_{k=1}^{n_g} x_{gk} = \frac{1}{N}\sum_{g=1}^{G} n_g \overline{x}_g$$

where n_g is the total number of observations of all kinds, i.e. $N = \sum_{g=1}^{G} n_g$.

Define:

$$\lambda = \beta(x)\Big/\omega(x) \quad (3)$$

Then generally, $\lambda \to Max$ is hoped. With a high value of λ, if all the observations are plotted on the same chart, the observations corresponding to the two different kinds (large or small) should be as far apart as possible while the observations corresponding to the same kind should be closely spaced. The factors with high λ values have the ability to greatly influence the structural damping ratio and they should be identified.

Having identified the influencing factors (M), it is possible to estimate the influence of each factor by solving an M-dimensional simultaneous linear equation. However, if the number of influential factors is equal or higher than 3 (M ≥ 3), it will be very difficult to obtain accurate solutions of the M-dimensional equations when there are close correlations between two or more factors according to the Crammer rule. We therefore look for a method to map from the M-space that is constituted by the influential factors (corresponding to large λ values) to a 2-dimensional space so the solutions can be accurately obtained. This mapping should not be affected by correlations between different factors and should maintain the characteristics of large λ values. To do this, the Nonlinear Mapping (NLM) method is used.

In general, the Euclidean distance is selected in NLM, giving[6]:

$$d_{ij} = \sqrt{(y_i - y_j)^T(y_i - y_j)} \quad (4)$$

where:

$$y_i = \begin{pmatrix} y_{i1} \\ y_{i2} \\ \vdots \\ y_{in} \end{pmatrix} \quad y_j = \begin{pmatrix} y_{j1} \\ y_{j2} \\ \vdots \\ y_{jn} \end{pmatrix}$$

However, since the Euclidean distance is dimensional and does not consider correlations between different factors, the results can be highly subjective. It is better to use the Mahalanobis distance (hereafter to be simply referred to as the M-distance), which overcomes the shortcomings of using the Euclidean distance. The M-distance is defined as[5]:

$$d_{ij} = \sqrt{(y_i - y_j)'v^{-1}(y_i - y_j)} \quad (5)$$

$$v = \begin{bmatrix} v_{11} & v_{12} & v_{13}\cdots & v_{1n} \\ v_{21} & v_{22} & v_{23} & v_{2n} \\ \vdots & \vdots & & \\ v_{n2} & v_{n2} & \cdots\cdots & v_{nn} \end{bmatrix}$$

Table 1. Summary of Full-Scale Experimental Data of Steel Building Structures.

NO	+	−	L (m)	H (m)	W (m)	Depth of Embedment (m)	Trans Disp. (mm)	Long Disp. (mm)	Trans-per. (S)	Long-per. (S)	Trans-damp (%)	Long. Damp (%)
1	48	3	108.8	243.4	44.8	20	10.6	25.4	4.31	3.67	0.5	0.5
2	43	3	67	180	65	24.4	500	300	2.7	2.7	0.4	0.5
3	47	3	79.8	169.8	25.9	19.8	1600	250	4.3	3.1	0.5	1.8
4	35	3	37.4	169.7	37.4	22.2	368.5	493	3.05	2.91	0.8	0.7
5	37	1	54	157	52	22	200	200	3.33	2.86	0.3	0.4
6	38	1	37	150	37	10	30	30	3.45	2.94	0.6	0.6
7	35	3	40	132	28	20.6	3000	2500	2.38	2.56	0.9	1
8	30	2	59	129	49.4	29	70	30	2.73	2.56	0.5	1
9	21	5	96	126	25.6	31.9	8	8	2.55	2.36	0.6	1.4
10	31	3	60	123.5	36	7	32.4	17.7	2.5	2.08	1.4	0.9
11	30	3	63.6	123	36	18.9	380	370	2.65	2.24	0.3	0.4
12	27	2	44.8	117.8	38.4	15.7	400	250	2.63	2.63	0.5	0.5
13	32	1	33.1	113	33	12.5	700	200	2.04	1.96	1.1	0.8
14	30	3	45	111.4	28.9	14.6	10000	400	3.2	2.71	0.9	1
15	29	4	36.6	109.5	36	24	2400	1800	2.43	2.35	1	1.1
16	25	3	59	103.5	35.2	21	200	300	2.33	2	0.5	0.5
17	28	2	36.9	100.4	36.3	21	120	120	1.85	1.85	1	1.1
18	23	3	44.8	100	32	21.7	8390	5420	1.98	1.76	0.9	0.9
19	24	3	64	99.2	40	14.5	600	450	1.79	1.61	0.7	0.6
20	21	3	64	99.2	40	14.5	650	700	2.04	1.72	0.5	1.1
21	26	3	42	97.3	33.8	14.9	1200	1500	1.82	1.54	0.7	1
22	26	3	17.5	95.2	8	16.4	1500	600	2.08	2	0.6	0.7
23	21	3	28.57	94.3	28.57	19.2	3000	3000	1.32	1.59	1.3	0.7
24	19	5	43.9	89	35.5	25	570	330	1.57	1.4	0.4	0.6
25	25	4	86.4	88.5	12.5	18.8	1000	510	1.87	1.48	0.7	1.3
26	21	3	34.8	83.4	32	23.5	800	400	2.08	2.04	0.6	0.7
27	20	4	41.4	80.9	20.6	18.6	77.2	45.1	1.82	1.72	2.4	4
28	19	4	90	80.7	22.4	21.1	200	200	1.58	1.52	1.2	1.2
29	23	2	47.4	79.1	25.9	10.7	440	500	1.82	1.41	1.4	1.5
30	18	2	48	76.8	32.85	18.1	160	110	1.93	1.75	0.9	0.8
31	17	2	35	76.5	27	6	563	368	1.83	1.71	1.1	0.9
32	21	3	37.5	69.2	22.5	15.6	950	850	1.91	1.89	2.1	1.9
33	18	2	44.8	68.2	33	14.4	1.1	0.92	1.8	1.66	1	0.7
34	12	2	38.85	61	28.8	12.9	300	850	1.48	1.43	1	1.2
35	15	4	32.4	60.5	24.3	20.2	750	980	1.56	1.52	1.1	1.1
36	14	2	38.4	59.7	23.3	12.6	100	100	1.4	1.3	1.1	1
37	13	1	30.2	59.5	30.2	8	69.6	15.5	1.23	1.19	0.3	0.6
38	13	3	79	58	40	21.7	100	80	0.89	0.75	0.9	1.2
39	12	1	90.4	58	79.2	9.8	560	480	0.97	0.92	2.2	2.3
40	14	1	28.8	56	23.4	12.2	1750	2000	1.17	1.14	0.5	0.6
41	10	1	88.4	51.4	46.6	9	250	180	0.63	0.54	1.7	2.5
42	14	1	87.5	50.7	87.5	8.3	250	180	1.32	1.26	2.5	2
43	10	1	48	48.8	30	7.5	850	80	0.91	0.4	1	2
44	12	1	28.4	46.6	24.8	7.1	25	25	1.19	1.11	1.9	1.5
45	9	1	29.3	31.2	14.5	4	2000	1000	0.94	0.8	1.9	1.4
46	9	1	17.9	30.9	5	2.2	2	2	0.79	0.56	1.2	1.2
47	11	0	59.4	30.5	10.7	1.5	2920	2800	0.54	0.74	2.2	2.2
48	8	0	8.6	30.4	5.9	2	200	200	0.76	0.65	0.8	1.1
49	7	1	17.5	21.4	5	3.2	15.2	11.5	0.73	0.42	1.7	2.8
50	5	0	24	19.1	12	3	620	370	0.39	0.36	2.5	3.9

Table 2. λ Value of different factors to transversal damping.

Factors	Number of floors above ground	Number of under- ground floors	Length	Height	Width	Foundation depth of embedment	Transverse amplitude	Longitudinal amplitude	Transversal period	Longitudinal period
λ	22.32	7.78	0.81	27.19	1.57	32.92	1.47	0.07	27.47	25.39

where:

$$v_{ij} = \frac{1}{m} \sum_{k=1}^{m} (y_{ki} - \overline{y}_i)(y_{kj} - \overline{y}_j)$$

in which m is number of observation times and:

$$\overline{y}_i = \frac{1}{m} \sum_{k=1}^{m} y_{ki} \qquad \overline{y}_j = \frac{1}{m} \sum_{k=1}^{m} y_{kj}$$

In equation (5), $\mathbf{V}$ is the covariance matrix of the influential factors and $\mathbf{V}^{-1}$ is reciprocal of $\mathbf{V}$. The M-distance is appropriate because it has been mathematically proven to be not affected by units and correlations among the different influential factors. Clearly when $\mathbf{V} = \mathbf{I}$ (where $\mathbf{I}$ is the unit matrix), formula (4) is identical to (5).

It should be pointed out that after using the M-distance, the mapping is no longer affected by the types of probability distribution of the different influential factors, their correlations and units of measurement. After mapping from a high dimensional space to a low (2) dimensional space, the two-dimensional space still retains the large λ values. This method has already been applied to weather forecasting with drastic improvement of results.

After mapping the measurement results of many observations of structural damping onto a plane, it is possible to predict the damping value of another structure. As the points that are closet to the point of damping value to be determined have the most influence on its value, the damping value D_e to be determined should be calculated using the following equation:

$$D_e = \sum_{i=1}^{m} \left(\frac{1}{d_i} \middle/ \sum_{i=1}^{m} \frac{1}{d_i} \right) D_i \qquad (6)$$

where m is the number of times of observations; d_i is the M-distance from the sample vector corresponding to the damping that is being determined in n-dimensional space to the observation vector of the ith sample; D_i is the damping of the ith observation. From equation (6), it can be seen that the nearer the distance between the sample point and the required point, the closer the damping values between these two points and vice versa.

3 APPLICATIONS

Table 1 presents the observation results for 50 full-scale steel framed buildings, giving values of 10 factors that may influence the damping ratio of a structure. Consider the damping ratio for the first mode of the structures. First, standardize the variance of the 10 factors. Then calculate the λ values of each factor for the transverse damping ratio, the results of which are given in table 2.

It can be seen that high λ values are associated with the total building height, number of storeys above ground, depth of embedment, transverse period and longitudinal period are bigger. Since the total building height and the number of storeys are closely related, only the total building height is included as an influential factor. Therefore, four factors can affect the transverse damping of a steel framed building, being total building height, depth of embedment, transverse period and longitudinal period. Figure 1 describes how the four factors affect the transverse damping ratio.

Figures 1(c) and 1(d) show that in general, the transverse damping ratio reduces as the frequency increases. The small variations within each figure indicate that other factors will also have some influence on the damping ratio. In common situations, the depth of foundation embedment is related to the building height (a higher building has a deeper embedment), so Figure 1(b) to some extent reflects the trend in Figure 1(a).

Since the influencing factors are correlated, using the M-distance can effectively deal with the correlations. After using the NLM method to map the observation results onto a 2 dimension space, structural damping may be predicted according to the distribution of damping values on the 2-dimensional space.

For example, figure 2 shows the mapping results for height, foundation depth of embedment, transverse frequency and longitudinal frequency. For a building which is 70 meters high, with a foundation depth of embedment of 10 meters, a transversal period of 2s and a longitudinal period of 2.3s, the damping value of this building structure may be calculated to be 1.066% using equation (6). Mapping this building structure according to NLM, its position is shown as the triangle in figure 2, which is near the centre of figure 2 on the horizontal axis.

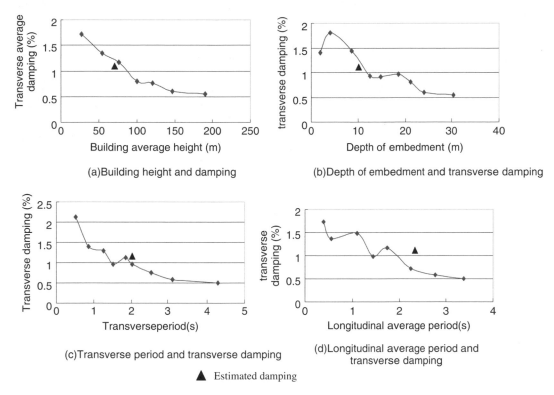

(a)Building height and damping

(b)Depth of embedment and transverse damping

(c)Transverse period and transverse damping

(d)Longitudinal average period and transverse damping

▲ Estimated damping

Figure 1.   Effects of different influencing factors on damping ratio.

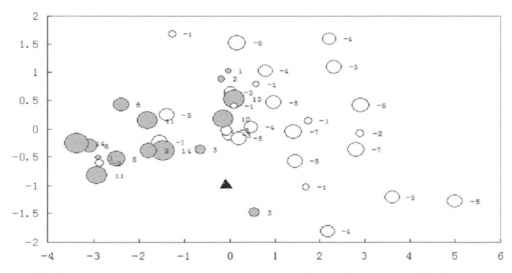

Note: Solid circles stand for large damping; Empty circles stand for small damping; Solid triangle stands for estimated damping value; The value on the right hand of a circle stands for the difference between the damping value and the average value (‰).

Figure 2.   Maping Results for Height, Foundation Depth of embedment, Transverse Period, Longitudinal Period.

4 DISCUSSION

Because the damping of a structure is affected by many factors, it is difficult to objectively and fully evaluate the damping of a building structure based on experimental data. Furthermore, all the influencing factors are likely to have some correlation between them to some extent. Using the Fisher rule and NLM method provides a feasible method to evaluate the damping ratio of a structure based on full scale experimental data. However, it should be pointed out that when selecting different experimental samples, it is important to define the conditions of the experiments. The full-scale experimental data in this paper were obtained in the condition of very small building amplitude such that its effect is small.

ACKNOWLEDGEMENTS

The work described in this paper was fully supported by a grant from the National Natural Science Foundation of China (50278074). The paper was written when the first author was a visiting scholar at the University of Manchester.

REFERENCES

[1] Y. Tamura and A. Jeary (1996), Forward, Journal of Wind Engineering and Industrial Aerodynamics, Vol. 59, pp. I–V.

[2] Q.S. Li, K. Yang, N. Zhang and C.K. Wong (2002), "Field Measurements of Amplitude-Dependent Damping in a 79-story Tall Building and Its Effects on the Structural Dynamic Response," The Structural Design of Tall Buildings, 11(2), pp.129–153.

[3] J.Q. Fang, A.P. Jeary, Q.S. Li and C.K. Wong (1999), "Random Damping in Buildings and Its AR Model," Journal of Wind Engineering and Industrial Aerodynamics, Vol. 79, pp. 159–167.

[4] G.Q. Li (1985), The Calculating Theory and Method of Earthquake-Resistant Structures [M], Earthquake Press, Beijing (in Chinese).

[5] J.W. Sammon (1969), "A Nonlinear Mapping for Data Structure Analysis", IEEE Transactions on Computers, 18(5), 401–408.

[6] R.T. Zhan, K.T. Fang (1997), Multivariate Statistical Analysis[M] Science Press, Beijing (in Chinese).

[7] F.X. Yang and Z.L. Lu (1985), Numerical Analysis[M], Tianjin University Press, Tianjin (in Chinese).

[8] X.L. Zhang (1990), Engineering Structural Wind Load Theory and Wind Resistance Manual, Tongji University Press, Shanghai.

Steel and Composite Structures – Wang & Choi (eds)
© *2007 Taylor & Francis Group, London, ISBN 978-0-415-45141-3*

Dynamic analysis of steel frameworks with joint damage after earthquake

Huaijin Song, Yongfeng Luo & Xue Li

College of Civil Engineering, Tongji University, Shanghai, China

ABSTRACT: This paper presents an analysis of cracks in weld seams existing in rigid welded flange and bolted web connections of steel framework. Based on the fracture mechanics, the effective length and effective depth of a crack are derived. The relationship between moment (M) and rotation (θ) of the damaged rigid connection is obtained. Numerical analysis is also conducted to ascertain the effects of joint stiffness on the natural frequencies of multi-story and high-rise damaged steel frames. The dynamic response of the damaged steel frames under earthquake is studied at the end of this paper. The results show that the joint damage has significant effects on aseismic behavior of the frames. Therefore, joint damage must be taken into account to analyze the aseismic behavior of steel frames under earthquake.

1 INTRODUCTION

The earthquake is a kind of natural calamity with great harmfulness. Destructive earthquakes have happened many times all over the world in the recent 20 years, such as the one in Loma Prieta, America, 1989, the one in Northridge, America, 1994 and the strong earthquake in Osaka and Kobe, Japan, 1995. During the violent earthquake, plastic deformations occurred on the beam-to-column connections of multi-story and high-rise steel frames. The damage on the weld seams may lead to the collapse of the whole structure. In order to prevent the damaged steel frame from being damaged again in subsequent earthquakes and to provide valuable references for the repairment, the moment-rotation ($M - \theta$) relationship of the beam-to-column connections with cracks in welded seams is deduced based on damage detection and the principle of the fracture mechanics. Finite element models are built and numerical analysis is conducted. The effects of joint stiffness on the natural frequencies of damaged steel frames and dynamic response of the damaged steel frames under earthquake is analyzed.

2 ROTATIONAL STIFFNESS OF A RIGID JOINT CONSIDERING THE DAMAGE

2.1 *Classification of the connections by the stiffness*

The stiffness of beam-to-column joints has been studied a lot in America and many European countries. Such connections are classified to pinned, rigid and semi-rigid according to the stiffness. The rotation θ is defined as the damage of the beam-to-column angle when loaded.

In EC3, When all the different parts of the joint are sufficiently stiff, there will be almost no difference between the respective rotations at the ends of the members connected at the joint. The joint can be considered as rigid. A rigid joint experiences a single global rigid-body rotation which is the nodal rotation in the commonly used analysis methods for framed structures. If the joint has a very small flexural stiffness, the beam will behave much as a simply supported beam, whatever the behavior of the other connected member. The joint can be considered to be a nominally pinned joint, the relative rotation between the beam and the column being almost the same as that for the simply supported beam end. For intermediate cases (non-zero finite joint stiffness), the transmitted moment will result in a difference between the absolute rotations of the two connected members. Then the joint is said to be "semi-rigid".

2.2 *The effective length of the crack*

The behavior of the beam-to-column connections has direct effects on the overall behavior of steel frames under loads. Bending deformation is dominant compared with axial, sheer and torsion deformations. The damaged rigid connections can not be treated as perfect rigid connections as usual in structure analysis. Instead, numerical simulation is employed to calculate the damaged rigid connections by using a rotation spring which has variable stiffness to model beam-to-column connection bending behavior as semi-rigid. The rotational stiffness of the welded flange-bolted

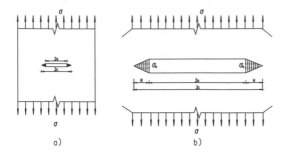

Figure 1. Banding plastic zone model.

by far-field stress σ and uniformly distributed stress σ_s on plastic zone respectively.

$$K_I^{(1)} = \sigma\sqrt{\pi c} \tag{1}$$

$$K_I^{(2)} = -\frac{2\sigma_s}{\pi}\sqrt{\pi c}\cos^{-1}\left(\frac{a}{c}\right) \tag{2}$$

K_I^c can be expressed as

$$K_I^c = K_I^{(1)} + K_I^{(2)} = \sigma\sqrt{\pi c} - \frac{2\sigma_s}{\pi}\cos^{-1}\left(\frac{a}{c}\right) \tag{3}$$

Since point c is the crack tip of plastic zone, the matrix should be not singular. That is to say $K_I^c = 0$. Thus

$$c = a\left/\cos\frac{\pi\sigma}{2\sigma_s}\right. = a\,\sec\frac{\pi\sigma}{2\sigma_s} \tag{4}$$

$\sec\frac{\pi\sigma}{2\sigma_s}$ can be expanded into series as follows

$$\sec\frac{\pi\sigma}{2\sigma_s} = 1 + \frac{1}{2}\left(\frac{\pi\sigma}{2\sigma_s}\right)^2 + \frac{5}{24}\left(\frac{\pi\sigma}{2\sigma_s}\right)^4 + \cdots\cdots \tag{5}$$

Omitting higher order terms, Equation (5) is simplified as

$$\sec\frac{\pi\sigma}{2\sigma_s} \approx 1 + \frac{1}{2}\left(\frac{\pi\sigma}{2\sigma_s}\right)^2 \tag{6}$$

Since $R = c - a$, it can be obtained from Equation. (4) and Equation. (6)

$$R = \frac{a}{2}\left(\frac{\pi\sigma}{2\sigma_s}\right)^2 \tag{7}$$

Then the effective length of crack can be written as

$$l^* = 2c = 2a + 2R = a\left(2 + \left(\frac{\pi\sigma}{2\sigma_s}\right)^2\right) \tag{8}$$

2.3 The Effective depth of the crack

In actual welding structures, cracks usually appear on the surface of the welding seams or are deeply embedded in the welding seams. They are penetrating cracks. The equivalent stress intensity factor K is needed. The

web connection depends mainly on the effective area of both top and bottom flanges of the beam. Therefore, the relationship of moment and rotation depends on the effective area of seams. Following assumptions are adopted for the deviation:

(1) Each crack is independent.
(2) All cracks are banding shape. The plastic zones of the crack tips stretch along two edges of the crack in wedge banding shape. The material of plastic zone is considered to be perfect plastic.
(3) The welding seams on beam flanges bear uniformly distributed loads.

Theoretical and experimental analyses show that the load on cracks leads to the further displacement of the surfaces of crack tips in plastic zone. The displacements are called crack opening displacement. Through a tensile test Dugdale puts forward the assumption that the plastic zone of crack tips appears to be wedge banding shape. This Dugdale's model is named as Model D-B, which is suitable for small-area and large-area yielding. It is assumed in Model D-B that crack tips appear to be wedge banding shape in plastic zone, as shown in Fig.1a, and the material is perfect plastic. The whole cracks and plastic zone are surrounded by large elastic zone. The assumptions also include that the yield stress σ_s acted on the interface of the two zones is uniformly distributed (Fig.1b).

Consequently, the crack length increases from $2a$ to $2c$ under far-field uniformly distributed tensile stress σ. The length of plastic zone can be expressed as $R = c - a$. Here c is the "crack tip" of banding plastic zone. The crack tip opening displacement δ is the opening displacement of the original crack ($2a$).

Taking out the plastic zones theoretically, the uniformly distributed tensile stress σ_s can be applied on the interface between elastic and plastic zones taking instead for simplification. By means of this method, the problem can be simplified as the crack with the length of $2c$ under interaction of far-filed stress σ and interfacial stress σ_s. The stress intensity factor K_I^c of crack tip c consists two parts, $K_I^{(1)}$ and $K_I^{(2)}$, generated

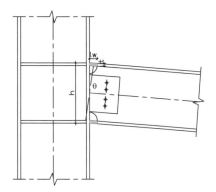

Figure 2. Rigid connection of beam-to-column.

stress intensity factor of a plate with limited width can be expressed as follows

$$K_I = \left[\frac{2b}{\pi a} tg \frac{\pi a}{2b} \right]^{1/2} \sigma \sqrt{\pi \left(\alpha^2 a \right)} \qquad (9)$$

When it is equivalent to a penetrating crack, the equivalent length is

$$a^* = \alpha^2 a \qquad (10)$$

Where α is a modified coefficient

$$\alpha = \frac{1.1}{\Phi_0} \qquad (11)$$

Φ_0 is the second kind elliptic integral.

Supposing that the measured crack depth is d, the effective crack depth can be expressed as

$$d^* = \left(\frac{1.1}{\Phi_0} \right)^2 d \qquad (12)$$

2.4 The rotation stiffness of the Connection

The effective area and the effective depth of the seam can be expressed as follows according to Equation.8 and Equation.12 (Fig. 2)

$$A^* = A_0 - \sum l_i^* d_i^* \qquad (13)$$

$$t^* = \frac{A^*}{b} \qquad (14)$$

Where, A_0 is the original area of the seams on beam flanges; A^* is the effective area of the seams on beam flanges; b is the width of the seams on beam flanges; t^* is the effective thickness of the seams on beam flanges

Plane cross-section assumption is adopted in the analysis of connection deformation since the deformation is small. It is also assumed that all moment is transferred by flange seams which are uniformly loaded and high strength bolt bears all sheer forces. The inertia moment of the damaged seams on the beam flanges is

$$I^* = 2A^* \left(\frac{h}{2} - \frac{t^*}{2} \right)^2 + \frac{b \left(t^* \right)^3}{6} \qquad (15)$$

The stress that the flange seams carries is

$$\sigma = \frac{Mh}{2I^*} \qquad (16)$$

While the strain is

$$\varepsilon = \frac{\sigma}{E^D} = \frac{Mh}{2E^D I^*} \qquad (17)$$

The elongation of the seams is

$$\Delta l_w = \frac{Mhl_w}{2E^D I^*} \qquad (18)$$

Where, l_w is the width of the seam on a beam flange; E^D is Elastic Modulus of damaged seam obtained from test.

Connection rotation is

$$\theta \approx tg\theta = \frac{\Delta l_w}{h/2} = \frac{Ml_w}{E^D I^*} \qquad (19)$$

Hence, the bending moment of the damaged rigid connection is

$$\begin{cases} M = k\theta & \theta < \theta_p \\ M = M_p & \theta \geq \theta_p \end{cases} \qquad (20)$$

Where,

$$\begin{cases} k = \frac{E^D}{l_w} \left[2A^* \left(\frac{h}{2} - \frac{t^*}{2} \right)^2 + \frac{b \left(t^* \right)^3}{6} \right] \\ M_p = \frac{2\sigma_s}{h} \left[2A^* \left(\frac{h}{2} - \frac{t^*}{2} \right)^2 + \frac{bh^3}{6} \right] \\ \theta_p = \frac{M_p}{k} \end{cases} \qquad (21)$$

σ_s is the yield strength of the seams on the beam flanges.

Figure 3. Plane steel frame with rigid connections.

3 NUMERICAL SIMULATION AND ANALYSIS

Substituting the stiffness of damaged connection for the initial stiffness of undamaged connection in the theoretical analysis, the influence of the damage on the dynamic behavior of the structure can be described by parameter k, which can be expressed as

$$\begin{cases} k = \dfrac{E^D I}{l_w}\left(\dfrac{\gamma_i}{1-\gamma_i}\right) \\ \gamma_i = 1 - \dfrac{E^D I^*}{E_0 I_0} \end{cases} \tag{22}$$

Where, γ_i is variable coefficient of the stiffness of the connection; I is the inertia moment of the seams of undamaged beam flanges. If $\gamma_i = 0$, then $k = 0$. The connection is pinned. If $\gamma_i \rightarrow 1$, then $k \rightarrow \infty$. The connection is rigid.

A 7-storey plane steel frame is taken as an example shown in Figure.3. Ansys is used for the dynamic analysis of the steel frames with different damaged joints. The cross-sections of beams are H300*150*8*12, and those of columns are H400*250*10*16. The member steel is Q345.

3.1 The influence of stiffness of the damaged joints on the natural frequencies of the frame

The eigenvalue analysis of the plane steel frame (Fig.3) is performed. The first 3 vibration modes are obtained and listed in Table.1. The curve of $\omega_s/\omega_r - \gamma_i$ is given in Figure.4. ω_s is the natural frequencies of the plane steel frame with semi-rigid beam-to-column connections. ω_r is the natural frequencies of the plane steel frame with rigid beam-to-column connections.

It can be concluded from Table.1 and Figure.4 that joint damage has significant effect on structural natural frequencies. If the joint damage is severer, the natural frequencies of the frame decrease more significantly. The dynamic behavior of the frame depends mainly on the first vibration mode. Hence, if the steel

Table 1. Natural frequencies of the frame with different joint stiffness (Hz)

Frequencies γ_i	ω_1	ω_2	ω_3
0	0.6109	3.7557	10.3235
0.1	1.8191	6.3323	13.06
0.2	2.0471	6.9946	13.8523
0.3	2.1526	7.2301	14.2327
0.4	2.3138	7.3961	14.4565
0.5	2.2538	7.5049	14.6041
0.6	2.282	7.5817	14.7087
0.7	2.303	7.6388	14.7867
0.8	2.3192	7.683	14.8472
0.9	2.3322	7.7182	14.8954
0.99	2.3417	7.7442	14.931

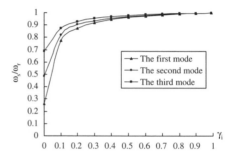

Figure 4. Influence of the joint stiffness on the natural frequency.

frame is calculated as rigid connections regardless of the damage, the error in dynamic analysis will be great.

3.2 Time-history analysis

Time-history analysis of the plane steel frame (Fig.3) is conducted. Raleigh damping model is adopted, and EI Centro seismic wave (1940) is used. The analysis lasts for 10 seconds with interval of 0.02s. The maximum acceleration of the wave is 220 cm/s².

The displacement of the top point and the base shear of the plane steel frame with different joint stiffness are shown in Figure.5 and Figure.6 respectively. Table.2 presents the maximum displacements and shears in time-history curve. It can be seen from Figure.5 and Table.2 that when the joint is almost perfect rigid, the frame has the largest stiffness and the top point displacement is the smallest. With the reduction of γ_i, the frame becomes weaker and the top point displacement increases. TPD reaches the maximum when the joint is nominal pinned. It can be obtained in Fig.6 and Tab.2 that the maximum base shear of damaged steel frame is between those of nominally pinned and perfect rigid frame. The severer the joint is damaged, the weaker the frame is. At the same time, the structural damping

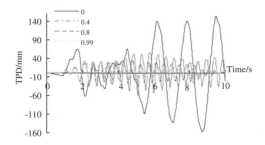

Figure 5. Displacements of the top point.

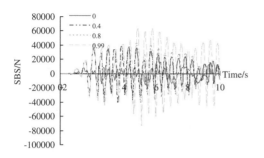

Figure 6. Base shear of the frame.

Table 2. Maximum displacements and shears of the frame with different stiffness

γ_i	TPD/mm	SBS/N
0	155.9	17742.8
0.4	59.7	40080.2
0.6	32.7	42998.1
0.99	31.1	73133.2

Note: TPD—top point displacement
SBS—structural base shear

increases. The structure will consume more seismic energy under earthquake. Therefore, with the degeneration of the joint stiffness the structural deformation will increase and the base shear will decrease. On evaluating the safety of steel frame after earthquake, the joint damage must be taken into account.

4 CONCLUSIONS

(1) The beam-to-column connection of steel frame with cracks in welded seams after earthquake is studied. The rotation stiffness of the damaged joint is deduced by employing the principle of fracture mechanics. It can be applied to the dynamic finite element analysis of steel frames with damaged joints.

(2) The results of numerical analysis show that joint damage has significant influence on natural frequencies, especial the structural basic frequency. The structural natural frequencies decrease significantly when the joint damage becomes severer. Since the dynamic behavior of the structure depends mainly on the basic vibration mode, the influence of joint damage on dynamic behavior should be taken into account in the seismic analysis of steel frame with damaged joint.

(3) According to the time-history analysis, the top point displacement increases while structural shear decreases if the joint stiffness degenerates. Thus, the joint damage must be taken into consideration in dynamic analysis of the structures.

REFERENCES

Bertero V V, Anderson J C,Krawinkler H.Performance of Steel Building Structures During the Northridge Earthquake.EERC Report.Berkeley:Univ.of Calif.,1994.

Investigation report of hanshin-earthquake[R]. Tokyo: JASC, 1995.

W.F. Chen, Zhou Suiping. Stability design of steel frame[M]. World publishing company, Shanghai, 2001.

Zhou Xuejun, Zhang Xianglong. The researches and applications of steel frames with semi-rigid connections[J]. Journal of Shandong University of architecture and engineering. Vol.18 (2003), pp85–88.

Ding Suidong, Sun Limin. Fracture mechanics[M]. China machine press, Beijing, 1997.

Dugdale D S. Yielding of steel containing slit[J].Journal of the Mechanics and Physics of Solids,1960,8:100~104.

Steel and Composite Structures – Wang & Choi (eds)
© 2007 Taylor & Francis Group, London, ISBN 978-0-415-45141-3

Evaluation of buckling-restrained braced frames overstrength factor

B. Asgarian & H.R. Shokrgozar

K.N.Toosi University of Technology, Tehran, Iran

ABSTRACT: Ordinary braces have defects such as: low ductility, unsymmetric hysterics response at tension and compression force and buckling of braces at compression force. To overcome the above mentioned problems, new type of braces as "Buckling-Restrained Braces" have been developed and subjected to investigation and examination. Buckling-restrained braces frames (BRBFs) have been shown to exhibit very favorable energy-dissipation characteristics. Analytical and experimental results showed that symmetric hysterics response, high ductility capacity and large drift resist were the preference of this bracing system compared to the other conventional bracing system Ordinary braces. The object of this paper is to evaluate the over-strength factor (Ω) of buckling-restrained brace frames. In this purpose, two dimensional frames with 4, 6, 8,10,12,14 stories and with variety configurations of bracing system (Diagonal, chevron Inverted-V, chevron V and X and split X type) were considered and designed based on seismic codes. In this paper overstrength factor of the BRBF were evaluated considering two methods. In first method static pushover analysis of the frames was performed and in the second method Incremental Dynamic Analysis (IDA) was performed using Opensees software. The overstrength factor of the frames were computed using results of the static pushover and Incremental Dynamic Analysis methods and as an average, overstrength factor of 1.7 was proposed for buckling-restrained brace frames.

1 INTRODUCTION

Braces are efficient structural elements for resisting lateral forces and much type of braces is used in buildings. Braces in concentrically braced frames (CBFs) have long been known to be prone to many non-ductile modes of behavior when subjected to large ductility demands. Such modes include connection and member fracture, severe loss of strength, stiffness due to beam ductility resulting from unbalanced tension and compression strengths and unable to dissipate energy (AISC, 2002) have been observed in concentrically braced frames. In recent years new generated of braces "Buckling-restrained braces" (BRBs) have been found that do not exhibit any of the unfavorable behavior characteristics of conventional braces. The buckling restrained brace is a brace whose core plate is covered with a restraining part to prevent buckling. An unbonded material or a clearance is provided between the core plate and restraining part so that the axial force borne by the core plate is not transmitted to the restraining part (Figure 1). The steel core member is designed to resists the axial forces with a full tension or compression yield capacity without the local or global flexural buckling failure. When the brace

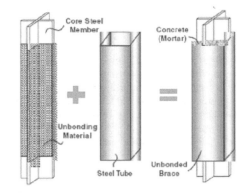

Figure 1. Buckling-Restrained Brace. [1].

is subjected to compressions, an unbonding material placed between the core member and the infill concrete is required to reduce the friction. Thus, a BRB basically consists of three components, including steel core member, buckling restraining part and the unbonding material. [1, 2]

Figure 2 shows a schematic of a commonly used BRBs type. The core is divided into three zones: the

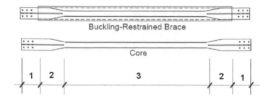

Figure 2. Schematic Diagram of Buckling-Restrained Brace [1].

yielding zone, a reduced section within the zone of lateral restraint provided by the sleeve (zone 3 in the figure); transition zones (of larger area than the yielding zone and similarly restrained) on either side of that yielding zone (zone 2); and connection zones that extend past the sleeve and connect to the frame, typically by means of gusset plates (zone 1). [3,4]

2 ADVANTAGES OF BUCKLING-RESTRAINED BRACE

BRBs are desirable for seismic design and rehabilitation for their superior ductile performance. Compared to conventional tension-compression CBFs, the BRBs system are capable of yielding in both tension and compression instead of buckling. Figure 3 shows typical hysteretic behavior of buckling-restrained. BRBFs system exhibits very favorable energy-dissipating characteristics, low post-yield stiffness of the braces, leave this system vulnerable to unfavorable behavioral characteristics such as maximum story drift and residual story drift. In a study on system performance for BRBFs, Sabelli et al. (2003) reported that residual story drifts were on average 40% to 60% of the maximum drifts. The other main significant difference of buckling-restrained brace frames to other seismic lateral force resisting systems high ductility capacity, resistance fatigue loading tolerance with repetition cycle more than twenty cycles, symmetric hysterics response of frame and cumulative plastic ductility more than 200. BRBFs have been reported to have 50% of the steel weight of SMRFs designed according to the UBC (ICBO 1997), while achieving maximum drifts of 50% to 70% of those reached by SMRFs in static pushover analyses (Clark et al. 1999).[5,6]

3 OBJECTIVE AND SCOPE

Structural engineers association of northern California (SEAONC) provides recommended provision for buckling-restrained brace frames. As established in this provision for BRBFs system design factors are: response modification coefficient, R = 8; system

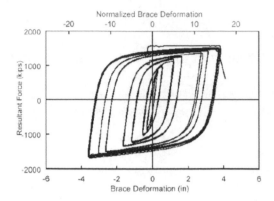

Figure 3. Typical hysteresis responses of buckling-restrained brace. [5].

overstrength factor, $\Omega = 2$; and deflection amplification factor, $C_d = 5.5$.

The object of this paper is calculating exact value of overstrength factor that were proposed the value of 2 for buckling-restrained braced frames in the SEAONC provision.

4 OVER STRENGTH FACTOR

In the development of seismic design provision for building structures, the most important part is the development of the force reduction factors and displacement factor. The force reduction factor, expressed as either a response modification factor, R, in National Earthquake Hazard Reduction Program (NERHRP 1988) or behavior factor, R_w in Iranian seismic code (Standard NO.2800) is used to reduce the linear elastic design response spectra.[8]

It is now accepted that force reduction factor accounts for the inherent ductility, overstrength. Last Definitions of force reduction factor suggested by *Uang,* sub-dividing R into two components mentioned above. Thus,

$$R = R_\mu \Omega \qquad (1)$$

Where R_μ is the ductility factor and Ω is the overstrength factor (Figure 4). The reserve strength that exist between the actual structural yield level (V_y) and first significant yield level (V_s) is defined in term of overstrength factor (Ω). [8]

$$\Omega = \frac{V_y}{V_s} \qquad (2)$$

Structural overstrength factor results from internal force redistribution (redundancy), higher material strength than those specified in designed, strain

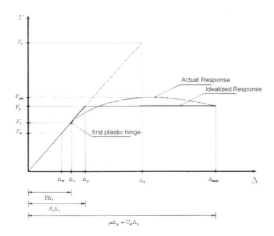

Figure 4. General structure response [8].

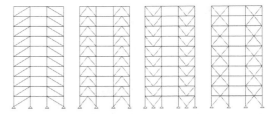

Figure 5. Prototype of frames.

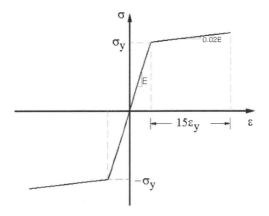

Figure 6. Steel01 Material for BRBs [9].

hardening, deflection constraints on system performance, member oversize, multiple loading combinations, effect of nonstructural elements, strain rate effects, and so on.

Overstrength factor can be obtained from analytical studies. The value of 2 was proposed to establish R for buckling restrained braced frames[6]. The overstrength factor shown in formula 2 is based on the use of nominal material properties. Denoting this overstrength factor as Ω_o, the actual over strength factor Ω that can be used to formulate R should consider the beneficial contribution of some other effects:

$$\Omega = \Omega_o F_1 F_2 ... F_n \quad (3)$$

Parameter F_1 may be used to account for difference between actual static yield strength and nominal static yield strength. For structural steel, a statistical study shows that the value of F_1 may be taken as 1.05. Parameter F_2 may be used to consider the increase in yield stress as a result of strain rate effect during an earthquake excitation. A value of 1.1, a 10 percent increase to account for the strain rate effect, could be used (Ellingwood et al. 1980). Other parameters can also be included when reliable data is available. These parameters include nonstructural component contributions, variation of lateral force profile, and so on.

5 STRUCTURAL DESIGN

Buildings with the $18\,\text{m} \times 18\,\text{m}$ dimensions in plan and $6\,\text{m} \times 6\,\text{m}$ grid, 3.2 m height having different bracing types (split X, chevron-V, chevron-Inverted V and diagonal Types) with different number of stories (4, 6, 8, 10, 12 and 14) are designed in accordance with the requirement of Iranian Earthquake Resistance Design Code [5] and Iranian National Building Code, part 10,

steel structure design [10]. The places for bracings were considered in 2 lateral bays (Figure 5).

6 MODELING THE STRUCTURE IN SOFTWARE OPENSEES

The software which was used in this paper is OPENSEES [9]. OPENSEES is finite element software which has been specifically designed in performance systems of soil and structure under earthquake and for this purpose it has been under study and development since 1990. For modeling the braces in this software to do nonlinear analysis following assumptions were considered.

6.1 Members' Modeling

To model braces nonlinear beam and columns element and behavioral modeling of materials, Steel01, have been used. Compressive and tensional strength were considered equal to steel yield strength. The used section for each member of brace is the fiber section. The strain hardening of 2% and the maximum ductility of 15 were considered. (Figure 6)

To model beams and columns, nonlinear beam-column element has been used. The used section for beams and columns is the fiber section.

7 THE METHOD OF OVERSTRENGTH FACTOR CALCULATIN

In this paper two methods have been used for calculating of overstrength factor.

In first method static pushover analysis of the frames was performed. Static pushover analysis is the one of most desirable and favorable method to estimate behavior of structure. In this method target displacement and lateral load pattern are considered as follow.

7.1 Target displacement

Maximum Inelastic (target displacement) Response Displacement, Δ_M, shall be computed as follows. [3]

$$\Delta_M = 0.7R\Delta_w \qquad (4)$$

Where Δ_w is the design level response displacement. Δ_w is used for elastic analysis can be defined as follow. [3]

$$\Delta_w = \frac{\overline{\Delta_M}}{R} \qquad (5)$$

Calculated story drift using $\overline{\Delta_M}$ shall not exceed 0.025 times the story height for structures having a fundamental period of less than 0.7 second. For structures having a fundamental period of 0.7 second or greater, the calculated story drift shall not exceed 0.020 times the story height. By setting this definition in equation 5 and 4 we have:

$$\Delta_M = 0.0175H \to T \le 0.7s$$
$$\Delta_M = 0.014H \quad \rightarrow T \ge 0.7s \qquad (6)$$

7.2 Lateral load pattern

The lateral distributed load pattern over the height of the structure is the most important thing in the analysis of structure. In this paper, the Formula that proposed at the Iranian standard code NO.2800 has been used.[12]

In the second method incremental nonlinear dynamic analysis is used. In this method by the use of time history of the earthquakes of Tabas, kobeh and Elcentro, their PGA with several try and errors have changed in a way that the gained time history resulted in the structure reaching to a one of the determined mechanism. The maximum nonlinear base shear of this time history is the inelastic shear base of structure. To gain the shear base like the first plastic hinge formation in structure, the nonlinear pushover static analysis, V_s, has been used. It means that the linear ultimate limit of structure in nonlinear static analysis and nonlinear dynamic analysis has been considered equally.[9]

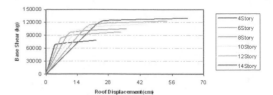

Figure 7. Roof Displacement-Base Shear curve for BRBFs that have Diagonal brace.

Table 1. Overstrength factor of BRBFs that have Diagonal braces.

NO. Story	V_s (kg)	V_y (kg)	Ω
4	58139	78000	1.34
6	74150	97616	1.32
8	81960	105235	1.28
10	92433	116519	1.26
12	98792	123019	1.24
14	109405	129979	1.19

7.3 The Relative Displacement between the floors

The relative displacement limitation between the floors was selected based on the Iranian standard code 2800.[3]

For the frames with the main period less than 0.7 seconds: $\overline{\Delta_M} \langle 0.025H$

For the frames with the main period more than or equal to 0.7 seconds: $\overline{\Delta_M} \langle 0.02H$

7.4 Forming Failure Mechanism and Frame Instability

To determine the ultimate limit which is defined by the maximum relative displacement it is necessary to be sure that the frame has kept its stability. In the case of story mechanism or overall mechanism happening in a frame under an earthquake even if the target displacement (maximum relative displacement) doesn't occur, the nonlinear dynamic analysis was stopped and the least scale factor of mechanism was introduced as ultimate limit. So to control structure stability forming plastic hinges was investigated in the final analysis of each frame under each earthquake. [9]

In the case that in each frame stories in each of columns sections the plastic hinge has been formed or all braces of that floor has reached to the ultimate ductility $\mu = 15$, it means the whole mechanism of structure and its instability.

Figures 7 through 10 show the charts of roof displacement-base shear gained by nonlinear static pushover analysis respectively for buckling restrained braced frames, diagonal braces, inverted V, V type and split X. In tables 1, 2, 3 and 4 the ultimate base shear

Table 2. Overstrength factor of BRBFs that have chevron Invert-V braces

NO. Story	V_s (kg)	V_y (kg)	Ω
4	63562	102984	1.62
6	78960	118525	1.5
8	94221	133764	1.42
10	104800	142631	1.36
12	115162	155768	1.35
14	120001	159927	1.33

Table 3. Overstrength factor of BRBFs that have chevron V braces

NO. Story	V_s (kg)	V_y (kg)	Ω
4	48401	85405	1.76
6	72606	123185	1.69
8	89897	135105	1.51
10	100471	144499	1.42
12	111093	153327	1.38
14	115308	157138	1.36

Table 4. Overstrength factor of BRBFs that have chevron V braces

NO. Story	V_s (kg)	V_y (kg)	Ω
4	55156	91154	1.65
6	80021	129422	1.62
8	95944	149992	1.56
10	108082	158538	1.47
12	120946	173607	1.44
14	135662	185922	1.42

Table 5. Nonlinear maximum Base Shear for BRBFs that have Diagonal brace under Elcentro, Kobe and Tabas ground motion

No.	V_y (ELC)	V_y (KOB)	V_y (TAB)	$\overline{V}_y$	Ω
4	78400	78350	78170	78307	1.35
6	103400	97650	97000	99350	1.34
8	109620	100500	109500	108040	1.32
10	120700	121000	120000	120567	1.3
12	126600	126400	130530	127843	1.3
14	131200	131000	130100	130767	1.2

Table 6. Nonlinear maximum Base Shear for BRBFs that have chevron invert-V brace under Elcentro, Kobe and Tabas ground motion

No.	V_y (ELC)	V_y (KOB)	V_y (TAB)	$\overline{V}_y$	Ω
4	89760	89280	89970	89670	1.85
6	121940	126750	122560	123750	1.71
8	132060	127500	135200	131587	1.46
10	145200	141440	148000	144880	1.44
12	151950	148700	161060	153903	1.38
14	159500	157940	166150	159500	1.38

Table 7. Nonlinear maximum Base Shear for BRBFs that have chevron V brace under Elcentro, Kobe and Tabas ground motion

No.	V_y (ELC)	V_y (KOB)	V_y (TAB)	$\overline{V}_y$	Ω
4	104800	104900	100500	103400	1.63
6	115350	120200	119000	118183	1.5
8	128670	128170	134400	130600	1.39
10	142520	137960	146200	142233	1.36
12	152520	153670	164300	156830	1.36
14	16080	159740	164110	161547	1.35

Table 8. Nonlinear maximum Base Shear for BRBFs that have chevron split brace under Elcentro, Kobe and Tabas ground motion

No.	V_y (ELC)	V_y (KOB)	V_y (TAB)	$\overline{V}_y$	Ω
4	93710	93750	94500	89670	1.7
6	123700	134500	125020	123750	1.6
8	152600	145980	147820	131587	1.55
10	155730	152370	161900	144880	1.45
12	168200	173500	183500	153903	1.45
14	185700	184690	192500	159500	1.44

V_y and V_s first yield that obtained from static pushover analysis have been shown.

In tables 5, 6, 7 and 8 the ultimate base shear V_y maximum acceleration and ultimate limit resulted by nonlinear dynamic analysis of frames are shown under Elcentro, Kobeh and Tabas time history for diagonal braces, chevron inverted V, chevron V and split X type.

According to the obtained overstrength factors at first method for each structure, overstrength factor calculated as follow:

The diagonal buckling restrained braces $\Omega_o = 1.27$ $\Omega = 1.47$

The inverted V buckling restrained braces $\Omega_o = 1.52$ $\Omega = 1.76$

The Chevron V type buckling restrained braces $\Omega_o = 1.43$ $\Omega = 1.65$

The split X buckling restrained braces $\Omega_o = 1.53$ $\Omega = 1.76$

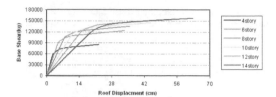

Figure 8. Roof Displacement-Base Shear curve for BRBFs that have Invert-V brace.

According to the obtained overstrength factors at second method for each structure, overstrength factor calculated as follow:

The diagonal buckling restrained braces $\Omega_o = 1.3$ $\Omega = 1.5$

The inverted V buckling restrained braces $\Omega_o = 1.53$ $\Omega = 1.77$

The Chevron V type buckling restrained braces $\Omega_o = 1.43$ $\Omega = 1.65$

The split X buckling restrained braces $\Omega_o = 1.52$ $\Omega = 1.76$

8 CONCLUSION

The results of obtained from static pushover analysis and incremental dynamic analysis as follow:

1. With comparing the two method of calculating overstrength factor (static pushover an incremental dynamic analyses) conclude that this two method have a difference less than five percent.
2. The obtained over strength factor for braced structures with buckling restrained braces in type V, inverted V, split X and diagonal are respectively, 1.65, 1.77, 1.76 and 1. 5.
3. In the general state, the braced structures with concentric buckling restrained braced frames (BRBF) over strength factor 1.7 have been suggested.

REFERENCES

[1] Black C., Makris N. & Aiken, I. 2002 "Component Testing, Stability Analysis and Characterization of Buckling-Restrained Braces" Report of Pacific Earthquake Engineering Research Center, University of California, Berkeley.
[2] Uang C. & Nakashima N. August 1991 "Steel buckling-Restrained Braced Frames", section 16 of Bozorgnia Handbook.
[3] "Iranian Code of Practice for Seismic Resistant Design of Buildings" No.2800, 3rd Edition, Building and Housing Research Center standard, 2005.
[4] Uang C. & Nakashima N. "Steel Buckling-Restrained Braced Frames", section 16 of Bozorgnia Handbook.
[5] Escu E.O July 2003 "Comparative Parametric Study on Normal on Buckling-Restrained Steel Brace" A Dissertation Submitted in Partial Fulfillment of Requirements for the Master degree in Earthquake Engineering, European School of Advance Studies in Reduction of Seismic Risk "Rose School".
[6] Lai J. & Tsai K. August 2004. "Research and Application of Buckling Restrained Braces in Taiwan" 13th World Conference on Earthquake Engineering, Paper No.2179, Vancouver Canada.
[7] Structural Engineers Association of Northern California October 2001 "Recommended Additions to the AISC Seismic for Structural Steel Buildings".
[8] Uang C.M. 1991. "Establishing R (or R_w) and C_d Factor Building Seismic Provision" Journal of Structure Engineering, Vol.117, No.1.
[9] Mwafy A.M. & Elnashai A.S. 2002. "Calibration of Force Reduction Factors of RC Buildings" Journal of Earthquake Engineering, Vol 6, Page 239–273,
[10] Central Research and building house 1996 "Iranian National Building Code, part 10, steel structure design"
[11] Mazzoni S., McKenna F., Scott M.H., Fenves G.L. & Jeremic B. December2004. "OpenSees Command Language Manual".
[12] Kravinkler H. November 1994 "Static Pushover Analysis" The Developing Structural Engineers of Northern California, Sanfrancisco.

Steel and Composite Structures – Wang & Choi (eds)
© 2007 Taylor & Francis Group, London, ISBN 978-0-415-45141-3

Experimental research on seismic behavior of two-sided composite steel plate walls

Fei-fei Sun, Guo-qiang Li & Hui Gao
Tongji University, Shanghai, P.R. China

ABSTRACT: A two-sided composite steel plate wall(CSPW) omits the connection between an ordinary CSPW to its boundary columns, in order to remove its adverse influence on seismic performance of those columns. Three scaled models of single-span, single-floor, two-sided CSPWs were tested. The CSPW specimens were pre-fabricated and built by connecting its concrete plate to its steel plate with bolts, providing the steel plate out-of-plane buckling resistance. In the test, each specimen was bolted to a hinged rectangular frame with rigid steel beams and columns and hence carried all of the horizontal force by itself. No vertical load was applied as CSPWs were not designed to withstand such a load. One specimen with single concrete plate was loaded monotonically, while the other two with bilateral concrete plates were tested under cyclic loading. The latter two specimens differed to each other in the thickness and reinforcement of the concrete plates. Experimental results showed that: (1) two sided CSPWs were sensitive to global out-of-plane instability; (2) two-sided CSPWs could have a ductility greater than 9, exhibiting excellent seismic performance, provided that there were reliable out-of-plane bracing as well as reliable restraint against local buckling by means of one- or bi-lateral concrete plates; (3) detailing of the connection between the steel plate and the boundary beams was essential for protecting two-sided CSPW from premature failure.

1 INTRODUCTION

Composite steel plate wall (CSPW) is a type of lateral member for tall buildings, composed of steel plate and concrete panel connected by shear connectors (See Figure 1). The concrete panel provides out-of-plane restraint preventing premature failure of the steel plate due to buckling. Both the shear and the energy dissipating capacity of the steel plate are thus significantly improved. Moreover, the concrete panel acts also as fire proof for the steel plate.

Past research has revealed the good seismic performance of the composite wall, including large initial stiffness, large plastic deformation capacity, stable hysteretic curve and so on. Li et al.(1995) initiated research on such member in China. Only very few buildings have incorporated the composite wall system, e.g. China World Trade Center (the third stage), which is still under construction as shown in Figure 2, as still a lot of issues need to be investigated.

(a) (b) (c)

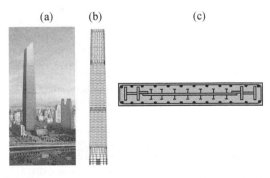

Figure 2. China World Trade Center (the third stage): (a) Rendered drawing; (b) Structural system (CSPWs are used in lower floors); (c) Typical configuration of the CSPWs.

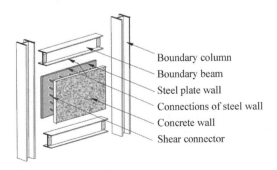

Boundary column
Boundary beam
Steel plate wall
Connections of steel wall
Concrete wall
Shear connector

Figure 1. Main components of a CSPW.

Table 1. Key information of the CSPW specimens.

Specimen	CW1	CW2	CW5
Concrete panel			
Number of panels	2	2*	1
Thickness of panels	30 mm	60 mm	60 mm
Reinforcement of panels	1 layer	2 layers	2 layers
Load scheme	Cyclic	Cyclic	Monotonic
Thickness of fish plate	4 mm	6 mm	6 mm

* One of the two panels in CW2 was replaced by two 30 mm-thick panels with single layer of reinforcement before the test due to the time limit of the test and the temporary change of the thickness from 30 mm to 60 mm.

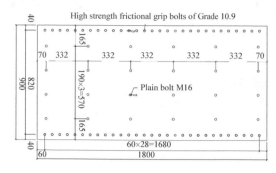

Figure 3. Details of the steel plate.

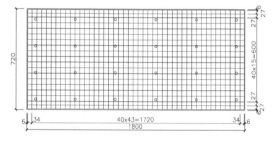

Figure 4. Details of the concrete panel.

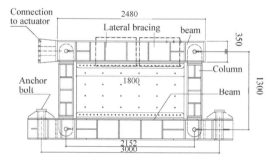

Figure 5. Test set-up and a specimen in it.

Most of the steel plate shear wall and composite wall studied were designed to connect with both beams and columns. By omitting the connection to columns, the adverse effect of the composite wall on the columns that the shear force of the columns is significantly enlarged can be avoided (Xu, Lu 1994) while the construction effort can be reduced. Up to now, most experimental researches focused on four-sided CSPW(Astaneh-Asl 2002, Zhao & Astaneh-Asl 2004). Hitaka and Matsui (2000) did tests on two-sided CSPWs with slits in the steel plate. The specimens by Li et al.(1995) were essentially two-sided CSPW without slits, the performance of which, however, seemed not satisfactory. The paper aims at validating the seismic behavior of two-sided CSPW specimens without slits by means of static tests.

2 TEST SCHEME

2.1 Test specimens

Three single-span, single-floor specimens were designed in a scale of 1/4 to 1/3. The bottom and top edges of the steel plate were connected to boundary beams by means of M16 × 50 high strength friction grip bolts. The concrete panel was pre-cast and connected with the steel plate by means of M16 plain bolts. 50 mm × 50 mm backup plates in the thickness of 6 mm were used to avoid local failure of concrete in compression.

Key information of the specimens is summarized in Table 1. The steel plate of all specimens had a common thickness of 2 mm. The concrete panels were designed according to AISC Seismic Provisions (AISC 2005): The concrete thickness shall be a minimum of 100 mm on each side when concrete is provided on both sides of the steel plate and 200 mm when concrete is provided on one side of the steel plate; the reinforcement ratio in both directions shall be not less than 0.0025; the maximum spacing between bars shall not exceed 450 mm.

As there is no design method provided for the concrete panel, the minimum requirement was adopted for Specimen CW1 and Specimen CW5, i.e. 30 mm and 60 mm, respectively, after considering the reduced scale. The thickness of the concrete panel for Specimen CW2 was changed to 60 mm after observing poor performance of CW1 due to premature failure of its concrete panel. Single and double layers of steel wire meshes were used for 30 mm and 60 mm concrete panels, respectively. Details of the steel plate and the concrete panel are shown in Figures 3 and 4.

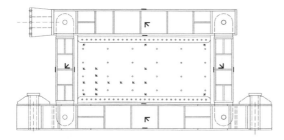

Figure 6. Location of the strain gauge rosettes.

Table 2. Properties of steel coupons.

Coupon Specimen	SPW CW1/2/5	Fish plate		
		CW1	CW2	CW5
Thickness (mm)	1.9	3.9	5.9	5.7
Elastic modulus (MPa)	174499	162124	187095	179587
Yield strength (MPa)	287.0	304.0	380.3	376.7
Tensile strength (MPa)	362.9	464.0	463.3	537.3
Elongation (%)	36.2	32.6	30.8	30.5

Table 3. Properties of concrete coupons.

Specimen	CW5	CW1	CW2
Size of Coupon (cm)	$10 \times 10 \times 10$	$15 \times 15 \times 15$	$15 \times 15 \times 15$
Compressive strength (Mpa)	31.7	27.7	32.3
Coefficient of variation	0.17	0.29	0.26

2.2 Test set-up

In order to obtain the resisting force of CSPW avoiding the contribution of the boundary frame, a hinged frame was designed for the boundary frame (See Figure 5). A large section was adopted for the boundary beams to provide rigid boundary for the specimen. The load was exerted by a 100-ton electro-hydraulic servo actuator to one end of the top boundary beam. No vertical load was employed in the test as the CSPW in a real structure do not assume vertical forces, which are usually transferred by boundary columns only. Two anchor bolts in diameter of 80 mm were used to fix the frame to the laboratory pedestal and two 50-ton jacks were installed to bear the horizontal reaction of the frame against the pulling and pushing action of the actuator. Horizontal bracing was applied on the flange of the top boundary beam to prevent out-of-plane movement.

2.3 Instrumentation and collection of data

Displacement gauges were installed to measure rigid body movement of the hinged frame as well as out-of-plane displacement of the specimens. As there was considerably large gap between the pin bolt and the pin hole of each hinge, additional displacement gauges were used to detect relative in-plane and out-of-plane rotation of the upper beam and the columns. Eighteen strain rosettes were placed to measure the strain distribution of the steel plate as shown in Figure 6. Additional strain gauges were laid to monitor the status of the pin frame. About 100 channels in the data acquisition system were recording data from the instruments.

3 TEST RESULTS

3.1 Material properties

The materials were tested in the Material Testing Laboratory of Tongji University. Steel coupons were tested in term of China National Standard "Tensile testing method for metal materials at room temperature" (GB/T 228-2002). Concrete coupons were tested in

term of China National Standard "Testing method for mechanical property of normal concrete" (GB/T 50081-2002). The results are listed in Tables 2 and 3 for steel and concrete, respectively.

3.2 Behavior of Specimen CW5

Specimen CW5 had a single concrete panel, which was 60 mm thick and had two layers of steel wire meshes. The steel plate was so thin that it got a large initial deformation during the ball spray treatment. Obvious out-of-plane deformation was still visible after the steel plate was bolted to the frame. Such deformation disappeared after the installation of the concrete panel. Specimen CW5 was monotonically loaded to failure. The force-displacement curve is shown in Figure 7.

During preloading, the composite wall of CW5 suffered global out-of-plane buckling at the load level of 114 kN. Distinct crease of the steel plate can be spotted right above the fish plate as shown in Figure 8. Remedy was taken by pushing the buckled plate back to its original position and then supplementing out-of-plane restraints to the concrete panel with 3 short channel steel welded to the flange of each boundary beam. No evidence of the crease could be found after the remedy and final failure of the specimen did not occur at this position, indicating that the above-mentioned buckling did not exert significant effect on the behavior of the specimen in the follow-up formal loading.

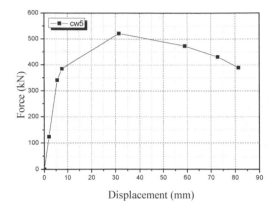

Figure 7. Force-displacement curve of CW5.

Figure 9. Tensile strips of CW5 at failure.

Figure 8. Out-of-plane buckling of CW5 under pre-loading.

Figure 10. Cracks in the concrete panel of CW5 at failure.

Local buckling was observed at the right edge of the steel plate between the two lower rows of bolts, accompanied with small noises. With the increase of the load, a tensile strip formed inclining to the upper right corner while the local buckling was growing up. Afterwards, more and more inclined strips appeared, while only some small cracks emerged on the concrete panel till the failure of the steel plate due to that the upper and lower ends of tensile strips were torn open. Out-of-plane deformation developed in the fish plates. The tensile strips formed in the steel plate and the cracks in the concrete panel at failure were shown in Figures 9 and 10, respectively.

3.3 Specimen CW1

Specimen CW1 had two concrete panels, which were 30 mm thick and had single layer of steel wire mesh. Out-of-plane restraints were installed to keep the concrete panels moving in their own plane.

The specimen was tested under cyclic loading. The hysteretic force-displacement curve is plotted in Figure 11. In the first stage, load control method was

used up to 400 kN, with single cycle for each load level and an increment of 50 kN for successive levels. Two vertical long cracks appeared in the right portion of the frontal panel at the load of 175 kN and the same panel cracked vertically through the whole section in the left portion at −170 kN. The cracking was due to that the panel had no sufficient bending stiffness and strength to resist the buckling action of the steel plate, as the panels were too thin and had only single layer of steel mesh.

An obvious change in slope angle appeared in the load-displacement hysteretic curve at -400 kN, with a significant increase of the oblique strain measured by the strain gauge rosette in the middle of the steel plate, indicating yield of the central portion in the inclined direction. Displacement control method was then activated and three cycles were exerted for each displacement level with an increment of 15 mm for successive levels. At the end of the first displacement level the rear panel cracked through the vertical section and the steel plate buckled. In the second displacement

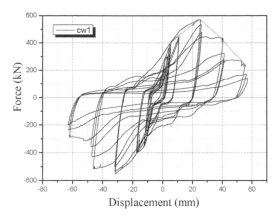

Figure 11.   Force-displacement hysteretic curve of CW1.

Figure 13.   Residual deformation of the steel plate of CW1 after removal of the panels at the end of the test.

Figure 12.   Cracks in the frontal panel of CW1 at failure.

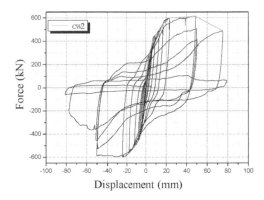

Figure 14.   Force-displacement hysteretic curve of CW2.

level, many cracks emerged on the rear panel and new cracks appeared on the middle of the frontal one. In the first cycle of the third level, left lower portion of the steel plate was torn open with a big noise. Global buckling occurred in the next cycle with a great strength degradation in the load-displacement curve. Failure modes of the steel plate and the concrete panel were shown in Figures 12 and 13, respectively.

3.4   Specimen CW2

Specimen CW2 had a 60 mm-thick concrete panel at the rear side and two 30 mm-thick panels at the frontal one. As in CW5 and CW1, the thicker and thinner panels had double and single layers of steel wire mesh, respectively. Screw rods (16 mm in diameter, Grade 4.8) were used to connect the panels.

Cyclic loading was applied. The hysteretic force-displacement curve is shown in Figure 14. In the first stage, a load control with single cycle for each level and a level increment of 50 kN was used. At the load

level of 250 kN, lower corners of the steel plate buckled. During the following incremental loading levels, the buckled portions were alternately straightened and buckled. Eventually the left and right lower corners cracked at −500 kN and 550 kN, respectively, exhibiting a typical form of fatigue rupture. At −550 kN, upper exposed portion of the steel plate developed several upper right strips while many cracks inclined in the same direction appeared on the frontal panel. The cracks extended through the whole panel at −600 kN.

In the second stage, an increment of 26 mm was used for displacement control method. Many upper left cracks occurred on the frontal panel at the first level, crossing previous cracks. Inclined cracks were also came forth in the rear panel. The bolt holes at the upper edge of the steel plate distorted significantly. In the second and third displacement level, the deformation of the bolt holes increased and finally led to cracking through some holes. However, the specimen failed due to that the lower part of the steel plate was torn through as a result of low cycle fatigue and tensile strip action.

Failure modes of the steel plate and the panel are shown in Figures 15–16.

4 EVALUATION

4.1 *Ductility*

In Figure 17, force-displacement skeleton curves are collected for all three specimens. As shown in Figure 18, ultimate displacement was defined as the displacement corresponding to 85% of the maximum load and the yield point D was determined following the criteria of equal energy, i.e. the energy covered by the simplified curve ODE is equal to that covered by the skeleton curve OABC. The calculated parameters of the characteristic points for Specimens CW1, CW2 and CW5 are shown in Table 4.

Figure 15. Cracks in the outer frontal panel of CW2 at failure.

Figure 16. Residual deformation of the steel plate of CW2 after removal of the panels at the end of the test.

4.2 *Energy dissipation*

Accumulative process of energy dissipation is shown in Figure 19, from which CW2 can be found much better than CW1 in energy dissipation.

Dissipative coefficient E is defined as follows,

$$E = \frac{S_{ABC} + S_{CDA}}{S_{OBE} + S_{ODF}} \qquad (1)$$

where, S_{ABC} and S_{CDA} are the areas of the curved triangles ABC and CDA while S_{OBE} and S_{ODF} are the

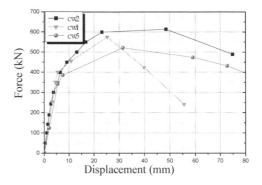

Figure 17. Force-displacement skeleton curves of all the specimens.

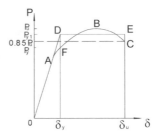

Figure 18. Determination of yield point.

Table 4. Calculated parameters of the characteristic points of equivalent elastic-ideally plastic model.

Specimen	CW1	CW2	CW5
Yield point			
Force (kN)	413.1	427.1	385.8
Displacement (mm)	6.5	7.2	7.6
Inter-story drift	1/200	1/181	1/171
Theoretical yield strength (kN)	504.1	568.1	472. 3
Point of ultimate strength			
Force (kN)	574.6	613.1	521.2
Displacement (mm)	25.3	48.6	31.5
Inter-story drift	1/51	1/27	1/41
Failure point			
Displacement (mm)	33.6	68.0	68.8
Inter-story drift	1/39	1/19	1/19
Ductility	5.2	9.5	9.1

areas of the triangles OBE and ODF, as shown in Figure 20. The calculated results of the first cycles of all levels are listed in Table 5 and 6 for Specimen CW1 and CW2.

5 CONCLUSIONS

(1) two-sided CSPWs are sensitive to out-of-plane global buckling. Out-of-plane restraints need to be provided for the concrete panels.

(2) minimum requirements for the panel thickness of CSPWs with two panels and single panel was shown insufficient and sufficient, respectively. Design method is in need for the former case.
(3) the capability of the concrete panel to prevent local buckling of the steel plate controls the energy dissipation and ductility of CSPW.
(4) the ductility of two-sided CSPWs with single and double panels can reach 9, exhibiting excellent seismic performance, provided that out-of-plane global restraints are strong enough.

ACKNOWLEDGEMENT

The research is supported by Project 50408036 sponsored by National Natural Science Foundation of China, the project-sponsored by SRF for ROCS, SEM, and the Key Project of Chinese Ministry of Education "Research on seismic performance of new SPSW-composite frame system".

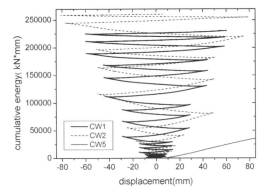

Figure 19. Accumulation of dissipated energy.

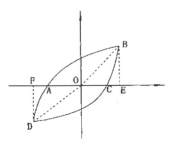

Figure 20. Energy dissipated in a hysteretic cycle.

Table 5. Energy dissipative coefficient E of the first cycle of each load level for CW1.

Level of cycle	350 kN	400 kN	15 mm	30 mm	45 mm
E	0.62	0.95	1.45	1.76	2.05

REFERENCES

AISC. 2005. *Seismic provisions for structural steel buildings*, Chicago: American Institute of Steel Construction, INC.
Astaneh-Asl, A. 2002. *Seismic behavior and design of composite steel plate shear walls*. Steel TIPS Report. Structural Steel Educational Council, Moraga, California.
Guo, J.Y., Guo, W.B., et al. 2006. Structural design for the main tower of the third-stage project of China World Trade Center (in Chinese), *Proc. of the Fourth Conference of Steel Structures for China Mainland, Taiwan and the Hong Kong Region*, pp. 188–195.
Hitaka, T., Matsui, C. 2000. Strength and behavior of steel-concrete composite bearing wall, *Proceedings of 6th ASCCS conference*, Los Angeles, USA, March 22–24, p. 777–784.
Li, G.Q., Zhang, X.G., Shen, Z.Y. 1995. Experimental research on hysteretic behavior of steel plate encased with RC shear panel (In Chinese), *Industrial Building*, vol. 25, p. 32–35.
Xue, M., Lu, Le-Wu. 1994. Interaction of Steel Plate Shear Panels with Surrounding Frame Members, *Proceedings of the Structure Stability Research Council Annual Technical Session*, Bethlehem, PA, pp. 339–354.
Zhao.Q and Astaneh-Asl. A. 2004. Cyclic behavior of traditional and an innovative composite shear wall, *Journal of Structural Engineering*, p. 271–284.

Table 6. Energy dissipative coefficient E of the first cycle of each load level for CW2.

Level of cycle	350 kN	400 kN	450 kN	500 kN	550 kN	26 mm	52 mm
E	0.85	0.78	1.02	1.30	1.48	1.67	2.56

Steel and Composite Structures – Wang & Choi (eds)
© *2007 Taylor & Francis Group, London, ISBN 978-0-415-45141-3*

Hysteretic behavior of thin-walled steel tubular columns with partially in-filled concrete

Yoshiaki Goto & Ghosh Prosenjit Kumar

Department of Civil Engineering, Nagoya Institute of Technology, Japan

ABSTRACT: Partially concrete-filled thin-walled steel tubular columns (PCFT columns) are often used for elevated highway bridge piers in Japan because of small cross-sectional areas and high earthquake resistance. Considering the lessons learned from the 1995 Kobe earthquake, the Japanese seismic design code for highway bridges adopted a new design concept to control the damage of the columns. Therefore, it is important to predict accurately the hysteretic behavior of PCFT columns. However, up to the present, no research has been conducted on the analysis of PCFT columns in a direct manner. In this analysis, steel tube and concrete are represented by nonlinear shell and solid elements, respectively. To express the material behavior, the 3-surface cyclic plasticity model is used for steel tube, while the concrete damaged plasticity model is used for in-filled concrete. Contact and friction behaviors are considered for interface modeling. The accuracy of the proposed FEM model is examined by comparing with experimental results.

1 INTRODUCTION

Partially concrete filled thin-walled steel tubular columns referred hereinafter as PCFT columns are often used as elevated highway bridge piers in the urban areas of Japan due to the advantages of small cross sectional areas, short construction period, improved ductility & strength and large energy dissipation capacity, compared with thin-walled hollow tubular columns. Since the concrete partially filled at the lower part of hollow tubes is confined by diaphragms, the ductility & strength of tubular columns are improved without so much increasing the seismic inertia force. Considering the lessons learned from the 1995 Kobe earthquake, the Japanese seismic design code for highway bridges adopted a new design concept where the damage of the columns is controlled such that their residual deformation under the in-put earthquake waves becomes within an allowable value. Therefore, it is important to predict accurately the ultimate behavior of PCFT columns during severe earthquakes.

In recent decades, several cyclic unidirectional experimental studies (Iura et al. 2002, Morishita et al. 2000, Hanbin Ge et al. 1996 and Public Work Research Institute, 1997–2000, among others) have been conducted to examine the hysteretic behavior of PCFT columns. Following these experimental studies, some analytical models based on beam theory (Susantha

et al. 2002, Varma et al. 2002) are proposed to predict the hysteretic behavior of PCFT or CFT columns. However, these models are not capable of considering properly the local buckling of thin-walled steel tube along with the interaction between steel tube and n-filled concrete. Matsumura et al. (2003) proposed a 3-D FEM model where the steel tube and in-filled concrete are respectively represented by geometrically and materially nonlinear shell elements and solid elements. Material steel is represented by a modified two-surface model to express its cyclic behavior accurately. Regarding the constitutive mode for in-filled concrete, a strain-softening plasticity model with the Drucker-Prager yield criterion is usually used for the compressive behavior, while the smeared cracking model is adopted for the tensile behavior. In their model, the contact behavior between steel tube and in-filled concrete is considered, although the interface friction in the tangential direction is ignored. Unfortunately, they succeeded only in analyzing the ultimate behavior of a CFT column under monotonic loading due to some numerical instability. Furthermore, the post-peak behavior of the CFT column predicted by the model differs significantly from the existing experimental results. Fujii et al. (2003) used a FEM model similar to Matsumura, although the constitutive model for in-filled concrete and the modeling of the interface action is simplified. That is, an elastic-perfectly plastic model is used for the in-filled concrete and a contact elastic spring model is used for the interface. The contact spring elements are inserted at the initial state and their locations are assumed not to change during the subsequent loading. Probably because of these simplified models, the pinching hysteretic loop characteristics of the CFT columns are not obtained in their analysis. Hsu et al. (2003) also presented a FEM model similar to Fujii et al. (2003). However, they disclose neither the constitutive models nor the interface model. The application of their model is limited only to the analysis of the local buckling pattern under monotonic loading.

In the PCFT columns, the cyclic local buckling of thin-walled steel tube that usually governs their ultimate behavior is prevented by the internal passive pressure due to the dilation of in-filled concrete. On the other hand, this pressure that acts as confinement to the in-filled concrete also increases the strength of concrete. Therefore, accurate modeling of the interface action together with the behavior of confined concrete and thin-walled steel tube may be an only versatile method to predict the hysteretic behavior of PCFT columns in a direct manner. The objective of the present paper is to propose a more accurate, yet numerically stable FEM model that considers all the factors quoted above. The accuracy of the proposed model is herein examined by comparing the computed results with those of the cyclic loading experiments conducted by the Public Work Research Institute (1997–2000).

2 HYSTERETIC CURVE OF PCFT COLUMN

According to the results of the unidirectional cyclic loading test (Public Work Research Institute, 1997–2000), PCFT columns exhibit a characteristic pinching hysteresis loop shown in Fig. 1, expressed in terms of horizontal load-displacement relationship. The typical hysteretic behavior of PCFT columns has two distinct features. First, the hysteresis loop shows a recovery of stiffness in the loading process. This may be explained by the opening and subsequent closing

behavior of concrete cracks transverse to the column axis. When the horizontal displacement approaches zero, the cracks that are closed in the compression side open due to the decrease of the compressive stress. This results in the stiffness degradation. However, when the horizontal displacement increases from zero in the opposite direction, the cracks again close on the compressive side and the column recovers its stiffness. Second, being different from the RC columns, the energy dissipation capacity of the hysteresis loop is rather stable, regardless of the magnitude of the amplitude. This is because the local buckling of steel tube is restrained by the confined concrete. To express the above-mentioned complicated behavior of PCFT columns, an accurate FEM model is proposed.

3 THREE-SURFACE CYCLIC PLASTICITY MODEL FOR STEEL

Hysteretic behavior of PCFT columns is strongly influenced by the behavior of the thin-walled hollow steel tube. Therefore, it is necessary to use an accurate constitutive model to express the cyclic behavior of the material steel. Herein, the modified 3-surface cyclic plasticity model developed by Goto et al. (1998, 2006) is adopted as a constitutive model. This model includes material parameters, such as Young's modulus E_s, Poisson's ratio υ_s, yield stress σ_y, ultimate tensile stress σ_u, original length of yield plateau ε_{yp}, plastic modulus H^p_{mon} obtained by tensile coupon test, minimum radius of elastic range f_b, elastic range reduction rate β, elastic range expansion coefficient ρ, discontinuous coefficient κ and curve-fitting parameter ξ for tensile coupon test. The parameters shown above are calibrated by tensile coupon test along with the cyclic loading experiments on thin-walled steel columns (Goto et al. 1998, 2006). The modified 3-surface model is implemented in ABAQUS ver.6.6 by user subroutine feature.

Table 1. 3-surface model parameters for SS400 steel.

No.16	No. 29 & 30
$E_s = 205.8(GPa)$	$E_s = 205.8(Gpa)$
$\sigma_y = 308(MPa)$	$\sigma_y = 308(MPa)$
$\sigma_u = 559.5(MPa)$	$\sigma_u = 534\,(MPa)$
$\upsilon_s = 0.3$	$\upsilon_s = 0.3$
$\varepsilon_{yp} = 0.0183$	$\varepsilon_{yp} = 0.0183$
$f_b/\sigma_y = 0.38$	$f_b/\sigma_y = 0.38$
$\beta = 100$	$\beta = 100$
$\rho = 2.0$	$\rho = 2.0$
$\kappa = 2.0$	$\kappa = 2.0$
$\xi = 0.25$	$\xi = 0.10$
$H^p_{mon}(*)$	$H^p_{mon}(*)$

(*) Multilinear curve is used to approximate the hardening behavior (Fig. 5).

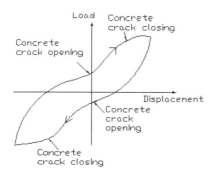

Figure 1. Typical hysteretic behavior of PCFT column.

4 CONCRETE DAMAGED PLASTICITY MODEL FOR IN-FILLED CONCRETE

Concrete that exhibits brittle behavior under tensile and small compressive stresses becomes ductile in the presence of high hydrostatic pressure. The brittle behavior of concrete disappears when the confining pressure is sufficiently high to prevent crack propagation. Under these circumstances, a failure driven by the collapse of concrete microporous structures exhibits a macroscopic response that resembles the behavior of a ductile material. Therefore, in the conventional triaxial concrete models, the behavior under compressive stresses is represented by plasticity model, while the behavior under tensile and small compressive stresses is expressed by smeared cracking model. This model, however, often encounters numerical difficulty under cyclic loading, when applied to FEM. To circumvent this situation, the concrete damaged plasticity model (Lee et al. 1998) implemented in ABAQUS is used herein. Although this is a thoroughly plastic model more approximate than the above-mentioned conventional model, better numerical stability is ensured. This model is based on the non-associated flow rule, where the Drucker-Prager hyperbolic flow potential G is used to calculate plastic strain increment $d\varepsilon^{pl}$ as shown below.

$$d\varepsilon^{pl} = d\lambda \frac{\partial G(\overline{\sigma})}{\partial \overline{\sigma}} \tag{1}$$

$$G = \sqrt{(\in \sigma_{t0} \tan \psi)^2 + \overline{q}^2} - \overline{p} \tan \psi \tag{2}$$

where, ψ = dilation angle, σ_{t0} = tensile strength, $\in$ = eccentricity of flow potential, $\overline{p} = -(1/3)trace(\overline{\sigma}) =$ effective hydrostatic pressure and $\overline{q} = \sqrt{1.5(\overline{s} : \overline{s})} =$ Mises equivalent effective stress; where $\overline{s}$ = effective stress deviators as given by $\overline{s} = \overline{\sigma} + \overline{p}I$.

The yield function F of the concrete damaged plasticity model is expressed by two hardening variables, $\tilde{\varepsilon}_t^{pl}$ and $\tilde{\varepsilon}_c^{pl}$, as

$$F = \frac{1}{1-\alpha} \left(\overline{q} - 3\alpha\overline{p} + \beta(\tilde{\varepsilon}_c^{pl})\langle\hat{\overline{\sigma}}_{max}\rangle - \gamma\langle-\hat{\overline{\sigma}}_{max}\rangle \right)$$
$$-\overline{\sigma}(\tilde{\varepsilon}_c^{pl}) = 0 \tag{3}$$

$\langle.\rangle$ is the Macauley bracket defined by $\langle x \rangle = (|x| + x)/2$. The coefficients of yield function are given by

$$\alpha = \frac{(\sigma_{b0}/\sigma_{c0}) - 1}{2(\sigma_{b0}/\sigma_{c0}) - 1}; \quad 0 \leq \alpha \leq 0.5$$
$$\beta = \frac{\overline{\sigma}_c(\tilde{\varepsilon}_c^{pl})}{\overline{\sigma}_t(\tilde{\varepsilon}_t^{pl})}(1-\alpha) - (1+\alpha); \quad \gamma = \frac{3(1-K_c)}{2K_c - 1} \tag{4}$$

where, K_c = shape factor of yield surface, $\hat{\overline{\sigma}}_{max}$ = maximum principal effective stress and

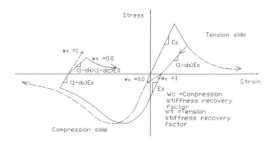

Figure 2. Concrete model under cyclic loading.

σ_{b0}/σ_{c0} = equibiaxial stress ratio. The degraded elastic stiffness for unloading case is defined as $D^{el} = (1 - d)D_0^{el}$, where D_0^{el} = initial elastic stiffness and d = damage variable. For loading case, consistent tangent elasto-plastic stiffness operator based on classical plasticity theory is expressed as

$$D^{ep} = D_0^{el} - \frac{(D_0^{el} g)(D_0^{el} f)^T}{\overline{H} + f^T D_0^{el} g} \tag{5}$$

where $f = \partial F/\partial \overline{\sigma}$, $g = \partial G/\partial \overline{\sigma}$ and $\overline{H}$ = plastic modulus.

Under uniaxial cyclic loading, the behavior of concrete is strongly influenced by the opening and closing of previously formed micro-cracks. There is some recovery of elastic stiffness when loading changes from tension to compression. This stiffness recovery concept is an important aspect under cyclic loading. The stiffness recovery is caused by the closing of the tension crack. In general, the compressive stiffness of concrete is fully recovered as loading changes from tension to compression, i.e., $w_c = 1.0$. On the other hand, the stiffness is not recovered as loading changes from compression to tension, which corresponds to $w_t = 0.0$. The present uniaxial stress-strain relation model under tension-compression-tension loading cycle is schematically shown by solid line in Fig. 2. For comparison, the stress-strain relations under monotonic loading are illustrated by broken lines.

5 INTERFACE MODELING BETWEEN STEEL TUBE AND IN-FILLED CONCRETE

Most of the complicated contact problems may be modeled by the surface-based interaction. The separated surfaces come into contact when clearance between them reduces to zero and contact pressure acts on the respective surfaces. The surfaces separate, if the contact pressure becomes zero. This contact behavior is simulated with "*hard contact*" model implemented in ABAQUS. In this model, master and slave surfaces are referred to as contact pair and the slave nodes are not allowed to penetrate into the master surface. When the slave node comes into contact with master surface, the contact pressure p occurs in the direction normal to the surface. At the same time, friction force occurs in the tangential direction. This friction behavior is simulated by "*Coulomb friction*" model, where the resultant shear stress $\overline{\tau} = \sqrt{(\tau_1^2 + \tau_2^2)}$ is computed from two orthogonal components of shear stresses τ_1 and τ_2, acting on the interface between two contact bodies. The friction model assumes that two contact surfaces can carry the resultant shear stress up to a certain magnitude at their interface before they start to slip. The magnitude of this shear stress is defined as critical shear stress τ_c, which is proportional to contact pressure p ($\tau_c = \mu p$, *where* μ = friction coefficient). The slip will occur along the interface when the shear stress reaches the critical stress ($\overline{\tau} = \tau_c$). In this case, constant shear stress ($\overline{\tau} = \tau_c$) acts at the interface during the slip. If $\overline{\tau} < \tau_c$, no relative motion occurs.

6 NUMERICAL EXAMPLE

A PCFT column model for the present analysis is illustrated in Fig. 3. This model is determined based on the PCFT column specimens used in the unidirectional cyclic loading experiment conducted by the Public Work Research Institute (1997–2000). The PCFT column specimens are fixed at the base. At the top of the specimen, alternating horizontal load under displacement control shown in Fig. 4 is applied with keeping the vertical load P constant. In the FEM model, only half of the model is discretized by virtue of its symmetry. Steel tube with diaphragms is modeled with the 4-node thick shell element with reduced integration (S4R). The lower part of the column is discretized from fine mesh to relatively coarse mesh, while the upper part is modeled by elastic beam element with pipe section (B31). The concrete core is represented by the 8-node solid element with reduced integration (C3D8R). The numerical analysis considering the geometrical and material nonlinearity is carried out by using the general-purpose finite element package program ABAQUS ver. 6.6.

The model specimens are made of carbon steel pipe (SS400) with 900 mm in diameter and 9 mm in thickness, and three diaphragms with 6 mm in thickness are welded to the inside of the pipe at an interval of 900 mm. The geometric and material parameters for the specimens are summarized in Table 2. Among these parameters, the radius-to-thickness ratio parameter R_t has a great influence on the local buckling behavior of PCFT column. Herein, the three model specimens (No. 16, No. 30 & No. 29) have the same radius-to-thickness ratio parameter of $R_t = 0.123$. No.16 and No.30 are PCFT columns with different axial force ratio ($P/\sigma_y A$). No. 29 has a hollow section without in-filled concrete. The

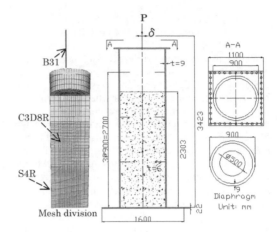

Figure 3. PCFT specimen and FEM model.

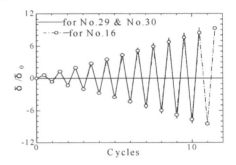

Figure 4. Cyclic loading program.

Table 2. Geometric and material property of specimens.

Specimen	No.16 (PCFT)	No.30 (PCFT)	No.29 (hollow)
Tube material	SS400	SS400	SS400
h(m) (height)	3.423	3.423	3.423
hc(m) (concrete height)	2.303	2.303	–
t(mm) (pipe thickness)	9	9	9
R(mm) (pipe radius)	450	450	450
$\overline{\lambda}$	0.268	0.268	0.268
R_t	0.123	0.123	0.123
$H_o(kN)$	443.94	400.82	400.82
$\delta_0(mm)$	11.53	10.5	10.5
p/σ_y^A	0.114	0.199	0.199
$f_c'(MPa)$	27.93	21.46	–

Note: $\overline{\lambda} = \frac{2h}{\pi\sqrt{I/A}}\sqrt{\frac{\sigma_y}{E_s}}$ (slenderness ratio parameter),
$\delta_0 = H_0 h^3/3EI$ (initial yield displacement),
$H_0 = (\sigma_y - P/A)Z/h$ (initial yield horizontal force),
$R_t = \frac{R}{t}\frac{\sigma_y}{E_s}\sqrt{3(1-v_s^2)}$ (radius to thickness ratio parameter).

experimental result of this specimen is used herein as a reference to identify the steel material parameters for the modified 3-surface model. The uniaxial stress-strain relations for the material steel are illustrated in

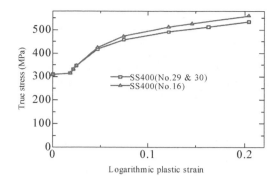

Figure 5. Uniaxial stress-strain relation of steel (SS400).

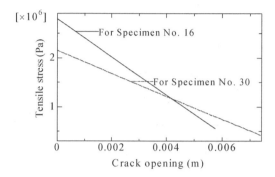

Figure 6. Tensile stress-crack opening relation of concrete.

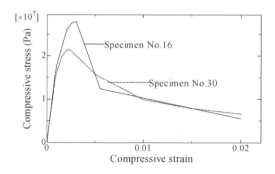

Figure 7. Uniaxial stress-strain relation of concrete.

Fig. 5 and the material parameters for the modified 3-surface model are summarized in Table 1. Regarding the in-filled concrete, tensile strength is assumed 10 percent of the compressive strength obtained by the compression test and the elastic stiffness calculated by using ACI guideline $E_0 = 4700\sqrt{f_c'} MPa$ is adopted up to the maximum point. In the post-peak range, the linear stress-crack opening relationship with the negative stiffness graphically shown in Fig. 6 is considered. For the compressive behavior, the uniaxial stress-strain curves shown in Fig. 7 are used. The shapes of these curves are determined considering the

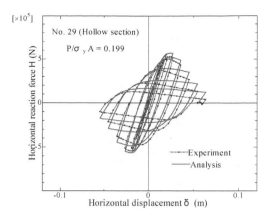

Figure 8. Hysteretic behavior of hollow steel column (No. 29).

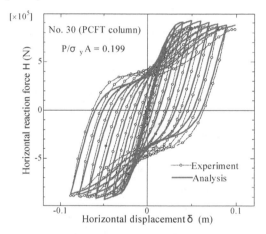

Figure 9. Hysteretic behavior of PCFT column (No. 30).

compressive strengths of the respective concrete such that these shapes are almost proportional to the stress-strain curves shown by Lee et al. (1998) and Sakino et al. (2004). The four parameters, such as $\psi = 33^0$, $\in = 0.99$, $\sigma_{b0}/\sigma_{c0} = 1.0$ and $K_c = 0.67$ are used to define the yield and potential surfaces of in-filled concrete of No.16 & 30 specimens. $\mu = 0.2$ is adopted as a interface friction parameter between steel tube and in-filled concrete. Among these parameters, eccentricity $\in$ is calculated by using William and Warnke three-parameter model (Chen et al. 1982). Remaining parameters including the friction coefficient are computed by applying trial and error method so that the numerical results best fit the experimental results.

7 DISCUSSIONS

The horizontal force H displacement δ relations obtained by numerical analysis are compared with the

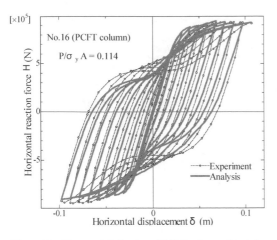

Figure 10. Hysteretic behavior of PCFT column (No. 16).

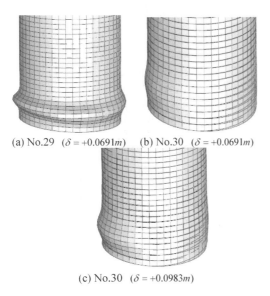

(a) No.29 ($\delta = +0.0691m$)   (b) No.30 ($\delta = +0.0691m$)

(c) No.30 ($\delta = +0.0983m$)

Figure 11. Deformed shape at the lower part of steel pipe ($\times 1.5$).

experimental results in Figs. 8~10 and the computed deformed shapes of the steel columns are illustrated in Fig. 11. It can be seen from Figs. 8~10 that the accuracy of the computed hysteretic behavior of hollow and PCFT columns is generally acceptable. Specifically, the computed results exhibit the pinching hysteretic loops observed in the experiment of CFT columns. From Fig. 11, it is observed that the cyclic local buckling of thin-walled steel tubular column is delayed by the interface action and dilation of in-filled concrete. This results in the improved ductility and higher energy dissipation capacity of PCFT columns of No.16 and No.30, comparing with the hollow column of No.29. If the computed hysteretic behavior is more precisely

examined, the last three hysteretic loops differ from the experimental results in terms of the stiffness degradation when the horizontal displacement approaches zero. This stiffness degradation is strongly influenced by the interface friction effect between steel and concrete. Therefore, more research is needed regarding the modeling of the interface friction.

8 SUMMARY AND CONCLUDING REMARKS

An accurate and numerically stable model for PCFT columns is presented to analyze the hysteretic behavior in a direct manner. In this model, the steel tube and the in-filled concrete are represented by nonlinear shell and solid elements, respectively. To express the material behavior, the 3-surface cyclic plasticity model is used for steel tube, while the concrete damaged plasticity model is used for the in-filled concrete. In the modeling of the interface action, both contact and friction behaviors are considered. In comparison with experimental results, the accuracy of the computed hysteretic behavior of PCFT columns is proved to be generally acceptable. Specifically, the computed results exhibit the pinching hysteretic loops observed in the experiment of CFT columns.

REFERENCES

Chen, W.F. 1982. Plasticity in reinforced concrete. McGraw-Hill, Inc, USA.
Fujii, K., Fujii, T., & Dai, H. 2003. An analysis on concrete filled steel tubular circular column under cyclic horizontal loads. J. of Struct. Engrg., JSCE, Vol. 49A, 1041–1050 (in Japanese).
Goto, Y., Wang, Q.Y. & Obata, M. 1998. FEM analysis for hysteretic behavior of thin-walled columns. J. Struct. Eng., ASCE, 124(11): 1290–1301.
Goto, Y., Jiang, K. & Obata, M. 2006. Stability and Ductility of Thin-Walled Circular Steel Columns under Cyclic bidirectional Loading. J. Struct. Eng., ASCE, 132(10): 1621–1631.
Hanbin Ge & Usami, T. 1996. Cyclic test of concrete filled steel box columns. J. of structural engineering, ASCE, 122(10): 1169–1177.
HKS. 2006. ABAQUS/Standard User's Manual, Version 6.6, Hibbit, Karlson &Sorensen, Inc.
Hsu, H.-L. & Yu, H.-L. 2003. Seismic performance of concretfilled tubes with restrained plastic hinge zones. J. of Construct. Steel Research. 59: 587–608.
Iura, M., Orino, A. & Ishizawa, T. 2002. Elasto-plastic behavior of concrete-filled steel tubular columns. J. Struct. Mech. and Earthquake Engrg., JSCE, No.696/I-58, 285–298 (in Japanese).
Lee, J., & Fenves, G.L. 1998. Plastic-Damage Model for Cyclic Loading of Concrete Structures. Journal of Engineering Mechanics, ASCE, 124(8): 892–900.
Matsumura, T. & Mizuno, E. 2003. 3-D FEM deformation analysis for confining effect on the behavior of CFT

column. 5th Japanese-German symposium on steel and composite bridges, Osaka, Japan. pp 337–344.

Morishita, M., Aoki, T. & Suzuki, M. 2000. Experimental Study on the Seismic resistance performance of Concrete-filled Steel Tubular Columns. J. of Struct. Engrg., JSCE, 46A: 75–83 (in Japanese).

Public Work Research Institute. 1997–2000. Report of cooperative research on limit state seismic design for bridge piers I–VIII and summary (in Japanese).

Sakino, K., Nakahara, H., Morino, S. & Nishiyama, I. 2004. Behavior of Centrally Loaded Concrete-Filled Steel-Tube short Columns. J. of Struct. Engrg., ASCE, 130(2): 180–188.

Susantha, K.A.S., Hanbin Ge & Usami, T. 2002. Cyclic analysis and capacity prediction of concrete-filled steel box columns. Earthquake Engrg. And Struct. Dynamics, 31: 195–216.

Varma, A.H., Ricles, J.M., Sause, R. & Lu, L.-W. 2002. Seismic behavior and modeling of high strength composite concrete-filled steel tube (CFT) beam-columns. J. of Construct. Steel Research, 58: 725–758.

Steel and Composite Structures – Wang & Choi (eds)
© *2007 Taylor & Francis Group, London, ISBN 978-0-415-45141-3*

Nonlinear dynamic analysis of Steel-Concrete Composite frame structures under earthquake excitation

A. Zona
Dipartimento di Progettazione e Costruzione dell'Ambiente, University of Camerino, Ascoli Piceno, Italy

M. Barbato & J.P. Conte
Department of Structural Engineering, University of California, San Diego, USA

ABSTRACT: Composite beams with deformable shear connection were specifically introduced as an extension of conventional monolithic beam models for the analysis of steel-concrete composite (SCC) structures in which the flexible shear connection allows development of partial composite action influencing structural deformation and distribution of stresses. The use of beams with deformable shear connection in the analysis of frame structures raises very specific modeling issues, such as the characterization of the cyclic behavior of the deformable shear connection and the assembly of composite beam elements with conventional beam-column elements. In addition, the effects on the dynamic response of SCC frame structures of various factors such as the shear connection boundary conditions are still not clear and deserve more investigation. The object of this work is to provide deeper insight into the nonlinear seismic response behavior of SCC frame structures and how it is influenced by various modeling assumptions.

1 INTRODUCTION

Among the various models available for the analysis of composite structures (Spacone & El-Tawil 2004), frame models allow to obtain significant information at reasonable computational cost compared to more sophisticated two-dimensional (plate/shell) and three-dimensional (solid) finite element (FE) models. Even if frame elements can only account approximately for local effects (e.g., shear lag, local instabilities in the compressed portion of the steel beam, cracking and crushing of concrete), good test-analysis correlation results were obtained by a number of researchers (e.g., Liew et al. 2001, Kim & Engelhardt 2005) for quasi-static tests and global response quantities. As an extension of conventional monolithic beam models, beams with deformable shear connection were specifically introduced and adopted for the analysis of SCC beams. Flexible shear connectors allow development of partial composite action (Viest et al. 1997, Oehlers & Bradford 2000), influencing structural deformation and distribution of stresses under service and ultimate load conditions. Furthermore, the shear connection can be responsible for collapse, e.g., when partial shear connection design is adopted, connectors fail due to limited ductility. Thus, a composite beam model with deformable shear connection has some important advantages over the common Euler-Bernoulli monolithic beam model, i.e., it allows a more

accurate modeling of the structural behavior, provides information on the slab-beam interface slip and shear force behavior, permits to evaluate the effects of the interface slip on stress distribution, and enables to model damage and failure of the connectors. Up to date, applications of beam elements with deformable shear connections to the analysis of SCC frames have mainly been limited to quasi-static behavior, e.g., recent work by Dissanayake et al. (2000), Ayoub & Filippou (2000), and Salari & Spacone (2001). On the other hand, there is limited experience on nonlinear dynamic analysis of SCC frames based on beam elements with deformable shear connection (Bursi et al. 2005). Furthermore, some modeling aspects deserve further investigation. In fact, the use of beams with deformable shear connection in the analysis of frame structures raises very specific modeling issues, such as the characterization of the cyclic behavior of the deformable shear connection and the assembly of composite beam elements with conventional beam-column elements. In addition, the influence of various factors (e.g., shear connection boundary conditions, mass distribution between the two components of the composite beam) on the dynamic response of SCC frame structures needs to be better understood through a systematic parametric study.

The objective of this work is to provide deeper insight into nonlinear dynamic analysis results of SCC structures and how different modeling assumptions affect these results. For this purpose, a materially-nonlinear-only FE formulation for static and dynamic analysis of SCC structures using displacement-based locking-free elements with deformable shear connection (Dall'Asta & Zona 2002) is employed. Realistic uni-axial cyclic constitutive laws are adopted for the steel and concrete materials of the beams and columns and for the shear connection. Nonlinear dynamic seismic analysis results of two-dimensional moment resisting frames made of steel columns and composite beams are provided. These results and their discussion focus on: (i) the influence of partial composite action on the dynamic nonlinear analysis of SCC frames; (ii) the effects of different modeling assumptions related to SCC structures.

2 MODELING AND ANALYSIS OF STEEL-CONCRETE COMPOSITE FRAMES

2.1 Beam model with deformable shear connection

The formulation for two-dimensional beams with deformable shear connection is based on the Newmark et al. (1951) model in which (i) Euler-Bernoulli beam theory (in small deformations) applies to both components of the composite beam, and (ii) the deformable shear connection is represented by an interface model with distributed bond allowing interlayer slip and enforcing contact between the steel and concrete components.

2.2 Finite Element Formulation

The present study makes use of a simple and effective two-dimensional 10 nodal degrees-of-freedom (DOFs) displacement-based SCC frame element with deformable shear connection (Dall'Asta & Zona 2002). This element was proven to produce accurate results even for local response quantities (e.g., section deformations, section stress resultants, shear force distribution at the steel-concrete interface, etc.) provided that a sufficiently refined mesh is adopted (Dall'Asta & Zona 2004c). The same element was employed for finite element response sensitivity analysis of SCC structures (Zona et al. 2005) under both monotonic and cyclic loading conditions. Its response results were validated through test-analysis comparison studies (Zona et al. 2005). Results and observations provided in the present paper are, however, not restricted to the specific finite element adopted.

In the finite element adopted, the concrete slab and steel beam components have their own user-defined reference system allowing proper selection of the position of the axial DOFs to be used in the FE assembly of beam and column elements. Typical choices are: reference systems located at the centroid of each component; reference systems both located at the interface

of the two components. Choice of the reference system does not influence the precision of the element, since the element shape functions in different reference systems are consistent so as to avoid the eccentricity issue and slip locking problems (Dall'Asta & Zona 2004b). Another useful feature of the FE formulation adopted is an internal constraint that can be introduced at each beam end independently. This internal constraint enforces zero slip at the beam end at which it is applied and can be used in modeling many real situations.

2.3 Modeling of inertia and damping properties

In this study, the inertia properties are modeled via translational (horizontal and vertical) masses lumped at the DOFs of the frame elements' external nodes. A lumped mass matrix formulation is preferred to a consistent mass matrix formulation for several reasons: (i) using a lumped mass matrix formulation, the inertia properties of the finite element model are independent of the finite element types employed, and thus the same structure mass matrix is obtained using displacement-based, force-based or mixed-formulation frame elements; (ii) a lumped mass matrix formulation yields a diagonal mass matrix at the structure level and entails significant computational advantages (over a fully populated consistent mass matrix) for calculations involving the inverse of the mass matrix such as in eigen-analysis and in explicit time integration methods; (iii) the use of consistent mass matrices at the element level presents some additional complications when internal DOFs need to be condensed out at the element level, as in the case of the locking free composite frame element used herein.

Information about damping properties of SCC frame structures inferred from experimental dynamic data is very limited (Bursi & Gramola 2000, Dall'Asta et al. 2005). This study assumes Rayleigh-type proportional viscous damping (Chopra 2001). The Rayleigh damping matrix used herein is proportional to the mass matrix and initial stiffness matrix.

2.4 Hysteretic Modeling of Structural Materials and Shear Connection

The selected constitutive model for the steel material (reinforcement and beam steel) is the uni-axial Menegotto & Pinto (1973) constitutive model, a computationally efficient nonlinear smooth law able to model both kinematic and isotropic hardening (Filippou et al. 1983) and Baushinger's effect, which are typical of the cyclic behavior of structural steel materials. Further details on the model and its numerical implementation can be found in (Barbato & Conte 2006), where the model is extended for finite element response sensitivity computation. A typical cyclic response is shown in Figure 1.

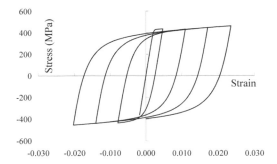

Figure 1. Menegotto-Pinto material constitutive model for structural steel: typical cyclic stress-strain response.

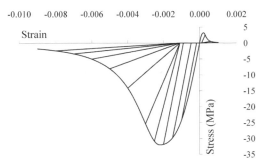

Figure 2. Hysteretic concrete material model: typical cyclic stress-strain response.

The selected constitutive law for the concrete material is a uniaxial cyclic law with monotonic envelope defined by the Popovics-Saenz law (Balan et al. 1997, 2001). The details of the formulation of this constitutive model and related material parameters can be found in (Zona et al. 2006). A typical cyclic response is shown in Figure 2.

The constitutive law used for the shear connectors is a slip-force cyclic law with monotonic envelope given by the Ollgaard et al. law (1971). The cyclic response of the shear connectors is a modified version of the model proposed by Eligenhausen et al. (1983). The details of the formulation of this constitutive model and related material parameters can be found in (Zona et al. 2006). A typical cyclic response is shown in Figure 3.

3 DYNAMIC RESPONSE SIMULATION OF SCC FRAME STRUCTURES

3.1 Description of the SCC frames analyzed

The basic testbed SCC frame structure considered in this section is a realistic 5-story 2-bay moment resisting frame made of steel columns and composite beams (Fig. 4). Each bay has a span of 5.00 m and each story has a height of 3.00 m. The steel columns are made of European HEB300 wide flange beams, while the

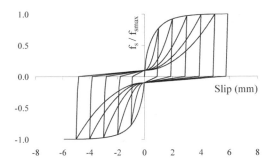

Figure 3. Hysteretic model of shear connection: typical cyclic shear force-slip response.

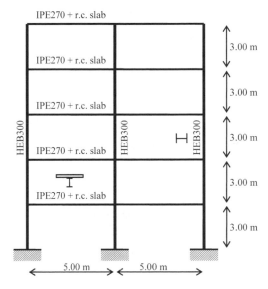

Figure 4. Configuration of testbed SCC frame structure analyzed.

composite beams are made of steel European IPE270 I-beams connected by means of stud connectors to a 100 mm thick concrete slab with an effective width estimated at 800 mm (kept constant along the beam), top and bottom reinforcements of 400 mm^2 and a concrete cover of 30 mm (Fig. 5).

This SCC frame was designed according to Eurocode 4 (CEN 2004a) to resist the static loads (composite cross section self weight = 2.36 kN/m, permanent load G = 16 kN/m, and live load Q = 8 kN/m with G and Q uniformly distributed along the composite beams) and seismic forces evaluated using response spectrum analysis with peak ground acceleration = 0.35 g, Type 1 spectrum of Eurocode 8 (CEN 2004b), modal damping ratio = 0.05, soil B, and behavior factor q = 3. A full shear connection (i.e., the ultimate strength of the composite section is not affected by the shear connectors) was designed using the plastic approach of Eurocode 4 (CEN 2004a). For

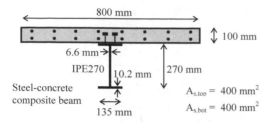

Figure 5. Composite beam cross-section definition of the testbed SCC frame structure analyzed.

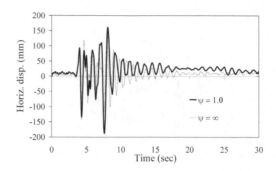

Figure 6. Horizontal displacement of left column at roof level: effect of deformable shear connection (Northridge seismic input).

the sake of simplicity, the shear connection strength was taken as constant along all composite beams.

All finite element models adopted in the response simulations illustrated use the same spatial discretization: four 10-DOF composite frame elements of equal length for each beam (between two adjacent columns) and two conventional displacement-based Euler-Bernoulli (monolithic) frame elements for each column (between two adjacent floors).

The inertia properties are modeled using lumped (horizontal and vertical) translational masses representing the mass of beams and columns, and the mass equivalent to both the permanent load G and the live load Q. The mass equivalent to the self-weight of the SCC beams is distributed between the slab and beam DOFs according to the actual mass of the steel beam and concrete slab components. The mass equivalent to the permanent load G is considered evenly divided between the slab and steel beam components. The mass equivalent to the live load Q is assigned entirely to the slab DOFs.

Three different slip boundary conditions are considered: (i) slip restrained at every beam-column joint (i.e., the relative slip between slab and steel beam is prevented at the face of every column); (ii) slip restrained at the central beam-column joints only (i.e., relative slip between slab and steel beam is prevented at the face of the central column only); (iii) slip free at each joint (i.e., due to sufficient space between columns and slabs, relative slip between slab and steel beam is not prevented). The slip constraint condition is not applied to the SCC beams at the roof level where the slab is free to slip. In this way, the present study aims at describing and analyzing the effects of different realistic modeling options (in terms of slip constraint conditions) at the joints between SCC beams and steel columns.

In addition, a finite element frame model with conventional Euler-Bernoulli monolithic beams (i.e., full shear interaction and full shear connection) was included in this study for comparison purposes.

3.2 Nonlinear earthquake response analysis

After quasi-static application of a vertical distributed load of 26.36 kN/m representing self weight,

permanent and live loads, four nonlinear dynamic analyses were carried out for each frame model considered by using two ground motion accelerograms and two different levels of viscous damping in the structure. The two historic earthquake accelerograms used as base excitation are: (i) the 1994 Northridge earthquake recorded at the Pacoima Dam station with a peak ground acceleration (PGA) of 1.585 g, corresponding to a return period of about 180 years (at the recording site), and (ii) the 1979 Imperial Valley earthquake recorded at the Bonds Corner station with PGA = 0.775 g, corresponding to a return period of about 40 years (at the recording site). Rayleigh damping was assumed with a specified damping ratio of ξ at both the first and third vibration modes of the structure. This study uses the constant average acceleration method and the corresponding set of nonlinear algebraic equations is solved iteratively using Newton's method (Chopra 2001). A constant time step $\Delta t = 0.005$ s was used for the numerical integration of the equations of motion in all the nonlinear dynamic analyses performed. Due to space limitation, in the sequel only selective results are presented.

The response of the frame model with rigid shear connection ($\psi = \infty$) is compared to the response of the frame model characterized by deformable shear connection ($\psi = 1.0$, full shear connection) and slip restrained at the central beam-column nodes only (frame model 10TC). In Fig. 6, time histories for the horizontal displacement of the left column at the roof level ($z = 15$ m) are reported. The differences between the two models are clearly noticeable. Taking the response of the monolithic frame as reference, the difference in magnitude of the positive and negative peaks is +0.2% and +31.5%, respectively. The maximum absolute difference over the duration of the earthquake is 129.9 mm, and the average absolute difference is 22.5 mm. It is observed that in the case of the monolithic frame, the displacement response oscillates around the static displacement response due to

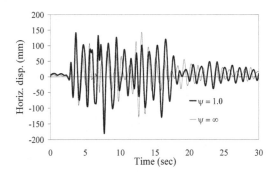

Figure 7. Horizontal displacement of left column at roof level: effect of deformable shear connection (Imperial Valley seismic input).

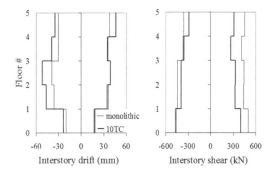

Figure 8. Interstory drift and shear demand: effect of deformable shear connection (Northridge seismic input).

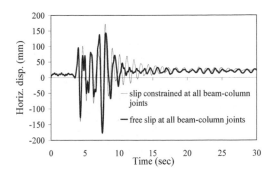

Figure 9. Horizontal displacement of left column at roof level: effect of slip constraint (Northridge seismic input).

floor displacements and interstory drifts. This conclusion is also corroborated by other analysis results (based on other combinations of modeling parameters/assumptions and the Imperial Valley earthquake input) not presented here due to space limitation.

Fig. 9 shows the effect of slip constraints on the roof horizontal displacement by considering two extreme cases, namely slip (i) restrained (frame model 10TA) and (ii) unrestrained (frame model 10TN) at all beam-column joints. Taking frame model 10TA as reference, the difference in magnitude of the positive and negative peaks is −16.5% and +5.7%, respectively. The maximum absolute difference over the duration of the earthquake is 95.8 mm, and the average absolute difference is 14.1 mm.

gravity loads only, which is not the case for the frame with deformable shear connection. To evaluate the effect of a different seismic excitation, Fig. 7 presents the roof displacement responses for the same models (monolithic frame and frame 10TC) as in Fig. 6 subjected to the (less intense) Imperial Valley earthquake. Again, the difference in the two model responses is evident (difference in magnitude of the positive and negative peaks = −1.5% and +40.8%, respectively, maximum absolute difference = 108.6 mm, average absolute difference = 51.2 mm).

Fig. 8 represents graphically the interstory drift and shear demands for the frame model with monolithic beams and frame model 10TC subjected to the Northridge earthquake input. It is found that response simulation of SCC frame structures modeled using frame elements with deformable shear connection (as compared to monolithic frame elements) leads to lower seismic demand in terms of interstory shears. As expected, the interstory shear demand increases with the overall stiffness of the frame model. Thus, response simulation of SCC frame structures modeled using frame elements with deformable shear connection (as compared to monolithic frame elements) leads to larger seismic demand in terms of

4 CONCLUSIONS

From the above results it can be concluded that the inelastic partial composite action in the SCC frame structures considered in this study plays an important role on their global seismic (dynamic) response behavior. The shear connection deformability has a significant effect on global seismic response, i.e., increase of floor displacements and interstory drifts, and decrease of interstory shear demand. These effects are amplified when slip constraints are not present at any beam-column joints. Thus, a proper representation of the slip boundary conditions for all composite beams is crucial for accurate response simulation. In addition, a frame model with deformable shear connection not only accounts for the effect of partial composite action in the global response prediction, but can also provide useful information on the response of the shear connection (in terms of interface slip and shear force). This local response prediction, that can be obtained at a relatively low additional computational cost (as compared to frame models with monolithic beams), allows to asses the shear connection behavior under dynamic/seismic loads.

ACKNOWLEDGEMENTS

Partial supports of this research by the National Science Foundation under Grant No. CMS-0010112, the Pacific Earthquake Engineering Research (PEER) Center through the Earthquake Engineering Research Centers Program of the National Science Foundation under Award No. EEC-9701568, and the National Center for Supercomputing Applications (NCSA) under Grant No. MSS040022N involving utilization of the IBM P690 computers are gratefully acknowledged.

REFERENCES

Ayoub, A. & Filippou, F.C. 2000. Mixed formulation of nonlinear steel-concrete composite beam element. *Journal of Structural Engineering ASCE* 126(3): 371–381.

Balan, T.A., Filippou, F.C. & Popov, E.P. 1997. Constitutive model for 3D cyclic analysis of concrete structures. *Journal of Engineering Mechanics ASCE* 123 (2): 143–153.

Balan, T.A., Spacone, E. & Kwon, M. 2001. A 3D hypoplastic model for cyclic analysis of concrete structures. *Engineering Structures* 23(4): 333–342.

Barbato, M. & Conte, J.P. 2006. Finite element structural response sensitivity and reliability analyses using smooth versus non-smooth material constitutive models. *International Journal of Reliability and Safety* 1(1–2): 3–39.

Bursi, O.S. & Gramola, G. 1999. Behaviour of headed stud shear connectors under low-cycle high amplitude displacements. *Material and Structures RILEM* 32: 290–297.

Bursi, O.S. & Gramola, G. 2000. Behaviour of composite substructures with full and partial shear connection under quasi-static cyclic and pseudo-dynamic displacements. *Material and Structures RILEM* 33: 154–163.

Bursi, O.S., Sun, F.F. & Postal, S. 2005. Non-linear analysis of steel-concrete composite frames with full and partial shear connection subjected to seismic loads. *Journal of Constructional Steel Research* 61(1): 67–92.

CEN, Comité Européen de Normalization. 2004a. *Eurocode 4: Design of composite steel and concrete structures – Part 1.1: General – General rules and rules for buildings*, EN 1994-1-1, Brussels.

CEN, Comité Européen de Normalization. 2004b. *Eurocode 8: Design of structures for earthquake resistance - Part 1: General rules, seismic actions and rules for buildings*, EN 1998-1-1, Brussels.

Chopra, A.K. 2001. *Dynamics of Structures: Theory and Applications to Earthquake Engineering*, Second Edition. Prentice Hall, Englewood Cliffs, NJ.

Dall'Asta, A., Dezi, L., Giacchetti, R., Leoni, G. & Ragni, L. 2005. Dynamic response of composite frames with rubber-based dissipating devices: experimental tests. *Proceedings, Fourth International Conference on Advances in Steel Structures*, Z.Y. Shen, G.Q. Li and S.L. Chan Editors, Elsevier Publisher, Oxford, UK, 741–746.

Dall'Asta, A. & Zona, A. 2002. Non-linear analysis of composite beams by a displacement approach. *Computers and Structures*, 80(27–30): 2217–2228.

Dall'Asta, A. & Zona, A. 2004a. Three-field mixed formulation for the non-linear analysis of composite beams with deformable shear connection. *Finite Elements in Analysis and Design* 40(4): 425–448.

Dall'Asta, A. & Zona, A. 2004b. Slip locking in finite elements for composite beams with deformable shear connection. *Finite Elements in Analysis and Design* 40(13–14): 1907–1930.

Dall'Asta, A. & Zona, A. 2004c. Comparison and validation of displacement and mixed elements for the non-linear analysis of continuous composite beams. *Computers and Structures* 82(23-26): 2117–2130.

Dissanayake, U.I., Burgess, I.W. & Davison, J.B. 2000. Modelling of plane composite frames in unpropped construction. *Engineering Structures* 22(4): 287–303.

Eligehausen, R., Popov, E.P. & Bertero, V.V. 1983. Local bond stress-slip relationships of deformed bars under generalized excitations. *Report No. 83/23, EERC Earthquake Engineering Research Center*, University of California, Berkeley, CA.

Filippou, F.C., Popov, E.P. & Bertero, V.V. 1983. Effects on bond deterioration on hysteretic behavior of reinforced concrete joints. *Report EERC 83–19*, Earthquake Engineering Research Center, University of California, Berkeley, CA

Kim, K.D. & Engelhardt, M.D. 2005. Composite beam element for nonlinear seismic analysis of steel frames. *Journal of Structural Engineering ASCE* 131(5): 715–724.

Liew, J.Y.R., Chen, H. & Shanmugam, N.E. 2001. Inelastic analysis of steel frames with composite beams. *Journal of Structural Engineering ASCE* 127(2): 194–202.

Menegotto, M. & Pinto, P. E. 1973. Method of analysis for cyclically loaded reinforced concrete plane frames including changes in geometry and nonelastic behavior of elements under combined normal force and bending. *Proceedings, IABSE Symposium on Resistance and Ultimate Deformability of Structures Acted on by Well-Defined Repeated Loads*, International Association for Bridge and Structural Engineering, Zurich, 112–123.

Newmark, N.M., Siess, C.P. & Viest, I.M. 1951. Tests and analysis of composite beams with incomplete interaction. *Proceeding of the Society for Experimental Stress Analysis* 9(1): 75–92.

Oehlers D.J. & Bradford M.A. 2000. *Elementary Behaviour of Composite Steel and Concrete Structural Members*. Butterworth-Heinemann, London.

Ollgaard, J.G., Slutter, R.G. & Fisher, J.W. 1971. Shear strength of stud connectors in lightweight and normal weight concrete. *AISC Engineering Journal* 2Q: 55–64.

Salari, M.R. & Spacone, E. 2001. Analysis of steel-concrete composite frames with bond-slip. *Journal of Structural Engineering ASCE* 127(11): 1243–1250.

Spacone, E. & El-Tawil, S. 2004. Nonlinear analysis of steel-concrete composite structures: state-of-the-art. *Journal of Structural Engineering ASCE* 130(2): 159–168.

Viest, I.M., Colaco, J.P., Furlong, R.W., Griffs, L.G., Leon, R.T. & Wyllie, L.A. 1997. *Composite Construction Design for Buildings*. McGraw-Hill, New York, NY.

Zona, A., Barbato, M. & Conte, J.P. 2005. Finite element response sensitivity analysis of steel-concrete composite beams with deformable shear connection. *Journal of Engineering Mechanics ASCE* 131(11): 1126–1139.

Zona, A., Barbato, M. & Conte, J.P. 2006. Finite element response sensitivity analysis of steel-concrete composite structures. *Report No. SSRP-04/02*, Department of Structural Engineering, University of California, San Diego, CA.

Steel and Composite Structures – Wang & Choi (eds)
© *2007 Taylor & Francis Group, London, ISBN 978-0-415-45141-3*

A new method of push-over analysis for evaluation of elasto-plastic seismic performance of long-span structures

MuWang Yang, Yongfeng Luo, Xue Li & Xi Zhou
College of Civil Engineering, Tongji University, Shanghai, China

ABSTRACT: Few researches about the static elasto-plastic seismic analysis method of long-span spatial structures have been found. A new static elasto-plastic computation method, named Push-down, is proposed in this paper. The fundamental principles of the method are described and approved, such as forming the elasto-plastic capacity spectrum, the demand spectrum and proposing some kinds of vertical load patterns. The seismic performance of the long-span spatial structures which fundamental mode is mainly vertical deformation can be simply and effectively evaluated by the method of Push-down. Meanwhile, it easily enables the designer to understand and evaluate the seismic performances of the long-span spatial structures under vertical excitation.

1 INTRODUCTION

The long-span structure is one of the most rapidly developing types of structures. It is widely used in large-scale and public buildings. Its seismic characteristics are paid more attention by the researchers. With the development of seismic resistance researches of structures, different computation methods for seismic analysis are correspondingly used in different type of structures for predicting the structures response accurately. There are usually base-shear method, response spectrum method, pseudo-dynamic method, time-history method and static elasto-plastic analysis method. The static elasto-plastic analysis method, named Push-over method, is only suitable for the regular tall-buildings. Then the Push-over method is not suitable for the long-span spatial structures, since the stiffness distribution and the dynamic characteristics of the spatial structures are different with the tall buildings. Therefore the time-history method is usually adopted in the seismic analysis of the spatial structures these days.

In generally speaking, few researches about the vertical seismic computation method of long-span spatial structures have been found till now. According to the dynamic characteristics of the long-span spatial structure and based on the principle of FEMA273 (Federal Emergency Management Agency) and ATC-40 (Applied Technology Council) about seismic resistance analysis of the tall-buildings, a new static elasto-plastic analysis method for long-span spatial structures, named "Push-down method", is proposed in this paper. It is convenient to use the new method evaluating the seismic performance of long-span spatial structures which fundamental mode is mainly vertical deformation. This research conducts a new branch in seismic resistance analysis of the long-span structures.

2 THE PRINCIPLE OF PUSH-DOWN METHOD

2.1 *Basic assumptions and applicability*

Basfic assumptions: ①Plane-section assumption used in structure members. ② The effect of shearing deformation is neglected. ③Tension stress and force are defined to be positive. Clockwise rotation is defined to be positive. ④No separation between steel and concrete in composite structures.

Applicability: The method is applicable to the long-span spatial structures which the fundamental mode is vertical deformation. Especially it is easy to evaluate seismic resistance performance of the long-span spatial structures modeled with the beam elements. In the dynamic analysis of the structures, the dynamic characteristics are mainly composed of the fundamental vertical vibration mode.

2.2 *The constitutive relationship of plastic hinge and limit plastic angle*

There are four kinds of plastic hinges, which are moment hinge (M), shear hinge (V), axial hinge (P), and axial-moment hinge (PMM). They are adopted in the corresponding elements. The constitutive relationships of plastic hinges are shown in figure 1.

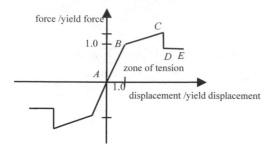

Figure 1. The constitutive relationship of plastic hinges.

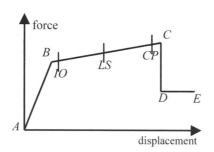

Figure 2. The limit displacements of the plastic hinge under different levels.

The curve in the figure is divided into four segments: elastic segments (AB), strengthening segments (BC), unloading segments (CD) and plastic segments (DE). The moment hinging (M) is adopted in beam elements and the axial-moment hinging (P-M-M) is adopted in column elements.

It is assumed that the rotation of the plastic hinge is concentrated on the section with the largest bending moment in the definite length range of plastic hinge. This section is named critical section. The limit rotation angle θ_p is shown as follows:

$$\theta_p = \left(\varphi_u - \varphi_y\right)\bigg/ l_p \tag{1}$$

The yielding rotation angle of a beam or a column in equation (1) is:

for beam $$\theta_y = \frac{f_y W_p l_b}{6EI_b} \tag{2}$$

for column $$\theta_y = \frac{f_y W_p l_c}{6EI_c}\left(1 - \frac{P}{f_y A}\right) \tag{3}$$

Where, φ_u = the limit plastic curvature of the critical section.

φ_y = the yielding curvature of the critical section.
l_p = the length of plastic hinge.
f_y = the yielding strength of material.
W_p = plastic inertial moment.
l_b = the actual length of the beam.
l_c = the actual length of the column.
P = the axial force of the element.

The node with the largest vertical displacement is taken as the control node in the seismic evaluation of long-span spatial structure in which the first vibration mode is mainly vertical deformation. When the displacement of control node reaches the preset limit displacement or the number of plastic hinges exceeds the limit number during the process of Push-down analysis, the structure will collapse. Then the analysis is done. There are two distinguished criterions about the structure collapse.

(1) Under a definite mode of loads, the vertical deformation of the structure becomes obviously larger than normal one. The structure becomes unstable and can't bear further load.
(2) According to the existent researches, the preset deformation in the vertical direction is defined. When the deformation of the control node reaches the preset value, the analysis is done. The structure meets the requirement of the code yet.

2.3 The pattern of vertical loads

The most importance of Push-down analysis is to choose the suitable pattern of vertical loads. The load pattern directly affects the numerical accuracy of the distribution of forces and deformations of the structure under the earthquake. During the process of linear analysis, the inertial force will be influenced by the frequency characteristics of the structure and dynamic characteristics of earthquake. During the process of nonlinear analysis, the distribution of inertial forces will change with the effect of the nonlinear deformations and lasting of the earthquake.

The vertical distribution mode of loads is assumed to be constant in the process of seismic analysis for the structure in which the effects of high order of vibration modes can be neglected and the only one certain yielding mode will occur. When the elasto-plastic analysis of the structure is conducted under the rare earthquake, the natural vibration periods and inertial forces will change. The inertial forces of structure can't be represented by one mode.

To the regular long-span spatial structure, three kinds of vertical load patterns are recommended in this paper.

(1) The load distribution mode is the same as the fundamental vibration mode. That is to say, the

distribution of vertical load is in proportion with the displacements of the first vibration mode.

(2) The combination coefficient method. That is to say, the response of the structure is the largest one by the load distribution in different spans.

(3) The combination mode of SRSS method. The inertial forces of the structure are the same as the forces of combination of response spectrums.

2.4 Vertical capacity spectrum

According to the principle of dynamics, the dynamic equations about multiple degrees of freedom under the earthquake are shown as follows:

$$[M]\{\ddot{X}\}+[C]\{\dot{X}\}+\{Q\}=-[M]\{I\}\ddot{X}_g \quad (4)$$

Where,

$[M]$ = mass matrix of the structure.
$[C]$ = the damping matrix.
$\{X\}$ = the vector of relative displacements.
$\{Q\}$ = the vector of restoring forces.
$\ddot{X}_g$ = acceleration of ground.
$\{I\}$ = the unit vector.

It is assumed that the displacements of a system with multiple degrees of freedom can be represented by a certain shape vector $\{\phi\}$. Then the displacement vector of the system with multiple degrees of freedom can be represented by $\{X\}=\{\phi\}x_c/\phi_c$. Where, x_c = the displacement of the control node of the structure. ϕ_c = the value of the shape vector in the control node.

The equation (4) can be expressed as:

$$[M]\{\phi\}\{\ddot{x}_c/\phi_c\}+[C]\{\phi\}\{\dot{x}_c/\phi_c\}+\{Q\}=-[M]\{I\}\ddot{X}_g \quad (5)$$

The two sides of the above equation are multiplied by $\{\phi\}^T$ in left, then:

$$\{\phi\}^T[M]\{\phi\}\{\ddot{x}_c/\phi_c\}+\{\phi\}^T[C]\{\phi\}\{\dot{x}_c/\phi_c\}+\{\phi\}^T\{Q\}$$
$$=-\{\phi\}^T[M]\{I\}\ddot{X}_g \quad (6)$$

let

$$\frac{x_c}{\phi_c}=\frac{\{\phi\}^T[M]\{I\}}{\{\phi\}^T[M]\{\phi\}}S_d \quad (7)$$

substituting the formula (7) into formula (6), the dynamic equation of system with single degree of freedom under earthquake is obtained:

$$M^*\ddot{S}_d+C^*\dot{S}_d+Q^*=-M^*\ddot{X}_g \quad (8)$$

where, S_d = the displacement of the equivalent structure with a single degree of freedom.

The mass of the equivalent structure:

$$M^*=\{\phi\}^T[M]\{I\} \quad (9)$$

The damping of equivalent structure:

$$C^*=\{\phi\}^T[C]\{\phi\}\frac{\{\phi\}^T[M]\{I\}}{\{\phi\}^T[M]\{\phi\}} \quad (10)$$

The restoring force of equivalent structure:

$$Q^*=\{\phi\}^T\{Q\}=\sum_{i=1}^n\phi_iQ_i=\sum_{i=1}^n\phi_im_i\frac{\phi_i}{\phi_c}\ddot{x}_c=\frac{\ddot{x}_c}{\phi_c}\sum_{i=1}^n m_i\phi_i^2 \quad (11)$$

The total lateral force of the structure:

$$V_v=\{I\}^T\{Q\}=\sum_{i=1}^n Q_i=\sum_{i=1}^n m_i\ddot{x}_i=\sum_{i=1}^n m_i\frac{\phi_i}{\phi_c}\ddot{x}_c=\frac{\ddot{x}_c}{\phi_c}\sum_{i=1}^n m_i\phi_i \quad (12)$$

Where, Q_i = restoring force of the node i.
m_i = the mass of the node i.
x_i = the displacement of the node i.
ϕ_i = the sharp value of the node i.

Comparing formula (11) with formula (12), the result is as follows:

$$Q^*=V_v\frac{\sum_{i=1}^n\phi_i^2 m_i}{\sum_{i=1}^n\phi_i m_i}=V_v\frac{\{\phi\}^T[M]\{\phi\}}{\{\phi\}^T[M]\{I\}} \quad (13)$$

The modal participation factor: $\Gamma=\dfrac{\{\phi\}^T[M]\{I\}}{\{\phi\}^T[M]\{\phi\}}$,

The effective modal mass: $M^*=\dfrac{(\{\phi\}^T[M]\{I\})^2}{\{\phi\}^T[M]\{\phi\}}$,

According to the equation (8) and the formula (13), the acceleration response spectrum S_a and the displacement response spectrum S_d can be described as follows:

$$S_d=\frac{x_c}{\Gamma\phi_c} \qquad S_a=\frac{Q^*}{M^*}=\frac{V_v}{\frac{(\{\phi\}^T[M]\{I\})^2}{\{\phi\}^T[M]\{\phi\}}}=\frac{V_v}{M^*} \quad (14)$$

Considering the long-span spatial structure that the first vibration mode is vertical deformation, the relationship curve between the vertical force excited by earthquake and the displacement of the control node can be transferred into the relationship curve between the acceleration spectrum and the displacement spectrum:

$$S_d=\frac{\Delta_v}{\Gamma_1\Delta_{v,1}} \qquad S_a=\frac{V_v}{M^*} \quad (15)$$

Where,
Γ_1 = participant coefficient of the vertical displacement of the fundamental vibration mode.
M^* = the modal participant mass.
V_v = the total of vertical shear.
Δ_v = the displacement of the control node with the largest dynamic response.
$\Delta_{v,1}$ = the displacement of the control node with the largest shape value in the first vibration mode.

2.5 Vertical demand spectrum

In general, the vertical acceleration is the 1/2~2/3 of the horizontal acceleration. Due to the random record of earthquake and based on existing researches about the vertical response spectrum, the curve of the vertical response spectrum can be described by a horizontal line in short periods and the hyperbolic curve in long periods:

$$\begin{cases} \alpha = \alpha_{v\max} & T \le T_g \\ \alpha = \alpha_{v\max}(T/T_g)^b & T > T_g \end{cases} \quad (16)$$

where, T_g = the characteristic period. T = the fundamental period. b = damped exponential.

The vertical earthquake coefficient k_v and the vertical earthquake effect coefficient α_v can be described as follows by the results of statistics:

$$\alpha_{v\max} = k_v \times \beta_{v\max} = \frac{S_{ac}}{g} \quad (17)$$

$$\begin{cases} k_v = 0.65k_H & \text{bedrock} \\ k_v = \begin{cases} 1.0k_H & T < 0.1 \\ [1.0 - 2.5(T-0.1)]k_H & 0.1 \le T < 0.3 \\ 0.5k_H & T > 0.3 \end{cases} & \text{soil layer} \end{cases} \quad (18)$$

where, k_H = horizontal seismic coefficient, $\beta_{v\max}$ = maximum value of dynamic coefficient.

The calculating parameters concerned with the vertical seismic influence coefficient α_v are listed in the table 1. The vertical response curves of intensity 7,8,9 on site I are shown in figure 3.

To obtain the elasto-plastic demand spectrum, the curve of the $S_{ae} - T$ can be converted into the curve of ADRS format of $S_{ae} - S_{de}$ with following formulas:

$$S_{ae} = u_{v\max} g \qquad S_{de} = \frac{T^2}{4\pi^2} S_{ae} g \quad (19)$$

Based on the elastic demand spectrum of the code, the elasto-plastic demand spectrum is established by the reduction coefficient R and the ductility coefficient μ. The steps are described as follows:

(1) The bilinear capacity spectrum and the elastic demand spectrum are drawn in the same figure. It is shown in figure 4. The acceleration spectrum S_a is displayed in the longitudinal coordinate, while displacement spectrum S_d is displayed in the horizontal coordinate.

(2) There is an intersection point in the figure of bilinear capacity spectrum and elastic demand response spectrum. The elastic spectrum displacement of this point is defined as S_{de}, while the spectrum acceleration of this point is defined as S_{ae}. S_{dy} is the yielding spectrum displacement. The ductility coefficient of the structure is defined as $\mu = S_{de}/S_{dy}$. The reduction coefficient of strength R_μ is defined by the formula (20).

Table 1. The calculating parameters of vertical seismic influence coefficient α_v.

Design earthquake intensity		7			8			9	
$k_H(g)$		0.10			0.20			0.40	
b		−1.0			−1.0			−1.0	
Type of ground	I	II	III/ I	II	III/ I	II	III/	I	III/ I
$T_g(s)$	0.2	0.3	0.6	0.2	0.3	0.6	0.2	0.3	0.6
$\beta_{v\max}$		2.30	2.0	2.30	2.0	2.30	2.0	2.30	2.0

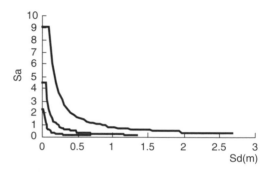

Figure 3. Vertical response curves of intensity 7,8,9 on site I.

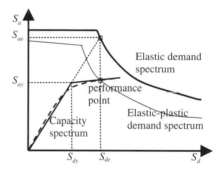

Figure 4. The calculating parameters of elasto-plastic demand spectrums.

$$\begin{cases} R_\mu = (\mu-1)\dfrac{T_{eq}}{T_g} + 1 & T \le T_g \\ R_\mu = \mu & T \ge T_g \end{cases} \quad (20)$$

Where, T_{eq}, the elastic period of the equivalent structure with a single degree of freedom, is calculated by the formula (21).

$$T_{eq} = 2\pi \sqrt{S_{dy}/S_{ay}} \quad (21)$$

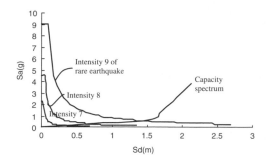

Figure 5. Example curve of elasto-plastic Push-down analysis.

The formula of transferring the elastic demand spectrum to the elasto-plastic demand spectrum is listed as follows.

$$\begin{cases} S_a = S_{ae} / R_\mu \\ S_d = S_{de} \mu / R_\mu \end{cases} \qquad (22)$$

2.6 *Calculation of the performance point*

When the curves of capacity spectrum and the certain intensity seismic demand spectrum are drawn in the same chart, the intersection point is named performance point. The corresponding displacement of performance point is the spectrum displacement of the equivalent structure with a single degree of freedom under the earthquake. This spectrum displacement can be converted into the displacement of control node of the structure by the formula (15). The distribution of plastic hinges, the vertical node deformations and the structure deflection can be obtained meanwhile. Then the seismic resistance evaluation of the structure can be achieved. If there is no intersection point, the analysis must be retried for satisfying seismic performance.

If the performance point is obtained, the structure can be evaluated in following:

(1) Whether the element stress of the structure meets the requirement of strength defined in the code.
(2) Whether the largest deflection of each span meets the requirement of the code. It is necessary to

satisfy the requirement of strength and stiffness specified in the code by the effects of earthquake.
(3) Whether the overall deformation or local deformation induces the structure into dynamic unstable situation which enables structure collapse.

Figure 5 is an example of using the Push-down method to evaluate the vertical seismic performance of a long-span structure.

3 CONCLUSION

In this paper, the fundamental principle of capacity spectrum and vertical demand spectrum is derived and established. The static elasto-plastic seismic evaluation method of long-span spatial structures, named Push-down method, is proposed and derived in detail. Then a calculating example is given. The validity and the feasibility of the method are verified. This new method can be simply used to evaluate the seismic performance of long-span spatial structures whose fundamental vibration mode is mainly vertical deformation.

REFERENCES

ShiZhao Shen. 1998. The development of long-span space structure. *China Civil Engineering Journal.*
MuWang Yang & YongFeng Luo. 2005. Research advances of antiseismic analysis methods about long-span steel structures. *Civil Engineer.*
FEMA273. 1997. NEHRP Guidelines for the seismic rehabilitation of buildings. Federal Emergency Management Agency. Washington, D. C.
ATC 40. 1996. Seismic evaluation and retrofit of concrete building. Applied Technology Council. California.
Xueye Xiong & ChunXiang Li. 2004. Nonlinear static procedure(pushover analysis) of large-span prestressed concrete structures, *Earthquake Engineering and Engineering Vibration.*24(1):101~107.
WeiXian Hu. 1988. Earthquake engineering. *Earthquake Publishing Company* (Beijing).
ShuWei Geng & XianXin. Tao 2004. The ratios of vertical to horizontal acceleration response spectra. *Earthquake Engineering and Engineering Vibration.*24(5):33~38.

Steel and Composite Structures – Wang & Choi (eds)
© 2007 Taylor & Francis Group, London, ISBN 978-0-415-45141-3

Seismic retrofit of r.c. frames with hysteretic bracing systems

D. Dubina, A. Stratan & S. Bordea

Politehnica University of Timisoara, Timisoara, Romania

ABSTRACT: Reinforced concrete structures built in seismic zones before 1960's were designed to resist mainly gravity loads and wind. The main deficiencies in reinforced concrete gravity frames are related to poor detailing and lack of capacity design, leading to reduced local and global ductility. At present, when such type of structures are subjected to structural evaluation, according to modern seismic design provisions, one finds out, that in almost all cases strengthening is needed. In present paper the strengthening of non-seismic r.c. frames with dissipative bracing system is examined. A detailed study case for a r.c. frame, designed according to provisions of 1950's and strengthened with steel Buckling Restrained Braces according to modern seismic provisions is presented.

1 INTRODUCTION

Romania is a country with a high seismic risk and before 1963, when first seismic design code was introduced, the r.c. (reinforced concrete) structures were designed only to resist gravitational loads mainly. Later, new codes were drafted (e.g. 1978, 1991, 2006) the last one being aligned with EN 1998-1. Practically almost all the buildings located in sever seismic zones, designed before 1970's must be evaluated and consolidated.

The aim of this study case is the structural seismic upgrade of a r.c. moment resisting frame (MRF)

designed for gravity loading only using steel buckling-restrained braces (BRB). The r.c. frame is shown in Figure 1.

Common materials used in the 1950's, as concrete B200 (corresponding to class C12/15 in Eurocode 2) and steel OB38 (with a characteristic yield strength of 235 N/mm^2) were considered used for the structure.

2 FRAME DESIGN

The gravity load design of the r.c. frame was done according to the old Romanian codes (original design). The building is located in Bucharest. The beam's effective width was considered only for sections at mid-span (Figure 1). The design strength was computed according to the modern code design. Detailing of reinforcement was characteristic for design practice used in Romania during 1950's:

 – poor anchorage length of bottom bars at the supports
 – inclined reinforcement used for shear force resistance
 – open stirrups, largely spaced (20–25 cm) in potential plastic zones.

An important remark regarding the structure is the existence on the outer frames of an infill masonry wall of 0.38 m thickness, and with a characteristic weight of 18 kN/m^3. In Table 1 are presented the loads and in Table 2 the combination of loads, both according to original design and modern code design. Live load was distributed in 3 ways: Live Load 1 (LL1) – distributed

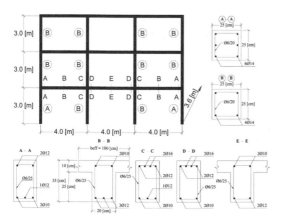

Figure 1.   Frame geometry and characteristic beam and column cross-sections.

Table 1. Type of loads.

Loads	Original design [KN/m²]	Modern code design [KN/m²]
Dead Load (DL)	3.08	3.08
Live Load* (LL) – roof level	3.00	2.00
Live Load (LL) – current level	1.50	1.50
Snow Load(SL)	1.00	1.60
Wind Load (WL)	0.70	0.52

Table 2. Load combinations.

Fundamental Combinations	Original design	Modern code design
1	1.3(DL + LL1 + SL)	1.35DL + 1.5LL1 + 1.05(SL + WL)
2	1.3(DL + LL2 + SL)	1.35DL + 1.5LL2 + 1.05(SL + WL)
3	1.3(DL + LL3 + SL)	1.35DL + 1.5LL3 + 1.05(SL + WL)
4	1.2 (DL + LL1 + SL + WL)	DL + LL1 + 0.7(WL + SL)

Table 3. Verification of beams.

Beam Sections	Internal actions symbol	Design strength	Internal actions Original code	Internal actions Modern code
A	M [KNm]	15.32	15.00	14.80
	Q [KN]	118.00	23.84	22.95
B	M [KNm]	23.00	22.83	21.47
C	M [KNm]	42.21	31.84	25.68
	Q [KN]	143.00	31.47	30.31
D	M [KNm]	42.36	27.58	26.02
	Q [KN]	144.30	28.40	26.02
E	M [KNm]	15.34	13.31	11.37

on all elements; Live Load 2 and 3 (LL2 and LL3) – as a chess distribution.

The frame geometry and the obtained cross sections are presented in Figure 1.

The results of design checking for beam and column sections under gravity loads only are summarised in Table 3 and Table 4.

Table 4. Verification of columns.

Beam Sections	Internal actions symbol	Design* strength	Internal actions Original code	Internal actions Modern code
A	M [KNm]	43	4.25	7.72
	N [KN]		399.35	390
B	M [KNm]	38	10.95	6.22
	N [KN]		371	368

*The design strengths were determined to gravity axial load of the each column element

3 STRENGTHENING SOLUTION

The different strengthening solutions were considered for seismic upgrade i.e. steel BRB's only; confinement of the ground and first floor columns using fiber reinforced polymers (FRP); and the combination of the previous two solutions.

The BRB's were introduced only in the middle span, as an inverted V braces, pinned at the ends. The design of the BRB's was accomplished according to Eurocode 3, following the procedure described in AISC 2005. Design seismic forces were obtained using spectral analysis with a reduction factor q equal to 6. BRB frames and eccentrically braced frames are expected to possess similar structural ductility and they are assigned same values of force reduction factor R in AISC 2005. Therefore behaviour factor q to be used for BRB system was considered equal to the one assigned by Eurocode 8 for eccentrically braced frames (q = 6).

The core of the buckling restrained brace was considered to be of rectangular shape. Cross-section areas of braces resulted from design are:

– braces of ground floor: A = 250 mm²
– braces of first floor: A = 230 mm²
– braces of second floor: A = 112 mm²

4 ANALYSIS PROCEDURE

Pushover analysis was applied in order to evaluate the differences between the original frame (MRF) and the retrofit ones. Displacement demand was estimated according to the N2 method implemented in Eurocode 8. Seismic action is characterised by the elastic response spectrum, shown in Figure 3 (peak ground acceleration $a_g = 0.24$ g, control period $T_c = 1.6$ s). Performance of the structure was evaluated in terms of inelastic deformation capacities corresponding to ultimate limit state. Development of plastic mechanism was also observed.

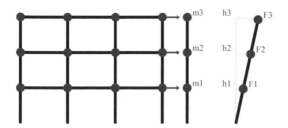

Figure 2.   Mass distribution.

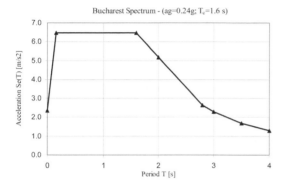

Figure 3. Elastic response spectrum for Bucharest (P100-1/2006).

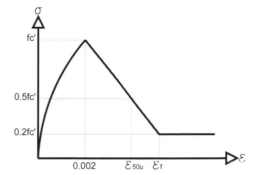

Figure 4.   Kent & Park unconfined concrete definition.

Table 5.   Equivalent yielding strength of the bars.

Element	Section	Diameter [mm]	L_{breq} [mm]	L_{bav} [mm]	$f_{y,eq}$ [N/mm²]
Beams	A	Φ12	505	225	104.70
		Φ10	421.2	225	125.53
	C	Φ10	421.2	250	139.48
		Φ12	505	250	116.34
	D	Φ10	421.2	250	139.48
Column	A, B	Φ14	589.7	560	223.16

The lateral forces for pushover analysis were considered of inverse triangular distribution (Figure 2), and were determined as in Equation 1 below:

$$F_i = \frac{m_i \cdot h_i}{\sum m_i \cdot h_i} \quad (1)$$

where h_i = the height of level i relative to the base of the frame and m_i = the mass at level i computed from the load combination DL + 0.4(LL1 + SL) and distributed in the main nodes.

Performance of structural elements at the ultimate limit state was defined in terms of:

– plastic rotations of beams and columns;
– axial plastic deformations for buckling restrained braces.

5   MODELLING FOR PUSHOVER ANALYSIS

5.1   Materials

Considering the inappropriate detailing of r.c. elements, concrete was taken as unconfined (FEMA356). The material model was considered according to Kent & Park, from Park & Paulay (1975) (Figure 4), as unconfined material with linear softening of the rigidity and no tension. Concrete compressive strength

was considered equal to $f'_c = 12.5\,\text{N/mm}^2$, while the ultimate strain $\varepsilon_f = 0.015$.

Due to the poor anchorage length of the bottom longitudinal reinforcement in the beams an equivalent yield strength of the steel has to be used (FEMA356) see Equation 2 below:

$$f_{y,eq} = f_y \times \frac{L_{b,av}}{L_{b,req}} \quad (2)$$

where, $f_{y,eq}$ = equivalent yield strength; f_y = steel reinforcement yield strength; $L_{b,av}$ = available anchorage length; $L_{b,req}$ = required anchorage length (according to Eurocode 2).

Table 5 presents the sections where insufficient anchorage was present, with the values of f_{yeq}. The reinforcing steel has the characteristic yield strength of 235 N/mm² and it was defined as a uniaxial bilinear material with strain hardening according to Eurocode 3.

5.2   Members modelling

5.2.1   Beams and columns

Unlike the beams, where original effective width was considered only for the sections along the span, a 72 cm

Table 6. Column stiffness reduction according to FEMA356.

Edge Columns	Floors	Stiffness reduction	Internal Columns	Floors	Stiffness reduction
	second	0.5		second	0.5
	first	0.5		first	0.525
	ground	0.7		ground	0.67

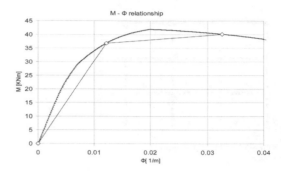

Figure 5. M-Φ relationship.

effective width, according to FEMA356, was considered. Four Φ 8 mm rebars at 18 cm spacing, have been considered for the slab in the effective width.

The effective stiffness of the members corresponding to cracked cross-section was reduced according to FEMA 356 as follows:

– beam flexural stiffness was reduced by 0.5;
– column flexural stiffness was reduced depending on level of axial force (Table 6).

For plastic analysis, beams and columns were modelled using concentrated plasticity at the ends, defined as rigid plastic bilinear moment-rotation relationship. The plastic hinge length (L_p) was computed according to Paulay & Priestley (1991) resulting $L_{p(column)} = 0.19$ m and $L_{p(beam)} = 0.21$ m. See for example Equation 3 below:

$$L_{pi} = 0.08 \times L_i + 0.022 \times d_i \times f_y \tag{3}$$

where, L_i = half of the span of the element, d_i = the diameter of the longitudinal reinforcement and f_y = characteristic strength for steel.

The bilinear idealization of moment curvature was obtained considering:

– the yield point occurred when rebars yield or concrete attains its compressive strength;
– ultimate curvature is calculated at the point when the materials reach their ultimate strains (e.g. 0.005 for concrete and 0.05 for steel);
– a 1% hardening applied to initial stiffness was considered (Figure 5).

The columns M-Φ relationships were obtained corresponding to the axial force from the gravitational loads in the earthquake combination.

5.2.2 Buckling Restrained Braces

Buckling restrained braces were considered pinned at the ends. Inelastic behaviour was modelled by concentrated plasticity. The material used for BRB was S235 grade steel and for a length of 3.6 m a yield displacement $\Delta_y = 4$ mm resulted. The ultimate displacement Δ_u was estimated based on experimental results presented in Newell (n.d.) tests. Based on these results, ductility ratios Δ_u/Δ_y were estimated for tension and

compression, amounting to 8.3 and 7.5 respectively. In order to obtain the adjustment of the design strengths (maximum compression strength C_{max} and maximum tension strength T_{max}) the AISC formulas were applied, see Equations 4 and 5 below:

$$T_{max} = \omega \cdot R_y \cdot f_y \cdot A \tag{4}$$

$$C_{max} = \omega \cdot \beta \cdot R_y \cdot f_y \cdot A \tag{5}$$

where, f_y is the yield strength; R_y is the ratio of the expected yield stress to the specified minimum yield stress f_y (considered equal to 1). Concerning the experimental values of the compression adjustment factor $\beta = 1.05$ and strain hardening adjustment factor $\omega = 1.25$ they were obtained in same manner as the coefficient Δ_u/Δ_y was found using AISC formulas, see Equation 6:

$$\beta = \frac{C_{max}}{T_{max}} \quad \text{and} \quad \omega = \frac{T_{max}}{f_{fysc} \cdot A} \tag{6}$$

where $f_{ysc} =$ is the measured yield strength of the steel core.

BRB member behaves according to a bilinear force-deformation relationship with hardening. In Figure 6 is presented BRB behaviour model for all 3 storeys.

5.2.3 Modelling of strengthening with FRP

In order to enhance ductility of reinforced concrete columns, their strengthening with FRP was considered. The fabric was applied in horizontal layers, its effect being confinement of concrete. The effect of confinement by FRP was determined according to FIB Bulletin 14/2001, and consisted in an increase of concrete compression strength (from 12.5 N/mm^2 to 40.8 N/mm^2) and ultimate strain (from 0.005 to 0.02). A more favourable behaviour of the confined columns is resulting (Figure 7). The design axial strength increases 3 times (from 987 kN to 2771 kN for column section A) and by about 20% for design moment resistance corresponding to an axial force of 389.6 kN in the seismic design situation (in the column section A).

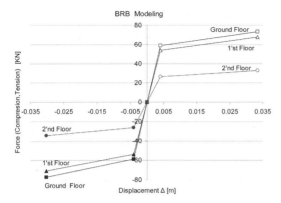

Figure 6. BRB behaviour model.

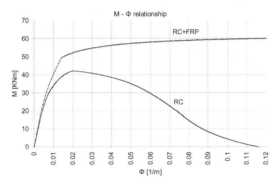

Figure 7. Effect of confinement by FRP on the moment-curvature relationship of column A (corresponding to an axial force of 389.6 KN from seismic combination).

6 PERFORMANCE ASSESSMENT

6.1 *Moment resisting frame (MRF)*

Analysis of the original MRF showed an unsatisfactory seismic response. First plastic hinge appears in the column. Plastic mechanism involves mostly columns from the first and second floors (Figure 8a), but also some beams from the first storey. Lateral drifts at the ultimate limit state also indicate concentration of damage in first two storeys (Figure 10). Ultimate rotations in plastic hinges corresponding to ultimate limit state are first reached in columns (Figure 9). It can be observed that the structure has a limited global ductility, because columns attain ultimate plastic deformations at a top displacement roughly four times smaller than the top displacement demand due to design earthquake action. Fundamental period of vibration and target displacements at the ultimate limit state for the original reinforced concrete frame and several alternative strengthening solutions are presented in Table 7.

Table 7. Fundamental period of vibration and target displacements for the considered structures.

Structure	Period T [s]	Target displacement d_t [m]
MRF + FRP + BRB	0.64	0.222
MRF + BRB	0.64	0.224
MRF + FRP	1.0	0.395
MRF	1.0	0.39

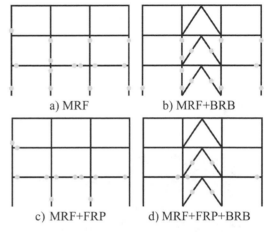

a) MRF b) MRF+BRB

c) MRF+FRP d) MRF+FRP+BRB

Figure 8. Plastic hinges with inelastic deformations larger than the ULS capacity at the target displacement.

6.2 *Strengthening with buckling restrained braces*

Strengthening by buckling-restrained braces increased considerably the strength and stiffness of the frame (Figure 9), decreasing by almost 50% the top displacement demand at the ultimate limit state. The first plastic hinges formed in column, followed by the ones in braces and beams. The plastic mechanism involved also the first two storeys (Figure 8b) and (Figure 10). This strengthening solution reduced the overall damage in the structure, as less plastic hinges formed in reinforced concrete elements at the target displacement (Figure 8b). However, seismic performance is still unsatisfactory, as inelastic deformations corresponding to ultimate limit state are recorded in columns, braces and beams before reaching the target displacement.

6.3 *Strengthening by fiber reinforced polymers*

As an alternative to strengthening by buckling restrained braces, the possibility to improve seismic performance by confining the columns with FRP was investigated. FRP was applied only for columns from the ground floor and first floor. The FRP fabric was

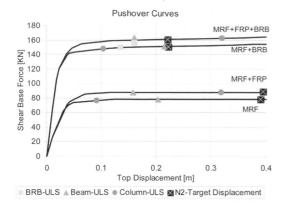

Figure 9. Pushover curves of the analysed frames.

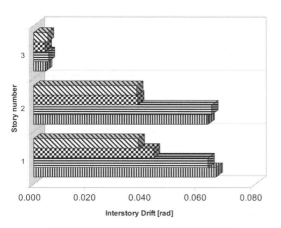

Figure 10. Interstorey drift demands at the target displacement.

considered applied in horizontal direction only, which ensures a confinement of concrete, but does not act supplement reinforcement. The effect of application of FRP was an increase of axial force capacity of the columns and ductility, but just a slight increase of bending moment capacity.

The overall structural response did not change significantly due to application of FRP (Figure 9) but, the ultimate deformation in columns (corresponding to ultimate limit state) was increased. Consequently the first plastic hinge forms in beam element and concerning the column the ultimate deformation is attained at larger top displacement demands than in the case of the original frame (Figure 8c and Figure 9). Also, the top displacement and interstorey drift demands at the ultimate limit state do not change significantly compared to initial frame.

6.4 BRB and FRP strengthening

Strengthening of the r.c. frame by means of BRB only did not eliminated failure of r.c. members. Therefore, a consolidation by both FRP and BRB systems was considered.

The main effect of the BRB system is improvement of global force-deformation characteristics (increased strength and stiffness), which results in decreased top displacement demands at the ultimate limit state (Figure 9). On the other hand, FRP technique enhances the local behaviour of columns by increasing their ductility, this being the reason of attaining ultimate deformation after the demand displacement. Also, it must be specified that the first plastic hinge in column elements is attained in the unconfined column from the second floor. Consequently, less damage is observed in columns (Figure 8d and Figure 9). Inelastic demands in beams and buckling restrained braces are still large. Ultimate plastic deformations in bracings and beams are attained at top displacements lower than the displacement demand at the ultimate limit state.

The structure was also studied as moderately dissipative using a behaviour factor q = 3 in order to design buckling restrained braces. Consequently the cross-section of BRB's was doubled, resulting a more rigid system for which the displacement demand was reduced with 35% with respect to ductile system with q = 6.

7 CONCLUSIONS

R.C. moment-resisting frames in severe, or even moderate seismic zones designed for gravity loads only in the past are in need for seismic rehabilitation in order to comply with modern seismic design requirements. Seismic upgrade of this type of structures using buckling restrained braces was investigated. The main effect of the dissipative bracing system is the improvement of overall strength and stiffness. However, application of the dissipative bracing system alone is not sufficient for an appropriate seismic performance. Additionally, the existing r.c. members should be strengthened. The most convenient solution seems to be application of FRP on beams and columns.

The analysis showed that seismic rehabilitation of nonseismic r.c. frames cannot be accomplished by means of very ductile dissipative bracing system without a proper strengthening of r.c members.

In present study only the columns have been confined (not strengthened!) with FRP. A better response capacity of BRB strengthened moment-resisting frame is expected if both r.c. columns and beams would be properly reinforced with FRP. In fact, if FRP reinforcement could be enough effective, the beams and columns will work mainly in elastic domain, while ductile steel BRB will be responsible for dissipative behaviour. Therefore a Performance Based Design approach could be applied on this line.

REFERENCES

AISC (2005). "Seismic Provisions for Structural Steel Buildings". American Institute of Steel Construction, Inc. Chicago, Illinois, USA.

CR 0-2005 (2006) *Cod de proiectare. Bazele proiectarii structurilor in constructii* (Romanian modern design code regarding the combination of loads)

CR 1-1-3-2005 (2006) *Cod de proiectare. Evaluarea actiunii zapezii asupra constructiilor* (Romanian modern design code regarding the snow load)

Eurocode 2 (December 2003) *Design of concrete structures – Part 1-1: General rules and rules for buildings* FINAL DRAFT prEN 1992-1-1. CEN – European Committee for Standardization

Eurocode 3 (2003). *Design of steel structures. Part 1-1: General Rules and Rules for Buildings.* CEN – European Committee for Standardization.

Eurocode 8 (January 2003) *Design of structures for earthquake resistance*, Part 1: General rules, seismic actions and rules for buildings, DRAFT No 6, Version for translation (Stage 49). CEN – European Committee for Standardization

Fajfar, P. (2000) A Nonlinear Analysis Method for Performance Based Seismic Design in Eurocode 8 *Annex B (Informative) Determination of the target displacement for nonlinear static (pushover) analysis.*

F. McKenna et al., (February 2005) *Open System for Earthquake Engineering Simulation User Manual*, OpenSees version 1.7.0

FEMA 356, (2000) *Prestandard and commentary for the seismic rehabilitation of buildings*, Federal Emergency Management Agency, Washington (DC).

FIB Bulletin 14 (2001) *Externally bonded FRP reinforcement for RC structures*

Newell, J.& Higgins, C. (n.d.) *Steel Confined Yielding Damper For Earthquake Resistant Design*, NHMJ Young Researchers Symposium June 21, 2003,http://cee.uiuc.edu/sstl/nhmj/ppt/Newell.ppt

NP-082-04 (2005) *Cod de proiectare. Bazele proiectarii si actiuni asupra constructiilor. Actiunea vantulu.* (Romanian modern design code regarding the wind load)

Park, R. & Paulay, T (1975) *Reinforced Concrete Structures*, New Zealand, John Wiley & Sons, Inc., New York.

Paulay, T. and Priestley, M.J.N., (1992) *Seismic Design of Reinforced Concrete and Masonry Buildings*, John Wiley & Sons, Inc., New York.

P100-1/2006 (2006). Cod de proiectare seismica – Partea I – Prevederi de proiectare pentru cladiri (Romanian modern design code regarding the seismic load).

Steel and Composite Structures – Wang & Choi (eds)
© 2007 Taylor & Francis Group, London, ISBN 978-0-415-45141-3

Practical design method for large span corridor structure between twin-tower tall building

J.J. Hou
South China University of Science and Technology, Guangdong Provincial Architectural Design &
Research Institute, Guangzhou, China

X.L. Han
South China University of Science and Technology, Guangzhou, China

B.S. Rong
Guangdong Provincial Architectural Design & Research Institute, Guangzhou, China

ABSTRACT: To investigate the vertical seismic response of space corridor built between twin-connected-tower tall buildings, firstly time history analysis is applied with the input of vertical seismic records, and the examples vary with towers' height and truss span. Secondly this type of structure is simplified as a 2-DOF system and analyzed in frequency domain, and the dynamic character of vertical seismic response is drawn out. The gravity coefficient method adopted by current China code for seismic design of building may not lead to safe result in the corridor being researched. A practical design formula considering interaction between towers and corridor is suggested. A project example of truss at the top of twin-tower super tall building is analyzed with method suggested and time history method, the results coincide well.

1 FORWARD

In recent years, more and more large span corridors are build between twin-tower or multi-tower tall buildings e.g. Petronas Towers in Kuala Lumpur, Malaysia and Grande Arche des La Defense in Paris, France. Usually truss is selected as the structure of the corridor. During the design, it is found that the vertical seismic response based on gravity coefficient method is possible smaller than the result based on time history analysis or shaking table experiment. The gravity coefficient method indicates the vertical seismic response of a truss equals to the gravity internal forces multiplies the maximum coefficient of vertical seismic spectrum. Since the concept of spectrum is based on system of single degree freedom (1-DOF), the gravity coefficient method is safe for large span truss supported on the ground or low story buildings only, which is directly excited by ground seismic motion. The practical vertical seismic design formula by Y.G Zhang leads to more precise result than the gravity coefficient method (Zhang, Y.G. 1985), but it is also based on ground motion excitation assumption. Unfortunately, under vertical seismic excitation, the large span truss built on twin-connected-tower tall buildings will interact with

the tower and work differently with ground-supported large span truss.

2 TIME DOMAIN ANALYSIS

This section will discuss seismic motion input, points in building finite element model and parameter analysis of large span truss seismic response.

2.1 *Input of seismic motion*

Researches show that although vertical seismic spectrum has similar shape with horizontal seismic spectrum of the same site, but they have different predominant period and attenuation principle. The vertical seismic spectrums vary with site condition, are relative with seismic intensity and function of period. In high and medium frequency range, the vertical to horizontal response spectra ratio is about 0.65, but in low frequency range, the ratio is close to 1.0. (Gao, Y.C. 2006., Geng, S. 2004.) The current China seismic design code of building doesn't provide vertical spectrum, thus, it is necessary to apply time history analysis with input of vertical seismic records to acquire accurate results.

Table 1. Statistic of 54 random selected seismic record.

Predominant period T_g (s)	Numbers
0.20 ~ 0.30	42
0.30 ~ 0.40	4
0.40 ~ 0.50	4
0.50 ~ 0.90	2
>0.95	2

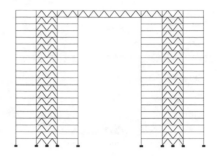

Figure 1. B4 model.

The site's vertical seismic predominant period is close to the vertical vibration period of 6~20 floors building, and the site's horizontal seismic predominant period is close to second or higher horizontal vibration periods of buildings(HU, Y.X. 2006.). The statistic of site predominant periods of 54 seismic records randomly selected from PEER's database is shown in Table 1.

Since most records fall in the 0.20~0.30s range of predominant period, in this paper the records among this range are selected as time history analysis input, and the mean results are used for further analysis. The acceleration of each record is scale to 8th Degree seismic of grade II site, that is $a_g = 455\,\text{mm/s}^2$. The gravity coefficient should be 10% according to China Building Seismic Design Code for this condition.

2.2 Analysis model of finite element model

The stiffness and equivalent mass of slabs and beams should reflect the character of the real structure. Shell element or plate element is recommended to simulate the slabs' out-of-plan deformation. The rigid slab assumption is not valid in vertical dynamic analysis.

The full mass matrix of beam is first choice. If lumped mass matrix is applied, at least one more joint should be added to the mid span of beam. To achieve 90% participation mass, the vertical vibration will consume vast computer resource due to numerous local vibration modes, even the effecian Ritz modes method is applied. But according statistic of 2 dimension frame examples in this paper, the first vibration mode will achieve about 70% participation mass. To guarantee the analysis result precise, 100 modes is calculated in each example.

2.3 Example design

Totally 11 examples are designed in line with China design code (Tables 1 & 2). They are large span steel truss built at the top of brace-frame twin-tower and classified in groups according to the variable parameter. The variables include tower floors and truss span, e.g. example B4 is illustrated in Figure 1, which tower is 20 floors high and truss span is 32 meters.

Table 2. Examples' characters and analysis result collection.

Character	ID of example	Tower floor	Truss span	Tower mass m_0(ton)	Truss mass m_1(ton)	Tower vertical frequency ω_0 (r/s)	Truss vertical frequency ω_1(r/s)	m_1/m_0	ω_1/ω_0	Acce' Amplify factor tower top to ground	Acce' Amplify factor truss mid span to ground	Acce' Amplify factor truss mid span to tower top
Single	A1	5	–	854	–	34.2	–	–	–	2.20	–	
Tower	A2	10	–	1830	–	32.6	–	–	–	2.25	–	
without	A3	15	–	2781	–	28.7	–	–	–	2.45	–	
truss	A4	20	–	3756	–	25.0	–	–	–	2.20	–	
Change	B1	5	32 m	1850	286	63.7	20.4	0.155	0.32	2.19	3.03	1.38
Tower	B2	10	32 m	3707	286	48.9	19.2	0.077	0.39	2.56	3.14	1.40
story	B3	15	32 m	5608	286	28.1	18.5	0.051	0.66	2.91	3.43	1.40
	B4	20	32 m	7560	286	24.6	17.9	0.038	0.73	2.58	3.85	1.75
	B5	1	32 m	0	286	–	17.9	0.038	0.73	2.58	3.85	1.75
Change	D1	20	10 m	7560	68	24.4	71.2	0.009	2.92	3.13	2.73	1.24
truss	D2	20	20 m	7560	177	27.2	24.1	0.023	0.89	2.74	4.25	1.93
span												

Table 3. Model material and member dimension

Member or material		Dimension or grade
Column	16–20 floor	500 × 500
Section	11–15 floor	650 × 650
(mm)	6–10 floor	750 × 750
	1–5 floor	850 × 850
Beam Section (mm)		300 × 700
Brace (mm)		Rec 350 × 350 × 20
Truss Section (mm)	Upper chord	Rec 350 × 350 × 20
	Lower chord	Rec 350 × 350 × 20
	Web member	Rec 300 × 300 × 20
Concrete		C35
Steel		Q235

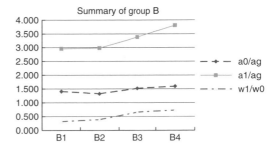

Figure 3. Analysis result of group B.

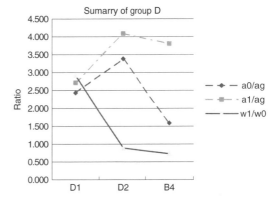

Figure 4. Analysis result of D group.

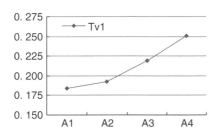

Figure 2a. Basic vibration periods of A group.

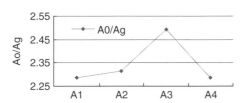

Figure 2b. Acceleration amplitude of group A.

2.4 *Analysis result*

2.4.1 *Example group A: Changing single tower stories*

The vertical vibration period and the tower floors increase synchronously. And the acceleration amplitude of tower top reaches peak value when the vertical period is close to site predominant period, this phenomenant may due to resonance occurs. (Figures 2a & 2b)

2.4.2 *Example group B: changing twin-tower floor*

In Figure 3, a_0/a_g is acceleration amplitude of tower top to ground, a_1/a_g is acceleration amplitude of truss mid span to ground, and w_1/w_0 is basic mode's frequency of truss to tower.

2.4.3 *D group: changing span of truss*

D1, D2 and B4 form a group of examples with fixed tower stories and various truss spans. This group can draw similar principle with group B, only difference is the frequencies range is larger.

Based on time domain FEM analysis, the basic principle can be drawn out:

1) Acceleration amplitude of tower with truss is smaller than same height tower without truss (Figure 2b & Figure 3);
2) The truss acceleration amplitude increases synchoronously with height of tower (Figure 3);
3) When frequency ratio between truss to tower is close to 1.0, the amplitude reach peak value both for tower and truss (Figure 4).

3 FREQUENCY DOMAIN ANALYSIS

3.1 *Simplified analysis model*

According to time history analysis, the sum of participation mass ratios of tower and truss's basic vibration modes is greater than 70%. This type of structure can be simplified into a 2 DOF system, and the 1st DOF is twin-tower and 2nd is truss (Figure 5a & b).

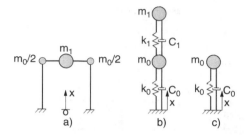

Figure 5. Simplified analysis model.

The dynamic equation is

$$\begin{bmatrix} m_0 & 0 \\ 0 & m_1 \end{bmatrix}\begin{Bmatrix} \ddot{x}_0 \\ \ddot{x}_1 \end{Bmatrix} + \begin{bmatrix} c_0+c_1 & -c_1 \\ -c_1 & c_1 \end{bmatrix}\begin{Bmatrix} \dot{x}_0 \\ \dot{x}_1 \end{Bmatrix} + \begin{bmatrix} k_0+k_1 & -k_1 \\ -k_1 & k_1 \end{bmatrix}\begin{Bmatrix} x_0 \\ x_1 \end{Bmatrix} = -\begin{bmatrix} m_0 & 0 \\ 0 & m_1 \end{bmatrix}\begin{Bmatrix} \ddot{x}_g \\ 0 \end{Bmatrix} \quad (1)$$

where $m_0 = $ the equivalent mass of towers; $m_1 = $ the equivalent mass of large span truss;

$k_0 = $ the equivalent axial compressive stiffness of towers; $k_1 = $ equivalent bending-shearing stiffness of large span truss respectively;

$c_0 = $ damper coefficients of towers; $c_1 = $ damper coefficients of truss;

$x_0 = $ the vertical displacement of tope of tower; $x_1 = $ the vertical displacement at mid-span of truss relative to the ground;

$x_g = $ ground vertical seismic motion acceleration.

3.2 Transfer function of towers and truss

Assumed the following parameters:

$$\omega_0^2 = k_0/m_0, \quad \omega_1^2 = k_1/m_1, \quad \xi_0 = c_0/2\omega_0 m_0,$$

$$\xi_1 = c_1/2\omega_1 m_1, \quad \lambda = m_1/m_0.$$

The equation (1) becomes:

$$\begin{cases} \ddot{x}_0 + 2\xi_0\omega_0\dot{x}_0 + 2\xi_1\omega_1\lambda\dot{x}_1 - 2\xi_1\omega_1\lambda\dot{x}_1 + \left(\omega_0^2 + \lambda\omega_1^2\right)x_1 - \lambda\omega_1^2 x_1 = -\ddot{x}_g \\ \ddot{x}_1 + 2\xi_1\omega_1\left(\dot{x}_1 - \dot{x}_0\right) + \omega_1^2\left(x_1 - x_0\right) = 0 \end{cases} \quad (2)$$

Apply Fourier transform to equation 2 (Zhuang, B.Z. et al. 1995):

$$[A]\begin{Bmatrix} X_0(\omega) \\ X_1(\omega) \end{Bmatrix} = \begin{Bmatrix} 1 \\ 1 \end{Bmatrix}X_g(\omega) \quad (3)$$

where, $X_0(\omega)$ is the Fourier transform of x_0; $X_1(\omega)$ is the Fourier transform of x_1. $X_g(\omega)$ is the Fourier transform of $\ddot{x}_g$ calculated as Equation (4);

$$[A] = \begin{bmatrix} -\omega^2 + i(D_0 + D_1\lambda)\omega + \omega_0^2 + \lambda\omega_1^2 & -i\omega D_1\lambda - \lambda\omega_1^2 \\ -i\omega D_1 - \omega_1^2 & -\omega^2 + i\omega D_1 + \omega_1^2 \end{bmatrix}$$

$$X_g(\omega) = -\int_{-\infty}^{\infty}\ddot{x}_g e^{-i\omega t}dt \quad (4)$$

$$\begin{Bmatrix} X_0(\omega)/X_g(\omega) \\ X_1(\omega)/X_g(\omega) \end{Bmatrix} = [A]^{-1}\begin{Bmatrix} 1 \\ 1 \end{Bmatrix} \quad (5)$$

The frequency domain transfer function describes the transfer relationship between system input and output variables in frequency domain, thus:

$$\{H(i\omega)\} = \begin{Bmatrix} H_{x_0}(i\omega) \\ H_{x_1}(i\omega) \end{Bmatrix} = [A]^{-1}\begin{Bmatrix} 1 \\ 1 \end{Bmatrix}$$

where, $H_{x_0}(i\omega) = $ the transfer function of x_0; $H_{x_1}(i\omega) = $ the transfer function of x_1.

Assuming $D_0 = 2\xi\omega_0$; $D_1 = 2\xi\omega_1$, then

$$\begin{cases} H_{x_0}(i\omega) = \dfrac{-\omega^2 + i\omega(D_1 + D_1\lambda) + \left(\omega_1^2 + \lambda\omega_1^2\right)}{Det(A)} \\ H_{x_1}(i\omega) = \dfrac{-\omega^2 + i\omega(D_1 + D_0 + D_1\lambda) + \left(\omega_1^2 + \omega_0 + \lambda\omega_1^2\right)}{Det(A)} \end{cases} \quad (6)$$

in which: $Det(A)$ is the determinant of matrix A,

$$Det(A) = \omega^4 - i\omega^3(D_1 + D_0 + D_1\lambda) - \omega^2(\omega_1^2 + \lambda\omega_1^2 + \omega_0^2 + D)$$
$$+ i\omega(\omega_1^2 D + D_1\omega_0^2) + \omega_0^2\omega_1^2$$

Seismic action $\ddot{x}_g$ is random load. If the ground motion acceleration is assumed to be white noise stochastic procedure with zero mean value, the acceleration's spectrum intensity $S_g(\omega)$ is:

$$S_g(\omega) = S_0 \quad (7)$$

According to random vibration process theory, the structure displacement mean square value is:

$$E[x^2(t)] = R_x(0) = \int_{-\infty}^{+\infty}S_x(\omega)d\omega$$

$$= \int_{-\infty}^{+\infty}|H(i\omega)|^2 S_g(\omega)d\omega = \int_{-\infty}^{+\infty}|H(i\omega)|^2 S_0 d\omega \quad (8)$$

in the above equation, $R_x(\tau)$ is the auto-correlation function of structure displacement $x(t)$, $S_x(\omega)$ is the spectrum density of structure displacement. S_0 is the spectrum density of ground motion acceleration, it is constant in frequency domain without relationship with frequency.

The transfer function $H(i\omega)$ in equation (8) is,

$$H(i\omega) = \frac{-i\omega^3 B_3 - \omega^2 B_2 + i\omega B_1 + B_0}{\omega^4 A_4 - i\omega^3 A_3 - \omega^2 A_2 + i\omega^3 A_1 + A_0}$$

$$\int_{-\infty}^{+\infty}|H(i\omega)|^2 d\omega = \pi \cdot HD \quad (9)$$

$$\left[\frac{B_0^2}{A_0}(A_2 A_3 - A_1 A_4) + \right.$$

$$A_3(B_1^2 - 2B_0 B_2) + A_1(B_2^2 - 2B_1 B_3)$$

$$\left. + \frac{B_3^2}{A_4}(A_1 A_2 - A_0 A_3) \right]$$

in which, $\quad HD = \dfrac{}{A_1(A_2 A_3 - A_1 A_4) - A_0 A_3^2}$

Thus the mean square value of tower's displacement $E[x_0^2]$ and the mean square value of truss's displacement $E[x_1^2]$ are:

$$\begin{cases} E[x_0^2] = S_0 \int_{-\infty}^{+\infty} |H_{x_0}(i\omega)|^2 \, d\omega \\ E[x_1^2] = S_0 \int_{-\infty}^{+\infty} |H_{x_1}(i\omega)|^2 \, d\omega \end{cases} \quad (10)$$

Substituting all factors in Eq.(6) of $H_{x_0}(i\omega)$ and $H_{x_1}(i\omega)$ who are relative to Eq.(8) into Eq.(9), the mean square value of structure displacement in Eq.(10) can be obtained.

Since the mean square values of Eq.(10) are relative with S_0, to change them into dimensionless format, divide them with $E[x^2]$, which is the mean square displacement of 1-DOF system excited with zero mean value white noise:

thus, we obtain:

$$E[x^2] = \frac{\pi S_0}{2\xi_0 \omega_0^3} \quad (11)$$

$$\begin{cases} \overline{E[x_0^2]} = \dfrac{E[x_0^2]}{E[x^2]} = \dfrac{2\xi_0 \omega_0^3}{\pi} \int_{-\infty}^{+\infty} |H_{x_0}(i\omega)|^2 \, d\omega \\ \overline{E[x_1^2]} = \dfrac{E[x_1^2]}{E[x^2]} = \dfrac{2\xi_0 \omega_0^3}{\pi} \int_{-\infty}^{+\infty} |H_{x_1}(i\omega)|^2 \, d\omega \end{cases} \quad (12)$$

Changing different parameters in Eq.(12) as variables, different curves to reflect relationship between dimensionless structure's mean square responses $E[x_0^2]$, $E[x_1^2]$ and each parameter variables can be drawn out. These parameter variables include vibration frequencies ratio between truss and tower $\mu = \omega_1/\omega_0$, masses ratio between truss and tower $\lambda = m_1/m_0$, tower damper ratio ξ_0 and truss damper ratio ξ_1.

In this paper, the visualization function of Matlab is applied to draw the relative curves between μ, λ, ξ and $\sqrt{E[x_0^2]}, \sqrt{E[x_1^2]}$. From these curves we can analyze the influence of each parameter.

If change ξ_0, ω_0 with ξ_1, ω_1 in Eq.(11) and substitute them into Eq.(12), we can obtain the amplified factor of seismic response in truss under different support conditions:

$$E[x^2] = \frac{\pi S_0}{2\xi_1 \omega_1^3} \quad (13)$$

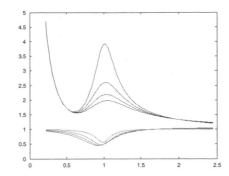

Figure 6. Relevant curves of mean square values of displacement response and frequency ratioswhen mass ratio λ is small (The upper group curve represent truss, and the lower group curve represent tower).

$$\sqrt{\overline{E[x_1^2]}} = \sqrt{\frac{E[x_1^2]}{E[x^2]}} = \sqrt{\frac{2\xi_0 \omega_0^3}{\pi} \int_{-\infty}^{+\infty} |H_{x_1}(i\omega)|^2 \, d\omega} \quad (14)$$

3.3 Mean square response of truss

When the mass ratio is small, the peak value of truss response occurs at frequency ratio $\mu = 0.98$, and the range of frequency of peak value is narrow; when the mass ratio is bigger than 10%, the peak displacement response of truss occur at smaller frequency ratio. The dimensionless mean square deviation truss displacement is always bigger than 1, that means the vibration of truss are always amplified (Figure 6).

4 PRACTICAL DESIGN METHOD

4.1 Gravity ratio method and old practical design method

The seismic code suggests at the site of seismic degree 8, 8.5 and 9, the standard value of vertical seismic is 10%, 15% and 20% of the gravity load's standard value. These percentage are exactly the maximum influence factor of horizontal seismic spectrum when damping ratio is 5% multiply 0.65, and 0.65 is the ratio between vertical and horizontal seismic acceleration spectrum peak value.

The large span truss practical design formula under vertical seismic suggested (Zhang, Y.G. 1985) is:

$$F_i^e = C \cdot (\mu_{min} + R \cdot l_i) \cdot F_i^s \quad (i=1,2\ldots n) \quad (15)$$

in which, F_i^e is the internal force of members under vertical seismic; F_i^s is the member force under gravity load;

C-seismic intensity factor, the values are 0.5, 1, 2 for seismic degree 7,8,9 respectively;

Table 4. Recommend value of R, μ_{min} in Eq.(15)

Site seismic grade	μ_{min}(%)	R(/m) Chord	R(/m) Web
I	3.0	0.15	1.4
II	4.0	0.30	1.5
III	5.6	0.40	1.7
IV	7.0	0.60	1.9

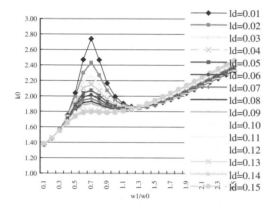

Figure 7. Acceleration Amplitude with different λ and μ (ld is mass ratio, $\xi_0 = 0.05$ 和 $\xi_1 = 0.02$).

l_i-the horizontal distance from mid point of ith member to the support, the unit is meter;
μ_i is provided in Chart.4.

4.2 The practical design method by this paper

A new practical design formula is suggested as follows, which include the tower and truss interaction factor k_0:

$$F_i^e = C \cdot k_0 \cdot \left(\mu_{min} + R \cdot l_i\right) \cdot F_i^s \ (i=1,2...n) \qquad (16)$$

The factor k_0 can be calculated with Eq.(14). For the convenience of application, k_0 is list in Figure 7,

in which λ and μ have the same meaning with that in Section 3.1. The rest variables are the same as Eq.(15).

5 CONCLUSION

For the response of large span truss at the top of tall buildings under vertical seismic excitation, the following conclusions are drawn:

1) The gravity coefficient method and the practical method by Zhang do not consider the interaction working of tower and upper truss, will probably lead to unsafe result;
2) With the increase of frequency ratio between truss and tower, the amplify factor of response of truss is bigger;
3) When the mass ratio between truss and tower is close to 1, the response of truss is amplified further due to resonance;
4) The formula suggested in this paper can calculated the truss internal force convenience and precisely enough for engineering design.

ACKNOWLEDGEMENT

My deepest gratitude goes to the sponsoring from Guangzhou Government Construction Techniques Development Fund.

The section of Project example, Discussion on some problems and Reference are omitted due to limitation of size, welcome contact authors for details.

Steel and Composite Structures – Wang & Choi (eds)
© *2007 Taylor & Francis Group, London, ISBN 978-0-415-45141-3*

Bolted links for eccentrically braced frames: Influence of link stiffness

A. Stratan, A. Dogariu & D. Dubina
Politehnica University of Timisoara, Timisoara, Romania

ABSTRACT: Eccentrically braced steel frames represent a suitable solution for multi-storey buildings located in seismic areas. A bolted connection between the dissipative element (link) and the beam is suggested to facilitate replacement of damaged dissipative elements (links) after a moderate to strong earthquake, which reduces repair costs. Influence of structural configuration (homogeneous/dual) and flexibility of bolted links on seismic performance of eccentrically braced frames incorporating removable links is investigated.

1 INTRODUCTION

Design of multi-storey structures in high-seismicity areas is usually based on dissipative structural response, which accepts significant structural damage under the design earthquake. However, capacity-based design, applied in modern seismic design codes promotes plastic deformations in predefined areas only, called dissipative zones. A bolted connection between dissipative zone and the rest of the structure would allow replacement of the dissipative elements damaged as a result of a moderate to strong earthquake, reducing the repair costs.

Application of this principle to eccentrically braced frames, where link elements serve as dissipative zones, is presented in Figure 1. The connection of the link to the beam is realized by a flush end-plate and high-strength bolts.

Extended end-plate bolted connections for eccentrically braced frames with link-column connection configuration were previously investigated experimentally by Ghobarah & Ramadan 1994. Their inelastic performance was found to be similar to fully-welded connections. Recently, Mansour et al. (2006) investigated replaceable shear links composed of two bolted back-to-back channels.

2 ANALYSIS PROCEDURE

2.1 Homogeneous/dual structural configuration

Eccentrically braced frames represent a convenient lateral-force resisting system as it offers a relatively high lateral stiffness and excellent ductility. When designing an eccentrically braced frame, it may be convenient to use pinned connections for beams not containing links (see Figure 2). This approach has the advantage of both simple and economic design, as pined connections are preferred due to lower fabrication and erection costs in comparison with moment-resisting connections.

Alternatively, dual structural configurations can be used. In this structural system strength and stiffness under lateral forces is provided by both eccentrically braced and moment-resisting frames (see Figure 2). It has several advantages over "pure" eccentrically braced frames in the context of seismic-resistant design. The first one consist in the fact that the

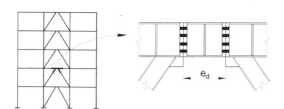

Figure 1. Bolted link concept.

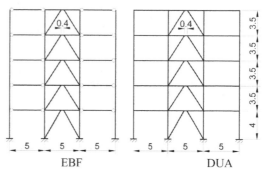

Figure 2. Eccentrically braced (EBF) and dual (DUA) frames.

secondary lateral-force resisting system (moment-resisting frames) may provide an alternative load path in the event of failure of the primary system lateral-force resisting system (eccentrically braced frames).

A second advantage of dual configurations is related to permanent deformations experienced after a seismic motion that drives the structure in the inelastic range of response. It is especially desirable to keep permanent deformations at minimum in case of bolted links, in order to make possible their easy removal and replacement. In dual configurations, moment-resisting frames are expected to experience limited yielding, providing the elastic restoring force necessary to reduce permanent lateral displacements.

In order to assess the influence of dual structural configuration on seismic performance of eccentrically braced frames, both homogeneous and dual configurations were considered in this study. In a first phase the EBF and DUA structures were analysed considering conventional links, obtained as part of the middle span beams, neglecting the influence of the link-to-beam connection on frame response. In a second phase, the influence of bolted link-to-beam connection on the seismic performance of the dual frame was evaluated.

2.2 Design of structures

The design was carried out for the dual structure (DUA), according to Eurocode 3 (1993) and Eurocode 8 (1994). A $4.75 \, \text{kN/m}^2$ dead load on the floors, $1.70 \, \text{kN/m}^2$ for exterior cladding and $3.0 \, \text{kN/m}^2$ live load were considered. Seismic design parameters were: $0.35 \, \text{g}$ peak ground acceleration, stiff soil conditions (class A), behaviour factor $q = 5.5$, and interstorey drift limitation of 0.006 of the storey height.

A short link whose behaviour is governed by shear was considered. Design of braces and beams segments outside links in the middle span was governed by capacity design requirements. Critical design forces on columns and beams in the outer spans were recorded under gravitational loading. Dimensions of structural elements were the same for both dual and homogeneous structural configurations. The following cross-section dimensions of structural elements were obtained: HEB260 grade S355 columns, IPE330 grade S355 beams in the outer bays, IPE240 grade S235 links and beams in the middle bay, and rectangular hollow sections $120 \times 120 \times (7.1 - 12.5)$ grade S235 braces. Fundamental period of vibration was $0.58 \, \text{s}$ for DUA and $0.64 \, \text{s}$ for EBF.

2.3 Seismic action

Seismic performance of the frames was analysed using pushover and non-linear time-history analysis. Two sets of ground motions each containing seven records were used. Both sets of records are representative for Vrancea seismic source in Romania, but different soils

conditions: stiff soil (corresponding to control period $T_C = 0.5 \, \text{s}$) and soft soil ($T_C = 1.4 \, \text{s}$). The two target spectra (for stiff and soft soils) were scaled to match the same spectral acceleration in the 0.58–0.64 second period range as the EC8 spectrum used in design. As a result, peak ground accelerations of $0.28 \, \text{g}$ and $0.23 \, \text{g}$ resulted for the $T_C = 0.5$ and $T_C = 1.4$ sets of records, respectively, corresponding to $0.35 \, \text{g}$ peak ground acceleration for the EC8 type A spectrum used in the design.

This procedure assures approximately the same design earthquake forces as the ones used in the initial design using the EC8 design spectrum. While the $T_C = 0.5$ set of records consisted in recorded accelerograms only, the $T_C = 1.4$ set comprised both recorded and semi-artificial accelerograms.

2.4 Performance levels

Three performance levels were considered: serviceability limit state (SLS), ultimate limit state (ULS), and collapse prevention (CP) limit state.

Intensity of earthquake action at the ultimate limit state was considered as corresponding to the design one. It was assigned a ground motion intensity factor $\lambda = 1.0$. Ground motion intensity at the serviceability limit state was obtained by scaling the ground motion by $\lambda = 0.5$ (corresponding to $\nu = 0.5$ in Eurocode 8), while the one for the collapse prevention limit state was considered for $\lambda = 1.5$ (FEMA 356).

Links in eccentrically braced frames have an excellent ductility (Kasai & Popov, 1986). Accepted plastic shear deformations range from 0.08 rad (AISC 2002) and 0.11 rad (FEMA 356, 2000). Ultimate link deformations of $\gamma_u = 0.1$ rad were used in this research. Ultimate plastic rotation capacities of flexural members (beams, columns, connections) were considered equal to $\theta_u = 0.03$ rad. Criterion for attainment of the ultimate limit state (ULS) was considered attainment of ultimate plastic deformations (γ_u, θ_u) in elements.

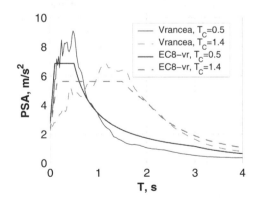

Figure 3. Target pseudo-acceleration spectra and average spectra of the two sets of accelerograms.

Performance level corresponding to serviceability limit state was considered as the attainment of 0.006 h interstorey drift, while the one for the collapse prevention – dynamic instability. Dynamic instability was defined as significant increase in lateral displacements at a small increase in intensity and was determined according to the FEMA 350, 2000 procedure.

3 SEISMIC PERFORMANCE: DUAL VS. HOMOGENEOUS CONFIGURATION

3.1 Element modelling

A series of pushover and time-history analyses were carried out using the Drain-3dx computer program. The inelastic shear link element model was based on the one proposed by Ricles & Popov (1994). As the original model consisted in four linear branches, it was adapted to the trilinear envelope curve available in Drain-3dx (see Figure 4).

Beams, columns and braces were modelled with fibre hinge beam-column elements, with plastic hinges located at the element ends. Nominal steel characteristics were used.

3.2 Pushover analysis

A pushover analysis was performed first (Figure 5), inverted triangular load distributions, while

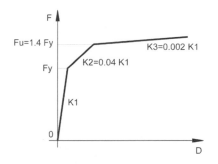

Figure 4. Shear force – deformation model for link element.

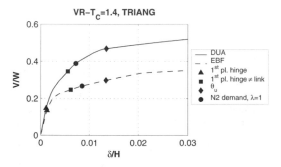

Figure 5. Normalised base shear force versus normalised top displacement for the DUA and EBF structures.

displacement demands were evaluated by the N2 method (Fajfar, 2000).

The EBF structure has a slightly lower stiffness than the dual one. However, both structures show similar base shears at the first plastic hinge, which indicate that design is governed by the properties of the eccentrically braced frames. The difference between the two structures is more evident in the post-elastic range, the DUA structure being characterised by larger overstrength. Displacement demands obtained using the N2 method for an earthquake intensity corresponding to the ultimate limit state ($\lambda = 1.0$) are larger for the EBF structure, especially in the case of the $T_C = 1.4$ spectrum.

3.3 Time-history analysis

Time history analyses were then performed, under increasing ground motion intensities (λ factor). Results presented in the following represent the average of the seven records in each group of accelerograms.

Interstorey drifts at SLS are less than the performance objective (0.006 rad), ranging between 0.0023 and 0.003 rad. Only links yielded at SLS.

Performance at the ULS ($\lambda = 1$) was satisfactory for both structures, plastic deformation in elements being lower than ultimate ones. Maximum link shear deformations are presented in Table 1. Inelastic deformation demands in elements other than links were minor. With the exception of the EBF structure subjected to the $T_C = 1.4$ set of ground motions, maximum interstorey drifts were below the 0.006 rad limit, indicating limitations of non-structural degradation even at the ULS. Ground motions from the $T_C = 1.4$ set generated higher transient and permanent lateral drifts and element deformation demands (see Table 1). Effect of structural configuration was minor in the case of $T_C = 0.5$ accelerogram set, but was important for $T_C = 1.4$, dual configuration reducing both top displacements and interstorey drifts. Distribution of interstorey drifts for $\lambda = 1$ is presented in Figure 6. Though in both structural configurations displacement demands concentrate in lower storeys, dual structure is characterised by more uniform distribution over the height of the building. The same trend is observed from local link deformation demands.

Accelerogram multiplier λ_u at the attainment of ULS criteria (generally governed by ultimate link deformations) was equal to 4.55 and 4.47 for the DUA and EBF structures under the $T_C = 0.5$ set of ground motions, and 2.15 and 1.72 for the DUA and EBF structures under the $T_C = 1.4$ set of ground motions. Safety level is high ($\lambda_u > 1$) in the case of $T_C = 0.5$, but decreases for the $T_C = 1.4$ set of accelerograms. Dual structural configuration improves structural performance, but this effect is important only for the $T_C = 1.4$ set of accelerograms.

Table 1. Average values of link deformations (γ_{link}), maximum (IDR_{max}) and permanent (IDR_{per}) interstorey drift demands for $\lambda = 1$.

Structure	γ_{link}, rad		IDR_{max}, rad		IDR_{per}, rad	
	$T_C = 0.5$	$T_C = 1.4$	$T_C = 0.5$	$T_C = 1.4$	$T_C = 0.5$	$T_C = 1.4$
DUA	0.020	0.024	0.0050	0.0059	0.0005	0.0011
EBF	0.018	0.034	0.0056	0.0091	0.0002	0.0032

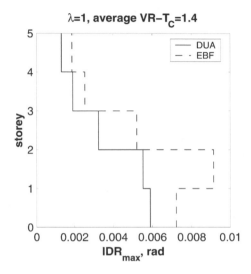

Figure 6. Distribution of interstorey drift demands IDR_{max} for the DUA and EBF structures $\lambda = 1$, $T_C = 1.4$ set of ground motions.

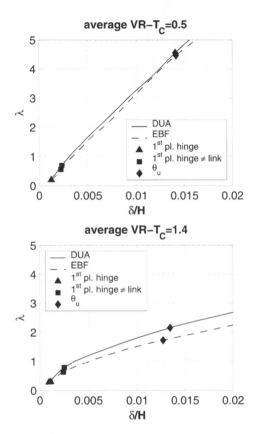

Figure 7. Incremental dynamic analysis curves for the DUA and EBF structures.

Dynamic instability was not detected for accelerogram multipliers less than $\lambda = 7$ for the $T_C = 0.5$ accelerogram set. In the case of $T_C = 1.4$ accelerogram set there is higher probability of dynamic instability, most of λ factors ranging between 1.2 and 5. Dual structural configuration has an enhanced performance from this point of view.

A comparison of top displacement demands in the form of incremental dynamic analysis (IDA) curves is presented in Figure 7. The dual configuration has two effects on the structural response. The first one is related to reduction of top displacements due to higher hardening of the global force-displacement curve. This effect is important for $T_C = 1.4$ accelerogram set only, for which fundamental period of vibration of the analysed structures is located in the constant acceleration region of the response spectrum ($T < T_C$). Due to larger post-yield stiffness of the global force-displacement curve, response of the dual structure is "closer" to the elastic one, resulting in lower displacement demands in comparison with EBF structure.

The second effect is the reduction of interstorey drift demands by the dual configuration which is mainly related to a more uniform distribution of interstorey displacements over the height of the structure. This effect is independent of the ground motion characteristics.

In addition to maximum transient drifts, the dual configuration is efficient in reducing permanent displacements of the structure, but only for displacement demands that do not cause yielding in structural elements outside links (beams, braces and columns). It is to be noted that inelastic deformations in beams, columns and braces were minor for $\lambda = 1.0$–1.5 in the case of the DUA frame.

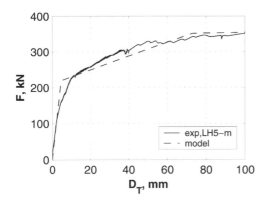

Figure 8. Shear force – deformation relationship for the bolted link: experiment vs. model.

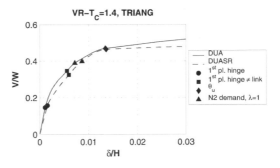

Figure 9. Normalised base shear force versus normalised top displacement for the DUA and DUASR structures.

4 SEISMIC PERFORMANCE: CONVENTIONAL VS. BOLTED LINKS

4.1 Bolted link model

In order to validate performance of bolted links, an experimental program was carried out. A detailed description and outcomes of the experimental investigation on removable bolted links is available elsewhere (Stratan & Dubina, 2004). It was shown that the bolted connection between the removable link and the beam is both partial strength and semi-rigid. While the effects of reduced moment resistance of the connection can be eliminated by using shorter links, influence of connection flexibility on the response of the removable link need be considered in global analysis of eccentrically braced frames.

Three main sources of deformation affecting the shear force – deformation relationship of bolted links were identified. These components are shear deformation of the link, rotation of the flush end-plate connection and slip in the connection. Independent implementation of all three components in a global model of the structure seems cumbersome. An alternative approach is to determine an equivalent shear force – deformation relationship, accounting for additional deformation due to bolted connection. The latter approach was adopted here.

A trilinear model of the link shear force – deformation relationship similar to the one shown in Figure 4 was adopted for the bolted link. The model was fit to the results of an experimental specimen (LH5-m) with the same characteristics as the ones of the links used in the numerical study. Initial stiffness K_1 of the bolted link model was considered equal to the experimental one, determined using a linear regression of the force-deformation relationship up to a force equal to 30% of the maximum one. Post-yielding stiffness values K_2 and K_3 were determined similarly to the ones in

Figure 4 ($K_2 = 0.04K_1$, $K_3 = 0.002K_1$). The yield force F_y was considered equal to the experimental one, while the F_u was assigned a value of $1.6F_y$ in order to fit to experimental data.

The most important difference between characteristics of the bolted and conventional links is the initial stiffness, the former being of the order of 15–20% of the latter. In order to investigate the influence of bolted link characteristics on the global frame response, a new frame was analysed. It was denoted by DUASR and was obtained by using the model of the bolted link described above in the DUA frame. In the final bolted link model used in global frame analysis, the nominal yield force was used, in order to allow comparison with the conventional link.

4.2 Pushover analysis

The pushover curves of the DUA and DUASR structures under an inverted triangular lateral load are shown in Figure 9. The effect of reduced stiffness of bolted links is reduction of initial global stiffness. However, neither base shear at first yield, nor the global strength of the structure is modified significantly in comparison with the DUA structure.

4.3 Time-history analysis

Consideration of bolted link characteristics did not affect significantly performance at the SLS. In the case of the $T_C = 0.5$ set of ground motions, interstorey drifts increased from 0.0027 rad (DUA structure) to 0.0034 rad (DUASR structure). As in the case of the DUA structure, only links yielded at the SLS for the DUASR structure.

At the ULS ($\lambda = 1$) the DUASR structure experienced slightly higher inelastic deformation demands in dissipative elements – links (see Table 2), but performance was satisfactory. Maximum interstorey drifts in DUASR structure were higher in comparison with DUA structure at this level of ground motion intensity, indicating higher damage in non-structural

Table 2. Average values of link deformations (γ_{link}), maximum (IDR_{max}) and permanent (IDR_{per}) interstorey drift demands for $\lambda = 1$.

Structure	γ_{link}, rad		IDR_{max}, rad		IDR_{per}, rad	
	$T_C = 0.5$	$T_C = 1.4$	$T_C = 0.5$	$T_C = 1.4$	$T_C = 0.5$	$T_C = 1.4$
DUA	0.020	0.024	0.0050	0.0059	0.0005	0.0011
DUASR	0.024	0.035	0.0060	0.0085	0.0002	0.0010

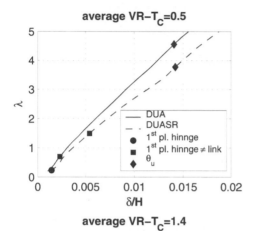

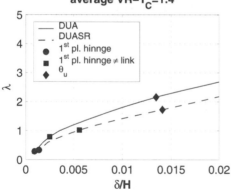

Figure 10. Incremental dynamic analysis curves for the DUA and DUASR structures.

at lower values of ground motion intensities in the case of the DUASR structure. However, non-dissipative elements (braces, columns and beam segments outside links) experience lower inelastic deformation demands for the DUASR structure. In the case of the $T_C = 0.5$ set of ground motions first yield in non-dissipative elements is attained at $\lambda = 0.7$ for the DUA structure and $\lambda = 1.49$ for the DUASR structure. Under the $T_C = 1.4$ set of ground motions non-dissipative elements experience first yield at $\lambda = 0.78$ for the DUA structure and $\lambda = 1.02$ for the DUASR structure. It indicates lower structural damage in non-dissipative (and non-removable) elements for the DUASR structure.

5 CONCLUSIONS

A removable link for eccentrically braced frames and dual frames (EBFs combined with MRFs) is suggested. Previously reported experimental investigations proved feasibility of this solution. In the present paper influence of structural configuration and reduced stiffness of bolted links on global seismic performance of eccentrically braced frames were investigated under two sets of ground motions.

It was shown that ground motion characteristics affect in a greater extent structural response than the homogeneous/dual configuration. However, dual structural configurations have beneficial effects on seismic response, including more uniform transient drift demands along the height of the structure and reduced permanent drifts. Dual structural configuration is to be preferred especially for buildings located on soft soils.

Bolted links are much more flexible in comparison with conventional links. The most convenient way to consider bolted link characteristics in the global frame analysis is through equivalent link characteristics, accounting for flexibility and slip in the link-to-beam connection. A trilinear shear force – deformation relationship based on experimental results was proposed in this paper. Further research will address analytical model for characteristics bolted links. Reduced stiffness of bolted links affects seismic performance of eccentrically braced frames by increasing lateral drifts and inelastic deformation demands in links, On the other hand, inelastic deformation demands in non-dissipative elements are reduced in the case of

elements. Distribution of interstorey drift demands over the height of the structure did not show important change due to bolted connections. Permanent drifts were slightly lower for the structure with bolted links (see Table 2).

Neither the DUA, nor the DUASR structure attained dynamic instability at the CP limit state ($\lambda = 1.5$).

Figure 10 presents the relationship between ground motions scale factor λ and the normalised top displacement δ/H (IDA curves) for the DUA and DUASR structures. The most evident consequence of bolted connections on seismic response of the eccentrically braced dual structure is increased displacement demands. Inelastic deformation demands are attained

frames with bolted links. Overall, global structural performance of dual eccentrically braced frames with bolted links was satisfactory, making them a promising structural solution in seismic regions.

ACKNOWLEDGEMENTS

Support of the Ministry of Education and Research of Romania and National Authority for Scientific Research through grant CEEX ET no. 1434/2006 is gratefully acknowledged.

REFERENCES

AISC 2002. Seismic Provisions for Structural Steel Buildings. *American Institute of Steel Construction, Inc.* Chicago, Illinois, USA.

Eurocode 3. 1993. Design of steel structures. Part 1-1: General Rules and Rules for Buildings. *CEN - European Committee for Standardization.*

Eurocode 8. 1994. Design provisions for earthquake resistance of structures. *CEN – European Committee for Standardisation.*

Fajfar, P. 2000. A nonlinear analysis method for performance-based seismic design. *Earthquake Spectra*, 16(3): 573–92.

FEMA 350, 2000. Recommended Seismic Design Criteria for New Steel Moment-Frame Buildings, SAC Joint Venture.

FEMA 356. 2000. Prestandard and commentary for the seismic rehabilitation of buildings; Washington (DC): Federal Emergency Management Agency.

Ghobarah, A. and Ramadan, T. 1994. Bolted link-column joints in eccentrically braced frames. *Engineering Structures*, Vol.16 No.1: 33–41.

Kasai, K., and Popov, E.P. 1986. General Behaviour of WF Steel Shear Link Beams, *ASCE, Journal of Structural Engineering*, Vol.112, No.2: 362–381.

Mansour, N., Christopoulos, C., Tremblay, R. 2006. Seismic design of EBF steel frames using replaceable nonlinear links. *STESSA 2006* – Mazzolani & Wada (eds), Taylor & Francis Group, London, p. 745–750.

Ricles J.M., Popov E.P. 1994. Inelastic link element for EBF seismic analysis. ASCE, Journal of Structural Engineering, 1994, Vol. 120, No. 2: 441–463.

Stratan, A. & Dubina, D. 2004. Bolted links for eccentrically braced steel frames. *Connections in Steel Structures V*, Ed. F.S.K. Bijlaard, A.M. Gresnigt, G.J. van der Vegte. Delft University of Technology, the Netherlands. pp. 223–232.

Dynamic behaviour of composite steel-concrete floor systems

B. Uy
University of Western Sydney, Sydney, Australia

F. Tahmasebinia
University of Wollongong, Wollongong, Australia

ABSTRACT: Composite steel-concrete beam systems are now fairly commonplace in the use of floor systems for steel framed multi-storey buildings. There is an increasing reliance for these systems in being able to achieve larger spans involving column free area. It is not uncommon for these floor systems to span up to about 20 metres for these structural forms whilst making use of pre-camber in the steel sections to overcome long term deflection problems. This increasing trend has thus made the serviceability limit state of vibration much more critical. This paper provides a study of some of the salient issues which affect the dynamic performance of composite steel-concrete beams. A finite element model is developed which is calibrated with known solutions for simple cases. The effects of adjacent beams, slab geometry and secondary beams on the natural frequency of floor panel systems is investigated in this paper and compared with existing theoretical models. A comprehensive parametric study is also conducted to look at the key influences in the assessment of the natural frequencies of composite steel-concrete floor systems and compared with existing theoretical models. Parametric studies are also carried out to provide designers with an opportunity to understand the key parameters involved in influencing the behaviour of the vibration performance of these systems.

1   INTRODUCTION

Composite steel-concrete beam systems as illustrated in Figure 1 are now commonplace in use for the construction of multi-storey steel buildings.

The serviceability limit states quite often govern the design of composite steel-concrete floor systems. This is particularly true in long span floor systems which are often used in multi-storey buildings to provide column free areas from core to perimeter frame. The use of semi-continuous behaviour and pre-cambering may be able to be used to overcome long term deflections which are primarily due to creep and shrinkage of concrete, however in the case of pre-cambering this then gives rise to the members becoming susceptible to vibrations controlling the design. Two significant examples of the design of long span composite floor systems in Australia include

- Grosvenor Place, Sydney (1988); and
- Latitude, Sydney (2005)

1.1   *Grosvenor Place, Sydney (1988)*

The Grosvenor Place building is a landmark building on George Street, Sydney which is on the border of the financial and rocks districts. The building has a total height of 180 metres with 44 storeys above ground as illustrated in Figure 2. The building was designed by the structural engineers, Ove Arup and Partners and construction was completed in 1988 by Concrete Constructions, (Gillett and Watson, 1987). The beams which span approximately 16 metres from reinforced concrete core to perimeter frame were designed for serviceability as semi-continuous, with a semi-rigid joint assumed between the beam and core as illustrated in Figure 3.

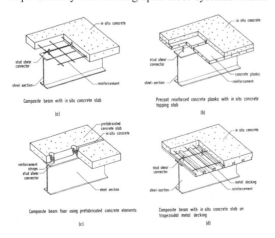

Composite beam with in situ concrete slab

(a)

Precast reinforced concrete planks with in situ concrete topping slab

(b)

Composite beam floor using prefabricated concrete elements

(c)

Composite beam with in situ concrete slab on trapezoidal metal decking

(d)

Figure 1.   Typical steel-concrete composite beams (Uy and Liew, 2002).

Figure 2.   Grosvenor Place, Sydney (1988).

(a)

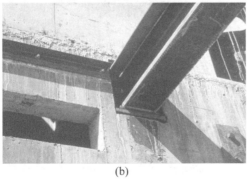

(b)

Figure 3.   (a) Plan view of steel beams with steel decking being placed; (b) typical beam to core connection used to achieve semi-continuous behaviour for serviceability.

Figure 4.   Latitude, Sydney (2005).

1.2   *Latitude, Sydney (2005)*

The Latitude building in Sydney has been completed in 2005 and exists on George Street on the World Square Site and is directly adjacent to Sydney's Chinatown at Haymarket as illustrated in Figure 4. The building is a landmark building which was designed by Hyder Consulting and constructed by Multiplex. The building has a total height of 222 m over 45 floors and has some very innovative features in it's design. The beams in the floor system span a total of 14 metres from core to perimeter frame and in order to achieve this the beam's were pre-cambered by 40 mm to overcome 60 mm long term deflections as shown in Figure 5, (Chaseling, 2004 and Australian Steel Institute, 2004).

1.3   *Previous research*

Uy and Belcour (1999) used a finite element approach to study the dynamic behaviour of composite steel-concrete beam systems to consider the effects of semi-continuous behaviour. Uy and Chan (2001) and Uy and Yap (2001) then expanded their studies to consider the effects of partial shear connection and the effects that column stiffness plays in the vibration performance of composite floors.

Sapountazkis (2004) developed a numerical analysis to consider the dynamic problem of a reinforced concrete slab stiffened by steel beams. However with this study the results and suggestions are based isolated beam only.

Hicks (2004) measured the fundamental frequencies and damping factors of 18 floors in different

Figure 5. View of underside of beam and slab showing primary span from core to perimeter frame.

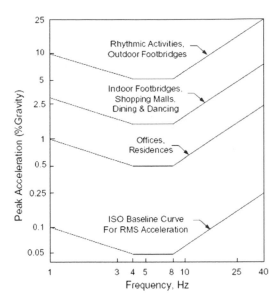

Figure 6. Peak accelerations for human comfort for vibrations due to human (ISO, 1989).

locations. He conducted a comprehensive study of the dynamic behaviour of composite steel-concrete floor systems. This paper also suggested that by assuming that primary beams have zero deflection, the secondary (joists) beams vibrate as simply supported beams. He added that slab flexibility is also affected by an equal deflection of the supports, as a result of this situation, and that slab frequency can be evaluated on the basis of fixed-end boundary conditions. In reality the boundary conditions for secondary beams (joists) are not simply supported; in this case it is just to suggest a simplified approach for the secondary beam. We cannot, under normal conditions, ignore the amount of deflection due to the primary beams which directly impacts the fundamental frequency of the system.

El-Dardiry and Tianjian (2005) conducted experimental and numerical analysis under human and machine excitation. Their study focused on the modes of composite slabs in a theoretical analysis.

Ebrahimpour and Ronald (2005) described some methods of controlling floor vibration. They discussed the tuned mass damper (TMD) for passively controlling floor vibration and mentioned that (TMD) it can provide artificial damping in composite steel-concrete beam systems.

2 ACCEPTABLE NATURAL FREQUENCIES

Prior to undertaking a structural analysis of composite beam systems it is important to ascertain the spectrum of natural frequencies which are generally acceptable for floor systems. The type of loading, which is present, usually determines this. In office buildings the most prevalent form of loading is live loading exhibited from persons walking, which can promote excitation of floor systems. The Australian Standard AS 2327.1-1980 (Standards Association of Australia 1980) provides guidance on the acceptable ranges of natural frequencies, which are outlined herein.

The potential problems caused by these vibrations is the annoyance resulting from the promotion of nausea and sound transmission arising from the oscillation of floors. People walking alone or together can create period forces in the frequency range of 1 to 4 Hz. However, for very repetitive activities such as dancing, it is possible to excite frequencies of around 10 Hz.

In general, natural frequencies less than 5 Hz should be avoided otherwise walking resonance may occur. However, it is recommended that for very repetitive activities such as dancing, the natural frequency of floors be set at 10 Hz or more. It is also suggested that for spans greater than 6 metres, especially for those without partitions and with natural frequencies smaller than 15 Hz, that a more rigorous analysis is undertaken.

Excessive continual vibrations could also lead to fatigue damage of structural components and thus issues associated with resonance should be avoided. In addition to controlling resonance effects through controlling natural frequencies, the peak acceleration also plays an important role in the vibration performance of floor systems as illustrated in the ISO curves in Figure 6.

857

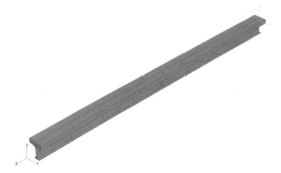

Figure 7. Single composite steel-concrete beam in ABAQUS.

3 FINITE ELEMENT MODEL

This section outlines the development of a finite element model used to consider isolated composite steel-concrete beams and composite steel-concrete beam systems to illustrate that the interaction between the beams and slabs plays a significant role in the vibration behaviour of the composite floor system.

3.1 Isolated steel-concrete composite beam

To consider the behaviour of isolated steel-concrete composite beams six beams were modelled under free vibration conditions as illustrated in Figure 7. Table 1 shows a summary of results from ABAQUS which have been compared with an analytical solution from AS2327.1-1980 (Standards Association of Australia, 1980) where f is the fundamental natural frequency of a simply supported composite beams using the relationship in Equation 1

$$f = K \sqrt{\frac{E_s I'_{cs}}{wL^4}} \qquad (1)$$

where $K = 155$ for simply supported beams; E_s = the modulus of elasticity of the steel joist (N/mm²); I'_{cs} = the second moment of area of the composite section (mm⁴); w = the dead load on the beam (kN/m); and L = the beam span length (mm).

Table 1 illustrates that the natural frequency of the isolated composite beams predicted by the FEM analysis is very close to that based on the model of AS2327.1-1980.

3.2 Composite steel-concrete beam systems

In order to consider the interaction effects of the primary, secondary beams and the slabs as illustrated in Figure 8, one must model a full panel as illustrated in Figure 9.

Figure 10 provides the results of the ABAQUS FEM for a number of spans with varying distances between

Table 1. Comparison between theoretical and analytical analysis for single composite beam.

L (m)	f AS2327.1-1980 (Hz)	f FEM (ABAQUS) (Hz)
6	34.2	35.6
8	20.2	20.0
10	12.9	12.8
12	8.9	8.9
14	6.5	6.5
16	4.9	5

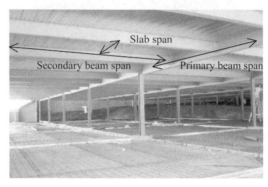

Figure 8. Typical primary and secondary composite beam floor system, to be considered in the paper.

Figure 9. System solid model in ABAQUS without slabs for clarity.

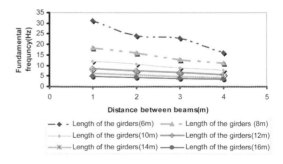

Figure 10. Parametric results highlighting slab span.

858

the secondary beams. The results highlight that by increasing the distance between beams, the fundamental frequency in the system tends towards the mode of slab but when the distance between beams is small, the frequency of the system depends on the frequency of girders.

4 COMPARISONS

The results of the FEM in the previous section highlight the importance of considering the influence of both the primary and secondary beams as well as the slab in the natural frequency of the system. Structural engineers may not always have access to such sophisticated software and thus it may be necessary to consider a simplified solution. A method known as Dunckerly's equation is used in the AISC guideline on vibrations, (Murray et al., 1997). According to the AISC guidelines, there is a special relationship between slab mode, girder mode and secondary beam mode. This relationship can be represented by Dunckerly's equation highlighted in equation 2

$$\frac{1}{f_n^2} = \frac{1}{f_g^2} + \frac{1}{f_j^2} + \frac{1}{f_s^2} \tag{2}$$

where:
f_n : Fundamental frequency for system
f_g : Girders mode (primary beam span as in Fig 9)
f_j : Joists mode (secondary beam span as in Fig 9)
f_s : Slab mode (slab span as in Fig 9)
   Furthermore, this equation can be expressed by higher orders in equations 3 and 4

$$\frac{1}{f_n^4} = \frac{1}{f_g^4} + \frac{1}{f_j^4} + \frac{1}{f_s^4} \tag{3}$$

$$\frac{1}{f_n^6} = \frac{1}{f_g^6} + \frac{1}{f_j^6} + \frac{1}{f_s^6} \tag{4}$$

A more general equation is given in equation 5

$$\frac{1}{f_n^N} = \frac{1}{f_g^N} + \frac{1}{f_j^N} + \frac{1}{f_s^N} \tag{5}$$

where N = 2, 4, 6, ...
   One can thus compare the results of the Dunckerly equation with those of the FEM to see the viability of the hand based method for design. Tables 2–5 indicate this comparison with respect to the different distances between girders. The comparisons are shown for spacing between girders of 1, 2, 3 and 4 metres in Tables 2–5 respectively.

Table 2. Comparison between Dunckerly's equation and FEM (Distance between girders = 1 m).

Model (N = 2)	Model (N = 4)	Model (N = 6)	FEM (Hz)	L (m)
32.3	33.2	33.2	31.2	6
18.5	18.6	18.7	18.5	8
11.9	11.9	11.9	12.1	10
8.2	8.3	8.3	8.5	12
6.09	6.1	6.1	6.3	14
4.7	4.7	4.69	4.9	16

Table 3. Comparison between Dunckerly's equation and FEM (Distance between girders = 2 m).

Model (N = 2)	Model (N = 4)	Model (N = 6)	FEM (Hz)	L (m)
21.5	24.9	25.9	24.1	6
14	15	15.1	16.1	8
9.3	9.69	9.7	10.6	10
6.6	6.7	6.71	7.56	12
4.89	4.93	4.9	5.6	14
3.75	3.73	3.71	4.3	16

Table 4. Comparison between Dunckerly's Equation and FEM (Distance between girders = 3 m).

Model (N = 2)	Model (N = 4)	Model (N = 6)	FEM (Hz)	L (m)
17.33	20.6	21.8	22.8	6
12.37	14	14.3	12.8	8
8.78	9.38	9.42	8.9	10
6.34	6.56	6.5	6.80	12
4.73	4.8	4.82	4.83	14
3.65	3.69	3.6	3.76	16

Table 5. Comparison between Dunckerly's Equation and FEM (Distance between girders = 4m).

Model (N = 2)	Model (N = 4)	Model (N = 6)	FEM (Hz)	L (m)
13.8	16.3	17.1	16	6
10.2	11.7	12.1	11.2	8
7.4	8	8.05	7.8	10
5.4	5.6	5.7	5.7	12
4.1	4.2	4.2	4.3	14
3.19	3.24	3.2	3.5	16

It is worth highlighting that Dunckerly's equation using the fourth or sixth power generally gives a result within ten percent of the FEM result. Thus the method which is amenable to calculation by hand once each of the individual fundamental frequencies have been established, appears to be suitable for design purposes.

5 CONCLUSIONS AND FURTHER RESEARCH

This paper has identified the dynamic performance of composite steel-concrete beam systems as being one of the key serviceability limit states which quite often control the design of these floor systems.

This paper has provide some case studies of composite steel-concrete floor systems where vibration has been a controlling criterion and worthy of further consideration.

A finite element method has been developed here to consider the isolated composite steel-concrete beam behaviour. This model was expanded to consider system behaviour to look at the interaction between primary and secondary beams and the supported slabs.

A simplified solution has been compared with the FEM which is shown to be suitable for use by structural engineers in assessing the vibration behaviour of floor systems.

Finally further research is necessary to look at system behaviour and incorporating the effects of partial interaction as well as the effects of damping. In long span systems quite often damping techniques have been used to justify vibration performance. In particular this was necessary on a recent UK project completed in 2002/2003, the More London project (see Figures 11 and 12) in London which employed 20 metre primary spans and required specialised damping procedures at the ends of the beams to control the vibration performance of the systems.

REFERENCES

Australian Steel Institute (2004) Latitude reaches skyward in steel: Construction technology at its best, *Steel Australia*, 17 (1), p. 10–13.

Chaseling, C. (2004) Star attraction, *Modern Steel Construction*, 37, 36–42.

Ebrahimpour, A. and Sack, R.L. (2005) A review of vibration serviceability criteria for floor structures. *Computers and Structures* 83, 2488–2494.

El-Dardiry, and Tianjian, J. (2005) Modelling of the dynamic behaviour of profiled composite floors. *The Structural Engineer. 28*, 576–579.

Gillett, D.G. and Watson, K.B. (1987) Developments in steel high rise construction in Australia, *Steel Construction, Journal of the Australian Institute of Steel Construction*, 21 (1), 2–8.

Hicks, S.J. (2004) Vibration characteristics of steel–concrete composite floor systems. *Steel Construction Institute*, Ascot, Great Britain.

International Standards Organization (1989) Evaluation of Human Exposure to Whole-Body Vibration – Part 2: Human Exposure to Continuous and Shock Induced Vibrations in Buildings *International Standard*, ISO 2631–2.

Murray, T. M., Allen, D. E. and Ungar, E. E. (1997). AISC Steel Design Guide Series 11:Floor Vibrations Due to Human Activity. *American Institute of Steel Construction*.

Sapountzakis, E. J. (2004) Dynamic analysis of composite steel–concrete structures with deformable connection. *Computers and Structures. 82*, 717–729.

Standards Association of Australia (1980) *AS2327.1–1980, SAA Composite Construction Code, Part 1: Simply supported beams.*

Uy, B., and Belcour, P.F.G. (1999) Dynamic response of semi-continuous composite beams, *Proceedings of the 16th Australasian Conference on the Mechanics of Structures and Materials, Sydney, Balkema*, 613–618.

Uy, B. and Chan, A. (2001) Effects of Partial Shear Connection on the Dynamic Response of Semi-Continuous Composite Beams, *First International Structural Engineering and Construction Conference, ISEC01, Honolulu, Hawaii, January, Balkema*, 801–805.

Uy, B. and Liew, J.Y.R. (2002) Composite steel-concrete structures, Chapter 51 *Civil Engineering Handbook, CRC Press*

Uy, B. and Yap, K.K (2001) Dynamic Response of Semi-Continuous Composite Beams in Typical Braced Frames, *Proceedings of the First International Conference on Advances in Steel & Composite Structures, ASCS01, Pusan, June*, 393–400.

Figure 11.   More London project, Thames London.

Figure 12.   Composite beams in More London project.

Steel and Composite Structures – Wang & Choi (eds)
© *2007 Taylor & Francis Group, London, ISBN 978-0-415-45141-3*

Pushover analysis and performance evaluation of a high-rise building structure

Zhaobo Wang & Yongfeng Luo

College of Civil Engineering, Tongji University, Shanghai, China

ABSTRACT: As an effective method to evaluate the performance of structures subjected to seismic hazards, the pushover analysis has played an important role in the performance-based design. The pushover analysis of the 3D-model of an actual complicated high-rise SRC/CFT structure is presented in this paper. According to the special function of the structure, the more critical performance demands are put forward based on the three aseismic-levels in Chinese code. The plastic hinge curves for SRC/CFT members are obtained through the numerical simulation. The pushover analysis of the high-rise structure is conducted on the basis of these curves. The target displacement is obtained by the capacity spectrum method and the corresponding performance state is gotten. The comprehensive evaluation of the seismic behavior of the structure is concluded. Finally, based upon the application status of pushover analysis in China, the problems of performance-based design that should be pay attention to in practice are summarized.

1 INSTRUCTION

Lately in the twenty-century, the researches on seismic theory and application based on seismic performance are widely developed in the America and Japan. As an effective method, pushover was firstly advocated in ATC40 and FEM273, and has been an important method in the performance-based design. Pushover analysis is an elasto-plastic static analysis method. In the elasto-plastic dynamic time history analysis, the estimations of kinetic parameters and the restoring force are inevitable. At the meantime, it is difficult to apply in practice project because of the more time-consuming, expensive calculation and large quantity of data. The pushover analysis can predict the whole loading process of the structure including elastic deformation, cracking, yielding, and collapse. The analysis results can indicate the occurring sequence and distribution of plastic hinges, deformation mode, weak stories and collapse mode. The engineer can grasp the detail information of the structure directly. Therefore, the non-linear pushover analysis is considered to be the practical analysis method based on seismic performance at present.

2 THE MECHANISM OF PUSHOVER ANALYSIS

The hypotheses of the pushover analysis are as follows:

1) The response of structure is relevant to an equivalent single degree of freedom system, and this means that the response of the structure is only under the control of the first mode;

2) Despite of the magnitude of displacement, the deformation shape of the structure along the height keeps invariable.

The steps in pushover analysis are shown in detail as follows:

1) Prepare the structure data, such as the model of structure, the physical parameters of members and the plastic hinge model, etc.;

2) Calculate the inner forces of the structure under the vertical loads (to be the preliminary step before the push load step);

3) Apply the horizontal loads on the mass center of each storey, which obeys certain shape vector along the height. The summation of the inner forces generated by horizontal forces and that obtained in step 2 will just result in the yielding or cracking of one or several members;

4) Update the stiffness of yielding or cracking members and increase the more load to make another or other members yield or crack;

5) Repeat step 3,4 until the displacement of the structure reaches to the target (for the ordinary pushover analysis) or the structure collapse eventually (for the capacity spectrum method).

The pure pushover analysis can't evaluate the seismic performance of the structure and it only presents the capacity of the structure under some load distribution. In order to evaluate the seismic performance, the performance point must be determined under the corresponding earthquake. The estimation of the performance point may be made by using the capacity spectrum method in ATC-40 (ATC 1996), inelastic spectra (Chopra and Goel 1999; Fajfar 1999), the displacement coefficient method in FEMA-356 (FEMA 2000), or yield point spectra (Aschheim and Black 2000). Other scholars propose to use the equivalent storey model to calculate the performance point by the time history analysis, Such as the inelastic static-dynamic method proposed by Zhiyong Yang.

Pushover analysis has been commended in Chinese code (2001) for seismic design of buildings. But few specifications about this method are contained in the code, and it is hard to instruct the design. In fact, the applications of pushover analysis are limited to the structures with simple arrangements or shapes, and only with the RC or S materials. With the development of the architecture, the structure systems become more complex and the applications of pushover analysis are far from the need of the complex structures design.

According to the design demand, the pushover analysis of a composite structure is conducted in this paper. The practicability of the method is proved. Some practical problems in the analysis process are summarized.

3 THE DESIGN CONDITIONS OF THE STRUCTURE

3.1 *The introduction of the structure*

The tower of Ping'an Finance Building has 38 storeys. The height of the tower is 164 m. The plane of structure has only one symmetric axis. The largest outer sizes in x and y direction are 73.5 m and 45.0 m respectively. The plans of the structure shrink twice in 29th and 37th storey respectively, i.e. the plane area of the structure along the height is variable. There is a large shared space in the building. The surroundings of the shared space are elevator shafts. The roof and indoor dome are located on the 38th storey and the 6th storey. The structural arrangement is shown in Figure 1.

The CFT columns are arranged around the center elevator shafts and the shared space. The vertical steel braces are located in the x- and y- direction. The center brace frame is composed of the braces, CFT columns and steel beams. The surrounding frame is made up of the circular SRC columns and SRC beams. The space between the outer SRC columns is 3 m. The standard floor arrangement of the tower is shown in Figure 2.

The member properties of the structure are shown in detail in Table 1.

In order to assure the sufficient ductility and bearing capacity of the structure under the strong earthquake and keep the limited and stable deformation, the static elasto-plastic pushover analysis is conducted

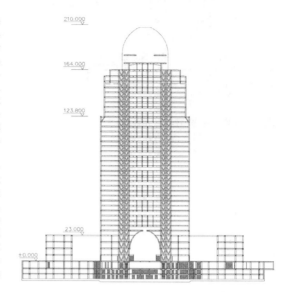

Figure 1. The arrangement of the structure.

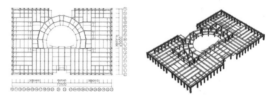

Figure 2. The standard floor with SRC beams.

in the optimum design of Ping'an Finance Building. The relevant analysis methods adopted are shown in reference.

3.2 *Parameters for asesmic design*

According to the relative Chinese codes, the significant coefficient of the structure is 1.0. The type of building height is B. The classification for earthquake-resistance of the structure is C-class.

The parameters of common seismic design are as followed. The seismic design intensity of the structure is 7 degree. The classification of the earthquake is I. The site soil is IV class.

The parameters of the strong earthquake are list as follows: the damping ratio of the structure is 0.05. The maximum seismic influence coefficient of horizontal earthquake is 0.45. The design characteristic period of ground motion is 1.1s.

3.3 *The requirement of seismic behavior and the performance indices*

The three aseismic-levels in Chinese code usually require that the structure can sustain a minor earthquake without any damage, a moderate earthquake

Table 1. The properties of the members of the structure.

Members	Material
Main members: column, girder, truss, brace	Q345 (steel)
Secondary members: beam, lateral brace of beam	Q235 (steel)
All CFT of the tower	C60
SRC beam	C30
Concrete of each storey slab	
SRC column	1~15storey	C60
	16~23storey	C50
	24~29storey	C40
	30~38storey	C30

Members	Size
CFT column	Ø1100~800
	T50~22 mm
SRC corner column	Ø1200~1000
	2H700 × 350 × 14 × 36
	~2H600 × 200 × 9 × 19
SRC edge column	Ø900~850
	2H550 × 250 × 14 × 32
	~2H500 × 200 × 9 × 19
SRC beam	900 × 600
	H600 × 250 × 12 × 25
	~H600 × 250 × 9 × 16
Steel braces	H500 × 300 × 28 × 40
	~H500 × 250 × 12 × 22

with repairable structural damage, and a strong earthquake without collapse. This structure is for an important finance organization, so the more critical performance demands are put forward out of the code. The demands in detail are as follows.

The elastic design should be conducted under the minor earthquake. The strength, stability and the deformation should meet the demand of the codes. The interstory drift limit is 1/400. The performance of the structure under the moderate earthquake should be ensured in elastic phase. The main members should keep elastic. The secondary members can turn into be plastic, and the displacement should be within the limit to keep the glass curtain safe. It means that the displacements under the moderate earthquake can be limited only 3 times as that under the minor earthquake. The elasto-plastic interstory drift under the strong earthquake should be within 1/70.

The deformations of the members in the elasto-plastic state should also be controlled under the strong earthquake. The rotation of the SRC beams should be limited within 0.02 rad and that of the columns should be within 0.012 rad. The tensile strain of the steel braces should be limited within 0.012, and the compressive strain should be controlled within 0.004

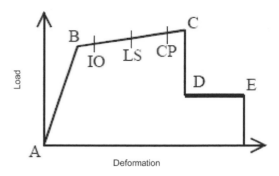

Figure 3. Generalized load-deformation relationship.

The occurring sequences and distribution of plastic hinges should follow the principles as follows:

1) The setback inclining columns, the braces and the members supporting the dome of the structure should keep elastic;
2) Though both ends of the main beams are allowable for plastic hinges, the structure must still keep stable bearing-capacity even if the plastic hinges occur.

4 THE SELECTION OF PLASTIC HINGE CURVE

The typical plastic hinge curve is shown in Figure 3. Point B represents the initial yielding point of members. Point C represents the buckling point of the structure. Point E represents the collapse of the structure. IO represents the immediate occupancy point. LS represents the life safety point. CP represents the collapse prevention point.

In the pushover analysis of the structure, the hinge properties of the steel member employ the default properties provided by the program. The hinges of steel columns are the PMM hinges by considering the coupling of axial force and moment. The hinges of steel braces are the P hinges by considering the axial loads. For the main members of CFT and SRC, the hinges are not the default ones provided by software but the PM hinges defined by user.

There are available researches on the plastic hinge curves of concrete or steel members. Few ready references or data about hinge curves of SRC or CFT members are found. The definition of the curves is based on the numerical simulation. The self-programming is adopted to calculate the hysteretic curves of the typical members. The backbone curves of the members can be obtained by the hysteretic curves. The constitutive relationship of steel material is assumed to be ideal elasto-plastic. Kent-Park model is used for the unconfined concrete in SRC sections. The constitutive relationship of the cored concrete in CFT sections recommended in Japan code is introduced, which takes consideration of the confined effect of the pipe.

4.1 The basic hypotheses of the ascertain of the backbone curves

The basic hypotheses for SRC and CFT members to ascertain the backbone curves are as follows:

1) Plane section hypothesis, that is, the sections of the members remain plane during the loading or unloading;
2) The deformation curve of the members obeys the cosine function;
3) The constitutive relationship of the steel is assumed to be ideal elasto-plastic, which is shown as Equation 1.

$$\sigma_s = \begin{cases} E_s \varepsilon_s & |\varepsilon_s| \leq \varepsilon_y \\ (f_y + 0.01E_s(\varepsilon_s - \varepsilon_y))\varepsilon_s / |\varepsilon_s| & \varepsilon_y < |\varepsilon_s| \leq \varepsilon_{sh} \end{cases} \quad (1)$$

Where σ_s is the stress; ε_s is the strain; E_s is Young's modulus; f_y is the yielding strength; ε_y is the yielding strain; ε_{sh} is the limited strain.

The constitutive relationship of the unconfined concrete in the SRC beams and the SRC columns is Kent-Park model, which is shown as Equation 2. The model takes consideration of the restriction effect of stirrups.

$$\sigma_c = \begin{cases} f_c \left[2\varepsilon_c / 0.002 - (2\varepsilon_c / 0.002)^2 \right], & \varepsilon_c \leq 0.002 \\ f_c \left[1 - Z(\varepsilon_c - 0.002) \right], & 0.002 \leq \varepsilon_c \leq \varepsilon_{20c} \\ 0.2 f_c, & \varepsilon_{20c} \leq \varepsilon_c \\ Z = 0.5 / (\varepsilon_{50c} - 0.002) \end{cases} \quad (2)$$

Where ε_{20c} is the strain corresponding to 20% of the ultimate stress; ε_{50} c is the strain corresponding to 50% of the ultimate stress.

The constitutive relationship of cored concrete of CFT is shown as Equation 3.

$$\sigma_c = \begin{cases} \sigma_0 \left[A(\varepsilon_c / \varepsilon_0) - B(\varepsilon_c / \varepsilon_0)^2 \right], & (\varepsilon_c \leq \varepsilon_0) \\ \sigma_0(1-q) + \sigma_0 q(\varepsilon_c / \varepsilon_0)^{0.1\xi}, & (\varepsilon_c \leq \varepsilon_0, \xi \geq 1.12) \\ \sigma_0(\varepsilon_c / \varepsilon_0) / \left[\beta(\varepsilon_c / \varepsilon_0 - 1)^2 + \varepsilon_c / \varepsilon_0 \right], & (\varepsilon_c \leq \varepsilon_0, \xi < 1.12) \end{cases} \quad (3)$$

Where

$$\sigma_0 = f_{ck} \left[1.194 + \left(\frac{13}{f_{ck}} \right)^{0.45} \left(-0.07485\xi^2 + 0.5789\xi \right) \right]$$

$$\varepsilon_0 = \varepsilon_{cc} + \left[1400 + 800 \left(\frac{f_{ck} - 20}{20} \right) \right] \xi^{0.2} \quad (\mu\varepsilon)$$

$$\varepsilon_{cc} = 1300 + 14.93 f_{ck} \quad (\mu\varepsilon)$$

$$k = 0.1\xi^{0.715}, A = 2 - k, B = 1 - k$$

$$q = \frac{k}{0.2 + 0.1\xi}, \beta = \left(2.36 \times 10^{-6} \right)^{\left[0.25 + (\xi - 0.5)^7 \right]} f_{ck}^2 \times 5 \times 10^{-4}$$

$$\xi = \left(A_s f_y \right) / \left(A_c f_{ck} \right)$$

Where ξ is the parameter that represents the hoop action of the pipe; A_s is the area of steel pipe; f_y is the yielding strength of steel; A_c is the area of concrete; f_{ck} is the characteristic strength of concrete.

4) The member with both ends fixed can be decomposed into two cantilevers. Thus the problem will turn into the solution of the M-θ relationship of the cantilevers.

The fiber model method is adopted to calculate the M-θ relationship of the section. From the hypothesis 2, the relationship of end curvature ($y''(0)$) and the relative displacement of two fixed ends (Δ) is shown as Equation 4:

$$\phi = \frac{\Delta \pi^2}{2L^2} \quad \text{or} \quad \Delta = \frac{2\phi L^2}{\pi^2} \quad (4)$$

The M-ϕ hysteretic curve will be inverted into the M-θ curve by Equation 4. By connecting the peak point of the hysteretic curve, the backbone curve can be formed.

4.2 Parameters of plastic hinge curve

Parts of the Parameters of plastic hinge curve are shown in table 2 and table 3.

5 THE PUSHOVER ANALYSIS AND PERFORMANCE EVALUATION

5.1 The Pushover curve

The selection of the horizontal load distribution should reflect the distribution characters of inertia forces of each storey under earthquake. In this paper the seismic forces calculated by the elastic spectrum method are taken as the lateral forces mode in pushover analysis. The representative gravity load is equal to the dead load and a half of the live load.

The Y-direction pushover curve of the structure is given in figure 4.

5.2 Target displacements

Transform the base shear-roof displacement curve of pushover analysis into ADRS (spectrum displacement-acceleration) capacity spectrum and transform the elastic 5% damped response spectrum of 7 degree strong earthquake from traditional acceleration-period form to the ADRS form. According to the control principles under strong earthquake, few members turn into plastic when the target displacement reaches except for the horizontal members. The performance point is obtained based on the elastic 5% damped response spectrum. Plotting the demand spectrum and capacity spectrum on the same chart, the crossing point of them is the performance point. Transform the spectrum displacement of the crossing point into the roof displacement. This roof displacement is the target displacement. The spectrum displacement in Y-direction in elasto-plastic pushover analysis is 780 mm, and the corresponding target displacement is 1080 mm.

Table 2. Parameters of SRC and CFT (Unit: kNm, rad).

Member	Type	Yielding rotation	moment Yielding	Limiting rotation	Limiting moment
C1F1F4	SRCCOL	0.006	3800	0.012	3800
C2F1F4	SRCCOL	0.006	9320	0.012	9320
C2F14F22	SRCCOL	0.006	6970	0.012	6970
C5F1F4	CFTCOL	0.01	9330	0.02	9330
C5F14F19	CFTCOL	0.01	7660	0.02	7660
G1F1F5	SRCBEAM	0.004	2600	0.02	2600

Table 3. Parameters of parts steel members (Unit: kN·m, rad).

Member	B		C		E		IO	LS	CP
	M	R	M	R	M	R			
B1277	2213	0.006	2644	0.039	885	0.051	0.009	0.03	0.039
B1596	2213	0.004	2644	0.025	885	0.033	0.006	0.019	0.025
B1283	1805	0.005	2157	0.032	722	0.041	0.007	0.024	0.032

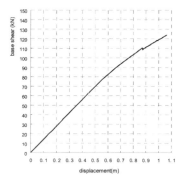

Figure 4. The capacity curve in Y-direction.

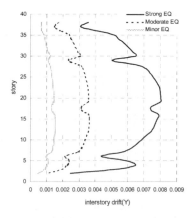

Figure 5. The interstory drift at three aseismic-levels.

5.3 The evaluation of seismic performance

5.3.1 The interstory drift at three aseismic-levels
According to the target displacements, the interstory drift at three aseismic-levels are draw in figure 5.

5.3.2 The performance evaluation of the structure under the moderate earthquake
The displacement in the Y-direction of the 38th storey under the minor earthquake response spectrum is 167 mm (interstory drift is about 1/1010). Due to few members under the moderate earthquake turn into plastic, the displacement that is 2.857 times as the above one is taken as the target displacement of the elasto-plastic pushover analysis under the moderate earthquake. When the displacement in Y-direction reaches to 470 mm, the distributions of plastic hinges are shown in Figure 6.

The number of SRC columns, CFT columns, SRC beams and steel beams is 12194 in all. No plastic hinges occur in the columns and braces when the pushover analysis reaches to the medium earthquake,

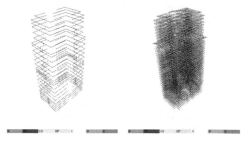

Figure 6. The distribution of plastic hinges of SRC/steel beams under moderate EQ.

Only few plastic hinges occur in SRC beams and steel beams. The results meet the former demand that main members keep in elastic, secondary members are permitted to be in plastic and the displacements are small enough to assure the safety of glass curtain.

865

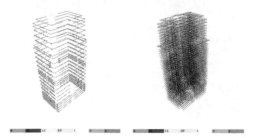

Figure 7. The distribution of plastic hinges of SRC/steel beams under strong EQ.

5.3.3 *The performance evaluation of the structure under the strong earthquake*

When reaching to the target displacement in the pushover analysis, the maximum of the storey drift in Y-direction is 1/125, which is within the acceptable limit 1/70.

When the displacement of Y direction reaches to the target in the pushover analysis, the main members such as columns and the braces all keep in elastic. The plastic hinges occur in about 700 SRC beams and steel beams. The distribution of the plastic hinges is shown in Figure 7. From the Figure 7 it can be drawn that most of the plastic hinges in steel main beams occur in IO phase, and only few occur in LS phase. The deformations of most members are in the plastic phase. From the results of the elasto-plastic pushover analysis it can be concluded that the structure keep in the beginning of the plastic under the strong earthquake, and the performance meets the predefined control demand.

From the distribution of the plastic hinges it can be drawn that most of the SRC beams and steel beams are in elastic and only a few beams are in plastic. The plastic rotations of the beams are within 0.02 rad. No plastic hinges occur in the columns. The member rotations are within 0.006 rad. No plastic hinges occur in the steel braces. The limited tensile strain and the limited compressive strain all meet the demand of the controlling requirement.

6 THE EXISTING PROBLEMS

This paper introduces a complicated high-rise SRC/CFT structure and it's seismic performance evaluation, to provide a reference for the similar case. The complicated three-dimensional pushover analysis is conducted, and capacity spectrum method is used to get the target displacements. The performance of the building at different aseismic-levels can satisfy the predefined demands. It can be concluded that the pushover analysis is also effect and necessary for the complicated structures.

Although the static non-linear analysis for aseismic design is recommended in Chinese code, the complement standard items for the application of the analysis are deficient. Such as:

1) The proper scope of its application is to be ascertained;
2) The research on the plastic hinge properties of SRC and CFT members is scarce, and the experiments or numerical simulations are necessary;
3) The research given in this paper is pushed only along its symmetry axis. The pushover analyses of asymmetric buildings should be studied further. The torsion will occur in the asymmetric structure, which may lead to the incorrect target displacement;
4) Too few specifications about this method are contained in Chinese code. It is hard to instruct the design. More detailed instruction should be provided.

REFERENCES

Nikken Sekkei Design CO. LTD. & East China Architecture Design & Research Institute CO. LTD. 2004. Report of China Ping'an Finance high-rise building design (R).

Steel structure research-lab, College of Civil Engineering, TongJi University. 2005. Optimal design report of Ping'an Finance high-rise building in China (R)

T.M. Murray, D.E. Allen & E.E. Ungar. 1997. Floor Vibrations Due to Human Activity. Steel Design Guide Series 11, AISC and CISC.

ATC. 1996. Seismic Evaluation and Retrofit of Concrete Buildings, Volume 1, ATC-40 Report. Applied Technology Council, Redwood City, California.

FEMA. 2000. NEHRP Guidelines for the Seismic Rehabilitation of Buildings. Developed by the Building Seismic Safety Council for the Federal Emergency Management Agency (Report No. FEMA 273). Washington, D.C.

LIU Da-hai, YANG Cui-ru. 2003. The Calculation and Conformation of the Profiled Bar (Steel Tube) Concrete Tower. Beijing: China Architecture and Building Press.

Chopra, A. K., and Goel, R. K. 1999. Capacity-demand diagram methods based on inelastic design spectrum. Earthquake Spectra. 15(4), 637–656.

Fajfar, P. 1999. Capacity spectrum method based on inelastic demand spectra. Earthquake Eng. Struct. Dyn., 28(9), 979–993.

Aschheim, M., and Black, E. 2000. Yield point spectra for seismic design and rehabilitation. Earthquake Spectra, 16(2), 317–335.

YANG Zhi-yong & HE Ruo-quan. 2003. Inelastic Static-dynamic Analysis Method of High-rise Steel Structures under Earthquake Action. Beijing: Journal of Building Structure. 24(3), 25–32.

Code for seismic design of buildings (GB50011-2001). 2001. Beijing: China Architecture and Building Press.

Code for seismic design of buildings (DGJ08-9-2003 J10284-2003). 2003. Shanghai: The construction and management committee of shanghai.

Code for design of steel structures (GB50017-2003). 2003. Beijing: China plan press.

Steel and Composite Structures – Wang & Choi (eds)
© *2007 Taylor & Francis Group, London, ISBN 978-0-415-45141-3*

Issues in the mitigation of progressive collapse through the tying force method for steel framed buildings with simple connections

M.P. Byfield & S. Paramasivam

School of Civil Engineering and the Environment, University of Southampton, Southampton, UK

ABSTRACT: After the removal of support to a column due to damage caused by impact or blast, loads can be redistributed to the remaining structure by for example either catenary actions in the floor frame or through double span beam actions. Steel framed buildings with flexible joints, such as fin plate, double angle web cleat and flexible end plate connections are routinely assumed to be compliant with the tying force design method, which aims to ensure the column loads are redistributed via catenary action in the event of progressive collapse. This is possible only if the joint possess sufficient ductility and strength to survive the performance demands imposed by this mechanism. This paper demonstrates that many industry standard joints lack rotational ductility. Moreover, a couple can develop between beam flange and columns which is shown to lead to early joint fracture, resulting in collapse. This paper demonstrates the issues involved in transferring load from the removed column to rest of the structure through catenary action. The factor of safety against collapse is determined for the typical steel frame.

1 INTRODUCTION

In recent years high-rise office buildings have increased risks of experiencing loads from vehicle borne improvised explosive devices. Using suicide bombers large amounts of explosives can be detonated at close range. As a result, the building is subjected to huge pressure, which can result in the failure of one or more load bearing elements. The loss of the elements can cause either the structural instability and/or overload the adjacent members, resulting in a chain reaction of failure to an extent that is disproportionate to the original localized damage. This is called as disproportionate collapse. Such collapse was well demonstrated in the Murrah building incident. In this incident, the direct blast pressure destroyed one column by brisance effect and another two columns by shear. The loss of these three columns led to a collapse that consumed ½ of the floor area of this nine storey building. Although it was correctly designed to the requirements at that time it was unable to redistribute the column loads. The building had no emergency means for redistributing loads, as it was lacking the strong internal partition walls or cladding. Such events can be avoided in three ways: event control, indirect design, and direct design. In event control, the minimum threat is measured for the building and/or the load bearing elements and the stand off distance is maintained. In case of indirect design, the load carried by the removed load bearing elements is transferred to the surviving structure through minimum strength, continuity and ductility. In direct design, the safety of the building is ensured by two methods, namely the specific local resistance method and the alternate load path method. In the alternate load path method, loads from the removed load bearing members are transferred to the rest of the structure through for example: catenary action (tying force approach in Figure 1(a)), beam action (the double span approach in Figure 1(b)), slab arching (Figure 1(c)) or through truss systems (Figure 1(d)), such as those incorporated into the top of the World Trade Centre Towers, and also used in other land mark structures such as the 2nd International Financial Centre in Hong Kong. This paper concentrates on the tying force approach as related to steel xframed buildings incorporating simple (shear) joints.

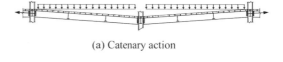

(a) Catenary action

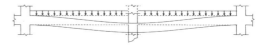

(b) Double span beam action

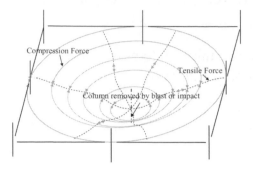

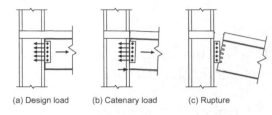

(a) Design load   (b) Catenary load   (c) Rupture

Figure 2. Prying and catenary action.

(c) Diaphragm action (arching)

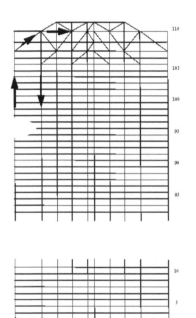

(d) Out-rigger truss system used in WTC 1

Figure 1. Mechanism of transferring the accidental load.

2 THE TYING FORCE APPROACH

The tying method was introduced into the UK Building Regulations after the Ronan Point collapse. In this approach, all floor members are required to be effectively tied together. The tying force approach assumes the accidental load, at the time of damage, is equal to 0.3 of the imposed load plus 1.05 of the dead load. When an intermediate column is removed due to impact or blast, adjacent beams on either side support the accidental load through catenary action. As the shear force due to accidental load (double span)

is less than the design shear force (i.e. using 1.4 dead + 1.6 imposed), the beams do not fail by shear and the tying force should comfortably exceed the shear force at the connection. In practise, designers must ensure that the tensile strength of connections is atleast equal to the design shear force, although most industry standard steel connections have no difficulty in meeting this requirement. UK Building Regulations place no requirement for joints to have sufficient ductility in order to accommodate the beam rotations that occur during catenary action. The tying capacity of the connections, as presented in the Steel Construction Institute "Joints in Steel Connections: Simple Connections" (SCI, 2002), are calculated in the absence of any rotation, see Figure 2(a). Horizontal equilibrium of the joint dictates that the prying action sketched in Figure 2(b) must reduce the tying force capacity of the joint, leading to a large reduction in the tying strength of the joint. This raises the possibility that joints can fail by early rupture as shown in Figure 2(c).

3 METHODOLOGY

3.1 Assumptions

To study the feasibility of this approach, a simple analysis method is developed based on basic structural mechanics. The following assumptions are considered in the analysis: (i) load from fully glazed cladding was removed during the accidental limit state, (ii) during the accidental limit state, 1.05 dead load + 0.3 imposed load was assumed, (iii) all columns are assumed to remain inline, (iv) the design tying force is maintained only until the rotation capacity of the connection is exceeded and the tying capacity of the connection is calculated by assuming zero rotation (SCI, 2002), and (v) the tying capacity of profiled sheet is calculated by considering the bearing failure of shear studs.

3.2 Dynamic Effect

When the column is removed and starts falling, potential energy is converted into the kinetic energy as well as strain energy. The maximum tying force generated during this process can be represented by the final at rest tying force, multiplied by the dynamic amplification factor (DAF). The magnitude of the DAF will depend on the amount of damping in the system, as well as the time duration of column removal. Fire would let a column down slowly and produce a DAF of 1.0. At the other extreme, a brisance failure in the column would represent an instantaneous column removal. The extent of damping is very difficult to estimate without full-scale experimental tests. If the system were undamped, then the DAF could be in the region of 2.0. Since blast is the trigger for catenary action of most interest to this study, a DAF of 1.5 is assumed in the absence of reliable data.

3.3 Method to estimate factor of safety

After removal of the column, the columns, above it, start to drop down and the beams on either side start to rotate about extreme joints. Let the beam-column joint, just above the removed column be C and the beam-column joints on either side of C in the plane of drop be A and B. The beams ACB shown in Figure 3(a) act like an inverted three hinged arch as shown in Figure 3(b). As the inverted arch is a determinate structure, the analysis is relatively simple. Let the rotation at 'A' be θ_{con}. The beam parallel to the beam 'ACB' marked as '$A_1C_1B_1$' is shown in Figure 3. To find the load resistance through catenary action of reinforcement mesh and profiled sheet, the panels are divided into one metre strips along AA_1 from $A_1C_1B_1$. A one metre wide strip XZY located a distance x from $A_1C_1B_1$ is considered, the mechanics of which are sketched in Figure 3. The upwards lift generated by T_{slab} is dependent on the end slope of the slab, which varies from zero at $A_1C_1B_1$ to θ at ACB. It is assumed that the full tensile capacity of the slab is mobilised along the entire width of the slab. Therefore, the upwards reactive force offered by catenary action in the slab is:

$$P_{up} = 2\,T_{slab}\sin\left(\frac{\theta x}{L}\right) \qquad\qquad - (2)$$

Where, P_{up} is the upward reactive force caused by slab tying capacity per metre; $\frac{\theta x}{L}$ provides the inclination of slab catenary to the horizontal, as shown in Figure 3(c). Figure 3(d) shows the final loading on beam CC_1. There will not be any reduction in the accidental load on the secondary beams SS_1 due to catenary action in the slab as no change in angles of tying force occurs on either side.

The factor of safety against collapse is defined as the ratio between the tying capacity of the joint, T_{con} and the tying force, T acting in the joint when the rotation limit of the connection is reached, T_{con}. The contribution of the reinforcement mesh and profiled sheet is very complex to predict because of high level of material non-linearity occurring in the structure. In

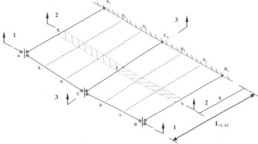

(a) Deflected shape after removal of column 'C'

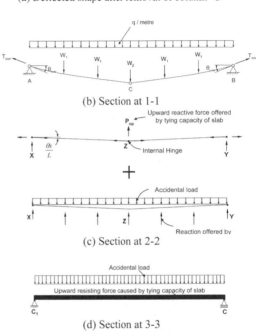

(b) Section at 1-1

(c) Section at 2-2

(d) Section at 3-3

Figure 3. Analysis for Catenary Action.

addition to that, its confidence in resisting the accidental load is less because of the chances of discontinuity in the reinforcement (if there is lapping, the reinforcement bar can not transfer the tying force to the beam) and the tensile crack in the slab. Hence, the lower and upper bound values of factor of safety are estimated by ignoring and considering the tensile strength of slab for a typical building frame.

3.4 Case study

The typical building frame, shown in Figure 4, is aimed to provide a reasonable match with the arrangement found in typical medium rise office developments, with 4 m floor to floor height. The floor slab is extended by 0.5 m beyond the edge beams to support a curtain wall system. The total dead and imposed loads were 5.1 kN/m² and 6.0 kN/m² (inc. partitions)

869

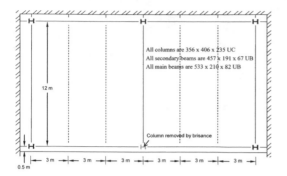

All columns are 356 x 406 x 235 UC
All secondary beams are 457 x 191 x 67 UB
All main beams are 533 x 210 x 82 UB

12 m

Column removed by brisance

3 m — 3 m — 3 m — 3 m — 3 m — 3 m

0.5 m

Figure 4.   Idealized frame for the analysis.

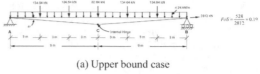

(a) Upper bound case

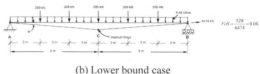

(b) Lower bound case

Figure 6.   Accidental loads on main beam 'ACB'.

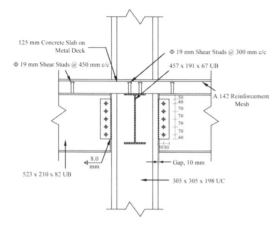

125 mm Concrete Slab on
Metal Deck

Φ 19 mm Shear Studs @ 450 mm c/c

Φ 19 mm Shear Studs @ 300 mm c/c

457 x 191 x 67 UB

A 142 Reinforcement
Mesh

523 x 210 x 82 UB

8.0
mm

Gap, 10 mm

305 x 305 x 198 UC

Figure 5.   Beam – column joint details.

respectively and this provides an accidental limit state load of 7.1 kN/m². All steel is S355 grade and all concrete is C35 normal weight.

In the lower bound case, the tensile strength of the slab has been ignored and the DAF has been taken as equal to 2.0, in accordance with US practice (GSA, 2003). This provides a factor of safety of 0.08. In upper bound case, the full tensile strength of the slab was included and the DAF was taken as equal to 1.0. These assumptions provided a factor of safety of 0.19. The accidental load and the tying force necessary to achieve equilibrium, together with the factor of safety (FoS) are shown in Figure 6 for both cases.

3.5 Discussions

Insufficient ductility is the primary reason for these low factors of safety, although it should be noted that this analysis assumed 4 degrees of maximum rotation in the joints, whereas ductility predictions based on the Liu *et al* (2004) method show a limit of 2 degrees on rotation. Using an iterative procedure it is possible to define what deformation is necessary in the example considered herein, in order to support

the damaged column via catenary action. Assuming that the beam column joints remain able to transfer their design tying force (regardless of the rotation), the connections would need to rotate by 24 degrees to the horizontal in order to support the damaged column by catenary action (assuming full tensile capacity of the slab was mobilised and assuming a dynamic amplification factor of 1.5). This would require a downwards movement of 3.84 m. Such deformations are not possible without fracturing the joints and the reinforcement in the slab. Therefore a progressive collapse will occur is other emergency load paths are absent.

4   CONCLUSIONS

The feasibility of supporting damaged columns using the tying force method is studied by way of a case study of a steel framed multi-storey building. Using a simplified method, the factors of safety against collapse were bounded within the range of 0.1 to 0.2. Insufficient joint ductility is the primary reason for these low values. Even if there was sufficient joint ductility present, the deflection required to support the damaged column via catenary action would be approximately one floor height. Such deformations are not possible without fracturing the joints and the reinforcement in the slab. The tying force method is therefore deemed to have failed to prevent progressive collapse in this case study. The case study is representative of "typical" office buildings constructed with nominally-pinned joints. In such situations designers should consider alternative means for providing robustness to buildings, if as part of a risk assessment column fracture is identified as a likely design scenario.

ACKNOWLEDGEMENTS

This paper extends work originally published in the workshop, "Urban habitat constructions under catastrobic events".

REFERENCES

BSI, Structural use of steelwork in building, Part 1: *Code of Practise for design – rolled and welded sections*, BS 5950–1, 2000.

BSI, Structural use of steelwork in building, Part 4: *Code of Practise for design of composite slabs with profiled steel sheeting*, BS 5950–4, 1994.

Byfield, M. P. & Paramasivam, S. 2007. The prevention of disproportionate collapse using catenary action. In Wald. et al. (ed.), *Urban Habitat Constructions under Catastrophic Events*; *Proc.of workshop.,Prague,30–31 March 2007*. Prague: Prazska technika.

General Services Administration (GSA). (2003). *Progressive collapse analysis and design guidelines for new federal office buildings and major modernization projects*. Office of Chief Architect, Washington, D.C.

Liu, J., and Astaneh-Asl, A (2004). Moment-rotation parameters for composite shear tab connections. *J. Struct. Eng.*, 130(9),1371–1380.

The Steel Construction Institute (SCI), Joints in Steel Construction: Simple Connections, P212, 2002.

Steel and Composite Structures – Wang & Choi (eds)
© *2007 Taylor & Francis Group, London, ISBN 978-0-415-45141-3*

Impact behavior of lightweight fiber-reinforced sandwich composites

J.Y. Richard Liew, K.M.A. Sohel, K.S. Chia & S.C. Lee
Department of Civil Engineering, National University of Singapore, Singapore

ABSTRACT: The results of an experimental study on the impact behavior of sandwich composite panels, in-filled with lightweight fiber-reinforced concrete between steel faceplates, are presented. In this study, the panels were impacted by a 58-kg drop-weight projectile with a hemispherical head from a height of 4.0 m. Parametric studies in these tests include concrete type and volume percent of fiber in the concrete core. From the tests, the impact force and indentation in the top steel plate were measured while visual damage in the core material was noted. The experimental results indicated that the plain concrete cores without any fiber reinforcement cracked into separate pieces upon impact. On the other hand, the concrete cores incorporating 1 to 2% fiber content remained intact after impact, with increasing reduction in the average crack width as well as number of crack formations as fiber content increased. In addition, the concrete core incorporating 2% fiber had no radial cracks, which are cracks propagated outwardly from point of impact, except for local indentation. Depth of local indentation was observed to decrease with the increase in core strength. From the impact shear test on lightweight concrete, it is clear that hooked-ends steel fiber performed better under impact.

1 INTRODUCTION

Today civil, marine, aeronautic, automotive, military and construction industries have acknowledged the benefits of sandwich construction in various applications. The smoothness of surfaces and potential for avoiding complex lateral stiffening systems are other favorable properties of sandwich system. In recent years new types of sandwich designs have been developed for ship and offshore applications. One such design has been proposed by Bergan and Bakken (2005), which is based on using sandwich plates made of steel plates with lightweight concrete core also known as Steel-Concrete-Steel (SCS) sandwich system. The performance of SCS sandwich system has shown its superiority over traditional engineering materials in application requiring high strength, high ductility, as well as high energy absorbing capability (Sohel *et al*. 2003, Oduyemi & Wright 1989). The apparent advantages of the system are that the external steel plates act as both primary reinforcement and permanent formwork, and also as impermeable, impact and blast resistant membranes.

Low velocity impact with dropping and floating objects or moorings can cause local indentation of the face sheet, permanent compression of the underlying core material, local damage of core and introducing of interfacial cracks between steel plate and core. These may extend under service loads, possibly causing a catastrophic failure of the sandwich plate. 'Low velocity impact' generally refers to impact range 1 to 10 m/s which are ordinarily introduce in the laboratory by mechanical (drop weight or pendulum) test machine (Richardson & Wisheart, 1996). During the impact of a projectile on a sandwich structure, overall structural deformation and local damage at the impact point are observed. Appropriate modeling of local indentation is important to predict the impact force history and overall response of the sandwich structure impacted by a foreign object.

This paper addresses the dynamic indentation behavior of SCS Sandwich panel with a concrete core, as well as the impact performance of concrete reinforced with different fiber types and volume fractions. The investigation focused on the behavior of fiber-reinforced lightweight concrete cores, such as foam concrete and lightweight aggregate.

2 EXPERIMENTAL DETAILS

The experimental investigation was focused on impact performance of SCS sandwich panels and fiber-reinforced LWA concrete panel subjected to low-velocity impact. In this experimental programme, the thickness of the steel plates and properties of the core materials are considered.

To investigate local impact behavior of the SCS sandwich panels, ten specimens were prepared. Three different types of concrete core were used in the

SCS panels, which were either normal-weight concrete (NC), foam concrete (FC) or lightweight aggregate concrete (LC). Polyvinyl alcohol (PVA) fiber of 30-mm length with an aspect ratio of 45 was added at various volume fractions ranging from 1 to 3% into some of the SCS sandwich specimens. All the SCS panels for local impact test were 300 mm square in size with a core thickness of 60 mm (Fig. 1a). Steel face plates thickness varied from 4 mm to 8 mm. In addition, LC panels (Fig.1b) with different types of fiber were also tested for impact performance. Details of the specimens are given in Tables 1 and 2.

The impact resistance of the SCS sandwich panels and LC panels was determined by an instrumented drop-weight impact test machine developed by Ong et al. (1999) as shown in Fig. 2. A steel test frame was designed and fabricated to accommodate a 300 × 300 mm specimen under impact. The frame was made by welding four 100 × 50 × 10 mm parallel flange channels, each of length 404 mm, together to a base plate as shown in Fig. 2. The projectile is allowed to slide freely up and down a guide fixed to an external self-supporting steel frame (Fig. 2). The impact was achieved by dropping a 58 kg cylindrical projectile with a hemispheric head of 90 mm diameter from a height of 4 m. Just before the projectile hit

the specimen, its speed was approximately 8.12 m/s. Dynamic load cells were instrumented in the projectile to record the signals related to its impact pressure and the observed impact load to which the projectile was subjected when it struck the panel.

In addition, for the impact test on the fiber-reinforced LC panels, a 200 mm diameter opening was provided in the center of the base plate (Fig. 3). This entire frame was bolted to the heavy base frame of the test rig. Due to the stiff configuration of the frame, minimum bending was expected when the specimen was loaded centrally. Thus, the mode that absorbs most of the energy from the impact would therefore be the local punching and shear failure mode.

Table 2. Properties of LC panels.

Panel ref.	% fiber	Fiber description (Aspect ratio)	Air (%)	S.G.	f_{cu} (MPa)
SF1-1	1	30-mm hooked-ends steel (80)	17	1.40	24.2
SF1-2	2		24	1.40	21.6
SF2-1	1	13-mm straight steel (65)	16	1.44	23.4
PVA1-1	1	30-mm PVA (45)	20	1.27	20.6
PVA1-2	2		23	1.23	23.0

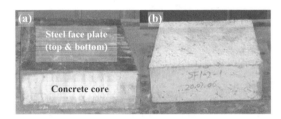

Figure 1. (a) SCS sandwich panel; (b) LC panel.

Table 1. Properties of SCS sandwich panels.

Sandwich ref. no.*	t_{steel} (mm)	Core material	S. gravity of core	f_{cu} (MPa)
FC4-60-0	4	FC	1.20	16.0
FC6-60-0	6	FC	1.20	16.0
FC8-60-0	8	FC	1.20	16.0
FC4-60-1	4	FC-1 % PVA	1.20	16.3
FC4-60-2	4	FC-2 % PVA	1.20	16.9
FC4-60-3	4	FC-3 % PVA	1.20	17.9
LC4-60-0	4	LC	1.44	27.4
LC4-60-1	4	LC-1% PVA	1.44	28.7
LC4-60-2	4	LC-2% PVA	1.44	28.2
NC6-60-0	6	NC	2.40	83.0

*FC4-60-0 = core type (FC = foam concrete; LC = light aggregate concrete; NC = normal concrete), steel plate thickness (mm), core thickness (mm), fiber volume fraction (%).

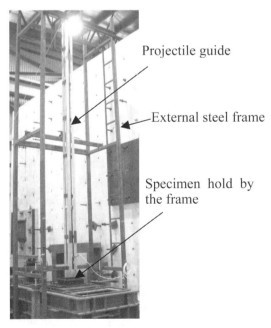

Figure 2. Experimental set-up for impact on SCS sandwiches.

3 IMPACT TEST RESULTS AND DISCUSSION

After each impact test, the damage level was evaluated based on the indentation depth, average dent diameter, and crack propagation in the concrete core. For the

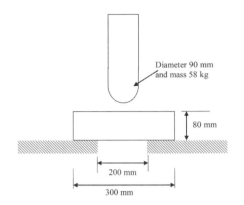

Figure 3. Schematic diagram of test set-up for LC panel.

Figure 4. Local impact damage (indentation) of SCS sandwich panel due to projectile impact.

SCS sandwich panels, the steel plates were dented by the impact (Fig. 4) regardless of the composition of the concrete core material. In the SCS panels, the concrete cores were cracked due to the impact with crack-lines propagated radially from the center (Fig. 5a & b). For those with plain concrete cores (without fiber), the cracking was comprehensive with several pieces of the core material separating entirely from each other along with numerous fragments (Fig. 5a). For those with FC core, the concrete pulverized below the point of impact between the two steel plates (Fig. 5c & d). For NC core, the concrete had less pulverization. In the case of fiber-reinforced cores, the fibers held the core material together after cracking (Fig.5c to f), and the concrete could therefore be lifted out in one piece after impact though deep radial cracks were formed. Dent was formed at the point of impact on the top surfaces of the cores similar to that formed on the steel plates.

For the impact test on LC panels, the aim was to evaluate the impact performance of the panels with different fiber types and dosages. The results revealed good impact resistance for the panels reinforced with steel fiber of higher aspect ratio and possessing hooked-ends anchorage geometry. Typical damage on the impact face after first impact for some panels is shown in Fig. 6.

3.1 Denting in SCS sandwich panel

The dent on the top steel plate of each test specimen was measured after the impact. A transducer was used to measure the depth of the permanent deformation at interval points of half-centimeter apart along a diameter on the circular area of exposure to impact. The

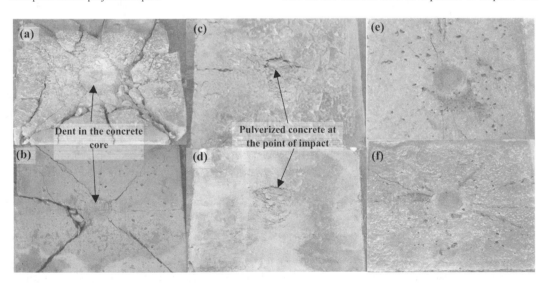

Figure 5. Drop-weight impact damage on concrete cores (a) FC4-60-0, (b) NC6-60-0, (c) FC4-60-2, (d) FC4-60-3, (e) LC4-60-1, and (f) LC4-60-2.

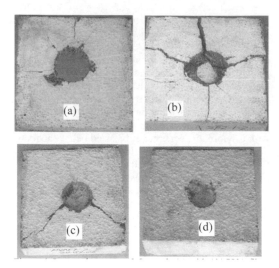

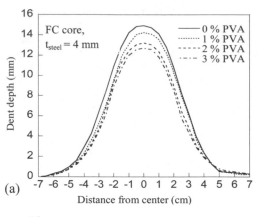

Figure 6. Impact damage on LC panels (a) with 1% PVA fiber (b) with 1% straight steel fiber (c) with 2% PVA fiber (d) with 2% hooked-ends steel fiber.

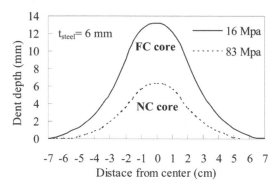

Figure 7. Permanent dent profile on face plate of SCS sandwich panels with different core compressive strength.

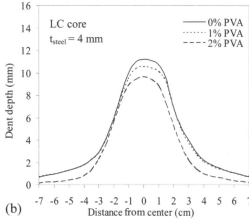

Figure 8. Permanent dent profile on face plate of SCS sandwich panels with different volume fractions of fiber.

Table 3. Impact test results of SCS sandwich panel.

Sandwich ref. no.	f_{cu} (MPa)	Max. impact force (kN)	Permanent indentation (mm)
NC6-60-0	83.0	358	6.2
FC4-60-0	16.0	195	15.0
FC6-60-0	16.0	215	13.2
FC8-60-0	16.0	241	11.0
LC4-60-0	27.4	264	11.2
LC4-60-1	28.7	—	10.5
LC4-60-2	28.2	278	9.6

process was repeated thrice to obtain an average dent profile. The dent profiles for the sandwich specimens, which were generally parabolic in shape, are plotted in Figs. 7 and 8 and given in Table 3.

The maximum dent depth reduced significantly with increase in compressive strength of the core material. In sandwich panels with FC (16 MPa) and NC (83 MPa) cores, the maximum dent depth of 13.2 and 6.3 mm were observed, respectively (Fig. 7). Besides this, increased volume fraction of fiber in the core material reduced the maximum dent depth (Fig.8). The dent depth was also reduced significantly with the increase of steel face plate thickness. As the steel face plate thickness increased from 4 to 8 mm, the maximum dent depth decreased from 15.0 mm to 11.0 mm (Table 3). The reason may be related to two effects: local bending and membrane action of the steel face

plate on the core which is directly related to the plate thickness and the ability of a thicker plate to distribute the applied load to a larger contact area since this distribution is also directly related to thickness.

Thus it can be concluded that for sandwich panel the dent depth is a function of the compressive strength of the core material, the volume fraction of fiber in the core, and the face plate thickness. The increase in any

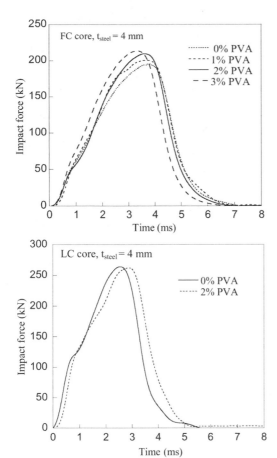

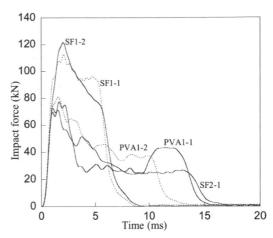

Figure 10. Impact force history of LC panels with different fiber dosage and type.

Figure 9. Effect of fiber (PVA) on the impact force history of SCS sandwich panels.

of these three parameters will lead to a reduction in the dent depth.

3.2 Impact force and time history

The observed impact reaction force versus time traces is shown in Figs. 9 and 10. Upon impact, the SCS sandwich panel experienced a sudden increase in impact force which rises to the maximum value. There are two phases, one is loading and another is unloading phase. The time to reach peak impact force was very short because of the hard contacts between the steel projectile and the steel plate. The peak loads may be influenced by various parameters. However, in the present study, the nose shape of the projectile, core thickness and the boundary conditions were kept constant and hence, the peak load was affected mainly by the drop height of the projectile, face plate thickness and type of concrete core in the specimen. Core strength had important effects on the loading response

of the sandwich panels. The impact test indicated that the maximum impact force generally increased with increasing core strength as indicated in Table 3.

This can generally be explained by a stiffness argument as a weaker core material deformed more easily upon impact and thus a lower reaction force was generated. The use of fiber in the core material also had little effect on the impact force history (Fig. 9) as the impact force increased slightly when the volume fraction of PVA fiber increased at similar impact velocity.

The impact force response for the sandwich specimens was affected by the steel face plate thickness. This is shown in Table 3 that the impact force increased as the face plate thickness increased at similar impact velocity. The peak impact force increased about 10% as the steel face plate thickness increased from 4 mm to 6 mm, and increased 12% when steel face plate thickness increased from 6 mm to 8 mm (Table 3). This can be attributed to the increase in bending and membrane stiffness when the steel face plate thickness increased.

The second series of impact test consisted of five LC panels that contained PVA or steel fibers as shown in Table 2. The aim was to evaluate the impact performance of the panels with different fiber types and dosages. The LC panels with steel fibers had higher unit weight of about 1400 kg/m^3 while those with PVA fiber had lower unit weight of about 1250 kg/m^3. Figure 10 shows the load–time curves for the LC panels under impact. From the figure, it is apparent that the specimens which contained PVA and straight steel fibers have near-identical performance. The significant increase in load capacity of LC panels with hooked-ends steel fiber is due to a stiffer panel indicated by higher maximum impact force generated upon impact (Fig.10), compared to the others.

Table 4. Impact test results of LC panels.

Panel ref.	% Fiber	v (m/s)	No. of cracks Top	No. of cracks Side	No. of cracks Bottom	Max. Impact Force (kN)
SF1-1	1	8.0	1	5	5	113
PVA1-1	1	8.0	12	15	16	77
SF2-1	1	8.0	Become 4 pieces			72
SF1-2	2	8.0	0	5	6	122
PVA1-2	2	8.0	4	7	14	81

Generally concrete panels with hooked-ends steel fiber are found to show significantly improved dynamic shear resistance. This may be due to a higher aspect (length-diameter) ratio and an effective hooked-ends geometry improving the fiber anchorage. Typical damage on the impact face after first impact for some panels is shown in Fig. 6. For panels reinforced with PVA fiber or straight steel fiber, the impact caused a through hole at the impact point. On the other hand, panels reinforced with hooked-ends steel fiber exhibits only dent on the top surfaces and some cracks in the bottom surface.

4 CONCLUSIONS

The behavior of SCS sandwich panels under low velocity impact was studied experimentally. In general, the generated impact reaction force varied inversely with the permanent indentation. For a given impact energy; the impact reaction force and permanent indentation are function of the steel plate thickness, core compressive strength, materials' elasticity and fiber volume fraction. From the results, it is observed that the impact reaction force increased and permanent indentation decreased when the steel plate thickness, core compressive strength or fiber volume fraction increased, or when the elasticity of the materials decreased.

Using 1 to 2% fiber in the concrete core reduced cracking in terms of the width, length and numbers. The fibers also ensured that the core remained as a single piece after the impact. For both foam and lightweight aggregate concrete cores, the use of 2–3% PVA fiber minimized radial cracks in the core.

The impact performance on lightweight aggregate concrete (LC) panels showed that under the drop weight hammer impact for punching-shear, the panel reinforced by 30-mm hooked-ends steel fiber of between 1–2% volume fractions exhibited improved resistance and reduced damage compared to the other LC panels, which were reinforced by either 13-mm straight steel fiber or 30-mm PVA fiber. This could be due to a higher aspect (length-diameter) ratio and an effective hooked-ends geometry improving the fiber anchorage.

REFERENCES

Bergan P.G. & Bakken K. 2005. Sandwich Design: a Solution for Marine Structures? *International Conference on Computational methods in Marine Engineering*, Marine 2005, Oslo, Norway, 27–29 June.

Kennedy S.J. & Kennedy J. L. 2004. Innovative use of sandwich plate system for civil and marine applications. *International Symposium on Innovation and Advances in steel structures*, 30–31 August 2004, Singapore: 175–185.

Ong K.C.G., Basheerkhan, M., & Paramasivam, P., 1999. Resistance of fibre concrete slabs to low velocity projectile impact. *Cement & Concrete Composites*, 21(5–6): 391–401.

Sohel K.M.A., Liew J.Y.Richard, Alwis W.A.M. & Paramasivam P. 2003. Experimental investigation of low-velocity impact characteristics of steel-concrete-steel sandwich beams. *Steel and Composite Structures*, 3(4): 289–306.

Steel and Composite Structures – Wang & Choi (eds)
© 2007 Taylor & Francis Group, London, ISBN 978-0-415-45141-3

Behaviour of different types of columns under vehicle impact

M. Yu & X.X. Zha
Shenzhen Graduate School, Harbin Institute of Technology, Shenzhen, China

Y.C. Wang
University of Manchester, UK

ABSTRACT: This paper presents the results of a numerical simulation study of the behaviour of vehicle impact on different types of columns. The simulations were performed using the finite element software LS-DYNA, which was verified to perform this type of analysis by comparing the simulation results against three impact tests on concrete filled tubular columns, the results of which show good agreement between the simulation and test results. A numerical parametric study was then performed to investigate the effects of three types of vehicles impacting on three types of commonly used columns (steel tubular columns, reinforced concrete columns and concrete filled tubular columns). The parametric study covered different vehicle approaching velocities, different column sizes and different boundary conditions. This paper will present the impact force results and discuss how the impact force may be affected by the aforementioned different parameters considered in the simulations.

Keywords: Car-pole impact, Concrete filled steel tube, Impact force, finite elements, digital filter

1   INSTRUCTION

Since the accident of Maraca Ibo Bridge disaster in May 1964, there have been many serious accidents of ships ramming bridge piers. In the Chernobyl Nuclear Power Plant disaster, the debris generated by the initial explosion fiercely impacted the surrounding concrete structures, and caused the structures to deform and finally collapse, which led to the disastrous leak of nuclear materials. The 9″ 11″ incident, which shocked the world, is the direct result of plane impact on the Twin Towers. The losses of these disasters are incalculable, making it imperative to devote efforts to understand the effect of impact on structures so that the engineering community can design safe structures for the protection of lives and properties. Meanwhile, with rapid development of the society, transport has become an indispensable part in modern lives. Whilst the safety of people under vehicle impact is one of the critical issues of car manufacturers, there is also a need to understand the effect of vehicle impact on structures when they are involved. However, in current codes, these impact loads are seldom considered in design, and only a few countries have some relevant design guides, such as the British Standard BS5400 [12]. However, these design guides generally give calculation methods of vehicle impact load based on experience and prescribe structural detailing rules to ensure

structural safety. The related calculation methods and designing requirements are very rough.

Columns are important structural components and their safety directly affects safety of the entire structure. This paper will present the results of a finite element simulation study of the effects of vehicle impact on structural columns, giving impact forces on steel tube columns, reinforced concrete columns and concrete-filled steel tubular (CFT) columns impacted by three representative vehicles with difference velocities.

2   COMPARISON BETWEEN RESULTS OF IMPACT TESTS ON CFT COLUMNS AND FINITE ELEMENT SIMULATIONS

2.1   *Description of tests and results*

The impact tests [5, 6] on CFT columns were conducted at Taiyuan Science and Engineering University, China. In total, the tests included impact tests on 26 circular CFT specimens and 3 hollow steel tube specimens, and 3 static load tests on CFT specimens. The tests were carried out by dropping falling hammers on the test specimens, from a variable height up to 13.5 m. The highest impact speed was 16 m/s. The hammer had a wedge flat impact head and was made of chromium 15 with a hardness of 64 HRC. The size of the impact

Figure 1. Typical deformed specimen.

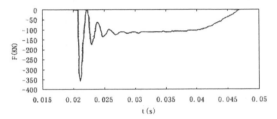

Figure 2. Typical impact force – time curve.

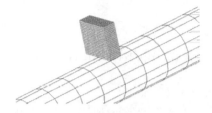

Figure 3. Finite element model of the drop hammer on a CFT member.

area was 30 mm × 80 mm and the weight of the hammer was 203 kg.

2.2 Test results

This paper will analyze 3 circular CFT column tests all having the following properties: external diameter 114 mm, overall specimen length 1300 mm, clear span 1200 mm, and steel tube thickness 3.5 mm. The average cube strength of the concrete was 47.5 N/m², the yield strength of steel was 2.41×10^8 N/m² and ultimate strength of steel was 3.44×10^8 N/m². The boundary condition was one end fixed and the other simply supported on a roller. The three tests analysed had three different hammer falling heights: 6 m, 8 m and 8.4 m. Figure 1 shows one deformed specimen after test. Figure 2 shows a typical recorded impact force versus time curve of the test specimens.

As can be seen from figure 2, after impact, the impact force to the specimen reached the peak value. This was then followed by vibration of test specimen causing fluctuation in the impact force. Afterwards, the impact force reached a plateau for a period of time before finally disappearing. The appearance of the plateau is a consequence of "plasticity hinge" which develops after the peak impact force and resists a constant moment.

2.3 Finite element model and results

The following properties of steel and concrete were used in numerical simulations:

Steel: density: 7850 kg/m³; elastic modulus: 206000 N/mm²; tangent modulus: 660 N/mm²; yield strength: 241 N/mm²; Poisson ratio: 0.3; failure strain: 0.2.

Concrete: According to Holmquist [4] with a compressive strength of 32 MPa, which can simulate well the dynamic effect of concrete under large strain and high strain rate.

Shell elements and solid elements were used for the steel tube and concrete core respectively. The drop hammer was modeled as a rigid body of cuboid with a size of 80 mm × 80 mm × 30 mm to ensure that the contact between the drop hammer and the steel tube is consistent with the test condition. Automatic Single Surface Contact was adopted between the drop hammer and the concrete tube to be consistent with later simulations of vehicle impact. The model used the same experimental boundary condition of one end being fixed and the other being simply supported. Figure 3 shows a view of the finite element model.

Under impact, the interface between the steel tube and the concrete core is considered to have negligible influence on the behaviour. Shared nodes between the steel tube and concrete core were adopted. However, this would cause initial penetrations, which would induce the inappropriate contact behaviour of negative slipping interface energy. To avoid the problem of initial permeation, viscous contact damping was introduced in the model with a viscous damping coefficient (VDC) of 40. Calculation of the contact stiffness was according to a segment-based contact algorithm in LS-DYNA, with the value of "SOFT = 2". This treatment was found to be very effective for handling the contact problem and eliminated mutual permeation at contact interface.

In impact analysis, the control of hourglass is very important. Generally hourglass cannot exceed 10% of the total energy. Through trial and error, it was found that hourglass control was best achieved with the hourglass No. 3 option in LS-DYNA. Figure 4 shows that the hourglass energy was only 0.78% of the total energy. Figure 5 compares the simulated and test final deformations and figure 6 presents the simulated impact force versus time curves for one test.

The simulated impact force time-deformation curve contains high frequency noise (which becomes more obvious in vehicle impact simulation), it is necessary

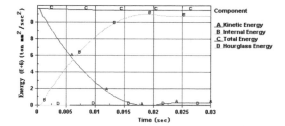

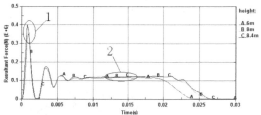

(a) Impact force – time curves after digital data filtering

Figure 4. Evolution of various energy quantities for drop hammer height of 6 m.

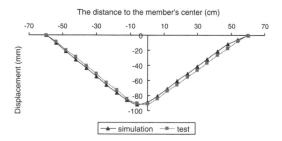

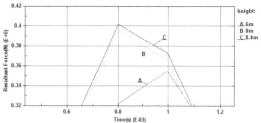

(b) Enlarged view of point 1 in (a)

Figure 5. Comparison between final deformations between simulation and test for drop hammer height of 6 m.

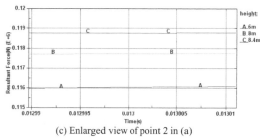

(c) Enlarged view of point 2 in (a)

Figure 7. Simulated results of impact force versus time curve filtering high frequency noise.

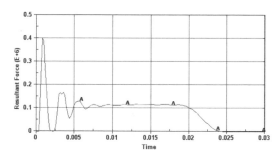

Table 1. Comparison of test results and finite element results of impact forces.

Drop height (m)		6	8	8.4
Peak	Test results	357	389	408
Value	FE results	401	496	485
(Kn)	FE results after filtering	355	395	402
Plateau	Test results	114	117	116
value	FE results	116	117	118
(kN)	FE results after filtering	116	118	119

Figure 6. Simulation result of impact force versus time curve for drop hammer height of 6 m.

to filter out the high frequency noise. By using the arithmetic average value method to digitally filter the raw data by setting Point Average being 2; it is possible to remove the effect of high frequency noise from the raw time-deformation curve to obtain more appropriate impact force – time curve. Figure 7 shows the effect on force – time curve after data filtering.

Table 1 compares the test and finite element (FE) simulation results of peak and plateau values of impact force for the three tests. The simulated peak impact force is affected by the high frequency noise so the FE results after filtering are much lower than the raw FE results. The FE results after filtering are very close to the test results. The plateau results occur in the low

frequency region so they are not affected by filtering. Again they agree with the test results very well.

To summarize, the software LS-DYNA can be used to simulate impact on columns. However, it is important that the hourglass energy and the contact viscous damping coefficient (VDC) are appropriate. For this study, the hourglass energy should be minimized and a value of VDC of 40 has been found to provide satisfactory results. Also digital filtering should be applied to eliminate the influence of high frequency noise.

3 PARAMETRIC STUDY OF VEHICLE COLLISION ON DIFFERENT TYPES OF COLUMNS

3.1 Finite element model of cars

To simulate the effect of vehicle impact on structures, it is important to use validated finite element models of the vehicles. This study will use three models, which have been extensively tested by others. These three models may be used to represent a mini-car, a light truck and a medium weight truck. They are:

1. Geo Metro, Reduced model (16,100 elements);
2. Chevrolet C2500 Pickup, Reduced model (10,500 elements);
3. Ford Single Unit Truck, Reduced model (21,400 elements).

The finite element models of these vehicles have been downloaded from the website of NCAC (National Crash Analysis Center), which may also be downloaded from the website of US Highway Traffic Safety Administration (NHTSA). Since the focus of this study is on the behaviour of the columns, the reduced models of the vehicles were used. Detailed FE models of these vehicles can be found from the same websites.

These vehicle models and others like them have been developed and validated by the US Federal Government for use in crashworthiness simulation. The models were created through a painstaking process, whereby the geometry of various parts of each vehicle was digitized and imported into finite element modeling software [1].

3.2 Finite element models of various columns

Simulations have been performed to compare the performance of the following three types of columns: steel tube columns, reinforced concrete columns and concrete-filled steel tube columns. As before, shell elements were used for the steel tube and solid elements for the concrete. Beam elements were used for the steel reinforcing bars. The steel tube thickness is 10 mm and the height of the column is 3600 mm. Eight 40 mm rebar were assumed to be in RC columns. Figures 8 and 9 show the steel and concrete material models and Tables 2 and 3 give the values of the various material properties as requested for input in LY-DYNA.

When conducting vehicle impact simulation using LS-DYNA, it is important to set the parameters in the software to their optimal values for efficient and appropriate simulation. Some guidance on selecting appropriate parameter values can be found Zhao[3].

A large number of simulations have been carried out. Each simulation case will be referred to by a four digit number XXXX, with the 1st to 4th digitals representing vehicle type (Geo Metro, Chevrolet, Ford

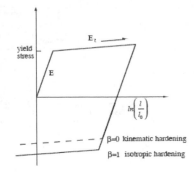

Figure 8.   Constitutive model of steel.

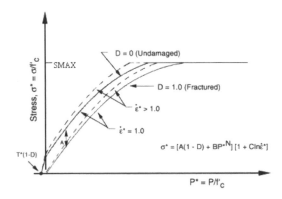

Figure 9.   Constitutive model of concrete.

Truck), speed (30 km/h, 50 km/h, 100 km/h), boundary condition (fixed-free, fixed-pinned, fixed-fixed) and column type (Steel, RC, CFT) respectively.

4 FINITE ELEMENT SIMULATION RESULTS

4.1 Effect of hourglass energy

Single point Gauss integration is employed to conduct nonlinear dynamic analysis in LS-DYNA, which can greatly save the computation time and also is efficient for large deformation analysis. However, single point integration can cause the simulation to become a zero energy model, also called hourglass, which makes the simulation results invalid. Therefore, it is important to the control hourglass energy. If the hourglass energy exceeds 10% of the total energy, the results may not be valid. Sometimes the hourglass energy should not even exceed 5% of the total energy. The validation exercise in section 2.2 indicates that the hourglass energy was very strictly controlled, which gave very good agreement between test and simulation results. Figure 10 presents a typical breakdown of the various energy terms and it indicates that very good control of the hourglass energy was achieved.

Table 2. Values for steel stress-strain curve (mm, kg, S, N)

RO	E	PR	SIGY	ETAN
7850	210000	0.30	300.0	1000.0

Table 3. Values for concrete stress-strain curve (mm, kg, S, N)

RO	G	A	B	C
2440	14860.0	0.79	1.6	0.0070
N	FC	T	EPS0	EFMIN
0.61	48.0	4.0	1.0	0.010
SFMAX	PC	UC	PL	UL
7.0	16.0	0.0010	800.0	0.10
D1	D2	K1	K2	K3
0.0400	1.0	85000.0	−171000.0	208000.0

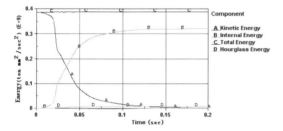

Component	
A	Kinetic Energy
B	Internal Energy
C	Total Energy
D	Hourglass Energy

Figure 10. Evolution of various energy quantities for Geo Metro with speed of 100 kph impacting on a CFT column with both ends fixed (No. 1333).

4.2 Deformations and impact forces

Figures 11–13 present the impact force – time curves after digital filtering with CFC 60 [8, 9]. It can be seen that the shape of the impact force – time curve is dependent on the type of vehicle and column. Under impact by a Geo Metro (figure 11) or a Chevrolet Truck (figure 12), all three types of column show similar behaviour (with different magnitudes of impact forces): the maximum peak impact force followed by gradual diminishing. Under impact by a Ford Truck (figure 13), the impact force – time curve of the steel column shows a prolonged period of low impact force, but the curves of RC and CFT columns develop two peaks. The main reasons are that the mass of a Geo Metro or a Chevrolet Truck is relatively uniformly distributed, which may be assumed to be a mass block. In contrast, the mass of a Ford Truck is concentrated in the head and back of the vehicle, which may be assumed to be two mass blocks connected by a spring. When a column is impacted by a Geo Metro or a Chevrolet Truck, the momentum is the highest when the vehicle initially contacts the column so that the first wave

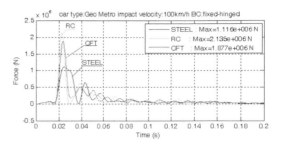

Figure 11. Impact force versus time curve for three columns impacted by Geo metro with speed of 100 kph.

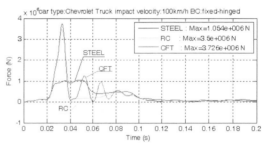

Figure 12. Impact force versus time curve for three columns impacted by Chevrolet Truck with speed of 100 kph.

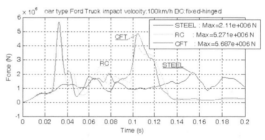

Figure 13. Impact force versus time curves for three column impacted by Ford Truck with speed of 100 kph.

crest first appear, but the impact force diminishes with their gradual contact action; for a Ford Truck, the two peaks simply reflect the two occasions of the head and back of the truck impact the column at different times. Under impact by a Ford Truck, the behaviour of the steel column is different because it is flexible (so the impact force is low) and when the head of the truck impacts the steel column, the column is flattened so the impact by the back of the truck cannot be felt by the column.

Table 4 presents the peak values of impact force for all the simulation cases. The following may be observed: (1) The higher the impact speed (parameter 2), the higher the peak impact force. However, the peak force does not vary with the impact speed linearly. (2) The type of column end supports has some influence

Table 4. Peak values of impact force after filtering with CFC60.

Parameter and number			Vehicle type					
			Geo Metro		Chevrolet Pickup		Ford Truck	
			Speed (km/h)					
			30	100	30	100	30	100
Connecting type (parameter 3) 1	Column type (parameter 4)	1	304	1077	398	1040	578	1923
		2	362	1989	398	3700	579	5306
		3	327	1655	397	3702	582	5354
2		1	354	1116	401	1054	639	2110
		2	349	2135	403	3500	594	5271
		3	402	1877	403	3726	614	5687
3		1	383	1140	401	1242	615	2823
		2	351	2210	405	3848	598	5357
		3	327	2099	397	3882	659	5441

Note: see table 4 for definition of different parameters.

on the peak impact force, but the effect is relatively minor. This comes about because the impact time is very short so the impact is felt locally by the column with very little effect by the end conditions. (3) Since the RC and CFT columns are much stiffer than the steel column, except at the low impact speed of 30 kph, the impact forces on the steel column are much lower in all cases than the impact forces on the RC and CFT columns, which are very similar. At the impact speed of 30 kph, the impact forces on all three types of column are similar. At this impact speed, the impact forces on all columns are low and the impact force is mainly affected by the vehicle type, rather than the column. (4) The impact force depends not only on the weight and stiffness of the vehicle, but also the distribution of the weight stiffness of the vehicle.

The above vehicle types represent a mini car, a small truck, and a medium truck respectively. For the CFT column used in the study with fixed-hinge end supports, the peak values of impact force are as follows: at 30 kN/h, they are 402 kN, 405 kN, and 659 kN for the three different types of vehicle; at 50 kN/h, they are 989 kN, 1305 kN, and 2065 kN respectively; at 100 kN/h, they are 2210 kN, 3882 kN, and 5687 kN respectively. For comparison, the car loading values currently recommended in the Chinese code JTG D60 – 2004 are: 1000 kN along the car running direction and 500 kN perpendicular to the car running direction and BS5400 are: 500 kN along the car running direction and 100 kN perpendicular to the car running direction, which can be unsafe.

5 CONCLUDING REMARKS AND ONGOING WORK

This paper has presented the results of a numerical simulation of vehicles impacting on steel, reinforced concrete and concrete filled tubular columns. The software LS-DYNA was used and it was verified for this application by comparing the simulation results with three impact tests on CFT columns under a dropping hammer. The following general conclusions may be drawn:

1. The mass, stiffness and their distributions of the impacting vehicle have significant influences on the peak impact force on the impacted columns;
2. The peak impact force increases with the impact speed, but the relationship is not linear;
3. The effect of the end support conditions of a column on the peak impact force is small;
4. The column type has major influence on the peak impact force, especially at high impact speeds;.
5. The Chinese code criterion for automobile load may not be unsafe.

This study is a preliminary investigation of vehicle impact on columns. A large amount of research studies are necessary before a practical design method to estimate vehicle impact forces on columns can be developed. Future research studies will consider column load carrying capacity and loading condition, vehicle characterization (stiffness, mass and their distributions), impact position and development of simplified theoretical models.

REFERENCES

[1] Sherif El-Tawil, Severino E and Fonseca P (2005), Vehicle Collision with Bridge Piers. JOURNAL OF BRIDGE ENGINEERING, ASCE, Vol. 10, pp.345–353.
[2] Fasanella E L and Jackson K E (2002), Best practices for crash modeling and simulation, NASA/TM–2002–211944.
[3] H O Zhao (2003), LS-DYNA dynamic analysis and guidance, Bing Qi Gong Ye Press (Armory Publication House), (in Chinese).
[4] Holmquist T J, Johnson G R and Cook W H (1993), A computational constitutive model for concrete subjected to large strains, high strain rates, and high pressures, Proceedings of 14th International Symposium on Ballistics, 591–600.
[5] Wang K Z (2005), Experimental Research on Concrete-Filled Steel Tubes under Lateral Impact Load, MSc thesis, Taiyuan University of Technology, (in Chinese).
[6] Gu D B (2005), Accident Research on Steel Tube-Confined Concrete under Lateral Impact, MSc thesis, Taiyuan University of Technology, (in Chinese).
[7] Li H J (2003), A Study of Side Crashworthiness of Passenger Car Body, MSc thesis, JiLin University, (in Chinese).

[8] International Standards Organization (1980), ISO 6487: Road vehicle-crash test measuring and testing techniques, Compiled by Chang Chun car institute (in Chinese).

[9] SAE (1984), J211 Instrumentation for impact test, SAE Handbook.

[10] Chen H (1998), Research on the Application of Digital Filter Technology in Auto Crash Test, Automotive Engineering (in Chinese).

[11] JTG(2004), JTG D60 – 2004: General Code for Design of Highway Bridges and Culverts, (in Chinese).

[12] BSI (1988), BS 5400: Steel, concrete and composite bridges, Part 2: specification for loads, British Standards Institution, London.

Steel and Composite Structures – Wang & Choi (eds)
© *2007 Taylor & Francis Group, London, ISBN 978-0-415-45141-3*

Impact analysis of a composite guardrail

M.M. Shokrieh, K. Daneshjoo & M. Esmaeili

Composite Research Laboratory, Mechanical Engineering Department, Iran University of Science and Technology, Tehran, Iran

ABSTRACT: Glass fiber reinforced plastics (GFRPs) are lightweight and non-corrosive materials and these characteristics are highly desirable for the highway guardrail applications. Standard pendulum impact testing of the composite guardrail prototypes was conducted at the federal outdoor impact laboratory (FOIL) at the Turner Fairbank Highway Research Center in McLean, Virginia, USA. To gain an insight about the pendulum impact test details, a finite element simulation study was performed using the explicit finite element code LS-DYNA. Accuracy of the simulation was verified using qualitative and quantitative comparisons. The results of the simulation showed that the failure of the composite guardrail prototype during the pendulum impact test was due to an insufficient transverse crushing capacity of the central portion of the prototype. In the light of this finding, the current composite guardrail prototype model was modified using additional tubes in the central portion. Subsequent simulation results predict that the improved model performs successfully according to requirements of occupant ridedown accelerations and has adequate ability to resist concentrated loads.

1 INTRODUCTION

Fiber reinforced polymer (FRP) composites continue to be the most popular for civil engineering and transportation industry applications. Bridge supports, bridge reinforcement, bridge decking, sign supports, lighting poles and guardrail systems are examples. Appropriately designed composite material structures can be highly energy-absorbent, lightweight, durable and cost effective alternatives to conventional metal, concrete or timber structures. The highway guardrail is a specific product that will be sold in mass quantities.

Beginning in 1990, the researchers began to study the possibility of using FRP composite materials for highway guardrail systems (Dutta 1998; Svenson 1994; Svenson et al. 1994; Svenson et al. 1995). Researchers have conducted fundamental and applied research and development studies to evaluate the feasibility of using pultruded composite materials to develop a composite material highway guardrail.

Impact tests were conducted to understand the behavior of glass fiber reinforced composite materials when subjected to dynamic loading (Svenson et al. 1992).

Design of connections and joints is an important part of the design process of the guardrails. Multi-bolted joints were tested in tension to determine the effect on tensile capacity of varying the width of the specimen, the side distance, the edge distance, the bolt pattern, and the number of bolts for pultruded E-glass and isophthalic polyester sheets (Hassan et al. 1997).

Static tests were conducted on FRP guardrail specimens with rectangular cross sections (Bank & Gentry 2001). The specimens tested were constructed from standard pultruded profiles produced with vinyl-ester resin and E-glass fiber reinforcement. The tests were conducted to investigate the effects of overall guardrail geometry, material properties and guardrail configuration.

Based on the successful results of the static tests on the guardrail specimens, a full scale pendulum impact test was performed at the Turner Fairbank Highway Research Center in McLean, Virginia. More information on testing setup and time histories from each of the experiments on composite rails can be found in (Brown 1998).

2 PROBLEM STATEMENT

According to the conditions of NCHRP Report 350 (Ross et al. 1993), for the safety performance evaluation of highway structures, three dynamic performance evaluation factors are given together with recommended evaluation criteria and applicable tests. The factors are: (1) structural adequacy; (2) occupant risk; and (3) post-impact vehicular response. By pendulum impact testing, the structural adequacy and occupant risk factors can be evaluated. The pendulum impact tests were conducted to compare the dynamic response from standard w-beam guardrail with the dynamic response of an FRP composite rail (Svenson & Brown 1998; Brown 1998). The tests used a 900 kg pendulum at impact velocities ranging from 10 to 35 km/h. Tests were conducted with end restraint and without end restraint to determine the effect on tensile capacity. An accelerometer was attached to the pendulum to obtain data. The acceleration data was converted into velocity and displacement data by integration. The specimens tested without end restraint were not capable of stopping the pendulum at high impact speeds. The connections to the test system failed. The specimens tested with end restraint demonstrated a behavior similar to the behavior of a steel guardrail. The FRP prototypes, specimen CGR10, were tested with each end rigidly fixed using standard 25-mm-diameter cable anchors tightened. Cable anchor brackets were attached to the backside of the FRP rail, one at each end. The anchor brackets were welded to a steel plate bolted to the FRP rail. The cables were passed between the rail anchor bracket and the anchor stanchions and were fastened with a 25 mm cable-nut and washer at each end. The blockouts affixed to the FRP rails were attached to two standard strong-posts (I-sections) using two standard blockout-to-post bolts. The FRP rail, specimen CGR10, was fabricated from several extruded FRP rectangular box-sections, bonded together and then bonded to a 6-mm-thick FRP sheet with an epoxy resin. In addition to the resin, small self-tapping screws were used to attach the FRP sheet to the FRP box-sections.

In this study to gain an insight about the pendulum impact test details, a finite element simulation study is performed using the explicit finite element code LS-DYNA. Accuracy of the simulation will be verified using qualitative and quantitative comparisons.

The intent for the analysis was:

- To accurately simulate the observed full-scale pendulum impact test behavior and further investigate the design shortcomings.
- To compare the effects of different design alternatives that result in acceptable pendulum impact test performance in a cost-effective manner.

The baseline model consisted of different components. These components include: steel posts and blockouts, standard steel cables, steel plates, multi-cellular pultruded FRP rail, specimen CGR10, and the pendulum. The nose of pendulum was a rigid 150 mm diameter steel tube and only this tube was modeled as a representative of the pendulum.

3 FINITE ELEMENT SIMULATION

3.1 Element properties

Four node thin shell elements were used throughout to model the surfaces of baseline model components.

Additionally, the LS-DYNA default Belytschko-Tsay shell element formulation was selected because of its computational efficiency. Bank used spot-welds to model the reduced strengths of the composite tubes at the corners (Bank & Gentry 2001). The composite tubes were modeled as four separate plates that were attached together using spot-welds. Results of the simulations showed the disagreement between the experimental and simulated impact test results coming from the fact that spot-welds tended to unzip too quickly in the initial stages of the simulation. The spot-weld is a rigid beam that connects the nodal points of the nodal pairs; thus, nodal rotations and displacements are coupled. Spot-weld failure parameters were calculated from the transverse tensile and in-plane shear strengths of the composite material. More information on details of spot-weld failure can be found in LS-DYNA theoretical manual (Hallquist 1998). It was determined that using spot-welds to connect the separate surfaces of multi-cellular pultruded FRP rail is not necessary. Beam elements were used to model the standard steel cables and truss beam element formulation was selected for these beam elements.

3.2 Material properties

Four different types of material models were used from the library of the LS-DYNA code in the development of finite element analysis described herein:

- Material Model 54 (mat-enhanced-composite-damage), which models damages in composite materials.
- Material Model 24 (mat-piecewise-linear-plasticity), which models a piecewise linear plasticity for metals.
- Material Model 20 (mat-rigid), which models rigid materials.
- Material Model 1 (mat-elastic), which models elastic materials.

For the multi-cellular pultruded FRP rail modeling, the Material Model 54, with the Chang and Chang failure criterion was used, which is applicable to such materials and more suitable than the other related available ones, for this case. The material properties used for the composite materials are shown in Table 1. For modeling of steel posts and blockouts, the Material Model 24 was used. By using this material model, an elasto-plastic material with an arbitrary stress versus strain curve and arbitrary strain rate dependency can be defined. Also, failure based on a plastic strain or a minimum time step size can be defined. The material properties used for the steel posts and blockouts are shown in Table 2. Moreover, for modeling of the nose of pendulum, the Material Model 20 was used, since its deformation was neither necessary nor desirable

Table 1. Typical material properties used for composite materials (Smith et al. 2000).

Parameter	Value		
	MPa	kg/mm³	mm/mm
ρ		1.939×10^{-6}	
ν_{ba}			0.10
E_a	20.69×10^3		
E_b	6.89×10^3		
G_{ab}	2.5×10^3		
G_{bc}	1.25×10^3		
G_{ca}	2.5×10^3		
X_C	0.207×10^3		
X_T	0.207×10^3		
Y_C	0.103×10^3		
Y_T	0.048×10^3		
S_C	0.069×10^3		

Table 2. Material properties of steel posts and blockouts (Atahan & Cansiz 2005).

Parameter	Value		
	MPa	mm/mm	kg/mm³
ρ			7.85×10^{-6}
ν		0.30	
E	200,000		
σ_y^*	336.80		
$(\sigma - \varepsilon_e^p)_1^{**}$	336.80	0.000	
$(\sigma - \varepsilon_e^p)_2$	337.10	0.026	
$(\sigma - \varepsilon_e^p)_3$	401.20	0.045	
$(\sigma - \varepsilon_e^p)_4$	490.90	0.108	
$(\sigma - \varepsilon_e^p)_5$	555.20	0.203	
$(\sigma - \varepsilon_e^p)_6$	604.60	0.254	
$(\sigma - \varepsilon_e^p)_7$	657.70	0.277	
$(\sigma - \varepsilon_e^p)_8$	677.40	0.300	

* Yield stress.
** Stress versus effective plastic strain.

in this investigation. A nose mass of 912 kg, which was similar to the mass of the pendulum used in the pendulum impact test, was utilized. Elements, which are rigid, are bypassed in the element processing and no storage is allocated for storing history variables (no stresses, no strains occur in a rigid body); consequently, the rigid material type is very cost efficient. And finally for modeling standard steel cables, the Material Model 1 was used.

3.3 Contact interfaces modeling

Apart from the geometry, meshing and assignment of element and material properties, equally important part of the model creation was the declaration of contact interface types between the independent interacting parts of the model and within each of them, as deformation proceeds. The treatment of sliding and impact along interfaces has always been very important in order to simulate a "physical" performance between interacting structural members being in contact, i.e. keeping geometric boundaries without model's independent parts penetrating each other. Two different contact interface types required for the appropriate processing of any possible contact interaction among all parts of the model are defined below:

- Contact-Automatic-Single-Surface-Title:
  This type of contact was used for the contact between the surfaces of multi-cellular pultruded FRP rail.
- Contact-Automatic-Surface-To-Surface-Title:
  This type of contact surface is fully symmetric and the choice of slave and master surfaces can be arbitrary. Furthermore, since this is an automatic contact surface, orientation of the segment normals for the contacting surfaces is unnecessary. It was determined that this type of contact sufficiently represented the contact between the pendulum and multi-cellular pultruded FRP rail as well as the contact between the other surfaces of model.

3.4 Boundary conditions

The boundary conditions refer to the pendulum. The moving mass is modeled as a rigid body with five constraints. There are no displacement along the model global axes x and y, and no rotations about the three global basic axes. Therefore, displacement along the z axis is only permitted, with initial velocity of 9.72 m/s (35 km/h). The lower nodes of the posts and the end of each cable had six restraints. There are no displacements along three global basic axes, and no rotations about them.

4 EXPERIMENTAL VALIDATION

The validation of the finite element model is made by a direct comparison to the available results of the pendulum impact tests (Bank & Gentry 2001). Figure 1 shows the comparison between the pendulum impact tests and the finite element simulation of the time histories acceleration, velocity and displacement. During simulation the model of composite guardrail resists the impact load as well as the behavior observed in pendulum impact test, by: (1) an initial flexural damage phase that is accompanied by significant tension in the guardrail; (2) an unloading/softening phase dominated by bending and twisting of the posts along with the loss of tension in the rail; and (3) a final stiffening phase where the cables and rail tension stop the pendulum,

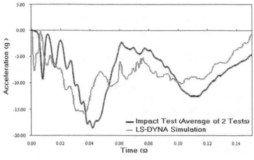

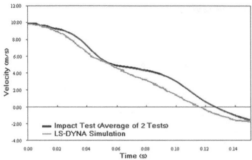

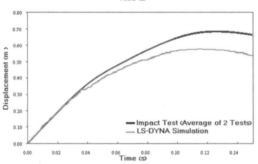

Figure 1. Comparison between impact tests (Bank & Gentry 2001) and finite element simulation results of specimen CGR10.

The LS-DYNA simulation predicts a deceleration peak of 16 g in the first phase whereas in the experiments, a deceleration of 19 g was observed. It is possible that the model of the steel posts and blockouts is too soft, allowing them to torque and translate rearward more early in the simulation than they do in the experiment. Thus in the simulation, pendulum is rebounded more early than its behavior in the experiment. Predicted velocity and displacement data were close to pendulum impact test time histories. A qualitative comparison between the test and simulation results is shown in Figure 2. As shown in this figure, a fairly similar behavior obtained from both the experimental and finite element methods.

Figure 2. Qualitative comparison between test (Brown 1998) and simulation results of specimen CGR10.

5 THE DEFICIENCY OF CGR10 COMPOSITE GUARDRAIL

The guardrail when a car crashes into it must have the ability to resist concentrated lateral forces in addition to the ability to develop tensile forces longitudinally. Similar to the previous pendulum impact test simulation, another simulation was done on pultruded composite guardrail prototype, specimen CGR10. The difference was that the height of pendulum to be modeled was lessened from both top and bottom that it reduced near to 50 percent, so that it was hitting the central portion of guardrail during impact simulation. In fact this modeling was a way to show the impact simulation of vehicle's bumper according to requirements for NCHRP Report 350 crash-test 3–11 , in which a heavy vehicle traveling 100 km/hr colliding into a guardrail at an angle of 25 degrees (Ross et al. 1993). The results showed the weakness of the central portion of the composite guardrail to stop the pendulum. In fact, the central portion of the composite guardrail is not able to tolerate the impact of the pendulum and fails. These results are shown in Figure 3. Moreover, regarding to full-scale crash test which has done according to the conditions of NCHRP Report 350 test 3–11 on this guardrail system in October 2001, penetration of vehicle to central portion of guardrail, no development of tensile forces longitudinally and at last failing the guardrail system were reported.

The reasons of why the requirements for structural adequacy of CGR10 composite guardrail are not satisfied are:

- The pultruded composite material's low resistance to concentrated loads.
- The lowest cross section of the guardrail in its central portion.
- Contact of the vehicle's bumper to the central portion of the guardrail during the collision.

These deficiencies were not found in viewpoint of standard pendulum impact testing of CGR10 composite guardrail.

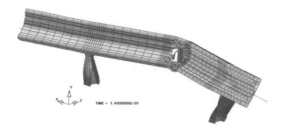

Figure 3. The weakness of the central portion of CGR10 composite guardrail to stop the pendulum.

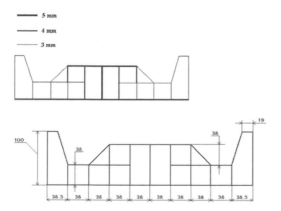

Figure 4. The new prototype composite guardrail profile (dimensions are in mm).

6 NEW PROTOTYPE COMPOSITE GUARDRAIL PROFILE

According to below factors, a new prototype composite guardrail profile was designed:

- Strengthening the central portion of guardrail profile using additional cells to avoid penetration of vehicle's bumper.
- Strengthening the front and back faces of the cross section in order to raise the concentrated load resisting capability and transferring of this load to the other part of the cross section and preventing the longitudinal splitting of the guardrail.
- Constructing the new prototype composite guardrail profile from same cells in order to provide an adequate tensile strength for splice connection.
- Considering the weight of new prototype in comparison with CGR10 composite guardrail.

Based on the above mentioned requirements a new composite guardrail is designed and shown in Figure 4.

A similar simulation that was done for composite guardrail, specimen CGR10, which showed the weakness of its central portion, was performed again for

Figure 5. The ability of the new composite guardrail prototype in stopping the pendulum.

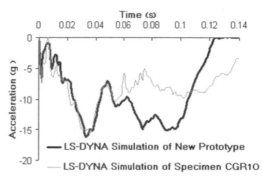

Figure 6. Comparison between impact simulation results of the specimen CGR10 and the new composite guardrail prototype.

the new prototype composite guardrail profile and the results showed the ability of new composite guardrail prototype in stopping the pendulum.

Also simulation of the standard pendulum impact test was performed on new composite guardrail prototype to estimate the deceleration peak in order to compare with the requirements for NCHRP Report 350. In comparison with the simulation of pendulum impact test of composite guardrail, specimen CGR10, it is obvious that the new composite guardrail prototype stops the pendulum almost 0.01 second earlier because of its reinforcement in the central portion. The new composite guardrail prototype exhibits a deceleration peak about 16.5 g in recent simulation which is less than its standard limit, 20 g.

In Figure 6 a comparison between the impact simulation results of the composite guardrail, specimen CGR10, and the new prototype composite guardrail is shown. It is obvious that as in both simulations the pendulum hits to the deepest tubes which are toward the edges of guardrails which are almost similar in both models, these two guardrails behave almost similarly until 0.06 second. But after this specific time, because of contact of the central portion of guardrails to the pendulum which are different in two models, they behave different. In unloading/softening phase of the pendulum impact test, simulation of the specimen CGR10 composite guardrail shows a deceleration peak about 5 g and the new composite guardrail shows a deceleration peak about 7 g which is because of the reinforcement of its central portion.

7 CONCLUSIONS

In this study by using the explicit finite element code LS-DYNA, a non-linear finite element analysis program, pendulum impact test on the cable-anchored composite guardrail, specimen CGR10, was simulated. The results presented an acceptable rang of accuracy in comparison with test results. The energy absorption of the composite rail comes not from the elasto-plastic behavior of the material, but rather from the controlled tearing and splitting of the composite material. After validating the simulation, the low resistance of the central portion of this composite guardrail to concentrated loads was assessed. Regarding to the effective factors, the new composite guardrail specimen was designed and the pendulum impact test was simulated. The results showed the ability of new prototype composite guardrail in stopping the pendulum in addition to satisfying the requirements for NCHRP Report 350.

REFERENCES

Atahan, A.O. & Cansiz, O.F. 2005. Impact analysis of a vertical flared back bridge rail-to-guardrail transition structure using simulation. *Finite Element in Analysis and Design* 41: 371–396.

Brown, C.M. 1998. *Pendulum testing of an FRP composite guardrail: FOIL test numbers 96P019 through 96P023, 97P001 and 97P002.* Report FHWA-RD-98-017, Federal Highway Administration.

Bank, L.C. & Gentry, T.R. 2001. Development of a pultruded composite material highway guardrail. *Composites* A(32): 1329–1338.

Dutta, P.K. 1998. *Investigations of plastic composites materials for highway safety structures.* CRREL Report 98–7, US Army Corps of Engineers.

Hallquist, J.O. 1998. *LS-DYNA Theoretical Manual.* California: Livermore Software Technology Corporation.

Hassan, N.K., Mohamedien, M.A., & Rizkalla, S.H. 1997. Multi bolted joints for GFRP structural members. *ASCE Journal of Composites for Construction* 1(1): 3–9.

Smith, J.R., Bank, L.C. & Plesha, M.E. 2000. Preliminary study of the behavior of composite material box beams subjected to impact. In *6th LS-DYNA users conference 2000-simulation 2000, Dearborn, MI, 9–11 April 2000:* 11-1–11-16.

Svenson, A.L. 1994. *Impact characteristics of glass fiber reinforced composite materials for use in roadside safety barriers.* Report FHWA-RD-92-090, FHWA, US.

Svenson, A.L. & Brown, C.M. 1998. *Pendulum impact testing of steel w-beam guardrail: FOIL test numbers 94P023-94P027, 94P030 and 94P031.* Report FHWA-RD-98-018, Federal Highway Administration.

Svenson, A.L., Hargrave, M.W. & Bank, L.C. 1992. Impact performance of glass fiber composite materials for roadside safety structures. *Advanced Composite Materials in Bridges and Structures:* 559–568.

Svenson, A.L., Hargrave, M.W., Bank, L.C. & Ye B.S. 1994. Data analysis techniques for impact tests of composite materials. *ASTM J Testing Evaluation* 22(5): 431–441.

Svenson, A.L., Hargrave, M.W., Tabiei, A., Bank, L.C. & Tang, Y. 1995. Design of pultruded beams for optimization of impact performance. In *50th Annual SPI Conference Proceedings, 10-E:* 1–7.

Ross, H.E., JR., Sicking, D.L., Zimmer, R.A., & Michie, J.D. 1993. *Recommended Procedures for the Safety Performance Evaluation of Highway Features: National Cooperative Highway Research Program Report 350.* Washington, D.C.: Transportation Research Board.

Fracture and fatigue

Steel and Composite Structures – Wang & Choi (eds)
© *2007 Taylor & Francis Group, London, ISBN 978-0-415-45141-3*

Cumulative damage model for steel weld under low-cycle fatigue

A. Wu, Z.Y. Shen & X. Zhao

College of Civil Engineering, Tongji University, Shanghai, P. R. China

ABSTRACT: A non-linear damage model based on continuum damage mechanics for weld material under low-cycle fatigue loading is presented in this paper. Deterioration of material properties including yield strength, Young's modulus as well as the hardening coefficient is considered in the hysteretic model. Monotonic tensile and strain-controlled low-cycle fatigue tests are carried out on a number of steel and all-weld specimens extracted from a welded T-joint. Material parameters defined in the damage model are obtained from regression analysis of the experimental results. Plastic strain amplitude and loading sequence are found to have more significant effect on the hysteretic behavior of the steel and weld material than that of the initial yield stress.

1 INTRODUCTION

Stiffness, strength and energy dissipation capacity of structures deteriorate when subjected to cyclic loading due to a damage cumulating process in the material, which has already micro-defects resulting from the construction and service period. The damage accumulated will inevitably affect the remaining strength or life of components or structures. Structural failure occurs when the damage accumulated to a critical level.

The reliability of a structure depends on that of its weakest part. Welded connections are widely used in steel structures and have, unfortunately, been proved often acting as the weakest part due to the inherent imperfections resulted from the welding process. Residual stresses with a magnitude of yield stress may exist in the weld area; the material properties of the weld metal, across the heat affected zone to the unaffected base metal varies greatly resulting "metallurgical stress concentration"; defects such as micro-cracks, inclusions, undercuts etc. are difficult to avoid completely, the effect of which on the structural response to cyclic loading is significant. These make the welded connection more susceptible to damage cumulation. About half of the brittle fracture occurred in the 1994 Northridge earthquake was in welded beam-column connections and originated from or near the weld area. Intensive studies thus are necessary for establishing a more reliable damage cumulation seismic analysis of welded connections.

Fracture mechanics has been successfully applied in predicting the failures of many engineering structures with existing macrocracks. However, the deterioration of stiffness and strength of metals occurs well before the initiation of a macrocrack due to the initiation of microcracks, intergranular debonding or decohesion between inclusions and the matrix [Lemaitre 1996]. This deterioration should be taken into account in a complete and reliable structural analysis. Continuum damage mechanics provides an efficient tool for the analysis of damage cumulation and has been successfully applied in pressure vessel and aeronautics engineering for creep, fatigue and ductile damage problems.

The deterioration of young's modulus, yield strength and hardening coefficient in the hysteretic model of a material subjected to cyclic loading can be considered having relation with the damage variable (or damage index), a basic concept in damage mechanics. To quantify the damage in a material, an appropriate damage variable, which is able to describe the microstructural damaged state in terms of measurable mechanical parameters, is first to be defined.

The common way of incorporating the damage cumulation model in structural analysis presented in the literature is to formulate a comprehensive equation, which finally relates the damage variable to the number of cycles, in the framework of thermodynamic based on the damage variable defined and the dissipation potential function assumed. The D-N relation proposed was validated by comparing the calculated D-N curves with measured ones. This brings serious disadvantages of the D-N method: the model is applicable only to the structure with the material and loading conditions tested, the experimental procedure is too complicated to apply in practice, and it is not possible to use in a subsequent numerical seismic response analysis unless the damage model is included in the constitutive equation and the hysteretic model.

Damage has also been defined directly on structural component level in terms of the following three aspects: (1) material property deterioration: Hearn & Testa [1991] incorporate the damage as the reduction of the stiffness in the element stiffness matrix; Lardner [1967] defines the damage in terms of tensile strength deterioration and Jesus et al [2001] in terms of mean stress. As strength and stiffness are two different properties of a material which should both be considered in a structural analysis, it is thus inadequate for the above models to consider only one of them; (2) deformation or strain: Powell & Allahabadi [1998] argued that the damage variables defined based on deformation were the most reasonable ones. Krawinkler & Zohrei [1983] simplified the non-linear deterioration curve obtained from cyclic testing of a cantilever steel member into three stages and applied linear damage cumulation for each stage. The model, though, did not consider the effect of load sequence, and the separation point for the first and second stages is difficult to define, thus the model is difficult to apply. (3) combination of strain and energy: Park and Ang [1985a, 1985b], Usami & Kumar [1998] and Kumar & Usami [1996] defined the damage variable as a linear combination of the damage caused by excessive cyclic loading effect in terms of deformation and hysteretic energy. Although these models consider the effect of both the maximum deformation and the loading history, they apply only to ideal elasto-plastic cases and are difficult to apply in practice due to the complication of determining the required material constants.

Compared to the D-N method, not much work has been found in the literature on application of damage cumulation model in structural seismic analysis due to the complexity of the problem. Considerable work has been done by Shen and co-researchers [Shen & Wu 2007, Shen & Shen 2002] on developing a reliable, systematic and practical analytical method to take into account the damage cumulation effect in seismic analysis of steel structures. Ding et al [2005] incorporated the damage model developed by Shen & Dong [1997] into the stiffness matrix and the structural dynamic equilibrium equation of spatial truss structures to study their structural response under seismic loading.

2 MATERIALS DAMAGE MODEL

Damage variable on material level has been defined based on various physical or mechanical parameters, from the net area reduction first proposed by Rabotnov[1969] in the 1960s to the variation of Young's modulus [Lemaitre 1996], reduction of sectional area [Cheng et al 1996], variation of HL hardness [Chen & Zhao 2005], and variation of the area of the stress-deformation curve [Chen & Jiang 2005], etc. However, none of these models enjoys universal acceptance due to the limitations of not being able to account for all the load-level dependence, load interaction,

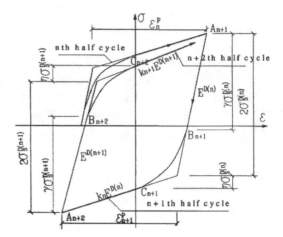

Figure 1. The proposed hysteretic model considering damage cumulation for structural steel [Shen & Dong 1997].

and load sequence effects, difficulties in defining the required constants and lack of standardised measuring method.

Compared to a hysteretic model for a structure component which is reliable only for the type of geometry and loading conditions tested, a hysteretic model at material level with a reliable numerical method has wider applicability as experiments are more expensive, time-consuming, and not always feasible.

Shen & Dong [1997] proposed a non-linear damage cumulation model based on plastic strain for structural steel which takes the complete loading history and energy dissipation as well as the effect of the maximum plastic strain into account. The model was developed based on the one proposed by Park and Ang [1985a] and the cyclic response of the material with the constants determined from regression analysis of a series of experimental results of simple standard tensile and cyclic tests. In this paper, Shen's damage model for steel was first used for weld material and experiments were then carried out for further refinement and determination of specific material parameters for weld material. Figure 1 shows the damage model adopted, where ε_n^p and ε_{n+1}^p are the plastic strain in the nth and $(n + 1)$th half cycle, respectively; $E^{D(n)}$ and $E^{D(n+1)}$ the Young's modulus at the nth and $(n + 1)$th half cycle; $\sigma_s^{D(n)}$ and $\sigma_s^{D(n+1)}$ the yield strength of the nth and $(n + 1)$th half cycle; $k^{(n)}$ and $k^{(n+1)}$ the hardening coefficient of the nth and $(n + 1)$th half cycle; and γ and η the constants.

The damage variable D is defined as

$$D = (1 - \beta) \frac{\varepsilon_m^p}{\varepsilon_u^p} + \sum_{i=1}^{N} \beta \frac{\varepsilon_i^p}{\varepsilon_u^p} \qquad (1)$$

where ε_m^p is the maximum plastic strain occurred during the cyclic loading; ε_u^p the ultimate plastic strain of the material obtained from standard tensile test; N the number of half cycle, and β a constant.

The deteriorated young's modulus, yield strength and hardening coefficient for half-cycle n is calculated by

$$E^{D(n)} = (1 - \xi_1 D_n) E^{D(n-1)} \qquad (2)$$

$$\sigma_s^{D(n)} = (1 - \xi_2 D_n) \sigma_s^{D(n-1)} \qquad (3)$$

$$k_n = k_0 + \xi_3 \sum_{i=1}^{n} \frac{\varepsilon_i^p}{\varepsilon_u^p} \qquad (4)$$

where ξ_1, ξ_2, and ξ_3 are material constants.

3 EXPERIMENTAL STUDY

Two steel plates of nominal dimensions $100 \times 1200 \times 30\,mm$ and $200 \times 1200 \times 30\,mm$ were joined by two fillet welds to form a T-joint. The steel plates used were GB/T700 Q235 structural steel, with a specified yield strength of not less than 225 MPa and a tensile strength of between 375 and 460 MPa for the specific thickness. Standard tensile and strain controlled low cycle fatigue tests were carried out on both the steel and weld materials.

3.1 Apparatus

The tensile test was performed using an INSTRON testing machine. The gauge length of the extensometer used is 15 mm with a 5 mm travel distance. The testing machine used for the low-cycle fatigue test is MTS 809 with a loading capacity of 5 t. The gauge length of the extensometer used is 25 mm, which is the smallest available.

3.2 Tensile test

Standard tensile tests conforming to GB/T 228-2002 on both steel and weld material from the welded T-joint were carried out to obtain the monotonic properties of these two materials. In total nine standard round bar tensile specimens were extracted from the steel plate, the right and left side weld.

3.3 Low-cycle fatigue test

Fully-reversed tension-compression low-cycle fatigue tests were carried out on both steel and weld material on specimens extracted from the welded T-joint. Full stress-strain response for each load cycle was recorded digitally by a computer. Regression analysis was then carried out on the hysteretic curves obtained from the tests to determine the material parameters defined in the proposed damage cumulation model as described in section 2.

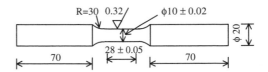

Figure 2. Low-cycle fatigue specimen, dimensions in mm.

3.3.1 Specimen

Care has to be taken in the designation of the low-cycle fatigue specimen for weld material. The ratio of the gauge length to diameter of the specimen should be minimized to avoid buckling as large compression load is involved; however, the minimum value of the gauge length of the specimen is limited by that of the available extensometer, which means the diameter of the specimen should be maximized. Unfortunately, the diameter of the specimen is at the same time limited by the capacity of the fatigue testing machine and the fact that the all-weld specimen is extracted from the welded T-joint. The final dimension of the specimen adopted is shown in Figure 2. In order to ensure the specimens extracted contain all weld material, a 20 mm thick specimen was first cut from the T-joint and polished and etched to give the exact area of the weld metal.

3.3.2 Loading mechanism

In total eight specimens were tested under axial loading for each material. Constant strain amplitude of 0.0075, 0.0125, 0.0150 and 0.0175 was applied to four specimens respectively. The other four specimens were subjected to continuously varying strain amplitudes in order to study its effect on the material parameters defined in the damage model. For two of them, 40 half cycles was applied at each strain amplitude in ascending manner from 0.0075 to 0.0125 and to 0.0150, the specimens were then tested under 0.0175 strain amplitude until failure occurred. Same strain amplitudes but varying in descending manner was applied to the rest two specimens.

Triangular strain waveform was used with a constant loading frequency of 0.2 Hz. Loading started with tension up to the designated strain amplitude, unloading and compression are then applied until the strain is fully reversed.

4 RESULTS AND DISCUSSION

4.1 Tensile test results

The average values of those obtained from test records of the yield/proof strength σ_y, ultimate tensile strength σ_u, yield strain ε_y and ultimate strain ε_u are listed in Table 1. Figure 1 shows some typical engineering stress-strain curves for the steel and weld material, respectively.

Table 1. Tensile test results for steel and weld materials.

Material	σ_y Mpa	σ_u MPa	ε_y	ε_u
Steel	316.7	515.6	0.0016	0.1574
Weld1	444.8	575.0	0.0020	0.1228
Weld2	456.0	591.7	0.0020	0.1259

Table 2. Number of half cycles to failure.

Specimen number	Strain amplitude	n Steel	n Weld	n_0 Weld
1	0.0075	1652	1352	7
2	0.0125	364	328	6
3	0.0150	392	194	4
4	0.0175	236	134	4
5	0.0075–0.0175	394	260	9
6	0.0075–0.0175	250	202	2
7	0.0175–0.0075	604	460	4
8	0.0175–0.0075	1180	313	2

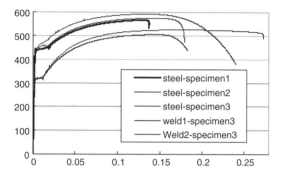

Figure 3. Typical stress-strain curves of the steel and weld materials obtained from the tensile test.

The results for the three specimens extracted from the same weld are consistent. However, slight difference in properties of the two fillet welds was seen as shown in Table 1. This may be explained from the effect of the welding process. There is no clear plateau in the stress-strain curves for weld materials as that for steel.

Figure 3 shows big variations in the properties of the steel material. This is because the specimens were extracted from plate areas close to the weld, where property variation from the weld to certain distance in the steel plate is significant due to the welding process. The low-cycle fatigue test results on steel show similar variation. The test results for steel plate obtained in this study are thus not representative for normal as-delivered steel plate material.

4.2 Cyclic test

The number of half-cycles to failure, n, for the steel and weld specimens are listed in Table 2. Steel specimens are seen to have better performance than weld under same cyclic loading conditions. This is expected due to the presence of the inherent micro-imperfections in the weld area resulted from the welding process. These micro defects make the weld more susceptive to damage cumulation effect under cyclic loading.

Figure 4 shows two typical hysteretic curves obtained from the cyclic test for weld material. Cyclic hardening is seen for the first few cycles. The deterioration of yield strength and young's modulus happens in the following cycles and becomes significant at the later stage of the cyclic loading, as shown by dotted lines in Figure 4.

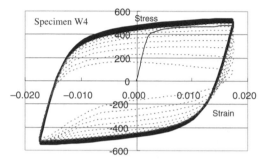

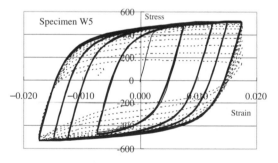

Figure 4. Hysteretic curves for weld specimen W4 and W5.

The yield strength, $\sigma_s^{D(n)}$, as defined in the damage cumulation model (see Fig. 1) was calculated from the experimental hysteretic curves for each specimen. Figure 5 shows the results for weld specimen W3. In addition to the hysteretic curves as shown in Figure 4, cyclic hardening is clearly seen from the $\sigma_s^{D(n)} - n$ curves for the first several cycles. The damage variable model was modified accordingly by including a damage initiating half-cycle number, n_0, in Equation 1, i.e.

$$D = (1-\beta)\frac{\varepsilon_m^p}{\varepsilon_u^p} + \sum_{i=n_0}^{N}\beta\frac{\varepsilon_i^p}{\varepsilon_u^p} \tag{5}$$

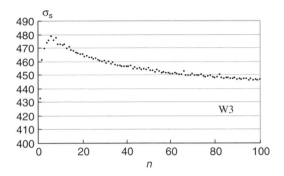

Figure 5. Variation of yield strength with number of half-cycles for weld specimen W3.

Table 3. Material parameters obtained from experimental results.

Specimen	β	ξ_1	ξ_2	ξ_3	k_0
W1	0.007494	0.162	0.533	0.00251	0.00871
W2	0.015645	0.195	0.162	0.00096	0.01353
W3	0.020456	0.217	0.175	0.00078	0.01379
W4	0.024146	0.301	0.131	0.00048	0.01402
W5	0.014904	0.271	0.038	0.00619	0.02689
W6	0.020014	0.123	0.033	0.00459	0.03188
W7	0.014232	0.162	0.387	0.00132	0.01296
W8	0.018546	0.318	0.398	0.00234	0.00607
S1	0.007756	0.351	0.0539	0.00108	0.02425
S2	0.018214	0.203	0.0750	0.00050	0.04040
S3	0.013407	0.270	0.0748	0.00027	0.01389
S4	0.017930	0.248	0.0704	0.00056	0.01084
S5	0.012018	0.239	0.0209	0.00039	0.04346
S6	0.020075	0.096	0.0927	0.00404	0.05872
S7	0.014927	0.243	0.1321	0.00019	0.02199
S8	0.008526	0.271	0.1386	0.00097	0.02362

The number of half-cycles showing cyclic hardening was determined from the $\sigma_s^{D(n)} - n$ curve for each weld specimen and is listed in Table 2. In general, for higher strain amplitude, fewer cycles are needed to initiate material property deterioration. The average number of half-cycles for the eight tested weld specimens is 4.

With the value of D at failure taken as 1, the constant β can be determined by

$$\beta = \frac{1 - \dfrac{\varepsilon_m^p}{\varepsilon_u^p}}{\displaystyle\sum_{i=n_0}^{N} \dfrac{\varepsilon_i^p}{\varepsilon_u^p} - \dfrac{\varepsilon_m^p}{\varepsilon_u^p}} \qquad (6)$$

With the calculated value of β, the value of D_i, $E^{D(i)}$, $E^{D(i-1)}$, $\sigma_s^{D(i)}$, $\sigma_s^{D(i-1)}$, k_i, ξ_1^i, ξ_2^i, and ξ_3^i for each half cycle, i, can be obtained from regression of experimental stress-strain curves and Equations 2–4. The value of ξ_1, ξ_2, and ξ_3 of a specimen is taken as the average of the obtained ξ_1^i, ξ_2^i, and ξ_3^i values. The final results of the material constants β, ξ_1, ξ_2, and ξ_3 for the tested specimens of steel and weld material are listed in Table 3.

The scatter of the data is demonstrated in Table 3. The values of the ultimate strain, ε_u^p, is expected to have big variation in the real cyclic specimens and will have more significant effect on the material parameters in the damage model. However, ε_u^p can only be taken as that obtained from tensile test for all the eight specimens for each material. This may be the reason attribute to the scatter of data.

The effect of various factors including the original yield strength, plastic strain amplitude, and loading sequence on the hysteretic behavior and material parameters was studied. The plot of yield stress obtained from the first tensile half-cycle versus various material parameters shows no clear relation between them.

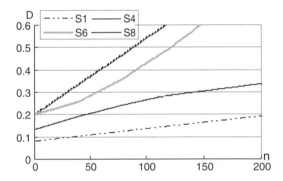

Figure 6. D–n curves for steel specimens.

The strain amplitude and loading sequence are seen to have more significant effect on the material behavior and parameters. Figure 6 shows the variation of the damage variable D with number of half cycles to failure obtained for some of the steel and weld specimens. For all the constant strain amplitude tests, damage cumulates linearly with numbers of cyclic loading. The rate of damage development increases with increasing strain amplitude, as demonstrated by curves S1 and S4 in Figure 6. The slope of the D-n curve for varying strain amplitude tests changes with the magnitude of the strain undertaken in the same way as that for the constant strain amplitude test, i.e. the curve is convex for decreasing strain amplitude and concave for increasing strain.

5 CONCLUSIONS AND FUTURE WORK

A damage cumulation model for steel weld material is presented in this paper. Tensile and strain-controlled

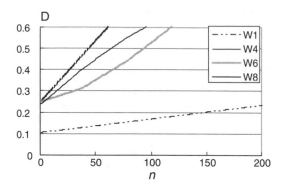

Figure 7. *D–n* curves for weld specimens.

low cycle fatigue tests were carried out on specimens extracted from a welded T-joint to study their behavior under cyclic loading. Material parameters defined in the damage cumulation model were obtained from regression analysis of test results.

To take into account the cyclic hardening for the first several cycles observed from the experimental hysteretic curves, the damage model originally proposed was modified by calculating the damage variable from an damage initiating number of cycle, as shown in Equation 5.

The damage cumulation process is found to be affected by the plastic strain amplitude and the varying and sequence of the cyclic loading.

Due to the effect of welding process, the pre-loading material properties of the specimens extracted from different positions in a welded T-joint show certain degree of variation. Scatter of material parameter results is seen and may be caused by using a same ε_u^p value for all the specimens.

More tests need to be done to obtain more accurate material parameters defined in the model and to study further the effect of various factors such as steel grade, welding process etc. on the model. Tests on specimen from homogenous structural steel plates with different specified grade will be carried out next to check the consistency of the material parameters and to exclude the effect of ε_u^p. Weld specimens will be produced from weldments using different welding process to study their effect on the model. With reliable material parameters determined for various commonly-used structural steels and weldments, engineers in practice would be able to take into account the damage cumulation effect in seismic analysis of new welded structures and residual-strength/life prediction for maintenance or repair of existing structures.

ACKNOWLEDGEMENT

This project was granted financial support from China Postdoctoral Science Foundation.

REFERENCES

Chen L. Jiang J. 2005. A new low cyclic fatigue damage model and its experimental verification. *ACTA Metallurgica Sinica* 41 (2): 157–160 (in Chinese)

Chen X. & Zhao S. 2005. Evaluation of fatigue damage at welded tube joint under cyclic pressure using surface hardness measurement. *Engineering Failure Analysis* 12: 616–622

Cheng G.X. & Zuo J.Z. & Lou Z.W. & Kuang Z.B. 1996. Continuum damage model of low-cycle fatigue and fatigue damage analysis of welded joint. *Engineering Fracture Mechanics* 55(1): 155–161

Ding Y. Guo & F. Li Z.X. 2005. Elasto-plastic analysis of spatial trusses under earthquake excitations considering damage accumulation effect. *Engineering Mechanics* 22(1): 54–58

Hearn G. & Testa R.B. 1991. Model analysis for damage detection in structures. ASCE ST10: 3042–3063

Jesus A.M.P. & Ribeiro A.S. & Fernandes A.A. 2005. Finite element modelling of fatigue damage using a continuum damage mechanics approach. *ASME Journal of Pressure Vessel Technology* 127:157–164

Krawinkler H. & Zohrei M. 1983. Cumulative damage in steel structures subjected to earthquake ground motions. *Computers and Structures* 16(1–4): 531–541

Kumar S. Usami T. 1996. An evolutionary-degrading hysteretic model for thin-walled steel structures. *Engineering Structures* 18(7): 504–514

Lardner R.W. 1967. A theory of random fatigue. *Journal of the Mechanics and Physics of Solids* 15(3): 205–221

Lemaitre J. 1996. A course on damage mechanics, 2nd edition. Berlin Heidelberg: Springer

Park Y.J. & Ang A.H.S.1985a. Mechanistic seismic damage model for reinforced concrete. *ASCE Journal of Structural Engineering* 111(4): 722–739

Park Y.J. & Ang A.H.S. 1985b. Seismic damage analysis of reinforced concrete buildings. *ASCE Journal of Structural Engineering* 111(4): 740–757

Powell G.H. & Allahabadi R. 1998. Seismic damage prediction by deterministic methods: concepts and procedures. *Earthquake Engineering and Structural Dynamics* 16: 719–734

Robotnov Y.N. 1969. Creep Problems in Structural Members. Amsterdam : North-Holland

Shen Zu-yan & Dong Bao. 1997. An experiment-based cumulative damage mechanics model of steel under cyclic loading. *Advances in Structural Engineering* 1(1): 39–46

Shen Z.Y. & Shen S. 2002. Seismic analysis of tall steel structures with damage cumulation and fracture effects. *Journal of Tongji University* 30 (4): 393–398 (in Chinese)

Shen Z.Y. & Wu A. 2007. Seismic analysis of steel structures considering damage cumulation. *Frontiers of Architecture and Civil Engineering in China* 2(1):1–11

Usami T. & Kumar S. 1998. Inelastic seismic design verification method for steel bridge piers using a damage index based hysteretic model. *Engineering Structures* 20(4–6): 472–480

Steel and Composite Structures – Wang & Choi (eds)
© 2007 Taylor & Francis Group, London, ISBN 978-0-415-45141-3

The fatigue strength of butt welds made of S690 and S1100

R.J.M. Pijpers
Delft University of Technology, Faculty of Civil Engineering and Geosciences, The Netherlands
Netherlands Institute for Metals Research

M.H. Kolstein, A. Romeijn & F.S.K. Bijlaard
Delft University of Technology, Faculty of Civil Engineering and Geosciences, The Netherlands

ABSTRACT: With modern steel manufacturing techniques it is possible to produce steel with nominal strengths up to 1100 MPa (very high strength steels, VHSS). For the successful application of higher strength steels material performance and structural behavior should be thoroughly understood. According to EN1993-1-9 (2005) the fatigue strength of the steel structure depends on the applied detail, plate thickness and machining condition. Several tests have been performed in order to study the influence of base material strength on the fatigue strength of a transverse butt welded joint in the as-welded condition. A constant amplitude load, R = 0.1, was generated by a Schenck servo hydraulic test rig. The base materials of the butt welds are Naxtra M70 and Weldox S1100E. The influence of the use of VHSS on fatigue strength has been shown by comparing the resulting S-N curves to EN1993-1-9 (2005).

1 INTRODUCTION

1.1 Research project VHSS

Modern steel manufacturing techniques make steel production possible with yield strengths up to 1100 MPa (very high strength steels, VHSS). The number of applications where VHSS can be used to economic advantage over lower strength steels is growing.

Although very high strength steels (VHSS, see Table 1) are already available they are not yet generally used by manufactures in the field of civil engineering. In 2006 TU Delft has started the research project 'Very High Strength Steels for Structural Applications'. This research project evaluates the suitability of the latest generation of steels in terms of structural behaviour, mechanical performance and weldability.

The main objective of the proposed project is to supply the steel construction industry with information relevant for the design and fabrication of VHSS structures.

The use of VHSS is especially cost-effective in situations where the weight of the structure itself forms an important part of the total load on the structure and/or when an optimization on dimension is a decisive parameter. The design stress in a VHSS structure is usually higher than in a structure made from conventional steel and the stress due to own weight will be

Table 1. Various types of steel and their applications.

Strength [MPa]	Description	Steel quality	Application
<300	Regular structural steel	S235	Buildings
300 400 500 600	Conventional high strength steel (CHSS)	S355/S460	Bridges/High rise buildings
700 800 900 1000 1100	Very high strength steel (VHSS)	S690/S960/ S1100	Cranes/ Bridges/High rise buildings

lower. This results in absolute and relative higher stress variation due to an external cyclic load. For structures with a high number of cycles during their lifetime (e.g. bridges) high cycle fatigue (HCF) is of importance.

1.2 Literature on fatigue of VHSS transverse butt welds

The first part of the research project focuses on the fatigue strength of a VHSS welded connection. From Maddox (1990) it is a well known fact that fatigue

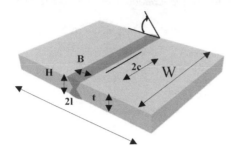

Figure 1. EN 1993-1-9 (2005) detail category 90.

strength of steel structures is independent of yield strength of the base material. This is due to the presence of welded joints, near which micro cracks are likely to give rise to fatigue cracks. As the fatigue strength is not proportional to the yield stress, fatigue is more often governing the design.

According to EN 1993-1-9 (2005), which gives rules for the design of steel structures up to S690, the fatigue strength of the steel structure depends on the applied detail (notch factor), plate thickness and machining condition.

However, tests on base materials and details with low stress concentration factor show increasing fatigue strength by increased yield strength. The geometry of a transverse butt weld detail, having a low notch factor compared to other welded details, appears to be most similar to the base material geometry. If the use of VHSS would increase the fatigue strength of a welded joint, than most likely this increase would be shown in the butt weld detail.

According to EN 1993-1-9 (2005) the fatigue design strength at 2 million cycles of the butt weld detail (see Fig. 1) is 90 MPa. The following criteria account for this detail class: transverse splices in plates or flats; weld run-on and run-off pieces to be used and subsequently removed; plate edges to be ground flush in direction of stress; welded from both sides in flat position; NDT applied.

The design fatigue curve of EN 1993-1-9 (2005) is only valid if the geometry of the butt welded detail is valid by the requirements for execution of steel work according to EN 1090-2 (2005). This code takes account of hot-rolled, structural steel products up to S960. The S1100 steel is thus out of the range of this code.

According to crane code CEN/TS 13001-3-1 (2004) the fatigue strength of the same quality butt weld is 140 MPa. Hamme et al. (2000) discuss the use of high strength steels in crane structures and show fatigue results on base materials, welded and post weld treated joints of S355, S690 and S960.

Demofonti G. et al (2005) have performed axial fatigue tests on 10 mm butt welds made of S355 up to S960. The results of the investigation showed that

Figure 2. Fatigue test specimen in test rig Stevin Laboratory.

under constant amplitude loading no significant differences in bearable local stresses. Advantages for the S960 steel can be noticed in case of variable amplitude loading Machining of welds, in order to achieve low notch factors is found to give an advantage for high-strength steels. In practice however machining critical details is time consuming, especially for large structures.

In the ECSC research program 'Efficient lifting equipment, with extra high strength steel', Puthli et al. (2005), performed fatigue experiments on various butt-welds details made of S690, S960 and S1100. Resulting S-N curves show a slope m = 5 in for S1100, which is higher than the value of m = 3 used in design codes.

In this research it is expected that fatigue strengths of VHSS welds with low SCF are higher than the values proposed in EN 1993-1-9 (2005). This would mean a shift to the right and/or a higher slope in S-N curves.

2 SPECIFICATION TEST SETUP

2.1 Test rig

The base plate materials for the experiments are Naxtra M70 (S690) and Weldox S1100E (S1100). All tensile static and fatigue specimens have been tested with a 600 kN dynamic servo hydraulic test rig (Schenck PCX 0001, Fig. 2) at the Stevin Laboratory of the Delft University of Technology.

Table 2. Tensile test results.

Measured	R_{eh} MPa	R_m MPa	R_{eh}/R_m –	A %	E MPa
S690	733	787	0.93	17	$2.04*10^5$
S1100	1055	1083	0.97	11	$2.00*10^5$
Data sheet					
S690	800	830	0.96	16	–
S1100	1197	1432	0.84	11	–

Table 3. Geometry data of fatigue specimens, see Figure 1.

Steel type	S690	S1100
Number of test specimens	6	7
Number of strain gauges	14	14
Thickness t [mm]	12	10
Length 2l [mm]	750	1000
Width W [mm]	120	100
Weld metal	Megafill 742M	Tenacito 75 & SH NI 2 K 140
Weld process	FCAW	SMAW
Number of weld layers	5	9

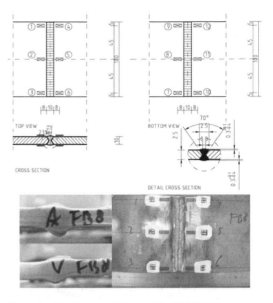

Figure 3. Geometry and pictures of a S1100 fatigue specimen incl. location of strain gauges.

2.2 Tensile tests

Table 2 shows the results of tensile tests on the plate material, based on average results of three test pieces per material.

The tensile test results of the S690 material are in line with the tests performed by Kolstein et al. (2006) on the same material and with the data sheet of the steel manufacturer. The S1100 results however are very different from the data sheets. In particular the S1100 material shows low ductility as the values of yield strength are close to tensile strength.

Probably while making the test pieces they have been influenced by heat treatment, which reduced the material strength. This resulted in differences in measured values and data sheet values up to 12% in yield strength and 25% in tensile strength.

At the time of writing this paper the exact cause of the lower strength values is not clear yet. The effect of manufacturing on material strength will be investigated in follow up research.

2.3 Welded specimens

The geometry of the strip test specimens has been chosen to qualify the EN 1993-1-9 (2005) class 90 butt weld detail in the as-welded condition. In total thirteen test specimens have been tested; six made of S690, seven made of S1100. Table 3 shows general dimensions of the test specimens.

Figure 3 shows the geometry of the S1100 butt welds. Apart form the width and thickness the S690 weld geometry is identical. The width of the test specimens is taken ten times the thickness of the plate material. The edges of the welded plates have been ground for reducing the chance of crack initiation at the sides. The stresses at the side however still remained larger at the side compared to the middle of the plate, confirmed by strain gauge measurements.

Each specimen contains 12 strain gauges (FLA-6-11) at 8 mm from the weld toe to monitor the strains on both sides of the welded connection and to see the influence of stress amplification near the weld and near the sides of the specimen. For local measurements special strain gauges (FXV-1-11-002LE) have been applied at 4 mm from the weld toe.

Before fatigue testing static measurements have been performed to set the strain range for the fatigue test by determining minimum and maximum force. The strain controlled fatigue tests have been performed at R = 0.1 at constant amplitude.

3 STATIC RESULTS

3.1 Static measurements

In order to define the stress levels for fatigue loading the specimens have been first statically loaded up to stress levels of 30% of the yield strength. Figure 4 shows the longitudinal stress pattern for various applied forces at six strain gauge positions.

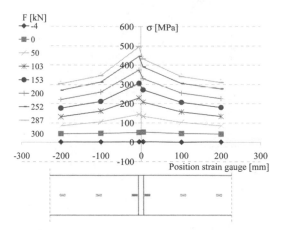

Figure 4. Longitudinal stress pattern.

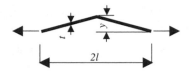

Figure 5. Misalignment parameters after Hobbacher (2004).

The measured values are the average tensile stresses at the top of the specimens near the weld, at 100 mm and 200 mm from the weld and at the middle of the plate.

3.2 Misalignment

Due to shrinkage after welding the test specimens are all bended along the axis of the weld. This causes a non-linear behaviour of the stresses as a result of increasing the tensile load when the specimens are clamped in the test rig. At high stress levels the test rig straightens the test specimens. Hobbacher (2004) gives calculation rules of equations 1 and 2 for a misalignment factor in transverse butt welds, K_m with the parameters given in Figure 5:

$$K_m = 1 + \frac{3y}{t} \cdot \frac{\tanh(\frac{\beta}{2})}{(\frac{\beta}{2})}$$ (1)

$$\beta = \frac{2l}{t} \cdot \sqrt{\frac{3 \cdot \sigma_m}{E}}$$ (2)

The formula shows the influence of the nominal stress level σ_m on the misalignment factor. At high stress levels the effect of secondary bending stresses is large compared to low stress levels.

Table 4. Local geometry fatigue specimens, see Figures 1 & 5.

	Specimen	B_{min}/B_{max} mm	H_{min}/H_{max} mm	H_{max}-t mm	y mm
S690	FA1	10/15	17/17	2.5	4
	FA2	10.3/14.1	16.9/17.7	2.85	–
	FA3	8/11	14.3/14.3	1.15	5
	FA4	9/13.5	14.5/14.5	1.25	2.5
	FA5	7/10	14/14.5	1.25	7
	FA9	9/14	15/15	1.5	4.5
S1100	FB1	10/16	16/16	3	6
	FB2	11/15	15.6/15.9	2.8	3.7
	FB3*	9/16	15.2/17	3.5	7
	FB4*	13/13	15.5/16	3	7
	FB5	12/16	15/15	2.5	8
	FB6	11.5/17.5	14.5/15.3	2.65	8
	FB9*	11/15.8	16.2/16.6	3.3	7.5

*Edges badly grounded at the side of the plate.

Table 4 summarizes local geometry details of all fatigue tested specimens.

The table gives minimum and maximum values of weld width B, weld height H, the excess of weld metal H-t (see Fig. 1) and the misalignment parameter y (see Fig. 5).

The yield and tensile strength of the weld material is lower than the plate material strength of S1100. The undermatched S1100 welds therefore have been made thicker than allowed according to EN1090. The code limits for S690 the excess of weld metal to H-$t = 0.1*B < 2$ mm. In several test pieces this value has been exceeded.

Measured average stress values at 4 mm from the weld toe are compared to calculated values of $K_m^* \sigma_m$ according to Hobbacher (2004). The stress measurements of both the S690 and the S1100 welds are in good correspondence with the calculated results at high stress levels.

4 FATIGUE RESULTS

4.1 Crack initiation and propagation

In general when determining the fatigue strength two main stages are important: crack initiation and crack propagation. Table 5 and figure 8 show the results of crack initiation and crack propagation at the various stress levels of the S690 and S1100 test specimens.

Crack initiation is visible when strain gauge data start to bend off the regular scheme (Fig. 6, point A). The number of cycles at this point A is defined as N_i. Figure 6 shows the monitoring of crack initiation by strain gauge measurements. In the graph the

Table 5. Fatigue results.

Specimen		$\Delta\sigma_{mean}$ MPa	$N_i(*10^5)$ cycles	$N_f(*10^5)$ cycles
S690	FA1*	110		47.0
	FA2	282	2.4	3.1
	FA3	285	2.4	3.4
	FA4	137	13.0	19.0
	FA5	206	2.2	6.0
	FA9	138	13.0	24.0
S1100	FB1*	200		80.0
	FB2	454	0.05	0.2
	FB3	296	3.0	4.1
	FB4	240	1.5	3.0
	FB5	236	19.0	21.0
	FB6	357	0.6	1.0
	FB9	240	0.9	2.1

*No crack initiation.

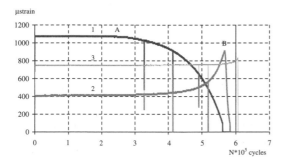

Figure 6. Strain gauge measurement results of specimen FA5.

strain interval values of three strain gauges are mentioned: strain gauge 1 (tension) near which a crack has initiated, strain gauge 2 on the back side (compression) and strain gauge 3 at 100 mm of the weld.

When a crack initiates the crack interval values of the tensile strain gauge near the crack decrease while the values of the compressive strain gauge at the opposite side of the plate increase.

If the crack reaches the full depth of the plate the strain interval values of the compressive strain gauge immediately decrease (Fig. 6, point B).

An alarm system has been used which immediately shuts off the system in case of 10% deviations, which also makes it possible to indicate the locations of crack initiation on the test specimens.

The crack propagation phase can be subdivided in three stages, characterised by the monitoring method.

The first stage of crack propagation shows small cracks monitored by use of a liquid (petrol). The second stage, defined as N_f, is the number of cycles when the crack reaches a length of 10 mm, which is easily visible in practise. The third stage is the moment the

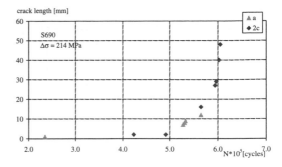

Figure 7. Crack growth curve in depth (a) and width (2c) of FA5 specimen (S690).

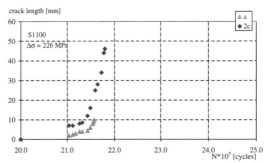

Figure 8. Crack growth curve in depth (a) and width (2c) of FB5 specimen (S1100).

cracks have critical length, which introduces brittle failure of the test specimen.

Crack propagation is monitored visually. Figures 7 and 8 show the crack propagation in the plate width direction and in the plate depth direction of a S690 and a S1100 specimen that are tested at rather similar stress levels. Figure 9 illustrates the crack propagation parameters.

When comparing the crack growth curves of the S690 and the S1100 test specimens large differences in crack initiation phase and crack propagation phase are found. Crack initiation of the S1100 specimens is much longer while crack propagation is shorter than the S690 specimens.

Except from the FA4 specimen of S690 in all specimens cracks have initiated from the side of the plate in the weld toe. In the FA4 specimen cracks occurred at multiple locations. In two specimens no cracks have occurred at high number of cycles; in one of the S690 specimens at a stress range of 110 MPa and in one of the S1100 specimens at a stress range of 200 MPa.

4.2 S–N curves

Hobbacher (2004) describes a statistical method for the evaluation of S-N curves. The characteristic stress

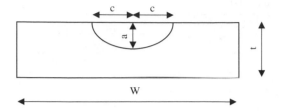

Figure 9. Crack growth parameters c and a.

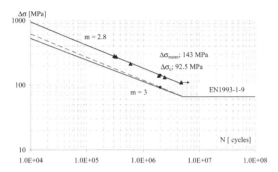

Figure 10. S–N curve of S690 specimens.

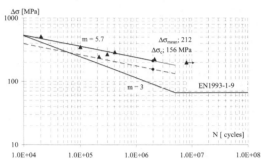

Figure 11. S–N curve of S1100 specimens.

range values at 2 million cycles, which are calculated for 95% survival probability on a two-sided confidence level of 75% of the mean, are based on Equation 3, described in Hobbacher (2004):

$$\log N_c = (a - k * Stdv) + b * \log \Delta \sigma_c \qquad (3)$$

The number of test pieces is small; therefore the value of the k is set to 3. Figures 10 and 11 show the resulting S–N curves of the S690 and S1100 butt welds compared to S-N curve of detail class 90 of the EN1993-1-9 (2006).

The fatigue behaviour of the S690 weld is identical to the values of EN 1993-1-9 (2005), although the actual fatigue strengths are a little higher than the values of this code, which are based on numerous test results on S235 up to S690 welds.

Figure 10 shows the S–N line S690. The mean stress level at 2 million cycles $\Delta \sigma_{mean}$ is 143 MPa, which is in good correspondence to a high quality butt weld of regular steel. The calculated characteristic value $\Delta \sigma_c$ is 92.5 MPa. The slope value $m(=-b)$ of the S–N line of 2.8 is lower than the slope value of 3 according to EN 1993-1-9 (2005).

The results of the S1100 material show a fatigue behaviour completely different from the S690 material (see Fig. 11). The slope m of 5.7 is much higher than the value according to EN 1993-1-9 (2005). The $\Delta \sigma_{mean}$ level at 2 million cycles is 212 MPa, which is much higher than the S690 steel results. The calculated characteristic value $\Delta \sigma_c$ is 156 MPa.

It should be noticed that the S–N curve shows two deviant values at the stress level of 240 MPa. This can be explained by differences in geometry. One specimen tested at 240 MPa stress range has not been ground at the side. Especially in the S1100 steel this results in a weaker fatigue strength.

5 EVALUATION

A fatigue test series on transverse butt welds is the first part of a PhD project on the applicability of VHSS in civil engineering structures.

Tensile tests on S690 and S1100 show a brittle behaviour, because of low ratio of yield to tensile strength. Especially for the S1100 material tensile strengths are close to the yield strengths.

Although literature tends to minimize a positive effect of increased base material yield strength on fatigue strength of welded connections this research shows that use of S1100 in transverse butt welds increases the fatigue strength, especially in the high cycle fatigue region. Fatigue strength results of S690 transverse butt welds corresponds well to well to the design rules of EN 1993-1-9 (2005), however showing slightly higher fatigue strength.

While the crack initiation phase is very long, crack propagation of S1100 butt welds is short compared to S690 butt welds.

REFERENCES

Demofonti, G., Riscuifuli, C.M. Sonsino, H. Kaufmann, G. Sedlacek, C. Müller, F. Hanus, H.G. Wegmann 2001. *High-strength steels in welded state for lightweight constructions under high and variable stress peaks.* Luxembourg: European Commission Directorate-General for Research.

EN 1090-2. 2005. *Execution of steel structures and aluminium structures – Part 2: Technical requirements for the execution of steel structures.* Brussels: European Committee for Standardization.

CEN/TS 13001-3-1 en. 2004. *Cranes – General design – Part 3.1:Limit states and proof of competence of steel structures.* Brussels: European Committee for Standardization.

EN 1993-1-9. 2005. *Design of steel structures – General – Part 1.9: Fatigue strength of steel structures.* Brussels: European Committee for Standardization.

Hamme, U, J. Hauser, A. Kern, U. Schriever 2000. Einsatz hochfester Baustähle im Mobilkranbau. In *Stahlbau 6*: 295–300. Berlin: Ernst & Sohn.

Hobbacher, A. 2004. *Recommendations for fatigue design of welded joints and components, IIW document XIII-1965-03 / XV-1127-03.* Paris: International Institute of Welding.

Kolstein, M.H. 2006. *Integrity of high strength steel structures; fatigue design aspects using hss – A literature study – Stevin Report 6-05-10.* Delft: TU Delft.

Maddox, S.J. 1991. *Fatigue strength of welded structures.* Cambridge: Abington publishing.

Puthli, R., Herion, S., Bergers, J. 2006. *Untersuchungen zum Ermüdungsverhalten von hochfesten Stählen im Rahmen von LIFTHIGH.* In *Stahlbau 75 Heft 11*: 916–924. Berlin: Ernst & Sohn.

Steel and Composite Structures – Wang & Choi (eds)
© *2007 Taylor & Francis Group, London, ISBN 978-0-415-45141-3*

Factors influencing the fatigue design of high-strength fine grained steels

J. Bergers, S. Herion & R. Puthli
Research Centre for Steel, Timber and Masonry, University of Karlsruhe, Karlsruhe, Germany

ABSTRACT: High-strength steels are used today in wide areas in structural engineering, although the standards do not yet adequately cover their use. To increase the knowledge on fatigue behaviour of high-strength steels, extensive tests have been carried out at the Research Centre for Steel, Timber and Masonry in Karlsruhe. The most important conclusions derived from investigations on longitudinal attachments are presented here.

1 INTRODUCTION

Modern high-strength steels have found increased application in many areas of structural engineering, such as bridges, crane structures and in the offshore industry, and they are becoming more and more popular. This is particularly so, since they now possess sufficient toughness and show good weldability. At present, fine grained steel with yield strengths up to $1100 \, \text{N/mm}^2$ are available and S1300 is in development. The technical potential for further development is not yet fully exhausted.

However, the use of high-strength steel is strongly restricted by current standards. In general, steels up to a yield strength of $460 \, \text{N/mm}^2$ can now be considered as mild steels with regard to fatigue, and they are covered by almost every modern design recommendation. Steels with strengths up to $690 \, \text{N/mm}^2$ are still partly represented, however often only in a qualified sense.

Higher steel grades are mostly not considered and conservative reduction factors are applied for these grades when permitted.

The advantages of high-strength steel are the more slender design resulting in light structures and the consequent decrease in weld volume. Although the high yield strength is capable of being adapted in static calculations, fatigue increasingly becomes the governing factor for design. Therefore, to maximise the use of high-strength steels, a suitable non-conservative fatigue design is indispensable. Influencing factors such as shape of a structure, particularly joints, the execution of welds and post weld treatment play an important role.

2 HIGH-STRENGTH FINE GRAIN STEELS

As the name already suggests, for high-strength fine grained steels, the increase of strength is obtained by a refinement of the grain structure. To also ensure a sufficient toughness in addition to the high strength, a fine martensitic structure is formed using a special heat treatment. It is also important that the steel is not produced scrap-based, but by using blast furnaces, whereby the number of undesired inclusions and foreign atoms can be minimised and a high quality can be achieved.

In principle, this method of manufacture has no advantage for fatigue resistance. But it has to be assumed that the risk of fatigue failure due to imperfections in the atomic bond is reduced by means of high steel quality and low degree of impurity of atoms. However, with smaller grain size a void extends over several grains and therefore contributes more readily to the crack propagation.

3 YIELD STRENGTH

In principle, the yield strength does not have any significance for the determination of fatigue strength (unless perhaps by the application of strength dependent reduction factors). This definitely makes sense for applications where several million load cycles are expected for middle and low stress ranges, the so called high-cycle fatigue. For special applications, however, such as crane structures, relatively low stress ranges occur, but they are nearly always close to yield strength.

Also, especially for overload occurrences, low-cycle fatigue plays a significant role.

Of course it could be argued whether such a load situation counts as fatigue at all. However, investigations on fatigue resistance of modern high-strength steels have shown, that the yield strength influences the slope of the S-N-curve.

In Figure 1 the test results from fatigue investigations of longitudinal attachments (length $l = 80 \, \text{mm}$)

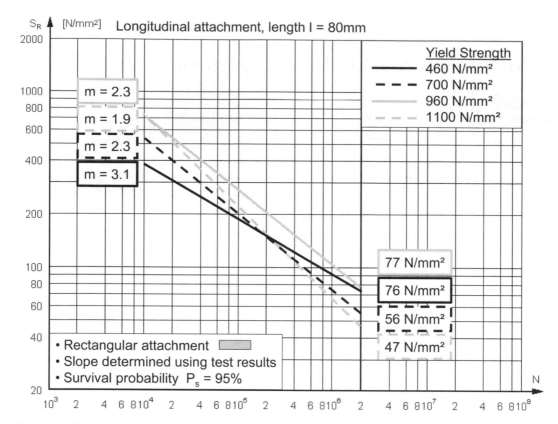

Figure 1.   S-N-curves for longitudinal attachments (80 mm).

made of high-strength steel are shown. In EN 1993–1–9 (2005), which is taken as a reference for standardization in this article, a slope of m = 3.0 is usually given for such S-N-curves. As can be seen in the Woehler-diagram, the slopes of the curves get steeper with an increasing yield strength. This results from the fact that in the area of low-cycle fatigue, the stresses reach the yield strength. Another effect is, that with steeper curves the fatigue resistance, usually taken at 2 million load cycles, declines.

The apparently lower fatigue strength for S1100 in comparison to S960 is due to the fact that for S1100 no weld metal with corresponding strength is as yet available. At present, the same weld metal is used for S1100 as for S960.

4   WALL THICKNESS

It is generally known that with larger wall thicknesses the fatigue strength decreases. In Figure 2, some of the available recommendations concerning the size effect of welded structures are illustrated. As can be seen, the wall thicknesses influence in the standards

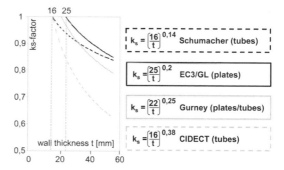

Figure 2.   Reduction-factor k_s for influence of wall thickness on the fatigue strength according to different standards.

differ, especially for large wall thicknesses, as used in bridges and offshore structures. For plates or tubes with thicknesses larger than 60 mm, very little data are available.

Figure 2 shows, that for thicknesses from 25 mm and greater, the discrepancy between the different guidelines is quite large. This raises the question, as to which database was used to establish the different formulae.

Schumacher (2003) arrived at her formula for the influence of wall thickness based on investigations on the fatigue strength of welded tubes only, especially for use in bridges. Numerous investigations by Maddox (1991) are the background of the recommendations given in EN 1993–1–9 (2005).

The fatigue design guidelines by Germanische Lloyd GmbH (2005) (GL), is close to that for EN 1993–1–9 (2005). The recommendation given by Gurney (1981) is limited up to a thickness of 50 mm, but is valid for both plates and tubes.

A large number of tests is taken into account within the CIDECT design guides (Zhao et al., 2001), but was originally meant for wall thicknesses smaller than 25 mm. This is only valid for tubular structures, but the investigations and test results of other authors are then evaluated together with the CIDECT research to provide an universal recommendation.

Altogether, much more literature is available on investigations concerning the size effect. More information about the existing data is given by Mashiri & Zhao (2005). But they finally conclude, that further research is necessary to cover the topic comprehensively.

In spite of different bases for the different formulae, the obvious aim should be to minimise wall thickness. This can be achieved by using high-strength steel.

Of course it is not always possible to use small wall thicknesses since mostly other influencing factors have to be considered e.g. stability. Nevertheless, the wall thickness can often be reduced by up to half by using high–strength steel. Taking the CIDECT design guide as an example, a reduction of about 30% in fatigue strength can be obtained by reducing the wall thickness from say 50 mm to 25 mm, which is possible by using S690 instead of S355.

This indirect influence on fatigue strength is often forgotten when weighing the advantages and disadvantages of high-strength steel.

5 POST WELD TREATMENT

As already mentioned at the beginning, a good fatigue design is indispensable for the use of high-strength steel. Otherwise, the high static strength reserve cannot be fully exploited. In the last few years, post weld treatment has proved to be a simple and effective method for increase fatigue resistance.

As an example of post weld treatment, two methods of treating the weld toes are given here: burr grinding and TIG dressing of the toes.

With the first method, burrs in the weld are ground off and an even transition from the attachment to the base plate is ensured, as is illustrated in Figure 3.

During TIG dressing, the existing weld metal is remelted by a tungsten inert gas electrode without adding filler metal. This toughens the weld seam and

Figure 3. Burr grinding of weld toe.

Figure 4. TIG dressing of weld toe.

so makes it more resistant to fatigue. A longitudinal attachment with a TIG dressed filler weld is shown in Figure 4.

In comparison to burr grinding, TIG dressing is the more cost intensive method, but is less prone to fault. Burr grinding sometimes results in defects such as pores in the weld, which where previously not there. In Figure 5, the S-N-curves of test results of specimen with TIG dressed welds and burr ground welds are illustrated. For a better estimation of the improvement, comparative tests have been carried out on as welded longitudinal attachments, which are also shown in Figure 5.

Within this test series, only a few specimens per steel grade were investigated, so that for the evaluation the steel grades S690, S960 and S1100 are collectively evaluated. To eliminate the influence of the slope of the curves, it has been fixed at m = 3.0 according to EN 1993–1–9 (2005).

What can clearly be observed in Figure 5 is the substantial advantage achieved by both methods of post weld treatment. Based on the evaluations carried out here, the specimens with TIG dressing and

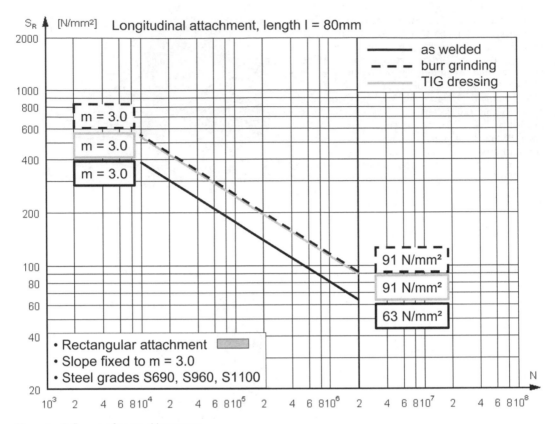

Figure 5. Influence of post weld treatment.

burr grinding both give an improvement of over 40% in stress range at 2 million load cycles in comparison to the untreated samples. Since burr grinding is more economical, this has to be the preferred method. But as mentioned before, special attention must be paid to a diligent execution of the grinding, to avoid pore defects. Other tests carried out in Karlsruhe have shown that even better results can be achieved with a professional execution of burr grinding.

The methods shown here are only two examples of many different possibilities of post weld treatment. Of course, similar improvements can be achieved e.g. by hammer or shot peening, heat or UIT-treatment, depending on the type of application.

6 DESIGN RECOMMENDATIONS

Fatigue resistant design here does not only refer to post weld treatment, but also to a proper overall design. This implies, for example, not providing weld seams in highly stressed zones and to avoid sharp edges. Nevertheless, attachments are sometimes necessary for different installations such as equipment, guy ropes or connections. But also here, the fatigue strength can

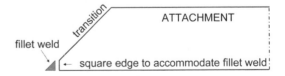

Figure 6. Attachment with square edge for a proper execution of the fillet weld.

be influenced positively by following some basic rules, which are not only valid for large structures in general, but also for small applications.

In principle, it is necessary to avoid sharp edges and hard transitions. The variant investigated here with a simple longitudinal rectangular attachment represents the most unfavourable solution. Attachments with gentle transitions to the structural member are preferable. So different variations of attachments with transition have been investigated.

The aim was first to investigate the influence of radial or linear transition, with different radii or angles. But for all attachments, for the proper execution of the fillet weld around the attachment, a square edge with a height equal to the weld leg is provided (see Figure 6).

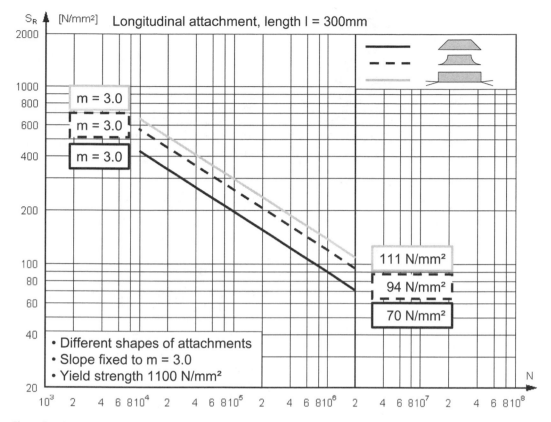

Figure 7. Comparison of different shapes of attachments.

The tests have shown that these attachments are no better than rectangular attachments for fatigue behaviour.

Therefore, the welds were treated for the subsequent test series. Here radius transitions have been investigated, whereby the weld was ground, so that an even and smooth transition was again restored. This results in an obvious advantage compared to untreated welds, as can be seen in the Woehler-diagram in Figure 7. But this method is based on burr grinding, which is more dominant than the influence of the gentle transition, due to its vicinity to the hot spot.

This poses the question of how to reduce the hot spot at the beginning of the attachment without treatment of the weld. Since the weld seam itself has the most severe influence, the solution is simple: The beginning of the welded seam must be located far enough away from the beginning if the attachment.

This idea led to the next test series: the inward-running welds (see Figure 7).

Here the weld seam is not continued around the attachment, but starts at approximately 5 cm in front of the attachment and is then continued along the side of the attachment, as illustrated in Figure 8. It is important

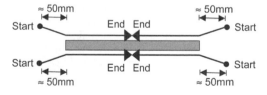

Figure 8. Execution of inward-running welds.

that the beginning of the weld is sufficiently far from the attachment. To research into this effect more precisely, finite element analyses have been performed. The distribution of the stresses along the attachment is illustrated in Figure 9.

The stress distribution for the attachment with the commonly used fillet weld seams around the attachment shows one large peak at the weld toe, as expected. For the inward-running welds two peaks occur, one at the weld toe and another at the begin of the attachment. The important conclusion is, that the stresses here are only half as high as in the case of the attachment with commonly executed fillet welds, so that fatigue strength can be expected to improve.

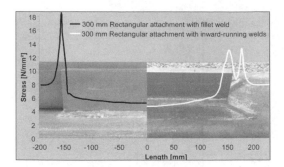

Figure 9. Stress distribution of inward-running welds.

The finite-element analyses are confirmed by the results of the fatigue tests. Figure 7 shows, that with inward-running welds, a 40% improvement in fatigue strength can be achieved. Also, this method is easy to use and cost effective as well.

7 CONCLUSIONS

High-strength steels should not automatically be avoided in structures on grounds of lack of adequate standards or by assuming that for design governed by fatigue, the yield strength does not play any role. The advantages of high-strength steels are a reduction of weight and weld volume with a consequent design and not least an economy of effort and energy. The challenge is posed in an implementation of a fatigue resistant design, which gets twice as important when using high-strength steel. Otherwise, the obvious benefits of using high-strength steel cannot be fully utilised. Together with post weld treatment, the fatigue strength of such structures can also be increased by optimisation of shape and by accurate welding. Also, the reduction of wall thickness and the corresponding favorable increase in reduction factors must be considered.

To keep pace with the forward-looking developments in ultra high-strength steels, further investigations by research institutions and standards committees are necessary.

8 ACKNOWLEDGEMENTS

The test results shown here only represent a small extract of the investigations carried out on longitudinal attachments. Further information can be found in the final reports of FOSTA-Project P512 (2005) and ECSCS-Project 4553 (2006). In this context the authors would like to thank FOSTA (Forschungsvereinigung Stahlanwendung e.V.) and the ECSC (European Community for Steel & Coal) for financial support to carry out the investigations and the accompanying working group of P512 "Beurteilung der Ermüdungsverhaltens bei Einsatz hoch- und ultrahochfester Stähle" and P4335 "Efficient Lifting Equipment With Extra High-Strength Steel". Furthermore the authors would like to thank Rautaruukki Metform, Finland, SSAB Oxelösund, Sweden, ThyssenKrupp Stahl, Germany for providing steel and Faun GmbH, Liebherr-Werk Biberach GmbH, Liebherr-Werk Ehingen GmbH and TEREX DEMAG GmbH for fabricating the specimens.

REFERENCES

Bergers, J, & Herion, S, & Höhler, S, & Müller, C, & Stötzel, J (2006) "Beurteilung des Ermüdungsverhaltens von Krankonstruktionen bei Einsatz hoch- und ultrahochfester Stähle", Stahlbau, Heft 11, Ernst & Sohn Verlag, Berlin, Germany

EN1993-1-9, (2005), "Design of steel structures – Fatigue", European Committee for Standardization, Brussels, Belgium

ECSC Project 4553 (2006), "Efficient Lifting Equipment With Extra High-Strength Steel", Final report, Brussels

FOSTA P512 (2005), "Beurteilung der Ermüdungsverhaltens bei Einsatz hoch- und ultrahochfester Stähle", Forschungsvereinigung Stahlanwendung e.V., Abschlussbericht zum Forschungsprojekt, Düsseldorf, Germany

Germanischer Lloyd Wind Energy GmbH (2005), "Guideline for the Certification of Offshore Wind Turbines", Hamburg, Germany

Gurney, TR (1981): "Some comments on fatigue design rules for offshore structures", 2nd Int Symp on Integrity of Offshore Structures, pp. 219–234, Applied Sciences Publishers, Barking, Essex, England

Hobbacher, A (1997), "Empfehlungen zur Schwingfestigkeit geschweißter Bauteile", IIW-Dokument XIII–1539–96/XV–845–96. DVS-Verlag, Düsseldorf, Germany

Maddox, SJ (1991) "Fatigue strength of welded structures", Abington Publishing, Cambridge, United Kingdom

Mashiri, FR, & Zhao, X-L (2005), "Thickness effect in Welded Joints – A Review", Proc 15th Int Offshore and Polar Eng Conf, ISOPE, Seoul, Vol 5, pp 325–332

Mecozzi, E, & Demofonit, G, & Quintanilla, H, & Izquierdo, A, & Cumino, G (2004), "Fatigue Behavior of Girth Welded Joints of High Grade Steel Risers", Proceedings of the Fourteenth International Offshore and Polar Engineering Conference. Toulon, France,

Puthli, R, & Herion, S, & Bergers, J (2006), "Untersuchungen zum Ermüdungsverhalten von hochfesten Stählen im Rahme von LIFTHIGH", Stahlbau, Heft 11, Ernst&Sohn Verlag, Berlin, Germany

Schumacher, A (2003), "Fatigue behaviour of welded circular hollow section joints in bridges", doctoral thesis, École Polytechnique Fédérale de Lausanne, Lausanne, Switzerland

Zhao, X-L, & Herion, S, & Packer, JA, & Puthli, RS, & Sedlacek, G, & Wardenier, J, & Weynand, K, & van Wingerde, AM, & Yeomans, NF (2001), "Design guide for CHS and RHS welded joints under fatigue loading," CIDECT (ed.) and Verlag TÜV Rheinland GmbH, Köln, Germany

Steel and Composite Structures – Wang & Choi (eds)
© *2007 Taylor & Francis Group, London, ISBN 978-0-415-45141-3*

Experimental study on a crack formation for railway composite girders with negative bending

N. Taniguchi & M. Ikeda
Railway Technical Research Institute (RTRI), Tokyo, Japan

T. Yoda
Waseda Univ., Tokyo, Japan

ABSTRACT: The continuous composite girders are frequently used for railway bridges. The features of the composite girders for railway bridges are characterized by Perfo-Bond Leisten and steel fiber reinforcement concrete. In this study, the behavior of a crack formation is investigated in comparison with experimental results of the test-specimens with Perfo-Bond Leisten and those of studs. In addition, the experimental results of the test-specimens with steel fiber reinforcement are examined with a view to the effect of the reinforcement on the crack. The validity of the tension stiffening theory was confirmed by comparison with experimental results and numerical results.

1 INSTRUCTION

In recent years, crack control has often been adopted in the design of the intermediate support points of continuous composite girders. The concept of this design is largely based on the tension stiffening effect theory proposed by Hanswille and other researchers in Germany, which is a theory widely discussed in relation to its application to road bridges in Japan. The design of composite girders in consideration of the tension stiffening effect assumes a certain degree of stress-bearing after cracking by the concrete between the cracks. This is different from the conventional technique to use "the steel girder plus reinforcement bar section" as the resistance section while neglecting the concrete subjected to the tensile force at the intermediate support points. When the tension stiffening effect is introduced into the design of continuous composite girders, positive bending tends to become smaller and negative bending larger than that of the conventional design method. The design section therefore changes depending on whether or not the tension stiffening effect is taken into account. This does not always mean the rationalization of design, but indicates that the stress borne by concrete should receive appropriate consideration in the design of continuous composite girders and other statically indeterminate structures, as pointed out by Hanswille and other researchers.

Railway bridges also commonly use continuous composite girders to reduce noise, improve economy and increase earthquake resistance. Their intermediate support points characteristically use steel fiber reinforced concrete and Perfo-Bond Leisten (PBL) dowels as shear connectors. Punching shear test results show that PBL dowels are advantageous from the viewpoint of fatigue strength, but there has been little research into the behavior of composite girders after cracking. It is also necessary to investigate the effects of PBL dowels on crack occurrence. Steel fiber reinforced concrete is said to effectively prevent cracks from dispersing and their widths from expanding, although this is mostly based on the results of concrete element tests. As a result, there is a need for experimental study of the phenomenon using composite girders.

To this end, the authors implemented negative bending load tests on the intermediate support points of simple composite girders, and confirmed the stress-bearing effects of concrete after cracking. The same tests were also repeated with specimens using PBL dowels as shear connectors with steel fiber reinforced concrete, and their effects were examined. The authors then compared the test results with the theory proposed by Hanswille and other researchers with due discussion.

2 SUMMARY OF TESTS

Figure 1 shows an outline of the specimen (with a span of 4 m, a floor-slab thickness of 250 mm, a reinforcement ratio of 2% and main reinforcement spacing of 150 mm). The authors used three types of specimen, A, B and C. Specimen A (often referred

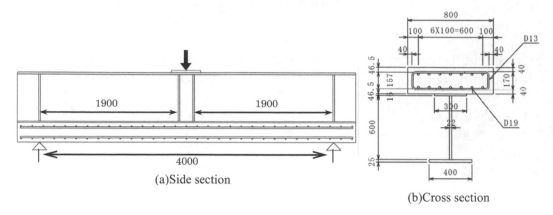

(a)Side section

(b)Cross section

Figure 1. Outline of the specimen (unit: mm).

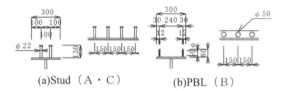

(a)Stud （A・C）　　　　(b)PBL （B）

Figure 2. Outline of shear connectors.

to as the standard specimen) was made of normal concrete with stud shear connectors. Specimen B (a perforated dowel specimen) was the same as specimen A, with the exception that the stud shear connectors were replaced with perforated steel plates. Specimen C (a specimen containing steel fibers) was made of steel fiber reinforced concrete with stud shear connectors. The shear connectors were designed on the basis of the conventional method for standard designs. The steel fiber reinforced concrete was prepared by mixing concrete with steel fibers (30-mm Shinko fiber at a volume ratio of 1%) and an expanding agent (Denka CSA100R) of a volume sufficient to cancel drying shrinkage according to the references.

The authors measured changes in the strain of the reinforcement bars caused by drying shrinkage. For this purpose, strain gauges were installed above and beneath the distribution reinforcement at the center of the girder to measure the strain in the distribution reinforcement before and after concrete casting, and an average of the measurements obtained from the two gauges was taken. When the concrete was cast and cured, the specimens were placed with the floor slab face-up and fitted with box-type forms at the sides. In this state, the girders were supported at plural points and constrained to minimize warping and deformation (Fig. 1 (b)).

Figure 3 shows the changes in the strain of each specimen. Specimen C demonstrates the effect of the expanding agent even before the removal of the forms,

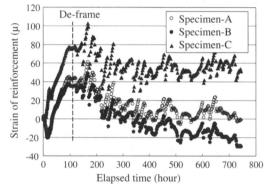

Figure 3. Measurements of the strain in reinforcement bars caused by drying shrinkage.

in that it has expanded about 40 μ as compared to specimens A and B that do not contain expanding agents. Specimen C also presents smaller changes caused by drying shrinkage after form removal, presumably because of the sustained effect of the expanding agent and the confining effect of the steel fibers. Comparison of specimens A and B indicates slight differences in the strain values even before form removal. This discrepancy, which seems to increase over time, may be attributed to the intrinsic difference that exists between concrete lots (even when made of exactly the same proportions) as well as the difference between the shrinkage-confining degrees caused by different shear connectors. In other words, it suggests that the confining force against drying shrinkage is stronger with the stud connectors of specimen A than with the PLB connectors of specimen B.

After the drying shrinkage had stabilized, negative bending load tests were conducted by static loading with unloading process once at the load levels 200, 400, 700 and 1,300 kN and continued loading thereafter until the top and bottom flange webs of the steel

916

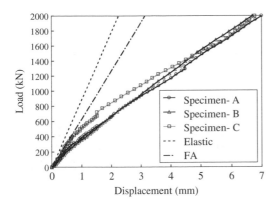

Figure 4. Relationship between load and displacement (at the center of the girder of each specimen, with the hysteresis at unloading omitted).

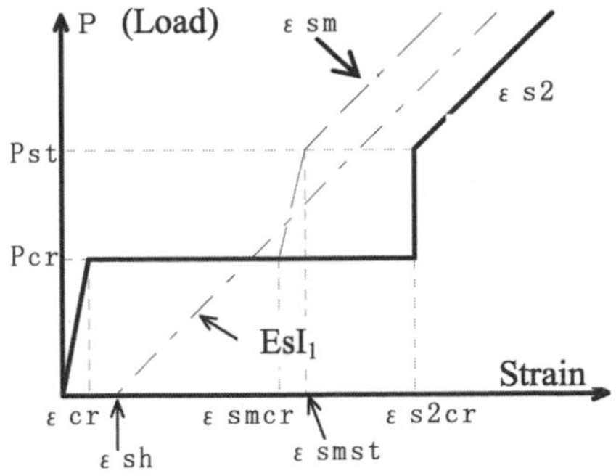

girder yielded. For unloading process in the tests, the load was removed at the calculated initial cracking load (200 kN) and at the stationary cracking load (700 kN) to check the cracking.

3 TEST RESULTS

3.1 Relationship between load and displacement

Figure 4 shows the relationship between load and displacement at the center of the girder of each specimen approximately until the distribution rein- forcement yielding load (1,800 kN). After the start of cracking, the displacement of specimen C tends to be smaller than that of specimens A and B, presumably because of the effects of the steel fibers and expanding agent. In the vicinity of the reinforcement bar (i.e. the distribution reinforcement) yielding load of 1,800 kN, however, the difference in dis- placement disappears, with all specimens presenting the same degree of displacement. This phenomenon is thought to indi- cate that the effects of the steel fibers and expanding agent almost completely disappear when cracking has progressed to a certain degree.

Comparison of specimens A and B shows that the relationships between load and displacement are almost the same immediately after the start of crack- ing, but the displacement of specimen B (perforated) becomes a little less at around 1,000 kN. This is pre- sumably because the rigidity of the PBL dowels affects the rigidity of the entire girder when cracking has pro- gressed to a certain degree. In Fig. 4, the "Elastic theoretical curves" are calculated on the basis of the beam theory and the bending rigidity when the entire cross section is still effective before the start of crack- ing. The "fiber analysis curves (FA)" are calculated using simplified linear analysis that is capable of reflecting the shear deformation when the entire cross section remains intact.

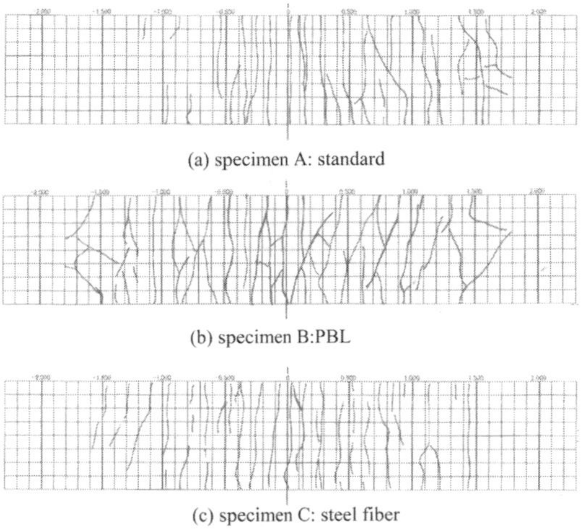

Figure 5. Approximate relationship between load and strain.

(a) specimen A: standard

(b) specimen B:PBL

(c) specimen C: steel fiber

Figure 6. Cracks on each specimen after the loading test (100-mm square meshes on the floor slab).

3.2 Occurrence of cracks

Figure 6 shows the cracks on the concrete floor slab after the loading test. The cracks run mostly in a direc- tion perpendicular to the bridge axis, which indicates that they were caused by the dominant bending behav- ior. With all specimens, crack spacing ranges from 100 to 200 mm, with an average value of about 150 mm. The value of 150 mm and the crack spacing coincide with the spacing between the main reinforcement bars, in a similar way to the values in the experimental results obtained by Nagai et al. The maximum crack spacing (L_{max}) is calculated as 217 mm using the equa- tion of The Japan Society of Civil Engineers, which is an appropriate value when compared with the test results.

Figure 6 (c) represents the case of specimen C, which is made of steel fiber reinforced concrete with a crack-dispersing effect. However, this specimen presents no significant difference from other speci- mens, except that slightly fewer penetrated cracks are

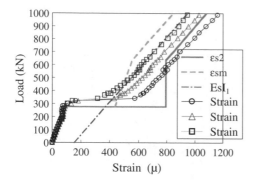

Figure 7. Relationship between load and strain in distributing bars (specimen A: standard).

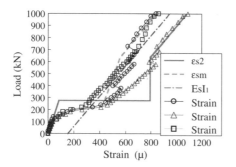

Figure 8. Relationship between load and strain in distributing bars (specimen B: PBL).

generated. This is presumably because the positioning of crack occurrence is largely governed by the main reinforcement bars, even if steel fiber reinforced concrete is used.

3.3 Relationship between load and strain in distribution reinforcement

Figures 7 to 9 show the relationship between load and strain in the vicinity of the center of the girder at each specimen, in which ε_{s2} and ε_{sm} are the maximum and average strain values calculated on the basis of the Hanswille theory, while the relationship between the load and the maximum/average strain is assumed as shown in Fig. 5. The line marked "EsI_1" shows the values calculated using the rigidity when the concrete is neglected. The values of ε_{s2} and ε_{sm} were calculated using the equations shown below.

(1) Initial cracking load P_{cr}

$$M_{cr} = \frac{f_t}{Z_0 + h_c/2} \cdot I_0 \cdot n \qquad (1)$$

$$P_{cr} = \frac{4M_{cr}}{l} \qquad (2)$$

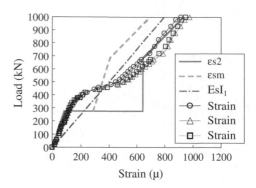

Figure 9. Relationship between load and strain in distributing bars (specimen C: steel fiber).

(2) Stationary crack-starting load P_{st}

$$N_{scr} = f_t \cdot k_c \cdot A_c \cdot (1 + n\,\rho_s) \qquad (3)$$

$$k_c = k_{co} + 0.3 = \frac{1}{1 + \dfrac{h_c}{2 \cdot Z_0}} + 0.3 \qquad (4)$$

$$M_{st} = \left(N_{scr} - \beta_m \frac{f_t \cdot A_s}{\rho_s \cdot \alpha}\right) \cdot \frac{I_1}{A_s \cdot Z_1} \qquad (5)$$

$$\alpha = \frac{A_1 \cdot I_1}{A_g \cdot I_g} \qquad (6)$$

$$P_{st} = \frac{4M_{st}}{l} \qquad (7)$$

(3) Strain immediately before the start of cracking ε_{cr}

$$\varepsilon_{cr} = \frac{f_t}{E_c} \qquad (8)$$

(4) Maximum strain after the start of cracking ε_{s2st} (at the start of stationary cracking)

$$\varepsilon_{s2st} = \frac{N_{scr}}{E_s \cdot A_s} + \varepsilon_{sh} \qquad (9)$$

(5) Average strain at the start of stationary cracking ε_{smst}

$$\varepsilon_{smst} = \varepsilon_{s2st} - \beta_m \frac{f_t}{E_s \cdot \rho_s} \qquad (10)$$

(6) Average strain immediately after the start of cracking ε_{smcr}

$$\varepsilon_{smcr} = \varepsilon_{s2st} - \beta \frac{f_t}{E_s \cdot \rho_s} \qquad (11)$$

where,
f_t: Concrete tensile strength
I_0, I_1, I_g: Geometrical moment of inertia; composite section, steel girder plus reinforcement bar section, and steel girder alone

Z_0, Z_1, Z_g: Distance from the neutral axis to the floor slab center; composite section, steel girder plus reinforcement bar section, and steel girder alone

A_s, A_c: Sectional area; reinforcement bar and concrete

E_s, E_c: Young's modulus; reinforcement bar and concrete

h_c, I: Floor slab thickness and span length of specimen

n: Ratio of Young's modulus, steel to concrete $(=7)$

ρ_s: Reinforcement ratio $(=0.02)$

ε_{sh}: Drying shrinkage strain

β, β_m: Factors to express the magnitude of tension stiffening at initial cracking and stationary cracking (set at 0.6 and 0.4 respectively)

For the drying shrinkage strain, the authors assumed $\varepsilon_{sh} = 150\,\mu$ (a normally used design value) for specimens A and B, and $\varepsilon_{sh} = 0\,\mu$ for specimen C to calculate strain, in view of considering the effect of the expanding agent and ignoring the effect of steel fibers. It is said that the figure "0.3" is included in equation (4) to consider the effect of drying shrinkage, but there is vivid argument about its validity. For the sake of comparing test results in this case, therefore, the authors used the equation (4) as it is. As a result, the effect of drying shrinkage represents only that of parallel shift in the x-axis direction of the strain in the stationary cracking state, as ε_{sh} is added to the maximum and average strain values in equations (9) to (11).

In these figures as a whole, the large strain values are approximately equivalent to the calculated maximum strain. It can therefore be thought that appropriate results are obtained using this method of calculation with β_m set as 0.4. Since the measured strain scatters to a large extent, however, both the maximum strain values as well as smaller ones are considered to be included in the measurement results. This indicates that it is difficult to correctly calculate the average strain based on the test results. However, an average strain value is required to ensure the validity of $\beta = 0.6$, in the sense that the results obtained in this study are insufficient. The authors therefore feel that further studies should be carried out on this issue in the future.

After the start of cracking, the strain near the point where a crack has occurred increases rapidly (a phenomenon referred to as *strain jump*). In the case of specimen C (with steel fibers) however, the strain jump is smaller (i.e. the strain increases more smoothly) than that of specimens A or B, presumably because strain is prevented from increasing quickly due to the bridging effect of the steel fibers. After cracking has progressed to a certain degree, however, the effect of the steel fibers diminishes, and the same degree of strain as in the other specimens is presented.

4 CONCLUSION

In this study, the authors conducted loading tests on continuous composite girders to simulate negative bending, and checked the effects of different shear connectors and concrete properties on the cracking and strain in distribution reinforcement. The test results were compared with the theory proposed by Hanswille et al. for theoretical verification, and the following conclusions were reached:

(1) In comparison of a specimen using PBL dowels as shear connectors and a specimen using studs, no significant difference was found in the relationship between load and displacement, the relationship between load and distribution reinforcement strain, and the crack positions.

(2) With a specimen using steel fiber reinforced concrete, an effect preventing strain jump in distribution reinforcement was observed when cracking occurred. After cracking had progressed to a certain extent, however, there were no significant differences in terms of strain or crack spacing between the specimen and that without steel fiber reinforcement.

(3) The maximum strain in distribution reinforcement obtained through the tests in this study approximately corresponded to the value calculated through application of the tension stiffening theory proposed by Hanswille et al. Static loading tests also provided appropriate results at $\beta_m = 0.4$ or a value proposed by these researchers. When $\beta = 0.6$, however, larger maximum strain values should be used for specimens made of steel fiber reinforced concrete than those without steel fibers, since the initial crack-suppressing effect can be expected.

Future subjects to be addressed on the theme discussed in this study include the assessment of cracking and verification of β_m setting when girders are subjected to fatigue loads.

REFERENCES

Yasukawa, Y. & Inaba, N. 2001. Design and Construction Work of Continuous Composite Bridges with Two Main Girders, *Proceedings of the 4th Symposium on Steel Structures and Bridges, Committee on Steel Structures, The Japan Society of Civil Engineers*: 11–2, Japan.

Hanswille, G. 1997. Cracking of concrete mechanical models of the design rules in EUROCODE4, *Conf.Report, Composite Construction in Steel and Concrete III, ASCE*: 420–433, America.

Roik, K. & Hanswille, G. (authors), Ito, K. & Hiragi, K. (translators) 1993. Restriction of Cracking Width of Composite Girders, *Bridge and Foundation, 93 – 2*: 33–40, Japan.

Nagai, M., Okui, Y. & Iwasaki, E. 2004. A Proposal on Methods to Calculate Cracking Widths and Reinforcement Bar

Strains of Continuous Composite Girders in Consideration of Initial Cracking, *Proceedings of The Japan Society of Civil Engineers, No. 759/I – 67*: 283–292, Japan.

Kurita, A., Ohyama, O. & Rutner, M. 2001.Present Status and Subjects of Dual Composite Continuous Box Girder Bridges, *Proceedings of the 4th Symposium on Steel Structures and Bridges, Committee on Steel Structures, The Japan Society of Civil Engineers* : 45–58, Japan.

Oka, A., Ohyama, O., Rutner M. & Kurita, A. 2002. Effect of Concrete Tension Stiffening of Dual Continuous Box Girder Bridges, *the 57th Annual Conference of the Japan Society of Civil Engineers, I – 345*: 689–690, Japan.

Nagai, M. & Iemura, T. 2000. An interview with Professor Hanswille, *Bridge and Foundation, Vol. 34, No. 11* :33–39, Japan.

Hosaka, T. & Sugimmoto, I. 2000.Recent Steel/Concrete Composite Bridges in Railways, *Bridge and Foundation, 2000 – 7*: 31–40, Japan.

Committee on Steel Fibers, Kozai Club.1998. A Guide to Steel Fiber Concrete (3rd edition), *THE KOZAI CLUB*, Japan.

Kamiya, T., Taniguchi, N., Irube, T., Ikariyama, H., Onozawa, T. & Yoda, T. 2004.An Experimental Study on the Cracking Behavior of Continuous Composite Girders at Intermediate Support Points (No. 1, PBL dowels), *the 59th Annual Conference of the Japan Society of Civil Engineers*, Japan.

Sato, K., Taniguchi, N., Irube, T., Ikariyama, H., Onozawa, T., & Yoda,T. 2004. An Experimental Study on the Cracking Behavior of Continuous Composite Girders at Intermediate Support Points (No. 2, Steel Fiber Reinforced Concrete), *the 59th Annual Conference of the Japan Society of Civil Engineers*, Japan.

Yajima, S., Ichikawa, A., Murata, K. & Kitazono, S. 2003. An Experimental Study on the Application of SRC Floor Slab Construction to Railway Trussed Steel Through-Bridges, *Proceedings of The Japan Society of Civil Engineers, No. 731/I - 63*:283–298, Japan.

Taniguchi, N. & Yoda, T. 2001. A Study on the Cracking of Composite Girders Subjected to Negative Bending, *Proceedings of The Japan Society of Civil Engineers, No. 668/I – 54*: 243–257, Japan.

Taniguchi, N. & Yoda, T. 1997. A Study on the Method of Simplified Bending Analysis for Composite Girders with Corrugated Steel Webs, *Proceedings of The Japan Society of Civil Engineers, No. 577/I – 41*: 107–120, Japan.

Nakamoto, K., Nagai, M., Okui, Y., Iwasaki, E. & Hosomi, M. 2003. An Experimental Study on the Cracking Behavior of Composite Girders Subjected to Negative Bending, *Proceedings of Structural Engineering, The Japan Society of Civil Engineers, Vol. 49A*: 1,143–1,152, Japan.

Taniguchi, N., Nishida, H., Murata, K., Yajima, S. & Yoda, T. 2002. Simplified Analysis on the Axial Tensile Behavior of Steel Fiber Reinforced Composite Floor Slabs, *Proceedings of Concrete Engineering, Vol. 13, No. 3*, Japan.

Steel and Composite Structures – Wang & Choi (eds)
© 2007 Taylor & Francis Group, London, ISBN 978-0-415-45141-3

Enhancing fatigue strength by ultrasonic impact treatment for welded joints of offshore structures

P. Schaumann & C. Keindorf

Institute for Steel Construction, Leibniz University, Hannover, Germany

ABSTRACT: This paper summarises fatigue tests on Y-joints to estimate the influence of a post weld treatment method called Ultrasonic Impact Treatment. With this method the fatigue resistance could be increased significantly. Furthermore tubular joints of tripods are analysed with numerical simulations to judge these welded joints with the hot-spot-concept. The stress concentration factor for the treated weld toe geometry was determined numerically and compared to experimental results.

1 INTRODUCTION

For the planned offshore wind farm "Kriegers Flak" in the Baltic Sea a fatigue design study was carried out including experimental and numerical investigations for welded joints of a tripod. A tripod is one kind of a supporting structure for wind energy converter as shown in Figure 1. The calculations were based on Baltic Sea conditions with 25 m water depth and for a 2 MW turbine. The welded joints were designed for a fatigue life of 20 years with numerical simulation based on the hot-spot-concept.

Furthermore experiments were carried out to estimate the fatigue resistance for such welded joints. Because of large dimensions for tripods the fatigue tests can't be performed in real scale.

Therefore 12 Y-joints were tested with $t_c = 90$ mm thickness for the chord and $t_b = 40$ mm for the brace welded in an angle of $\theta = 60°$C. The plate thicknesses are comparable with those of tripods. The objective of these tests was to check the fatigue resistance for welded joints with thick plates. Additionally the influence of post weld treatment by Ultrasonic Impact Treatment (UIT) should be estimated. This method introduces compressive stresses and plastic deformations at the weld toe to reduce residual stresses and stress concentrations. Because of these effects the fatigue strength increases significantly compared to as welded conditions.

Furthermore different types of tubular joints for offshore structures were investigated with numerical simulations to estimate the fatigue limit state for both conditions, as welded and treated by UIT. The stress concentration factors (SCF) for the treated weld toe geometry were determined numerically using submodel analysis and compared to experimental results.

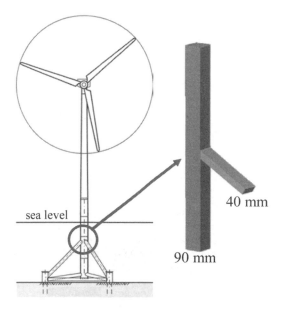

Figure 1. Offshore wind energy converter with tripod as support structure and Y-joint for experiments.

Finally a comparison between welded and cast iron joints was carried out in a fatigue design study under consideration of UIT-effects.

2 EXPERIMENTS WITH Y-JOINTS AND UIT

2.1 Test specimen

The test specimens are Y-joints with $t_c = 90$ mm thickness for the chord and $t_b = 40$ mm for the brace. Chord and brace were welded in an angle of 60° with fillet

Table 1. Chemical composition of material S355 J2G3 in [%].

C	Si	Mn	P	S	N	Cr	Ni
0.16	0.34	1.43	0.014	0.004	0.006	0.07	0.06

Table 2. Mechanical properties of material S355 J2G3.

Yield strength $R_{e,H}$ [N/mm²]	Ultimate strength R_m [N/mm²]	Elongation at failure A_5 [%]	Impact ductility KV [J, −40°C]
386	537	23.5	106

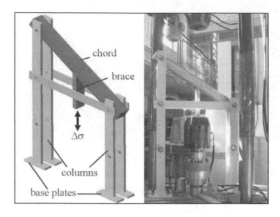

Figure 3. Test setup for Y-joints.

strain gauges were also installed to evaluate the local stress state in the near of weld toe and weld root.

2.3 Post weld treatment by UIT

Due to the plate thickness effect the fatigue resistance must reduce according to offshore-guideline (Germanischer Lloyd 2004). For example the reduced fatigue resistance at the chord is:

$$\sigma_{c,red} = \left(\frac{t_{ref}}{t_c} \right)^{0.25} \cdot \sigma_c = 0.73 \cdot \sigma_c \qquad (1)$$

where $t_{ref} = 25$ mm as reference plate thickness and $\sigma_c =$ fatigue resistance at 2 million cycles.

The reduction in fatigue resistance due to thickness effect (e.g. for the chord thickness 27%) has to consider in design studies, which is mostly limiting for the dimensions of tubular joints for offshore structures.

But the fatigue resistance of welded joints can be enhanced by post weld treatment. One method for this is Ultrasonic Impact Treatment (UIT). It is a proprietary technology developed originally in the Soviet Union for use on naval ships to reduce welding stresses (Statnikov et al. 1977). The equipment comprises a handheld tool and an electronic control box (Fig. 4). The tool is easy to handle during application. It operates at the head movement with a mechanic frequency of 200 Hz overlain by an ultrasonic frequency of 27000 Hz. The noise is negligible compared to other peening devices. Several kinds of heads and pins are available and can be chosen on the basis of the surface condition of the weld details to be treated. The method involves post-weld deformation treatment of weld toe by impacts from single or multiple indenting needles excited at ultrasonic frequency, generating mechanic impulses on the work surface (Statnikov et al. 1997).

The objective of the treatment is to introduce beneficial compressive residual stresses at the weld toe

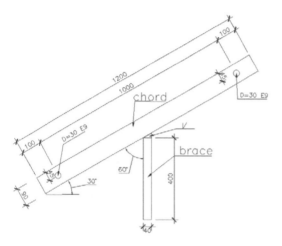

Figure 2. Test specimen.

welds. The Y-joints were fabricated of a steel S 355 J2G3. Chemical and mechanical properties of this steel are presented in Tables 1 and 2.

The specimen geometry with detailed dimensions of these welded joints is shown in Figure 2.

2.2 Experimental setup for fatigue testing

Fatigue testing was performed using a 600 kN servo-hydraulic test frame. The experiments were carried out up to 2 million cycles with test frequencies between $f_P = 3 - 5$ Hz depending on the value of the dynamic force. The test setup consisted of four columns with base plates and two horizontal bracing members. The Y-joint was supported by bolted connections at both ends of the chord. The position of the brace was vertically to fix the end of the brace in the testing machine.

The dynamic force loaded at the end of the brace simulated a stress range $\Delta\sigma$. During the tests the dynamic force was measured with a load cell and the deformation were recorded online by inductive displacement transducers. For some test specimens

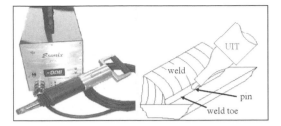

Figure 4. Ultrasonic Impact Treatment (UIT) [Esonix].

zones and to reduce stress concentration by improving the weld toe profile. Furthermore the area being treated is highly plastically deformed which has the effect of work hardening.

Compared to other impact treatment methods such as air hammer peening, shot peening or needle peening, UIT is claimed to be more efficient involving a complex effect of strain hardening, reduction in weld strain, relaxation in residual stresses reduction in stress concentration and thereby achieving a deeper cold worked metal layer (Statnikov et al. 1997).

2.4 Test program

The test program comprises two test series on Y-joints with and without post weld treatment by UIT (Table 3). The first test series without post weld treatment (Y_1 – Y_6) was carried out to get a reference S-N-curve for Y-joints for the as-welded condition. The dynamic loads were recorded as nominal stress ranges $\Delta\sigma_n$ which were varied for the test specimens. The ratio between minimum and maximum stresses was $R = \sigma_{min}/\sigma_{max} = 0.08 - 0.16$ depending on the servo-hydraulic system of the testing machine.

The second test series (Y_7 – Y_12) have been performed with the same procedure like the first test series but additionally with post weld treatment by UIT. The treatment was carried out according to the manufacturer's procedure document (Applied Ultrasonics 2000). The indenter consisted of three 3 mm diameter pins, fitted in a single holder. The treatment was carried out in short multiple passes.

3 FATIGUE TEST RESULTS

3.1 Failure modes

During the tests fatigue cracks occurred at mainly two positions – at the toe or at the root of the weld. For test specimens without additional fillet welds at the root the crack was detected always at the root because of higher notch effects.

But for tests with additional fillet welds at the root the position of cracks changed to the toe. It can be noticed that welding with additional fillet welds at the root has a great influence on the place where the fatigue

Table 3. Test program.

Test series	Test No.	weld toe [–]	$\Delta\sigma_n$ [N/mm^2]	R [–]	N [·10^6]
1	Y_1	as welded	28.8	0.13	0.12
	Y_2	as welded	25.6	0.08	0.15
	Y_3	as welded	15.0	0.15	1.10
	Y_4	as welded	19.8	0.13	0.24
	Y_5	as welded	10.2	0.16	3.39
	Y_6	as welded	16.4	0.13	0.45
2	Y_7	UIT	27.1	0.15	0.82
	Y_8	UIT	22.7	0.12	3.75
	Y_9	UIT	34.5	0.12	0.16
	Y_10	UIT	26.1	0.11	0.75
	Y_11	UIT	32.0	0.10	0.18
	Y_12	UIT	28.5	0.10	0.51

Figure 5. Test specimen Y_2 with fatigue crack at the weld toe.

crack will begin. For a good performance of post weld treatment by UIT, it is therefore desirable to match the fatigue crack growth life from root defects to the fatigue life of the treated toe. In this way larger size fillet weld reduced stress concentration adjacent to the weld root, contributing to increased fatigue life. An example for a fatigue crack is shown in Figure 5.

After fatigue tests several weld details and crack surfaces were cut out from the Y-joints and were examined for origins of fatigue cracks. Figure 6 presents the weld toe of test specimen Y_5 for as welded conditions. In the near of weld toe there can be three zones identified: 1. weld metal, 2. heat affected zone and 3. base material. The notch radius of the weld toe is $r_{as\ welded} = 0.4$ mm.

Figure 7 shows a photomicrograph of a typically treated weld toe at 50X magnification. Due to the post weld treatment by UIT the surface of the weld toe was highly plastically deformed and the notch was rounded. The weld toe of test specimen Y_9 is shown in Figure 7 after post weld treatment and fatigue testing. Analog to the test specimen Y_5 in Figure 6 the three zones are visible. The notch radius of the treated weld

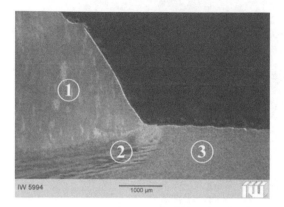

Figure 6.   Weld toe of Y_5 (as welded).

Figure 7.   Treated weld toe of Y_9 with fatigue crack (UIT).

toe increased to $r_{uit} = 1.8$ mm. Furthermore the fatigue crack can be observed starting from surface between the weld metal and the heat affected zone. Fatigue Cracks in welded joints are often detected at this position because the heat affected zone has a high degree of hardness.

3.2   S-N curves

All test results are summarized in two S-N-curves (Fig. 8). The S-N curve of the joints with UIT shows a significant increase in fatigue resistance compared to as welded joints. The as welded joints can be classified in FAT 90. This result corresponds with recommendations for tubular joints according to the offshore-guideline (Germanischer Lloyd 2004) based on the hot-spot-concept.

With $\Delta\sigma_c = 204.5$ N/mm^2 for 2 million cycles the fatigue strength after post weld treatment by UIT was double compared to as-welded ($\Delta\sigma_c = 95.5$ N/mm^2). The slope of the first test series with as-welded joints is m = 3.47. This value can be compared with recommendations of design guidelines for fatigue limit state (m = 3 for N $< 5 \cdot 10^6$). But for the second

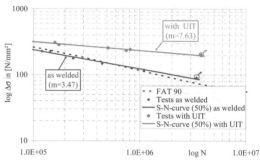

Figure 8.   S-N-curves of both test series (as welded and UIT).

Table 4.   Parameters of S-N-curves.

Test series	$\Delta\sigma_c$ [N/mm^2]	slope m [–]	N_R for $\Delta\sigma_c = 100$ [$\times 10^6$ cycles]
As welded	95.5	3.47	1.7
UIT	204.5	7.63	475.1

test series with UIT the slope is with m = 7.63 significantly higher. With the experimental results the validity of the thickness effect could be confirmed also for plate thickness of 90 mm. The parameters of both S-N-curves are shown in Table 4.

4   NUMERICAL SIMULATIONS OF Y-JOINT

In the first level a 3D-model of the whole test specimen was analyzed using commercially available finite element code ANSYS (Fig. 9). This model contained all boundary conditions, but excluded the actual weld notch effects. The weld profile was modeled as a notch having 60° flank angle with a theoretically zero toe radius. With this model the stress concentrations at the weld can be observed using the hot spot concept.

The stress concentration factors (SCF) were determined for the toe and root of the weld with the following equation:

$$SCF = \frac{\sigma_S}{\sigma_N} \qquad (2)$$

where σ_S = local stress at the hot spot and σ_N = nominal stress at the end of the brace.

To determine the hot spot stress two extrapolation points are necessary. The first point is located at the chord surface in a distance of $0.4 \cdot t_C$ from the hot spot the second in a distance of $1.0 \cdot t_C$. The calculation is comparable to the calculation for strain concentration factors if the stresses keep in linear elastic range. The experimental SCF's at the chord were measured with strain gauges in the near of toe and root of the

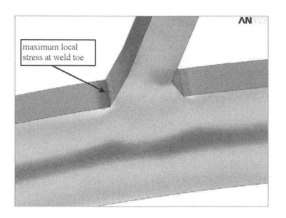

Figure 9. Stress concentrations at toe and root of the weld.

Table 5. Comparison of strains and SCF's at the chord.

Hot spot	Method	ε_S (0.4 t_C) [μm/m]	ε_S (1.0 t_C) [μm/m]	SCF [–]
Weld Toe	FEM	725	632	6.61
Weld Toe	Experiment	707	614	6.46
Weld Root	FEM	598	509	5.53
Weld Root	Experiment	609	524	5.59

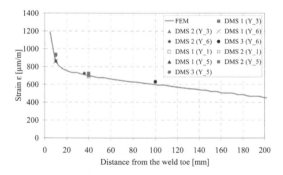

Figure 10. Numerical stress curve perpendicular to the weld.

weld. The SCF's derived with numerical simulations are compared to experimental results. The comparison is presented in Table 5. The experimental and numerical SCF's for weld toe and also for weld root show a very good agreement.

Figure 10 shows a curve of the stress path perpendicular to the weld at the chord. The curve was estimated with the FE-model. The stress increases nonlinear in the near of the weld toe. Furthermore the diagram includes all SCF values measured with strain gauges at the chord for the same load level. The test results agree very well with the numerical stress curve. So it can be noticed that the stress concentration due to the local geometry of the Y-joint is correctly comprised in the 3D-model.

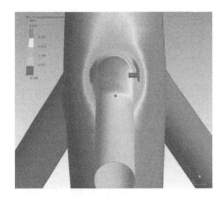

Figure 11. Geometry of tripod and stress concentration.

5 COMPARISON BETWEEN WELDED AND CAST JOINTS

5.1 Geometry of offshore structure (Tripod)

Welded and cast joints for Tripods are analyzed for the Baltic Sea conditions of the planned wind farm "Kriegers Flak". The water depth is assumed with 25 m for a 2 MW turbine. The diameter of the central tube (chord) is $D_C = 4.0$ m and for the braces $D_B = 2.0$ m which are jointed to the chord in an angle of 45°. The objective of the design study is the optimization of thicknesses t_C and t_B for fatigue resistance.

The loads of wind and wave calculated with deterministic concept are summarized in a rainflow count and load classes. The joints are designed for fatigue limit state with hot-spot-concept. With techniques of sub-modelling the hot-spot-stresses of the joints are calculated, to estimate finally the linear cumulative fatigue damage by Palmgren-Miner. The stress concentration factors for cast joints can be optimized by variable fillet radiuses. Thus the cumulative fatigue damage for cast joints is decreased and the thicknesses of chord and brace can be reduced significantly compared to welded joints. Weight saving between 20% and 40% are possible for different types of joints for tripods. The consideration of wave spreading allows further reduction for plate thicknesses, but this is possible for both variants. During the fatigue tests a nonlinear lost of stiffness for the joint could be monitored with strain gauges. This has to be taken into account for the dynamic behaviour of support structures.

5.2 Comparison of fatigue strength

The experimental results are taken into account in a reanalysis for tubular joints of tripods. In this way the fatigue resistance of welded joints can significantly be increased by post weld treatment with UIT. The estimated value in Table 4 for the treated condition by UIT is higher than fatigue class (FAT) for the unnotched base material thus the maximum fatigue resistance

Table 6. Parameters of S-N-curves.

Fatigue classes (FAT)	FAT $\Delta\sigma_c$ [N/mm^2]	slope m_1 [-]	slope m_2 [-]	Damage D [-]
As welded	100	3	5	27.2
UIT	160	7	7	1.4

Table 7. Comparison of plate thicknesses.

type of joint	Upper tripod-joint		Lower tripod-joint	
	t_C [mm]	t_B [mm]	t_C [mm]	t_B [mm]
Welded without UIT	200	100	120	50
Welded with UIT	90	60	80	50
Cast iron	90	60	80	50

is assumed to $\Delta\sigma_c = 160\,N/mm^2$. The parameter of the fatigue classes for both condition are presented in Table 6:

Two effects of UIT cause a better fatigue performance in the treated condition. At first the higher value for $\Delta\sigma_c$ and second the lower slope of the S-N curve. The lower cumulative fatigue damage for the same life time estimated in Table 6 allows weight savings which are comparable with savings by cast iron joints. The optimized thicknesses for different type of joints are compared in Table 7. For welded joints of future offshore wind farms the fatigue design would be more competitive if the effects of post weld treatment by UIT will be considered.

But for detailed numerical analyses of UIT-effects a sub-model is necessary. In this way a second level of numerical study have to carry out with simulations for a single weld seam including the welding process and afterwards the UIT-process.

6 CONCLUSIONS

Fatigue tests on welded joints were carried out to estimate the influence of post weld treatment by Ultrasonic Impact Treatment (UIT). This method introduces compressive stresses and plastic deformations at the weld toe thus residual stresses and stress concentration factors can be reduced. Because of these effects the fatigue strength increased significantly up to $\Delta\sigma_c = 204.5\,N/mm^2$ compared to as welded condition with $\Delta\sigma_c = 95.5\,N/mm^2$. This result corresponds with recommendations for tubular joints according to actual offshore-guidelines based on the hot-spot-concept. A second effect is observed for the slope of S-N curve which changed from 3.47 to 7.63 after post weld treatment by UIT.

The experimental results were compared with numerical solutions on a 3D-model of the whole test specimen using commercially available finite element code ANSYS. This model contained all boundary conditions, but excluded the actual weld notch effects. With this model the stress concentration at the weld could be analysed using the hot spot concept. The stress concentrations factors derived in experiments and numerical simulations have a very good agreement for the weld toe and also for the weld root.

Furthermore different types of tubular joints for offshore structures were investigated with numerical simulations to estimate the fatigue limit state for both conditions, as welded and treated by UIT. The stress concentration factor for the treated weld toe geometry was determined numerically using sub-model analysis. Finally a comparison between welded and cast iron joints was carried out in a fatigue design study under consideration of experimental results. The lower cumulative fatigue damage for the treated condition allows weight savings which are comparable with savings by cast iron joints. For welded joints of future offshore wind farms the fatigue design would be more competitive if the effects of post weld treatment by UIT will be considered.

ACKNOWLEDGEMENTS

The study was conducted at the Institute for Steel Construction of the Leibniz University, Hannover, Germany. Special acknowledgement is due to Warnow Design GmbH, Rostock, Germany, the engineering consult which supported the work. The authors will also like to thank Peter Gerster, Esonix, for supplying the UIT equipment.

REFERENCES

Applied Ultrasonics. 2002. Esonix Ultrasonic Impact Treatment: Technical Procedure Document, Applied Ultrasonics, Birmingham, AL, USA.

Bäumel, A., Seeger, T. 1990. Materials Data for Cyclic Loading, Suppl 1, Amsterdam, Elsevier Science.

Germanischer Lloyd 2004. Rules and Regulations IV, Non-marine Technology, Part 2: Offshore Wind Energy, Hamburg, Germanischer Lloyd Industrial Services.

Roy, S., Fisher, J.W. 2005. Enhancing Fatigue Strength by Ultrasonic Impact Treatment, Steel Structures 5, 241–252, Korean Society of Steel Construction (KSSC), Korea.

Schaumann, P., Kleineidam, P., Wilke, F. 2004. Fatigue Design bei Offshore-Windenergieanlagen, Stahlbau Vol. 73(9), 716–726, Ernst&Sohn-Verlag, Hannover, Germany

Statnikov, E.S. et al. 1977. Ultrasonic impact tool for strengthening welds and reducing residual stresses, New Physical Methods of Intensification of Technological Processes.

Statnikov, E.S. et al. 1997. Applications of operational ultrasonic impact treatment (UIT) technologies in production of welded joints, IIW, Doc. XIII-1668-97, International Institute of Welding, Paris, France.

Steel and Composite Structures – Wang & Choi (eds)
© 2007 Taylor & Francis Group, London, ISBN 978-0-415-45141-3

Buckling of centre-cracked shear panels

M.M. Alinia, H.R. Habashi & S.A.A. Hosseinzadeh
Department of Civil Engineering, Amirkabir University of Technology, Tehran, Iran

ABSTRACT: Shear panels are utilized as thin steel plate shear walls (TSPSW) or panels created within the web of link beams in eccentrically braced frame (EBF) structures. They buckle under relatively small forces and their buckling behavior is to some extent influenced by crack defects that may occur during assembling procedure, transition operations, fatigue, or oxidations. In this paper, the result of a large number of numerical analyses carried out to investigate the shear buckling of thin steel panels subjected to in-plane shear loading is presented. Sensitivity analyses based on different geometrical and mechanical properties, such as the crack length and its angle of inclination, the panel thickness, the Poisson's ratio, the modulus of elasticity and boundary conditions are carried out. The results are then utilized to introduce a stiffness reduction factor for panels defected by central cracks.

1 INTRODUCTION

A thin shear panel subjected to shear force, utilized, for instance, in aeronautic and naval industries, in thin steel plate shear walls, in plate and box girders, and in frame bracing systems, experiences equal tensile and compressive principle stresses in the linear elastic stage. With the increase of shear force, the panel starts to buckle due to compressive stresses and undergoes large out-of-plane deformations in the diagonal direction before yielding.

Panels may suffer from various types of crack defects, and therefore, it is imperative to estimate the residual strength of damaged structures. The effect of cracks on the buckling behavior, load carrying capacity, and safety assessment of shear panels should be investigated and considered both in the design procedure and during service life.

According to the literature review on cracked plates, there are no specific reports on the buckling of cracked shear panels and no guidelines are available for predicting the sensitivity of crack characteristics on the amount of reduction of their load carrying capacity. However, there are a few reports on the buckling of cracked plates under tension or compression and on shells subjected to internal pressure and axial compression.

Riks et al. (1992) used the finite element method to study the buckling behavior of cracked plates under tension. Brighenti (2005) investigated the effect of crack lengths and locations on the buckling of cracked elastic rectangular plates under tension or compression. Paik and Satish Kumar (2005) experimentally and numerically investigated the behavior of cracked plates under axial compression or tension.

Also, numerous tests and numerical studies have indicated the formation of several types of fatigue cracks in the webs of slender plate girders subjected to combinations of bending and shear (2002).

In this paper, thin shear panels having cracks at their centers are analyzed. The aim is to determine their buckling behavior and to show how much their strength could be degraded by various types of central cracks.

2 METHOD OF STUDY

The analyses were carried out by means of the finite element method. The numerical approach seems to be the most promising method, since the analytical formulation of the aforementioned problem is noticeably complicated. As even in the case of numerical analysis, the large number of interacting parameters such as the angle of inclination and the length of crack, the slenderness and the aspect ratio of panels, plus the complicated buckling behavior makes it quite difficult to ascertain the common applicable conclusions.

The general purpose FE program, ANSYS was utilized for the modeling and buckling analysis of cracked shear panels having different mesh refinements near cracks.

Cracks were assumed to be in the centre of plates, having different lengths and angles of inclination. Figure 1 illustrates the crack characteristics considered in this study. Cracks were presumed to be through thickness, having no friction between their edges and no propagation was allowed.

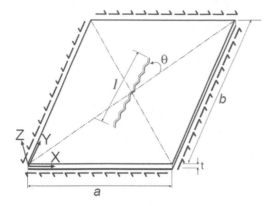

Figure 1. Panel and crack characteristics.

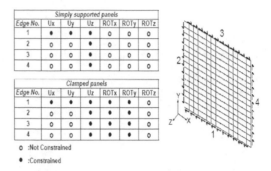

Simply supported panels						
Edge No.	Ux	Uy	Uz	ROTx	ROTy	ROTz
1	●	●	●	o	o	o
2	o	o	●	o	o	o
3	o	o	●	o	o	o
4	o	o	●	o	o	o

Clamped panels						
Edge No.	Ux	Uy	Uz	ROTx	ROTy	ROTz
1	●	●	●	●	●	o
2	o	o	●	●	●	o
3	o	o	●	●	●	o
4	o	o	●	●	●	o

o :Not Constrained

● :Constrained

Figure 2. Modeling and boundary conditions.

The material, for most parts of the work, was assumed to be mild steel with properties equivalent to: ($E = 210\,GPa$, and $\nu = 0.3$).

Other material properties were also considered in the parametric studies, but unless otherwise stated, the default material is mild steel. Either simply supported or clamped boundary conditions were assumed throughout the work.

The following geometrical and material parameters were also considered:

- *Aspect ratio of panels* (φ): Different aspect ratios were considered to evaluate the effects of cracks on short and long panels. However, when other parameters were concerned, the aspect ratio was taken equal to one. According to Figure 1, the length (a) and width (b) of panels were respectively along the x- and y-axes.
- *Slenderness ratio of panels* (λ): The thickness to width ratio of panels varied according to the practical purposes from thin to very thin panels. The ratios considered were: $\lambda = 0.002{\sim}0.006$.
- *The relative crack length* (l'): Different crack lengths were considered. In order to relate the length of cracks to the dimensions of panels, the relative crack length was defined as the ratio of the actual crack length (l) to the length of panel (a). The different relative crack lengths were: $l' = 0.1{\sim}0.5$.
- *The angle of crack* (θ): The inclination angle of crack relative to the equivalent tension diagonal of a square panel varied between 0 to 90 degrees as illustrated in Figure 1. Therefore, $\theta = 45°$ characterizes a crack which is parallel to the y-axis.
- *Young's modulus* (E): In order to inspect the effect of Young's modulus on the behavior of cracked panels, apart from mild steel properties mentioned earlier, two other moduli, i.e. $E_1 = 140\,GPa$ and $70\,GPa$, were also considered. The latter corresponds to aluminum.
- *Poisson's ratio* (υ): To investigate the effects of material compressibility, υ was considered to vary

in the range: $\upsilon = 0.1{\sim}0.49$. The ratio $\upsilon = 0.49$ corresponds to a nearly incompressible material.
- *Critical stress multiplier* (ψ): The ratio of elastic buckling stress of a cracked panel to its corresponding uncracked panel is defined as the critical stress multiplier ($\tau_{cr}^{cracked} = \psi \cdot \tau_{cr}$). Therefore, $\psi = 1$ refers to a perfect uncracked panel. The critical stress reduction factor is thus calculated as $(1 - \psi)$.

3 MESH SENSITIVITY FOR BUCKLING ANALYSIS

The linear analysis option of the FE program ANSYS was incorporated to predict the elastic shear buckling stress of perfect uncracked panels having a uniform mesh distribution.

The Eigen buckling method of ANSYS and its "Shell 63" element was used as the basic concept. This four-node quadrilateral shell element is capable of modeling elastic behavior and can simulate both membrane and flexural behaviors. In addition, it has three rotational and three translational degrees of freedom per node.

Regarding the convergence study and verification of results, panels were divided into sufficient number of elements to allow for the development of shear buckling modes and displacements.

Figure 2 shows a typical modeling system in which uniformly distributed shear loads were applied along the middle plane of all edge nodes, while the corner nodes were given half those values.

The rotational and translational constrains applied to different edges are also illustrated in Figure 2. In this way, the panels were under pure shear stress action.

The elastic buckling shear stress values derived from numerical analyses were then compared to those obtained from the theoretical Equation 1:

$$\tau_{cr} = \frac{c_s \pi^2 E}{12(1-\upsilon^2)}\left(\frac{t}{b}\right)^2 \tag{1}$$

928

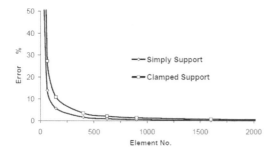

Figure 3. Results of FE convergence studies.

Figure 3 shows the variation of percentage errors obtained by comparing the FE analysis results to the theoretical value for different number of incorporated elements. The presumed panel dimensions considered for convergence study were: ($a = b = 1000$ mm, $t = 3$ mm) According to the results presented in Figure 3, the models with a mesh refinement of 30×30 (900 elements) produced results that were in good agreement with the theory and was therefore used as the minimum requirement.

The FE modeling and analysis of cracked panels involve maximizing the precision associated with the calculation of stresses and displacements near cracks and the local effects imposed by the cracks on the overall response of panels. The first obvious solution to increase precision was to use denser FE meshes near cracks. Therefore, for meshing purposes, the panels were divided into three zones, namely the "crack tips," along the "crack sides," and "away" from the cracks. These divisions were suggested for assigning different mesh densities for zones relative to the position of cracks.

A comprehensive study on the FE meshing techniques of cracked panels was previously carried out by the authors (unpubl.) and it was concluded that the mesh density at the crack tips plays a dominant role in the accuracy of analysis in an exceptional manner, but the regions around crack sides may have mesh refinements similar to uncracked panels.

Hence, a suitable mesh pattern as shown in Figure 4 was selected and incorporated throughout this work. According to this figure, due to stress concentration, very dense meshing was utilized near the crack tips. The element sizes were then gradually increased to the optimum size of an uncracked panel as they parted away from cracks.

4 DISCUSSION OF RESULTS

4.1 Elastic buckling of cracked panels

The effects of different parameters (described in Section 2) on the bifurcation point and critical stresses

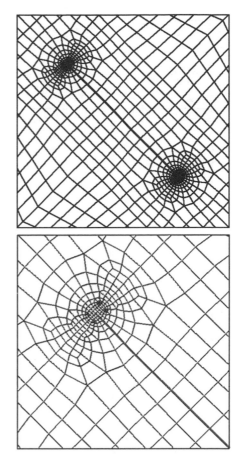

Figure 4. FE Meshing details. Top: the crack length. Bottom, the crack tip.

of central cracked shear panels are discussed in this section. Cracked panels with different slenderness ratios, having various crack angles, crack lengths and boundary conditions were analyzed.

Figure 5 presents a summary of typical results obtained from the FE analyses. The curves represent the reduction in critical stress multipliers of cracked panels (ψ) against different crack lengths (l'), various crack angles (θ), and panel slenderness ratios (λ). The first obvious conclusion is that the panel slenderness ratio does not have any effect on the bifurcation point of cracked panels. On the other hand, the length of crack, as expected, reduces the buckling stress. However, the most important conclusion is that the effect of crack length is very much influenced by the crack's angle of inclination, especially when the crack is laid along compressive principle stresses ($\theta = 90°$).

In order to have a better understanding of the importance of the combinational effects of the length and the angle of cracks, the typical curves of Figure 6 are presented. These curves correspond to a

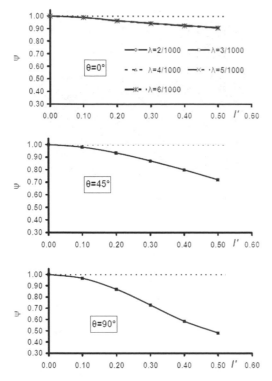

Figure 5. The effect of panel slenderness ratio on panels having different crack lengths and angles.

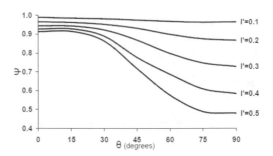

Figure 6. Critical stress multiplier vs. different crack angles and lengths.

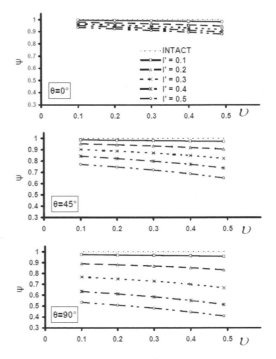

Figure 7. The effect of Poisson's ratio on panels having different crack lengths and angles.

simply supported steel panel with the dimensions of $1000 \times 1000 \times 3$ mm. Similar curves have been obtained for panels with different material and geometrical properties, and are thus omitted for brevity. In these curves, the critical stress multiplier (ψ) is drawn against various crack angles (θ) and lengths (l').

It can be seen that between the crack angles of 30 to 75 degrees, sharp reductions in the buckling stresses of panels having crack lengths $l' \geq 0.3$ were obtained. It is observed that for crack angles $0 \leq \theta \leq 30$, whatever the crack length may be; or for the crack lengths of $l' \leq 0.2$ whatever the angle may be, the reduction in critical stresses was very little ($1 - \psi < 10\%$). The maximum reduction in critical stresses was recorded at $\theta = 90°$; which amounted to 27% for $l' = 0.3$ and 52% for $l' = 0.5$.

The above procedure was then repeated for panels having clamped edges and similar curves were obtained. The results showed that the boundary condition did not really affect the buckling stresses. In fact a maximum difference of 5% was recorded in the case of ($\theta = 90°$). These curves are omitted for brevity.

In order to study the effect of the Poisson's ratio, the results of analyses on panels with various υ are presented in Figure 7. It is shown that the relative buckling stress (ψ) of cracked panels decreases with the increase

in the Poisson's ratio. This behavior is due to the diagonal contraction of the plate responsible for the local buckling phenomenon near cracks.

However, the decrease in the buckling stress was very much influenced by both the angle and the length of cracks. The biggest decrease in critical stress (about 60%) took place in the panel made of nearly incompressible material ($\upsilon = 0.49$), having a relative crack length of $l' = 0.5$ and an angle of inclination of $\theta = 90°$.

The effect of aspect ratio (φ) was considered by analyzing panels having different lengths and similar widths. It is noted that the buckling mode shape of a shear panel is very sensitive to its aspect ratio, and in some instances it is difficult to foresee the first

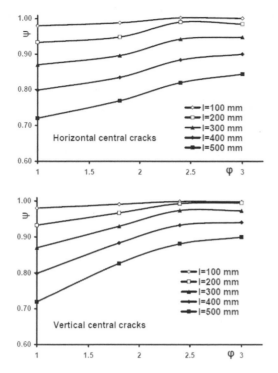

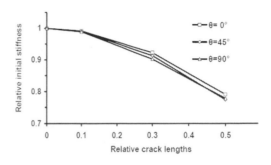

Figure 9. The decrease in the initial stiffness of shear panels.

Figure 8. The effect of aspect ratio having different crack lengths. Top: horizontal cracks. Bottom: vertical cracks.

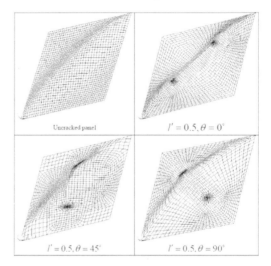

Figure 10. Buckling shapes of uncracked and cracked panels.

buckling mode shape and the angle at which tension field builds up. Therefore, two typical crack angles, i.e. vertical cracks with $\theta = 45°$ and horizontal cracks with $\theta = -45°$ were considered for this part.

Figure 8 presents the effect of aspect ratio on the buckling strength of cracked panels having unrelated crack lengths. The curves show the variation of relative critical stresses (ψ) against different aspect ratios (φ) for various absolute crack lengths ($l = 100 \sim 500$ mm). These results were obtained for typical steel panels with the dimensions of $a = 1000 \sim 3000$ mm, $b = 1000$ mm and $t = 3$ mm. The figure also compares the results of panels having either vertical or horizontal cracks.

It is observed that in both crack types, ψ increases with φ. This increase is directly related to the position of the crack with respect to the tension field of panels. In a square panel the tension field is formed along the diagonal line covering the central cracked area; whereas, in a rectangular panel, the tension field may develop in areas away from the center-cracks.

Finally, the effect of the modulus of elasticity was considered. Many analyses were carried out; varying the E while other parameters, including υ, were kept constant. The results showed that the rates of change of critical shear stresses in panels with different Young's modulus were very similar. In other words, regardless

of the effects of Poisson's ratio, the conclusions drawn earlier for steel panels can be equally extended to other materials.

4.2 Stiffness degradation of cracked panels

The effects of crack lengths and their angles of inclination on the initial stiffness of shear panels are shown in Figure 9. Initial stiffness of panels was obtained from their initial linear portion of $p - \Delta$ curves (applied shear force vs. in-plane displacements). The relative stiffness was then calculated by dividing the stiffness of cracked panel (k^c) to the stiffness of corresponding uncracked panel (k).

Figure 9 shows the relative reduction in the initial stiffness of panels against their relative crack length (l') and different values of θ. It is observed that the initial stiffness of shear panels decreases as the crack length increases; and that this decrease is almost insensitive to the angle of inclination of the crack. The maximum reduction was evaluated about 22%, which corresponds to $l' = 0.5$. A very similar set of results were obtained for various panels having different slenderness ratios and boundary conditions; and as a result, the following simple relationship may be defined:

$$\frac{k_c}{k} = 1 - 0.9\,l'^2, \quad l' \le 0.5, \quad \text{for all values of } \theta \qquad (2)$$

Finally, Figure 10 shows the deformed shapes of some typical cracked panels having different angles of inclination and compares them with that of an intact panel. This figure is presented for a graphical demonstration of the buckling behavior of centrally cracked shear panels.

5 CONCLUSIONS

The buckling behavior of centrally cracked thin shear panels was investigated. The influence of several geometrical and mechanical characteristics on the behavior of cracked shear panels was examined by utilizing the numerical FEM. The results showed that:

– When the relative crack length is $l' \ge 0.2$, considerable reduction of the critical stress is expected; smaller cracks may be ignored.
– The influence of a crack greatly depends on its angle of inclination. If the crack is laid normal to the tension field ($\theta = 90°$), it can impose significant effects.
– There is a combinational effect between the length and the angle of cracks. It is shown that beside their individual degradation effects, they can amplify each other and result in extensive loss of the load bearing capacity of shear panels.
– The initial stiffness of a shear panel decreases with the length of crack. This change is insensitive to the angle of crack.
– Poisson's ratio can have some effects; a maximum reduction of 13% was recorded in the case of incompressible materials.

REFERENCES

Alinia MM, Hosseinzadeh SAA. & Habashi HR. 2007. Numerical modeling for buckling analysis of cracked shear panels. *Thin-Walled Structures;* submitted for possible publication.
ANSYS, user manual, version 5.4, *Canonsburgh,* PA, USA.
Brighenti R. 2005. Buckling of cracked thin-plates under tension or compression. *Thin-Walled Structures*; 43; 209–224.
Paik JK., Satish Kumar YV. & Lee JM. 2005. Ultimate strength of cracked plate elements under axial compression or tension. *Thin-Walled Struct*; 43: 237–272.
Riks E., Rankin CC. & Bargon FA. 1992. Buckling behaviour of a central crack in a plate under tension. *Eng. Frac. Mech.*; 43: 529–48.
Roberts TM. &. Davies AW, 2002. Fatigue induced by plate breathing. *J Const. Steel Res*; 58, 1495–1508.

NOTATIONS:

a	Length of panel
b	Width of panel
c_s	Shear buckling coefficient
E	Modulus of elasticity
k^c	Initial stiffness of a cracked panel
k	Initial stiffness of an uncracked panel
l	Length of crack
l'	Relative crack length
t	Thickness of panel
φ	Aspect ratio of panel
λ	Slenderness ratio of panel
θ	Angle of inclination of crack
τ_{cr}	Shear buckling stress of an intact panel
$\tau_{cr}^{cracked}$	Shear buckling stress of a cracked panel
υ	Poisson's ratio
ψ	Critical stress multiplier

Thin-walled structures

Steel and Composite Structures – Wang & Choi (eds)
© 2007 Taylor & Francis Group, London, ISBN 978-0-415-45141-3

Elastic buckling of elliptical hollow sections

A.M. Ruiz-Teran

University of Castilla-La Mancha, Spain. Academic Visitor at Imperial College London, UK

L. Gardner

Department of Civil and Environmental Engineering, Imperial College London, UK

ABSTRACT: Hot-rolled and cold-formed structural steel tubular members of elliptical cross-section have recently been introduced into the construction sector. However, there is currently limited knowledge of their structural behaviour and stability and comprehensive design guidance is not yet available. This paper examines the elastic buckling response of elliptical hollow sections in compression, which has been shown to be inter-mediate between that of circular hollow sections and flat plates. The transition between these two boundaries is dependant upon both the aspect ratio and relative thickness of the section. Based on numerical results, formulae to accurately predict the elastic buckling stress of elliptical sections have been proposed, and shortcomings of existing expressions have been highlighted. Length effects have also been investigated. The findings form the basis for the development of effective section properties for slender elliptical hollow sections.

1 INTRODUCTION

The use of elliptical structural forms dates back to the mid-nineteenth century. In the initial designs of the Britannia Bridge developed in 1845, elliptical sections were considered for the compression flange of the main box girder (Ryall 1999), whilst the primary arched compression elements of the Royal Albert Bridge, designed by Brunel and constructed in 1859, were of elliptical form and fabricated from wrought iron plates (Binding 1997). Elliptical hollow sections (EHS) are now available as hot-rolled and cold-formed structural sections and have recently begun to appear in steel construction. Examples include several canopies and buildings, such as the coach station at Terminal 3 and the main building at Terminal 5 of Heathrow Airport in London (McKechnie 2006) and the main building at Terminal 4 of Barajas Airport in Madrid (Viñuela & Martinez 2006), as well as bridges, such as the Highland Society Bridge in Braemar, Scotland (Corus 2006). From an architectural perspective, these new sections offer an interesting and unusual appearance, whilst from the structural standpoint, they possess differing flexural rigidities about each of the principal axes (allowing the sections to be orientated to most efficiently resist the applied loading) as well as high torsional stiffness.

Towards the establishment of structural design rules, this paper examines the influence of aspect ratio, relative thickness and length on the elastic buckling of EHS and presents formulae to predict accurately the elastic buckling stress in compression. The described research underpins the development of effective section properties for slender EHS.

2 BACKGROUND

Previous research into EHS has included analytical studies, conducted primarily in the 1960s and 1970s, and more recent experimental and numerical studies.

Kempner (1962) obtained the elastic buckling stress for an oval hollow section (OHS) and established that it is equivalent to the buckling stress of an axially compressed circular hollow section (CHS) with a radius equal to the maximum radius of curvature of the OHS. Kempner & Chen (1964) analysed the post buckling behaviour of OHS, finding that the higher the aspect ratio of the OHS, the more stable post-buckling behaviour (approaching a plate-like response) and, the lower the aspect ratio, the more unstable the post-buckling behaviour (approaching a shell-like response). They also found that the higher the relative thickness, the less stable the post-buckling response. Kempner & Chen (1967) showed that load carrying capacities above the bifurcation load would be attained for OHS with high aspect ratios, due to the redistribution of stresses to the stiff major axis regions of the section of high curvature. The study of Tvergaard (1976), based on the application of Hutchinson's (1968) asymptotic theory to elastic-plastic OHS, showed that this extra load carrying capacity above the bifurcation load may not achieved when elastic-plastic material behaviour is considered, due to premature yielding of the major axis regions. High aspect ratio elastic-plastic OHS were found to be moderately imperfection sensitive, whilst the lower the aspect ratio, the greater the imperfection sensitivity.

With the emergence of elliptical hollow sections of structural proportions, a number of recent studies have been performed (Gardner 2005; Gardner & Ministro 2005). Experimental and numerical studies of EHS under axial compression and bending have been described by Chan & Gardner (2006) and a system of cross-section slenderness limits has been established by Gardner & Chan (2006). In order to derive effective properties for sections deemed to be slender (Class 4), the present research was initiated and formulations to achieve accurate prediction of the elastic buckling stress of EHS have been developed and are described herein.

3 BOUNDARY BEHAVIOUR OF EHS

EHS are geometrically defined by the length of the two principal axes: the major ($2a$) and minor ($2b$) axes. When both axes are of equal length, the section has an aspect ratio of unity and becomes a CHS, whilst when the minor axis length is negligible in comparison to the major axis length, the aspect ratio approaches infinity, and the case of two parallel plates results (Figure 1). Therefore, the elastic buckling stress for an EHS should be a function of the aspect ratio and be bounded by the elastic buckling stress of a CHS (when the aspect ratio is equal to unity) and the elastic buckling stress of a flat plate (when the aspect ratio approaches infinity).

The elastic buckling stress of a CHS under compression is given by:

$$\sigma_{CHS} = \frac{E}{\sqrt{3(1-v^2)}} \frac{t}{r} C_x \tag{1}$$

where r = radius of curvature, t = thickness, E = Young's modulus, v = Poisson's ratio and C_x = a coefficient defined by the relative length of the CHS (for medium-length cylinders $C_x = 1$).

The elastic buckling stress of a compressed plate is given by:

$$\sigma_{Plate} = \frac{K\pi^2 E}{12(1-v^2)} \left(\frac{t}{w}\right)^2 \tag{2}$$

where w = width of the plate and K = a coefficient dependant on the relative length and the boundary conditions of the plate.

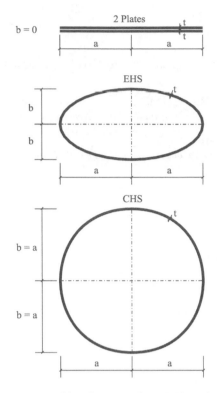

Figure 1. Transition between plate, EHS and CHS geometry.

Previous research has been based on the assumption that the elastic buckling stress of an EHS under compression is the same as that of a CHS with a radius equal to the radius of curvature at the ends of the minor axis of the EHS, where $r = a^2/b$. Therefore, the assumed elastic buckling stress of a compressed EHS (Kempner, 1962) is given by:

$$\sigma_{EHS}^K = \frac{E}{\sqrt{3(1-v^2)}} \frac{t}{a^2/b} \tag{3}$$

An alternative proposal (Corus 2004) gives the elastic buckling stress of a compressed EHS as:

$$\sigma_{EHS}^C = \frac{E}{\sqrt{3(1-v^2)}} \frac{t}{a\sqrt{a/b}} \tag{4}$$

For an EHS with an aspect ratio of unity, Equations 3 and 4 provide the same result, which is equal to that of Equation 1 with $C_x = 1$. Therefore, both equations fit the CHS bound. However, for an EHS of high aspect ratio, the elastic buckling stress of the EHS should tend to that of a flat plate, which neither Equation 3 nor Equation 4 predict.

A formulation for the elastic buckling stress of an EHS that complies with both boundaries (CHS and

Table 1. Results of models with low relative thickness. $L/2a = 2.5$.

Geometry of the models			Elastic buckling load MN	Relative equivalent diameter
2a mm	2b mm	t mm		
400	400	0.1	8.04×10^{-3}	0.99
400	300	0.1	5.44×10^{-3}	1.29
400	200	0.1	3.20×10^{-3}	1.92
400	135	0.1	1.98×10^{-3}	2.87
400	100	0.1	1.42×10^{-3}	3.84
400	65	0.1	8.79×10^{-4}	5.99
400	40	0.1	5.40×10^{-4}	9.56
400	20	0.1	2.69×10^{-4}	18.97
400	0	0.1	7.14×10^{-6}	711.87

Table 2. Results of models with relative thickness less than 0.02. $L/2a = 2.5$.

Geometry of the models			Elastic buckling load MN	Relative equivalent diameter
2a mm	2b mm	t mm		
400	400	0.1	8.04×10^{-3}	0.99
400	300	0.1	5.44×10^{-3}	1.29
400	200	0.1	3.20×10^{-3}	1.92
400	135	0.1	1.98×10^{-3}	2.87
400	100	0.1	1.42×10^{-3}	3.84
400	65	0.1	8.79×10^{-4}	5.99
400	40	0.1	5.40×10^{-4}	9.56
400	20	0.1	2.69×10^{-4}	18.97
400	0	0.1	7.14×10^{-6}	711.87
400	400	2	3.22	0.99
400	300	2	2.19	1.27
400	200	2	1.31	1.87
400	135	2	0.82	2.74
400	100	2	0.60	3.61
400	65	2	0.38	5.46
400	40	2	0.24	8.37
400	20	2	0.13	16.26
400	0	2	0.06	35.61
400	400	5	19.96	0.99
400	300	5	14.00	1.24
400	200	5	8.49	1.78
400	135	5	5.44	2.56
400	100	5	4.02	3.33
400	65	5	2.61	4.95
400	40	5	1.61	7.95
400	20	5	1.07	11.98
400	0	5	0.89	14.28

plates) is therefore sought. Defining the equivalent diameter $D_{Eq,EHS}$ of an EHS as the diameter of a CHS with the same elastic buckling stress, and the relative equivalent diameter as the ratio between the equivalent diameter and the major axis ($2a$) of the EHS leads to Equations 5 and 6 for the CHS and plate boundaries, respectively.

$$\frac{D_{Eq,EHS}\left(\frac{a}{b}=1\right)}{2a} = 1 \tag{5}$$

$$\frac{D_{Eq,EHS}\left(\frac{a}{b}\to\infty\right)}{2a} = \frac{8\sqrt{3(1-v^2)}}{K\pi^2}\frac{2a}{t} \tag{6}$$

From Equations 5 and 6 it may be observed that the equivalent diameter is a function of both the aspect ratio a/b and relative thickness $t/2a$, and that the diameters proposed by Kempner and Corus do not fit the exact value for EHS of high aspect ratios.

4 NUMERICAL MODELLING

In order to investigate how the equivalent diameter of an EHS varies with aspect ratio a/b and the relative thickness $t/2a$, numerical solutions (Tables 1–4) of elastic buckling loads for a set of EHS has been obtained by means of the FE package ABAQUS.

The models had a fixed major axis dimension of 400 mm and a fixed length of 1000 mm. Nine variations of the minor axis dimension were taken: 400, 300, 200, 135, 100, 65, 40, 20 and 0 mm (plate). Ten variations of thickness were considered: 40, 32, 25, 20, 16, 10, 8, 5, 2 and 0.1 mm. This provided aspect ratios a/b ranging between unity and infinity relative thickness $t/2a$ ranging between 1/10 and 1/4000. A Young's modulus of 210 GPa and a Poisson's ratio of 0.3 were assumed. The members had fixed boundary conditions

and the lowest local buckling modes are presented in Tables 1–4.

5 ANALYSIS OF RESULTS

The relative equivalent diameters obtained from the models have been presented in Figure 2, together with the relative equivalent diameter proposed by Kempner and Corus, given by:

$$\frac{D_{Eq,EHS}^K}{2a} = \frac{a}{b} \tag{7}$$

$$\frac{D_{Eq,EHS}^C}{2a} = \sqrt{\frac{a}{b}} \tag{8}$$

The results show that the equivalent diameter based on Kempner's assumption fits well for very thin tubes (more slender than the practical range produced), whilst the equivalent diameter proposed by Corus fits well for very stocky tubes (stockier than the practical range produced).

Table 3. Results of models with relative thickness between 0.02 and 0.04. $L/2a = 2.5$.

Geometry of the models

2a mm	2b mm	t mm	Elastic buckling load MN	Relative equivalent diameter
400	400	8	49.92	1.00
400	300	8	36.50	1.20
400	200	8	22.44	1.71
400	135	8	14.55	2.43
400	100	8	10.82	3.13
400	65	8	6.72	4.89
400	40	8	4.78	6.84
400	20	8	3.91	8.37
400	0	8	3.63	8.97
400	400	10	77.38	1.01
400	300	10	57.62	1.18
400	200	10	35.68	1.67
400	135	10	23.34	2.35
400	100	10	17.01	3.09
400	65	10	10.96	4.67
400	40	10	8.49	6.03
400	20	10	7.41	6.91
400	0	10	7.06	7.20
400	400	16	190.66	1.03
400	300	16	151.35	1.13
400	200	16	95.37	1.57
400	135	16	60.42	2.27
400	100	16	45.29	2.91
400	65	16	34.59	3.79
400	40	16	30.62	4.31
400	20	16	29.06	4.54
400	0	16	28.38	4.59

Table 4. Results of models with relative thickness greater than 0.04. $L/2a = 2.5$.

Geometry of the models

2a mm	2b mm	t mm	Elastic buckling load MN	Relative equivalent diameter
400	400	20	297.07	1.02
400	300	20	239.44	1.11
400	200	20	150.31	1.53
400	135	20	96.14	2.19
400	100	20	76.03	2.70
400	65	20	62.28	3.29
400	40	20	57.41	3.57
400	20	20	55.71	3.68
400	0	20	54.51	3.73
400	400	25	473.90	0.99
400	300	25	375.57	1.09
400	200	25	232.76	1.52
400	135	25	157.95	2.05
400	100	25	131.48	2.43
400	65	25	113.85	2.80
400	40	25	108.02	2.95
400	20	25	106.43	3.00
400	0	25	103.81	3.06
400	400	32	720.07	1.04
400	300	32	588.40	1.11
400	200	32	386.68	1.46
400	135	32	282.87	1.85
400	100	32	246.72	2.12
400	65	32	223.38	2.37
400	40	32	216.55	2.45
400	20	32	215.75	2.46
400	0	32	208.52	2.50
400	400	40	1066.30	1.08
400	300	40	898.21	1.11
400	200	40	628.64	1.36
400	135	40	488.37	1.76
400	100	40	439.65	1.86
400	65	40	409.10	2.00
400	40	40	401.48	2.03
400	20	40	402.42	2.03
400	0	40	383.86	2.12

Figures 3 and 4 show the ratios between the elastic buckling stress obtained based on the equivalent diameters of Kempner and Corus and that obtained from the FE models. Kempner's assumption consistently leads to conservative predictions, with errors between 14 and 22% for the practical commercial range ($a/b = 2$ and thicknesses between 8 mm and 16 mm). Corus' proposal results in unconservative predictions with errors between 11 and 21% for commercial geometries. The higher the aspect ratio, the greater the errors associated with both existing proposals.

Both the aspect ratio a/b and relative thickness $t/2a$ define the manner of buckling of an EHS. The resistance to buckling of an EHS is given by the transverse curvature as well as the local bending stiffness. The lower the relative thickness, the greater the contribution of the transverse curvature to the buckling resistance (closer to CHS behaviour), whereas the higher the relative thickness, the larger the contribution of the local bending stiffness to the buckling resistance (closer to plate behaviour). Also, the lower the aspect ratio, the greater the contribution of the transverse curvature to the buckling resistance (closer

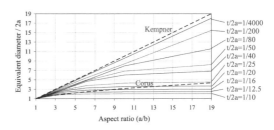

Figure 2. Relative equivalent diameters versus aspect ratio.

to CHS behaviour), whereas the higher the aspect ratio, the larger contribution of the local bending stiffness to the buckling resistance (closer to plate behaviour).

For high aspect ratios, the relative equivalent diameter reaches a plateau (Figure 2), the value of which

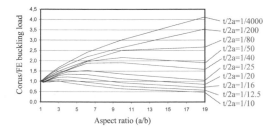

Figure 3. Ratio between the elastic buckling load obtained with Kempner assumption and that obtained by FE versus the aspect ratio of the EHS.

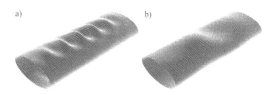

Figure 4. Ratio between the elastic buckling load obtained with Corus assumption and that obtained by FE versus the aspect ratio of the EHS.

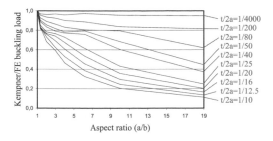

Figure 5. Lowest buckling mode for (a) EHS 400 × 200 × 5 ($L = 1000$ mm) and (b) EHS 400 × 200 × 25 ($L = 1000$ mm).

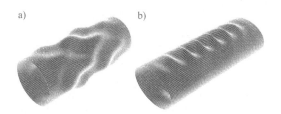

Figure 6. Lowest buckling mode for (a) CHS 400 × 400 × 5 ($L = 1000$ mm) and (b) EHS 400 × 300 × 5 ($L = 1000$ mm).

is governed by the relative thickness (Equation 6). Therefore, the higher the relative thickness, the lower the aspect ratio required to achieve plate behaviour (Figure 2) and the longer the buckling wavelengths (Figure 5).

The lower the aspect ratio, the larger the area over which the buckling waves spread (Figure 6). In fact,

Table 5. Summary of results of models with $L/2a = 7.5$.

2a	t	L	Elastic buckling load (MN)		
mm	mm	mm	a/b = 1	a/b = 4/3	a/b = 2
400	40	3000	894.22	820.66	605.71
400	32	3000	570.97	518.98	374.78
400	25	3000	351.04	315.86	225.04
400	20	3000	229.21	204.49	143.55
400	16	3000	152.16	132.18	92.40
400	10	3000	64.25	55.24	35.44
400	8	3000	44.80	36.23	22.28
400	5	3000	18.64	13.92	8.44
400	2	3000	3.11	2.19	1.30

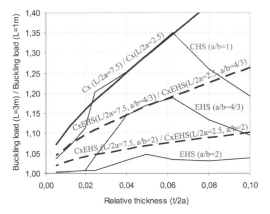

Figure 7. Influence of length on elastic buckling load of EHS.

for an aspect ratio of unity (i.e. for CHS), the buckling waves spread fully around the section.

6 LENGTH EFFECTS

The elastic buckling load of CHS and EHS is influenced by member length. Buckling loads for members of three times the length of members in Section 4 have been obtained (Table 5). The ratio between these elastic buckling loads has been plotted in Figure 7, to which the ratio of length factors (C_x/C_x^*) proposed by EC3-1-6 and given by Equations 10 and 11 has been added. Based on the results, Equation 9 has been proposed for EHS.

$$C_{x,EHS} = \begin{cases} 1 + \left(\dfrac{C_x}{C_x^*} - 1 \right) \left(1 - 1.4 \left(1 - \dfrac{b}{a} \right) \right) & a/b < 3.5 \\ 1 & a/b \geq 3.5 \end{cases} \qquad (9)$$

$$C_x = 1 + \frac{0.2}{6} \left(1 - 4\sqrt{2} \frac{L}{2a} \sqrt{\frac{t}{2a}} \right) \leq 1 \qquad (10)$$

939

$$C_x^* = C_x\left(\frac{L}{2a}=2.5\right)=1+\frac{0.2}{6}\left(1-10\sqrt{2}\sqrt{\frac{t}{2a}}\right)\leq 1 \qquad (11)$$

Figure 7 shows that, as expected, EHS are less sensitive to length effects than CHS, and that application of the EC3-1-6 (for CHS) is conservative for EHS. However, improved agreement is achieved with the proposed expression (Equation 9).

7 PROPOSED FORMULATIONS

It is proposed that the elastic buckling stress of an EHS under compression be given by:

$$\sigma_{EHS}=\frac{E}{\sqrt{3(1-v^2)}}\frac{2t}{D_{Eq.EHS}}C_{x,EHS}>\sigma_{Plate}^* \qquad (12)$$

The elastic buckling stress of a plate with the same boundary conditions as the EHS, taking into consideration shear deformation (Mindlin plates), can be approximated by:

$$\sigma_{Plate}^*=\frac{K\pi^2 E}{12(1-v^2)}\left(\frac{t}{w}\right)^2\varphi_1\varphi_2 \qquad (13)$$

where w is the width of the plate ($w=2a$), $K=7$, φ_1 accounts for the boundary conditions at the loaded ends (Equation 14) and φ_2 accounts for shear deformation (Equation 15).

$$\varphi_1=\begin{cases}1+0.1\left(\dfrac{L/w-5}{2.5}\right)^2 & w\leq L<5w \\ 1 & 5w\leq L\end{cases} \qquad (14)$$

$$\varphi_2=1-16\left(\frac{t}{w}\right)^2 \qquad (15)$$

Consequently, the relative equivalent diameter for plates is given by:

$$\frac{D_{Eq.Plate}}{2a}=\frac{8\sqrt{3(1-v^2)}}{K\pi^2}\frac{2a}{t}\frac{1}{\varphi_1\varphi_2} \qquad (16)$$

7.1 Proposed relative equivalent diameter for EHS of any aspect ratio

The proposed relative equivalent diameter for EHS of any aspect ratio and relative thickness $t/2a$ up to 0.1 is given by Equation 17, where coefficients may be determined from Equations 18–21.

$$\frac{D_{Eq.EHS}}{2a}=c_1^*\left(\frac{a}{b}\right)^{c_2}<\frac{D_{Eq.Plate}}{2a} \qquad (17)$$

$$c_1^*=\begin{cases}c_1 & a/b\geq 4/3 \\ 1+3(c_1-1)(a/b-1) & a/b<4/3\end{cases} \qquad (18)$$

$$\begin{cases}c_1=c_{11}\ c_2=c_{21} & \text{for } 0.00<\dfrac{t}{2a}\leq 0.02 \\ c_1=c_{12}\ c_2=c_{22} & \text{for } 0.02<\dfrac{t}{2a}\leq 0.05 \\ c_1=c_{13}\ c_2=c_{23} & \text{for } 0.05<\dfrac{t}{2a}\leq 0.1\end{cases} \qquad (19)$$

where $c_{11}, c_{12}, c_{13}, c_{21}, c_{22}$ and c_{23} are coefficients that depend on the relative thickness and are given by:

$$\begin{bmatrix}c_{11}\\ c_{12}\\ c_{13}\end{bmatrix}=\begin{bmatrix}-1.848 & 1.659 & -0.690 & 1.000\\ -0.637 & 0.777 & -0.478 & 0.983\\ -3.718 & 8.111 & -5.515 & 2.053\end{bmatrix}\begin{bmatrix}(5t/a)^3\\ (5t/a)^2\\ (5t/a)\\ 1\end{bmatrix} \qquad (20)$$

$$\begin{bmatrix}c_{21}\\ c_{22}\\ c_{23}\end{bmatrix}=\begin{bmatrix}-13.156 & 6.756 & -1.283 & 1.000\\ -0.240 & 0.591 & -0.491 & 0.985\\ 5.859 & -12.676 & 8.197 & -0.805\end{bmatrix}\begin{bmatrix}(5t/a)^3\\ (5t/a)^2\\ (5t/a)\\ 1\end{bmatrix} \qquad (21)$$

With this proposed equivalent diameter, the errors in the predicted elastic buckling stresses are less than 14% for all aspect ratios. These errors are markedly lower than those of Kempner (90%) and Corus (324%).

7.2 Proposed relative equivalent diameter for EHS of aspect ratio less than 4

A simpler formula for the relative equivalent diameter of EHS of practical aspect ratios (a/b < 4) is given by:

$$\frac{D_{Eq.EHS}}{2a}=1+f\left(\frac{a}{b}-1\right)<\frac{D_{Eq.Plate}}{2a} \qquad (22)$$

where $f=\left(1-10\dfrac{t}{2a}\right)>0.65$ $\qquad (23)$

With this proposed equivalent diameter the errors in the evaluation of the elastic buckling stress of EHS with aspect ratios less than 4 and relative thickness $t/2a$ less than 0.05 are below 10%, and the predictions are always conservative. This represents a considerable reduction in errors as compared to Kempner (33%) and Corus (92%).

8 CONCLUSIONS

Tubular members of elliptical cross-section have recently been introduced into the construction market, offering a new alternative to structural engineers and architects. As part of the development of structural design rules for elliptical hollow sections, this paper has focused on accurate prediction of their elastic buckling response, being intermediate between that of circular sections and flat plates. The transition between these two boundaries is dependant upon both the aspect ratio and relative thickness of the section. Based on numerical results, formulae to determine the elastic buckling stress have been proposed. These lead to safe predictions and considerably reduced errors in comparison to existing approaches.

REFERENCES

Binding, J. 1997. Brunel's Royal Albert Bridge: A Study of the Design and Construction of His 'Gateway to Cornwall' at Saltash. Twelveheads Press.

Chan, T.M. & Gardner, L. 2006. Experimental and numerical studies of Elliptical Hollow Sections under axial compression and bending. *11th International Symposium on Tubular Structures* – ISTS11. 31 August – 2 September 2006, Quebec City, Canada.

Corus. 2004. Celsius 355® Ovals – Size and resistances. Structural & Conveyance Business.

Corus. 2006. Celsius 355® Ovals. Structural & Conveyance Business.

Gardner, L. 2005. Structural behaviour of oval hollow sections. International *Journal of Advanced Steel Construction* 1(2):26–50.

Gardner, L. & Chan, T.M. 2006. Cross-section classification of Elliptical Hollow Sections. *11th International Symposium on Tubular Structures* – ISTS11. 31 August – 2 September 2006, Quebec City, Canada.

Gardner, L. & Ministro, A. 2005. Structural steel oval hollow sections. *The Structural Engineer* 83(21): 32–36.

Hutchinson, J.W. 1968. Buckling and initial post-buckling behaviour of oval cylindrical shells under axial compression. *Journal of Applied Mechanics*, March 1968: 66–72.

Kempner, J. 1962. Some results on buckling and postbuckling of cylindrical shells. *Collected papers on instability of shell structures*. NASA TND-1510, Dec 1962: 173–186.

Kempner, J. & Chen, Y.N. 1964. Large deflections of an axially compressed oval cylindrical shell. Proceedings of the 11th International Congress of applied mechanics. Munich, Springer-Verlag, Berlin, 299–305.

Kempner, J. & Chen, Y.N. 1967. Buckling and Post-buckling of an axially compressed oval cylindrical shell. *Symposium on the Theory of shells*. Houston, Texas.

McKechnie, S. 2006. Terminal 5. London Heathrow: The main terminal building envelope. *The Arup Journal* 41(2): 36–43.

Ryall, M.J. 1999. Britannia bridge: from concept to construction. *Proc. Inst. Civ. Engrs, Civ. Engng* 132: 132–143.

Tvergaard, V. 1976. Buckling of elastic-plastic oval cylindrical shells under axial compression. *International Journal of Solid and Structures* 12: 683–691.

Viñuela, L. & Martinez, J. 2006. Steel structure and prestressed façade of the new Terminal Building. *Hormigón y Acero* 239: 71–84.

Steel and Composite Structures – Wang & Choi (eds)
© *2007 Taylor & Francis Group, London, ISBN 978-0-415-45141-3*

Local buckling of thin-walled elliptical tubes containing an elastic infill

A. Roufegarinejad & M.A. Bradford
Centre for Infrastructure Engineering & Safety, School of Civil & Environmental Engineering,
The University of New South Wales, Sydney, Australia

ABSTRACT: This paper addresses the local buckling of thin-walled elliptical tubes subjected to uniform axial compression and which contain an elastic infill that inhibits the local buckling into the tube. It makes use of an energy-based technique in which the strain energy stored in both the thin-walled elliptical tube and in the elastic restraint during buckling is calculated, as well as the work done by the external compression during buckling. The displacement function chosen is based on the physical observation that the buckling in an elliptical tube is localised, and the critical stress for local buckling is determined by seeking the extent of the localisation that minimises the buckling stress. The buckling problem is presented in analytic form based on fundamental mechanics, although simple numerical techniques are needed to derive explicit solutions. The accuracy of the approach is discussed, as well as that of an approximation in which the elliptic profile is transformed to that of an equivalent circle.

1 INTRODUCTION

Thin-walled circular steel tubes subjected to axial compression find widespread application in many branches of engineering and their buckling behaviour has been researched fairly extensively. On the other hand, the structural behaviour of elliptical tubes has been far less studied, despite their growing use in engineering structures (Bortolotti *et al.* 2003, Corus 2004, Chan & Gardner 2006, Gardner & Chan 2006, Zhu & Wilkinson 2006). While it is widely known that imperfection sensitivity is dominant in the buckling of thin-walled circular tubes and leads to sudden failures which must be analysed by Donnell shell theory or the like (Teng 1996), the buckling of elliptical tubes is less explosive insofar as failure may occur beyond the initial buckling load and the postbuckling range is not necessarily accompanied by rapid strain softening (Hutchinson 1968).

It is widely known in structural engineering that the strength enhancement of infilling a hollow steel tube with concrete arises not only because of the strength of the concrete infill itself, but because it also increased the local buckling capacity of the tube by inhibiting the buckling of the tube into the infill (Uy 2001). This latter effect was considered for circular tubes by Bradford *et al.* (2002), who derived an analytical expression for the local buckling of a thin-walled circular tube with a rigid, concrete infill. The possibility of the infill being elastic rather than rigid, and its influence on buckling, was considered by Bradford & Vrcelj (2004) for

square tubes and Bradford & Roufegarinejad (2006) and Bradford *et al.* (2006) for circular tubes. Hitherto, it appears the influence of an elastic infill (and even a rigid infill of infinite stiffness) on the buckling of thin-walled elliptical tubes has not been reported.

Sustained research on the stability of elliptical tubes appears to date from the work of Marguerre (1951), who proposed a "mean value" of the curvature of the ellipse as a basis for an equivalent radius of a circular tube, but the use of this concept led to erroneous predictions of the buckling stress. This concept was explored further by Kempner (1962). Kempner's work recognised the 'localisation' of the buckle of an elliptical tube, as shown in Figure 1, where the buckle is localised in a region adjacent to the position in the profile where the curvature is smallest. This concept is used in the present paper. Several other researchers have addressed the hollow elliptical tube buckling problem, including Myers & Hyer (1999) and Hyer and Vogl (2001), but research of the topic has been far from extensive.

This paper considers the local buckling of thin-walled elliptical tubes containing an elastic infill under uniform compression. It is based on a consideration of the similar problem for circular tubes reported elsewhere (Bradford *et al.* 2006), with a consideration of the observed localisation of the buckled shape in order to simplify the analysis. The problem is stated in analytical form, from which numerical solutions may be derived, and that are compared with finite element results obtained using ABAQUS (2006). The use

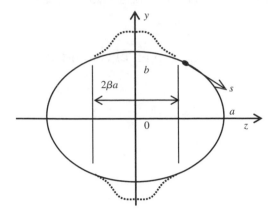

Figure 1. Localisation of buckling of elliptical tube.

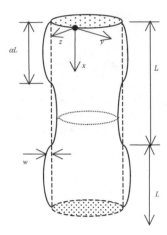

Figure 2. Axes, buckled shape and buckling wavelength.

of an approximate solution, again in closed form, is discussed.

2 ENERGY FORMULATION

The buckling mode shown in Figure 2 is assumed to be infinitesimal and of magnitude w in the local normal (s) direction, that is produced when a constant uniaxial strain ε_0 applied in the x-direction reaches its critical value $\varepsilon_{0\ell}$ at which elastic bifurcation buckling takes place. One local buckling cell of wavelength L in the x-direction is considered here, being one of a number of such cells assumed to form lengthwise. The energy formulation (Bradford et al. 2006) requires statements of the strain energy stored due to bending only (U_b), the membrane strain energy due to stretching (U_m), the strain energy stored in the elastic infill (U_r) as well as the work done during buckling (V).

The middle or reference surface strains for the ellipse (where u and v are the axial and circumferential displacements respectively) are

$$\varepsilon_x = \partial u/\partial x;\ \varepsilon_s = \partial v/\partial s + w/\rho;\ \gamma_{xs} = \partial u/\partial s + \partial v/\partial x\ ;$$

$$\kappa_x = -\partial^2 w/\partial x^2;\ \kappa_s = -\partial^2 w/\partial s^2;\ \kappa_{xs} = -2\partial^2 w/\partial x \partial s \quad (1)$$

in which ρ is the local radius of curvature of the undeformed tube that varies around its elliptic profile. The generalised curvatures in Equation 1 lead to the well-known strain energy stored due to bending only as

$$U_b = \frac{D}{2}\int_0^L \oint_C \left[\kappa_x^2 + \kappa_s^2 - 2(1-v)\left(\kappa_x\kappa_s - \kappa_{xs}^2\right)\right]\mathrm{d}s\,\mathrm{d}x \quad (2)$$

in which $D = Et^3/[12(1 - v^2)]$, E is Young's modulus, v is Poisson's ratio and C denotes the elliptic contour.

From elementary elasticity theory the stresses through the thickness ($z \in [-t/2, t/2]$) corresponding to ε_x and ε_s are

$$\sigma_x(z) = \left(\frac{E}{1-v^2}\right)\left[\left(\varepsilon_x - z\kappa_x\right) + v\left(\varepsilon_s - z\kappa_s\right)\right] \quad (3)$$

and

$$\sigma_s(z) = \left(\frac{E}{1-v^2}\right)\left[\left(\varepsilon_s - z\kappa_s\right) + v\left(\varepsilon_x - z\kappa_x\right)\right] \quad (4)$$

which lead to edge forces per unit length of

$$N_x = \int_{-t/2}^{t/2}\sigma_x\,\mathrm{d}z = Et\left(\varepsilon_x + v\varepsilon_s\right)/\left(1-v^2\right); \quad (5)$$

$$N_s = \int_{-t/2}^{t/2}\sigma_s\,\mathrm{d}z = Et\left(\varepsilon_s + v\varepsilon_x\right)/\left(1-v^2\right); \quad (6)$$

as well as

$$N_{xs} = \int_{-t/2}^{t/2}\tau_{xs}\,\mathrm{d}z = \tfrac{1}{2}Et\gamma_{xs}/\left(1+v\right), \quad (7)$$

where γ_{xs} is the shear strain at the middle surface of the ellipse. At buckling, the axial force intensity is $(\varepsilon_0 E)t$, which must be equal to N_x in Equation 5, which results in

$$\varepsilon_0 = \left(\varepsilon_x + v\varepsilon_s\right)/\left(1-v^2\right), \quad (8)$$

while the strain $-v\varepsilon_0$ in the s direction because of Poisson's effect must be augmented by the membrane buckling strain w/ρ, producing

$$\varepsilon_s = w/\rho - v\varepsilon_0, \quad (9)$$

944

and solving Equations 8 and 9 simultaneously leads to

$$\varepsilon_x = \varepsilon_0 - v\,w/\rho\,. \tag{10}$$

The strain energy due to membrane stretching during buckling is

$$U_s = \tfrac{1}{2}\oint_C \int_{-t/2}^{t/2}(N_x\varepsilon_x + N_s\varepsilon_s + N_{xs}\gamma_{xs})\,\mathrm{d}z\,\mathrm{d}s \tag{11}$$

which leads to

$$U_s = \frac{\tfrac{1}{2}Et}{1-v^2}\oint_C \int_{-t/2}^{t/2}\left[(\varepsilon_x + \varepsilon_s)^2 - 2(1-v)\left(\varepsilon_x\varepsilon_s - \frac{\gamma_{xs}^2}{4}\right)\right]\mathrm{d}z\,\mathrm{d}s \tag{12}$$

The work done by the external compressive forces prior during buckling is equal to the end load multiplied by the axial shortening $1/2\int(\partial w/\partial x)^2\mathrm{d}x$ and by the change in axial length caused by the change in strain $\varepsilon_x-\varepsilon_0$ (Equation 10), so that

$$V = Et\varepsilon_0\int_0^L\oint_C\left[(\varepsilon_x - \varepsilon_0) + \frac{1}{2}\left(\frac{\partial w}{\partial x}\right)^2\right]\mathrm{d}s\,\mathrm{d}x\,. \tag{13}$$

Figure 3 shows a meridional local buckle cell, with the region $x\in\Gamma=[0,\alpha L]$ representing that for which the buckle penetrates the medium of constant stiffness k; clearly the penetration parameter α that defines the extent for which the buckle penetrates the buckle is not known *a priori*. For the general case. Because strain energy is stored in the medium only in the region $\Gamma\in[0,\alpha L]$, the strain energy stored during buckling associated with the stiffness of the infill is

$$U_r = \tfrac{1}{2}\int_0^{\alpha L}\oint_C kw^2\,\mathrm{d}s\,\mathrm{d}x\,. \tag{14}$$

Finally, prior to buckling, the strain energy stored in the tube due to axial compression is

$$U_0 = \tfrac{1}{2}\int_0^L\oint_C(E\varepsilon_0)\varepsilon_0 t\,\mathrm{d}s\,\mathrm{d}x = \tfrac{1}{2}Et\int_0^L\oint_C\varepsilon_0^2\,\mathrm{d}s\,\mathrm{d}x\,, \tag{15}$$

and so the total change in potential during buckling is

$$\Pi = U_b + U_m + U_r - V - U_0\,. \tag{16}$$

The circumference S of an ellipse with semi-major axis a and semi-minor axis b is

$$S = \oint_C \mathrm{d}s = \oint_C[1 + (\mathrm{d}y/\mathrm{d}z)^2]^{\frac{1}{2}}\,\mathrm{d}z\,, \tag{17}$$

which when using the change of variable

$$\mathrm{d}s = \frac{1}{a}\sqrt{\frac{m}{n}}\,\mathrm{d}z \tag{18}$$

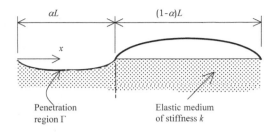

Figure 3. Meridional view of buckle and penetration region Γ.

with

$$m = a^4 - a^2 z^2 + B^2 z^2 \quad\text{and}\quad n = a^2 - z^2 \tag{19}$$

produces

$$S = \frac{4}{a}\int_0^a\sqrt{\frac{m}{n}}\,\mathrm{d}z\,. \tag{20}$$

Equation 20 can be stated using Ramanujan's first approximation as

$$S \cong \pi\left[3(a+b) - \sqrt{(a+3b)(3a+b)}\right]\,. \tag{21}$$

The total change in potential in Equation 16 therefore becomes

$$\begin{aligned}
\Pi =\ & \frac{Et^3}{6(1-v^2)\,a}\int_0^L\int_0^a\left(\frac{\partial^2 w}{\partial x^2}\right)^2\sqrt{\frac{m}{n}}\,\mathrm{d}z\,\mathrm{d}x \\
& + \frac{Et^3 a^3}{6(1-v^2)}\int_0^L\int_0^a\left(\frac{\partial^2 w}{\partial z^2}\right)^2\sqrt{\frac{n}{m}}\,\mathrm{d}z\,\mathrm{d}x + \frac{Et^3 a^7 b^4}{6(1-v^2)}\int_0^L\int_0^a\left(\frac{\partial w}{\partial z}\right)^2\frac{z^2\,\mathrm{d}z\,\mathrm{d}x}{m^3\sqrt{mn}} \\
& - \frac{Et^3 a^5 b^2}{3(1-v^2)}\int_0^L\int_0^a\frac{z}{m^2}\sqrt{\frac{n}{m}}\frac{\partial w}{\partial z}\frac{\partial^2 w}{\partial z^2}\,\mathrm{d}z\,\mathrm{d}x \\
& + \frac{vEt^3 a}{3(1-v^2)}\int_0^L\int_0^a\frac{\partial^2 w}{\partial x^2}\frac{\partial^2 w}{\partial z^2}\sqrt{\frac{n}{m}}\,\mathrm{d}z\,\mathrm{d}x - \frac{vEt^3 a^3 b^2}{3(1-v^2)}\int_0^L\int_0^a\frac{\partial^2 w}{\partial x^2}\frac{\partial w}{\partial z}\frac{z\,\mathrm{d}z\,\mathrm{d}x}{\sqrt{m^3 n}} \\
& + \frac{Et^3 a}{3(1+v)}\int_0^L\int_0^a\sqrt{\frac{n}{m}}\left(\frac{\partial^2 w}{\partial x\partial z}\right)^2\mathrm{d}z\,\mathrm{d}x + 2Etb^2 a^7\int_0^L\int_0^a\frac{w^2\,\mathrm{d}z\,\mathrm{d}x}{\sqrt{nm^5}} \\
& + \frac{2k}{a}\int_0^{\alpha L}\int_0^a\sqrt{\frac{m}{n}}w^2\,\mathrm{d}z\,\mathrm{d}x - \frac{2Et\varepsilon_0}{a}\int_0^L\int_0^a\sqrt{\frac{m}{n}}\left(\frac{\partial w}{\partial x}\right)^2\mathrm{d}z\,\mathrm{d}x. \tag{22}
\end{aligned}$$

3 RITZ-BASED SOLUTION TECHNIQUE

In implementing the Rayleigh-Ritz technique, the axial strain at elastic buckling ε_0 in Equation 22 is determined by making the change in potential Π stationary with respect to all variations of the buckled shape w which satisfies the kinematic boundary conditions for buckling of the elliptical cylinder. The function $^x w(x)$ in the domain $x\in[0,L]$ (or $^x w(\xi)$ where $\xi=x/L\in[0,1]$) given by

$$^x w = {}^x\!\Delta\cdot\sin[\pi(\alpha - \xi)]\cdot\sin[\pi(\xi - 1)] \tag{23}$$

945

in which $^x\Delta$ is a deflection parameter, satisfies the boundary conditions that

$$^xw(\xi = 0) = {}^xw(\xi = \alpha) = {}^xw(\xi = 1) = 0. \tag{24}$$

In addition, because

$$\partial^xw/\partial x = \left(\pi\,{}^x\Delta/L\right)\cdot\sin[\pi(\alpha + 1 - 2\xi)], \tag{25}$$

the function ^xw is periodic along the wavelength L of the cylinder and is symmetric and because

$$\left(\frac{\partial^xw}{\partial x}\right)_{\xi=0} = \begin{cases} 0 & \alpha = 0 \\ -\pi\,{}^x\Delta/L & \alpha = \frac{1}{2} \end{cases} \quad\text{and} \tag{26}$$

$$\left(\frac{\partial^xw}{\partial x}\right)_{\xi=1} = \begin{cases} 0 & \alpha = 0 \\ -\pi\,{}^x\Delta/L & \alpha = \frac{1}{2}, \end{cases} \tag{27}$$

the buckling mode is antisymmetric in the interval $\xi \in [0, 1]$ and the deformation in Equation 23 satisfies the required kinematic boundary conditions for a tube with no infill ($k = 0$) or with a rigid infill ($k \to \infty$).

The displacement function $^sw(s)$ chosen in the meridional direction should also satisfy the boundary conditions. By a transformation of variables to $\eta = z/a \in [0, 1]$ (where $z \in [0, a]$ represents the right hand portion of the ellipse), the function

$$^sw = {}^s\Delta\cdot\left\{2\left[\frac{1-(b/a)}{\beta^3}\right]\eta^3 - 3\left[\frac{1-(b/a)}{\beta^2}\right]\eta^2 + 1\right\} \tag{28}$$

where $^s\Delta$ is a deflection parameter, satisfies the kinematic condition $^sw(\eta = 0) = 1$ so that the maximum buckling displacement takes place at the point of minimum curvature on the ellipse. In Equation 28, β is the localisation parameter discussed in Section 1 that is relevant to the local buckling of elliptical tubes, so that $2a\beta$ represents the projection of the buckled width in the cross-section onto the z-axis. Equation 28, which satisfies the additional boundary conditions that

$$\left(\partial^sw/\partial z\right)_{\eta=0} = \left(\partial^sw/\partial z\right)_{\eta=\beta} = {}^sw_{\eta=\beta} = 0, \tag{29}$$

is a generic form chosen to be valid for all $a \geq b$ in modelling the meridional buckling displacement. For a circular tube for which $a = b$, Equation 29 produces $^sw = {}^s\Delta$ which represents an axisymmetric ring buckling mode and which is the same as that assumed by Bradford *et al.* (2000, 2006), while as $b/a \to 0$ and the oval tube tends towards a flat plate fixed at the edges $z = \pm a$, it produces

$$^sw = {}^s\Delta\cdot\left[2(\eta/\beta)^3 - 3(\eta/\beta)^2 + 1\right]. \tag{30}$$

For a flat plate which is built-in at its edges, the plate may buckle across its entire width and so the

localisation parameter is $\beta = 1$, and using this in Equation 30 reduces the term in square brackets to the cubic interpolation function for a plate built-in along its edge.

Combining the buckling deformations over the domains $\xi \in [0, 1] \in \Re^1$ and $\eta \in [0, 1] \in \Re^1$, the buckling deformation $w \in \Re^2$ that satisfies the buckling boundary conditions for the elliptic tube can be written as

$$w = {}^xw\cdot{}^sw = q\cdot\left[2\left(1-\frac{b}{a}\right)\left(\frac{\eta}{\beta}\right)^3 - 3\left(1-\frac{b}{a}\right)\left(\frac{\eta}{\beta}\right)^2 + 1\right] \\ \times\sin[\pi(\alpha - \xi)]\sin[\pi(\xi - 1)] \tag{31}$$

or

$$w = q\cdot F, \tag{32}$$

in which q is the maximum magnitude of the buckling displacement and the function $F = \mathrm{fn}(a, b, \beta, \alpha, \xi, \eta)$.

By substituting Equation 32 and its appropriate derivatives into Equation 22 and minimising by setting $\partial\Pi/\partial q = 0$ for arbitrary q, the local buckling stress can be written in the form

$$\sigma_{0\ell} = \omega\cdot\frac{E}{\sqrt{1-\nu^2}}\frac{t}{(d/2)} \tag{33}$$

in which $d = 2a^2/b$ is the diameter of the curvature of the end of the minor axis (where the localised buckling initiates) and ω is the local buckling coefficient. This takes the explicit form

$$\omega = \frac{a^2t}{12L^2b\sqrt{1-\nu^2}}f_1 + \frac{tb^2L^2}{12\sqrt{1-\nu^2}}f_2 - \frac{\nu tab}{6\sqrt{1-\nu^2}}f_3 \\ + \frac{ta^2}{6b}\sqrt{\frac{1-\nu}{1+\nu}}f_4 + \frac{a^4bL^2\sqrt{1-\nu^2}}{t}f_5 + \frac{k\sqrt{1-\nu^2}a^2L^2}{Et^2b}f_6 \tag{34}$$

with the terms f_1 to f_9 being given by

$$gf_1 = \int_0^1\int_0^\beta\sqrt{\frac{\Theta}{\Gamma}}\left(\frac{\partial^2F}{\partial\xi^2}\right)^2\,d\eta\,d\xi\,;\;\; gf_2 = \int_0^1\int_0^\beta\frac{\eta^2}{\sqrt{\Theta^7\Gamma}}\left(\frac{\partial^2F}{\partial\eta^2}\right)^2\,d\eta\,d\xi\,;$$

$$gf_3 = \int_0^1\int_0^\beta\frac{\eta}{\sqrt{\Theta^5\Gamma}}\frac{\partial^2F}{\partial\xi^2}\frac{\partial^2F}{\partial\eta^2}\,d\eta\,d\xi\,;\,gf_4 = \int_0^1\int_0^\beta\sqrt{\frac{\Gamma}{\Theta}}\left(\frac{\partial^2F}{\partial\xi\partial\eta}\right)^2\,d\eta\,d\xi\,;$$

$$gf_5 = \int_0^1\int_0^\beta\frac{F^2\,d\eta\,d\xi}{\sqrt{\Theta^5\Gamma}}\,;\;\; gf_6 = \int_0^\alpha\int_0^\beta\sqrt{\frac{\Theta}{\Gamma}}F^2\,d\eta\,d\xi \tag{35}$$

with

$$g = \int_0^1\int_0^\beta\sqrt{\frac{\Theta}{\Gamma}}\left(\frac{\partial F}{\partial\xi}\right)^2\,d\eta\,d\xi \quad\text{and} \tag{36}$$

$$\Gamma = 1 - \eta^2 \quad\text{and}\quad \Theta = a^2\Gamma + b^2\eta^2. \tag{37}$$

946